Informatik-Fachberichte 293

Herausgeber: W. Brauer
im Auftrag der Gesellschaft für Informatik (GI)

W0260123

J. Encarnação (Hrsg.)

Telekommunikation und multimediale Anwendungen der Informatik

GI-21. Jahrestagung
Darmstadt, 14.-18. Oktober 1991

Proceedings

Springer-Verlag
Berlin Heidelberg New York London Paris
Tokyo Hong Kong Barcelona Budapest

Herausgeber
José L. Encarnação
Technische Hochschule Darmstadt
Fachbereich Informatik
Fachgebiet Graphisch-Interaktive Systeme
Wilhelminenstr. 7, W-6100 Darmstadt

CR Subject Classification (1991): A.0, B.4, C.2.0-3, D.0, J.7, K.3.0

ISBN-13: 978-3-540-54755-6 e-ISBN-13: 978-3-642-77060-9
DOI-13: 978-3-642-77060-9

Dieses Werk ist urheberrechtlich geschützt. Die dadurch begründeten Rechte, insbesondere die der Übersetzung, des Nachdrucks, des Vortrags, der Entnahme von Abbildungen und Tabellen, der Funksendung, der Mikroverfilmung oder der Vervielfältigung auf anderen Wegen und der Speicherung in Datenverarbeitungsanlagen, bleiben, bei auch nur auszugsweiser Verwertung, vorbehalten. Eine Vervielfältigung dieses Werkes oder von Teilen dieses Werkes ist auch im Einzelfall nur in den Grenzen der gesetzlichen Bestimmungen des Urheberrechtsgesetzes der Bundesrepublik Deutschland vom 9. September 1965 in der jeweils geltenden Fassung zulässig. Sie ist grundsätzlich vergütungspflichtig. Zuwiderhandlungen unterliegen den Strafbestimmungen des Urheberrechtsgesetzes.

© Springer-Verlag Berlin Heidelberg 1991

Satz: Reproduktionsfertige Vorlage vom Autor

33/3140-543210 – Gedruckt auf säurefreiem Papier

Vorwort

Telekommunikation und Multimediale Anwendungen der Informatik

Wir stehen heute am Übergang von der Industriegesellschaft zur Kommunikationsgesellschaft. Diese ist geprägt von der dezentralen Nutzung und dem kooperativen Bearbeiten multimedialer Information. Die Basis dazu bietet die Kommunikationstechnologie. Sie ist zu realisieren als ein Verbund aus Informatik, Elektronik, Telekommunikation sowie Übertragungs- und Vermittlungstechnik integriert mit benutzergerechten, multimedialen Nutzungsmöglichkeiten.

Die breite Erschließung der Möglichkeiten der Kommunikationstechnik für die multimediale Nutzung steht noch am Anfang ihrer Entwicklung. In Deutschland müßten insbesondere die Telekommunikationsindustrie und die Anbieter von Kommunikationsdiensten eine (auch internationale) Vorreiterrolle spielen, wenn sie künftig im Wettbewerb bestehen wollen, denn die strukturelle und industrielle Entwicklung der Bundesrepublik ist davon stark abhängig. Es müssen Pilot- und Prototyp-Realisierungen durchgeführt werden, die Lösungsalternativen erproben und analysieren und den Weg für Produkte und neue Strukturen vorbereiten und einleiten. Hierzu müssen Methoden, Werkzeuge und Modelle sowie methodische Vorgehensweisen entwickelt, angeboten und verwendet werden, die auch sehr stark von der Informatik geprägt werden bzw. herrühren.

Die 21. Jahrestagung der Gesellschaft für Informatik hat sich in Darmstadt vom 14. bis 18. Oktober 1991 mit dieser Thematik beschäftigt. Die behandelten Themen waren in folgende Klassen unterteilt:

<u>Grundlagenthemen:</u>

In diesem Zusammenhang werden Themen behandelt, die Grundlagencharakter haben bzw. Werkzeuge sind für die Entwicklung von Telekommunikationssystemen und ihre multimedialen Anwendungen.

<u>Systemarchitekturen:</u>

Hierbei stehen die Telekommunikationssysteme, die dazugehörigen Rechnerarchitekturen und Endgeräte, die notwendigen Programm- und Dienstentwicklungsumgebungen sowie Anwendungsschnittstellen im Vordergrund.

<u>Anwendungen:</u>

Eingehende Behandlung von Anwendungen und praxisbezogene Erfahrungsberichte.

Dabei stehen folgende Schwerpunkte im Vordergrund:

- Multimedia, Kommunikation und Interaktion
- Systemarchitekturen, Protokolle und Integration von Multimedia
- Netze, Dienste und Anwendungen von Multimedia und Telekommunikation

In diesem Tagungsband werden die Hauptvorträge von Experten von internationalem Rang aus Deutschland, England, Japan und USA sowie die Fachbeiträge zu acht Fachgesprächen des technisch-wissenschaftlichen Programms veröffentlicht.

Wir glauben, daß es uns gelungen ist, in einer sinnvollen Breite und ausgezeichneter technisch-wissenschaftlicher Qualität, das höchstaktuelle Thema "Telekommunikation und multimediale Anwendungen der Informatik" in der Tagung zu behandeln und in Buchform als Tagungsband einem breiten Publikum zugänglich zu machen. Dafür werden sich sicherlich nicht nur Fachleute aus der Informatik und aus der Telekommunikationstechnik interessieren, sondern auch die Experten, die darauf aufbauende Systeme, Dienste und Anwendungen realisieren, anbieten oder betreiben.

Eine solche Tagung wäre nicht möglich ohne die Hilfe Vieler. Obwohl sicherlich nicht ausschöpfend, möchte ich folgende Danksagungen aussprechen:

- den Programmkomitees für die konstruktive und engagierte Mitarbeit beim Aufstellen des technisch-wissenschaftlichen Programmes,

- allen Mitgliedern des Komitees, die die verschiedenen Teile der GI'91 getragen und durch ihre persönliche Beratung geprägt haben, für ihre Unterstützung,

- allen Sponsoren, die geholfen haben, die GI'91 zu finanzieren,

- der Technischen Hochschule Darmstadt und derem Fachbereich Informatik für die personelle und technische Unterstützung sowie für die Bereitstellung der Räumlichkeiten,

- dem Springer-Verlag für das Zustandekommen dieses interessanten Tagungsbandes,

- den Kollegen Henhapl (lokale Organisation), Hoffmann (Tutorien), Kammerer (Industrieprogramm) und Waldschmidt (Finanzen) für die geleistete Zusammenarbeit; ohne sie wäre die GI'91 nicht möglich gewesen,

und "last but not least!..."

- Herrn Dr. Rolf Lindner vom Fachgebiet Graphisch-Interaktive Systeme der TH Darmstadt und insbesondere Frau Beatrice Barth, meiner Sekretärin, für das große Engagement, die viele Mühe und für die ausgezeichnete Leistung, die sie zu meiner Unterstützung bei der Planung, Organisation und Durchführung der GI'91 geleistet haben.

Darmstadt, 14.08.1991

J. Encarnaçao

Wissenschaftliche Programmkomitees

Professor Dr. J. Encarnaçao, TH Darmstadt (Vorsitzender)
Professor Dr. W.Henhapl, TH Darmstadt
Dipl.-Inform. U. Claussen, AITEC GmbH & Co KG, Dortmund
Dr. W. Glatthaar, IBM Deutschland GmbH, Stuttgart
Professor G. Hommel, TU Berlin
Dr. Ch. Hornung, FhG-AGD, Darmstadt
Professor Dr. F. Hoßfeld, Forschungszentrum Jülich GmbH
Professor Dr. P. Jensch, Universität Oldenburg
Professor Dr. E. Jessen, TU München
Dr. P. Kohlhammer, Loewe-Opta GmbH, Kronach
Professor Dr. G. Müller, Universität Freiburg
Professor Dr. E. J. Neuhold, GMD-IPSI, Darmstadt
Professor Dr. R. Popescu-Zeletin, TU Berlin
Professor Dr. E. Raubold, GMD, Darmstadt
Professor Dr. W. Sammer, Siemens AG, München
Dr. M. J. Schachter-Radig, Neue Technologie u. Entwicklungs GmbH, München
Professor Dr. G. Schlageter, Fernuniversität Hagen
Professor Dr. K. Waldschmidt, Johann Wolfgang Goethe Universität
Dr.-Ing. K.-W. Westphal, Telenorma GmbH, Frankfurt
Professor Dr. W. Zschunke, Technische Hochschule Darmstadt

Professor Dr. H.-J. Hoffmann, TH Darmstadt
Professor Dr. R. Hofmann, TH Darmstadt
Professor Dr. R. Reischuk, TH Darmstadt
Professor Dr. P. Kammerer, TH Darmstadt
Professor Dr. H. Tzschach, TH Darmstadt
Professor Dr. K.- W. Wente, FH Darmstadt

Organisationskomitee

Komitee: **Koordination**

Professor Dr. J. Encarnaçao, TH Darmstadt (Sprecher)
Professor Dr. H. Waldschmidt, TH Darmstadt
Professor Dr. W. Henhapl, TH Darmstadt
Professor Dr. P. Kammerer, TH Darmstadt
Professor Dr. H.-J. Hoffmann, TH Darmstadt
Professor Dr. R. Piloty, TH Darmstadt

Komitee: **Finanzen**

Professor Dr. H. Waldschmidt, TH Darmstadt (Sprecher)
K. Schmidt, ZGDV, Darmstadt
Professor Dr. G. Lustig, TH Darmstadt
Professor Dr. H. Walter, TH Darmstadt
Professor Dr. W. Bibel, TH Darmstadt

Komitee: **lokale Organisation**

Professor Dr. W. Henhapl, TH Darmstadt (Sprecher)
Dr. R. Lindner, TH Darmstadt
M. Derkinderen, TH Darmstadt

Komitee: **Industrieprogramm**

Professor Dr. P. Kammerer, TH Darmstadt (Sprecher)
Professor Dr. G. Lustig, TH Darmstadt
Professor Dr. R. Hoffmann, TH Darmstadt
Professor Dr. H. Waldschmidt, TH Darmstadt
H. Kuhlmann, ZGDV Darmstadt
Professor Dr. E. Raubold, GMD Darmstadt
Dr. U. Kling, GMD Darmstadt
U. Bleimann, FH Darmstadt

Komitee: **Tutorien**

Prof. Dr. H.-J. Hoffmann, TH Darmstadt (Sprecher)
Dr. G. Snelting, TH Darmstadt
Dr. J. Schönhut, FhG-AGD Darmstadt
Dipl.-Ing. Kracker, GMD Darmstadt
Prof. Dr. H. Schneider, FH Darmstadt
Prof. Dr. G. Weber, FH Darmstadt

Inhaltsverzeichnis

Eingeladene Vorträge

Fachgespräche

Eingeladene Vorträge

USER INTERFACE SOFTWARE TOOLS

James D. Foley
Graphics, Visualization and Usability Center
College of Computing, Georgia Institute of Technology
Atlanta, GA 30332-0280

Developing high-quality user interfaces is becoming the critical step in bringing many different computer applications to end users. Ease of learning and speed of use typically must be combined in an attractively-designed interface which appeals to application-oriented (not computer-oriented) end users. This is a complex undertaking, requiring skills of computer scientists, application specialists, graphic designers, human factors experts, and psychologists.

User interface software is the foundation upon which the interface is built. The quality of the building blocks provided by the software establishes the framework within which an interface designer works. The tools should allow the designer to quickly experiment with different design approaches, and should be accessible to the non-programmer designer.

In this paper we discuss important directions in software tools for building user interfaces:

- Unified representation serving multiple purposes;
- Integration with software engineering tools:
- Interactive programming and by-example creation of interfaces and interface components.

Most of our focus is on the first two areas.

1. Background on User Interface Software Tools

Figure 1 shows the various levels of user-interface software, and suggests the roles for each. The application program has access to all software levels; programmers can exploit the services provided by each level, albeit with care, because calls made to one level may affect the behavior of another level. In this paper we discuss just the interaction technique toolkit and user interface management system layers. See [FOLE90] for discussions of the window manager and graphics layers.

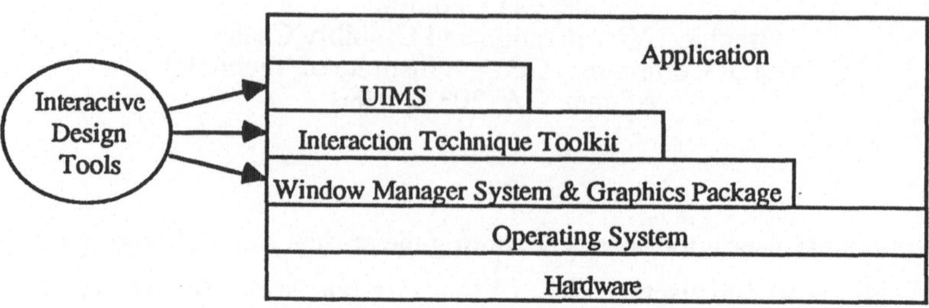

Fig. 1 User interface software. The application program has access to the operating system, window manager system and graphics package, toolkit, and UIMS. The interactive design tools allow non-programmers to design windows, menus, dialogue boxes, and dialogue sequences.

1.1. INTERACTION–TECHNIQUE TOOLKITS

Interaction techniques are the means by which users interactively input information to a computer system. A typical set of interaction techniques includes a dialogue box, file–selection box, alert box, help box, list box, message box, radio–button bank, radio button, choice–button bank, choice button, toggle–button bank, toggle button, fixed menu, pop-up menu, text input, and scroll bar. Interaction–technique toolkits are subroutine libraries of interaction techniques which are made available for use by application programmers. Widely used toolkits include the Andrew window–management system's toolkit [PALA88], the Macintosh toolkit [APPL85], OSF/Motif [OPEN89] and InterViews [LINT89] for use with X Windows [SCHE86], several toolkits that implement OPEN LOOK [SUN89] on both X Windows and NeWS, and Presentation Manager [MICR89]. In the X Window system, interaction techniques are called *widgets*, and we will often use this term.

The look and feel of a user–computer interface is determined largely by the collection of interaction techniques provided for it. Designing and implementing a good set of interaction techniques is time consuming, requiring experimentation with users to ensure ease of learning and speed of use. This toolkit approach, which helps to ensure a consistent look and feel among application programs, is clearly a sound and well–accepted software engineering practice. Considerable programmer productivity is gained by using an existing toolkit. The interaction technique toolkit is used in the user interface of both applications programs and the window manager itself.

Notice that the previous list of widgets includes both high- and low-level items, some of which are composites of others. For example, a dialogue box might contain several radio–button banks, toggle–button banks, and text input areas. Hence toolkits include a means of composing widgets together, typically via subroutine calls.

Creating composites by programming is tedious. Interactive editors allow composites to be created and modified quickly, facilitating easy changes to user interface details based on user feedback to the design team. Commercially available dialogue box editors include Developer's Guide from Sun Microsystems [SUN90] and UIMX from Visual Edge [VISU90].

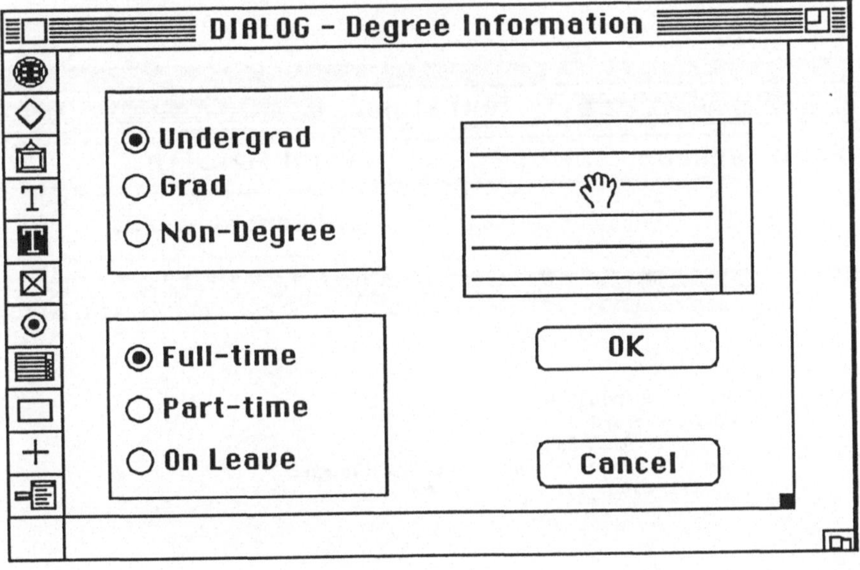

Fig. 2 The Now Software (formerly SmethersBarnes) Prototyper dialogue-box editor for the Macintosh. A scrolling-list box is being dragged into position. The menu to the left shows the widgets that can be created; from top to bottom, they are buttons, icons, pictures, static text, text input, check boxes, radio buttons, scrolling lists, rectangles (for visual grouping, as with the radio-button banks), lines (for visual separation), pop-up menus, and scroll bars. (Courtesy Now Software.).

The output of these editors is a representation of the composite, either as data structures that can be translated into code, as code, or as compiled code. In any case, mechanisms are provided for linking the composite into the application program. Programming skills are not needed to use the editors, so the editors are available to user–interface designers and even to sophisticated end users. These editors are typical of the interactive design tools. Figure 2 shows an example of such an editor.

Another approach to creating menus and dialogue boxes is to use a higher-level programming–language description. In MICKY [OLSE89], an extended Pascal for the Macintosh, a dialogue box is defined by a record declaration. The data type of each record item is used to determine the type of widget used in the dialogue box: enumerated types become radio–button banks, character strings become text inputs, Booleans become check boxes, and so on. Figure 3 shows a dialogue box and the code that creates it. An interactive dialogue–box editor can be used to change the placement of widgets.

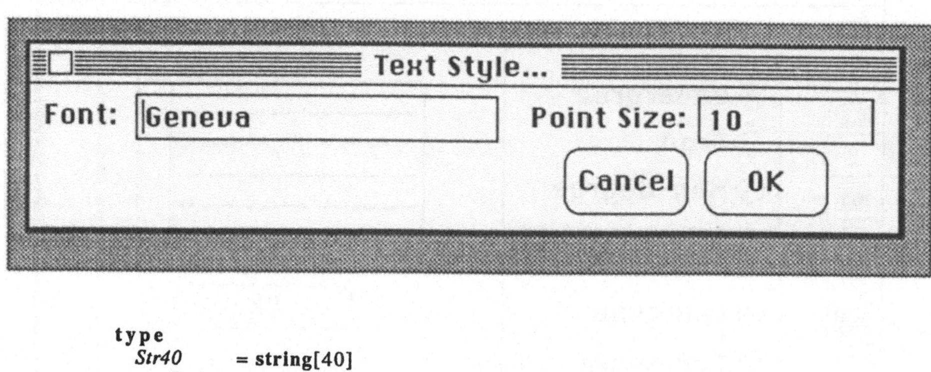


```
type
  Str40     = string[40]
  textStyle = record
                font : Str40;
                points (*Name = 'Point Size'): integer
              end
```

Fig. 3 A dialogue box created automatically by MICKY from the extended Pascal record declaration. (Courtesy Dan Olsen, Jr., Brigham Young University.)

Peridot [MYER86; MYER88] takes a radically different approach to toolkits. The interface designer creates widgets and composite widgets interactively, by example. Rather than starting with a base set of widgets, the designer works with an interactive editor to create a certain look and feel. Examples of the desired widgets are drawn, and Peridot infers relationships that allow instances of the widget to adapt to a specific situation. For instance, a menu widget infers that its size is to be proportional to the number of items in the menu choice set. To specify the behavior of a widget, such as the type of feedback to be given in response to a user action on a menu item, the designer selects the type of feedback from a Peridot menu, and Peridot generalizes the example to all menu items.

1.2. User Interface Management Systems

A user–interface management system (UIMS) is built on top of an interaction technique toolkit, and provides additional functionality in implementing a user interface. All UIMSs provide some means of defining admissible sequences of user actions and may in addition support overall screen design, help and error messages, macro definition, undo, and user profiles. Some recent UIMSs also manage the data associated with the application.

UIMSs, like toolkits, can increase programmer productivity, speed up the development process, and facilitate iterative refinement of a user interface as experience is gained in its use. The more powerful the UIMS, the less the need for the application program to interact directly with the operating system, window system, and interaction-technique toolkit.

In some UIMSs, user–interface elements are specified in a programming language that has specialized operators and data types. In others, the specification is done via interactive graphical editors, thus making the UIMS accessible to non-programmer interface designers. The former approach tends to be more powerful; the latter, more accessible.

Applications developed on top of a UIMS are typically written as a set of subroutines or, in contemporary object-oriented environments, as *methods*. The UIMS is responsible for calling appropriate methods in response to user inputs. In turn, the methods influence the dialogue—for instance, by modifying what the user can do next on the basis of the outcome of a computation. Thus, the UIMS and the application share control of the dialogue—this is called the *shared-control* model. The key concept here is that of a user–interface *state* and associated user actions that can be performed from that state. The state can be affected by user actions and by methods.

If a context-sensitive user interface is to be created, the system responses to user actions must depend on the current state of the interface. System responses to user actions can include invocation of one or several methods, state changes, and enabling, disabling, or modifying interaction techniques or menu items in preparation for the next user action. Help can also be made dependent on the current state.

Sequencing and state dependencies can be specified using a variety of linguistic approaches, such as state diagrams, augmented transition networks, and event languages. Green [GREE87] surveys event languages and all the other sequence-specification methods we have mentioned, and shows that general event languages are more powerful than are transition networks, recursive transition networks, and grammars. He also provides algorithms for converting these forms into an event language. ATNs that have general computations associated with their arcs are also equivalent to event languages.

A quite different way to define sequencing is by example. Here, the user interface designer places the UIMS into a "learning" mode, and then steps through all acceptable sequences of actions (a tedious process in complex applications, unless the UIMS can infer general rules from the examples). The designer might start with a main menu, select an item from it, and then go through a directory to locate the submenu, dialogue box, or application-specific object to be presented to the user in response to the main menu selection. The object appears on the screen, and the designer can indicate the position, size, or other attributes that the object should have when the application is actually executed. The designer goes on to perform some operation on the displayed object and again shows what object should appear next, or how the displayed object is to respond to the operation; the designer repeats this process until all actions on all objects have been defined. This technique works for sequencing through items that have already been defined by the interface designer, but is not sufficiently general to handle arbitrary application functionality. User-interface software tools with some degree of by-example sequencing specification include Menulay [BUXT83], TAE Plus [MILL88] and the SmethersBarnes Prototyper [COSS89]. Peridot, mentioned earlier, builds interaction techniques (i.e. hardware bindings) by example. Another way to define user interfaces consisting of interconnected processing modules is with data-flow diagrams. For instance, the NeXT Interface Builder [NEXT90] allows objects to be interconnected so that output messages from one object are input to another object. Type checking is used to ensure that only compatible messages are sent and received.

Further background on UIMSs can be found in [HART89; MYER89; OLSE87].

2. The User Interface Management System – Data Base Management System Analogy

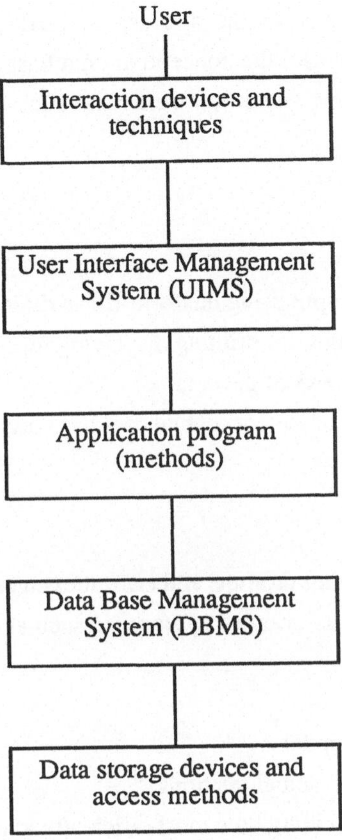

Figure 4. Symmetry between User Interface Management Systems (UIMS) and Data Base Management Systems (DBMS).

There is a useful analogy and symmetry between UIMS concepts and those found in data base management systems (DBMS) as suggested in Figure 4. Just as a DBMS manages data, a UIMS manages the interface. A DBMS hides the user from details of storage organizations, just as a UIMS hides the programmer from details of interaction devices and techniques. A DBMS has one or more specialized languages with which the programmer defines views, report formats, and queries. So too a UIMS may have specialized languages for specifying sequencing, state dependencies and changes, screen organizations, and menu designs.

Driving this analogy further, a key to the success of DBMSs is the use of a unifying representation. In the relational world, the conceptual data base design serves this role. This representation is used at design time and at run time in a variety of important ways, which we list here without elaboration.

Design time uses:

- Form creation for data input using the form-fill-in dialogue style (Form Builder)
- Report template generation for printing/displaying reports (Report Formatter)
- Definition of multiple views of the data
- Generation of diagrams of the conceptual data base organization

Run time uses:

- Helping user by displaying prompts and error messages as the user specifies a query or an update, and in forms-oriented approaches such as Query-by-Example (QBE)
- Integrity checking
- Alerting/triggering
- Processing queries and updates
- Mapping from views into actual relations
- Query optimization -- deciding how most efficiently to access information
- Generating reports

3. Unified Representation for User Interfaces

In UIMS research, the values of a unified representation are only now coming to be recognized. Some of the types of information found in such a representation are typified by UIDE, the User Interface Design Environment [FOLE88, FOLE89, FOLE91A]. The representation provides:

- The class hierarchy of objects which exist in the application
- Properties of the objects
- Actions which can be performed on the objects

- Units of information (parameters) required by the actions
- Pre- and postconditions for the actions.

UIDE is implemented in an expert system shell, using seven types of schemata: object schema, action schema, parameter schema, pre-condition schema, post-condition schema, attribute schema, and attribute type schema. (In the following description, slots, or information contained with a schema, are given in italics).

Instances of the Object Schema represent the objects defined as part of the user interface design. Within this schema, a *Description* slot is a textual description of the object (created by the user interface designer) which can be provided to the user at run time in response to a help request. *Actions on object* is a relation linking an object schema instance to the action schema instances for those actions which can be applied to the object class. The object schemata makes the UIMS aware of the data model assumed by the application.

There is an instance of the Action Schema for each action defined in the user interface. Actions can affect an attribute, an object, or an object class action. *Description* is available for run-time help. The *Action routine name* provides the link to the run-time procedure which actually carries out the action. (In object-oriented programming, this would be the method name.) The *Actions mutually-exclusive with* slot refers to all actions which cannot be available at the same time as this action. This slot can be used to organize menus: mutually exclusive commands, such as "turn x on" and "turn x off", can be assigned the same menu slot. *Inverse action* is the name of the action which is the inverse of this action, if such an action exists. This assists in implementing an undo command. *Parameters*, *Pre-conditions*, and *Post-conditions*, refer to schema instances which further describe the action. For each action there is a Parameter Schema instance for each parameter, or unit of information, required by the action.

Pre-conditions are predicates which must be true in order for an action to be enabled and thus available to the user for selection. Arbitrary predicates can be used, along with pre-defined predicates having to do with the number of instances of different types of objects. Post-conditions, implemented as add-lists and drop-lists, change the values of predicates. Pre- and post-conditions specify just enough of the application's semantics to encode the state of the interface, to allow context-sensitivity in presenting menus, to give context-sensitive help, and for the very limited understanding of the semantics which is necessary for transformations of interface styles.

A Pre-condition Schema is instantiated for each action. A *Description* explains what the pre-condition means. The actual *Expression* of the pre-condition is represented in a form which can be conveniently evaluated at run-time, to determine whether the pre-condition is true or false. A Post-condition Schema contains the *Add-list* and *Drop-list* of predicates.

The types of information encoded by predicates and used in the pre- and post-conditions include:

- The number of objects of a particular object class which are in existence. A creation command increments this number as a post-condition; a deletion command decrements the number.
- The number of objects of different object classes which have been selected as belonging to the Currently Selected Set.
- The existence of a currently selected command.
- The existence of a value for a parameter.
- Any predicates established by the interface designer as being important to defining the context of the interface.

Each attribute of an object is described by an instance of the Attribute Schema, which includes an optional *Default value* for the attribute. The Attribute Type Schema records the attribute *Data type* as being integer, real, enumerated, etc., and gives other information which is specific to the data type.

This unified representation is evolving with time; in its present form, it is not sufficiently robust to serve all of the purposes we have in mind, and it continues to evolve [FOLE91b]. However, it is already quite useful. The following list indicates things that we would like to be able to do with the representation; items indicated with an * have already been developed:

Design time uses:

* Test properties of the interface, such as functional completeness, consistency and reachability [FOLE89, BRAU90]
* Automatically organize menus and dialogue boxes [KIM90, DEBA91]
* Support application of correctness-preserving transformations to the interface specification, allowing the designer to quickly explore a space of design alternatives [FOLE87]

* Automatically create an interface to the application, using menus, dialogue boxes, and direct manipulation [FOLE91a]
* Evaluate the interface design with respect to speed of use, using a key-stroke model type of analysis [CARD80, SENA89]
- Optimize the interface design for speed of use
- Predict learning times and error rates using cognitive models [KIER85] integrated with the representation
* Generate a parser which accepts natural-language commands to the application [KOVA90]

Run time uses:

* Explain why a command is disabled [FOLE89]
* Explain (partially) what a command does [FOLE89]
- Explain what information is needed before a command can be performed
* Provide procedural help, via animation, taking into account the current application context [SUKA88, SUKA90]. Also show, via animation, what sequence of commands must be performed before a currently-disabled command can be performed.
- Generate a guided tour of the application for new users
- Adapt to individual users based on their demonstrated knowledge of and usage of the application
- Provide an undo capability
* Control actual execution of the application, including enabling and disabling of menus items, as well as display of menus, dialogue boxes, and windows [FOLE91a, GIES91]

Because the unified representation includes aspects of data base schemata as well as those of traditional user interface representations, the situation of Figure 5 is more evocative of the state of affairs which becomes possible with this approach.

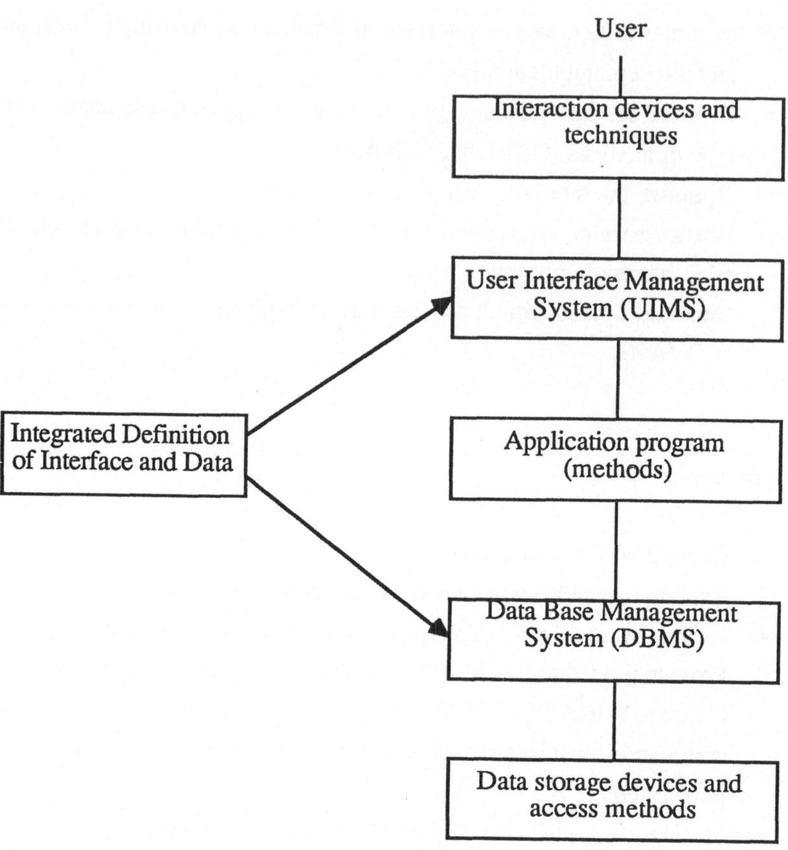

Figure 5. Unified definition driving/controlling UIMS and DBMS.

Because the unified representation includes aspects of data base schemata as well as those of traditional user interface representations, the situation of Figure 5 is more evocative of the state of affairs which becomes possible with this approach.

4. Integration with Software Engineering / Data Engineering Tools

The software and data engineering community has already developed a variety of tools for describing information about applications in general, without any real focus on the user interface. In the meantime, the user interface software community has been developing tools, as discussed above, which include some of the very same information, plus more specialized user interface information. This means that some information has to be specified twice, once by the software engineer and once by the user interface designer. These tools should clearly be merged to speed up the design process, avoid duplication of efforts, and avoid potential inconsistencies between the dual specifications.

As a start in this direction, we have coupled the data modeler component, called D2M2 Edit of the Delft Direct Manipulation Manager (D2M2) [BEEK90] with Sun's DevGuide [SUN90]. Once an object-oriented data model has been created using D2M2's interactive design tool, the user interface designer selects attributes or actions (i.e. methods) of one or more objects, and drags them into DevGuide. We have augmented DevGuide with a rule base developed from the Open Look Style Guide [SUN89] utilizing application semantic information in the specifications, so that a control panel, complete with menus and possibly other widgets, is automatically created [DEBA91]. The designer can then use DevGuide's interactive design capabilities to fine-tune the design, changing either the specific selection of controls or the positioning of controls in the window.

5. Conclusions

We have presented the concept of a unified user interface representation which can be used for a variety of run time and design time purposes, and have drawn analogies to data base management systems. A specific unified model has been overviewed, and current and projected uses have been described. Much work is needed to realize the full potential of this concept; in the meantime, some of the promises of the approach have already been realized.

Acknowledgements: The Graphics and User Interface Research Group at George Washington University and later the Graphics, Visualization, and Usability Center at Georgia Tech provided the intellectual environment supporting this work. Direct contributors to UIDE include D. DeBaar, M. Frank, C. Gibbs, D. Gieskens, M. Gray, W. Kim, S. Kovacevic, L. Moran, K. Murray, H. Senay, J. Sibert and P. Sukaviriya. Financial support has been provided by the National Science Foundation, Sun Microsystems, Inference Corporation, Software Productivity Consortium, Siemens Corporation, the GWU Dept. of EE & CS Industrial Liaison Program, and Georgia Institute of Technology. Thanks to Martin Frank, Mark Gray, Srdjan Kovacevic, Joan Morton, and Piyawadee Sukaviriya for technical assistance in preparing this paper.

REFERENCES

APPL85 Apple Computer, *Inside Macintosh*, Addison-Wesley, Reading, MA, 1985.

BEEK90 Beekman, W., *D2m2edit,* Master's Thesis, Delft University of Technology, July 1990.

BRAU90 Braudes, R., *A Framework for Conceptual Consistency Verification,* DSc dissertation, Dept. of Electrical Engineering and Computer Science, The George Washington University, 1990.

BUXT83 Buxton, W. et al., Towards a Comprehensive User Interface Management System, Proceedings 1982 SIGGRAPH Conference, published as *Computer Graphics*, 17(3), July 1982, pp. 35–42.

CARD80 Card, S., T. Moran, and A. Newell, The Keystroke-Level Model for User Performance Time with Interactive Systems, *Communications of the ACM*, 23(7), July 1980, pp. 398–410.

COSS89 Cossey, G., *Prototyper*, Now Software, Portland, Oregon, 1989.

DEBA91 DeBaar, D. and J. Foley, Coupling Application Design and User Interface Design, *Report GIT-GVU-91-10,* Graphics, Visualization and Usability Center, Georgia Institute of Technology, Atlanta GA 30332-0280, 1991.

FOLE87 Foley, J., W. Kim, and C. Gibbs, Algorithms to Transform the Formal Specification of a User-Computer Interface, *Proceedings INTERACT '87, 2nd IFIP Conference on Human-Computer Interaction*, Elsivier Science Publishers, Amsterdam, pp. 1001–1006.

FOLE88 Foley, J., C. Gibbs, W. Kim, and S. Kovacevic, A Knowledge Base for a User Interface Management System, *Proceedings CHI '88 - 1988 SIGCHI Computer-Human Interaction Conference*, ACM, New York, 1988, pp. 67–72.

FOLE89 Foley, J., W. Kim, S. Kovacevic, and K. Murray, Designing Interfaces at a High Level of Abstraction, *IEEE Software*, 6(1), January 1989, pp. 25–32.

FOLE90 Foley, J., A. van Dam, S. Feiner, and J. Hughes, *Computer Graphics – Principles and Practice*, Addison-Wesley, Reading, MA, 1990.

FOLE91a Foley, J., W. Kim, S. Kovacevic and K. Murray, UIDE - An Intelligent User Interface Design Environment, in J. Sullivan and S. Tyler (eds.) *Intelligent User Interfaces,* Addison-Wesley, Reading MA, 1991.

FOLE91b Foley, J., D. Gieskens, W. Kim, S. Kovacevic, L. Moran and P. Sukaviriya, A Second-Generation Knowledge Base for the User Interface Design Environment, *GWI-IIST-91-13,* Dept. of Electrical Engineering and Computer Science, The George Washington University, 1991.

GIES91 Gieskens, D. and J. Foley, Controlling Interface Objects Through Pre- and Postconditions, *GIT-GVU-91-09,* Graphics, Visualization and Usability Center, Georgia Institute of Technology, Atlanta GA 30332-0280, 1991.

GREE87 Green, M. A Survey of Three Dialog Models, *ACM Transactions on Graphics*, 5(3), July 1987, pp. 244–275.

HART89 Hartson, R. and D. Hix, Human–Computer Interface Development: Concepts and Software, *ACM Computing Surveys*, 21(1), March 1989, pp. 5–92.

KIER85 Kieras, D. and P. Polson, An Approach to the Formal Analysis of User Complexity, *International Journal of Man-Machine Studies*, 22, 1985, pp. 365–394.

KIM90 Kim, W. and J. Foley, DON: User Interface Presentation Design Assistant, *Proceedings of the SIGGRAPH Symposium on User Interface Software and Technology*, Snowbird, Utah, October 1990.

KOVA90 Kovacevic, S., A Compositional Model of Human-Computer Dialogues, *GWU-IIST-90-36,* Department of Electrical Engineering and Computer Science, The George Washington University, Washington DC 20052, 1990.

LINT89 Linton, M., J. Vlissides, and P. Calder, Composing User Interfaces with InterViews, *IEEE Computer*, 22(2), February 1989, pp. 8–22.

MICR89 Microsoft Corporation, *Presentation Manager*, Microsoft Corporation, Bellevue, WA, 1989.

MILL88 Miller, P., and M. Szczur, Transportable Application Environment (TAE) Plus Experiences in 'Object'ively Modernizing a User Interface Environment, *Proceedings OOPSLA'88*, pp. 58–70.

MYER86 Myers, B., Creating Highly-Interactive and Graphical User Interfaces by Demonstration, Proceedings 1986 SIGGRAPH Conference, published as *Computer Graphics*, 20(4), August 1986, pp. 249–257.

MYER88 Myers, B., *Creating User Interfaces by Demonstration*, Academic Press, New York, 1988.

MYER89 Myers, B., User-Interface Tools: Introduction and Survey, *IEEE Software*, 6(1), January 1989, pp. 15–23.

NEXT90 NeXT, *Interface Builder*, NeXT Inc, Sunnyvale, CA 1990.

OLSE87 Olsen, D., ed., ACM SIGGRAPH Workshop on Software Tools for User Interface Management, *Computer Graphics*, 21(2), April 1987, pp. 71–147.

OLSE89 Olsen, D., A Programming Language Basis for User Interface Management, Proceedings of CHI, 1989, ACM, New York, 1989, pp. 171-176.

OPEN89 Open Software Foundation OSF/MOTIF manual, Cambridge, MA, 1989.

PALA88 Palay, A. et al., The Andrew Toolkit: An Overview, *Proceedings 1988 Winter USENIX*, Feb. 1988, pp. 9–21.

SCHE86 Scheifler, B., and J. Gettys, The X Window System, *Transactions on Graphics*, 5(2), April 1986, pp 79–109.

SENA89 Senay, H., P. Sukaviriya, L. Moran, Planning for Automatic Help Generation, *Proceedings of Working Conference on Engineering for Human Computer Interactions*, Napa Valley, California, August 1989.

SUKA88 Sukaviriya, P., Dynamic Construction of Animated Help from Application Context, *Proceedings of ACM SIGGRAPH 1988 Symposium on User Interface Software and Technology (UIST '88)*, 1988, ACM, New York, NY.

SUKA90 Sukaviriya, P and J. Foley, Coupling a UI Framework with Automatic Generation of Context-Sensitive Animated Help, *Proceedings of ACM SIGGRAPH 1990 Symposium on User Interface Software and Technology (UIST '90)*, Snowbird, Utah, October, 1990, in press.

SUN89 Sun Microsystems, Inc., *Open Look Graphical User Interface Application Style Guidelines*, Mountain View Ca, December 1989.

SUN90 Sun Microsystems, Inc., *Open Windows Developer's Guide 1.1, Reference Manual*, Part No l. 800-5380-10, Revision A of June 1990.

VISU90 Visual Edge, *UIMX User's Manual*, Ville St. Laurent, Quebec, Canada, 1990.

Visualization Activities in the UK

F.R.A. Hopgood

Informatics Department, Rutherford Appleton Laboratory, UK

The Advisory Group On Computer Graphics (AGOCG) for the UK academic community organized a workshop on Scientific Visualization, held at Cosener's House, Abingdon, UK, in February 1991. The workshop attracted 35 attendees from the UK, both from academia and industry. Todd Elvins from the San Diego Supercomputer Centre was invited to the workshop, to give a USA perspective. The workshop followed the now traditional pattern of presentations followed by working group discussions. The discussions ranged from trying to understand what "visualization" really is, to discussing the practical requirements for visualization software.

Two publications arising from the workshop are in preparation, Brodlie et al.[1] and Earnshaw and Wiseman.[2] The first volume is a digest of the whole workshop, the second is an introductory guide to the current approaches and techniques in scientific visualization. Both contain many illustrations of results achieved with existing systems.

One of the major remits of the workshop was to define a strategy for software support for visualization in the UK. Opinion was divided at the workshop as to the form this should take. Some felt that the field is now sufficiently mature that AGOCG should recommend one of the existing visualization systems *now*. Others felt that such a course would be premature and that there was a need to evaluate and assess existing systems before making a recommendation. In the event the latter view prevailed and the workshop recommendations to AGOCG included a programme of evaluation and assessment of existing systems.

AGOCG subsequently endorsed this recommendation and an evaluation programme is now underway, involving a number of UK universities and research institutes (including the author's Department). The systems being considered are AVS, apE and KHOROS. These systems differ from other so-called visualization systems in that they are highly flexible and configurable and allow users to add their own modules, they can be thought of as 'application builders' rather than application programs *per se*. They are applicable across a broad spectrum of application domains, other systems are, in general, domain specific.

The evaluation is taking the form of 'pencil and paper' studies together with case studies on specific problems. The case studies will include the fields of computational fluid dynamics (CFD), analysis of satellite image data, and the solution of partial and ordinary differential equations.

In August 1990, a survey of superworkstation hardware, commissioned by AGOCG, was completed, and one of the issues being explored in the visualization systems assessment is the extent to which superworkstation hardware is a prerequisite for visualization software, and the extent to which useful work can be accomplished using more modest, lower cost, hardware.

Interim results of the evaluation studies will be available by October 1991, and the talk will present and discuss the current situation then.

References

1. K.W. Brodlie et al., *Scientific Visualization Techniques and Applications,* Springer-Verlag, in press.
2. R.A. Earnshaw and N. Wiseman, *Introductory Guide to Scientific Visualization,* Springer-Verlag, in press.

Neue Entwicklungen der Gerätetechnik für Graphiksysteme

Wolfgang Straßer
Wilhelm-Schickard-Institut für Informatik
GRIS
Universität Tübingen
Auf der Morgenstelle 10 C9
7400 Tübingen

Die Leistungsfähigkeit heutiger Graphiksysteme reicht aus, um Szenen beschränkter Komplexität wie sie z.B. im Maschinenbau oder bei Flugsimulationen auftreten, in Echtzeit darzustellen. Die von der Graphikhardware unterstützten lokalen Beleuchtungsverfahren vermitteln bereits ein sehr realitätsnahes Bild. Die Zahl der auf Farbmonitoren darstellbaren Punkte und Farben erlaubt, hohe Ansprüche an Bildqualität zu befriedigen. Die hierdurch gegebene Kompabilität zum geplanten hochauflösenden Fernsehen muss für den Einsatz in multimedialen Systemen gerätetechnisch durch Prozessoren zur Videosignalverarbeitung für effiziente Datenübertragung, Speicherung und Druckerausgabe ergänzt werden. Die Mischung von Video und Computergraphik steigert die Ansprüche an Visualisierungstechniken und Bildqualität, die nur durch den Übergang auf globale Beleuchtungsverfahren erfüllt werden können, was gleichzeitig neue Anwendungsgebiete erschliesst. Die Echtzeitfähigkeit hierfür verlangt völlig neue Architekturen für Graphiksysteme. Die technologischen Möglichkeiten sind vorhanden. Sie werden im Vortrag kritisch durchleuchtet und analysiert; Mega-Trends werden aufgezeigt und erläutert.

Heutige Graphiksysteme haben durch diese technologischen Möglichkeiten aber auch neue Anwendungen wie "Visualisierung für wissenschaftliches Rechnen", "Virtuelle Realität", "Multimedia" u.a. erschlossen und in einem vernünftigen Preis-Leistungsverhältnis möglich gemacht. Von diesen Anwendungen kommen weitergehende Anforderungen, die sich in vier Klassen einteilen lassen:

- Photorealismus oder Ausdruckstärke der erzeugten Bilder

- Echtzeit (25 Bilder/sec) für die Bilderzeugung

- Kompatibilität zu anderen Systemen und Diensten

- Neue Ein-/Ausgabegeräte

Die hiermit gestellten technischen Probleme und ihre Lösungsmöglichkeiten werden im Vortrag dargestellt.

Betriebserfahrungen und Messungen an einem großen FDDI-Netz ..... und sich daraus ergebende Konsequenzen

Otto Spaniol
Lehrstuhl für Informatik IV
RWTH Aachen

Kurzfassung:

Lokale Netze sind im letzten Jahrzehnt nach einer recht zähen Einführungsphase zum Allgemeingut geworden. Nachdem zunächst vorwiegend geographisch eng begrenzte Netze im 10 Megabit/sec.-Bereich (Ethernet, Token Ring,...) erfolgreich erprobt wurden, sind heute einigermaßen marktfähige Produkte der 100 Megabit-Klasse verfügbar. Netze mit einer nominell im Gigabitbereich liegenden Datenrate sind im Entwicklungsstadium.

Es hat sich bei der Einführung von Netzen gezeigt, daß - etwa bei Ethernet - zwischen den Phasen "Konzeption" (1973), "erste halbwegs marktfähige Produkte" (1981) und "jeder hat's" (1990) jeweils knapp 10 Jahre lagen. Ähnliches ist bzgl. der 100-Megabit/sec.-Netzklasse bzw. erst recht für die noch futuristischeren Netze zu erwarten. Um diese Zeiten vielleicht in Zukunft etwas zu verkürzen, ist es sehr wichtig, frühzeitig Betriebserfahrungen zu sammeln und konkrete Aussagen zum wirklichen Leistungsverhalten von Hochgeschwindigkeitsnetzen wie FDDI zu machen - nicht zuletzt auch zur Behebung von Schwachstellen, welche die Akzeptanz neuer Netzformen behindern.

Entscheidende Frage für die Verwendbarkeit von Hochgeschwindigkeitsnetzen ist - neben der profanen Kostenproblematik: Was bleibt für den Anwender von der nominell sehr hohen Kapazität übrig? Mindestens drei Faktoren können dazu führen, daß die "am Ende herauskommende" Leistung im Vergleich zur nominellen Datenrate als sehr niedrig erscheinen kann:

- *unzureichende Architektur bzw. mangelnde Rechenkapazität der Netzcontroller*
- *überflüssige Protokollfeinheiten sowie aufwendige Protokollstacks*
- *ineffiziente Implementierungen bzw. für andere Netztypen konzipierte und von dort in (allzu) vereinfachender und daher unzulässiger Weise übernommene Protokolle wie etwa TCP/IP.*

Es werden Erfahrungen und Messungen an einer der weltweit größten FDDI-Installationen vorgestellt, welche zeigen, daß alle diese Effekte in der Praxis auftreten.

"Abfallprodukt" der Untersuchungen ist eine Meßmethodik, die für angewandte Mathematiker möglicherweise trivial, aber in der Informatik kaum gebräuchlich ist. Die Methodik ist universell einsetzbar und einfach zu verstehen ist. Sie benötigt nur einen vergleichsweise minimalen Rechenaufwand.

1. Präliminarien

An der RWTH Aachen wird eines der weltweit größten Hochgeschwindigkeitsnetze von Typ FDDI (Fiber Distributed Data Interface) betrieben. Die ringförmige Topologie von FDDI ist aus historischen und auch aus betrieblichen Gründen in zwei 'Halbringe' aufgeteilt, welche im Rechenzentrum über eine Ethernetstrecke miteinander verbunden sind (siehe Abb. 1). Nach wie vor ist der Datenverkehr innerhalb der Hochschule weitgehend auf Verbindungen von und zum Rechenzentrum konzentriert - eine Folge der bisherigen sternförmigen Anschaltungen -, und dies wird sich so rasch nicht ändern. Daß die Ethernetverbindung beider Ringe zum Engpaß werden könnte, hat sich bisher nicht feststellen lassen. In diesem Fall wäre eine Umstellung der Topologie - d.h. eine Zusammenschaltung beider Halbringe zu einem Gesamtring - leicht und schnell durchführbar.

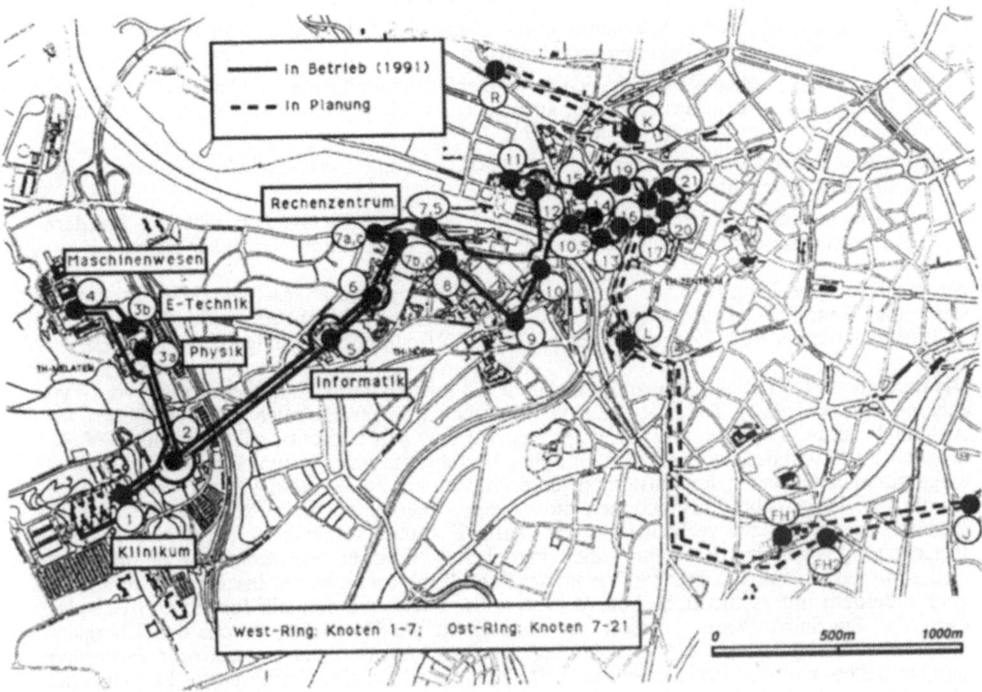

Abb. 1: Die Topologie des FDDI-Rings an der RWTH Aachen

Die Halbring-Strategie hat Auswirkungen auf die Antwortzeiten, welche den Verkehr zwischen beiden Ringen betreffen; dies wird aus den im weiteren Verlauf beschriebenen Messungen evident. Während man dies als Nachteil ansehen muß, spricht umgekehrt die betriebliche Seite zugunsten einer Zerlegung in Teilnetze, denn kleinere Einheiten sind einfacher zu überwachen als größere - und außerdem wird das Risiko eines Totalausfalls stark vermindert.

Das Netz besteht (zur Zeit) aus 23 postalischen Endverteilern. Die Gesamtausdehung des aus Kostengründen fast überall 'flachgedrückt' verlegten Rings (d.h. Hin- und Rückrichtung in derselben Trasse und sogar in einem gemeinsamen Kabel; siehe Abb. 1) beträgt etwa 16 Kilometer, wenn man die hausinternen Strecken zwischen postalischem Endverteiler und FDDI-Knoten mitzählt.

Der Betrieb des Netzes erfolgt als Backbone für lokale Subnetze sowie für die beiden neuen Großrechner des Rechenzentrums (IBM 3090 - 600E und Siemens Fujitsu S - 400/10). Die Chance zur Realisierung eines derart großen Hochgeschwindigkeitsnetzes ergab sich übrigens nur aufgrund dieser Großrechnerbeschaffungen. Obwohl frühzeitige Installationen immer sehr teuer und daher häufig umstritten sind (zum Vergleich kann etwa darauf verwiesen werden, daß der Preis für einen Ethernet-Transceiver zwischen 1980 und 1990 von ca. DM 8.000,-- auf weniger als DM 400,-- gefallen ist), betrugen die Gesamtkosten für das Netz nur einen Bruchteil der Aufwendungen für die Großrechner.

Für die Entscheidung, keine Workstations bzw. sonstige Rechner direkt an den FDDI-Ring anzuschließen, sind mehrere Faktoren verantwortlich:

- die (noch allzu) hohen Preise für FDDI-Adapter, welche einen zu großen relativen Anteil der Netzkosten an den Gesamtkosten ausmachen würden ("der Fernsehgerätestecker darf nicht halb so viel kosten wie das Gerät selbst"),
- das bisher verwendete Verfahren (Encapsulation Bridging), welches keine Stationen außer den FDDI-Brücken am Ring direkt zuläßt,
- die Tatsache, daß bei Inbetriebnahme der beiden Halbringe (im Oktober 1989 bzw. im März 1990) noch keine Produkte zur Direktanbindung von Workstations verfügbar waren.

An den Backbone-Ring sind zur Zeit mehr als 800 Endgeräte angeschlossen - mit ständig steigender Tendenz. Als Protokolle oberhalb der Netzzugangsebene werden vorwiegend TCP/IP (von mehr als 600 Endgeräten) bzw. DECNET (mehr als 100 Endgeräte) eingesetzt.

Die durch den FDDI-Ring miteinander verbundenen lokalen Netze sind überwiegend vom Typ Ethernet (ca. 60 Netze). Token-Ringe sind wesentlich weniger zahlreich (etwa 10 Netze), ihr Anschluß wurde erst Mitte 1991 in einer aus betrieblichen Gesichtspunkten stabilen Weise möglich. Aufgrund des Encapsulation-Bridging-Verfahrens, welches immer noch in einer allzu eingeschränkten Version im Einsatz ist (eine Umstellung auf Custom Filtering bzw. Source Routing ist längst überfällig) können diese beiden Netztypen nicht "miteinander reden". Dadurch entstehen bzw. verbleiben zwangsläufig Kommunikationsinseln, was eine gewisse Ernüchterung gegenüber dem hehren Ziel einer ebenso einfachen wie offenen Kommunikation "jeder mit jedem" mit sich brachte. Verantwortlich für die nach wie vor inselartigen Strukturen sind außerdem unterschiedliche höhere Protokolle. Diese Problematik führte zum Beispiel dazu, daß Spezialhardware und -software angeschafft werden mußte, welche den Übergang zwischen DECNET und SNA ermöglichte. Es lassen sich schöne theoretische Festreden darüber halten, wie wunderbar denn die Welt durch die allgemeine Einführung, Akzeptanz und Bezahlbarkeit der OSI-Protokolle werden wird. Die Realität sieht leider völlig anders aus.

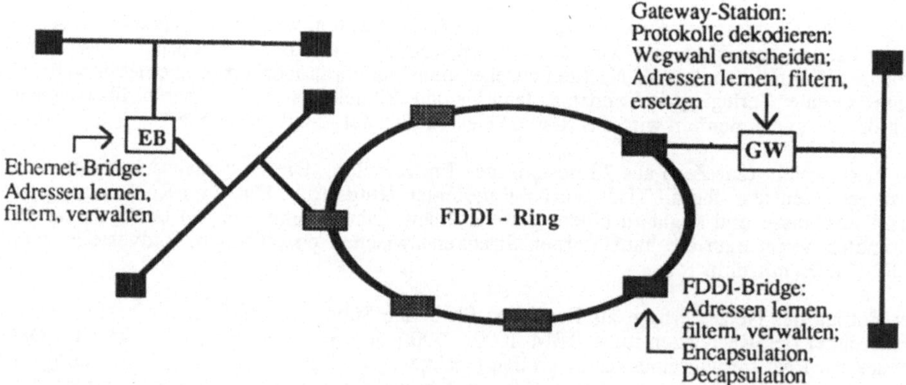

Abb. 2: Kommunikationsstruktur und Aufgabe der verschiedenen Komponenten

2. Betriebserfahrungen (vorwiegend nichttechnischer Art)

Wenn das Netz 'funktioniert' (und zum Glück ist die Zahl der Störungen nach Behebung einer Vielzahl kleinerer, aber lästiger 'Kinderkrankheiten' sehr stark zurückgegangen), dann läuft es stabil - und es gibt keinerlei auf das Netz zurückzuführende Lastprobleme. Problemsituationen sind sie auf sehr unterschiedliche Ursachen zurückzuführen. Es beginnt damit, daß Hausmeister oder Schlüssel nicht auffindbar sind, es geht weiter mit fehlerhaft konfigurierten Gateways (welche den Gesamtdurchsatz auf nahe Null reduzieren und dafür in sehr vereinfachender Weise den FDDI-Ring verantwortlich machen), und es endet mit scheinbar unbegreiflichen Effekten, welche zu gegenseitigen Schuldzuweisungen zwischen Hersteller, Betreiber und Anwender führen.

Es kommt uns so vor, als ob der bei Betriebssystemen wohlbekannte Thrashing-Effekt (dramatisches Absinken der Leistung bei Überschreiten eines gewissen Schwellwerts für die Zahl gleichzeitig in Bearbeitung befindlicher Prozesse) sich auf Netzinstallation und -betrieb übertragen würde: Die Wahrscheinlichkeit für das Auftreten einer Komplikation scheint stärker als proportional mit der Anzahl der Netzknoten zu steigen.

Das für den Netzbetrieb verfügbare Personal kann sich nur um die dringendsten Dinge kümmern. Anwenderberatung kann deshalb nicht annähernd im wünschenswerten Umfang druchgeführt werden, dies hat zwangsläufig auch Konsequenzen für die Akzeptanz des Netzes.

Ein für Universitäten wohl typisches Problem ist Dezentralisierung der Verantwortlichkeiten: Die durch den FDDI-Ring miteinander verbundenen Netze werden zum überwiegenden Teil von Mitarbeitern der betroffenen Institute verwaltet. Der Kenntnisstand dieser Mitarbeiter ist naturgemäß sehr unterschiedlich - nicht zuletzt eine Folge der Tatsache, daß sich die meisten davon auf Zeitstellen befinden, weshalb sehr häufig Neubesetzungen erfolgen.

Das Engagement und die Begeisterungsfähigkeit dieser Mitarbeiter ist außerordentlich hoch, aber dieser Umstand hat - so begrüßenswert er auch ist - eine Reihe von negativen Auswirkungen für den Betrieb des Gesamtnetzes. Nicht selten hat ein solcher Mitarbeiter eine Reihe von Ideen, die nicht in allen Fällen ausgereift sind. Wenn er versucht, sie 'probeweise' in die Tat umzusetzen, kann er damit evtl. nicht nur den Betrieb des eigenen lokalen Netzes, sondern auch die Operation des Gesamtnetzes negativ beeinflussen. Es ist auch vorstellbar (um es vorsichtig auszudrücken), daß besonders preiswerte - weil ungeprüfte oder gar 'schwarze' - Software Verwendung findet. Selbst wenn dies nur eine irreale Befürchtung sein sollte, wird doch die pure Vorstellung dieser Möglichkeit eine Reihe von potentiellen Netzbenutzern abschrecken. Es gibt Institute der RWTH Aachen, die wegen solcher Gerüchte bisher den Anschluß an das FDDI-Netz nicht vollzogen haben ("wenn ich meine Straße zu anderen hin verbreitere, kommen die anderen auch einfacher zu mir - und darunter sind vielleicht einige, die ich lieber draußen lassen würde").

Es muß (bzw. sollte) gewährleistet sein, daß die Stationen des Backbone-Netzes (also die FDDI-Adapter) eine unkorrekte Arbeitsweise der dahinterliegenden Subnetze sehr schnelle erkennen und daß sie diese Netze daran hindern, ihr fehlerhaftes Arbeiten auf den Gesamtring zu übertragen. Die heute im Einsatz befindlichen FDDI-Komponenten erfüllen diese Forderungen noch nicht im wünschenswerten Maß; beispielsweise kann ein ungeprüftes Einwickeln und Transportieren von Nachrichten (so wie es beim 'encapsulation bridging' üblich ist) zu ebenso merkwürdigen wie lästigen Störungen führen. Fordert man höchste zeitliche Verfügbarkeit und Zuverlässigkeit oder setzt man das Netz gar für 'lebenswichtige' Echtzeitanwendungen ein, dann ist bei der Verwendung von heute auf dem Markt befindlichen FDDI-Komponenten doch Vorsicht oder Zurückhaltung zu empfehlen.

3. Messungen

Die entscheidende Frage für den Anwender ist: "Was bringt mir das Hochgeschwindigkeitsnetz über die Funktionalität hinaus, die mir bisher verfügbar war?". Erwartet wird in erster Linie einfacherer Zugang und beschleunigte Abwicklung des Transports von großen Datenmengen.

Der 'naive' Anwender wird auch eine Beschleunigung des Transports von geringen Datenmengen erhoffen, weil ihm der Name 'Hochgeschwindigkeitsnetze' das suggeriert. Allerdings wird er sehr schnell erkennen, daß die reine End-End-Verzögerungszeit durch ein 'Hochgeschwindigkeitsnetz' nicht verkürzt wird - denn die Übertragungszeit ist an die Lichtgeschwindigkeit gebunden, und deren Steigerung wurde bisher vergeblich versucht.

Es wird sogar mit einiger Sicherheit der umgekehrte Effekt eintreten, d.h. für kleine Datenmengen wird der FDDI-Ring 'langsamer' sein als Netze mit wesentlich geringerer Datenkapazität, denn die FDDI-Knoten müssen diese Daten aufwendiger verarbeiten (man betrachte etwa den zusätzlichen Aufwand für 'encapsulation' bzw. 'decapsulation' im Falle der von uns eingesetzen Komponenten gegenüber einer Ethernet-Brücke, welche diese Maßnahmen nicht erfordert; siehe Abb. 2). Durch erhöhten Einsatz von Parallelverarbeitung und durch Verlagerung von Protokollfunktionen so weit wie möglich in Hardware sollte sich der Geschwindigkeitsverlust in engen Grenzen halten; das ist aber bei unseren FDDI-Brücken leider noch nicht der Fall. Wirkliche Einsparungen durch FDDI können sich höchstens bei sehr großen Datenmengen ergeben, die im Bereich File Transfer und erst recht bei Bildverarbeitung anfallen. Analogie: Ein Lastwagen transportiert große Frachtmengen "schneller" als ein Kleinwagen, obwohl er sogar langsamer fährt. Dagegen ist für den Transport von geringen Frachtmengen der Kleinwagen vorzuziehen. Ein Letztes zu dieser Analogie: Die Antwortzeit wird irrelevant, wenn es auf einige Minuten oder Stunden mehr oder weniger nicht ankommt.

Trotz dieser Vorbemerkungen über die zu erwartenden Ergebnisse und über Sinn bzw. Unsinn des Transports kleiner Datenmengen über den FDDI-Ring ist festzustellen, daß in allen Anwendungsformen zahlreiche kurze Nachrichtensequenzen über den Ring übertragen werden - und sei es auch nur zum Verbindungsaufbau. Es wird sich ferner herausstellen, daß durch spezielle Organisationsformen höherer Protokolle (die einst für ein anderes Umfeld - d.h. andere Kabeltypen, Störraten, Entfernungen, ... - entwickelt wurden!) große Datenmengen künstlich in kleine Teile zerlegt werden können, weshalb dann der zuvor geschilderte für FDDI sehr unangenehme Effekt auftreten kann - und auch tatsächlich auftritt!

Bei den an unserem Ring durchgeführten Messungen (von denen hier nur über erste Ergebnisse berichtet werden kann) wurden folgende Parameter untersucht:

- Antwortzeit für kurze Nachrichten (das kann man wie bereits erwähnt als ein dem Ring gegenüber 'unfaires' Szenario ansehen, aber die Ergebnisse sind trotzdem von grundsätzlichem Interesse).
- Maximal erreichbarer Durchsatz für File Transfer.

Beides wurde untersucht für einzelne Verbindungen, die zusätzlich zum sonstigen Netzverkehr geschaltet wurden. Eine großer Lasttest, bei dem alle Knoten gemeinsam den Ring zu belasten versuchen (und bei dem dann festzustellen sein wird, ob und wie stark sich diese Anforderungen gegenseitig behindern) steht noch aus.

3.1. Meßmethodik

Bisher werden (wegen der Nichtverfügbarkeit entsprechender Geräten) die Meßwerte nicht direkt auf dem FDDI-Ring 'abgegriffen', sondern auf einem Meßgerät (Lanalyzer), das an ein zu Knoten 5 (Informatik; siehe Abb. 1) gehörendes Ethernet angeschlossen wird.

Um mit einem einzigen Lanalyzer auszukommen, werden Round-Trip-Verzögerungszeiten gemessen, d.h. dasselbe Meßgerät mißt die Gesamtverzögerung vom Absenden einer Nachricht (bzw. genauer gesagt vom Zeitpunkt des Passierens dieser Nachricht beim Lanalyzer) bis zur Rückkehr der Antwort an die Quellstation (bzw. Lanalyzer). Abb. 3 erläutert das Meßprinzip.

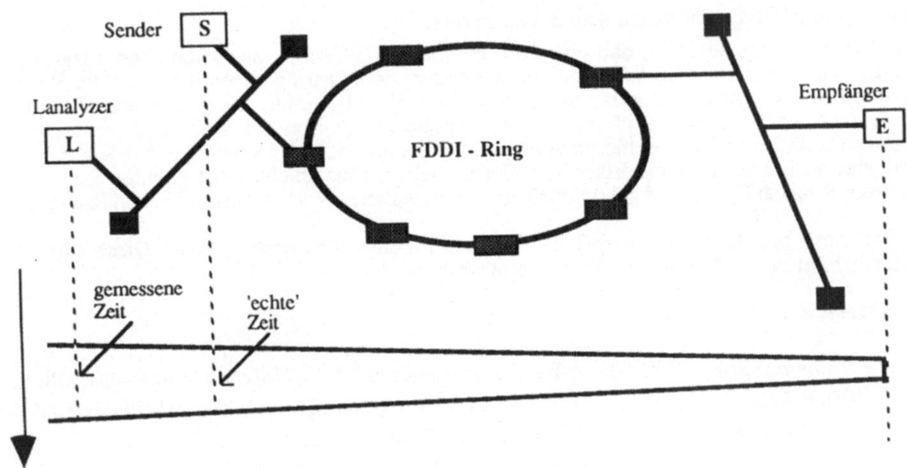

Abb. 3: Meßprinzip und Lage der Meßpunkte

Die erhaltenen Meßergebnisse schwanken aus verschiedenen Gründen:

- unterschiedliche aktuelle Belastung der FDDI-Knoten und des FDDI-Backbone
- nicht konstante Reaktionszeiten der angeschlossenen Endgeräte
- ungenau gehende bzw. nur sehr grob auflösende Uhren des Meßgeräts.

Vernachlässigen wir für den Moment einmal diese statistischen Schwankungen, dann können wir in Abhängigkeit von der Lokation der betroffenen Endgeräte eine Anzahl von Faktoren ausmachen, die für die Gesamtverzögerungszeit verantwortlich sind. Dazu zählen:

- Die End-End-Signallaufzeit; diese ist sowohl von der geographischen Entfernung als auch von der Gesamtzahl zu passierender FDDI-Knoten abhängig. Dabei kann man sich auf diejenigen FDDI-Knoten beschränken, welche einen Encapsulation- oder Decapsulation-Vorgang ausführen (die anderen reichen die Pakete mit vernachlässigbarer Verzögerungszeit weiter). Auch die unterschiedlichen geographischen Entfernungen sind gemessen an den anderen Faktoren von untergeordneter Bedeutung für die Gesamtverzögerungszeit.
- Verzögerungszeiten durch informationsverarbeitende Komponenten. Diese wiederum lassen sich unterteilen in mehrere Typen:
  - Bearbeitungszeiten in den Endgeräten.
  - Verzögerungszeiten durch Komponenten, die zu lokalen Netzen gehören; hierzu gehören z.B. Wartezeiten, die durch filternde Ethernet-Brücken verursacht werden.
  - Verzögerungszeiten durch FDDI-Adapter, diese entstehen durch Ausführung von Encapsulation- bzw. von Decapsulation-Vorgängen bzw. durch zu seltenes Leeren der Sendepuffer der angeschlossenen Lokalen Netze.

Eine ausführlichere Liste von Verzögerungszeitparametern wird im nächsten Abschnitt bei der Beschreibung und Auswertung der Meßszenarien gegeben.

Zur Meßmethodik:

Jede der durchgeführten Messungen enthält - abhängig vom Meßszenario (d.h. von der Verkehrsart und von der Lokation der miteinander kommunizierenden Geräte) die für die beobachtete Gesamtverzögerungszeit verantwortlichen Faktoren in unterschiedlicher Anzahl. Daher kann jedes Meßergebnis als fehlerbehaftete lineare Gleichung mit unbekannten Parametern interpretiert werden. Mit wachsender Zahl und mit wachsender Diversität von Tests erhalten wir ein größer werdendes Gleichungssystem, dessen Lösung (sofern sie existiert) die uns interessierenden unbekannten Parameter zu bestimmen gestattet.

Eine 'typische' Gleichung erhält man etwa wie folgt:

Das Szenario bestehe darin, daß eine Kommunikation zwischen zwei Stationen ausgeführt werde, von denen eine dem Westring und die andere dem Ostring zugeordnet sei. Die Station des Ostrings sei direkt über das am zugehörigen FDDI-Knoten angeschlossene Ethernet erreichbar; bei der ebenfalls über Ethernet erreichbaren Station des Westrings werde angenommen, daß dieses Ethernet hierarchisch strukturiert sei und daß zwischen FDDI-Knoten und Endgerät noch zwei filternde Ethernet-Brücken stehen. Eine solche Konfiguration kommt in unserem Ring häufig vor und gehört auch zu den im weiteren Verlauf beschriebenen Szenarien.

Die Messung bestehe nun darin, daß die Wartezeit für ein Echo ermittelt werde. Diese Zeit setzt sich wie folgt zusammen (Reze = Rechenzentrum):

Gemessene Zeit + Meßfehler

$$\begin{aligned} = \; & 2 \times u_{Ethernet\text{-}Bridge} + u_{FDDI\text{-}A\text{-}West\text{-}Encapsulation} + u_{FDDI\text{-}Reze\text{-}West\text{-}Decapsulation} \\ & + u_{FDDI\text{-}Reze\text{-}Ost\text{-}Encapsulation} + u_{FDDI\text{-}B\text{-}Ost\text{-}Decapsulation} + u_{Verarbeitungszeit\text{-}B} \\ & + u_{FDDI\text{-}B\text{-}Ost\text{-}Encapsulation} + u_{FDDI\text{-}Reze\text{-}Ost\text{-}Decapsulation} \\ & + u_{FDDI\text{-}Reze\text{-}West\text{-}Encapsulation} + u_{FDDI\text{-}A\text{-}West\text{-}Decapsulation} \\ & + 2\, u_{Ethernet\text{-}Bridge} + u_{Verarbeitungszeit\text{-}A} \,. \end{aligned}$$

Wenn wir annehmen, daß der Aufwand für Encapsulation und für Decapsulation etwa gleichgroß ist (diese Annahme ist nicht gravierend, denn Verzögerungszeiten treten sowieso immer paarweise auf, weil jede Encapsulation eine Decapsulation bedingt), dann ergibt sich folgende kürzere Gleichung:

$$u_{echo} + 4\, u_{EB} + 8\, u_{FDDI} = m + \varepsilon.$$

Dabei ist:

u_{echo} die Gesamtbearbeitungszeit in den Endgeräten.
u_{FDDI} die Verzögerungszeit durch einen Encapsulation- oder Decapsulation-Vorgang.
u_{EB} die Verzögerungszeit, welche auf eine filternde Ethernet-Brücke zurückzuführen ist.
m der Meßwert.
ε der Meßfehler.

Durch insgesamt n solche Tests mit unterschiedlichen Szenarien erhalten wir auf diese Wesie ein Gleichungssystem folgender Art:

$$\begin{aligned} a_{11} \cdot u_1 + \ldots\ldots\ldots + a_{1n} \cdot u_n &= m_1 + \varepsilon_1 \\ a_{21} \cdot u_1 + \ldots\ldots\ldots + a_{2n} \cdot u_n &= m_2 + \varepsilon_2 \\ \ldots\ldots\ldots \qquad & \ldots\ldots\ldots \\ a_{p1} \cdot u_1 + \ldots\ldots\ldots + a_{pn} \cdot u_n &= m_p + \varepsilon_p \end{aligned}$$

bzw. in Matrixform: $\mathbf{A} \cdot \mathbf{u} = \mathbf{m} + \varepsilon.$

Dabei sind die Koeffizienten a_{ij} (d.h. die Anzahlwerte des Auftretens der betreffenden Komponenten) ganzzahlig, und die Matrix ist dünnbesetzt, wenn sehr unterschiedliche Testkonfigurationen einbezogen werden.

Es gilt nun [siehe z.B. Magnus, J.R., Neudecker, H.: Matrix Differential Calculus with Applications in Statistics and Ecoonomics. Wiley, Chichester 1988, S. 257ff.]:

Sei wie oben $\mathbf{A} \cdot \mathbf{u} = \mathbf{m} + \varepsilon$,
wobei ε ein Zufallsvektor mit $E(\varepsilon) = 0$ und $Cov(\varepsilon) = \sigma^2 \cdot \mathbf{I_p}$ (I_p = p-zeilige Einheitsmatrix). Wenn Rang($\mathbf{A}$) = n ist, d.h. sofern genügend Tests zur Bestimmung der unbekannten Parameter gemacht wurden, dann gilt:

1. $(\mathbf{A}^T\mathbf{A})^{-1}$ existiert.

2. $\hat{\mathbf{u}} = (\mathbf{A}^T\mathbf{A})^{-1}\mathbf{A}^T\mathbf{m}$ ist der beste lineare erwartungstreue Schätzer für den Vektor $\mathbf{u}$.

   Man bezeichnet dies auch als Gauß-Markov-Schätzung bzw. als BLUE (Best Linear Unbiased Estimator).

   Besonders interessant ist, daß der Meßfehler hierbei keinerlei Rolle mehr spielt; es muß nur sichergestellt (oder vorausgesetzt!) werden, daß der Erwartungswert der Meßfehler den Wert Null hat.

3. Für die Covarianzmatrix gilt:
   $Cov(\hat{\mathbf{u}}) = \sigma^2 \cdot (\mathbf{A}^T\mathbf{A})^{-1}$.

3.2. Meßszenarien und -ergebnisse

3.2.1. Das Szenario 'Ping'

Die einfachste - und für die Komponenten des FDDI-Rings auch 'unfairste' - Messung besteht in einer Analyse des Ping-Szenarios. 'Ping' ist im wesentlichen gekennzechnet durch die Gesamtwartezeit für eine einzelne Nachricht, die an der Zielstation per Echo an den Sender zurückgespielt wird.

Der FDDI-Ring wurde nicht installiert, um überwiegend oder ausschließlich Ping-Nachrichten zu verarbeiten. Andererseits treten kurze Nachrichtensequenzen dieser Art etwa zu Beginn eines File Transfers immer auf. Schließlich wird es sich (siehe unten) zeigen, daß durch spezifische Eigenschaften von höheren Protokollen auch lange Nachrichtentransfers (bei denen FDDI entscheidend besser abschneiden sollte als bei einem Ping-Szenario) in letzter Konsequenz in eine Folge von kurzen Blöcken zerlegt werden, die einzeln als Ping interpretierbar sind.

Die untersuchten Szenarien für Ping betreffen eine Quellstation im Subnetz 'Informatik' (FDDI-Knoten 5) und eine Reihe von verschiedenen Zielstationen, wobei möglichst viele unterschiedliche Situationen berücksichtigt werden sollten. Lage, Name und Adresse der in die Messungen einbezogenen Stationen gehen aus Abb. 4 hervor.

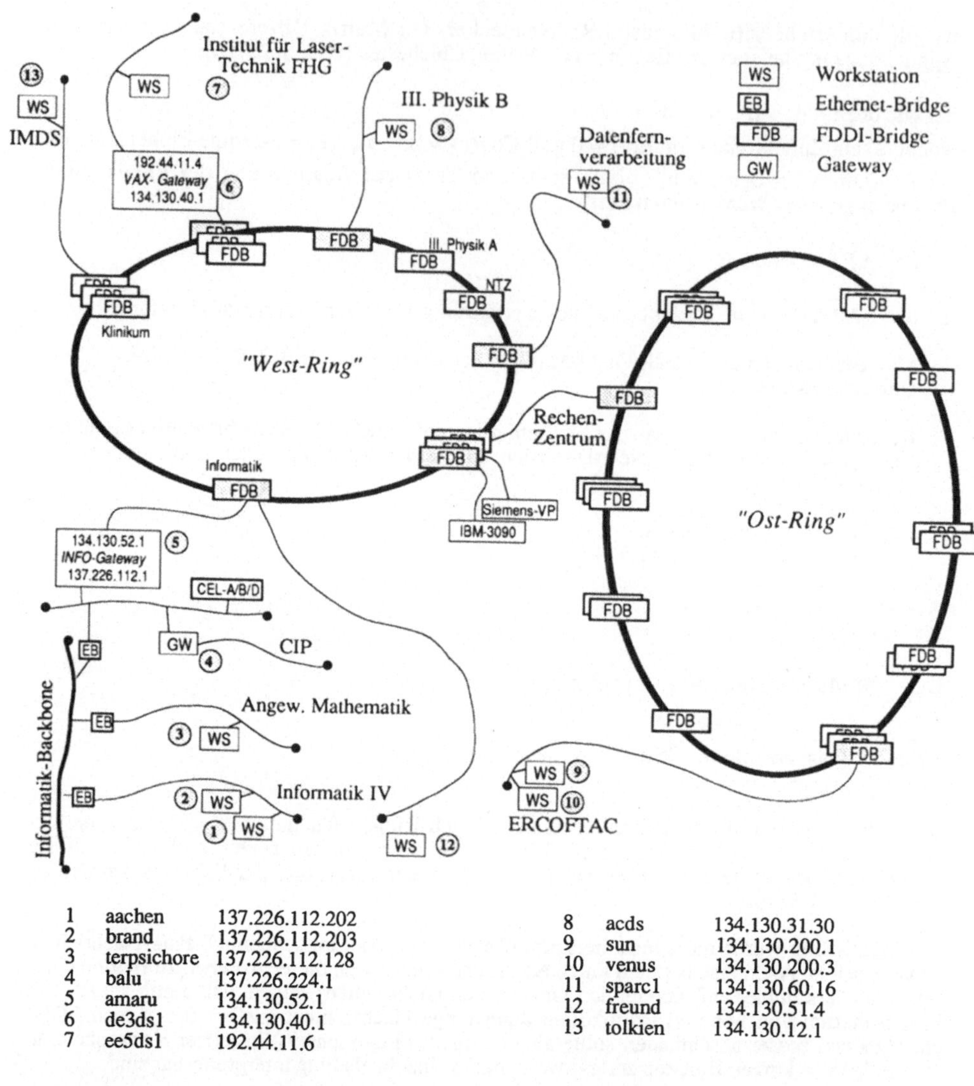

Nr.	Name der Workstation	TCP/IP-Adresse	Nr.	Name der Workstation	TCP/IP-Adresse
1	*aachen*	*137.226.112.202*	*8*	*acds*	*134.130.31.30*
2	*brand*	*137.226.112.203*	*9*	*sun*	*134.130.200.1*
3	*terpsichore*	*137.226.112.128*	*10*	*sun*	*134.130.200.3*
4	*balu*	*137.226.224.1*	*11*	*sparc1*	*134.130.60.16*
5	*amaru*	*134.130.52.1*	*12*	*freund*	*134.130.51.4*
6	*de3ds1*	*134.130.40.1*	*13*	*tolkien*	*134.130.12.1*
7	*ee5ds1*	*192.44.11.6*			

Abb. 4: Konfiguration des FDDI-Rings und Lage der beteiligten Stationen

Die Ergebnisse der Messungen für das 'Ping'-Szenario sind:

Stationen	Charakteristik	Meßergebnis
aachen ↔ brand	*Ethernet-intern*	$u_{echo} = 0{,}8\ ms$
aachen ↔ terpsichore	*Ethernet-Backbone*	$u_{echo} + 4\, u_{EB} = 1{,}5\ ms$
freund ↔ acds	*FDDI-Westring*	$u_{echo} + 4\, u_{FDDI} = 11{,}2\ ms$
freund ↔ venus	*Beide FDDI-Ringe*	$u_{echo} + 8\, u_{FDDI} = 19{,}1\ ms$
aachen ↔ venus	*Ethernet-Backbone + Gateway + beide Ringe*	$u_{echo} + 4\, u_{EB} + 8\, u_{FDDI} + 2\, u_{GW} = 25{,}2\ ms$
amaru ↔ freund	*FDDI-'rein-raus'; (ohne Kapselung; FDDI-Knoten agiert fast wie eine Ethernet-Brücke)*	$u_{echo} + 2\, u_{F\text{-}local} = 4{,}3\ ms$
aachen ↔ freund	*FDDI-'rein-raus' + Gateway + Ethernet-Bridge*	$u_{echo} + 4\, u_{EB} + 2\, u_{F\text{-}local} + 2\, u_{GW} = 8{,}1\ ms$

Die Auswertung dieses Gleichungssystems mit der zuvor beschriebenen Methode liefert folgendes auf den ersten Blick völlig überraschende - und für den FDDI-Ring zunächst deprimierend scheinende - Ergebnis:

$\hat{u}_{echo}$ = **0,96 ms** $\hat{u}_{EB}$ = **0,13 ms** $\hat{u}_{GW}$ = **2,37 ms**

$\hat{u}_{FDDI}$ = **2,11 ms** $\hat{u}_{F\text{-}local}$ = **1,42 ms.**

Bevor wir im einzelnen auf Erklärungen zu und auf Konsequenzen aus diesen Meßergebnissen eingehen, wollen wir ein weiteres - für den FDDI-Einsatz relevanteres - Szenario vorstellen.

3.2.2. Das Szenario 'FTP (File Transfer Protocol)'

Ermittelt wurde hier der maximal erreichbare Durchsatz zwischen zwei Stationen und zwar unter Verwendung folgender Protokolle der Ebene 3/4 (durch die beteiligten Quell- bzw. Zielstationen wird implizit vorgegeben, welche Protokolle zum Einsatz kommen können):

- TCP/IP mit relativ kleinem Quittungsfenster (ACK-Zwang nach ca. 32 Kilobit an Daten).
- TCP/IP mit größerem Quittungsfenster (so groß, daß es für den erreichbaren Durchsatz keine Engpaßfunktion mehr darstellt).
- UDP (Unacknowledged Datagram Protocol mit 70.400 Bits pro Paket).

Für das FTP-Szenario wurden folgende Werte gemessen:

Stationen	Charakteristik	Maximaler Durchsatz (Megabit/sec.)	
aachen -> brand	*Ethernet-intern*	*2,48*	*mit TCP/IP (kleines Fenster)*
aachen -> terpsichore	*Zwei Ethernet-Brücken*	*2,24*	"
freund -> acds	*Zwei FDDI-Brücken*	*1,44*	"
freund -> venus	*Vier FDDI-Brücken*	*0,96*	"
aachen -> acds	*Zwei Ethernet-Brücken, zwei FDDI-Brücken, ein Gateway*	*0,74*	"
aachen -> freund	*Zwei Ethernet-Brücken, eine FDDI-'rein-raus'-Brücke, ein Gateway*	*1,68*	"
aachen -> venus	*Zwei Ethernet-Brücken, vier FDDI-Brücken, ein Gateway*	*0,72* *1,64*	*mit TCP/IP (kleines Fenster)* *mit UDP*
amaru -> sparc1	*Zwei FDDI-Brücken*	*3,04*	*mit TCP/IP (großes Fenster)*

Auch aus diesen Werten ergibt sich, daß 100 Megabit pro Sekunde an nomineller Kapazität nicht annähernd für eine einzelne Verbindung erreicht werden können. Aufgrund des zu kleinen TCP/IP-Quittungsfensters, welches bei der Mehrzahl der unter TCP/IP arbeitenden Endgeräte eingesetzt wird, ergibt sich als unerwünschte Konsequenz, daß große Datenbestände, die eigentlich wie bei einer Pipeline (also ohne weitere Verzögerungszeiten mit Ausnahme der unvermeidlichen Verzögerungszeit des Startpakets) durchgereicht werden sollten, in 'Pings' der Größe 32 Kilobit zerstückelt werden. Das hat entscheidende Auswirkungen sowohl auf die Gesamtübertragungszeit als auch auf den erzielbaren Durchsatz.

3.2.2. Das Szenario 'Spray'

Dieses Szenario ist das bestmögliche für FDDI (unter Berücksichtigung der Beschränkungen, welche durch unzureichende höhere Protokolle impliziert werden).

Bei 'Spray' werden Pakete ohne Quittungsmechanismus zwischen einer Quell- und einer Zielstation übertragen. Die maximale Leistungsfähigkeit des FDDI-Rings für diese Anwednung kann dadurch ermittelt werden, daß man die Spray-Rate so hoch dreht, daß jede weitere Steigerung zu Datenverlusten am Ziel führen würde (wobei wir mit unseren Meßmitteln nicht feststellen können, ob solche Verluste in den FDDI-Knoten oder - was uns wahrscheinlicher scheint - in der Zielstation erfolgen).

Das Spray-Szenario war rechnerbedingt bei unseren Messungen nur auf eine einzige Kommunikationsbeziehung anwendbar und ergab:

Stationen	Charakteristik	Maximaler Durchsatz (Megabit/sec.)
amaru -> sparc1	*Zwei FDDI-Brücken*	*5,76 Mbit/sec.*

Dieser Wert ist (als kleines 'Trostpflaster' für die bisher so niedrigen FDDI-Werte) sogar höher als bei Abwicklung desselben Szenarios in derselben Workstation (also z.B. *amaru -> amaru*). Eine Erklärung für diesen Effekt könnte sein, daß die Speicherkapazität des FDDI-Rings die 'entfernte' Kommunikationsbeziehung bevorteilt, d.h die Zielstation kann offenbar höhere Datenmengen verarbeiten, wenn zwischen ihr und der Quelle der FDDI-Ring als Puffer steht.

4. Konsequenzen

Auf den ersten Blick sind die gemessenen Leistungsdaten des FDDI-Rings schockierend niedrig. Eine evtl. daraus gezogene Schlußfolgerung, der FDDI-Ring sei ein ungeeignetes Netzkonzept, wäre aber vorschnell und unkorrekt. Einige Gründe dafür:

- 'Ping' (siehe 3.2.1) ist ein für FDDI denkbar schlechtes bzw. ungeeignetes Szenario.
- Die Ergebnisse für 'Spray' zeigen, daß nicht zuletzt auch die Endgeräte für eventuelle Engpässe verantwortlich sind. Die Erfahrung zeigt leider, daß bei jeglichem auftretendem Problem zunächst das für den Anwender 'mysteriöse' Netz als Prügelknabe herhalten muß.
- Die Leistungsfähigkeit pro Verbindung ist noch kein vollständiges Charakteristikum für die Gesamtleistungsfähigkeit. Über FDDI lassen sich viele Verbindungen gleichzeitig abwickeln, ohne daß sie sich signifikant gegenseitig behindern. Diesbezügliche Messungen sind in Vorbereitung.

  Allerdings: den Anwender interessiert es wenig, ob viele andere Anwender gemeinsam mit ihm aktiv sein können; er ist ausschließlich an seinen eigenen Leistungsdaten interessiert. Wenn diese für ihn zu schlecht ausfallen - aus welchen Gründen auch immer -, wird er das Netz nicht akzeptieren.
- Die in manchen Fällen hoch scheinenden Antwortzeiten sind nicht so extrem als daß sie der 'normale Anwender' nicht akzeptieren würde.
- Die FDDI-Adapter haben eine Reihe von Tätigkeiten durchzuführen (in unserem Falle etwa Encapsulation oder Decapsulation), die nicht 'zeitlos' abgewickelt werden können (siehe dazu auch Abb. 2). Die nur filternde Ethernet-Brücke ist hier viel besser dran - und deshalb sind auch ihre Leistungsdaten besser als diejenigen der FDDI-Adapter.
  Es muß außerdem zur Ehrenrettung von FDDI gesagt werden, daß einerseits Ethernet über eine so große Distanz wie in unserer Konfiguration weder möglich noch zweckmäßig wäre und daß zum anderen FDDI gleichzeitig viele Transportvorgänge in Bearbeitung haben kann, während auf Ethernet zu einer gegebenen Zeit maximal eine Nachricht kollisionsfrei präsent sein kann.
- Ein realistischerer Vergleich betrifft 'FDDI-Bridge' und 'Gateway-Station' (anstelle Ethernet-Bridge'). Hier sind die Verzögerungszeiten in etwa gleichgroß, was zugunsten der wie ein Router arbeitenden Gatewaystation spricht, zumal sie viele Funktionen in Software durchführt, während die FDDI-Brücken mögliche Beschleunigungen durch Hardware-realisierungen nur unzureichend ausnutzen. Eine plausible - aber kaum akzeptable - Erklärung wäre, daß die FDDI-Brücke zu stark mit der Beobachtung des FDDI-Verkehrs beschäftigt ist und daß sie deshalb die angeschlossenen Ethernets 'vernachlässigen' muß.
- Für die unzureichend scheinenden Durchsatzraten des FDDI-Rings ist nicht allein die Rechnerleistung der FDDI-Adapter verantwortlich (obwohl hier sehr bald eine deutliche Verbesserung durch Einsatz leistungsfähigerer Komponenten erfolgen muß - das gilt jedenfalls für die bei uns derzeit eingesetzten FDDI-Adapter).
  Eher noch wichtiger ist aber die Anpassung bzw. geeignete Abstimmung des gesamten Protokollstacks! Es nützt kaum etwas (wenn überhaupt), die Parameter der Ebene 2 durch Einsatz höherratiger Netze wie FDDI zu verbessern, wenn nicht im Gegenzug die Protokolle der höheren Ebenen an diese Gegebenheiten angeglichen werden. In unserer Netzinstallation 'bremsen' die unzureichenden (weil für andere Zwecke entworfenen und von dort unzulässig übernommenen) TCP/IP-Protokollrealisierungen den FDDI-Ring ganz erheblich. Abbildung 5 soll diese Zusammenhänge plakativ darstellen.

  Es bestätigt sich wieder einmal die wohlbekannte - aber leider nur selten beachtete - Erfahrung, daß nur eine globale Anpassung des Protokollstacks wirkliche Leistungs-verbesserungen bringt. Eine Kette ist immer nur so stark wie ihr schwächstes Glied - was bedeutet, daß eine Verstärkung der sowieso schon relativ starken Glieder nur wenig nutzt. In Abbildung 6 ist die weitere Beschleunigung des 'Rennpferdes FDDI' wertlos, wenn die 'Schnecke TCP/IP' mitberücksichtigt werden muß. Diese Binsenweisheit bleibt nach meiner Meinung auch deshalb unberücksichtigt, weil die Mittel fehlen, mehrere Protokollebenen gleichzeitig realitätsnah zu modellieren und auf diese Weise wirklich verstehen zu lernen.

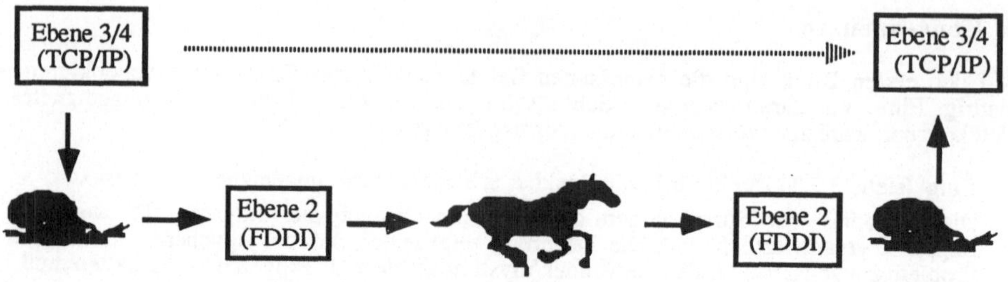

Abb. 5: Der bremsende Einfluß von TCP/IP auf das Leistungsverhalten des FDDI-Netzes

5. Zusammenfassung

Erfahrungen aus und Messungen an unserer FDDI-Installation zeigen, daß Realisierung und Betrieb eines solch großen Netzes eine Vielzahl von Überraschungen mit sich bringen.

- Der Anwender sollte nicht erwarten, daß die magische Zahl 100 Megabit pro Sekunde (mancher Anwender kennt von FDDI wenig mehr als eben diese Zahl) für ihn auch wirklich erreichbar ist. Die real erzielbaren Werte liegen sehr viel niedriger. Es gibt Gründe dafür, aber die Hersteller sollten doch gehalten sein, die Diskrepanz zwischen Fiktion und Realität nicht gar zu kraß ausfallen zu lassen.
- Es wird nach meiner Auffassung zuviel mit absurd hohen Vorstellungen über die benötigte Netzkapazität um sich geworfen. Man bedenke, daß der Inhalt großer Bibliotheken in kürzester Zeit über eine 100 Megabit/sec.-Verbindung übertragen werden könnte. Für die überwiegende Zahl von heute dringlichen Anwendungen selbst bei großen Institutionen ist die nach Abzug aller Verlusteffekte verbleibende Datenrate bei FDDI mehr als ausreichend.
- Völlig neue Probleme werden sich einstellen, wenn wirklich Bildverarbeitung in großem Stil eine weite Verbreitung und Akzeptanz gefunden haben wird, wenn also etwa Bildschirminhalte in hoher Qualität (d.h. mit höchster Pixelauflösung und zahlreichen Grauwert- bzw. Farbstufen) über das Netz übertragen werden. Hierzu werden Netzformen wie FDDI nicht mehr in der Lage sein.

  Eine überspitzte Folgerung daraus wäre, daß FDDI für 'normale Daten' zuviel Kapazität anbietet, dagegen für Bildverarbeitung zuwenig. Man könnte deshalb sogar schlußfolgern, daß FDDI in ein gewisses Niemandsland bzgl. sinnvoller Anwendungen fällt. Soweit möchte ich allerdings keinesfalls gehen. Nach meiner Meinung macht FDDI sehr viel Sinn, weil es eine zweckmäßige Backbone-Vernetzung (und damit eine geeignete Hierarchisierung) für zahlreiche Institutionen ermöglicht. In absehbarer Zeit werden auch kleinere FDDI-Workstation-Ringe als Anhängsel zu FDDI-Backbone-Ringen verfügbar und bezahlbar sein. Wenn das Netz nicht permanent bis an die Halskrause belastet ist, dann halte ich auch das besser als den umgekehrten Zustand. Schließlich ist es keine besonders wünschenswerte Vorstellung für den Autofahrer, daß alle Spuren einer Autobahn Stoßstange an Stoßstange mit Autos gefüllt werden müßten. Bei Autobahnen versteht man sofort, daß dies nicht gut ist; von Netzen wird es nur allzu oft verlangt. Warum eigentlich?

Dank gilt meinen Mitarbeitern Peter Davids und Thomas Meuser für die Durchführung der Messungen - auch wenn bzw. gerade weil sie sich sehr stark um eine sehr kritische Wertung der Produkte des in unserem FDDI-Ring eingesetzten Herstellers (nicht zuletzt auch durch für FDDI besonders "geeignete" Auswahl von Testszenarien) bemüht haben.

Besonderer Dank gilt Herrn Dr. Rudolf Mathar (Lehr- und Forschungsgebiet "Stochastik, insbesondere Anwendungen in der Informatik" der RWTH Aachen), der uns die in Abschnitt 3 geschilderte ebenso einfache wie effziente Auswertetechnik vermittelte.

FROM BROADBAND ISDN TO MULTIMEDIA COMPUTER NETWORKS

Prof. Dr. R. Popescu-Zeletin
GMD-FOKUS/DETECON/TU-Berlin
Hardenbergplatz 2
D-1000 Berlin 12, F.R. Germany

Abstract

The introduction of fiber optics and new switching technologies allows the development of new services and applications based on the integration of different information types (voice, data, video, images). The paper outlines the architecture and the major solutions adopted for the development of a broadband, multifunctional computer network in F.R.Germany under the acronym BERKOM Berliner Kommunikationssystem. BERKOM is a pilot project of the Deutsches Bundespost and aims at an integrated broadband communication system. A reference model was developed for the global system focussing on new multifunctional end-system architectures.

Key words:

Broadband communication, multimedia data structures, reference model, cooperative applications

1. Introduction

During the last decade one can witness a great deal of effort in the development and introduction of wide-area networks (WANs). Parallel to the availability of WANs, conceptual work and resulting products have been developed for the hardware and software architecture of the end-systems.

The variety of technologies in the domain of WANs covers networks based on different switching technologies (circuit switching, packet switching), for different applications (telephony, telex, data communication, telematic services, TV distribution), using different transmission media (cables, fiber-optics, satellites).

Presently, the major direction of development is the integration of the networks already in operation into one network which can satisfy all the different requirements with respect to application profiles and their required qualities of services. This is motivated not only by the maintenance costs of different existing networks operating in parallel, but also by the variety of applications one may imagine in a multifunctional environment.

The digitization of the traditional applications based on analogue transmission signals (voice, TV etc.) is the necessary premise for their integration into one universal network. This is not only necessary for the transmission of data through the multi-purpose network, but also for the development of multimedia end-systems coping with digital information.
The first step in this integration process was the ISDN technology, based on data and voice integration in one network employing transmission channels of up to 64 kbit/s.
The introduction of new transmission media allows a new step in this area of development where transmission speeds ranging from 2 Mbit/s to Gbit/s are available. A major step in this development is the introduction of the Broadband-ISDN with transmission channels of up to 140 Mbit/s.
Broadband ISDN is the premise to a universal IBCN network (Integrated Broadband Communication Network). The different steps of integration must be coherent in order to ensure the integration of developments already existing in new environments. It is therefore mandatory, due to the financial and intellectual efforts invested in each step of this development, that a smooth transition from one step to the next is achieved.
Based on the same basic switching principles as ISDN, B-ISDN provides new capabilities for the next step of integration for applications which require transmission rates of up to 140 Mbit/s.
Note that the ongoing discussion on the switching principles and techniques (ATM, STM) within the B-ISDN WANs is quite irrelevant when viewed from the perspective of a global system.
Nevertheless, some major points have to be mentioned regarding the controversial discussion of circuit switching versus packet switching within the B-ISDN network:

- the switching technique employed within the network is irrelevant, if the subscriber interface is of the same type as that of ISDN (D-channel signalling, B- and H-channels). This is an important aspect, since the future B-ISDN network should incorporate products already developed for an ISDN environment,
- the technique used internally should provide end-to-end channels of up to 140 Mbit/s (needed for cheap TV integration) with guaranteed transmission rates.
- the switching technique used internally should be oriented to the requirements of the majority of applications in the future universal network. Most end-systems will require circuit switching transmission (telephone, TV sets etc.).

Also, it should be noted that most data communication applications have a connection-oriented profile (X.25 networks provide virtual circuits based on an underlying packet switching technology).
The same development takes place in metropolitan/local area networks where the introduction of fiber optics as transmission media and suitable network access technologies (FDDI, DQDB) follow the same integration idea of services. The standardization of protocol formats in WAN and MAN allows a smooth integration and interconnection of network technologies in a global network.
The introduction of optic technology in transmission and storage will influence the development of optical processing as last bastion of electronics in the chain storage/transmission/processing. The present bottle-neck are the end-systems and not the networks.
New software and hardware architectures for multimedia end-systems are required for the integration of the different communication types (data, voice, video) and the support of the high speed applications. Special attention is required for new concepts and development of high speed protocols and their imple-

mentation. In line with the above considerations a pilot project (BERKOM) was started in Berlin in 1986 by the Deutsches Bundespost and the Berlin Senate. The aim of BERKOM is to design and develop a B-ISDN global network. In this context a global network implies a network architecture not only for layer 1-4 in the ISO/OSI Reference Model [8], but also new applications, techniques and end-systems architectures in this new environment.

2. The BERKOM Test Network

The prerequisite for a B-ISDN network is a fiber-optic infrastructure. Berlin has a fiber-optic network of approximately 17,000 km interconnecting the major research and development institutions and computer manufacturers. The switching nodes must provide the functionality necessary for the future B-ISDN. This means that they have to present to the subscriber channels at 64 kbit/s, 2 Mbit/s and 140 Mbit/s, conforming to the PCM hierarchy and the H_X-channel structure [1].
The D-channel protocol, as defined for the ISDN by CCITT, can be easily enhanced to support the establishment of the circuit switched channels with the above mentioned characteristics.
The channel structure provided on a subscriber line is:

$S_B = 1^*D + 2^*B + 4^*H_1 + 1^*H_4$

- D = 16 kbit/s
- B = 64 kbit/s
- H_1 = 920 kbit/s
- H_4 = 135,168 Mbit/s at a total transmission speed of 153,6 Mbit/s

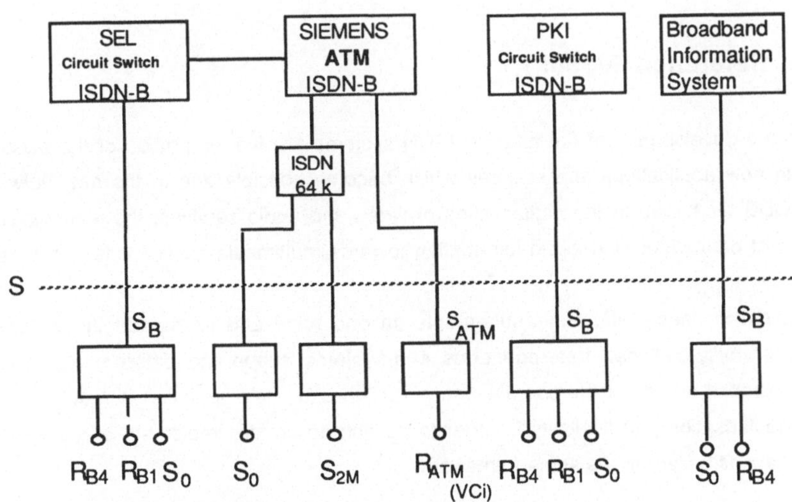

Figure 1: BERKOM Test Network

The switching nodes for the BERKOM test network are provided by Standard Elektrik Lorenz, Philips Kommunikations Industrie and SIEMENS AG, the switching techniques cover STM (SEL,PKI) and ATM (SIEMENS) (figure 1). As the standardization of the H_4-channels has not yet been finalized and discussion concerning line speeds and channel structures is still in progress, a terminal adapter (TA) with an R-interface point for each end-system provides the end-system decoupling from the internal switching technolog. The switching nodes are pilot developments and each serves up to 30 end-systems.

There are two classes of end-systems (fig.2):

- end-systems directly connected to the B-ISDN network
- end-systems which are indirectly connected to the network via local networks or PABXs.

These two classes identify the two endsystem architectures in the lower layers. The first one has to comply in all layers (1-4) with the choice of standards for interworking, the second one only in layers 3 and 4 [10]. The profiles defined above have an immediate impact on the architecture for B- and H-channels in the two classes of end-systems specified above (directly and indirectly connected to B-ISDN) and for the B-ISDN - LAN gateways.

For interworking of LANs and a B-ISDN network a "Modular Gateway System" (BERGATE) was designed and developed. The supported LAN-types are CSMA-CD, Token Ring, Token Bus, which follows the ISO/IEEE P802 - Standard and the protocol profiles which are given by MAP (Manufacturing Automation Protocols), TOP (Technical and Office Protocols) and SPAG (Standards Promotion and Application Group) [12, 14, 16]. The interconnection of the LANs and the B-ISDN network is realized in the ISO Layer 3 or 4, using the appropriate protocol profiles. Additionally, FDDI and DQDB technologies are supported by the gateway.

3. The Berkom Reference Model

The main directions in the development of the future B-ISDN systems are the integration of the existing telematic services, with new applications and services which become possible due to the new network technologies (ATM, DQDB etc.). Due to the digitalization of video and audio services, the technological base for the integration of data, video and audio information towards multimedia applications becomes a reality.

In order to be able to satisfy these different requirements on one hand and to support in the future global systems interoperability between heterogeneous end-systems supporting different classes of applications a global reference model for the coordination in design and implementation is required. The model should allow modularisation and flexibility in order to be able to create, implement and introduce new services and applications based on the same framework.

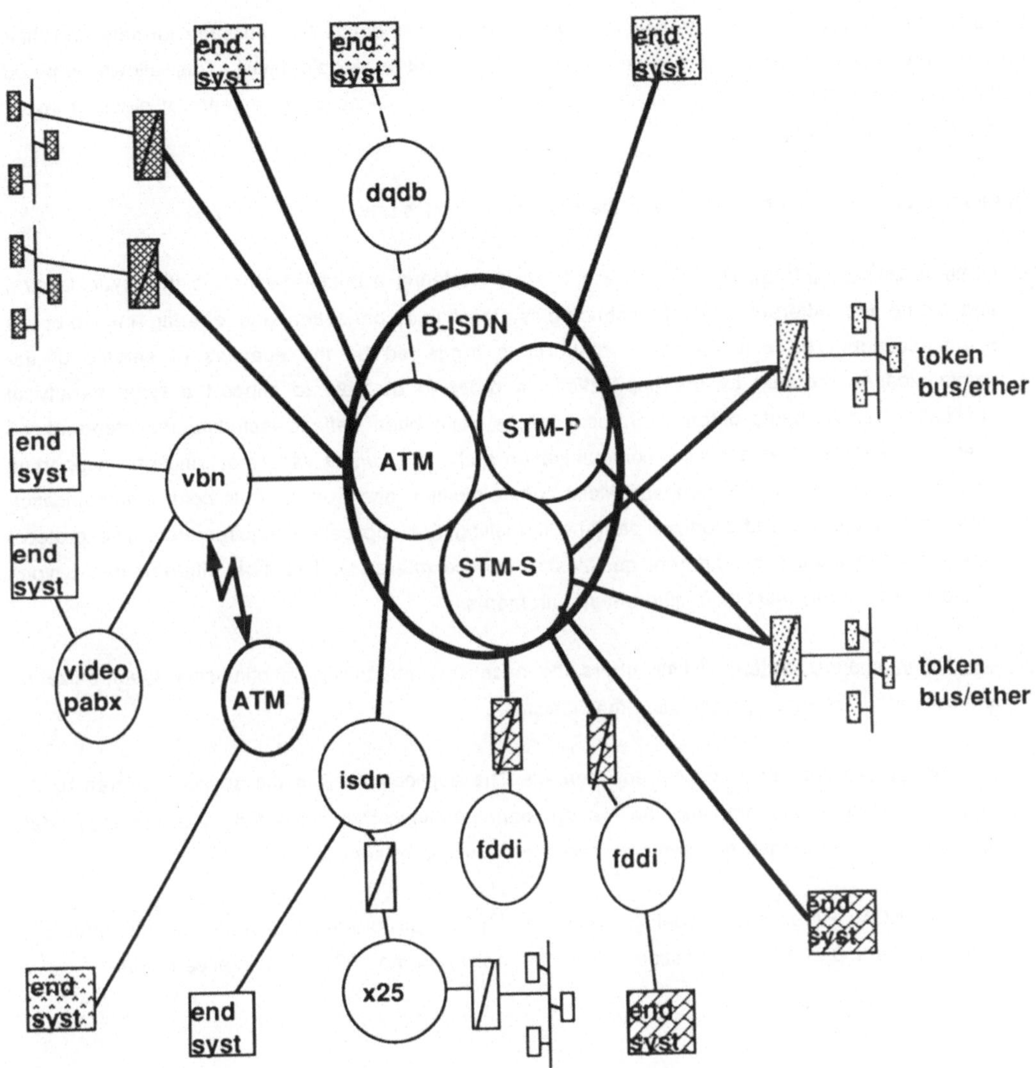

Fig.2

The BERKOM Reference Model as shown in fig. 3 is based on enhancements of the ISO/OSI Reference Model. It has the same basic structuring mechanisms as the ISO/OSI one: layers, services and protocols. The motivation for the adoption of the same technique and discipline is on one hand the easy integration of the existing telematic services and on the other hand the flexibility offered which is necessary for the development of mulitmedia end-systems and applications.

Note that the term "service" used below is defined by the capabilities offered by each layer in the model. In other words the functionality of each layer is defined completely by the service offered to the layer above, hiding all layers below. The protocols within each layer define the message formats and their allowed sequences in order to provide the functionality of a certain layer. This allows required modularization and flexibility of the communication architecture since developments at different layers can be done independently.

Three main platforms can be identified in the BERKOM architecture:

- communication platform including layers 1 - 4 are offering a unique service to the layers above, and hiding the different network technologies, their interconnection and offering inter-process communication. The communication platform is trigge-red by the qualities of service of the communication required by the application. In order to be able to support a large variety of application requirements different protocol stacks have been defined including the standardized ones. The service offered by the communication platform offers a variety of functions supporting from connection oriented to connectionless communication and from point-to-point communication patterns to multicast and arbitrary patterns modelling the application requirements. The protocol stacks chosen modell the different qualities of service offered and support different traffic types (video, audio, data) with their different requirements.

- application support platform which offers the necessary structuring mechanisms synchronization and common information presentation mechanisms.

- the applications serving a specific environment. The application using the services offered by the application support platform and the management functions may instantiate the necessary data structures and model their requirements based on qualities of service.

  The BERKOM Reference Model provides additional structuring mechanisms for functional distribution on top of a communication architecture based on the ISO-OSI Reference Model.

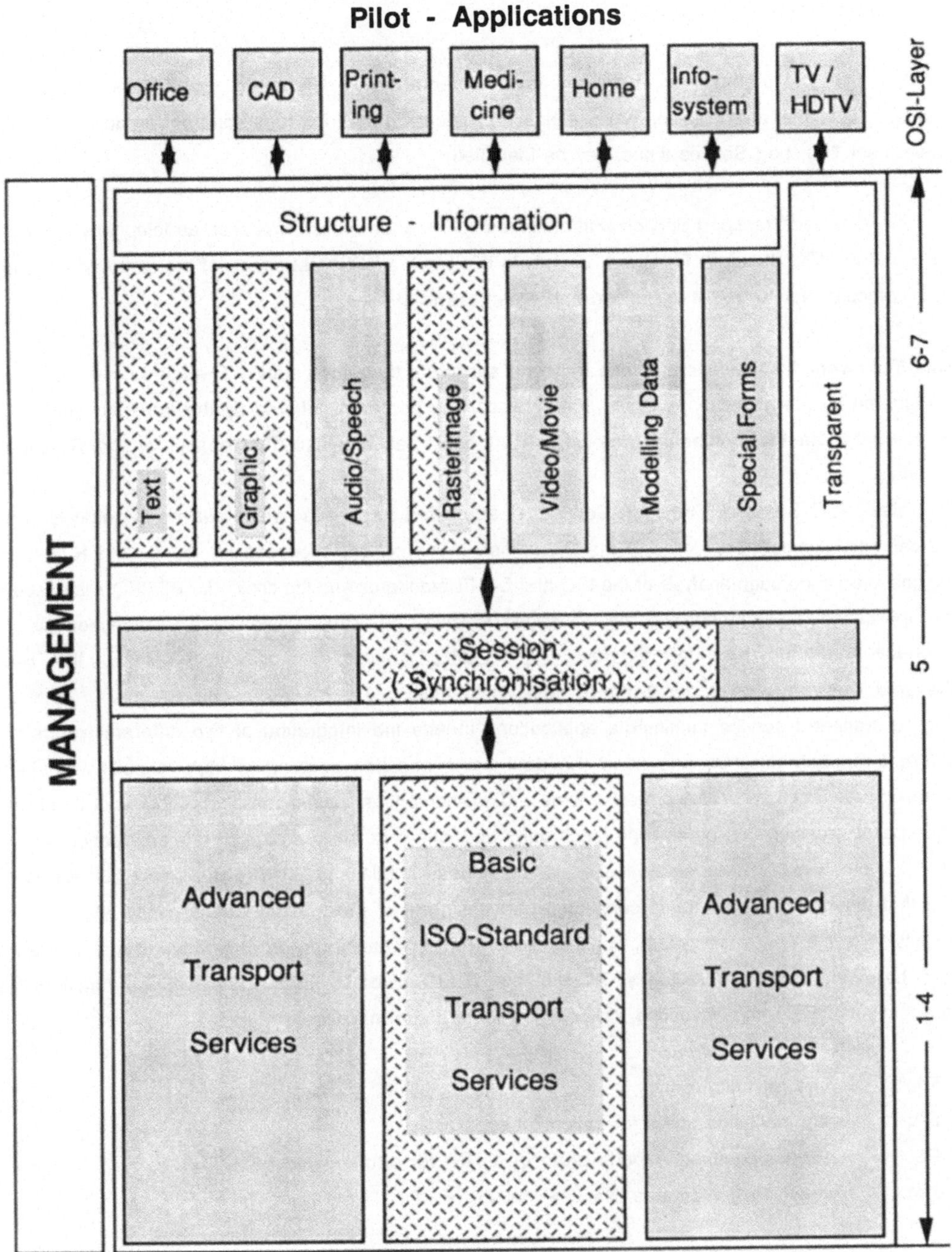

BERKOM - Reference Model

- ISO/OSI RM (existing standards)

3.1 Layers 1-4

A major technical problem for layers 1-4 is the integration of different characteristics and existing standards for different application types (voice, TV, telematic etc.) into one coherent model.
Three main Transport Service types may be identified:

1) Advanced Transport Service with minimal overhead for applications such as telephony, TV etc..
2) Basic ISO Standard Transport Service. In this block a BERKOM choice of the standard - services and protocols - for layer 1-4 is defined and organized in profiles.

Note that layers 1-4 provides a unique transport service to the layers above. The use of a certain block is triggered by the application via the quality of service parameter offered by the transport service. B-ISDN offers, via the D-channel protocol, an out-band call establishment phase for the B- and H-channels.
The "Basic ISO Standard Transport Service" is structured in profiles, which can be negotiated during the call establishment procedure of the D-channel. The profiles proposed for the B- and H-channels are based on a thorough analysis of the ISO and CCITT standards and the choice for BERKOM is based on the application requirements and the envisaged topology [6]. They provide both "Connectionless and Connection-oriented Transport Services" [9]. Table 1 outlines the possible profiles in the "Basic Standard Transport Service" for B- and H- and V channels.
For the transport service multimedia applications means the integration of two different worlds with different requirements. On one hand the data communication demanding high reliability requiering protocols based on hand-shaking mechanisms and anisochronous communication and the other hand video and speach transmission demanding isochronity and where reliability is not of primarily concern.
The OSI profiles are not optimally suited for continuous media and B-ISDN characteristics due to the fact that they include procedures consuming processing time and transmission capacity which are not necessary in the new environment. For this reason protocol profiles for an Advanced Transport Service must be defined which guarantee an efficient Transport Service by reducing the protocol functionalities to a minimum. Following Advanced Protocol Profiles are currently focussed on:

- R1: Frame definition only
- R2: Frame definition and error indication by CRC
- R3: R2 and a checkpoint mode protocol (CMP) for error/flow control [3]
- R4: Light-weight protocol as like XTP, VMTP or LTP [4, 5, 12]

Profil - Nr. / ISO-Layer		Connectionless -		Connection - oriented -				
		1 . 1	1 . 2	2 . 1	2 . 2	2 . 3	2 . 4	2 . 5
4 Transport	Service	ISO 8072 DAD1 CLTS	ISO 8072 DAD1 CLTS	ISO 8072 COTS	ISO 8072 COTS	ISO 8072 COTS	ISO 8072 COTS	ISO 8072 COTS
	Protocol	ISO 8602 CLTP	ISO 8602 CLTP	ISO 8073 Class 4 COTP	ISO 8073 Class 4 COTP	ISO 8073 Class 0 COTP	ISO 8073 Class 2 COTP	ISO 8073 Class 0 COTP
3 Network	Service	ISO 8348 DAD1 CLNS	ISO 8348 DAD1 CLNS	ISO 8348 DAD1 CLNS	ISO 8348 DAD1 CLNS	ISO 8348 CONS	ISO 8348 CONS	ISO 8348 CONS
	Protocol	ISO 8473 Inactive Subset CLNP	ISO 8473 Full Protocol CLNP	ISO 8473 Inactive Subset CLNP	ISO 8473 Full Protocol CLNP	T.70 Minimum Netw. Layer CONP	T.70 Minimum Netw. Layer CONP	ISO 8208 ≙ X.25 PLP CONP
2 Data Link	Service	ISO 8886 CLLS	ISO 8886 CLLS	ISO 8886 CLLS	ISO 8886 CLLS	ISO 8886 COLS	ISO 8886 COLS	ISO 8886 COLS
	Protocol	ISO 8802/2 + Framing + CRC CLLP	ISO 8802/2 + Framing + CRC CLLP	ISO 8802/2 + Framing + CRC CLLP	ISO 8802/2 + Framing + CRC CLLP	ISO 7776 ≙ X.25 LAPB COLP	ISO 7776 ≙ X.25 LAPB COLP	ISO 7776 ≙ X.25 LAPB COLP

CLTS - Connectionless Transport Service
CLTP - Connectionless Transport Protocol
COTS - Connection-oriented Transport Service
COTP - Connection-oriented Transport Protocol
CLNS - Connectionless Network Service
CLNP - Connectionless Network Protocol
CONS - Connection-oriented Network Service
CONP - Connection-oriented Network Protocol
CLLS - Connectionless Link Service
CLLP - Connectionless Link Protocol
COLS - Connection-oriented Link Service
COLP - Connection-oriented Link Protocol

Table 1

The Advanced Transport Service should support new applications and distributed communication services, where the transaction, the distribution and conferencing functions are integrated in the Transport Service. Those functions consist of unrealiable and reliable multicasting and broadcasting and could be provided in connectionless or connection-oriented protocols. Protocol profiles to support this functionality are currently under discussion.

For the BERKOM transport profiles a common frame format have been defined:

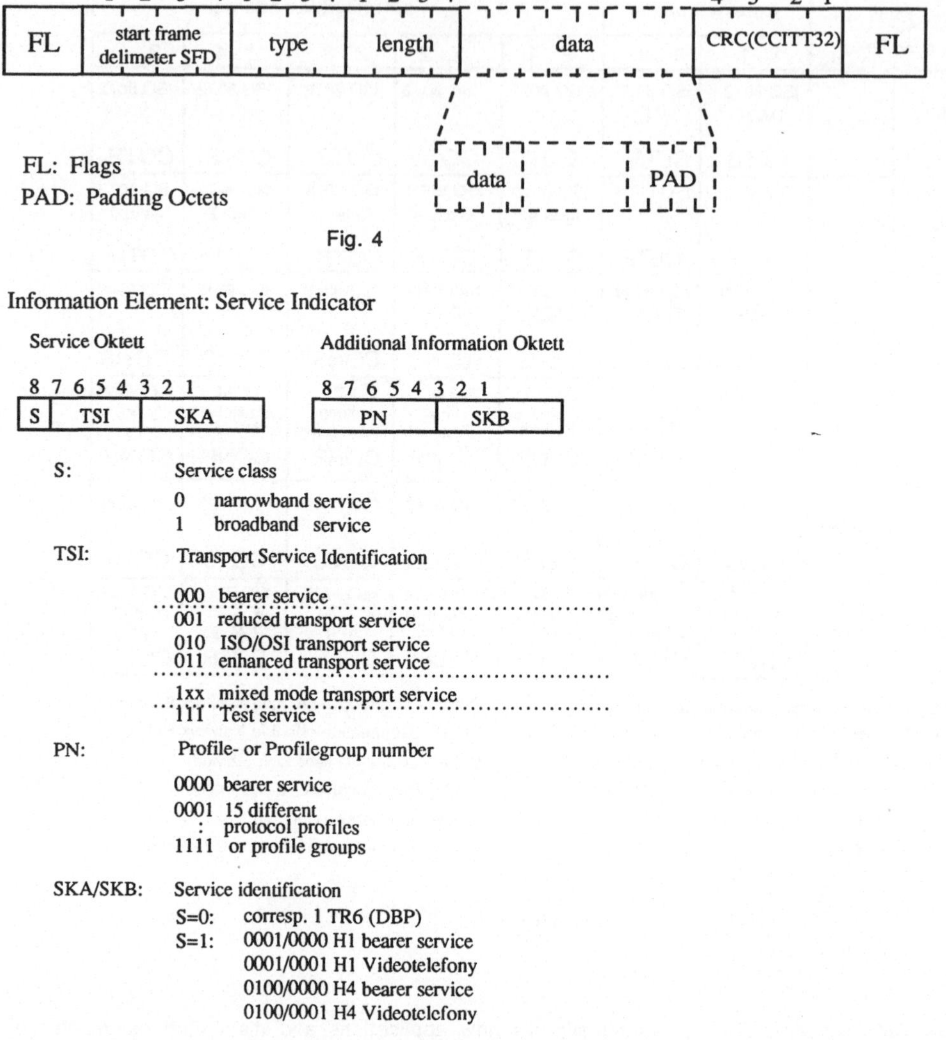

Fig. 4

Information Element: Service Indicator

Service Oktett

8 7 6 5 4 3 2 1

S	TSI	SKA

Additional Information Oktett

8 7 6 5 4 3 2 1

PN	SKB

S: Service class
- 0 narrowband service
- 1 broadband service

TSI: Transport Service Identification
- 000 bearer service
- 001 reduced transport service
- 010 ISO/OSI transport service
- 011 enhanced transport service
- 1xx mixed mode transport service
- 111 Test service

PN: Profile- or Profilegroup number
- 0000 bearer service
- 0001 15 different
- : protocol profiles
- 1111 or profile groups

SKA/SKB: Service identification
- S=0: corresp. 1 TR6 (DBP)
- S=1: 0001/0000 H1 bearer service
- 0001/0001 H1 Videotelefony
- 0100/0000 H4 bearer service
- 0100/0001 H4 Videotelefony

Table 2: Transport Service Selection by D-channel SET UP

B-ISDN offers - comparable to the narrowband ISDN out-band signalling. Connections (B-, H- or VC-channels) are established in an outband call setup phase. Different parameters of the call can be passed to the peer endsystem by "information elements" and define on successful call "CONNECT" the quality of the connection. This feature of outband signalling makes possible to inform the peer endsystem on the required Transport Service (TS) functions, the communication between the two endsystems should/must have. If the called endsystems can provide the requested TS-functions, the connection can be established. If not, an alternative TS-functionality can be signalled back or the call can be cleared by either endsystems.

The required Transport Service and the selected protocol profile can be signalled by coding this in the "Service Indicator" element. The possible alternatives are summarized in Table 2. Applications may exchange different types of information. An example can be the transfer of multimedia document, on one high speed connection where text sequences are followed by video frames, images and then by audio information and so on. For performance and synchronization reasons, is desirable to switch from a protocol profile to onather on an established connection instead of using a dedicated channel for each form of information type. In the context of the BERKOM project, we decided to support connections with "static profiles" and "dynamic profiles" (or mixed profiles fig.5): The requested behaviour of a connection is signalled via the "Transport Service Indicator" (fig.6) at call setup time. At this time, one specific behaviour for the connection is selected. In the case of "dynamic profile selection" the actual protocol profile must be indicated in Layer 2a, by the "type field" of each transfered frame. This means that the endsystems must be able to support parallel protocol machines:

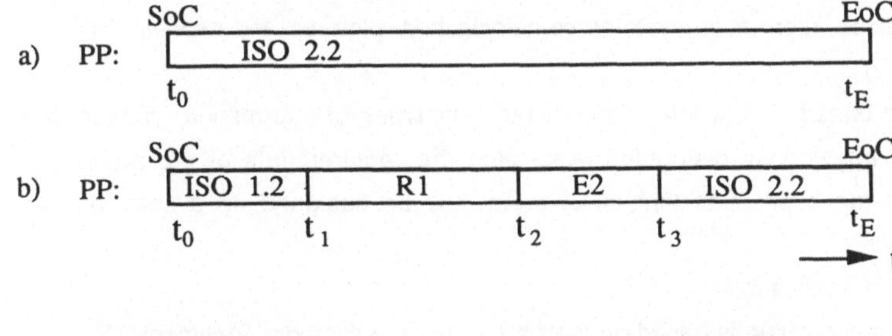

a) static protocol profile
b) dynamic protocol profile (mixed mode)

PP: Protocol Profile
SoC: Start of Connection
EoC: End of Connection

Fig. 5

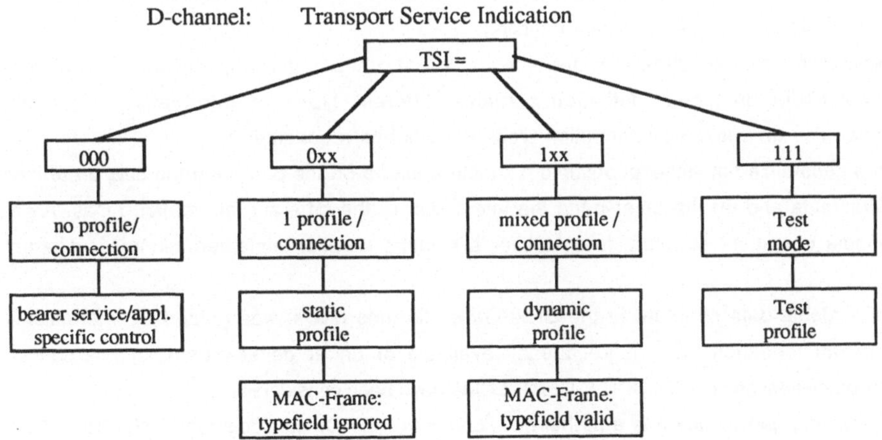

Fig. 6

3.2 The Upper Layers of the BERKOM Reference Model

New applications may be designed based on the new qualities offered by the broadband environment. The application fields shown in figure 6 on which BERKOM focusses are:

Office: Multimedia and document (text,voice, video, raster image) processing in an office en vironment

CAD: Interactive working in the field of Computer Aided Design, Computer Aided and/or Inte grated Manufacturing

Publishing: High quality previewing and printing in a publishing environment

Medicine: Analysis of X-ray or other raster images combined with textual or/and voice diagnostics

Home: Using high quality image and audio information in the private area.
Distribution of (high quality) television programs to one or more receivers

Infosystem: Access to centralized information systems containing text, pictures and movies (for example TV/HDTV)

The applications envisaged are based on the integration of different types of information (speech, text, video, graphics, etc.) and are functionally distributed. Analysing the requirements of the above pilot applications one may arrive at the conclusion that all of them may be designed on a common data structure and common behaviour.

There are two basic categories of applications:

- Applications in which the communication is based on the exchange of multimedia documents. Here the main problem is the definition of the structure of the multimedia document.
- Applications which are based on real-time interaction between users via audio-visual information exchange and working on a common object, a (multimedia) document. Main problems to be solved are again the common multimedia datastrucure and additionaly the integration of isochronous transmission necessary for the real time components with the anisochronous, high reliable communication necessary for the document handling.

The common generalized data structure provides a model for all of them, the model of a multimedia document. The multimedia document definition contains different types of information. A certain application field can use one or more information types to model its requirements.

The definition of a generalized multimedia document structure allows on the one hand the support of new application requirements and on the other hand the integration of the different information types. A generalized multimedia document structure is a tool for alleviating parallel implementations for different applications.

The definition of a global data structure in the BERKOM Reference Model was guided by international standards and recommendations which are already available or under development. In this context standards and Recommendations of ISO, CCITT, CEPT, ECMA have been considered.

The global data structure can be seen as a further development of the Office Document Architecture ODA and should serve as a multimedia document data structure including substructures for text, graphic, speech, raster, movie, modelling data and special forms. Each data type is represented by a specific module in the multimedia document.

In the BERKOM Reference Model an application in an office environment (text, graphic, speech) and an application in the field of medicine (raster, speech, text) for example, uses subsets of the same global data structure via the common interface between the applications and the BERKOM communication modules located in layer 6 and 7.

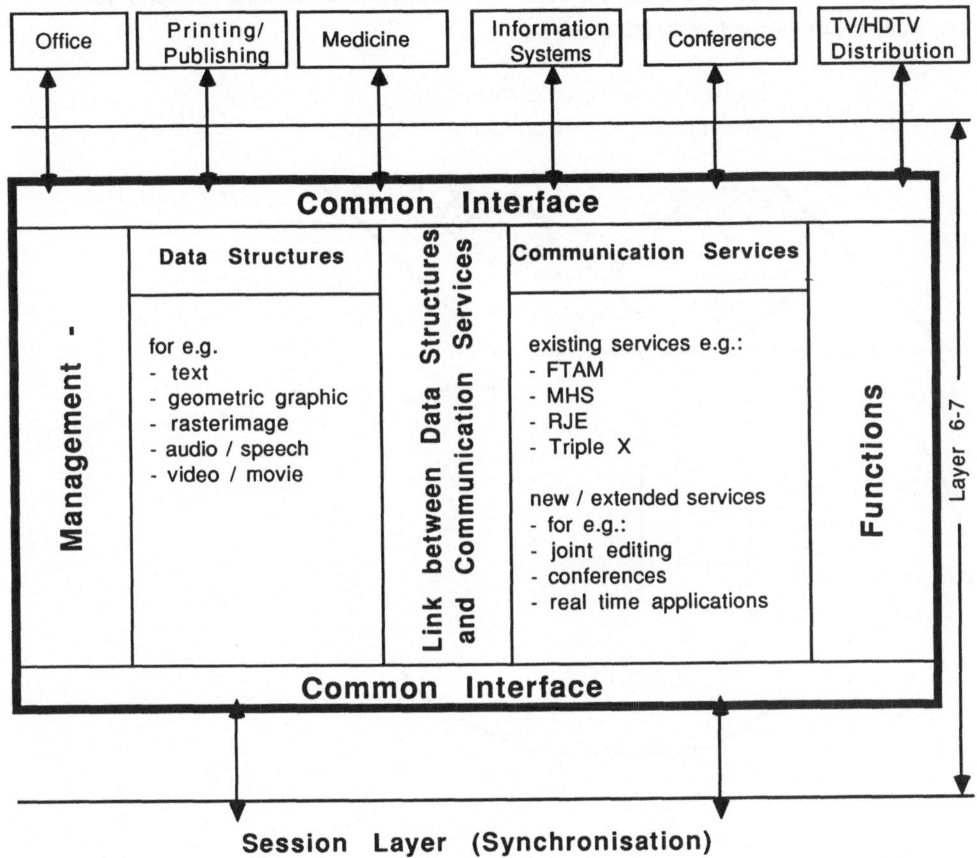

Fig. 7

The modules in the multimedia document are independent of each other. Each module can be developed separately and modified internally, as long as the functionality of the modules is preserved. In order to be able to integrate these independent modules, structural information for the multimedia document is required. The structural information module is divided in two parts:

- specific to each data structure,
- generally serving as a semantic bracket between particular modules.

Hierarchical relationships are allowed between the basic data structures (e.g. the geometric graphic data structure might be a subset of the modelling data structure etc). There are two alternatives for handling a multifunctional document:

use of the existing ISO/OSI services and protocols (MHS, FTAM, etc.). This alternative implies certain restrictions in the operations allowed on a multifunctional document due to the functionality provided by these services and protocols,

definition of new services and protocols for the layers 6 and 7. In this case these services and protocols should be specific to the operations on the multi-functional data structure. Both alternatives are followed up in BERKOM (fig 7).

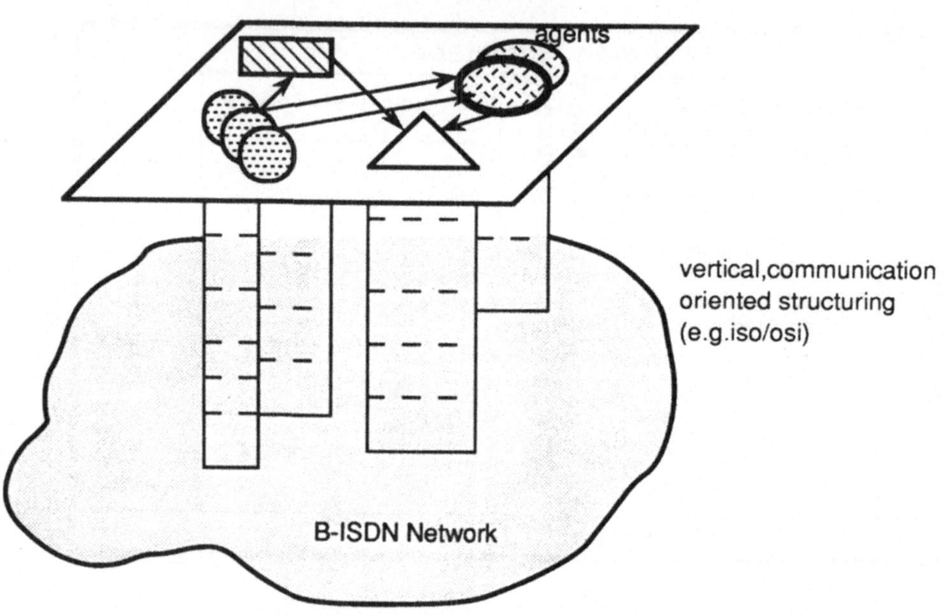

Fig.8

As already mentioned the applications envisaged in BERKOM are functionally distributed. If the ISO-OSI Reference Model focusses mainly on point-to-point communication between entities functionally equivalent, the BERKOM Reference Model develops concepts and mechanisms for the cooperation of diverse functional entities to provide a framework for the development of distributed applications. The roles identified in each application field (fig. 8) are supported by a common layered communication architecture (vertical/communication-oriented structuring). The need for an Open Distributed Processing (ODP) reference architecture for the distributed application fields is to be satisfied with the help of BERKOM. The BERKOM Reference Model provides a management function to support these characteristics (figure 7).

4. Conclusions

The paper presents the main rationales and technical solutions adopted by BERKOM. Several projects have been started for the pilot implementation of the different modules identified in the BERKOM Reference Model. Due to the precise definition of the services offered by each module the integration work will be eased and a global system will be operated and demonstrated at the end of 1989. The participants in the BERKOM projects represent over 30 computer manufacturers and research institutions.

5. References

1. /CEP 86/ Studies on Broadband Aspects of ISDN-Status Report, GSCB-CEPT, DOC T/6SCB,86
2. /BER 87/ BERKOM-Referenzmodell Version 01 4/6/87 - BERKOM-Dokument 0075/06/87
3. /BUX 88/ W. Bux, P. Kermany, W. Kleinoder, Performance of an improved Data Link Control Proc. 9th ICCC , Tel Aviv Israel, 1988
4. /CCITT 88/ CCITT Blue Book, Integrated Services I.451, Digital Network (ISDN), Recommen dations of the series I, Recommendation I.451, ISDN, Usernetwork interface layer 3 specification, 1988
5. /CHE 86/ O. R. Cheriton: "VMTP: A Transport Protocol for the Next Generation of Communication Systemes", Proceddings of SIGCOMM '86, ACM, August 1986
6. /HEN 87/ BERKOM/AK1-Dokument: BERKOM-Architekturmodell für die ISO/OSI-Schichten 1-4, L. Henckel, G. Goldacker, B. Henckel, Forschungszentrum für Offene Kommunikationssysteme FOKUS, GMD Berlin, F.R. Germany
7. /HUI88/ C. Huitema, W. Dabbous, Real Time Communication in LAN Proceedings EFOC-LAN 88, Amsterdam, The Netherlands, June 1988
8. /ISO 7498/ International Standardization Organization (ISO): Information Processing Systems-Open Systems Interconnection: Basic Reference Model, IS 7498,
9. /ISO 8072/ International Standardization Organization (ISO): Information Processing Systems-Open Systems Interconnection: Transport Service Definition IS 8072,1984
10. /ISO 8348/ International Standardization Organization (ISO): Information Processing Systems-Data Communications - Network Service Definition, DIS 8348, 1984
11. /ISO 8348/ Information Processing Systems - Data Communications DAD2 - Network Service Definition, Addendum 2, Network layer addressing
12. /MAP 88/ Manufacturing Automation Protocols: Specification Version 3.0, Gen. Motors,88
13. /PEI88/ Protocol Engin. Inc. XTP Protocol Definition Rev 3.2, Santa Barbara, USA
14. /SPA 86/ Standard Promotion and Application Group: Guide to the Use of Standards, Rev3/86
15. /SWO 86/ "Communication Support for Distributed Processing: Design and Implementation Issues", L. Svoboda, IBM Europe Seminar 1986, Oberlech, Austria
16. /TOP 88/ Technical and Office Protocols: Specification, Boeing, 1985

Constructing Image Databases for the Collection of Art Museums

Tsuyoshi Teshima

Multimedia Communication Center
DAI NIPPON Printing Co., Ltd.
Ichigaya-Kagacho, Shinjuku-ku, TOKYO, 203 Japan

1. Introduction

Art Museums, possessing a large number of art works, can open only a small portion of them to the public due to limited exhibition space and preservation considerations. Thus, for museums today, image databases of art collections for electronic image presentation systems are extremely useful. Indeed, such systems have already been implemented with great success in several museums. High-definition television (HDTV) technology applied to high-quality still image presentation systems further improves the image quality and integrates various types of information including text, graphics, and audio.

In recent years, HDTV image database systems have been introduced to several public art museums in Japan. The image databases are created using high-quality image processing techniques based on printing technologies.

2. Basic Technology

Printing, the original medium, has also taken a place in the expansion of multimedia systems. Indeed, recently there has been a growing demand for the true reproduction of electronic images such as video, computer graphics, and HDTV images onto printed materials.

2.1 Printing Video Images

Printing of conventional video images (NTSC, PAL, or SECAM) was achieved early on through the Video Printing System (VPS). The VPS converts a digitized video image into the high-quality image data required for printing through an image quality improvement process, pixel interpolation between scan lines, and adjustment of color and gradation. The final image is exposed onto a color transparency film with a film recorder and the film then used for printing in the conventional way. The quality improvement is, however, limited due to the small amount of data from the original video signal, therefore, the process is unsuitable for large-sized printing.

2.2 Printing Computer Graphics Images

Direct printing of the digital image data of computer graphics was realized several years ago with the Computer Graphics Printing System (CGPS). The CGPS is a digital image processing system for creating high-quality color separation film for printing. RGB image data are converted into four-color (Yellow, Magenta, Cyan, blacK) screened-dot ink density data. The reproduction process includes color and gradation adjustment, pixel interpolations, size conversion, antialiasing, and other image processing. A color scanner is then used to output the four-color separation films for high-quality printing. This non-photographic process yields sharp and vivid images suitable for printing of large-sized materials such as posters or calendars. (Fig.1)

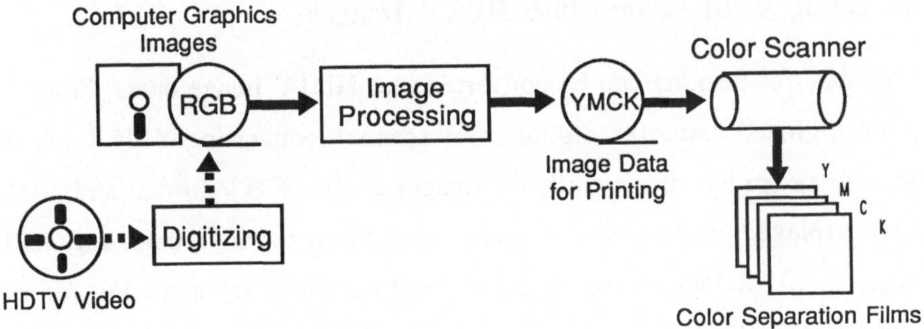

Fig.1: Printing Process of CG images and HDTV images

2.3 Printing HDTV Images

CGPS technology is applied to the printing of HDTV images. The amount of data contained in HDTV images is several times that of conventional video data and can be converted into print-quality images through digitizing and an image quality improvement process.

This technique has led us to HDTV publishing, and several books containing color images from HDTV broadcasting programs have already been published. (Fig. 2)

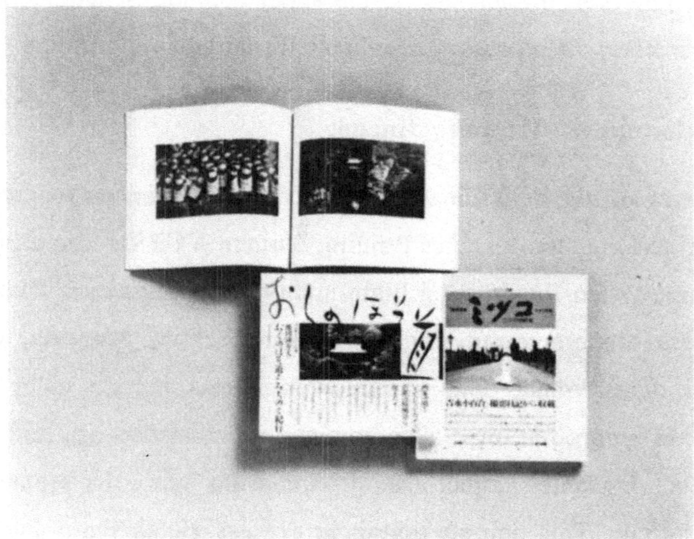

Fig. 2: Examples of the HDTV publishing books

2.4 Converting Print Images into HDTV Images

Image data for printing can be converted into HDTV image data. Color films for printing are digitized using the digital color scanner, converting YMCK ink data into RGB data. Image processing records the images in the HDTV format and print-quality images are displayed on the screen. Today, in addition to video tape, various kinds of media such as CD-ROMs, video discs, and optical discs are available for recording HDTV still images. (Fig. 3)

We are promoting this application called Hi-Vision Graphics which boasts the following features:

1) Displays various kinds of information including text and graphics (display capacity: 3,000 to 4,000 characters per screen)
2) Makes the most of graphic design through use of a wide screen
3) Enables economical software production through reuse of print data
4) Provides effective presentation of voice information and music

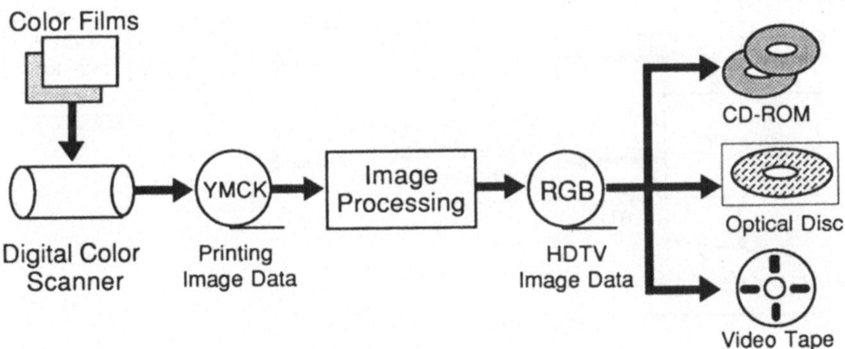

Fig.3: Process of HiVision Graphics

3. HDTV Image Database in an Art Museum

In April 1989, the first HDTV image database system called "Hi-Vision Gallery" was introduced at the "Museum of Fine Arts, Gifu" which is located in Gifu city, 400 kilometers west of Tokyo. The museum is well known for its collection of works by Odilon Redon, the French Symbolist painter .

3.1 System Overview

Fig.4 shows the overview of the system. A 32bit personal computer (PC) controls the whole system. Image data are recorded on CD-ROMs of the HDTV Still Picture Player System and output as HDTV signals. The character fonts for screen display (including kanji, or Chinese characters) are created by the HDTV Character Font Generator, which has 40x40 dot bit map fonts of our traditional "Shuei" typefaces. The

PC graphics for operating menus are converted into HDTV signals and then all the information is displayed on a single HDTV monitor.

The text information and the database index are stored on the hard disk of the PC. The digital audio for narration and music is also recorded on CD-ROM.

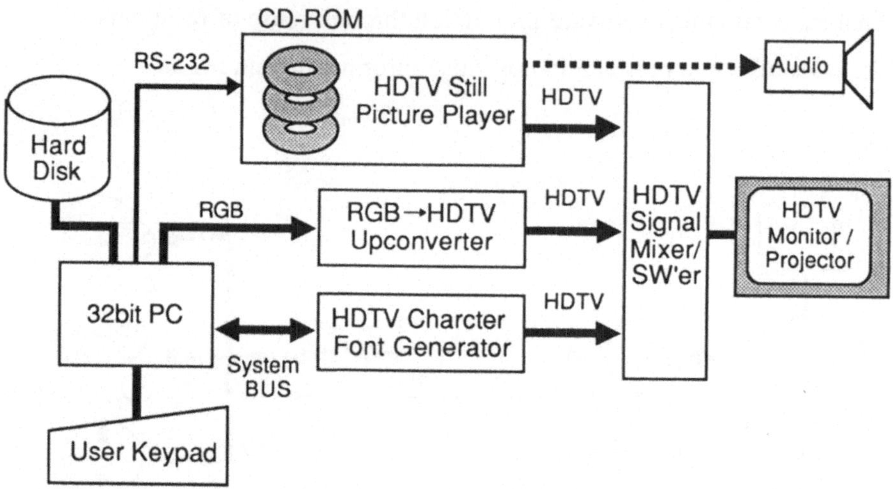

Fig.4: Overview of HDTV image database

3.2 User Interfaces

Both program screening services and database services are available to the gallery's visitors. In the program screening galleries, they can choose one of 30 titles of five-minute long HDTV still image programs, with narration and music, which are focused on the artists and their works. Visual effects like scrolling or wiping are programmed for effective presentation.

In the database booths, visitors can retrieve the image and text of any work from the collection database which is classified by artist, medium, and production date, etc. Two kinds of retrieval interfaces are provided: the general retrieval for general visitors and the specialized retrieval for researchers or curators. Fifty items of specialized information are registered such as artistic commentaries, artist biographies, and relevant literary materials. As at March 1991, the database contained information on 1,100 works of art.

3.3 Creation of the Image Database

The image database is created through the following procedure. The works of art are photographed on 4"x5" color films. The films are then digitized using the digital color scanner for printing. The HDTV paint system performs image processing, YMCK to RGB conversion, color and tone adjustment, trimming, and size conversion. Filtering is also carried out to reduce the flickers caused by the interlacing. True reproduction of art works requires frame-by-frame retouching by the operators. After composing the image, text, and other aspects of the design, HDTV format image data are created. These become the master data for the CD-ROM production. The text database is compiled through the CTS (computerized typesetting) printing process.

4. Further Applications

HDTV image database systems have already been installed in several art museums in Japan, and more are scheduled for installation.

Just as the text processing techniques in the printing industry promoted the creation of text databases, the image processing techniques of today will improve the quality of image databases. Moreover, the technical basis of the multimedia databases which integrate text, image, and audio, is already firmly established.

Fig.5: The Program Screening Gelleries

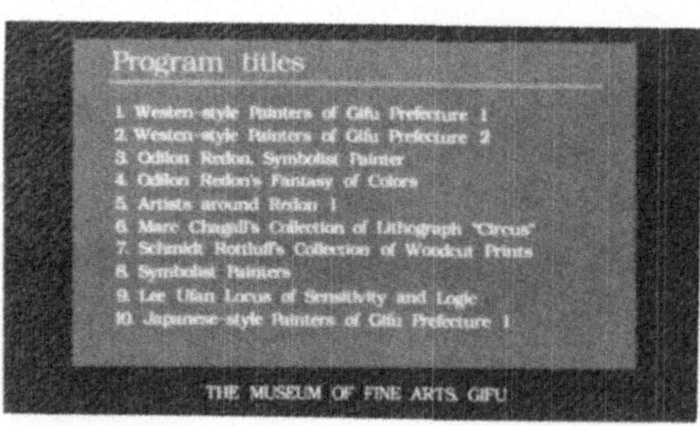

Fig. 6: The Program Selection Menu

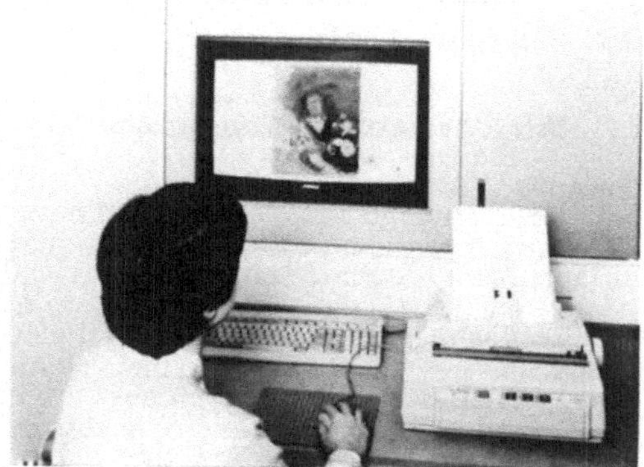

Fig. 7: The Database Booth

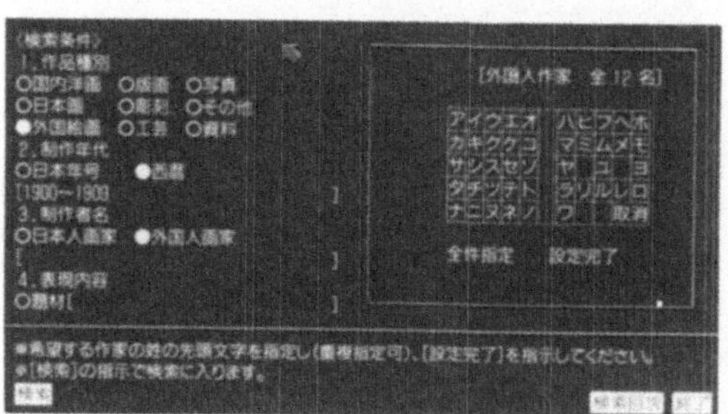

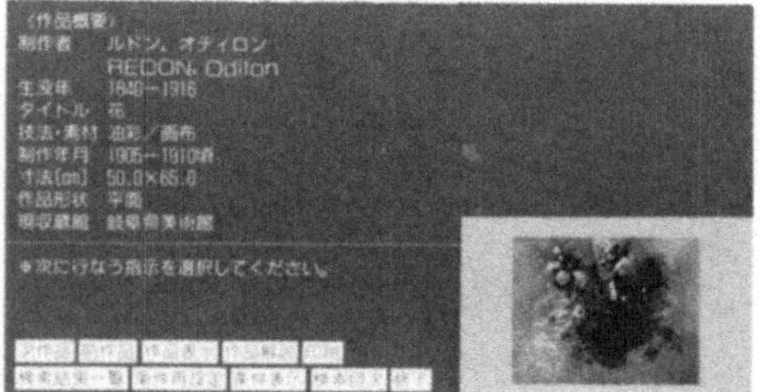

Fig.8 Example of The Database Retrieving

HDTV-Technologie in Europa

Statusreport und Neue Anwendungen

Michael Hausdörfer, BTS Darmstadt

Vereinte Anstrengungen der europäischen Industrie haben in kurzer Zeit einen Rückstand gegenüber der japanischen Technik aufgeholt und einen beachtlichen Stand der HDTV-Technologie erreicht.

Seit 1986 arbeiten zahlreiche Firmen im Eureka-Projekt EU95 zusammen an dem Ziel, eine Übertragungskette für hochauflösendes Fernsehen HDTV zu realisieren.
Diese umfaßt die Bildaufnahme und Produktionsbearbeitung ebenso wie datenratenreduzierte Verfahren zur Übertragung zum Teilnehmer, einschließlich der Endgeräte, im Sinne eines HDTV-Rundfunkdienstes.

Eine erste Präsentation anläßlich der IBC in Brighton 1988 konnte das Potential des neuen Mediums an Hand erster Produktionen der Öffentlichkeit vermitteln. In 1992 werden während der Olympischen Spiele (Albertville und Barcelona) sowie während der Weltausstellung in Sevilla Übertragungen des HDTV in HD-MAC-Format vorgenommen werden, die an ausgewählten Plätzen in Europa empfangen werden können; hierzu wird zunächst ein erstes Los von etwa 1000 Empfängern zur Verfügung stehen.

Die Standardisierung des HDTV-Signals hat hierbei besonderes Gewicht. Im Gegensatz zu dem japanischen HDTV-Standard auf der Basis 1125Z/60,2:1, dessen Kennwerte in keinem harmonischen Verhältnis zu den Abtastparametern der traditionellen Fernsehsysteme stehen, stehen die Parameter des europäischen Lösungsansatzes im diskreten Verhältnis zu den bestehenden Fernsehsystemen, um eine Abwärtskompatibilität zwischen den Systemen mit einfachen Mitteln zu erzielen. Der Versuch, progressive Abtastung gegenüber der heute durchgängig angewandten Interlace-Abtastung durchzusetzen, mußte vor den technologischen Realitäten zurückstehen, da die Verarbeitung der dabei anfallenden hohen Datenrate heute mit ökonomischen Mitteln und der erforderlichen betrieblichen Sicherheit noch nicht möglich ist.

So verwendet der europäische HDTV-Standard derzeit 1250Z/50 mit Zeilensprung in der HDI-Konfiguration. Die Anwendung progressiver Abtastung in der HDP-Konfiguration wird als Ziel weiter beibehalten.

HDTV ist ein Sammelbegriff für eine Familie höherzeiliger Fernsehsysteme; die Anzahl der Zeilen ist wesentlich höher (x2), die Anzahl der Bildelemente in Zeilenrichtung ist mehr als doppelt so hoch wie bei den herkömmlichen Systemen, wenn man die Breitbilddarstellung im Seitenverhältnis 16:9 berücksichtigt. Der Vorteil des HDTV, dem Betrachter bei kleinerem Betrachtungsabstand (E/H=3) einen größeren Bildausschnitt etwa um den Faktor 5,3, gegenüber dem heutigen Standardfernsehen anbieten zu können, ist für das Realitätsempfinden des Betrachters relevant.

HDTV ist demnach auch für eine Großbildanwendung geeignet; allerdings sind Zweifel angebracht, ob sich hierfür eine breite Akzeptanz finden wird, wenn im Heimbereich Bilddiagonalen >1,50m propagiert werden. Auf dem Gebiet des HDTV-Displays besteht gegenwärtig ein technologischer Rückstand.

Die Realisierung der Studiogeräte zur Bildaufnahme und -verarbeitung machte zahlreiche technologische Neuentwicklungen erforderlich. Für die Farbkamera z.B. wurden hochauflösende Aufnahmeröhren erforderlich. Das Auflösungsvermögen der bisher verwendeten Aufnahmeröhren reichte für HDTV nicht aus, da die MTF wegen der wesentlich kleineren Bildelemente zu früh abfiel. HDTV-taugliche Neuentwicklungen von Plumbicon- und Saticon-Aufnahmeröhren konnten hier Abhilfe schaffen. Die Auflösung herkömmlicher Kameraobjektive war ebenfalls nicht genügend hoch, was in der Folge zur Verwendung neuer Gläser und Bearbeitungsverfahren bei der Herstellung speziell hochauflösender Objektive für HDTV führte.

Die Aufzeichnung der HDTV-Signale muß erheblich größere Signalbandbreiten verarbeiten. Genügten noch ca. 5,5 MHz Bandbreite für das konventionelle Fernsehen, so muß bei HDTV zunächst eine Bandbreite von ca. 44 MHz (später 60 MHz), d.h. um den Faktor 8 - 10 mal größer aufgezeichnet werden. Auch hierfür wurden Lösungen durch Steigerung der Aufzeichnungsdichte und der benötigten Bandfläche entwickelt, die sich in der Praxis bewährt haben.

Für den Bildmischer sind Breitbandkoppelfelder zur Signalanwahl, Misch- und Effekteinheiten sowie ein Chromakey-Prozessor entstanden, der den hohen Anforderungen des HDTV voll entspricht.

Diese und weitere notwendige Systemkomponenten bilden heute den technologischen Stand des HDTV in Basisbandkonfiguration.
Mit ihnen werden gegenwärtig mobile und ortsfeste Versuchsstudios ausgestattet, die Produzenten den praktischen Umgang mit HDTV näher bringen sollen; hierzu gehören z.B. die Gesellschaft "Vision 1250" und das Technologiezentrum in Oberhausen, HDO, die gegenwärtig ausgerüstet werden.

Die technologische Evolution von HDTV ist aber noch nicht abgeschlossen. Bereits während der Entwicklung der Phase 1 zeigte sich, daß die Systemreserven hinsichtlich des Signal-Rauschabstands bereits im HDTV-Basisband erheblich geringer sind, als dies bei Standardfernsehsystemen der Fall ist. Dies war zu erwarten.
Dies betrifft neben der Bildaufnahme insbesondere die Signalaufzeichnung, bei der eine systematische Reduktion des Störabstands bei Überspielungen wesentlich früher als beim Standardfernsehen eintritt.

Hieraus erwächst im praktischen Einsatz eine Beschränkgung der Produktionsbearbeitung, die im Versuchsbetrieb hingenommen, im praktischen Einsatz jedoch nicht akzeptiert wird.

Ein Ausweg aus dieser Situation bietet die digitale Aufzeichnung des HDTV-Signals. Ein digitales Interface-Format mit 8- bzw. 10-Bit-Quantisierung und 72-MHz-Abtastfrequenz ist bereits erarbeitet und in der Beschlußphase.

Untersuchungen über die Verteilmedien Koaxialkabel oder Glasfaser werden gegenwärtig betrieben.

An der Entwicklung eines digitalen Recorders für HDTV wird gegenwärtig intensiv gearbeitet. Die hohe Datenrate - sie beträgt mehr als 1,2 Gbit/s - stellt außerordentlich hohe Anforderungen. Dies bedeutet in letzter Konsequenz, daß alle übrigen Studiokomponenten mit digitaler Signalverarbeitung verfügbar sein müssen, um die Systemkomponenten des Studios in digitalem Format zusammenfügen können.

HDTV ist mehr als ein erweitertes Fernsehsystem. Mit der Großbilddarstellung, ohne Beschränkungen durch Zeilenstruktur und hoher spatialer Auflösung bieten sich zahlreiche andere Anwendungsfelder des HDTV, wie z.B. in der Medizin, der technischen Konstruktion, der Ausbildung und nicht zuletzt der Werbung.

Es ist nur zu wünschen, daß HDTV und die Computertechnik zum gegenseitigen Nutzen künftig stärker als bisher in Verbindung treten.

Literatur

H. Schönfelder: Probleme eines HDTV-Studios, NTZ Bd.42 (1989) H.9 544-554

G.M. Walker: HDTV Part 2, World Broadcast New, H.11 (1989) S. 60-66

U. Reimers: From 4:3 to 16:9, Aspect Ratio - a Big Change in TV Production and Broadcasting, Conv.Rec. 17th International Television Symposium and Technical Exhibition, Montreux, June 1991, S. 15-23

Multimedia Telecommunications toward the 21st Century

Takahiko Kamae
NTT Human Interface Laboratories
1-2356 Take, Yokosuka-shi, 238-03 Japan

Abstract: Telecommunication networks are moving toward multimedia services. NTT made ISDN service commercially available in June, 1988, in Japan, and now provides a complete set of ISDN service including H_0 and H_1, user to user information, and B and D packet services.

Multimedia applications of ISDN are being developed. Video conferencing based on CCITT H.261 is available. ISDN facsimiles are being widely spread over the country along with the decrease of their cost. Standard color picture transmission based on the JPEG standard is drawing attention in various fields. ISDN interface boards are stimulating personal computer applications to ISDN.

The boardband ISDN (B-ISDN), which is believed to be the infrastructure in the 21st century, is being developed in parellel with its standardization at CCITT. Based on B-ISDN, NTT announced the service vision toward the 21st century by the name VI & P (Visual, Intelligent and Personal communications system). Personal pocket telephones, text mails and visual telephones are expected to be three basic telecommunication services in the future. The development of various applications has been started toward the in-house VI & P experiments at NTT laboratories. ISDN and B-ISDN will enhance multimedia telecommunications at offices and gradually at home.

1. Introduction

NTT inaugurated ISDN commercial service in 1988. The number of ISDN customers is over 30,000 in March, 1991. In parallel with ISDN implementation, NTT is developing the broadban ISDN (B-ISDN) toward the 21st century.

The applications of ISDN are also being developed. Video conferencing and video telephone, which are believed to be among the most important services using ISDN, were standardized by CCITT. NTT developed a one board codec to visualize the standard video conferencing. The standardization of color picture transmission codec, which is generally called a JPEG codec, is exploring another application of ISDN to visual information. An ISDN pay telephone, which was installed in 1990, enables the general public to become familiar with ISDN. ISDN is also promoting multimedia personal computers (PCs).

B-ISDN is expected to push telecommunications more toward multimedia services. Asynchronous transfer mode (ATM) technologies facilitate B-ISDN to support a much wider variety of information media from data to a high definition TV (HDTV).

NTT announced a telecommunication service vision toward the 21st century by the name VI & P (visual, intelligent and personal telecommunication services). One of the most important keywords associated with B-ISDN is Fiber-to-the-Home (FTTH). FTTH will enhance the "visualization" of telecommunications. The picture quality of visual telephone service in B-ISDN will be made much better than in the present ISDN (called N-ISDN hereafter), and so the applications will be of wide variety.

In this paper described are the Japanese status of N-ISDN service and its multimedia applications, and the B-ISDN effort toward the 21st century.

2. N-ISDN and multimedia services

2.1 N-ISDN

The N-ISDN service NTT is providing covers the basic rate (BRI) and the primary rate (PRI) interfaces, both B channel and D channel packet services, user to user information (UUI) service, and switched H_0 (384K b/s) and H_{11}(1.5M b/s) services. In 1990, NTT started ISDN pay telephone service in Japanese major cities, which enables customers to transmit information at a 64K b/s rate at corners of streets.

The number of N-ISDN customers in Japan is shown in Table 1. About a half of the customers use N-ISDN with terminal adapters (TA). About 30% of the customers is using digital telephones. Just recently the standard video conferencing service was launched by NTT, and it starts contributing to expanding ISDN utilization.

Service	Number of customers
BRI	27313
PRI	560
Packet	4405 (BRI) 24 (PRI)

Table 1 ISDN Customers (at the end of March, 1991)

2.2 Video telephone and video conferencing through N-ISDN

Fig. 1 shows the standard model of video telephone/video conferencing system. For standardization of video codec, the common interface format (CIF) for TV signal was defined to enable the interconnection of the 525/60 (North American, Japan) and 625/50 (Europe) TV systems, as shown in Fig. 2. The standard codec encodes TV signal with CIF. The interchange between a national TV standard and the CIF can be done in each country. For low cost video codec, a quarter CIF (QCIF) was also defined.

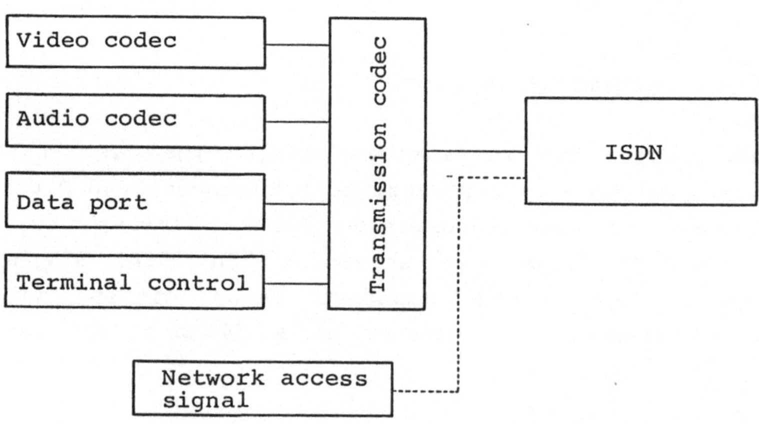

Fig. 1 Standard video telephone/video conferencing system

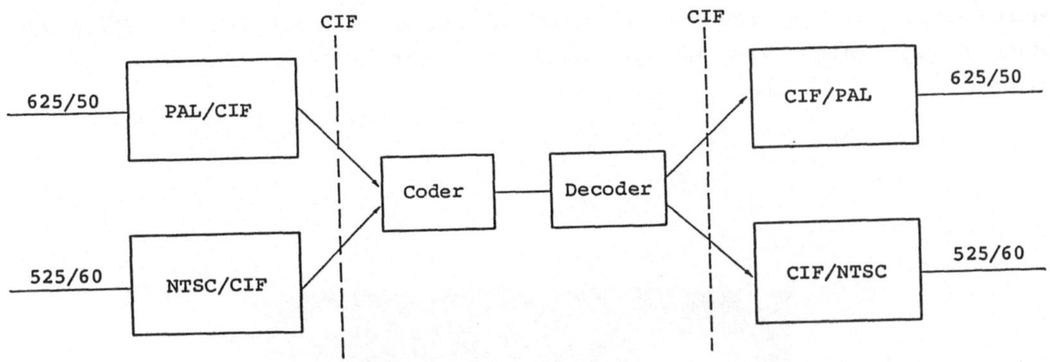

Fig. 2 International Connection through the Common Interface Format (CIF)

Table 2 shows typical usages of ISDN channels to video telephone/video conferencing service. The use of B or 2B is likely suitable to video telephone, and H_0 or H_1 to video conferencing. The combination of QCIF, 48K b/s video and 16K b/s audio will facilitate offering low-cost video telephone service.

Channel	Division	Coding
H_{11}	B (audio) + 23B (video)	audio: SB-ADPCM PCM video: hybrid coding
H_0	B (audio) + 5B (video)	
2B	B (audio) + B (video)	
	16k (audio) + 112k (video)	audio: being standardized video: hybrid coding
B	16k (audio) + 48k (video)	

Table 2 Bitrates for video telephone/video conferencing

Fig. 3 shows a one-board video codec developed by NTT. This codec can be used to B and 2B in Table 2. Its functionalities cover NTSC/CIF conversion, hybrid video codec and transmission codec. A 16K b/s audio codec can be mounted as a child board on this board. Four 240 MOPS DSPs, which was newly developed, encodes and decodes CIF TV signal up to 128K b/s.

Fig. 3 Standard video codec (28 cm x 28 cm)

2.3 Color picture transmission

The standard color picture codec, based on the JPEG standard, was developed as shown in Fig. 4. This codec has an interface to the VME bus, and can encode a picture having maximally 8192 pixels in one direction. By assigning bitrates suitably to brightness and color difference signals, the data compression rate can come down to 1/20 without noticeable degradation.

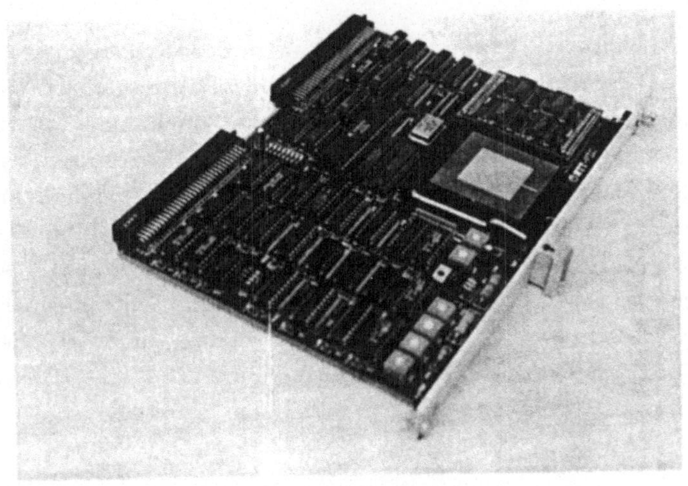

Fig. 4 JPEG coded

An optical disk memory and the JPEG code board are attached to a UNIX workstation to constitute a color picture filing system, as shown in Fig. 5. A color copier attached to the workstation can be used as an I/O terminal. Color pictures are transmitted through ISDN. A JPEG codec is useful to save transmission time and storage capacity.

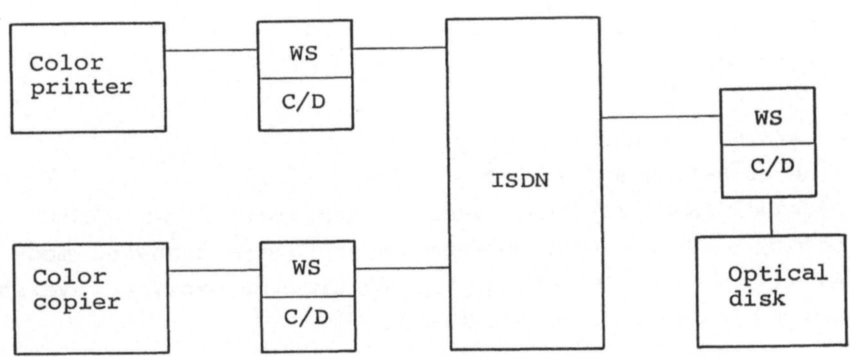

C/D : JPEG codec
WS : Workstation

Fig. 5 Still Color Picture Filing and Transmission

2.4 Multimedia PC communication

The addition of ISDN interface board to a personal computer (PC) turns PC to an ISDN terminal. Using two 64K b/s (B) channel in the ISDN basic rate interface, voice and text/image data can be transmitted simultaneously. The ISDN interface board developed by NTT has telephone capability using a B channel and 64K b/s data transmission capability using the other B channel. It can be applied to simultaneous transmission of voice and data. If a 32K b/s ADPCM voice codec is attached, the simultaneous voice and data transmission can be done through one B channel. The voice and image data may be stored in the hard disk of PC in advance, and transmitted as a 64K b/s data stream; then a multimedia electronic mail can be realized.

2.5 N-ISDN and consumer electronics

The above-mentioned digital technologies are giving an impact on consumer electronics. For example, a digital electronic camera is likely to have a JPEG codec, and images taken by the camera are digitally encoded and stored in an IC card. NTT is marketing an ISDN terminal called a telephotographic system. An IC card is set on this system and then the stored data is transmitted through ISDN. Thus photographic pictures can be transmitted to a remote spot instantaneously.

3. B-ISDN and VI & P service vision

3.1 Outline of B-ISDN and VI & P

The broadband ISDN (B-ISDN) being standardized at CCITT can be understood by such keywords as the asynchronous transfer mode (ATM), the fiber-to-the-home (FTTH), and a multimedia network covering data up through a high definition TV (HDTV).

As shown in Fig. 6 transmitted data stream is divided into sequences of a small amount of data called a cell. Each cell has a 5 octets header and a 48 octets information field. At the transmission line, cells are packed without any gap, as shown in Fig. 6. Various identifiers in the header specify a cell, and contain sufficient information to deliver a cell. When a high bitrate is necessary, more cells are captured. In other words, network customers can define the capacity of a channel at their own will. This is called a virtual channel. A customer can take as many vitual channels as it desires in its subscriber's line.

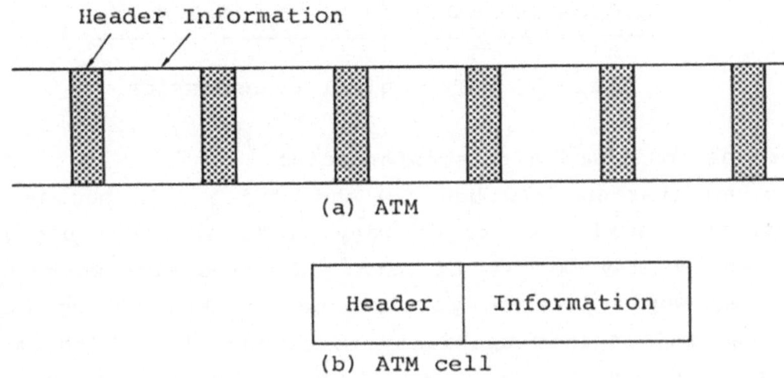

Fig. 6 Asynchronous Transfer Mode (ATM)

B-ISDN is widely believed to be the telecommunication infrastructure in the 21st century. On the basis of B-ISDN NTT announced the 21st century telecommunication service vision by the name VI & P, which is an acronym of the visual, intelligent and personal communications service.

3.2 Multimedia service example

3.2.1 Visual telephone

Visual telephone is believed to be one of the major terminals in the 21st century. Its picture quality wil be much better compared with the CIF based picture in the N-ISDN environment.

3.2.2 HDTV

Depending on the application of the high definition TV (HDTV), necessary bitrates will become versatile, as shown in Table 3. FTTH in B-ISDN will make a wide range of HDTV application easy.

Usage	Bitrate
Studio grade	620M b/s
Program distribution grade	100-150M b/s
High quality entertainment	40-50M b/s
Application specific	10-20M b/s

Table 3 HDTV digital transmission

3.2.3 Personal multimedia teleconferencing

A desktop workstation (WS) used in the office will become multimedia. Office workers would like to do many kinds of works using their WS. Teleconferencing may be one of them. Since office workers has their own WS, they would like to participate in the meeting at their own desk, and so multipoint capability is essential. At the meeting each participant would like to see one another, and so multi-motion-picture-windowing is important. It may be desirable to distinguish who is speaking audibly.

Fig. 7 shows a typical screen image of such a multimedia teleconferencing system developed by us. From the WS a participant can send text, scanned image, telewriting information, his own voice and image, and stored video in color. It has audio window capability. The voice of a person whose window is in the left comes from left, and the voice in the right from right. Another feature is private conversation capability. Any two can talk confidentially in the meeting, and this private conversation voice comes from extreme right.

Fig. 7 Personal multimedia teleconferencing system

3.2.4 Networkcasting

Since B-ISDN transmits a good quality picture, TV programs can be distributed through B-ISDN. This kind of TV casting may be called networkcasting. People living in a big city may be able to watch a baseball game held in their home town at any time.

4. Multimedia computing

4.1 Difference between telecommunication and computer fields

In the telecommunication field "multimedia" suggests "non-voice". On the contrary, texts and numerics have been performing a major part in the computer field. A multimedia computer suggests capabilities of new media such as voice. At the WS used in the personal multimedia teleconferencing system mentioned in 3.2.3, motion picture signal goes to a window subsystem directly, and the only thing the WS does is to control windows. Codec for motion picture signals and control circuits for voice signals are installed outside the WS.

4.2 Multimedia computing

The demand for multimedia computers is rising. However there are various aspects and levels in multimedia computing. Multimedia computing technologies may augment capabilities of the workstation mentioned in 4.2. Motion picture display and audio windowing capabilities could be added to the present window system.

A more enhanced multimedia computer might be able to process motion pictures and audio signals. For example, a multimedia workstation may take every frame of motion picture into its main memory, and obscure the background to make a human image conspicuous. It may be able to detect a human image in each frame and adjust a TV camera so that a human image may always be in the center. At an audio side the workstation may distinguish a human voice from background noise and make the voice clearer. A multimedia database will increase the application of detabases. The progress of multimedia computing technologies is a key factor for N-ISDN and B-ISDN to become widespread.

4.3 Multimedia society

B-ISDN and VI & P will give a great impact on the 21st century society. A portion of newspapers and books might become electronic, and be distributed through B-ISDN. The distance between broadcasting and telecommunications will become closer. In other words, information may be represented using various media, and so people can choose media depending on the information they want. People may be able to access any information at any time, and to access any people at any place.

Fachgespräche

Betriebsaspekte eines großen OSI-Netzes

Fragen des Netzbetriebes erlangen im Rahmen des Managements von großen und gekoppelten Netzen erhebliche technische und wissenschaftliche Bedeutung. Es soll versucht werden, einen Überblick über die Betriebsfunktionen innerhalb des Deutschen Forschungsnetzes (DFN) zu geben. Eine vorrangige Rolle wird dabei das Wissenschaftsnetz (WIN) spielen, ein von der DBP Telekom im Auftrag des DFN-Vereins bereit gestelltes X.25 - Netz für 9.6 kb/s bis 1,92 Mb/s. Zu den Betriebsfunktionen gehören nicht nur Übergänge in andere OSI-Wissenschaftsnetze sondern auch die Gateway-Funktion zu Bitnet und den IP-Netzen für verschiedene Anwendungsdienste, sowie Informationsdienste, wobei der anlaufende Betrieb von X.500-Directory-Systems eine grundsätzlich wichtige Rolle spielt.

Koordinator: Prof. Dr. E. Jessen, TU München

Anforderungen an das Deutsche Forschungsnetz (DFN)

- Ein Abriß der Entwicklung des DFN -

K. Ullmann, DFN-Verein Berlin

1. Einführung

In den frühen achtziger Jahren begannen die Entwicklungen zum Deutschen Forschungsnetz (DFN). Organisatorisch unter dem Dach des "Vereins zur Förderung eines Deutschen Forschungsnetzes" (DFN-Verein) zusammengefaßt, entwickelte eine Gruppe von Wissenschaftlern 1983 ein Konzept zur Entwicklung des DFN, die in /1/ dokumentiert ist. Dieses Konzept unterlegt ein Anwenderszenario, das im ersten Teil von Kapitel 2 beschrieben wird. Dieses Anwenderszenario mußte später korrigiert werden, da durch den rapiden Fortschritt der anzuschließenden Rechentechnik weitere Anforderungen berücksichtigt werden mußten. Für den technischen Entwicklungsprozeß des DFN galten (und gelten) einige Prinzipien, die in Kapitel 3 beschrieben werden.

Anwenderszenarien und Konstruktionsprinzipien ergeben Konstruktionsszenarien, die jeweils Grundlage der Entwicklungsprojekte waren. Sie werden im Kapitel 4 diskutiert. Im Kapitel 5 werden dann einige Teile der Realisierung des DFN dargestellt. Im Kapitel 6 werden die wichtigsten Erkenntnisse zusammengefaßt, wobei die Realisierungen im Kontext der Anforderungen bewertet werden.

Auf wichtige Aspekte wird in dieser Darstellung nur am Rande oder gar nicht eingegangen: Alle betrieblichen Aufgabenstellungen, die organisatorischen Aspekte sowie die Einbettung der Arbeiten des DFN-Vereins in das europäische und internationale Umfeld werden nicht beschrieben. Ferner wird nicht auf neuere (nach 1987 entwickelte) Pläne zum Ausbau des DFN eingegangen.

2. Anforderungen

2.1 Allgemeines

Ein Rechnernetz stellt sich einem Nutzer als eine Menge von Diensten dar, die er zum Zwecke der Datenübermittlung nutzen kann. Die klassischen Netzdienste sind

- der Dialog zum interaktiven Zugriff auf Rechenanlagen
- der Filetransfer zum Senden von Files in ein Filesystem eines entfernten Rechners

- die Elektronische Post zum Versenden von Nachrichten von Personen untereinander

- der Remote-Job-Entry zur Nutzung entfernter Rechenanlagen.

Diese Dienste sind die Basisdienste im DFN, die von Wissenschaftlern genutzt werden können.

In der Bundesrepublik Deutschland wird die Zahl der Wissenschaftler, die Kommunikationsdienste nutzen oder nutzen werden, mit etwa 100.000 geschätzt (siehe /2/, /3/). Diese Zahl kennzeichnet das technische System, das für diese Nutzer entwickelt werden sollte, in mehrfacher Hinsicht:

- Es ist kein System, das ausschließlich von DFN-Experten genutzt wird.

- Es ist ein System, das in der Dimensionierung unter einem technischen Großsystem wie dem Telefonnetz steht, aber deutlich über den klassischen DV-Infrastrukturen (Rechenzentren).

Ferner ist für die Nutzung des DFN charakteristisch, daß die nutzenden Wissenschaftler neben nationalen auch internationale Kommunikationsverbindungen haben.

Eine weitere durch die Erfahrung gestützte Annahme besagt, daß Nutzer den Zugang zu Kommunikationsdiensten über die ihnen zugänglichen DV-Infrastrukturen finden. Eine Darstellung dieser Strukturen findet man u.a. in /2/.

Im weiteren werden Abstraktionen der Szenarien aus /2/ dargestellt, die im jeweils zeitlichen Kontext der DFN-Entwicklungen gültig waren.

2.2 Frühes Anwenderszenario

Zum Beginn der achtziger Jahre waren DV-Infrastrukturen in wissenschaftlichen Einrichtungen gekennzeichnet durch Rechenzentren, deren Aufgabe es war, Rechenleistung für die gesamte Einrichtung bereitzustellen. Die Rechenzentrumsanlagen wurden durch Abteilungs/Fachbereichsrechner ergänzt.

Die Aufgabe, ein DFN zu entwickeln, bestand also darin, auf der Basis einer geeigneten Kommunikationstechnik die DFN-Dienste auf allen wesentlichen Rechenanlagen des Wissenschaftsbereiches zu realisieren (Anschluß des Rechnertyps x an das DFN).

2.3 Weiterentwickeltes Anwendungsszenario

Im Verlauf der letzten 5 - 6 Jahre wurden die Möglichkeiten der Rechnernutzung entscheidend verbessert. Durch die Einführung von PC's und Arbeitsplatzrechnern (sog. Workstations) steht Rechenleistung direkt am Arbeitsplatz des Wissenschaftlers zur Verfügung.

Die lokale Vernetzung - ebenso wie PC und Arbeitsplatzrechner - ist kostengünstig verfügbar. Groß angelegte Beschaffungsmaßnahmen (siehe z.B. /2/) trugen dazu bei, daß sich das Anwendungsszenario für DFN-Dienste damit wesentlich veränderte.

3. Konstruktionsprinzipien

3.1 Allgemeine Prinzipien

Es war die Aufgabe gestellt, vor dem Hintergrund des (sich wandelnden) Anwendungsszenarios umfangreiche Implementierungen der DFN-Dienste zu beginnen. Bei der sehr heterogenen technischen Basis - es waren Implementierungen in ca. 10 unterschiedlichen Systemumgebungen vorzunehmen - war es notwendig, allgemeine Konstruktionsprinzipien zu formulieren. Diese Prinzipien waren:

- Der DFN-Dienst sollte weitgehend auf Systemsoftware des Herstellers (des Systems) aufbauen und soweit wie möglich vom Hersteller selbst realisiert sein.

- In den Fällen, wo dies nicht möglich war, sollte eine professionelle Erstellung der Software (Pflichtenheft, Abnahmeprozeduren etc.) dafür garantieren, daß stabile Implementierungen entstehen.

3.2 Protokollarchitektur

Zur Konstruktion eines Kommunikationssystems müssen dessen Architektur und dessen Kommunikationsverhalten definiert werden. Der DFN-Verein hat sich sehr frühzeitig auf die Verwendung von Standards (ISO/OSI bzw. CCITT) festgelegt. Als Rahmen dient das OSI-Referenzmodell; die Protokolle wurden entsprechend definiert. Im Anhang 1 sind die Definitionen für das DFN bezogen auf das erweiterte Anwendungsszenario zusammengefaßt dargestellt.

4. Konstruktionsszenarien

Die Konstruktionsaufgabe für das DFN zerfällt in zwei große Teile (siehe auch Abbildung 1 im Anhang 2).

a) Bereitstellung eines Weitverkehrsnetzes und

b) Anschluß eines Infrastrukturkomplexes (Einzelrechner, Rechenzentrum, lokales Netz)

Für das Weitverkehrsnetz war die X.25-Empfehlung der CCITT die verbindliche Festlegung für das DFN. Die ersten Realisierungen, die organisatorisch nur unzureichend zusammengefaßt waren, bestanden aus regionalen (X.25-) Netzen und dem öffentlichen Datex-P-Netz (Abb. 2a im Anhang 2).

Beim Anschluß der Infrastrukturkomplexe wurden die im Anhang 2 Abb. 3a, b (vereinfacht) dargestellten Szenarien unterlegt.

5. Realisierungen

5.1 Angeschlossene Rechentechnik

Die im Anhang 4 aufgeführten Realisierungen der DFN-Dienste in Endsystemen wurden mit einem Aufwand von ca. 100 Mannjahren durchgeführt. Für die wesentlich kompliziertere Realisierung des in Abb. 3b Anhang 2 dargestellten Szenarios gibt es ebenfalls Realisierungen. Allerdings stehen einer flächendeckenden Einführung die Produkte der US-INTERNET-Dienste, die dem Protokollnamen folgend auch TCP/IP-Dienste genannt werden, entgegen. Diese Produkte werden für die Mehrzahl der UNIX-Realisierungen kostenlos geliefert. In den meisten Fällen wird der Anschluß an die DFN-Architektur durch sogenannte Kommunikationsserver (s. Abb. 3b Anhang 2) geliefert. Diese Server sind (Applikations-) Gateways - sie sind leider an der Benutzerschnittstelle nicht zu verbergen.

Die Dienstrealisierungen sind gut angenommen, wie die Nutzungszahlen (s. Anhang 5) für die Elektronische Post belegen.

5.2 Das X.25-Wissenschaftsnetz

Die Realisierung des Weitverkehrsnetzes seit 1989 umfaßt ein X.25-Wissenschaftsnetz (WIN) sowie DATEX-P und eine Anbindung an das europäische X.25-Netz IXI (International-X.25-Interconnect), so daß ein Nutzer von seinem Rechner jeden Rechner, der an dieser Struktur angeschlossen ist, direkt erreichen kann. Die genaue Zahl dieser Rechner ist nicht zu bestimmen, da Wissenschaftseinrichtungen Adreßräume jeweils selbst verwalten. Direkt angeschlossen am WIN sind z. Z. ca. 250 Einrichtungen. Die derzeitige Nutzung des WIN ist im Anhang 3 zusammengestellt.

Das WIN wird nicht nur für DFN-Dienste genutzt, sondern auch für TCP/IP-Nutzung, für DECNET usw. Das WIN ist insofern ein Multiprotokollnetz.

6. Zusammenfassung

Es ist im Zeitraum der letzten 6 Jahre gelungen, im deutschen Wissenschaftsbereich die Weitverkehrskommunikation technisch und organisatorisch stabil zu gestalten. Neben den Entwicklungen, die dazu erfolgreich durchgeführt wurden, konnte auch der Betrieb angemessen organisiert werden.

All dies wäre ohne massive Unterstützung durch das Bundesministerium für Forschung und Technologie nicht möglich gewesen.

7. Literatur

/1/ Deutsches Forschungsnetz
Vorschlag für die erste Projektphase
Berlin, Januar 193

/2/ Zur Ausstattung der Hochschulen in der Bundesrepublik Deutschland mit Datenverarbeitungskapazität für die Jahre 1988 - 1991
Kommission für Rechenanlagen der DFG
Bonn, 1988

/3/ COSINE-Studie

Anhang 1: **Erweitertes Protokollszenario des DFN**

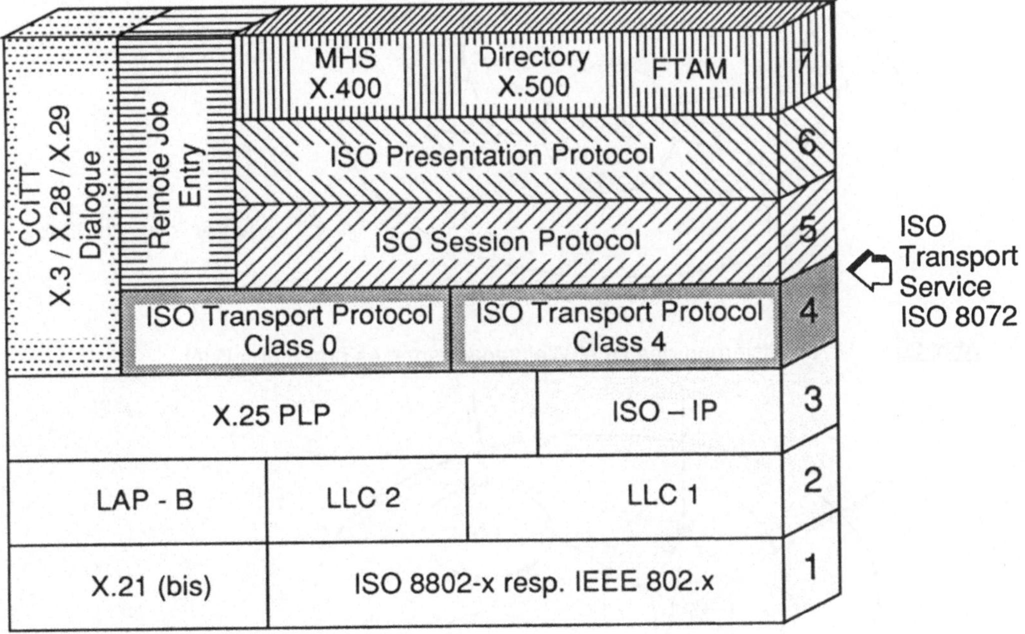

Anhang 2: **Konstruktionsszenarien des DFN**

Abb. 1 Gliederung des DFN

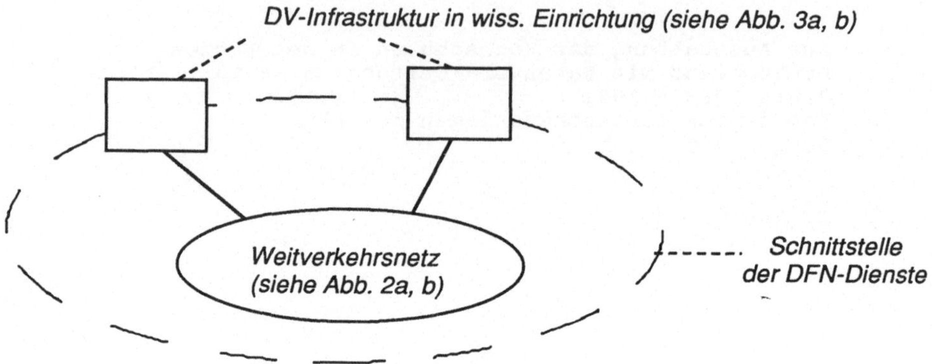

Abb. 2a Frühe Realisierung des DFN-Weitverkehrsnetzes

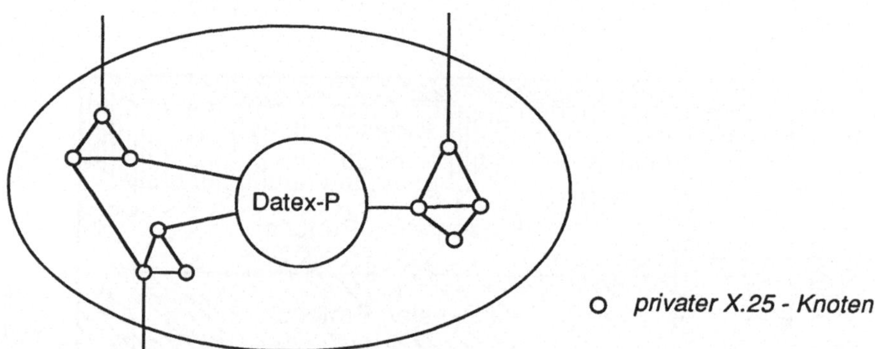

Abb. 2b Realisierung des DFN-Weitverkehrsnetzes durch das WIN

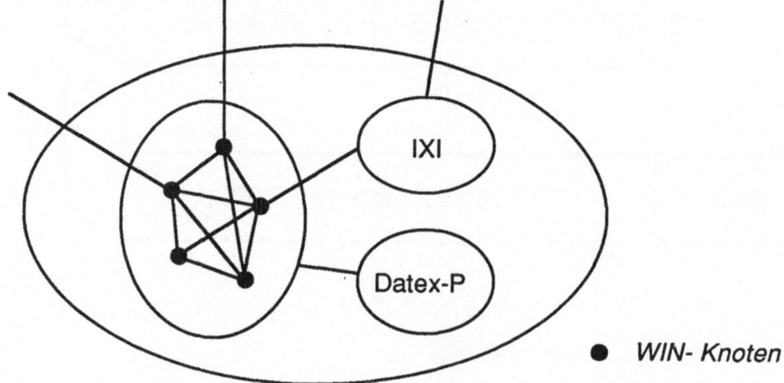

Abb. 3a Frühes DFN-Konstruktionsszenario
DV-Infrastruktur = Rechenanlage

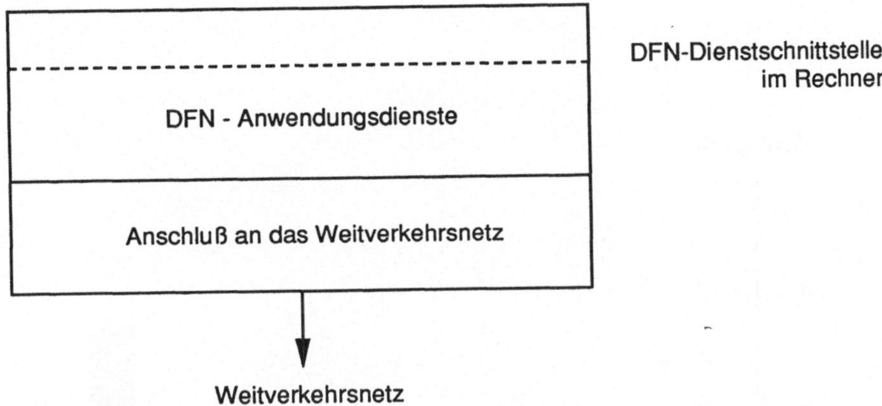

Abb. 3b Weiterentwickeltes DFN-Konstruktionsszenario

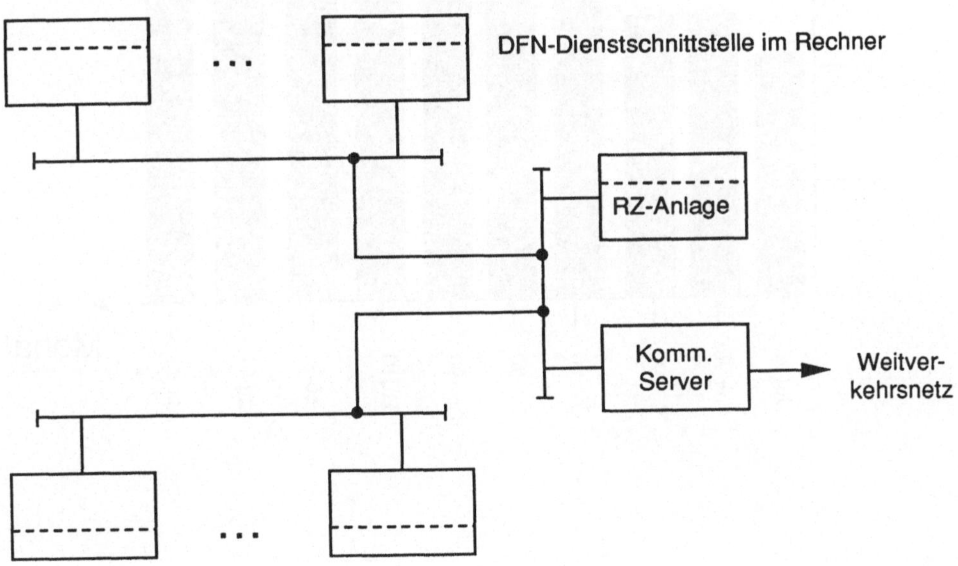

Anhang 3: **WIN - Statistik 1990/91**

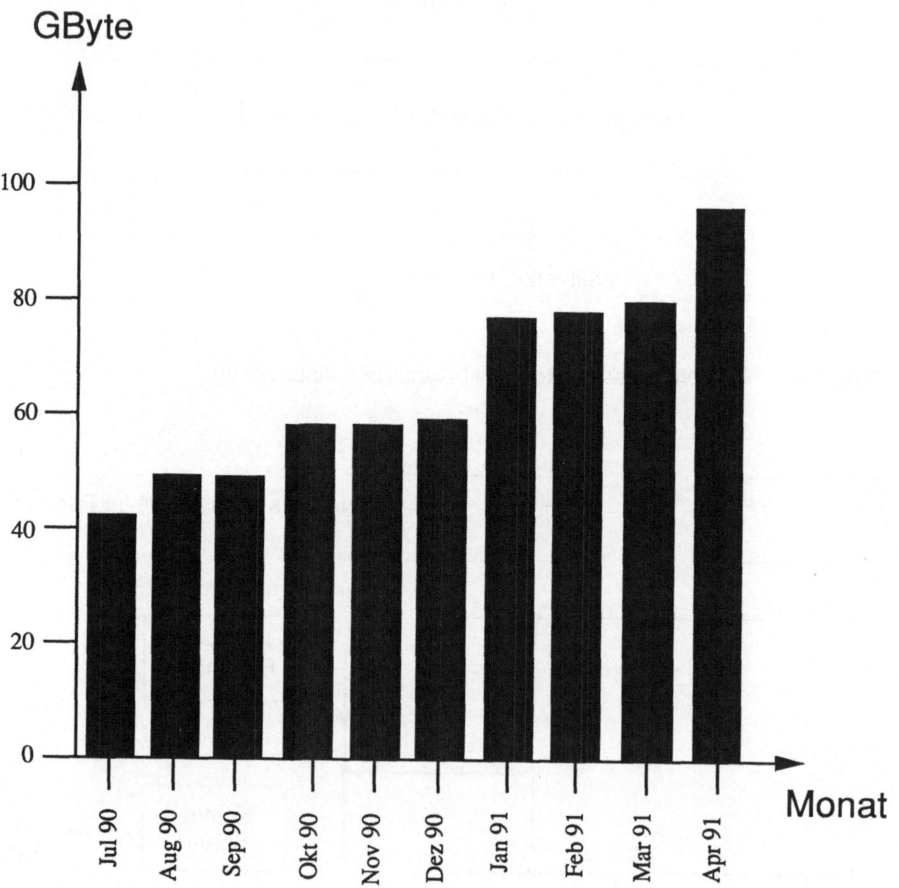

Anhang 4: Verfügbarkeit von DFN-Diensten

Rechnertyp / Betriebssystem

DFN Dienste

	AIX	AT&T UNIX V	Bull UNIX V	Bruker UNIX V	CDC NOS/ VE	Convex C1	Data General AOS/ VS	DEC UNIX/ ULTRIX	GTC XENIX	HP UX 9000	IBM MS-DOS
X.25	H	H	H		H	H	H	H	H	H	H
X.29	H	H	H		H		H	H		H	H
PAD	H	H	H	H	H	H	H	H		H	H
RJE					DFN			DFN			
FTAM	H	H			H		H	H		H	DFN
MHS	H	DFN			H		H	H/DFN	H	H/DFN	

	IBM MVS	IBM VM	Nixdorf Targon UNIX V	PCS MUNIX	Prime Primos	Siemens BS 2000	Siemens SICOMP WS 30	Siemens MX 2	Stollmann	SUN SUNOS	
X.25	H	H	H	H	H	H	H	H	H	H	
X.29	H	H	H	H	H	H	H	H	H	H	
PAD	H	H	H	H	H	H	H	H	H	H	
RJE	DFN	DFN	DFN	DFN	DFN	DFN				DFN	
FTAM	H	H	H	DFN	H	H		H		ISODE	
MHS	H/DFN	H/DFN	H	DFN	H	H		H		H/DFN	

H: Verfügbar Hersteller - Produkt
DFN: Verfügbar als DFN - Produkt

Anhang 5: **Nutzungszahlen für den X.400 Dienst des DFN**

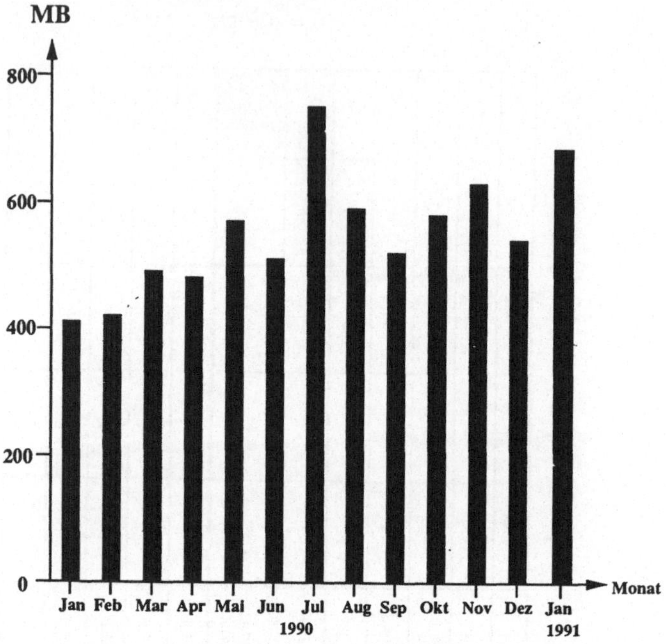

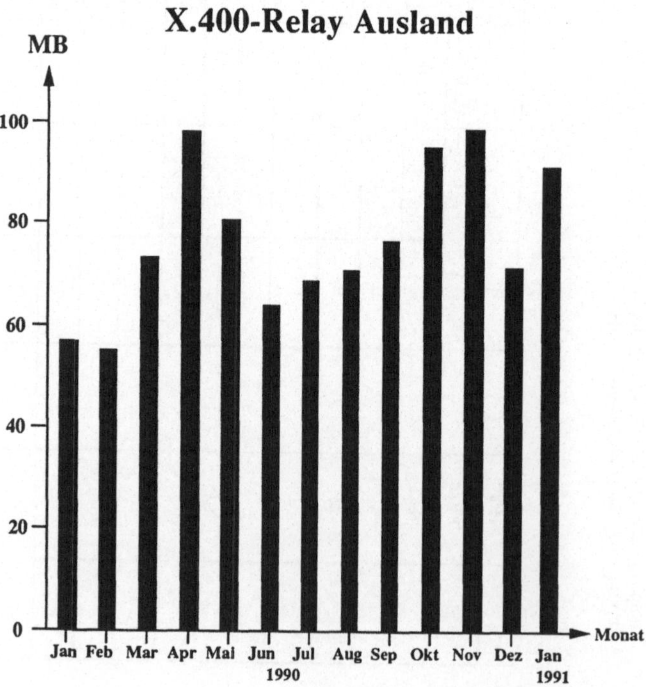

X.400-BITNET-Gateway

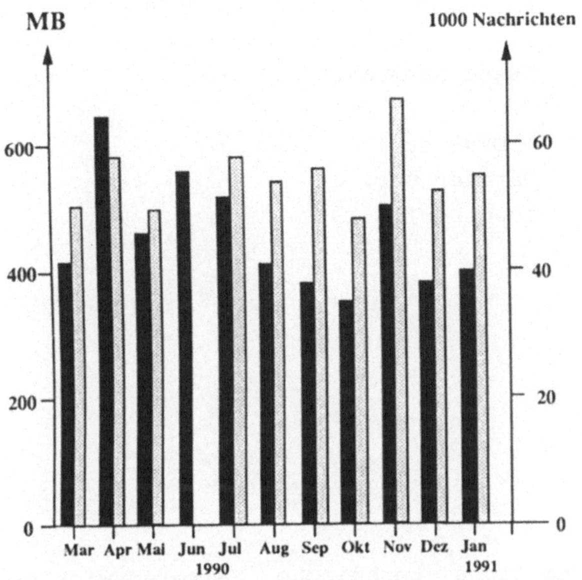

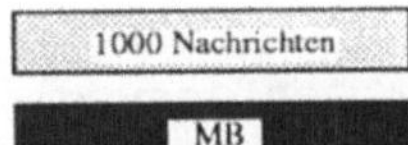

X.400-SMTP Gateway

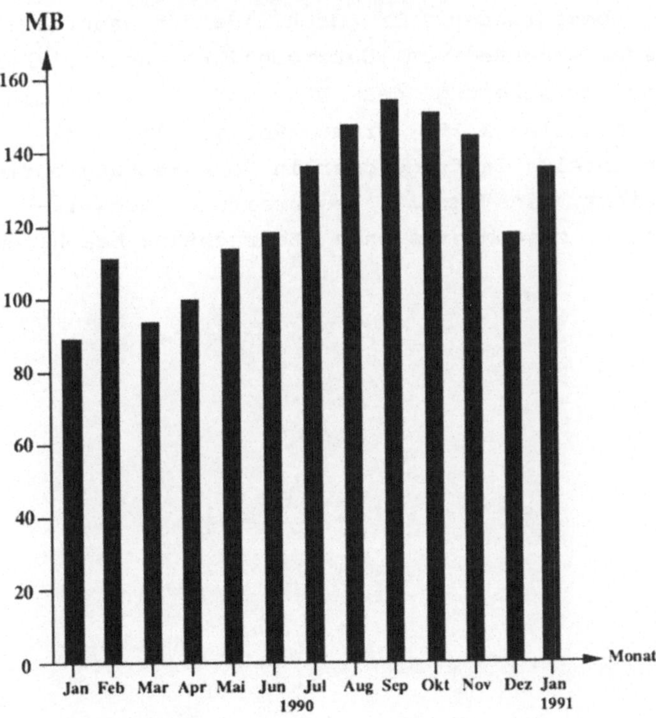

PERSPEKTIVEN FÜR DIE WEITERENTWICKLUNG DES WISSENSCHAFTSNETZES (WIN)

Peter Kaufmann

DFN-Verein
Pariser Str. 44
1000 Berlin 15

1) Zusammenfassung

Der Beitrag diskutiert mögliche technische Weiterentwicklungen für das Wissenschaftsnetz (WIN) des DFN-Vereins. Zu Beginn werden der augenblickliche Stand sowie der Pilotbetrieb für die Erweiterung auf Zugänge mit 2Mb/s beschrieben. Im anschließenden Hauptteil werden wesentliche technische Aspekte von Netztechniken beschrieben, die zur Zeit für eine Erweiterung des WIN auf Zugänge mit 34Mb/s untersucht werden: Schnelles X.25, Frame Relay, ATM, MAN (DQDB). Dabei werden auch bereits laufende oder in Vorbereitung befindliche Pilotprojekte skizziert. Eine kurze Beschreibung des T3-Backbones des NSFNET und eine zusammenfassende Betrachtung beschließen den Beitrag.

2) Stand des Wissenschaftsnetzes

Im Sommer 1989 vereinbarten der DFN-Verein und die Deutsche Bundespost/Telekom (DBP/T), daß DBP/T im Auftrag des DFN-Vereins ein X.25-Paketnetz (Wissenschaftsnetz, WIN) betreiben sollte. Der DFN-Verein als Auftraggeber kauft den Dienst zu einem jährlichen Festbetrag von DBP/T, vergibt die Anschlüsse ebenfalls zu Festpreisen an seine Mitglieder, und ermöglicht ihnen dadurch eine Datenkommunikation zu volumenunabhängigen Gebühren.

Die Aufnahme des WIN-Regelbetriebes erfolgte im Frühjahr 1990. In der ersten Stufe werden vereinbarungsgemäß Zugangsleitungen für wahlweise 9,6Kb/s oder 64Kb/s angeboten. An den fünf WIN-Netzknoten waren im Mai 1991 120 Leitungen mit 64Kb/s-Anschlüssen und 115 Leitungen mit 9,6Kb/s-Anschlüssen geschaltet. Der Gesamtdurchsatz im WIN betrug im April 1991 ca 100 Gigabyte. Einrichtungen mit hohem Kommunikationsbedarf haben sich mehrere 64Kb-Anschlüsse schalten lassen.

Die Verbindung des WIN mit dem europäischen X.25-Netz IXI (z.Z. 64Kb/s) ermöglicht den WIN-Teilnehmern eine kostenfreie Datenkommunikation mit den Kollegen in anderen europäischen Forschungsnetzen.

3) Erweiterung des Wissenschaftsnetzes auf Zugänge mit 2Mb/s

Bei Vertragsabschluß war ein wesentlicher Teil der Vereinbarung, daß das WIN gemäß verfügbarer moderner Technik weiterentwickelt und somit stets auf dem neuest möglichen Stand gehalten werden sollte.

Eine Schlußfolgerung aus dieser Absichtserklärung ist die Arbeit an der Bereitstellung von 2Mb/s-Anschlüssen im WIN am lokalen Anschlußpunkt. Solche Fortentwicklungen des Dienstangebotes sind möglich, weil seit 1990 Produkte von Herstellern von X.25-Untervermittlungen zur Verfügung stehen, die diese Leistungsklasse anbieten.

Der vereinbarte Pilotbetrieb für das WIN-Upgrade umfaßt die folgenden Stufen:

Dezember 1990: Beginn des Testes mit einem abgesetzten WIN-Knoten in Augsburg und zwei Zugangsleitungen zum LRZ München und zur Universität Nürnberg/Erlangen. In beiden Einrichtungen waren frühzeitig lokale "schnelle" X.25-Untervermittlungen installiert worden, die die 2Mb/s-X.25-Anschlüsse in der lokalen Umgebung verteilen können und eine gute Test- und Meßumgebung für den Pilottest bieten.

Juni 1991: Nach den Tests an einem WIN-Knoten erfolgt die Zuschaltung eines weiteren WIN-Knotens in Hannover mit zwei Zugangsleitungen zur Universität Hannover und zum DESY (Hamburg).

In dieser Testphase wird das Zusammenspiel zweier WIN-Knoten auf dem Weg zwischen zwei Endteilnehmern geprüft.

Dezember 1991: Sofern alle Tests erfolgreich verlaufen, soll Ende des Jahres die Abnahme des 2Mb/s-Zuganges erfolgen, so daß ab Anfang 1992 alle WIN-Teilnehmer diesen Dienst regulär in Anspruch nehmen können.

Der DFN-Verein schätzt die Kosten pro Anschluß auf 250.000 DM pro Jahr.

4) Perspektiven jenseits von 2Mb/s

Angesichts der laufenden Einführung eines 2Mb/s-Zuganges im WIN besteht natürlich die Frage der weiteren Entwicklung hin zu Anschlußkapazitäten mit 34Mb/s oder 140Mb/s am lokalen Anschlußpunkt.

Die Deutsche Bundespost/TELEKOM hat in Gesprächen mit dem DFN-Verein ihre grundsätzliche Bereitschaft für eine Erweiterung des WIN auf zunächst 34Mb/s erklärt. Vom DFN-Verein werden daher z.Z. die nächsten Schritte zur Erweiterung auf 34Mb/s (und später gegebenenfalls 140Mb/s) in Angriff genommen. Zunächst werden Informationen über den Stand und die Entwicklung unterschiedlicher technischer Lösungen gesammelt und ausgewertet.

Unter der Randbedingung einer herstellerunabhängigen, an ISO orientierten Lösung werden technische Eigenschaften, Stabilität der Normung und (voraussichtliche) Entwicklung der Produktverfügbarkeit analysiert. Eine möglichst bruchlose Einpassung in das gegenwärtige WIN (bis 2Mb/s) ist dabei ebenso wichtig wie die Weiterentwicklung der nächsten WIN-Stufe zur voraussichtlich übernächsten WIN-Stufe, nämlich der späteren Integration in das Breitband-ISDN (B-ISDN).

Die folgenden Kandidaten für die Hochgeschwindigkeitserweiterung des WIN zeichnen sich zum jetzigen Zeitpunkt ab: X.25 mit LAP-B und G.703, Frame Relay mit LAB-D, DQDB als MAN-Standard, Asynchronous Transfer Mode (ATM).

In den folgenden Abschnitten werden die wichtigsten Eigenschaften unter der Fragestellung "WIN-Upgrade" diskutiert. Dabei werden v.a. die WAN-Eigenschaften diskutiert, während die mögliche Einbettung in LANs nicht Gegenstand der folgenden Betrachtungen ist.

Weitergehende technische Beschreibungen finden sich jeweils in der angeführten Literatur.

4.1) Schnelles X.25

Unter diesem Terminus ist eine Realisierung des X.25 PLP mit den im wesentlichen gleichen Eigenschaften wie im Bereich bis zu 2Mb/s zu verstehen. Statt X.21 als unter X.25/LAP-B liegende Schicht zur Anbindung an 2-Draht-Kupernetze wird aber G.703 mit 34Mb/s zur Anbindung an Glasfaserkabel verwendet.

Zu klärende Fragen befassen sich u.a. mit den End-zu-End Durchlaufzeiten der Pakete, den erreichbaren Durchsätzen in Abhängigkeit von Paket- und Fenstergröße, den Quittierungsmechanismen (End-zu-End oder Knoten-zu-Knoten) sowie den notwendigen bzw. verfügbaren Prozessorleistungen für die

Kommunikation. In <1,2,3> werden die entsprechenden Fragestellungen ausführlich und mit unterschiedlichen Gewichten diskutiert. Insgesamt zeigt sich aber, daß X.25 im WAN-Bereich nicht notwendig auf den Bereich bis 2Mb/s beschränkt sein muß und daß mit entsprechenden Parametereinstellungen, die heute allerdings erst von wenigen Produkten unterstützt werden, durchaus auch über 2Mb/s liegende Durchsätze erzielt werden können.

Gegenwärtig wird ein DFN-Projekt vorbereitet, in dem viele der obigen Fragen in der Praxis getested werden. Zwischen den Universitäten Erlangen/Nürnberg, Würzburg und dem LRZ München werden 34-Mb-Glasfaserleitungen mit herkömmlicher TDM-Technik auf 4 Kanäle zu je 8Mb/s herunter geschaltet. Die 8Mb-Kanäle sind an X.25-Untervermittlungen angeschlossen, die in den beteiligten Einrichtungen bereits verfügbar sind und über X.25-Anschlüsse mit bis zu 10Mb/s verfügen. Im Rahmen dieses Projektes wird erstmals X.25 mit 8Mb/s im WAN-Bereich praktisch getestet und angewendet.

Neben der praktischen Erprobung wird der DFN-Verein eine Simulationsstudie zum Thema "Schnelles X.25" in Auftrag geben, die die praktischen Erfahrungen durch theoretische Untersuchungen begleiten wird.

4.2) Frame Relay

Frame Relay ist ein DTE-DCE-Interface-Protokoll in der OSI-Ebene 2 mit einer V.35-Schnittstelle. Zur Zeit wird es bei CCITT bis zum Bereich 2Mb/s standardisiert (I.122). Der Herkunft nach aus dem ISDN-Bereich stammend (I-Serie), verwendet es LAP-D (Q.921) und Outbandsignalisierung (Q.931) <9,10,11>. Es ist ein auf OSI-Ebene 2 verbindungsorientiertes Protokoll, das keine Fehlerkorrektur zur Verfügung stellt. Wie bei X.25 werden auf beiden Seiten des Netzes permanente logische Kanäle für eine Kommunikation aufgebaut (DLCI: Data Link Connection Identifier).

Die Standardisierung ist weit fortgeschritten, aber noch nicht in allen Einzelheiten abgeschlossen. Es fehlen u.a. Festlegungen für die Kommunikation der Frame-Relay-Switches untereinander oder für die Bereitstellung von Switched-Virtual-Channels (SVC); bisher sind nur Permanent-Virtual-Channels (PVC) spezifiziert.

Zu den wesentlichen Eigenschaften von Frame Relay zählen:

- Variable Framelänge (Minimum 262 Byte);
- Die maximale Anzahl der PVCs beträgt 910;
- Keine Fehlerkorrekturbehandlung;
- Die Reihenfolge der Frames bleibt erhalten;
- Geringe End-zu-End Verzögerung (angestrebt sind ca. 2ms pro Knoten);
- Keine Congestion-Control (sie muß von der DTE geleistet werden);
- Multiplexing in Schicht 2;
- Auftretende Frameverluste müssen von höheren Schichten behoben werden.

Noch zu entwickelnde zukünftige Optionen sind vorgesehen in den Bereichen:

- Globale Adressierung;
- Multicast-Fähigkeit;
- End-zu-End Flußkontrolle.

Zu den noch zu untersuchenden Fragestellungen des Frame-Relay-Protokolls gehören u.a.:

- Welche Auswirkungen hat die fehlende "Congestion Control" auf die Dienstgüte?
- Frame Relay ist ein DTE-DCE Interface Standard (wie X.25); das Zusammenspiel verschiedener Frame-Relay-Teilnetze ist noch nicht festgelegt.
- Welche Performanceverbesserungen etwa gegenüber üblichen X.25-Netzen sind in der Praxis erreichbar?

Im Grundsatz können oberhalb von Frame Relay beliebige andere Netzprotokolle (OSI-Schicht 3: X.25, ISO-IP, DoD-IP, SNA, ...) verwendet werden. Unterhalb von Frame Relay kann im Prinzip ATM verwendet werden. Die jetzigen Implementierungen verwenden aber herstellerabhängige, z.T. dem Cell-Relaying von ATM ähnliche,

Verfahren. Festlegungen in den Standards für die unterhalb und oberhalb von Frame Relay nutzbaren Protokolle sind noch nicht abgeschlossen.

Ein Gruppe von Herstellern hat die verfügbaren Festlegungen zur gemeinsamen Definition einer Produktschnittstelle verwendet <4,6,7>. Die ersten Produkte (sowohl Netzknoten als auch Nutzerequipment) gemäß diesen Spezifikationen befinden sich am Markt. Die für die verfügbaren und angekündigten Frame-Relay-Netzknoten angebotenen Anschlußkapazitäten betragen bis zu 2Mb/s; Erweiterungen an den Netzknoten auf bis zu 34Mb/s sind aller Wahrscheinlichkeit nach Ende 1993 zu erwarten. Demgegenüber ist bereits für die nahe Zukunft Anschlußequipment (z.B. Router) mit 34Mb/s angekündigt.

DBP/T und der DFN-Verein sind grundsätzlich übereingekommen, im Rahmen des WIN einen Pilotversuch mit Frame Relay durchzuführen. Es ist angestrebt, den Versuch gegen Ende 1991 beginnen zu lassen. Die technischen und organisatorischen Einzelfragen werden zur Zeit diskutiert.

Fazit: Frame Relay kann eine erfolgversprechende Schicht-2-Technik werden. Viele Fragen sind aber noch nicht ausreichend geklärt. Zur Zeit sind auch die Standardisierungen noch nicht abgeschlossen. Die jetzt am Markt erscheinenden Produkte sollten für die weiteren Untersuchungen der Frame-Relay-Eigenschaften in Pilottests genutzt werden; die Beschränkung der gegenwärtigen Netzprodukte auf 2Mb/s führt aber über das verfügbare 2Mb/s-WIN noch nicht hinaus.

4.3) DQDB

DQDB (Distributed Queue Dual Bus) ist von IEEE 802.6 seit 1987 als Grundlage für die Entwicklung eines MAN-Standards (Metropolitan Area Network) verwendet worden.

DQDB ist als Doppelbussystem realisiert. Jede Station ist an beide Bussysteme angeschlossen. Über beide Anschlüsse werden Daten empfangen bzw. versendet oder Reservierungen für den Versand vorgenommen (unidirektionales Lesen und Schreiben). Von beiden

Kopfstationen des Doppelbusses wird eine ständige Folge von Slots erzeugt (Abbildung 1). Die Enden des Doppelbusses können in einem Gerät miteinander verbunden werden, so daß ein logischer Doppelbus in Form eines physischen Doppelringes entsteht.

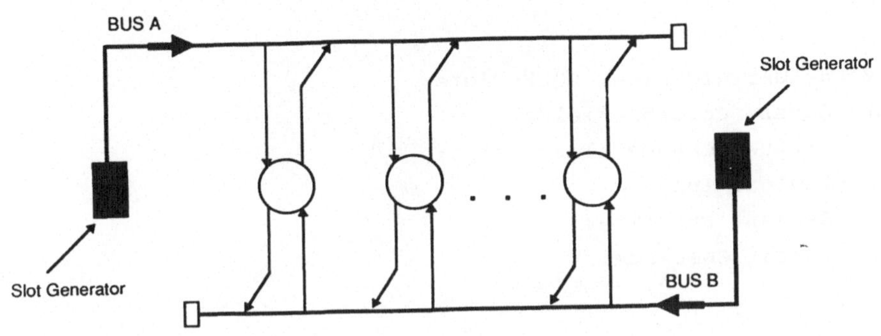

Abbildung 1: DQDB-Doppelbussystem

Die Slotgröße ist seit 1989 in Übereinstimmung mit der Festlegung für ATM auf 53 Byte (5 Byte Header + 48 Byte Daten) festgelegt. Die Zusammensetzung des Headerfeldes eines DQDB-Slots ist in Abbildung 2 dargestellt. Das ACF-Feld liegt außerhalb des Bereiches für die Header-Checksum (HCS). Dadurch ist eine Umsetzung von DQDB-Slots auf ATM-Zellen nicht ganz so trivial wie die Verwendung der gleichen Länge vermuten läßt (siehe Diskussion weiter unten).

In DQDB sind zwei wesentliche Zugriffsmodi auf die Slots vorgesehen: Queue-Arbitrated (QA) für den asynchronen in Schicht 2 verbindungslosen Datentransfer und Pre-Arbitrated (PA) für die isochrone in Schicht 2 verbindungsorientierte Übertragung. Insofern sollen im zukünftigen MAN-Netzwerk neben Datenkommunikation auch Audio- und Videodienste bereit gestellt werden können. Die ersten Implementierungen werden allerdings nur den QA-Zugriff (ohne Retransmission im Verlustfall) realisieren. Flußkontrolle oder Congestion-Control werden von DQDB nicht angeboten.

Die Durchsatzraten sind im Grundsatz in DQDB verschieden implementierbar. Die anlaufenden Pilottests sind auf 34 Mb/s (bzw. 45 Mb/s in den USA) ausgelegt; höhere Raten sind in Zukunft möglich.

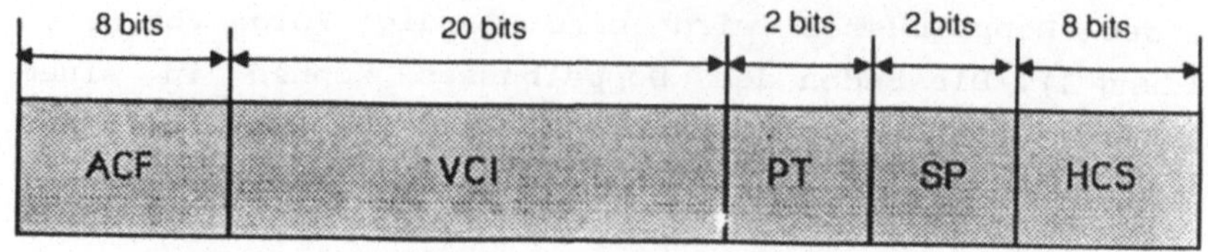

Abbildung 2: Headerfeld eines DQDB-Slots.
ACF: Access Control Field;
VCI: Virtual Channel Identifier;
PT: Payload Type;
SP: Segment Priority;
HCS: Header Checksum;

DQDB bietet zwar in weiten Bereichen allen Stationen einen fairen Zugriff für asynchronen Transfer. In Hochlastfällen oder bei wenigen weit auseinanderliegenden Stationen wird der Zugriff aber zunehmend "unfair". Randständige Stationen des Bussystems erhalten im Vergleich zu zentralen Stationen einen größeren Anteil an den zur Verfügung stehenden Slots. Ein großer Anteil von isochronen Daten erhöht die Fairness für die asynchronen Zugriffe ebenso wie eine unter dem maximal Möglichen liegende Lastausnutzung <12>.

In Abbildung 3 ist die typische Topologie einer DQDB-Installation gezeigt. Der innere Kern wird von den sogenannten Edge-Gateways (EGW) gebildet, die typischerweise bei den öffentlichen Betreibern stehen. Die Verbindung zum Nutzer wird durch die Customer-Gateways (CGW) hergestellt, wobei ein EGW mehere CGWs bedienen kann. Die CGWs stehen üblicherweise beim Kunden und bieten dort verschiedene Interfaces an: anfangs nur Ethernet, später sollen Token Ring, FDDI, X.21 und G.703 hinzukommen. DQDB kann im Bridge-Mode als auch im Router-Mode zur Verknüpfung der lokalen Netze eingesetzt werden.

Mit Hilfe von MAN-Switching-System Routers (MSS Routers) können mehrere DQDB Teilsysteme verbunden werden oder es kann eine Verbindung zu einem ATM-Switch hergestellt werden.

DQDB ist für die Anwendung in öffentlichen Netzen konzipiert worden. Dementsprechend nehmen Sicherheitsaspekte einen wichtigen Raum ein. Die Adreßauthentifizierung des Senders beim Eingang von Datenpaketen in eines der EGWs und des Empfängers beim Verlassen von Datenpaketen eines EGWs sollen zuverlässig verhindern, daß Daten falsch ausgeliefert werden. Die Einrichtung von Closed-User-Groups ermöglicht die Nutzung eines DQDB-Systems durch völlig unabhängige und disjunKte Nutzergruppen.

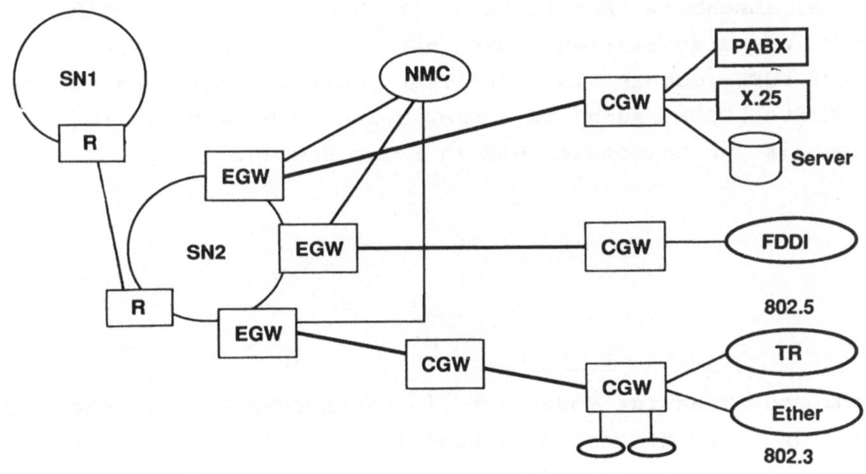

Abbildung 3: Typische DQDB-Konfiguration
EGW: Edge Gateway
CGW: Customer Gateway
R : MAN-Switching-System Router
SNx: MAN Subnetwork
NMC: Network Management Center

Oftmals wird SMDS (Switches Multi-Megabit Data Service) im Zusammenhang mit DQDB (MAN) genannt. SMDS ist für den öffentlichen verbindungslosen (keine Flow Control, keine Congestion Control) Paketdienst konzipiert, der im Einzelfall technisch durch ein DQDB-Netzwerk realisiert werden kann, der aber auch auf anderen Netztechologien (herstellerspezifische Netze, ATM-Netze) basieren kann. Mit einer Paketgröße von bis zu 9188 Bytes ist SMDS v.a. für die LAN-LAN-Verbindung gedacht.

DBP/T bereitet zwei Pilotversuche mit DQDB-Technik vor. Im Sommer 1991 sollen in München und in Stuttgart jeweils einige Testnutzer mit Ethernet-Interfaces an die von Siemens bzw. SEL bereitgestellten DQDB-Installationen angeschlossen werden. Ein Antrag des LRZ München an den DFN-Verein zur Förderung eines DQDB-Pilotanschlusses befindet sich in der Vorbereitung.

Fazit: DQDB bzw. MAN ist von seiner technischen Konzeption her auf eine Verwendung im regionalen Bereich ausgelegt. Wenngleich im Prinzip auch auf große Entfernungen verwendbar (anders als FDDI), macht es die absehbare Entwicklung im breitbandigen Netzbereich (ATM) eher unwahrscheinlich, daß nationale oder internationale WAN-Netze auf DQDB-Technik als zentralem Baustein aufbauen werden. Insofern dürfte DQDB auch im DFN-Bereich wohl nur als regionale Zubringertechnik zum nationalen WIN in Frage kommen.

4.4) ATM

ATM (Asynchronous Transfer Mode) ist die Basistechnik, auf die sich CCITT für ein zukünftiges Breitband-ISDN (I.121) <18> vor einiger Zeit geeinigt hat. Zuvor war STM (Synchronous Transfer Mode) als mögliche Alternative diskutiert worden.

In Schicht 1 ist ein verbindungsorientiertes, netzinternes Protokoll für die Kommunikation zwischen den Netzknoten definiert (ATM Layer). Interfaceprotokolle zu möglichen Nutzern von ATM sind im wesentlichen im AAL (ATM Adaptation Layer) in Schicht 2 festgelegt. Im ATM Layer werden Zellen konstanter Länge ausgetauscht; die Zellänge ist im Sommer 1990 auf 53 Byte (5 Byte Header, 48 Byte Informationen) festgesetzt worden. Das ATM Layer bietet zwei Arten der logischen Verbindung: Virtual Channel und Virtual Path. Ein Virtual Path ist die Zusammenfassung mehrerer Virtual Channels, die die gleichen Nutzer miteinander verbinden. So kann eine komplexe Anwendung, die mehrere Virtual Channels parallel verwendet (z.B. Video, Audio, Datentransfer) über einen Virtual Path sehr effektiv geroutet werden.

Die Zusammensetzung des Headerfeldes einer ATM-Zelle ist in Abbildung 4 dargestellt.

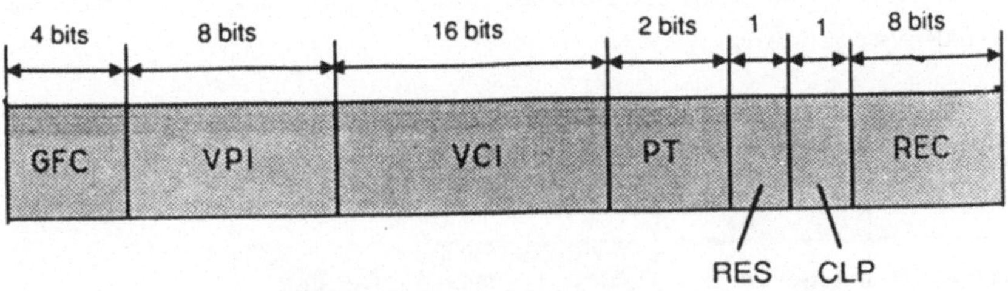

Abbildung 4: Headerfeld einer ATM-Zelle am User-Network-Interface
GFC: Generic Flow Control;
VPI: Virtual Path Identifier;
VCI: Virtual Channel Identifier;
PT: Payload Type;
CLP: Cell Loss Priority;
HEC: Header Error Control;
RES: Reserve Field

Wichtige Eigenschaften von ATM sind:

- Zellmultiplexing;
- Flußkontrolle;
- Flexible Bitrate für die Nutzung;
- Asymmetrische Verbindung: Von A nach B kann eine andere Bitrate angefordert werden, wie von B nach A;
- Die Signalisierung läuft über gewöhnliche virtuelle ATM-Kanäle.

ATM ist in der Lage, sehr unterschiedliche Kommunikationsarten zu bedienen, von asynchronem Datentransfer bis zu real-time isochronen Anwendungen. Die möglichen Dienstmodi werden durch vier verschiedene Dienste im AAL angeboten (siehe Abbildung 5). Die Kennzeichnung der vier Dienste erfolgt im PT-Feld des Headers einer ATM Zelle.

Class A bedeutet die Emulation einer Leitungsverbindung mit fester Bitrate, z.B. für Audio oder Video; Class B könnte Audio/Video mit variabler Bitrate unterstützen; Class C und D wären für den verbindungsorientierten bzw. verbindungslosen Datentransfer geeignet. Insgesamt ist ATM also u.a. für die Integration ganz unterschiedlicher Datennetzdienste geeignet: X.25, Frame Relay, SMDS, LAN-LAN-Verbindung.

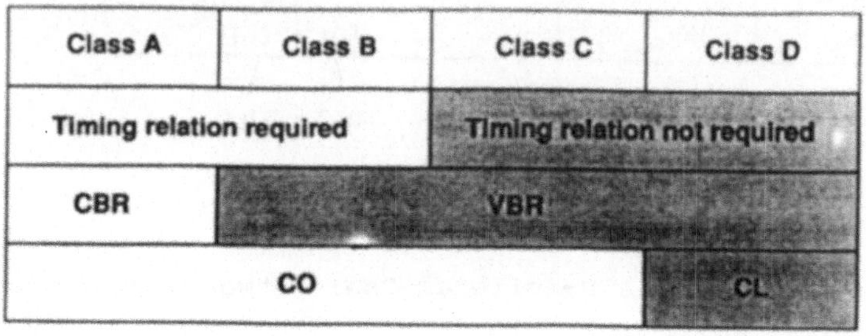

<table>
<tr><td>Class A</td><td>Class B</td><td>Class C</td><td>Class D</td></tr>
<tr><td colspan="2">Timing relation required</td><td colspan="2">Timing relation not required</td></tr>
<tr><td>CBR</td><td colspan="3">VBR</td></tr>
<tr><td colspan="3">CO</td><td>CL</td></tr>
</table>

Abbildung 5: Dienste des ATM Adaptation Layers (I.362)
CBR: Constant Bit Rate
VBR: Variable Bit Rate
CO : Connection Oriented
CL : Connectionless

Der Piloteinsatz eines ATM-Vermittlers findet seit 1989 im Rahmen des BERKOM-Projektes statt. Sollte demnächst eine neue Generation von ATM-Vermittlern zum Einsatz kommen, wäre der DFN-Verein an einer Beteiligung im Rahmen eines Pilotprojektes interessiert.

Interworking DQDB/ATM

Die einheitliche Festlegung von DQDB-Slots und ATM-Zellen auf 53 Byte läßt eine einfache Verknüpfung beider Netztechniken vermuten. Dieser erste Eindruck trügt allerdings. Bereits die verschiedene Berechnung der Header Checksum in DQDB (ohne ACF-Feld) und der Header Error Control (mit GCF-Feld) macht beim Übergang jeweils eine Neuberechnung notwendig. Daneben ist die Adressierung in beiden Systemen anders strukturiert.

Trotzdem liegt es nahe, die vier AAL-Dienste (siehe Abb. 5) mit entsprechenden Diensten im MAN zu verknüpfen. So wird die Dienstklasse A des AAL mit den isochronen PA-Slots vom DQDB verknüpft werden. Allerdings zeigen sich gerade in diesem Beispiel noch beträchtliche Lösungsprobleme. Im DQDB werden isochrone Kanäle auf die einzelnen Bytes eines PA-Slots abgebildet, während im ATM isochrone Anwendungen stets ganzen Zellen zugeordnet werden.

Am problemlosesten zeichnet sich zur Zeit das Zusammenspiel zwischen ATM mit AAL Class D (CL) und QA-Slots (CL) ab. In das ATM-Netz könnte ein Server für den verbindungslosen Datenverkehr integriert werden, dessen Dienste vom MAN aber auch von direkt an das ATM-Netz angeschlossenen verbindungslosen LANs genutzt werden könnte (siehe Abbildung 6). In Anlehnung an die Integration von X.25 in ISDN würde man so eine "minimale Integration" eines CL-Servers in ATM realisieren.

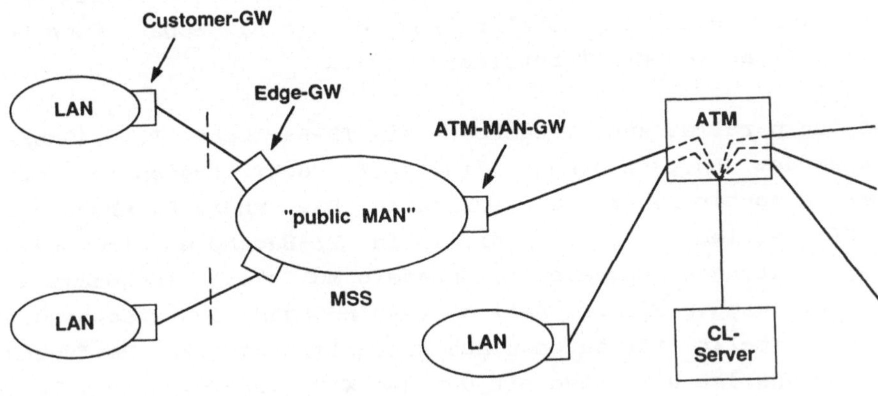

Abbildung 6: Integration eines verbindungslosen (CL) DQDB-Dienstes in ein ATM-Netz
MSS: MAN Switching System

Eine weitergehende Diskussion zum Thema Interworking von ATM und DQDB findet man in <13>.

5) Situation im NSFNET

Das NSFNET ist ein von der National Science Foundation (NSF) finanziertes Kommunikationsnetz für die Belange von Forschung und Lehre in den USA. Wesentliche finanzielle Unterstützung in Form von Ausrüstung, Personal und Leitungskapazität werden zusätzlich von IBM und MCI geleistet. Der Betrieb des NSFNET wird von MERIT (Bundesstaat Michigan) durchgeführt. Neben NSFNET gibt es weitere große US-weite Datennetze, die von anderen US-Agencies finanziert und betrieben werden (ESNET, NASA, DARPA, CSNET). Alle Netze sind über zwei Knotenpunkte (FIX East, FIX West, FIX = Federal Internet Exchange) miteinander verbunden.

Im NSFNET ist 1989 ein sogenannter T1-Backbone (T1 = 1,5Mb/s) in Betrieb genommen worden. Dieser Backbone verbindet seitdem eine große Zahl regionaler Netze in den gesamtem USA miteinander. Er ist als ein teilweise vermaschtes Netz aus Mietleitungen realisiert, wobei die T1-Switches in lokalen Zentren (z.B. Super Computer Zentren) im Auftrag von MERIT betrieben werden.

Seit 1990 ist parallel zum T1-Backbone ein T3-Backbone (T3 = 45Mb/s) eingeführt worden. Am T3-Backbone sind z.Z. acht regionale Netze bzw. Super Computer Zentren angeschlossen. Das innere Backbone-Netz wird von MCI betrieben. Anders als beim T1-Backbone sind keine dedizierten Leitungen angemietet, sondern MCI stellt insgesamt die verabredete Dienstgüte sicher. Fallen Verbindungen aus technischen Gründen aus, stellt MCI automatisch Ersatzleitungen zur Verfügung. Der Betriebsmodus ist hier also mit dem des WIN vergleichbar. Wegen dieser gegenüber anderen US-Backbones geänderten Dienstart, findet man in Darstellungen den T3-Backbone nicht mehr als ein Leitungsnetz sondern als eine Dienstwolke beschrieben.

Die T3-Switches (ENSS) werden von IBM gestellt. Es sind Prototypen, die bisher kommerziell nicht verfügbar sind. Sie basieren auf der RS6000-Workstation und enthalten zusätzliche Hardwareinterfaces. Für den Anschluß an den T3-Backbone ist jeweils ein T3-Switch beim Nutzer und beim nächstgelegenen Anschlußpunkt zum T3-Backbone, also bei MCI, aufgestellt.

6) Schlußfolgerungen

Für die nächste WIN-Ausbaustufe mit 34Mb/s-Zugang lassen sich zur Zeit noch keine endgültigen Festlegungen treffen. Mehrere technisch unterschiedliche Kandidaten stehen zur Auswahl.

Zum Teil stehen diese Techniken einander gegenüber, zum Teil sind sie aber auch evolutionär ergänzend. Die oben angesprochenen Techniken sind auf unterschiedlichen OSI-Schichten angesiedelt. Dadurch lassen sich verschiedene Szenarien gemeinsamer Implementierungen denken <19>. Während eine mögliche Verbindung von DQDB und ATM trotz der oben angesprochenen Integrationsprobleme als "horizontale" Integration denkbar ist, können Schnelles X.25, Frame Relay und ATM in "vertikaler" Form miteinander verbunden werden. Inwiefern Schnelles X.25 allein, auf Frame Relay oder direkt auf ATM aufsetzend realisiert werden kann und sollte, muß sich in naher Zukunft auf Grund technischer Eigenschaften und Produktverfügbarkeit zeigen. Dabei ist insbesondere zu klären, ob X.25-Interfaces (oder andere Netzprotokolle) in weiterer Zukunft über ein Frame-Relay-Netz an das ATM-Netz gekoppelt sind, oder ob diese Netzprotokolle direkte Zugänge zum ATM-Netz haben werden.

Weitere Fragestellungen betreffen die Entwicklung neuer Protokolle im "Gigabit"-Bereich. Zwar zeigen Untersuchungen, daß das eigentliche Netzprotokoll nur in verhältnismäßig geringen Umfang für die Performance und die oftmals auftretenden Durchsatzbegrenzungen verantwortlich ist. Den überwiegenden Ausschlag für den Durchsatz geben die Güte der Implementierung, die Hardware, das Kommunikationsinterface und die Kommunikationsarchitektur (Bus, Prozessor) (siehe u.a. <16>). Insofern können verbesserte Implementierungen und Parameteradaptionen der verfügbaren Protokolle sicherlich noch manche Bandbreitenerhöhung an die Anwendungen durchreichen.

Gleichwohl kann man davon ausgehen, daß alle zur Zeit gängigen Netzprotokolle (X.25, DoD-IP, ISO-IP, SNA, ...) in Bereichen sehr hoher Bandbreite keine adäquaten Kommunikationsprotokolle mehr darstellen werden. Die Entwicklung neuer Netzprotokolle (Light Weight Protocols: Express Transfer Protocol (XTP), Versatile Message Transaction Protocol (VMTP), ...) und deren international vereinbarte Durchsetzung werden zunehmend an Bedeutung gewinnen.

Ein anderer Protokollaspekt wird im BERKOM-Referenzmodell <17> behandelt. In zukünftigen Multi-Media-Anwendungen wird es sinnvoll sein, in einer Globalkommunikation mehrere Teilanwendungen parallel und in Reihenfolge ablaufen zu lassen. Diese Teilanwendungen werden völlig unterschiedliche Anforderungen an das darunter liegende Transportsystem stellen. Dadurch wird es notwendig werden, zwischen Diensten und Protokollen in einer Globalanwendung beliebig zu wechseln und je nach Bedarf der Anwendung Dienstgüten vom Transportsystem anzufordern, wie z.B. Synchronität oder Asynchronität, verbindungslose oder verbindungsorientierte Kommunikation, diverse Quality-of-Service Parametrisierungen (Fehlerkorrekturen, Flußkontrolle, ...) usw.

Gerade der letzte Aspekt der Wahl verschiedener Transportsysteme wird sich mittel- und langfristig auch auf die Architektur des Kommunikationsnetzes, also auf die des WIN, auswirken.

Zurück zur Ausgangsfragestellung bedeutet das, daß eine für die nächste WIN-Stufe auszuwählende technische Lösung in jedem Fall den Ausblick auf die übernächste Stufe berücksichtigen sollte: Viele technische und organisatorische Anhaltspunkte sprechen dafür, daß in der zweiten Hälfte der Neunziger Jahre ATM in Form des B-ISDN <18> <u>die</u> Hochgeschwindigkeits-WAN-Vermittlungstechnik sein wird. Diese Technik kommt für den jetzt zu planenden nächsten Ausbau des WIN wahrscheinlich zu spät - gleichwohl kann erwartet werden, daß zu gegebener Zeit auch im WIN die ATM-Technik zum Einsatz kommen wird und daß der DFN-Verein in Form von Piloteinsätzen zu den frühen Anwendern von ATM gehören wird.

Die Evaluation der nächsten WIN-Stufe sollte bis 1992 soweit gediehen sein, daß eine Entscheidung über die technische Umsetzung erfolgen kann. Ein Pilotversuch mit dieser Technik sollte 1992/93 beginnen und eine stabile betriebliche Einführung sollte für 1993/94 angestrebt werden.

Neben der technischen Umsetzung muß in den Jahren 1992/93 die organisatorische und v.a. finanzielle Umsetzung vorbereitet werden. Die Mittel für einen 34Mb/s-Zugang werden erheblich sein und bedürfen der vorbereitenden Planung in den Haushalten der interessierten Forschungseinrichtungen.

7) Literaturverzeichnis

<1> Integration von ISO-CONS in UNIX bsd 4.4, Husemann, D., Diplomarbeit an der FAU Erlangen-Nürnberg, 1991.

<2> 10 Mbps X.25!, Wells, D., Computer Networks and ISDN Systems 19, 1990;

<3> A Profile for Wide Area X.25 Operating at 2 Mbps, Sales, B., Helios-B Internal Report, Feie Universität Brüssel, März 1991

<4> Frame Relay specification with extensions, Doc. Nr. 001-208966, 1990, Digital Equipment, Northern Telecom, StrataCom.

<5> Broadband networks, McQuillan, J.M., Data Communication International, June 1990, pp 62-74.

<6> Frame Relay white paper, Button, B., StrataCom, May 1990.

<7> The Frame Relay solution, US Sprint, Telenotes Vol. 1, Nr. 2, Sept. 1990.

<8> Frame Relay networks: Not as simple as they seem, Opderbeck, H.,Data Communications International, Dec. 1990, pp 89-91.

<9> I.122, Framework for providing additional packet mode bearer service, CCITT, Melbourne, Nov. 1988.

<10> Q.921, ISDN user-network interface - Data link layer specification, CCITT, Melbourne, Nov. 1988.

<11> Q.931, ISDN user-network interface - layer 3 specification for basic call control, CCITT, Melbourne, Nov. 1988.

<12> Performance Analysis of DQDB, Davids, P., Martini, P., IEEE, IPCCC 1990, Scottsdale, Arizona, Mar. 1990.

<13> DQDB/ATM Interworking, Burak, M., Luckenbach, Th., Third IFIP High Speed Networking Conference, Berlin, Mar. 1991.

<14> New Proposal extends the reach of metro area nets, Evans, P., Hullet, J. L., Data Communication International, Feb. 1988.

<15> Metropolitan Area Network, Alcatel, Oct. 1990

<16> High-Speed Transport Components, Zitterbart, M., IEEE Network Magazin, Jan. 1991.

<17> BERKOM Referenzmodell Lower Layers, DETECON, Berlin, Version 2.1, Feb. 1991.

<18> I.121, Broadband Aspects of ISDN, CCITT, Geneva, 1989

<19> Heutige und zukünftige Paketdienste, Wurzenberger, J., Datacom, pp. 49-51, Apr. 1991.

GeNeRIC: German Networking Research and Information Center

Manfred Bogen [1]

Gesellschaft für Mathematik und Datenverarbeitung (GMD)
Institut für Informationstechnische Infrastrukturen (Z1)
Abteilung für Wissenschaftsnetze (WN)
Riemenschneiderstraße 11
D–5300 Bonn 2
Tel.: +49 228 81996 50
E-Mail: MABOGEN@DEARN

In diesem Papier werden die Dienste, die die Gesellschaft für Mathematik und Datenverarbeitung (GMD) im Auftrag des Vereines zur Förderung eines Deutschen Forschungsnetzes (DFN) für DFN-Nutzer, für EARN-Nutzer und für IP-Kunden in der Bundesrepublik Deutschland anbietet, vorgestellt. Die Dienste beinhalten Electronic-Mail, File-Transfer, Remote Logon, Paketvermittlung und Auskunftsdienste. Faktoren wie die Vielzahl der eingesetzten Protokolle, die hohen Leitungsgeschwindigkeiten und die hohen Datenaufkommen stellen besondere Anforderungen an das Netzmanagement und die Dienstgüte. Es wird beschrieben, wie innerhalb der GMD diese komplexen Probleme bearbeitet werden. Darüberhinaus wird versucht, ein Verständnis für diese Probleme und für die Tatsache, daß jeder zum erfolgreichen Austausch von Daten beitragen kann, zu wecken.

1. Wissenschaftsnetze

Die in den letzten Jahrzehnten entstandenen Rechnernetze bieten für die Wissenschaft neue Möglichkeiten der Kommunikation und Kooperation. Die hohen Leitungsgeschwindigkeiten, die unterschiedlichen Verbindungstechniken und die heterogenen Kommunikationsprotokolle machten jedoch auch schnell deutlich, daß es nicht genügte, zwei Rechner über ein Kabel miteinander zu verbinden und dann die Stecker in die Wand zu stecken. Komplexe Probleme beim Aufbau und Betrieb von Rechnernetzen wurden identifiziert und es wurde versucht, diese unter dem Oberbegriff "Netzmanagement" zu lösen.

Die Entwicklung ist auch in den 90er Jahren noch nicht abgeschlossen. Es wird an der Realisierung von noch schnelleren Datenleitungen im GBit-Bereich gearbeitet, es wird erwartet, daß sich ein Metanetz von Rechnern, das die ganze Welt umspannt und die Kommunikation zwischen beliebigen Personen und Anwendungen ermöglicht, entwickelt, neue Kommunikationsdienste und -anwendungen sind im Entstehen [Bir91].

---

[1] Manfred Bogen ist seit 1983 wissenschaftlicher Mitarbeiter der GMD. Er studierte Informatik und Physik an der Universität Bonn. Seit 1987 arbeitet er im Institut für Informationstechnische Infrastrukturen mit den Schwerpunkten Gruppenkommunikation, OSI-Dienste und Netzkoordination. Herr Bogen ist Projektleiter des GeNeRIC-Projektes.

Alle Nutzer im Netz werden in einem globalen Auskunftsdienst (Directory) registriert und somit auch auffindbar sein.

Damit werden auch die Anforderungen an das Netzmanagement und die Koordination der Netze noch steigen: es wird nicht mehr genügen, sein eigenes Netz unter Kontrolle zu haben, sondern es wird erforderlich werden, Informationen über das eigene Netz in allgemein verständlicher Form zur Verfügung zu stellen und auf Störungen und Änderungen in anderen Netzen zu reagieren. Netzsicherheit, die beim autorisierten Zugang zu Rechnernetzen beginnt und beim Entdecken von "Computerviren" noch lange nicht aufhört, wird spätestens dann zum Thema werden, wenn ernsthafte Schäden durch den Mißbrauch von Rechnernetzen zu Tage treten. Darüberhinaus werden sowohl die einzelnen Wissenschaftler selbst, als auch die genutzten Netze vor der Informationsflut geschützt werden müssen, die sich über eben diese leistungsfähigen Netze ergießen kann. Vor diesem Hintergrund wird deutlich, daß Netze nicht "so nebenbei" betrieben werden können, sondern daß die Koordination und das Management Personen erfordern, die diesen Aufgaben als Arbeitsinhalte "full time" nachgehen.

Als "Wissenschaftsnetze" werden nun solche Netze bezeichnet, die ausschließlich wissenschaftlichen Zwecken, im Gegensatz zur kommerziellen oder firmeninternen Nutzung, dienen und die Wissenschaftlern weltweit die Nutzung zahlreicher Kommunikationsdienste ermöglichen. Sie sind heutzutage ein gängiges Arbeitsmittel für die Wissenschaft geworden und befinden sich in der Tradition von Telefonnetzen, Postämtern und Bibliotheken.

Der Wissenschaftler als Nutzer der Wissenschaftsnetze ist zunächst einmal nicht frei in der Auswahl der ihm gebotenen Kommunikationsdienste. In den meisten Fällen betreibt er die Netze nicht selbst, sondern er kann nur aus dem bestehenden Angebot in seiner Institution auswählen. Dieses Angebot ist häufig erst nach und nach von den lokalen Netzbetreibern entsprechend ihrer Möglichkeiten zusammengestellt worden und richtet sich wiederum danach, um welche Organisation es sich handelt und welche Betriebssysteme oder Rechner im Laufe der Zeit angeschafft worden sind. Der Wissenschaftler als Endnutzer ist nicht interessiert an den verschiedenen Verbindungstechniken, an den unterschiedlichen Kommunikationsprotokollen, an den gezogenen Leitungen oder an der Zugehörigkeit "seines" Rechners zu internationalen Netzen. In den meisten Fällen steht für ihn in seinen Kommunikationswünschen nicht einmal der Kostenaspekt im Vordergrund. Er ist und bleibt interessiert an seinen Arbeitsinhalten, an seiner Erreichbarkeit und Anbindung, an der Zuverlässigkeit und Transparenz der genutzten Kommunikationsdienste, an einer synchronisierten Zeit und an einer einzigen, guten Benutzeroberfläche. Selbst die Netzbetreiber vor Ort sind häufig überfordert, wenn Netzprobleme den lokalen Bereich überschreiten. Sie sind froh, wenn sich ihr Verantwortungsbereich auf die lokalen Belange von Wissenschaftsnetzen und die Benutzerbetreuung beschränken kann, da sie ihre Netztätigkeit häufig nur neben anderen Arbeiten verrichten können. Die globalen Aufgaben müssen also von anderer Stelle wahrgenommen werden.

In der Bundesrepublik Deutschland gibt es mehrere Anbieter von Kommunikationsdiensten für die Wissenschaft und mehrere Betreiber von Wissenschaftsnetzen, die die Endnutzer und die lokalen Rechner- und Netzbetreiber in ihrer Kommunikation verantwortlich unterstützen. Der Verein zur Förderung eines deutschen Forschungsnetzes (DFN) bietet im Auftrag des Bundesministeriums für Forschung und Technologie (BMFT) Kommunikationsdienste für seine Mitglieder mit dem Schwerpunkt "Open Systems Interconnection (OSI)" an. Die Gesellschaft für Mathematik und Datenverarbeitung (GMD) betreibt Wissenschaftsnetze für DFN, IBM und die firmeninterne Kommunikationsinfrastruktur auf nationaler und internationaler Ebene. Das Angebot der GMD umfaßt Zentralknoten-Dienste, Anwendungs-Gateways, Auskunftsdienste und netzübergreifendes Netzmanagement, die für den Auftraggeber DFN im gemeinsamen Projekt "German Networking, Research, and Information Center (GeNeRIC)" zusammengefaßt sind. Dieses Projekt wird im folgenden mit seinen Diensten, seinen Problemen und seiner Zielsetzung (Motto: "Wir binden Sie an!") vorgestellt.

Abbildung 1 zeigt die Dienstmaschinen (Server), auf denen die von der GMD angebotenen Kommunikationsdienste zur Verfügung gestellt werden.

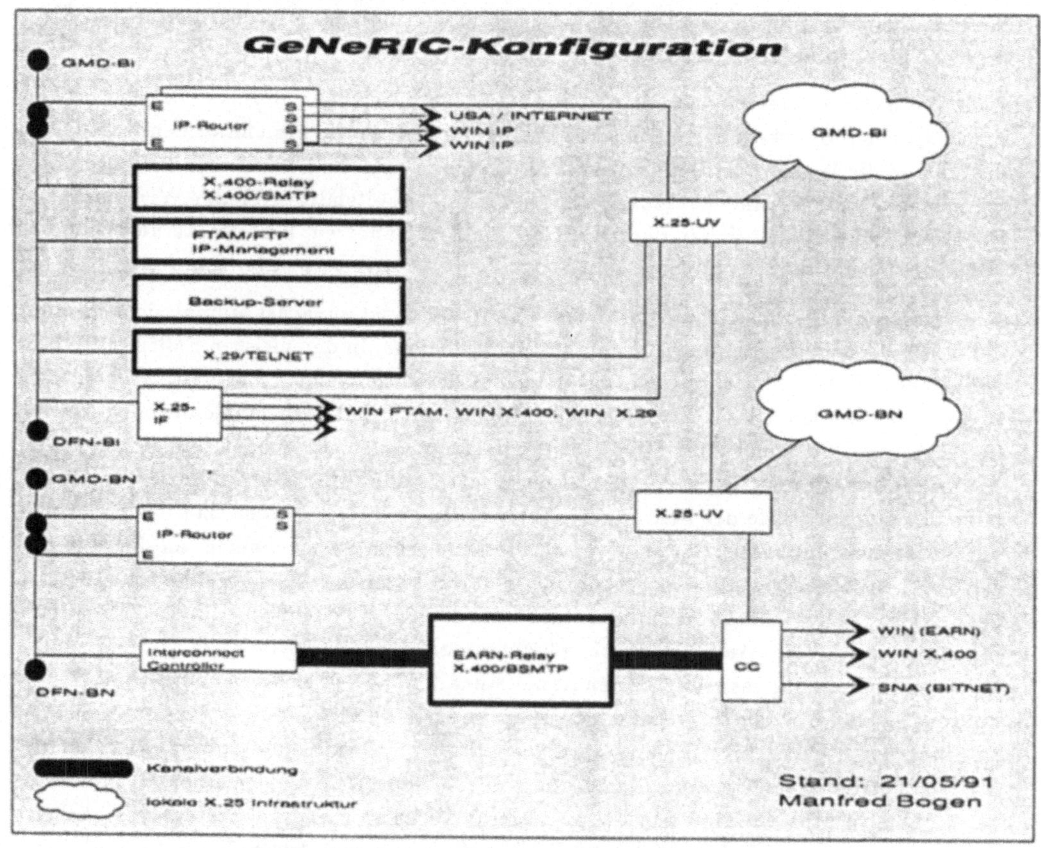

Abbildung 1: GeNeRIC-Konfiguration

2. Koordination und Management

Trotz verteilter Anwendungen ist es bisher immer noch so, daß Netze an zentraler Stelle koordiniert und geplant werden. Die in den einzelnen Netzzentren beschäftigten Personen gehen dieser Tätigkeit zumeist "hauptamtlich" nach. Ein hohes Maß an Expertentum und Erfahrung ist hierzu erforderlich.

Die GMD als Auftragnehmer von DFN ist Netzzentrum für

— den nationalen DFN Message-Handling-System-Verbund (DFN MHS-Verbund),
— das European Academic and Research Network (EARN) und
— den deutschen Teil des INTERNET.

2.1 X.400-Relay (DFNRELAY)

Das Herzstück des Kommunikationsverbundes der DFN-Mitglieder auf der Basis der Protokolle für "Open Systems Interconnection" ist der Message-Handling-Verbund nach X.400 [BBD*91]. X.400 ist die 1984 zum ersten Mal vorliegende Spezifikation eines OSI-Electronic-Mail-Protokolls, das von CCITT erarbeitet wurde [X.400-84]. Inzwischen liegt schon eine von CCITT und ISO gemeinsam getragene Verfeinerung dieser Spezifikation seit 1988 vor [X.400-88]. Die X.400-Installationen in der Bundesrepublik sind im DFN Message-Handling-System-Verbund (DFN MHS-Verbund) zusammengefaßt. Dabei gibt es eine ausgezeichnete X.400-Installation, den DFNRELAY, eine Art X.400-Zentralknoten, der von der GMD im GeNeRIC-Projekt betrieben wird. Dieser Zentralknoten ist erste Anlaufstelle für X.400-Installationen, die zum DFN-MHS-Verbund hinzukommen wollen. Hier können die ersten Verbindungstests gemacht werden und die Namen für die neuen Teilnehmer im Verbund festgelegt werden. Hier werden die Informationen über alle deutschen X.400-Installationen gesammelt, verwaltet und verteilt, hier werden die neuen Installationen bekannt gemacht und hier werden die Nachrichten für Neuinstallationen und Installationen, die keine Direktverbindungen betreiben wollen, weitergeleitet. Im Prinzip kann sich im DFN MHS-Verbund jede X.400-Installation mit jeder anderen X.400-Installation direkt "unterhalten". Dies erfordert jedoch ein hohes Maß an Verkehrsüberwachung und an Management, das häufig nicht im rechten Verhältnis zum Nutzen steht angesichts der Tatsache, daß gerade in kleineren Institutionen das "Mail-Geschäft" nebenamtlich betrieben wird.

Neben diesen betriebsorientierten Tätigkeiten werden auch die nationalen Postmaster am DFNRELAY beraten.

Wie die nationalen X.400-Installationen im DFN MHS-Verbund zusammengefaßt sind, so sind auch die europäischen X.400-Systeme in den RARE [2] MHS-Verbund integriert [RARE91]. Im europäischen Verbund gibt es pro Mitgliedsland einen sogenannten "well-known entry point (WEP)", wiederum eine auf internationaler Ebene ausgezeichnete X.400-Installation, über die alle Nachrichten von und zu Nutzern dieses Landes laufen. Diese Eingangspunkte kennen sich gegenseitig und bauen bei Bedarf Verbindungen zueinander auf. Der WEP für Deutschland ist DFNRELAY, so daß mit den Betreibern der ausländischen MHS-Relays kooperiert werden muß.

Neben dem Store-and-Forward auf nationaler und internationaler Ebene werden am DFNRELAY auch die für das Funktionieren des X.400-Verbundes unbedingt notwen-

[2] Reseaux Associes pour la Recherche Europeenne

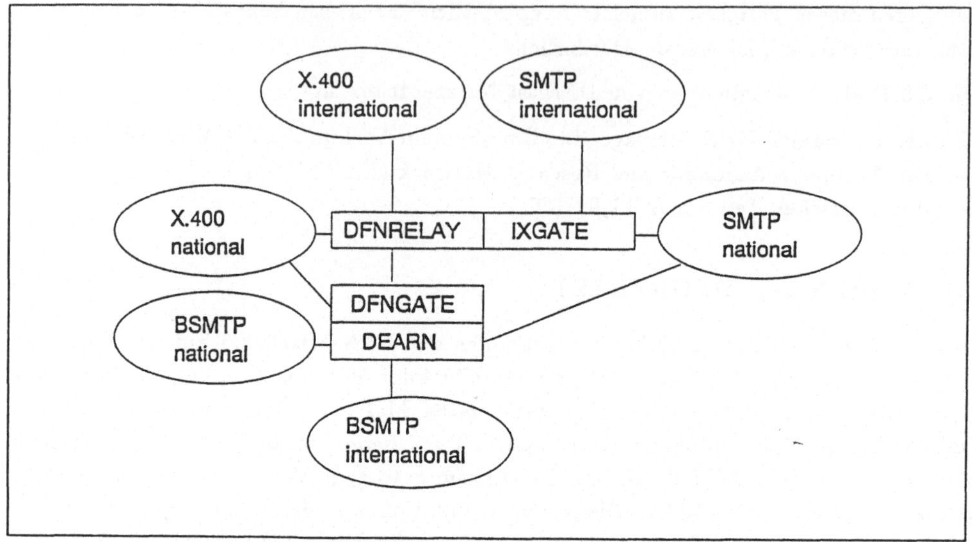

Abbildung 2: Die GeNeRIC-Server im Mail-Verbund

digen Auskunftsdienste angeboten: Verteilerlisten nicht nur für DFN-Mitglieder, Infoblöcke für die X.400-Installationen und Verzeichnisdienste in Form des DFN Nameservers und, derzeit noch geplant, des X.500-Directory-Dienstes [X.500].

2.2 EARN-Relay (DEARN)

In der Bundesrepublik gibt es jedoch nicht nur OSI-Dienste, sondern auch Kommunikationsdienste, die in Netzen angeboten werden, die keine standardisierten OSI-Protokolle verwenden. Ein Beispiel hierzu ist EARN [EARNa].

EARN ist der europäische Ableger des nordamerikanischen BITNET (Because It's Time Network), das 1981 für die nicht-kommerzielle Nutzung gegründet wurde und das technisch dem IBM firmeninternen VNET (1972) entspricht. EARN wurde 1983 auf Initiative von IBM gestartet und expandiert derzeit von Mitteleuropa aus stark in Richtung Osteuropa und Nordafrika [Qua90].

EARN gab die Management-Struktur für andere Netze dahingehend vor, daß es pro Mitgliedsland mindestens einen internationalen Knoten gibt, der spezielle Anforderungen hinsichtlich Hard- und Software und Management erfüllen muß, bei dem die internationalen Leitungen enden und der als Zugangspunkt für den EARN-Verkehr von und zu Nutzern des jeweiligen Landes fungiert. Dieser EARN-Zentralknoten für die Bundesrepublik steht seit 1988 bei der GMD in Bonn [BSWW89].

Neben dem eigentlichen Betrieb gehört es zu den Aufgaben am deutschen EARN-Zentralknoten

— die EARN-Dienste wie LISTSERV, NETSERV, MAILER und RELAY der Wissenschaft, speziell den DFN-Nutzern, am Zentralknoten anzubieten,
— den Einsatz von Netzanwendungen innerhalb des deutschen EARN zu koordinieren,

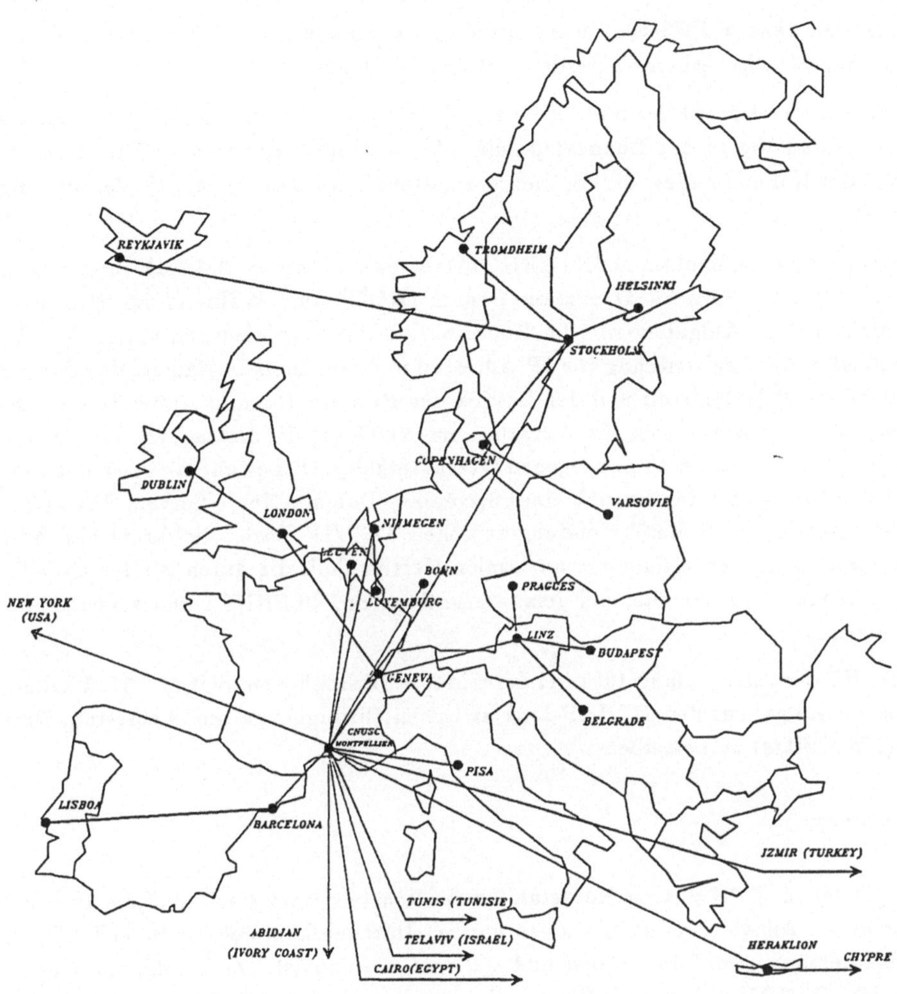

Abbildung 3: EARN

- die Migrationsstrategien für die EARN-Dienste zu OSI-Diensten in nationaler und internationaler Kooperation zu erarbeiten und umzusetzen,
- das Netzmanagement für EARN Deutschland in Kooperation mit den Koordinatoren der anderen Mitgliedsländern durchzuführen,
- die EARN-Nutzer beim Übergang von den herstellerabhängigen zu den OSI-Anwendungsprotokollen und beim Übergang von den EARN-Diensten zu den DFN-Diensten zu beraten und diesen Übergang zu koordinieren.

2.3 IP-Relay und NOC

Ein weiteres Beispiel für ein existierendes Netz ohne OSI-Protokolle ist das TCP/IP-basierte INTERNET. Dieses weltweite Netz entstand aus dem nordamerikanischen ARPANET, das schon 1968 vom amerikanischen Verteidigungsministerium begonnen wurde. In den Jahren 1973-1981 wurden die heute verwendeten TCP/ IP-Protokolle, die weiterhin ständig verbessert werden, entwickelt [Qua90].

Abhängig von der Anzahl der internationalen Leitungen, über die das IP-Protokoll gefahren wird und die in der Bundesrepublik oder in Nachbarländern wie der Schweiz (CERN) oder Holland enden, gibt es mehrere nationale Knoten, die die IP-Vermittlung betreiben.

Das Management im deutschen INTERNET teilt sich zwischen Network Information Center (NIC) und Network Operation Center (NOC) auf. Während im NIC eher die administrativen Aufgaben wie die Koordination der verschiedenen nationalen IP-Dienstanbieter, die Registrierung von IP-Adressen und von Domain Namen, das Führen der nationalen IP-Datenbank und der Betrieb des Primary Domain Name Servers angesiedelt ist, beschränken sich die Aufgaben im NOC auf die Betreuung der eigenen IP-Kunden, das Management der eigenen internationalen IP-Leitung und der nationalen IP-Vermittlung und den Betrieb eines Secondary Domain Name Servers. Die nationale IP-Vermittlung soll die Verbindung zwischen TCP/IP-Hosts, die direkt am WiN angeschlossen sind oder Teilnehmer an lokalen Netzen sind, die durch Router an WiN angeschlossenen sind einerseits, und dem internationalen INTERNET andererseits, hergestellt werden.

Das DFN NOC ist seit Anfang 1991 bei der GMD in Birlinghoven. Seit Mai 1991 gehört es zu seinen Aufgaben, den DFN IP-Link zwischen Birlinghoven und Princeton, New Jersey (128 KBit/s) zu betreiben.

3. Gateways

Die Mitglieder im DFN nutzen Kommunikationsdienste derart stark, daß die Bundesrepublik in der Anzahl der Installationen und im Datenaufkommen im Bereich "Open Systems Interconnection" in Europa und weltweit führend ist. Darüberhinaus gibt es jedoch viele DFN-Mitgliedsinstitutionen, die aus verschiedenen Gründen, z.T. historischer Art, zu Netzen gehören und mit Kommunikationspartnern weltweit verkehren, die nicht OSI-Protokolle verwenden. Um diese Kommunikationsbedürfnisse zu befriedigen, bietet DFN nicht nur die Koordination und Kommunikation im OSI-Verbund an, sondern es werden auch Netzübergänge (Gateways) zu Nicht-OSI-Netzen wie EARN und INTERNET betrieben. Diese Gateways ermöglichen nicht nur den Übergang zwischen verschiedenen Netzen mit unterschiedlichen Kommunikationsprotokollen, sondern auch das "Durchdringen" verschiedener Adreßbereiche. Je weniger der Endnutzer von einem Gateway sieht, um so besser ist es. Die GMD versucht die Gateways so zu betreiben, daß jeder Nutzer sie nur als Fortführung seines eigenen Netzes sieht. Neben diesen primären Aufgaben dienen die Gateways auch der Migrationsunterstützung. Nutzer können Kommunikationsdienste wie z.B. Dateiserver nutzen, obwohl diese nicht im eigenen Netz lokalisiert sind. Die Kommunikation läuft dann über einen Gateway, so daß auch Institutionen, die noch keine OSI-Protokolle verwenden, am DFN-Verbund, der auf OSI-Diensten basiert, teilnehmen und ihre Benutzer schrittweise an die OSI-Dienste heranführen können.

Es gibt eine große Anzahl von Anwendungs-Gateways zwischen den verschiedenen Protokollwelten. Die GMD betreibt nur solche Gateways als zentralen Dienst, die mit einem Bein in der OSI-Welt beheimatet sind, also z.B. einen X.400-BSMTP-Gateway (OSI/ EARN), nicht aber einen SMTP-BSMTP-Gateway (INTERNET/ EARN). Die möglichen Kommunikationspfade werden dadurch nicht eingeschränkt, im Gegenteil: durch den Betrieb von Gateways lediglich von der OSI-Welt aus lassen sich Personaleinsparungen im Bereich des Hard- und Software-Supports von 20 bis 30 Prozent, im Bereich der Wartung und Modifikation sogar 70 bis 80 Prozent erzielen. Darüberhinaus kommt man mit maximal 2 Netzübergängen von einem Netz zu einem beliebigen anderen. Dies ist wichtig angesichts der Tatsache, daß Gateways Engpässe bilden und Qualitäts- und Durchsatzeinschränkungen bewirken können.

An den Anwendungs-Gateways müssen komplexe Aufgaben wie Protokollkonvertierung, Adreßumsetzung, Dienstanpassung, Informationserhaltung, Wegewahl, Flußkontrolle und auch Netzmanagement wahrgenommen werden. Immer wieder gibt es deshalb Zweifel an der Existenzberechtigung und der Leistungsfähigkeit von zentralen Gateways, zumal es im lokalen Bereich auch immer mehr Anwendungs-Gateways gibt.

Zentrale Gateways haben jedoch auch Vorteile wie z.B. die einfachere Koordination und die eindeutigen Verantwortlichkeiten. Außerdem bieten sie die einzige Möglichkeit für Nutzer ohne lokale Gateways, andere Netze zu erreichen. Es bleibt der Entscheidung der lokalen Postmaster überlassen, eigene oder zentrale Gateways zu nutzen. Lokale Gateways erfordern einen gewissen Installations- und Verwaltungsaufwand, der im Verhältnis zu den Vorteilen wie die Vermeidung von Engpässen und die eigene Routing-Entscheidung gesehen werden müssen. Lokale Gateways werden umso attraktiver, je größer die sie betreibende Institution ist.

Gateways verbinden Kommunikationsinseln miteinander. Alle im folgenden beschriebenen Gateways werden als bi-direktionale Netzübergänge betrieben.

3.1 X.400-SMTP (IXGATE)

Der X.400-SMTP-Gateway ermöglicht die Kommunikation von Benutzern, die X.400 als Electronic-Mail-Protokoll verwenden, mit Benutzern, die das im INTERNET gebräuchliche Simple Mail Transfer Protocol (SMTP) [RFC821] einsetzen. Dieser Gateway wird auf derselben Hardware-Konfiguration betrieben wie DFNRELAY und nutzt auch dieselbe Software.

Mit zu den Aufgaben der Gateway-Betreiber gehört neben dem Betrieb und der Verkehrsüberwachung die Wartung der internationalen Adreßabbildungstabellen [RFC987].

3.2 X.400-BSMTP (DFNGATE)

Der X.400-BITNET-Gateway ermöglicht die Kommunikation per Electronic Mail zwischen Benutzern in Netzen, die CCITT X.400-Protokolle (1984, Red Book) und Benutzern in Netzen, die BSMTP-Protokolle mit RFC-Adressierung [RFC821], [RFC822] benutzen. Dabei ist das "Batch Simple Mail Transfer Protocol (BSMTP)" eine Batch-Version des im INTERNET genutzten SMTP-Protokolls [BSMTP].

Dieser Gateway wird auf derselben Hardware-Konfiguration betrieben wie der EARN-Relay DEARN. Die eingesetzte Software-Konfiguration ermöglicht darüberhinaus in Zusammenarbeit mit der TCP/ IP-Software den Betrieb eines örtlich getrennten Backup-Dienstes für den X.400-SMTP-Gateway und ggfs. den Betrieb eines örtlich getrennten Backup-Dienstes für den X.400-Relay.

3.3 FTAM-FTP

Auch der FTAM-FTP-Gateway ermöglicht die Kommunikation zwischen OSI- und TCP/ IP-Netzen: FTAM-Nutzer können durch ihn auf Dateien in einem FTP-Server, FTP-Nutzer können durch ihn auf Dateien in einem FTAM-Server zugreifen. Dabei ist FTP das im INTERNET gebräuchliche File Transfer Protocol [RFC959], FTAM ist der OSI-Standard für "File Transfer Access and Management" [ISO8571].

Diese Gateway-Funktionalität als zentrales Dienstangebot der GMD ist neu. Zunächst soll der Betrieb des Gateways mit Public Domain Software (ISODE) unter Berkeley UNIX/ Ultrix aufgezogen und erbracht werden. Zeitgleich dazu wird der Einsatz von Hersteller-Software geprüft.

Der FTAM-FTP-Gateway wird auf derselben Hardware-Konfiguration betrieben wie auch der IP-Relay und der Secondary Domain Name Server.

3.4 X.29-TELNET

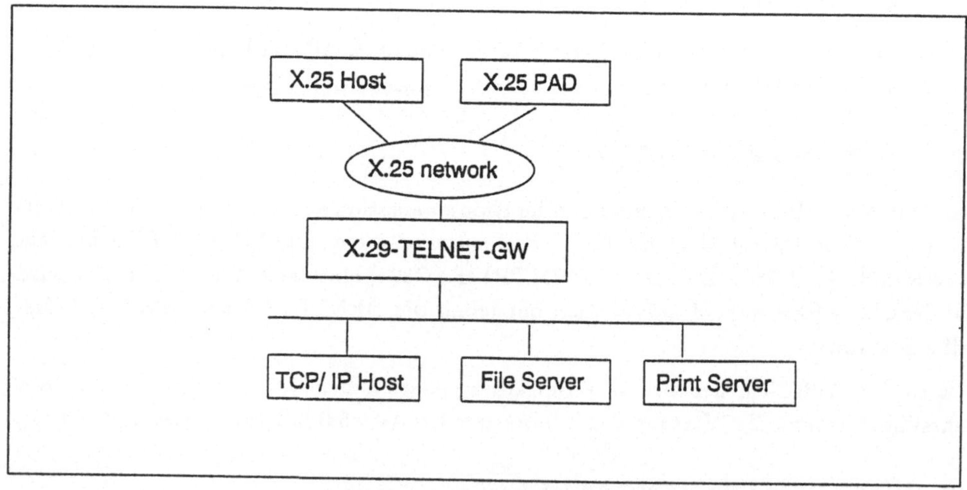

Abbildung 4: X.29-TELNET

Der X.29-TELNET-Gateway ermöglicht einerseits Nutzern in paketvermittelnden Netzen nach dem X.25-Standard den Dialogzugang zu Rechnern, die als TCP/ IP-Hosts zum INTERNET gehören, andererseits den TELNET-Nutzern im INTERNET den Dialog-Zugang zu Rechnern in Netzen mit X.25-Kommunikationsinfrastruktur [RFC854].

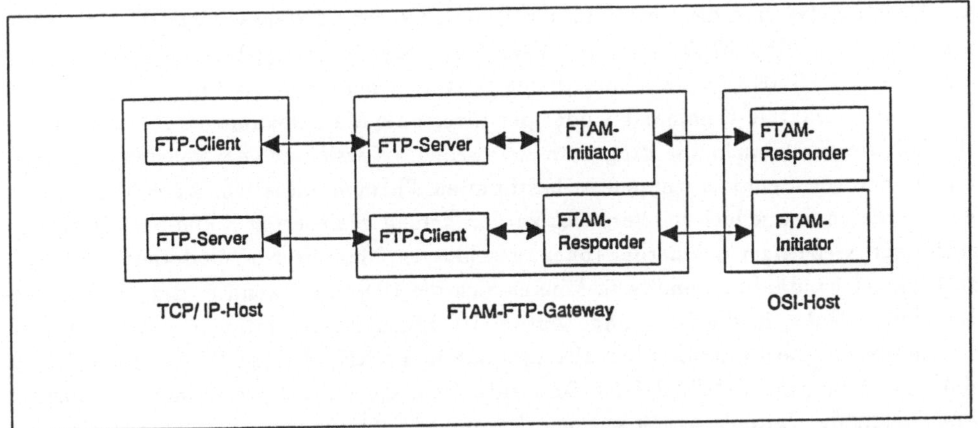

Abbildung 5: FTAM-FTP

Im Gegensatz zu den bisher beschriebenen Gateways, die jeweils als spezielle Anwendungs-Software auf einem Universalrechner laufen, handelt es sich beim X.29-TELNET-Gateway um eine fertige Box, in der alles enthalten ist, die mit einem X.25-Netz einerseits und einem Ethernet andererseits verbunden und nur noch in den Bereichen Netzadressen und Zugangsbeschränkungen den lokalen Begebenheiten angepaßt werden muß.

Auch dieser Dienst ist neu als zentraler Dienst der GMD.

4. Auskunftsdienste

Zum richtigen Funktionieren eines verteilten Kommunikationsverbundes gehört nicht nur das Angebot der eigentlichen Kommunikationsdienste, sondern auch die Bereitstellung von Information. Diese Information kann Aussagen über den Kommunikationsverbund selbst, über seine Nutzer, über die Arbeitsinhalte und Interessen der Nutzer und speziell über die Erreichbarkeit im Kommunikationsverbund enthalten. Auch Programme und Tabellen, die z.B. zum Management der Netze gebraucht werden, gehören zu den Informationen, die angeboten werden müssen.

An den jeweiligen Zentralknoten im Netzzentrum der GMD werden diese Informationen für die direkten Netznutzer, aber auch für die Benutzer jenseits der Gateways bereitgehalten.

Am X.400-Relay werden die Informationen über die im DFN MHS-Verbund beteiligten X.400-Systeme derzeit noch in Form von sogenannten "Infoblöcken" bereitgehalten. Sie werden regelmäßig verschickt und können auch per File Transfer abgeholt werden. Benutzergruppen sind hier zu sogenannten "Verteilerlisten" zusammengefaßt und können somit über eine Adresse angesprochen werden. Als Benutzerverzeichnis dient der DFN Nameserver, der jedoch durch den standardisierten Auskunftsdienst nach X.500 [X.500] abgelöst wird, der derzeit im Pilotprojekt der GMD namens "VERDI" eingeführt wird [VERDI90].

Am EARN-Relay sind die Nutzer in der "User Directory Database (UDD)" bei einem Server namens LISTSERV oder im "User Directory System (UDS)" bei einem Server namens NETSERV verzeichnet. Beide Server dienen auch als File Server, so daß dort über interaktive Kommandos oder per Electronic Mail Dokumente abonniert oder abgeholt werden können, die Programme, Tabellen, Übersichten oder vollständige Konversationen von Benutzergruppen zu bestimmten Themen enthalten. Auch hier gibt es Verteilerlisten, die jedoch im Vergleich zu den derzeit realisierten X.400-Verteilerlisten mächtiger, vielfältiger definierbar (offen, geschlossen, abonnierbar, moderiert etc.) und auch besser handhabbar sind, z.B. können sich die Benutzer, wenn es der Eigentümer der Liste erlaubt, in die Liste ohne Zutun des Eigentümers eintragen oder auch aus ihr austragen. Die Kenntnis über alle Systeme in EARN/ BITNET/ NetNorth ist in einer Datei namens BITEARN NODES enthalten, die einmal pro Monat aktualisiert wird, indem die Veränderungen seit dem Vormonat netzweit verschickt werden. Für die unterschiedlichen Betriebssysteme im EARN-Verbund gibt es Benutzerschnittstellen zu dieser Datei zum Abfragen und Aktualisieren von Knoteneinträgen.

Am IP-Relay wird der Secondary Domain Name Server für die im IP-Verbund involvierten Netze betrieben. Er enthält Daten wie IP-Adressen, Domain Namen und auch Routen, nicht aber Informationen über einzelne Nutzer. Weiterhin gibt es auch hier einen FTP-Server, bei dem eine Anzahl von Dokumenten, die für den Netzbetrieb nötig sind, abgeholt werden können.

5. Projektziele

Primäres Ziel des Projektes "GeNeRIC" (German Networking, Research and Information Center) ist es also, im Auftrag des Vereins zur Förderung eines deutschen Forschungsnetzes (DFN) Kommunikationsdienste für die deutsche Wissenschaft und Forschung zur Verfügung zu stellen.

Daneben gibt es jedoch noch weitere Ziele. Während der Projektlaufzeit

- soll die Dienstgüte der derzeit angebotenen Dienste wesentlich verbessert werden.
- sollen Werkzeuge zur Überwachung und zum Management der beteiligten Netze entwickelt werden, die es ermöglichen, automatisch auf Fehlersituationen und Störungen im Netz zu reagieren.
- sollen im Hinblick auf die stetig wachsenden Benutzerzahlen Maßnahmen ergriffen werden, um einen kostengünstigen Betrieb durch die konsequente Entwicklung und den Einsatz von automatisch ablaufenden Prozeduren und sonstigen Werkzeugen sicherstellen zu können. Diese sollten einen Beitrag zur Verbesserung der Kosten/ Leistungsrelation erbringen.
- soll der Übergang zu Herstellerprodukten, wo möglich, vollzogen werden. Durch den Einsatz von Hersteller-Software soll die bei der derzeit eingesetzten Software

ungeklärte und oft unbefriedigende Wartungssituation deutlich verbessert und der Betreuungsaufwand für die eingesetzte Software mittelfristig sehr reduziert werden.[3]

— sollen die Konzepte und Protokolle der "Open Systems Interconnection (OSI)" durch das Angebot von hochwertigen OSI-Diensten und leistungsfähigen Gateways zu anderen Netzen weiterhin propagiert werden.
— sollen speziell die im Betrieb und Management der verschiedenen Netze erworbenen Kenntnisse und Erfahrungen mit Projekten und Aktivitäten zu den Themen "Network Management" und "Directory" ausgetauscht werden.

6. Probleme und Maßnahmen

Zum Projektbeginn von GeNeRIC konnten schon zahlreiche Maßnahmen zur Verbesserung der Dienstqualität realsiert werden. Die Hardware-Konfigurationen für DFN-RELAY/ IXGATE und DFNGATE/ DEARN wurden gegen erweiterbare, leistungsfähigere ausgetauscht. Die Software für DFNRELAY und IXGATE wurde neukonfiguriert und auf einer Dienstmaschine zusammengelegt. Am DFNGATE wurde neue Software eingesetzt, die aufgrund einer Kooperation zwischen der "Universidade Federal do Rio de Janeiro (UFRJ)" und der GMD in Bonn für tauglich befunden und angepaßt wurde [Pizz91]. Beim eingesetzten IBM-Rechner wurde der Übergang auf ein leistungsfähigeres Betriebssystem (VM/ XA) vollzogen.

Die Kommunikationsinfrastruktur wurde durch zahlreiche projektdezidierte WiN-Anschlüsse verbessert. Der Personalaufwand zur Betreuung der Dienstmaschinen und ihrer Nutzer, zur Betriebsüberwachung und zur Bearbeitung der betriebs- und problemorientierten Forschungsinhalte wurde gesteigert.

Alle diese Maßnahmen wurden weitgehend transparent für die Nutzer vollzogen. Seither ist die Akzeptanz der angebotenen Dienste stark gestiegen. Einige Probleme verbleiben jedoch.

Zunächst einmal sind solche Probleme zu nennen, die durch bestimmte Randbedingungen im Kommunikationsverbund verursacht werden und die somit nicht durch Maßnahmen der GMD allein gelöst werden können.

6.1 Auskunftsdienste

Bei allen Arten von Auskunftsdiensten gibt es ein allgemeines Problem: die Akzeptanz dieser Dienste steht und fällt mit der Vollständigkeit und Zuverlässigkeit der gespeicherten Daten. Die häufigsten Fragen in einem Kommunikationsverbund sind:

— Kann ich meinen Kommunikationspartner über das Netz erreichen?
— Wie lautet seine Anschrift?
— Sind meine Informationen bei ihm angekommen?
— Warum dauert dies denn so lange?

[3] Um den angestrebten Übergang auf Hersteller-Software zur Erbringung der Dienste zu erleichtern, wurde eine Kooperation mit der Firma Digital (München/ Köln) eingegangen, um in beidseitigem Interesse Industrieprodukte im täglichen Betrieb in komplexer Netzumgebung zu überprüfen.

Auf die ersten beiden Fragen muß ein Auskunftsdienst Antwort geben können. Es gibt jedoch einen Kreislauf, da derzeit das Eintragen in die Auskunftsdienste freiwillig ist: nur wenige Nutzer registrieren sich, nur wenige Nutzer sind im Auskunftsdienst auffindbar, die Akzeptanz bei den Nutzern ist entsprechend schlecht, nur wenige Nutzer registrieren sich, ...

Unter Berücksichtigung der Datenschutzgesetze müssen deshalb Maßnahmen ergriffen werden, die Endnutzer und auch die Systeme systematisch in die Auskunftsdienste zu bringen. In einer englischen Universität beispielsweise wird die Telefonliste nur aus den Einträgen im X.500-Directory erzeugt. Ähnliches geschieht im EARN: die Routing Tabellen werden aus der Datei BITEARN NODES, die alle Knoten weltweit enthält, erzeugt. Auch für den OSI-Verbund ist eine solche Vorgehensweise denkbar und wünschenswert.

Im Vorfeld des verteilten X.500-Auskunftsdienst werden derzeit noch Hilfsmittel wie zentrale Dateien genutzt, um die Informationen über die Kommunikationspartner im X.400-Verbund zu verwalten und zu verteilen. Diese Informationen sind, ähnlich wie Telefonbücher, zum Zeitpunkt ihres Erscheinens schon wieder veraltet. Gerade während des Übergangs vom öffentlichen Paketvermittlungsnetz (Datex-P) zum privaten Wissenschaftsnetz (WiN) macht sich dies in verschlechterter Erreichbarkeit und in unnötig längeren Durchlaufzeiten bemerkbar. Benutzerschnittstellen zu den X.400-Infoblöcken zum Abfragen oder Aktualisieren, wie sie im EARN bekannt sind, gibt es nicht und auch die Disziplin, jede kommunikationsrelevante Änderung sofort zu melden, ist noch nicht sehr verbreitet. Derzeit ist es leider immer noch so, daß die Verwaltung von 4000 Knoten in EARN/ BITNET/ NetNorth weltweit besser funktioniert als die Verwaltung von 250 X.400-Installationen in Deutschland. Es wird jedoch erwartet, daß X.500 hier Abhilfe schafft, da die Zugangsprotokolle schon im Standard festgelegt sind.

6.2 Quo vadis EARN?

Die Kommunikationsdienste, die im europäischen Rechnernetz EARN schon seit Jahren zuverlässig angeboten werden, basieren auf dem von IBM entwickelten "Network-Job-Entry (NJE)-Protokoll" und werden deshalb auch "NJE-Dienste" genannt [NJE]. NJE ist ein Anwendungsprotokoll, das zahlreiche Möglichkeiten für die eigentlichen Netzprotokolle zuläßt: zunächst einmal sind dies wieder die IBM-Protokolle wie BSC (Binary-Synchronous Communication) oder SNA (Systems Network Architecture), dann Protokolle der Firma Digital, die im Rahmen von "Decnet" angeboten werden, auch OSI-Protokolle wie X.25 und nicht zuletzt TCP/ IP.

Während in der Bundesrepublik die NJE-Dienste fast ausschließlich über X.25 angeboten werden, ist die Situation im internationalen EARN-Backbone wesentlich heterogener. Da die Randbedingungen und Entscheidungen der internationalen Kommunikationspartner in einem Rechnerverbund toleriert und akzeptiert werden müssen, bleibt es einem internationalen Netzknoten wie dem DEARN nicht erspart, alle üblichen Netzprotokolle auch zu unterstützen, damit die Entscheidungen, welche Knoten direkt ohne unnötiges "Store-and-Forward" miteinander verbunden werden, nur auf der Basis von Faktoren wie Datenaufkommen und Dienstgüte, nicht aber aufgrund von technischen Restriktionen gefällt werden können.

Dies wurde bei den Planungen für den deutschen Zentralknoten DEARN berücksichtigt, zumal Deutschland derzeit ein Drittel des internationalen EARN-Verkehrs produziert oder entgegennimmt. Bei der GMD werden derzeit folgende Netzprotokolle für NJE unterstützt: NJE/ SNA/ X.25, NJE/ OSI (ohne SNA) und NJE/ TCP/ IP. Es steht derzeit leider noch nicht fest, ob sich EARN auf dem internationalen Backbone auf ein Netzprotokoll beschränken will.

Die Finanzierung von EARN in der Bundesrepublik ist gesichert, seitdem die Kosten auf die Mitglieder umgelegt werden und die DFN-Geschäftsstelle diese Mittel uneigennützig verwaltet. Dies ist jedoch nicht in allen europäischen Ländern so geregelt, so daß auch hier die Entwicklung abgewartet werden muß.

6.3 Die Systeme im Verbund

Eine Grundidee im DFN MHS-Verbund ist es, Kommunikation zu ermöglichen. In einer Zeit, in der X.400-Systeme, die dem 1984er Standard entsprechen, nach wie vor in der Entwicklung stehen ,– an den 1988er Standard für X.400 denken derzeit nur wenige –, ist es sinnvoll, auch Systeme, die nicht hundertprozentig dem X.400-Standard entsprechen, zum Verbund zuzulassen. Der Schwerpunkt wird nicht auf die Konformität, sondern auf die Interoperabilität gelegt. Dennoch: einige Implementierungen im Verbund sind sehr inflexibel, wenn sie beispielsweise mit Systemen kommunizieren sollen, die mehrere DTE-Adressen (X.25) oder mehrere Transport Service Access Points (TSAPs) verwenden. Oder aber sie nutzen nicht die volle Leistungsfähigkeit von X.400 wie z.B. das Verschicken von Dokumenten in beiden Richtungen über eine einmal etablierte Verbindung aus. Darüberhinaus gibt es leider viele "Dialekte" von X.400, die aufgrund von Inkonsistenzen im Standard, die auch der 1988 herausgegebene Implementor's Guide [X.400-IG] nicht ausräumen konnte, und von unterschiedlichen Interpretationen bei der Implementierung entstanden sind. Ein X.400-System, das am DFNRELAY eingesetzt wird, muß mit allen diesen Umständen fertig werden. Der Betrieb und die Betreuung werden hierdurch jedoch erschwert.

6.4 Die Postmaster

In vielen kleineren Institutionen wird der Kommunikationsbetrieb von einer Person "nebenamtlich" betreut. Dies hat zur Folge, daß es im Verhinderungsfall für diese Person wie z.B. durch Urlaub, Krankheit, Dienstreise oder starke inhaltliche Inanspruchnahme durch die eigentlichen Arbeitsinhalte keine Vertreterregelung gibt. Dies macht sich zuerst an einer zentralen Stelle wie im Netzzentrum der GMD bemerkbar: Warteschlangen bilden sich, Verzögerungen in der Zustellung von Nachrichten treten auf, Beschwerden häufen sich und die Versuche der GMD, einen Ansprechpartner zur Behebung der Probleme zu finden, scheitern, da sich niemand sonst mit dem Kommunikationsbetrieb auskennt.

Um dieser Tatsache Rechnung zu tragen, werden die Teilnehmer am DFN-Verbund in Zukunft aufgefordert werden, sogenannte "Betriebsvereinbarungen" abzuschließen, in denen sie ebenfalls eine bestimmte Dienstqualität vor Ort zusichern müssen.

6.5 Die Endnutzer

Zu den Aufgaben der GMD in ihrem Netzzentrum gehört auch die Beratung der Dienstnutzer. Die Dienstnutzer sind dabei jedoch nicht die Endnutzer, sondern Postmaster, Knotenadministratoren oder Systemadministratoren, also eher die technisch-orientierten Leute. Diese werden ständig beispielsweise über Verteilerlisten über Ausfälle, Störungen oder Änderungen informiert. Die Anzahl dieser Kunden ist beschränkt. Kritisch wird es für die Beratung jedoch, wenn sich die Endnutzer in ihren Nöten direkt an die GMD wenden, ohne vorher ihre lokalen Kollegen konsultiert zu haben. Oft handelt es sich um ein lokales Problem, oft weiß der Postmaster vor Ort schon Bescheid. In den Fällen von nicht-lokalen Problemen wird er mit Sicherheit informiert oder seine Kontaktaufnahme mit der GMD bringt den gewünschten Erfolg bzw. die Information, die er an seine Kollegen in der Institution vor Ort weitergeben kann.

6.6 Public Domain Software im Netzzentrum

Public Domain Software hat zunächst einmal den Vorteil, daß sie frei erhältlich ist. Sie wurde zumeist auf der Basis von individueller Eigeninitiative entwickelt und dann an Interessenten weltweit verteilt. Ein weiterer Vorteil ist es, daß diese Software mit allen Quellen verteilt wird und somit geändert und angepaßt werden kann. Dies ist bei dem derzeitigen Entwicklungsstand von Kommunikations-Software sehr wichtig. Der wesentliche Nachteil bei Public Domain Software ist die ungeklärte Wartungssituation: kein Hersteller steht hinter dieser Software, an den man sich, wenn man einen entsprechenden Wartungsvertrag abgeschlossen hat, mit der Bitte um Behebung des Problems wenden kann. Es ist zwar bei bestimmten Software-Paketen wie z.B. TCP/ IP schon vorgekommen, daß Hersteller diese aufgrund ihrer Marktpräsenz und ihrer Akzeptanz als De-Facto-Standard adaptiert haben, doch dies ist eher die Ausnahme. Speziell für Anwendungs-Gateways findet man häufig keine Software bei den Herstellern. Ausnahmen gibt es jedoch auch hier im Bereich X.29-TELNET (z.B. Cisco) und FTAM-FTP (z.B. Digital und IBM).

Im GeNeRIC-Projekt wird Public Domain Software zur Zeit noch an vielen Stellen eingesetzt: DFNRELAY, DFNGATE, IXGATE, Domain Name Server, FTAM-FTP, DEARN und speziell im Bereich Netzmanagement, in dem aber wiederum das IP-Management auszuklammern ist, da es hier bereits kommerzielle Software zu kaufen gibt.

Die Wartungsproblematik wird dadurch entschärft, daß es fast zu jeder Public Domain Software internationale Verteilerlisten als Diskussionsforum gibt, auf denen die Autoren, die Betreiber und die Nutzer dieser Software miteinander im Dialog stehen. Auch gibt es internationale "Selbsthilfegruppen" wie z.B. das im Rahmen von COSINE gebildete EAN Support Team, das zur Aufgabe hat, die im internationalen X.400-Verbund häufig eingesetzte EAN-Software wartungsmäßig abzusichern und in Richtung volle X.400-Konformität zu entwickeln [EAN].

Es wird für die Zukunft erwartet, daß im GMD-Netzzentrum die X.400- und die X.400-SMTP-Gateway-Software durch Hersteller-Software oder durch Software, für die Firmen Wartungsverträge anbieten, abgelöst werden kann. Auch am FTAM-FTP-Gateway zeichnet sich eine ähnliche Entwicklung schon ab. Zu diesem Zweck gibt es eine von der Firma Digital zur Verfügung gestellte Hardware-Konfiguration, die einerseits wie eben beschrieben als Backup-Maschine dient und andererseits die Umgebung für die Tests alternativer Software ist.

6.7 Verfügbarkeit der Dienste

In der Vergangenheit ist es leider schon vorgekommen, daß ein Dienstort durch Störungen wie Stromausfall oder einen Bagger, der die Netzleitungen zerstörte, vollkommen außer Gefecht gesetzt wurde. Auch der Ansatz, alle Dienste auf einem Universalrechner anzubieten, hat Nachteile.

Die zuvor beschriebenen Dienste werden folgerichtig auf verschiedenen, örtlich getrennten "Service-Maschinen" angeboten. Einzelne Service-Maschinen beherbergen dabei mehrere Dienstleistungen. Dies ist abhängig von der Software, die zur Erbringung der Dienstleistung für das jeweilige Betriebssystem verfügbar ist. Jede Komponente ist autonom und somit unabhängig von Störungen der anderen Komponenten. Die Last, die durch die elektronische Kommunikation der DFN-Nutzer erzeugt wird, wird auf mehrere Maschinen verteilt. Dies hat erhebliche Vorteile gegenüber einer zentralen Lösung, da Engpässe, die in Spitzenzeiten an einer Gateway-Relay-Komponente entstehen können, die anderen Komponenten, da verteilt, nicht beeinträchtigen. Dadurch wird die Dienstgüte, in diesem Fall die Verfügbarkeit, die Antwortzeiten und der Durchsatz, wesentlich gesteigert.

Für die angebotenen Dienste X.400-Relay, X.400-SMTP-Gateway, FTAM-FTP-Gateway und IP-Routing ist mindestens eine Backup-Lösung vorgesehen, die es den Kommunikationspartnern bei geplanten oder unvorhergesehenen Systemzusammenbrüchen erlaubt, ihr Routing kurzfristig für die Dauer der Störung zu ändern. Die Verfügbarkeit der Dienste wird somit stark erhöht. In Notsituationen besteht sogar die Möglichkeit, über derartige Backup-Maschinen Lastausgleich für den Nachrichtenverkehr zu machen.

Zur Erbringung des X.400-Relay- und des X.400-SMTP-Gateway-Dienstes wird nicht nur eine leistungsstarke, erweiterbare Hardware-Konfiguration eingesetzt, sondern es wird auch eine durch die Industriekooperation ermöglichte Backup-Hardware-Konfiguration bereitgehalten, die ein sofortiges Umschalten bei lokalen Problemen ermöglicht. Gleiches gilt für die zwei IP-Router, die gleichzeitig zu "Backup-Zwecken" und zur Kapazitätserweiterung betrieben werden.

Die GMD hat darüberhinaus angeboten, eigenes, projektfremdes Equipment zu Backup-Zwecken einzusetzen.

Am X.400-BSMTP-Gateway besteht auch noch die Möglichkeit, Backup für den X.400-SMTP-Gateway zu spielen, so daß Nachrichten von deutschen DFN-Nutzern an Nutzer im INTERNET den Weg X.400 (DFN) - BSMTP (BITNET) - SMTP (INTERNET) oder umgekehrt nehmen können.

Für den X.400-BSMTP-Gateway selbst gibt es keinen direkten Backup. Nur ein Umweg über SMTP unter Nutzung eines SMTP-BSMTP-Gateways kann den Kommunikationsfluß zwischen X.400-Nutzern und EARN-Nutzern aufrechterhalten.

Zur Zeit gibt es keinen Backup-Dienst für den EARN-Zentralknoten DEARN. Im internationalen EARN wird jedoch daran gearbeitet, EARN, ähnlich der Entwicklung im BITNET, in sogenannte "Regionen" aufzuteilen. In jeder Region gibt es dann zwei Dienstzugangspunkte, sogenannte "Core Sites", die mit allen anderen Core Sites verbunden sind. Diese beiden Knoten einer Region stehen sich gegenseitig zu Backup-Diensten zur Verfügung. Ein Zugangspunkt für eine Region ist auf jeden Fall DEARN.

Diese Regionalisierung von EARN basiert auf der Tatsache, daß die EARN-Dienste auch oberhalb von TCP/ IP angeboten werden können. In jeder EARN-Region soll es jeweils eine IP-Verbindung in die USA, die für NJE oberhalb von TCP/ IP (VMNET) genutzt werden soll, geben. Ziel ist es, die derzeitigen Engpässe in der Kommunikation mit Nordamerika und darüber versorgten Bereichen, die durch die einzige transatlantische EARN-Leitung zwischen Montpellier und New York zwangsläufig entstehen, zu vermeiden [EARNb]. Die vollzogene IP-Anbindung des DEARN erlaubt es erst, an diesem Konzept aktiv teilzunehmen und die Dienstgüte für die EARN-Nutzer, auch im Backup-Fall, zu verbessern.

6.8 Betriebsstörungen

In einem verteilten Kommunikationsverbund gibt es eigentlich nie Pausen, in denen die Dienste nicht genutzt werden. Dies bedeutet aber auch, daß zu jeder Tages- und Nachtzeit eine Störung physikalischer Art oder auch von Dienstnutzern verursacht auftreten kann. Eine Betreuung der Dienstmaschinen rund um die Uhr wäre sinnvoll. Dies ist jedoch im Netzzentrum der GMD aus Kostengründen nicht realisierbar.

Es geht nicht, daß wie z.B. in ähnlichen Netzzentren der Firma IBM 8 Personen pro Arbeitsplatz aufgeboten werden, um einen 24-Stunden-Betrieb das ganze Jahr über anzubieten. Die Betreuung der Dienste in der GMD durch Personen kann nur in der normalen Dienstzeit realisiert werden. Ansonsten werden die Service-Maschinen durch speziell entwickelte Monitor-Programme überwacht, die auf Systemzusammenbrüche reagieren, die benötigte Betriebsmittel nachstarten und die fehlerhafte Informationsobjekte wie Nachrichten oder Dateien aus dem Verkehr ziehen.

Aus personellen und betrieblichen Gründen ist es auch nicht vorgesehen, eine 24-Stunden-Rufbereitschaft für den Relay- bzw. Gateway-Betrieb zu realisieren. Rufbereitschaft ist lediglich für Notfälle wie Feuer, Stromausfall und Klimaausfall realisiert. Ansonsten ist für alle Betriebsorte je ein Anrufbeantworter mit Fernabfrage vorgesehen.

Für stichprobenartige Überprüfungen der Dienstgüte außerhalb der Dienstzeit wird pro Dienstort ein Mitarbeiter mit einer "Remote Control Station" ausgestattet, die ihm Zugriff auf die Service-Maschinen von "zu Hause aus" ermöglichen.

Für die im Projekt eingesetzte Hardware wurden entsprechende Wartungsverträge abgeschlossen, – der IBM-Rechner ist sogar in der Lage, bei Störungen diese selbst per Telefon an den Wartungsdienst zu melden –, und für die Kommunikationsleitungen wurde mit der Bundespost auch Entstörung außerhalb der Dienstzeit vereinbart.

6.9 Personalausfall

Trotz der verbesserten allgemeinen Personalsituation kann es vorkommen, daß MitarbeiternInnen, die einen Kommunikationsdienst im Projekt betreuen, gleichzeitig verhindert sind. Um dieser Tatsache Rechnung zu tragen, wurden innerhalb der GMD, wo möglich, projektübergreifende Arbeitsgruppen etabliert, in den MitarbeiterInnen aus unterschiedlichen Projekten mit verwandten Arbeitsinhalten zusammenarbeiten, um in solchen Fällen Synergie-Effekte zu nutzen.

6.10 Benutzerbetreuung

Die Zufriedenheit der Benutzer mit einem Dienst wächst mit der Zuverlässigkeit des Dienstes und auch mit der Menge an Informationen, die die Benutzer über den Dienst erhalten. Dies bedeutet nicht nur: Wie kann ich den Dienst nutzen? Sondern auch: was ist denn jetzt schon wieder los? Wo kann ich mal nachfragen? Für die Nutzer ist es nicht akzeptabel, wenn Nachrichten an einem Tag zum Empfänger kommen, an einem anderen Tag nicht, und er kennt den Grund nicht.

Die GMD hat deshalb einige Maßnahmen eingeleitet, um eine "offene Informationspolitik" zu realisieren. Zunächst einmal werden die Kunden, die Postmaster und die Administratoren über die Verteilerlisten über Störungen, Ausfälle und Änderungen informiert. Es werden Logbücher geführt, in denen man derartige Ereignisse nachlesen kann. Es wurden spezielle Benutzerkennungen für Fehlermeldungen eingerichtet, der Absender einer Fehlermeldung bekommt eine Empfangsbestätigung und wird über den Stand der Fehlerbearbeitung informiert. Bei schriftlich oder elektronisch gemeldeten Fehlern durch die Postmaster oder Administratoren laufen feste Fehlerbearbeitungsprozeduren ab. Anfragen von Endnutzern werden an die zuständigen Postmaster weitergeleitet. Außerhalb der Dienstzeit stehen Anrufbeantworter zur Verfügung. Die umfangreichen Statistiken werden dem Auftraggeber vorgelegt und werden bei Bedarf veröffentlicht.

6.11 Qualitätskontrolle

An den GMD Service-Maschinen werden Daten über deren Nutzung und die Auslastung der Kommunikationsleitungen gesammelt und ausgewertet. Umfangreiche Statistiken werden erstellt. Diese Zahlen werden regelmäßig veröffentlicht und dienen dem Management als Überblick über die Nutzung ihrer Dienste und als Basis für Entscheidungen wie z.B. ob neue Leitungen anzumieten sind. Sie dienen den Betreibern als Kontrolle und können auch zur Kostenrückerstattung verwendet werden.

Diese Zahlen geben allerdings keinen Aufschluß über die Akzeptanz der Dienste, über die Zufriedenheit der Nutzer und über die Dienstgüte, da in diesen Statistiken Angaben, wie lange Nachrichten unterwegs waren und wie lange die Service-Maschinen ohne Ausfall den Dienstnutzern zur Verfügung standen, nicht enthalten sind.

Um diesem Mißstand abzuhelfen, hat die GMD Testzyklen definiert, die stündlich durchlaufen werden und die die Gateways nach X.400 (DFNGATE, IXGATE), den EARN- und X.400-Relay und die NJE-Server wie LISTSERV, MAILER und RSCS (Remote

Spooling Communication Subsystem) einbeziehen. Eine Nachricht durchschnittlicher Länge wird unter Nutzung dieser Server von einem EARN-Nutzer zu einem SMTP-Nutzer und wieder zurück geschickt. Da die Server-Maschinen an unterschiedlichen Dienstorten stehen, wird auch die zugrundeliegende X.25-Infrastruktur in den Test mit einbezogen.

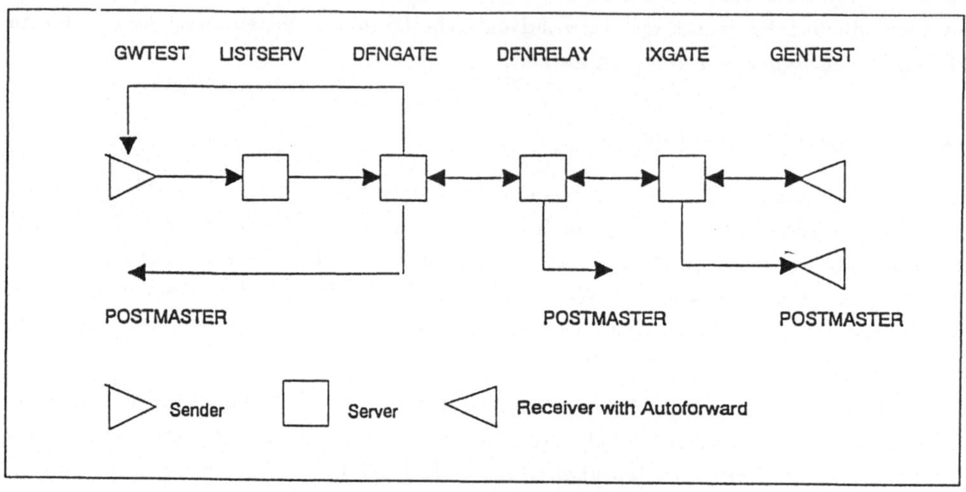

Abbildung 6: Der Testzyklus

Die ausgewerteten Zahlen geben am Monatsende Auskunft darüber, wie groß die Bearbeitungszeit innerhalb der GMD war und wie lange die Service-Maschinen ohne Ausfall genutzt werden konnten. Zusätzlich wird dieser Test noch zur Betriebsüberwachung genutzt, indem die Postmaster jeweils diesseits und jenseits der Gateways diese Testnachricht auch erhalten.

Der IP-Relay, der FTAM-FTP-Gateway und der X.29-TELNET-Gateway ist in diesen Testzyklus für Electronic Mail natürlich nicht integriert. Der IP-Relay wird per Selbsttest (Ping) alle 10 Minuten überprüft, die beiden anderen Gateways befinden sich noch in der Aufbauphase.

7. Teamwork

Trotz den soeben dargestellten Maßnahmen und Aktivitäten der GMD bleiben noch einige Aufgaben für die Kunden des GMD-Netzzentrums. Die Postmaster sollen wie die GMD ständig die Erreichbarkeit ihrer Installation überprüfen und die Informationen über die eigene Installation und die Kommunikationspartner aktualisieren. Meldungen von Betriebsstörungen, Ausfällen und Änderungen sollen an die Endnutzer weitergegeben werden. Die im DFN MHS-Verbund definierten Verteilerlisten und Dokumente, die zum Kommunikationsbetrieb gebraucht werden, sollten abonniert werden. Es sollte pro Installation mindestens 2 Leute geben, die sich mit dem Alltag im Kommunikationsbetrieb auskennen. Im DFN MHS-Verbund sollten die Institutionen die Betriebsvereinbarungen mit dem DFN eingehen, in denen sie sich verpflichten, selbst eine bestimmte

Dienstqualität für ihre Kommunikationspartner, zu denen auch die GMD gehört, zu gewährleisten.

Auch die Endnutzer können zum Gelingen des Kommunikationsverbundes beitragen, indem sie den Postmaster vor Ort in seiner Arbeit unterstützen und seine Position innerhalb der Institution stärken, indem sie ihn bei Kommunikationsproblemen als ersten konsultieren und indem sie auf "Experimente" an Sonn- und Feiertagen verzichten.

8. Einige Zahlen zum Schluß

Einige Zahlen zum Schluß sollen dazu dienen, die bisher beschriebene Problematik abzurunden.

Tabelle 1 bezieht sich auf den Monat Juni 1991 und gibt Aufschluß über die Datenmengen, die ausschließlich bei der GMD gemessen wurden. Für den IP-Relay, den FTAM-FTP-Gateway und den X.29-TELNET-Gateway liegen noch keine Zahlen vor, da sich diese Dienste noch in der Aufbauphase befinden.

Juni '91	**Daten in MB**	**Nachrichten**
X.400 national	726,5	145.300
X.400 international	114,5	22.900
DEARN	24.021,5	1.484.760
IP-Vermittlung	–	–
X.400-SMTP-GW	298	59.600
X.400-BSMTP-GW	651,5	96.960
FTAM-FTP-GW	–	–
X.29-TELNET-GW	–	–

Tabelle 1: GeNeRIC-Verkehrsaufkommen

	9.6 KBit/s	**64 KBit/s**
WiN (X.25)	DM 18.000	DM 60.000
EARN	DM 7.000	DM 13.000
	DM 1.000	DM 1.500
IP	DM 10.000	DM 18.000
OSI	–	–

DFN DM 500, 5.000, 2.000, 5.000, 10.000

Tabelle 2: DFN-Tarife p.a., Stand Juni '91

Tabelle 2 zeigt die derzeit gültigen Tarife. Die Anschlußgebühr für WiN, EARN und die IP-Vermittlung richten sich nach der Geschwindigkeit, mit der die Institution an das Wissenschaftsnetz angeschlossen ist. Die zweite Zeile für die EARN-Beiträge gilt für

Knoten, die als sogenannte "Pseudo-Knoten" geführt werden und für die es eine Weiterleitung für Electronic Mail von EARN nach X.400 gibt. Die DFN-Mitgliedsbeiträge richten sich nach den Status der Institution (Hochschule, Großforschungseinrichtung, Wirtschaftsunternehmen etc.). Die Preise für den OSI-Verbund einschließlich der Nutzung der Anwendungs-Gateways werden für Ende 1992 oder Ende 1993 erwartet.

Im vorliegenden Papier wurde gezeigt, daß die GMD in ihrem GeNeRIC-Projekt ein Netzzentrum für die offene Kommunikation in heterogenen Netzen mit dem Schwerpunkt "Open Systems Interconnection" anbietet. Die gestellte Aufgabe ist nicht einfach und bedarf unbedingt der Kooperation mit nationalen und internationalen Kommunikationsdienstanbietern. Insellösungen sind nicht gefragt. Die GMD hat schon zahlreichen Maßnahmen ergriffen, um die Dienstgüte zu verbessern und die Zuverlässigkeit zu steigern. Doch weitere Aufgaben für die kommenden Jahre warten schon. Die Postmaster und Administratoren und sogar die Endnutzer können die GMD in der Lösung dieser Aufgaben zum Wohle des internationalen Kommunikationsverbundes unterstützen.

Im GeNeRIC-Projekt der GMD arbeiten derzeit 8 Wissenschaftler und 5 mathematisch-technische Assistenten/-innen. Darüberhinaus gibt es noch weitere MitarbeiterInnen der GMD, die dieses Projekt direkt oder indirekt unterstützen. All diesen soll an dieser Stelle für ihre Mitarbeit und Kooperation gedankt werden.

REFERENCES

[Bir91] Klaus Birkenbihl. Wissenschaftsnetze. In: Jahresbericht 1990, Gesellschaft für Mathematik und Datenverarbeitung, St. Augustin, 1991

[BBD*91] Bogen, Büttner, Dillmann, Eckert, Engelke, Grape, Hennings, Klever, Melcher, Mayerhofer, Siebert, Spirk, Sylvester, Bonacker. PRMD-Manager-Handbuch. DFN-Bericht Nr. 62, Deutsches Forschungsnetz, Berlin, Januar 1991

[BSMTP] E. Alan Crosswell. Batch Simple Mail Transfer Protocol, Columbia University Center for Computing Activities, September 1982

[BSWW89] Manfred Bogen, Peter Sylvester, Clemens Wermelskirchen, Peter Wunderling. Kommunikationsunterstützung für die Wissenschaft. In: Jahresbericht 1988. Gesellschaft für Mathematik und Datenverarbeitung, St. Augustion 1989.

[CCITT80] CCITT, Data Communication Networks - Services and Facilities, Terminal Equipment and Interfaces, Recommendations X.1 - X.29, Yellow Book Vol. VIII Fascicle VIII.2, VII Plenary Assembly, Geneva, 1980, International Telecommunication Union (ITU), Geneva, 1981

[CCITT84] CCITT, Data Communication Networks Interfaces - Recommendations X.20 - X.32, Red Book Vol. VIII Fascicle VIII.3, VIII Plenary Assembly, Malaga-Torremolinos, 1984, International Telecommunication Union (ITU), Geneva, 1985

[EAN] G. Neufeld. The EAN Diustributed Message System, User's Manual, Version 2.1. University of British Columbia, Vancouver, B.C. Canada, July 1987

[EARNa] P. Sylvester. EARN - ein europäisches Computer-Netzwerk für Wissenschaft und Forschung. Der GMD-Spiegel, 85(1): 71-74, März 1985

[EARNb] D. Bovio, H.U. Giese, H. Nussbacher. Proposal for the regionalization of EARN. Document EXEC121 90. Draft. März 1991

[ISO8571] ISO. Information Processing Systems - Open Systems Interconnection - File Transfer Access adn Management, Part 1-5

[LaQ90] Tracy L. LaQuey. Editor. The User's Directory of Computer Networks. Digital Press, Bedford, Massachusetts, 1990

[NJE] John M. Hutchinson, David A. Stamper. Network Job Entry Concepts and Protocols Overview. Technical Bulletin. Washington Systems Center, Gaithersburg, MD, November 1985

[Pizz91] Juan Altmayer Pizzorno. Gateway X.400/ BITNET de Queens Em Uso Em Producao Na Alemanha (DEARN) E Na UFRJ: Um Trabalho A Quatro Maos. 90. Simposio Brasileiro de Redes de Computadores (SBRC), Florianopolis, Santa Catarina, Brazil, Mai 1991

[Qua90] John S. Quaterman. The Matrix: Computer Networks and Conferencing Systems Worldwide. Digital Press, Bedford, Massachusetts, 1990

[RARE91] RARE. Annual Report 1990. Amsterdam, 1991

[X.400-IG] CCITT, X.400–Series Implementor's Guide, Question 33/ VII, Special Rapporteur's Group on Message Handling Systems, Version 6, November 1987

[X.400-84] CCITT, Data Communication Networks, Message Handling Systems, Recommendations X.400-X.430, Malaga–Torremolinos, Juni 1984,

[X.400-88] CCITT, Data Communication Networks, Message Handling Systems, Recommendations X.400-X.420, Melbourne, November 1988

[X.500] CCITT, Data Communication Networks, Directory, Recommendations X.500-X.521, Melbourne, November 1988

[RFC821] Jon Postel, Simple Mail Transfer Protocol. RFC 821, USC Information Science Institute, August, 1982.

[RFC822] David H. Crocker, Standard for the Format of ARPA Internet Text Messages. RFC 822, University of Delaware, August, 1982.

[RFC854] J. Postel, J. Reynolds, TELNET Protocol Specification, RFC 854, USC Information Science Institute, May 1983

[RFC959] J. Postel, J. Reynolds, File Transfer Protocol (FTP), RFC 959, USC Information Science Institute, October, 1985

[RFC987] S. E. Kille. Mapping between X.400 and RFC822. RFC987, University College London, June, 1986

[VERDI-90] K. H. Bonacker, A. Jerusalem, S. Kolvenbach, M. Tschichholz, VERDI III - DFN-Directory - Verwaltung von MHS-Daten, Draft Version, DFN, November 1990

Anschluß von lokalen Netzen an ein X.25-Paketvermittlungsnetz

Dr. P. Holleczek
Universität Erlangen-Nürnberg
Regionales Rechenzentrum Erlangen

Zusammenfassung

Das von DBP-Telekom im Auftrag des DFN-Vereins betriebene X.25-Paketvermittlungsnetz WiN erlebt einen unerwartet hohen Anstieg des Verkehrsaufkommens, der am einleuchtendsten durch die Vielzahl der am WiN angeschlossenen lokalen Netze erklärt werden kann. Es wird dargelegt, welche Methoden es zum Anschluß von lokalen Netzen an ein X.25-Netz gibt, welche Protokolle und welche Produkte zum Einsatz kamen und welche Erfahrungen damit gesammelt werden konnten. Besonderer Augenmerk gilt der zukünftigen Ausgestaltung mit 2 Mbps-Übertragungswegen und dem zu erwartenden Durchsatz. Messungen in Laborumgebung zeigen, daß ein 2 Mbps-X.25-Netz zusammen mit heutigen LAN-Vermittlungseinrichtungen eine volle Ausnutzung der Bandbreite ermöglicht. Angesichts des zu erwartenden EC-GOSIP, das auch in lokalen Netzen X.25 als Netzwerkprotokoll vorsieht, stellt ein X.25-Weitverkehrsnetz, das sich höheren Geschwindigkeiten öffnet, ein strategisches Instrument dar.

1. Einleitung

Das WiN (<u>Wi</u>ssenschafts<u>N</u>etz) als X.25-Weitverkehrsnetz (WAN), im Auftrag des DFN-Vereins von der DBP-Telekom betrieben und konzipiert für eine monatliche Summendatenrate von 50 Gigabyte, ist seit etwas mehr als einem Jahr in Betrieb. Die Summendatenrate lag Mitte 1991 bereits bei 100 Gigabyte. Die Steigerung ist durch die erfreuliche Ausweitung der Anschlußzahlen nicht allein zu erklären. Während bei den WiN-Vorbildern, z.B. den Landesnetzen in Bayern und Nordrhein-Westfalen, X.25-basierte Endsysteme im Vordergrund standen, sind es mittlerweile die Endsysteme an lokalen Netzen (LAN), die den stärksten Verkehrszuwachs am WiN verursachen. Das ist Grund genug, sich mit dem Anschluß von LANs an das WiN auseinanderzusetzen.

2. Problemstellung

Das WiN erfüllt zunehmend einen doppelten Zweck. Es soll eine transparente Verbindung von lokalen Netzen mit ihren Endsystemen ermöglichen und eine Übergangsmöglichkeit von Endsystemen an LANs zu Endsystemen am WAN bieten. Während in lokalen Netzen eine Vielfalt von Protokollen vorherrscht, wird in Endsystemen mit direkten WAN-Anschlüssen hauptsächlich der klassische ISO-Protokollstack mit Transportklasse Ø und dem verbindungsorientierten Netzwerkdienst (X.25) eingesetzt. Für eine Protokollwandlung ist dieser Protokollstack das Hauptziel.

Beim Anschluß von lokalen Netzen mit ihren Endsystemen an ein X.25-Weitverkehrsnetz (wie das WiN) stellen sich folgende Fragen:

- Wie verträglich sind die in lokalen Netzen eingesetzten Protokolle mit X.25; welche Standards gibt es?
- Was kann über das WiN transportiert werden; welche Gateways gibt es?
- Welche Produkte gibt es generell und wie sind die Erfahrungen damit?

Da die Übertragungsgeschwindigkeiten in lokalen Netzen in der Regel höher sind als in Weitverkehrsnetzen, ist auch interessant, wie der Datendurchsatz beim Übergang zwischen LAN und WAN bzw. zwischen LANs, die über ein WAN verbunden sind, beeinflußt wird. Neben den zur Zeit üblichen Übertragungsraten von 64 kbps sind besonders die im WiN angestrebten 2 Mbps von Interesse.

Aufgrund der Protokollvielfalt in lokalen Netzen muß für diese Erörterung eine gewisse Auswahl getroffen werden, wobei das Hauptinteresse bei der Anwendungsschicht oder bei den mittleren Ebenen wie Netzwerk oder Transportschicht liegt.

Folgende LAN-Protokolle werden untersucht:

- Die Protokolle mit dem höchsten Verbindlichkeitsgrad, die ISO-Protokolle mit den zwei äquivalenten Varianten /BH89/,
  - TPØ/CONS, der einfachen Transportschicht mit dem mächtigen, verbindungsorientierten Netzwerkdienst, obligatorischer Bestandteil des kommenden EC-GOSIP /CAF90/,
  - TP4/CNLS, der mächtigen Transportschicht mit dem einfachen, verbindungslosen Netzwerkdienst, obligatorischer Bestandteil des US-GOSIP /NG90/,
- die US-DoD-Protokollsuite als herstellerunabhängiger Quasistandard mit TCP/IP auf den niederen Ebenen,
- als Beispiel für eine LAN-Herstellerlösung, das Netzwerkprotokoll für PC-Netze der Firma Novell mit IPX auf der Netzwerkebene.

Protokolle, die für lokale Netze nicht typisch sind (wie SNA) oder Protokolle, die ohnehin klar in Richtung ISO tendieren (wie DECNet) bleiben hier unberücksichtigt. Außerdem sollte festgehalten werden, daß LAN-Architekturen, die typisch für die Forschungslandschaft sind, wie Ethernet und FDDI, im Vordergrund stehen, und nicht z.B. Token Ring.

Erfahrungen, die Grundlage für die Erörterungen bieten, konnten nicht nur im ersten WiN-Betriebsjahr, sondern vor allem anhand eines der Vorläufer des WiN, dem aus Standleitungen mit eigenen X.25-Knoten aufgebauten Bayerischen Hochschulnetzes (BHN), gewonnen werden. Das BHN besteht als organisatorischer Zusammenschluß der Universitäten und Fachhochschulen des Landes Bayern fort.

3. Das WiN als transparentes Transportmittel

In diesem Abschnitt soll untersucht werden, welche Möglichkeiten es zur Abbildung der verschiedenen LAN-Protokolle auf das X.25 des WiN gibt. Das eleganteste Verfahren besteht darin, in LAN und WAN die gleichen Protokolle (also CONS) einzusetzen. Bei dieser homogenen Lösung kann man WiN und angeschlossene LANs als ein Netz mit gemeinsamem Adreßraum betrachten.

Passen die Protokolle nicht zueinander, müssen die Dateneinheiten aus dem LAN auf X.25-Pakete abgebildet werden. Da die LAN-Protokolle in der Regel auch nach Art eines Schichtenmodells gegliedert sind, ist es von Bedeutung, welcher Schicht des LAN-Protokolls die Dateneinheiten entnommen werden. Werden sie als 'Frames' der Link-Ebene (2) entnommen, ist die äquivalente Konstruktion zur LAN-Strukturierung eine 'remote Bridge' und das X.25-Netz ist Ersatz für eine Punkt-zu-Punkt-Leitung. Werden sie als 'Pakete' der Netzwerkebene (3) entnommen, wird quasi das LAN mit seiner Netzstruktur auf das X.25-Netz abgebildet. Die äquivalente Konstruktion zur LAN-Strukturierung ist ein 'remote Router'. Da lokale Netze zunehmend durch Router strukturiert werden, wird in weiteren hauptsächlich dieser Ansatz verfolgt, während die Bridge-Lösung allenfalls einen Notbehelf darstellt.

3.1 ISO-CONS

ISO-CONS erlaubt es, den gleichen verbindungsorientierten Netzwerkdienst in Weitverkehrsnetzen und lokalen Netzen /ISO8881/ einzusetzen. X.25 (ab Version '84) kann ISO-CONS-Dienste erbringen. Die Adressierung erfolgt in WAN und LAN durch NSAP-Adressen. Eine einheitliche, wohlstrukturierte Adressierung im gesamten Netzbereich ist möglich.

Auf Ebene 2 werden in LAN und WAN verwandte Protokolle eingesetzt (LLC2 vs HDLC, s. Figur 1). Aufgrund der Ähnlichkeit ist die Umsetzung in Vermittlungseinrichtungen relativ einfach bzw. laufzeiteffizient.

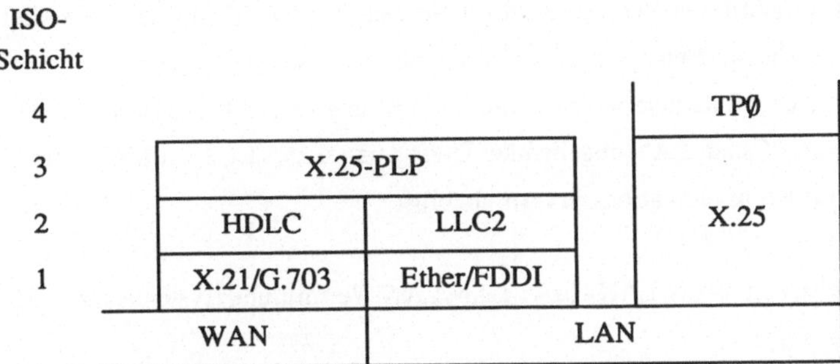

Figur 1: Vermittlungseinrichtung für ISO-CONS in WAN und LAN

ISO-CONS in lokalen Netzen wurde erstmals im englischen Forschungsnetz JANET eingesetzt, beruhte allerdings auf Spezialentwicklungen, die nicht in die Betriebssysteme/Netzwerksysteme der Rechnerhersteller integriert waren. Seit sich abzeichnet, daß ISO-CONS auf den unteren Ebenen auch für lokale Netze Bestandteil des zu erwartenden EC-GOSIPs wird, wird dieser Protokollstack zunehmend durch US-amerikanische Computerhersteller implementiert. Tabelle 1 enthält eine Zusammenstellung der in der Regel erst kürzlich verfügbaren Host-Implementierungen. Eine entsprechende Zusammenstellung für Vermittlungseinrichtungen enthält Tabelle 2.

Tabelle 1: Stand Verfügbarkeit X.25/LLC2-Produkte in Endsystemen

Rechner / Netzwerk / Betriebssystem	**Dienst**	**Status**
PC / BICC-Rainbow / DOS	X^3-Dialog FTAM	verfügbar in Vorbereitung
SUN / SUNLINK V.7 / UNIX	X^3-Dialog FTAM X.400	verfügbar verfügbar verfügbar
VAX / PSI V.5 / VMS	X^3-Dialog FTAM X.400	verfügbar verfügbar verfügbar
CYBER / CDCNet / NOS/VE 1.5.3	X^3-Dialog FTAM X.400	verfügbar verfügbar verfügbar
UNIX 4.4 BSD	X^3-Dialog VT FTAM X.400 X.500 X-Windows	in Vorbereitung in Vorbereitung in Vorbereitung in Vorbereitung in Vorbereitung in Vorbereitung

Alle in Tabelle 1 aufgeführten Rechner sind im Bereich des RRZE mit der Hersteller-Implementierung von X.25/ LLC2 bereits am Netz. Betriebserfahrungen in größerem Stil liegen hauptsächlich für X^3-PADs und PC-Cluster (mit ca. 50 Rechnern) sowie mit Vermittlungseinrichtungen der Firmen Netcomm und Spider vor. Für den von PCs am LAN abgehenden Dialog hat X.25/LLC2 am RRZE inzwischen die konventionelle Lösung mit Kommunikationsservern abgelöst.

Tabelle 2: Verfügbarkeit WAN/LAN- bzw. LAN/LAN-Vermittlungssysteme für X.25 (bzw. CONS)

Hersteller	**WAN (HDLC)**		**LAN (LLC2)**		**Status**
	X.21	**G.703**	**Ethernet**	**FDDI**	
Spider	X		X		verfügbar
CAMTEC	X		X		verfügbar
Netcomm SW2000	X	X (2 Mbps)	X		verfügbar
Netcomm SW3500	X	X (8 Mbps)	X	X	in Vorbereitung
CISCO	X		X		in Vorbereitung
Wellfleet	X		X	X	in Vorbereitung

Interessant sind für den bislang noch nicht verbreiteten Protokollstack die Durchsatzfragen. Während sie bei 64 kbps noch keine größere Rolle spielen, stellt sich die Frage, ob bei Übertragungsraten von 2 Mbps Vermittlungseinrichtungen oder Endeinrichtungen eine Grenze für den Durchsatz darstellen. Da ein 2 Mbps-WiN noch nicht verfügbar ist, wurde am RRZE eine Testkonfiguration aufgestellt, die es erlaubt, Durchsatzmessungen an einem modellhaften 2 Mbps-X.25 mit Ethernet-LANs vorzunehmen. In Konfiguration 1 hängen Sender und Empfänger am gleichen Ethernet-Segment. In Konfiguration 2 wird die Verbindung lokal durch das Ethernet-Board eines X.25-Vermittlers geleitet, um dessen Einfluß abzuschätzen. In Konfiguration 3 hängen Sender und Empfänger an verschiedenen Segmenten, die über Vermittlungssysteme direkt (über eine 2 Mbps-X.25-Strecke) miteinander verbunden sind. In Konfiguration 4 sind die Vermittlungssysteme nicht direkt, sondern über eine X.25-Untervermittlung mit 2 Mbps-Anschlüssen, also quasi über ein minimales Netz verbunden. Figur 2 zeigt die Meßwerte für ein einfaches Memory-Memory-Transportprotokoll für die verschiedenen Konfigurationen. Ersichtlich ist, daß bei Paketlänge 1024 zwischen zwei PCs von den möglichen 200 Paketen/sec. immerhin noch 180 Pakete/sec. über ein (minimales) 2 Mbps-Netz transportiert werden können. Der limitierende Faktor ist offensichtlich die PC-Leistung und ein konstanter Verlust von ca. 10 % durch die Behandlung von Ethernet-Paketen in der Vermittlungseinrichtung. Dieser Verlust kommt vermutlich von einer nicht sehr effizienten Implementierung der Quittungsbearbeitung der Ethernet-Karte der Vermittlungseinrichtung. Es ist zu erwarten, daß dieser Effekt in kürze behoben sein wird. Bei zwei Paaren von PCs und Simplex-Betrieb (gleiche Richtung) erreicht man dann bei Paketlänge 1024 mit 250 KByte/sec. genau die theoretische Grenzleistung. Bei Duplex-Betrieb von 4 PCs (unterschiedliche Richtung) erreicht man in der Summe ca. 335 kByte/sec., also fast die Summe der möglichen Einzelleistungen von 2 x 180 kByte/sec. Der limitierende Einfluß von Kollisionen auf dem Ethernet bleibt unberücksichtigt.

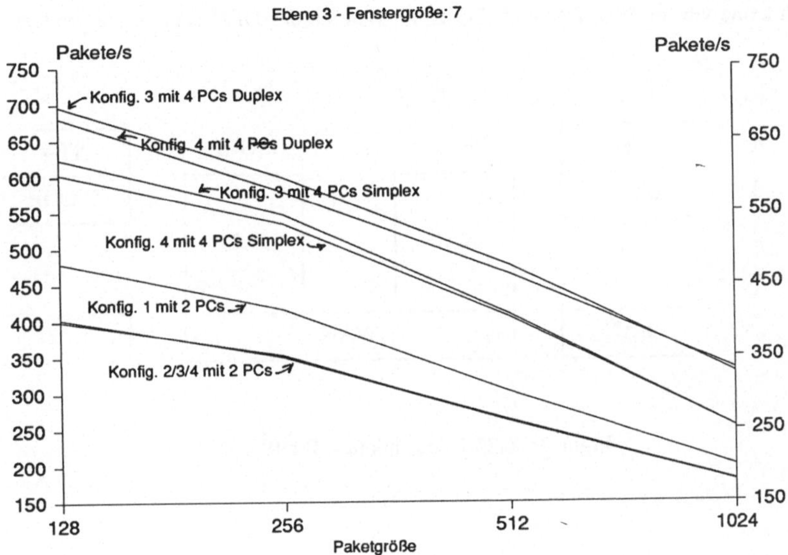

Figur 2: X.25-Durchsatzmessung für PCs mit CONS im WAN/LAN-Übergang

Bei großen Paketen kann also die verfügbare Bandbreite des Übertragungsmediums (bis 2 Mbps) voll ausgenutzt werden. Das kommt dem Transfer von großen Datenmengen, wie z.B. beim Filetransfer, voll zugute. Bei kürzeren Paketen, wie z.B. beim Dialog, ist die Ausnutzung nicht so gut, allerdings hauptsächlich durch die Endsysteme begrenzt.

3.2 ISO-CNLS

Der US-GOSIP stützt sich in lokalen Netzen hauptsächlich auf die mächtige Transportklasse 4 und den verbindungslosen Netzwerkdienst (ISO-CNLS oder kürzer ISO-IP). Für eine Abbildung auf X.25 sieht US-GOSIP zwei Möglichkeiten vor:

- Abbildung von TP4 auf X.25,
- Abbildung von ISO-IP auf X.25.

Während die erste Variante hauptsächlich für Endsysteme gedacht ist, die direkt an einem X.25-WAN hängen, bietet sich die zweite Möglichkeit für lokale Netze an. In diesem Fall müssen ISO-IP-Pakete in X.25-Pakete verpackt werden ('Encapsulation', s. Figur 3). Da hier ein verbindungsloser Dienst auf einem verbindungsorientierten Dienst abgebildet wird, gibt es die Möglichkeit, einen oder mehrere virtuelle Kanäle des verbindungsorientierten Dienstes zu benutzen. Sollte die für einen virtuellen Kanal verfügbare Bandbreite begrenzt sein, bietet die Verwendung mehrerer virtueller Kanäle die Möglichkeit, die verfügbare Bandbreite voll auszunutzen. Die meisten X.25-Router beherrschen diese Technik. Erfahrungen liegen allerdings noch keine vor, da ISO-TP4 und ISO-IP im Bereich des DFN noch keine größere Verbreitung haben.

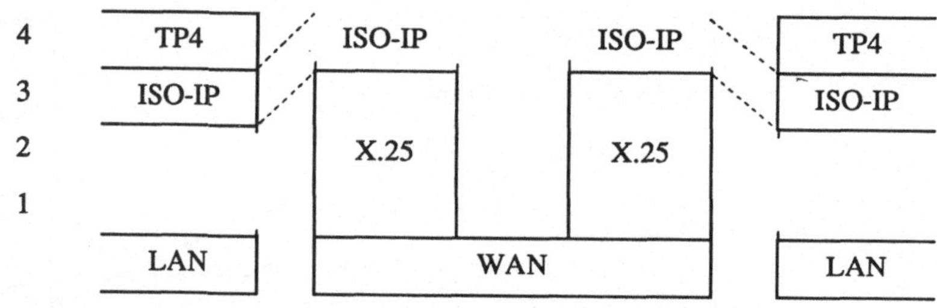

Figur 3: X.25-Encapsulation für ISO-IP

3.3 TCP/IP

Das derzeit in der Forschungsszene am häufigsten vertretene LAN-Protokoll ist die Protokollsuite des US-DoD, meist abgekürzt TCP/IP. TCP ist die Abkürzung für das Transportprotokoll, IP die Bezeichnung für das Internet- (Netzwerk-) Protokoll. Die Abbildung auf X.25 gestaltet sich ähnlich wie bei ISO-CNLP. Die Abbildung von IP auf X.25 wird durch den RFC877 (Request For Comment) beschrieben und beruht wieder auf Encapsulation, die derjenigen von ISO-IP (Figur 3) entspricht. Diese Technik hat sich soweit durchgesetzt, daß man sie in allen Routern, sofern sie überhaupt X.25 berücksichtigen, vorfindet.

Erfahrungen mit dem 'Routen' von IP über X.25 liegen regional im Bayerischen Hochschulnetz schon seit 1988 vor. So zeigte sich bald, daß diese Technik zu einem hohen Verbrauch an Ressourcen führen kann:

- Da jedes Internet-Paket Adreßinformation enthält, aber auf feste logische Verbindungen abgebildet wird, geht statisch bereits eine gewisse Bandbreite verloren.
- Der Austausch von dynamischer Routinginformation belastet das Netz. (Für X.25 bzw. ISO-CONS liegen darüber noch keine Aussagen vor.)
- Stimmen die TCP/IP-Verkehrsbeziehungen nicht mit der X.25-Netzstruktur überein, muß Information ggf. mehrfach über das Netz transportiert werden /NI91/.

Während die ersten beiden Effekte bereits im BHN offen zu Tage traten, machte sich der Mehrfachtransport erst im WiN bemerkbar. Der Grund ist relativ einfach zu nennen: Im BHN war die X.25-Netzstruktur an den Verkehrsbeziehungen zwischen den Universitäten ausgerichtet. Im WiN war und ist dies aus erfindlichen Gründen nicht mehr möglich. Die globalen Verkehrsbeziehungen sind zu heterogen. Um die WiN-Last in Grenzen zu halten, empfiehlt sich eine sorgsame Planung der IP-Routingkonzepte.

Inwieweit die Laufzeiteffizienz von TCP-IP durch Encapsulation in auf X.25 und die damit verbundene 'doppelte' Verarbeitung der Netzwerkschicht beeinträchtigt wird, war lange Zeit umstritten.

Messungen mit IP-Routern, die über eine mit X.25 betriebene 2 Mbps-Strecke miteinander verbunden sind, zeigen bei maximaler Paketlänge von 1024 Bytes in der Summe beim (FTP-) Filetransfer einen Durchsatz von 230 kByte/ sec. Da bei maximaler Paketlänge von 1.5 kBytes auf dem Ethernet X.25-Pakete von 1 kBytes Länge im Wechsel mit kürzeren Paketen erzeugt werden, entspricht der obige Wert der theoretischen Grenzleistung des Übertragungsmediums abzüglich des statischen Overheads für den IP-Header.

Ein anderes Prinzip scheint sich, so die Messungen, bewahrheitet zu haben: Die Verwendung mehrerer virtueller Kanäle hat durchaus Vorteile, wenn es (was auch im 2 Mbps-Bereich möglich ist) zu Störungen kommt. Ist ein virtueller Kanal kurzfristig gestört, kann die Übertragung auf anderen virtuellen Kanälen weiterlaufen. Das führt im Bereich ≥ 2 Mbps im Störungsfall zu einem deutlich besseren Durchsatz als bei Verwendung nur eines virtuellen Kanals.

Man kann also sagen, daß für die heutigen Router das Transportieren von IP über X.25 für 2 Mbps (zumindest für Filetransfer-ähnliche Anwendungen) keine Beeinträchtigung darstellt.

3.4 IPX

Ähnlich wie die hauptsächlich mit TCP/IP (lokal) vernetzten UNIX-Workstations beginnen sich die lokalen PC-Netze über das WiN zusammenzuschließen. Gemeinsame, oft landesweit organisierte Ressourcenverwaltung (wie z.B. Softwareverteilung) ist die Haupttriebfeder.

Der vom Marktführer, der Firma Novell, eingesetzte Protokollstack umfaßt die Protokolle PSX/IPX auf Transport/ Netzwerkebene. Das Routing von IPX über X.25 ist möglich, aber herstellerabhängig. Entsprechende herstellerabhängige Lösungen finden sich in PC-basierten Kommunikationsservern oder auch schon in Routern. Herstellerunabhängige Standards fehlen.

IPX-Routing ist seit kurzem im Bereich des BHN sowohl mit Kommunikationsservern als auch mit Routern im Einsatz. Größere Erfahrungen liegen noch nicht vor. Die Abbildung von IPX auf X.25 scheint aber nicht besonders laufzeiteffizient zu sein, vermutlich begründet durch den Fenster-Mechanismus von IPX.

3.5 Pseudo-CONS

Will man ISO-Protokolle auf TPØ-Basis (d.h. klassische WiN-Protokolle, die üblicherweise auf dem CONS-Stack laufen) in lokalen Netzen einsetzen, führt das dann zu Problemen, wenn solche lokalen Netze nur TCP/IP-Verkehr zulassen. Man kann diese TCP/IP-Struktur aber durchaus beibehalten und TPØ etc. über TCP einsetzen, benutzt man entsprechende Funktionen aus der von Marshal T. Rose entwickelten UNIX-Public-Domain-Software ISODE (ISO Development Environment, /RC86/). Sie bietet mit TCP eine sichere Pseudo-Netzwerkverbindung (/RFC983/, s. Figur 4) zu entsprechenden ISO-CONS-Gatewayrechnern. Diese ISO-CONS-fähigen Gatewayrechner sind UNIX-Workstationa, die sowohl zum X.25-Netz als auch zum TCP/IP-Netz Zugang haben. Auf allen TCP/IP-Stationen mit ISODE-Software im so erschlossenen Subnetz können dann die klassischen ISO-Anwendungen angeboten werden. Diese Technik bietet gleichzeitig eine einfache Migrationsstrategie von TCP/IP zu ISO-Protokollen.

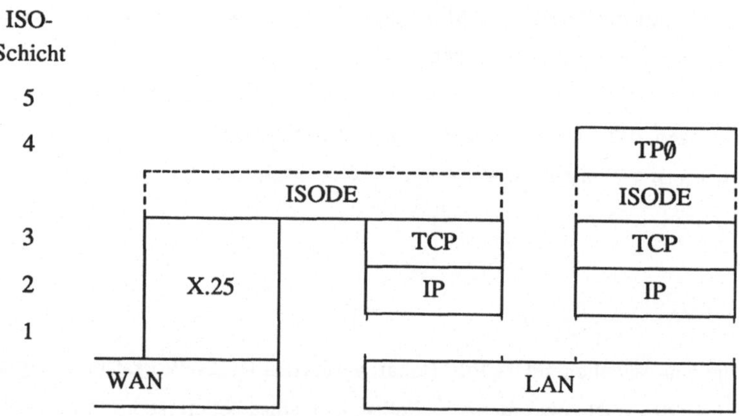

Figur 4: TCP als Pseudo-CONS in ISODE

Bereits 1988 wurde ISODE mit den entsprechenden Funktionen im BHN in Betrieb genommen, so daß damit z.B. der DFN RJE-Dienst in TCP/IP-Umgebung verfübar wurde. Rundum positive Erfahrungen liegen auch am LRZ /ST91/ vor. Fragen der Laufzeiteffizienz stehen bei diesem Produkt allerdings nicht im Vordergrund, da es hier wesentlich um das Anbieten einer Funktionalität geht.

4. WiN und Gateways

Das WiN erlaubt als transparentes Transportmedium die Zusammenarbeit von LAN-Inseln innerhalb der einzelnen Protokollwelten durch Routing über X.25. Das ist eine notwendige, aber nicht hinreichende Funktion. Je größer die Protokollvielfalt um das WiN wird, desto öfter geht es gerade darum, eine Zusammenarbeit zwischen Rechnern aus unterschiedlichen Protokollwelten bewerkstelligen zu müssen. Dabei sind besonders die Fälle interessant, die eine Wandlung von bzw. zu den klassischen TP∅-basierten DFN/OSI-Protokollen im WiN erfordern. Es geht dabei in der Regel um die Anpassung von Protokollen der Anwendungsschicht, wie Dialog, Filetransfer, Elektronische Post etc. Auch unterschiedliche Transportprotokolle können eine Hürde sein. Geeignete globale Gateways am WiN stehen zum Teil bereits zur Verfügung, doch erhebt sich die Frage, ob im Sinne einer sparsamen Ressourcenverwaltung am WiN und eines besseren Durchsatzes lokale, dezentrale Gateways nicht eine sinnvolle Ergänzung sind.

4.1. TP∅/TP4-Gateway

Auch wenn sich in Europa eine Ausrichtung auf ISO-TP∅/X.25 abzeichnet, wird es sicher ISO-TP4/IP-basierte Anwendungen geben - u.a. für die Anbindung an die USA. Die Kluft zwischen unterschiedlichen Transportprotokollen muß (durch Gateways) überbrückt werden (s. Figur 5). Lösungen hierfür gibt es auch im bereits erwähnten ISODE. Betriebserfahrungen liegen im DFN noch keine vor.

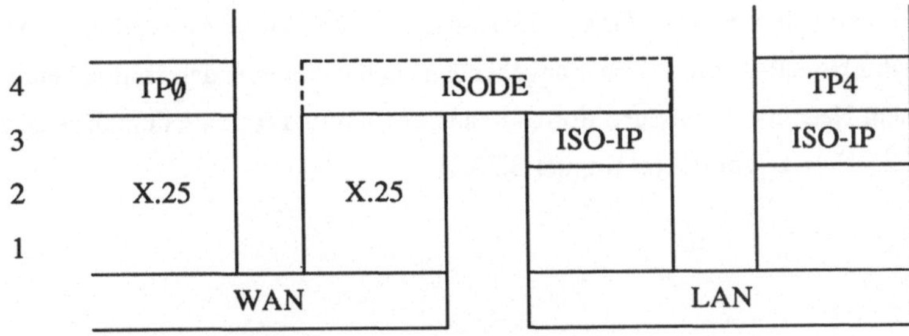

Figur 5: ISO-TP∅/TP4-Gateway

4.2 ISO/TP∅-TCP/IP-Gateways

Die zahlenmäßig größten Anforderungen an Gateways an Wandlungen von/zu ISO/TP∅ liegen im Bereich TCP/IP. Folgende Wandlungen müssen bewerkstelligt werden:

- FTAM - FTP,
- X.400 - SMTP,
- X^3 - Telnet.

FTAM-FTP-Gatewayfunktionen gibt es in der bereits öfter erwähnten UNIX-Software ISODE. Erfahrungen im größeren Stil gibt es (noch) nicht.

Zur Wandlung zwischen X.400 und SMTP wird oft die ehemalige RARE-Entwicklung 'Mailway' eingesetzt. Auf X.400-Seite bietet der Mailway einen MTA (Message Transfer Agent), keinen UA (User Agent) an. Dies ist Sache von lokalen X.400-Systemen. Im BHN ist der Mailway seit 1988 im Dauereinsatz. Die Erfahrungen sind rundum zufriedenstellend.

Die Wandlung zwischen X^3 und Telnet wird heute oft schon von IP-X.25-Routern quasi 'mit'erledigt. Am RRZE übernimmt diese Aufgabe seit ca. 1/2 Jahr das CDC-Netzwerksystem CDCNet.

4.3. IPX

Voll inhaltliche Übergänge zwischen lokalen PC-Netzen mit Novell-Betriebssystem und Rechnern an WiN mit TPØ/ X.25-Basis sind noch nicht in breiter Front verfügbar. Eine effiziente Anbindung sollte so erfolgen, daß der Fileserver über entsprechende Funktionen verfügt und so selbst als Gateway arbeitet. Mit der kürzlich erschienen Version 3.11 von (Novell-) Netware 386 ist dies erstmals für FTAM der Fall. Anwendungserfahrungen liegen noch keine vor.

Für eine Anbindung eines ganzen PC-Netzes an X.400 gibt es zwar bereits Lösungen, doch sind sie noch nicht in den Fileserver integrierbar. Fragen der Anpassung von Novell-Anwendungen an X^3-Dialog stellen sich nicht, da das Novell-Konzept weder abgehenden noch ankommenden Dialog kennt. So bleibt nur, jeden PC in einem Netz mit der nötigen Software auszurüsten und für eine durchlässige Anbindung an das öffentliche X.25-Netz zu sorgen (s. Kapitel 3.3.).

5. Ergebnis

Das WiN als X.25-Netz ist in der Lage, nicht nur die klassischen ISO/TPØ-basierten Protokolle zu transportieren, sondern praktisch auch alle anderen bekannten LAN-basierten Protokolle. Das erfolgt am besten durch 'Router', die verschiedene Protokolle gleichzeitig bearbeiten können. Auch die im LAN-Bereich üblichen höheren Geschwindigkeiten können über ein entsprechend 'schnelleres' WiN transportiert werden. Durchsatzeinbußen durch 'Encapsulation' anderer Protokolle sind auch für höhere Geschwindigkeiten (wie 2 Mbps) nicht zu erkennen. In Anbetracht des zu erwartenden EC-GOSIP mit X.25 in lokalen Netzen stellt ein X.25-orientiertes WiN darüber hinaus einen effizienten und strategischen Netzansatz dar.

Für Protokollanpassungen zwischen klassischen WiN-Protokollen und LAN-typischen Protokollen (z.B. TCP/IP) steht erprobte 'public domain'-Software zur Verfügung. Eine besondere Rolle spielt dabei das Produkt ISODE, insbesondere dadurch, daß es Bestandteil von UNIX 4.4BSD sein wird. Es liefert nicht nur volle ISO-Implementierungen und Gateways, sondern verkörpert eine elegante Migrationsstrategie von TCP/IP-Protokollen zu ISO-Protokollen.

Danksagung

Mein besonderer Dank gilt meinen Kollegen U. Hillmer, J. Dannenberg und T. Eckert für das unermüdliche Aufstellen immer neuer Meßreihen und Herrn E. Hergenröder dafür, die Messungen in eine ordentliche Form zu bringen.

Literatur

/BH89/ Bauerfeld, W., Holleczek, P.: Global Connectivity; Computer Networks and ISDN Systems 17 (1989), 300-304

/RC86/ Rose, M.T., Cass, D.E.: OSI Transport Services on top of the TCP; Computer Networks and ISDN Systems 12 (3), 1986

/CAF90/ Caffrey, L.: EPHOS: Towards a European GOSIP; Computer Networks and ISDN Systems 19 (1990), 265-269

/NG90/ Nitzan, R., Gross, P.: The Role of U.S. GOSIP; Computer Networks and ISDN Systems 19 (1990), 270-274

/ST91/ Storz, M.: ISODE Experience Panel; European Control Data Users Group (ECODU) 51, Kopenhagen 1991

/ISO8881/ ISO8881: Information processing systems - Data Communications: Use of the X.25 packet level protocol in local area networks, 1989

/NI91/ Nipper, A.: IP über X.25-Netze (WiN-IP); Netzwerkmanagement-Tutorium, GI Deutsche Informatik-Akademie, Bonn, Mai 1991

Betriebserfahrungen der Endsystembetreiber und Endnutzer am Wissenschaftsnetz

W. Held, G. Richter, D. Schulze

Westfälische Wilhelms-Universität Münster, Universitätsrechenzentrum

Das Endsystem

Erste Experimente zum Aufbau eines lokalen Rechnernetzes (LAN) an der Universität Münster, das wir als Endsystem sehen, gehen in das Jahr 1984 zurück. Von Anfang an war für alle Fachbereiche ein gemeinsames Netz vorgesehen, das zur Leistungssteigerung leicht in homogene Teilnetze zerlegt und über ein Backbone-Netz wieder zusammengefügt werden kann. Im Frühjahr 1991 waren ca. 800 Rechner vernetzt, am Ende dieses Jahres werden es infolge einer größeren DV-Beschaffungsmaßnahme über 1.000 Geräte sein.

In unserem LAN auf der Basis von ISO 8802/3 (Ethernet) setzen wir die Leitungskontrollprotokolle LLC1 und LLC2 (siehe auch nachfolgende Abbildung) gemäß der internationalen Norm ISO 8802/2 ein, die auf Datagramm-Basis bzw. verbindungsorientiert arbeiten. LLC1 nutzen wir für verbindungslose Netzdienste CLNS (Connectionless Network Service) mit dem OSI-Transportprotokoll ISO 8073 Class 4 (TP4). CLNS ist gemäß ISO 8473 ein Datagramm-Netzwerkdienst. CLNS tritt in Ebene 3 des OSI-Modells an die Stelle von X.25, ähnlich dem IP- Protokoll im ARPA-INTERNET. Daneben kennen wir das verbindungsorientierte Protokoll CONS (Connection oriented Network Service), das in Form von X.25 realisiert ist. TP4 ist neben TP0 eine wichtige Variante des Protokolls der Ebene 4 im OSI-Modell, denn TP4 ist der Transportdienst für eine fehlerkorrigierte Übertragung zwischen Endsystemen. Man findet TP4-Protokolle vor allem in LANs. TP0 war ursprünglich vorgesehen für die Nutzung einfacher Teletex-Terminals ohne Fehlerkorrektur, insbesondere ohne Endsystem-zu-Endsystem-Kontrolle. Der Einsatz von TP0 auf der Basis von X.25 ist eigentlich nur deshalb zu tolerieren, weil der Netzwerkdienst X.25 in Deutschland mit hoher Dienstgüte angeboten wird und weil Endsysteme bisher im wesentlichen direkt an die X.25-Netze angeschlossenen und LANs als Endsystem weniger betrachtet werden. Das verbindungsoriente Protokoll CONS und das OSI-Transportprotokoll TP0 setzen wir im LAN nicht ein.

Auf der Basis von LLC2 benutzen wir Netbios und SNA. Netbios ist die häufig verwendete Netz-Schnittstelle für MS-DOS-Systeme, SNA ist die Netz-Architektur der Firma IBM.

Auf der Basis von TP4 haben wir im Auftrag des Vereins zur Förderung eines Deutschen Forschungsnetzes e. V. (DFN) dessen Protokolle für Filetransfer (FT) und Remote Job Entry (RJE) auf Unix- und VM-Systeme (IBM) portiert. Von den OSI-Protokollen haben wir FTAM-Filetransfer für MS-DOS-Rechner in Zusammenarbeit mit der Humboldt-Universität in Berlin und der Firma Coconet angepaßt. Damit sind für DOS-Rechner auch bereits die Grundlagen für Mail-Anbindungen auf Basis des X.400-Message-Handling-Systems gelegt.

In einem RARE-Pilot-Projekt (RARE - Réseaux Associés pour la Recherche Européenne -, eine

Fördereinrichtung der Europäischen Gemeinschaft für den Wissenschaftsbereich) sind an anderer Stelle für UNIX FTAM- und X.400-Anbindungen ebenfalls im CLNS-Stapel mit den ISODE-Protokollen bereitgestellt (ISODE - ISO Development Environment -, eine Software-Entwicklung der Firma Wollongong, die als Public Domain Software zur Verfügung steht).

Schicht		
4	ISO 8073 Transport Class 4 (Error Detection and Recovery)	Class 0 (Simple Class)
3	ISO 8473 ConnectionLess Network Service	ISO 8208 X.25 Packet Level Protocol
2b	Logical Link Control ISO 8802/2 Class 1 (Connectionless Service)	Class 2 (Connection oriented Service)
2a	ISO 8802/3	
1	LAN ("Ethernet")	

Abbildung: Ausschnitt aus dem ISO-OSI-Protokollstapel

Damit ist für DOS-Rechner unabhängig von den Netzkarten und für viele Unix-Systeme die Basis für FTAM und X.400 gelegt worden. Was wir nicht als Widerspruch dazu verstanden wissen wollen, daß z. B. FTAM in den funktionalen Ausprägungen selbst verbesserungsbedürftig ist, daß Adressierungsfragen noch zügig und flexibel zu lösen sind oder daß die Routing-Verwaltungsmechanismen weiterentwickelt werden müssen.

Der Zugang vom Endsystem zum Wissenschaftsnetz (WIN) und zum europäischen Verbundnetz (IXI) erfolgt über einen Router (Firma Cisco) und einzelne Rechner. Über den Router werden im Rahmen des RARE CLNS-Pilotprojektes FTAM- und X.400-Anwendungen zwischen unserem LAN und den WIN/IXI-Netzen möglich. Dabei wird X.25 als Subnetzprotokoll unter CLNS verwendet. Über die einzelnen, direkt angeschlossenen Rechner werden die Protokolle X.3/X.28/X.29 genutzt. Diese legen den Zugang zu einem X.25-Netz für asynchrone Geräte oder Geräte fest, die nicht X.25-fähig sind. Da diese einzelnen Rechner auch an das LAN angeschlossen sind, dienen sie dort als Gateway für andere Systeme. Die übrigen Kommunikationsprotokolle und -dienste, die nicht auf OSI beruhen (z. B. EARN/BITNET, INTERNET und DECNET), sollen hier nicht betrachtet werden.

Durchsatz, Transportzeit, Verfügbarkeit usw.

Erste positive Erfahrungen im angesprochenen RARE-Pilot-Projekt zeigen, daß die Durchsätze im CLNS/TP4-Protokollstapel selbst dann mit dem CONS/TP0-Stapel vergleichbar sind, wenn X.25 als Subnetz-Protokoll verwendet wird. Auch daraus folgern wir, daß die OSI-Protokoll-Stapel dual implementiert werden müssen (CLNS und CONS), da andernfalls immer Gateways erforderlich sind, die leider nicht allgemein genutzt werden können. Daß der gleichzeitige Ablauf der dualen Protokollstapel im Netz machbar ist, haben wir durch parallelen Einsatz von INTERNET- und OSI-Protokollen in unserem LAN gezeigt. Wegen der Einschränkungen, die ein Gateway stets mit sich bringt, sind wir der Meinung, daß nicht nur im LAN sondern auch in Weitverkehrsnetzen Datentransfer zusätzlich und parallel zu X.25 auf der Basis von CLNS (und TP4) eingesetzt werden müßte. Unsere Erfahrungen mit LANs zeigen übrigens auch, daß die Protokolle LLC1, CLNS und TP4 bezüglich der Zuverlässigkeit dem INTERNET-Protokolls (TCP/IP) ebenbürtig sind.

Systematische Meßergebnisse über Durchsatz, Transportzeiten, Übertragungsraten usw. im OSI-WIN liegen in Münster weniger vor. Aus unserer Sicht stellt sich das Netz i. a. als stabil dar. Die Verfügbarkeit des WIN beträgt nahezu 100 %. Der Ausfall von 1 oder 2 Minuten/Woche sollte tolerabel sein. Aus unserer Sicht ist die Fehlerrate auf Ebene 2 gleich 0.

Die Qualität der Fehlerbehandlung ist aufgrund der geringen Fehlerrate kein Thema mehr. Beratung und Betreuung durch die Telekom-Mitarbeiter an dem für uns zuständigen Knoten sind gut, was wohl auch auf die wechselseitige Unterstützung zwischen Telekom-Mitarbeitern und Mitarbeitern der Universitätsrechenzentren zurückgeführt werden kann. Als gleichermaßen gut empfinden wir die Unterstützung und Beratung durch die Mitarbeiter der Geschäftsstelle des DFN-Vereins.

Der eigene Personalaufwand zur Betreuung unseres Knotens erfordert etwa 0.1 bis 0.2 Mannjahre. Darin enthalten sind u. a. Arbeiten in Verbindung mit der Nutzer-Abrechnung, der Überprüfung der Funktionalität der Endgeräte, die Umstellung auf neue Netz-Standards sowie Zeiten für die Weiterbildung sowie Vorbereitung und Abhaltung von Vorträgen.

Zum Jahreswechsel 1990/91 verschlechterte sich im WIN die Verweilzeit von X.400-Nachrichten, die über Relay-Rechner bzw. aus der X.400-Welt über Gateway-Rechner in die INTERNET- bzw. EARN/BITNET-Welt liefen, durch Verzögerungen in den Umsetzungsrechnern dramatisch. Ein Gateway-Rechner hat die Aufgabe, Protokolle verschiedener Netze umzusetzen, damit Verbindungen möglich werden. Er wird z. B. genutzt, um Nachrichten aus X.400-Netzen in das INTERNET oder in die EARN-Welt und umgekehrt zu transportieren. Relay-Rechner werden in einer einheitlichen Protokoll-Umgebung (z. B. X.400) genutzt, um Nachrichten zwischenzuspeichern und weiterzuleiten oder zu verteilen (Store and Foreward). Es gingen damals vereinzelt Daten verloren, manchmal wurden sie auf wundersame Weise vermehrt. Uns erreichte z. B. eine zweizeilige Nachricht, die sich auf ca. 50 Seiten hochgeschaukelt hatte. Durch den Einsatz leistungsfähigerer Gateway-Rechner und neuer Umsetzsoftware verbesserte sich die Situation sehr. In Testmessungen der Geschäftsstelle des DFN-Vereins wurden in der Zeit vom 08.05. bis 15.05.91 folgende Laufzeiten gemessen:

X.400 Relay: bei 40 Testnachrichten
$37 \leq 2$ min, $3 \leq 12$ min

X.400-EARN-Gateway: bei 32 Testnachrichten
$30 \leq 2$ min, $2 \leq 5$ min

X.400-INTERNET-Gateway: bei 36 Testnachrichten
$26 \leq 2$ min, $7 \leq 5$ min, $1 \leq 8$ min, $2 = 5$ h

Bis auf die beiden offensichtlich auf Fehler zurückführbaren Ausreißer im X.400-INTERNET-Gateway (die Adresse war fehlerhaft), scheinen diese Ergebnisse zufriedenstellend zu sein.

Da sich die Laufzeit L einer Nachricht bei einem Übergang in eine andere Netzwelt im wesentlichen aus der Bearbeitungszeit B im Gateway/Relay-Rechner und der Wartezeit W (wegen Nichterreichbarkeit eines Endsystems) ergibt, d. h. $L = B + W$, ist neben der Größe B auch die Größe W zu beachten, die letztlich zu Rückstaus im Gateway/Relay-Rechner führt. Durch Verträge mit den Endnutzern wird der DFN-Verein sicherstellen, daß die Endsysteme ständig betriebsbereit gehalten werden.

Beim Übergang von X.400 nach EARN/BITNET bzw. INTERNET treten bisher allerdings auch noch unerfreuliche Fehlersituationen bei der Nutzung der relativ beliebten Listserver auf. Die meisten weltweit nutzbaren Listserver sind von X.400 aus nicht oder nicht immer erreichbar, was natürlich die Ablösung von EARN-Bitnet sehr erschwert. Der Fehler ist den Verantwortlichen bekannt, aber offensichtlich nicht einfach behebbar.

Notwendige Nachbesserungen

Wir wollen noch auf einige verbesserungswürdige Punkte aufmerksam machen. Das fehlende durchgängige Netzmanagement für gekoppelte Netze (WIN-IXI, aber auch X.400-EARN) hat negative Auswirkungen auf die Nutzer im Endsystem, da Fehler z. B. im Verbindungsaufbau kaum nachvollziehbar sind. Die Benutzeroberflächen (z. B. für Mail geeignete Menues zur Bearbeitung der vorhandenen elektronischen Post) bedürfen noch der weiteren Verbesserung, wenn man in Zukunft stärker als bisher auch gelegentliche DV-Nutzer an die Kommunikation heranführen will. Erforderlich sind auch eine Benutzer-Unterstützung schon vor der Aufnahme von Kommunikationsverbindungen durch eine verteilte Informationsdatenbank (X.500) sowie eine Postverteilung für die Arbeitsplatzrechner z. B. über die bisher nur ansatzweise implementierten Standards Message Store und Remote User Agent aus der X.400-Reihe.

Ein schmorendes Problem, das demnächst gelöst werden muß, ergibt sich aus einer fehlenden Möglichkeit zur Kontingentierung im WIN. Wenn extreme WIN-Nutzer nicht in ihre bezahlten Schranken gewiesen werden können, droht das einfache pauschalierte Abrechnungsverfahren im DFN-Verein mit jährlichen Fixkosten über kurz oder lang auseinanderzubrechen. In die Kontingentierungsregelung könnte eingehen, ob die Belastung das gesamte WIN oder nur den regionalen Postknoten berührt. Wenn sich z. B. München und Erlangen erhebliche Datenmengen zusenden, stört mich das solange nicht, wie ich von Münster aus beide erreichen kann.

ISO-OSI-Akzeptanz

Aus der Sicht des Endnutzers müssen noch größere Anstrengungen als bisher unternommen werden, um OSI breit einzuführen. Während in den USA erst im vergangenen Jahr eine „User Alliance for

Open Systems - Houston 30“ eingeführt wurde, die inzwischen diverse OSI-Barrieren ausgemacht und Aktionen zur Überwindung dieser Barrieren vorgeschlagen hat, sind wir in Deutschland mit dem DFN-Verein eigentlich in der besseren Situation. Jedoch scheinen auch bei uns größere Anstrengungen notwendig zu sein. Leider kosten noch größere Anstrengungen als bisher noch mehr Personal und Geld.

Unserer Meinung nach muß das strategische ISO-OSI-Ziel noch stärker herausgearbeitet werden, damit es dem Endnutzer deutlicher bewußt wird. Dazu gehören die Darstellung des Nutzens von OSI sowie die Fortschritte gegenüber der verbreiteten INTERNET-Welt. Zur Zeit hat man den Eindruck, daß die Entwicklung von OSI und INTERNET eher zugunsten von INTERNET auseinanderdriftet, da auf INTERNET-Basis immer mehr Dienste und Tools bereitstehen. Wann wird es z. B. einen OSI-Standard geben, der X-Windows entspricht? Für jedes einzelne Universitätsrechenzentrum ist es mühsam und aufwendig, dies strategische OSI-Ziel herauszustellen. Diese OSI-Diskussion ist sowohl für die INTERNET-Welt als auch für die EARN/Bitnet-Welt zu führen. Ist z. B. mit dem X.400-Standard inzwischen wirklich der Stand erreicht, um EARN/Bitnet abzulösen? Oder entwickelt sich auch die EARN/Bitnet-Umgebung sogar fast noch schneller als die X.400-Umgebung. Faktisch mag es nicht so sein. Die Unsicherheit ist dennoch groß. Oben hatte ich die Notwendigkeit von X.500-Directories angesprochen. Wie man hört, sollen diese Ende 1992 implementierbar sein. Welchen Stand wird die INTERNET-Welt zu diesem Zeitpunkt erreicht haben?

Wie kann man die OSI-Akzeptanz beim Benutzer erhöhen? Wie kann der Anwender zu OSI bewegt werden, wenn herstellerspezifische Netze mehr Komfort bieten oder mehr Komfort zu bieten scheinen? In unserer Universität müssen wir z. Z. zusehen, wie ein Fachbereich die herstellerunabhängige Netzwelt durch die herstellerspezifische Welt eines größeren Anbieters austauschen will. Ein Gesichtspunkt ist dabei die Erreichbarkeit von Partnern in großen Universitäten und Forschungseinrichtungen, die oftmals diese herstellerspezifischen Netze einsetzen und sie für den Zugang zu ihnen nahelegen. Durch Vorgabe standardisierter OSI-Protokolle gerade in den großen Universitäten und Forschungseinrichtungen könnten normende Schritte vorgegeben werden.

Ich denke, daß es dringend ist, die zur Einführung der OSI-Dienste erforderlichen Kraftanstrengungen, wenn möglich über das bisherige hinaus, fortzusetzen. Fehlende Mitarbeiter und fehlendes Geld könnte man dazu vielleicht auch in den neuen Bundesländern finden, wo man u. a. über staatliche Beschäftigungsgesellschaften nachdenkt. Beschäftigungsgesellschaften hätten in diesem Falle den großen Vorteil, daß sie eine Verbesserung der Infrastruktur bewirken würden. Die Bereitstellung von Infrastrukturen ist schließlich Hauptaufgabe von Bund und Ländern.

Diese Anstrengungen sind vermehrt auch auf Hochgeschwindigkeits-Übertragungen auszudehnen, wo wir durch die Verzögerungen der Deutschen Bundespost Telekom bereits zwei wertvolle Jahre im Wettbewerb mit anderen Ländern verloren haben. Wenn man von verstopften Straßen und einer fortschreitenden Ausweitung des Kfz-Verkehrs ausgeht, müßte man wenigstens einmal ausprobiert haben, ob nicht die viel preiswertere und umweltfreundlichere Verkehrs-Infrastruktur der Kommunikationswege eine Entlastung bringen könnte. Könnten nicht in vielen Fällen Bits anstelle von Menschen reisen! Für Wissenschaftler unserer Universität, vor allem der Medizin, ist dies ein dringendes Thema. Mit der Datenübertragung über das WIN allein ist ihnen schon heute nicht mehr geholfen.

FINANZIELLE UND VERTRAGLICHE REGELUNGEN DES NETZBETRIEBES

K.-E. Maass
Verein zur Förderung eines Deutschen Forschungsnetzes e. V. (DFN-Verein)

Berlin

1. Datenkommunikation als Grunddienst

Wissenschaft lebt vom Gedanken- und Meinungsaustausch. Technischer Träger dieses Kommunikationsbedürfnisses ist ein gut ausgebautes Telekommunikationsnetz. Da die Nutzung von Datenverarbeitungskapazität ein wichtiger Bestandteil der Arbeit von Wissenschaftlern ist, besteht Konsens, daß Datenkommunikation (als Teil der Telekommunikation in der Wissenschaft) ein Grunddienst für Forschung und Lehre ist, der allen Mitgliedern einer Hochschule oder Forschungseinrichtung, einschließlich Studenten und Verwaltung, angeboten werden muß. Grunddienste wie Telefon, Licht, Heizung, Wasser, - und Datenkommunikation zeichnen sich dadurch aus, daß sie generell zur Verfügung gestellt werden und ihr Bedarf nicht im Einzelfall nachgewiesen werden muß.

Datenkommunikation als Grundversorgung muß sich an folgenden Anforderungen orientieren:

- Datenkommunikation muß berücksichtigen, daß Wissenschaft in einem Netz von weltweiten Informationsbeziehungen lebt,

- Datenkommunikationsnetze sind Hilfsmittel für wissenschaftliche Arbeiten, die der Verbindung zwischen Teilnehmern bei der Nutzung von Rechnern, bei Zugriff auf Daten- und Softwarebanken sowie beim Austausch von Daten und Nachrichten dienen.

- Datenkommunikationsdienste müssen als integrierte Werkzeuge am Arbeitsplatz des Mitarbeiters angeboten werden.

- Die Kommunikation zwischen Teilnehmern muß auch bei unterschiedlichen technischen oder organisatorischen Bedingungen ungehindert möglich sein.

- Kein Mitarbeiter darf durch seinen Standort innerhalb einer Hochschule bei der Versorgung im Grunddienst benachteiligt sein.

Die Aufwendungen für den Grunddienst Datenkommunikation lassen sich wie folgt gliedern:

1 Aufwendungen, die innerhalb einer Wissenschaftseinrichtung (Hochschule) anfallen: Dies sind insbesondere Personalkosten sowie Kosten für das interne Netz (Hardware und Software für die Kommunikation, Leitungen)

2 Aufwendungen, die außerhalb einer Wissenschaftseinrichtung anfallen: Dies sind insbesondere Kosten für Datenfernübertragung und -vermittlung über das X.25-Wissenschaftsnetz WIN (Anschlußkosten pro WIN-Port, Adressendienste) sowie Kostenanteile für die Vermittlung in andere Netze (z. B. in den allgemeinen Datenübermittlungsdienst Datex-P oder in ausländische Netze wie EARN/BITNET oder INTERNETs)

Dieser Vortrag befaßt sich mit den unter Ziffer 2 genannten Aufwendungen. Hierfür hat der DFN-Verein Verantwortung übernommen. Zur begrifflichen Klarstellung: WIN ist die Kurzbezeichnung für das auf X.25 basierende Datenübermittlungsnetz des DFN-Vereins (bis Ebene 3 des OSI-Schichtenmodells). WIN ist Teil des Datenkommunikationssystems Deutsches Forschungsnetz DFN, das desweiteren auch die Kommunikationsanwendungsdienste wie Dialog, FT, RJE, MHS sowie Vermittlungsdienste in andere Netze beinhaltet.

2. Aufwendungen, die innerhalb einer Wissenschaftseinrichtung anfallen

Über die Aufwendungen zu Ziffer 1 hat die Kommission für Rechenanlagen der Deutschen Forschungsgemeinschaft mit ihrem Netzmemorandum - Notwendigkeit und Kosten der modernen Telekommunikationstechnik im Hochschulbereich; Bonn, 1987, einen Orientierungsrahmen gesetzt. Die wichtigen Aussagen sind hier kurz zusammengefaßt:

Die Deutsche Forschungsgemeinschaft empfiehlt in ihrem Netzmemorandum (1985) zur Unterstützung der Anwender innerhalb einer Hochschule eine Wissenschaftlerstelle pro 1.000 Netzanschlüsse, wobei bei DV-intensiven Fachbereichen auf 10 Hochschulangehörige 1 Netzanschluß und in Fachbereichen mit bisher geringerer DV-Durchdringung auf 50 Hochschulangehörige ein Netzanschluß kommt. Ferner empfiehlt die Deutsche Forschungsgemeinschaft je nach Art und Umfang der überregionalen Aktivitäten 1 bis 3 Stellen pro Hochschule für die Betreuung der überregionalen Netzdienste.

Für Investitionen je nutzerbezogenem Netzanschluß nennt die DFG einen Durchschnittswert von 2.500,-- DM. Hinzu kommen jährliche Wartungskosten von 6 bis 8 % der Investitionssumme. Die für interne Netze anfallenden Kosten hängen sehr von den lokalen Gegebenheiten ab und entziehen sich daher einer Veranschlagung durch den DFN-Verein.

3. Aufwendungen für Datenfernübertragung und -vermittlung über das WIN

3.1 Anforderungen an ein Datennetz

An ein Datennetz für die Wissenschaft werden folgende Anforderungen gestellt:

- Kommunikation muß technisch machbar sein!
  Dies heißt, daß Software- und Hardwareprodukte für die Kommunikation verfügbar sind und daß beim Aufbau eines Datennetzes auch neue Kommunikationsformen, die z. B. breitbandige Übertragungswege benötigen, berücksichtigt werden müssen.

- Kommunikation muß bezahlbar sein!
  Das heißt, die Kosten müssen sich - wenn schon nicht an Sonderpreisen für die Wissenschaft - an Selbstkosten und nicht an Gebühren oder Marktpreisen orientieren. Dies heißt auch, daß die budgetären Möglichkeiten öffentlicher Einrichtungen, denen eine kameralistische Haushaltsführung vorgeschrieben ist, berücksichtigt werden (Etatansätze für mindestens 2 Jahre im voraus planbar).

- Kommunikationswege müssen für genormte Kommunikationsprotokolle offen sein!
  Das heißt, langfristig kann eine herstellerübergreifende Kommunikation ökonomisch und technisch nur sichergestellt werden, wenn international normierte Bausteine für die Datenkommunikation verwendet werden; dies schreiben auch die Beschaffungsrichtlinien der öffentlichen Hand vor.

Wie können die Anforderungen an das Datennetz für die Wissenschaft erfüllt werden?

Kommunmikationsdienste zählen nicht zum Monopolbereich der DBP TELEKOM, sondern unterliegen dem Wettbewerb. Daher gab es vor der Einrichtung des Wissenschaftsnetzes verschiedene Denkansätze zur Realisierung eines Wissenschaftsnetzes:

<u>Entweder</u> Nutzung des allgemeinen Datenübermittlungsdienstes DATEX-P der DBP TELEKOM.

Vorteil: gute Qualität, keine zentrale Organisation nötig.

Nachteil: Vom übertragenen Datenvolumen abhängige Gebühren, die haushaltsmäßig nicht vorausplanbar sind; kein Einfluß auf Weiterentwicklungen.

oder DFN-eigener Netzbetrieb

Vorteil: Selbststeuerung, volle Eigenverantwortung
Nachteil: Keine Erfahrungen lagen beim DFN-Verein vor, es fehlte Fachpersonal, kein Zugang zu öffentlichen Netzen möglich.

oder Auftrag an "jedermann", der Erfahrung beim Aufbau und Betrieb eines Spezialnetzes hat (sog. Outsourcing). Dieser "jedermann" kann eine Privatfirma sein oder auch ein öffentliches Unternehmen wie die DBP TELEKOM.

Vorteil: Einfluß auf Netzgestaltung und Weiterentwicklung des Netzes ohne eine eigene Betriebsmannschaft aufbauen zu müssen; Freiheit bei der Gestaltung der Preise für die Anwender aus der Wissenschaft im Rahmen globaler Kostendeckung.
Nachteil: Aufbau einer zentralen Organisation zur Koordinierung der Netznutzung; bei einem privaten Anbieter als "jedermann" fehlender Zugang zu öffentlichen Netzen.

Als Einstiegsbedarf für das zu übertragende Datenvolumen in der Wissenschaft wurden bei einer Umfrage unter den Mitgliedern des DFN-Vereins im Jahre 1988 ca. 50 GigaByte pro Monat ermittelt. Eine Studie des DFN-Vereins vom August 1988 (DFN-Bericht Nr. 52: Datenkommunikation in Lehre und Forschung - Bedarf der Wissenschaft und Anforderungen an die Deutsche Bundespost; Berlin 1988) kam zu einem auf das Jahr 1995 extrapolierten Bedarf von 500 GigaByte pro Monat, wenn keine Möglichkeit zur Bilddatenübertragung vorhanden ist, und auf einen Bedarf von 3.500 GigaByte pro Monat, wenn Bilddatenübertragung möglich ist (Voraussetzung hierfür ist das Vorhandensein eines Hochgeschwindigkeitsdatenkommunikationsnetzes für die Wissenschaft). 50 GigaByte entsprechen ungefähr einem Datenvolumen von 15 Mio. Seiten DIN A 4.

3.2 Kostenvergleich der verschiedenen Realisierungsmöglichkeiten für einen Netzbetrieb

Vor der Erteilung eines Auftrages zur Errichtung eines Wissenschaftsnetzes wurden die verschiedenen Realisierungsmöglichkeiten in ihrem Preis-/Leistungsverhältnis verglichen. Dies basierte auf der Annahme von 100 Anschlüssen für eine Übertragungsgeschwindigkeit mit 64 kbit/s sowie 100 Anschlüssen für eine Übertragungsgeschwindigkeit von 9,6 kbit/s sowie einer Auslastung (Volumen) des gesamten Netzes von 50 GigaByte pro Monat.

Die Nutzung des allgemeinen Datenübermittlungsdienstes Datex-P zur Erfüllung der Anforderungen würde Kosten in Höhe von ca. 15 bis 20 Mio. DM pro Jahr ergeben je nach Mengenstaffelung, Tag- oder Nachttarifzuordnung und Ansatz von Zeitgebühren.

Ein DFN-eigener Netzbetrieb, der den Anforderungen nach einem sicheren Netz unter quasi kommerziell üblichen Bedingungen entspricht, würde ca. 25 Mio. DM pro Jahr kosten (angenommene durchschnittliche Entfernung von 50 km zum nächsten Verbindungsknoten und Kostenverhältnis von Anschlußleitungen zu Verbindungsleitungen 50 : 50 sowie Abschreibungen auf Investitionen und Personalkosten von 2 Mio. DM pro Jahr).

Angebote von "jedermann" mit der notwendigen Erfahrung im Netzgeschäft lagen zwischen ca. 10 und 20 Mio. DM. Die Bandbreite ergibt sich aus Unsicherheiten bei der Rückrechnung von Angeboten auf der Basis einer höheren Anzahl von Anschlüssen und einer höheren Auslastung des Netzes. Nach Maßgabe des günstigsten Preis/Leistungsverhältnisses wurde der Auftrag zur Errichtung und zum Betrieb des Wissenschaftsnetzes an das öffentliche Unternehmen DBP TELEKOM erteilt. Bei der Leistungsfähigkeit sind besonders hervorzuheben die bewährte, den internationalen Normen entsprechende X.25-Technik, die Möglichkeit des Ausbaus zur schnellen Datenkommunikation, ein uneingeschränkter Übergang ins Datex-P-Netz und ins europäische Wissenschaftsnetz IXI. Der Vertrag mit der DBP TELEKOM zu einem Preis unterhalb 10 Mio. DM pro Jahr wurde im August 1989 unterzeichnet.

Ein Vergleich der Leistungen des Wissenschaftsnetzes mit den gestellten Anforderungen ergibt:

- Die Kommunikation über das X.25-Wissenschaftsnetz WIN ist technisch in guter Qualität machbar. Die Weiterentwicklung zu höheren Geschwindigkeiten ist vertraglich vorgesehen.

- Die Kosten des X.25-Wissenschaftsnetzes sind bezahlbar, wenn sich solidarisch alle Wissenschaftseinrichtungen daran beteiligen. Eine Umfrage unter den Mitgliedern des DFN-Vereins aus dem Jahre 1988 ergab, daß diese Einrichtungen für Postleitungen (Datex-P, Datex-L, HfD, private Leitungen) ca. 12 Mio. DM für die Weitverkehrskommunikation und nochmals ca. 3 Mio. DM für die interne Kommunikation an die Deutsche Bundespost gezahlt haben.

- Das X.25-Wissenschaftsnetz ist für genormte Kommunikationsprotokolle offen. Es unterstützt damit auch die Ziele des DFN-Vereins und die Beschaffungsrichtlinien der Öffentlichen Hand.

3.3 Preis- und Vertragsgestaltung durch den DFN-Verein

Das X.25-Wissenschaftsnetz wurde in dem strategischen Dreieck Anwender, DFN-Verein und DBP TELEKOM realisiert (s. Abb. 1: Strategisches Dreieck). Der Erbringer der Leistung gegenüber dem Anwender (Hochschule, F- und E-Einrichtung) ist die DBP TELEKOM. Der Koordinator der Leistung ist der DFN-Verein. Der Anwender zahlt an ihn für die Leistung ein Entgelt. Der DFN-Verein finanziert wiederum die Erbringung der Leistung durch die DBP TELEKOM insgesamt.

Strategisches Dreieck

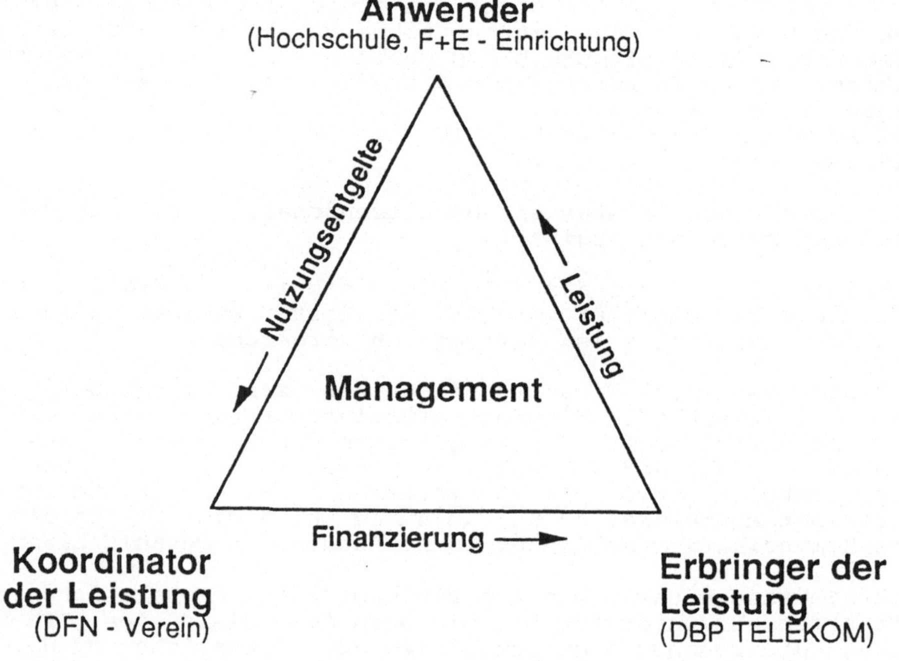

Abbildung 1

Ein wesentlicher Vorteil des X.25-Wissenschaftsnetzes WIN liegt darin, daß der DFN-Verein den Anwendern Einzelanschlußkosten pauschal und volumenunabhängig im Rahmen der vertraglich vorgegebenen Grenzwerte für das Gesamtnetz in Rechnung stellen kann. Die Preise für Einzelanschlüsse wurden vom DFN-Verein nach seinen Preis/Leistungsvorstellungen festgelegt und nicht von der DBP TELEKOM vorgegeben, wie oftmals angenommen wird. Die pauschalen Einzelanschlußpreise geben den Hochschulen die Möglichkeit, mit einem Zeitvorlauf von 2 Jahren ihre Etats für Datenkommunikation anzusetzen. Der einzelne Nutzer (Wissenschaftler, Student, Mitarbeiter) wird dadurch in die Lage versetzt, praktisch uneingeschränkt von seinem Rechner mit entfernten Rechnern zu kommunizieren.

Bei voller Ausschöpfung der maximalen Anschlußzahlen, derzeit 235 Anschlüsse, decken sich Einnahmen und Ausgaben des DFN-Vereins für das Gesamtnetz (ohne Overheadkosten). Der Vertrag mit der DBP TELEKOM enthält deswei teren eine Preisanpassungsklausel, die sich an den Preisen für das Datex-P orientiert.

Der Vertrag mit der DBP TELEKOM enthält Grenzwerte für eine maximale Anzahl von Anschlüssen für 64 kbit/s und 9,6 kbit/s und für ein maximales Datenübertragungsvolumen (50 GigaByte pro Monat). Wenn diese Grenzwerte überschritten werden, gibt dies der DBP TELEKOM das Recht zu Nachverhandlungen mit dem DFN-Verein über den Preis des Gesamtnetzes. Mehrkosten aufgrund der Nachverhandlungen müssen sich nicht unbedingt auf die Einzelanschlußpreise auswirken. Dies liegt im Entscheidungsbereich des DFN-Vereins, wie er die Gesamtkosten durch Einnahmen decken kann. Die Steigerung der Übertragungsleistung auf 2 Mbit/s rechtfertigt allerdings einen höheren Preis für diese neue Anschlußkategorie. Der DFN-Verein hält hierfür einen gegenüber 64 kbit/s um das vierfache erhöhten Preis für angemessen und hat als Jahresentgelt für den 2 Mbit/s-Anschluß an das X.25-Wissenschaftsnetz den Preis von 250 TDM pro Jahr vor einem Jahr angekündigt. Bei der Erweiterung der Obergrenze für das insgesamt zu übertragende Datenvolumen sollte überlegt werden, ob hier nicht zu allererst die Hauptnutzer zu einer Finanzierung der Mehrkosten herangezogen werden können, um den Einstiegspreis in die Datenkommunikation so niedrig wie möglich zu halten.

3.4 Zubringernetzregelung

Eingangs wurde als eine Anforderung an die Grundversorgung mit Datenkommunikation unter anderen genannt, daß kein Wissenschaftler durch seinen Standort innerhalb einer Hochschule benachteiligt werden darf. Dieser Nachteil ergibt sich insbesondere für die Hochschulen, die innerhalb eines Ortsnetzes sehr verstreute Institute besitzen, im Vergleich zu den Hochschulen, die auf einem Campus liegen. Es war daher Ziel der Verhandlungen des DFN-Vereins mit der DBP TELEKOM, die unterschiedliche Gebührenbelastung von Wissenschaftseinrichtungen, die auf einem Campus liegen, und solchen, die in Stadtgebieten verstreut sind, abzubauen. Dies wurde durch die Zubringernetzregelung teilweise realisiert. Die Regelung sieht vor, daß Gebühren für interne Leitungen, die Daten an den X.25-Wissenschaftsnetzanschluß bringen (sogenannte Zubringerverbindungen), um 50 % gegenüber dem Listenpreis reduziert werden. Ferner besteht - vertraglich abgesichert - die Möglichkeit, neue Hochgeschwindigkeitsstrecken innerhalb einer Hochschule preisgünstig außerhalb der Listenpreise der DBP TELEKOM durch die TELEKOM bauen zu lassen (sogenanntes "Aachener Modell", weil diese Sonderregelung erstmals von der RWTH Aachen angewandt wurde).

3.5 Finanzierung des Netzbetriebes

Das X.25-Wissenschaftsnetz wurde von der DBP TELEKOM ab 1990 für den Wirkbetrieb zur Verfügung gestellt. Die DFN-Mitgliedseinrichtungen konnten das Netzangebot aber erst nach und nach nutzen, da sie die Haushaltsvoraussetzungen zur Finanzierung schaffen mußten. Für das Jahr 1991 kündigt sich bereits eine in etwa ausgeglichene Bilanz der Einnahmen und Ausgaben an.

Mit der Erweiterung des X.25-Wissenschaftsnetzes um 2 Mbit/s-Anschlüsse ab 1992 ist allerdings eine Neuberechnung der Kosten für das Gesamtnetz verbunden, über deren Höhe und Auswirkungen auf die Anschlußpreise derzeit noch nichts gesagt werden kann. Die Verhandlungen mit der DBP TELEKOM hierüber haben begonnen.

3.6 Finanzierung von Wissenschaftsnetzen in anderen Ländern

In der Wissenschaftslandschaft der Bundesrepublik Deutschland besteht Konsens, daß die Bereitstellung und der Betrieb von Netzen und von Zusatzdiensten ganz oder überwiegend aus Einnahmen der Anwender finanziert werden. Die Dienste müssen jeder für sich kostendeckend betrieben werden. Die Preise sollen pauschaliert erhoben werden.

Ähnliche Vorstellungen wurden auch in der Schweiz von der dortigen akademischen Netzorganisation SWITCH realisiert. Auch die holländische Netzorganisation SURFnet gleicht seit Juli 1989 ihre Ausgaben durch Einnahmen von den teilnehmenden Einrichtungen aus.

In Großbritannien wird das dortige Wissenschaftsnetz JANET derzeit voll über eine Organisation, die in etwa der Deutschen Forschungsgemeinschaft vergleichbar ist, subventioniert. Das JANET ist allerdings auch offen für Industrieforschung, dann mit Vollkostenverrechnung.

Auch in Frankreich wird die Datenkommunikation in der Wissenschaft voll von öffentlicher Seite finanziert.

In den USA betreibt die National Science Foundation (NSF) den NSFNET Backbone. Er wird von der NSF finanziert. Zuschüsse kommen u. a. auch vom Staate Michigan, von IBM und von MCI, die die Weitverkehrsleitungen finanziert. Die NSFNET Regionals, die in der Regel eine größere Stadt oder einen Staat umfassen und vom NSFNet Backbone verbunden werden, finanzieren sich nach Starthilfe durch die NSF jetzt selbst (Quelle: Matrix News, Vol. 1, No. 1, April 1991).

4. Aufwendungen für die Vermittlung in andere Netze

Die Vermittlung von Daten und Informationen über das X.25-Wissenschaftsnetz hinaus zu anderen Netzen ist eine Aufgabe, die der DFN-Verein zur Sicherstellung weltweiter Kommunikation für die deutsche Wissenschaft übernommen hat. Die Kosten für diesen Service übernehmen derzeit weitgehend noch Zuwendungsgeber; langfristig sind sie von den Anwendern zu tragen.

Als zusätzlichen Service werden derzeit vom DFN-Verein angeboten:

a) der Übergang in das öffentliche Datenübermittlungsnetz Datex-P
b) die Nutzung des europäischen X.25-Kernnetzes IXI
c) die Vermittlung von DFN-OSI-Kommunikationsdiensten zu ausländischen Wissenschaftsnetzen
d) die Vermittlung im deutschen EARN und von dort zum europäischen EARN und BITNET
e) die Vermittlung innerhalb der INTERNET-Protokoll-Welt (TCP/IP).

Die Leistungen werden vergleichbar dem WIN-Konzept derzeit zumindest durch Outsourcing oder im Rahmen internationaler Zusammenarbeit erbracht.

Die Kosten für die Nutzung des Datex-P aus dem Wissenschaftsnetz heraus werden von der DBP TELEKOM direkt mit den Anwendern verrechnet, wobei nur die Verbindungsgebühren in Rechnung gestellt werden.

Die Kosten für die Nutzung des IXI trägt derzeit die Kommission der Europäischen Gemeinschaft. Eine spätere Finanzierung des deutschen Anteils durch Umlage auf die deutschen Anwender ist vorgesehen. Es wird derzeit mit einer Belastung der deutschen Anwendergemeinschaft in Höhe von ca. 700 TDM pro Jahr gerechnet.

Die Kosten für die Vermittlung von DFN-OSI-Kommunikationsdiensten in ausländische Wissenchaftsnetze trägt derzeit in der Piloterprobungsphase der BMFT. Sie liegen bei ca. 2 Mio. DM pro Jahr. Auch hier sind ab ca. 1993 die Kosten auf die Anwender umzulegen.

Für die Vermittlung im deutschen EARN und von dort in das EARN/ BITNET tragen bereits ab 1991 die Anwender die vollen Kosten in Höhe von ca. 560 TDM. Der Anwender zahlt pauschal einen Preis von 13 TDM pro Jahr bei einem Zugang mit 64 kbit/s in das WIN und von 7 TDM pro Jahr bei einem Zugang mit 9,6 kbit/s in das WIN. Das Risiko der Finanzierung trägt der DFN-Verein für den Fall, daß die Erlöse den Aufwand nicht decken.

Auch für die Vermittlung innerhalb der TCP/IP-Welt tragen ab 1991 bereits die Anwender die Kosten. Die Preise liegen bei 18 TDM pro Jahr, wenn der Anwender einen 64 kbit/s-Anschluß an das WIN hat bzw. bei 10 TDM pro Jahr bei einem Zugang mit 9,6 kbit/s an das WIN. Die Gesamtkosten liegen bei 625 TDM pro Jahr einschließlich des Domain Name Server-Betriebes. Sie sind bei ca. 45 Nutzerinstitutionen gedeckt. Das Risiko trägt auch hier der DFN-Verein.

5. Weitverkehrskosten, bezogen auf eine einzelne Hochschule

Vor der Hochschulrektorenkonferenz hat Anfang des Jahres 1991 der DFN-Verein folgende finanziellen Belastungen von Hochschulen für externe Datenkommunikation als Grunddienst der Hochschulen genannt:

Art der Hochschule	WIN-Anschluß in kbit/s	WIN-Kosten in TDM/Jahr	Internationale Verbindungen in TDM/Jahr	Sonstige Dienste (DFN-Verein) TDM/Jahr	Gesamtsumme ir TDM/Jahr
Kleine, geisteswissenschaftliche Hochschule					
heute	9,6	18	---	---	18
in zwei Jahren	64	60	20	10	90
in fünf Jahren	2 x 64	120	30	20	170
Normale Hochschule					
heute	64	60	20	---	80
in zwei Jahren	2 x 64	120	40	10	170
in fünf Jahren	2.000	250	60	20	330
Naturwissenschaftlich/technische Hochschule					
heute	2 x 64	120	20	---	140
in zwei Jahren	2.000	250	60	10	320
in fünf Jahren	?	400	60	20	480

Die Ausgabenprognose basiert auf folgenden Überlegungen:

Eine wissenschaftliche Hochschule benötigt mindestens einen 64 kbit/s-Anschluß. Die naturwissenschaftlich/technisch orientierten Hochschulen benötigen einen höheren Anschlußwert für die Nutzung von Workstations und großen Gemeinschaftsrechnern oder Datenspeichern. Gegenüber dem heutigen Zustand ergeben sich in den kommenden 2 bis 5 Jahren erhebliche Kostensteigerungen aus folgenden Gründen:

- technische Verbesserungen gegenüber den heutigen Möglichkeiten,
- Anwachsen der Forderungen der Wissenschaft aufgrund der Tatsache, daß auf Datenkommunikation beruhende wissenschaftliche Arbeitsweisen national und international selbstverständlich werden,
- Fortfall von Zuwendungen des BMFT, mit denen im Rahmen der Piloterprobung Teile der bisherigen Dienste finanziert werden konnten (der BMFT kann betriebliche Leistungen für Landeseinrichtungen wie z. B. Hochschulen aus haushaltsrechtlichen Gründen nicht auf Dauer finanzieren),
- stärkere Inanspruchnahme von Kommunikationsdiensten, um im internationalen Rahmen den Informations- und Konferenzverkehr elektronisch zu erleichtern.

Die Kosten für die Datenkommunikation liegen in der Größenordnung von ca. 1 bis 2 % des Sachhaushalts einer Hochschule.

6. Aufwendungen für die Weiterentwicklung des Deutschen Forschungsnetzes

Das Wissenschaftsnetz ist kein statisches Gebilde. Es muß als Grunddienst für die Versorgung der Wissenschaftseinrichtungen mit Datenkommunikationsleistungen ständig weiterentwickelt und künftigem Bedarf angepaßt werden. Erinnert sei daran, daß in den USA bereits Wissenschaftsnetze mit 45 Mbit/s realisiert werden. In Deutschland ist als Ziel für das Jahr 2000 das Gigabit/s-Netz im Gespräch.

Grundlage für die Weiterenwicklung des Deutschen Forschungsnetzes ist das "Entwicklungsprogramm" des DFN-Vereins, das derzeit bis 1993 fortgeschrieben ist. Dieses Programm enthält Ausgaben für 1991 in Höhe von 17 Mio. DM, in 1992 in Höhe von 13 Mio. DM und in 1993 in Höhe von 10 Mio. DM.

Während die Betriebskosten von den Anwendern aufgebracht werden sollen, besteht weitgehend Konsens, daß die Entwicklungsarbeiten auch in Zukunft aus Zuwendungen des Bundes finanziert werden. Für die Weiterentwicklung wird der DFN-Verein wie in der Vergangenheit Aufträge an Dritte vergeben.

Der DFN-Verein hat für Entwicklungsaufgaben ca. 100 Mio. DM in den Jahren 1983 bis 1990 ausgegeben, die er als Zuwendungen vom BMFT für die Errichtung des auf internationalen Normen basierenden Deutschen Forschungsnetzes (DFN) erhielt. Die Ausgaben verteilen sich wie folgt:

- 30 Mio. DM für Softwareentwicklungen von DFN-Basisdiensten (Dialog, Dateitransfer, Remote Job Entry, Electronic Mail, Directory u. a.)
- 10 Mio. DM für Hochgeschwindigkeitsdatenkommunikation
- 5 Mio. DM für sichere Netzdienste
- 15 Mio. DM für Beschaffungen von Kommunikations-Hard- und Software zur Nutzung des DFN
- 20 Mio. DM für DFN-Vereins-eigene Betriebseinrichtungen zur Softwarepflege und zur Sicherstellung des Zuganges zu ausländischen Netzen
- 20 Mio. DM für Anwendungen innerhalb ausgewählter Nutzergruppen.

Für diese Entwicklungsarbeit flossen
45 Mio. DM über 175 Verträge in die Hochschulen
35 Mio. DM in die außeruniversitären staatlichen Forschungseinrichtungen und
16 Mio. DM in private Wirtschaftsunternehmen.

7. Ausblick

Der DFN-Verein versteht sich als Selbsthilfeorganisation der Wissenschaft für Belange der Datenkommunikation. Er hat das Deutsche Forschungsnetz als Kommunikationssystem für die Wissenschaft aufgebaut und seine Erprobung gefördert. Er berät die Anwender und vertritt ihre Interessen gegenüber staatlichen Stellen, gegenüber Zuwendungsgebern, gegenüber Herstellerfirmen und im internationalen Rahmen.

Er wird diese Funktionen auch weiterhin ausüben. Hinzu kommt in immer stärkerem Maße seine Rolle als Anbieter von Mehrwertdiensten im Wissenschaftsbereich, als Koordinator der Interessen und als ordnender Faktor für einen normengerechten, zugleich wirtschaftlichen und Anwenderansprüchen genügenden Kommunikationsservice. Er dient zugleich auch als Anwenderkreis für neuartige Kommunikationsformen mit Vorreiterrolle für die Wirtschaft.

Der DFN-Verein betreibt seine Dienste nicht im Monopol, sondern im Wettbewerb mit Dritten, z. B. mit dem Datex-P-Dienst der Deutschen Bundespost oder mit den Diensten des EUNET (Universität Dortmund) oder des XLINK (Universität Karlsruhe).

Der DFN-Verein wird von sich aus alles Denkbare unternehmen, um die Finanzierbarkeit der Datenkommunikation für die Wissenschaft auch in Zukunft sicherzustellen. Hierzu gehören

- enge Abstimmung mit der Hochschulrektorenkonferenz, der DFG, der AGF und anderen Wissenschaftsorganisationen sowie mit dem BMFT und BMBW im nationalen Rahmen und mit RARE, der EG und anderen Partnern im internationalen Rahmen,

- Gespräche mit den DV-Länderreferenten in den jeweiligen Wissenschaftsministerien,

- Einfluß auf die DBP TELEKOM, der Wissenschaft spezielle günstige Preise anzubieten im Gegenzug für die Bereitschaft der Wissenschaft, an der Weiterentwicklung der Datenkommunikation aktiv teilzunehmen und diese Erfahrungen der DBP TELEKOM zur Verfügung zu stellen

- und Überlegungen, bestimmte Teilbereiche ggf. zu nationalen Aufgaben zu erklären, die dann über Bund/Länderfinanzierung oder über andere Wege grundfinanziert werden könnten.

Sichere und authentische Kommunikation in Netzwerken

Patrick Horster
Hans-Joachim Knobloch

E. I. S. S.
Europäisches Institut für Systemsicherheit
Universität Karlsruhe
Am Fasanengarten 5
D-7500 Karlsruhe 1

Zusammenfassung

Beim Einsatz moderner Kommunikationsnetze treten die Forderungen nach zuverlässiger Identifikation und Authentifikation potentieller Kommunikationspartner zunehmend in den Vordergrund. Die meist ebenfalls geforderte Vertraulichkeit der übertragenen Daten, die außerdem so schnell wie technisch möglich (online) erfolgen soll, wird durch Verwendung symmetrischer Kryptosysteme erreicht. Ein fundamentales Problem beim Einsatz symmetrischer Kryptosysteme, bei denen jeweils zwei Kommunikationspartner über einen gemeinsamen Schlüssel verfügen müssen, ist der erforderliche sichere Schlüsselaustausch. Dieses Problem wird um so größer, je mehr Schlüssel ausgetauscht, verwaltet und bei Bedarf aktualisiert werden müssen. Eine notwendige Forderung ist weiterhin, daß die Schlüssel authentisch sein müssen, also dem jeweiligen Benutzer (Rechner) nachweisbar zugeordnet werden können. Eine Möglichkeit zur Lösung dieses Schlüsselaustauschproblems bietet die Verwendung von asymmetrischen Kryptosystemen. Auf der Basis des diskreten Logarithmusproblems werden ausgehend vom Diffie-Hellman Public-Key-Distribution-System und vom Unterschriftensystem von ElGamal Protokolle zum Austausch authentischer Schlüssel in Netzwerken vorgestellt. Werden solche authentischen Schlüssel bei einer verschlüsselten Kommunikationssitzung verwendet, so kann infolge der untrennbaren Verknüpfung von Authentifikation und Schlüsselaustausch die Aufrechterhaltung einer authentischen Verbindung sichergestellt werden. Im Gegensatz zu anderen Verfahren sind an der Authentifikation nur die betroffenen Parteien beteiligt – insbesondere wird kein "Online Security Server" benötigt.

1. Grundlagen und Nomenklaturen

Soweit nicht ausdrücklich darauf hingewiesen wird, werden Berechnungen im endlichen Körper $GF(p)$ durchgeführt. Mit $\mathbb{Z}_n^*$ wird die multiplikative Einheitengruppe modulo n bezeichnet.

1.1 Das diskrete Logarithmusproblem

Eine der bedeutendsten Einwegfunktionen der Kryptologie basiert auf dem "diskreten Logarithmusproblem" im $GF(p)$. Sind eine große Primzahl p und eine Primitivwurzel α modulo p gegeben, so ist es zwar für alle x relativ leicht, den Wert $y := \alpha^x$ zu berechnen (hierzu sind $O(\log x)$ Langzahlmultiplikationen mit anschließender Moduloreduktion erforderlich). Bei Kenntnis von y, α und p ist es jedoch (insbesondere dann, wenn $p-1$ einen großen Primfaktor enthält) i. allg. unmöglich, den zugehörigen Wert x zu ermitteln [PoHe78]. Zur Lösung dieses diskreten Logarithmusproblems sind $O(\exp(const \cdot \sqrt{\log p \log\log p}))$ Langzahloperationen erforderlich [Odly85].

1.2 Das Diffie-Hellman Public-Key-Distribution-System

Das Verfahren von Diffie und Hellman [DiHe76] nutzt diese Einwegeigenschaften aus, um zwischen jeweils zwei Benutzern auf einem öffentlichen Kanal Schlüssel zu vereinbaren. Dieses als Public-Key-Distribution-System bekannte Verfahren kann für zwei Benutzer A und B folgendermaßen dargestellt werden. Jeder Benutzer i wählt eine geheime Zufallszahl x_i und veröffentlicht den Wert $y_i := \alpha^{x_i}$, wobei die Parameter p und α allen Benutzern bekannt sind. Auf diese Art können A und B nun einen gemeinsamen Schlüssel K_{AB} generieren: $y_A^{x_B} = K_{AB} = y_B^{x_A}$.

Sind die öffentlichen Werte nicht nachweisbar authentisch, so kann sich unter Umständen ein "Benutzer" C gegenüber A als B und/oder gegenüber B als A ausgeben – die gewünschte Kommunikationssicherheit kann dann nicht gewährleistet werden.

1.3 Das Unterschriftensystem von ElGamal

Die beim Diffie-Hellman-System verwendeten Parameter werden im Verfahren von ElGamal [ElGa85] genutzt, um ein System zur Erzeugung digitaler Unterschriften zu realisieren. Ist $m \in GF(p)$ ein zu signierendes Dokument (gesendet von A nach B), so wählt A eine (genau einmal zu verwendende) Zufallszahl $k \in \mathbb{Z}_{p-1}^*$ und berechnet $r := \alpha^k$. Hierdurch ist A in der Lage, die Kongruenz $m \equiv x_A \cdot r + k \cdot s \pmod{p-1}$ eindeutig zu lösen, womit $\alpha^m = \alpha^{x_A \cdot r} \cdot \alpha^{k \cdot s}$ gilt. Das so ermittelte Paar $u = (r, s)$ stellt eine Unterschrift von m bezüglich der Primitivwurzel α dar. Sind m, r und s gegeben, so kann der Empfänger B die Unterschrift u leicht auf Korrektheit überprüfen, denn es muß $\alpha^m = y_A^r \cdot r^s$ gelten.

Vertauscht man die Rollen von r und s, so erhält man eine modifizierte ElGamal-Unterschrift [AgMV90]. Hierbei löst A die Kongruenz $m \equiv x_A \cdot s + k \cdot r \pmod{p-1}$, womit $\alpha^m = \alpha^{x_A \cdot s} \cdot \alpha^{k \cdot r}$ gilt. Das Paar $u' = (s, r)$ stellt eine modifizierte Unterschrift von m dar. Die Unterschrift u' kann wiederum leicht vom Empfänger B auf Korrektheit überprüft werden, denn es muß $\alpha^m = y_A^s \cdot r^r$ gelten.

2. Das System Kerberos

Um dem Authentifikationsproblem (Nachweis der Identität eines Benutzers bzw. Rechners) bei der Anwendung von Kryptosystemen in Rechnernetzen gerecht zu werden, wurde im Rahmen des am MIT durchgeführten "Athena Project" das System "Kerberos" entwickelt [MNSS87, StNS88]. Die Mechanismen des Kerberos Systems benutzen hierzu modifizierte Needham-Schroeder Protokolle [NeSc78], die auf symmetrischen Kryptosystemen basieren.

Verwendet man in einem Netzwerk zur Authentifikation reine Paßwortmechanismen, so existieren zahlreiche Attacken, um solche Mechanismen außer Kraft zu setzen. Wird beispielsweise das Paßwort im Klartext oder in verschlüsselter Form übertragen, so können entsprechende Informationen bereits durch einfaches Abhören gewonnen werden.

Kerberos vermeidet die Übertragung von Paßwörtern, setzt allerdings die Existenz eines zentralen (sicheren, vertrauenswürdigen) Kerberos-Servers voraus, in dem alle Authentifikationsdaten (insbesondere die verschlüsselten Paßwörter) der potentiellen Benutzer gespeichert sind. Die Arbeitsweise wird hier vereinfacht dargestellt, wobei $\{m\}k$ die mit dem Schlüssel k verschlüsselte Nachricht m bezeichnet. Das verwendete Kryptosystem ist systemweit bekannt, zumeist kommt der Data Encryption Standard (DES) zur Anwendung.

2.1 Der Authentifizierungsdienst von Kerberos

Möchte ein Benutzer B mit einem Server S im System kommunizieren, so meldet er sich mit seinem netzweit bekannten Benutzernamen und dem Verbindungswunsch (Name des Servers S) beim Kerberos-Server an. Der Kerberos-Server antwortet mit einer Nachricht $\{K_{BS}, \{T_{BS}\}K_S\}K_B$, die mit dem Paßwort K_B von B verschlüsselt ist und somit lediglich von B entschlüsselt werden kann. Die Nachricht enthält einen Sitzungsschlüssel K_{BS} und ein Ticket T_{BS}, das mit dem Schlüssel K_S des Servers S verschlüsselt ist. Der Sitzungsschlüssel dient zur Kommunikation zwischen B und dem Server S. Das Ticket enthält u. a. den Benutzernamen I_B von B, den Kommunikationsschlüssel K_{BS}, einen Zeitstempel sowie Informationen über die Gültigkeitsdauer des Tickets.

Fordert der Benutzer B vom Server S einen Dienst an, so verwendet er diese Informationen, um sich gegenüber dem Server zu identifizieren. Hierzu sendet B die Nachricht $\{A_B\}K_{BS}, \{T_{BS}\}K_S$ an S; A_B bezeichnet dabei den Kerberos-Authentifikator des Benutzers B, der u. a. den Benutzernamen I_B von B und einen Zeitstempel enthält. Der Server ist somit in der Lage, die Identität von B zweifelsfrei zu verifizieren. Da B und S den Schlüssel K_{BS} für eine verschlüsselte Kommunikation verwenden, kann auch der Benutzer B zweifelsfrei die Identität des Servers überprüfen.

Wesentlicher Bestandteil des Kerberos Systems ist der sogenannte Ticket Granting Server G, der sicherheitskritische Funktionen innerhalb des Kerberos Systems übernehmen kann. Eine Aufgabe dieses Servers ist, den Austausch authentischer Schlüssel zwischen zwei Benutzern B und C zu realisieren.

Wünscht B mit dem Benutzer C eine Kommunikationsbeziehung, so fordert er von C das Ticket $\{T_{CG}\}K_G$ an. B sendet (z.B. unter Verwendung des Schlüssels K_{BG}) die Tickets $\{T_{CG}\}K_G$ und $\{T_{BG}\}K_G$ an den Server G. Der Server G kann somit die Identitäten von B und C überprüfen, er generiert einen Sitzungsschlüssel K_{BC} und sendet $\{K_{BC}, \{I_B, K_{BC}\}K_{CG}\}K_{BG}$ an B. B sendet $\{I_B, K_{BC}\}K_{CG}$ an C, wodurch auch C in den Besitz des Schlüssels K_{BC} gelangt. Vertrauen die Benutzer dem Server G und verwenden sie bei einer verschlüsselten Kommunikation (mit einem geeigneten Kryptosystem) den Schlüssel K_{BC}, so können sich die Benutzer gegenseitig identifizieren, wodurch eine sichere und authentische Kommunikation gewährleistet werden kann.

2.2 Schwächen des Systems

Kerberos besitzt einige offensichtliche Schwächen, insbesondere ist die Verwendung von Paßwörtern und der Einsatz eines zentralen Servers kritisch zu betrachten.

- Die Verwendung von Paßwörtern, wie sie im Kerberos System benutzt werden, beseitigt nicht die generellen Probleme der Wahl und Handhabung von Paßwörtern.
- Da sämtliche Paßwörter (wenn auch verschlüsselt) gespeichert sind, sind besondere Maßnahmen zu ergreifen, um diese gegen Angriffe zu schützen.
- Der zentrale Kerberos-Server kann zum Engpaß werden und ist daher für "Denial of Service"-Angriffe anfällig.
- Die Frage der Gültigkeitsdauer der Tickets ist ein ungeklärtes Problem – es ist ein Kompromiß zwischen Sicherheit (gegen Replay) und Benutzerkomfort zu finden.
- Da der Server alle Paßwörter und somit sämtliche Schlüssel kennt, besitzt er die "Big Brother"-Eigenschaft. Er ist somit in der Lage, verschlüsselte Kommunikationen abzuhören und zu entschlüsseln.

3. Das Konzept SELANE

Das am E.I.S.S. entwickelte Konzept eines sicheren lokalen Netzwerks SELANE (SEcure Local Area Network Environment) basiert auf dem diskreten Logarithmusproblem [BaKn89, Günt89, Baus90, HoK·91] und verfolgt dieselben Ziele wie das Kerberos System. Das SELANE Konzept basiert jedoch auf einem asymmetrischen Kryptosystem, wodurch der beim Kerberos System erforderliche (Online Authentication) Kerberos-Server nicht benötigt wird.

Die bei SELANE benutzten KATHY-Protokolle (K-ATH-Y: K Y für Key Exchange und ATH für die darin eingebettete Authentifikation) eignen sich zur Lösung einer Vielzahl von Problemen, von denen hier schwerpunktmäßig das Problem des Austauschs authentischer (geheimer) Schlüssel und der damit verbundenen authentischen Kommunikation in Netzen betrachtet wird.

Für die vorgestellten Protokolle ist die Existenz einer übergeordneten vertrauenswürdigen Institution, der sogenannten SKIA (Secure Key Issuing Authority), erforderlich. Die SKIA ist lediglich für die Ausgabe von Benutzeridentifikationen (Zertifikate) zuständig, an der eigentlichen Authentifikation ist sie jedoch nicht aktiv beteiligt. Die von der SKIA ausgegebenen Identifikationen enthalten eine nur von der SKIA berechenbare Signatur von Benutzermerkmalen. Nur die Besitzer einer von der SKIA unterschriebenen Merkmalsliste können eine funktionsfähige authentische Verbindung aufbauen.

Das Konzept geht davon aus, daß für eine Authentifikation bereits authentische Daten vorhanden sein müssen – Vertrauen kann nur übertragen, nicht aber erzeugt werden. Hierzu ist eine vertrauenswürdige Instanz (Zertifizierungsinstanz) erforderlich, die das in sie gesetzte Vertrauen weitergeben kann. Als authentische Basisdaten sind hier lediglich Daten der SKIA notwendig. Die eigentliche Authentifikation von Kommunikationspartnern kann nach einer geeigneten Initialisierung bzw. Registrierung unabhängig von der SKIA erfolgen.

4. Die KATHY-Protokolle

Die KATHY-Protokolle gliedern sich in drei Phasen:

1. Initialisierung der SKIA; diese muß genau einmal bei Inbetriebnahme des Systems durchgeführt werde.
2. Registrierung eines Benutzers durch die SKIA; jeder Benutzer erhält einmalig (evtl. für eine gewisse Geltungsdauer) seine Benutzeridentifikation.
3. Authentifikation eines Benutzers; während dieser Phase des Protokolls werden authentische Schlüssel vereinbart.

4.1 Das KATHY-Basisprotokoll

Die Initialisierung der SKIA und die für die Registrierung eines Benutzers erforderliche Festlegung einer Liste von Benutzermerkmalen sind bei den vorgestellten Protokollen identisch. Das Basisprotokoll dient als Grundlage für die Modifikationen des Protokolls.

4.1.1 Initialisierung der SKIA

Als Basisparameter dienen eine große Primzahl p und eine Primitivwurzel α modulo p. Die SKIA wählt eine geheime Zufallszahl $X \in GF(p)$ und berechnet $Y := \alpha^X$. Die Parameter p, α und Y werden allen Benutzern bekanntgegeben. Den Wert X hält die SKIA geheim.

4.1.2 Registrierung eines Benutzers durch die SKIA

Die Registrierung eines Benutzers i geschieht interaktiv zwischen der SKIA und dem Benutzer. Zunächst wird eine (codierte) Merkmalsliste $m_i \in GF(p)$ (z.B. Name, Geburtsdatum, Anschrift, Privilegien, Verfallsdatum) festgelegt.

Beim Basisprotokoll wählt die SKIA ein zufälliges geheimes $k_i \in \mathbb{Z}^*_{p-1}$ und berechnet zu $r_i := \alpha^{k_i}$ ein eindeutig bestimmtes s_i derart, daß die Kongruenz $X \cdot r_i + s_i \cdot k_i \equiv m_i \pmod{p-1}$ erfüllt ist. Die Parameter m_i, r_i und s_i werden dem Benutzer in einer geeigneten Form (z.B. Chipkarte) ausgehändigt und bilden die Benutzeridentifikation. Der Wert k_i wird nicht weiter benötigt. Das Paar (r_i, s_i) stellt eine ElGamal-Unterschrift der Merkmalsliste m_i dar. Hierbei ist insbesondere darauf zu achten, daß der Parameter s_i vom jeweiligen Benutzer geheimgehalten werden kann. Werden die Parameter in eine Chipkarte gespeichert, so kann diese Anforderung leicht erfüllt werden.

4.1.3 Authentifikation eines Benutzers

Bei der Authentifikation eines Benutzers A gegenüber einem Benutzer B wird nun folgendermaßen verfahren. A teilt B seine Parameter m_A und r_A mit. B wählt eine zufällige Zahl z_B und sendet den Wert $v_B := r_A^{z_B}$ an A. Hierdurch sind beide Benutzer in der Lage, den (Schlüssel-) Wert K_A zu berechnen. A berechnet $K_A := v_B^{s_A}$, und B berechnet $K_A = (\alpha^{m_A} \cdot Y^{-r_A})^{z_B}$.

Beide Benutzer verfügen nun über einen gemeinsamen Schlüssel K_A. Wird dieser Schlüssel beim Chiffrieren verwendet, dann erfolgt die Überprüfung der Authentifikation dadurch, daß sich A und B "verstehen". Hiermit hat sich aber lediglich A gegenüber B authentifiziert; zur gegenseitigen Authentifikation muß das Protokoll nochmals mit vertauschten Rollen durchgeführt werden.

Da die (in den Abbildungen dargestellten) Protokolle zur gegenseitigen Authentifikation unabhängig voneinander sind, können sie auch überlappend bzw. gleichzeitig ausgeführt werden.

Benutzer A	Kanal	Benutzer B
A sendet (m_A, r_A) an B	$\longrightarrow$	(m_A, r_A)
		B wählt $z_B \in \mathbb{Z}_{p-1}^*$
v_B	$\longleftarrow$	$v_B := r_A^{z_B}$
$K_A = v_B^{s_A}$		$K_A = (\alpha^{m_A} \cdot Y^{-r_A})^{z_B}$
(m_B, r_B)	$\longleftarrow$	B sendet (m_B, r_B) an A
A wählt $z_A \in \mathbb{Z}_{p-1}^*$		
$v_A := r_B^{z_A}$	$\longrightarrow$	v_A
$K_B = (\alpha^{m_B} \cdot Y^{-r_B})^{z_A}$		$K_B = v_A^{s_B}$

Abb. 1: Austausch authentischer Schlüssel

Die Überprüfung der Authentifikation der Kommunikationspartner A und B geschieht dadurch, daß die Authentizität der (Schlüssel-) Werte K_A und K_B sichergestellt wird, d.h. sowohl A als auch B besitzen nachweislich die Werte K_A und K_B. Einigen sich A und B auf ein geeignetes Verschlüsselungssystem, so kann dies durch die verschlüsselte Übertragung von Zufallszahlen R_A und R_B überprüft werden. Bezeichnet $E(M, K)$ die Ver- und $D(M, K)$ die Entschlüsselung eines Datums M mit dem Schlüssel K, so kann diese Überprüfung folgendermaßen vollzogen werden.

Benutzer A	Kanal	Benutzer B
A wählt R_A zufällig		
$c_A := E(R_A, K_A)$		
A sendet c_A an B	$\longrightarrow$	c_A
		$C_A := E(D(c_A, K_A), K_B)$
C_A	$\longleftarrow$	B sendet C_A an A
A testet $D(C_A, K_B) = R_A$?		
		B wählt R_B zufällig
		$c_B := E(R_B, K_B)$
c_B	$\longleftarrow$	B sendet c_B an A
$C_B := E(D(c_B, K_B), K_A)$		
A sendet C_B an B	$\longrightarrow$	C_B
		B testet $D(C_B, K_A) = R_B$?

Abb. 2: Authentifikation mittels Zufallsfolgen

Wurden die Benutzer erfolgreich identifiziert, so können ihre authentischen Schlüssel für eine verschlüsselte authentische Kommunikationssitzung genutzt werden. Der Benutzer A verschlüsselt seine Daten mittels des Schlüssels K_A, und der Benutzer B verwendet zum Verschlüsseln seinen Schlüssel K_B. Einigen sich die Benutzer auf eine geeignete Funktion f, so können sie aus den Schlüsselwerten K_A und K_B auch einen gemeinsamen Schlüssel $K := f(K_A, K_B)$ ermitteln.

Wird eine Kommunikationssitzung unverschlüsselt durchgeführt, so ist es denkbar, daß nach einer erfolgreichen Authentifikation eine dritte Partei unbemerkt die Rolle eines

Kommunikationspartners annimmt. Für die gesicherte Aufrechterhaltung einer authentischen Verbindung ist daher eine verschlüsselte Kommunikation zwingend erforderlich.

4.2 Verdeckte Unterschriften

Archiviert die SKIA bei der Registrierung eines Benutzers A den "geheimen" Wert s_A und beobachtet den öffentlichen Kanal, so ist sie in der Lage, den authentischen Schlüssel K_A nachzubilden. Die SKIA muß also das absolute Vertrauen aller Benutzer genießen. Dieser Sachverhalt liegt darin begründet, daß die SKIA beim Basisprotokoll den vom jeweiligen Benutzer geheimgehaltenen Schlüssel s kennt und somit bei Bedarf den (Schlüssel-) Wert v^s berechnen kann. Zur Vermeidung dieser Schwäche kann ein Benutzer seine Merkmalsliste "verdeckt" unterschreiben lassen. Hierzu wählt der Benutzer eine geheime Zahl $a_i \in \mathbb{Z}_{p-1}^*$ und berechnet die Primitivwurzel $\beta_i := \alpha^{a_i}$, die er der SKIA mitteilt. Die SKIA verwendet nun anstelle von α den Wert β_i und berechnet Werte $r_i := \beta_i^{k_i}$ sowie b_i derart, daß die Kongruenz $X \cdot r_i + b_i \cdot k_i \equiv m_i \pmod{p-1}$ erfüllt ist. Die Parameter m_i, r_i und b_i werden dem Benutzer ausgehändigt. Das Paar (r_i, b_i) stellt eine modifizierte ElGamal-Unterschrift der Merkmalsliste m_i dar. Der Benutzer löst die Kongruenz $s_i \equiv b_i \cdot a_i^{-1} \pmod{p-1}$. Der Wert s_i kann von der SKIA nicht ermittelt werden.

Die Authentifikation eines Benutzers A gegenüber einem Benutzer B verläuft analog zum Basisprotokoll. A berechnet $K_A = v_B^{s_A} = r_A^{z_B \cdot s_A} = \alpha^{a_A \cdot k_A \cdot z_B \cdot b_A \cdot a_A^{-1}} = \alpha^{k_A \cdot z_B \cdot b_A}$, und B berechnet $K_A = (\alpha^{m_A} \cdot Y^{-r_A})^{z_B} = \alpha^{b_A \cdot k_A \cdot z_B}$. Die SKIA kann den Wert K_A nicht ermitteln, da sie weder a_A noch z_B kennt.

4.3 Die r^r-Variante

Wählt man die folgende Modifikation des Basisprotokolls, so können Benutzer einerseits Werte v vorberechnen und andererseits gemeinsame feste authentische Schlüssel vereinbaren. Die Modifikation besteht darin, daß die SKIA für einen Benutzer i die Kongruenz $X \cdot s_i + r_i \cdot k_i \equiv m_i \pmod{p-1}$ löst, womit auch $\alpha^{m_i} = Y^{s_i} \cdot r_i^{r_i}$ gilt. Das Paar (s_i, r_i) stellt eine modifizierte ElGamal-Unterschrift der Merkmalsliste m_i dar. Die Authentifikation eines Benutzers A gegenüber einem Benutzer B verläuft analog zum Basisprotokoll.

Benutzer A	Kanal	Benutzer B
A sendet (m_A, r_A) an B	$\longrightarrow$	(m_A, r_A)
		B wählt $z_B \in \mathbb{Z}_{p-1}^*$
v_B	$\longleftarrow$	$v_B := Y^{z_B}$
$K_A = v_B^{s_A}$		$K_A = (\alpha^{m_A} \cdot r_A^{-r_A})^{z_B}$

Abb. 3: Schlüsselaustausch der r^r-Variante

Benutzt man das folgende Protokoll, so können die Benutzer A und B einen gemeinsamen festen authentischen Schlüssel $K_{AB} = K_{BA}$ vereinbaren.

Benutzer A	Kanal	Benutzer B
A sendet (m_A, r_A) an B	$\longrightarrow$	(m_A, r_A)
(m_B, r_B)	$\longleftarrow$	B sendet (m_B, r_B) an A
$K_{AB} = (\alpha^{m_B} \cdot r_B^{-r_B})^{s_A}$		$K_{BA} = (\alpha^{m_A} \cdot r_A^{-r_A})^{s_B}$

Abb. 4: Austausch eines gemeinsamen Schlüssels

4.4. Hierarchische Systeme

Zur Authentifikation eines Benutzers muß lediglich sichergestellt werden, daß der Parameter Y der SKIA authentisch ist. Berücksichtigt man diesen Sachverhalt, so können sich Benutzer auch dann gegenseitig authentifizieren, wenn sie von unterschiedlichen, unabhängigen SKIAs registriert worden sind. Wurde der Benutzer A von der SKIA1 und der Benutzer B von der SKIA2 registriert, und sind Y_1 und Y_2 die öffentlichen Parameter der zugehörigen SKIAs, so ergeben sich die Schlüssel K_A und K_B (vgl. Abb. 1) durch $K_A = (\alpha^{m_A} \cdot Y_1^{-r_A})^{z_B}$ und $K_B = (\alpha^{m_B} \cdot Y_2^{-r_B})^{z_A}$.

Die vorgestellten Verfahren lassen sich auch auf beliebige Netzwerkhierarchien übertragen. Dabei werden sogenannte Sub-SKIAs eingerichtet, die von der (Top-Level-) SKIA wie Benutzer behandelt werden, ihrerseits aber auch Benutzeridentifikationen vergeben können. Hierzu verwendet der zur Sub-SKIA werdende Benutzer anstelle von (α, Y, X) das Tripel (r, r^s, s), wobei r (infolge des zu $p-1$ teilerfremden Parameters k) ebenfalls eine Primitivwurzel modulo p ist.

Zur Authentifikation eines Benutzers sind dann allerdings neben dem Parameter Y der Top-Level-SKIA auch die öffentlichen Parameter sämtlicher SKIAs erforderlich, die zwischen dem Benutzer und der Top-Level-SKIA liegen. Vertrauen muß aber nur der Authentizität von Y entgegengebracht werden.

5. Bemerkungen

Wird in einem Netzwerk das Verfahren von Diffie und Hellman eingesetzt, so muß sichergestellt werden, daß die veröffentlichten Parameter y_i authentisch sind, also tatsächlich zum Benutzer i gehören. Diese Forderung ist jedoch insbesondere dann, wenn die Anzahl der Benutzer groß ist und die Parameter aus Gründen der Sicherheit häufig geändert werden müssen, nur sehr schwer zu realisieren. Fordert man darüber hinaus noch, daß nach jeder Kommunikation der Schüssel gewechselt werden muß, so kann dies praktisch nicht mehr verwirklicht werden.

Bei den vorgestellten KATHY-Protokollen werden bei jeder Kommunikation neue authentische Schlüssel vereinbart. Hiervon ausgenommen ist der Austausch gemeinsamer Schlüssel bei der r^r-Variante.

Verwendet man in einem Netzwerk Kerberos-Protokolle, so ist das Vorhandensein eines (Online Authentication) Kerberos-Server für die Authentifikation von Benutzern zwingend erforderlich. Da ein solcher Server beim SELANE Konzept nicht benötigt wird, entfällt auch die Notwendigkeit, sich über die Auslastung, Ausfallrate und Verletzbarkeit einer solchen zentralen Einrichtung Gedanken zu machen.

Geht man davon aus, daß zur Verschlüsselung ein standardisiertes Verfahren (z.B. DES) verwendet wird und die Authentifikation mittels des Basisprotokolls oder der r^r-Variante durchgeführt wird, so ist die SKIA in der Lage, verschlüsselte Kommunikationen der beschriebenen Art zu entschlüsseln (Big Brother Eigenschaft). Wird die Modifikation der verdeckten Unterschriften eingesetzt, so hat die SKIA keinen Vorteil gegenüber einem externen Lauscher. Die Auswahl des zu verwendenden Verfahrens richtet sich also stark nach den Bedürfnissen der Benutzer sowie der Zertifizierungsinstanz. Für die r^r-Variante ist uns derzeit keine Möglichkeit zur Erzeugung verdeckter Unterschriften bekannt.

Jeder registrierte Benutzer ist mit Hilfe seiner Benutzeridentifikation (m, r und s) auch in der Lage, ein beliebiges Dokument D mit einer ElGamal-Unterschrift zu versehen. Der Empfänger einer solchen Unterschrift kann bei der Verifikation der Unterschrift überprüfen, daß die Unterschrift von der Benutzeridentifikation abgeleitet wurde und somit authentisch ist. Um ein Dokument D zu unterschreiben, wählt der Benutzer ein zufälliges (geheimes) $K \in \mathbb{Z}_{p-1}^*$, berechnet den Wert $R := r^K$ und löst die Kongruenz $D \equiv s \cdot R + K \cdot S \pmod{p-1}$. Das Paar $U = (R, S)$ ist die zu D gehörende Unterschrift. Zur Verifikation wird überprüft, ob $r^D = (\alpha^m \cdot Y^{-r})^R \cdot R^S$ gilt.

Die Verwendung von SELANE als Schwellwertschema, bei dem lediglich bestimmte Koalitionen von Benutzern gemeinsame Aktionen durchführen können, ist ebenfalls denkbar. Der einfachste Ansatz hierzu wäre, einen Schwellwert-Parameter in der Merkmalsliste eines Benutzers vorzusehen. Benutzer könnten beispielsweise genau dann eine gemeinsame Aktion durchführen, wenn die Summe der Parameter einen gewissen Schwellwert überschreitet.

Sind die Parameter m_A und r_A eines Benutzers A bekannt (z.B. durch ein netzweites Verzeichnis), so kann das Verfahren auch zur E-Mail-Verschlüsselung verwendet werden. Dazu berechnet B seinen Schlüssel K_A und sendet sowohl den Parameter v_B als auch die mittels K_A verschlüsselten Nachrichten an A. Lediglich der Benutzer A kann unter Benutzung seines geheimen Parameters s_A aus der Information v_A den Schlüssel K_A berechnen.

KATHY-Protokolle wurden am E. I. S. S. unter Beteiligung von Y.Arman, F.Bauspieß, C.Otto und S.Stempel als “SELANE-Pilot-Implementationen” in verschiedenen Rechnerumgebungen integriert.

Rechner	Zahlenlänge (Bit)				
	256	512	640	768	1024
SUN 3/60	0.27	1.0	1.9	2.8	6.1
VAX 3400	0.23	1.2	2.1	3.5	7.5
80386, 25 MHz	0.18	0.97	1.7	2.8	6.2

Abb. 5: Zeiten der Modulo-Potenzierung in Sekunden

Derzeit stehen Terminalprogramme mit integriertem Authentifikationssystem für die Systeme SUN 3 und Macintosh sowie IBM-AT kompatible Rechner zur Verfügung, wobei RS 232 oder TCP/IP als Trägermedium dient. Terminalprogramme für weitere 680X0 und 80X86 Systeme befinden sich in Bearbeitung. Die Host-Software steht auf SUN 3 (UNIX) für RS 232 und TCP/IP zur Verfügung. Um die Authentifikationszeiten klein zu halten, wurden die benötigten Modulo-Potenzierungen in der jeweiligen Assemblersprache realisiert. Die in Abb. 6 angegebenen Zeiten beziehen sich auf eine Authentifikation zwischen zwei SUN 3/60.

Zahlenlänge (Bit)	256	512	640	768	1024
Sekunden	1.3	3.8	6.5	10.4	23.1

Abb. 6: Authentifikationszeiten

Literatur

[AgMV90] G.B.Agnew, R.C.Mullin, S.A.Vanstone: Improved Digital Signature Scheme Based on Discrete Exponentiation, Electronics Letters, Vol. 26 (1990) 1024.

[Baus90] F.Bauspieß: SELANE – An Approach to Secure Networks, Abstracts of SECURICOM '90, Paris (1990) 159-164.

[BaKn89] F.Bauspieß, H.-J.Knobloch: How to Keep Authenticity Alive in a Computer Network, Advances in Cryptology, Proceedings of EUROCRYPT '89, Springer LNCS 434 (1990) 38-46.

[DiHe76] W.Diffie, M.E.Hellman: New Directions in Cryptography, IEEE Transactions on Information Theory, Vol. IT-22 (1976) 644-654.

[ElGa85] T.ElGamal: A Public-Key Cryptosystem and a Signature Scheme Based on Discrete Logarithms, IEEE Transactions on Information Theory, Vol. IT-31 (1985) 469-472.

[Günt89] C.G.Günther: Diffie-Hellman and ElGamal Protocols With One Single Authentication Key, Advances in Cryptology, Proceedings of EUROCRYPT '89, Springer LNCS 434 (1990) 29-37.

[HoK191] P.Horster, H.-J.Knobloch: Protocols for Secure Networks, Abstracts of SECURICOM '91, Paris (1991) 27-35.

[HoK291] P.Horster, H.-J.Knobloch: Protokolle zum Austausch authentischer Schlüssel, GI-Fachtagung VIS '91, Verläßliche Informationssysteme, Springer Informatik-Fachberichte 271 (1991) 321-328.

[HoK391] P.Horster, H.-J.Knobloch: Discrete Logarithm Based Protocols, Advances in Cryptology, Proceedings of EUROCRYPT '91, to appear.

[MNSS87] S.P.Miller, B.C.Neumann, J.I.Schiller, J.H.Saltzer, Section E.2.1: Kerberos Authentication and Authorization System, MIT Project Athena, Cambridge, Ma. (1987).

[NeSc78] R.M.Needham, M.D.Schroeder: Using Encryption for Authentication in Large Networks of Computers, Comm. of the ACM 21 (1978) 993-999.

[Odly85] A.M.Odlyzko: Discrete Logarithms in Finite Fields and their Cryptographic Significance, Advances in Cryptology, Proceedings of EUROCRYPT '84, Springer LNCS 209 (1985) 224-314.

[PoHe78] S.C.Pohlig, M.E.Hellman: An Improved Algorithm for Computing Logarithms Over GF(p) and its Cryptographic Significance, IEEE Transactions on Information Theory, Vol. IT-24 (1978) 106-110.

[StNS88] J.G.Steiner, B.C.Neumann, J.I.Schiller: Kerberos – An Authentication Service in Open Network Systems, Usenix Workshop Proceedings, UNIX Security Workshop, Portland, Or. (1988).

The BERKOM Administration Infrastructure

Jane Hall, Michael Tschichholz
GMD-FOKUS, Hardenbergplatz 2, D–1000 Berlin 12
E-Mail: hall / tschichholz@fokus.berlin.gmd.dbp.de

Abstract

OSI Management and Directory standards are gradually becoming stable and are gaining support from a variety of interests. This paper discusses one particular architecture for an administration system based on OSI concepts, the BERKOM Administration Infrastructure. The BERKOM Administration Infrastructure offers administration support to distributed applications in an open services environment in which diverse networking capabilities are being interconnected by a B-ISDN backbone network. In the paper particular emphasis is given to the architecture of the management system.

1 Introduction

The concepts of the BERKOM Administration Infrastructure have been developed within the BERKOM Project (**Ber**liner **Kom**munikationssystem) which was started in 1986 in order to provide experience in using the high-speed 140 mbit/s broadband ISDN network that the German PTT, the Deutsche Bundespost Telekom, was installing in Berlin. Various projects have been initiated within the framework of BERKOM in order to evaluate its potential, with particular attention being paid to new protocols for high-speed bulk transfer over reliable links, group communication, inter-organisational networking, and cooperation in an open services environment [POPES-88]. Within the BERKOM framework new areas of application are being developed which are based on work investigating the design of a multimedia document model [MOEL-90]. Various projects are concerned with the possibilities offered by the high-speed network for transmitting video pictures, text and speech and with the innovative applications that can result from this. All these projects need to be concerned with the management of their particular application and it was recognized that a unified management concept should be developed for the BERKOM integrated services [BERK-REF]. In order to prevent each project developing its own management solutions that would then be incompatible with the other BERKOM projects, the BERMAN project was started in the autumn of 1988. The aim of the project is to develop a model for managing distributed applications in an open broadband ISDN environment and to provide basic management components that can be used by other BERKOM projects. Individual management solutions developed by each BERKOM project can thus be prevented and a "BERKOM-wide" management system supported.

2 The BERKOM Administration Infrastructure

The BERMAN project is currently developing the BERKOM Administration Infrastructure (BAI) which is a management platform providing generic management functions to BERKOM applications. The services it provides are part of the BERKOM infrastructure which consists of a service pool of generally useful services intended to support distributed applications in the BERKOM environment. The advantage of such an infrastructure is that it provides basic services that can be used by all applications and which do not have to be designed anew by each application designer. In the case of the BAI, the services provided are related to administrative support and they can be selected and configured according to a distributed application's requirements.

The BAI consists of two services at present, the *Management Support Service* (MSS) and the *Directory Service* (DS) (see Figure 1). A third service, the *Administration Information Service* (AIS) is still being designed and is intended to provide functions that extend beyond the MSS and DS. Further services will be specified and incorporated into the BERKOM Administration Infrastructure as and when required.

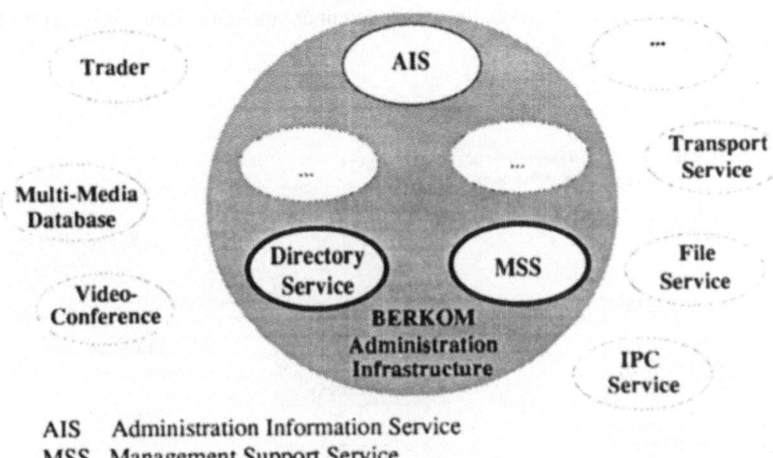

Figure 1: The BERKOM Administration Infrastructure in the BERKOM Service Pool

Two standardised systems are being used for the administration of distributed applications in an open services environment in order to reflect the different characteristics of the information being maintained. Selection of a system is determined by which can best do justice to the characteristics of any individual piece of information as well as conformance to existing standards:

The Directory Service is based on an implementation conforming to the X.500/ISO-9594 Standard [ISO-9594]. The BERKOM Directory can store longer-lasting information (for example, information about services, servers, communication addresses, etc.) and can be used to administer such information on behalf of the integrated services. This can obviate the need to provide separate information storage for the services and it also enables locally administered information to be published globally or to a large sector of the user community, thus contributing to the interoperability and cooperation between autonomous services in a heterogeneous environment.

The Management Support Service (MSS) is provided by the BERKOM Management Platform. It is based on OSI systems management concepts [ISO-7498-4] that have been extended to support distributed applications in the BERKOM environment and it uses the object-oriented modelling approach of OSI management. The MSS is particularly concerned with administering management data that is liable to change rapidly and so cannot be stored using more permanent information storage tools. A concept has been developed in BERMAN for the implementation of a Management Information Base (MIB) which also supports the different characteristics of management information.

3 The BERKOM Management Platform

The BERKOM Management Platform shown in Figure 2 is called the Basic Management Support Service (BMSS), because it permits any service or distributed application to perform management operations on generic or application specific MOs, via standardized interaction mechanisms, unified interfaces, and generic management functions [TSCHI-91a]. The BMSS is based upon the OSI management architecture in which a manager (manager role) can perform operations upon managed objects (MOs) of a managed open system using standardized communication services or protocols (CMIS/CMIP) via an agent (agent role). Events concerning the MOs (e.g. error reports) can be sent asynchronously to a manager (via notifications). The communication between the management components occurs via the OSI Management services or protocols.

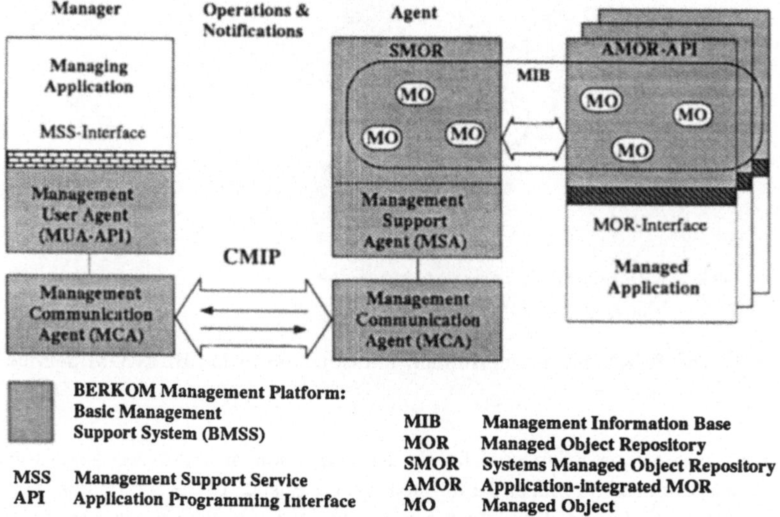

Figure 2: Abstract Architecture of the Basic Management Support System

The BMSS has adapted the OSI Management Architecture [ISO-10040]. Based upon the OSI communication infrastructure (CMIS/CMIP) and an agent realized by BERMAN, two programming interfaces are provided to implementors of the management platform:

- By integrating the Management User Agent application programming interface into a managing application, or manager, the Management Support Service (MSS) functions are made available to the application.
- In order that an application component can be managed by an OSI manager, it must both offer a suitable interface and realize the standardized management information model (MIM) [ISO-10040]. A programming interface to the Application-Integrated Managed Object Repository (AMOR) is provided which enables managed applications to implement their MOs easily.

The *Management Communication Agent* is the central switching and communication component of an end system for the management context. It permits management activities between local Management User Agents and a remote Management Support Agent on the one hand, and between remote Management User Agents and the local Management Support Agent on the other. Both synchronous and asynchronous activities are supported.

A *Management Support Agent* is the component of an end system responsible for carrying out all management operations which are to be performed locally. It serves as an access point for the local Management Information Base (MIB), which is represented by the additional system components *Systems Managed Object Repository* and *Application-oriented Managed Object Repositories.* The Management Support Agent and its Repositories together with the local Management Communication Agent perform the tasks of an OSI agent. OSI Managed Objects represent the resources in a system that are to be managed.

The following sections describe in more detail the architectural concepts of these BMSS components.

3.1 Management User Agent (MUA)

The Management Support Service (MSS) provides an interface which permits an application to assume the manager role. The MSS interface is provided by the MUA as an application programming interface (MUA-API) which can be integrated in managing applications. Included in this set of functions are, for example, the OSI Object Management Function [ISO-10164-1] and the OSI State Management Function [ISO-10164-2].

In addition, support for handling management connections is provided. This enables a managing application to determine for itself when the MUA should establish or terminate connections to MSAs on its behalf, although if the connections are not being used the MUA can independently take the necessary steps to terminate them.

Using such management connections an MUA can address the MSA of any end system and so access any managed application component residing there. This idea goes beyond OSI management where it is currently not possible to address all the components of a distributed application in a single operation because each component local to an end system is represented individually by an MO and each operation is end-system specific.

Applications which are to be managed using the BMSS can be monitored and controlled using the MSS management functions as required. This applies not only to local and remote BMSS-managed applications, but also to those which are integrated in a remote management system if this management system supports OSI management.

The MUA hides the interaction with the Management Communication System so that the location of the desired management information is transparent to the managing application. The concept of global managed object (MO) names makes such transparency possible. A local MO name receives a prefix (global Directory distinguished name) from which the system and the local context can be identified. For management operations on local MOs, direct connections can be established between the MUA and MSA using interprocess communication mechanisms. For remote operations, the Management User Agent must map the MSS operations onto the communications functionality provided by a Management Communication Agent (MCA).

3.2 Management Communication Agent (MCA)

Each Management Communication Agent (MCA) is responsible for exactly one end system, and therefore responsible for the switching and establishment of connections between MUAs and MSAs. For local connections, the MCA fulfills only the switching role. For remote management operations, the MCA is responsible for the establishment of connections with the remote MSA and MUAs.

When performing an operation a Management User Agent sends to the MCA the name of the relevant managed object. The MCA then determines whether a local or a remote MO has been specified. If the MO is local, the Management Communication Agent gives the MUA the local MSA address so that the MUA can use interprocess communication mechanisms to establish its own connection to the MSA. If a remote MO is involved, the MCA first uses the Directory to determine the remote MSAs address, based on the global MO name.

It then establishes a connection with the MSA of the remote system and informs the MUA of the connection identifier so that it can initiate management transactions directly with this MSA. The corresponding procedures are required for the reverse case, if notifications are to be sent by the MSA to the local or remote MUAs.

In order for the MCA to assume this switching role, all MUAs and the MSA must inform the local MCA of their existence. Each MCA must be able to establish a connection with every other MCA. For this reason, it has been decided to store connection data for each Management Support Agent and each managing application (including the address of the local MCA) in the Directory System so that it is globally available and can thus be accessed by remote systems. This means that the Basic Management Support System cannot effectively support management in an open systems environment without using the Directory, which is why a close link has been established between the two systems.

The interchange of management information between MCAs and the execution of management operations is ensured by a set of protocols called *Management Support Protocols*. Each Management Communication Agent supports at least the *Common Management Information Protocol* (CMIP) [ISO-9596]. Further Management Support Protocols may include, for example, the *File Transfer Protocol* or the *Transaction Protocol.*

3.3 Management Support Agent (MSA)

The Management Support Agent is the central access point to the MIB of an end system and it communicates with the MCA, or in the case of local management with the local MUA. In the BMSS, the MIB is represented by a set of Managed Object Repositories which are composed of Systems Management Object Repositories (SMORs) and Application-integrated Management Object Repositories (AMORs) (see Section 3.4).

The responsibilities of the MSA include access control of the incoming management jobs, selection of the Managed Object Repositories which correspond to the addressed MOs, forwarding the jobs to the respective MORs and returning the results. In the reverse direction, the MSA must send the notifications received from the MORs to the corresponding managing applications (or to their MUAs).

3.4 The Concept of Managed Object Repositories (MORs)

A concept has been developed in BERMAN for the implementation of the Management Information Base (MIB) [ISO-10165] as shown in Figure 3 which also supports the different characteristics of management information:

- For management data (MOs) that is more relevant to a specific end system and more static, there is a "Systems Managed Object Repository" (**SMOR**) available on every end system (Managed Open System).

- Management data that can change very rapidly (e.g. a byte counter) is maintained by the object, or process, in which the information is created. When required, these variables can be accessed, and if need be their values modified, using management services and protocols. An application programming interface is provided which can be used to integrate managed objects into application and network components ("Application-integrated Managed Object Repository" (**AMOR**)).

This distribution was introduced because the OSI Management Model does not extend to the modelling of relationships between the managed object and the actual resource. All the MOs comprising the MIB of an end system are maintained by the SMOR and the AMORs of the end system. Both types of Managed Object Repositories (**MORs**) are being realised using a generic MOR object and are distinguished only by the process to which they belong and by the MOs that they are responsible for. A MOR encompasses not only information aspects of the MIB but also functional ones, such as: creation and deletion of MOs, selecting specified MOs by means of scoping and filtering, access control at the application level, administering the containment relations, etc.

The MOR interface is similar that of the Common Management Information Service (CMIS) [ISO-9595]. As in CMIS there are operations for reading and altering attributes, creating and deleting MOs, initiating actions, and sending notifications.

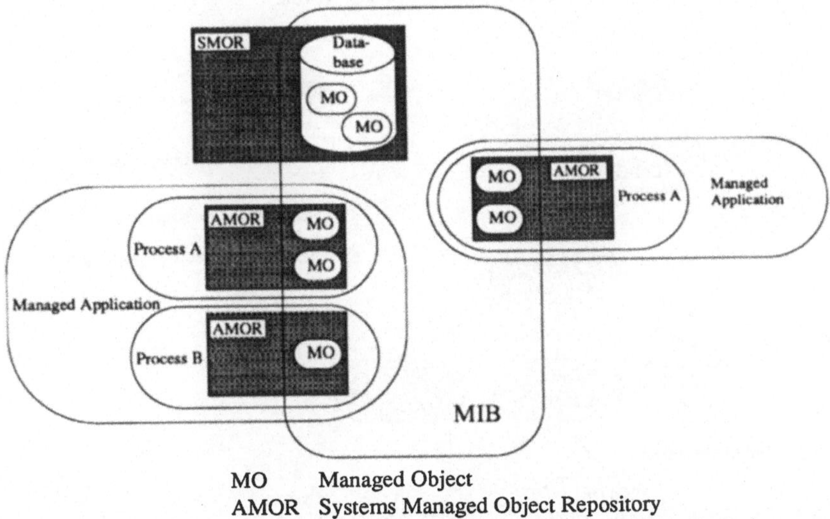

MO Managed Object
AMOR Systems Managed Object Repository
SMOR Application-integrated Managed Object Repository

Figure 3: The Relationship between the MORs, the MIB and Managed Applications

SMOR and AMOR

The *Systems Managed Object Repository* (SMOR) is introduced to store information relevant to a managed open system. The SMOR can use a database to store its MOs so that they will not be deleted when the agent process (MSA) terminates. The functionality of a SMOR otherwise resembles that of the MOR described above. Only one SMOR exists per end system and it forms the root of the containment tree of this system. The MOs stored in the SMOR are called *Systems Managed Objects* (SMOs)

In addition to the SMOR there is an *Application-integrated Managed Object Repository* (AMOR), which is integrated into an application process. *Application-integrated Managed Objects* (AMOs) contain attributes with rapidly changing values, which are not suitable for separation from the actual process. One reason for this can be the unacceptable updating of copies in a MIB database (SMOR) for such frequently changing information (e.g. packet counter).

Composite Managed Objects (CMOs)

Because an MO models an abstract view of any resource or application entity, data of different characteristics may be contained in such an abstract definition [DITT-91]. To enable management information to be stored in an optimal manner, as shown above, so-called *composite managed objects* (CMOs) will be introduced in the next project phase. CMOs consist of a master part, which is located in the SMOR, and one or more subaltern parts, which are located in one or more AMORs. This implementation concept best suits the contradictory requirements of abstract MO definitions and an optimal implementation of management data with varying characteristics.

4 The BERKOM Directory System

The BERMAN project is using the standardised X.500/ISO-9594 Directory for its Directory Service [TSCHI-91b]. The Directory enables entries to be created, read, modified and deleted and allows locally managed information to be available globally. Searches can be undertaken on the information to retrieve entries relevant to a given keyword or combination of keywords.

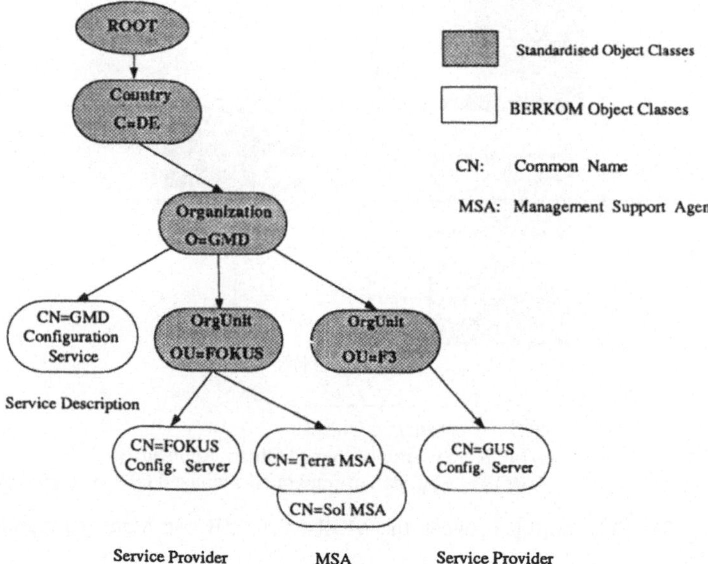

Figure 4: Example of BMSS-specific Directory objects

The service thus provides a nameserver functionality as the information stored locally about a component can be accessed using a globally unique name. The constraint on the information stored by the Directory is that it should be retrieved more often than it is modified and so it is used to store longer-lasting information, such as addresses, names, and information about services, users and other components within the distributed processing system that is of global interest.

A Directory Entry contains the data relevant to an object structured according to a particular schema. An initial schema is defined in the X.500/ISO-9594 standard, but it can be extended within Private Directory Management Domains by new object classes and attribute types. In the BERMAN project the Directory schema has been extended to store information relevant both to the BERKOM environment generally and to BERMAN management requirements in particular.

In order to support BERMAN management requirements new object classes have been defined. Information about both management services and other services is stored using the *service description* object class which defines the abstract service specification and the semantics of a service. The *service provider* object class contains information on individual servers and how they provide the service they offer. The servers are distributed applications that need to be managed and so this object class provides a link to the MSS in that the names of the MOs that represent the server are stored in the server's Directory entry. The object class *management support agent* provides information on the management functions and managed object classes supported by the MSA, and the object class *systems management application entity* is used to store information about the Common Management Information Services that are supported. Figure 4 shows where BERKOM management components can be placed in the directory information tree structure and the kind of objects that can be created using these new object classes.

The BERMAN project is thus using the X.500/ISO-9594 Directory to support its management objectives and has established a link between the Directory Service and the Management Support Service by using the new object classes mentioned above and by storing the identifiers of the MOs representing the server in the server's Directory entry. In addition, work is being pursued on how to represent an OSI managed system in the Directory so that it is uniquely identifiable. The OSI standards specify a MIB located on only one end system. However, the MOs that it contains need to be globally available and this requires a uniform naming context. Such a naming context can be achieved by collecting together the individual hierarchically ordered MIBs under a common global root. As the Directory provides such a naming structure, various methods of representing the MIBs in the Directory are being investigated [WASS-90]. One of the favoured methods is to bind each MIB to the Directory System by means of a specific Relative Distinguished Name. This enables the MIB to be bound to the DIT naming structure and the MOs to be named according to this structure.

5 The Use of the BERKOM Administration Infrastructure

Within the BERKOM framework, the first project to make use of the BAI is the BERCIM project in connection with its approach to service trading [TSCHA-90]. The trader allows users to state their requirements on a service and it selects the most appropriate server. The task of the trader is therefore to use its knowledge about servers and their attributes to match service requirements with service offers and to establish a binding between requestor and server based on an agreed quality-of-service contract.

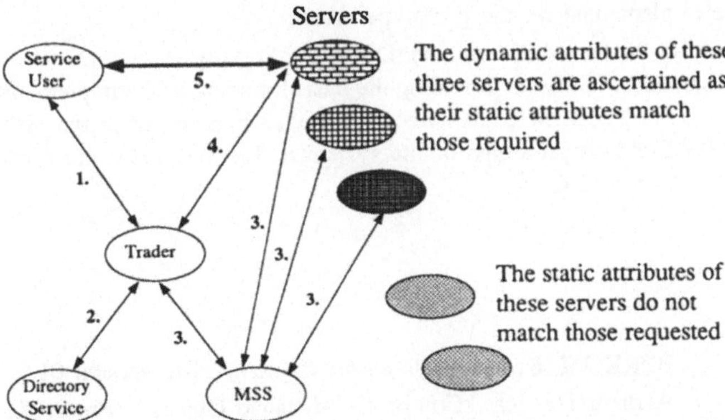

Figure 5: Use of the BAI in the BERKOM Trading Scenario

It has similar functions to that of an intelligent trader currently under discussion in standardisation work, and in ODP terminology it provides a search service, a selection service and an invocation service [ODP-N312].

The trader uses the BERKOM Administration Infrastructure to obtain information on the requested service and the servers offering the service [HALL-91]. The object class service description provides the correct invocation of the service as well as the domains for which the service is defined. The service description also informs the trader about the service attributes defined for the service and which are static and which are dynamic.

After receiving a request from the service user (step 1 in Figure 5) the trader invokes a search operation on the Directory Service to obtain those servers providing the service in the domains of the service requestor (or in the search area defined by the requestor) whose static service attribute values match those requested by the user (step 2 in Figure 5). The trader now requests up-to-date values for the dynamic service attributes of these servers via the Management Support Service (step 3 in Figure 5). As the values of these attributes change rapidly, the trader negotiates with the servers until it obtains one that is prepared to provide the service with the requested values (step 4 in Figure 5). Once a server has been selected the trader links the service requestor with the server (step 5 in Figure 5).

6 Summary

An architecture for an administration system providing generic functions to distributed applications in an open services environment has been described. The BERKOM Administration Infrastructure has been designed as an administration platform for a large heterogeneous environment independent of the underlying communication service. The architecture is based on standards where they are available and applicable. When this has not been the case, development work has been undertaken within the project itself. A close link has been established between the X.500/ISO-9594 Directory and OSI management, and the OSI information model has been refined to meet the needs of implementors. A building block approach has been adopted in providing common generic functions so that applications can import the particular functionality they require.

In order to support the definition of managed objects a tool called DAMOCLES is currently being developed. This tool supports the definition of managed objects which conform to standardized MO classes and is based on an extensive MO library. It offers an X-interface (OSF/Motif) to human users and also allows the BMSS components to access the MO definitions contained in its library.

The trading example shows how components based on OSI/CCITT Management and Directory standards can be used to support trading. This work is being undertaken in close cooperation with the BERCIM project and will be implemented on the prototype system.

A pilot implementation of the MSS and the Directory Service is currently in progress. The ISO Development Environment (ISODE) is providing the basis for some of the implementation, which is being written using C and C++. The BERKOM Directory Service is based upon the ISODE X.500 Directory System QUIPU [KILLE-90]. A prototype of the system is due mid 1991 and further development will take place during 1991 and 1992.

References

[BERK-REF] BERKOM. *Berkom-Referenzmodell*, 04.06.87. Version 01.

[DITT-91] Andreas Dittrich. "Composite Managed Objects". *Proceedings of IFIP 6.6 - 2nd International Symposium on Integrated Network Management*, pages 789–800. North-Holland, ISBN 0 444 89028 9, April 1991.

[HALL-91] Jane Hall and Michael Tschichholz. "Management in a Heterogeneous Broadband Environment". *Proceedings of IFIP 6.6 - 2nd International Symposium on Integrated Network Management*, pages 613–624. North-Holland, ISBN 0 444 89028 9, April 1991.

[ISO-10040] ISO/DIS 10040. "Information processing systems — Open Systems Interconnection (OSI) — Systems Management Overview". June 1990.

[ISO-10164-1] ISO/DIS 10164-1. "Information processing systems — Open Systems Interconnection (OSI) — Systems Management Functions". *Part 1: Object Management Function*, June 1990.

[ISO-10164-2] ISO/DIS 10164-2. "Information processing systems — Open Systems Interconnection (OSI) — Systems Management Functions". *Part 2: State Management Function*, June 1990.

[ISO-10165] ISO/DIS 10165. "Information processing systems — Open Systems Interconnection (OSI) — Structure of Management Information". June 1990.

[ISO-7498-4] ISO 7498-4. "Information processing systems – Open Systems Interconnection (OSI)". *Part 4: Management Framework*, 1989.

[ISO-9594] ISO 9594. "Information processing systems – Open Systems Interconnection (OSI) – The Directory". 1989.

[ISO-9595] ISO 9595. "Information processing systems — Open Systems Interconnection (OSI) — Common Management Information Service Definition". January 1990.

[ISO-9596] ISO 9596. "Information processing systems — Open Systems Interconnection (OSI) — Common Management Information Protocol Specification". January 1990.

[KILLE-90] Stephen E. Kille, Colin J. Robbins, and Michael Roe. "QUIPU". *The ISO Development Environment: User's Manual*, volume 5, University College London, January 31 1990. Department of Computer Science.

[MOEL-90] E. Moeller, A. Scheller, and G. Schürmann. "Distributed Multimedia Information Handling". *Computer Communications*, 13, 1990.

[ODP-N312] ISO/TC97/SC21/WG7. "Working Document on the Trader". October 1990.

[POPES-88] R. Popescu-Zeletin, B. Butscher, P. Egloff, and J. Kanzow. "A Global Architecture for Broadband Communication Systems: The BERKOM Approach". *in Proceedings of the Workshop on Future Trends of Distributed Computing in the 1990's*, pages 366–373. Hong Kong, IEEE Computer Society Press., September 1988.

[TSCHA-90] V. Tschammer, A. Wolisz, and J. Hall. "Support for cooperation and coherence in an open services environment". *In Proceedings on 2nd Workshop on the Future Trends of Distributed Computing in the 1990's*. Cairo, Egypt, IEEE Computer Society Press, ISBN 0 8186 2088 9., September 1990.

[TSCHI-91a] M. Tschichholz, M. Behrendt, A. Consael, A. Dittrich, J. Hall, N. Jantzen, O. Schittko, S. Waßerroth, and M. Wittig. "Guide to the Basic Management Support System (BMSS)". *Arbeitspapiere der GMD, No. 509*, February 1991. ISSN 0723-0508.

[TSCHI-91b] M. Tschichholz et. al. "Guide to the BERKOM Directory". *Arbeitspapiere der GMD, No. 510*, February 1991. ISSN 0723-0508.

[WASS-90] S. Waßerroth and M. Tschichholz. *Konzept für die globale eindeutige Identifizierung von Managed Objects in offenen Systemen*. GMD FOKUS, Berlin, 1990.

Ein modellbasiertes Expertensystem für die Wartung von Telekommunikationsnetzwerken

Carla Decker, Susanne Lösken

Danet GmbH
GS Kommunikation und Anwendungen
Pallaswiesenstr. 201
D-6100 Darmstadt
GERMANY
Tel. +49 6151 8097 332
Fax +49 6151 828 17

Walter Kehl, Heiner Hopfmüller

SEL Alcatel
Research Centre
Dept. ZFZ/SW2
Lorenzstr. 10
D-7000 Stuttgart 80
GERMANY
Tel. +49 711 869 2106
Fax +49 711 869 2185

Zusammenfassung

Der folgende Beitrag beschreibt eine Anwendung von Expertensystemtechniken für die Wartung von Telekommunikationsnetzwerken. Diese Techniken, wie explizites Modellieren der Netzwerke und die Verwendung eines ATMS-basierten (assumption based truth maintenance system) Inferenzmechanismus, werden in verschiedenen Teilen des Wartungsprozesses angewandt. Die Architektur des Wartungssystem ist so allgemein wie möglich gehalten, mit einer strikten Unterteilung des generischen Wissens und des spezifischen Wissens. Das Wissen wird nach Struktur- und Verhaltenswissen unterschieden, wobei strukturelles Wissen in Form eines funktionalen und eines physikalischen Modells repräsentiert wird. Regeln beschreiben das Verhalten von lokalen Netzwerkkomponenten sowie auch des gesamten Systems. Der Inferenzmechanismus benutzt dieses Wissen für abduktive und deduktive Schlußfolgerungen, um die Fehlerhypothesen, Tests und Reparaturen der Komponenten herleiten zu können. Prototypische Anwendungen sind entwickelt worden, die unter anderem das BERKOM Netzwerk in CLOS modelliert haben.

Die besonderen Vorteile dieses modellunterstützten Ansatzes im Bereich der Telekommunikation werden in diesem Beitrag erläutert.

Schlüsselwörter:

Telekommunikationssysteme, objektorientierter Ansatz, modellbasierter Inferenzmechanismus (Schließen über "first principles"), Wartung (Maintenance), Korrelation, Diagnose, ATMS-Mechanismus, generische Wissensrepräsentation, Wissensbasen mit strukturellem und Verhaltenswissen

Modellbasiertes Expertensystem zur Wartung von Telekommunikationsnetzwerken

1 Einleitung

Dieser Beitrag beschreibt die Anwendung von wissensbasierten Techniken für die Wartung (Maintenance) von Telekommunikationssystemen. Das Wartungssystem welches hier beispielhaft beschrieben wird, ist eine spezielle Anwendung eines generischen Wartungssystems, das für zukünftige IBCNs (Integrated Broadband Communication Networks) im Rahmen des RACE[1] Projekts AIM[2] entwickelt wurde. Das gemeinsame Ziel von AIM und anderen RACE Projekten ist die Entwicklung eines integrierten Breitband Telekommunikationsnetzwerk Managementsystems (TMN Telecommunication Management Network), um durch automatische Dienstleistungen hochkomplexe Telekommunikationsnetze zu handhaben. Eine dieser Dienstleistungen ist die Wartung des Netzes, mit der sich das AIM Projekt beschäftigt.

Bei dem Wartungsproblem, das hier beschrieben wird, handelt es sich um die Fehlerbehandlung von Hardwarefehlern im laufenden Netzbetrieb (online corrective maintenance). Wartung heißt hier mehr als nur die Lösung des Diagnoseproblems von elektronischen Komponenten. Gemeint ist hier das Problem, eine Vielzahl von Fehlermeldungen zu verarbeiten und entsprechende Fehlerhypothesen aufzustellen und zu verifizieren, damit entsprechende Reparaturmaßnahmen und gegebenenfalls vorbeugende Maßnahmen generiert werden können. Dabei ist es vorrangig, das Problem so schnell wie möglich einzugrenzen und Folgefehler der eigentlichen Fehlerursache so schnell wie möglich als solche zu erkennen und somit den Suchraum einzugrenzen.

Die Arbeiten von de Kleer, Williams (de Kleer, J: (1986), de Kleer, J.; Williams, B.C.:(1987)) und die von Davis (Davis, R.:(1984)) sind bei der Behandlung dieser Themenstellung Ausgangspunkt. Der Inferenzmechanismus, der in Kapitel 4 ausführlich behandelt wird, ist innerhalb des AIM Projektes entwickelt worden (Bigham, J.; Pang, D.; Chau, T. (1990). Er basiert auf dem Dempster-Shafer Ansatz (Spillman, R.:(1990)).

---

[1] RACE: Research and Development for Advanced Communication Technologies in Europe

[2] AIM: RACE Project R1006, AIP (Advanced Information Processing) Application to IBCN Maintenance

2 Das BERKOM-System als Anwendungsbeispiel

2.1 Architektur des BERKOM-Systems

Das BERliner KOMmunikationssystem ist ein im probeweisen Einsatz befindliches Breitbandkommunikationssystem. Derzeit sind 34 Pilotanwender an dieses System angeschlossen. Zur Verfügung gestellt werden 32 Dialogdienste und 2 Verteildienste. Die Übertragungsrate für Dialogverbindungen beträgt maximal 150 Mbit/s und 600 Mbit/s für Verteildienste.

2.2 Wissensakquisition für BERKOM

Um das zur Erstellung des hier behandelten Prototyps erforderliche Wissen zu akquirieren, wurden Interviews mit Experten geführt sowie Dokumentationen über das BERKOM-System herangezogen.

Es hat sich als Vorteil erwiesen, daß spezielle Szenarios von möglichen Fehlern erstellt wurden. Dadurch wurde sowohl strukturelles Wissen über Netzkomponenten als auch ihr Verhalten im Netz akquiriert. Die Wissenserhebung von BERKOM und von System X führte dann zur Erstellung einer generischen Wissensbasis, die in die Wissensakquisitionskomponente eingegeangen ist.

3 Die Modellierung des BERKOM-Systems

3.1 Die Wissensbasen

Die Wissensbasen stellen das erforderliche Wissen über BERKOM zur Verfügung, das zur Fehlerbehandlung notwendig ist. Es wird logisch unterschieden zwischen strukturellem Wissen einerseits und Wissen über die Beziehungen zwischen den verschiedenen strukturellen Einheiten andererseits. Das heißt sowohl für das funktionale als auch für das physikalische Modell wird Wissen über die funktionale, respektive die physikalische Struktur und Wissen über das Verhalten einzelner funktionaler, respektive physikalischer Komponenten abgebildet. Auf diese Weise ist es möglich, Schlußfolgerungen zu ziehen, die auf einem detaillierten kausalen Modell des BERKOM-Systems basieren.

Dieser Ansatz wurde gewählt, da davon ausgegangen wird, daß im Hinblick auf ein Telecommunication Management Network (TMN) die so gewonnen Erfahrungen nützlicher sind als rein heuristisches Wissen über ein spezielles System. Heuristisches Wissen kann vielmehr, falls erforderlich, ergänzend herangezogen werden.

Die folgenden Wissensarten werden unterschieden:

1) **Funktionales Wissen;** hierzu gehören Informationen über den funktionalen Aufbau des Systems sowie über die zwischen diesen Einheiten bestehenden funktionalen Zusammenhänge.

2) **Diagnosewissen;** im Zusammenwirken mit dem funktionalen Wissen wird es möglich, gemeinsam zu wartende Einheiten (maintainable entities; MEs) zu definieren. Hier sind Informationen über verschiedene Tests und unter welchen Gegebenheiten diese angewendet werden sollen, gespeichert. Weiterhin gehört hierzu Wissen über erwartetes Testverhalten bei korrektem Verhalten des Systems sowie auch bei bestimmten Fehlersituationen.

3) **Physikalisches Wissen;** hierzu gehören Informationen über die einzelnen physikalischen Komponenten des Systems und die Verknüpfungen zwischen diesen.

4) **Reparaturwissen;** im Zusammenwirken mit dem physikalischen Wissen ergeben sich Informationen darüber, welche physikalischen Einheiten gemeinsam ausgetauscht werden müssen.

Im folgenden wird näher auf das funktionale Modell, seine Struktur und das Verhalten der funktionalen Komponenten eingegangen.

3.2 Das funktionale Modell

Das funktionale Modell spielt in dem hier dargestellten Zusammenhang eine zentrale Rolle. Die Trennung in ein funktionales und physikalisches Modell ermöglicht es, das Problem der Fehlerdiagnose und Fehlerbehandlung unabhängig von der spezifischen Konfiguration des physikalischen Systems zu betrachten. Das funktionale Wissen nach strukturellem Wissen und nach Verhaltenswissen zwischen den Komponenten zu unterscheiden, ist Voraussetzung dafür, daß man auf sehr fundierte Weise über das System Rückschlüsse ziehen kann. Im weiteren wird deshalb nur das funktionale Modell ausführlich erläutert.

Das System läßt sich vollständig beschreiben durch die Beschreibung seiner Struktur und des Verhaltens jeder einzelnen Komponente. Dieses Verhalten wird durch Regeln repräsentiert, die beschreiben, wie sich die einzelnen Systemkomponenten zueinander verhalten. Ein hierauf basierender Inferenzmechanismus kann Rückschlüsse von "first principles" ziehen und von daher bis ins Detail zurückverfolgen, wie verschiedene Arten von Inputverhalten zu einem bestimmten Outputverhalten geführt haben. Dies führt zu einer qualitativ besseren Absicherung des Problemlösungsprozesses.

3.2.1 Das strukturelle Modell

Den funktionalen Aufbau des strukturellen Modells bilden die funktionalen Einheiten (functional entities). Eine funktionale Einheit (FE) ist definiert als ein Objekt, das eine bestimmte Funktion des BERKOM-Systems modelliert. Es liegt auf der Hand, daß sich solche Funktionen auf verschiedenen Ebenen betrachten lassen, wobei eine Funktion einer höheren Ebene in mehrere Funktionen auf einer tieferen Ebene aufgeteilt ist. Diese Ebenen sind untereinander durch die Beziehungen has-part und is-part-of verbunden. Wie fein abgestuft eine solche Einteilung funktionaler Einheiten sein sollte, ist eine Designentscheidung. Diese ist im Hinblick auf die untersuchte Problemstellung getroffen worden. Es werden vier Ebenen unterschieden, die Funktionen bis auf die Ebene der funktionalen Baugruppe und der funktionalen Unterbaugruppe unterscheiden. Diese Anlehnung an das physikalische Modell entsteht dadurch, daß eine Baugruppe eine in sich geschlossene Funktion wahrnimmt, z.B. Multiplexergruppe, die die Funktion des Multiplexens übernimmt. Durch diesen Detaillierungsgrad ist es möglich, im Modell die funktionalen Einheiten als Ursache von Fehlersymptomen zu bezeichnen, die auch im BERKOM-System von der dafür verantwortlichen Funktion ausgelöst werden. Das entscheidende Kriterium hierbei ist das der Isomorphie zwischen Zielsystem und seiner Abbildung im Modell.

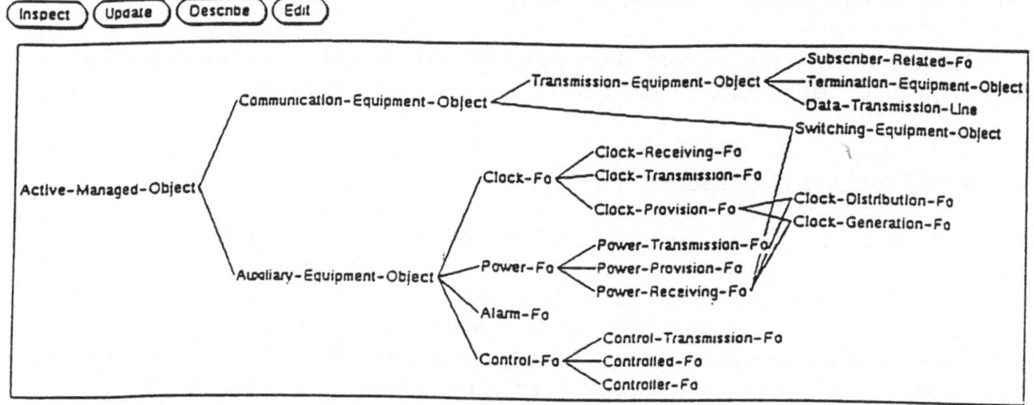

Bild 1: Ausschnitt aus dem strukturellen Modell

Durch Eigenschaften, die an die funktionalen Einheiten angehängt sind, läßt sich feststellen, welche physikalischen Einheiten dieser Funktion zuzuordnen sind. Diese Korrespondenzen sind sehr wichtig, da sie es ermöglichen, nach Lokalisierung der Funktion, die zu einem fehlerhaften Verhalten des Systems geführt hat, die entsprechenden physikalischen Einheiten auszutauschen. Dabei können die Beziehungen zwischen funktionalen und zugehörigen physikalischen Einheiten 1:1, 1:n, n:1 sein.

Eine funktionale Einheit wird charakterisiert durch ihre Eigenschaften (Attribute wie z.B. korrespondierende physikalische Komponenten) sowie durch die Art der Beziehungen, die sie mit anderen funktionalen Einheiten verbinden. Diese Beziehungen zwischen funktionalen Einheiten werden durch **Ports** dargestellt, d.h. zwei funktionale Einheiten sind dann miteinander verbunden, wenn ein output-port der einen Einheit mit einem input-port der anderen Einheit verbunden ist (siehe Bild 1).

Eine Taxonomie dieser funktionalen Einheiten bildet die funktionale Struktur des Systems ab. Die dargestellte funktionale Struktur enthält einerseits BERKOM-spezifische Klassen, andererseits sind die höheren Ebenen so gewählt, daß sie allgemein genug sind, um für andere Telekommunikationssysteme gelten zu können, z.B. die Unterteilung in Kommunikationskomponenten und Hilfskomponenten. Diese wiederum sind unterteilt in Vermittlungs- und Übertragungskomponenten einerseits und Stromversorgungs-, Steuerungs- und Taktkomponenten andererseits (siehe Bild 2).

Die Grenze festzulegen, ab welcher Ebene die funktionalen Einheiten nicht länger für alle Telekommunikationssysteme gelten, d.h. nicht mehr generisch sondern speziell sind, wird innerhalb des AIM Projektes in Zusammenarbeit mit den AIM Mitarbeitern von British Telecom und dem von ihnen untersuchten DSSS diskutiert.

Hier zeigen sich die Vorteile einer von der physikalischen Sicht getrennten funktionalen Betrachtung. Ein Teil des Wissens ist als generisch anzusehen und kann von daher für die Wartung anderer Telekommunikationssysteme verwendet werden.

3.2.2 Wissen über das Verhalten der Netzkomponenten

Dieses Verhalten ist für jede Netzkomponente beschrieben. Dadurch läßt sich mittelbar über den Inferenzmechanismus auch das Verhalten des Gesamtsystems bestimmen.

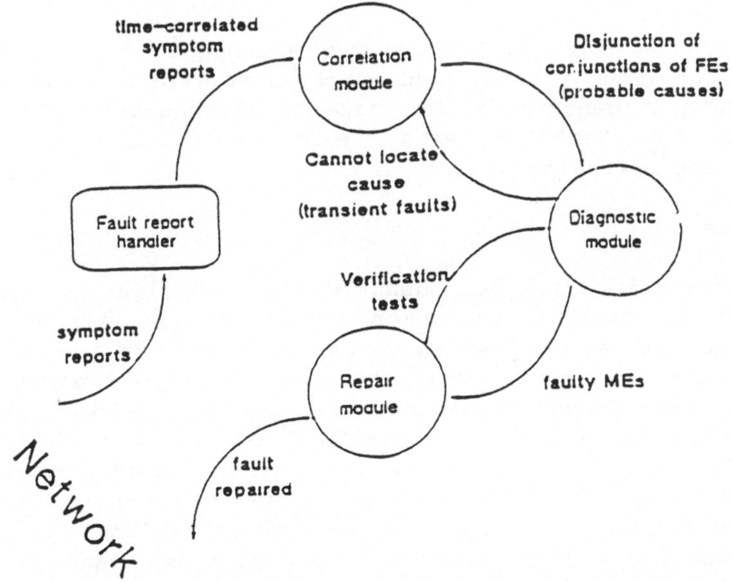

Bild 2: Generische Objekte der Klassenhierarchie

Der zugrundeliegende Inferenzprozeß betrachtet sowohl den im System vorliegenden Fluß des Verhaltens der einzelnen Komponenten, als auch den bei korrektem Systemverhalten erwarteten Ablauf. Danach wird durch das Zusammenwirken von deduktiven und abduktiven Schlußfolgerungen die Fehlerursache lokalisiert. Zum Inferenzmechanismus siehe auch Kapitel 4.

Die Abbildung des Verhaltens erfolgt entlang der folgenden Arten von Informationen:

1) **internal states:** dieses Attribut ist jeder funktionalen Einheit zugeordnet. Komponentenzustände (internal states) wie z.B. working/not-working sind die grundlegende Ursache eines spezifischen Verhaltens des Systems. Dementsprechend bilden Aussagen über den Zustand der Komponenten den Inhalt der Hypothesen über mögliche Fehlerursachen.

2) **port states:** ports als Träger der Informationen über Beziehungszusammenhänge (connectivity information) zwischen den funktionalen Einheiten beinhalten Informationen über den jeweiligen Status der Beziehungen zwischen funktionalen Einheiten.

3) **rules:** diese Regeln geben Auskunft über den Zustand der output-ports in Abhängigkeit vom Zustand der zugehörigen input-ports sowie der internal states.

Der Vorteil, in dieser Weise Regeln für jede einzelne Komponente zu schreiben, hat mehrere Aspekte. Zum einem ist es möglich für jede einzelne Komponente recht einfache Regeln zu formulieren, die lokal auch sehr leicht geändert werden können, ohne als Folge davon im System weitere Änderungen vornehmen zu müssen. Zum anderen wird es bei Betrachtung des Gesamtbeziehungsgefüges möglich, ein sehr komplexes Beziehungsgefüge zu betrachten, das gleichzeitig übersichtlich und leicht modifizierbar ist.
Da sämtliche Regeln den zugehörigen Komponenten angehängt sind, sind sie entlang des strukturellen Systemaufbaus gegliedert. Folglich lassen sich auch die Regeln in generische, für alle Telekommunikationssysteme geltende Regeln, z.B. Verhalten einer Verbindung (line) und in BERKOM-spezifische Regeln unterscheiden.

3.3 Der Wartungszyklus

Der Wartungszyklus basiert auf der oben beschriebenen Modellierung des BERKOM Systems. Der Wartungszyklus wird unter dem Blickwinkel gesehen, daß nach Auftreten eines Fehlers Maßnahmen zu dessen Erkennung und Behebung vorgenommen werden. Dieser Wartungszyklus ist in Bild 3 dargestellt. Er besteht aus vier Hauptkomponenten, die jeweils eine in sich geschlossene Teilfunktion im Hinblick auf das Ziel der Fehlerbehebung erfüllen. Diese Module sind die Fehlerreportbehandlung (event report handler), das Korrelationsmodul, das Diagnosemodul und die eigentliche Fehlerbehebung. Die Beziehungen zwischen diesen Teilfunktionen sind logisch sequentiell, wobei die Ergebnisse des Vorgängers den Input der nachfolgenden Teilfunktionen darstellen. Rückkopplungen bestehen zwischen Korrelation und Diagnose und Fehlerbehebung. In besonders schwer zu ermittelnden Fehlerursachen kann es notwendig sein, erneut auf die Ergebnisse des Korrelationsmoduls zurückzugreifen.

Nach Behebung eines Fehlers wird wieder auf die Diagnose zurückgegriffen, um festzustellen, ob der Fehler auch tatsächlich nicht mehr auftritt, d.h. die Diagnose hat die richtige Ursache ermittelt.

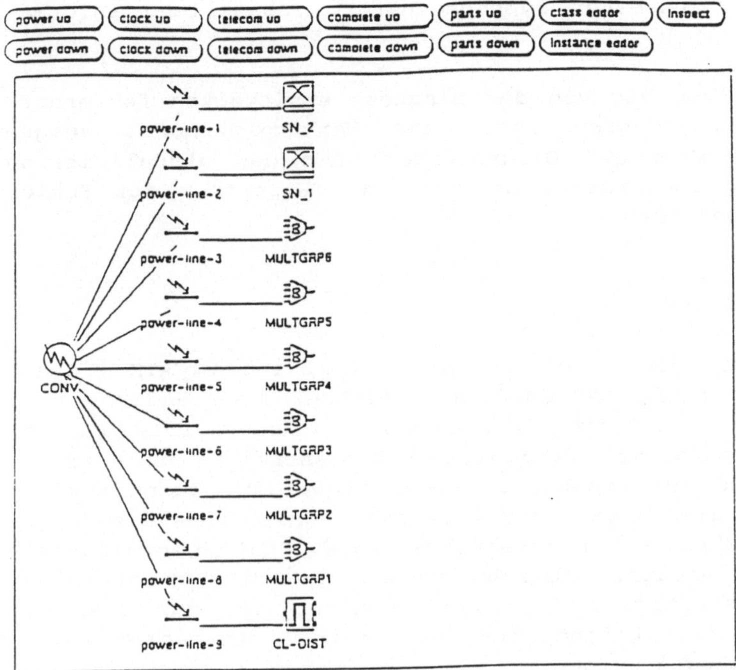

Bild 3: Der Wartungszyklus

Nachfolgend eine kurze Beschreibung der Aufgabe der einzelnen Komponenten:

1) **Fehlerreportbehandlung:** hier werden vom Netzwerk ankommende Symptome gesammelt. Diese Symptome werden in einfacher Form zu Gruppen zusammengefaßt und an das Korrelationsmodul weitergeleitet. Nach erfolgreicher Fehlerbehebung werden die zugehörigen Symptome wieder gelöscht.

2) **Korrelationsmodul:** Ziel dieses Moduls ist es zur Eingrenzung der Diagnose Hypothesen darüber aufzustellen, welcher Fehler den einlaufenden Symptomen zugrunde liegen könnte. Hauptkriterium der Zusammengehörigkeit von Symptomen ist deren zeitliches Zusammentreffen. Dabei sollen soviele Symptome wie möglich durch eine zugrundeliegende Ursache erklärt werden. Als Ergebnis liefert der Korrelationsmodul eine Disjunktion von Konjunktionen funktionaler Einheiten. Jede dieser Konjunktionen ist eine Fehlerhypothese, die an das Diagnosemodul weitergeleitet wird.

3) **Diagnosemodul:** Diagnose hat die Aufgabe, die von der Korrelation aufgestellten Fehlerhypothesen so einzugrenzen, daß im Ergebnis die physikalischen Komponenten bekannt sind, die den Fehler verursacht haben. Hierzu werden sowohl automatische als auch von einem Operator manuell durchzuführende Tests eingesetzt. Geräte bzw. funktionsspezifische Tests werden zur genaueren Lokalisierung der Fehlerursache und

der zugehörigen physikalischen Komponenten eingesetzt. Sind die durchgeführten Tests nicht in der Lage, eine Fehlerursache einzugrenzen, so wird wieder an das Korrelationsmodul zurückgegeben, um von dort aus, unter Einbeziehung der neuesten Symptomemuster, zusätzliche Informationen zu gewinnen.

4) **Reparatur:** hier werden die von der Diagnose ermittelten fehlerhaften physikalischen Komponenten durch das Wartungspersonal ausgetauscht. Anschließend wird das Diagnosemodul mit den aktualisierten Daten versorgt, um zu überprüfen, ob durch die Reparatur der Fehler auch tatsächlich behoben ist.

4 Der Inferenzmechanismus

Der Inferenzmechanismus, der hier verwendet wird, ist im AIM Projekt entwickelt worden. Er heißt GMS-GMS[1] und zeichnet sich dadurch aus, daß er nicht an eine spezifische Wissensbasis gebunden ist, d.h. er kann für Wissensbasen aus dem Telekommunikationsbereich zur Korrelation, zur Diagnose und zur Reparatur verwendet werden. Zudem basiert er auf dem Dempster-Shafer Ansatz und kann daher unsicheres Wissen gemäß der Wahrscheinlichkeitstheorie berücksichtigen, ist aber allgemeiner als der Bayesian Ansatz. Dadurch ist es möglich auch fehlendes Wissen mit zu berücksichtigen, d.h. die Tatsache, daß für eine bestimmte Annahme weder bestätigende noch ablehnende Informationen vorliegen, kann mit einbezogen werden.

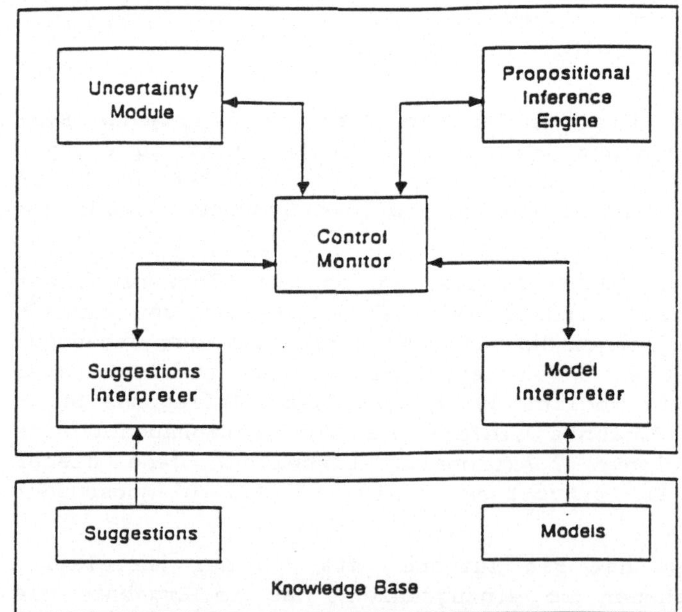

Bild 4: Aufbau der Inferenzmaschine

[1] GMS-GMS: Generic Managed Searcher of the Generic Maintenance System

4.1 Die Komponenten der Inferenzmaschine

Die hier beschriebene Inferenzmaschine besteht aus den folgenden Komponenten (siehe Bild 4), die jeweils andere Arten von Inferenzen ermöglichen. Dadurch kann der Wartungsprozeß (maintenance process) wirksam unterstützt werden. Dies wird im folgenden näher erläutert.

1) **Hypotheseninterpretierer** (suggestion interpreter); die Wissensbasis enthält Annahmen über das BERKOM-System. Diese werden vom Hypotheseninterpretierer als Grundlage herangezogen, um dem Kontrollmonitor Vorschläge zu machen, welche Annahmen in der gegebenen Situation untersucht werden sollen. Es handelt sich um einen abduktiven Inferenzmechanismus,d.h. if-cause-then-effect Regeln werden in if-effect-then-cause Regeln umgewandelt. Die Regeln, die dann mit den vorliegenden Symptomen übereinstimmen, werden angewendet.

2) **Kontrollmonitor** (control monitor); hier wird das Problem in Teilprobleme aufgegliedert, unter Verwendung anderer Schlußfolgerungskomponenten im GMS-GMS. Auf diese Weise soll der sehr umfangreiche Suchraum eingegrenzt werden. Hauptaufgabe ist es, Steuerungsinformationen für diese anderen Komponenten bereitzustellen. Dabei jede Teilaufgabe, wie Korrelation oder Diagnose, getrennt gesteuert. Für den notwendigen Austausch von Informationen zwischen diesen Steuerinformationen ist der Kontrollmonitor zuständig.

3) **Modellinterpretierer** (model interpreter); ein Inferenzmechanismus, der das Verhalten der Komponenten deduktiv interpretiert. Grundlage dafür bilden die Wissensbasen, einschließlich Wissen darüber, welche Deduktionen sinnvoll sind, z.B. kann es an dieser Stelle sinnvoll sein, daß zugrundeliegende tiefe Modell (deep model) durch heuristisches Wissen zu ergänzen. In diesem Fall würde das Erfahrungswissen der Experten, über das, was als sinnvolle Deduktion anzusehen ist, mit einfließen. Die gefundenen Deduktionen werden von PIE (siehe unter 4)) verwendet.

4) **Propositionaler Inferenzmechanismus** (propositional inference engine); PIE ist ein Mechanismus, der temporäre Veränderungen berücksichtigen kann (truth maintenance system) und auf dem ATMS[1] basiert. PIE fungiert als Cache für die vom Modellinterpretierer gemachten Deduktionen. Wie oben beschrieben, entscheidet der Modellinterpretierer darüber, welche Deduktionen abgeleitet werden. PIE baut ein Netz auf, daß die Annahmen (propositions), die die gemachten Deduktionen unterstützen, in den Knoten abspeichert. Gleichzeitig werden Informationen darüber abgespeichert, welche Kombinationen von diesen Annahmen inkonsistent sind. Dies dient dazu die Anzahl der Knoten zu reduzieren, indem solche, die inkonistent sind, herausgenommen werden (pruning). In den Kanten werden Informationen über die Beziehungen zwischen den Annahmen (propositions) gespeichert.

---

[1] ATMS: Assumption Based Truth Maintenance System

5) **Vages Schlußfolgern** (uncertainty module); diese Komponente ist als eine Ergänzung zu PIE anzusehen und kann dazu benutzt werden, vage Annahmen bei der Auswahl vermuteter Fehlerursachen in geeigneter Weise mitzuberücksichtigen. Der zugrundeliegende Ansatz ist der Dempster-Shafer Ansatz. Ein Hauptvorteil dieses Ansatzes ist es, nicht nur Angaben machen zu können, wie sicher man etwas annimmt, sondern auch mit welcher Sicherheit man etwas nicht weiß.

Im folgenden Abschnitt wird beispielhaft auf den Inferenzmechanismus für Korrelation eingegangen.

4.2 Inferenzmechanismus für Korrelation

Wie oben gesagt, besteht die Aufgabe der Korrelation darin, mögliche Fehlerursachen, die möglichst viele der auftretenden Symptome erklären können, zu identifizieren. Zu diesem Zwecke werden verschiedene funktionale Beziehungszusammenhänge untersucht.

Nachfolgend wird kurz beschrieben wie diese Angaben vom Inferenzmechanismus verwendet werden. Der Hypotheseninterpretierer benutzt die Regeln über das Verhalten einzelner Komponenten abduktiv, um so mögliche Ursachen für die auftretenden Symptome zu finden. Jede dieser Hypothesen hat ein mit ihr verbundes Maß, daß angibt, wie stark oder schwach diese Hypothese aufgrund der vorliegenden Tatsachen abgesichert ist. Zwei Aufgaben werden durch Anwendung des Inferenzmechanismus erreicht:
1) Herausfiltern einer weitreichenden, d.h. möglichst viele Symptome erfassenden, konsistenten Erklärung für die auftretenden Symptome.

2) Herausfiltern einer qualitativ guten, d.h. einer in vielen Fällen mit der tatsächlichen Ursache übereinstimmenden, konsistenten Erklärung für die vorliegenden Symptome. Hierzu werden nach bestimmten, meist heuristischen Kriterien, Schwerpunkte (focus assumptions) im Inferenzprozeß gesetzt. Das derzeitig einzige Kriterium, das verwendet wird, ist das der Glaubwürdigkeit (credibility).

5 Resümee

Wie aus der obigen Darstellung der Inferenzkomponente ersichtlich ist, wird für die Problemlösung nicht nur der modellbasierte Ansatz gewählt, sondern ein Teil des Wissens ist auch in heuristischem und Erfahrungswissen repräsentiert. Wir sind der Ansicht, daß diese Techniken besonders für das Wartungsproblem von Telekommunikationsnetzen, im Gegensatz zu einem rein regelbasierten Ansatz, die folgenden Vorteile haben:

1) Telekommunikationsnetzwerke werden sehr oft als Varianten eines Basissystems installiert. Die Modellierung dieser Varianten wird mit Hilfe des modellgestützten Ansatzes wesentlich vereinfacht, da eine eindeutige Trennung von generischem und netzspezifischem Wissen unterstützt wird.

2) Die Wissensbasis, auf die hauptsächlich die Inferenzmaschine zugreift, kann auch von anderen Systemkomponenten benutzt werden. Dadurch kann jederzeit der aktuelle Zustand des Systems abgefragt werden und das dynamische Wissen (z.B. die aktuellen Verbindungen der Teilnehmer) ebenfalls zu Schlußfolgerungen zur Verfügung gestellt werden.

Danksagung

Die Autoren danken allen Partnern des AIM Projekts, die durch ihre Arbeiten diesen Beitrag unterstützt haben.

Diese Arbeiten werden im Rahmen des RACE Programms von der Europäischen Gemeinschaft gefördert.

6 Literaturverzeichnis

Azmoodeh, M.: Representation of Generic Structure and Behaviour of Networks, Proceedings of the fourth RACE TMN Conference, 1990, 244-254

Bigham, J.; Pang, D.; Chau, T.: Inference in a Generic Maintenance System for Integrated Broadband Communication Networks, Proceedings of the fourth RACE TMN Conference, 1990, 199-206

Bigham, J.: Computing Beliefs according to Dempster-Shafer and Possibilistic Logic, Third International Conference on Information Processing and Management of Uncertainty in Knowledge Based Systems, 1990, Paris

Davis, R.: Diagnostic Reasoning Based on Structure and Behavior, Artificial Intelligence Vol. 24 (1984) 347-410

Decker, C.; Loesener, C.R.; Loesken, S.: Basic Knowledge Representation Tools and their Functional Specification, 1990, AIM Deliverable 06/DAN/WBS/DR/C/037/a2

Decker, C.; Chau, T.; Torres, A., Refined Knowledge Representation/Knowledge Acquisition Tools, 1990, AIM Deliverable 06/DAN/WBS/DR/I/066/a1

de Kleer, J.: An assumption based TMS, Artificial Intelligence Vol. 28 (1986), 127-162

de Kleer, J.; Williams, B.C.: Diagnosing Multiple Faults, Artificial Intelligence, Vol. 32 (1987), S. 97 - 130

Laskey, K.B.; Lehner, P.E., Belief maintenance: an integrated approach to uncertainty management, Proceedings of the AAAI, August 1988, 210-214

Newstead, M.; Kehl, W.; Loesken, S.: Prototype for Digital Network I: Representation of Existing Digital Network, Knowledge Base for Online Diagnosis and Knowledge Base for Repair Planning, 1990, AIM Deliverable 06/SEL/FZS/DR/X/061/a1

Spillman, R.: Managing with Belief, AI Expert, Mai 1990, S. 44 - 49

Stahl, B.; Azmoodeh, M., Knowledge Representation of Networks in the RACE Project AIM, Proceedings of the fourth RACE TMN Conference, 1990, 64-74

Kommunikation und neue Dienste

Der technische Fortschritt und die Digitalisierung der Nachrichtentechnik ermöglicht die Bereitstellung von Netzen mit sehr hohen Kapazitäten. Soll diese Technik Chancen auf allgemeine Realisierung haben, wird nachzuweisen sein, daß neben den überzeugenden technischen Eigenschaften vor allem Anwendungen und Nachfrage im privaten und öffentlichen Bereich vorhanden sind, die die hohen Investitionen rechtfertigen. Dabei ist die Heterogenität als zentrale Idee des OSI zu wahren, wenn auch die Wirksamkeitsanforderungen an die Datenverarbeitung die gegenwärtige Nutzung von OSI nur eingeschränkt verwendbar macht.

Koordinatoren: Prof. Dr. G. Müller, Universität Freiburg
Dr. L. Mackert, IBM ENC, Heidelberg

Mehrwertdienste in Rechnernetzen - Technik und Anwendungen

Werner Zorn

Universität Karlsruhe
Kaiserstr. 12
7500 Karlsruhe

0 Kurzfassung

Mehrwertdienste (VAS) sind alle Netzdienste, die dem zugrundeliegenden Transportnetzwerk wesentliche Leistungsmerkmale hinzufügen. Hierzu zählen einmal die klassischen Basis-Dienste wie Electronic Mail, News, Directory Service, zum andern darauf aufsetzende Dienste für den Austausch von Dokumenten, Faksimilen, Graphik, den Dialogzugriff auf Datenbanken u.v.a.m. Unter der Bezeichnung CSTA (Computer Supported Telefonic Application) wachsen derzeit die rechner- und telefon-orientierten Mehrwertdienste zusammen. Der Beitrag behandelt die Techniken, inkl. Administration, stellt Nutzenbetrachtungen an, und versucht, einen Überblick über die Marktentwicklung zu geben.

1 Einführung

Während man in den offenen Rechnernetzen, z.B. im Wissenschaftsbereich bei Netzdiensten an die "klassischen" Basisdienste

- Electronic Mail
- File Transfer
- Dialog
- RJE

denkt, geht der Begriff "Mehrwertdienste" (VAS = Value Added Services) weit darüber hinaus. In [Gut90] wird folgende Definition vorgenommen:

VAS sind

- auf Netzen angebotene Dienstleistungen, die
- einen über die Netzleistung hinausgehenden Nutzen bieten und
- durch einen Betreiber gegen Verrechnung zugänglich gemacht werden

Sie gliedern sich in:

- Basic Services	(z.B. E-Mail, EDI, ...)
- Information Services	(z.B. Börsen- und Auskunftsdienste)
- Processing Services	(z.B. DATEV, Reservierungsdienste)

im Gegensatz zu

VAN sind

- Netze mit Dienstleistungen, die eine durchgängige Kommunikation zwischen Teilnehmern ermöglichen.
  (Dies sind Funktionen wie vermitteln, umsetzen, speichern, sichern, etc.)

An weiteren Begriffen soll unterschieden werden

- RN - Rechnernetz, d.h. Verbund von Rechnern, welche Nachrichten austauschen können
- ÖN - Öffentlich vermitteltes Netz (z.B. Telefon, DATEX-P, ISDN)
- MS - Managementsystem für die Dienste
- IN - Öffentlich vermitteltes Mehrwertnetz

So kann man folgende Zusammenhänge herstellen:

RN + VAS => VAN

VAN + ÖN + MS => IN

Die folgende Abb. 1 soll die Struktur eines Intelligenten Netzes veranschaulichen

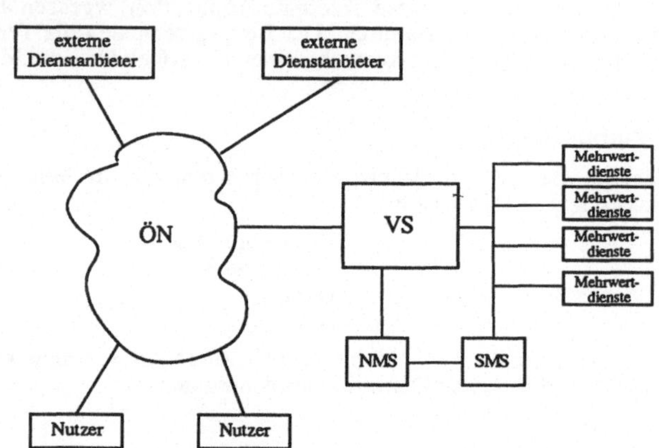

Abbildung 1:

Mehrwertdienste mit Ihrer Einbettung in ein öffentlich vermitteltes Intelligentes Netz

ÖN: Öffentliches Netz
VS Vermittlungssystem
SMS: Service Management System
NMS: Network Management System

Die in Abb. 1 dargestellte Struktur eine Intelligenten Netzes stellt die Vereinigung von zwei völlig unterschiedlichen Netzwelten dar, nämlich den

- offenen Rechnernetzen privater Betreiber (DV, LAN, WAN)
- Öffentlich vermittelten Netzen der Postgesellschaften (Sprache, Daten)

2 Architektur von Rechnernetzen

Betrachtet man den Aufbau der offenen Rechnernetze, so stellt man unabhängig von der jeweiligen Protokollarchitektur die folgenden Entwicklungsphasen fest, wie sie in Abb. 2 veranschaulicht sind [Zo 2.91]

zu a): In der ersten Phase (ca. 1970-79) griffen die Nutzer auf die angeschlossenen Hosts zu, zumeist in (Fern-)Dialog zur Inanspruchnahme von Rechenleistung (Dialog, File Transfer, RJE).

zu b): in der zweiten Phase (ca. 80-89) siedelten sich weitere Dienste an den Netzrändern an für E-Mail, DB, Conferencing-Systeme

zu c): in der dritten Phase (ca. seit 90) siedelten sich zunehmend Dienste innerhalb des Netzes an, wie z.B. Mail-Relaying, News-Distribution, Name Server sowie allgemeine Netzwerk-Managementfunktionen.

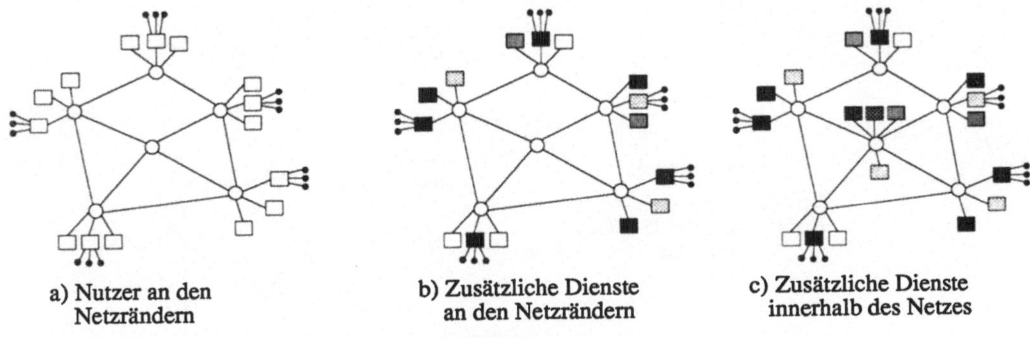

a) Nutzer an den Netzrändern

b) Zusätzliche Dienste an den Netzrändern

c) Zusätzliche Dienste innerhalb des Netzes

Abbildung 2:

Die Entwicklung von Netzdiensten in offenen Rechnernetzen

Die Architektur der offenen Rechnernetze wird seit ca. 10 Jahren bestimmt durch das von der ISO entwickelte OSI-Referenzmodell (Abb. 3)

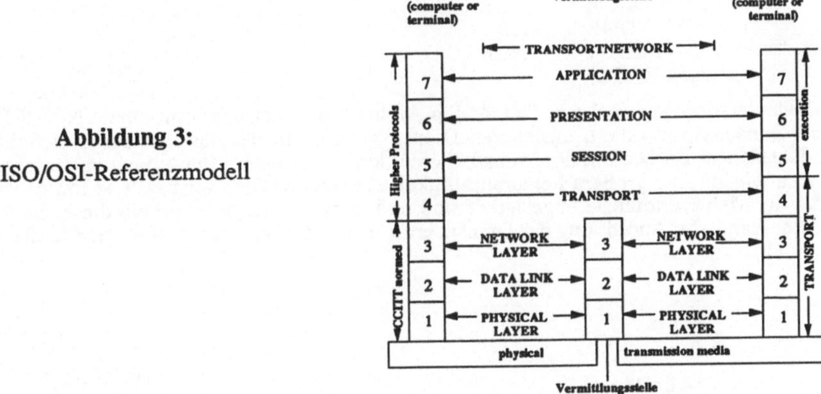

Abbildung 3:

ISO/OSI-Referenzmodell

Dieses ist weltweit inzwischen so selbstverständlich akzeptiert, daß man der damit verbundenen Gesamtprobleme erst allmählich gewahr wird. Zu den Grundproblemen zählen dabei in erster Linie

- Adresskontrolle
- Flußkontrolle
- Fehlerkontrolle

Verfolgt man diese Problematik durch sämtliche Schichten hindurch über sämtliche Knoten hinweg, wobei einem Abb. 4 eine Vorstellung von der Komplexität des Systems vermitteln soll

Abbildung 4:

Zur Veranschaulichung der Komplexität des ISO/OSI-Modells

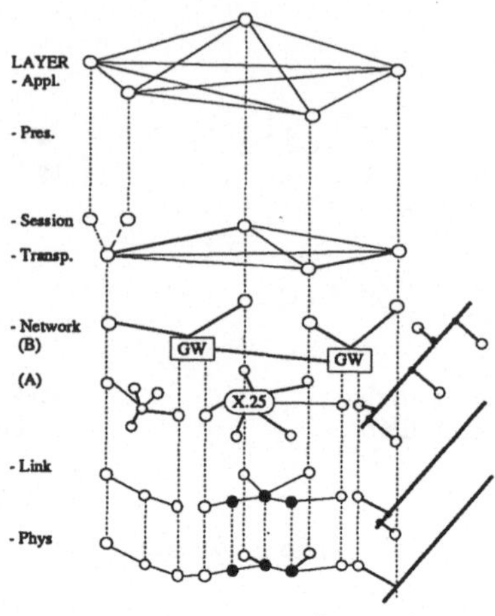

so stellt man folgende Eigenschaft fest [Zo 5.91]

Schicht	(1)	X.21:	NCR (Not Completely Reliable)
"	(2)	HDLC:	CR (Completely Reliable)
"	(3)	X.25:	NCR
"	(4)	TP4:	CR
"	(5-7)	HLP:	NCR
	z.B. E-Mail		

Dies bedeutet, daß man real eine "Sandwich-Architektur" vorfindet mit einem NCR/CR-Wechsel: man beginnt zunächst mit einem unsicheren Dienst, welcher in der darüberliegenden Schicht abgesichert wird. Die sichere Basis dient der nächsthöheren Schicht zu einem Funktionalitätssprung, durch welchen man sich wieder eine größere Fehleranfälligkeit einhandelt. Dies setzt sich so fort bis in die Schicht 7, auf der die Mehrwertdienste angesiedelt sind und setzt sich dann innerhalb dieser Schicht fort. Abb. 5 zeigt eine mögliche Anordnung der Dienste entsprechend ihrem Integrationsgrad in die Applikationen.

Abbildung 5:

Stufen der Dienstintegration

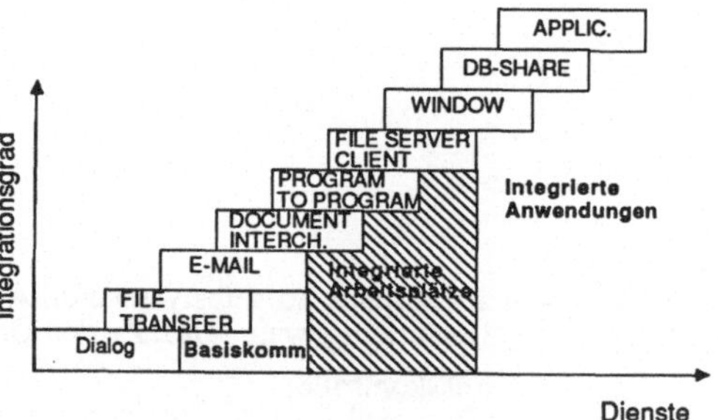

Die NCR/CR-Sandwich-Struktur läßt sich dabei sehr plausibel an dem E-Mail/EDIFACT-Dienst veranschaulichen: während der E-Mail-Dienst weltweit funktioniert und eine hohe Komplexität aufweist, ist er gleichzeitig noch viel zu unzuverlässig für kritische Anwendungen. EDIFACT muß also weitere END-TO-END-Sicherungsmechanismen bereitstellen, um die erforderliche CR-Funktionalität zu garantieren.

Abb. 6 zeigt dabei eine typische Protokollarchitektur

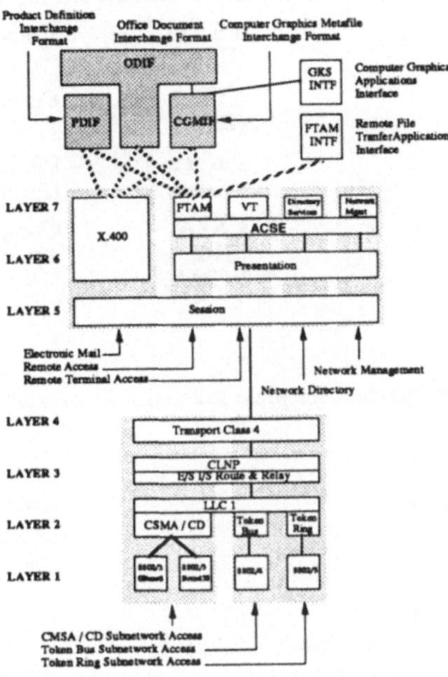

Abbildung 6:

Typische Protokollarchitektur einer Arbeitsplatzstation

3 Integrierte Rechner- und Telekommunikationsnetze

Durch die in Abb. 7 veranschaulichte Integration der öffentlichen Postdienste auf Basis ISDN und deren "Zusammenprallen" mit der Welt der Rechnernetze wird es einerseits ein konkurrierendes Angebot geben, andererseits völlig neuartige "Synergetische" Lösungen

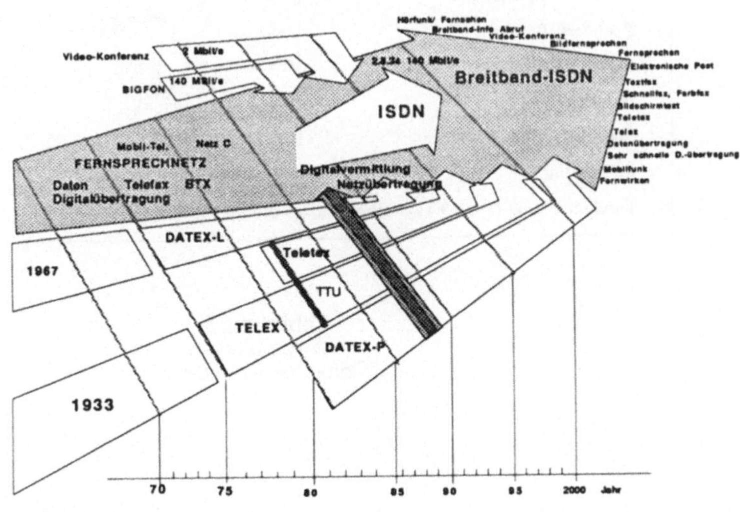

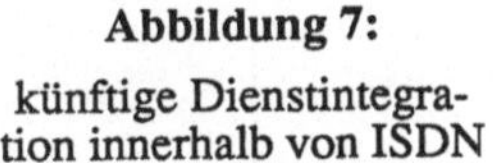

Abbildung 7:

künftige Dienstintegration innerhalb von ISDN

Folgende Bezeichnungen werden derzeit für solche "Telefon + DV"-Systeme verwendet:

VEA: (IBM)	Voice Enhanced Application
CIT: (DEC)	Computer Integrated Telephony
PaCT: (Siemens)	PBX and Computer Teaming (Private Branch Exchange)
CSTA: (SNI, ECMA)	Computer Supported Telephony Application

Die folgenden Beispiele sollen die Anwendungsmöglichkeiten exemplarisch veranschaulichen [Zo2.91]

<u>Telefon + -Anwendungen</u>

(Quelle: ENATOR)
"Computer Help Desk"

<u>Anwendung:</u>

Hilfesuchender Nutzer braucht Unterstützung durch Spezialisten

<u>Ablauf:</u>

- Nutzer ruft CSTA-System an
- Anruf wird weitergeleitet an Spezialisten
- Rechnergeführter Dialog klärt Nutzerproblem
- Spezialist erhält Diagnoseinformation
- Direkter Dialog (VOICE) zwischen Nutzer und Spezialist

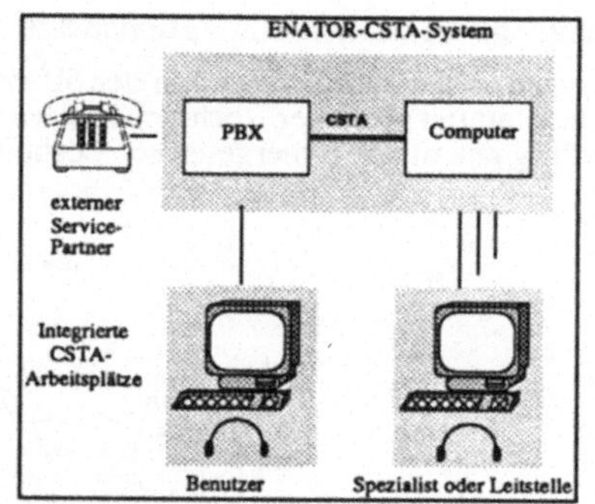

Abbildung 8:
Anwendung
"Computer Help Desk"

Weitere Telefon + -Anwendungen

(Quelle: ENATOR)

Anwendung:

Info-System von Fluggesellschaften

Ablauf:

- Kunde ruft bei AIRLINE an
- AIRLINE nimmt Anfrage entgegen über
  - Spracherkennungseinheit
  - Tasteneingabe am Telefon (MF)
  - "natürlicher" Mitarbeiter
- AIRLINE antwortet über
  - synthetisches Ausgabesystem
- Kunde kann sich zu
  - Buchungssystem
  oder
  - "natürlichem" Mitarbeiter
  weitervermitteln lassen
- Kunde bucht Flug

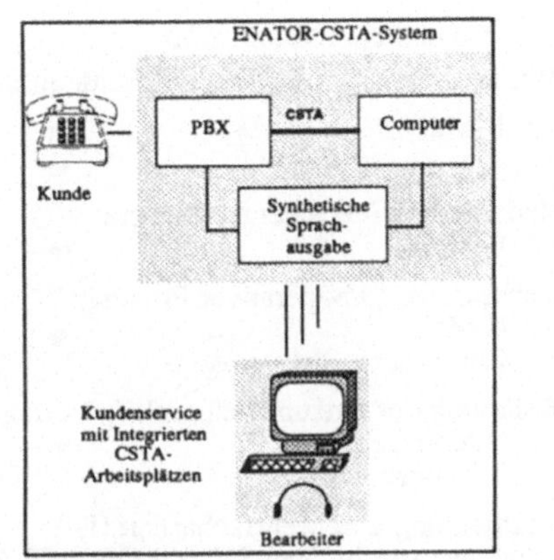

Abbildung 9:
"Info-System von Fluggesellschaften"

Beispiele für weitere Anwendungen sind

- Auskunftsdienste der DBP
- Branchenauskünfte
- Tele-Marketing (evtl. KI-gestützt)
- Mahnwesen
- Anrufbeantwortung
- Telefon-Operator

4 Marktentwicklungen

Marktforschungen zufolge handelt es sich bei dem VAS-Markt um eine der am stärksten wachsenden Sparten der DV-Branche mit geschätzten 20 Mrd US$-Umsatz in 92 [Gut90]. Dementsprechend hart umkämpft sind die Marktanteile.

Abb. 10 veranschaulicht die strategischen Vorstöße

Abbildung 10:
Marktentwicklungen im VAS-Bereich

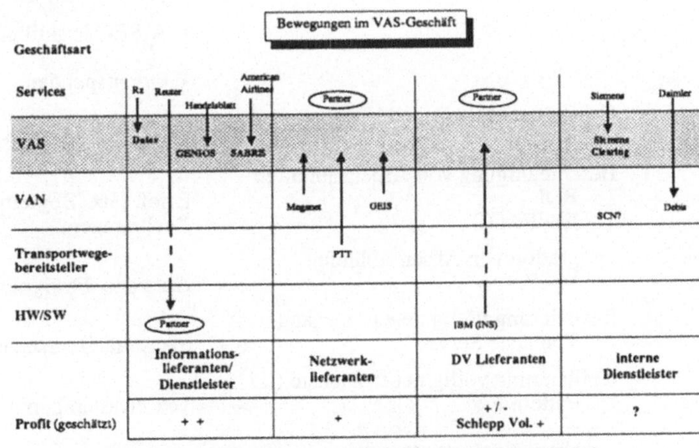

5 Nutzen

Die folgende stichwortartige Übersicht soll die potentiellen Nutzenkategorien für VAS den möglichen Gefahren gegenüberstellen [Zo2.91]

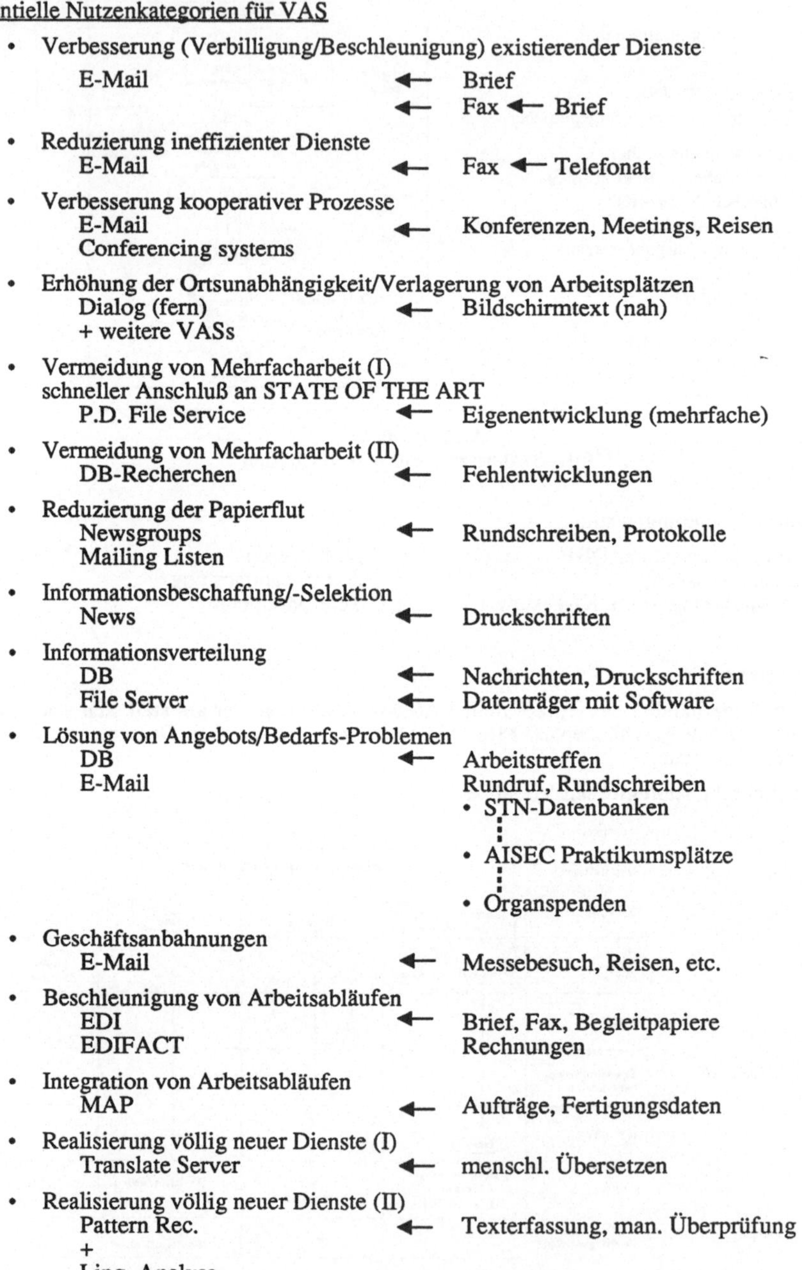

Potentielle Nutzenkategorien für VAS

- Verbesserung (Verbilligung/Beschleunigung) existierender Dienste
  E-Mail ← Brief
  ← Fax ← Brief
- Reduzierung ineffizienter Dienste
  E-Mail ← Fax ← Telefonat
- Verbesserung kooperativer Prozesse
  E-Mail ← Konferenzen, Meetings, Reisen
  Conferencing systems
- Erhöhung der Ortsunabhängigkeit/Verlagerung von Arbeitsplätzen
  Dialog (fern) ← Bildschirmtext (nah)
  + weitere VASs
- Vermeidung von Mehrfacharbeit (I)
  schneller Anschluß an STATE OF THE ART
  P.D. File Service ← Eigenentwicklung (mehrfache)
- Vermeidung von Mehrfacharbeit (II)
  DB-Recherchen ← Fehlentwicklungen
- Reduzierung der Papierflut
  Newsgroups ← Rundschreiben, Protokolle
  Mailing Listen
- Informationsbeschaffung/-Selektion
  News ← Druckschriften
- Informationsverteilung
  DB ← Nachrichten, Druckschriften
  File Server ← Datenträger mit Software
- Lösung von Angebots/Bedarfs-Problemen
  DB ← Arbeitstreffen
  E-Mail   Rundruf, Rundschreiben
  - STN-Datenbanken
  - AISEC Praktikumsplätze
  - Organspenden
- Geschäftsanbahnungen
  E-Mail ← Messebesuch, Reisen, etc.
- Beschleunigung von Arbeitsabläufen
  EDI ← Brief, Fax, Begleitpapiere
  EDIFACT   Rechnungen
- Integration von Arbeitsabläufen
  MAP ← Aufträge, Fertigungsdaten
- Realisierung völlig neuer Dienste (I)
  Translate Server ← menschl. Übersetzen
- Realisierung völlig neuer Dienste (II)
  Pattern Rec. ← Texterfassung, man. Überprüfung
  \+
  Ling. Analyse

- Realisierung völlig neuer Dienste (III)
  rechnergeführter Dialog ⋮← menschl. Auskunftssystem
  (X)

- VAS als strategische Waffe (I)
  rechnergeführte Auftragsabw. ← Lieferantenneutrale Auftragsabwicklung

- VAS als strategische Waffe (II)
  rechnergestütze Werbung ← tradit. Werbung Marketing

- VAS als kriminelle Waffe (I)
  Bargeldautom. ← Banküberfall
  elektronische Transfers ← Bargeldtransfers (schwarz/weiß)

- VAS als kriminelle Waffen (II)
  Ausspähen von Daten ← Industriespionage
  Vernichten/Man. von Daten ← Sabotage

- VAS als kriminelle Waffen (III)
  Verletzen des Datenschutzes ← Bespitzelung

- VAS als kriminelle Waffe (IV)
  SW-Diebstahl über Netz ← SW-Diebstahl (phys.)

6 Zusammenfassung

Angesichts der aufgezeigten technischen Entwicklungen, der Marktentwicklungen sowie der auf Anwenderseite vorhandenen Nutzenpotentiale läßt sich eine sinnvolle Vorgehensweise für den Anwender wie folgt angeben

- Unternehmerische Strategie
- VAS-Szenario '95
- Ist-Analyse
- Kosten/Nutzen-Analyse
- DV-Migrationskonzept

Besondere Probleme bestehen dabei in folgenden Bereichen

- OSI versus TCP/IP
- Wettbewerbssituation
- eigener KNOW HOW-Aufbau
- Gesamtentwurf

sowie insbesondere

- Netzwerk-Management
  - Konfiguration
  - Leistung
  - Fehler
  - Abrechnung
  - Sicherheit

Das Hauptproblem wird jedoch in der Vereinigung der "unterschiedlichen Welten" von Rechnernetzen und Telekommunikationsnetzen liegen, wie es die folgende Tabelle veranschaulicht:

Tabelle 1.:
Rechnernetze vs. Postnetze

	Rechnernetze		Postnetze	
	TCP/IP	OSI	ÖN	PN
• Standards	RFC	ISO	CCITT	
• Adreßvergabe				
• USER	lokal	?		
• Netz	"NIC"		zentral	
• Plugs				
• Netzimplem.	dezentr.	dezentr.	zentral	
• Netzadmin.	dezentr. + zentral "NOC"		zentral	
• VAS-Anbieter	dezentr.	dezentr.	zentr./dezentral	

Hierbei wird die Problematik im Bereich "Netzwerkmanagement" die größte Herausforderung darstellen.

Literatur

|Gut90| Guttmann, Jürgen "Value Added Services
VAS-Projekt, Siemens AG,
München 1990

|Zo2.91| Zorn, Werner "Mehrwertdienste in Netzen"
Tutoriumsbeitrag auf der
KIVS'91, Universität Mannheim, 19.2.91

|Zo5.91| Zorn, Werner "Offene Rechnernetze-Aufbau und Beherrschbarkeit"
Vortragsfolien, Universität Hagen, 28.5.91

Verteilte Dokumente in offenen Netzen

H. Bunz, H. Fanderl, J. Kämper, B. Paul[1]

IBM Deutschland
European Networking Center
Tiergartenstr. 8
D-6900 Heidelberg
e-mail: { bunz, fanderl, kaemper, paul } at DHDIBM1.bitnet

Kurzfassung

Die Büroautomatisierung in heterogenen Systemen basiert auf internationalen Standards wie ODA und DFR. Diese ermöglichen ein gemeinsames Verständnis von Applikationen über den Inhalt und die Form von Dokumenten beziehungsweise über Operationen auf derartigen Objekten wie den entfernten Zugriff auf Dokumentenspeicher. Viele Anwendungsszenarien motivieren Konzepte und Dienste, die es erlauben, Dokumente auf Netzwerke zu verteilen und auf einzelne Teile der Dokumente entfernt zugreifen zu können. In dieser Hinsicht ist die Funktionalität heutiger Normen unzureichend - diese erlauben nur die Handhabung vollständiger Dokumente und ermöglichen nicht, daß die Verteilung eines Dokuments auf verschiedene Netzwerk-Knoten für Anwendungen transparent gehalten wird. Erfahrungen aus Prototyp-Entwicklungen im Rahmen von europäischen Kooperationsprojekten werden in diesem Beitrag vorgestellt. Für die daraus abgeleiteten notwendigen Erweiterungen der involvierten Normen werden Konzepte vorgeschlagen.

Abstract

Office automation in open systems is based on international standards like ODA and DFR. These standards provide heterogeneous applications with a common understanding of the format and content as well as a definition of operations on documents, e.g. remote access to a document store. Current standardization covers only the processing of complete documents. However, in many office scenarios the need arises to distribute documents over a network and to access and process parts of these documents. Prototyping work in the framework of European projects results in necessary extensions to the current standards.

1. Einleitung

Im Mittelpunkt vieler Büroanwendungen stehen elektronisch verarbeitbare Dokumente. Dokumente im Sinne dieses Beitrages sind Mengen strukturierter Informationen, die der menschlichen Wahrnehmung zugänglich sind. Beispiele sind Geschäftsbriefe, Bilanzen, Rechnungen, Normen, Patente und Gesetzestexte. Ursprünglich wurden elektronische Dokumente lediglich von lokalen Anwendungen erstellt und bearbeitet. Der Austausch geschah in Form von Papierdokumenten.

---

1 Dieser Beitrag wurde teilweise durch das ESPRIT Projekt Nr. 2374 der Europäischen Gemeinschaft gefördert.

Neuere Anwendungsszenarien wie die elektronische Post oder Bildschirmtext sind dadurch gekennzeichnet, daß elektronische Dokumenta zwischen verschiedenen Anwendungen ausgetauscht werden, oder daß von verschiedenen Anwendungen auf sie zugegriffen wird. Grundlagen derartiger Anwendungen sind ein gemeinsames Verständnis über die behandelten Dokumente und geeignete Kommunikationsdienste etwa zum Versenden dieser Dokumente. Die internationalen Standardisierungsgremien ISO und CCITT haben entsprechende Normen entwikkelt. DOAM (Distributed Office Applications Model) [1] beschreibt eine grundlegende Struktur für verteilte Büroanwendungen. X.400 [2] stellt einen Dienst zum Versenden elektronischer Dokumente dar. Der sich entwickelnde Standard für *Document Filing and Retrieval* (DFR) [3] ermöglicht den Zugriff auf entfernt gespeicherte Dokumente. Neben derartigen Kommunikationsdiensten gibt es auch Normen, die ein gemeinsames Verständnis über die mit Hilfe dieser Dienste übermittelten Informationen ermöglichen. Mit Bezug auf Dokumente ist hier insbesondere die Architektur ODA (Open Document Architecture) [4] zu nennen.

Die obigen Normen fassen Dokumente als isolierte, vollständige Informationseinheiten auf. Für zukünftige Anwendungen wird es notwendig sein, Dokumente logisch in Teile zu zerlegen und diese auf verschiedenen Knoten in einem Netzwerk zu speichern. Diese Beobachtung soll anhand von zwei Beispielen näher erläutert werden.

- Ein Hypermediasystem, das zur Verwaltung sehr umfangreicher Dokumente wie zum Beispiel eines Lexikons geeignet ist, muß die Möglichkeit beinhalten, die Dokumente auf mehrere Knoten in einem Netzwerk zu verteilen, und den Anwendungen eine von dieser Verteilung unabhängige Sicht auf die Dokumente gestatten.
- Das verteilte Bearbeiten von Dokumenten (*joint editing*) kann nur dann sinnvoll realisiert werden, wenn verschiedene Teile eines Dokuments für Anwendungen gesperrt werden können, und diese Teile lokal bei den entsprechenden Anwendungen gespeichert werden.

Diese und eine Vielzahl weiterer Anwendungen motivieren also neue Konzepte, die es gestatten, Dokumente logisch zu zerlegen und dann auf verschiedene Knoten eines Netzwerkes zu verteilen. In heterogenen Umgebungen sollten diese Konzepte im Rahmen internationaler Standards entwickelt werden.

Im Mittelpunkt dieses Beitrages stehen Erweiterungen der existierenden Kommunikationsdienste und Dokumentenarchitekturen, die zur Realisierung verteilter Dokumente geeignet sind. Zum einen werden Ansätze untersucht, die es erlauben, ODA Dokumente in Teile zu zerlegen und die Beziehungen zwischen diesen Teilen zu beschreiben. Zum anderen wird ein Mechanismus zur verteilten Speicherung von Dokumenten entwickelt, der auf dem DFR Standard basiert. Dieser Standard wird konform zu DOAM durch Server-Server Protokolle derart erweitert, daß Teile eines Dokuments auf unterschiedlichen Knoten eines Netzwerkes verwaltet werden können. Die DFR Funktionalität in Bezug auf den Dokumentenzugriff und das Dokumenten-Retrieval bleibt dabei erhalten, die Verteilung der Dokumente erfolgt transparent für die Anwendungen. Die beschriebenen Untersuchungen, aber auch Überlegungen zur Integration multimedialer Informationen wie Audio und Video machen deutlich, daß ein neues, sehr flexibles Konzept zur Informationsmodellierung notwendig ist. Diese Aussage wird in dem Beitrag näher motiviert. Außerdem wird ein derartiges Konzept vorgestellt.

Im Rahmen des durch die Europäische Gemeinschaft geförderten ESPRIT Projektes PODA-2[2] wurden bereits einige der in dieser Arbeit vorgestellten Konzepte als Prototypen realisiert. So wurden innerhalb von PODA-2 neben Prototypen, die den Austausch von Dokumenten in offenen Netzen unterstützen [5], auch eine Implementierung des DFR Standards realisiert. Über die daraus resultierenden Erfahrungen und zukünftige Projektpläne wird in diesem Beitrag ebenfalls berichtet.

In dem folgenden Abschnitt behandeln wir zunächst die oben erwähnte DFR Implementierung. Die Abschnitte 3 und 4 beinhalten die Modellierung verteilter Dokumente und die Erweiterun-

---

2 PODA-2 (Piloting the Open Document Architecture) beschäftigt sich mit der Verbreitung und den Erweiterungen von ODA. Die Partner dieses Projektes sind: British Telecom, BULL, IBM, ICL, Nixdorf, Océ, Olivetti, Siemens, TITN, University College London.

gen des DFR Standards, die notwendig sind, um Dokumententeile auf verschiedenen Knoten eines Netzwerkes zu verwalten. Ein neues Konzept zur Informationsmodellierung wird in Abschnitt 5 motiviert und vorgestellt. Eine Zusammenfassung, die auch einen Ausblick auf zukünftige Prototypentwicklungen enthält, rundet den Beitrag ab.

2. *DFR: Entfernter Zugriff auf komplette Dokumente*

Wie bereits in der Einleitung ausgeführt wurde, entsteht aufgrund der Anforderungen verschiedenster Applikationen die Notwendigkeit, Informationen über verschiedene Netzwerkknoten zu verteilen. Dies kann u.a. aus folgenden Gründen geschehen:

- Verbindung von aus organisatorischen Gründen verteilt gespeicherter Information
- Einbindung spezialisierter Server (z.B. Video-, Audioserver, Datenbanken)
- Redundanz identischer Information für höhere Verfügbarkeit oder zur Minimierung der Kommunikationskosten
- Modularität für gleichmäßiges Wachstum
- Paralleles Abarbeiten unabhängiger Teiloperationen (z.B. Suchfunktion)

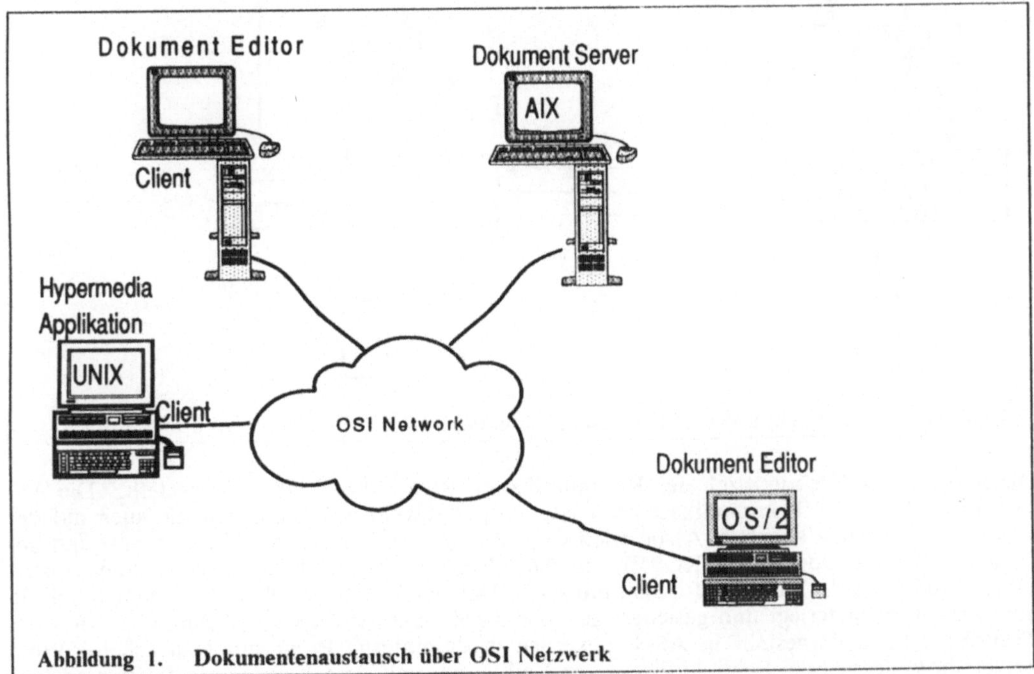

Abbildung 1. Dokumentenaustausch über OSI Netzwerk

Daraus ergibt sich die Anforderung, daß auf Informationen in einem offenen, heterogenen Netzwerk zugegriffen werden kann. Dazu müssen Dokumente in entfernten Dokument-Speichern abgelegt und über das Netzwerk ausgetauscht, sowie deren Inhalt an verschiedenen Knoten im Netzwerk verarbeitet werden können. Um die dabei entstehenden Probleme untersuchen und weitergehenden Anforderungen gerecht werden zu können, wird im Rahmen des bereits erwähnten ESPRIT-Projektes PODA-2 ein Prototyp implementiert, der auf den ISO-Normen DOAM, DFR und ODA basiert.

Die Hardware Architektur der Prototyp-Entwicklung ist in Abbildung 1 dargestellt. Eine Hypermedia-Applikation sowie verschiedene andere Applikationen verwenden DFR Clients, um die Dokumente in einem Dokumenten-Speicher abzulegen und wiederzufinden. Die Kommunikation zwischen diesen Clients und dem Server, wie der Austausch von Dokumenten, basiert auf OSI Diensten. Die OSI Software wird auf dem Entwicklungssystem ISODE [6] realisiert und läuft unter AIX.

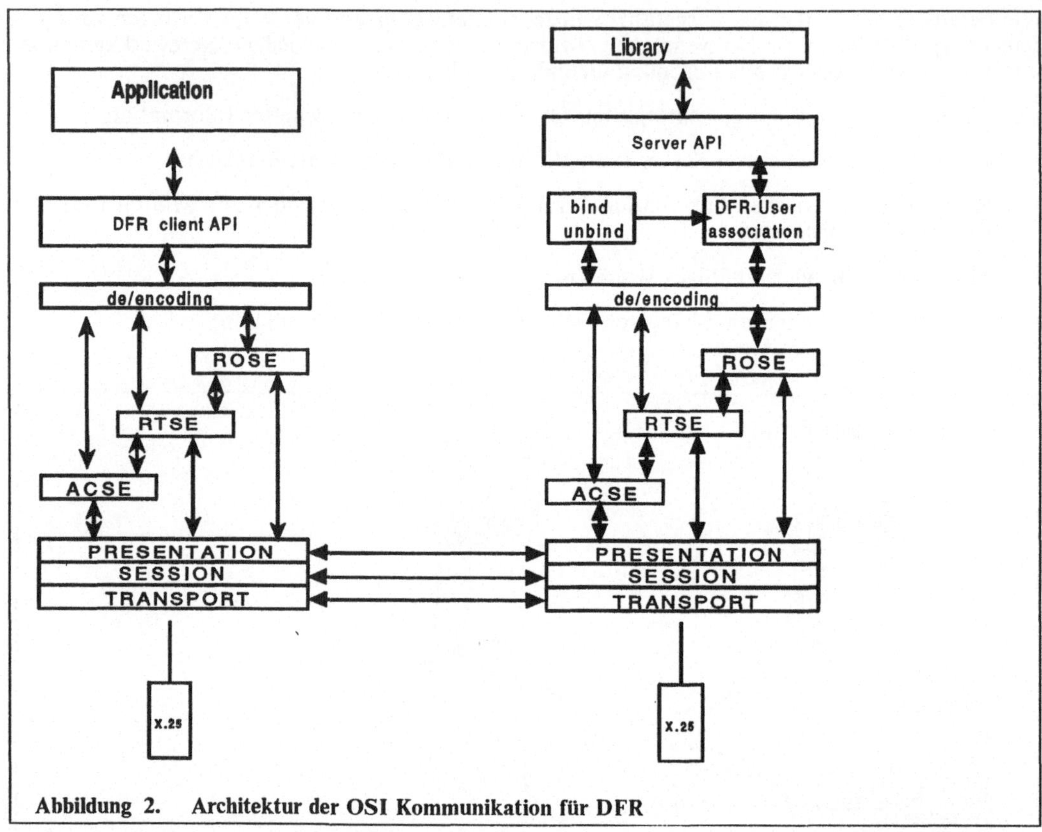

Abbildung 2. Architektur der OSI Kommunikation für DFR

Die Software-Architektur bzgl. der Kommunikation ist in Abbildung 2 dargestellt. Die Anwendungsschicht der Kommunikationsdienste enthält sowohl auf der Client als auch auf der Server-Seite folgende Dienste: Association Control (ACSE), um eine Verbindung auf- und abzubauen, Remote Operations (ROSE), um die DFR-Operationen beim Server-Knoten auszuführen, und Reliable Transfer (RTSE), um große Dateneinheiten, wie den Dokument-Inhalt, in mehreren Blöcken zerlegt und gesichert zu übertragen. Die Anfragen und Antworten aus der Transfer Syntax, dargestellt in ASN.1, werden in das interne Programmiersprachen-Format kodiert bzw. dekodiert. Beim Aufbau einer Verbindung mit dem Kommando DFR-*bind* über ACSE wird auf der Server-Seite zunächst ein Authentisierungs-Test durchgeführt. Dabei wird festgestellt, ob der Benutzer authorisiert ist, Operationen auszuführen. Diese Prüfung wird von einem Authentication & Security Service ausgeführt, der seine Informationen mit Hilfe eines Directory Service (X.500) abruft. Anschließend kann eine DFR-Benutzer-Assoziation etabliert werden, so daß nachfolgende DFR-Kommandos über ROSE direkt an das Server API weitergeleitet werden. Sie werden dann auf die Operationen eines konkret zur Verfügung stehenden (lokalen) Dokumenten-Verwaltungssystems (*library*) abgebildet. Die Möglichkeiten, die für die Architektur eines Dokumenten-Verwaltungssystems gegeben sind, sowie die Frage, welche bestehenden Systeme dafür als Basis verwendet werden können, sollen in diesem Papier nicht erörtert werden. Ein Beispiel, wie man die DFR-Operationen auf der Basis eines Non-Standard-Datenbanksystems realisieren kann, findet man in [7].

Ein wichtiger Ansatz für die Implementierung der DFR-Norm ist es, ein allgemeines Programmiersprachen API für den DFR Client zu entwickeln. Es soll leicht handhabbar sein und die Anwendungsprogramme von der sehr allgemein gehaltenen und umfangreichen Schnittstelle, wie sie in der Norm angegeben ist, befreien. Ziel ist es, daß die wichtigen Anwendungen mit den häufigen Aufruf-Folgen die DFR-Funktionalität einfach benutzen können. Das Client API wird dazu in mehrere Funktionsgruppen aufgeteilt, die die verschiedenen Phasen der Bearbeitung widerspiegeln (s. Abbildung 3):

- Initialisierung und Bearbeitung einer Anwendungsumgebung
- Auf- und Abbau der DFR-Benutzer Assoziation
- Ausführung der DFR-Operationen (Daten-Austausch)
- Abruf der Ergebnisse von asynchron angestoßenen Operationen

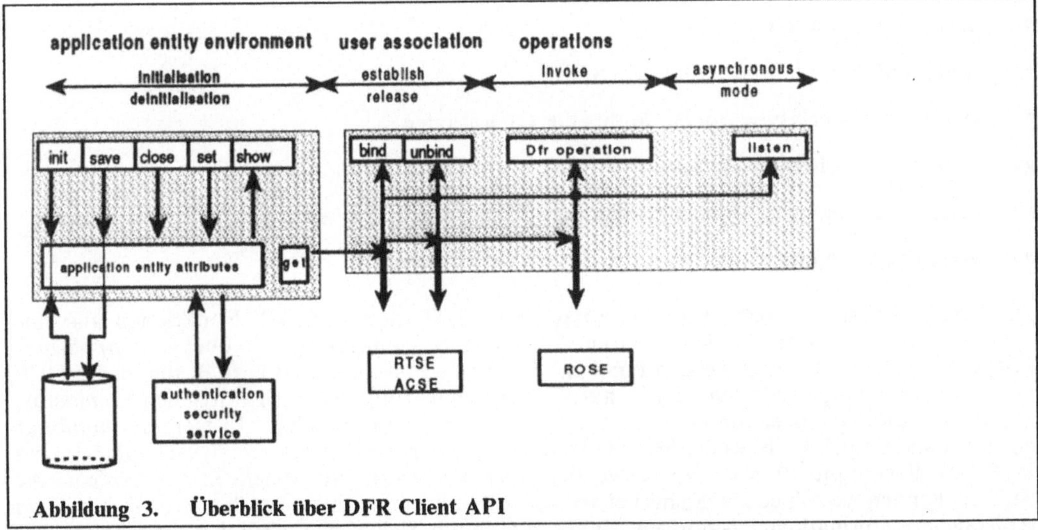

Abbildung 3. Überblick über DFR Client API

In dem *Application Entity Environment* werden die Parameter gesetzt und verwaltet, die häufig von verschiedenen Operationen benutzt werden, die in Operationen im Regel-Fall nicht benutzt werden oder die für eine Folge von Aufrufen unverändert bleiben sollen. Beim Aufruf einer DFR-Operation werden die fehlenden Parameter aus dem *Application Entity Environment* gelesen und der dann komplette Satz von Parametern kodiert und übertragen. Die Anwendungsumgebung kann für spätere Wiederverwendung permanent gespeichert (*save*) und zu Beginn einer Sitzung eingelesen (*init*) werden. Außerdem können die Attribute gesetzt (*set*), gelesen (*show*) und freigegeben (*close*) werden.

Das gesamte Spektrum an Möglichkeiten, die der DFR Service bietet, wird in mehrere Funktionsstufen (*functional profiles*) zerlegt. Damit wird einerseits eine größere Flexibilität erreicht, andererseits können Systeme mit unterschiedlichen Eigenschaften und Anforderungen in einem Gesamtnetz verwendet werden. Im DFR Standard sind bereits 3 Stufen festgelegt worden, die im wesentlichen das Informationsmodell betreffen: es gibt einen flachen Dokumenten-Speicher (nur eine Gruppe für den ganzen Dokumenten-Speicher), das Schema der Gruppenhierarchie ist von einem Administrator vordefiniert oder die Benutzer können beliebig Gruppen anlegen und manipulieren.

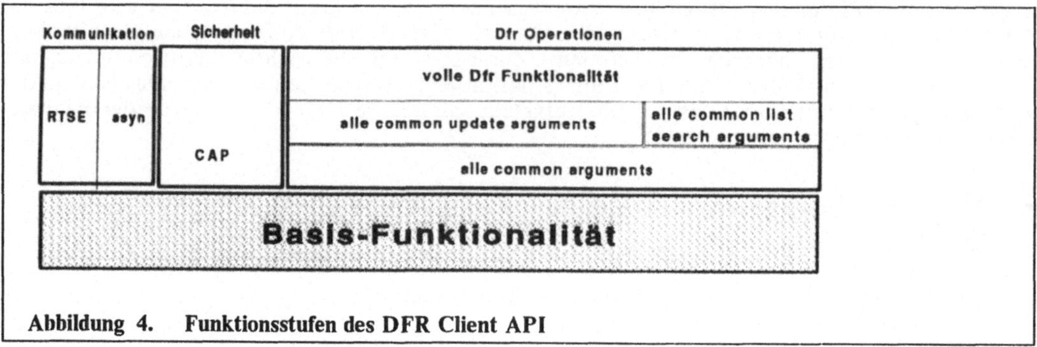

Abbildung 4. Funktionsstufen des DFR Client API

In dem Prototyp werden weitere Stufen eingeführt (s. Abbildung 4):

- Basis-Funktionalität
- Kommunikation mit oder ohne RTSE
- Synchroner oder asynchroner Modus der Operationen
- Zusätzliche Sicherheitseigenschaften, z.B. *control access package* (CAP)
- Benutzung von verschiedenen Teilen der allgemeinen Argumente der DFR-Operationen
- Volle Funktionalität

Die Möglichkeiten der Basis-Funktionalität sind direkt als Client-API-Funktionen aus einer Programm-Bibliothek zugänglich, während die weiteren Funktionen nur über das *Application Entity Environment* benutzt werden können. Die Basis-Funktionalität enthält die in der DFR-Norm als minimal geforderten Eigenschaften. So können nur die obligatorischen (*mandatory*) Argumente und ein sogenanntes *basic attribute set* verarbeitet werden. In weiteren, unabhängigen Funktionsstufen gibt es die Möglichkeit, den in der Norm als optional bezeichneten Dienst RTSE im Kommunikationsteil zu verwenden oder die Operation asynchron aufzurufen. Außerdem können verschiedene Sicherheitsstufen eingeführt werden, die den Austausch unterschiedlicher Informationen benötigen, damit bestimmte Sicherheitsrichtlinien eingehalten werden. Schließlich können neben den obligatorischen auch die optionalen Argumente der DFR-Operationen bearbeitet werden. Die Argumente können unterschieden werden in allgemeine Argumente für alle Operationen, die Argumente der Suchoperationen und die Argumente der Änderungsoperationen. Eine genaue Beschreibung des DFR-Client-API findet sich in [8].

Die DFR-Prototyp-Implementierung dient als Grundlage verschiedener Erweiterungen. Mit der bisher beschriebenen Funktionalität des Prototyps können nur komplette ODA Dokumente ausgetauscht werden, außerdem besteht zwischen den Servern keine direkte Verbindung. Um den Anforderungen der Applikationen - siehe u.a. die Szenarios Hypertextsysteme und Joint Editing in der Einleitung - zu genügen, müssen Dokumente jedoch auch zerlegt und verteilt gespeichert werden können und es muß Verweise (Referenzen) zwischen ihnen geben. Um partielle Dokumente sinnvoll zu definieren sind neben der Erweiterung der Informationsarchitektur ODA um den Begriff der partiellen Dokumente, auch die entsprechenden Austauschformate, Kommunikationsprotokolle und Speicherungsmodelle zu erweitern. Um eine Zusammenarbeit zu gewährleisten, sind diese Erweiterungen konsistent durchzuführen (z.B. gemeinsamer Referenzierungsmechanismus). Diese Problematik und die daraus folgenden Erweiterungen werden in den beiden folgenden Kapiteln behandelt: zunächst unter dem Blickwinkel der Dokumenten-Architektur und anschließend aus der Sicht der Kommunikationsdienste.

3. Partielle Dokumente

Im Mittelpunkt dieses Abschnitts steht die Modellierung von partiellen Dokumenten und Referenzen zwischen Dokumententeilen. Zunächst wollen wir aber einen Überblick über weitere Aktivitäten geben, die ebenfalls die Erweiterung des ODA Standards zum Ziele haben, und verdeutlichen, daß partielle Dokumente für eine Vielzahl dieser Erweiterungen ein notwendiges Basiskonzept darstellen.

Motiviert durch unterschiedliche Anwendungen werden augenblicklich sehr viele potentielle Erweiterungen zu ODA in den Standardisierungsgremien diskutiert [9]. In naher Zukunft steht eine Erweiterung der ODA Norm an, die unter anderem Farbe in Dokumenten und Sicherheitsaspekte betrifft. Bezüglich des letzten Punktes wird es möglich sein, auf Teile von Dokumenten unterschiedliche Funktionalitäten festzulegen, wie beispielsweise

- das Verdecken von Teilen des Inhalts eines empfangenen Dokuments (*confidentiality*),
- die Überprüfung auf Konsistenz von Inhaltsteilen (*integrity*) oder
- die Überprüfung auf die Authentizität der Urheber eines Dokuments oder eines Dokumententeils (*authenticity*).

Weitere Beispiele zur Erweiterung von ODA, bei welchen sich im Augenblick schon konkrete Diskussionsgrundlagen herausgebildet haben und die teilweise schon als *working draft* den Gremien vorliegen, sind

- die Normierung von Daten in Dokumenten, welche als Grundlage für Spreadsheats oder Geschäftsgraphiken dienen und auf die von mehreren Stellen eines Dokuments zum Zweck der Präsentation aber auch für diverse Berechnungen zugegriffen werden kann,
- die Normierung von zeitabhängigen Medien, wie vor allem Sprache aber auch Bewegtbilder,
- die Erweiterung der in ODA darstellbaren Tabellen und
- die Normierung von Dokumententeilen und Referenzen zwischen Dokumententeilen, auch verschiedener Dokumente.

Wie bereits erwähnt wurde, soll im folgenden der letzte Punkt dieser Aufzählung näher betrachtet werden. Die Fähigkeit, Dokumententeile und Referenzen zwischen diesen Teilen auszeichnen, austauschen und präsentieren zu können, ist eine Voraussetzung für viele der anderen obigen Erweiterungen. Dabei kann beispielsweise an Zugriffsbeschränkungen gedacht werden, die sich auf Teile eines Dokuments beziehen, ebenso an Daten, die wie externe Dokumententeile referenziert werden. Aber auch die Verwendung von Audio und Bewegtbildern kann die Referenzierung und Übertragung von Dokumententeilen erforderlich machen, insbesondere dann, wenn diese Teile auf spezialisierten Servern ausgelagert sind und zu Darstellungszwecken übertragen werden.

Die existierende ODA Norm stellt bereits ein Konzept für Teile von Dokumenten und externe Referenzen zur Verfügung: das *Resource* Dokument. Dabei ist es allerdings nur möglich, ausschließlich Teile der generischen Strukturen aus genau einem *Resource* Dokument in ein gegebenes Dokument einzufügen. Dieses Konzept ist geeignet, um bei der Übertragung von Dokumenten auf häufig benutzte, unveränderliche Teile zu verzichten und diese aus vorher übertragenen Dokumenten zu extrahieren. Beispiele hierfür sind Firmenlogos oder Briefköpfe. Sicherlich kann das Konzept des *Resource* Dokuments als ein erster Schritt zur Definition von Dokumententeilen dienen. Aus den angeführten Gründen, insbesondere aufgrund der Beschränkung auf Teile der generischen Strukturen, ist allerdings offensichtlich, daß ein verallgemeinertes Extraktionskonzept entwickelt werden muß.

Als Grundlage für ein derartiges, verallgemeinertes Konzept kann ein *Working Draft* zur Darstellung von ODA Erweiterungen zur Realisierung von „Partiellen Dokumenten und externen Referenzen zwischen Dokumenten" von C. und U. Bormann [10] dienen. In diesem Vorschlag zur ODA Erweiterung wird die Möglichkeit geschaffen, Unterbäume der logischen Struktur oder der Layout Struktur eines Dokuments nicht mit zu übertragen, sondern diese aus referenzierten Dokumenten zu importieren. Notwendige Voraussetzung dazu ist, daß der Empfänger diese referenzierten Dokumente besitzt oder zumindest auf diese zugreifen kann und daß darin genau die referenzierten Teile zur Verfügung gestellt werden. Bei der Auflösung einer Referenz (Resolution) wird diese durch die entsprechenden Teile des referenzierten Dokuments ersetzt, wobei stets ein gesamter Unterbaum dieses Dokuments zusammen mit allen angefügten Inhaltsteilen übernommen wird. Da möglicherweise auch in diesem Unterbaum weitere Referenzen vorkommen können, kann diese Auflösung von Referenzen als ein Prozeß angesehen werden, der möglicherweise erst nach mehreren Schritten zum Ziel, einem kompletten Dokument, führt.

Die Referenz im unvollständigen Dokument besteht aus einem eindeutigen Identifikator des externen Dokumentes und aus einem Bezeichner für das betreffende Objekt, welches, mitsamt seines Unterbaums, anstatt der Referenz in das Dokument eingebaut werden soll. Im Gegensatz dazu findet sich im referenzierten Dokument eine Tabelle, die diese Bezeichner auf die realen ODA Identifikatoren abbildet und somit eine Liste der exportierten Dokumententeile darstellt.

Dieses Konzept von partiellen Dokumenten und externen Referenzen auf Dokumente ist damit eine Erweiterung des vorher beschriebenen *Resource* Dokuments auf spezifische Dokumentenstrukturen. Ob nun diese ODA Erweiterung die Anforderungen von verschiedenen Anwendungen befriedigen kann, soll anhand der Szenarien Hypermediaserver und Joint Editing System untersucht werden.

Für die Anwendung Hypermediaserver ist die Einführung von Dokumententeilen eine notwendige Voraussetzung, um den Aufwand beim Dokumententransfer in akzeptablen Grenzen zu halten. Hier ist es vor allem notwendig, lediglich die angeforderten Teile des auf einem Server vorliegenden Dokuments, möglicherweise auch unvollständige Unterbäume, in effizienter Art zu übertragen. Die Definition eines partiellen Dokuments als vollständiger Unterbaum der spezifischen, logischen Struktur oder der Layout Struktur schränkt die Verwendbarkeit des Konzepts somit ein, ist für viele Hypermedia-Anwendungen aber ausreichend. Wesentlich schwerer wiegt die Tatsache, daß der obige Vorschlag nicht die Übertragung von Dokumententeilen vorsieht, sondern nur die Versendung vollständiger Dokumente und die spätere, lokale Extraktion von Teilen erlaubt.

Bei einer Anwendung aus dem *Joint Editing* Bereich sollten genau diejenigen Teile von Dokumenten ausgetauscht werden, die sich bei einem Editiervorgang verändert haben. Hinderlich für die Verwendbarkeit des oben beschriebenen Ansatzes erweist sich in diesem Zusammenhang die notwendige Deklaration von zu exportierenden Dokumententeilen, da Austausch und Modifikation bei einer *Joint Editing* Anwendung beliebige Dokumententeile, die im übrigen nicht unbedingt mit Unterbäumen der spezifischen Strukturen identisch sein müssen, betreffen.

Für die beiden beschriebenen Applikationen sind somit folgende, über das oben beschriebene Konzept hinausgehende Forderungen an ein erweitertes Konzept für Teile von Dokumenten zu stellen:

1. Ein Teil eines Dokuments ist ein beliebiger Ausschnitt aus einer der Strukturen dieses Dokuments.

2. Ein Dokumententeil kann entweder durch einen Bezeichner (*label*), wie in dem beschriebenen Normierungsvorschlag dargestellt, oder durch die explizite Angabe der Identifikatoren aus der internen ODA Struktur referenziert werden. In diesem zweiten Fall müssen Teile eines Dokuments, die von außen referenziert werden können, nicht explizit als solche deklariert werden.

3. Die Übertragung von Dokumententeilen ist möglich.

Diese zusätzlichen Anforderungen an Dokumententeile werden augenblicklich in ein verallgemeinertes Konzept für partielle Dokumente umgesetzt und anhand eines Prototyps für ein Hypermediasystem auf Praktikabilität und Vollständigkeit erprobt.

4. *Zugriff auf verteilte Dokumente in offenen Netzen*

Der im Kapitel 2 beschriebene DFR Prototyp bietet bisher den Zugriff auf komplette Dokumente in einem lokalisierten Dokumentenspeicher. Als Verweise auf extern gespeicherte DFR-Objekte werden im Sinne von DOAM Referenzen verwendet. Diese Referenzen werden jedoch nicht vom Server verfolgt, sondern müssen vom Client aufgelöst werden. Ebenso werden Referenzen bei einer Veränderung (z.B. *move, delete*) des referenzierten Objektes nicht geändert. Für die Verteilung von Dokumenten ergeben sich aus den betrachteten Szenarien folgende Anforderungen hinsichtlich der Kommunikationsdienste:

- transparenter Zugriff auf Dokumente oder Dokumententeile im Serversystem
- mehrere Kopien eines Dokumentes oder Dokumententeiles können auf verschiedenen Servern liegen
- ein Dokument kann auf verschiedene Server verteilt sein

Im folgenden wird ein diese Anforderungen erfüllendes Konzept vorgestellt. Konform zu DOAM werden die Erweiterungen für ein verteiltes DFR durch ein Client-Server Modell mit geeigneten Server-Server Protokollen modelliert. Dies ist in Abbildung 5 dargestellt.

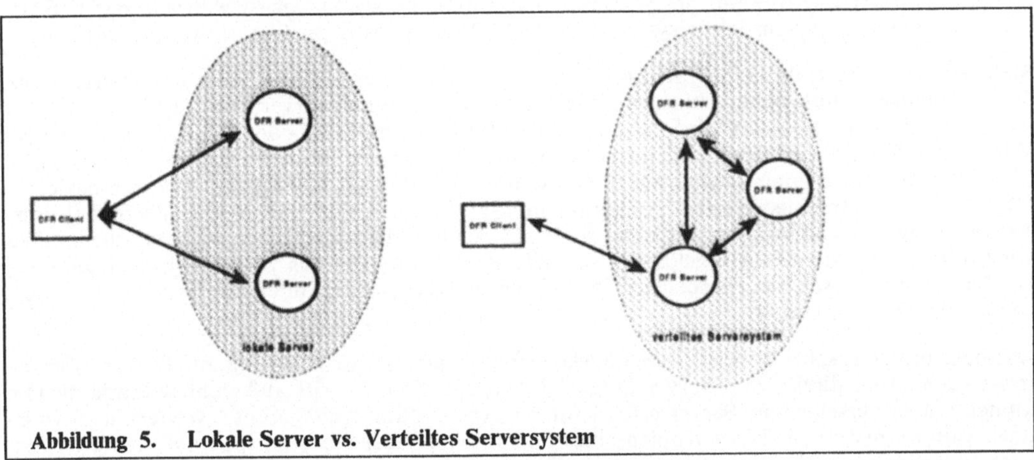

Abbildung 5. Lokale Server vs. Verteiltes Serversystem

Für die Auswahl der Server-Server Protokolle sind die Anforderungen der Applikationen an die Konsistenz der Informationsobjekte entscheidend. Für eine Vielzahl von DFR-Anwendungen z.B. beim lesenden Zugriff auf gespeicherte Dokumente in räumlich getrennten Abteilungen, oder bei der Verteilung von *Resource Dokumenten* auf mehrere Server, gelten die folgenden Annahmen:

- Leseoperationen sind wesentlich häufiger als Schreiboperationen
- zeitweilige Inkonsistenz duplizierter Daten ist akzeptabel
- konkurrierende Zugriffe auf dasselbe Objekt sind selten

Für diesen Bereich der **schwachen Konsistenz** kann der Aufwand für die aufwendigeren Transaktionsprotokolle vermieden werden. Die Server-Server Protokolle für eine verteilte Speicherung werden im folgenden ohne die Duplikation von Objekten entwickelt und dann auf den Bereich der schwachen Konsistenz ausgedehnt.

Im Serversystem werden unabhängige Objekte oder Teilobjekte abgespeichert. Alle DFR-Operationen auf einem DFR-Eintrag werden dabei als atomare Einheit ausgeführt und entsprechend den Regeln von DFR im Fehlerfall lokal, auf einem Server, zurückgesetzt. Der beauftragende Prozeß erhält dabei immer ein Ergebnis der Operation. Das Server-Server Protokoll basiert auf dem Client-Server Protokoll und es werden keine neuen Protokollelemente und Operationen eingeführt. Das Server-Server Protokoll ist symmetrisch, d.h beide Server können über eine Assoziation alle Operationen ausführen oder anstoßen. Die Operationen mehrerer DFR-Benutzer können simultan über eine Server-Server Assoziation ausgetauscht werden. Um die Sicherheit beim Zugriff sicherzustellen, werden die Zugriffsprivilegien des jeweiligen Benutzers bei jeder Operation ausgetauscht. DFR ist so konzipiert, daß der Zugriff auf ein Serversystem ohne wesentliche Änderungen des Client-Server Protokolls möglich ist. Um die Rückwärtskompatibilität zu erhalten werden diese Änderungen als neue, optionale Datenfelder eingefügt. Wenn diese Parameter nicht vorhanden sind, dann wirkt der Server wie ein lokaler Dokumentenspeicher.

Der Zugriff auf Dokumente kann über den Pfadnamen oder direkt über einen eindeutigen Identifikator (UPI, Unique Permanent Identifier) erfolgen. Für diese beiden Zugriffsmethoden werden im folgenden Konzepte für die Verteilung vorgestellt. Beim Zugriff über den Pfadnamen werden analoge Konzepte wie bei X.500 [11] verwendet. Enthält ein Pfadname eine Referenz auf einen externen DFR Server, dann wird diese Referenz verfolgt, und die DFR Operation auf dem referenzierten Objekt ausgeführt. Beim direkten Zugriff über den UPI muß der UPI im Serversystem eindeutig sein. Dies wird durch eine Namenskonvention gegeben. Bei der *create* Operation wird lokal ein eindeutiger Identifikator vergeben. Um die globale Eindeutigkeit zu erhalten wird der Identifikator des Servers hinzugefügt. Dieser Server ist während der Lebensdauer des Dokumentes im Serversystem für die Auflösung des Identifikators verantwortlich.

Stellt ein Server fest, daß eine Operation entfernt auszuführen ist, dann wird die Operation im Server solange unterbrochen, bis das Ergebnis vom entfernten Server zur Verfügung gestellt wird. Das gesamte Resultat wird dann an den anfordernden Server oder Client zurückgegeben. Wird das referenzierte Objekt verändert, gelöscht oder bewegt, dann werden in der bisherigen Norm die Referenzen nicht aktualisiert. Mit diesem Ansatz sind Referenzen vom referenzierten Objekt separiert. In einem verteilten Dokumentenspeicher ist dies jedoch nicht akzeptabel. Eine Veränderung des referenzierten Objektes sollte sich auch in den Referenzen wiederspiegeln. Dies wird durch eine Attributliste gelöst, in der alle Referenzen auf das DFR Objekt eingetragen werden. Die in dieser Liste gespeicherten Referenzen werden bei Änderung des DFR Eintrages durch den Server gewartet.

Mit dem bisher beschriebenen Verteilungskonzept können als Erweiterung zur DFR-Norm externe Referenzen direkt vom Server aufgelöst werden. Der Zugriff auf nicht redundante Dokumente über verschiedene Server erfolgt für die Anwendung transparent. Weiterhin ist es jedoch wünschenswert, mehrere Kopien eines Dokumentes oder einer Gruppe auf mehreren Servern zu speichern. Die Verwaltung der Kopien wird durch einen Attributsatz analog zur *user-reference* [3] ermöglicht. Die Kopien eines DFR-Eintrages bilden einen Baum und damit kann für die verteilt gespeicherte Information ein hierarchisches Sperrkonzept verwendet werden. Es existiert ein eindeutiges Original eines DFR-Eintrages im Serversystem. Eine Operation auf diesem Objekt wird an allen Kopien ausgeführt. Es liegt in der Verantwortung des Servers, die Veränderung auf allen Kopien des betreffenden Objektes auszuführen. Alle kritischen Operationen wie *reserve, unreserve, modify* müssen über das Original ausgeführt werden. Bei den Operationen *delete, move* werden die Referenzen in den Attributlisten aktualisiert oder der ganze Baum gelöscht. Die unkritischen Operationen *read, copy, list, search* können auf jede Kopie angewandt werden. Die Applikation kann jedoch immer einen Zugriff auf das Original verlangen.

Ein analoger Mechanismus wie bei den Kopien ermöglicht es auch, Teile eines Dokumentes als unabhängige Teilobjekte auf verschiedenen Servern abzulegen. Eine für die Applikation transparente Verteilung eines Dokumentes ist damit nicht möglich - der Server unterstützt nur die Speicherung der Teilobjekte, indem die entsprechenden Attribute verwaltet werden. Die Aufteilung eines Dokumentes in Teilobjekte ebenso wie das Zusammenfügen eines Dokumentes bleibt aber in der Verantwortung der Applikation. Alle Änderungen am Inhalt des Dokumentes muß die Applikation an den entsprechenden Teildokumenten ausführen. Für spezielle Informationsarchitekturen wie z.B. ODA ist es möglich, diese Prozesse im Server zu implementieren. Eine allgemeine Informationsmodellierung, die diese Beschränkung auf einzelne Architekturen überwindet, wird im nächsten Abschnitt vorgestellt.

5. *Integrierte Informationsmodellierung*

Der bereits betrachtete, internationale Standard ODA ist eine Architektur, die die Modellierung von Dokumenten und insbesondere der zur Präsentation eines Dokuments notwendigen Informationen betrifft. Ein weiteres Beispiel für eine derartige Informationsarchitektur ist EDI. Im Gegensatz zu ODA, das die Beschreibung von Dokumenten ermöglicht, bezieht sich EDI auf die Modellierung formatierter Daten. Die Notwendigkeit, neue, andere Aspekte beleuchtende Architekturen zu definieren, wird mit der Komplexität verteilter Anwendungen wachsen. In diesem Abschnitt werden wir einen generellen Ansatz für die Informationsmodellierung vorstellen. Zunächst sollen drei wesentliche Anforderungen abgeleitet werden, die ein derartiger Ansatz erfüllen muß, die Integration der Informationsmodellierung mit Kommunikationsdiensten, die Erweiterbarkeit der Informationsmodellierung und die Verknüpfung existierender Architekturen.

Integration mit Kommunikationsdiensten: Wie bereits in der Einleitung dieses Beitrags ausgeführt wurde, basieren eine Vielzahl von verteilten Anwendungen auf zwei Klassen von Normen. Auf der einen Seite benötigen sie Kommunikationsdienste wie X.400 oder DFR, die es ermöglichen, Datenströme auszutauschen. Die mit Hilfe dieser Protokolle ausgetauschten Datenströme tragen zunächst keine Information. Erst durch die oben erwähnten Informationsarchitekturen wie ODA oder EDI erzielen die beteiligten Anwendungen ein gemeinsames Verständnis über die ausgetauschten Datenströme - diese werden dadurch zu Informationsträgern.

In der Standardisierung wurden diese beiden Bausteine, Kommunikationsdienste und Informationsarchitekturen, bisher weitgehend getrennt behandelt. Sobald Kommunikationsdienste aber derart zu spezifizieren sind, daß sie in ihrem Verhalten von der zu übertragenden Information abhängig sind, kann diese starre Trennung nicht länger aufrecht erhalten bleiben. Der DFR Standard enthält zum Beispiel erste Ansätze zur Integration der beiden Bausteine. Er definiert Attribute, die etwa die Speicherung und das Retrieval von Dokumenten betreffen. Neben diesem und anderen Aspekten der Informationsmodellierung beschreibt der Standard insbesondere Kommunikationsdienste, die den entfernten Zugriff auf Dokumenten-Server gestatten. Einige der DFR Attribute betreffen Zugriffsrechte für das jeweilige Dokument. Bei dem entfernten Zugriff auf Dokumente werden diese Attribute von den entsprechenden DFR Kommunikationsdiensten beachtet. Somit ist es möglich, unterschiedlichen Applikationen einen Zugriff unterschiedlicher Funktionalität (Schreib-Zugriff, Lese-Zugriff, ...) auf einzelne Dokumente oder Dokumentengruppen zu gestatten.

Man beachte, daß der beschriebene Ansatz nicht DFR spezifisch ist. Die Realisierung von Zugriffsrechten wird immer aus zwei Teilen bestehen, die Modellierung der Zugriffsrechte und die von dieser Information in ihrem Verhalten abhängigen Kommunikationsdienste. Der Zwang, Informationsarchitekturen und Kommunikationsdienste miteinander zu verknüpfen, wird mit der Komplexität der betrachteten Informationen und mit der Funktionalität der verwendeten Dienste wachsen. Daraus folgt die Notwendigkeit einer Schnittstelle zwischen Informationsarchitekturen und den Kommunikationsdiensten der Anwendungsschicht (OSI Schicht 7). Diese Schnittstelle hat Teile der durch die betrachteten Architekturen modellierten Informationen auszuzeichnen, auf die die Dienste zugreifen können. Die Spezifikation der Dienste muß deren Verhalten abhängig von diesen Informationsausschnitten definieren.

Erweiterbarkeit: Im Abschnitt 2 wurden bereits einige Aktivitäten aufgezeigt, die Erweiterungen des ODA Standards betreffen. Dieses war lediglich ein kleiner Ausschnitt aus den gegenwärtigen Standardisierungsbemühungen. Die angestrebten Erweiterungen sind durch eine Vielzahl von Anwendungsszenarien motiviert und somit im einzelnen sinnvoll. In ihrer Gesamtheit gefährden sie jedoch den eigentlichen Zweck von ODA, eine Basis für den Austausch von Dokumenten in offenen Netzen zu etablieren. Entwickelt sich der Standard tatsächlich in derart viele verschiedene Richtungen fort, so wird keine Applikation mehr in der Lage sein, ihn im vollen Umfang zu unterstützen. Es gibt kein ausreichendes Konzept, mit dessen Hilfe Anwendungen eine zu verwendende Teilmenge der ODA Architektur auszeichnen können. Somit ist es nicht möglich, sich über die Funktionalität der auszutauschenden Dokumente zu verständigen. Auch die Definition von *Document Application Profiles* (DAPs) (für einen Überblick siehe [12]), die bereits heute Teilmengen des ODA Standards auszeichnen, ist in dieser Beziehung keinesfalls ausreichend. Diese DAPs sind starr und können von den jeweiligen Applikationen nicht beeinflußt werden.

Es ist ein Konzept zu fordern, das es den beteiligten Applikationen gestattet, den von ihnen benötigten Ausschnitt aus den Informationsarchitekturen auszuhandeln. So ist es denkbar, daß eine Anwendung die von einem der heutigen DAPs definierte Funktionalität und zusätzlich ein Konzept für Dokumententeile benötigt. Die verwendeten Kommunikationsdienste müssen es den an dieser Anwendung beteiligten Kommunikationspartnern gestatten, diesen Ausschnitt aus der ODA Architektur zu verhandeln. Erst im Anschluß daran kann der beabsichtigte Informationsaustausch stattfinden. Informationsarchitekturen müssen somit die Auszeichnung von funktionalen Teilmengen durch Kommunikationsdienste geeignet unterstützen. Nur so sind zukünftige Erweiterungen existierender Standards handhabbar. Man beachte, daß der bereits erwähnte DFR Standard erste Ansätze in dieser Richtung realisiert. Bei dem Verbindungsaufbau zwischen einem Client und einem Dokumenten-Server wird gemäß DFR die Menge der Attribute festgelegt, die für den weiteren Kommunikationsvorgang von Bedeutung sind.

Integration von Architekturen: Es gibt eine Vielzahl von Informationsarchitekturen oder augenblicklichen Entwicklungen, die unterschiedliche Aspekte von Anwendungen betreffen. ODA beschreibt Dokumente und beschränkt sich in seiner bisherigen Fassung auf deren Präsentation. DFR modelliert Informationen, die für die Speicherung von Dokumenten, für den Zugriff auf Dokumente und für das Retrieval von Dokumenten notwendig sind. EDI spezifiziert die Modellierung formatierter Daten. Neuere Ansätze wie HyTime [13] oder MHEG [14] betreffen ebenso wie ODA die Präsentation von Informationen, beziehen aber Referenzierungs- und Synchronisationsmechanismen, die für Hypertext- und Multimedia-Anwendungen wesentlich sind, mit ein. Viele Anwendungen werden in Zukunft diese Architekturen nicht isoliert verwenden, sondern deren Integration voraussetzen. Die Verknüpfung von ODA mit den MHEG oder HyTime Arbeiten wird zum Beispiel die Verwendung von Audio- und Videoteilen in ODA-Dokumenten ermöglichen. Dieses ist das Ziel der entsprechenden Standardisierungsgruppen. Die zusätzliche Verwendung von DFR Attributen würde etwa die Unterstützung von Retrieval Anwendungen gestatten. Ein Konzept zur Informationsmodellierung muß demnach die Integration existierender und zukünftiger Architekturen anstreben.

Um die abgeleiteten Anforderungen an die Informationsmodellierung befriedigen zu können, ist eine gemeinsame Basis aller in Betracht kommender Architekturen zu fordern. Diese gemeinsame Basis muß existierende Architekturen verknüpfen, um neue Aspekte der Informationsmodellierung erweiterbar sein und es Kommunikationsdiensten ermöglichen, Teilmengen der insgesamt zur Verfügung stehenden Architekturen auszuzeichnen. Die gemeinsame Basis sollte die Teile der Informationsmodellierung beinhalten, die heutigen Architekturen und den sich abzeichnenden neueren Entwicklungen gemeinsam sind. Dieser größte gemeinsame Nenner wird sehr genau durch den objektorientierten Ansatz wiedergegeben. Bei allen neueren Ansätzen (ODA, DFR, HyTime, MHEG) werden Objekte und Klassen unterschieden, diese lassen sich zu Hierarchien anordnen, und es gibt Vererbungsmechanismen. Dabei werden Objekte durch eine Menge von Attributen beziehungsweise deren Werte beschrieben. DFR definiert zudem Methoden, die auf Objekte anwendbar sind, zum Beispiel das Löschen oder das Modifizieren von Objekten. Es sei an dieser Stelle auch erwähnt, daß ODA keine derartig klar definierten Methoden für Objekte unter der Dokument-Ebene umfaßt. Der Layout Prozeß ist etwa nur für vollständige Dokumente definiert und nicht auf einzelne Objekte bezogen. Dieses ist unter anderem der Grund, warum eine Definition partieller Dokumente notwendig ist.

Als eine geeignete Basisarchitektur, die exakt den objektorientierten Ansatz wiederspiegelt, ist SGML anzusehen. Die Standard Generalized Markup Language (SGML) wurde 1986 von der

ISO als internationaler Standard verabschiedet [15]. Sie ist eine Dokumentenarchitektur, definiert aber im Gegensatz zu ODA nicht die Präsentation von Dokumenten, sondern bezieht sich lediglich auf deren logische Struktur. Die Festlegung der Präsentation erfolgt durch Verabredungen von Benutzergruppen. So wird augenblicklich in einer Arbeitsgruppe der ISO ein Präsentationsmodell für SGML, das als DSSSL bezeichnet wird, entwickelt [16]. Im folgenden werden einige Gründe angegeben, warum SGML als geeignete Basisarchitektur anzusehen ist.

- SGML beinhaltet eine objektorientierte Informationsmodellierung. Sie unterscheidet Objekte und Klassen, erlaubt Hierarchien von Objekten und Klassen und definiert geeignete Vererbungsmechanismen. Außerdem enthält SGML ein Konzept, um Methoden an Objekte und Klassen zu binden, die *processing instructions.*
- SGML erlaubt partielle und zerlegte Dokumente. Ein SGML Dokument kann somit in Teile zerlegt werden, die dann etwa auf verschiedenen Knoten in einem Netzwerk gespeichert werden können.
- SGML ist ein weithin akzeptierter, internationaler Standard.
- Es gibt bereits sehr viele Werkzeuge zur Unterstützung von SGML Anwendungen wie zum Beispiel SGML Parser.
- Für einige Informationsarchitekturen existieren bereits SGML Repräsentationen. ODL ist neben ODIF das zweite Austauschformat für ODA Dokumente, es basiert auf SGML. HyTime repräsentiert die zu definierenden Synchronisations- und Referenzierungskonzepte unmittelbar mit Hilfe der SGML Methodik.

Aufgrund dieser Eigenschaften von SGML bietet diese Architektur die geeignete Plattform, um gegebene und zukünftige Informationsarchitekturen zu integrieren. Im weiteren skizzieren wir das sich daraus ergebende Modell zur Informationsmodellierung. Die folgende Abbildung gibt die Struktur eines SGML Objekts wieder, das verschiedene Aspekte besitzt, deren Bedeutung durch unterschiedliche Informationsarchitekturen beschrieben wird.

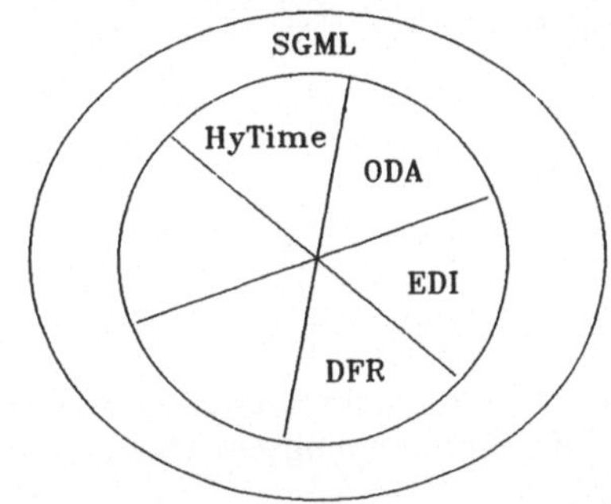

Die Semantik einiger Attribute des obigen Objektes bezieht sich auf die SGML Architektur. Dieses ist zum Beispiel mit Hinblick auf eindeutige Objektnamen oder Referenzen zwischen Objekten und Klassen notwendig. Die Bedeutung der anderen Attribute ergibt sich aus den angegebenen Informationsarchitekturen. Einzelne Anwendungen benutzen lediglich gewisse Aspekte eines derartigen Objektes. Eine ODA Applikation wird lediglich die ODA Attribute und zusätzlich die durch SGML definierten Strukturinformationen verwenden.

Nicht jedes Objekt muß alle zur Verfügung stehenden Aspekte umfassen. Falls ein Dokument als eine Einheit gespeichert wird, so ist etwa nur ein Satz von DFR Attributen sinnvoll. HyTime Attribute sind lediglich für solche Objekte zu vergeben, die als Teilobjekte zeitabhängige Informationen (Audio, Video) oder Referenzen zwischen Objekten umfassen. Die folgende Abbildung erläutert diesen Zusammenhang an einem Beispiel.

Man beachte, daß die Inhaltsteile eines Dokumentes analog zu ODA von den Blättern der obigen Struktur referenziert werden. (In der Abbildung wurden Inhaltsteile als Rechtecke dargestellt.) Die Attribute der einzelnen Inhaltsarchitekturen (Text, Audio, Video) werden gemäß der obigen Abbildung ebenfalls auf der Basis von SGML repräsentiert. Nähere Aussagen über spezielle Inhaltsarchitekturen sollen hier nicht getroffen werden.

Für eine Realisierung des geschilderten Ansatzes sind folgende Schritte notwendig:

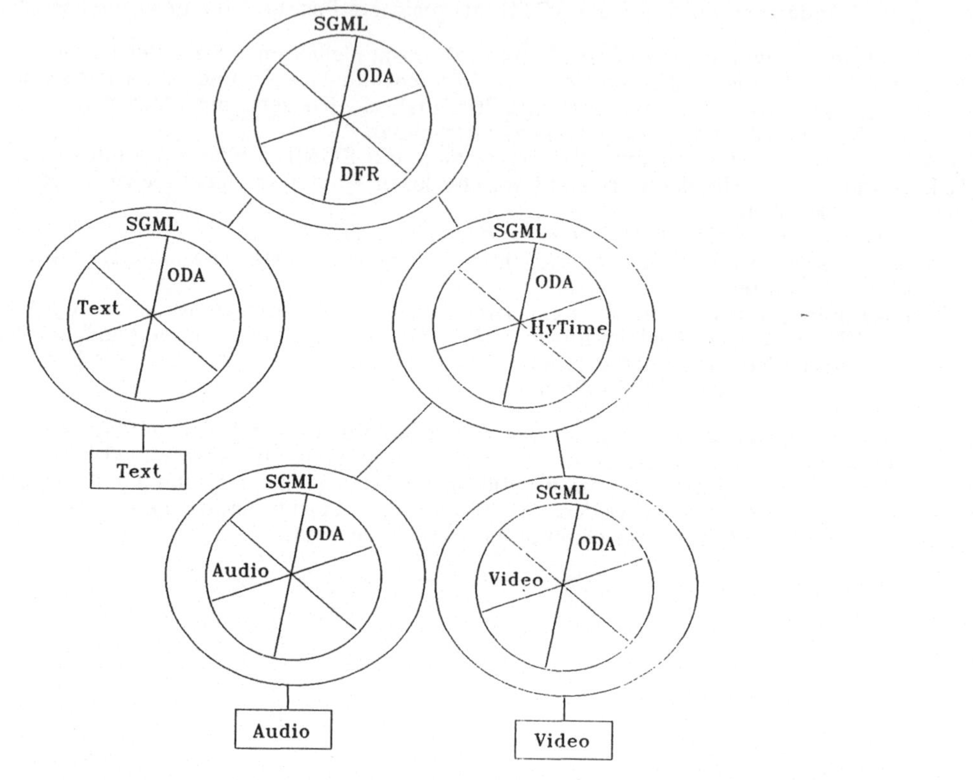

- Auszeichnung von geeigneten Aspekten eines Objektes. Diese Aspekte betreffen zum Beispiel die Speicherung von Objekten, die Präsentation von zeitunabhängigen Dokumententeilen und Synchronisationsmechanismen.
- Auswahl von gegebenen Architekturen, die die ausgezeichneten Aspekte beschreiben. Für die obigen Aspekte könnten dieses etwa ODA, DFR und HyTime sein.
- Spezifikation einer SGML basierten Repräsentation für die ausgewählten Architekturen, falls derartige Repräsentationen nicht bereits wie im Falle von ODA (ODL) verfügbar sind.
- Einführung von eindeutigen Identifikatoren für die einzelnen Architekturen wie ODA und Hytime, mit denen diese in SGML Dokumenten aber auch durch Kommunikationsdienste adressiert werden können.
- Aufbau eines lokalen API, das Dienste zur Analyse eines SGML Dokuments und zur Verarbeitung der einzelnen Aspekte eines solchen Dokuments zur Verfügung stellt. Dabei sollen die APIs, die für einzelne Architekturen wie ODA bereits vorliegen oder in der Entwicklung befindlich sind, geeignet eingebunden werden. Das lokale API ist derart zu modularisieren, daß es mit neuen Aspekten der Informationsmodellierung wachsen kann.
- Anbindung des API an Kommunikationsdienste wie die im Abschnitt 2 geschilderte DFR Implementierung oder die im Abschnitt 4 entwickelten Erweiterungen des DFR Dienstes. Dabei sind den Kommunikationsdiensten geeignete Konzepte hinzuzufügen, die das Aushandeln der für einen bestimmten Kommunikationsvorgang relevanten Aspekte gestatten.

Man beachte, daß bei jeder späteren Erweiterung der Informationsmodellierung diese Schritte wiederholt auszuführen sind.

Es folgt unmittelbar, daß der geschilderte Ansatz eine Integration existierender Informationsarchitekturen beziehungsweise deren Erweiterbarkeit unterstützt. Auch die dritte zu Beginn dieses Abschnitts abgeleitete Anforderung wird erfüllt. Der Ansatz erlaubt die Integration von Informationsarchitekturen und Kommunikationsprotokollen. Zum Beispiel ist es möglich, die bereits erwähnte, in dem DFR Standard enthaltene Modellierung von Zugriffsrechten zu isolieren und als einen weiteren Aspekt von Objekten zu integrieren. Kommunikationsdienste, die die so modellierten Zugriffsrechte zu beachten haben, müssen dann auf diesen genau abgegrenzten Informationsausschnitt zugreifen. Die SGML Architektur kann derart genutzt werden, daß Informationsausschnitte, auf die Kommunikationsdienste zugreifen müssen, durch geeignete Parser sehr schnell extrahiert werden können.

6. *Zusammenfassung*

Eine Vielzahl von Anwendungen benötigen Konzepte zur Verteilung von Dokumenten auf unterschiedliche Netzwerk-Knoten und Dienste, die den Zugriff auf derartige Dokumente gestatten. In einem heterogenen Umfeld muß diese Basis in Form von internationalen Standards gegeben sein. Die in dieser Hinsicht relevanten Normen, insbesondere DFR und ODA, besitzen zwei Einschränkungen, die sie für die betrachteten Szenarien ungeeignet machen: es werden nur vollständige Dokumente betrachtet und die Kommunikation beschränkt sich auf Client-Server Verbindungen.

In diesem Beitrag wurden Erweiterungen der Architektur ODA in Richtung auf partielle Dokumente sowie der durch DFR beschriebenen Kommunikationsdienste untersucht, die für die Verteilung von Dokumenten auf mehrere Server notwendig sind. Es wurde deutlich, daß für diese Konzepte eine Integration der Kommunikationsdienste und der Informationsmodellierung unabdingbar ist. Eine neues Konzept zur Informationsmodellierung, das diese Integration erlaubt und zudem erweiterbar ist, wurde vorgestellt. Für einige der in diesem Beitrag entwickelten Konzepte existieren bereits Prototypen. Erfahrungen bezüglich dieser Implementierungsarbeit wurden ebenfalls wiedergegeben.

Die zukünftige Arbeit wird die Weiterentwicklung der vorgestellten Konzepte und insbesondere des DFR Prototyps betreffen. Zunächst sollen die im Abschnitt 4 vorgestellten Konzepte derart in dieses DFR System integriert werden, daß mehrere Server ein logisches Server-System bilden. Anwendungen können dann auf ein Netz von Servern mit der in der DFR Norm definierten Funktionalität zugreifen, ohne die Verteilung der Dokumente auf verschiedene Server beachten zu müssen. Bis hierhin werden nur vollständige ODA Dokumente als DFR Einträge betrachtet. Die Verwendung partieller ODA Dokumente und ein geeignetes Konzept, um Referenzen zwischen Dokumententeile zu beschreiben, bilden dann die Grundlage, um Dokumente in Teile zu zerlegen und auf ein Netzwerk zu verteilen.

Parallel zu den beiden obigen Erweiterungen des DFR Prototyps wird das vorgeschlagene Konzept zur Informationsmodellierung auf lokale Applikationen angewendet. Anschließend soll eine auf DFR und DOAM basierende Speicherung von Informationsobjekten realisiert werden, die bezüglich der Informationsmodellierung auf den im Abschnitt 5 vorgestellten Konzepten beruht. Die vorgestellten Arbeiten stehen in Zusammenhang mit weitergehenden Überlegungen wie Sicherheitsaspekte in offenen Netzen und multimediale Dokumente in Hochgeschwindigkeitsnetzen.

Literatur

[1] **ISO/IEC** *Information technology, - Text and office systems - Distributed-office-applications-model, Part 1: General model,* Draft final for International standard 10031-1, 1990

[2] **CCITT** *Message Handling: System and Service Overview* CCITT Recommendation X.400, 1988

[3] **ISO/IEC** *Information technology, Document filing and retrieval (DFR), Part 1: Abstract service definition and procedures,* Draft International standard 10166-1, 1989

[4] **ISO/IEC** *Information Processing - Text and Office Systems - Office Document Architecture (ODA) and Interchange Format,* ISO 8613 (part 1,2,4,5,6,7,8), 1989

[5] **W. Filip, J. Kämper, W. Knobloch** *Conversion between the Open Document Architecture ODA and proprietary Architectures - A case study* to appear in: Proc. GI/ITG Fachtagung Kommunikation in verteilten Systemen, Mannheim, Informatik Fachberichte, Springer Verlag, 1991

[6] **S. E. Kille, C. J. Robbins, M. Rose, A. Turland** *The ISO Development Environment: User's Manual* Department of Computer Science, University College London, 1990

[7] **H.-B. Paul, A. Söder, H.-J. Schek, G. Weikum** *Unterstützung des Büro-Ablage-Service durch ein Datenbankkernsystem,* Proc. GI-Fachtagung Datenbanksysteme in Büro, Technik und Wissenschaft, Darmstadt, Informatik Fachberichte 136, Springer Verlag, 1987, pp. 196-211

[8] **H. Bunz** *Interface Specification for Remote Access to a Document Store* Technical Report 57, ESPRIT PROJECT 2374, Piloting of the Office Document Architecture 2

[9] *Framework for future extensions to ODA* ISO/IEC JTC1/SC18 SWG - ODA Design Directions and Extensions, Editor I.R. Campbell-Grant, February 1989

[10] **C. Bormann, U. Bormann, TU Berlin** *Proposal for a First Working Draft on an Addendum to ISO 8613 on Partial Documents and External References between Documents*

[11] **CCITT** *The Directory - Overview of Concepts, Models, and Service* CCITT Recommendation X.500, 1988

[12] **A. Spiceley** *(unknown title)* to appear in: Computer Networks and ISDN Systems

[13] *Hypermedia/Time-based Structuring Language (HyTime)* ANSI Project X3.749-D, 1990

[14] **ISO/IEC** *Coded Representation of Multimedia and Hypermedia Information* MHEG Working Document, 1990

[15] **ISO** *Information Processing - Text and Office Systems - Standard Generalized Markup Language (SGML),* ISO 8879, 1988

[16] **ISO** *Document Style Semantics and Specification Language* ISO SC 18 / WG 5

[17] **ISO/IEC** *Information technology, - Text and office systems - Distributed-office-applications-model, Part 2: Distinguished-object-reference and associated procedures,* Draft final for International standard 10031-2, 1990

[18] **ISO/IEC** *Information technology, Document filing and retrieval (DFR), Part 2: Protocol specification,* Draft International standard 10166-2, 1989

[19] **G. Krönert** *Genormte Austauschformate für Dokumente,* Informatik Spektrum Bd. 11, No. 2 (April 88), pp. 71 - 84

[20] **W. Appelt** *Dokumentenaustausch in Offenen Systemen.* Berlin, Heidelberg, New York: Springer 1990

[21] **ISO** *Abstract Syntax Notation 1 (ASN.1)* ISO 8824/8825, 1988

Experimentelle Analyse kooperativer Entscheidungsprozesse in Computerkonferenzen

B. Freisleben, B. Rüttinger†, A. Sourisseaux† und S. Schramme†*

*Technische Hochschule Darmstadt
Fachbereich Informatik
Alexanderstr. 10
D–6100 Darmstadt

†Technische Hochschule Darmstadt
Institut für Psychologie
Hochschulstr. 1
D–6100 Darmstadt

Kurzfassung

Die zunehmende Nutzung der Kommunikationsdienste von Rechnernetzen hat nicht nur quantitative, sondern auch qualitative Auswirkungen auf die Strukturen der Arbeitsorganisation. In diesem Beitrag werden die Ergebnisse eines interdisziplinären Forschungsprojektes zwischen Informatikern und Psychologen vorgestellt, dessen Zielsetzung die Analyse des Einflusses von Computerkonferenzen auf das Entscheidungs- und Problemlöseverhalten innerhalb einer Gruppe ist. Anhand einer experimentellen Untersuchung, an der insgesamt 12 Gruppen mit jeweils 5 Testpersonen teilnahmen, werden die Unterschiede der aus einer persönlichen Sitzung und einer Computerkonferenz resultierenden Entscheidungsprozesse aufgezeigt.

1 Einleitung

Entscheidungen in Organisationen sind in der überwiegenden Zahl Gruppenentscheidungen oder werden zumindest nach Beratung in Gruppen gefällt. Dies beruht nicht nur auf gesellschaftlichen Überzeugungen und organisationspolitischen Gründen, sondern auch auf sachlichen Erwägungen. Zum einen wird angenommen, daß in Gruppen mehr Ideen und bessere Bewertungskriterien produziert werden, zum anderen wird erwartet, daß die Akzeptanz von Entscheidungen, an denen die Betroffenen beteiligt sind, zunimmt.

Es ist jedoch nicht zu übersehen, daß Gruppenentscheidungen – unabhängig davon, ob die erwarteten positiven Effekte eintreten – im Vergleich zu Individualentscheidungen Nachteile aufweisen. Zu nennen ist vor allem der erhöhte Zeitaufwand, der mehrere Quellen hat:

- die Vorbereitung und Koordination der Sitzungen;
- die Wegezeit, die bei räumlich sehr distanten Sitzungsteilnehmern sehr gravierend ist;
- die Angleichung des Informationsstandes zu Beginn der Sitzung, ein Mehraufwand, der bei Befragungen von Führungskräften als besonders gewichtiger Zeitverlustfaktor beschrieben wird;
- und schließlich die verlängerte Sitzungsdauer aufgrund von Konflikten und Beziehungsproblemen.

Diesem vermehrten Zeitaufwand wird auf verschiedene Weise zu begegnen versucht:

- durch Planungstechniken;
- durch Telefonkonferenzen,
- durch gezieltes Informationsmanagement;
- durch schriftliche Entscheidungstechniken (z.B. Metaplantechnik, Delphi-Methode [8]), die zu einer Versachlichung der Diskussionen führen sollen.

Ein Medium, welches die Vorteile aller genannten Maßnahmen in sich zu vereinen scheint, sind die von heutigen Rechnernetzen angebotenen elektronischen Kommunikationssysteme. Sie ermöglichen die automatische Festlegung von Sitzungszeiten (*electronic calendaring* [7]), eine rasche und präzise Informationsübermittlung (z.B. durch *E-Mail* [9]), den Einsatz schriftlicher Entscheidungs- und Problemlösetechniken über Computer [12] und die Durchführung von Konferenzen mit räumlich distanten Personen (*Computerkonferenzen* [6]).

Es verwundert daher nicht, daß das Interesse für den Einsatz elektronischer Kommunikationssysteme zur Unterstützung von Gruppenentscheidungen in Organisationen sehr gewachsen ist und daß in den letzten Jahren mit dem Entstehen leistungsfähiger Netzwerke große Anstrengungen unternommen wurden, solche Systeme zu entwickeln und zu erproben.

Schon die ersten Anwendungen zeigten jedoch, daß der Einsatz solcher Systeme das Entscheidungs- und Problemlöseverhalten unter vielen Aspekten verändert, weil sich die computerunterstützte Koordination und Kommunikation in vielfältiger Weise von der direkten und persönlichen Interaktion unterscheidet [3, 4, 5, 10]. Dieser Unterschied wird häufig unter dem Stichwort "Depersonalisation" zusammengefaßt, womit ausgedrückt werden soll, daß wesentliche Aspekte der natürlichen Kommunikation ausgefiltert werden. Zu nennen ist vor allem das Wegfallen nonverbaler und paraverbaler Kommunikationsaspekte sowie der unmittelbaren und spontanen personenorientierten Abstimmung und Sequenzierung der Beiträge.

Die bei Gruppenentscheidungsprozessen in Computerkonferenzen auftretenden Phänomene wurden bislang keiner systematischen Analyse unterzogen. Es ist deswegen das Ziel der vorliegenden Arbeit, im Rahmen eines interdisziplinären Projektes zwischen Psychologen und Informatikern sowohl das Kommunikationsverhalten als auch das Problemlöse- und Entscheidungsverhalten umfassend zu beschreiben sowie zu klassifizieren und im Zusammenhang mit der Qualität, der Effizienz und der Akzeptanz der Entschlüsse oder Problemlösungen zu analysieren. Die Grundlage der Analyse ist eine vergleichende experimentelle Untersuchung der in einer Computerkonferenz und einer persönlichen Sitzung auftretenden Gruppenentscheidungsprozesse.

Der Beitrag ist folgendermaßen gegliedert: in Abschnitt 2 wird die Entwurfsphilosophie und Implementierung des in den Experimenten benutzten Groupware–Systems präsentiert. Abschnitt 3 beschreibt den Versuchsplan und die Ergebnisse der experimentellen Analyse. Abschnitt 4 faßt die Resultate des Projektes zusammen und liefert einen Ausblick auf zukünftige Forschungsaktivitäten.

2 Entwicklung eines Groupware–Systems

Um eine Computerkonferenz im Sinne der Zielsetzung des Projektes durchführen und analysieren zu können, muß die verwendete Kommunikationssoftware bestimmten Anforderungen genügen. Es soll beispielsweise möglich sein, die Kommunikation synchron/asynchron zu gestalten, unterschiedliche Verbindungen (1:n, 1:1) pro Teilnehmer aufzubauen und die Computerkonferenz moderiert bzw. nicht moderiert ablaufen zu lassen. Außerdem müssen Möglichkeiten zur automatischen Protokollierung verschiedener Bewertungskriterien (z.B. Inhalt und chronologische Reihenfolge der ausgetauschten Nachrichten, Anzahl der Länge der Beiträge pro Teilnehmer, Fehlerhäufigkeit) vorhanden sein, um die Ergebnisse entsprechend evaluieren zu können.

Da für die experimentellen Untersuchungen mehrere durch das Ethernet des Fachbereiches Informatik der TH Darmstadt miteinander vernetzte UNIX-Workstations unterschiedlichen Typs zur Verfügung standen und die rudimentären UNIX-Kommunikationsdienste (*talk, mail, news*) die Anforderungen nicht oder nur in beschränktem Umfang erfüllen konnten, wurde eigens für die Experimente am Rechner ein einfaches Groupware-System entwickelt. In den folgenden beiden Abschnitten wird auf dessen Funktionalität und Implementierung eingegangen.

2.1 Entwurfsphilosophie

Um insbesondere Benutzern mit wenig oder keinen Erfahrungen am Rechner einen leichten Einstieg zu ermöglichen, wurde beim Entwurf des Programmes besonderer Wert auf eine möglichst übersichtliche und komfortable Benutzerschnittstelle gelegt, die stark von der Fenstertechnik Gebrauch macht. Jeder Teilnehmer hat deshalb mindestens ein Fenster auf dem Bildschirm, in dem die noch nicht gelesenen Nachrichten, dargestellt durch jeweils eine Zeile mit Absender und Thema, angezeigt werden. Wählt der Benutzer eine Nachricht zum Lesen aus, so wird ein neues Fenster geöffnet, in dem der Inhalt der Nachricht mit Hilfe eines Editors begutachtet werden kann. Gleichzeitig wird die Nachricht aus der Liste der ungelesenen Nachrichten entfernt und in die Liste der gelesenen Nachrichten aufgenommen, deren Einträge ebenfalls in einem Fenster angezeigt und jederzeit per Editor inspiziert werden können. Alle Fenster bleiben immer so lange erhalten, bis sie vom Benutzer explizit geschlossen werden.

Neue Nachrichten werden in gleicher Weise durch den innerhalb eines Fensters aktivierten Editor erstellt, wobei Textausschnitte aus einer beliebigen anderen Nachricht per Maus einkopiert werden können. Die Nachricht kann anschließend an einen oder mehrere Empfänger gesandt werden und wird darauf in die Liste der abgeschickten Nachrichten aufgenommen. Da die einzelnen Arbeitsgänge in eigenen Fenstern durchgeführt werden, kann jederzeit zwischen ihnen umgeschaltet werden.

Alle Aktionen der Benutzer werden automatisch protokolliert, wobei unter einer Aktion nicht nur das Abschicken/Empfangen von Nachrichten verstanden wird, sondern auch die Auswahl einer Nachricht zum Bearbeiten oder die Erstellung neuer Nachrichten und der Abbruch einer solchen Erstellung. Bei jeder Aktion wird der auslösende Teilnehmer mit Uhrzeit und Art der Aktion sowie gegebenenfalls der Inhalt einer Nachricht notiert.

Zusätzlich kann für jedes Experiment eingestellt werden, ob eine Nachricht immer an alle Teilnehmer einer Computerkonferenz gehen muß, oder ob die Teilnehmer die gewünschten Empfänger auswählen können. Ebenso ist es möglich, einen Moderator zu bestimmen, der alle Nachrichten "zensieren" kann.

2.2 Implementierung

Das beschriebene Groupware-System wurde auf unterschiedlichen Rechnertypen (SUN Sparcstation, SUN 3/60, PCS Cadmus 9600, Hewlett Packard 300) mit den jeweiligen herstellerspezifischen Varianten des UNIX-Betriebssystems implementiert. Das dabei zugrunde liegende Entwurfsprinzip war das Client/Server-Paradigma der verteilten Programmierung [13], also die Wechselwirkung eines Servers mit mehreren Clients, wobei der Server die von den Clients an ihn erteilten Aufträge bearbeitet. Da die Clients nicht direkt, sondern ausschließlich über den Server miteinander kommunizieren, erschien dieser Ansatz ideal geeignet, um die komplette Client-Client Interaktion während eines Experiments für Analysezwecke zu protokollieren. Die Clients und der Server wurden jeweils durch mehrere Betriebssystemprozesse realisiert, die über die in UNIX üblichen Kommunikationsprimitive (Sockets, TCP/IP) Nachrichten miteinander austauschen. Die Funktionsweise beider Einheiten läßt sich wie folgt charakterisieren:

Der Server verharrt in einem Wartezustand, bis er von einem Client aufgefordert wird, eine Nachricht an die anderen Clients weiterzureichen. Wenn er eine solche Nachricht empfängt, wird ein Prozeß erzeugt, der ihren Inhalt in eine Protokolldatei schreibt und an alle Clients schickt.

Die Clients repräsentieren die Abstraktion der an einem Experiment teilnehmenden Testpersonen innerhalb der Computersysteme. Als komfortable Benutzeroberfläche wird das auf allen Rechnern vorhandene X-Windows Softwarepaket verwendet, mit dem Ein- bzw. Ausgaben in mehreren Fenstern auf dem Bildschirm dargestellt werden können. Innerhalb jedes Clients ist ein Prozeß für die Eingaben des Benutzers verantwortlich. Dieser eröffnet ein Fenster, innerhalb dessen der UNIX-Editor Micro-Emacs gestartet wird, um eine Nachricht über die Tastatur eingeben zu können. Sobald der Editor verlassen wird, schickt der Prozeß die Eingabe an den Server und ruft den Editor erneut auf. Ein zweiter Prozeß wartet auf Nachrichten vom Server und zeigt sie nach ihrer Ankunft ebenfalls mit Hilfe der Fenstertechnik auf dem Bildschirm an. Die üblichen systemseitigen "Aufräumungsfunktionen" bei Beendigung des Programmes (Abmelden des Clients beim Server, Schließen von Dateien, Löschen von Prozessen etc.) werden von einem dritten Client-Prozeß übernommen.

Es ist geplant, das Groupware-Paket auf weitere Workstations (Apple Macintosh, DEC VAXstation) zu portieren und seine Funktionalität aufgrund der in den Experimenten gemachten Erfahrungen zu erweitern, insbesondere im Hinblick auf ergonomische Aspekte und die Produkte der Open Software Foundation (OSF-Motif).

3 Experimentelle Untersuchung

In diesem Abschnitt werden die einzelnen Phasen der experimentellen Untersuchung beschrieben und die aus einer persönlichen Sitzung bzw. einer Computerkonferenz gewonnenen Ergebnisse ausführlich diskutiert.

3.1 Versuchsplan

Der Versuchsplan ist ein Meßwiederholungsdesign. 12 Gruppen mit jeweils 5 Teilnehmern hatten 2 Entscheidungsfälle zu lösen. Um die Variable "Entscheidungsfall" auszubalancieren, lösten 6 Gruppen zuerst Fall 1 und dann Fall 2, die anderen 6 Gruppen gingen in umgekehrter Reihenfolge vor.

Als unabhängige Variable wurde das "Medium" (Computerkonferenz vs. persönliche Entscheidungssitzung) berücksichtigt. Auch hier wurde die Reihenfolge ausbalanciert, indem 6 Gruppen zuerst an einer Face-to-Face Sitzung und anschließend an einer Computerkonferenz teilnahmen. Bei den anderen 6 Gruppen war die Reihenfolge umgekehrt.

Folgende Einflußgrößen wurden als abhängige Variablen berücksichtigt:

- Kommunikationskategorien

- Phasen des Entscheidungsprozesses
- Qualität der Entscheidung
- Effizienz der Entscheidung
- Akzeptanz der Entscheidung
- kritische Ereignisse.

Die Entscheidungsfälle sind zwei Übungen, die in ihrer Struktur der bekannten NASA-Übung [1] entsprechen und in ihrer Komplexität wichtigen Alltags–Entscheidungen ähnlich sind. Im ersten Fall geht um einen Schiffbruch, im zweiten Fall um einen Expeditionsunfall in der Antarktis. In beiden Fällen kann die verunglückte Gruppe von mehreren geretteten Gegenständen nur eine bestimmte Anzahl mitnehmen. Ihr Überleben hängt davon ab, daß die richtigen Gegenstände ausgewählt werden. In den Entscheidungssitzungen müssen deswegen die vorgegebenen Gegenstände nach ihrer Wichtigkeit in eine Reihenfolge gebracht werden. Diese beiden Entscheidungsfälle wurden ausgewählt, weil sie einerseits aus einem so entfernten Lebensbereich stammen, daß Teilnehmer verschiedener Studien– bzw. Berufsgruppen eine ähnlich gute Ausgangsposition haben und weil sie andererseits jeweils eine eindeutige Richtiglösung haben, so daß die Qualität der erarbeiteten Lösungen quantifiziert werden kann.

Die Qualität der Lösung ergibt sich aus einem Vergleich der von den Gruppen vorgeschlagenen Rangfolge mit der von Experten angefertigten, eindeutigen Richtiglösung für die Reihenfolge der Gegenstände.

Die Effizienz wird durch die Zeitdauer der Sitzung, die Akzeptanz der Lösung sowie der Entscheidungsfindung mittels eines Fragebogens am Ende der Sitzungen ermittelt.

Die Face-to-Face Sitzungen wurden mit Video aufgenommen und anschließend nach den 12 Kategorien der Interaktions-Prozeß-Analyse von Bales [2] klassifiziert. Nach dem gleichen Klassifikationsschema wurden die ausgedruckten Beiträge der Computerkonferenzen bewertet.

Unter dem Aspekt des Entscheidens wurden die Beiträge den heuristischen Phasen des Problemlöseprozesses – Strukturierung, Problemdefinition, Alternativengenerierung, Bewertung und Beschlußfassung – zugeordnet.

Schließlich wurden die Diskussionen auf kritische Ereignisse hin ausgewertet. Als solche Ereignisse werden Beiträge oder kurze Diskussionsphasen bezeichnet, die für die Qualität und Akzeptanz der Gruppenlösungen besonders fördernd bzw. behindernd waren.

Die am Experiment beteiligten Versuchspersonen waren Studenten der TH Darmstadt, die freiwillig an der Untersuchung teilnahmen. Da bei den Entscheidungsfällen teilweise auch technisches Wissen wichtig war und die Ausgangspositionen der Teilnehmer möglichst identisch sein sollten, wurden die Gruppen so zusammengesetzt, daß alle Teilnehmer entweder nur aus dem ingenieurwissenschaftlichen oder nur aus dem sozialwissenschaftlichen Bereich kamen. Die Studierenden waren sich vor der Untersuchung nicht näher bekannt.

Bei den Gruppensitzungen stellten sich die Teilnehmer zunächst kurz mit ihrem Namen und ihrem Studiengang vor. Nach einer kurzen Einweisung bearbeiteten die Versuchspersonen den Entscheidungsfall zunächst allein und stellten eine individuelle Rangreihe der Gegenstände auf. Diese Individuallösungen wurden nicht ausgetauscht.

Anschließend wurde in einer Gruppensitzung eine gemeinsame Lösung erarbeitet und beschlossen. Dabei wurde kein offizieller Moderator oder Gruppenleiter bestimmt. Die Face-to-Face Gruppen wurden mit Wissen und Einverständnis der Teilnehmer mit Video aufgenommen. Auch bei den Computerkonferenzen wurde jeweils eine Person gefilmt.

Am Ende der Sitzung fand eine Nachbefragung über die Entscheidungsfindung und die Akzeptanz der Lösungen statt.

3.2 Ergebnisse

Eine erste Analyse der gewonnenen Daten zeigt folgende Ergebnisse:

- Qualität:
  Die Lösungen der Computerkonferenzen liegen zwar näher bei den Expertenlösungen als die Lösungen der Face-to-Face Sitzungen, doch ist der Unterschied nicht bedeutend.

- Effizienz:
  Die Dauer der Computerkonferenzen ist bedeutend länger als die der Face-to-Face Sitzungen (56 vs. 36 Minuten).

- Akzeptanz:
  Die Akzeptanz der Gruppenlösungen ist unter beiden Medium–Bedingungen im Durchschnitt gleich hoch. Was die Bewertung des Entscheidungsprozesses betrifft, so findet sich bei den Computerkonferenzen eine stärkere Polarisierung in Befürworter und Gegner.

- Einfluß:
  Die Gruppenlösungen der Computerkonferenzen weichen im Durchschnitt weniger stark von den anfänglichen Individuallösungen der einzelnen Gruppenmitglieder ab. Die einzelnen Teilnehmer nahmen also eher gleichmäßig auf die Gruppenlösung Einfluß.

- Kommunikation:
  Zwischen den beiden Medium–Bedingungen zeigen sich große Unterschiede in der Partizipationsrate, in den Kommunikationskategorien und in der Länge der Beiträge. Die Partizipationsrate ist bei den Computerkonferenzen bedeutend ausgeglichener als bei den persönlichen Sitzungen. Die Beiträge in den Computerkonferenzen sind sehr kurz und gegen Ende der Sitzungen zunehmend im Telegrammstil abgefaßt.

  Die meisten Beiträge der Face-to-Face Sitzungen lassen sich der Kategorie "gibt Meinung" zuordnen, d.h. die Teilnehmer tauschen vor allem Vermutungen und Bewertungen aus. In den Computerkonferenzen ist die Hauptkategorie "gibt Information", d.h. es werden vor allem Fakten und Tatbestände mitgeteilt. Die Teilnehmer von Computerkonferenzen sind relativ stark mit der Diskussions–Koordination beschäftigt, was sich in einer erhöhten Besetzung der Kategorien "fragt nach Strukturierung" und "gibt Strukturierung" ausdrückt. Schließlich fällt bei den Computerkonferenzen die große Anzahl von Beiträgen auf, die der Kategorie "dramatisiert" zuzuordnen sind, d.h. Beiträge, die Gefühle und Stimmungen ausdrücken, welche sich direkt auf den Problemlöseprozeß beziehen.

  Zusammenfassend läßt sich zur Kommunikation feststellen, daß die Beiträge in Computerkonferenzen gleichzeitig viel sachlicher (Information, Strukturierung) und sozial-emotionaler ("Dramatisierung") sind als die Beiträge in Face-to-Face Gruppen. Es findet also eine stärkere Trennung zwischen der sachlichen und sozial–emotionalen Ebene statt. Beide Aspekte werden nicht wie bei den Face-to-Face Gruppen verbal/nonberbal oder in der Kategorie "gibt Meinung" vermischt, sondern deutlich, und zwar in voneinander abgehobenen Diskussionsphasen, getrennt.

- Entscheidungsphasen:
  Die Phase der Beschlußfassung wird bei Computerkonferenzen viel ausführlicher durchgeführt als bei Face-to-Face Sitzungen. Sie wird explizit und formal vorgenommen und ist für alle Teilnehmer in ihren Einzelschritten nachvollziehbar. Bei Face-to-Face Gruppen werden die Beschlüsse häufig unvermittelt und aufgrund der Wortführerschaft einzelner Gruppenmitglieder herbeigeführt. Die Planung und Strukturierung des Vorgehens wird bei Computerkonferenzen ausführlicher und detaillierter besprochen. Die Situationsanalyse wird bei beiden Medium-Bedingungen gleich kurz und unzureichend vorgenommen. Das Generieren von Vorschlägen ist bei den Computerkonferenzen leicht erhöht. Die Bewertung ist bei beiden Medien die Hauptphase und relativ gesehen ungefähr gleich lang.

- Kritische Ereignisse:
  Schlechte Entscheidungen bei Computerkonferenzen sind häufig darauf zurückzuführen, daß wichtige Beiträge einzelner Teilnehmer nicht beachtet und diskutiert werden.

.3 Diskussion

Vie die Ergebnisse zeigen, sind Computerkonferenzen nicht allgemein besser oder schlechter als atürliche Konferenzen. Beide Konferenzarten unterscheiden sich aber in vielfältiger Weise.

unächst bestätigt die Untersuchung den bisherigen Befund, daß Computerkonferenzen nicht, wie ei natürlichen Diskussionen häufig beobachtet, von wenigen Teilnehmern dominiert werden, sondern aß sich die Partizipationsraten stark angleichen.

)iese "Nivellierung" oder "Demokratisierung" hängt nach einer Schlußbefragung der Teilnehmer und ›eobachtungen während der Konferenzen mit folgenden Bedingungen zusammen:

- Zunächst können alle Teilnehmer unabhängig voneinander gleichzeitig schreiben (wenn auch die Beiträge nacheinander am Bildschirm erscheinen), "Vielredner" können sich also nicht mehr gegen andere durchsetzen.
- Auf ähnliche Weise können die Teilnehmer unabhängig voneinander entscheiden, auf welche Beiträge sie eingehen. Auch dadurch können Vielschreiber aus der Diskussion gedrängt werden.
- Da alle Teilnehmer kontinuierlich Beiträge abschicken und diese Bemerkungen gelesen werden müssen, wird ein "Zuhören" erzwungen. Gleichzeitig bremst das relativ langsame Schreiben – im Gegensatz zum raschen und spontanen Sprechen – die Tendenz, lange Beiträge zu formulieren. Es daher nicht verwunderlich, daß die Beiträge bei Computerkonferenzen viel kürzer sind als bei natürlichen Diskussionen.
- Diese Kürze ergibt sich damit zusammenhängend auch daraus, daß man nur mit kurzen und prägnanten Nachrichten Aufmerksamkeit gewinnen kann und daß man durch das Formulieren langer Beiträge den Anschluß an die aktuell diskutierte Fragestellung verliert. Die Teilnehmer an den Computerkonferenzen lernen im Laufe der Diskussion immer besser mit dieser Schwierigkeit umzugehen, was die deutliche Zunahme von Nachrichten im Telegrammstil gegen Ende der Sitzung belegt.

- Weiterhin werden eher unsichere Teilnehmer nicht mehr durch unmittelbare verbale oder nonverbale Reaktionen verunsichert. Sie sind deswegen eher bereit, an der Diskussion teilzunehmen.
- Schließlich ist die isolierte Situation und das Medium Computer auch für eher passive Personen ein Anreiz, den Entscheidungsprozeß aktiv mit zu beeinflussen.

Die bisherigen Befunde, daß Computerkonferenzen aggressiver und sozial-emotionaler als natürliche Diskussionen sind [11], müssen differenzierter betrachtet werden. Zwar steigt der Anteil sozial-emotionaler Beiträge an, gleichzeitig findet jedoch eine Versachlichung der Diskussion statt. Beide Tendenzen haben teilweise gleiche Ursachen. Die Distanz zu den anderen Teilnehmern führt zum einen dazu, daß Gefühle freier ausgedrückt und Gereiztheit eher formuliert wird. Die Teilnehmer sind sozusagen "Schreibtischtäter", die keinen direkten Schlagabtausch befürchten müssen. Der Schutz des Computers läßt aber auch gleichzeitig viele Vorsichtsfloskeln und Rückversicherungen wegfallen. Man formuliert direkter, was zu einer Erhöhung der Kategorie "gibt Information" führt.

Ein weiterer Grund für die Zunahme sozial-emotionaler Beiträge ist das Fehlen der nonverbalen Ausdrucksweise. Gefühle und Stimmungen müssen verbalisiert werden. Dabei ist allerdings zu beachten, daß nur reflektierte und bewußte Gefühle verbalisiert und explizit mitgeteilt werden.

Die stärkere Versachlichung schließlich beruht auch darauf, daß man nur – wie zuvor ausgeführt – mit kurzen und prägnanten, aber auch überprüfbaren Informationen die Diskussion beeinflussen kann.

Neben dem veränderten Kommunikationsverhalten sind die Computerkonferenzen vor allem durch eine starke Formalisierung gekennzeichnet. Die Entscheidungsgruppen legen starken Wert auf eine genaue Festlegung des Vorgehens und die Strukturierung der Problemlösung. Gleichzeitig wird die Beschlußfassung stark betont. Im Gegensatz zu den natürlichen Sitzungen, in denen spontan und frei diskutiert wird, kann man bei einigen Computerkonferenzen eher von einem formalisierten Verhandlungsprozeß sprechen, in dem Schritt für Schritt über den Rang der einzelnen Gegenstände entschieden wird. Alle Teilnehmer nehmen zu einem Gegenstand so lange Stellung, bis ein expliziter Konsens erreicht ist. Dann wird zum nächsten Gegenstand übergegangen. Auch unter diesem Aspekt kann man von einer "Nivellierung" oder "Demokratisierung" der Entscheidung sprechen. Es sind nicht mehr einzelne Wortführer, die eine Entscheidung herbeiführen, auch Mehrheits- oder Pluralitätsentscheidungen sind nicht mehr durchsetzbar, sondern konsensuale Beschlüsse, die formal durch Befragen jedes Teilnehmers und explizit durch die schriftliche Zustimmung herbeigeführt werden. Der Vorteil dieses Vorgehens ist, daß systematisch entschieden wird. Gleichzeitig sind jedoch spätere Revisionen der Einzel-Beschlüsse nur schwer durchsetzbar. Dieser Nachteil gewinnt dadurch an Gewicht, daß die Computerkonferenzen ebenso wie die natürlichen Sitzungen durch eine mangelnde Situationsanalyse des Entscheidungsfalles gekennzeichnet sind. Es ist zu vermuten, daß eine Groupware–Applikation "Situationsanalyse" die Qualität der Computerkonferenzen erheblich verbessern würde.

Die bisherigen Ausführungen machen deutlich, warum Computerkonferenzen länger dauern als Face-to-Face Sitzungen. Es ist nicht so sehr der verlangsamte Informationsaustausch durch Schreiben, der durch kürzere Beiträge kompensiert wird, als vielmehr die hohe Beteiligung aller Sitzungsteilnehmer und vor allem die formalisierte und konsensuale Entscheidungsfindung, die eine explizite positive Stellungnahme aller Diskussionsteilnehmer zu allen Teilfragen erfordert.

Was die Qualität der Lösungen betrifft, so bestehen zwischen den beiden Konferenzarten nur tendenzielle Unterschiede. Eine Analyse derjenigen kritischen Ereignisse, die starke Fehleinschätzungen verursachten, weist auf, daß große Fehler vor allem deswegen gemacht wurden, weil einzelne

Beiträge nicht in angemessener Weise in die Diskussion aufgenommen wurden. Einzelne Teilnehmer brachten zwar die richtigen Argumente, wurden aber von den anderen "übersehen". Diese Analyse zeigt, daß bei Computerkonferenzen nicht nur die Teilnehmer "nivelliert" werden, sondern auch die Beiträge. Zwar weist dieser Befund auch darauf hin, daß in Computerkonferenzen das Potential der Mitglieder stärker eingebracht wird, doch ist nicht gewährleistet, daß die kompetenten Beiträge kompetenter Mitglieder mehr Einfluß auf die Entscheidung haben als andere Beiträge. Dies zeigt sich auch in dem Befund, daß die Gruppenlösung bei Computerkonferenzen im Durchschnitt weniger von den individuellen Anfangslösungen abweichen als bei persönlichen Sitzungen. Bei guten Entscheidungen ist eine große durchschnittliche Abweichung zu erwarten, da sich dann die Argumente des besten Teilnehmern durchsetzen müßten. Eine Verbesserung der Computerkonferenzen könnte deswegen eventuell durch ein Groupwareprodukt erreicht werden, das den Teilnehmern erlaubt, wichtige Informationen zu markieren und damit eine besondere Aufmerksamkeit für diese Nachrichten zu gewährleisten.

4 Zusammenfassung und Ausblick

In dieser Arbeit wurde im Rahmen eines interdisziplinären Projektes zwischen Informatikern und Psychologen untersucht, wie sich das Problemlöse- und Entscheidungsverhalten einer Gruppe in einer persönlichen Sitzung von dem in einer Computerkonferenz unterscheidet. Die Analyse basierte auf einem an der TH Darmstadt durchgeführten Experiment, an dem mehrere Gruppen von Testpersonen teilnahmen und zwei Entscheidungsfälle unter verschiedenen Medienbedingungen gemeinsam zu lösen hatten. Dabei zeigten sich Unterschiede vor allem in der individuellen Beeinflussung der Entscheidung, in dem Engagement der individuellen Teilnahme am Problemlöseprozeß und im Stil der Kommunikation. Für die Experimente am Rechner wurde eigens ein einfaches Groupware-System entwickelt.

In zukünftigen Forschungsarbeiten sollen die gewonnenen Ergebnisse durch zusätzliche Experimente validiert und ergänzt werden. Es ist beabsichtigt, die daraus resultierenden Erfahrungen in die Entwicklung eines funktional erweiterten Groupware-Paketes einfließen zu lassen, das alle in Gruppenentscheidungsprozessen relevanten Faktoren im Sinne der Benutzer in angemessener Weise berücksichtigt.

Literatur

[1] K. Antons. *Praxis der Gruppendynamik.* Hogrefe, Göttingen, 1973.

[2] R.F. Bales. How People Interact in Conferences. *Scientific American*, 3:3–7, 1955.

[3] T.X. Bui. *CO-oP: A Group Decision Support System for Cooperative Multiple Criteria Group Decision Making*, volume 290 of *Lecture Notes in Computer Science.* Springer-Verlag, 1987.

[4] J.D. Eveland and T.K. Bikson. Work Group Structures and Computer Support: A Field Experiment. *ACM Transactions on Office Information Systems*, 6(4):354–379, 1988.

[5] F. Flores, M. Graves, B. Hartfield, and T. Winograd. Computer Systems and the Design of Organizational Interactions. *ACM Transactions on Office Information Systems*, 6(2):153–172, 1988.

[6] I. Greif, editor. *Computer Supported Cooperative Work: A Book of Readings.* Morgan Kaufmann Publishers, 1988.

[7] R. Johansen, J. Vallee, and K. Collins. *Electronic Meetings: Technical Alternatives and Social Choices.* Addison–Wesley, Reading, Mass., 1979.

[8] H.A. Lindstone and M. Turoff. *The Delphi Method: Techniques and Applications.* Addison–Wesley, Reading, Mass., 1975.

[9] W.E. Mackay. Diversity in the Use of Electronic Mail: A Preliminary Inquiry. *ACM Transactions on Office Information Systems*, 6(4):380–397, 1988.

[10] M.H. Olson. Remote Office Work: Changing Patterns in Space and Time. *Communications of the ACM*, 26(3), 1983.

[11] L. Sproull and S. Kiesler. Reducing Social Context Cues: The Case of Electronic Mail. *Management Science*, 32:1492–1512, 1986.

[12] M. Stefik, G. Foster, D. Borrow, K. Kahn, S. Lanning, and L. Suchman. Beyond the Chalkboard: Computer Support for Collaboration and Problem Solving in Meetings. *Communications of the ACM*, 30(1), 1987.

[13] A. Tanenbaum. *Computer Networks.* Prentice Hall, Englewood Cliffs, NJ, 1981.

Aufgaben und Stellung von EURESCOM

Horst A. Besier

EURESCOM, Schloßwolfsbrunnenweg 35

6900 Heidelberg

1. Einführung

EURESCOM (European Institute for Research and Strategic Studies in Telecommunications) geht auf eine Entscheidung der CEPT (Kommission der europäischen Post- und Fernmeldeverwaltungen) und eine Tagung des EG-Postministerrates vom September 1989 zurück, wo die Generaldirektion der Deutschen Bundespost beauftragt wurde, eine Gruppe hochrangiger Experten zur Erarbeitung eines Memorandum of Understanding (MoU) zu benennen. Dieses Memorandum of Understanding wurde Anfang des Jahres 1990 von 23 öffentlichen Netzbetreibern aus 18 europäischen Staaten unterzeichnet, die sämtlich der CEPT angehören.

In der Folge wurde ein Vorbereitungskomitee gebildet, das seine erste Sitzung am 16. März 1990 in München hatte, und das mit seinen Arbeitsgruppen die für die Gründung und die spätere Arbeit von EURESCOM notwendigen Grundlagen erarbeitete:

- Rechts- und Finanzstruktur,
- geistige Eigentumsrechte,
- Arbeitsprogramm für kurzfristige und für längerfristig angelegte Projekte sowie
- Programm- und Projektmanagement.

Die Deutschen Bundespost TELEKOM hatte sich bereits in der Vorbereitungsphase sehr intensiv für EURESCOM engagiert und stellte den Vorsitzenden des Vorbereitungskommittees. Der von der DBP TELEKOM vorgeschlagene Sitz des Institutes in Heidelberg konnte sich gegen seine beiden Mitbewerber Barcelona und Lissabon mit überwältigender Mehrheit durchsetzen.

2. Rechtsform und Kreis der Gesellschafter

Am 14. März 1991 wurde EURESCOM von 20 der Unterzeichner des MoU als GmbH gegründet, 2 Unterzeichner kamen etwas später als Gesellschafter hinzu.

Die Gesellschafter von EURESCOM bestehen aus den öffentlichen Netzbetreibern von 18 westeuropäischen Staaten: alle 12 EG-Staaten, die EFTA-Staaten Finnland, Island, Norwegen, Schweden und Schweiz sowie Jugoslawien.

Gesellschafter können nur solche Betreiber öffentlicher europäischer Telekommunikationsnetze sein, die nationale und/oder internationale öffentliche feste Netze zur Verfügung stellen und denen im Zusammenhang mit diesen Netzen oder Dienstleistungen besondere oder ausschließliche Rechte oder Verpflichtungen übertragen sind. Der Kreis der Gesellschafter ist auf den Geltungsbereich der CEPT, d.h. auf Europa, beschränkt. Eine weitere Voraussetzung ist, daß die Gesellschafter ein besonderes Interesse daran haben sollen

- die Konvergenz und Integrität der Netzinfrastruktur,
- die Interoperabilität der Dienste,
- die Entwicklung europaweiter Telekommunikations-Dienstleistungen sowie
- die Unterstützung moderner Forschungs- und Entwicklungstätigkeiten des Instituts

sicherzustellen.

3. Aufgaben von EURESCOM

EURESCOM hat es sich zur Aufgabe gemacht, die Entwicklung und Bereitstellung harmonisierter, europaweiter öffentlicher fester Telekommunikationsnetze und -dienste zu unterstützen. Dabei sollen Wissenschaft und Forschung in der Telekommunikation gefördert werden, die für die Gesellschafter eine grundsätzliche Bedeutung und große Anwendungsbreite haben. EURESCOM sieht sich hier als eine bedeutende Triebkraft, die entscheidend dazu beiträgt, den Prozess der europäischen Integration voranzutreiben.

Die Hauptzielrichtungen, in denen sich EURESCOM für seine Gesellschafter engagiert, sind

- Erarbeitung abgestimmter Strategien für die Planung und Bereitstellung zukünftiger europäischer öffentlicher Telekommunikationsinfrastrukturen für Netze und Dienste;

- Initialisierung und Koordinierung gemeinsamer Forschungsprojekte, die im vornormativen und im vorwettbewerblichen Bereich liegen;

- Initialisierung und Koordinierung von Pilotprojekten;

- Mitarbeit und Vorarbeit bei der europäischen und weltweiten Standardisierung in der Telekommunikation (ETSI, CCITT und dergl.);

- Gemeinsame Aktionen und Projekte mit anderen relevanten europäischen Forschungseinrichtungen, wie z.B. in Forschungs- und Entwicklungsprogrammen der EG (RACE II), der CEPT und dergleichen.

Es ist dabei besonders hervorzuheben, daß EURESCOM die Stelle ist, die solche Projekte initiiert und koordiniert, für die Durchführung aber bis auf wenige Ausnahmen auf die von den Gesellschaftern eingebrachten Ressourcen zurückgreift. Ein weiterer wichtiger Punkt ist, daß durch die Aktivitäten von EURESCOM der Wettbewerb zwischen den einzelnen öffentlichen Netzbetreibern natürlich nicht eingeschränkt werden soll und darf.

4. Durchführung von Projekten

Alle Aktivitäten von EURESCOM werden in Form von Projekten durchgeführt, die im Hinblick auf die angestrebten Ergebnisse, die benötigten Ressourcen sowie auf die Termine möglichst detailliert beschrieben und in einen Mehrjahresarbeitsplan aufgenommen werden, aus dem dann wiederum der jährliche kostenbewertete Arbeitsplan entsteht. Für jedes Projekt wird eine Ausschreibung (call for tenders) unter den Gesellschaftern durchgeführt und - bei genügendem Interesse - das Projekt unter Leitung eines Projektmanagers durchgeführt; die Koordinierung und regelmäßige Bewertung und Überprüfung eines Projekts ist Aufgabe des ständigen Mitarbeiterstabes von EURESCOM.

Für das Projektmanagement und das Berichtswesen werden moderne, IV-gestützte Methoden angewandt, wobei besonderer Wert auf eine effiziente, termingerechte Durchführung jedes einzelnen Projektes und eine gute Koordinierung zwischen verschiedenen Projekten gelegt wird. Bei der jährlichen Überprüfung und Bewertung durch eine unabhängige Bewertungsgruppe wird über Fortsetzung, Modifizierung oder Abbruch jedes Projektes entschieden.

Die Projekte lassen sich in zwei Klassen einteilen:

- Projekte allgemeinen Interesses (General Interest Projects) erfordern die Zustimmung einer qualifizierten Mehrheit und werden von allen Gesellschaftern sowohl hinsichtlich der vom ständigen Mitarbeiterstab zu erbringenden Koordinierungs- und Überprüfungsfunktion als auch durch aktive Mitarbeit von Experten in der Projektdurchführung getragen.

- Projekte speziellen Interesses (Special Interest Projects) werden nicht von einer qualifizierten Mehrheit getragen; hier schließen sich einige Gesellschafter zusammen, um solche Projekte gemeinsam unter der Koordinierung von EURESCOM durchzuführen.

Ergebnisse der Projekte sollen in folgender Form erzielt werden:

- Schutzrechte (Patente, gewerbliche Schutzrechte) sollen einerseits den Gesellschaftern einen technologischen Vorsprung verschaffen und andererseits Lizenzgebühren erzielen.

  Über ein IPR-Agreement (Intellectual Property Rights) ist die Vereinbarung zwischen den Mitgliedern hinsichtlich des Austausches von Patent- und Schutzrechten auf Produkte und Technologien der Telekommunikation auf dem Wege des Lizenzverfahrens geregelt. Patente für Produkte, die im Rahmen von EURESCOM-Projekten erarbeitet wurden, und Rechte an Technologien, deren Nutzung für die Umsetzung von EURESCOM-Projekten zwingend erforderlich sind, sollen offengelegt werden. Gesellschafter haben jedoch das Recht, bestimmte Patente und Schutzrechte von diesem Austausch auszunehmen.

- Beiträge zu europäischen und weltweiten Standardisierungsgremien (z. B. ETSI, CCITT) sollen die Standardisierungsarbeit vorantreiben, aber auch die Interessen der öffentlichen Netzbetreiber in diesen Gremien angemessen berücksichtigen helfen.

- Studien und Empfehlungen an die Netzbetreiber bilden die Grundlage, die Netz- und Diensteentwicklung in harmonisierter Form voranzutreiben.

- Pilotprojekte und Feldversuche sollen die Machbarkeit und Akzeptanz von Technologien und Produkten aufzeigen und ggf. Kristallisationspunkte für eine flächendeckende Einführung sein.

5. Organisationsstruktur von EURESCOM

Oberstes Gremium von EURESCOM ist die Generalversammlung, in der alle Gesellschafter vertreten sind. Der Stimmanteil bei Abstimmungen richtet sich nach dem Anteil (von 1 Stimme als Minimum bis zu 64 Stimmen als Maximum), der aus dem Jahresumsatz der jeweiligen Gesellschaften abgeleitet ist.

Für die Führung der Geschäfte - zusammen mit dem Direktor des Instituts - ist der neunköpfige Aufsichtsrat zuständig.

Das Organigramm zeigt die funktionale Organisationsstruktur des ständigen Mitarbeiterstabes EURESCOM (Permanent Staff), mit dem Direktor an der Spitze, und folgenden Organisationseinheiten:

- Abteilung für Programm-Management (Programme Management Office) unterteilt in die Bereiche für

  - Strategische Studien (und Netzaspekte) und
  - Forschung,

  zuständig für die Projektabwicklung, wobei im Hinblick auf Projektleiter und Projektteams, die üblicherweise nicht zum ständigen Mitarbeiterstab gehören, bei einigen Projekten eine andere Regelung getroffen werden kann;

- Verwaltungsabteilung;

- Büro für Rechtsangelegenheiten, dem ein Gremium für Patent- und Schutzrechte angeschlossen ist.

Für den ständigen Mitarbeiterstab ist bis Ende 1991 ein Personalkörper von ca. 15 Mitarbeitern vorgesehen, der dann entsprechend dem Fortgang der Projektarbeit auf 40 bis 80 erhöht werden soll.

6. Aktuelle Projekte von EURESCOM

Für 1991 (Startphase) und 1992 wurden Projekte ausgewählt, die mit begrenzter Laufzeit und relativ geringen Ressourcen bereits eine Reihe dringend benötigter Ergebnisse erzielen können. Die in diesem Zeitraum initiierten Projekte sind folgenden Schwerpunkten zuzuordnen:

- Infrastruktur und vermittelte (Breitband-) Netze, u.a. mit
  - Studien zu einem europäischen ATM-Pilotnetz (Asynchronous Transfer Mode), basierend auf den aktuellen Standards von B-ISDN in CCITT und ETSI,
  - Vorbereitung eines Pilotversuches mit integrierten 1-Mbit/s-Multimediadiensten,
  - Kooperation bei METRAN (Managed European Transmission Network), ein Netz elektronischer Highways, basierend auf Übertragungssystemen der SDH-Hierarchie (Synchronous Digital Hierarchy),
- Intelligente Netze (IN), u.a. mit
  - Virtuellen privaten Netzen
  - Universal Personal Telecommunication
  - Evolution der IN-Architektur

- Telekommunikations-Management-Netz (TMN), u.a. mit
  - Aspekten der Netzsicherheit
  - Management-Dienste
  - TMN-Organisationsmodell
  - Evolutionszenarien für TMN

- Strategische Studien, die einerseits die Langzeitevolution von Telekommunikationsdiensten und -netzen, einschließlich der dazu benötigten Technologien, in Form von Studien aufzeigen sollen und andererseits daraus resultierend Projektvorschläge erarbeiten sollen. Solche Projekte könnten auch außerhalb von EURESCOM organisiert werden.

Die eingeplanten Ressourcen für die Projekte betragen für die Startphase in 1991 etwa 200 Mannmonate, werden jedoch für 1992 mindestens auf das Fünffache anwachsen.

Zu den vorstehend genannten Projekten werden noch weitere Projekte hinzukommen, die dann auch ein stärkeres Gewicht auf die in der Anfangsphase etwas zu kurz gekommenen Forschungsaspekte legen werden.

7. Zusammenfassung

EURESCOM wurde im März 1991 als ein gemeinsames Institut von 22 europäischen Betreibern öffentlicher Telekommunikationsnetze mit ständigem Sitz in Heidelberg gegründet. Es hat die Aufgabe, Forschung und strategische Studien der Netzbetreiber in Form von Projekten zu initiieren und zu koordinieren, wobei besonderes Gewicht auf die Evolution von Telekommunikationsdiensten und -netzen gelegt wird.

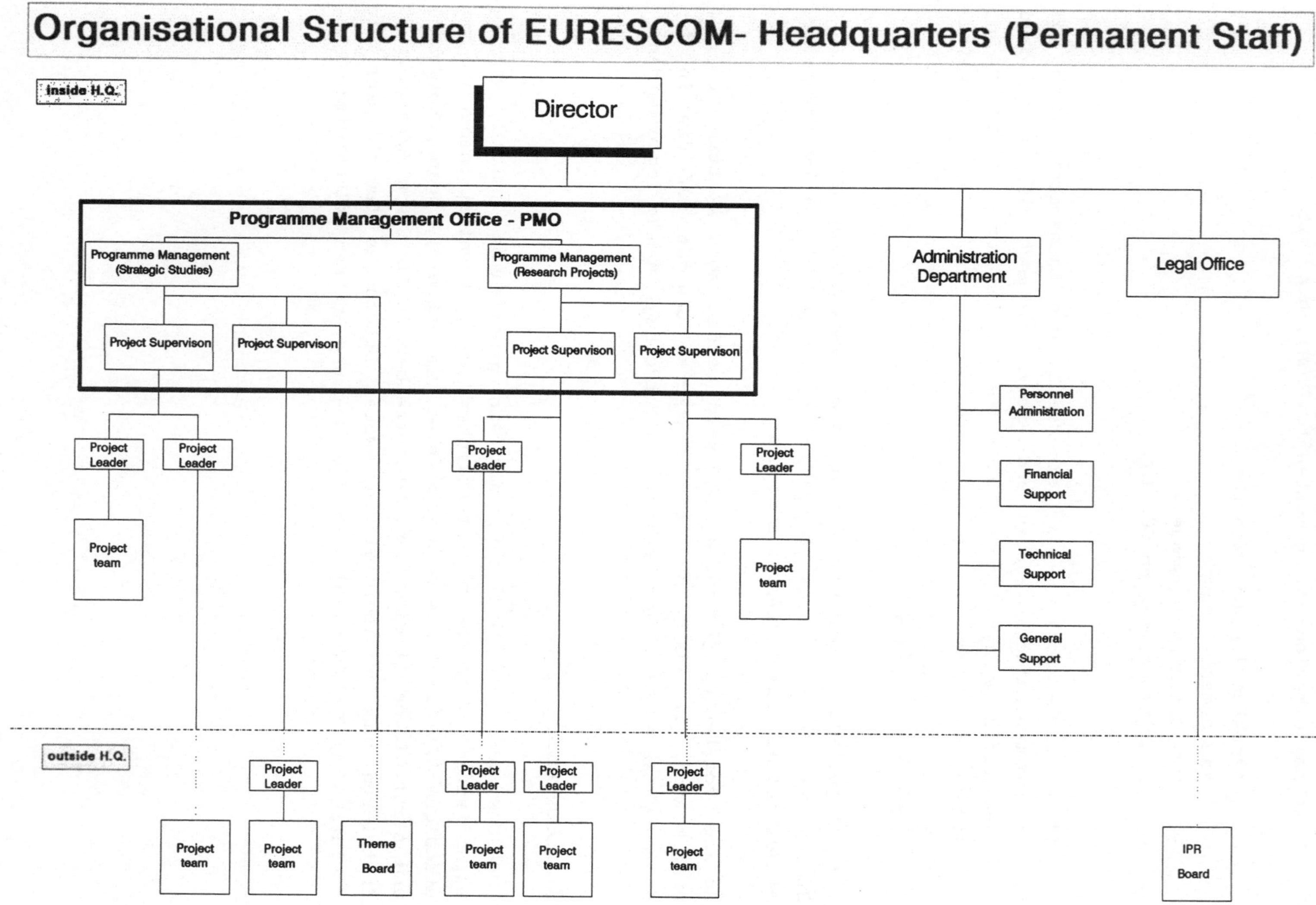
Organisational Structure of EURESCOM- Headquarters (Permanent Staff)
inside H.Q.
outside H.Q.
Director
Programme Management Office - PMO
Programme Management (Strategic Studies)
Programme Management (Research Projects)
Project Supervison
Project Supervison
Project Supervison
Project Supervison
Administration Department
Legal Office
Personnel Administration
Financial Support
Technical Support
General Support
Project Leader
Project team
Theme Board
IPR Board

Integration und Bedienung breitbandiger multimedialer Dienste auf zukünftigen Endgeräten

Das RACE Projekt MCPR[1]

M. Weiss[i] A. Lesch[i] R. Cordes[ii]
K.-H. Jerke[i] H. Rößler[i] P. Szabo[i]

(i) SEL - Alcatel
Research Center
Ostendstr. 3
7530 Pforzheim

(ii) Telenorma (Bosch Telecom)
Vorentwicklung
Mainzer Landstr. 128-146
6000 Frankfurt/Main 1

Kurzfassung

Im RACE[2]-Projekt MCPR (Multimedia Communication Processing and Representation) wird eine System-Architektur eines zukünftigen breitbandigen Endgerätes entwickelt. Hierbei werden relevante Technologien für die Kommunikation, der Verarbeitung sowie der Repräsentation und Präsentation multimedialer Dienste und Informationen eingesetzt. Zur Unterstützung der Bedienung sowie als Basiskonzept der Benutzerschnittstellen-Entwicklung werden Hypermediatechniken benutzt. Dieses Endgerät, das die heutige Workstation- und Videotechnologie als Basis für ein zukünftiges System integriert, ist als Terminal für das europäische Breitbandnetz (IBCN) zu sehen. Ein erster Prototyp bestehend aus zwei direkt verbundenen MCPR-Systemen wurde im Oktober 1990 der EG in Brüssel vorgestellt.

Das Konsortium dieses Projektes bestand in den Jahren 1988-1990 aus SEL-Alcatel (D; Prime Contractor), Alcatel-SESA (E), TELENORMA (Bosch Telecom) (D), Tecnopolis CSATA Novus Artus (I), Alcatel-FACE (I), Alcatel-STK (N)

---

[1]Die Arbeit entstand im Rahmen des RACE Projekts R1038 *Multimedia Communication Processing and Representation*

[2]RACE steht für Research and Development in Advanced Communications technologies in Europe

1 Einleitung

Systeme zur integrierten Bearbeitung von Text, Bild, Graphik, Sprache und in manchen Fällen auch Video oder Animation befinden sich derzeitig als Einplatzsysteme in der Entwicklung bzw. sind als erste Produkte verfügbar. Schon frühzeitig erkannte man, daß eine wichtige Anwendung der Austausch solcher **multimedialer Informationen** [24] ist. In der modernen Bürokommunikation widmete man sich sowohl im Bereich des Information Retrievals [3, 4] als auch im Bereich der Bürodokumente [14, 13] dieser Problematik. Heutzutage sind neben den Entwicklungen im Konsumgütermarkt für **Multimedia** [8, 9, 16], wesentliche neue Entwicklungen im Bereich neuartiger Netzstrukturen für **multimediale Kommunikation** [2, 20, 23, 26] sowie verteilter Anwendungen [10, 15, 18, 21] und deren Unterstützung durch Datenbanken [1, 7] zu sehen.

Ziel des RACE-Projektes R1038 *Multimedia Communication Processing and Representation* ist es, eine Architektur eines zukünftigen Breitbandendgerätes, das für Multimediale Dienste ausgelegt ist, zu definieren und zu realisieren. Ein wesentlicher Gesichtspunkt bei der Entwicklung eines derartigen multimedialen Arbeitsplatzsystems ist die Nutzung der Multimedia Technologie durch normale - gelegentliche - Benutzer (casual user). Durch die Integration unterschiedlicher Dienste wie Videotelefon, Videokonferenz, multimediale Dokumentenübertragung sowie Dienstkomponenten der Multimedia-Kommunikation werden spezielle Anforderungen an die Benutzerschnittstelle gerichtet. Die unterschiedlichen Arten der Diensteintegration sowie die diversen Qualitätsanforderungen werden im Rahmen des RACE Programms in Form von Feldversuchen untersucht, wobei das MCPR Terminal als Endsystem eingesetzt wird.

Aus diesen Aufgabenstellungen ergeben sich folgende Einflußfaktoren der Entwicklung:

- Breitbandtechnologie (ATM[3]),
- Medien (Text, Bild, Grafik, Audio und Video),
- Workstation Technologie und
- Zukünftige (Breitband-)Informationsdienste.

Im nächsten Abschnitt werden wir kurz die Bedeutung der Begriffe

- Kommunikation (communication)

[3]ATM = Asynchronous Transfer Mode

- Verarbeitung (processing) und
- Repräsentation/Präsentation (representation)

für unser Multimedia Endgerät Projekt erläutern. Daran anschließend werden die wesentlichen Konzepte der Bedienoberfläche präsentiert, um abschließend den ersten Prototypen zu erläutern und wesentliche Meilensteine für die verbleibende Projektlaufzeit zu präsentieren.

2 Multimedia Communication Processing and Representation

2.1 Multimediale Informationstypen

Für die Entwicklung unseres ersten Demonstrators gingen wir von den folgenden fünf Basisdatentypen aus:

- **Text** als beliebige Folge von Zeichen, die als ASCII Text codiert sind und durch unterschiedliche Attributierung (Font, Orientierung, Größe, ...) darstellbar sind.
- **Grafik** als Menge geometrischer Figuren, die durch Graphikeditoren erstellbar sind. Sie können in unterschiedlichen Formaten wie GKS3D, PHIGS++ oder ähnliches dargestellt werden.
- **Festbilder**, die im Gegensatz zum Datentyp Graphik aus bildpunktorientierten Informationen aufgebaut sind. Ein Bild, das zumeist in mehrdimensionalen Matrizen den einzelnen Bildpunkten Farb- oder Grauwerte zuordnet, kann z. B. in TIFF, CIFF oder nach dem JPEG Standard codiert bzw. komprimiert werden.
- **Video** oder **Moving Pictures** können als eine Folge von Einzelbilder (frames) angesehen werden, die mit einer Geschwindigkeit von ca. 25 frames pro Sekunde dargestellt werden müssen, um ein angemessenes Bild auf dem Monitor zu erhalten. Eine einfache Manipulation auf einem Videostrom ist das Grabbing (separieren eines einzelnen frames und Weiterbearbeitung als Festbild). Das Videosignal kann analog in das Display eines Arbeitsplatzrechners eingespeist werden, es gibt aber auch unterschiedliche Ansätze zur Digitalisierung von Video; z. B. innerhalb der ISO als MPEG-Standard [27].
- **Audio** kann sowohl Sprache als auch HiFi Musik in digitalisierter Form sein. Audioinformationen können einerseits als Erweiterung einer multimodalen Benutzerschnittstelle dienen, zum anderen auch als eigenständige Informationsträger z. B. als Annotationen eingesetzt werden.

Somit war eine wichtige Anforderung für den ersten Prototypen (Initial Demonstrator) die Integration der o. g. zeitabhängigen und zeitunabhängigen Datentypen auf einem Arbeitsplatzsystem als Kommunikationsendgerät.

2.2 Kommunikation

Die zentrale Forderung seitens der Kommunikation ist der Anschluß eines MCPR-Systems an ein ATM (Asynchronous Transfer Mode) basiertes Breitbandnetz wie z. B. das IBCN. Hierbei wird momentan die Einbeziehung eines sog. Terminal-Adapters verfolgt (s. Bild 1).

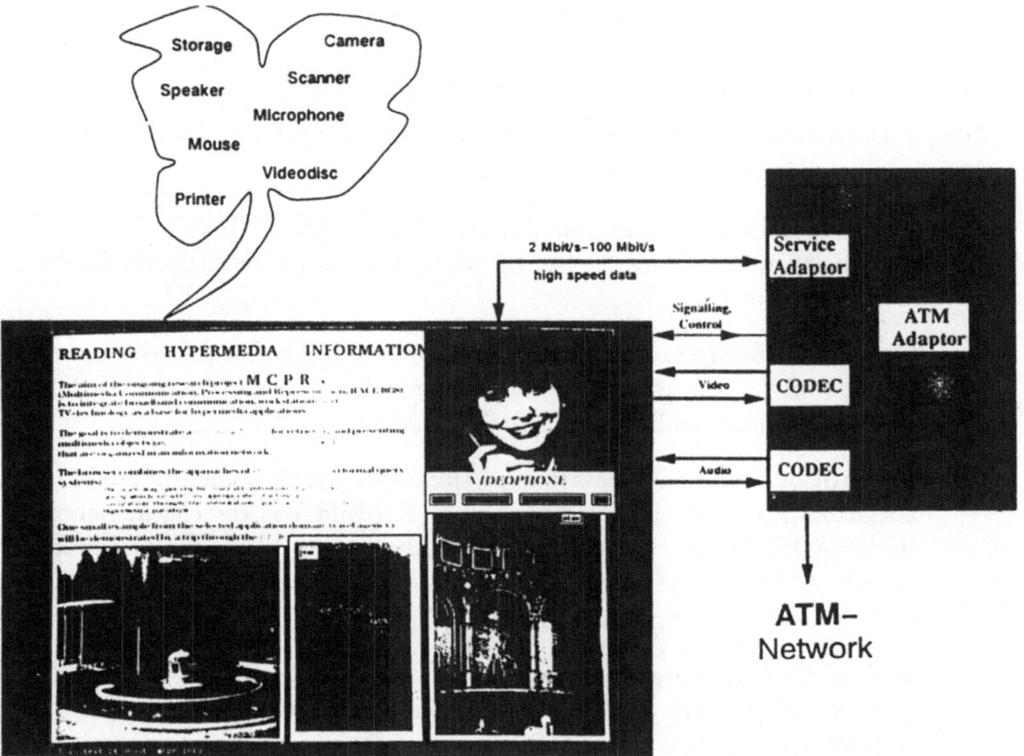

Bild 1: Kommunikationumfeld des Projektes MCPR

Ein derartiger Adapter beinhaltet neben der ATM-Umsetzung (Cell Multiplexing / Demultiplexing) auch mehrere Codecs für CBO (Continous Bitstream Oriented) Informationstypen wie Audio/Video sowie eine mögliche Diensteumsetzung oder einen Anschluß für andere physikalische Übertragungsarten wie 2 MB Kanäle oder 64 kbit ISDN.

Für den Initial Demonstrator sind als Kommunikationsdienste in erster Linie Multimedia Abrufdienste und Videotelefonie in Betracht gezogen worden.

Synchronisationsaspekte von Medien sind in diesem Rahmen nur am Rande betrachtet worden. Die Synchronisationsanforderungen werden durch ein entwickeltes Signalling Protokoll abgedeckt [17].

2.3 Repräsentation und Präsentation

Zur Repräsentation/Speichermodell (storage layer) sowie zur Präsentation (presentation layer) der multimedialen Anwendungen dient eine objektorientierte Architektur rsp. ein einfaches objektorientiertes Modell. Die Basisklassen werden durch folgende Beziehungen charakterisiert:

- is-a
- depends-on
- part-of

Bild 2 zeigt einen Ausschnitt des MCPR-Objektmodells. Es ist ein pragmatischer Ansatz zur objektorientierten Strukturierung von multimedialen Daten sowie zur Integration multimedialer Dienstelemente und weist somit nicht den Facettenreichtum auf, wie z. B. in [1, 3, 5, 10, 12, 15, 19, 21, 25] beschrieben.

Eine zentrale Klasse ist das monomediale *Particle*, aus denen komplexe, multimediale Objekte (composites) wie *Pages* aufgebaut sind. Durch subclassing erreicht man eine Unterteilung der Klasse *Particle* gemäß zeitabhängigkeit rsp. -unabhängigkeit in die Klassen *Permanent Particle* und *AV Particle*. Beide lassen sich dann medienspezifisch noch weiter verfeinern. Andererseits kann jedes *Particle* als direkt manipulierbares Objekt präsentiert werden und/oder gespeichert sein. Hierzu gibt es die Klassen *DM Objects* und *Storage Description*. Durch die Klasse *Storage Description* kann somit ein beliebiges Objekt persistent gemacht werden. Die Hypertextfunktionalitäten werden derzeit durch die Klassen *Reference Relation* sowie *List of Links* für jedes *Particle* realisiert.

Der Zusammenhang zwischen der Präsentationsschicht und dem Speichermodell wird durch das VIEW-Konzept realisiert (s. Bild 3). Durch das VIEW Konzept werden Objekte des Speichermodells auf dem Bildschirm als sog. *Pages* präsentiert. Diese sich überlappenden multimedialen Informationseinheiten bestehen aus sich nicht überlappenden *Particles* der Speicherschicht. In dieses Konzept sind ebenfalls die Ankerpositionen der unterschiedlichen Links integriert.

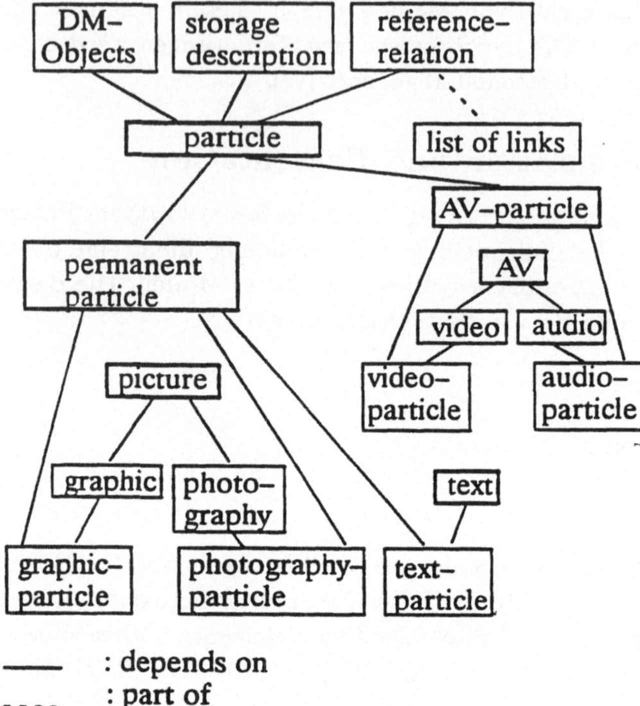

Bild 2: Objektorientiertes Datenmodell des Projektes MCPR

2.4 Be- und Verarbeitung

Durch die objektorientierte Strukturierung der multimedialen Informationseinheiten gelingt es, sowohl medienspezifische Bearbeitungsfunktionen - wie *rewind, play, record* für Audio oder Video bzw. *loudness* für Audio oder *contrast, brightness* für Video mit dem Medienobjekt gemeinsam zu implementieren, als auch medienübergreifende Funktion wie u.a der *Linking-Mechanismus* zur interaktiven Handhabung des multimedialen Informationsgeflechts zu realisieren. Einige einfache Medienkonvertierungen wie z. B. Videograbbing sind ebenfalls realisiert.

3 Benutzerschnittstelle

Aus logischer Sicht basiert die Benutzerschnittstelle auf einem Ereignismodell, in welchem der Nutzer über Eingabegeräte Ereignisse auslöst. Aus Anwendungssicht haben wir eine Hypermedia basierte Benutzerschnittstelle, in der Informationseinheiten in Knoten (nodes) und Beziehungen in Verweistypen (links) dargestellt werden.

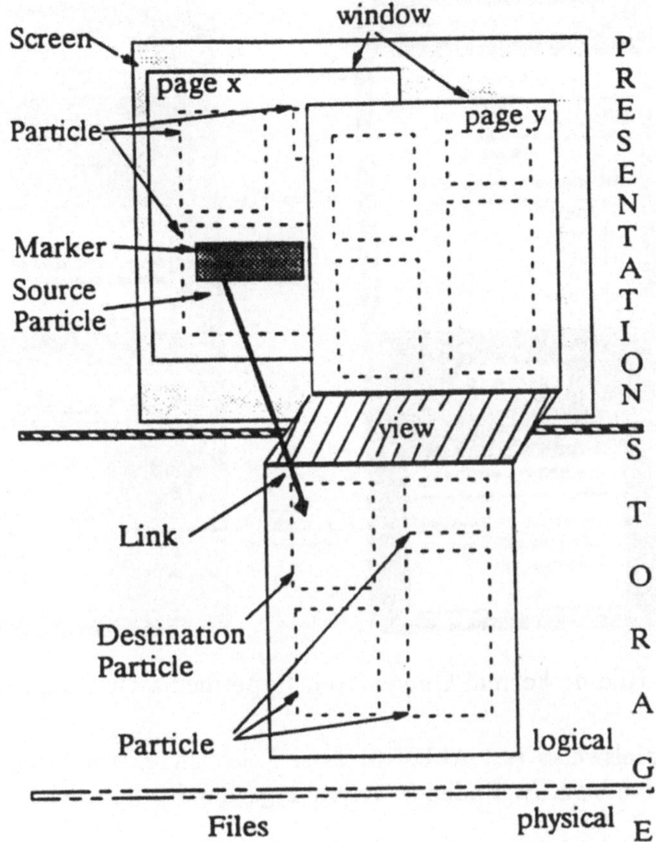

Bild 3: Zusammenhang zwichen Präsentations- und Speichermodell

Ereignisbasierte Mechanismen sind bei der Entwicklung asynchroner Dialoge bereits eingeführt. Physikalische Aktionen werden dabei von verschiedenen Geräten erfaßt und als *events* an die Objekte in der Präsentationsschicht der Softwarearchitektur weitergeleitet. Dort werden sie von den zugehörigen Objekten empfangen; somit sind diese Objekte direktmanipulativ.

Als Erweiterung der Hypermediafähigkeit der Benutzerschnittstelle und des Browsings ist eine Abfragekomponente für *formal queries* entwickelt worden. Die Architektur und die Integration in das MCPR-Konzept zeigt die Abbildung 4; in Abbildung 5 ist eine zweistufige Beispielanfrage dargestellt.

Die *Formal Query*-Komponente [11] basiert auf einer hierarchischen Konzeptstruktur der Anwendung. Die Möglichkeit, während einer Anfrage Modifikationen (Präzisierungen) durchzuführen, wird durch die Attribute der Objektinstanzen des Informationsgeflechts unterstützt. Somit kann ein inkrementelles Auswerten der Anfrage interaktiv unter Benutzung von Metainformationen gewährleistet werden.

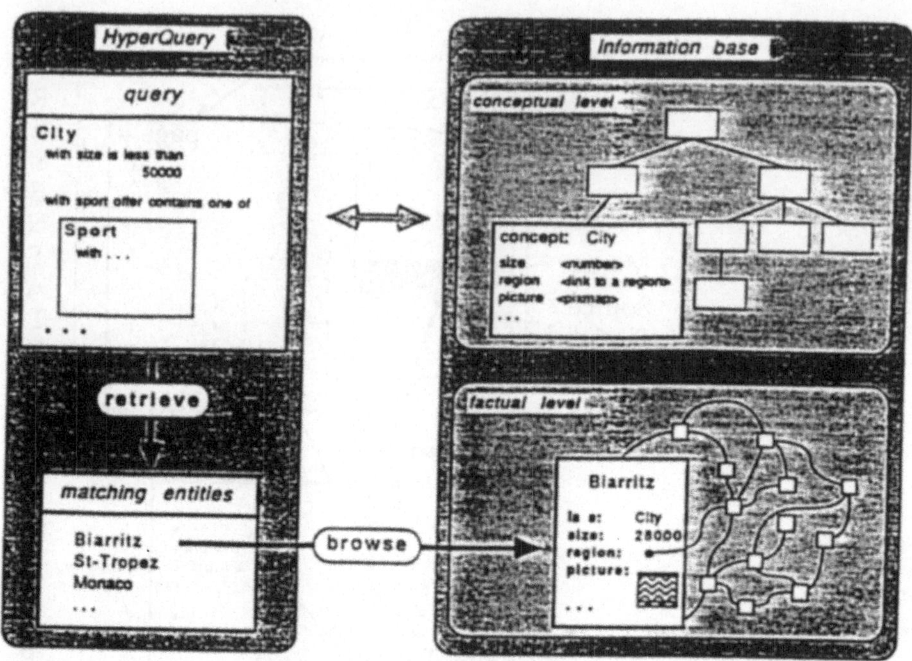

Bild 4: Formal Queries und Hypermedia Browsing in MCPR

Beispielsweise (s. Abb. 5) lassen sich zu einem geographischen Gebiet Deutschlands, z. B. Baden-Würtemberg, alle Städte mit einem Statement

FIND ALL *City*

auswählen, um sowohl die zugehörigen Information zu bearbeiten als auch gleichzeitig die Lage der Städte auf einer zugehörigen Landkarte anzuzeigen. Diese Landkarte kann als Ausgangspunkt für ein weiteres Browsing benutzt werden. Andereseits läßt sich auch die Anfrage durch eine gezielte Präzisierung so formulieren, daß nur die Städte betrachtet werden sollen, die ein Spielcasino und gute Restaurants besitzen. Als Ergebnis erscheint dann die Touristeninformation über Baden-Baden sowie ein zugehöriger Stadtplan für ein mögliches Browsing.

4 Stand der Arbeiten und Ausblick

Die Entwicklungsumgebung besteht z.Zt. aus SUN-Workstations und SYMBOLICS-Lispmaschinen, einem systemunabhängigen Videomixer sowie Videoboards, die Realzeit-Video in den hochauflösenden Farbbildschirm integrieren. Verschiedenste Videoquellen wie Videodisk, Videotape und Videokamera werden vom System (teilweise remote) gesteuert. Ein integrierter FDDI-Controller

Find all

City

with ***entertainment offer*** **contains** ***all–of***

Find all

entertainment

casino

restaurant

Bild 5: Zweistufige Beispielanfrage

ermöglicht das Experimentieren mit Datenraten wie sie bei der Breitbandkommunikation vorgesehen sind. Eines der wichtigen Ziele bei der Implementierung ist es, ein möglichst hohes Maß an Portabilität zu erreichen. Aus diesem Grund sind Standardsystemkomponenten wie UNIX, C, Common-Lisp oder das X-Window-System ausgewählt worden.

Der Demonstrator, der im Oktober 1990 der EG in Brüsssel vorgestellt wurde, basiert auf zwei direkt verbundenen SUN SparcStation zwischen denen parallele Verbindungen für Audio, Video, Daten und Signalling installiert worden. Die Applikation die dort präsentiert wurde - ein Reisebüro Szenario - integriert Bildkommunikationsdienste in Verbindung mit multimedialen Informationsdiensten.

In der Zukunft wird der Kommunikationsteil des MCPR Systems um eine Videokonferenzkomponente erweitert. Hiermit kann der Benutzer auch eine Bildkommunikationsverbindung mit mehreren Gesprächspartnern aufbauen, wobei die Videobilder der Partner auf dem Bildschirm des MCPR-Systems erscheinen. Diese Funktionalität kann lokal in Deutschland am VBN getestet werden, was für 1991 geplant ist. Für den Austausch der Multimedia Dokumente sowie einem darauf durchzuführenden Joint-Editing werden z. Zt. High-Level Protokolle untersucht. Ebenso wird das MCPR-System 1992 in mehreren paneuropäischen Breitband-Feldversuchen eingesetzt. Das MCPR System dient in diesen Feldversuchen dazu, adäquate Bedienoberflächen und neuartige Telekommunikationsdienste im Bereich multimedialer Anwendungen zu implementieren und zu validieren.

Literatur

[1] P. B. Berra et al.; *Architecture for distributed multimedia database systems*, computer communication, vol. 13, no. 4, May 1990

[2] S. Bulick et al., *The US West Advanced Technology Prototype multimedia Communication System*, GLOBECOM89, Texas, Nov. 1989

[3] S. Christodoulakis et al., *Multimedia Document Presentation, Information Extraction, and Document Formation in MINOS: A Model and a System*, ACM ToOIS, Vol.4, No.4, Oct 1986, pp. 345-388

[4] R. Cordes, R. Buck-Emden, H. Langendörfer, *Multimedia Information Management and Optical Disk Technologies as a Basis for Advanced Information Retrieval*; Proc. RIAO88, MIT, Cambridge (MA), pp. 65-80, März 1988

[5] R. Cordes, M. Hofmann, H. Langendörfer, R. Buck-Emden, *The Use of Decomposition in an Object-oriented Approach to Present and Represent Multimedia Documents*, Proc. HICSS-22, Kona (Hawaii), Jan. 1989, pp. 820-828

[6] R. Cordes, *On the Way to hypermedia and multimedia Terminals and Services*, Proc. Interactive Communication Tools, Paris, May 1990

[7] R.Cordes, T.Kummerow; *Multimedia Communication and Information Management based on Available and Emerging Standards*, accepted paper: IEEE-COMSOC, Interntional Workshop on Telematics 91

[8] *Digital Video Interactive*, Byte, May 1989, pp. 283-289

[9] *Digital Video Interactive*, CACM Special Issue, July 1989

[10] M.H. Hodges, R.M. Sannett, M.S. Ackermann, *A Construction Set for Multimedia Applications*, IEEE Software, Jan. 1989

[11] K.-H. Jerke et al., *Combining Hypermedia Brosing with Formal Queries*, Proc. Interact90, Cambridge, pp. 593-598

[12] W. Klas, E. J. Neuhold, M. Schrefl, *Using an object-oriented approach to model multimedia data*, computer communication, vol. 13, no. 4, May 1990

[13] Y. Komachi, K. Yamada, M. Aizawa, *Electronic Document and Publication Bank*, Proc. 5. IEEE Workshop on Telematics, Denver (CO), Sept. 1989

[14] C. Kou, O. Moriel, *The ODA - Toolkit - An efficient Approach to Office Application development*, Proc. 5. IEEE Workshop on Telematics, Denver (CO), Sept. 1989

[15] T. D. Little, A. Ghafoor, *Syncronization amd Storage Models for Multimedia Objects*, IEEE Journal on Selected Areas in Communication, Vol. 3, No.3, April 1990

[16] A.C. Luther, *Digital Video in the PC Environment*, McGraw Hill, 1989

[17] Race Project R1038 *Multimedia Communication Processing and Representation*, Deliverable D, *Initial Demonstrator Description*, November 1989

[18] E. Möller, A. Scheller, G. Schürmann, *Distributed processing of multimedia information*, Proc. ICDSC-10, Paris, May 1990

[19] M. Palaniappan, N. Yankelovich, M. Sawtelle, *Linking active anchors : a stage in the evolution of hypermedia*, Hypermedia, Vol.2, No.1, pp.47-66, 1990

[20] R. Popescu-Zeletin et al., *A Global Architecture for Broadband Communication Systems : The BerKom Approach*, Proc. Future Trends in Distributed Computing Systems in the 1990s, Hong Kong, 1988, IEEE Computer Society.

[21] K.V.B. Rao, A. Gafni, G. Raeder, *Dynamo: A Model for a Distributed Multimedia Information Processing Environment*, Proc. HICSS-22, Kona (Hawaii), Jan 1989, pp. 800-809

[22] K. Süllow, R. Cordes, *Einbeziehung von Hypermediatechniken in die multimediale Kommunikation* Proc. Workshop Hypertext/Hypermedia 90, Darmstadt, April 1990

[23] R. Steinmetz, *Synchronization Properties in Multimedia Systems*, IEEE Journal on selected Areas in Communication, Vol.3, No.3, April 1990

[24] D. Tsichritzis et al., *A Multimedia Office Filing System*, Proc. VLDB83, Florence, 1983

[25] D.Woelk, W.Kim, *Multimedia Information Management in an Object-oriented Database System*, Proc. VLDB87, Brighton, pp. 319-329

[26] D. Wybranietz, R. Cordes, F.-J. Stamen, *Support for Multimedia Communication in Future Private Networks*, Proc. 1st Int. Workshop on Network and Operating System Support for Digital Audio Video, Berkeley (CA), Nov. 1990

[27] H. Yasuda, *Standardization activities on multimedia coding in ISO*, Signal Processing - Image Communication, Vol.1, No.1, June 1989

Internet Communication with End-to-End Performance Guarantees

David P. Anderson
University of California at Berkeley
Department of Electrical Engineering and Computer Science
Berkeley, CA 94720, USA

Ralf Guido Herrtwich
IBM European Networking Center
Tiergartenstr. 8
D-6900 Heidelberg 1

Abstract. This paper describes the *Session Reservation Protocol* (SRP). SRP is defined in the DARPA Internet family of protocols. It allows communicating clients to reserve the resources, such as CPU and network bandwidth, necessary to achieve given performance (delay and throughput) objectives. The immediate goal of SRP is to enable IP-based distributed systems to handle "continuous media" (digital audio and video). However, SRP is applicable to any application that requires guaranteed-performance communication. The design goals of SRP include (1) independence from transport protocols (SRP can be used with standard protocols such as TCP or with specialized protocols); (2) compatibility with IP (data packet formats are not modified); (3) a host implementing SRP can benefit from its use even when communicating with hosts not supporting SRP. SRP is based on a workload and scheduling model called the *DASH Resource Model.* This model defines a parameterization of client workload, an abstract interface for hardware resources, and an end-to-end algorithm for negotiated resource reservation based on cost minimization. SRP implements this end-to-end algorithm, handling those resources related to network communication. The approach can easily be applied to other network environments such as OSI communication.

1.0 INTRODUCTION

The work described in this paper was motivated by the goal of incorporating *continuous media* (digital audio and video) into a general-purpose distributed system, using standard networks and IP-based protocols. It will soon be possible to equip average workstations with the hardware to handle digital continuous media [1] and to connect these workstations by high-speed wide-area networks capable of handling continuous-media data [2]. This hardware base can support distributed continuous-media applications like video conferencing systems and the real-time display of video data stored on a remote file server [3].

Such applications transmit continuous data streams between hosts, and impose end-to-end constraints on the performance (delay and throughput) of this communication. For example, an audio/video conversation using CD-quality sound and compressed NTSC-quality video might require a sustained throughput of about 3 Megabits per second and a maximum delay of a few hundred milliseconds. The performance of current computer hardware, such as CPUs, networks, and disks, is generally sufficient to handle several streams of continuous-media data. However, the scheduling policies used in hosts and gateways are designed primarily for fairness and simplicity. Therefore the performance of a given connection may fail to meet the application's requirements, especially during periods of heavy system load [4].

To allow end-to-end performance guarantees in the presence of concurrency and contention, a new approach to resource scheduling with the following properties is needed:

- Resources (CPU, network, disk, *etc.*) that can potentially become bottlenecks must be scheduled in a way that allows "reservations" (with associated performance guarantees) to be made to individual clients.
- In making a reservation, clients must specify their workload. This is only possible if the workload is known when the reservation is made, *i.e.*, if the software generating the workload operates in a predictable fashion.
- Since data may traverse resources on several hosts, a protocol for distributed resource reservation is needed. This protocol in turn requires a uniform interface to the various resources involved.

In this paper we describe a resource reservation protocol called SRP (*Session Reservation Protocol*) that allows performance guarantees to be made for communication based on IP [5]. SRP can be viewed as a management protocol operating at the internetwork (IP) layer. SRP is directly responsible for reserving only network resources. However, it is designed to function as part of a larger framework in which other system resources (disks, DSP chips, *etc.*) are reserved and scheduled together with network resources. The design goals of SRP also include the following:

- Independence from transport protocols. SRP can be used with standard protocols such as TCP or with specialized protocols for continuous-media transport.
- Compatibility with IP. Header fields of IP packets are not added or modified. On hosts that implement SRP, however, the IP module must be modified so that the relative priority of incoming packets can be established.
- A host implementing SRP can benefit from its use even when communicating with hosts not supporting SRP.

The TCP/IP framework provides both advantages and disadvantages for continuous media communication. IP has the advantages of providing widespread connectivity and of using an unreliable datagram service (continuous media applications often do not require reliability and could not, in fact, tolerate the delays caused by retransmissions in reliable link-layer protocols). However, the dynamic routing generally used by IP implementations poses problems for real-time communication.

For continuous-media data, regular delivery is at least as important as reliable delivery, and sometimes more important. TCP [6], the dominant stream transport protocol in IP networks, is not especially well-suited to continuous media applications. While it is possible to use TCP for continuous-media data, its main features (flow control and error recovery) are not useful in this context. They may actually interfere with timing, and increase workload, in ways that are detrimental to real-time performance. In the long run it will likely be necessary to develop new transport protocols for continuous media; in the meantime it is important to accommodate standard protocols.

The remainder of this paper is organized as follows. Section 2 deals with the workload and scheduling model on which SRP is based. Section 3 specifies the actual SRP protocol. The interaction between IP and SRP is described in Section 4. Section 5 describes an extension of SRP that accommodates hosts in the communication path that do not implement SRP.

2.0 THE DASH RESOURCE MODEL

To formalize the reservation of resource capacity, a model for expressing workload and processing is needed. The model used in SRP is called the *DASH Resource Model* [7]. In this model, the set of system components that handle continuous-media data is decomposed into a set of *resources*. In general, a resource corresponds to a schedulable hardware device and its accompanying software driver. For example, a CPU and its scheduler might comprise a resource. Resources may also be more complex: a local area network (which includes multiple interface devices, concurrent operation, and multiple scheduling mechanisms) might be treated as a single resource.

The DASH Resource Model assumes that work is assigned to resources in discrete units called *messages*, typically representing a segment of continuous-media data. Each message has a well-defined *arrival time* at which it is available for handling by a resource and *completion time* at which the handling is finished.

The flow of continuous-media data is considered to consist of linear simplex streams of messages that pass through one or more resources. Data is generated by a *source resource* (a disk or ADC), is then processed by a sequence of *handler resources* (networks and CPUs) and finally is consumed by a *sink resource* (disk or DAC). A message's completion time in one resource is its arrival time at the next resource. Many of these simplex data streams may exist concurrently, even within a single application. Therefore this scheme encompasses many continuous-media applications: playback of continuous media from disk, storage on disk, live conversations between human users, and others.

2.1 Linear Bounded Arrival Processes

Each data stream flowing across an interface defines an *arrival process* into the downstream resource. To describe the message arrival, the DASH Resource Model uses *linear bounded arrival processes* (LBAPs), an abstraction introduced by Cruz [8]. An LBAP has the following parameters:

- *maximum message size* S_{max} (bytes)
- *maximum message rate* R_{max} (messages/second)
- *maximum burst size* B_{max} (messages)

In any time interval of length T, the number of messages arriving at the interface may not exceed $B_{max} + TR_{max}$. The long-term data rate of the LBAP is $S_{max}R_{max}$ bytes per second. The burst parameter B_{max} allows short-term violations of this rate constraint, modeling programs and devices that generate "bursts" of messages that would otherwise exceed the rate constraint. Note that this model is not intended for bursty traffic; it merely takes irregularities in periodic data communication into account.

We define a function $b(m)$ representing the *logical backlog* of the arrival process. This is the number of messages by which the arrival process is "ahead of schedule" (relative to its long-term rate) when message m arrives. The logical backlog is not necessarily the number of queued messages since a resource may process messages upon their actual arrival if it is fast enough. $b(m)$ is defined by

$$b(m_0) = 0$$

$$b(m_i) = \max(0, b(m_{i-1}) - (t_i - t_{i-1})R_{max} + 1)$$

where t_i is the arrival time of message m_i. Using $b(m)$, we define the *logical arrival time*, $l(m)$, of a message m as

$$l(m_i) = t_i + b(m_i)/R_{max}$$

Intuitively, $l(m)$ is the time m would have arrived if the LBAP strictly obeyed its maximum message rate.

2.2 Sessions

The use of a resource by a particular data stream is called a *session*. A session represents a reservation of part of the capacity of the resource. Clients must request sessions with all of the resources they need (using a scheme defined below) prior to sending messages. As part of the reservation, the client must specify its workload; in return, the resource provides a bound on the delay it will impose. The client can then decide if this delay is sufficient for its purpose.

Each session has associated sets of LBAP parameters for its input and/or output interfaces. A handler resource accepts LBAPs, producing output LBAPs. The client of the resource must enforce the input LBAP parameters; the scheduler of the resource must enforce the output parameters. We assume that handlers do not modify the message stream, *i.e.*, that they do not lose messages, change the size of messages, speed up or slow down the message stream. Therefore, the incoming and outgoing LBAP for handler resources must have the same values for S_{max} and R_{max}. On the other hand, incoming and outgoing LBAP may have different burst sizes, depending on the overall workload and scheduling policy of the resource.

In addition to their LBAP specifications, handler sessions also have the following parameters:

- *maximum logical delay* L_{max}
- *minimum actual delay* A_{min}
- *maximum buffered delay* M_{max}

The *actual delay* of a message *m* in a handler resource is the time interval between its arrival at the input interface and its arrival at the output interface. The *logical delay* of *m* is the interval between *m*'s logical arrival time and its logical arrival time at the output interface. Logical delay, rather than actual delay, determines end-to-end delay bounds. The *buffered delay* is the portion of actual delay during which the message is stored in host memory. In resources such as wide-area networks, M_{max} will be smaller than L_{max} due to message propagation time. A_{min} and M_{max} are used to calculate buffer space needs.

When data traverses a sequence of resources, the *basic sessions* within the resources are said to form an *end-to-end session*. An end-to-end session represents a unidirectional point-to-point communication path that traverses several resources, either within a single host or across a network. The output interface of each resource in an end-to-end session is the input interface of the next resource. Each resource must be prepared to handle the burst size generated by the previous resource. The *end-to-end logical delay* of a message is the interval between its logical arrival time at the source output and its logical arrival time at the sink input. The maximum logical delay of an end-to-end session is the sum of the maximum logical delays of its handler resources.

2.3 Session Establishment Decision Algorithms

When establishing a new basic session, the session manager has to ensure that the workload of this session does not lead to a violation of previously given guarantees. How this "non-interference" can be determined is highly resource-specific, but any policy for which an upper bound on delay can be derived from given input LBAP parameters can be used. For example, *round-robin*, *FIFO*, *rate-monotonic* and *earliest-deadline-first* scheduling all have this bounded-delay property and the resource interface can be successfully implemented on top of them.

Many existing results in real-time scheduling can be applied [9], especially if we restrict our attention to resources that consist of a single hardware device (*e.g.*, a CPU). For example, if guaranteed delay can always equal the interarrival time of messages on a session, a simple test for preemptive *rate-monotonic* scheduling of a singular resource would be

$$\sum_{s \in E} R_{max}(s) T_{max}(s) \leq |E|(2^{1/|E|} - 1)$$

where E is the set of all established sessions (including the new one) and T_{max} is the *maximum service time* for each message. Under the same conditions,

$$\sum_{s \in E} R_{max}(s) T_{max}(s) \leq 1$$

is a sufficient non-interference condition for preemptive *earliest-deadline-first* scheduling [10]. Worst-case simulation provides a more general decision procedure [11].

For resources that encapsulate more than a single device, the management procedures are more complicated. For example, the sources of delay in an FDDI network (viewed as a single resource) include queuing, media access, and propagation, and many scheduling policies are possible. Establishing a session in such a resource may involve using network management protocols to reserve network bandwidth in "synchronous" or "isochronous" channels. The question of how to support sessions in such a resource is a subject of ongoing investigation.

2.4 An Economic Approach to Delay Allocation

It may be possible for a resource to guarantee a maximum delay anywhere within a certain range. The shorter the delay is, the more costly it is because the resource has less freedom to schedule other service requests – and the fewer additional sessions it can support. To divide the delay between resources in an end-to-end session the DASH Resource Model takes an approach based on economics. This approach has also been used for problems such as routing and load-balancing [12].

When a client reserves a session with a resource, the resource makes reservations for the smallest possible maximum delay. In addition, the resource provides a *cost function* indicating, for each larger maximum delay, the associated cost to the client. The client may relax the maximum resource reservation to minimize cost. Cost can either be real money, to be later billed to the client, or some metric reflecting the resource's current load. The cost function may be a function of the workload of a resource or of parameters like the time of day or the identity of the user (*e.g.*, so that frequent users pay less). For tractability, the DASH Resource Model requires that every cost function be (1) continuous and piecewise linear; (2) strictly monotonic decreasing, and (3) convex. The first value for which a cost is given is the smallest achievable maximum delay. We assume that at some point more delay will not lead to lower costs because the cost of buffering messages over the delay period will exceed the cost saved by the larger delay.

The cost of an end-to-end-session is the sum of the costs of its component sessions. Since we have defined cost functions to be convex piecewise linear functions, they can be combined by the following procedure: The segments of the functions are sorted in order of decreasing (more negative) slope. They are placed end-to-end, starting at the point which is the sum of the initial endpoints of the functions.

2.5 Buffer Reservation

The DASH Resource Model is designed to prevent message loss due to buffer overflow. For a given resource, this requires reserving enough buffer space to accommodate the input burst size plus messages being processed in the host. In bytes, this amount of buffer space is given by the expression

$$S_{max}(B_{max} + R_{max} M_{max})$$

Let $\overline{M}_{max}$ be the maximum buffered delay summed over the chained basic sessions within one host and let B_{max} be the input burst size of the first session. Then the number of buffers needed for all sessions in the host is

$$S_{max}(B_{max} + R_{max} \overline{M}_{max})$$

A second need for buffer space arises when the receiving end of an application must deliver messages at a constant rate to the output device (*e.g.*, audio or video converters). Suppose the first message of a stream arrives with minimum delay. If the application outputs the first message immediately, and the second arrives with maximum delay, there will be an unacceptable pause in the output between the two. The application must therefore buffer messages to ensure that there is no "jitter" in the output.

Assuming that the source resource generates messages fast enough to maintain a nonzero backlog, jitter can be avoided as follows. The receiver waits until

$$R_{max}(\overline{L}_{max} - \overline{A}_{min})$$

messages have been received (where $\overline{L}_{max}$ and $\overline{A}_{min}$ are the sums of L_{max} and A_{min} over all sessions in the end-to-end session), and then waits until the logical arrival time of the last of these messages. If O_{max} is the output burst size of the last session, the number of bytes needed for buffering is

$$S_{max}(O_{max} + R_{max}(\overline{L}_{max} - \overline{A}_{min}))$$

2.6 End-to-End Session Establishment

The DASH Resource Model defines an establishment protocol for end-to-end sessions. Using this protocol, the application's allowable end-to-end delay is divided between the resources, and burst sizes are established. The establishment protocol is carried out by *host resource managers* (HRMs).

Initially, the HRM at the source host is given a client request that specifies the resources involved, the message size and rate, and the end-to-end delay requirements. These requirements are given by a *target* and *maximum* value, denoted E_{target} and E_{max}. We do not specify how clients learn about their delay requirements. They could negotiate them immediately prior to establishing the session.

The protocol has two phases:

1. The first phase traverses the hosts from the source to the sink. A *request* message is exchanged between HRMs. The request message contains the data message size and rate, the client delays E_{target} and E_{max}, the burst size from the previous host, the cumulative sums of L_{max} and A_{min}, and the cumulative cost functions. Maximum reservations are made for each resource, and corresponding buffer space is reserved.
2. The second phase proceeds in the reverse direction. The receiving client evaluates the end-to-end session parameters and decides on a delay for the session. A *reply* message containing the remaining excess delay (see below) and the burst size into the next host is passed back towards the source. For each resource, the session parameters are relaxed appropriately. The delay may be increased and additional buffers may be reserved both for this purpose and to accommodate larger input bursts. Delays are relaxed only up to the amount of buffer space available.

Let E_{actual} be the actual end-to-end logical delay obtained in the first phase of the session establishment. If this delay is less than E_{target}, some *excess delay* E_{excess} defined as

$$E_{excess} = E_{target} - E_{actual}$$

can be distributed among the resources. This should be done as economically as possible, *i.e.*, in a way which saves the largest amount of money.

In the second phase of the protocol, each HRM hands the remaining excess delay to the previous one. Each HRM knows the outgoing accumulated cost function and the cost functions of all local resources. It shifts the outgoing cost function to the left, so that the first cost value is given for 0. From this cost function the segments from 0 to E_{excess} are examined. If any of these correspond to segments of local cost functions, the corresponding resources are relaxed by the time-extent of the segment, provided there is enough buffer space available. If E_{excess} lies in the middle of a segment, the amount of the relaxation is the part of the segment that lies to the left of E_{excess}. Any excess delay that is not returned to local resources is passed back to the previous host.

3.0 THE SESSION RESERVATION PROTOCOL

SRP is a mechanism to achieve performance guarantees for communication in the Internet. It is used in the process of establishing an end-to-end session with resources involved in an IP-based communication between a sending and a receiving client. This end-to-end session is associated with a connection of a particular IP-based protocol (for example, a TCP connection). The performance guarantees of the end-to-end session apply to the data traffic from the sending to the receiving client on the associated connection.

SRP is responsible for reserving (and relaxing) the following resources:

- On the sending host, SRP reserves the network resource through which messages leave the host.
- On a gateway, SRP reserves the CPU resource needed for protocol operation and the network resource through which messages leave the gateway.
- On the receiving host, SRP makes no reservations, and simply conveys the session request to the receiving client. (The incoming network resource was reserved by the previous host in the path.)

The clients of SRP are responsible for reserving all other resources that handle the stream of continuous-media messages (for example, the reservation of an input or output device). Thus, on the sending and receiving hosts the HRM function is divided between the sending client and SRP. We assume that the set of resources involved in a connection is always fixed, *i.e.*, that the connection is based on a static route through the network. How this can be achieved for IP-based communication will be described in Section 4.

Every resource reservation by SRP includes a corresponding reservation of buffer space. SRP reserves no buffer space to avoid jitter in the output, but it provides the receiving client with information about the delay so that the client can implement jitter avoidance.

3.1 Operation of SRP

A typical scenario for establishing an end-to-end session using SRP is the following:

1. The sending and receiving clients set up a transport-level connection. The performance goals for the connection are determined, and a globally unique end-to-end session ID is obtained.
2. The sending client reserves all resources related to the data source (disk, camera, VCR, *etc.*) and the CPU resource. The service times of the CPU reservation must include the requirements of all software modules used by the application, *i.e.*, not only the application itself, but also the file service, the network protocols, *etc.* The client can learn about these requirements, *e.g.*, by calling a corresponding function of the protocol module it uses (which in turn can call a function of the protocol *it* uses, and so on). The sending client then calls the local SRP module.
3. The SRP module on the sending host examines the address of the receiving client and consults the IP routing table to determine the network a message on the associated connection is sent on and the next node within the connection. This node may either be the receiving client itself or a gateway. The SRP module then reserves the appropriate network resource and the necessary buffer space, and forwards the end-to-end session request to the next hop.

4. On a gateway, the SRP module reserves both CPU and network resources and the corresponding buffer space. To make a CPU reservation, it first determines the maximum service time of the gateway software, which depends on the incoming and outgoing network of a connection. It then makes a reservation for the outgoing network just as described in the previous paragraph.
5. The SRP module at the receiving host notifies the receiving client, which reserves the CPU resource and the remaining resources related to the data sink (conversion hardware, *etc.*).
6. The receiving client compares its own requirements and those of the sending client with the obtained maximum delay. If the obtained guarantees exceed the requirements, the receiving client relaxes the reservations of local resources, and returns the remaining excess delay to the SRP module.
7. SRP completes the backward pass of the end-to-end session establishment through the gateways, relaxing reservations according to cost functions and remaining excess delay. It also adjusts burst sizes.
8. The SRP module on the sending host relaxes the network resource, then returns to the sending client. The client relaxes its local resource reservations, completing the session establishment.

Once a session has been established, the sending client can begin sending data. As IP datagrams arrive at each host (gateways and receiver) the IP modules at these host associate these datagrams with the corresponding end-to-end session, and pass this information to the schedulers of the resources involved (CPU, network, etc.). Thus the end-to-end performance guarantees are achieved.

3.2 Protocol Elements of SRP

SRP modules communicate using the Sun RPC protocol [13]. This protocol makes use of either the UDP or TCP protocols. Sun RPC offers reliable communication and has a convenient programming interface. It provides authentication mechanisms which could be used to prevent unauthorized users from reserving resources.

Each SRP module implements an RPC program with three procedures: a call to establish sessions, another to delete them, and a mandatory null call taking no parameters and returning no results (for determining round-trip time). The SRP module itself gets an identification number (which has to be assigned by a network authority in order to make it unique). In the Sun RPC interface specification language, an extension of the XDR External Data Representation Standard [14], the SRP program is as follows:

```
program SRP {
   version Original {
      void null_function (void) = 0;
      E_rep_data session_establish (E_req_data) = 1;
      void session_delete (E_session) = 2;
   } = 1;                            // first protocol version
} = ?;                               // to be assigned
```

The format of a request message is the following:

```
struct E_req_data {
   E_session     esid;             // end-to-end session ID
   int           destination;      // receiving host (Internet address)
   int           net;              // incoming network (Internet address)
   Connection    connection<>;     // associated connection
   float         rate;             // rate
   int           in_burst;         // input burst limit
   int           size;             // message size
   float         tgt_ete_delay;    // target end-to-end delay
   float         max_ete_delay;    // maximum end-to-end delay
   float         acc_max_delay;    // maximum logical delay so far
   float         acc_min_delay;    // minimum actual delay so far
   Point         acc_cost<>;       // cost function so far
};
struct E_session {
   int           source;           // sending host (Internet address)
   int           id;               // unique number
};
```

Each end-to-end session is associated with an upper-level connection. These connections can be defined at any layer in the protocol hierarchy, at the transport level or above. Each connection is therefore identified by a set of data items (connection numbers, addresses, etc.), one per protocol layer. Each IP datagram sent on the connection contains these data items in its various headers. The length of the items, and their position within the IP datagram, depend on the upper-level protocols. The sending client obtains this protocol-specific "association data" from the protocol layers it uses and passes it to SRP as part of the session establishment request.

```
enum Protocol { UDP, TCP, RPC, NFS, ...};
union Connection switch (Protocol proto) {
    case UDP:                  UDP_data udp_data;    // UDP datagram ID
    case TCP:                  TCP_data tcp_data;    // TCP connection ID
    ...
};
```

Reply messages have the following format:

```
enum Rep_code { ESTABLISHED, NOT_ESTABLISHED };
struct E_rep_data {
   E_session          esid;          // end-to-end session id
   union Reply switch (Rep_code rep_code) {
      case ESTABLISHED:      struct {
                                float    excess_delay; // remaining excess delay
                                int      out_burst;    // acceptable burst
                             } pos_info;
      case NOT_ESTABLISHED:  struct {
                                int      failure;      // failure code
                                int      culprit;      // failing host
                                char     msg<>;        // name of failing resource
                             } neg_info;
}; };
```

A failure can either be due to SRP or to an RPC failure further down the chain. In case of a failure the address of the host signaling the failure and a failure message (perhaps the name of the failing resource) are given.

4.0 SESSION ASSOCIATION IN IP

We want to achieve performance guarantees for conventional IP-based communication without changing the communication protocols involved. This means that data messages cannot contain an explicit identification of the session to which they belong. To correctly schedule guaranteed messages, hosts need to be able to distinguish them from other network traffic and know the parameters of the parent session.

4.1 Session Registration

To be able to identify the sessions of incoming IP packets, the IP module at a node must be informed of sessions through the node. Therefore, IP contains a *registration procedure* that is called by SRP when a session is established; it is passed the session ID and the association data obtained from the sending client. To achieve static routing for each same session, the registration call specifies the next machine to which all IP messages of the respective session will be sent, and the network to be used.

In practice, a registration procedure adds new entries to per-protocol "lookup tables" (hash tables keyed by protocol-specific ID fields) for each of the protocols in its association-data array. Each entry in a lookup table corresponds to a connection of the corresponding protocol. Its value is either a session ID (if the protocol is the last element of the association-data array) or identifies the next protocol up in the sequence.

4.2 Session Identification

When an IP packet arrives at a host, the IP module must find what session (if any) the packet is associated with. While various implementation approaches are possible, the following organization is suggested. For every upper-level protocol, there is an *identification procedure* that, given an incoming packet of that protocol, identifies the SRP session, if any, to which the packet belongs.

These procedures encapsulate the "lookup tables" maintained for each protocol (see above). They work as follows: The protocol-specific identifiers are extracted from the IP packet and used to hash into the lookup table. If no entry is found, a failure code is returned. If the entry contains a session ID, this ID is returned. Otherwise the number of the next-higher protocol is obtained from the packet, the identification procedure for that protocol is called, and its value is returned.

For each incoming packet, then, IP extracts the protocol field from the IP header and calls the corresponding identification routine. This is normally done at the interrupt level. If a session ID is found, the scheduling of subsequent handling of the packet (*e.g.*, the deadline of the process that handles it) is based on the logical arrival time of the packet and the session delay. Otherwise scheduling can be done by conventional means, *e.g.*, FIFO.

Outgoing packets on the sending host or on a gateway leave the node through the network resource. They are handled by IP in the following way: We assume that the SRP session ID is known. On the sending host, it can be passed down from the client, on gateways, it was obtained as above. Again, the session ID is used to calculate the logical arrival time of the message into the network resource, which in turn is used to determine the queuing order of the packet and perhaps the manner in which it is transmitted on the medium.

If higher-layer packets are fragmented, an IP packet may not contain association data. In this case session identification works as follows: Because of static routing all fragments arrive in order. The first fragment has the *more fragments* field set. We assume that this fragment contains the ID of the associated connection (a reasonable assumption since (1) IP packets are sufficiently large, and (2) in all Internet protocols the address field is contained in the packet header). IP calls the appropriate identification procedure for this packet and stores the obtained session ID together with the *packet ID* of the IP packet. Future fragments can be identified by having the same packet ID. Once the last fragment arrives (without the *more fragments* field set), IP can delete the fragment identification information.

In order not to have to call the identification procedure for *every* incoming IP packet, we request that all IP packets belonging to SRP sessions are sent with the *low delay* field set. This field is a hint to IP that a packet may have performance guarantees associated with it.

5.0 INTERACTING WITH NODES THAT DO NOT IMPLEMENT SRP

If one of the nodes in the route of an IP-based connection does not implement SRP, an end-to-end session is not possible. In some cases, a *partial session* among nodes supporting SRP is a useful alternative to no session at all. Of course, strict performance guarantees cannot be made for a partial session, but if the performance bottlenecks are within the sites involved in the partial session, the resulting performance may be adequate.

To establish a partial session, at least one of the client nodes must have an SRP module. We call this module the *anchor module*. Let us assume the anchor module is located on the sending host and the complete route of the session is known. In this case, after making local reservations, the anchor module sends RPCs to other SRP modules, starting with the second node in the route, until the call is successful. The session parameters and the remaining route are forwarded in this RPC. The contacted SRP module tries to continue the session establishment by contacting the next node of the route which implements SRP.

The accumulation of delay parameters and cost functions takes place just as in the establishment of complete sessions. Burst parameters and incoming network specifications, however, may not be valid if a previous node did not participate in the session establishment. The values of these parameters are merely hints, and needed buffer space and service time must be estimated. In the second phase, output burst relaxations should only be made if the resource offering to accept a larger burst is the immediate successor in the session. If not all immediate successors of a node implement SRP, the *low delay* field can no longer be used as a hint that a packet may belong to an SRP session.

Clients involved in partial sessions must be able to enforce the route that data packets will follow. In some cases there may be only one possible route. If this is not the case, the sender can send a "probe packet" with the IP *record route* option to find a route, and subsequently send data packets with the IP *strict source routing* option (this may involve modifications to transport protocol implementations). If the anchor module is located on the receiving host, a similar approach can be used (we assume that the route is reversible). The anchor module contacts the first SRP module in the route, which then initiates the reservation procedure.

The establishment protocol for complete sessions requires the receiving client to be contacted after the first phase of the protocol. In partial sessions, however, the receiving client may have no access to SRP. The sending client will then play the role of the receiving client in deciding about the final delay parameters. If the last SRP module on the route is not located on the receiving host, it calls back the anchor module on the sending site where an indication is delivered to the sending client.

6.0 CONCLUSION

We have presented a scheme to achieve communication with end-to-end performance guarantees in the framework of an existing and widely used protocol architecture. The mechanisms described in this paper, though implemented for IP-based network communication, are by no means restricted to that framework. This choice, however, enables a large number of systems to participate in guaranteed-performance communication, requiring a minimum degree of modification to existing systems. Performance guarantees could also be associated with OSI connections following the same approach.

The most fundamental decision in the design of SRP was the use of the DASH Resource Model and the model of data traffic associated with it. A variety of other models, both statistical and deterministic, could have been used instead, probably offering the possibility to describe a broader range of traffic properties (for example, average message loss). The model we chose has the advantage of being simple and useful for deriving performance guarantees for a variety of scheduling disciplines. The primary application area of SRP, continuous media, is well supported by our model. Some potential drawbacks are:

- During the establishment of an end-to-end session other establishment requests may be rejected needlessly because maximum reservations are made for each resource. This problem can be avoided easily by implementing a lock for each HRM, blocking each new request until pending requests are completed.
- SRP only establishes and deletes end-to-end sessions. No means to relax the requirements on such sessions are provided. However, should the requirements on a connection change (rare for continuous-media applications), a new end-to-end session can be associated with it.
- The algorithms we have presented are conservative. Optimistic approaches (analogous to those used in transaction processing) may be suitable for the real-time application domain, especially in regard to throughput reservation [15].
- The protocol in its present form (just like the established Internet protocols) deals with one-to-one communication only. Since continuous-media systems will be used to a large extent for groupware applications like conferencing, multi-point connections are an important issue. [16] presents an extension of end-to-end sessions for multiple receivers.

The protocol was implemented on Sun 3 workstations running the Mach operating system. A discussion of protocol implementation alternatives can be found in [17]. Not surprisingly, it turned out that not the implementation of SRP, but the implementation of the underlying resources was most difficult. Today's CPU schedulers and networks do not provide any support for continuous-media workload and require significant modification. With future operating systems providing real-time scheduling options and high-speed networks with bounded transit times, the realization of SRP will become a lot easier.

Acknowledgements

Martin Andrews, Robert Wahbe, Shin-Yuan Tzou, Ramesh Govindan contributed to the development of the DASH Resource Model. Carl Schaefer and Luca Delgrossi participated in the design of SRP. Thanks also to Domenico Ferrari for helpful comments and discussions.

References

1. A.C. Luther: Digital Video in the PC Environment, McGraw-Hill, 1989.
2. S. Newman: The Communications Highway of the Future, IEEE Communications Magazine 26, 10 (Oct. 1988), 45 – 50.
3. K.A. Frenkel: The Next Generation of Interactive Technologies, Comm. of the ACM 32, 7 (July 1989), 872 – 881.
4. R.G. Herrtwich: The Role of Performance, Scheduling, and Resource Reservation in Multimedia Systems, Proc. of Operating Systems of the Nineties and Beyond, Dagstuhl Seminar, July 1991.
5. J. Postel: Internet Protocol, DARPA Internet RFC 791, Sep. 1981.
6. J. Postel: Transmission Control Protocol, DARPA Internet RFC 793, Sep. 1981.
7. D.P. Anderson, S. Tzou, R. Wahbe, R. Govindan, M. Andrews: Support for Continuous Media in the DASH System, Proc. of the 10th International Conference on Distributed Computing Systems, Paris, May 1990.

8. R.L. Cruz: A Calculus for Network Delay and a Note on Topologies of Interconnection Networks, Ph.D. Dissertation, Report No. UILU-ENG-87-2246, University of Illinois, July 1987.
9. S. Cheng: Scheduling Algorithms for Hard Real-Time Systems – A Brief Survey, in Hard Real-Time Systems, J.A. Stankovic, K. Ramamritham (Eds.), IEEE Computer Society, 1988, 150–173.
10. C.L. Liu, J.W. Layland: Scheduling Algorithms for Multiprogramming in a Hard Real-Time Environment, J. ACM 20, 1 (Jan. 1973), 47–61.
11. M. Andrews: Guaranteed Performance for Continuous Media in a General Purpose Distributed System, Master's Thesis, UC Berkeley, Oct. 1989.
12. D. Ferguson, Y. Yemini, C. Nikolaou: Microeconomic Algorithms for Load Balancing in Distributed Computer Systems, Proc. of the 8th International Conference on Distributed Computing Systems, San Jose, June 1988.
13. RPC – Remote Procedure Call Specification, Version 2, DARPA Internet RFC 1057, Sun Microsystems, June 1988.
14. XDR – External Data Representation Standard, DARPA Internet RFC 1014, Sun Microsystems, June 1987.
15. R.G. Herrtwich: The DASH Resource Model Revisited or Pessimism Considered Harmful, Proc. of the 1st International Workshop on Network and Operating System Support for Digital Audio and Video, Report No. TR-90-062, International Computer Science Institute, Berkeley, Nov. 1990.
16. D.P. Anderson: Meta-Scheduling for Distributed Continuous Media, Report No. UCB/CSD 90/599, UC Berkeley, Oct. 1990.
17. D.P. Anderson, L. Delgrossi, R.G. Herrtwich: Process Structure and Scheduling in Real-Time Protocol Implementations, Proc. of Kommunikation in Verteilten Systemen, Mannheim, Informatik-Fachberichte 267, Springer-Verlag, Feb. 1991.

EASInet - Ein Beitrag zu einem europäischen Forschungsnetz

Von KLAUS BIRKENBIHL

Gesellschaft für Mathematik und Datenverarbeitung mbH (GMD), Bonn

---

Als IBM den Partnern ihres Supercomputerprogramms EASI den Aufbau eines Computernetzes zusagte, war klar, daß dies nur in Kooperation mit anderen Anbietern von Netzdiensten erfolgen konnte. So kam es frühzeitig zu Vereinbarungen mit nationalen und internationalen Netzanbietern. Wie richtig dieser Ansatz auch im Hinblick auf eine künftige europäische Netzinfrastruktur war, zeigt sich daran, daß sich EASInet heute als ein bedeutender Bestandteil eines künftigen europäischen Backbones für ein Forschungsnetz zeigt.

EASInet - die Vorgeschichte

EASI (European Academic Supercomputer Initiative) ist ein Angebot der IBM an wissenschaftliche Einrichtungen, das u.a. folgende Punkte umfaßt:

- Verkauf eines Supercomputers der Reihe IBM 3090 oder IBM /390 zu Vorzugskonditionen,
- Unterstützung von Kooperationen mit Supercomputeranwendungen auf IBM-Systemen,
- Aufbau von Netzverbindungen unter den EASI-Installationen mit einer Bandbreite von mindestens 64Kb,
- Bereitstellung einer Verbindung nach USA.

Im Rahmen der Überlegungen zur Realisierung des vorletzten Punkts, wurde schnell klar, daß keine EASI-Installation daran interessiert sein konnte, die Zahl ihrer schon bestehenden Netzverbindungen durch eine weitere zu vergrößern - und damit die Komplexität und den Aufwand für den Betrieb der Netzinfrastruktur zu erhöhen. Gefragt war vielmehr die Erhöhung der Bandbreite existierender Netzverbindungen - oder zumindest die Integration in neue Planungen.

Bemühungen von IBM, Postgesellschaften zu finden, die mit EASInet und ggf. anderen

---

Klaus Birkenbihl studierte Mathematik an der Universität Bonn, seit 1972 arbeitet er in der GMD als Software-Technologe, Berater und RZ-Leiter. Zur Zeit baut er die neue Abteilung für Wissenschaftsnetze auf. EASInet ist eines der Projekte dieser Abteilung.

Netzanbietern ein gemeinsames Pilotprojekt "europäische Breitbandvernetzung" gestartet hätten, scheiterten. So blieben die europäischen Netzanbieter in dem Bemühen eine einfachere, kostengünstigere und leistungsfähigere Netzinfrastruktur aufzubauen unter sich.

Von vornherein war klar, daß es nicht ausreichen würde, die Installationen mit Leitungen auszustatten. Vielmehr wurde zur Sicherstellung der Konnektivität ein zentrales Management und eine koordinierte Planung als Angebot an die EASI-Installationen für unerläßlich angesehen. Der Auftrag hierfür ging an die GMD.

EASI-Installationen		
Land	*Ort*	*Institution*
Belgien	Leuven	KUL
Deutschland	Aachen	RWTH
Deutschland	Braunschweig	U-Bs
Deutschland	Darmstadt	GSI
Deutschland	Hamburg	DESY
Deutschland	Karlsruhe	KfK
Frankreich	Lyon	IN2P3
Frankreich	Montpellier	CNUSC
Frankreich	Paris	CEA
Frankreich	Paris	CIRCE
Holland	Amsterdam	SARA
Italien	Bologna	CINECA
Italien	Rom	U-Rom
Österreich	Wien	U-Wien
Schweden	Skelleftea	UMEA
Schweiz	Genf	CERN
Schweiz	Zürich	ETH-Z
Spanien	Barcelona	FCR

Tabelle 1: Liste der EASI-Installationen

EASInet - die Technik

Eine Anforderung an EASInet war von Anfang an die Unterstützung unterschiedlicher Anwendungen auf unterschiedlichen Protokollen. Nur so konnte man der Situation in Europa, wo sich verschiedene Protokollwelten und Anwendungen nebeneinander etabliert haben, gerecht werden. Die für das EASInet wichtigen Protokolle sind TCP/IP, SNA und X.25/OSI.

Das EASInet-Management kümmert sich hauptsächlich um die Konnektivität und das Routing innerhalb des Netzes (also die Ebenen 2 bis 4 des OSI-Referenz-Modells). Zur Zeit gibt es also keine Anwendungsgateways, die im Auftrag von EASInet betrieben werden. Dies ist ein Ansatz, mit dem die EASI-Installationen ganz zufrieden sein können - handelt es sich doch um große Installationen, die in der Lage sind, ihre Daten in dem jeweils erforderlichen Protokoll zu erzeugen und deren Hauptinteresse zuverlässige und leistungsfähige Konnektivität ist.

Die Multiprotokollbasis

EASInet unterstützt zwei Techniken um eine Basis für Multiprotokollanwendungen zu schaffen: im internationalen Backbone mit Geschwindigkeiten von 256Kb bis 1,5Mb werden IDNX-Multiplexer eingesetzt. Auf nationalen Verbindungen mit geringerer Geschwindigkeit (64Kb) wird X.25 als Träger für die anderen Protokolle benutzt. In Deutschland ist das Wissenschaftsnetz des DFN die X.25-Basis. Beide Techniken haben ihre Vor- und Nachteile.

Die IDNX-Box unterstützt neben dynamischem Multiplexen die Überwachung von Leitungen und Verbindungen. Es ist sehr einfach, Statistiken und Zeitdiagramme über Verbindungsqualitäten, kurze Ausfälle, Schwankungen im *clocking* u. Ä. zu erhalten und frühzeitig Maßnahmen

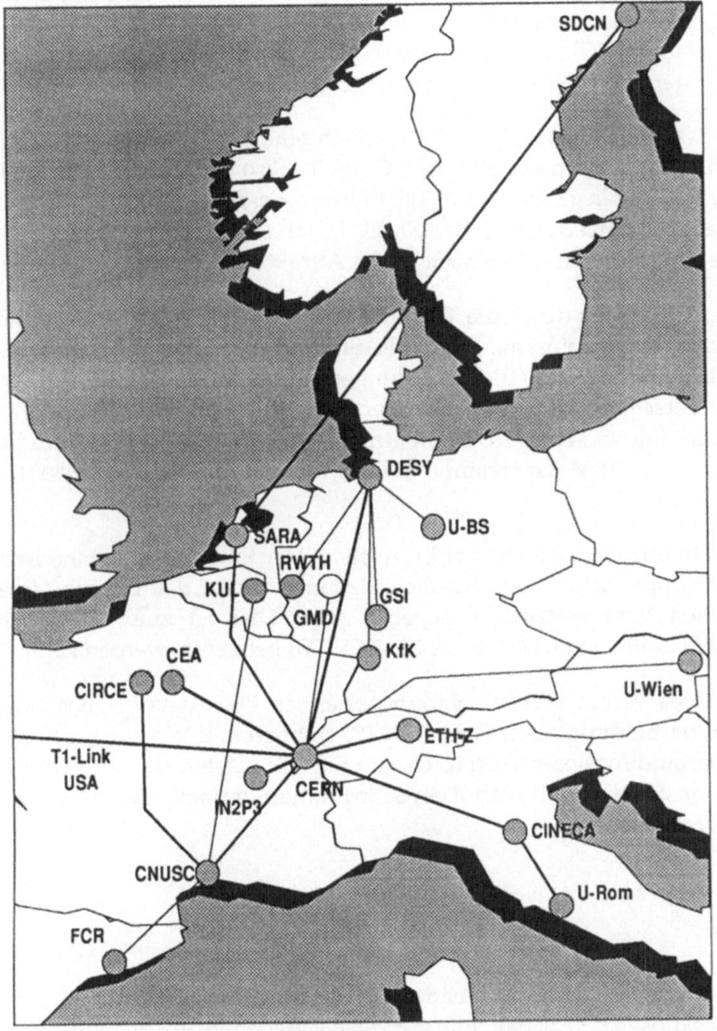

Abbildung 1: EASInet

zur Stabilisierung der Situation zu ergreifen. Untereinander fahren die IDNX-Boxen zu Netzmanagementzwecken ein eigenes Protokoll, das eine zentrale Steuerung des Netzes ermöglicht.

Ein Nachteil der IDNX-Box ist in vielen Fällen sicher die feste Bandbreitenverteilung für die verschiedenen Protokolle. Zwar kann von dem Netzzentrum aus diese Bandbreitenzuordnung dynamisch verändert werden - es ist aber nicht möglich, ein statistisches Multiplexen, bei dem je nach Verkehrsaufkommen Bandbreiten - nach entsprechenden Vorgaben - dynamisch zugeordnet werden zu definieren.

X.25 als Multiprotokollbasis wird im EASInet innerhalb Deutschlands eingesetzt, wo mit dem Wissenschaftsnetz des DFN (WIN) ein 64Kb-Netz zur Verfügung steht, das von den meisten deutschen Einrichtungen genutzt wird. X.25 Pakete werden im WIN derzeit alle gleichbehandelt (84er Standard). Das Netz stellt sich also dynamisch auf unterschiedliche Anforderungen der aufgesetzten anderen Protokollfamilien ein. Allerdings ist es nicht möglich - wie z.B. bei dynamischen Multiplexern - Durchsatzgarantien zu geben oder Prioritäten für bestimmte Betriebsarten festzulegen.

Unterstützte Protokolle

EASInet unterstützt die Protokolle TCP/IP, SNA und X.25/OSI. Wie in den anderen europäischen Netzen ist dabei TCP/IP das am meisten nachgefragte Protokoll. Verschiedene Ursachen sind hierfür ausschlaggebend: TCP/IP ist billig (gehört zur Standardausstattung eines UNIX Betriebssystems), TCP/IP ist - deshalb - verbreitet (insbesondere in den USA), TCP/IP wird auf den meisten LANs sowieso gefahren, TCP/IP ist bei weitem nicht so komplex wie fortgeschrittenere Protokolle (z. B. die meisten ISO/OSI-Protokolle oder aber auch SNA).

EASInet stellt direkte TCP/IP-Konnektivität zu den USA durch eine 1,5Mb Leitung (T1 nach US-Standard) her. Die Verbindung geht von CERN nach Cornell - also direkt ins NSFnet. Die Verantwortung für diese Verbindung liegt bei dem NSFnet Operationszentrum MERIT in Ann Arbor und dem EASInet Managementzentrum in der GMD. 1,5Mb das ist eine Kapazität von mehr als 23 mal eine 64Kb Leitung - entsprechend hoch ist die Attraktivität dieser Verbindung.

SNA erhält seine Bedeutung für EASInet durch die Tatsache, daß SNA sehr gut in EARN integriert ist und so eine gute Basis für gemeinsame EARN/EASInet-Verbindungen bietet. EARN ist - trotz eines starken Wachstums von TCP/IP-Verbindungen in den letzten Jahren - immer noch das größte europäische Forschungsnetz. SNA hat aber auch erhebliche funktionale Vorteile in der homogenen Verbindung von IBM Großrechnern. Dies hat für EASInet insofern Bedeutung, als das EASI-Programm ja IBM-Supercomputeranwendungen europaweit unterstützen soll.

OSI auf der Basis von X.25 ist in Europa nach wie vor im Aufbau. Der Entwicklungsstand ist in den verschiedenen Ländern Europas sehr unterschiedlich. EASInet fördert den Einsatz dieser Protokolle, wo die technischen Voraussetzungen gegeben sind. Dies ist zum Beispiel in Deutschland der Fall, wo WIN als eine natürliche Basis für ISO/OSI betrachtet werden kann.

Im Gegensatz zu dem in Europa verbreiteten verbindungsorientien OSI-Asatz (CONP) wird sich OSI in den USA in seiner verbindungslosen Form (CLNP) etablieren. EASInet hat gemeinsam mit NSFnet den ersten verbindungslosen Austausch von ISO/OSI Datenpaketen über den Atlantik im August 1990 demonstriert. Damit wurde ein Schritt unternommen, die europäische und die US- ISO/OSI-Welt zu verbinden.

Zentrales Netzmanagement

EASInet ist ein zentral verwaltetes Netz. Die Anwender von EASInet haben die Möglichkeit, Netzdienste quasi aus der Steckdose zu beziehen. Für die verschiedenen Protokolle gilt dies in Abhängigkeit von den unterschiedlichen technischen Gegebenheiten in unterschiedlichem Maße.

Für TCP/IP Konnektivität benutzt EASInet die NSFnet Technik der *nodal switching systems* in vereinfachter Form. Als Router kommen hier IBM PC's in RISC-Technologie zum Einsatz (RS/6000 oder RT/6150). Je nach Anzahl der angeschlossenen Leitungen werden eines oder mehrere dieser Systeme zu einem sogenannten *nodal switch* verbunden. Alle *nodal switches* werden zentral von der GMD verwaltet. Die Installation und Konfiguration aller Software und aller Daten und die Überwachung der Funktion dieser *switches* erfolgt zentral über die GMD. Für die EASI-Installation, die diesen Service in Anspruch nimmt, bedeutet das - nachdem sie ihre lokale Netze bei der GMD angemeldet hat - weltweite TCP/IP-Konnektivität für ihre lokalen Netze durch Anschluß an den *nodal switch*. Keine weiteren Aktionen sind erforderlich.

Das TCP/IP Netzmanagement erfolgt mit dem Netzmanagementwerkzeug XGMON. Dieses Werkzeug verfügt über umfangreiche Möglichkeiten zur Überwachung und Steuerung des Netzes.

Nicht ganz so weit geht der Service bei SNA. Hier sind die Vermittler üblicherweise Vorrechner vom Typ IBM 37xx und versehen mehr Dienste als nur WAN-Konnektivität herzustellen. Traditionell werden diese Systeme durch die lokalen Rechenzentren betreut und konfiguriert. Die GMD organisiert für SNA die internationale Konnektivität und stellt den Installationen die entsprechenden Dateien für die Konfiguration ihrer Vorrechner zur Verfügung. Netzmanagement-Informationen werden im Netview System der GMD zentral gesammelt.

Die Dienste für X.25 entsprechen in etwa denen für SNA. Bei WIN liegt allerdings das X.25 Netzmanagement bei der Post, so daß die GMD hier nicht involviert ist. Andererseits sind die Netzmanagementinformationen, die von den X.25-Vermittlern verschiedener Hersteller bereitgestellt werden, nicht einheitlich und deshalb schwieriger zu verarbeiten.

Wichtige Leitungen des EASInet-Backbone			
Leitung von	*Leitung nach*	*Bandbreite*	*Partner*
Stockholm	Amsterdam	194Kb	NORDUnet
Amsterdam	Genf	256Kb	HEPnet, SURFnet, EUNET
Hamburg	Genf	768Kb	HEPnet, DESY
Montpellier	Genf	256Kb	CNUSC, EARN
Lausanne	Genf	2048Kb	SWITCH
Paris	Genf	256Kb	CEA
Lyon	Genf	1024Kb	IN2P3

Tabelle 2: EASInet Kooperationen

Innerhalb von EASInet führt die GMD Untersuchungen und Entwicklungen durch, um ein Netzmanagement für Multiprotokollnetze auf Basis von CMIS/CMIP (ISO/OSI) aufzubauen. Dieses Netzmanagement-System wird Informationen von IDNX-Boxen, IP-Routern, SNA- und X.25-*switches* sammeln und aktuell aufbereiten.

Kooperation mit anderen Netzen

Wie bereits erwähnt, hat EASInet von Anfang an eine kooperative Haltung zu anderen Anbietern von Netzdiensten angenommen. Heute betreibt EASInet in Kooperation mit anderen Anbietern ein Netz, in dem europaweit Verbindungen mit einer Bandbreite, die ein Vielfaches von 64Kb ist zur Verfügung steht.

Die EASInet-Philosophie

Als Basis für solche Kooperationen dient das *EASInet Acceptable Use Statement*, das festlegt, daß das *EASInet Project Committee* (EPC) anderen nicht kommerziellen Forschungsnetzen die Nutzung der EASInet-Leitungen gestatten kann. Dies geschieht in der Regel dadurch, daß bestehende Leitungen zusammengelegt werden, um von den relativ günstigeren Leitungstarifen der Postgesellschaften für höhere Bandbreiten zu profitieren.

Sicher ist vielfach der Wunsch, die T1-Verbindung ins NSFnet nutzen zu können, der Auslöser für solche Kooperationen. Der Effekt ist aber, daß den EASI-Installationen und den mit Ihnen verbundenen Netzen eine Infrastruktur zur Verfügung steht, die oft weit oberhalb der von EASInet garantierten 64Kb liegt.

Beispiele für Kooperationen

EASInet hat auf dieser Basis Kooperationen mit nationalen Netzorganisationen wie GARRnet (Italien), SURFnet (Holland) und NORDUnet (Skandinavien). Mit einigen Netzen bestehen Absprachen über gegenseitiges Backup (z.B. DFN). Auch internationale Netzanbieter wie EARN haben gemeinsame Leitungen mit EASInet.

EASInet und die weitere europäische Vernetzung

EASInet hat sicher nicht nur den EASI-Installationen geholfen, ihre nationalen und internationalen Netzverbindungen zu verbessern. Andere Netze und ihre Nutzer haben hiervon ebenso profitiert, wie natürlich auch EASInet durch die Kooperation mit diesen Netzen. EASInet hat mit dazu beigetragen, einen Prozeß in Gang zu bringen, der lange überfällig war: die Integration der europäischen Forschungsnetze. RARE hat 1990 eine *European Engineering Planning Group* (EEPG) gebildet, deren Aufgabe es war, die Situation in Europa (Leitungen, Verkehrsaufkommen ...) zu analysieren und Vorschläge für eine künftige Struktur zu machen. Der Report dieser Gruppe weist verschiede Möglichkeiten zur Konsolidierung der Situation auf: die Vorschläge reichen von einem Multiplexer-Netz für Mehrprotokollbetrieb bis zu einem X.25-Netz (IXI-Nachfolger), das auch die Basis für andere Protokolle bilden soll. Welche Lösung sich letztendlich auch durchsetzt: EASInet hat hier wichtige Vorarbeit geleistet: durch Bündeln von Leitungen, durch praktischen Einsatz von Multiplexertechnik und X.25 für Mehrprotokollbetrieb und durch Erfahrung beim Einsatz eines zentralen Netzmanagements.

... und ein Dankeschön

an IBM (Herb Budd, Harry Clasper, Michael Nies, Wolfgang Schröter und Peter Streibelt), die EASInet initiierte, finanziert und technisch unterstützt, an die Kollegen bei unseren Partnernetzen in Europa und USA, an die EASI-Installationen und das *EASInet Project Committee*, und nicht zuletzt an meine Kollegen in der GMD (insbes. Stefan Faßbender, Willi Porten, Detlef Straeten und Peter Wunderling), die mit Geschick und Sachkenntnis EASInet zu einem beachtlichen Erfolg verholfen haben.

Praktischer Einsatz und Weiterentwicklung von Estelle

C. Andrae, J. Bredereke, C. Hille, D. Peter, T. Reimer, U. Schüler, R. Gotzhein, F. H. Vogt

Estelle AG, Arbeitsbereich Rechnerorganisation,

Universität Hamburg, Bodenstedtstr. 16, 2000 Hamburg 50

Kurzfassung. Estelle ist eine formale Beschreibungstechnik für die Spezifikation von verteilten informationsverarbeitenden Systemen, insbesondere von Kommunikationsdiensten und -protokollen des OSI Basisreferenzmodells. Estelle wurde von der International Standardization Organization (ISO) entwickelt und besitzt seit 1989 den Status eines internationalen Standards. Gegenwärtige Aktivitäten betreffen vor allem den praktischen Einsatz und die Weiterentwicklung von Estelle sowie die Bereitstellung von rechnergestützten Werkzeugen.

Der Beitrag stellt überblickartig mehrere Arbeiten auf der Grundlage von Estelle sowie dabei gewonnene Erfahrungen vor. Die Arbeiten wurden am Arbeitsbereich Rechnerorganisation der Universität Hamburg durchgeführt, teilweise in Kooperation mit industriellen Partnern. Dabei geht es zunächst um den praktischen Einsatz von Estelle als Spezifikationstechnik einerseits und als Implementierungstechnik andererseits. Weitere Arbeiten behandeln die Realisierung verteilter Applikationen mit Estelle sowie die Schaffung von Benutzerschnittstellen. Es wird ferner untersucht, inwieweit objektorientierte Konzepte in Estelle bereits verwirklicht sind. Schließlich wird ein Ansatz zur formalen Verifikation von Spezifikationen in Estelle diskutiert.

1 Einleitung

Estelle ist eine formale Beschreibungstechnik für die Spezifikation von verteilten informationsverarbeitenden Systemen, insbesondere von Kommunikationsdiensten und -protokollen des OSI Basisreferenzmodells ([ISO81]). Estelle wurde von der International Standardization Organization (ISO) entwickelt und besitzt seit 1989 den Status eines internationalen Standards ([ISO89-1]). Gegenwärtige Aktivitäten betreffen vor allem den praktischen Einsatz und die Weiterentwicklung von Estelle sowie die Bereitstellung von rechnergestützten Werkzeugen.

Eine Estelle-Spezifikation beschreibt ein hierarchisch aufgebautes System von nicht-deterministischen, sequentiellen Komponenten, den *Modulinstanzen* (kurz *Module*). Modulinstanzen können über bidirektionale *Kanäle*, die jeweils zwei *Interaktionspunkte* miteinander verbinden, kommunizieren. Sowohl die Systemstruktur als auch die Verbindungsstruktur werden durch die Spezifikation festgelegt und können sich dynamisch verändern.

Jede Modulinstanz besitzt eine externe Schnittstelle, gegeben durch eine endliche Menge von Interaktionspunkten mit zugehörigen FIFO-Warteschlangen unbegrenzter Kapazität. Nachrichten, die über einen Interaktionspunkt an eine Modulinstanz gerichtet sind, werden zunächst in die jeweilige Warteschlange eingereiht. Außerdem kann eine Modulinstanz über ihre Interaktionspunkte Nachrichten an andere Modulinstanzen senden. Der jeweilige Adressat ist durch die aktuelle Verbindungsstruktur eindeutig definiert.

Neben der externen Schnittstelle ist jede Modulinstanz durch die Menge ihrer (internen) Zustände, einen Anfangszustand sowie die Menge möglicher Zustandsübergänge charakterisiert. Unter den Zustandskomponenten ist der *Hauptzustand* hervorzuheben, welcher eine endliche Wertemenge besitzt. Damit eignet sich das Modell des erweiterten endlichen Automaten als Basis für die Festlegung der Estelle-Semantik. Die Menge möglicher Zustandsübergänge wird durch eine Anzahl von Transitionen definiert. Eine *Transition* kann schalten, wenn eine im Bedingungsteil spezifizierte Restriktion erfüllt ist. Darin kann auf den Hauptzustand, den sonstigen Zustand, die Nachrichten in den Warteschlangen sowie deren Parameterwerte Bezug genommen werden. Das Schalten einer Transition bewirkt einen atomaren Zustandsübergang, welcher als Folge PASCAL-ähnlicher Anweisungen im Transitionsblock beschrieben ist.

Das Gesamtverhalten eines in Estelle spezifizierten Systems ist näherungsweise durch die parallele Ausführung aller Modulinstanzen gegeben. Dieser maximale Parallelitätsgrad wird nach einem festgelegten Synchronisationsschema reduziert, was u.a. durch geeignete Attributierung von Modulen geschieht.

Der vorliegende Beitrag stellt überblickartig mehrere Arbeiten auf der Grundlage von Estelle sowie dabei gewonnene Erfahrungen vor. Die Arbeiten wurden am Arbeitsbereich Rechnerorganisation der Universität Hamburg durchgeführt, teilweise in Kooperation mit industriellen Partnern. Sie stellen einen repräsentativen Querschnitt durch das Spektrum der gegenwärtigen Estelle-Aktivitäten dar.

2 ESTELLE ALS SPEZIFIKATIONSTECHNIK

Bei der Normung des Basisreferenzmodells für offene Systeme wurden die Standards bisher in natürlichsprachlicher Form, teilweise ergänzt durch Zustandstabellen, verfaßt. In letzter Zeit hat man jedoch erkannt, daß als Grundlage für die Präzisierung und Weiterentwicklung eine genauere, formale Beschreibungsform benötigt wird. Daher begann man bei der Normung der Dienstelemente für die Anwendungsschicht (z.B. bei OSI Transaction Processing), gleichzeitig mit den bisher üblichen Beschreibungsformen auch formale Spezifikationen in Estelle zu erstellen. Eine formale Beschreibung hat eine eindeutige Interpretation und ist daher sowohl als Diskussionsgrundlage als auch für die Konsensbildung wesentlich geeigneter.

Damit zeichnet sich zugleich ein Wandel in der Methodik ab: Es gibt nun eine Technik, die die Entwicklung vom Standard über die Spezifikation zur Implementierung unterstützt. Dabei erlauben die Abstraktionsmöglichkeiten die isolierte Beschreibung eines offenen Systems, das später in eine Umgebung eingebettet werden kann.

Durch die intensive Auseinandersetzung mit dem Standarddokument und die Umsetzung in die Estelle-Strukturen entsteht beim Bearbeiter ein tieferes Verständnis für das spezifizierte System, Ungenauigkeiten und Widersprüche der informellen Beschreibung lassen sich aufdecken.

Um Vorteile und Methodik des Estelle-Einsatzes an einem aktuellen Beispiel zu verdeutlichen, wurde im Rahmen einer Studienarbeit ([And90]) eine Estelle-Spezifikation des CCR-Anwendungsbausteins ([ISO88-1, ISO88-2]) erstellt. CCR (Commitment, Concurrency and Recovery) ermöglicht die atomare Ausführung von Aktionen, an denen mehrere verteilte Knoten beteiligt sind. Es ist unter anderem wesentliche Grundlage für OSI Transaction Processing.

Im ersten Schritt der Spezifikation wurde die Systemstruktur festgelegt. Durch geeignete Definition der Umgebung und der Schnittstellen konnte der Schwerpunkt auf die Beschreibung des eigentlichen CCR-Anwendungsbausteins gelegt werden. Estelle unterstützt die Technik der Modularisierung; dabei wurden zunächst einzelne Estelle-Module der obersten Ebene festgelegt, die dann stufenweise weiter verfeinert wurden. Die gewählte Systemstruktur ist in Abbildung 1 dargestellt. Der Modul CCR-ASE repräsentiert den CCR-Anwendungsbaustein, die dunkel unterlegten Module gehören zur Umgebung und sind als EXTERNAL deklariert.

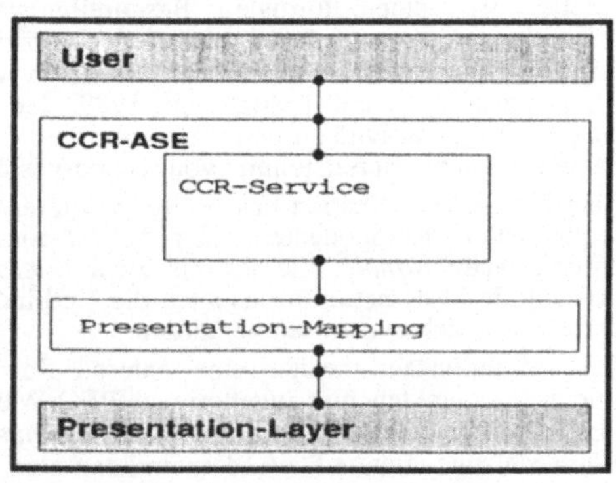

Abbildung 1: Systemstruktur für die Spezifikation des CCR-Anwendungsbausteins

Es ergaben sich zwangsläufig klare Schnittstellendefinitionen, da Estelle bestimmte Konstruktionen, wie z.B. globale Variablen, gar nicht unterstützt. Mit der Aufteilung der Funktionen auf die Module wurden auch die benötigten Kanäle definiert, dazu die Datentypen für die Nachrichten, die den Kanälen zugeordnet sind. Durch die Untersuchung der möglichen und zulässigen Nebenläufigkeiten ergibt sich die Attributierung der einzelnen Module.

Das Verhalten der einzelnen Module, das in den Transitionsteilen beschrieben wird, wurde im darauffolgenden Schritt festgelegt. Die benötigten Zustände und Transitionen ergaben sich dabei durch Vereinfachung und Zusammmenfassung der Zustandstabellen aus der ISO-Protokollspezifikation sowie durch Analyse der möglichen Kollisionsfälle ('gleichzeitiges' Eintreffen von Nachrichten an verschiedenen Interaktionspunkten). Neben der Einführung von zusätzlichen Zuständen kann die Bearbeitungsreihenfolge solcher 'gleichzeitiger' Nachrichten vor allem auch durch Prioritäten gesteuert werden.

Mit diesem durch die Anwendung von Estelle nahegelegten Vorgehen konnte das Verständnis für die Abläufe und Zusammenhänge des zu spezifizierenden Protokolls schrittweise vertieft werden. Wichtig sind in diesem Zusammenhang vor allem die Abstraktionsmechanismen der Sprache Estelle, die die Konzentration auf die wichtigen Aspekte fördern und die formale Beschreibung von Implementierungsdetails entlasten.

Anhand dieser Arbeit konnte gezeigt werden, daß der Einsatz von Estelle sinnvoll ist, ohne daß eine spätere Implementierung dabei im Vordergrund steht. Obwohl die Arbeit sich auf die Spezifikation beschränkt, steht zu erwarten, daß das entstandene umfassende Verständnis eine spätere Implementierung in jedem Fall wesentlich erleichtern wird.

3 ESTELLE ALS IMPLEMENTIERUNGSTECHNIK

Eine eindeutige formale Beschreibung ist gleichzeitig geeignet, um als Eingabe für weitergehende Werkzeuge zu dienen. Für Estelle wurden z.B. Compiler entwickelt, die eine Estelle-Spezifikation in Programmcode umsetzen. Dieser Programmcode ist anschließend von Hand zu ergänzen.

Interessante Fragestellungen ergeben sich dabei betreffend der Verwendbarkeit solcher Systeme in einer herstellerspezifischen Umgebung (Betriebssystem und Software) und Aussagen über den erreichbaren Grad an Unterstützung der Programmentwicklung (Verhältnis automatisch generierter Code/von Hand erstellter Code). Neben diesen Fragen zu einer rationellen Programmerstellung betreffen weitere Fragen die Effizienz des erstellten Programms selbst. Beide Aspekte stellen die Grundlage einer Abschätzung dieser weitergehenden Verwendung von Estelle auch für die kommerzielle Softwareerstellung dar.

Ergebnisse zu diesen Fragen wurden in einer Diplomarbeit ([Rei89]) an der Universität Hamburg in Zusammenarbeit mit der SIEMENS AG erzielt. In der Arbeit wurde die Spezifikation und Implementierung eines Reliable Transfer Service Elements (RTSE, [ISO88-3, ISO88-4], X.410-1984-Modus) im Rahmen einer UNIX-Umgebung und unter Verwendung herstellerspezifischer Kommunikationssoftware durchgeführt. Dabei fand ein Estelle-Compiler des National Institute of Standards and Technologies (NIST) ([NBS87]) Verwendung, der Code in der Programmiersprache C generiert.

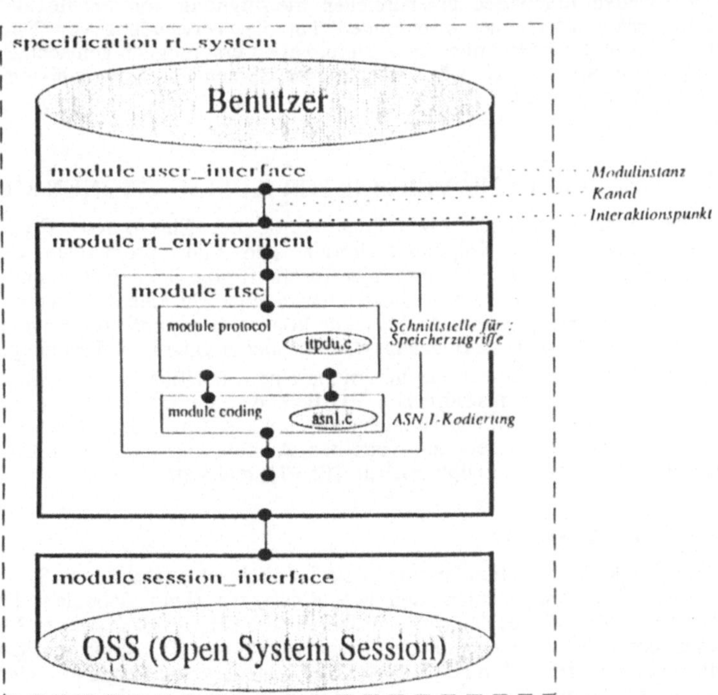

Abbildung 2: Systemstruktur einer Estelle-RTSE-Spezifikation als Grundlage einer Implementierung

Analog zum Vorgehen in Abschnitt 2 bestand auch in dieser Arbeit der erste Schritt in der Erstellung einer Estelle-Spezifikation. Dabei wurde die Systemstruktur festgelegt und das Verhalten der einzelnen Module in Form von Automaten beschrieben. Die Transitionsteile dieser Automaten sind entsprechend der natürlichsprachlichen Standardbeschreibung ergänzt. Dazu gehören unter anderem Prozeduren und Funktionen, die erst auf C-Programmebene ausgefüllt werden. Die Implementierung dieser C-Codeteile wird je nach vorhandener Zielmaschine und vorhandenen

Funktionspaketen vorgenommen; dadurch wird ein hohes Maß an Portabilität erreicht. Gleichzeitig kann die Übersichtlichkeit der Spezifikation erhalten bleiben, ohne das Laufzeitverhalten der späteren Implementierung zu verschlechtern.

Zur Verdeutlichung dient die Abbildung 2, welche die gewählte Systemstruktur der Estelle-Spezifikation wiedergibt und die Anbindung an Funktionspakete zeigt. Neben der vorgestellten Strategie der Anbindung an existierende Funktionspakete über Schnittstellenmodule (Modul 'user_interface' und Modul 'session_interface' in Abbildung 2), lassen sich vom Estelle-Compiler generierte Protokollmaschinen (Modul 'protocol machine') mit Hilfe der Bibliotheksfunktionen des Laufzeitsystems in eine beliebige Programmumgebung integrieren. Dieser Ansatz könnte schon bald für die Industrie an Bedeutung gewinnen, wenn compilergenerierte Protokollmaschinen von Hand erstellte ersetzen.

Die Verwendung von Estelle erleichterte zum Zeitpunkt der Entwicklung Umstellungen der Systemstruktur (Anzahl und Aufgabenbereiche von Modulen sowie deren Nachrichten und Verbindungsstruktur). Zum Zeitpunkt des Testens konnten Fehler leicht ermittelt werden, da der Ablauf der Spezifikation durch Nachrichten und Verhalten der Module als Automaten einfach nachvollziehbar ist.

Die Arbeiten zur Spezifikation und Implementierung nahmen in etwa ein Jahr in Anspruch. Dabei ist zu beachten, daß zunächst die Anwendung von Estelle erlernt und das Verständnis für RTSE, auch im Zusammenhang mit potentiellen Benutzern, erworben werden mußte. Es ergibt sich insgesamt eine Spezifikation von ca. 150 Seiten, die Grundlage für ein Programm von ca. 600 Seiten ist. Das Estelle-RTSE ist in Verbindung mit einem normkonformen SIEMENS-RTS verwendbar. Es konnte aber aufgrund eines unterschiedlichen Funktionsumfangs (Estelle-RTSE: keine Mehrfachverbindungen, keine Sicherheit gegen Systemabstürze) kein direkter Leistungsvergleich angestellt werden. Für Prototypimplementierungen kann das Leistungsverhalten bereits als ausreichend angesehen werden.

Die erzielten Ergebnisse unterstreichen die Eignung von Estelle als Implementierungstechnik für Kommunikationssysteme. Insbesondere hat die Verwendung von Estelle die Implementierung von RTSE im zeitlichen Rahmen einer Diplomarbeit erst möglich gemacht. Eine direkte Realisierung in der Programmiersprache C wäre wegen der fehlenden Kommunikationskonzepte schwieriger und langwieriger gewesen.

4 Realisierung verteilter Applikationen mit Estelle

In den vorangegangenen Abschnitten wurde der Einsatz von Estelle für die Spezifikation und Implementation von Kommunikationsdiensten und -protokollen des OSI Basisreferenzmodells beschrieben. Darüberhinaus eignet sich Estelle für die Spezifikation und Implementation von verteilten *Anwendungen*. Verteilte Anwendungen haben einerseits eine Reihe von Vorteilen, wie größere Leistung und erhöhte Verfügbarkeit. Andererseits können bestimmte Anwendungen nur verteilt implementiert werden; man denke da an Buchungssysteme oder automatische Fertigungsprozesse.

Estelle bietet Sprachkonstrukte an, um die einer Spezifikation innewohnende Parallelität und damit die maximale Verteilung auszudrücken. Existierende Estelle-Compiler generieren Code, der sequentiell auf einem Prozessor ablauffähig ist (siehe Abschnitt 3). Im Rahmen einer Studienarbeit ([Pet89]) wurde ein verallgemeinertes Konzept zur verteilten Ausführung von Estelle-Code erarbeitet. Dieses Konzept wurde im Rahmen einer Diplomarbeit ([Pet91]) realisiert.

4.1 Parallelität

Eine Estelle-Spezifikation beschreibt ein hierarchisch strukturiertes System von nichtdeterministischen, sequentiellen Komponenten, den Modulinstanzen (kurz 'Module'). Das Verhalten jedes einzelnen Moduls ergibt sich aus den Zustandsübergängen (Transitionen) dieses Moduls, oder aus dem Verhalten untergeordneter Module, wenn der Modul selbst keine Transition schalten kann. In diesem Fall schalten die untergeordneten Module parallel. Da untergeordnete Module nur dann Transitionen schalten dürfen, wenn der übergeordnete Modul keine schalten kann, sind die hierarchisch angeordneten Module auch gleichzeitig synchronisiert.

Weiterhin kann Parallelität von Transitionen über eine Strukturierung der Spezifikation in mehrere *Systemmodule* zustande kommen. Systemmodule bilden mit ihren untergeordneten Modulen voneinander unabhängige Hierarchien. Module in einer Hierarchie schalten Transitionen unabhängig und ohne Synchronisation zu Modulen in anderen Hierarchien. Diese zweite Möglichkeit läßt sich zwar sicher einfach implementieren, sie ist aber nicht so allgemein wie die erste Möglichkeit, da dann nur auf oberster Ebene Parallelität möglich ist.

4.2 VERTEILUNG

Damit Estelle-Code parallel ausgeführt werden kann, müssen die Komponenten, die auf ein reales System verteilt werden sollen, identifiziert werden. Dabei entstehen zwei Teilprobleme:

1. *Was* kann verteilt werden? (logische Verteilung)
2. *Wie* soll es verteilt werden? (physikalische Verteilung)

Die logische Verteilung legt die maximale Verteilung fest, die später bei Abbildung auf ein reales System reduziert werden darf. Dabei werden die kleinsten zu verteilenden Komponenten einer Estelle-Spezifikation bestimmt. Die Implementation einer Spezifikation erfolgt durch Reduktion der Verteilung, es können also nicht kleinere als die bestimmten Komponenten parallel ausgeführt werden. Gleichzeitig wird dem Benutzer von Estelle durch die logische Verteilung ein Modell angeboten, wie eine dem Problem gerecht werdende Verteilung in der Estelle-Spezifikation zum Ausdruck kommen kann.

Die Möglichkeiten, was als kleinste Komponente der Verteilung bestimmt werden kann, reichen von einzelnen Anweisungen bis zu Systemmodulen. Allerdings scheinen diese beiden Extreme nicht geeignet zu sein. Systemmodule erlauben nur eine Verteilung auf oberster Ebene einer Spezifikation. Die Verteilung einzelner Anweisungen setzt komplizierte Nebenläufigkeitsanalysen voraus.

Als Komponenten der logischen Verteilung wurden die Modulinstanzen bestimmt. Dieses Modell ist für Benutzer einfach zu verstehen, da es sich an der Struktur einer Estelle-Spezifikation orientiert. Außerdem sind, nach der Semantik von Estelle, Modulinstanzen sequentielle Komponenten, so daß etwa eine Verteilung kleinerer Komponenten, wie z.B. Transitionen, den Parallelitätsgrad nicht erhöhen würde.

Ziel der Implementierung einer Spezifikation ist es u.a., die Komponenten der logischen Verteilung - die Modulinstanzen (Module) - auf ein reales, verteiltes System von Prozessoren abzubilden. Dazu müssen den Modulen physikalische Prozessoren zugeordnet werden. Die Abbildung wird durch die Angabe von Paaren (Modul, Prozessor) definiert. Damit stellt sich das Problem der Benennung von Modulinstanzen.

Als Kompromiß zwischen Flexibilität und einfacher Benennung wurden *Modulvariablen* bestimmt. Modulvariablen dienen dazu, Modulinstanzen zu identifizieren. Sie werden benutzt, um die System- und Verbindungsstruktur festzulegen und dynamisch zu ändern.

Der Name einer Modulinstanz wird durch den Namen der Modulvariablen bestimmt, die benutzt wurde, um diese Modulinstanz zu kreieren. Wird eine Modulinstanz kreiert, so ist in der entsprechenden Estelle-Anweisung eine Modulvariable anzugeben, die nach der Kreierung auf diese Modulinstanz verweist. Dieses Modell der Benennung ist einfach zu verstehen, läßt effektiv eine Verteilung der Module auf ein reales System zu und schränkt die möglichen physikalischen Verteilungen nicht zu sehr ein.

Allerdings ist der Name einer Modulvariablen, wegen der Blockstruktur einer Estelle-Spezifikation, nicht immer eindeutig. Daher wird (rekursiv) der Name der Modulinstanz, in der die Modulvariable definiert ist, ermittelt. Dieser Name qualifiziert dann den Namen der Modulvariablen und macht ihn so eindeutig.

4.3 SYNCHRONISATION

Eine Estelle-Spezifikation ist ein hierarchisch strukturiertes System von Modulen, die entsprechend ihrer Hierarchie synchronisiert werden müssen (siehe Abschnitt 4.1). Verteilt man Module auf ein reales System, so muß sichergestellt werden, daß die Module bei einer beliebigen Verteilung synchronisiert werden können.

An den Algorithmus, der die Synchronisation der verteilten Module implementiert, sind aus praktischen Gründen einige Anforderungen zu stellen. So sollen wenig und nur kurze Nachrichten zwischen den Prozessoren, auf denen die Module ablaufen, ausgetauscht werden.

In der Studienarbeit wurde ein Algorithmus vorgestellt, bei dem pro Zustandsübergang in einem Modul zwei Nachrichten gesendet werden. Diese Nachrichten besitzen zudem keine Parameter, haben also lediglich eine Signalfunktion. Damit ist die effiziente Ausführung einer verteilten Estelle-Spezifikation möglich.

4.4 REALISIERUNG

Im Rahmen einer Diplomarbeit ([Pet91]) wurde das hier beschriebene Konzept realisiert. Ein Estelle-Compiler, der sequentiell ablauffähigen Code generiert ([NBS87]), wurde so modifiziert, daß der generierte Code verteilt ausgeführt werden kann. Gleichzeitig wurde das Laufzeitsystem neu erstellt. Der vom Compiler erzeugte Code benutzt das Laufzeitsystem, um die Estelle-Spezifikation gemäß der Semantik von Estelle zu implementieren.

Die Diplomarbeit ist damit gleichzeitig eine Studie über die Integration neuer Konzepte in ein bestehendes System. Als besonders schwierig stellte sich die Anpassung des Estelle-Compilers an das neue Laufzeitsystem heraus, da der vom Ausgangscompiler erzeugte Code Repräsentationen von Estelle-Konzepten (z. B. Modulinstanzen) mit ihrer Adresse im Hauptspeicher adressierte. Bei einer verteilten Ausführung sind solche Speicheradressen nutzlos, da sie nur auf einem bestimmten Prozessor gültig sind. Es mußte daher ein Adressierungsschema entwickelt werden, daß unabhängig von Speicheradressen ist, und es galt, den Estelle-Compiler an dieses Schema anzupassen.

Mit dem neuen Laufzeitsystem ist es möglich, verteilte Applikationen in einer Unix-Umgebung auf mehreren, über ein Netz verbundenen Prozessoren auszuführen. Das Laufzeitsystem benutzt dazu den Kommunikationsmechanismus des BSD-Unix (*Sockets*). Da der Estelle-Compiler Code in der Sprache *C* generiert und auch das Laufzeitsystem in dieser Sprache geschrieben ist, muß lediglich ein C-Compiler für jeden Prozessor vorhanden sein. Die einzig zur Zeit noch relevante Einschränkung besteht darin, daß alle Prozessoren Datentypen gleich repräsentieren müssen.

4.5 BEWERTUNG UND AUSBLICK

Das Konzept der Verteilung wurde ohne Änderung der Syntax und der Semantik von Estelle integriert. Damit wurde gezeigt, daß es möglich ist, eine Estelle-Spezifikation in beliebig verteilten realen Systemen zur Ausführung zu bringen. Dieses ist insbesondere unter dem Gesichtspunkt des Testens bemerkenswert, denn es ist damit möglich, eine Estelle-Spezifikation in fast allen Belangen lokal, d.h. auf einem Prozessor, zu testen. Nach dem Test kann die Estelle-Spezifikation dann auf einem realen System zur Ausführung gebracht werden.

Betrachtet man die Möglichkeiten von Estelle, wie sie im Rahmen der Studienarbeit deutlich geworden sind, so bietet sich die Spezifikationssprache Estelle auch als Programmiersprache für verteilte Anwendungen an. Estelle ist eine standardisierte Sprache mit Ausdrucksmöglichkeiten von Parallelität, Verteilung und Kommunikation. Das Modulkonzept von Estelle eignet sich sehr gut für funktionale Abstraktionen. Zudem läßt sich ein Estelle-Programm lokal, d.h. auf einem Prozessor, testen, ohne das semantische Probleme mit der gewünschten Verteilung auftreten.

In der Studienarbeit wurde weiter gezeigt, daß die *separate* Übersetzung von Teilen einer Estelle-Spezifikation, ähnlich wie in Modula-2 oder Ada, möglich ist. Damit kann eine umfangreiche Anwendung von mehreren Personen bei voller Schnittstellenüberprüfung spezifiziert und implementiert werden.

Es ist geplant eine Arbeitsumgebung zu erstellen, die die Schritte von der Erstellung der Anwendung über lokales Testen bis zur automatischen Verteilung der Komponenten auf die Prozessoren unterstützt.

5 BENUTZERSCHNITTSTELLEN FÜR ESTELLE

Estelle wird für die Spezifikation und Implementierung von verteilten Systemen eingesetzt. Die Überprüfung der Korrektheit ist wegen der Komplexität solcher Systeme, die sich aus der Parallelität interner Abläufe ergibt, oft sehr schwierig. Neben der formalen Verifikation (siehe Abschnitt 7) hat hier vor allem das systematische Testen eine besondere Bedeutung. Im Kontext verteilter Systeme bedeutet Testen, ein System wiederholt zu stimulieren, das dabei auftretende externe Verhalten zu beobachten ('black box' Testen) und mit dem erwarteten Verhalten zu vergleichen. Für weitergehende Untersuchungen ist es zweckmäßig, einzelne Systemkomponenten zu stimulieren und ihre Reaktionen sowie ihren internen Zustand zu betrachten ('white box' Testen). Dafür ist eine Benutzerschnittstelle erforderlich, wie sie von Estelle nicht bereitgestellt wird.

Eine Möglichkeit, ein in Estelle spezifiziertes System zu stimulieren, besteht darin, eine mögliche Systemumgebung durch Angabe weiterer Modulinstanzen festzulegen. System und Umgebung können dann mit dem Estelle-Compiler übersetzt und ausgeführt werden, ein interaktiver Eingriff ist nicht möglich. Das externe Systemverhalten läßt sich durch geeignet festzulegende Protokollierungsprozeduren, die in die Estelle-Transitionen oder das Laufzeitsystem eingefügt werden, aufzeichnen. In gleicher Weise lassen sich auch Informationen über interne Abläufe und Zustände gewinnen, wobei z. B.

- die Namen der aktiven Modulinstanzen,
- das Eintragen von Nachrichten in Warteschlangen,
- die Namen schaltender Transitionen,
- die aktuellen Hauptzustände,
- Änderungen der Verbindungsstruktur,
- Modulinstanziierungen

aufgezeichnet werden können. Neben der Möglichkeit, Protokollierung problemorientiert zu realisieren, bietet der NIST-Compiler ([NBS87]) die Option, mehrere der o. g. Informationen aufzuzeichnen.

Für interaktive Eingriffe in das Ablaufverhalten eines Systems ist eine weitergehende Benutzerschnittstelle erforderlich. Für das interaktive Testen des externen Systemverhaltens sind Eingaben in Form von Nachrichten an die externen Interaktionspunkte zu richten und außerdem die Nachrichten des Systems an seine Umgebung zu visualisieren und entgegenzunehmen. Eingriffe in das interne Systemverhalten können in der Auswahl der nächsten Transition oder der Modifikation von Nachrichtenwarteschlangen bestehen.

Im Rahmen einer Studienarbeit ([Sch91]) wurde die in Abbildung 3 gezeigte graphische Benutzerschnittstelle zu interaktiven Steuerung des Ablaufverhaltens der Estelle - Spezifikation 'game' realisiert. Das System besteht aus der Modulinstanz 'Spielmanager', seine Umgebung aus drei Spielern und einem Störenfried. Jede Modulinstanz wird durch ein eigenes Fenster repräsentiert. In diesem Fenster werden der aktuelle Hauptzustand sowie die Werte einiger lokaler Variablen dargestellt. Die Spieler können Anfragen an den Spielmanager richten, was interaktiv durch Anwahl sogenannter 'Buttons' durch den Benutzer mit Hilfe einer Maus geschieht. Dies hat zur Folge, daß der Spielmanager Nachrichten erhält und darauf in wohldefinierter Weise reagiert. Sämtliche Nachrichten an den Spielmanager werden in derselben Warteschlange gesammelt, und zwar in der Reihenfolge ihres Eintreffens. Diese Warteschlange kann in der hier beschriebenen Benutzerschnittstelle ebenfalls visualisiert werden.

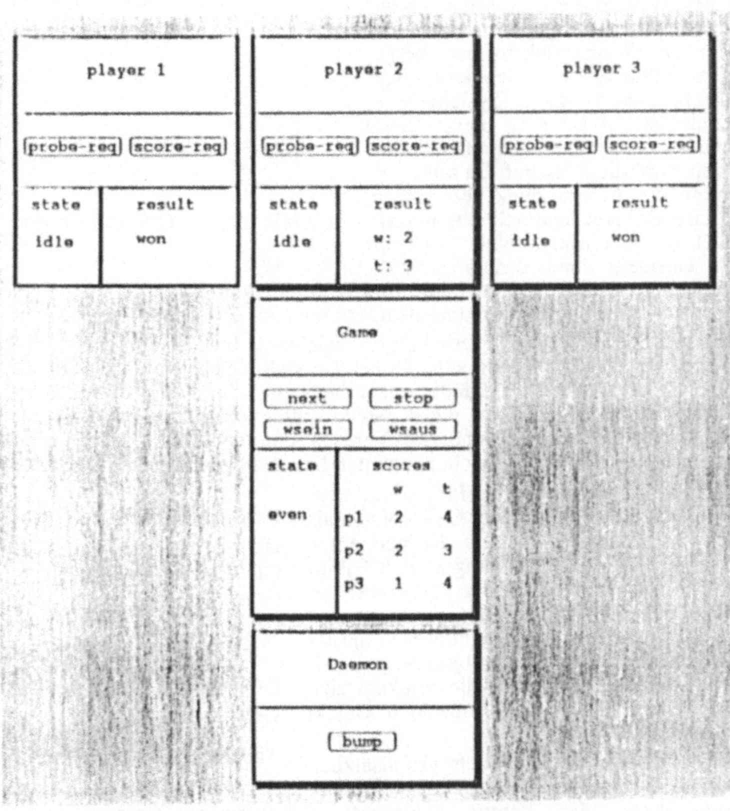

Abbildung 3: Graphische Benutzerschnittstelle für das System 'game'

Die gezeigte Benutzerschnittstelle wurde auf einer SUN4 unter SUN/OS 4.0 und Verwendung von Sunview in C implementiert. Es ist geplant, weitere Informationen wie z. B. die Verbindungsstruktur, Zustandsgraphen und Transitionsdeklarationen zu visualisieren, ggf. unter Einsatz von OPEN LOOK oder OSF/MOTIF.

6 ESTELLE, EINE OBJEKTORIENTIERTE SPRACHE?

Manufacturing Message Specification (MMS, [ISO89-2]) ist ein Kommunikationsstandard der Anwendungsschicht, dessen Beschreibung auf Konzepten der Objektorientierung beruht. Im Rahmen einer Diplomarbeit ([Hil90]), die in Kooperation mit der SIEMENS AG München entstand, wurde MMS in Estelle spezifiziert. Dabei wurde besonderer Wert darauf gelegt, die objektorientierten Aspekte des informellen Standards in der formalen Beschreibung auszudrücken. So ergab sich die Frage, inwieweit Estelle eine objektorientierte Darstellung unterstützt.

In objektorientierten Programmiersystemen (OOPs) wird der Versuch unternommen, die bisher getrennten Bereiche - Daten und Programmcode - zusammenzuführen. Das daraus resultierende Gebilde wird als *Objekt* bezeichnet. Es besteht die Möglichkeit, Objekte zu definieren, die alle Eigenschaften ihrer Vorfahren *erben*. Weiterhin besteht die Möglichkeit, über *Botschaften* von anderen Objekten kontrolliert die Fähigkeiten eines Objektes zu nutzen. Der kontrollierte Zugriff wird über eindeutig definierte Schnittstellen (*Methoden*) zu jedem Objekt realisiert.

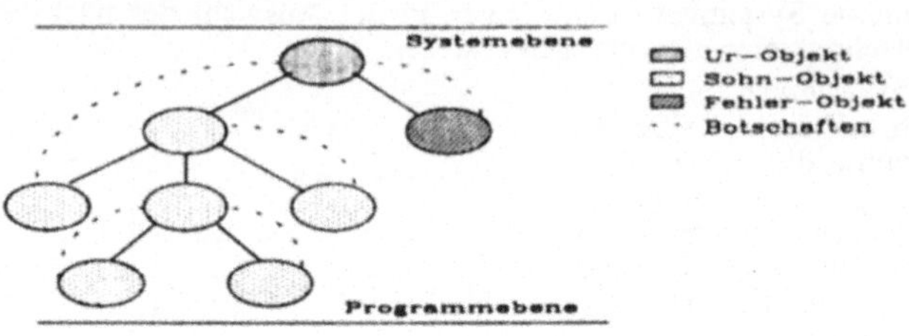

Abbildung 4: Objektorientierte Systemkonfiguration

Sollte eine Botschaft vom angesprochenen Objekt nicht bearbeitet werden können, so wird diese Botschaft an das Vater-Objekt weitergegeben. Das Vorgehen des Weiterreichens an übergeordnete Objekte wird solange durchgeführt, bis das Ur-Objekt (Systemebene) erreicht ist. Ist auch an dieser Stelle keine entsprechende Schnittstelle zu finden, wird die Botschaft im allgemeinen an ein Fehler-Objekt übergeben, wo eine entsprechende Reaktion stattfindet. Das Übermitteln einer Botschaft wird einerseits durch das umgebende Programmiersystem (z.B. Interprozeßkommunikation) und andererseits durch entsprechende Sprachkonstrukte (z.B. Prozeduraufrufe) unterstützt und ermöglicht.

Ein erster Schritt bei den Überlegungen, Estelle als objektorientierte Spezifikationstechnik einsetzen zu können, ist eine konzeptionelle Gegenüberstellung beider Ansätze (siehe Tabelle 1). Bei dieser Übersicht ist zu erkennen, daß die wesentlichsten Aspekte objektorientierter Methoden schon implizit in Estelle vorhanden sind. An dieser Stelle sollen einige der wichtigsten Konzepte objektorientierter Systeme näher betrachtet und der Versuch unternommen werden, mögliche Ähnlichkeiten zu den entsprechenden in der Übersicht genannten Estelle-Konzepten herauszuarbeiten.

Objekte in objektorientierten Programmiersystemen wie Smalltalk80 oder C++ können als nach außen abgeschlossene Komponenten aufgefaßt werden. Ein Zugang zu den Fähigkeiten eines Objektes ist nur über dafür bereitgestellte Schnittstellen (Methoden) möglich. Um das Ziel des kontrollierten Zugriffs auf die Fähigkeiten eines Objektes zu erreichen, ist es notwendig, daß die objektspezifischen Schnittstellen dem nutzenden Objekt bekannt sind.

OOP-Botschaften	<->	Estelle-Ereignisse
OOP-Instanz	<->	Estelle-Modul
OOP-Instanzvariable	<->	Estelle-Modulvariable
OOP-Kapselung	<->	leer
OOP-Methoden	<->	Estelle-Transitionsteil
OOP-Objekt	<->	Estelle-Modul
OOP-Polymorphie	<->	leer
OOP-Vererbung	<->	Estelle-Datentypvererbung

Tabelle 1: Objektorientierte Konzepte *vs* Estellekonzepte

Die Vorgehensweise entspricht grundsätzlich dem Modulkonzept von Estelle. Ein *Estelle-Modul* kann über bereits deklarierte Schnittstellen (Interaktionspunkte/Kanäle) mit anderen Modulinstanzen kommunizieren. Ein Estelle-Modul kennt lediglich seine eigenen Schnittstellen, i. allg. aber nicht die seiner Umgebung. Dies weicht vom entsprechenden objektorientierten Konzept signifikant ab.

Methoden gestatten den kontrollierten Zugriff auf die Daten und Fähigkeiten eines Objektes. Die Ausführung einer Methode hat die Modifikation von Daten und ggf. den Aufruf weiterer Methoden anderer Objekte über entsprechende Botschaften zur Folge. Ähnliches passiert beim Schalten einer *Estelle-Transition*. Auch hier ändert sich der interne Zustand, und es können ggf. Nachrichten an andere Module gesendet werden.

Anders als beim objektorientierten Ansatz ist jedoch in Estelle die Zuordnung zwischen Nachricht und Transition nach außen nicht sichtbar. Außerdem ist eine entsprechende Zuordnung i. allg. nicht eindeutig. Daher ist es nicht möglich, gezielt von außen Transitionen eines Moduls zu aktivieren.

Objekte in objektorientierten Programmiersystemen können ihre *Eigenschaften vererben*. Ein direktes Sohn-Objekt erbt alle Eigenschaften seines Vater-Objektes. Dabei wird die Notwendigkeit der Erzeugung und Freigabe von Sohn-Objekten deutlich. Dieses wird in den unterschiedlichen OOPs verschiedenartig gelöst. Grundsätzlich aber werden zu diesem Zweck Konstruktoren und Destruktoren angeboten.

Estelle bietet ähnliche Mechanismen zur Erzeugung und Freigabe von Sohn-Modulen. Bei der Instanzierung eines Estelle-Moduls werden aber lediglich die moduleigenen Datentypen an die Sohn-Module weitergegeben. Soll Estelle als objektorientierte Sprache Verwendung finden, wäre es u.a.

notwendig, die Fähigkeiten eines Moduls an seine Sohn-Module weiterzugeben. Estelle erlaubt eine solche Vorgehensweise auf Spezifikationsebene nicht.

Die konzeptionelle Gegenüberstellung objektorientierter Ansätze und der entsprechenden Estelle-Sprachkonstrukte ergab, daß die typischen objektorientierten Konzepte in fast allen Fällen durch Estelle-Sprachmittel ausgedrückt werden können. Einer der wichtigsten objektorientierten Aspekte, die Vererbung (inheritance), kann jedoch nur durch eine Spracherweiterung eingebracht werden. Die in Estelle fehlenden Konzepte waren für die eingangs genannte Anwendung von untergeordneter Bedeutung. Abschließend läßt sich feststellen, daß auch objektorientierte Spezifikationen von komplexen ISO-Standards wie MMS in Estelle formuliert werden können. Dieses ist für eine umfassende Verständnisausbildung vor der Implementierung von großem Nutzen.

7 VERIFIKATION VON ESTELLE-SPEZIFIKATIONEN

Verifikation im Kontext der Software-Entwicklung umfaßt alle Maßnahmen, die der vollständigen formalen Überprüfung von Eigenschaften, welche von Spezifikationen und Implementierungen gefordert werden, dienen. Beispiele solcher Eigenschaften sind Konsistenz, Verklemmungsfreiheit, Terminierung, Korrektheit und Konformität.

Ein System ist im anschaulichen Sinne korrekt, wenn es sich immer genau so verhält, wie es von ihm erwartet wird. Damit dieser intuitive Korrektheitsbegriff formal überprüfbar wird, sind mehrere Voraussetzungen zu erfüllen. Erstens muß die vorliegende Systembeschreibung eindeutig sein, was eine formale Semantik erfordert. Zweitens müssen die Erwartungen an das System ebenfalls eindeutig - das heißt unter Verwendung einer formalen Beschreibung - festgelegt werden. Schließlich ist drittens präzise zu definieren, wann die formale Systembeschreibung als korrekt bezüglich der formal spezifizierten Erwartungen bezeichnet werden darf.

Es existieren verschiedene Verfahren, um Korrektheit unter diesen Voraussetzungen zu überprüfen, zum Beispiel Programmverifikation (für sequentielle und nebenläufige Programme), Erreichbarkeitsanalyse, sowie algebraische Transformationen. Testen zählt nicht zu diesen Verfahren, da es im allgemeinen keine *vollständige* Überprüfung zuläßt und damit lediglich die Anwesenheit, nicht aber die Abwesenheit von Fehlern zeigen kann.

Wurde ein System mit verschiedenen Beschreibungstechniken spezifiziert und die wechselseitige Korrektheit dieser Beschreibungen nachgewiesen, so kann man sich wesentlich sicherer sein, daß diese Beschreibungen dem entsprechen, was gewünscht ist. Die Gefahr von Mißverständnissen beim Abfassen der Spezifikationen wird kleiner, und die Gefahr von einfachen Fehlern wird ganz gebannt.

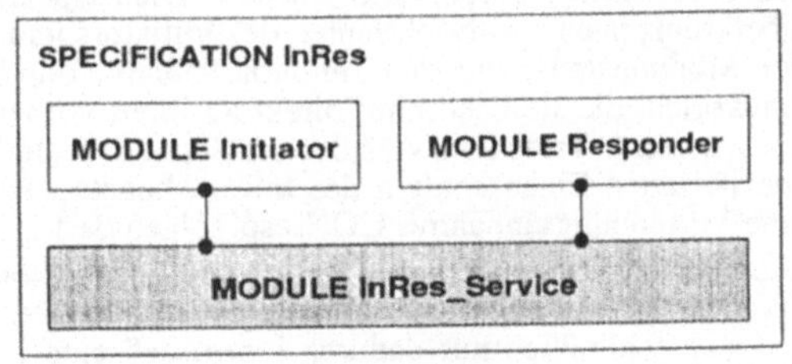

Abbildung 5: Systemarchitektur

Für Spezifikationen in Estelle ist ein Ausführungsmodell definiert ([ISO89-1]), mit dem den Spezifikationen eine formale Bedeutung gegeben wird. Damit kann eine Spezifikation in Estelle gegen andere formale Beschreibungen verifiziert werden, so daß die Vorteile der Verifikation sich für Estelle nutzen lassen.

Im Rahmen einer Studienarbeit ([Bre90]) wurde untersucht, inwieweit die Korrektheit von Estelle-Spezifikationen in Bezug auf eigenschaftsorientierte Spezifikationen in temporaler Logik definiert und verifiziert werden kann. Diese Untersuchung fand anhand eines einfachen Kommunikationsprotokolls, das in Estelle spezifiziert wurde, sowie einer Beschreibung des zu erbringenden Kommunikationsdienstes in temporaler Logik statt.

Der Nachweis der Korrektheit wurde manuell durchgeführt. Für jede in temporaler Logik spezifizierte Eigenschaft wurde mit Hilfe des Estelle-Ausführungsmodelles der jeweils interessierende Teil des Baums der möglichen Zustandsfolgen konstruiert und gezeigt, daß der Baum die betreffende Eigenschaft erfüllt.

Die Verifikation der Estelle-Spezifikation führte zu einigen unerwarteten Ergebnissen. Obwohl die erste Version auch nach erfolgtem Prüflesen durch mehrere Personen korrekt zu sein schien, stellte sich bei dem Versuch, ihre Korrektheit methodisch nachzuweisen, heraus, daß einige Eigenschaften nicht erfüllt waren. Dies lag daran, daß es durch die Nebenläufigkeit der Estelle-Modulinstanzen zu nicht bedachten Situationen kommen konnte. Um zu demonstrieren, wie auch in einem sehr einfachen verteilten System durch Nebenläufigkeit überraschend schnell eine schwer überschaubare Komplexität entsteht, soll eine dieser Situationen hier exemplarisch vorgestellt werden.

Bei dem erwähnten einfachen Protokoll handelt es sich um das 'InRes'-Protokoll (aus [Got90]). Es gewährt zwei Instanzen den Dienst, verbindungsorientiert zu kommunizieren. Der Dienst ist asymmetrisch, eine der Instanzen übernimmt die Rolle des 'Initiators', die andere die Rolle des 'Responders'. In die Verantwortung des Initiators fällt es, eine Verbindung aufzubauen; er kann Daten zum Responder senden, falls eine Verbindung existiert. Der Responder kann einen Verbindungsaufbauwunsch annehmen oder auch ablehnen, er kann Daten empfangen, und er kann eine Verbindung abbauen.

Abbildung 5 zeigt den Aufbau des Systems. Das Zeit-Signal-Diagramm in Abbildung 6 zeigt einen erfolgreichen Verbindungsaufbau, die Übertragung eines Nutzdatums und einen Verbindungsabbau.

Als der InRes-Diensterbringer in Estelle spezifiziert wurde, wurde er ganz natürlich in drei Estelle-Modulinstanzen verfeinert: Jeweils eine Modulinstanz, die einen der Interaktionspunkte zum Initiator beziehungsweise Responder bedient, und eine Modulinstanz, die ein Medium repräsentiert und die anderen beiden, möglicherweise verteilt realisierten Modulinstanzen verbindet. Damit ergab sich die Aufgabe, daß alle drei Modulinstanzen zusammen den InRes-Dienst erbringen sollen, dabei aber keine den aktuellen Systemgesamtzustand kennt.

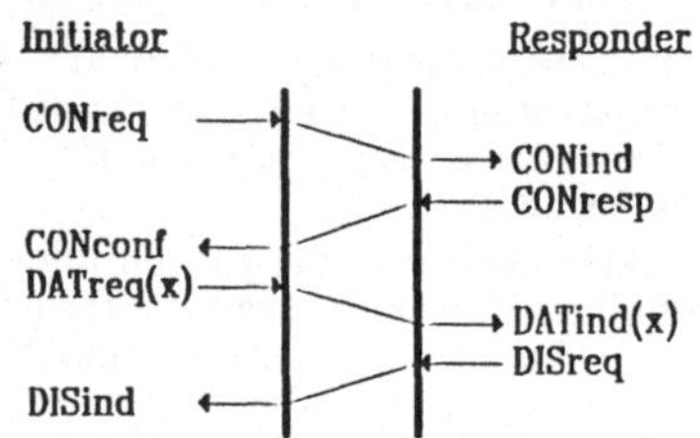

Abbildung 6: Ein Beispiel für die Nutzung des InRes-Dienstes

Hieraus resultierte eine jener Situationen, welche bei der ersten InRes-Spezifikation in Estelle nicht bedacht worden waren: Die formale Dienstspezifikation in temporaler Logik läßt zu, daß es zu Überkreuzungen von Nachrichten des Initiators und des Responders kommen darf. In Abbildung 7 wird der Modulinstanz, die den Initiator bedient, durch mehrere Überkreuzungen vorgegaukelt, daß die Antworten, die sie bekommt, direkt zu ihren vorherigen Anfragen gehören. Aus ihrer lokalen Sicht ist alles korrekt. Trotzdem verstößt die letzte Verbindungsbestätigung 'CONconf' gegen eine der formal spezifizierten Eigenschaften des InRes-Dienstes, nach welcher höchstens je ein 'CONconf' auf jeweils eine Verbindungsannahme 'CONresp' folgen darf.

Die Probleme, die aus dieser und anderen, noch komplexeren nicht erkannten Überkreuzungen resultierten, ließen sich schließlich nur dadurch lösen, daß eine interne Quittung eingeführt wurde. Diese beantwortet alle Nachrichten, welche die Modulinstanz, die den Responder bedient, an die Modulinstanz, die den Initiator bedient, sendet. Damit können alle Überkreuzungen erkannt und geeignet behandelt werden.

Die zusätzliche Quittung demonstriert gleichzeitig ein weiteres Ergebnis der Arbeit: Die resultierende Protokollspezifikation in Estelle ist deutlich anders geworden, als wenn sie wie üblich ohne eine fertig vorliegende Dienstspezifikation entworfen worden wäre. Insgesamt führte die Verifikation zu einem wesentlich tieferen Verständnis der Eigenschaften und Probleme des betrachteten Protokolls.

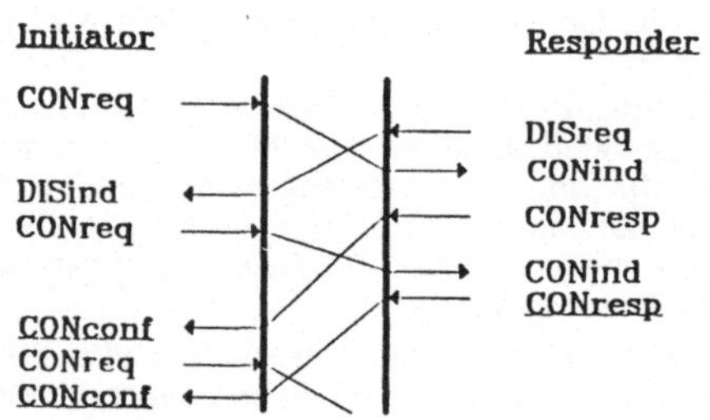

Abbildung 7: Nicht erkannte Überkreuzungen stiften Verwirrung

Die Untersuchungen im Rahmen der Studienarbeit dienten in erster Linie dazu, den Zugang zur Verifikation von Estelle-Spezifikationen herzustellen und praktische Erfahrungen zu sammeln. Um eine Verifikation vollständig formal durchführen zu können, wäre ein Formalismus notwendig, mit dem sich sowohl die Eigenschaften eines Dienstes als auch alle relevanten Eigenschaften des dazugehörigen Estelle-Protokolls formal beschreiben lassen, und der trotz dieser großen Ausdruckskraft übersichtlich bleibt. Eine durchgängig formale Behandlung der Problematik aus der Studienarbeit könnte beispielsweise durch Festlegung einer axiomatischen Semantik für Estelle erreicht werden. Hierzu existieren derzeit lediglich erste Ansätze.

8 SCHLUßWORT

Formale Beschreibungssprachen im OSI Umfeld sollen sowohl der präzisen Beschreibung von Standards als auch der Realisierung von standardisierten Protokollen dienen. Der zweite Aspekt ist oftmals unterschätzt worden, da der Realisierungaspekt nicht unmittelbar mit der Normungsabsicht vereinbar schien. Für die Protokolle der Anwendungsschicht gilt dies nicht, da hierfür die - zugegeben abstrakte - Benutzersicht die entscheidende Motivation für Entwicklung und Normung begründet. Da darüber hinaus für die flexible Gestaltung der unterschiedlichen Anwendungsanforderungen viele Protokollbausteine notwendig sein können, ist sowohl die prozedurale Abstraktion bezüglich dieser Bausteine als auch deren Zusammenfügung zu einem Ganzen zu unterstützen. Estelle verfügt über Konzepte zur Erfüllung dieser Forderungen und stellt deshalb ein geeignetes Instrument für die Normungsentwicklung dar. Darüber hinaus erlauben Estelle-Compiler die Übersetzung der Spezifikation in eine prozedurale Programmiersprache. Daß die Anwendung von Estelle nicht nur in der Normungsentwicklung, sondern auch in der industriellen Praxis in Deutschland zunehmende Verbreitung findet, ist vermutlich diesem Umstand zuzuschreiben. Für die Weiterentwicklung von Estelle bzw. Estelle-Werkzeugen hat sich die Kooperation zwischen Industrie und Universität bewährt. Es wäre wünschenswert, wenn diese Entwicklung auch von den zuständigen deutschen Normungsgremien gesehen werden würde, um Estelle in eine DIN-ISO-Norm überführen zu können.

LITERATURHINWEISE

[And90] Andrae, C.: Eine Spezifikation von CCR in Estelle, Studienarbeit, Fachbereich Informatik, Universität Hamburg, 1990

[Bre90] Bredereke, J.: Spezifikation und Verifikation des InRes-Protokolls unter Verwendung von Estelle und temporaler Logik, Studienarbeit Nr. 700, Fachbereich Informatik, Universität Hamburg,1990

[DeBu89] Dembinski, P., Budkowski, S.: Specification Language Estelle, in: M. Diaz et al. (eds.), The Formal Description Technique Estelle, North-Holland, 1989, S. 35-75

[Got90] Gotzhein, R.: The Formal Definition of the Architectural Concept 'Interaction Point', in: S. T. Vuong (ed.), Formal Description Techniques, II, North-Holland, 1990, S. 67-81

[Hil90] Hille, C.: Spezifikation eines Manufacturing Message Specification-Subsets (MMS) in Estelle mit einer nachfolgenden Implementierung in einer UNIX-Umgebung, Diplomarbeit, Fachbereich Informatik, Universität Hamburg, 1990

[Hog89] Estelle, LOTOS und SDL. Standard-Spezifikationssprachen für verteilte Systeme, Springer-Verlag, Berlin, 1989, 188 S.

[ISO81] ISO/TC97/SC16: Data Processing - Open Systems Interconnection - Basic Reference Model, Computer Networks 5, 1981, S.81-118

[ISO88-1] ISO DIS 9804.2 Service Definition for the Commitment, Concurrency and Recovery Service Element, 1988

[ISO88-2] ISO DIS 9805.2 Protocol Specification for the Commitment, Concurrency and Recovery Service Element, 1988

[ISO88-3] ISO IS 9066-1 Information processing systems- Text Communication Reliable Transfer Part 1: Model and Service Definition, 1988

[ISO88-4] ISO IS 9066-2 Information processing systems - Text Communication Reliable Transfer Part 2: Protocol Specification, 1988

[ISO89-1] ISO IS 9074, Estelle - A Formal Description Technique Based on an Extended State Transition Model, 1989, 179 S.

[ISO89-2] ISO DIS 9506 Industrial Automation Systems-System Integration and Communication-Manufacturing Message Specification / Part 1 Service / Part 2 Protocol, 1989

[NBS87] User Guide for the NBS Prototype Compiler for Estelle - Final Report, National Institute of Standards and Technologies, Report No. ICST/SNA-87/3, Okt. 1987, 73 S.

[Pet89] Peter, D.: Konzeption eines Environments zur Generierung und verteilten Ausführung von Estelle-Code, Studienarbeit Nr. 379, Fachbereich Informatik, Universität Hamburg, 1989

[Pet91] Peter, D.: Entwurf, Realisierung und Integration eines Protokolls zur verteilten Ausführung von Estelle-Spezifikationen, Diplomarbeit Nr. 791, Fachbereich Informatik, Universität Hamburg, 1991

[Rei89] Reimer T.: Implementierung eines Reliable Transfer Service Elements (RTSE) mittels der formalen Beschreibungstechnik Estelle, Diplomarbeit Nr. 593, Fachbereich Informatik, Universität Hamburg, 1989

[Sch91] Schüler, U.: Entwurf und Realisierung einer Benutzeroberfläche für die interaktive Simulation einer Estelle-Spezifikation, Studienarbeit Nr. 688, Fachbereich Informatik, Universität Hamburg, 1991

Die offene Dokumentenarchitektur ODA als Basis der Bürokommunikation

U. Einig, J. Kämper, W. Knobloch[1]
IBM, European Networking Center
Tiergartenstraße 8
D-6900 Heidelberg
Germany

Kurzfassung

Die Dokumentenarchitektur ODA, die als ein internationaler Standard den Dokumentenaustausch in offenen Netzen ermöglichen soll, gewinnt gegenwärtig einen immer stärkeren Einfluß auf den Bereich der Bürokommunikation. Sowohl neue Applikationen, die unmittelbar auf dieser Architektur basieren, als auch Konverter, die die Brücke zu gegebenen Anwendungen schlagen, werden im Rahmen von europäischen Kooperationsprojekten entwickelt. Zusätzlich besitzt ODA auch einen Einfluß auf Herstellerarchitekturen und zukunftsorientierte Anwendungen wie Hypertextsysteme. Insbesondere in Bezug auf derartige, neue Anwendungsbereiche zeigen sich aber auch die Grenzen der Architektur. In diesem Beitrag werden Erfahrungen aus der Mitarbeit an europäischen Kooperationsprojekten wiedergegeben, die die obigen Aspekte der Verwendung von ODA als Basis für die Bürokommunikation beleuchten.

1. Einleitung

Dokumente sind für viele Büroanwendungen von zentralem Interesse. Geschäftsbriefe, Bilanzen, Rechnungen und vieles mehr können als Dokumente aufgefaßt werden. Um Büroanwendungen automatisieren zu können, ist somit die Handhabung elektronischer Dokumente von besonderer Bedeutung. Lokale Büroanwendungen werden immer häufiger miteinander verbunden, die Unterstützung von Büroabläufen, die verschiedene Arbeitsschritte und -plätze betreffen, tritt in den Vordergrund. Eine derartige Verknüpfung verschiedener Arbeitsschritte macht ein gemeinsames Verständnis dieser Schritte über die zu verarbeitenden Dokumente notwendig. Dokumentenarchitekturen sollen diese Grundlage der Bürokommunikation liefern.

In heutigen Büros sind sehr unterschiedliche Anwendungen anzutreffen, es gibt eine Vielzahl heterogener Systeme etwa zur Textverarbeitung oder zur elektronischen Ablage von Dokumenten. Diese Beobachtung wird dadurch verstärkt, daß der Kommunikationsumfang zwischen verschiedenen Büros wächst und diese

---

[1] Dieser Beitrag wurde teilweise durch das ESPRIT-Projekt Nr. 2374 der Europäischen Gemeinschaft gefördert.

Kommunikation zunehmend elektronisch erfolgt. In einem derartigen, heterogenen Umfeld bilden internationale Standards die Brücke, über die die Kommunikation erfolgen kann. Der Standard ODA (Open Document Architecture) [2][3] übernimmt diese Brückenfunktion in Bezug auf Dokumente.

ODA beschreibt ein Modell zur Dokumentenverarbeitung, das verschiedene Schritte wie das Editieren, das Formatieren und das Präsentieren von Dokumenten umfaßt. Basierend auf diesem Dokumentenverarbeitungsmodell stellt ODA eine Sprache zur Verfügung, mit deren Hilfe Dokumente geeignet beschrieben werden können. Die Beschreibung eines Dokumentes kann dann zwischen Applikationen mit Hilfe von Kommunikationsdiensten ausgetauscht werden. Ein kurze Einführung in die wesentlichen Konzepte der Architektur ODA befindet sich im Abschnitt 2 dieses Beitrags.

Seit der Verabschiedung des ODA-Standards werden eine Vielzahl von Projekten durchgeführt, die die Umsetzung dieser Architektur in Anwendungen zum Ziel haben. Grundsätzlich können dabei zwei Ansätze unterschieden werden. Anwendungen (zum Beispiel Editoren), die ODA als internes Dokumentenformat verwenden, werden als ODA basierte Anwendungen (ODA „native" Anwendungen) bezeichnet. Sie ermöglichen die direkte Verarbeitung von Dokumenten, die in ihrer elektronischen Repräsentation ODA entsprechen. Zusätze zu existierenden Anwendungen, die deren internes Format in das ODA-Format übersetzen, werden als Konverter bezeichnet. Sie erlauben eine Kommunikation zwischen im Markt befindlichen Produkten.

Das ESPRIT Projekt PODA-2[2] , das von der europäischen Gemeinschaft gefördert wird, hat Ansätze in beiden der obigen Richtungen verfolgt. Die resultierenden Prototypen wurden auf verschiedenen Messen gezeigt. Das vorwiegende Anliegen dieses Beitrages ist es, wesentliche Erfahrungen aus dieser Projektarbeit wiederzugeben und dabei Vor- und Nachteile der oben skizzierten Alternativen zur Verwendung von ODA aufzuzeigen. Dieses geschieht in den Abschnitten 3 und 4.

Prototypen für zukunftsorientierte Anwendungen wie das verteilte Editieren (joint editing) oder Hypertextsysteme werden augenblicklich ebenfalls basierend auf dem Dokumentenverarbeitungsmodell von ODA entwickelt. Hier stoßen aber die Architektur selbst und insbesondere der Konverteransatz an ihre Grenzen. Die Beschreibung einiger Erfahrungen im Zusammenhang mit derartigen, zukunftsweisenden Anwendungen und der sich daraus als notwendig ergebenden Erweiterungen der ODA-Architektur befinden sich im Abschnitt 5. Dort wird außerdem auf den Einfluß eingegangen, den der Standard ODA auf andere Dokumentenarchitekturen ausübt.

2. Die offene Dokumentenarchitektur ODA

Die offene Dokumentenarchitektur ODA wurde von internationalen Standardi-

---

2 PODA-2 ist das Nachfolgeprojekt des PODA (Piloting the Open Document Architecture) Projektes. Die Partner dieses Projektes sind: British Telecom, BULL, IBM, ICL, Nixdorf, Océ, Olivetti, Siemens, TITN, University College London.

sierungsgruppen wie der ISO [3] und der CCITT[4] entwickelt, um ein einheitliches Modell für die Dokumentenverarbeitung zu etablieren. Basierend auf diesem Modell können verschiedene Anwendungen wie zum Beispiel der Austausch elektronischer Dokumente in offenen Netzen unterstützt werden. ODA wurde 1989 als internationaler Standard verabschiedet [2][3]. Für eine detailierte Einführung siehe [4]oder [5]. An dieser Stelle sollen nur einige wesentliche Konzepte vorgestellt werden.

ODA erlaubt die Beschreibung strukturierter Dokumente, die aus Textteilen, geometrischen Graphiken und Rastergraphiken bestehen. Dabei werden formatierte und editierbare Dokumente unterschieden. Außerdem gibt es Dokumente, die sowohl editierbar als auch formatiert sind. Es gibt verschiedene Schritte der Dokumentenverarbeitung, die im ODA-Standard beschrieben sind. Der Editierprozeß (editing process), der auf editierbare Dokumente anwendbar ist, dient der Erzeugung und der Veränderung von Dokumenten. ODA sagt nichts über die Art aus, wie Dokumente einem Benutzer zur Änderung anzubieten sind, der Standard bezeichnet lediglich die Teile eines Dokumentes, die durch den Editierprozeß verändert werden dürfen. Durch den Layoutprozeß (formatting process) wird ein editierbares Dokument in ein formatiertes Dokument überführt, das dann durch den Präsentationsprozeß (imaging process) auf einem physikalischen Gerät zum Beispiel einem Drucker oder einem Bildschirm dargestellt werden kann. Dieser Präsentationsprozeß ist von dem jeweiligen herstellerspezifischen Zielgerät abhängig und deshalb nur zum Teil in dem ODA-Standard beschrieben.

Selbstverständlich unterscheidet sich die Beschreibung eines editierbaren Dokumentes von der eines formatierten Dokumentes. Die wesentlichen Teile eines editierbaren Dokumentes beziehen sich auf seine Inhaltsteile, seine logische Struktur und einige Regeln, die den Layoutprozeß steuern. Die logische Struktur unterteilt ein Dokument etwa in Kapitel, Paragraphen und Abschnitte und zeichnet Fußnoten und Listen aus. Zusätzlich zu den Inhaltsteilen enthält ein formatiertes Dokument eine Layout-Struktur, durch die ein Dokument zum Beispiel in Seitenklassen, einzelne Seiten und rechteckige Bereiche auf diesen Seiten unterteilt wird. Den rechteckigen Bereichen, die als „Frames" bezeichnet werden, sind die Inhaltsteile zugeordnet.

Beide Strukturen, sowohl die logische Struktur als auch die Layout-Struktur, bestehen aus zwei Teilstrukturen, einer generischen und einer spezifischen Struktur. Die generischen Strukturen beschreiben Eigenschaften eines Dokuments, die einer Dokumentenklasse gemeinsam sind, die spezifischen Strukturen beschreiben zusätzliche Eigenschaften des speziellen Dokuments, also die Teile, die dieses Dokument von anderen Dokumenten der gleichen Klasse unterscheidet. Oft besitzen zum Beispiel alle Geschäftsbriefe einer Firma den gleichen Aufbau und den gleichen Seitenkopf. Diese Informationen können dann durch die generischen Strukturen für alle Geschäftsbriefe gemeinsam dargestellt werden. Die jeweilige Anschrift, das Datum, der Text und alle weiteren Teile, die nicht einheitlich sind, sind durch die spezifischen Strukturen zu erfassen.

Alle oben erwähnten Strukturen eines Dokuments sind hierarchisch aufgebaut. Die spezifischen Strukturen bestehen aus Objekten, die durch Attribute bezie-

---

[3] International Organization for Standardization

[4] Comité Consultatif International Télégraphic et Téléphonique

hungsweise Werte dieser Attribute beschrieben werden. Zum Beispiel werden der eindeutige Name eines Objekts oder die Liste der untergeordneten Objekte als Attribute dargestellt. Die generischen Strukturen bestehen aus Klassen, Objekte der spezifischen Strukturen referenzieren diese Klassen. Somit kann ein Dokument etwa mehrere Objekte enthalten, die je einen Paragraphen darstellen. Diese Objekte können alle die gleiche Klasse „Paragraph" referenzieren. Die Klassen der generischen Strukturen beschreiben Regeln, nach denen die spezifischen Strukturen aufgebaut sein müssen und geben Standardwerte („defaults") für einige Attribute vor. So kann etwa beschrieben werden, daß ein Paragraph in einem gewissen Dokument immer aus einer nicht leeren Folge von Textteilen und Graphiken bestehen muß und daß die Textteile immer mit einer gewissen Schriftgröße darzustellen sind, sofern das spezielle Objekt keine eigenen Angaben bezüglich der Schriftgröße enthält.

Stile (styles) werden in einem ODA-Dokument verwendet, um eine Faktorisierung von Informationen zu ermöglichen. Sie fassen einige Attributwerte zusammen und können von Objekten oder Klassen referenziert werden, mit der Folge, daß die referenzierenden Objekte und Klassen die Attributwerte übernehmen. Attribute, die die Darstellung gewisser Textteile betreffen, wie etwa die Schriftgröße, die Art der Hervorhebung und der aktuelle „Font", lassen sich etwa zu derartigen Stilen zusammenfassen.

Neben den bisher erläuterten Teilen umfaßt die Beschreibung eines ODA-Dokuments noch das „Document Profile", das Informationen enthält, die das gesamte Dokument betreffen wie den Autor, das Erstellungsdatum und Angaben über Vervielfältigungsrechte.

Wie bereits erwähnt wurde, kann ein ODA-Dokument Textteile, geometrische Graphiken und Rastergraphiken enthalten. Auf die entsprechenden Inhaltsarchitekturen soll hier nicht näher eingegangen werden.

Der Austausch von ODA Dokumenten geschieht mit Hilfe einer der beiden Austauschformate, die Teile des ODA-Standards sind. ODIF (Open Document Interchange Format) basiert auf ASN.1 ([1]), ODL (Open Document Language) dagegen auf SGML ([15]).

ODA stellt insgesamt sehr umfangreiche Konzepte zur Beschreibung von Dokumenten zur Verfügung. Der Standard macht aber keine Aussagen darüber, wie diese Konzepte zur Beschreibung bestimmter Dokumententeile wie Fußnoten, numerierte Segmente oder Inhaltsverzeichnisse einzusetzen sind. Insgesamt gestattet ODA eine solche Vielfalt von Möglichkeiten zur Beschreibung derartiger Dokumententeile, daß der Standard von heutigen Anwendungen nicht in vollem Umfang unterstützt werden kann. Aus diesem Grunde werden von internationalen Standardisierungsgremien sogenannte Dokumentenanwendungsprofile (Document Application Profiles, DAPs) definiert (für einen Überblick siehe [13]), die Teilmengen des ODA-Standards auszeichnen. In Europa wurde durch die EWOS[5] ein Satz von in ihrer Funktionalität hierarchisch geordneten DAPs entwickelt, Q111, Q112 und Q113. Dabei ist Q111 in der Lage, einfache Dokumente, die nur Text enthalten, darzustellen, Q112 ist vergleichbar zu den Fähigkeiten moderner Texteditoren (komplexe Strukturen, Graphiken, Bilder) und Q113 soll zukünftige Entwicklungen unterstützen, indem es zum Beispiel die Darstellung von Tabellen und automatischen Referenzierungen beschreibt.

---

[5] European Workshop for Open Systems

3. Konvertertechnologie

Der Austausch multimedialer Dokumente ist eines der Hauptziele, die mit der Entwicklung von ODA verfolgt wurden. Wie bereits erwähnt, stellen ODIF und ODL dabei allgemein anerkannte Austauschformate dar. Für den Übergang vom lokalen System zum offenen Netz muß nur ein einziger Transformationtyp zur Vefügung gestellt werden, unabhängig davon, welches Dokumentenverarbeitungssystem vom Kommunikationspartner benutzt und über welche verschiedenen Netztypen kommuniziert wird. In den bisherigen Implementierungen von ODA wird für den Austausch ausschließlich ODIF verwendet. Aus diesem Grunde beschränken sich unsere folgenden Betrachtungen auf dieses Austauschformat.

Verschiedene Hersteller von Hard- und Software haben sich bereits entschlossen, ODA zu unterstützen, und erste Transformationen für ihre eigenen Dokumentenverarbeitungssysteme entwickelt. Dabei ist es das Ziel, bereits im Markt befindliche Systeme miteinander kommunizieren zu lassen. Der allgemeine Ansatz dieser Konvertertechnologie besteht aus drei Teilen:

1. Basis ist ein existierendes System zur Dokumentenverarbeitung.
2. Zu diesem System wird eine Transformation entwickelt, die die interne Darstellung der Dokumente auf ODIF abbildet.
3. Zur Übertragung der ODIF Datenströme werden öffentlich verfügbare Telekommunikationsdienste verwendet, zum Beispiel X.25 Datex-P, und darauf basierende Implementierungen von X.400 [10].

Ein wesentlicher Vorteil dieses Ansatzes ist die Benutzung existierender Editorsysteme, die bereits im Markt befindlich und damit den Benutzern vertraut sind. Diese Systeme beinhalten im allgemeinen einen eigenen Formatierungsprozeß, der einen spezifischen Druckerdatenstrom erzeugt und die editierbare Darstellung eines Dokuments zur Eingabe hat. Aufgrund dieser Architektur sind solche Systeme nur in der Lage, editierbare Dokumente zu bearbeiten. Dies hat zur Folge, daß nur die editierbare Form eines ODA-Dokuments (Processable Document Architecture) ausgetauscht werden kann.

Der Konvertierungsprozeß für editierbare Dokumente verläuft dabei in Bezug auf die verschiedenen Inhalte und Eigenschaften eines Dokumentes in sehr unterschiedlicher Weise. Die Abbildbarkeit einzelner Eigenschaften und Inhalte läßt sich in drei Kategorien unterteilen, die hier am Beispiel der in [11] beschriebenen Konvertierung von ODA nach RFT:DCA[6] erläutert werden sollen:

- Die erste Kategorie enthält solche Inhalte und Eigenschaften, die sich editierbar abbilden lassen, ohne daß ein Formatierungsprozeß durchgeführt werden muß. Als Beispiele dafür seien der Textinhalt eines Dokuments oder die automatische Segment- und Seitennumerierung genannt.
- Die zweite Kategorie enthält solche Inhalte und Eigenschaften, die sich nur mit Hilfe eines zumindest teilweise durchgeführten Formatierungsprozesses editierbar abbilden lassen. Als Beispiel dafür eignet sich unter anderen die Zeilenlänge eines bestimmten Paragraphen. Während der Abbildung von

---

[6] Revisable Form Text Document Content Architecture

ODA nach RFT:DCA muß der Formatierungsprozeß teilweise simuliert werden, um die horizontale Dimension einer Textzeile zu errechnen, welche dann im resultierenden RFT:DCA-Dokument zu Beginn des Paragraphen als Kontrollinformation angegeben wird, wo sie bis zur nächsten gleichartigen Kontrollinformation gültig bleibt. Bei der Umkehrabbildung können nun alle Textzeilen zwischen solchen Kontrollinformationen als Paragraph interpretiert und als ein entsprechendes Objekt im entstehenden ODA-Dokument dargestellt werden.

- Die letzte Kategorie beschreibt solche Inhalte und Eigenschaften, die nicht editierbar abgebildet werden können. Ein Beispiel für eine solche Information ist das Inhaltsverzeichnis eines RFT:DCA-Dokuments, das dynamisch während der Formatierung aus den vorhandenen Segmentüberschriften zusammengestellt wird. Ein solches Konstrukt ist im Anwendungsprofil Q112 nicht verfügbar, so daß es nur als formatierter Text in den erzeugten ODA Dokumenten dargestellt werden kann.

Dem Konverteransatz sind einige Probleme immanent, die wir im folgenden kurz darstellen wollen, bevor wir am Ende dieses Abschnitts bestehende oder angestrebte Lösungen beschreiben. Eine detailliertere Untersuchung zu diesem Bereich, die auf den Fall des von den Autoren entwickelten Konverters zwischen ODA und RFT:DCA beschränkt ist, findet sich in [11].

Konvertierungen von Informationstypen sind sehr oft mit einem Informationsverlust verbunden. In dem hier betrachteten Fall sind für den Austausch von Dokumenten unter Verwendung von ODIF zwei Konvertierungen notwendig - von der internen Darstellung des Senders nach ODIF und von ODIF zum internen Format des Empfängers - , die beide zu einem Informationsverlust führen. Dieses Problem ist grundsätzlich unabhängig davon, ob es sich bei dem Sender und dem Empfänger um gleichartige Systeme handelt oder nicht, da der Versuch, die Abbildungsqualität für ein bestimmtes Partnersystem zu optimieren, die Gefahr des Verlustes der Austauschbarkeit mit anderen mit sich bringt.

Ein weiterer Problembereich ist die Durchschaubarkeit der Abbildung für den Endbenutzer, der in der Lage sein muß, die Menge der durch die Konvertierung abbildbaren Informationen zu erkennen. Dies ist von besonderer Bedeutung in Szenarien, in denen ein Dokument mehrmals zum Zweck der Revision zwischen verschiedenen Knoten ausgetauscht wird. In diesem Fall haben alle Benutzer der verschiedenen Endsysteme zu wissen, welche Dokumenteneigenschaften unter den gegebenen Transformationen erhalten bleiben. Dieses ist notwendig, damit die Eigenschaften, die nicht erhalten bleiben, in Dokumenten vermieden werden können. Eine auf Beachtung von Spezialfällen bedachte Spezifikation der Abbildung würde in dieser Hinsicht hinderlich sein. Die Bereiche Abbildungsqualität und Benutzerfreundlichkeit können demnach nicht unabhängig voneinander optimiert werden, sie sind negativ korrelliert.

Moderne Dokumentenarchitekturen wie ODA definieren ein Dokumentenmodell. Neben der Definition der Konstituenten eines Dokumentes und des benutzten Datenstromes werden auch die Verarbeitungprozesse für das Editieren, das Formatieren und das Präsentieren spezifiziert. Bei der Konvertierung veränderbarer Dokumente muß also nicht nur der aktuelle Inhalt eines Dokumentes, sondern auch sein Verhalten unter diesen Prozessen berücksichtigt werden, wobei die Abbildungsqualität für die verschiedenen Prozesse nicht unabhängig voneinander optimiert werden kann. (Beispiele für derartige Abhängigkeiten findet man in [11].)

Hier wirkt sich das Fehlen eines umfassenden Begriffes der Konformität zu ODA besonders aus. Die existierenden Definitionen von Konformität zu einem Dokumentenanwendungsprofil (Document Application Profile, DAP) beschränken sich im Wesentlichen auf die des ausgetauschten Datenstromes. Die Konformität einer Implementierung, wie zum Beispiel eines existierenden Textsystems und eines dazugehörigen ODA-Konverters, ist noch nicht definiert. In diesem Bereich werden zur Zeit Untersuchungen von den entsprechenden Gremien (EWOS, NIST[7] , PAGODA[8]) durchgeführt. Es ist geplant, daß ein Satz von „International Standardisierten Applikationsprofilen" (ISP), die in der Funktionalität Q111 bis Q113 entsprechen, Anforderungen an derartige ODA-Implementierungen definieren.

Ein wesentliches Merkmal der Architektur ODA ist die große Variabilität in der Darstellung eines Dokumentes und damit der ausgetauschten Datenströme. Beispiele hierfür sind der Ausnutzungsgrad der Vererbungsregeln (generisch zu spezifisch, über- zu untergeordnet) oder die Art der Positionierung von Text auf einer Seite (feste oder variable Rahmen (frame), Benutzung des „Offset" Attributes in der Rahmenspezifikation oder im Layout Stil). Obwohl solche verschiedenen Darstellungsweisen auch verschiedene Semantiken tragen, sind die Unterschiede zum Teil so gering, daß es für einen Konverter sehr viele verschiedene Möglichkeiten gibt, ein einziges Dokument darzustellen. Falls die Konvertierung ohne Benutzerinteraktion durchgeführt werden soll, so muß aber eine dieser Alternativen ausgewählt werden. Da jeder Konverter nur den seinem zugrundeliegenden Textsystem entsprechenden Teil des verwendeten Anwendungsprofils erzeugt, variieren die ausgetauschten Dokumente also stark in der Art der Darstellung. Daher muß der empfangsseitige Teil eines Konverters eine möglichst große Variabilität besitzen, was unter Umständen im Gegensatz zu den Möglichkeiten des internen Dokumentenformats steht.

Im folgenden werden wir Ansätze zeigen, die uns geeignet scheinen, die Abbildungsqualität eines Konverters zu beurteilen und die Austauschbarkeit der erzeugten ODA-Datenströme zu erhöhen. Sie bildeten die Grundlage der bereits erwähnten Prototypentwicklung im Rahmen des Projektes PODA-2.

Aus Sicht eines Konverters für existierende Textsysteme enthält ein ODA-Dokument verschiedene Ebenen von Information. Da ist zunächst der logische Inhalt des Dokumentes, der an den Empfänger möglichst unverzerrt weitergegeben werden muß. Dazu gehört nicht nur die Transformation der einzelnen Text- und Graphikteile eines Dokuments, sondern auch ihre logische Anordnung und ihre Beziehungen untereinander (z.B. Fußnotenreferenz zu Fußnotentext). In ODA DAP Q112 entspricht dies im Wesentlichen der Information, die in der logischen Struktur und den Inhaltsteilen beschrieben ist.

Als zweite Ebene der Information erscheint das Verhalten eines Dokumentes unter den Editier-, Formatierungs- und Präsentationprozessen, wie sie in ODA definiert sind. Da bei den existierenden Textverarbeitungssystemen Formatierung und Präsentation häufig gekoppelt sind, zerfällt diese Ebene aus Sicht eines Konverters in zwei Teile, die Editierbarkeit eines Dokuments einerseits und sein Layout andererseits. Das Verhalten eines Q112 Dokumentes unter einem Editierprozeß wird dabei im Wesentlichen von der generisch logischen Struktur und

---

7 National Institute for Standardisation, USA

8 Profile Alignment Group for ODA

der Definition des Editierprozesses in ODA bestimmt. Als Beispiele seien hier nur die Konzepte für automatisch generierte Segment- und Fußnotennummern, das Einfügen oder Löschen von Inhaltsteilen ohne Seiteneffekte auf die restliche logische Struktur oder die Restriktion von strukturellen Erweiterungen durch den „Generator for Subordinates" genannt, die geeignet in das jeweilige interne Dokumentenformat zu übersetzen sind.

Da ODA einen Layoutprozeß definiert und ein veränderbares Dokument laut Q112 alle notwendigen Informationen zur Anwendung dieses Prozesses enthält, trägt jedes solche Dokument implizit eine genaue Beschreibung seines Layouts mit sich.

Wie schon beschrieben, ist für die Konvertertechnologie hauptsächlich der Austausch von veränderbaren Dokumenten von Interesse. In diesem Falle ist die Optimierung der Abbildung des Editierverhaltens wichtiger als die exakte Übermittlung des Layouts. Für die Konvertierung editierbarer Dokumente ergibt sich also folgende Liste von zu erhaltender Information, geordnet nach sinkender Priorität:

- Logischer Dokumenteninhalt
- Verhalten unter Editierprozessen
- Dokument-Layout

Die Beschreibung einer aus diesem Ansatz resultierenden Abbildungsspezifikation für den Fall des von uns entwickelten Prototypen findet sich in [12].

Sowohl aus Gründen der Kosteneinsparung, als auch um die Austauschbarkeit von Dokumenten in höherem Maße gewährleisten zu können, wurde im Rahmen des Projektes PODA-2 ein Toolkit entwickelt, das den wahlfreien Zugriff auf ODA-Konstrukte ermöglicht und einige Diagnosewerkzeuge beinhaltet, die den Ausdruck von Teilen eines ODA-Dokumentes in lesbarer Form gestatten. Dies führte zu einer Vereinheitlichung der Architekturen der im PODA-2 Projekt entwickelten Konverter. Diese gemeinsame Architektur wird in der folgenden Graphik dargestellt.

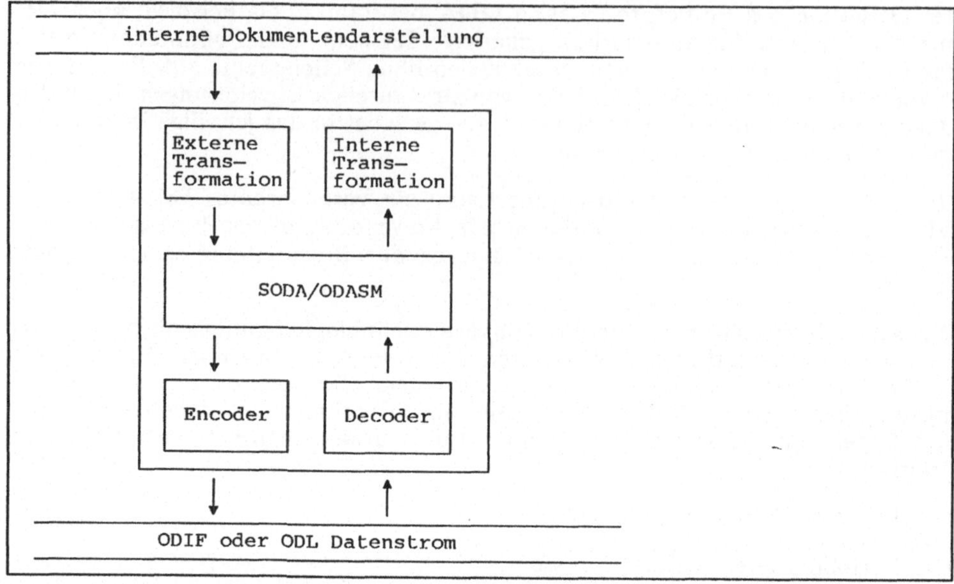

Im Zentrum des Konverters steht das SODA (Stored ODA) und ODASM (ODA Storage Manager) Toolkit [18], das einen wahlfreien Zugriff und gleichzeitig eine Speicherverwaltung für ODA-Konstrukte bereitstellt. Die Abbildung von und nach der internen Darstellung sowie von und nach der standardisierten Darstellung (ODIF oder ODL) werden von darauf aufbauenden, unabhängigen Modulen durchgeführt. Dies bietet unter anderem die Möglichkeit, die Konvertierungsabbildung zur internen Darstellung unabhängig vom verwendeten Austauschformat zu erhalten.

Da das API des Toolkits eine exakte Abbildung der ODA Konstituenten beschreibt, konnten jedoch sehr verschiedene Stile der Benutzung entstehen. Die gemeinsame Benutzung des Toolkits führte also nicht unmittelbar zu einer besseren Austauschbarkeit der Dokumente. Um dies zu ermöglichen, muß man das API um Begriffe, die durch das verwendete DAP definiert werden, erweitern. Derartige Begriffe wie etwa „numeriertes Segment" (numbered Segment) oder „Fußnote" (footnote) legen die Ausprägung eines oder mehrerer ODA Konstituenten fest. Die einheitliche Darstellung dieser Begriffe in ODA Konstrukten als Resultat der Benutzung eines solchen erweiterten API's durch alle am Austausch beteiligten Applikationen verbessert die Qualität der zur Verfügung gestellten Transformationsdienste. Dieser Ansatz eines DAP orientierten API's wurde im PODA-2 Projekt ebenfalls verfolgt. Die Firma Apple entwickelte im Rahmen dieses Projekts mit WOPODA [8] einen entsprechenden Prototypen zur Unterstützung von Wortprozessoren.

Zur Zeit gibt es, entstanden aus den Erfahrungen des PODA-2 Projektes, den Versuch, mittels eines Konsortiums[9] verschiedener Firmen einen ODA-Toolkit zur Produktreife zu entwickeln und dem Markt zur Verfügung zu stellen.

---

[9] Ziel dieses Konsortiums ist die gemeinsame Entwicklung eines ODA-Toolkits und die Verbreitung des ODA Standards. Der Formationsprozess ist noch nicht abgeschlossen, potentielle Partner sind Bull S.A., Digital Equipment Corporation (DEC), Inter-

4. ODA-basierte Anwendungen

Obwohl mit ODA/ODIF ein gemeinsames, standardisiertes Austauschformat definiert wurde, sind die heutigen Textsysteme bei der Unterstützung dieses Formates oft überfordert. Welche Probleme beim Konvertieren der internen Speicherstruktur hin zum ODIF-Format entstehen, wurde im letzten Kapitel eingehend erläutert. Anders stellt sich die Situation für Textsysteme und Editoren dar, die intern auf dem ODA-Dokumentenmodell basieren und eine entsprechende Speicherrepräsentation besitzen. Dokumente, die mit derartigen Systemen bearbeitet werden, können vollständig und ohne Informationsverlust in das ODIF-Format konvertiert werden. Solche Textsysteme werden als ODA-basierte („native") Editoren bezeichnet. Grundsätzlich lassen sich zwei Modelle für diesen Ansatz unterscheiden:

1. Alle ODA-Prozesse (Editier-, Formatier- und Präsentationsprozeß) werden von dem System unterstützt. Alle Prozesse arbeiten auf einer ODA basierenden Speicherstruktur.
2. Nur der Formatier- und der Präsentationsprozeß arbeiten direkt auf ODA-Speicherstrukturen. Änderungen lassen sich in diesem Format nicht vornehmen. Diese Lösung genügt, um ein empfangenes ODA-Dokument zu formatieren und auszudrucken.

Wir konzentrieren uns im folgenden auf Systeme, die dem ersten Modell entsprechen. Derartige Systeme können lokale Funktionalitäten besitzen, die zur Zeit über den ODA-Standard hinausgehen. Diese Erweiterungen können jedoch nicht mit anderen ODA-Systemen ausgetauscht werden.

Beispiele für ODA-basierte Editoren sind ISOTEXT[10] und PAPYRUS[11] . Im folgenden werden einige Vorteile dieser Editoren, aber auch Probleme solcher Implementierungen angesprochen.

Um die Probleme des Informationsverlustes bei der Konvertierung in das ODIF-Format so gering wie möglich zu halten, wurden bestimmte Teilmengen des ODA-Standards (DAPs) definiert (s. 2 . Kapitel). ODA-basierte Editoren sind unabhängig von diesen DAPs. Sie unterstützen wesentlich mehr Variations- und Beschreibungsmöglichkeiten in ihren Dokumentendefinitionen. Einschränkungen in der Wahl von logischen Objekten und Layoutstrukturen, wie sie die DAPs vorschreiben, sind überflüssig. Beliebige Dokumentenstrukturen können somit ausgetauscht werden.

---

national Computers Ltd. (ICL), International Business Machines Coporation (IBM), Siemens Nixdorf Informationssysteme AG und Unisys.

[10] Der ISOTEXT Editor wurde im Rahmen des ESPRIT Projektes PODA von der Firma Nixdorf in enger Zusammenarbeit mit der Firma TELES GmbH Berlin realisiert. Dieser Editor arbeitet direkt auf ODA-Strukturen. Er ist unabhängig von „Document Application Profiles" und unterstützt den gesamten ODA-Standard. Der Editor wurde auf der CeBIT '89 und der CeBIT '90 vorgeführt.

[11] PAPYRUS ist ein Textsystem der Philips Electronic Ldt. Der Editor wurde auf der CeBIT 1990 zum ersten Mal vorgestellt. Der Editor unterstützt die Funktionalität des „Document Application Profiles" Q.112.

Mit Hilfe eines ODA-basierten Editors lassen sich eine Menge bestimmter Dokumentenklassen erzeugen. Diese sind vergleichbar mit Anwendungsprofilen (DAPs), jedoch sehr viel restriktiver. Eine Dokumentenklasse besteht aus einer generischen, logischen Struktur, einer generischen Layoutstruktur, Stilen, sowie generischem Inhalt. Sie können sehr gut als Ausgangsdokumente zum weiteren Editieren benutzt werden. Beispiele solcher Dokumentenklassen sind Formulare, Artikel, Benutzerhandbücher, etc. ODA-basierte Editoren können eine auf solche Dokumentenklassen abgestimmte, kontextgesteuerte Benutzerschnittstelle realisieren, die den Benutzern die Arbeit sehr erleichtert. Am Bespiel eines Geschäftsbriefes soll dies näher erklärt werden: Eine Dokumentenklasse „Geschäftsbrief" kann neben einem bestimmten generischen Layout und dem Firmenlogo etwa eine einfache logische Struktur enthalten. Beim Aufruf des Editors und dieser Dokumentenklasse erhält der Benutzer einen vorgefertigten Brief auf dem Bildschirm, in den er nun seinen individuellen Text eintragen kann. Dabei kann er nur logische Objekte anlegen, die die Dokumentenklasse erlaubt, z.B. Adresse, Absender und einfache Paragraphen. Diese Objekte, die in jeder Dokumentenklasse verschieden sein können, werden dem Benutzer in einem Menü zur Auswahl angeboten. Dieses Menü ändert sich während der Editiersitzung; so kann etwa nur einmal das Objekt „Adresse" ausgewählt werden. Adresse und Absender werden an bestimmten Stellen im Layout positioniert. Diese Angaben sind schon in der Dokumentenklasse festgelegt. Alle Dokumente, die durch einen Editiervorgang aus einer Dokumentenklassen entstehen, lassen sich ohne Probleme als ODIF Datenstrom austauschen. Um bestimmte DAPs für den Austausch mit anderen Systemen zu unterstützen, werden Dokumentenklassen derart angelegt, daß sie den DAP Spezifikationen genügen. Ein Beispiel für einen ODA-Editor, der ein derartige, „klassengetriebene" Benutzerschnittstelle realisiert, ist ISOTEXT [9].

Im ODA-Dokumentenmodell wird deutlich zwischen der Dokumentenarchitektur und den Inhaltsarchitekturen unterschieden. Die Inhaltsarchitekturen werden als eigenständige Teile beschrieben. Dies hat in ODA-basierten Editoren, die analog zu diesem Architekturkonzept implementiert wurden, zweierlei Vorteile.

- Die Modulstruktur eines ODA-Textsystems ist klar gegliedert. Es gibt jeweils Editiermodule für die Dokumentenstruktur und die einzelnen Inhaltsarchitekturen. Das gleiche gilt für den Formatierprozeß, es gibt einen Dokumentenformatierer und drei Inhaltsformatierer. Dies erleichert den Austausch der einzelnen Module wesentlich, und bietet gleichzeitig die Möglichkeit, Teile des Systems, z.B. Inhaltseditoren durch bereits existierende Editoren zu ersetzen. Es genügt dazu, die Benutzerschnittstelle und die internen Modulschnittstellen anzupassen.
- Der zweite Vorteil dieser klaren Trennung besteht in der Erweiterbarkeit sowohl des Standards, als auch der Implementierung. Neue Inhaltsarchitekturen z.B. Sprachanmerkungen, Bewegtbilder, etc. lassen sich durch Hinzunahme neuer Module (Editier-, Formatier- und Präsentationsmodule) unterstützen.

Die Möglichkeit des „joint editing", d.h. das gleichzeitige, gemeinsame Editieren mehrerer Benutzer auf einem Dokument, ist nur möglich, wenn alle Beteiligten ein einheitliches Verständnis von der Dokumentenstruktur besitzen. Durch Konvertierungen zwischen verschiedenen Formaten wird eine Lokalisierung der veränderten Dokumententeile und ein Austausch dieser Teile sehr erschwert (s.a. Abschnitt 5).

Trotz der erwähnten Vorteile, die die ODA-basierten Textsysteme bieten, gibt es eine Reihe von Problemen, die ihre Realisierung behindern bzw. verzögern. Darauf soll im folgenden näher eingegangen werden.

Der Standard ODA unterstützt sehr viele Funktionen heutiger Textsysteme nicht oder nur in einem geringen Umfang. Viele dieser Funktionen und Attribute werden jedoch in Erweiterungen zum Standard gegenwärtig diskutiert. Hier sind insbesondere Tabellen, Formeln, Farben, Sicherheitsaspekte, der Austausch von Dokumententeilen sowie der Zugriff auf externe Daten zu nennen. Außerdem läßt der Standard viele Frage offen, die jede Implementierung individuell lösen muß. Eine solche Frage ist die Gestaltung der Benutzerschnittstelle und damit zusammenhängend die Manipulation der internen Strukturen und Attribute.

Neue Textsysteme verwenden meist eine WYSIWYG- (What You See Is What You Get) -Darstellung der Dokumente. Der Benutzer kann also Änderungen des in der Bearbeitung befindlichen Dokumentes sofort am Bildschirm erkennen. Die Formatierung des Dokumentes muß dazu sehr schnell erfolgen, um die Antwortzeiten des Systems möglichst gering zu halten. Deshalb wird oft ein Formatierungsprozeß realisiert, der sich nur auf Teile des Dokumentes bezieht (partielles Formatieren). Das in ODA definierte Dokumentenmodell beschreibt hingegen lediglich, wie das gesamte Dokument zu formatieren ist. Kleine Änderungen in der logischen Struktur können eine vollständige Neuberechnung des Layouts bedingen. Partielles Formatieren kann demnach zu Inkonsistenzen zwischen den Strukturen eines ODA-Dokuments führen. Insgesamt folgt, daß auf ODA basierende Editoren ein eigenes Konzept zur partiellen Formatierung beinhalten müssen und daß dieses Konzept einen geeigneten Kompromiß zwischen kurzen Antwortzeiten und der Konsistenz der Dokumentenstrukturen darstellen muß.

Der Einsatz von ODA-basierten Textsystemen wird nur dann Erfolg haben, wenn er in einer integrierten Umgebung z.B. einem Büro erfolgt. Dies bedeutet jedoch, daß das Textsystem definierte Schnittstellen zu anderen Anwendungen (Datenbanken, Serienbrieferstellung, Geschäftsgraphiken, Aktenverfolgungssysteme, elektronische Post, etc.) benötigt. Eine solche integrierte Umgebung existiert bis heute nicht.

5. ODAs zukünftiger Einfluß auf die Bürokommunikation

In den beiden vorherigen Abschnitten haben wir zwei wesentliche Möglichkeiten vorgestellt, wie ODA durch Anwendungen unterstützt werden kann, Editoren, deren internes Dokumentenformat auf ODA basiert, und Konverter, die das Dokumentenformat existierender Editoren auf ODA abbilden. Hier wollen wir einige weitere Aspekte des Einflusses dieser Architektur auf kommerzielle Bürosysteme, aber auch Grenzen dieses Einflusses erläutern. Dabei zeigen wir Tendenzen auf, die für zukünftige Entwicklungen im Bereich der Bürokommunikation wesentlich sein werden.

Moderne Dokumentenarchitekturen, die von verschiedenen Herstellern als zukünftige Basis ihrer Bürosysteme entwickelt werden, sind stark durch den ODA-Standard beeinflußt. Dieses gilt etwa für DECs Compound Document Architecture [17] und für IBMs Information Interchange Architecture [16]. Beide Architekturen enthalten etwa einen objektorientierten Ansatz für die Modellierung von Informationen und trennen die Beschreibung der Inhaltsteile von der Beschreibung der Dokumentenstruktur. Sie definieren ähnlich wie ODA ein logische Sicht auf Dokumente und eine auf das Layout bezogene Sicht und erlauben sowohl spezifische Dokumententeile als auch generische Teile der Beschreibung, die auf eine Klasse von Dokumenten zutreffen. Herstellerarchitekturen bewegen sich also auf den ODA-Standard zu, sie stellen Weiterentwicklungen gegebener Architekturen wie IBMs Document Content Architecture dar, die ODA-Konzepte enthalten. Die Konvertierung von Dokumentenbeschreibungen zwischen den Herstellerarchitekturen und dem ODA-Standard wird dadurch vereinfacht. Der im Abschnitt 3 beschriebene, bei der Konvertierung von Dokumentenarchitekturen auftretende Informationsverlust wird geringer. Die Editoren, die diese neuen, durch ODA beeinflußten Herstellerarchitekturen als internes Dokumentenformat verwenden, müssen für diese Architekturen ein geeignetes Manipulationsmodell realisieren. Dabei treten die im letzten Abschnitt beschriebenen Probleme erneut auf. Die in ODA enthaltenen Konzepte sind auf Dauer aber nur dann als adäquat zu bezeichnen, wenn es gelingt, diese Probleme zu lösen, und die Konzepte direkt in Anwendungen umzusetzen. Dieser Zusammenhang soll im folgenden am Beispiel von Tabulatoren verdeutlicht werden.

ODA spezifiziert einen Tabulator durch seinen Namen, seine Position und einige weitere Eigenschaften, die hier nicht von Interesse sind. Einem Objekt, das einen Textausschnitt repräsentiert, kann eine Liste derartiger Tabulatoren zugewiesen werden. Innerhalb des Textes können dann Sprünge erfolgen, die einen der angegebenen Tabulatoren durch dessen Namen adressieren. Die Position dieses Tabulators wird dadurch zur aktuellen Position, an der die Präsentation des Textes fortzusetzen ist. Andere Dokumentenarchitekturen wie zum Beispiel RFT:DCA enthalten ein „gewachsenes" Tabulatorkonzept, das dem einer Schreibmaschine weitgehend entspricht. Die Tabulatoren werden im wesentlichen durch ihre Position spezifiziert. In dem jeweiligen Textteil kann immer nur der nächste Tabulator adressiert werden, es ist nicht möglich, Tabulatoren zu überspringen oder einen Sprung entgegen der Schreibrichtung vorzunehmen. Wie in [11] näher ausgeführt wird, kann eine vollständige Abbildung zwischen diesen Konzepten nur dann realisiert werden, wenn der betrachtete Text vollständig formatiert wird. Dieses ist für editierbare Dokumente aber nicht sinnvoll. Aus diesem Grunde verlangen viele Konverter eine eingeschränkte Verwendung von Tabulatoren und garantieren nur dann eine korrekte Abbildung, wenn diese Einschränkungen eingehalten werden.

Falls ein Dokument zwischen zwei Editoren ausgetauscht werden soll, deren interne Formate beide das herkömmliche Tabulatorkonzept aufweisen, so können die obigen Schwierigkeiten durch eine direkte Abbildung zwischen diesen Formaten vermieden werden. In diesem Fall ist die Idee eines einheitlichen Austauschformates also mit dem Nachteil einer geringeren Abbildungsqualität behaftet. Diese Beobachtung gilt analog für andere Konzepte von Dokumentenarchitekturen. Um diesen Nachteil möglichst gering zu halten, sollte eine Architektur wie ODA Konzepte enthalten, die denen heutiger oder zukünftiger Anwendungen entsprechen. Die Frage, ob für ODA in Bezug auf Tabulatoren ein geeignetes Konzept gewählt wurde, kann also nur durch zukünftige Anwendungen beantwortet werden. Falls diese das herkömmliche Tabulatorkonzept weiter verwenden, so wurde für ODA die falsche Auswahl getroffen. Im anderen Falle müssen die zukünftigen Anwendungen das in ODA enthaltene Tabulatorkonzept geeignet für den Benutzer umsetzen, sie müssen für diesen Teil eine ODA-basierte Lösung anbieten. Dieses Beispiel verdeutlicht, daß ODA nur dann als geeignetes Austauschformat anzusehen ist, wenn ihre wesentlichen Konzepte zukünftig direkt in Anwendungen umgesetzt werden können, und der Benutzer geeignet bei der Handhabung dieser Konzepte unterstützt werden kann.

In Zukunft wird sich der Inhalt von Dokumenten nicht nur auf Texte und Graphiken erstrecken, sondern es werden auch zeitabhängige Informationstypen wie Audio und Video integriert werden. Dazu sind neue Synchronisationsmechanismen notwendig, die augenblicklich von verschiedenen Gremien entwickelt werden. Es sei an dieser Stelle lediglich die Normierungsaktivität HyTime [14] genannt, die basierend auf SGML Konzepte zur Synchronisation und zur Referenzierung von Dokumententeilen anstrebt. Diese Konzepte sollen derart definiert werden, daß sie ODA hinzugefügt werden können.

Neben diesen Entwicklungen in Richtung Multimedia, die die Komplexität der in einem Dokument enthaltenen Information erhöht, wird auch die Funktionalität der Anwendungen wachsen. In diesem Zusammenhang sind zum Beispiel Applikationen wie das verteilte Editieren (joint editing) und Hypertextsysteme zu nennen. Wie bereits im letzen Abschnitt verdeutlicht wurde, ist der Konverteransatz zur Unterstützung des verteilten Editierens nur schlecht geeignet. Die Einzelanwendungen müssen für diese verteilte Applikation exakt das gleiche Verständnis über die behandelten Dokumente besitzen. Sie müssen etwa in der Lage sein, die Teile, die sie manipulieren wollen, durch Sperren vor dem Zugriff anderer zu schützen. Die Identifizierung der Dokumententeile, auf die derartige Sperren wirken sollen, kann aber nur dann einheitlich vorgenommen werden, wenn die beteiligten Anwendungen ein gemeinsames Verständnis über diese Teile besitzen. Sofern Konverter verwendet werden, ist die mögliche Granularität der Sperren demnach abhängig von dem Maß, mit dem die Konverter ein derartiges, gemeinsames Verständnis gestatten. Der Einsatz von auf ODA basierenden Editoren wirft diese Probleme nicht auf.

Für ein Hypertextsystem, aber auch für viele andere Anwendungen ist es unverzichtbar, Teile eines Dokumentes getrennt beschreiben und speichern oder präsentieren zu können. Außerdem müssen Beziehungen zwischen den einzelnen Teilen beschrieben werden können. Die hier notwendigen Konzepte werden oft als „partial documents“ und „links“ bezeichnet. In Bezug auf derartige Anforderungen erweist es sich als hinderlich, daß ODA den objektorientierten Ansatz, den diese Architektur in Bezug auf die Informationsstrukturierung verfolgt, nicht konsequent auf Operationen wie das Editieren oder das Präsentieren fortsetzt. ODA spezifiziert den Präsentationsprozeß etwa nur für vollständige Dokumente und nicht für einzelne Objekte. Gemäß dem objektorientierten Ansatz sollte dieser Prozeß aber als eine Methode beschrieben werden, die sich für strukturierte Objekte aus den entsprechenden Methoden der Subobjekte zusammensetzt. Eine derartige Weiterentwicklung bedeutet eine konzeptionell weitreichende Änderung des Dokumentenverarbeitungsmodell von ODA. Sie ist aber für viele Anwendungen von zentraler Bedeutung (vergleiche auch die Ausführungen im Abschnitt 4 bezüglich des partiellen Formatierens). Man beachte, daß einige Herstellerarchitekturen bereits entsprechende Konzepte enthalten wie zum Beispiel IBMs „Object Method Architecture“ [16].

Es gibt sehr viele verschiedene Aktivitäten, die die Weiterentwicklung von ODA betreffen [6], die obigen Ausführungen stellen lediglich einige der zentralen Richtungen dar. Die entsprechenden Vorschläge sind durch Anwendungen motiviert und somit im einzelnen sinnvoll. In ihrer Gesamtheit aber stellen die Erweiterungen die Verwendbarkeit von ODA als ein Austauschformat in offenen Netzen in Frage. Es gibt kein Konzept, das es den einzelnen Anwendungen erlaubt, den Ausschnitt aus der ODA-Architektur zu verabreden, der für ein gewisses Dokument benötigt wird. Das bereits erwähnte Konzept der DAPs ist in dieser Hinsicht nicht verwendbar, da es keine flexible Verabredungen zum Zeitpunkt der eigentlichen Kommunikation gestattet. Die Funktionalität der versendeten Dokumente kann durch den Absender somit nicht genügend auf die Möglichkeiten des Empfänger abgestimmt werden. Notwendig sind geeignete Erweiterungen der Kommunikationsdienste (X.400), die Verabredungen über die zu verwendenden Teile der ODA Architektur erlauben. Dazu ist es aber auch notwendig, Kommunikationsdienste und die mit ihrer Hilfe übertragenen Informationen stärker miteinander zu verknüpfen. Diese Integration wird eine der wesentlichen Tendenzen der zukünftigen Entwicklungen innerhalb der Bürokommunikation bilden.

Abschließend wollen wir noch zwei verschiedene Möglichkeiten zum Austausch von Dokumenten betrachten. Zum einen ist es möglich, daß der Absender den Austausch initiiert, zum anderen ist es aber auch denkbar, daß der Empfänger das Dokument anfordert. Beispiele für den ersten Fall sind das Versenden von Briefen oder die Zustellung der Gehaltsabrechnungen an die Mitarbeiter einer Firma. Die bereits erwähnten X.400 Dienste bieten hier die geeignete Funktionalität. Der Zugriff auf eine elektronische Bibliothek oder die Suche eines bestimmten Eintrages in einem elektronischen Lexikon, das auf einem entfernten Knoten im Netz verwaltet wird, sind Szenarien, in denen der Empfänger der Information den Austausch initiiert. Falls der Austausch komplette Dokumente betrifft, so definiert der sich entwickelnde Standard für „Document Filing and Retrieval" (DFR) [7] die für derartige Szenarien geeigneten Kommunikationsdienste. Innerhalb des bereits erwähnten Projektes PODA-2 werden augenblicklich Prototypen entwickelt, die diese Form der Kommunikationsdienste für ODA-Applikationen realisieren. Falls aber lediglich auf Fragmente sehr umfangreicher Dokumente zugegriffen werden soll (elektronisches Lexikon), so sind Kommunikationsdienste zu fordern, die einen entfernten Zugriff auf Schnittstellen gestatten, die die Manipulation beliebiger Dokumententeile ermöglichen. Derartige Schnittstellen sind durch die im Abschnitt 3 beschriebenen ODA Toolkits gegeben. Zukünftige Aktivitäten werden demnach auch die Anbindung dieser lokalen Toolkits an Kommunikationsdienste betreffen.

6. Zusammenfassung

Der Standard ODA gewinnt gegenwärtig einen großen Einfluß auf den Bereich der Bürokommunikation. Dabei gibt es zwei wesentliche Ansätze, die eine Anbindung lokaler Systeme an dieses Austauschformat ermöglichen. Konvertoren realisieren Übersetzungen zwischen dem internen Dokumentenformat von gegebenen Editoren und dem ODA-Format. Der wesentliche Vorteil dieses Ansatzes ist die Weiterverwendbarkeit von Produkten und Konzepten, die dem Anwender vertraut sind. Der Einsatz von Konvertern wird aber immer zu einem teilweisen Informationsverlust bei der Dokumentenübertragung führen. Der zweite Ansatz verfolgt die direkte Umsetzung der durch ODA definierten Konzepte in neuen Büroanwendungen. Er erlaubt eine vollständige Verknüpfung heterogener Systeme, setzt aber die Einführung neuer Anwendungen voraus. Insgesamt sind die beiden geschilderten Ansätze nicht als konkurrierend, sondern als sich ergänzend anzusehen. Zur Durchsetzung des ODA-Standards als die gemeinsame Basis für die zukünftige Bürokommunikation ist es notwendig, sowohl gegebene Anwendungen durch Konvertoren anzubinden als auch neue, auf ODA basierende Anwendungen zu entwickeln, die insbesondere auch neue Anwendungsfelder wie „joint editing" oder Hypertext zu erschließen haben. Für derartige, zukunftsorientierte Anwendungen wird es aber auch notwendig sein, den ODA-Standard geeignet zu erweitern. Dieses bezieht sich vor allem auf eine stärkere Integration der Dokumentenarchitektur ODA und der Kommunikationsdienste wie X.400 und DFR.

Literatur

[1] *Abstract Syntax Notation 1 (ASN.1).* ISO 8824/8825, 1988.

[2] *CCITT Ninth Plenary Assembly: Recommendations T.411, T.412, T.414 - T.418. Open Document Architecture and Interchange Format,* CCITT Blue Book (1989), Volume VII.6, Melbourne, November 1988.

[3] *Information Processing - Text and Office Systems - Office Document Architecture (ODA) and Interchange Format,* ISO 8613 (part 1,2,4,5,6,7,8), 1989.

[4] **W. Appelt.** *Dokumentenaustausch in Offenen Systemen.* Berlin, Heidelberg, New York: Springer 1990.

[5] **R. Hunter, P. Kaijser, F. Nielsen.** *ODA: a document architecture for open systems.* Computer Communication 12, 69 - 79, 1989.

[6] *Framework for future extensions to ODA.* ISO/IEC JTC1/SC18 SWG - ODA Design Directions and Extensions, Editor I.R. Campbell-Grant, February 1989.

[7] *Information technology, Document filing and retrieval (DFR), Part 1: Abstract service definition and procedures,* ISO/IEC DIS 10166, 1989.

[8] **S. Friedrich, G. Holzhauer, V. Lubet.** *WOPODA - A High Level Word Processor's Interface to ODA.* to appear.

[9] **U. Bormann, C. Bormann.** *ISOTEXT - A WYSIWYG Editing and Formatting System for ODA and SGML Documents.* Technical University of Berlin, 1989.

[10] CCITT series of Recommendations X.400 ff., 1984-1987.

[11] **W. Filip, J. Kämper, W. Knobloch.** *Conversion between the Open Document Architecture ODA and proprietary Architectures - A case study.* to appear in: Proc. GI Fachtagung Kommunikation in verteilten Systemen, Mannheim, 1991.

[12] **W. Filip, J. Kämper, W. Knobloch.** *Conversion between ODA Level 2 and RFT:DCA - Experiences with a prototype.* to appear in: Computer Networks and ISDN Systems.

[13] **A. Spiceley.** *(unknown title).* to appear in: Computer Networks and ISDN Systems.

[14] *Hypermedia/Time-based Structuring Language (HyTime).* ANSI Project X3.749-D, 1990.

[15] *Standard Generalized Markup Language.* ISO 8879, 1988.

[16] *Information Interchange Architecture.* GG24-3503-0, IBM Corporation, 1990.

[17] *Compound Document Architecture Manual.* Digital Equipment Corporation (DEC), 1988.

[18] **S. Feather, P. Fraatz.** *SODA: Stored ODA (SODA) Interface.* ESPRIT project 2374, Piloting the Open Document Architecture, 1989.

Möglichkeiten und Grenzen von 'Cooperative Work'

Neue Perspektiven gruppenorientierter Büroarbeit

König, Rainer und Zoche, Peter
Fraunhofer-Institut für Systemtechnik und Innovationsforschung (ISI),
Abteilung Telematik,
Breslauer Straße 48, D-7500 Karlsruhe 1

Zusammenfassung

Unter dem Begriff *Cooperative Work* werden verschiedene Ansätze der computerunterstützten Gruppenarbeit subsummiert, die zunehmend ihren Experimentierstatus verlassen und in breitere Anwendungsfelder eindringen. Die Anfänge dieser Konzepte gehen zurück auf die sechziger Jahre. Angesichts leistungsfähigerer Arbeitsstationen und Übertragungsnetze, die eine simultane, multimediale Kommunikation ermöglichen, erlangen diese Konzepte neue Qualität; dies wird am Beispiel eines Pilotvorhabens zur Multimedia-Breitbandkommunikation im Büro diskutiert.

Die vorherrschende Arbeitsteilung einer weitgehend verrichtungsorientierten Büroarbeit fördert ungünstige Arbeitsbedingungen und erweist sich zudem als produktivitätshemmend. Informations- und Kommunikationstechniken eröffnen u.a. durch eine Vernetzung kooperierender Arbeitsplätze neue Möglichkeiten der Arbeitsstrukturierung. Allerdings steht ein Nachweis des Nutzens bisheriger Ansätze von Cooperative Work aus; hemmende und fördernde Faktoren der Einführung computerunterstützter Büroarbeit werden exemplarisch aufgezeigt.

1. Der Stellenwert der Kooperation in Arbeitsprozessen

Arbeitsteilung ist zwangsläufig auf Kooperation, das heißt auf das Zusammenwirken der Beteiligten eines Arbeitsprozesses angewiesen. Kooperationen sind auf gemeinsam verfolgte (Teil-)Ziele angelegt, unabhängig davon, ob dies den Mitgliedern des Kooperationsprozesses selbst bewußt ist. Funktionierende Kooperationsprozesse erfordern eine Organisation, mit der bewußt geplante, in funktionalem Verhältnis zueinander stehende Arbeitsaktivitäten ermöglicht werden. Die Formen der Kooperation sind im Hinblick auf die Komplexität des arbeitsteiligen Prozesses und in Bezug auf das Verhältnis der einzelnen Arbeitsaktivitäten untereinander (z.B. gleichartig, ergänzend, kombinierend) zu unterscheiden. Dabei kann Kommunikation als ein Bindeglied verstanden werden, das die Regelung und Steuerung kooperativer Prozesse ermöglicht und effektiviert. Als Ziele kooperativer Arbeitsprozesse werden oft genannt:

- ein rationeller und kostensparender Personaleinsatz,
- die ökonomische Verwendung von Werkzeugen, Maschinen und anderer Produktionsmittel,
- eine Zusammenführung räumlich getrennter Produktionsprozesse oder deren räumliche "Verengung",
- eine Qualitätssteigerung,
- eine Erhöhung der Produktivität
- eine Zeitersparnis.

Kooperative Arbeitsprozesse müssen jedoch nicht zwangsläufig solchen Bewertungsmaßstäben folgen, die fast ausschließlich an ökonomischen Effizienzkriterien orientiert sind. Es können gleichfalls kooperative Arbeitsprozesse gestaltet werden, die einen Beitrag zur Humanisierung der Arbeit leisten, indem sie das "Ideal" eines ganzheitlichen, nur eingeschränkt durch repetitive Tätigkeiten geprägten Arbeitsablaufes anstreben.

1.1 Gruppenorientiertes Arbeiten in der Produktion

Zunehmende Arbeitsbelastung und eine wachsende Unzufriedenheit der Mitarbeiter, hohe Fluktuationsraten unter den Belegschaften und eine geringe Produktionsflexibilität kennzeichneten in den siebziger Jahren die Grenzen einer an tayloristischen Prinzipien orientierten industriellen Produktion.

In Folge dieser Entwicklung kam es - zunächst im skandinavischen Raum und in der Automobilindustrie - zu Arbeitsstukturierungsmaßnahmen und zur Einrichtung "**teilautonomer Arbeitsgruppen**", die ihre Arbeitsaufgaben einschließlich der zugehörigen Organisations-, Steuerungs- und Kontrollaufgaben in gemeinsamer Verantwortung übernahmen. Die Kontrolle durch den Vorgesetzten wurde durch eine ergebnisorientierte Kontrolle der zuvor vereinbarten Gruppenleistung ersetzt (vgl. Fickert 1988, S. 75). Gruppenarbeitskonzepte bzw. autonomieorientiertes Arbeiten als betriebliches Gestaltungskonzept vereinigen demnach die Elemente Dezentralisierung und Aufgabenerweiterung in sich.

Betriebliche Erfahrungen mit Konzepten der sogenannten "**Inselfertigung**" belegen, daß die "Aktivierung der dispositiven und planerischen Fähigkeiten der Mitarbeiter nicht nur von diesen als eine Bereicherung ihrer Fähigkeiten empfunden wird, sondern auch dem Betrieb sehr konkreten Nutzen bringen kann" (Strötgen 1991, S.1). Gleichwohl setzen sich veränderte organisatorische Konzepte nur langsam durch. In einer neueren Untersuchung über den Einsatz computerintegrierter Fertigungskonzepte (CIM), der "Schlüsseltechnologie der *Fabrik der Zukunft*", wird denn auch das Fazit gezogen, daß viele CIM-Konzepte "auf halbem Wege stecken bleiben", weil sich bisher nur in wenigen Unternehmen die Auffassung durchgesetzt habe, "daß Arbeitsanreicherung, Gruppenarbeit und andere Formen arbeitsorganisatorischer Gestaltung auch ökonomische Zielsetzungen zur Verminderung von Produktivitätsverlusten unterstützen" (Esser 1991, S.168).

1.2 Gruppenorientiertes Arbeiten im Büro

Wie angeführt, kann durch eine humane Gestaltung teilautonomer Arbeitsgruppen eine zusätzliche Produktivitätsressource erschlossen werden. In dieser Erkenntnis liegt eine Attraktivität, die in zunehmendem Maße dazu führt, aus der Produktion stammende Überlegungen in den Bürobereich zu übertragen.

Im Zusammenhang mit der Verbreitung integrierter Bürosysteme werden Konzepte wie "integrierte Vorgangs- bzw. Dokumentenbearbeitung" (z.B. bei Banken, Versicherungen) entwickelt, die der Vorstellung folgen, sowohl individuelle, als auch kollektive Dispositions- und Handlungsspielräume bei der Aufgabenbewältigung zu erweitern (vgl. Rolf et al. 1990, S. 242 ff.), die ihrerseits wiederum zu einer Verkürzung der Durchlaufzeiten und zu einer größeren Kundennähe führen können (vgl. Bauer/Lindenthal 1986, S.234 ff.).

Daneben gibt es jedoch auch strukturelle Faktoren, die mit dem Wandel von gemeinsamen Rahmenbedingungen der industriellen Gesellschaften verwoben sind. Als solche "Triebfedern" der Entwicklung technikgestützter Kooperationsformen sind beispielsweise zu nennen (vgl. z.B. Johansen 1988, S. 67 ff.):

- eine Globalisierung der Volkswirtschaften und Märkte,
- die damit einhergehende Zunahme der Kooperation zwischen Unternehmen,
- eine zunehmende zwischenbetriebliche nationale und internationale Arbeitsteilung,
- ein daraus resultierender Trend zur Auslagerung von Aufgaben,
- die Entwicklung zu flexibleren, projektbezogenen und/oder zeitlich befristeten Arbeitsgruppen bis hin zu matrixorientierten Organisationsformen.

Vor diesem Hintergrund konstatiert der Organisationstheoretiker Bleicher einen Bedarf an "produktorientierten teilautonomen Geschäftseinheiten", die eine verstärkte Orientierung der Organisation am Kunden- und Marktdenken gewährleisten könnten. Diese marktnahen, unternehmerischen Einheiten greifen auf "logistische, funktionale Kerneinheiten" zurück, die mit Unterstützung der Systemtechnik interne Dienstleistungen für die Marktpartner erbringen (vgl. Bleicher 1988, S. 10 ff.).

Die Möglichkeit der Funktionsintegration beispielsweise in einer vorgangsorientierten Sachbearbeitung oder der Automatisierung von Kooperationsprozessen in Gruppen wird auch vor dem Hintergrund technischer Entwicklungen, wie der zunehmenden Integration von Daten und Anwendungen sowie über- und zwischenbetrieblicher Vernetzung, verstärkt diskutiert (vgl. Kubicek 1990, S. 365 ff.; Bellmann/Wittmann 1991, S. 495 ff.).

2. Computerunterstützung in der Gruppenarbeit

Heutige Büroarbeit ist durch eine "verrichtungsorientierte Arbeitsteilung" charakterisiert, die sich nicht nur als "ungünstig für die Arbeitsbedingungen, sondern auch als außerordentlich produktivitätshemmend" erweist. Dafür angeführt werden beispielsweise Intransparenz, Doppelarbeit, Medienbrüche und lange Durchlaufzeiten. Informations- und Kommunikationstechniken eröffnen über den Zugang zu Informationsquellen, die Vernetzung kooperierender Arbeitsplätze und die Bereitstellung von Werkzeugen neue Möglichkeiten der Arbeitsstrukturierung (vgl. Bellmann/Wittmann, 1991, S. 382).

2.1 Entwicklung von computerunterstützter Gruppenarbeit

Der Ansatz computerunterstützter Gruppenarbeit geht zurück auf die 60er-Jahre, als Douglas Engelbart am Stanfort Research Institute den Prototyp heutiger Forschung auf diesem Feld entwickelte. Das On-Line-System (NLS) kann als frühes Hypertext-System für Arbeitsgruppen bezeichnet werden, das Filter für selektives Editieren von Informationen, Computerkonferenzen, automatische Dialogprotokollierung, gemeinsames Editieren von Texten und Datenbanken sowie Multimediakommunikation bereitstellte (vgl. Engelbart/Lehtman 1988, S. 245 ff.). Viele der dort entwickelten Techniken, wie Fenstertechnik, Maus oder hypertextverbundene Dokumente wurden zum Standard in vielen Personal Computern. Während jedoch die damalige Technologie an die Leistungsgrenzen von Geräten und Vernetzungsinfrastruktur stieß, erscheinen diese Ansätze heute in einem neuen Licht (vgl. Johansen 1988. S. 3f.; Ellis et al. 1991, S. 41).

Parallel zu diesen überwiegend auf die Benutzerschnittstellen konzentrierten Pionierarbeiten von Engelbart kann eine Entwicklung verteilter Systeme bzw. verteilten Problemlösens beobachtet werden, die zur Metapher der "Blackboard" geführt hat (vgl. Krallmann 1990, S. 3ff.). Dabei schauen alle an der Problemlösung Beteiligten gleichzeitig "auf ein und dieselbe Tafel". Hinter diesem Modell steht eine Situation menschlichen Problemlösens, bei der verschiedene Spezialisten ein gemeinsamens Ziel verfolgen, von denen jeder alleine nicht in der Lage wäre das komplexe Problem zu lösen (bspw. der Bau eines Hauses).

Durch den zunehmenden Bedarf nach audio-visueller Kommunikation und Datenkommunikation im Bürobereich werden Informations- und Kommunikationstechniken integriert und in einem Arbeitsplatz zusammengeführt. Mit dem Einsatz von Multifunktionsgeräten für Telefax, PC-Anwendungen, gemeinsames Editieren von Dokumenten und Bewegtbildkommunikation wird eine effektivere Arbeitsplatzkommunikation und eine Verbesserung von Abstimmungsprozessen angestrebt.

2.2 Klassifikation verschiedener Ansätze in der computerunterstützten Gruppenarbeit

In bezug auf unterschiedliche Aufgabenstellungen entstanden unterschiedliche Ansätze einer Computerunterstützung von Gruppenarbeit. Diese reichen von E-Mail über Projekt-/Kalender-Management bis hin zur Unterstützung von Gruppenentscheidungsprozessen sowie persönlichen und Video-Konferenzen, um nur einige Beispiele zu nennen. Die jeweils unterschiedlichen Zwecksetzungen führten auch zu einer unübersichtlichen Vielfalt von Begriffen, wie beispielsweise *Computer-Supported Cooperative Work (CSCW), Workgroup Computing, Coordination Technology, Computer Conferencing* oder *Group Decision Support Systems* (Johansen 1988, S. 10ff.); im deutschen Sprachraum tritt auch der Begriff *Telekooperation* auf. Bisher hat sich kein allgemeingültiger Oberbegriff für computerunterstützte Gruppenarbeit herausgebildet, wir verwenden ***Cooperative Work*** in diesem Sinne.

Um die Gemeinsamkeiten der unterschiedlichen Entwicklungsstränge zusammenzuführen wurden verschiedene Ansätze zur Klassifikation entwickelt. Dabei wird meist nach der Art der Gruppenunterstützung (persönliche Treffen, elektronische Treffen, zwischen den Treffen), nach der Gruppengröße und nach der räumlichen und zeitlichen Verteilung der Gruppenmitglieder unterschieden (vgl. Johansen 1988, S. 42 ff.).

Darüber hinaus haben Rüdebusch/Mühlhäuser 1991 (S. 466 ff.) eine sehr differenzierte und systematische Klassifikation vorgeschlagen:

- nach der Art der **Interaktion**. Die Partner können explizit kommunizieren oder implizit durch das Lesen von Nachrichten oder die Kenntnisnahme von Veränderungen durch andere Benutzer. Dies kann zeitlich synchron oder asynchron erfolgen.
- nach der Unterstützung der **Koordination**, die im Hinblick auf die Lösung gemeinsamer Aufgaben unabdingbar ist. Zum Teil sind keine oder nur mündliche Absprachen vorhanden. Elementare Koordination ist beispielsweise die Vergabe von Zugriffsrechten auf Dokumente. Komplexe Koodination berücksichtigt die Einhaltung verschiedener Rahmenbedingungen, wie bspw. die Barbeitungsreihenfolge und kann sowohl überprüfend, als auch führend sein.
- nach der (räumlichen) **Verteilung** der Gruppenmitglieder. Zentral, wie etwa bei der Unterstützung eines "face-to-face-meetings", lokal oder global.
- nach **Anwendungsklassen**. Von der einfachsten Stufe der Mehrbenutzerspiele mit impliziter Interaktion und keiner Koordination, über Konferenz-Systeme, Group Decision Support Systems, Hypertext-Strukturen, Mehrbenutzer-Editoren bis hin zu Koordinationssystemen, die komplexe Arbeitsabläufe über Muster und Regeln koordinieren.

3. Breitbandtechnologie als Basis neuer Cooperative Work-Konzepte

Nach Rüdebusch/Mühlhäuser wird der künftige Einsatz von Cooperative Work zunehmend in multimedialen verteilten Systemen realisiert werden (vgl. Rüdebusch/Mühlhäuser, 1991, S.465 f.). Von der Einführung digitaler Breitbandnetze sind somit wichtige Impulse für neue und breitere Anwendungsfelder computerunterstützter Gruppenarbeit zu erwarten. Aus diesem Grunde ist es interessant, den Blick auf aktuelle Untersuchungen zum Nutzungspotential digitaler Breitbandnetze zu richten.

3.1 Nutzungsmöglichkeiten digitaler Breitbandnetze

Im Rahmen der Studien zum Marktpotential zur zukünftigen Nutzung digitaler Breitbandnetze (vgl. Bierhals et al. 1991) wurde aufgezeigt, daß Breitbandanwendungen vor allem in den Funktionsbereichen Forschung und Entwicklung, Produktentwicklung, Projektmanagement, Management dezentral organisierter Unternehmen, Absatz und Weiterbildung auftreten werden. Durch die Übertragung von Bewegtbildern und Dokumenten können räumlich verteilte Teilnehmer besser miteinander kommunizieren und es kann vorhandenes Wissen besser zugänglich gemacht werden; die bisherigen Grenzen technischer Kommunikation werden somit wesentlich erweitert. Wie in den Untersuchungsergebnissen dargelegt ist, werden insbesondere Anwendungen im Bereich kooperativer Aufgaben und Entscheidungen erschlossen. "Vor allem die gemeinsame Bearbeitung von Multi-Media-Dokumenten von verteilten Arbeitsplätzen lassen die Vorteile für das einzelne Unternehmen deutlich werden. Damit wird ein neues Kapitel im Bereich des Produktionsfaktors 'Kommunikation' aufgeschlagen. Durch gezielte, überschaubare Pilotanwendungen, die u.a. die Hemmschwellen und Probleme lokalisieren und abbauen helfen sollen, können hier Reallösungen 'zum Anfassen', echte Nutzenerfahrung und damit letztlich auch verbesserte Ausbreitungschancen geschaffen werden" (Klein/Nippa, 1991, S. 65).

3.2 Pilotvorhaben Office Broadband Communication (OBC)

Im Rahmen des BERKOM-Programms der Deutschen Bundespost Telekom werden solche Anwendungsprojekte in den verschiedensten Anwendungs- und Unternehmensbereichen durchgeführt; hierfür steht das vermittelnde Breitband-Netz (VBN) und das BERKOM-Testnetzt zur Verfügung. Eines dieser Anwendungspilotprojekte ist das 'Office Broadband Communication (OBC)' Projekt, das die Integration von Breitband-Diensten an verteilten Büroarbeitsplätzen im Bereich des Management erprobt (vgl. Deutsche Bundespost Telekom 1991, S. 6 f.; vgl. Abbildung 1). Alle Möglichkeiten der Breitbandkommunikation, wie Video, Audio, Text, Grafik und schnelle Datenübertragung, werden direkt in einen "Standard-Arbeitsplatz-Computer" integriert (vgl. Spohn, 1991). Dies ermöglicht eine direkte, unmittelbare Kommunikation zwischen räumlich entfernten Kooperationspartnern.

Abbildung 1: Multimedia-Breitbandkommunikation im Büro
(Quelle: Faltblatt zur CeBit '91, Kooperation: DETECON, SEL ALCATEL, Apple Computer)

Multimedia- Breitbandkommunikation im Büro.

Das Ergebnis des BERKOM- Projektes OBC mit der Deutschen Bundespost TELEKOM.

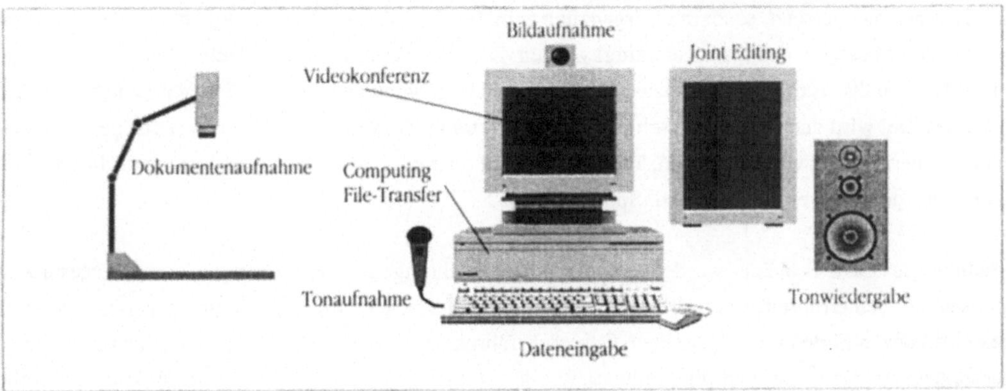

Die Hardware-Bestandteile der OBC-Workstation sind:
- Video-Karte,
- Kamera,
- Telefon (Lauthören, Freisprechen),
- AV-Modul (Bildmischung, Mikrophon, Lautsprecher),
- Ethernet-Karte.

Als Software werden eine integrierte Bedieneroberfläche, Video-Applikationssoftware und Cooperative-Working-Software mit den Funktionen Joint Viewing/Joint Editing (z.B. Aspects, Timbuktu) sowie Verarbeitungsprogrammen eingesetzt.

Dadurch werden folgende Grundfunktionen der OBC-Workstation bereitgestellt (vgl. Spohn, 1991):
- Desk-to-Desk-Kommunikation (Video, Ton, Bild, Text, Grafik, Daten),
- Integration von Video und Audio in Echtzeit,
- Schnelle Datentransferrate (2 Mbit),
- Joint Viewing eines Dokumentes durch die Gesprächspartner,

- Joint Editing an einem Dokument durch die Gesprächspartner,
- Verarbeitung des Dokuments,
- Interaktiver Zugriff auf Multimedia-Quellen,
- Videokonferenzen mit den beteiligten Arbeitsplätzen und auch mit öffentlichen Studios.

Das Neue an dieser Konfiguration ist die direkte Kommunikation ohne Umwege beim Informationsaustausch, der Entscheidungsfindung und der Erläuterung von Sachverhalten oder Objekten. Als Anwendungsmöglichkeiten bieten sich beispielsweise bei Projektmanagement und -dokumentation, bei der Abstimmung von Entwürfen, beim Erzeugen von Folien/Charts sowie bei der Integration neuer Softwarekomponenten die breitbandvermittelte Kooperation der Gesprächspartner an (vgl. Deutsche Bundespost Telekom 1991, S. 6 f.). Durch diese quasi-direkte Mensch-zu-Mensch Kommunikation soll das persönliche Gespräch nicht ersetzt, sondern in Ergänzung zum Telefonieren durch die Möglichkeit einer größeren Nähe und eines besseren Feedbacks erweitert werden. Die Videokommunikation bietet aber auch die Möglichkeit der Visualisierung von Gesprächsgrundlagen und trägt damit zu einer sachlicheren Kommunikation bei. Dieses Ziel wird auch durch die schnelle Datenübertragung erreicht, die Grundlage für gesprächsbegleitende Übermittlung von Entwürfen, Tabellen usw. ist oder bei Bedarf zum gemeinsamen Editieren von Dokumenten genutzt werden kann (vgl. Spohn, 1991).

Im Rahmen des OBC-Projektes werden die praktischen Erfahrungen mit kooperativen, computergestützten Arbeitssituationen ermittelt. Erste Erfahrungen zeigen, daß Besprechungen im Rahmen von Cooperative Work eine konzentrierte und zielgerichtete Gesprächsführung erfordern. Beim Umgang mit Dokumenten ist insbesondere der Ursprung und die Authentizität des Dokumentes durch eine genaue Versionskontrolle sicherzustellen (vgl. Spohn, 1991). Da es sich beim OBC-Projekt um eine Mehrbenutzerkommunikation handelt, bei der nicht sequentielle sondern simultane Bearbeitungen vorgenommen werden und darüber hinaus innovative Multi-Media-Anwendungen in nicht experimentelle Arbeitssituationen integriert sind, wird der Verlauf des Projektes klären helfen, die Veränderungen von Arbeitssituationen und Arbeitsinhalten durch Cooperative Work aufzuzeigen.

4. Hemmende und fördernde Faktoren von Cooperative Work

Bis heute erscheint der empirische Nachweis des Nutzens bisheriger Ansätze von Cooperative Work nicht - beziehungsweise nur teilweise - erbracht (vgl. Jarke 1987, S. 48). Aufgrund des veränderten Charakters der durch neue Ansätze von Cooperative Work entstehenden Arbeitssituationen - gekennzeichnet z.B. durch simultane, multimediale Kommunikationsmöglichkeiten -, kann auch nur in beschränktem Umfang auf Erfahrungen herkömmlich computerunterstützter Arbeit zurückgegriffen werden. Zur Weiterentwicklung der vorhandenen Ansätze ist deshalb eine Systematisierung von Einflußfaktoren auf Cooperative Work angebracht.

Mit den folgenden Ausführungen sollen einige der hemmenden und fördernden Faktoren von Cooperative Work aufgezeigt werden, die Eingang in eine vollständige und systematische Erhebung von Einflußfaktoren finden müßten, um über lose gekoppelte Technikinseln im Bürobereich hinauszuweisen (vgl. Kubicek, 1990, S.368).

Die Komplexität des Informationsaustausches und -zugriffs sowie die mögliche Einbeziehung mehrerer Teilnehmer in Cooperative Work erfordern eine wirksame **Koordination** der Kommunikationspartner, ihrer Aktivitäten und des Kommunikationsprozesses selbst. Eindeutige **Strukturen** aber auch hinreichend **flexible Nutzungsmöglichkeiten** der technischen Optionen tragen dazu bei, daß die objektive Leistungsfähigkeit der Systeme ausgeschöpft werden kann und die Gefahr, daß die subjektive Bewertung von Cooperative Work zur Unzufriedenheit der Teilnehmer führt, gering gehalten wird (vgl. Jarke 1987, S. 48). Vorhandene **Organisationsstrukturen** sind im Hinblick auf solche Anforderungen, die sich aus der Einführung von Cooperative Work ergeben, anzupassen. Dabei ist zu beachten, daß die Fähigkeit und Bereitschaft zur Kooperation bei Mitarbeitern, die zu sehr an eine "Koordination durch Hierarchie, Regeln und Programme gewöhnt sind, oft nicht mehr als selbstverständlich vorausgesetzt werden" kann (Staehle 1987, S. 459, zit. nach Straßburger 1990, S. 86). Dies setzt beispielsweise Organisationsstrukturen voraus, die jedem Kommunikationspartner eine problemlose Rückkoppelung im **Kommunikationsnetz** ermöglichen.

Der Einsatz von Cooperative Work auf Basis der Breitbandtechnologie kommt solchen Anforderungen entgegen, indem eine zeitgleiche Zusammenführung verschiedener Kommunikationsformen und -mittel und somit eine individuelle **Wahlmöglichkeit** am Arbeitsplatz gegeben ist. Diese Wahlmöglichkeit stellt andererseits hohe Anforderungen an die **Benutzerschnittstelle,** die Qualifikation der Benutzer und den **Einführungsprozeß** von Cooperative Work.

Arbeits- und Vorgangsbearbeitung in der computerunterstützten Kooperation erfordern in praktischen Situationen die Überprüfung und gegebenenfalls eine **Modifikation bestehender Arbeitstechniken**. Beispielsweise muß beim Zeigen personenbezogener und geschäftswichtiger Informationen oder von vertraulichen Dokumenten größere Sorgfalt als im persönlichen Gespräch angewandt werden, da die Möglichkeit des unbemerkten Kopierens besteht. Auch die Rechtsverbindlichkeit bei der Bearbeitung von Dokumenten (z.B. Unterschrift) stellt Anforderungen nicht nur an die Gestaltung der Systeme sondern auch die der Arbeitsorganisation, die heute noch nicht selbstverständlich sind. Ebenso erfordert der virtuelle Charakter der Dokumente bei der verteilten Bearbeitung besondere Vorkehrungen und Schutzmaßnahmen (z.B. Versionskontrolle, Zugriffsschutz) um Doppelarbeiten oder Manipulationen zu vermeiden.

Die Möglichkeiten paralleler Bearbeitung bei nicht zeitversetzter Kommunikation stellt an die Kooperationspartner ein **hohes Maß an Disziplin und Konzentration**. Die Unmittelbarkeit der erforderten Reaktion unterbricht andererseits den persönlichen Verarbeitungsprozeß und behindert das Nachdenken über mögliche Handlungsalternativen; auf diese Weise wird die Konzentration gestört und die sorgfältige Bewertung der Handlungsoptionen verhindert.

Es ist anzunehmen, daß die "Zahl der Kommunikationsprozesse in künftigen integrierten Netzen exponentiell ansteigen wird und Kommunikationsbeziehungen entstehen, die nur noch schwer zu durchschauen sind". Dadurch entsteht tendenziell ein Risiko mangelnder Transparenz und Kontrolle bei den Betroffenen und letztlich auch eine Erschwernis der Arbeit von Datenschutzbeauftragten und Aufsichtsbehörden (Roßnagel, 1989, S. 3).

5. Literatur

Bauer, G./Lindenthal, P., 1986, "Entwicklungsaussichten des Bankgewerbes in Hessen", Wiesbaden: HLT Gesellschaft für Forschung Planung Entwicklung mbH, 1986

Bellmann, K./Wittmann, E., 1991, "Modelle der organisatorischen Arbeitsstrukturierung - Ökonomische und humane Effekte", in: Bullinger, H.-J. (Hrsg.): Handbuch des Informationsmanagements im Unternehmen, München 1991, S. 487-516

Bierhals, R. et al., 1991, "Marktpotential für die zukünftige Nutzung digitaler Breitbandnetze", Karlsruhe: Fraunhofer-Institut für Systemtechnik und Innovationsforschung, 1991

Bleicher, J., 1988, "Die Organisation mit Zukunft", in: IBM-Nachrichten 38 (1988) Heft 292, S. 7-13

Deutsche Bundespost Telekom (Hrsg.), 1991, "Das BERKOM-Projekt 'Office Broadband Communication'", in: Visuell 2/91, S. 6ff.

Ellis, C.A. et al., 1991, "Groupeware. Some Issues and Experiences", in: Communications of the ACM, Jan. 1991, S. 39-58

Engelbart, D./Lehtman, H., 1988, "Working together", in: Byte December 1988, S. 245-252

Esser, U., 1991, "Auf dem Weg zu CIM - Technische, organisatorische und soziale Aspekte des Einsatzes integrierter EDV-Systeme II", in: Sozialwissenschaften und Berufspraxis 1/1991, S. 159-168

Fickert, J. et al., 1988, "Vernetzung und Integration von EDV-Systemen. Auswirkungen auf Beschäftigte - Handlungsmöglichkeiten für Betriebs- und Personalräte", in: Technologieberatungsstelle beim DGB Landesbezirk NRW, Heft 10, Oberhausen 1988

Jarke, M., 1987, "Informations- und Kommunikationsdienste zur Unterstützung von Gruppenentscheidungen", in: Handbuch der modernen Datenverarbeitung, 138/1987, S. 39-51

Johansen, R., 1988, "Groupware. Computer Support for Business Teams", New York, London 1988

Klein, G./Nippa,M., 1991, "Vermittelte Breitbandkommunikation im geschäftlichen Umfeld. Erste Ergebnisse einer Marktuntersuchung", in: OFFICE MANAGEMENT 3/1991, S. 56 ff.

Krallmann, H., 1991, "Verteilte wissensbasierte Systeme in der Fertigung", in: Forschung aktuell TU Berlin,1991, S.3-8

Kubicek, H., 1990, "Sozial- und ökologieorientierte Technikfolgenforschung. Probleme und Perspektiven am Beispiel der Büro- und Telekommunikation", in: Biervert, B./Monse,K. (Hrsg.): Wandel durch Technik?, Opladen 1990, S.353-385

Roßnagel, A., 1989, "Offene Kommunikation - geschützte Daten? Probleme des Datenschutzes", in: Fachtagung "Das Büro der Zukunft", 30. November 1989 in Münster

Rüdebusch, T./Mühlhäuser, M., 1991, "Ein Unterstützungssystem für Gruppenarbeit in verteilten Systemen", in: Effelsberg, W., Meuer, H.W., Müller, G. (Hrsg.): Kommunikation in verteilten Systemen, Berlin u.a. 1991, S. 464-478

Spohn, P., 1991, "Das BERKOM-Projekt 'Office Broadband Communication'", Vortrag zur Online 1991

Straßburger, F.X., 1990, "Chancen und Risiken eines integrierten Telekommunikationskonzeptes aus betriebswirtschaftlicher Sicht", München 1990

Strötgen, J., 1991, "Produktive Arbeitsorganisation", in: Fachtagung Sozialverträgliche Technik - Gestaltung und Bewertung am 12./13. März 1991 in Essen

Ein zukünftiges Kommunikationssystem für den privaten Netzbereich

Eine Übersicht über das DAMS-Projekt (ESPRIT 2146)

D. Wybranietz und F.-J. Stamen

TELENORMA GmbH - Bosch-Telecom
Vorentwicklung EVO 4
Mainzer Landstraße 128-146, D-6000 Frankfurt am Main 1
Tel.:(069) 266-3609 E-Mail: wybranie/stamen@tnevo.uucp

ZUSAMMENFASSUNG

Im ESPRIT-Projekt DAMS (Dynamically Adaptable Multi-Service System) wird ein Prototyp eines zukünftigen Kommunikationssystems für den privaten Netzbereich realisiert. Die Hauptziele des Projektes bestehen in der Integration von Audio-, Video- und Datendiensten sowie einer flexiblen, dynamischen Ausnutzung der zur Verfügung stehenden Übertragungsbandbreite. Weitere Anforderungen stellen eine erhöhte Zuverlässigkeit als die heutigen Systeme und die Anpaßbarkeit an veränderte Benutzerprofile und sich neu entwickelnde Standards dar. Die bereits existierenden Planungen für Breitbandkommunikationsdienste in Öffentlichen Netzen erfordern eine Migrationsstrategie zum ATM-Netz. Dieser Aufsatz gibt einen Überblick über die im DAMS-Projekt entwickelten Konzepte.

I. Einleitung

Die traditionellen Telefondienste und die Datenkommunikation stellen sich heute als nahezu völlig getrennte Welten dar: In den Büros existieren Nebenstellenvermittlungsanlagen, die die privaten Telefone untereinander und mit dem öffentlichen Netz verbinden, und lokale Netze, die den Datenaustausch zwischen Arbeitsplatzrechnern und Peripheriegeräten ermöglichen. Die Digitalisierung der weltweiten Telefonnetze und die Einführung von ISDN bieten zunehmend eine Infrastruktur für neue Dienste, die beide Welten zusammenwachsen lassen. Als Beispiel seien hier Multimedia-Systeme genannt, die Sprach-, Bild- und Datenkommunikation vereinigen.

Zukünftige Nebenstellenanlagen im geschäftlichen Anwendungsbereich sind von flexiblen Systemkonzepten geprägt, die hinsichtlich Ausbaugröße aber auch der zu vermittelnden Dienste variabel gestaltet sind. Die kostengünstige Integration bereits existierender Endgeräte - ISDN-fähig oder noch analog - sowie zukünftiger ATM-orientierter Endgeräte (ATM - Asynchronous Transfer Mode [Sch90]) in einem B-ISPBX-Vermittlungssystem (Broadband Integrated Services Private Branch Exchange) und die Abwicklung des Datenaustausches zwischen Endgeräten (z.B. Multimedia-Terminals) stellen wesentliche Leistungsmerkmale dar. Da Dienste noch nicht endgültig festgelegt oder noch nicht absehbar sind, ist ein flexibles Systemkonzept von entscheidender Bedeutung, das eine nachträgliche Einbeziehung neuer Dienste auf einfache Weise ermöglicht.

Während bei reinen Telefonsystemen gesicherte Untersuchungsergebnisse über das Verkehrsaufkommen vorhanden sind, fehlen derartige Informationen für die neuen Dienste. Bei der Konzeption zukünftiger Kommunikationssysteme ist daher eine Anpassung der Vermittlungs- und Übertragungskapazitäten an ein verändertes Verkehrsaufkommen vorzusehen, das durch die Nutzung neuer Dienste und Endgeräte herbeigeführt wird.

Private Kommunikationssysteme müssen alle im öffentlichen Bereich verfügbaren Standards- und Leistungsmerkmale unterstützen. Um bei dem hohen Wettbewerbsdruck im Telekommunikationsbereich auch zukünftig bestehen zu können, werden zusätzliche Leistungsmerkmale und Dienste die Kaufentscheidung des Kunden entscheidend beeinflussen. Die Konfigurationsfähigkeit und Modifizierbarkeit der Software und die hierfür erforderlichen Konzepte und Entwicklungsumgebungen werden dabei zu wesentlichen Anteilen den Markterfolg neuer Kommunikationssysteme bestimmen.

Die Hauptziele des DAMS-Projektes lassen sich folgendermaßen zusammenfassen:

- **Flexible Nutzung der Übertragungsbandbreite:** Die dynamische Zuweisung und Modifikation von Übertragungsbandbreite gestattet eine effiziente Ausnutzung der gesamten Übertragungskapazität und trägt somit zur Erhöhung der Systemleistungen bei. Hierbei sind ebenfalls variable Bitraten zu berücksichtigen, wie sie beispielsweise bei Datenkompressionsverfahren entstehen.

- **Integration von Diensten:** Die Integration von Sprach-, Bild- und Datenkommunikation und damit verbunden die Integration von paket- und leitungsvermittelten Übertragungsverfahren ist von großer Bedeutung.

- **Anpaßbarkeit an veränderte Benutzeranforderungen:** Die Einführung neuer, zur Zeit noch nicht endgültig festgelegter oder unbekannter Dienste muß nach der Installation eines Systems möglich sein. Ebenso muß das Entfernen von Diensten, eine Rekonfiguration des Systems und eine Erweiterung der Vermittlungs- und Übertragungskapazitäten durchführbar sein.

- **Erhöhte Zuverlässigkeit und Verfügbarkeit:** Beim Auftreten von Fehlern haben System- und Netzwerkmanagement den weiteren Betrieb des Systems sicherzustellen, notfalls mit einer verminderten Gesamtleistung des Systems (graceful degradation principle).

- **Erweiterbarkeit:** Die Systemkonzepte sollen einen weiten Bereich von einer kleinen bis hin zu einer großen Anzahl von Anschlüssen abdecken. Auch eine räumliche Verteilung des Systems ist in Betracht zu ziehen.

- **Offenheit:** Zur Nutzung von Software und Geräten fremder Hersteller ist die Berücksichtigung von Standards unbedingt erforderlich.

Zur Erreichung dieser Ziele ist ein äußerst modularer Aufbau des Systems zwingende Voraussetzung. Abbildung 1 gibt die Struktur eines verteilten Kommunikationssystems wieder. Über ein Verbindungsnetz (hier ein einfacher Ring) werden einzelne Knoten miteinander verbunden, die über lokale Vermittlungskapazität verfügen und Schnittstellen zu verschiedenen Endgeräten anbieten können. Auch bereits bestehende Kommunikationsinfrastrukturen, wie beispielsweise ein Ethernet, können integriert werden. Wir werden im nächsten Kapitel ausführlicher auf die Systemarchitektur eingehen.

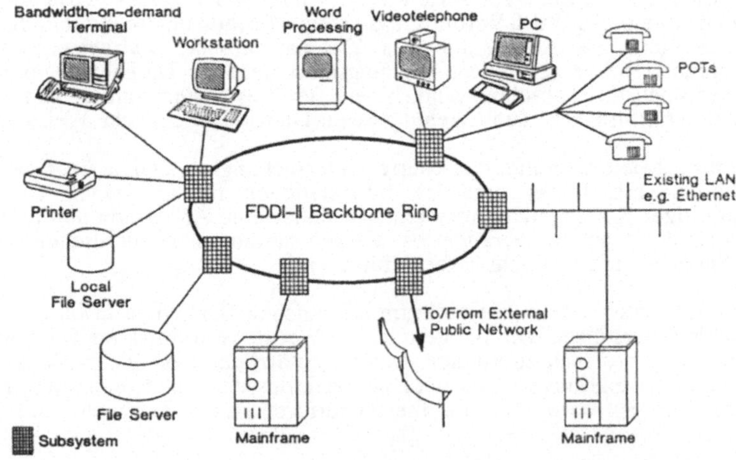

Abb. 1: Architektur des Kommunikationssystems

Der Prototyp wird einen ISDN-Basisanschluß für leitungsvermittelte Anwendungen, einen Ethernet-Anschluß (IEEE 802.3) für Paketdaten und einen Terminalzugang nach dem noch nicht endgültig festgelegten IEEE 802.9-Standard anbieten, der einen flexiblen Zugang für leitungs- und paketvermittelte Dienste mit wählbarer Bandbreite bereitstellt.

Das Projektkonsortium besteht aus den Partnern JS Telecom (Frankreich), BNR Europe (Großbritannien) und TELENORMA Bosch Telecom. Telenorma ist "Koordinierender Partner" und hat die Gesamtleitung des Projektes. DIT (Spanien), INESC (Portugal) und L-Cube (Griechenland) sind Assoziierte Partner, die Universitäten Aachen, Kaiserslautern, Stuttgart und Patras (Griechenland) sind Subkontraktoren.

In den beiden nächsten Kapiteln wird die Struktur der Hardware und der Steuersoftware beschrieben. Aufgrund der Größe des Projektes können die nachfolgenden Erläuterungen nur Überblickscharakter haben, auf eine Vielzahl interessanter Details mußte verzichtet werden.

II. Architektur der DAMS-Hardware

Die an das System gestellten Anforderungen werden in erster Linie durch eine modulare Hardware- und Software-Struktur erfüllt. Auf der Hardware-Ebene spiegelt sich dies in der Zerlegung eines vollständigen Systems in Untersysteme und speziellen Zwecken dienenden Komponenten wider. In Abbildung 1 wurde bereits die Gesamtarchitektur als ein aus mehreren Knoten bestehendes verteiltes System dargestellt. Den Aufbau eines einzelnen Knotens gibt Abbildung 2 wieder. Die zentrale Komponente ist die lokale Vermittlungseinheit (Local Switch Unit - LSU). Über eine knoteninterne Schnittstelle (Standard Internal Interface - SII) werden Port Units zur Adaption unterschiedlicher Anschlußarten, Backbone Units (BBU) zur Ankopplung an das Verbindungsnetzwerk und ein Steuerrechner (Subsystem Control Unit - SSCU) mit der LSU verbunden.

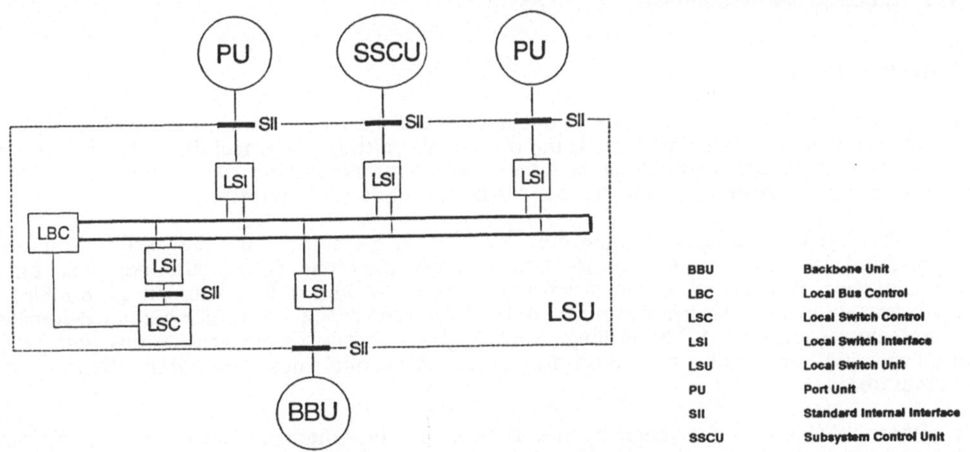

Abb. 2: Aufbau eines DAMS-Knotens

Dieses modulare Konzept gestattet eine flexible und kostengünstige Anpassung an verschiedenste Anwendungsbereiche. Bei Kleinsystemen kann vollständig auf die LSU verzichtet werden. Die Steuersoftware ist auch auf dem Prozessor einer Port-Unit ablauffähig. Je nach Art der verwendeten Schnittstellentypen wird die Obergrenze der anschließbaren Port-Units durch die Übertragungskapazität eines Knotens vorgegeben. Zur Erhöhung der Rechenleistung können mehrere Steuerrechner eingesetzt werden, die sich Vermittlungsaufgaben teilen oder Wartungs- oder Meßaufgaben wahrnehmen. Dabei ist es denkbar, daß auf einem solchen Steuerrechner ein völlig anderes Betriebssystem abläuft, wie etwa UNIX, das mit der Vermittlungs-Software kommunizieren kann. Ein einzelner Knoten ist autonom ablauffähig, d.h. bei fehlendem Bedarf kann auf das Backbone-Netz verzichtet werden. Zur Erhöhung der Gesamtübertragungskapazität können auch mehrere Backbone-Netze mit unterschiedlichen Topologien gleichzeitig eingesetzt werden. Alle Knoten des Systems können völlig unterschiedlich konfiguriert sein.

II.1 Die lokale Vermittlungseinheit LSU

Die LSU hat die Aufgabe, Verbindungswege zwischen an den Port-Units angeschlossenen Endgeräten einzurichten. Dies kann sowohl lokal innerhalb einer LSU als auch zwischen zwei LSUs über das Backbone-Netz geschehen. Paketdaten sind dabei gemäß den im Kopf angegebenen Adressen weiterzuleiten. Bei isochronen Verbindungen, wie etwa im Fall einer Sprachkommunikation, werden die dafür benötigten Übertragungskapazitäten reserviert. Der Verbindungsaufbau innerhalb eines Knotens oder über mehrere Knoten hinweg wird von einer Steuersoftware vorgenommen.

Die LSU beruht auf einem hybriden Vermittlungsprinzip. Den Kern stellt ein gefalteter Bus dar, der isochrone und nicht-isochrone Daten transportiert. Der isochrone Verkehr wird dabei nach dem Zeitmultiplexverfahren weitergeleitet. An der internen Schnittstelle SII besteht die Multiplexstruktur aus einem 125-μs-Rahmen mit einer wählbaren Anzahl von Slots. Der Zugang zum Bus und die Busarbitrierung werden von den SII-Schnittstelleneinheiten (Local Switch Interface - LSI) ohne zentrale Kontrolle durchgeführt. Die lokale Buskontrolleinheit (Local Bus Control - LBC) generiert alle Kontrollsignale für den lokalen Bus. Weiterhin werden von ihr Busoperationen überwacht und Fehlfunktionen der lokalen LSC-Steuerung (Local Switch Control - LSC) gemeldet. Die LSC ist ein Mikroprozessor, der neben dem Verbindungsaufbau auch Überwachungs- und Managementfunktionen ausführt. Innerhalb der LSU wurde bereits eine ATM-Vermittlungsstruktur berücksichtigt.

Für den Prototyp ist eine Übertragungskapazität von 262 Mbit/s geplant, eine Erhöhung bis auf über 800 Mbit/s wurde beim Entwurf berücksichtigt. In Verbindung mit einem Backbone-System bietet die gewählte Systemstruktur auch aus verkehrstechnischen Überlegungen heraus eine optimale Ausnutzung der Systemressourcen [KSS90].

II.2 Die Port Units

Jede Port-Unit stellt ein oder mehrere Zugänge des gleichen Typs bereit. Falls mehrere Zugänge vorhanden sind, wird in der Port-Unit keine direkte Vermittlung durchgeführt. Eine LSI ist jedoch in der Lage, eine direkte Verbindung zwischen zwei Endgeräten, die an der gleichen Port-Unit angeschlossen sind, herzustellen, ohne daß der lokale Bus der LSU belastet wird.

Die ISDN-Port-Unit adaptiert mehrere ISDN-Basisanschlüsse an die LSU. Ein ISDN-Basisanschluß besteht aus einem 16 kbit/s Signalisierkanal (D-Kanal), der u.a. für den Verbindungsauf- und abbau benötigt wird, und zwei 64 kbit/s Nutzkanälen (B-Kanäle). Mit entsprechenden Protokollkonvertierungen werden die isochronen Daten der beiden B-Kanäle des Basisanschlusses auf feste SII-Slots abgebildet. Nicht-isochrone Daten werden aus dem D-Kanal extrahiert und innerhalb des Systems unter Kontrolle des verteilten Betriebssystems weiterverarbeitet.

Die Ethernet-Schnittstelle ermöglicht die Einbindung bestehender Netze in das Kommunikationssystem. Dabei soll sowohl der Datenaustausch zwischen Geräten, die an unterschiedlichen, über das DAMS-System miteinander verbundenen Ethernets angeschlossen sind, als auch der direkte Zugriff über das Ethernet auf mit dem Backbone verbundene Geräte möglich sein. Hierfür sind eine eindeutige Identifikation von Quell- und Zielstationen, eine Pufferverwaltung zur Behandlung von unterschiedlichen Datenraten, eine Fehlerbehandlung auf den OSI-Ebenen Data Link und Transport, Routing, Paketisierung/Depaketisierung von Daten zur Behandlung unterschiedlicher Paketgrößen in den beteiligten Netzen und Relaisfunktionen zur Transformation von Adressen erforderlich.

Die IVDTE-Port-Unit umfaßt die Basisfunktionen der beiden zuvorgenannten Schnittstellen. Die Arbeiten orientieren sich an dem noch nicht endgültig festgelegten IEEE-802.9-Standard für integrierte Sprach- und Datennetze (Integrated Voice and Data Local Area Network - IVDLAN). Dieser Schnittstellentyp wurde in das Projekt mit aufgenommen, weil er neue Anforderungen an die

Steuerung stellt, insbesondere in Bezug auf die Integration der beiden Datentypen und die dynamische Veränderung der Bandbreite.

II.3 Die Schnittstelle zum Backbone-Netz

Die im Projekt implementierte Schnittstelle basiert auf dem FDDI-II-Standard (Fiber Distributed Data Interface, ANSI X3T9.5). FDDI-II besitzt eine Token-Ring-Architektur mit einer Glasfaser als Übertragungsmedium. Die Netzwerktopologie, ein doppelter Ring, kann bis zu 500 Stationen mit einer maximalen gesamten Glasfaserlänge von 100 km unterstützen. Benachbarte Ringstationen dürfen dabei bis zu 2 km voneinander entfernt sein. Die Übertragungsbandbreite beträgt 100 Mbit/s.

FDDI-II ist abwärtskompatibel mit FDDI: Bitrate, Übertragungscode, optische Eigenschaften, die Topologie etc. sind identisch zu FDDI. Beide Standards benutzen verteilte Kontrolle für die Initialisierung, das Management und den Wiederanlauf im Fehlerfall. Vereinfacht ausgedrückt, kann FDDI-II als eine Erweiterung von FDDI angesehen werden, die es erlaubt, über die Glasfaser sowohl isochrone Daten als auch Paketdaten zu übertragen. Auf eine detaillierte Darstellung der Arbeitsweise von FDDI-II kann hier nicht eingegangen werden. An dieser Stelle sei nur bemerkt, daß in FDDI-II ebenfalls ein 125-μs-Rahmen erzeugt wird, der in einzelne Slots unterteilt ist. Neben den FDDI-II-Funktionen regelt die Backbone-Unit die Abbildung der SII-Slots auf die FDDI-II- Slots für den isochronen Verkehr sowie die Übertragung von Paketen. Ähnlich wie beim lokalen Bus der LSU sind auch bei FDDI-II die Anteile von isochronem und nicht-isochronem Verkehr dynamisch einstellbar. Über das Netzwerkmanagement können z.B. Mindestkapazitäten für die beiden Verkehrsarten reserviert werden.

Für einen Knoten können grundsätzlich mehrere Backbone-Units und unterschiedliche Topologien gewählt werden. Es ist durchaus vorstellbar, daß in einem aus mehreren Knoten bestehenden Ring zwei Knoten, die besonders häufig mit hohen Datenraten miteinander kommunizieren, zusätzlich direkt verbunden werden. In derartigen Fällen werden jedoch erhöhte Anforderungen an die Steuersoftware gestellt, die in DAMS konzeptionell untersucht, jedoch nicht implementiert werden. Alternative Backbone-Netze wie DQDB oder zukünftige, höher-bitratige Systeme können ebenfalls als Knotenverbindungsnetz eingesetzt werden.

III. Die Steuersoftware

Die Leistungsfähigkeit eines verteilten Kommunikationssytems wie DAMS wird entscheidend durch die Betriebsorganisation und damit durch die Software bestimmt. Jeder einzelne Knoten stellt bereits ein eigenes verteiltes System dar: neben den beiden Steuerrechnern LSC und SSCU werden Teile der Betriebssoftware auf den Port Units und den Backbone Units ausgeführt. Zur Lösung dieser Problematik ist ausschließlich eine hochgradig verteilte Betriebsorganisation im Sinne der sehr strengen Definition von Enslow [Ens78] geeignet, die eine Verteilung von System- und Kontrollfunktionen vorsieht.

Kommunikationssysteme mit Vermittlungsaufgaben sind durch "weiche" Echtzeitanforderungen charakterisiert. Anwender erwarten beispielsweise für den Verbindungsaufbau Anwortzeiten, die sich an den CCITT-Empfehlungen (Q.514) orientieren und auch bei hohem Verkehrsaufkommen gewährleistet sein müssen. Für Audio- und Video-Mailsysteme müssen Sprach- und Bildinformationen unter der Kontrolle der Betriebssoftware von einem Speichermedium zu einem Endgerät transportiert werden. Hierzu ist es erforderlich, daß für solche Dienste bestimmte Nachrichtenverarbeitungsraten zur Beibehaltung spezifizierter Dienstqualitäten garantiert werden [Ste90, STW90]. Ein Benutzer wird sich nicht mit einem System zufrieden geben, wenn er die in seiner Audio-Mailbox gespeicherten Informationen zum Teil in verzerrter Form erhält.

Die Hardware-Konfiguration eines Kommunikationssystems hängt sehr stark von den einzelnen Einsatzgebieten ab. Die Software muß aus diesem Grunde ebenfalls flexibel konfigurierbar sein, insbesondere muß die Abbildung der logischen Software-Struktur auf die physische Hardware-Konfiguration in weiten Grenzen veränderbar sein. Die Integration neuer Leistungsmerkmale und Software-Versionen sowie Wartungsmaßnahmen dürfen keine Betriebsunterbrechungen bewirken, lediglich eine kurzzeitige Leistungsminderung ist akzeptabel. Von der Betriebssoftware werden daher Mechanismen erwartet, die das dynamische Nachladen und Auswechseln von Softwaremodulen unterstützen.

Im Fehlerfall muß sich die Software einer veränderten Hardware-Konfiguration anpassen können. Systemdienste, die auf nicht mehr betriebsbereiten Hardwarekomponenten installiert waren, sind an anderer Stelle neu zu starten. Um die Echtzeitfähigkeit des Systems auch bei Ausfällen von Komponenten zu erhalten, werden Hot-Stand-By-Verfahren erforderlich, bei denen jede aktive Komponente über einen oder mehrere redundante "Schatten" verfügt.

Die Einbeziehung von Fremdsoftware ist ebenfalls eine wichtige Anforderung. Gerade im Hinblick auf zukünftige Dienstleistungen, die sich durch die Schlagworte "Intelligente Netze" und "Mehrwertdienste" charakterisieren lassen, werden Datenbanksysteme, Editoren, Graphikpakete etc. benötigt, die nicht DAMS-konforme Schnittstellen besitzen.

Zur Erfüllung dieser Anforderungen werden im DAMS-Projekt ein verteiltes Betriebssystem und eine darauf aufbauende, verteilte Steuersoftware konzipiert und implementiert. In den folgenden Abschnitten werden das verteilte Betriebssystem und das Netzwerkmanagement charakterisiert.

III.1 Das verteilte Betriebssystem

Das Betriebssystem besteht aus einem *Betriebssystemkern* und darauf ablaufenden *Betriebssystemdiensten*, die als Prozesse oder Prozeßgruppen realisiert sind. Der Betriebssystemkern orientiert sich an Konzepten, die dem heutigen Stand der Technik entsprechen. Wesentliche Strukturierungseinheit ist das Team. In Form leichtgewichtiger Prozesse können beliebig viele Kontrollflüsse dynamisch erzeugt und vernichtet werden. Die durch ein Team implementierten Funktionen werden über dynamisch verwaltete Ports angesprochen. Der Kern unterstützt das Client-Server-Programmierparadigma durch eine sichere auftragsorientierte Kommunikation. Daneben existiert auch eine mitteilungsorientierte Kommunikation, die u.a. für den Verbindungsauf- und -abbau benötigt wird. Die Unterstützung eines einzigen Kommunikationsmodells, wie etwa des Remote-Procedure-Calls, wäre hier u.a. aus den in [TaR88] genannten Gründen hinderlich. Speziell beim Verbindungsaufbau wird eine Vielzahl von Einstellaufträgen für die Hardware abgesetzt, für die eine Auftragsbeziehung nicht wünschenswert ist. Der korrekte Aufbau der Verbindung wird über eine End-zu-End-Kontrolle durch übergeordnete Protokolle sichergestellt. Der Kern bietet darüber hinaus eine Multicast-Kommunikationsmöglichkeit an, über die mit einer Nachricht mehrere Empfänger-Teams erreicht werden können [Wyb90]. Die gesamte Kommunikations-Hardware des DAMS-Systems ist ebenfalls multicast-fähig.

Teams lassen sich während ihres Betriebs auf andere Prozessoren verlagern. Alle aktiven Prozesse innerhalb eines Teams können in konsistenter Weise angehalten, aus ihrer lokalen Umgebung herausgelöst und auf einem anderen Prozessor neu gestartet werden. Die hierzu vom Kern angebotenen Funktionen können auch für Checkpointing-Verfahren eingesetzt werden. Die Multicast-Kommunikation und die Verlagerung von Teams sind wichtige Voraussetzungen zur Erreichung von Fehlertoleranz in dynamischen Echtzeitsystemen. Die Konzeption des Kerns und seiner Basismechanismen stellt eine gute Grundlage für die Implementierung höherer Konzepte zur Strukturierung komplexer Softwaresysteme dar, wie sie beispielsweise in [WyB90] und [DKM87] beschrieben werden.

Nach der Untersuchung vieler, kommerziell verfügbarer oder in der Entwicklung bereits weit fortgeschrittener Systeme aus dem Forschungsbereich [ABB86, Che84, MuT86, RAF88] wurde entschieden, den an der Universität Kaiserlautern in der Realisierung befindlichen Betriebssystemkern PORDOS (Portable Real Time Distributed Operating System [NeG90]) einzusetzen. Die zuvor genannten Systeme genügten einzeln betrachtet nicht allen Anforderungen und boten auf Grund eingeschränkter Verfügbarkeit und Dokumentation keine geeignete Basis, eigene Konzepte einzubringen. Neue Anwendungen, wie etwa Multimedia-Arbeitsplätze, stellen neue Anforderungen an die Betriebssoftware, die zum jetzigen Zeitpunkt noch untersucht werden [WCS90]. Für den Vergleich und die Bewertung von alternativen Lösungsansätzen wird jedoch bereits jetzt ein Experimentierfeld benötigt. PORDOS bietet hierfür alle Voraussetzungen.

Die höheren Betriebssystemdienste sind als eine Menge von Server-Teams realisiert, die im Prinzip beliebig auf die Verarbeitungseinheiten des Systems verteilt werden können. Bis jetzt wurden ein

Name-Server, ein File-Server, ein Printer- und Terminal-Server, ein Boot- und Load-Server sowie ein Debug-Server realisiert.

Im allgemeinen ist in einem DAMS-System nicht an jedem Knoten Hintergrundspeicher verfügbar. Es muß also möglich sein, Software über den knotenlokalen Bus und/oder über den Backbone-Ring zu laden. Der *Boot-Server* lokalisiert die angeforderte Software-Komponente auf einem Hintergrundspeicher und lädt sie auf die angegebene Verarbeitungseinheit.

Bei der Systeminitialisierung wird vom Boot-Server in Kooperation mit der Konfigurationsverwaltung die aktuelle Struktur der Hardware ermittelt. Ausgehend von dieser Hardware-Beschreibung wird eine Plazierung der zur Verfügung stehenden Software durchgeführt [Sta86]. Für jedes Team existiert eine Attribut-Liste, die Angaben über die benötigten Hardware-Ressourcen (Hauptspeicher, E/A-Geräte, u.s.w.) und eventuell zusätzliche Plazierungsdirektiven enthält. So kann z.B. ausgeschlossen werden, daß aus Fehlertoleranzgründen zwei Teams auf die gleiche Verarbeitungseinheit oder den gleichen Knoten geladen werden. Nach einer erfolgreichen Plazierung wird der Ladevorgang gestartet.

Steht im gesamten System nur ein Hintergrundspeicher zur Verfügung, so wird vor dem eigentlichen Laden jeder Verarbeitungseinheit eine Liste der auf diese Einheit zu ladenden Software geschickt. Die komplette Systemsoftware wird anschließend per Broadcast transferiert. Jede Verarbeitungseinheit selektiert aus dem Strom von Lademodulen die lokal benötigte Software. Sind hingegen mehrere Hintergrundspeicher vorhanden, kann das Laden effizienter durchgeführt werden, indem einerseits der Ladevorgang parallelisiert wird und andererseits durch die Nutzung von Multicast-Mechanismen die Software nur an die Verarbeitungseinheiten verschickt wird, an denen sie auch benötigt wird.

Der *Debug-Server* ermöglicht die Fehlersuche in Teams zur Laufzeit auf Quellsprachenebene. Jedes Team kann von einer zentralen Station unabhängig von seinem Ausführungsort getestet werden. Das zu untersuchende Team (Target-Team) wird mit einem zusätzlichem Prozeß, dem Debug-Assistant, instrumentiert. Dieser Prozeß hat Zugriff auf sämtliche Speicherbereiche aller Prozesse des Teams. Der Debug-Server führt den Dialog mit dem Benutzer und verwaltet die Symboltabellen aller zum Target-Team gehörenden Prozesse. In Kooperation mit dem Debug-Assistant hat er die Möglichkeit, auf die Speicherbereiche der Prozesse zuzugreifen und ihren Ablauf zu beeinflussen.

Neben der Entwicklung von Diensten des verteilten Betriebssystems werden Fehlertoleranzmechanismen untersucht, die zur Verbesserung der Verfügbarkeit der Dienste eingesetzt werden können. An der Universität Kaiserslautern wurde ein Konzept für eine fehlertolerante Version des Boot-Servers entworfen [NeB90]. Ziel war es, durch redundante Speicherung der Lademoduln den Ausfall von Komponenten des Boot-Servers zu tolerieren. Der fehlertolerante Lade-Dienst besteht aus einer Menge von kooperierenden Boot-Servern, die verteilt ablaufen. Jeder Boot-Server verwaltet und speichert lokal (im Hauptspeicher oder auf dem Hintergrundspeicher) eine Untermenge aller zur Verfügung stehenden Lademoduln. Ein verteilter Kontrollalgorithmus stellt sicher, daß mindestens N Kopien eines Lademoduls im System vorhanden sind und daß kein Server mehr als 1 Exemplar eines Moduls verwaltet. Somit können also N-1 Ausfälle von Verarbeitungseinheiten toleriert werden. N ist frei wählbar, so daß der Grad der Fehlertoleranz den Kundenbedürfnissen angepaßt werden kann.

III.2 Netzwerkmanagement

Die Aufgabe des Netzwerk-Managements besteht in der effizienten Verwaltung, Steuerung und Beobachtung (Monitoring) eines Kommunikations-Netzwerkes. U.a. fallen die Aufgaben Konfigurations-Management, Leistungs-Management und Fehler-Management an. Für das DAMS-System wurde ein dezentraler, hierarchischer Ansatz zur Strukturierung des Netzwerkmanagements gewählt (Abb. 3). Jede Einheit des verteilten DAMS-Systems besitzt hierbei ihre eigene Managementfunktionalität.

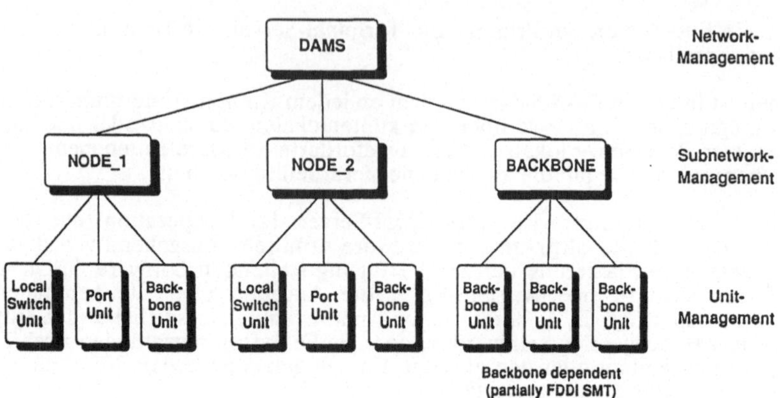

Abb. 3: Die DAMS-Management-Hierarchie

Die Nachteile dieser Lösung sind die größere Komplexität der einzelnen Managementfunktionen und das erhöhte Kommunikationsaufkommen zwischen den Funktionseinheiten. Dem gegenüber stehen aber eine Reihe von Vorteilen, die insbesondere die Anforderungen an die Systemsoftware erfüllen. Zum einen führt die Verteilung von Management-Funktionen automatisch zu einer höheren Verfügbarkeit des Managements. Zum anderen ermöglicht die Aufteilung der Management-Funktionen in kleinere Einheiten eine Migration von Funktionen im Fehlerfall. Die Kopplung von Management-Funktionen zu den Einheiten des DAMS-Systems erlaubt weiterhin eine einfache Erweiterung des Systems um neue Einheiten oder den Aufbau eines Systems aus heterogenen Komponenten.

Dieser dezentralen Struktur ist eine hierarchische Struktur wie folgt übergeordnet:

- Jede Hardware-Einheit des DAMS-Systems wird durch eine lokale Managementkomponente verwaltet (Unit-Management).

- Die Monitor- und Kontrollfunktionen für ein komplettes Subsystem werden von einer eigenständigen Management-Komponente ausgeführt, dem Subnetwork-Management. Im DAMS-System werden zwei Typen des Subnetwork-Managements unterschieden: das knotenspezifische und das backbone-orientierte Subnetwork-Management. Die Software des Subnetwork-Managements wird im jeweiligen Subnetwork ablaufen, um eine autonome Arbeitsweise des Subsystems zu gewährleisten.

- Eine spezielle Instanz des Managements koordiniert schließlich die Aktivitäten der einzelnen Subnetwork-Managements und stellt die Schnittstelle des Management nach außen dar. Dieses globale Network-Management kann an einer beliebigen Stelle des Systems ablaufen. Um die Verfügbarkeit des Managements zu erhöhen, wird diese Komponente verteilt realisiert.

Die wichtigste Aufgabe des Konfigurations-Managements ist die kontrollierte Initialisierung des gesamten Systems nach dem Einschalten. Die Initialisierung wird durch die folgenden Randbedingungen wesentlich beeinflußt:

- Die benötigte Software ist nicht an jeder Einheit resident verfügbar.

- Hintergrundspeicher ist nicht an jedem Subnetwork verfügbar; im Extremfall existiert nur ein Hintergrundspeicher im gesamten System.

- Die Systemkonfiguration wird nicht statisch festgelegt.

- Die Reihenfolge, in der die einzelnen Komponenten des Systems eingeschaltet werden, ist nicht bekannt.

- Jede Komponente kennt lediglich ihre Adresse und nicht ihre Plazierung im System.

Die Initialisierung wird in einem Bottom-Up-Verfahren durchgeführt, indem jede Einheit nach dem Einschalten ihre Verfügbarkeit dem lokalen Subnetwork-Management meldet. Das Subnetwork-Management meldet anschließend die Verfügbarkeit des jeweiligen Subnetworks an das Network-Management. Nach dem Laden der einzelnen Einheiten wird die Information über die aktuelle Systemkonfiguration im Top-Down-Verfahren wieder an die Subnetwork-Management-Ebene und von dort an die Unit-Management-Ebene zurückgemeldet.

Nach dem Erreichen der Betriebsbereitschaft stellt das Management eine komplettes Abbild der Hardware-Struktur des verteilten Systems bereit, das laufend aktualisiert wird. Änderungen oder Ausfälle im System werden dynamisch eingetragen. Das Konfigurations-Management nutzt die Management-Hierarchie, um das Hardware-Abbild als verteilte Datenstruktur zu implementieren. Hierfür wird auf die dezentral verwalteten Subnetwork-Strukturen und die auf den Einheiten vorhandenen lokalen Datenstrukturen zugegriffen.

Das Performance-Management bewertet und optimiert den Betriebszustand des Systems. Bewertungskriterien sind Paket-Durchsatz, Antwortzeiten, Fehlerhäufigkeiten, Verbindungsaufbau- und Abbauzeiten sowie Blockierungshäufigkeiten für leitungsvermittelten Verkehr. Die Beeinflussung der Leistungsfähigkeit des Systems erfolgt über das Setzen von Protokollparametern und über die Verwaltung der zur Verfügung stehenden Übertragungsbandbreite. Das Erstellen von Langzeitstatistiken ist eine weitere Aufgabe des Performance-Managements.

Die Aufgabe des Fehler-Managements ist es, fehlerhafte Systemteile zu entdecken und zu lokalisieren. Im Rahmen des Projektes wird zunächst lediglich das Erkennen von Fehlern implementiert. Hierfür werden analog zur Systemhierarchie Filtermechanismen genutzt, die die Sicht auf das System mit unterschiedlicher Abstraktion erlauben. Durch geeignete Darstellungsweisen kann auf einer zentralen Bedienstation der Systemzustand in einem frei wählbaren Detaillierungsgrad dargestellt werden. Über "Zooming"-Techniken ist das Bedienpersonal in der Lage, Teilsysteme oder einzelne Einheiten gezielt zu untersuchen. In [HaW90] werden Beispiele für die benutzergerechte Präsentation komplexer Leistungsdaten gegeben, die aus verteilten Systemen extrahiert wurden.

IV. Zusammenfassung

Zukünftige Kommunikationssysteme werden durch ein hohes Maß an Konfigurierbarkeit und Modifizierbarkeit sowohl im Bereich der Hardware als auch der Steuersoftware gekennzeichnet sein. Im vorliegenden Aufsatz wurden Konzepte eines fortgeschrittenen Kommunikationssystems vorgestellt, die diese geforderte Flexibilität wesentlich unterstützen. Der Prototyp dieses Systems wird gegenwärtig im ESPRIT-Projekt DAMS realisiert. Zum Abschluß des Projektes sind umfangreiche Messungen geplant, die mit Hilfe eines speziellen Verkehrsgenerators [CAR90] die Eignung der Konzepte nachweisen sollen.

Wir danken allen am DAMS-Projekt beteiligten Personen und Institutionen, ohne deren Einsatz und Kooperationsbereitschaft ein internationales Projekt dieser Größenordnung nicht durchführbar wäre. Unser besonderer Dank gilt Herrn M. Siegel für seine Mitwirkung bei einer früheren Version des Netzwerk-Management-Abschnittes und Frau G. Zäh für die editorische Bearbeitung dieses Aufsatzes.

V. Literaturverzeichnis

[ABB86] M. Accetta, R. Baron, W. Bolosky et al.; Mach: A New Kernel Foundation for Unix Development; in: Proc. USENIX Technical Conf. Exhibition; June 1986

[Bec90] T. Becker; Achieving High Availability in a Distributed ISDN Control System; ZRI-Bericht Nr. 7/90; Universität Kaiserslautern; August 1990

[CAR90] E. Carrapatoso, C. Aguiar, M. Ricardo; Traffic Generator/Detector System Concept; Proc. EFOC LAN; Munich; June 1990

[CCIT90] CCITT Recommendations I.150; B-ISDN ATM Aspects; Genf 1990

[Che84] D. Cheriton; The V Kernel - A Software Base for Distributed Systems; IEEE Software; April 1984

[DKM87] N. Dulay, J. Kramer, J. Magee et al.; Distributed System Construction: Experience with the CONIC Toolkit; in: Proc. Int. Workshop on Exp. with Distributed Systems; Springer LNCS; 1987

[Ens78] P.N. Enslow Jr.; What is a Distributed Data Processing System? Computer; Jan. 1978

[HaW90] D. Haban, D. Wybranietz; A Hybrid Monitor for Behavior and Performance Analysis of Distributed Systems; IEEE Trans. on Software Eng.; Feb. 1990

[KSS90] W. Krautkrämer, K. Sauer, D. Schlichthärle, R. Knobling; Flexibles Systemkonzept zur Einführung von BK-Diensten im privaten Netzbereich; ntz, Dez. 1990

[MuT86] S. Mullender, A. Tanenbaum; The Design of a Capability-Based Distributed Operating System; Computer Journal; 1986

[NeB90] J. Nehmer, T. Becker; A Fault Tolerance Approach for Distributed ISDN Control Systems; Proc. Fourth ACM SIGOPS Workshop an Fault Tolerance Support in Distributed Systems; Bologna; September 1990; ACM-OSR, Vol. 25, No. 1

[NeG90] J. Nehmer, T. Gauweiler; Design Rationale for the MOSKITO Kernel; Informatik-Fachberichte 264; Springer-Verlag 1990

[PoB89] R. Pohlit, M. Bleichrodt; Dynamically Adaptable Multi-Service Switch; Proc. of 6th Ann. ESPRIT Conf.; Brussels; Nov./Dec. 1989

[RAF88] M. Rozier, V. Abrossimov, F. Armand et al.; CHORUS Distributed Operating Systems; Chorus Systemes; Technical Report CS/TR-88-7.6; Nov. 1988

[Sch90] B. Schaffer; ATM Switching in the Developing Telecommunication Networks; Proc. ISS'90; Stockholm 1990

[Sta86] F.J. Stamen; Konfigurationsmanagement in verteilten Betriebssystemen; Proc. NTG/GI-Fachtagung Architektur und Betrieb von Rechensystemen; März 1986

[Ste90] R. Steinmetz; Synchronization Properties in Multimedia Systems; IEEE Journal on Selected Areas in Communication; Vol. 3; No. 3; April 1990

[STW90] D.P. Anderson, S.Y. Tzon, R. Wahbe, R. Govindan, M. Andrews; Support for Continuous Media in the DASH System; Proc. 10th ICDCS; Paris; Mai/Juni 1990

[Tan87] A.S. Tanenbaum; Operating Systems - Design and Implementation; Prentice-Hall; 1987

[TaR88] A. Tanenbaum, R. van Renesse; A Critique of the Remote Procedure Call Paradigm; in: Proc. EUTECO'88; 1988

[WCS90] D. Wybranietz, R. Cordes, F.-J- Stamen; Support for Multimedia Communication in Future Private Networks; in: Proc. 1st Int. Workshop on Networking and Operating System Support for Digital Audio and Video; Berkeley, Nov. 1990

[WyB90] D. Wybranietz, P. Buhler; The LADY Programming Environment for Distributed Operating Systems; Future Generation Computer Systems Journal; Vol. 6, Iss. 3; Dez. 1990

[Wyb90] D. Wybranietz; Multicast-Kommunikation in verteilten Systemen; Informatik-Fachberichte, Bd. 242; Springer Verlag; 1990

Kooperative Graphische Anwendungen in Hochgeschwindigkeitsnetzen

M. Bever[†], S. Noll[††], J. Rix[††], C. Schottmüller[†]

[†] IBM European Networking Center
Tiergartenstr. 8, D-6900 Heidelberg

[††] Fraunhofer-Arbeitsgruppe für Graphische Datenverarbeitung (FhG-AGD)
Wilhelminenstr. 7, D-6100 Darmstadt

Kurzfassung

In geographisch verteilten Entwurfsumgebungen, wie sie beispielsweise in der Europäischen Automobilindustrie anzutreffen sind, wird zunehmend der Entwurf eines Automobils durch die Kooperation von verschiedenen Entwurfsteams erreicht. Hierzu werden Entwurfsdaten in Form von komplexen Zeichnungen, Skizzen und Graphiken ausgetauscht. Die bei diesen Anwendungen anfallenden Datenvolumina stellen besondere Anforderungen an die Kommunikationsdienste und Netze. Am Beispiel kooperativer graphischer Anwendungen werden neuartige Anforderungen an einen Hochgeschwindigkeitsdatentransfer diskutiert und eine Realisierung auf der Basis eines Hochgeschwindigkeits-Transportsystems vorgestellt.

1. Kooperative graphische Anwendungen

Kooperatives Arbeiten in geographisch verteilten Entwurfsumgebungen, wie sie zum Beispiel in der Automobilindustrie, im Schiffs- und Flugzeugbau vorherrschen, erfordern den Austausch von Konstruktionsdaten zwischen Entwurfsinseln. Beispielsweise tauschen in einer Diskussion über Entwurfsalternativen eines neuen Automobils Konstrukteure verschiedener Europäischer Lokationen üblicherweise eine ganze Reihe von graphischen Objekten, wie Zeichnungen, Skizzen und Graphiken, aus. Das kooperative Arbeiten auf solchen graphischen Objekten kann durch geeignete Werkzeuge unterstützt werden.

Die Erzeugung und Manipulation von graphischen Objekte wird mit Hilfe von sogenannten graphischen Editoren durchgeführt. Sind diese verteilt, so erlauben sie den geographisch verteilten Entwurfsinseln das Editieren auf gemeinsamen (verteilten) graphischen Objekten.
Für die Realisierung eines verteilten graphischen Editors sind Kommunikationsdienste erforderlich, die den Austausch der graphischen Objekte unterstützen. Dabei muß die Heterogenität heutiger Graphikarbeitsplatzrechner berücksichtigt werden, die sich neben unterschiedlichen Rechnerarchitekturen und Betriebssystemen auch in unterschiedlichen Bilddarstellungsformaten, Farbdarstellungen und Fenstersystemen ausdrücken kann. Eine Lösungsmöglichkeit bietet die Verwendung standardisierter Kommunikationsprotokolle und Graphikaustauschformate, wie sie beispielsweise von der ISO (International Organization for Standardization) genormt sind.

Um den Benutzern eines verteilten graphischen Editors akzeptable Antwortzeiten zu gewährleisten, müssen die Objekte möglichst in Realzeit ausgetauscht werden können. Neue Lösungen für verteilte graphische Editoren ergeben sich dann, wenn der Austausch von komplexen Graphikobjekten über Unternehmensgrenzen hinweg durch schnelle Datentransferdienste unterstützt wird. Die großen Datenvolumen komplexer graphischer Objekte und die hohen Anforderungen an die Übertragungsgeschwindigkeit machen Kommunikationsnetze mit hohen Übertragungsraten erforderlich.

Mit der Einführung von Hochgeschwindigkeitsnetzen in Europa wird neben der Vernetzung von unterschiedlichen Entwurfsinseln auch die technische Grundlage für die Entwicklung solcher schnellen Dateitransferdienste geschaffen. So kann beispielsweise die Übertragungsrate für heute existierende Datentransferdienste bereits dann gesteigert werden, wenn die mit Hochgeschwindigkeitsnetzen verfügbaren Bandbreiten für die Anwendung beziehungsweise für anwendungsnahe Kommunikationsdienste zur Verfügung stehen.

In Kapitel 2 wird eine kooperative graphische Anwendung beschrieben und daraus in Kapitel 3 spezielle Kommunikationsanforderungen an einen schnellen Datentransfer in einem Hochgeschwindigkeitsnetz abgeleitet. Die Realisierung dieser Anforderungen wird exemplarisch auf der Basis eines Hochgeschwindigkeits-Transportsystems vorgestellt. Hierzu gibt Kapitel 4 zunächst ein Überblick über den derzeitigen Entwicklungsstand der Hochgeschwindigkeitsnetze. Darauf aufbauend wird in Kapitel 5 ein Hochgeschwindigkeits-Transportsystem eingeführt und ein existierender Dateitransferdienst als Beispiel eines Datentransferdienst darauf abgebildet. Kapitel 6 faßt die Ergebnisse zusammen und gibt einen Ausblick auf offene Fragestellungen.

2. Verteiltes graphisches Editieren

Graphische Editoren unterstützen die Erzeugung, Manipulation und die Darstellung von graphischen Objekten. Ein Anwendungsgebiet ist die rechnergestützte Konstruktion (CAD). CAD Systeme stellen dem Benutzer u.a. auf der Graphikebene Operationen zur Verfügung, die neben der Erzeugung und Manipulation von Konstruktionsdaten auch die Darstellung des Entwurfsobjektes betreffen, wie beispielsweise das Drehen, Vergrößern und Verkleinern von Objekten.

Neben dem Einsatz in der rechnerunterstützten Konstruktion werden graphische Editoren überall dort eingesetzt, wo die Darstellung und rechnerunterstütze Bearbeitung von graphischen Informationen notwendig ist. Besonders interessant ist für viele Anwendungen die Verfügbarkeit von graphischen Funktionen zur Erzeugung und Manipulation von Freihandzeichnungen mit einem elektronischen Schreibstift (Tablett, Maus).

In kooperativen Anwendungen wie beispielsweise einer technischen Konferenz kann ein graphischer Editor dazu benutzt werden, Freihandskizzen oder auch bereits digitalisierte Zeichnungen mittels graphischer Operationen weiter zu bearbeiten. Dabei ist es prinzipiell möglich, daß jeder Diskussionsteilnehmer graphische Operationen auf der allen Konferenzteilnehmern gemeinsamen Skizze durchführt. Das jeweilige Ergebnis der Änderungsoperationen ist allen Teilnehmern sichtbar.

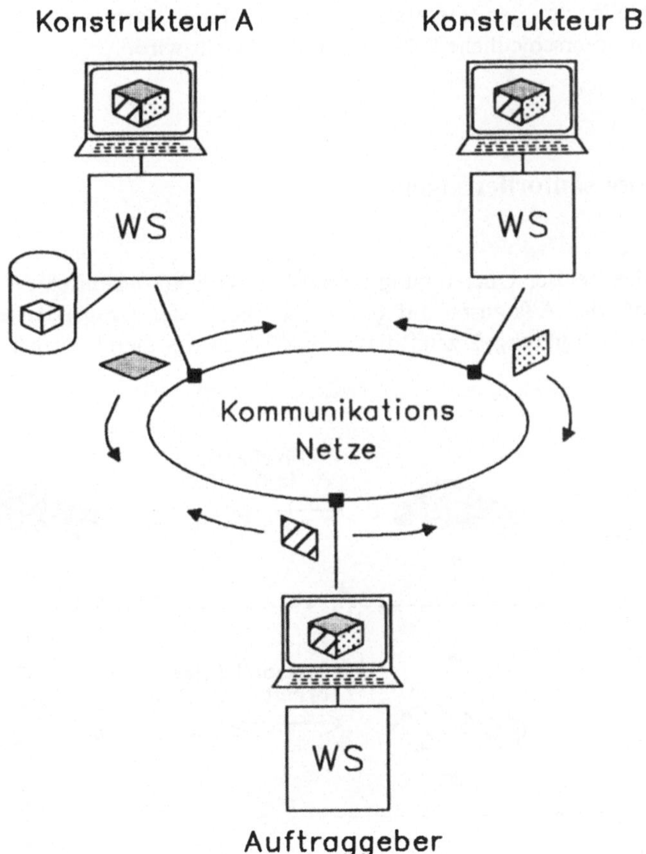

Abbildung 1. Kooperatives graphisches Szenario

In dem in Abbildung 1 dargestellten Szenario diskutieren zwei Konstrukteure mit einem Auftraggeber eine Entwurfsalternative, die hier vereinfacht als Würfel dargestellt wird. Sie benut-

zen dazu einen verteilten graphischen Editor (Distributed SketchPad), der speziell für die schnelle Erzeugung von Skizzen entwickelt wurde [17; 9].
Der noch unbearbeitete Würfel wurde durch den Konstrukteur A entworfen und lokal in einer Entwurfsdatenbank gespeichert. Während einer technischen Konferenz macht der Konstrukteur A dem Konstrukteur B und dem Auftraggeber möglicherweise verschiedene Versionen eines Entwurfsobjekt sichtbar. Hierzu muß das Entwurfsobjekt über einen geeigneten Kommunikationsdienst in Realzeit ausgetauscht werden. Das ausgetauschte graphische Objekt dient als Hintergrund für die nachfolgenden graphischen Operationen der einzelnen Partner. Dabei arbeitet jeder Benutzer konzeptionell auf einer ihm zugeordneten Ebene. Um allen Teilnehmern zu jedem Zeitpunkt ein konsistentes Bild der gemeinsamen Skizze geben zu können, müssen Änderungen allen Partnern in Realzeit zur Verfügung stehen.

Legt man die Ebenen aller Partner wie Klarsichtfolien übereinander, so ergibt sich unter Einbezug des Hintergrundes die gemeinsam erarbeitete Gesamtskizze. Die von einem Partner eingebrachten Anteile an der Gesamtskizze sind dabei farbig unterschieden, was in Abbildung 1 durch unterschiedliche Schraffuren dargestellt wird.

3. Kommunikationsanforderungen

Prinzipiell lassen sich bei der Übertragung graphischer Objekte technisch drei unterschiedliche Ebenen betrachten: der Austausch auf der **Pixel** Ebene, der **Graphikdaten** Ebene und der **Modelldaten** Ebene. Abbildung 2 zeigt diese drei Ebenen und deren Beziehungen untereinander:

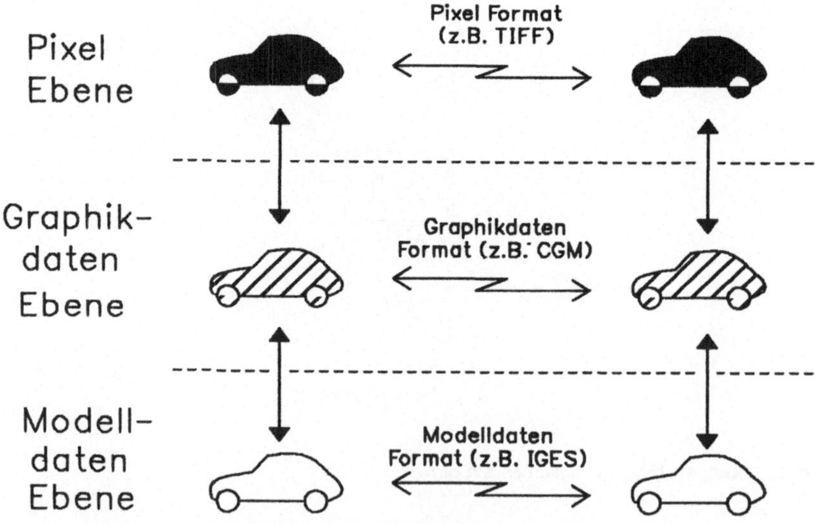

Abbildung 2. Ebenen der Graphikdatenübertragung

Beispielsweise werden aus Modelldaten, wie sie in einem CAD-System vorliegen, zur Interaktion mit dem Benutzer Graphikdaten generiert. Zur Bildausgabe auf Rasterbildschirmen erfolgt die Berechnung eines Pixelbildes aus diesen Graphikdaten.

Die Übertragung von graphischen Objekten ist auf jeder der drei Ebenen möglich. Der Austausch auf der Pixelebene bedeutet, daß alle Bildpunkte einer Graphik über einen geeigneten Kommunikationsdienst ausgestauscht werden müssen. Das zu übertragende Datenvolumen wird von der Größe, der Auflösung und der Farbinformationen der Graphik bestimmt. Bei einem Pixelbild mit einer typischen Auflösung von 1280 mal 1024 Bildpunkten (Pixel) und einer Echtfarbendarstellung von 24 Bit pro Pixel ergibt sich eine Größe von etwa 30MBit.

In unserem Szenario würde die Übertragung des Hintergrundbildes oder der einzelnen Teilskizzen als Pixelbild die Verfügbarkeit eines Datentransferdienstes erfordern, der die Übertragung des Datenvolumens in Realzeit garantiert. Die Anforderungen an einen solchen Dienst verschärfen sich dann, wenn beispielsweise dynamisch die Ansicht auf ein Objekt in einer 3D Präsentation geändert werden soll. Hierbei wäre die Übertragung von 10-20 Pixelbilder pro Sekunde notwendig.

Neben den Anforderungen an die Übertragungszeiten muß unter Umständen der Kommunikationsdienst auch unterschiedlichen Darstellungsformaten der Pixelbilder bei Sender und Empfänger Rechnung tragen. Übliche Pixelformate sind beispielsweise TIFF [18], GIF [5] oder X Bitmaps [12].

Beim Graphikdatenaustausch werden nicht die Pixelbilder, sondern alle Informationen zu ihrer Erzeugung übertragen. Die Darstellung und der Austausch der Graphikdaten erfolgt gemäß einem der üblichen Standards wie CGM, GKS, PHIGS [2; 14] oder X [10]. Die Datenvolumina liegen bei Graphikdaten im Bereich zwischen einigen Kilobyte und mehreren Megabyte. Bei einfachen Bildern (Skizzen, einfache Graphiken) bietet diese Austauschebene den Vorteil, daß die Datenvolumina im allgemeinen sehr viel geringer sind als bei der Übertragung auf der Pixelebene. Jedoch werden besondere Anforderungen an die Zuverlässigkeit der Übertragung gestellt. So ist die Verfälschung einer technischen Zeichnung beispielsweise durch die fehlerhafte Übertragung der Koordinaten einer Linie nicht tolerierbar.

Neben den beschriebenen Austauschebenen für graphische Objekte können in Entwurfsanwendungen Objekte auch bereits auf der Modelldatenebene ausgetauscht werden. Die Modelldaten beinhalten die Geometrie- und Präsentationsspezifikation eines Objektes. Austauschformate in diesem Bereich sind IGES, VDAFS und neuerdings auch STEP [4]. Die Modelldaten eines komplexen Entwurfsobjektes wie beispielsweise die eines Automobils, haben üblicherweise einen Umfang von mehreren Megabyte. Wird in unserem Beispielszenario der Austausch eines neuen Hintergrundbildes auf der Modelldatenebene durchgeführt, so muß ein Kommunikationsdienst dieses Datenvolumen möglichst schnell übertragen, um das kooperative Arbeiten mit dem verteilten Editor nicht durch lange Übertragungszeiten zu verzögern. Zusätzlich zu den hohen Übertragungsraten stellen sich auch auf dieser Ebene hohe Anforderungen an die Zuverlässigkeit der Übertragung.

Zusammenfassend läßt sich sagen, daß der Austausch von graphischen Objekten abhängig von der Ebene und des Objekttyps, unterschiedliche Anforderungen an einen Kommunikationsdienst stellt. Bestimmend sind dabei die Forderung nach hohen Bandbreiten und auf der Graphik- und Modelldatenebene auch die Forderung nach einer zuverlässigen Übertragung.

4. Hochgeschwindigkeitsnetze

Heutige öffentliche Netze wie das Datex-P Netz oder das ISDN-Netz der Deutschen Bundespost stellen Übertragungsraten im Bereich von 100 KByte zur Verfügung (144 KBit/s bei ISDN). Die Übertragungsraten heutiger lokaler Netzwerke liegen bereits im Bereich von 10 MBit/s. Diese Übertragungsraten insbesondere die der Weitverkehrsnetze sind für zukünftige kooperative graphische Anwendungen wie sie in den Kapiteln 1 und 2 beschrieben werden nicht ausreichend. Beispielsweise dauert die Übertragung eines Pixelbildes mit einer Größe von 30 MBit über das heute verfügbare ISDN-Netz der Deutschen Bundespost mit einer Übertragungsrate von 64 KBit/s ungefähr 8 Minuten. Die Übertragung von CAD-Modelldaten mit einer Größe von beispielsweise 10 MByte über das ISDN-Netz würde ungefähr 20min dauern.

Der Einsatz von Glasfaser als physikalisches Übertragungsmedium ermöglicht die Entwicklung von Hochgeschwindigkeitsnetzen. Im Gegensatz zu herkömmlichen Netzen, zeichnen sich Hochgeschwindigkeitsnetze durch Übertragungsraten von 100 MBit/s und mehr aus. Beispiel für ein lokales Hochgeschwindigkeitsnetz ist FDDI (Fibre Distributed Data Interface) mit einer Übertragungsrate von 100 MBit/s [15]. B-ISDN mit einer Übertragungsrate von etwa 150 MBit/s ist ein Beispiel für ein öffentliches Hochgeschwindigkeitsweitverkehrsnetz.

Lokale FDDI-Netzwerke sind bereits als Produkte erhältlich und nationale Hochgeschwindigkeitsweitverkehrsnetze wie beispielsweise das Vorläufer Breitband Netz der Deutschen Bundespost (VBN) sind kommerziell verfügbar. Im internationalen Bereich treibt die Europäische Gemeinschaft zusammen mit den Europäischen Postgesellschaften die Entwicklung eines Europäischen Hochgeschwindigkeitsnetzes voran.

5. Kommunikationsdienste für Hochgeschwindigkeitsnetze

5.1. Transportdienst

Der Einsatz einer Vielzahl unterschiedlicher miteinander verbundener Netze führt zu komplexen Netzinfrastrukturen zwischen kommunizierenden Rechnerknoten.
Gemäß dem ISO OSI Basisreferenzmodell besteht die Aufgabe des Transportdienst darin, dem Benutzer eine netzunabhängige Schnittstelle für die Übertragung von Daten bereitzustellen. Dabei werden die Netzinfrastruktur und die unterschiedlichen Dienste und Eigenschaften der Netzwerke verborgen (Abbildung 3). Beispielsweise bietet ein Transportdienst ein einheitliches Adressierungsschema, das von den Adressierungsschemas der unterschiedlichen Netzwerke abstrahiert, und das die eindeutige Adressierung von Transportdienstbenutzern ermöglicht.

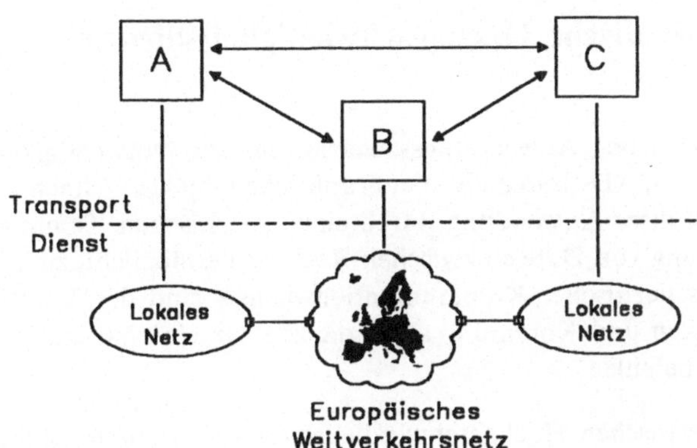

Abbildung 3. Transportdienst

Ein Hochgeschwindigkeits-Transportsystem realisiert einen Transportdienst für die Übertragung von Daten über Hochgeschwindigkeitsnetze.

Am Europäischen Zentrum für Netzwerktechnik wird derzeit ein Hochgeschwindigkeits-Transportsystem entwickelt [6]. Dieses Hochgeschwindigkeits-Transportsystem beinhaltet neben verbindungslosen Transportdiensten einen verbindungsorientierten Transportdienst, auf den im weiteren detailliert eingegangen wird.

Der verbindungsorientierte Transportdienst ermöglicht den Aufbau und den Abbau von logischen Transportverbindungen zwischen Transportsystembenutzern, über die Daten ausgetauscht werden können. Für jede Transportverbindung werden beim Verbindungsaufbau in Abhängigkeit von den Anforderungen, einer diese Transportverbindung benutzende Anwendung, bestimmte Dienstgüten mit dem Transportsystem verhandelt. Dienstgüten sind beispielweise die Übertragungsrate, die Fehlerrate oder die Übertragungsverzögerung. Die aushandelbaren Dienstgüten sind dabei von den Fähigkeiten des unterliegenden Netzwerkes abhängig. Zum Beispiel bestimmt die maximale Übertragungsrate eines Netzwerkes die maximal verhandelbare Übertragungsrate einer Transportverbindung. Für die Verhandlung der Dienstgüte verfügen die Dienstprimitive zum Verbindungsaufbau (T_Connect) einen Dienstgüteparameter (Quality of Service Parameter).

Je nach Anforderung der Anwendung muß die Datenübertragung zuverlässig erfolgen, d.h. die Daten werden fehlerfrei, ohne Verlust und ohne Vertauschung von Dateneinheiten übertragen. Beispielsweise müssen Modelldaten zwischen zwei CAD-Anwendungen zuverlässig übertragen werden. Im Gegensatz dazu sind bei der Übertragung von Bildsequenzen aus Rasterbildern Fehler tolerierbar, wenn sie bei der Anzeige der Bildsequenz vom Betrachter nicht wahrgenommen werden.

5.3 Anwendungsspezifische Hochgeschwindigkeitsdienste

Das in Kapitel 2 beschrieben Anwendungsszenario zeigt die Notwendigkeit des schnellen Datenaustauschs zwischen CAD-Systemen. Da graphische Objekte vielfach als Dateien verwaltet werden, ist es sinnvoll diese Anwendung durch einen spezifischen Kommunikationsdienst, der die schnelle Übertragung von Dateien zwischen Rechnern ermöglicht, zu unterstützen. Weitere Beispiele anwendungsspezifischer Kommunikationsdienste sind der Transfer von Bildsequenzen aus Rastergraphiken und Kommunikationsdienste zur Unterstützung der entfernten Ausführung von Graphikbefehlen.

Diese anwendungsspezifischen Hochgeschwindigkeitsdienste werden gemäß anwendungsspezifischen Kommunikationsprotokollen auf der Basis von Hochgeschwindigkeits-Transportsystemen realisiert (Abbildung 4).

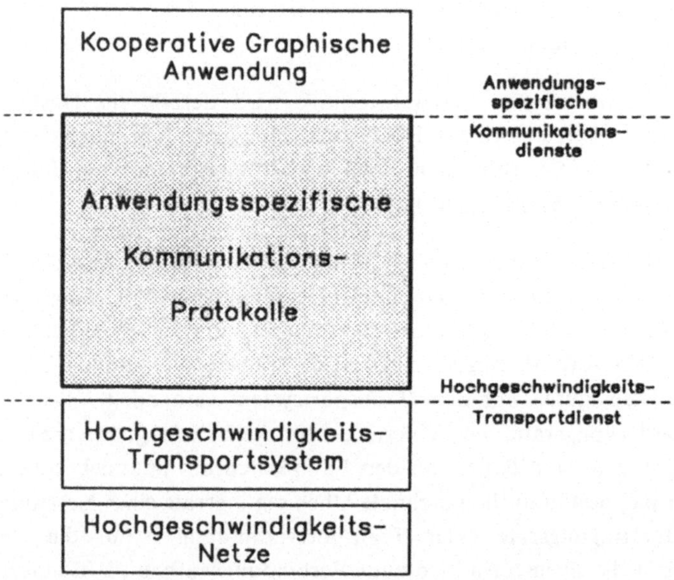

Abbildung 4. Anwendungsspezifische Protokolle

Die Realisierung eines anwendungsspezifischen Hochgeschwindigkeitsdienstes für kooperative graphische Anwendungen soll am Beispiel eines Hochgeschwindigkeits-Dateitransferdienstes erläutert werden. Als Ausgangspunkt dient dabei der ISO OSI Dateitransferdienst FTAM

Der ISO OSI Dateitransferdienst FTAM (File Transfer, Access and Management) [8; 1; 15; 19] unterstützt den Dateitransfer, den entfernten Dateizugriff und die Verwaltung von entfernten Dateien in offenen Systemen. Der FTAM Dienst abstrahiert von den Eigenschaften konkreter Dateisysteme durch ein virtuelles Dateikonzept, das komplexe Dateistrukturen, Typen von Dateiinhalten, Zugriffsrechte, etc. global beschreibbar macht. Dem FTAM Dienst liegt ihm ein Client/Server Modell zugrunde, wobei dem Server ein virtuelles Dateisystem zugeordnet ist (Abbildung 5). Der FTAM Dienst umfaßt Dienstprimitive zum Initiieren und Terminieren einer FTAM Verbindung und zum Ausführen von entfernten Dateioperationen, wie das Lesen oder Schreiben von Dateien bzw. Teilen von Dateien oder das Erzeugen oder Löschen von Dateien.

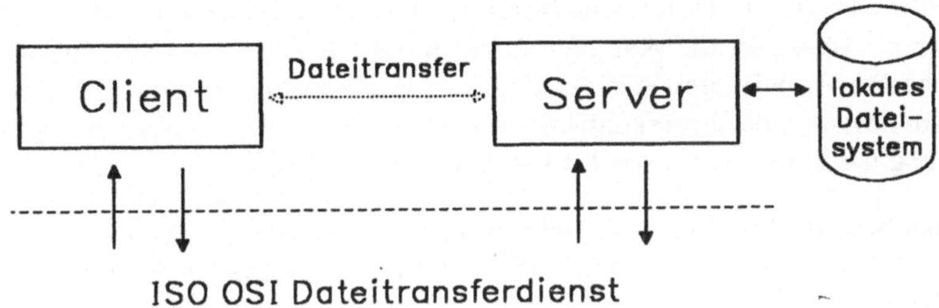

Abbildung 5. FTAM Client/Server Modell

5.4 Hochgeschwindigkeits-Protokollarchitektur

Der ISO OSI Dateitransferdienst wird gemäß dem ISO OSI Basisreferenzmodell [7; 1; 15; 19] durch eine Protokollarchitektur realisiert, die in sieben Schichten unterteilt ist. Jede Schicht erbringt einen Dienst durch ein schichtspezifisches Protokoll und benutzt dabei den Dienst der darunterliegenden Schicht.

Die Schichten 5 bis 7 erbringen den ISO OSI Dateitransferdienst FTAM auf der Basis des ISO OSI Transportdienstes. Das Protokoll der Schicht 7, der Anwendungsschicht setzt sich aus den Protokollen des Anwendungsdienstelementes ACSE (Application Control Service Element), und FTAM (File Transfer, Access and Management) zusammen. Das Protokoll des Anwendungsdienstelementes FTAM realisiert den ISO OSI Dateitransferdienst. Es benutzt dabei, den Dienst des Anwendungsdienstelementes ACSE das den Aufbau und den Abbau von Anwendungsverbindungen unterstützt und die Dienste der Schichten 5 und 6. Die Schicht 6, die Darstellungsschicht unterstützt die Aushandlung von globalen Datendarstellungen und globalen Datencodierschemata für die FTAM Protokolldateneinheiten und die zu übertragenden Dateiinhalte. Die Schicht 5, die Kommunikationssteuerungsschicht bietet Dienste zur Resynchronisation der Kommunikationspartner nach einem fehlerbedingten Abbruch des Dateitransfers. Abbildung 6 (links) zeigt die Protokollarchitektur des ISO OSI Dateitransferdienstes.

Die Übertragungsraten des ISO OSI Dateitransferdienstes, der durch die ISO OSI Protokollarchitektur realisiert ist, befriedigt nicht die Anforderungen zukünftiger kooperativer graphischer Anwendungen. War bei Kommunikationsdiensten auf der Basis herkömmlicher Netze die Übertragungsgeschwindigkeit des Netzes der Engpaß bei der Datenübertragung, so muß bei der Realisierung eines Hochgeschwindigkeits-Kommunikationsdienst berücksichtigt werden, daß nun die Protokollbearbeitung in den Rechnerknoten eines Hochgeschwindigkeitsnetzes den Engpaß der Datenübertragung darstellt. Ziel beim Entwurf einer Hochgeschwindigkeits-Protokollarchitektur ist deshalb die Minimierung des Protokollbearbeitungsaufwandes. Dieser ist abhängig zum einen von der Leistungsfähigkeit des Rechnerknotens, von der konkreten Implementierung der Protokollarchitektur auf einem Rechnerknoten und von der Komplexität der Protokollarchitektur selbst. Ein Ansatz zur Minimierung des Protokollaufwandes ist deshalb die Steigerung der Leistungsfähigkeit eines Rechnerknotens durch die Realisierung von Protokollfunktionen in Hardwarebausteinen oder durch parallele Rechnerarchitekturen [21]. Ein weiterer Ansatz ist die Nutzung effizienter Betriebssystem- und Interprozeßkommunikationsmechanismen, wie zum Beispiel effiziente Pufferstrategien [20] oder Upcalls [3]. Der Ansatz zur Minimierung des Protokollbearbeitungsaufwandes der im weiteren verfolgt wird, ist die Reduzierung der Komplexität der Protokollarchitektur.

In einem ersten Schritt wird derzeit im Rahmen einer prototypischen Implmentierung das ENC Hochgeschwindigkeits-Transportsystem, das in Kapitel 5.1 vorgestellt wurde, in die Protokollarchitektur des ISO OSI Dateitransferdienstes integriert (Abbildung 6).

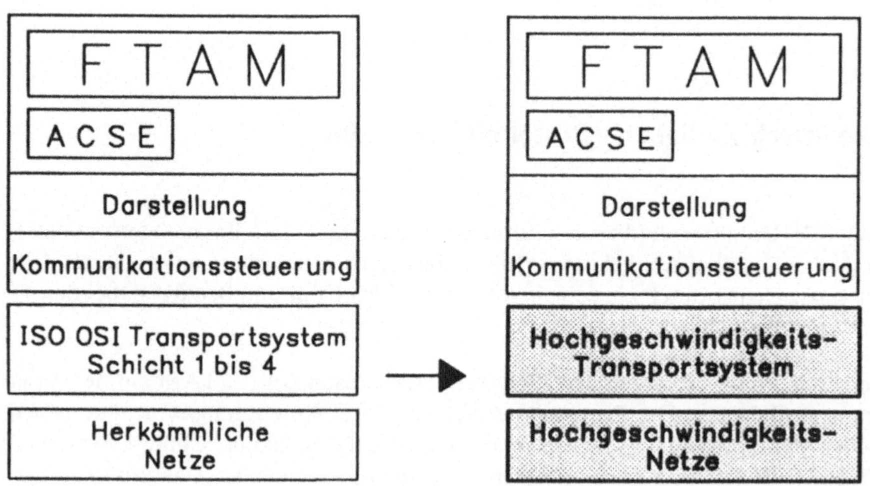

Abbildung 6. Integration des Hochgeschwindigkeits-Transportsystems

6. Zusammenfassung und Ausblick

Anhand eines konkreten kooperativen graphischen Anwendungsszenarios wurde der Austausch von graphischen Objekten auf den drei Ebenen der Pixel, der Graphikdaten und der Modelldaten diskutiert. Dabei sind die speziellen Hochgeschwindigkeits- und Zuverlässigkeitsanforderungen an Übertragungsdienste für graphische Objekte dargestellt und die neuen Hochgeschwindigkeitsnetze mit ihren hohen Übertragungsraten als technische Vorraussetzung kooperativer graphischer Anwendungen identifiziert worden. Hochgeschwindigkeits-Transportsysteme stellen einen netzunabhängigen Transportdienst zur Verfügung und bilden die Basis für die Entwicklung von anwendungsspezifischen Kommunikationsdiensten, die die Übertragung von graphischen Objekten in Realzeit unterstützen. Am Beispiel des ISO OSI Dateitransferdienstes wurde exemplarisch die Integration dieses Hochgeschwindigkeits-Transportsystems in die bestehende Protokollarchitektur diskutiert.

Ein weiteres Ziel ist die Durchführung von Messungen am Prototypen, die Aufschluß über Engpässe in der Protokollarchitektur geben können. Hierbei soll insbesondere das Protokollverhalten bei unterschiedlichen Lasten, d.h verschiedenen Dateigrößen und Dateioperationen, untersucht werden. Um den Protokollbearbeitungsaufwand des Dateitransferdienstes weiter zu reduzieren, sind die in Kapitel 5.4 beschriebenen Ansätze zur Minimierung des Protokollbearbeitungsaufwandes auch für die anwendungsspezifischen Protokolle des Dateitransferdienstes zu diskutieren. Ein Schwerpunkt bildet dabei die Reduzierung der Protokollkomplexität und auf den Einsatz effizienter Implementierungstechniken.

Danksagung

Wesentliche Teile dieser Arbeit sind im Rahmen der Kooperation zwischen dem Europaen Networking Center (ENC) der IBM in Heidelberg und der Fraunhofer Arbeitsgruppe Graphische Datenverarbeitung (FhG-AGD) entstanden. Allen Mitarbeitern sei an dieser Stelle für die interessanten Diskussionen gedankt.

Einige Arbeiten wurden im Rahmen des Europäischen Projektes RACE CAR (R1079) entwickelt und finanziert. Auch den beteiligten RACE CAR Projektpartnern gilt unser Dank.

Literatur

[1] **M.Bever:** *OSI: Application Layer - Entwicklungsstand und Tendenzen;* Tutorium: Kommunikation in Verteilten Systemen, GI/ITG Fachtagung, Februar 1989.

[2] **P.Bono, ed.:** *Special Issue on Graphics Standards;* in IEEE - Computer Graphics and Applications, Vol.6, No.8, August 1986.

[3] **D.D.Clark:** *The Structuring of Systems using Upcalls;* in Proc. of the 10th Symposium on Operating Systems Principles, ACM, New York, 1985, pp.171-180.

[4] **J.L.Encarnacao, R.Schuster, E.Vöge:** *Data Interfaces in CAD/CAM Applications;* Springer Verlag, 1986.

[5] *Graphics Interchange Format (GIF) Version 89a;* CompuServe Incorporated, 1990.

[6] **D.B.Hehmann, M.G.Salmony, H.J.Stüttgen:** *High-Speed Transport Systmes for Multi-Media Applications;* in Protocols for High-Speed Networks, North-Holland 1990, pp.303-321.

[7] **ISO:** *Information Processing Systems - Open Systems Interconnection - Basic Reference Model* International Standard, ISO 7498, 1984.

[8] **ISO:** *Information Processing Systems - Open Systems Interconnection - File Transfer, Access and Management* International Standard, ISO 8571, 1988.

[9] **J.J.Lee:** *Xsketch: a multi-user sketching tool for X11;* Conf. on Office Inf. Syst., ACM, 1990, pp.169-173.

[10] **M.Muth:** *Das X Window System;* Informatik Spectrum, Band 14, Heft 1, pp.34-36, Springer Verlag, Februar 1991.

[11] **L.H.Ngoh, T.P.Hopkins:** *Transport Protocol Requirements for Distributed Multimedia Information Systems;* The Computer Journal, Vol.32, No.3, 1989, pp.252-261.

[12] **A.Nye:** *Xlib Programming Manual;* O'Reilly & Associates, Inc., 1988.

[13] **G.M.Parulkar, J.S.Turner:** *Towards a Framework for High Speed Communication in a Heterogeneous Networking Environment;* in IEEE Network Magazine, Vol.4, No.2, March 1990, pp.19-27.

[14] **J.Rix, ed.:** *Special Issues on 3D Standards;* in Computers & Graphics, Vol.12, No.2, pp.145-190, April 1988.

[15] **M.T.Rose:** *The Open Book: A Practical Perspective on OSI;* Prentice Hall, 1990.

[16] **F.E.Ross:** *An Overview of FDDI: The Fiber Distributed Data Interface;* in IEEE Journal on Selected Areas in Communication, Vol.7, No.7, Sept. 1989, pp.1043-1051.

[17] **M.Schendel, S.Noll, J.Rix:** *Distributed SketchPad System - A tool for cooperative sketching in a network environment;* Contribution to the COMICS - Workshop in June 1991 in Tolouse, to be published in the Workshop Proceedings.

[18] *Tag Image File Format (TIFF) Specification Revision 5.0;* Aldus Corporation & Microsoft Corporation, 1988.

[19] **A.S.Tanenbaum:** *Computer Networks;* Prentice Hall, 1989.

[20] **C.M.Woodside, J.R.Montealegre:** *The Effect of Buffering Strategies on Protocol Execution Performance;* in IEEE - Trans. on Communications, Vol.37, No.6, June 1989.

[21] **M.Zitterbart:** *A Parallel Architecture for Transport Systems and Gateways;* in Kommunikation in verteilten Systen, Informatik Fachberichte 205, P.Kühn (ed.), Springer 1989, pp.774-757.

Basismechanismen für Multimedia-Systeme

Ziel dieses Fachgespräches ist es, die Datenverwaltungs- und Kommunikationsdienste zu identifizieren, die in einer Vielzahl von Multimedia-Anwendungen benötigt werden. Weiterhin werden Architekturvorschläge gesucht, die diese Dienste integrieren und dabei bekannte DB-Konzepte und Kommunikationsprotokolle verwenden und erweitern.

Koordinator: Prof. Dr. Th. Härder, Universität Kaiserslautern

Towards Integrated Multimedia Systems: Why and How

Ralf Guido Herrtwich
Ralf Steinmetz
IBM European Networking Center
Tiergartenstr. 8
D-6900 Heidelberg 1

Abstract. A multimedia system is characterized by the computer-controlled generation, manipulation, presentation, storage, and communication of independent discrete media such as text and graphics and continuous media such as audio and video. The innovation multimedia systems provide is flexibility through integration of different media into a single system. Such a system can unify the methods of information distribution, personalize information services through interactive access and individual information selection, enlarge the bandwidth of perception at the user interface, make information presentation more effective, and provide flexible media processing and transformation. The major obstacles to integrating continuous media with the discrete media in computer systems are limitations of today's digital solutions: In addition to high-capacity and high-speed hardware, system software is needed that meets the real-time demands of audio and video, and a multimedia application interface is compulsory. This paper elaborates on the key issues for providing multimedia support.

1. Introduction

The term "multimedia" has become one of the major buzzwords in computing and telecommunications for the 1990's [48; 50]. While everybody agrees that future systems should provide multimedia functionality, there is some uncertainty about what a multimedia system actually is and what its particular functions should be. Sometimes a product is called a "multimedia system" if it allows users to combine text and graphics. More ambitious people would not use this attribute before they can watch full-motion video in HDTV quality on their workstation screen while they are busy editing text files. The right answer is between these two extremes.

The purpose of this paper is to reasonably define the term and to examine the implications of this definition on future computer usage and computer system structures. The major challenges for building multimedia systems are identified. We investigate how issues that we consider crucial for multimedia systems are reflected in today's products or prototype systems and point out potential future developments. One assumption on which our discussion is based is the *distribution* of multimedia systems. Network attachment is a main feature of most computers today so that there is a strong requirement for multimedia functions to be applicable in a distributed environment as well.

Our discussion is based on the personal experiences we gained by working in two different multimedia research projects, DASH and DiME:

- DASH is a project at the International Computer Science Institute and the University of California at Berkeley concerned with providing operating system support for digital audio and video [3]. The goal of the DASH project is to design and implement system software that allows audio and video to be handled just as any other kind of data in a typical workstation environment.

- DiME is a project at the IBM European Networking Center Heidelberg dealing with distribution transparent access of multimedia resources like cameras and stored video sequences [52]. DiME aims to provide an "easy, but rich" communication service as part of an application programming interface to manipulate data streams by controlling their sources and sinks within a heterogeneous computing environment.

Apart from our own work, we take findings from recent multimedia conferences and workshops such as [5; 25; 40; 41; 49] into account.

The remainder of this text is organized in three main parts: Section 2 gives our definition of a multimedia system and provides the starting point of our discussion. Section 3 elaborates on the advantages of multimedia systems, identifying their potential for computing in general and for new application areas in particular. Based on this perspective, Section 4 as the core of the paper discusses the key issues faced in the design and realization of multimedia systems.

2. What is "Multimedia"?

Any information appealing to human senses is transported through some *medium*. From a computing perspective, media are means of communication between humans and computers - and between humans using computers as communication tools. Media represent information in particular ways: Whereas the medium "text" represents a formatted sentence visually as a sequence of characters, the medium "audio" represents it acoustically by means of pressure waves.

Each medium defines *presentation values* in a *presentation space* [19]: We are familiar with two-dimensional visual presentation space with color pixels on paper and on computer displays. But presentation spaces are not limited to the sense of vision: Stereophonic and quadrophonic sound define acoustic presentation spaces, and the sensors of our skin provide the presentation space for the sense of touch.

An additional presentation dimension for each medium is time. Some media, such as text and graphics, have time-independent values. These media are called *discrete*. Other media, such as audio and video, have values that change over time and these changes contribute to the media semantics. These media are called *continuous*. The terms "discrete" and "continuous" do not refer to the internal data representation, but to the users' view of the data. Continuous-media data often consist of a sequence of discrete values which replace each other as time progresses.

Media flexibility is the major requirement on a multimedia system. To be flexible enough to handle all kinds of media, a multimedia system should *combine* both discrete and continuous media. With this requirement, neither a VCR nor a desktop publishing system handling text and graphics are multimedia systems, whereas an editor with voice annotation is. Yet, the mere incorporation of different media is not enough to achieve flexibility: A multimedia system should also be able to handle each type of media *independently*, providing the opportunity to combine them in arbitrary ways. A VCR, *e.g.*, stores audio and video together, prohibiting to access each information stream separately as it would be possible if both audio and video where contained in different files on a disk. To flexibly separate and combine different media, the computer is the ideal tool.

In summary, we arrive at the following definition [55]: *A multimedia system is characterized by the* ***computer-controlled*** *generation, manipulation, presentation, storage, and communication of* ***independent discrete*** *and* ***continuous*** *media.*

The individual components of multimedia systems do not necessarily have to be new and should not ignore existing techniques; we now have more than four decades of experience with discrete media in computing, and know how to handle voice and video communication through global networks. The innovation multimedia systems provide is in the ***integration*** of all kinds of media into a single system, obscuring the lines between computing, telecommunications, and even mass media. From a computing perspective, which is the natural starting point of our discussion, the major challenge of building a multimedia system is the introduction of continuous media into today's computer systems.

3. Potential of Integrated Multimedia Systems

The fairly recent integration of graphics into computing makes the importance and impact of new media evident: Media determine *how* and *for what purpose* computers are used. With graphics, computers took over the function of traditional design tools, extending the capabilities of existing tools by applying the versatility of the computer to the new application domain. Beyond this, graphics changed the method of interaction with the computer. Window system paradigms evolved together with the new media, and improved the way man and machine could communicate. The integration of audio and video will lead to corresponding paradigms. Therefore, the implications of multimedia systems are not just in opening new application domains, but in changing the view of computing in general.

Integrating continuous media in computer systems can offer a variety of advantages such as

- the reduction of costs for existing services,
- the improvement in quality of existing services,
- the broader acceptance of existing services,
- the possible interaction between existing services, or
- the availability of new services.

With the availability of continuous media in distributed computer systems, potential advantages result from new possibilities for information distribution, selection, presentation, and processing. Since we should not design multimedia systems without these advantages in mind, we review the major potentials of integrated multimedia systems in the following subsections.

3.1 Unified Information Distribution

The telecommunications industry worldwide is moving rapidly towards the establishment of *integrated services digital networks* (ISDNs) [8], realizing that once all data has a common digital representation there is no need to maintain the existing multitude of channels for information distribution. The emerging optical technology together with advanced satellite communication leads to *integrated broadband communication networks* (IBCNs) [17; 6]: A single global multi-purpose network can carry not only text, data, and voice, but also video and high-fidelity audio, making it possible to offer high-quality video-conferencing and, on the long run, to provide television and radio services as well.

The high bandwidth and pervasiveness of future IBCNs make it possible not only to integrate existing communication networks, but also to replace and enhance traditional non-electronic channels of information distribution. The following example is typical for how the network can improve overall service quality: Instead of renting videos from a local video store, users can download them from a remote file server. This increases the availability of a particular video program (the cassette may have been rented by somebody else, but the master video on the file server is always accessible) and reduces the access time. The same considerations apply to the distribution of music albums and, eventually, books. Technically, traditional physical media carriers could be substituted by mass-storage servers that take over the role of private CD collection and public library. IBCNs will eventually allow information producers (such as film studios or record companies) to deliver their products directly to the client. This does not correspond to today's infrastructure for information distribution and implies severe economic changes.

In today's home electronics and communications equipment we can identify a significant duplication of hardware due to disjoint ways of information handling and encoding. Even in the future, there will not be only one universal audio or video format (HDTV, *e.g.*, seems to arise with three different standards). Format conversion services, however, will allow for the cooperation of various computer systems working with distinct data representations. Here, network integration pays off for the consumer: Avoiding redundant hardware for both the service provider and the consumer makes not only the provision of, but also the participation in information services less costly than today. Furthermore, having a universal "media platform" provides more flexibility: New services do not require new equipment, but merely new software. Their establishment can proceed more rapidly - especially since the software can be obtained through the network as well. With respect to the information distribution, users can electronically locate the best deal; there is no geographical limitation of a market, again implying economic change.

3.2 Individual Information Selection

Once information receivers have computer functions available at their end of the information distribution channel, they have a high degree of control about the information they obtain [15]. Today newspaper agencies and TV stations choose which information they present to their customers and when they broadcast it. With computers and telecommunication equipment, users can select information themselves, at a time of their own choosing: Videotex as a primitive example of such systems is also evolving towards the handling of multiple media [43]. Broadcast can be restricted to sending live information that is of interest to a large number of users at the same time, *e.g.*, a parliament session or a tennis match. All other information can be transmitted on demand; users stop a presentation, make a "detour" through background information, and return later to the interrupted program. Of course, not all users would like to change their current habits of information consumption and become involved in information services. The advantage of the computer lies in the adjustable degree of interaction; "couch potato mode" is always an option.

The powerful information retrieval abilities of the computer are essential for individual information access. Involvement of the user is not even needed: Computers can be programmed to automatically filter out news in which the user has no interest. Knowing the user's preferences, the computer can also customize the information presentation, *e.g.*, when reporting the daily news to a sports fan, football results would go first, while for another political events will be the prime items. This will also make more "personalized" information services possible [7]: Many people are more interested in news they receive through letters or electronic mail than in general news. A personalized information service would arrange all kinds of news items according to their importance to the user. Customized information services such as these are a major application field within artificial intelligence [47].

Distributed multimedia systems will not replace books. Apart from constituting a cultural asset, printed material has the advantage of not requiring an electronic device for reading. Paper will always be attractive for entertaining reading. For news, its may loose some importance as electronic services can be more up-to-date. Scientific publishing is likely to be an area where distributed multimedia systems have the most severe impact. Both search and composition are facilitated by a computer system hooked to an information network.

3.3 Enlarged Bandwidth of Perception

Media determine not just *how*, but also *how good* man and machine can communicate. In the early days of computing, humans had to adapt to the computer for input and output of data. The introduction of graphical user interfaces and pointing devices has improved man-machine communication significantly. But still, today's forms of I/O are neither a very natural way for humans to communicate, nor are they very efficient: To speak is faster than to write, to listen is easier than to read, and to show is better than to describe. Audio and video increase the bandwidth of perception at the user interface of a computer system [33]: According to the idea transfer model [61] the *idea expression spectrum* (amount and range of information required to express a thought) is transformed with less loss of information from and to the *media spectrum* (amount of information expressed using different media).

Choosing the appropriate medium to present information is determined by whom and in what way the information is used. With a variety of media available, it becomes possible to adjust the "look and feel" of computer-based work to familiar human operations. Proof for the importance of such adjustment is the success of the desktop paradigm used in today's window systems. To model sheets of paper on a computer screen is tightly coupled with the availability of appropriate media (graphics, in this case). Corresponding paradigms will evolve for systems incorporating audio and video.

The more media a system is able to support, the better it can be adjusted to the needs of users and applications. An application area where the flexible combination of media is particularly promising is self-guided learning [39]. Students in a history class can, *e.g.*, browse through databases containing documents, maps, newsreels, TV documentaries, recordings of speeches, *etc.*, at their own convenience, guided by a *hypermedia system* [11; 42]. Another application field for multimedia systems is *computer-supported cooperative work* (CSCW) [16]. CSCW systems (colloquially termed "groupware") are tools for teams of users that work on a common project. They do not require the users to be physically present in the same place or at the same time, but allow them to use the same media they use for their cooperation now. Video conferencing [53] or voice mail [60] are today's primitive examples of such systems. Their expansion and integration to include different kinds of media will result in tools that support applications such as co-authoring of multimedia documents or joint project management.

3.4 Flexible Media Processing and Transformation

Not only information access, but also information production is easier with a distributed multimedia system. While this is beneficial at an individual level, e.g., for the author of a paper, it can be questioned whether it is good for the society as a whole. The easier it is to put together a piece, the less important it is whether the content is worth the effort. Everything can be put together so fast that the amount of worthless information increases. The ratio of good versus bad pieces can become worse than today.

The ability to record and play back information coded in different kinds of media is the key element in supporting multimedia applications. The application range of a system, however, increases even further when these media can also be processed by the computer. A system's ability to "understand" audio, *e.g.*, makes it possible to issue spoken commands to it. This allows computers to be used by physically handicapped people or in situations where textual command input is impossible, *e.g.*, in a driving car.

One way in which a computer can process data is by transforming one information representation into another. This makes a flexible transition between different kinds of media possible: A movie can be generated from a textual description, a piece of music can be played from its score, *etc.* Today, media transformation is mainly used in the production of video images for flight and car simulators, but the same computer animation techniques that imitate the real world as closely as possible can be used to create a "virtual reality" that follows different laws of nature. Perhaps there is no other area where the generality of the computer, its boundless abilities to handle and modify information, is as stimulating as in audio and video production [35]. While the computer cannot turn the average user into George Gershwin or Walt Disney, it can at least remove technical and financial hurdles of creativity.

4. Key Issues in Integrated Multimedia System Design

We have identified *flexibility through integration* as the major feature of a multimedia system. This integration needs to take place throughout the entire system, *i.e.*, at the *hardware adaptation*, the *system management* and the *application interface level.* The integration issues, however, are different on each level as discussed in the following subsections.

4.1 Hardware Adaptation Level

Continuous and discrete media are traditionally handled in completely separate environments. As shown in Figure 1, exchange of continuous media takes place in analog form, involving very little processing. Even if in today's TV and radio studios many components handling continuous media become digital, analog signals are still switched between the various sources and sinks. Discrete media, on the contrary, are handled in a completely digital environment, both for processing and exchange.

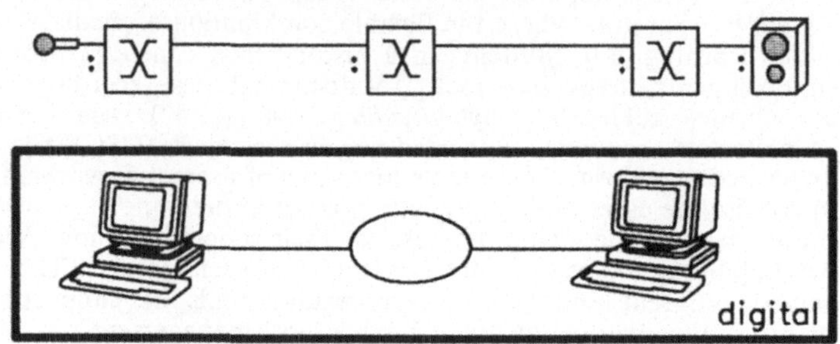

Figure 1. Disjoint analog and digital data paths

4.1.1 Analog Technology as a Starting Point

Using today's technology, the solution most readily available to integrate continuous and discrete media is not to abandon existing continuous-media equipment such as CD players or VCRs, but to connect this equipment to the computer via some interface (e.g., RS-232, RS-424, SCSI). Control functions are then executed rather from some software module in the computer than from the operating panel of these devices. This is shown in Figure 2. As a consequence, the representation of continuous media in such a system is determined by the existing devices. While for audio devices digital encoding can be used, video in such system is usually available in analog form only.

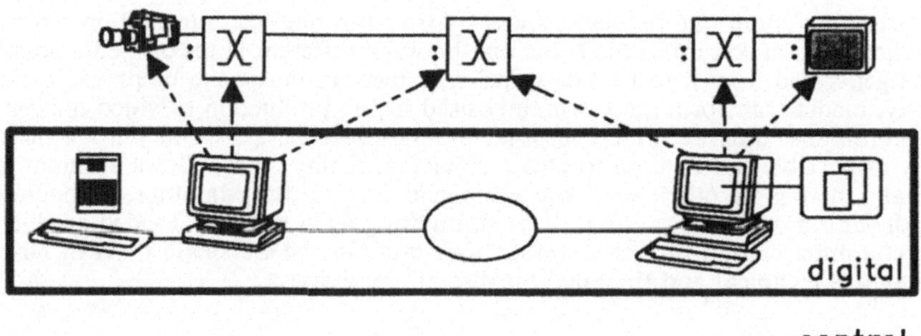

Figure 2. Computer controls all continuous-media devices

An example of a system using this approach is the Integrated Media Architecture Laboratory (IMAL) conceived at Bell Communications Research in Red Bank [32]. IMAL is an experiment in coordinating the provision of a variety media services offered through different communication utilities. Video services in the Muse and Pygmalion system of MIT's Project

Athena [22; 10; 38] function in the same way: Each video workstation is connected both to an Ethernet for traditional data communication and to a cable TV network to receive video signals. Through devices such as the Parallax video card [45], digital computer output and analog video information is combined so that it appears on the same screen.

The first experiments within the DiME project [52] at the IBM European Networking Center were performed on a similar basis using PS/2 workstations with AVC and M-Motion adapters [4; 36]. Devices generating or processing continuous media can be controlled from remote workstations. DiME, however, aims to use fully digital communication between various workstations as shown in Figure 3. Since devices receiving or generating analog signals may be used as local attachments, some digitization needs to take place before data can enter the digital network. A similar experiment took place at US West in Boulder [12] where a DS3 optical fiber interconnects two sites with their multimedia labs. Within the sites all continuous-media signals are interconnected using analog switches.

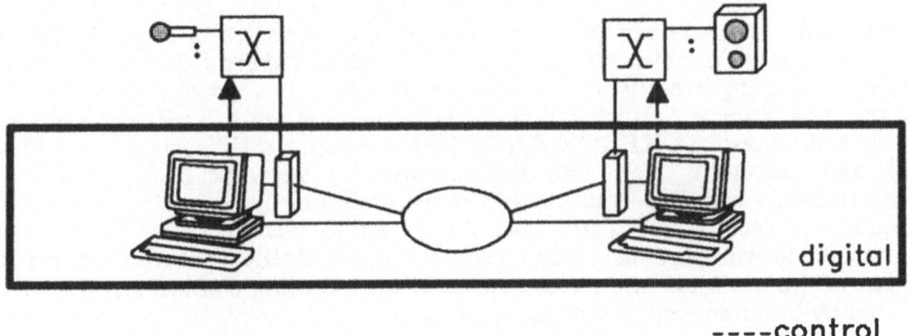

Figure 3. Local continuous-media devices attachments, nonintegrated digital continuous-media communication

4.1.2 Fully Digital Systems as the Final Goal

A main advantage of approaches utilizing analog technology is their feasibility today: Devices are available to handle audio and video in real time. They provide a testbed for experiments such as user interface and application studies for which the underlying technology is of minor importance. Demands for integrated solutions as described in Section 3 call for digital data representations. Digital encoding also allows for quality improvements since data can be stored, copied, and exchanged without loss of signal quality.

The above-mentioned approaches have the problem that the computer handles continuous-media device rather than the continuous-media data. This data does not enter the computer system after being generated; it passes through separate devices and its own communication lines. It cannot be manipulated directly by the computer; functions such as speech recognition are only possible in this approach if the signals are fed somewhere into the computer. Furthermore, the granularity of control over continuous media (*e.g.*, for synchronization purposes) is limited. The structure of such a system is similar to a process automation system in which the sensors and actors that connect the computer with the technical process impose uncertainties and delays.

Intermediate solutions operate with analog devices and some analog data streams for continuous media within the local environment (see Figure 4). The hybrid data handling remains; it imposes functional restrictions and adds complexity. Data from local analog devices is routed through a special local switch. It is treated differently from a digital media stream derived, *e.g.*, from the fixed disk and delivered to a video window.

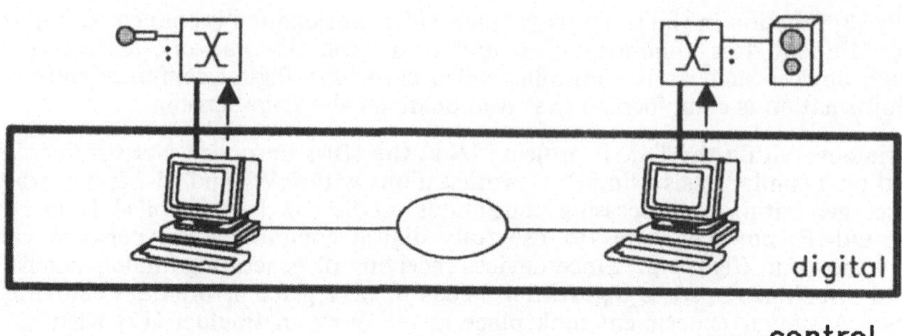

Figure 4. Local continuous-media devices attachments, digital continuous-media communication

The control of continuous media can be more direct if all data passes through the computer system itself. This is only possible with digital data encoding (where it is of secondary importance whether data is digitized in the workstation as shown in Figure 5 or in the I/O device. One of the first systems featuring digital audio in a computer environment was the Etherphone system developed at XEROX PARC [59] in which an Ethernet is used for data communication and telephony. Yet, with the exception of network communication, the Etherphone system keeps voice information and other data strictly separate. A similar approach was used in a project by AT&T in Naperville [27; 28]: A fast packet-switching network was directly attached to workstations. Enhancements to the UNIX operating system introduced the notion of "connectors" and "active devices" for handling continuous media. This was demonstrated by building a conferencing application.

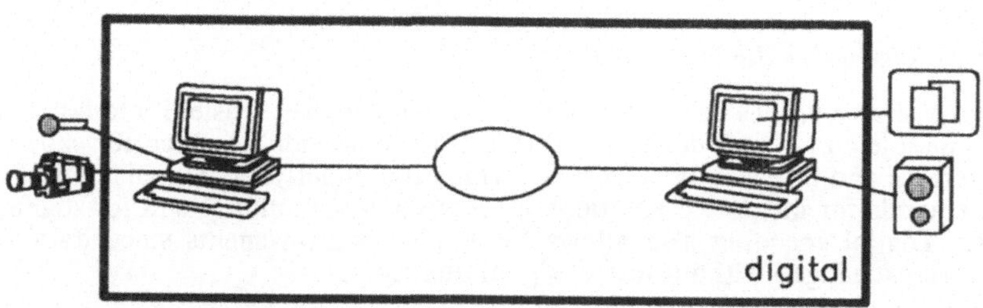

Figure 5. Unified approach

Conceptually, digital continuous-media I/O devices could be interfaced directly to the workstation bus once they have digital interfaces. At least for video, however, the sheer amount of data constitutes the major obstacle to transporting and storing it in raw digital form: A digital TV signal in traditional studio quality encoded according to CCIR 601, *e.g.*, requires 216 Mbit/s, an HDTV signal (e.g., HD-MAC) requires 216 Mbit/s * 5.33 $\simeq$ 1.152 Gbit/s [57]. Even higher data rates occur at the production of these HDTV signals.

Video data needs to be compressed before it can enter the computer system. Compression techniques, such as the one used with Intel's Digital Video Interactive (DVI) system [34], achieve a compression factor of appr. 150, albeit by tolerating some loss of video quality. In the DVI system, continuous media can be stored in computer memory and can pass through the standard system bus. The same is possible with Digidesign's audio system for the

Macintosh [31]. With algorithms such as MPEG developed at ISO/IEC JTC1/SC2 [63] or DVI low TV quality is possible at data rates around 1.4 Mbit/s.

Digital systems provide the additional advantage of increased flexibility. Today an audio mixer is typically a piece of hardware, but its functions could also be implemented in software running on a digital signal processor. The advantage of the software solution is that it can easily be adapted to changing requirements. The DVI system provides a good example: Its video display processor VDP1 (the pixel processor) is microcoded. Its code, which is part of the video compression/decompression chain, is loaded during the initialization phase, leaving the door open for future improvements of the compression algorithm. While performance requirements may not always allow software solutions, digital systems offer the possibility of continuously adjusting the border between hardware and software implementations based on the ratio of performance over processing power, cost and flexibility.

4.2 System Management Level

Available systems using DVI or the Compact Disc Interactive system as a whole (CD-I was developed mainly by Philips and published as the "Green Book" standard in a joint effort of Philips and Sony) [9] are intended for stand-alone local applications. In these systems, contention for hardware resources is ruled out by design. In multi-process/multi-user or even networked systems, where multiple concurrent applications on the same workstation are supported, contention may arise and will conflict with the performance requirements of continuous media. To avoid such problems, Olivetti's Pandora's Box [21], developed together with the University of Cambridge, keeps compressed digital video data out of the workstation. This, of course, cures only the symptoms, not the cause. To allow compressed digital video to share standard system resources is not so much a question of capacity, but a question of resource administration. It is a system management rather than a hardware problem.

4.2.1 Real-Time Scheduling

If concurrent processes handling continuous and discrete media share one machine, the operating system has to provide them with the system resources they need and to resolve resource conflicts. In traditional multitasking systems such as UNIX, "fairness" is the main criterion for resource administration. This criterion is insufficient for handling continuous media. Apart from high *throughput* requirements, continuous media impose *timing* demands on computer systems that result from the periodically changing value of continuous-media data: Each single value in an audio or video stream represents stream information for some fraction of time. Changes in the times at which values are played or recorded result in a modification of the original data semantics and must not happen unintentionally. To ensure correct timing, *delay* and *jitter* for the handling of continuous media have to be bounded if some I/O equipment (and, obviously, some human user sitting in front of it) is involved in the application [13]. Without I/O (*e.g.*, when copying a video file), the handling of continuous media is not time-critical.

To fulfill the timing requirements of continuous media, the operating system must use *real-time scheduling* techniques. These techniques have to be applied to all system resources through which continuous-media data passes, *i.e.*, to the entire *end-to-end* data path, not just the CPU. Networks and disks can contribute more to delay and jitter than processors. With DMA capabilities of controllers, continuous-media data may not even pass through the CPU. To support the function of these schedulers, the deterministic behavior of the operating system has to be ensured. Unpredictable effects of caching, process switches or page faults of a virtual memory system, *e.g.*, can ruin any carefully planned schedule.

Unfortunately, existing real-time systems are not well suited to support continuous media. Real-time scheduling is traditionally used for *command and control systems* in application areas such as factory automation or aircraft piloting. For these applications, a large variety of real-time tasks, a plethora of I/O devices to interface with the technical process to be controlled, and high fault-tolerance requirements (that somewhat counteract to real-time scheduling efforts) are typical. Continuous media have different (in fact, more favorable) real-time requirements:

- A sequence of digital continuous-media data results from periodically sampling a sound or image signal. Hence, in processing the items of such a data sequence, all time-critical operations are periodic. Schedulability considerations for periodic tasks are much easier than for sporadic ones [37].

- For many applications missing a deadline in a multimedia system is - although it should be avoided - not a severe failure. It may even be unnoticed: If an uncompressed video frame (or parts of it) are not available on time it can simply be dropped. The human viewer will hardly notice it, provided this does not happen for a contiguous sequence of frames. For audio, requirements are higher because the human ear is more sensitive to audio gaps than the human eye is to video jitter.

- The fault-tolerance requirements of continuous-media systems are usually less strict than for those real-time systems that have physical impact. The failure of a continuous-media system will not directly lead to the destruction of technical equipment or constitute a threat to human life. Of course, multimedia systems should be reliable, but not more or less than traditional data processing systems.

- The bandwidth demand of continuous media is not always *that* stringent. As some compression algorithms are capable of using different compression ratios - leading to different qualities - the required bandwidth can be negotiated. If not enough capacity for full quality is available the application may also accept a reduced quality (instead of no service at all). The quality may also be adjusted dynamically to the available bandwidth, *e.g.*, by changing encoding parameters.

In a traditional real-time system, timing requirements result from the physical characteristics of the technical process to be controlled, *i.e.*, they are given externally. Some continuous-media applications have to meet external requirements, too. Distributed music rehearsal is an example: Music played by one musician on an instrument connected to his workstation has to be made available to all other members of the orchestra within a few milliseconds, otherwise they cannot keep a common time. If human users are involved in only the input or only the output of continuous media, delay bounds are flexible. Consider the play-back of a video from a remote disk. How long it takes for a single video frame to be transferred from the disk to the monitor is unimportant to the user as long as frames arrive in a regular fashion. The user will notice any difference in delay only in the time it takes for the first video frame to be displayed. While the traditional real-time scheduling problem is to find a schedule for a set of processes with given delay bounds, often the problem in multimedia systems is to find reasonable delay bounds so that a set of processes is schedulable.

Continuous media are an addition to - not a substitute for - the discrete media already available. In future multimedia systems, time-critical continuous-media tasks and non-critical discrete-media processes will run concurrently. Such a mixed operation is a new demand on scheduling; traditional systems usually have to support only one class of processes. The operating system must fulfill two conflicting goals:

- Time-critical processes must never be subjected to *priority inversion* (*i.e.*, be kept from running by non-critical processes for an indefinite time) [51].

- Uncritical processes should not suffer from *starvation* because time-critical processes are executed.

A solution to this conflict is possible if multimedia systems have control over the time-critical workload they accept. A fraction of the overall resource bandwidth can be set aside to serve non-critical processes. A scheduling system based on these considerations is used in the DASH system [2] developed at the International Computer Science Institute and the University of California at Berkeley.

4.2.2 Quality of Service Management

An appropriate way to reconcile an application's specific needs with a system's current possibility of accommodating work items is to let both entities negotiate a *"quality of service"* immediately before this service is used [18]:

- Applications specify the workload they will impose on system resources and their performance requirements (if they have any) for the handling of this workload.
- In return, the operating system checks whether it can meet the requirements and, if so, provides performance guarantees (for throughput, loss and delay) and ensures meeting them as long as no hardware or software failure occurs and the application does not violate its workload specification.

Using this model, the operating system has control over the workload it accepts. It can refuse to service a new application if this application creates a workload that endangers the timing guarantees established for current users. Traditional real-time systems contain no mechanisms to turn down requests from the technical process to be controlled. Instead, they have to take into account exception handling procedures for alarm situations.

To guarantee a certain quality of service to an application, it is common that the system reserves a fraction of the overall system capacity exclusively for this application. Usually such a reservation is *pessimistic* [2]: It is made for the worst case, *i.e.*, for the largest workload. If the actual workload can differ significantly from the worst case (as, *e.g.*, in variable bit-rate encoding schemes) this may lead to an underutilization of resources. Service requests may also be turned down needlessly. An alternative is *optimistic* reservation [20] where requests are only turned down if they cannot even be tolerated in the best-possible case. Unlike the pessimistic scheme, optimistic reservation cannot avoid that violations of service quality occur. It, therefore, has to provide methods for *quality-of-service monitoring* (*e.g.*, detecting the violation of deadlines for the delivery of continuous-media messages) and *conflict resolution* (*e.g.*, by aborting the least important application). Both schemes have their applications: Pessimistic reservation is needed for video production and high-fidelity audio playback; optimistic methods are appropriate for voice communication, video conferencing and television services.

Quality-of-service negotiation is usually thought of as transferring a request with a fixed set of parameters from the application to the system. These parameters will then either be accepted or refused. This *imperative* approach can be extended to a more *cooperative* negotiation scheme: The application can define its preferred qualities and a range of other qualities that are also acceptable; the system can then flexibly reconcile the application's needs with its own possibilities. Such a quality-of-service negotiation can become arbitrarily complex, in particular if different parameters depend on each other. In deciding on a quality-of-service, not only the usefulness of a service quality (indicated, *e.g.* by a *service metric* as described in [18]), but also the *service cost* could be used. Again, the DASH project has addressed these issues [1].

4.3 Application Interface Level

The benefits of letting continuous media media pass through standard system resources cannot be exploited by the user if continuous media cannot be handled in the same software framework as other data types. Such a framework in today's distributed computer systems includes not only the services of the operating and communication system, but also the window system, the programming toolkit, and the data model. Well-established user interface paradigms such as I/O redirection and typical application tools such as mail should be applicable to continuous media as they are to text. When graphics was introduced to computing, not much care was spent on this aspect of integration: It is still almost impossible today to send graphics mail and rely on its correct presentation to the receiver.

In addition to new mechanisms needed at the application interface level (discussed below), existing problems in any distributed programming environment become more severe in an integrated multimedia system. These problems include heterogeneity (caused by connecting a large number of different I/O devices to the system), transparency (resulting from the need to handle multimedia objects at different locations) and protection (required to prevent unauthorized media usage).

4.3.1 Program and Data Model

Object-oriented approaches seem to be suited best to model the integration of different media. Almost all program models for multimedia systems use them (*e.g.*, [14; 29; 44; 54; 56; 58]). Often multimedia devices are represented through a class hierarchy with common operations on a generic device as the basic class. Subclasses may be input, output, in/out or storage devices. A camera and a microphone are, *e.g.*, subclasses of input devices. Other approaches try to establish a media hierarchy with operations common to the specific medium. A more application-oriented solution can focus on a representation of, *e.g.*, a book as the paradigm for documents and build subclasses for types or components of books such as video clips.

Generic features and inheritance mechanisms in object-oriented models make it feasible to apply standard system functions and application tools to a variety of data types. Yet, they take into account that different operations are applicable to each medium. In particular, they can handle the various presentation functions by a unified form. For discrete media, presentation involves static data display. For continuous media, it requires a dynamic reproduction of the data sequence. In addition, the common graphical interface of object-oriented systems provides a uniform "look-and-feel" that makes these systems easy to use.

Continuous media – by definition – elude the common *event-feedback loop* of user input and system output. While a video is displayed the user needs to be able to issue commands to switch between channels or to change the volume. This is reflected best by a *multi-threaded* application structure where sporadic events and each periodically recurring operation are represented by their own threads. Hence, not only are multimedia applications used in a concurrent environment, they are inherently concurrent themselves.

Inter-thread communication will look different for discrete and continuous media. Whereas for discrete media single data values are transferred, for continuous media the application should just need to define sources and sinks once and then continuously transfer data. In addition to "read" and "write", "connect/start" and "disconnect/stop" operations are required [27].

In regard to the data model, enhancements and/or adaptions of existing models such as SGML [24] or ODA [23] are desirable. SGML provides the basis for multimedia specific enhancements like SDML and HyTime, or may even be used as a framework for other information architectures [26]. In contrast to SGML, ODA has predefined semantics to specify the structure and the layout of documents which must be enhanced towards multimedia capabilities [19].

4.3.2 Synchronization

The temporal aspects of continuous-media data contribute to its semantics and are, therefore, not only important at the system management level, but also at the application interface. They result in a need for *synchronization* of threads that present continuous-media and discrete-media data (to users or user processes). The following reasons for synchronization can be distinguished:

- *Between different continuous-media streams:* If several continuous-media streams are semantically connected, their values have to be presented together. A movie and its soundtrack, *e.g.*, have to be displayed in a way that synchronizes the spoken voice with the movement of the speaker's lips. Another example is the synchronization of two stereo audio channels.

- *Between continuous-media streams and discrete-media data items:* If discrete-media data items are incorporated in continuous-media streams (as subtitles are in a movie) their processing or output has to occur when predefined events or time stamps of the continuous-media stream are reached. If embedding works the other way round (as in voice-over text) one can either start its presentation automatically together with the discrete-media data display or let the user initiate the presentation explicitly.

Whether the user can influence the synchronization that shall take place depends on the application. Consider the following two examples:

1. A camera and a microphone are attached to a workstation on which a joint editing application runs as part of a video conference. While speaking, the owner of a shared window points to various graphical objects. At the remote workstation the audio and video data must be presented simultaneously and synchronized with the pointer.

2. Surrogate travel applications such as the Aspen Movie Map of the MIT or a tour through the Palenque ruins [62] allow users to move electronically through a new building: The user defines direction and actions to be performed like "turn right and open door" by pressing appropriate buttons on the graphical user interface. Then the sound of the opening door and a video showing the walk through the room are presented. Additional information may appear as overlay text from time to time.

The first example involves *live synchronization*: Synchronous input of data shall result in synchronous output. Live synchronization should happen automatically; the user is not involved. In the second example, data does not evolve on the fly, but is available beforehand. In this case, which is called *synthetic synchronization* [30], users can freely order the various data entities in the time domain. For this purpose, they need to be able to express their synchronization requirements through corresponding language constructs. Notions like "present data entity A simultaneously/after/independently from data entity B" are needed [46]. These constructs apply either to the whole information entity or may refer to time or event stamps within the entity [56]. Within an integrated multimedia system both types of synchronization are required.

5. Final Remark

In the previous sections, we have identified the main criteria for the development of integrated multimedia systems. Looking at today's products, we find that they are still far away from having accomplished the goal of integration. If we, *e.g.*, look at CD-I, one of the most advanced technologies for multimedia retrieval, we notice that it is only provided within closed systems intended for this single application. If we look at the NeXT station which has pioneered audio in the workstation environment, we find that parallel audio output and mouse movements interfere with each other, causing annoying sound glitches. Yet, such systems are essential to understand the requirements of multimedia applications and to experiment with different solutions. Today's systems are beneficial as long as we keep in mind that they do not yet represent the future of information processing and exchange – they are, however, leading the way.

Acknowledgement

We would like to thank our colleagues within the DASH and DiME projects, in particular David Anderson of the University of California at Berkeley and Johannes Rückert, Bernd Schöner and Hermann Schmutz of the IBM European Networking Center Heidelberg, who have contributed indirectly to this paper by many - often very controversial – discussions.

References

[1] *David P. Anderson, Ralf Guido Herrtwich, Carl Schaefer;* **SRP: A Resource Reservation Protocol for Guaranteed-Performance Communication in the Internet;** International Computer Science Institute, Berkeley, Technical Report 90-006, Feb. 1990.

[2] *David P. Anderson, Ralf G. Herrtwich;* **Resource Management for Digital Audio and Video;** IEEE Workshop on Real-Time Operating Systems and Software, Charlottesville, May 1990, pp. 99-103.

[3] *D.P. Anderson, S. Tzou, R. Wahbe, R. Govindan, M. Andrews;* **Support for Continuous Media in the DASH System;** Proc. of the 10th International Conference on Distributed Computer Systems, Paris, May 1990.

[4] **IBM Audio Visual Connection, Product Documentation;** IBM 1990.

[5] **International Workshop on Network and Operating System Support for Digital Audio and Video;** International Computer Science Institute (ICSI), Berkeley, CA, Nov. 8-9, 1990.

[6] *William R. Byrne, Toomas A. Kilm, Bruce L. Nelson, Marius D. Soneru;* **Broadband ISDN Technology and Architecture;** IEEE Network Magazine, vol.3, no.1, Jan. 1989, pp. 23-28.

[7] *Steward Brand;* **The Media Lab, Inventing the Future at MIT;** Viking Penguin, 1987.

[8] *The International Telegraph and Telephone Consultative Committee.* **I-Series Recommendations;** VIIIth Plenary Assembly, Melbourne 1988.

[9] *Philips International B.V.;* **Compact Disc-Interactive - A Designers Overview;** Kluwen, Technische Boeken B.V., Deventer, Netherlands, 1989.

[10] *Georg Champine, Daniel Geer, William Ruh;* **Project Athena as a Distributed Computer System;** IEEE Computer, vol.23 no.9, September 1990, pp.40-51.

[11] *J. Conklin.* **Hypertext: A Survey and Introduction.** IEEE Computer, Sep. 1987, pp. 17-41.

[12] *Doug Corey, Jill Schmidt, Mark Abel, Stephen Bulick, Steve Coffin;* **Multimedia Communications: The US West Advanced Technologies Prototype Telecollaboration System;** 5th IEEE International Workshop on Telematics, Denver, Colorado, USA, Sept. 17-21, 1989.

[13] *Domenico Ferrari;* **Client Requirements for Real-Time Communication Services;** International Computer Science Institute, Technical Report 90-007, Berkeley, March 1990.

[14] *E. Fiume, D. Tsichritzis;* **Multimedia Objects;** in: Active Object Environments, D. Tsichritzis (Ed.), University of Geneva, June 1988, 121-128.

[15] *Karen A. Frenkel;* **The Next Generation of Interactive Technologies;** Communications of the ACM, vol.32, no.7, Jul. 1989, pp. 872-881.

[16] *Irene Greif;* **Computer-Supported Cooperative Work: A Book Of Readings;** Morgan Kaufmann Publishers, 1988.

[17] *Rainer Händel;* **Evolution of ISDN Toward Broadband ISDN;** IEEE Network Magazine, vol. 3, no.1, Jan. 1989, pp. 7-13.

[18] *Ralf Guido Herrtwich, Uwe Wolfgang Brandenburg;* **Accessing and Customizing Services in Distributed Systems;** International Computer Science Institute, Technical Report 89-059, Berkeley, Oct. 1989.

[19] *Ralf Guido Herrtwich, Luca Delgrossi* **ODA-Based Data Modeling in Multimedia Systems** International Computer Science Institute, Technical Report 90-043, Berkeley, 1990.

[20] *Ralf Guido Herrtwich;* **The DASH Resource Model Revisited or Pessimism Considered Harmful;** International Workshop on Network and Operating System Support for Digital Audio and Video; International Computer Science Institute, Berkeley, Nov. 8-9, 1990.

[21] *Andy Hopper;* **Pandora - An Experimental System for Multimedia Applications;** ACM, Operating Systems Review, Apr. 1990, pp. 19-34.

[22] *Matthew E. Hodges, Russel M. Susnett, Mark S. Ackerman;* **A Construction Set for Multimedia Applications;** IEEE Software Magazine, Jan. 1989, pp. 37-43.

[23] *ISO;* **Information Processing - Office Document Architecture and Interchange Format;** ISO, Genf 1989.

[24] *ISO;* **Information Processing - Standard Generalized Markup Language;** ISO, Genf 1986.

[25] **5th IEEE International Workshop on Telematics;** Denver, Colorado, USA, September 17-21, 1989.

[26] *Jürgen Kämper;* **Towards Extensible Multimedia Document Architectures;** IEEE Multimedia '90, Bordeaux, Nov. 15-17., 1990.

[27] *Wu-Hon F. Leung, Gottfried W. R. Luderer;* **The Network Operating System Concept for Future Services;** AT & T Technical Journal, vol.68, no.2, Apr. 1989, pp. 23-35.

[28] *W.H. Leung, T.J. Baumgartner, Y.H. Hwang, M.J. Morgan, S. C. Tu.* **A Software Architecture for Workstation Supporting Multimedia Conferencing in Packet Switching Networks.** IEEE Journal on Selected Areas in Communication, vol.8, no.3, Apr. 1990, pp. 380-390.

[29] *T.D.C. Little, A. Ghafoor;* **Synchronization and Storage Models for Multimedia Objects;** IEEE Journal on Selected Areas in Communication, vol.8, no.3, Apr. 1990, pp. 413-427.

[30] *Thomas D.C. Little, Arif Ghafoor;* **Network Considerations for Distributed Multimedia Objects Composition and Communication;** IEEE Network Magazine, vol.4 no.6, Nov. 1990, pp. 32-49.

[31] *W. Lowe, R. Currie* **Digidesign's Sound Accelarator: Leasons Lived and Learned** Computer Music Journal 13, 1, 36-46.

[32] *L.F.Ludwig, D.F.Dunn;* **Laboratory for Emulation and Study of Integrated and Coordinated Media Communication;** Frontiers in Computer Technology, Proc. of the ACM SIGCOMM '87 Workshop, Aug. 11-13,1987.

[33] *V.Y. Lum, K. Meyer-Wegener;* **An Architecture for a Multimedia Database Management System Supporting Content Search;** in: Proc. Int. Conf. on Computing and Information, Niagara Falls, Ontario, Canada, May 23-26, 1990.

[34] *Arch C. Luther.* **Digital Video in the PC Environment.** Intertext Publications McGraw-Hill Book Company New York.

[35] *Wendy E. Mackay, Glorianna Davenport;* **Virtual Video Editing in Interactive Multimedia Applications;** Communications of the ACM, vol.32, no.7, Jul. 1989, pp. 802-810.

[36] **M-Motion Video Adapter/A, User's Guide, Product Description;** IBM 1990.

[37] *Al K. Mok;* **The Design of Real-Time Programming Systems Based on Process Models;** IEEE Real-Time Systems Symposium, Austin, Dec. 1984, pp. 5-17.

[38] *W.E. Mackay, W. Treese, D. Applebaum, B. Gardner, B. Michon, E. Schlusselberg, M. Ackermann, D. Davis* **Pygmalion: An Experiment in Multimedia Communication** Proceedings of SIGGRAPH 89, Boston, July 1989.

[39] *M. Mühlhäuser;* **Requirements and Concepts for Networked Multimedia Courseware Engineering;** International Conference on Computer Aided Learning 89, 1989, Dallas, USA.

[40] **2nd IEEE COMSOC International Multimedia Communications Workshop;** Montebello, Quebec, Canada, Apr. 1989.

[41] **3rd IEEE COMSOC International Multimedia Communications Workshop;** Bordeaux, France, Nov. 15-17, 1989.

[42] *Jakob Nielsen;* **Hypertext and Hypermedia;** Academic Press, 1990.

[43] *F.Oguet, C.Schwarts, F.Kretz, M.Quere;* **RAVI, A Proposed Standard for the Exchange of Audio/Visual Interactive Applications;** IEEE Journal on Selected Areas in Communication, vol.8, no.3, April 1990, pp.428-436.

[44] *A.J. Palay* **The Andrew Toolkit: An Overview** Proceedings of the 1988 Winter USENIX Conference, Dallas, Feb. 1988, 9-21.

[45] *Parallax Graphics;* **The Parallax 1280 Series Videographic Processor;** 1987.
[46] *Jonathan B. Postel, Gregory G.Finn, Alan R. Katz, Joyce K.Reynolds.* **An Experimental Multimedia Mail System.** ACM Transactions on Ofiice Information Systems, vol.6, no.1, Jan. 1988, pp. 63-81.
[47] *Stephen Pollock;* **A Rule-Based Message Filtering Sysytem;** ACM Transactions on Office Information Systems, vol. 6, no. 3, Jul. 1988, pp. 232-254.
[48] *Larry Press;* **Computer or Teleputer?;** Communications of the ACM, vol. 33, no. 9, September 1990, pp. 29-36.
[49] **Real-Time Systems Symposium;** Orlando, December 5-7, 1990.
[50] *Phillip Robinson;* **The Four Multimedia Gospels;** Byte, vol.15, no. 2, Feb. 1990, pp. 203-212.
[51] *R. Rajkumar, L. Sha, J. P. Lehoczky;* **Real-Time Synchronization Protocols for Multiprocessors;** IEEE Real-Time Systems Symposium, Huntsville, December 6-8, 1988, pp. 259-269.
[52] *Johannes Rückert, Hermann Schmutz, Bernd Schöner, Ralf Steinmetz;* **A Distributed Multimedia Environment for Advanced CSCW Applications;** IEEE Multimedia '90, Bordeaux, Nov. 15-17, 1990.
[53] *Sunil Sarin, Irene Greif;* **Computer-Based Real-Time Conferencing Systems;** IEEE Computer, vol. 18, no. 10, Oct. 1985, pp. 33-45.
[54] *Ralf Steinmetz, Reinhard Heite, Johannes Rückert, Bernd Schöner;* **Compound Multimedia Objects - Integration into Network and Operating Systems;** International Workshop on Network and Operating System Support for Digital Audio and Video, International Computer Science Institute, Berkeley, Nov. 8-9, 1990
[55] *Ralf Steinmetz, Johannes Rückert, Wilfried Racke;* **Multimedia-Systeme;** Informatik Spektrum, Springer Verlag, vol.13, no.5, 1990, pp.280-282.
[56] *Ralf Steinmetz;* **Synchronization Properties in Multimedia Systems;** IEEE Journal on Selected Areas in Communication, vol. 8, no. 3, April 1990, pp. 401-412.
[57] *L. Stenger;* **Digital Coding of Television Signals-CCIR Activities for Standardization;** Signal Processing Image Comm. vol. 1 nr. 1, Jun. 1989, pp. 29-43.
[58] *J.S.Sventek;* **An Architecture for Supporting Multimedia Integration;** IEEE Computer Society Office Automatition Symposium, Apr. 1987, pp.46-56.
[59] *Daniel C. Swinehart;* **Telephone Management in the Etherphone System;** IEEE Globecom'87, 1987, pp. 30.3.1-30.3.5.
[60] *Robert H. Thomas, Harry C. Forsdick, Terrence R. Crowley, Richard W. Schaaf, Raymond S. Tomlinson, Virginia M. Travers;* **Diamond: A Multimedia Message System Built on a Distributed Architecture;** IEEE Computer, Dec. 1985, pp. 65-78.
[61] *Hitoshi Watanabe;* **Integrated Office Systems: 1995 and Beyond;** IEEE Communication Magazine, vol. 25, no. 12, Dec. 1987, pp. 74-80.
[62] *K. S. Wilson;* **Palenque: An Interactive Multimedia Optical Disk Prototype for Children;** Bank Street College of Education, NY, Technical Report 2, 1987.
[63] *Hiroshi Yasuda;* **Standardization Activities on Multimedia Coding in ISO;** Signal Processing: Image Communication, vol. 1, no. 1, 1989, pp.3-16.

X-MOVIE:
Digitale Filmübertragung und Darstellung im X-Window-System

Bernd Lamparter Wolfgang Effelsberg
Lehrstuhl für Praktische Informatik IV
Universität Mannheim

Kurzfassung

Diese Arbeit beschreibt ein System zur Speicherung, Übertragung und Darstellung von Bewegtbildsequenzen in einem digitalen Rechnernetz. Dabei kommen ausschließlich Hardwarekomponenten zum Einsatz, die in typischen Netzen aus Arbeitsplatzrechnern und Servern vorhanden sind; spezielle Hardware ist nicht erforderlich. Die Darstellung der Filme erfolgt in Fenstern des X-Window-Systems, um eine weitgehende Integration der Bewegtbildsequenzen mit den klassischen Bestandteilen einer EDV-Anwendung, wie zum Beispiel Text, Farbgraphik, Menüs und Symbole (Icons), zu erreichen. Der Schwerpunkt der Arbeit liegt bei den Darstellungsformaten und Übertragungsprotokollen für digitale Bewegtbildsequenzen.

1 Einleitung

Moderne Arbeitsplatzrechner sind heute mit schnellen Prozessoren, hochauflösenden Farbgraphikschirmen und leistungsfähigen Netzanschlüssen ausgestattet. Dabei ist für alle drei Komponenten mit einer weiteren erheblichen Leistungssteigerung in den nächsten Jahren zu rechnen (RISC-Prozessoren, VLSI-Technik, Hochgeschwindigkeitsnetze).

In einem solchen Szenario werden neuartige Anwendungen möglich, die den Rahmen der klassischen EDV sprengen. Eine solche Anwendung ist die Integration von Bewegtbilddarstellungen in die fensterorientierte Oberfläche eines Arbeitsplatzrechners.

Bereits seit einigen Jahren werden Computer zur Bewegtbildverarbeitung eingesetzt. Die wichtigsten Forschungs- und Entwicklungsrichtungen lassen sich in die drei Kategorien Computeranimation [LMS91], Computersimulation und Digitales Video [LD87, MD89] einteilen.

Diese Arbeit konzentriert sich auf die Übertragung und Darstellung von digitalen Bildsequenzen in einem verteilten System aus klassischen Workstations und Servern. Die Anforderungen an Übertragung und Darstellung werden am Beispiel des Systems X-MOVIE erläutert, das an der Universität Mannheim entwickelt wird. Ziel ist dabei die Ausgabe der digitalen Bewegtbildsequenzen in einem Fenster des X-Window-Systems [Jon88, SGN88] unter Unix. Das gesamte System kommt ohne spezielle Hardware aus. Im Gegensatz beispielsweise zu dem kommerziell verfügbaren DVI-System [Lut89] von Intel, das spezielle Hardware zur Darstellung und vor allem zur Kompression [Tin89] der Video- und Audiodaten benötigt.

In Kapitel 2 wird die Architektur des experimentellen X-MOVIE-Systems beschrieben. Kapitel 3 geht auf die Darstellungsformate digitaler Bilder und Bildsequenzen ein. Als ein weiterer Schwerpunkt werden dann in Kapitel 4 die Anforderungen der Bewegtbildübertragung an das Netz erläutert. Die Integration des Filmfensters in das X-Window-System wird in Kapitel 5 beschrieben. Über erste Implementierungserfahrungen berichtet Kapitel 6. Eine kurze Zusammenfassung beschließt die Arbeit.

2 Architektur des X-MOVIE-Systems

Das System X-MOVIE ist ein verteiltes System aus Unix-Workstations zur Erprobung integrierter digitaler Bewegtbildübertragung. Die Workstations sind in der ersten Phase über Ethernet verbunden; eine Umstellung auf einen FDDI-Ring ist in Vorbereitung. Die Gesamtarchitektur des Systems zeigt Abb. 1.

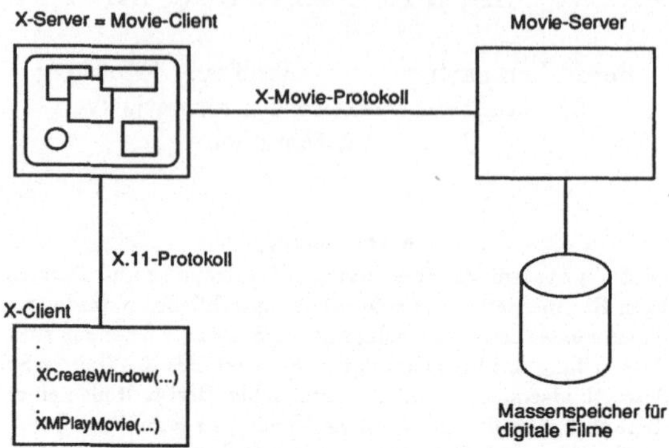

Abb. 1: Architektur des Systems X-MOVIE

Die wichtigsten Komponenten von X-MOVIE sind der Movie-Server und der Movie-Client. Wie auch im X-Window-System üblich, können Server und Client auf demselben oder auf verschiedenen Rechnern ablaufen.

Die Aufgabe des Movie-Servers ist die Speicherung und Wiedergabe der Bewegtbildsequenzen (digitale Filme). Er verwaltet ein Filmverzeichnis. Auf Abruf wird die Bildfolge eines Films über das Netz an den Movie-Client gesendet. Der Movie-Client ist eine Erweiterung des X-Servers. Da der Schwerpunkt der hier beschriebenen Arbeiten auf der Übertragung und Darstellung der digitalen Filme liegt, wird hier auf eine Diskussion der Erzeugung solcher Filme nicht näher eingegangen. Bisher wurden für das X-MOVIE-System nur Filme eingesetzt, die offline generiert und offline auf einer Standard-Workstation in das X-MOVIE-Format transformiert wurden. Die Originalfilme wurden an der Universität Karlsruhe als Computeranimation erstellt und uns von dort zur dankenswerterweise Verfügung gestellt.

3 Formate digitaler Bilder und Filme

3.1 Formate digitaler Bilder

Ein mit hoher Qualität digitalisiertes Bild benötigt pro Pixel 24 Bits Speicherplatz (je ein Byte pro Farbkomponente). Damit können ca. 16 Millionen unterschiedliche Farbtöne dargestellt werden, bei jeder einzelnen Farbe können 256 Abstufungen unterschieden werden. Diese Farbtiefe verhindert die Entstehung von Streifen in Bildern mit langsamen Farbübergängen. Das menschliche Auge kann kleinere Farbunterschiede praktisch nicht mehr wahrnehmen (siehe [Lut89]). Diese Darstellung benötigt allerdings sehr viel Speicher (720 kB für ein 600×400 Pixel großes Bild). Eine Kompression ist daher unumgänglich. Falls die Bilder von Graphikprogrammen stammen, z.B. CAD-Paketen, wird der pixelweisen Darstellung eine Speicherung der graphischen Primitive (Linien, Kreise, ...) vorgezogen.

Leider können heute übliche Workstations meist nur 256 Farben gleichzeitig darstellen. Deshalb braucht nur ein Byte pro Pixel abgespeichert bzw. übertragen zu werden, ohne daß bei der Darstellung ein Verlust an Bildqualität auftritt. Dieses Byte ist dann nur ein Index in eine Farbtafel, in der die eigentliche Farbe steht (siehe Abb. 2).

Die Farben selbst können je nach Graphikadapter wieder aus der oben erwähnten Menge von 16 Millionen Farben ausgewählt werden. Damit können zwar nur 256 Farben gleichzeitig auf einem Bildschirm dargestellt werden, aber die Auswahl der Farben erfolgt aus einer sehr großen Menge. Die Farbtafel muß zu jedem Bild gespeichert werden. Damit verringert sich die Datenmenge auf fast ein Drittel, ohne zusätzliche Einschränkungen an die Qualität der Einzelbilder.

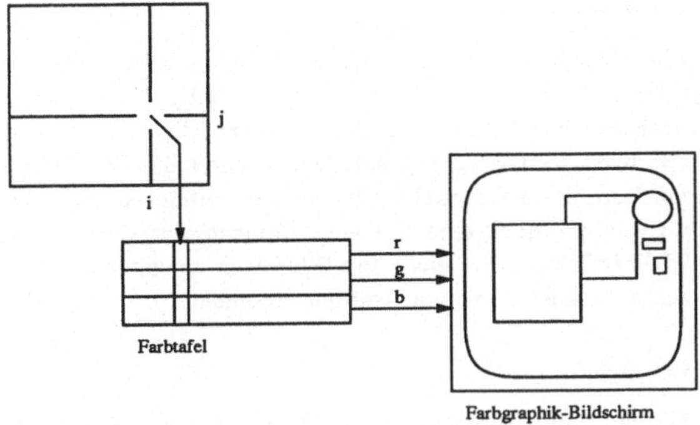

Abb. 2: Graphikadapter mit Farbtafel und Bildschirm

3.2 Kompressionsverfahren für Bilder

Wegen der immens großen Bilddateien müssen Bilder komprimiert werden. Dabei werden verlustfreie und verlustbehaftete Verfahren unterschieden. Bei der verlustfreien Kompression werden die Daten so komprimiert, daß Bilder korrekt dekomprimiert werden können; sie sind nach der Dekompression mit dem Original identisch. Dagegen werden bei der verlustbehafteten Kompression kleinere Abweichungen hingenommen.

Eine zweite Unterscheidung verschiedener Kompressionsalgorithmen ist das Zeitverhalten. Symmetrische Verfahren benötigen ungefähr die gleiche Zeit zur Kompression wie zur Dekompression. Bei asymmetrischen Verfahren hingegen ist der Kompressionsaufwand wesentlich höher als der zur Dekompression. Man erhofft sich dabei eine Verlagerung des Gesamtaufwandes zur Kompression hin, um die Dekompression in Echtzeit ausführen zu können. Im übrigen kann in vielen Anwendungen die komprimierte Version gespeichert und vielfach abgerufen werden, so daß durch das asymmetrische Verfahren der Gesamtaufwand minimiert wird.

Die Bilddarstellung mit einer Farbtabelle ist bereits eine einfache, verlustbehaftete Kompression. Eine weitere Kompression erfordert dagegen auch einen höheren Aufwand an Rechenleistung zur Dekompression auf der Workstation. Dabei sind von der Hardwareseite her folgende Varianten denkbar:

- Der Hauptprozessor wird mit dieser Aufgabe belastet.
- Die Workstation wird mit Spezialhardware zur Dekompression ausgerüstet oder solche Hardware wird direkt in die Graphikkarte eingebaut.
- Die Workstation wird mit einem oder mehreren zusätzlichen Prozessoren ausgerüstet, z.B. mit Transputern.

Von der Softwareseite her ist vor allem das CCC-Verfahren (Color-Cell-Coding, [CDF+86]) interessant. Eine Erweiterung dieses Verfahrens findet sich in [Pin90]. Die Kompression nach dem CCC-Verfahren erreicht 2 Bit/Pixel. Kompression und Dekompression lassen sich sehr gut parallelisieren, da die Kompression blockweise erfolgt. Das Verfahren ist verlustbehaftet und symmetrisch.

Zur Zeit werden außerdem von der ISO Verfahren zur Kompression für Einzelbilder und Bewegtbilder genormt [Gal91]: JPEG (Joint Picture Expert Group) und MPEG (Motion Photographic Expert Group). Beide Verfahren sind verlustbehaftet und symmetrisch. Durch die vorangestellte Cosinustransformation werden sehr gute Kompressionraten erzielt, allerdings ist ein hoher Aufwand an Rechenzeit notwendig. Aus diesem Grund sind für JPEG bereits spezielle Chips erhältlich und für MPEG angekündigt.

3.3 Formate digitaler Filme

Genauso wie bei Einzelbildern können digitale Filmdaten mit 24 Bits pro Pixel oder als Indizes auf eine Farbtabelle gespeichert werden. Pro Sekunde müssen mindestens 25 Bilder gezeigt werden, damit der Eindruck einer kontinuierlichen Bewegung entsteht. In der amerikanischen Fernsehnorm NTSC werden 30 Bilder pro Sekunde verwendet.

3.3.1 Kompression bei Filmen

Schon bei großen Einzelbildern kann die Datenmenge ein Problem sein, umso mehr bei Filmen. Da sich während des Ablaufs eines Films zwischen zwei Bildern meist wenig ändert, bietet es sich an, zweidimensionale Kompressionsverfahren zu dreidimensionalen zu erweitern, die dritte Dimension entspricht dann der Zeitachse. Ein häufig angewandtes Verfahren besteht darin, zuerst die Differenz zweier aufeinanderfolgender Bilder zu berechnen und dann diese Differenzbilder zu komprimieren. Dies ermöglicht, solange sich im Filmverlauf von Bild zu Bild wenig ändert, eine effiziente Kompression ohne Verlust.

Das oben erwähnte CCC-Verfahren kann leicht für Bildfolgen erweitert werden. Die Rechtecke werden dann zu Quadern die weiterhin parallel berechnet werden können.

3.3.2 Filme und Farbtabelle

Bei der Übertragung von Filmen, deren Einzelbilder mit einer Farbtabelle kodiert sind, zeigt sich folgendes Problem: Werden zwei aufeinanderfolgende Bilder unabhängig voneinander kodiert, so muß vor dem Laden des zweiten Bildes die zugehörige Farbtabelle geladen werden. Dadurch ist kurzzeitig das erste Bild in Falschfarben sichtbar. Dieser Effekt ist selbst dann unangenehm, wenn die beiden Bilder ähnliche Farben besitzen, da die jeweils ähnlichen Farben nicht den gleichen Index besitzen müssen. Dieses Problem kann z.B. behoben werden, wenn die Bildfolge im gesamten auf der Basis derselben Farbtabelle kodiert wird, genauer gesagt, wenn die Farbtafel für die gesamte Bildfolge gemeinsam festgelegt wird. Der Nachteil dieser Technik liegt nun darin, daß die größeren Farbunterschiede einer längeren Bildfolge nicht berücksichtigt werden.

Eine Möglichkeit, beide Ziele zu vereinen, d.h. die Farbtafel nicht nach jedem Bild auszutauschen und einer Bildsequenz trotzdem mehr als 256 Farben zur Verfügung zu stellen, ist folgende: Für jedes einzelne Bild werden nicht alle möglichen 256 Farbindizes ausgenützt, sondern z.B. 32 Indizes freigelassen. Vor dem Laden des nächsten Bildes können dann die Farben für diese 32 Indizes in die Farbtabelle der Workstation eingetragen werden, ohne das momentane Bild zu verfälschen. Das zweite Bild benützt diese Indizes und läßt seinerseits andere Plätze für das nächste Bild frei (siehe Bild 3, CLT_i steht für die Farbtafel (Color-Lookup-Table) des i-ten Bildes).

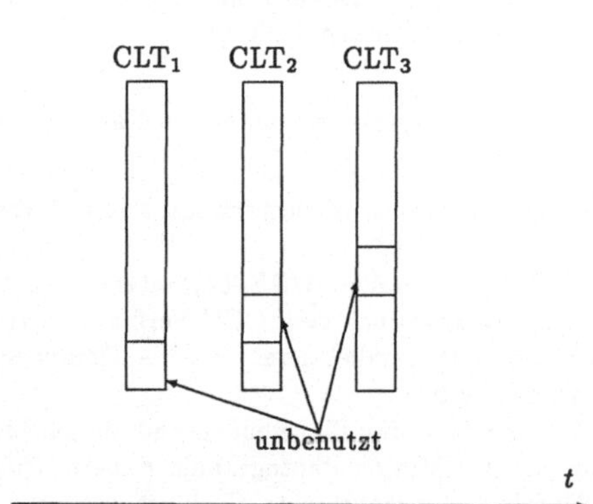

Abb. 3: Freilassen einiger Farbtabelleneinträge

$Bild_1$	$Bild_2$	CLT_1	CLT_2
0 0	0 0	c_{11}	c_{21}
0 0	0 1	c_{11}	c_{22}
0 0	1 0	c_{11}	c_{23}
0 0	1 1	c_{11}	c_{24}
0 1	0 0	c_{12}	c_{21}
0 1	0 1	c_{12}	c_{22}
0 1	1 0	c_{12}	c_{23}
0 1	1 1	c_{12}	c_{24}
1 0	0 0	c_{13}	c_{21}
1 0	0 1	c_{13}	c_{22}
1 0	1 0	c_{13}	c_{23}
1 0	1 1	c_{13}	c_{24}
1 1	0 0	c_{14}	c_{21}
1 1	0 1	c_{14}	c_{22}
1 1	1 0	c_{14}	c_{23}
1 1	1 1	c_{14}	c_{24}

Abb. 4: Mehrere Bilder zugleich im Framebuffer

Die Anzahl freigelassener Indizes kann von Bild zu Bild variieren, so daß schon nach wenigen Bildern die Farbtafel vollständig erneuert sein kann. Hier arbeiten wir noch an Algorithmen, die ein Optimum zwischen schnellem Wechseln der Farbtabelle und möglichst wenig unbenützten Indizes finden. Liegt die

Bildfolge beim Kodieren in Originalfarben (24 Bit RGB) vor, so kann die Anzahl freigelassener Indizes an Hand des Farbunterschieds zum nächsten Bild entschieden werden. Je größer der Farbunterschied, desto mehr Indizes sollten frei bleiben. Wenn nun aber Bilder einer Videokamera direkt übertragen werden sollen, dann bedeutet ein solcher Blick in die Zukunft eine Übertragungsverzögerung um ein oder mehrere Bilder.

Eine andere, ähnliche Möglichkeit zeigt Abb. 4 für eine exemplarisch vereinfachte Graphikkarte mit vier Bit Farbtiefe und zwei Bildern mit je zwei Bit Farbtiefe, also vier Farben. Der Inhalt des Bildspeichers ist in den ersten zwei Spalten dargestellt, die dritte Spalte ist die Farbtafel zum Darstellen des ersten Bildes, die vierte Spalte die Farbtafel zum Darstellen des zweiten Bildes (c_{ij} bezeichnet den j-ten Eintrag der zum Bild i gehörenden Farbtafel). Durch geschickte Wahl der Farbtabellen kann abwechselnd das eine oder das andere Bild gezeigt werden. Währenddessen kann das jeweils unsichtbare Bild ausgetauscht werden, ohne das aktuell dargestellte Bild zu stören (Wechselpuffer). Dies funktioniert natürlich auch für Graphikadapter mit acht Bits Farbtiefe und bis zu acht Bildern mit dann nur noch je zwei Farben.

4 Übertragungsprotokolle für Bilder und Filme

Als bevorzugte Netzprotokolle auf UNIX-Workstations haben sich die Internet-Protokolle weitgehend durchgesetzt; sie werden häufig auch als TCP/IP-Protokolle bezeichnet [Com88].

IP ist das Protokoll der Netzwerkschicht (Schicht 3); es bietet einen verbindungslosen Pakettransport zwischen zwei Rechnern. Dabei kann es sowohl zu Paketverlusten als auch zu Sequenzfehlern kommen. Bitfehler werden dagegen erkannt, und das entsprechende Paket wird verworfen.

TCP bietet eine zuverlässige, verbindungsorientierte Transportschicht. Es benutzt dabei IP zum Versenden einzelner Pakete und behebt Fehler durch Wiederholung verlorener Pakete nach einem Time-Out. Außerdem wird mittels eines Sliding-Window-Protokolls eine Flußkontrolle durchgeführt.

Ein drittes Protokoll ist das User-Datagram-Protocol (UDP). Es ist im wesentlichen eine Benutzerschnittstelle für IP. Die meisten Applikationen im Internet benützen TCP als Transportschicht, einige wenige UDP (siehe Abb. 5).

<table>
<tr><td colspan="2">telnet, ftp, mail, rlogin, X-Windows, ...</td><td colspan="2">nfs, X-MOVIE</td></tr>
<tr><td colspan="2">TCP</td><td colspan="2">UDP</td></tr>
<tr><td colspan="4">IP</td></tr>
<tr><td>Ethernet</td><td colspan="2">Tokenring</td><td>...</td></tr>
</table>

Abb. 5: Protokollhierarchie des Internet (Auszug)

4.1 Anforderungen an Protokolle zur Filmübertragung

Eine Bildsequenz enthält außer den einzelnen Bildern einen Header und eine Farbtabelle, sowie Farbtabellenänderungen. Auf dieser Datenstruktur operiert der Bildserver. Er generiert Antworten auf Anfragen eines Clients. Dabei bietet die Anfrageschnittstelle folgende Funktionalität:

- Anzeigen eines Inhaltsverzeichnisses
- Informationen über einen Film
- „Play"
- „Stop"
- „Step Forward/ Step Backward"
- „Show picture $< n >$"
- „Slower/Faster"

Vor allem die Filmübertragung ist dabei auf einer Transportschicht wie TCP nur mit Abstrichen realisierbar. Dies liegt vor allem an der mangelnden Effizienz aufgrund des Aufwandes für Flußkontrolle und

Fehlererkennung und -behebung. Das folgende Kapitel beschreibt kurz, welche Anforderungen zukünftige Transportschichten erfüllen sollten, bzw. welche Dienste angeboten werden sollten (siehe auch [HS91]).

4.2 Isochroner Datenfluß

Die Übertragung von Filmen stellt eine neue Anforderung an Übertragungsprotokolle: die isochrone Übertragung eines Datenstromes. Dazu ist nicht nur eine Obergrenze für die Verzögerung einzelner Datenpakete und eine Mindestübertragungsrate erforderlich, sondern auch eine konstante Verzögerung.

Ein isochrones Übertragungsprotokoll muß auf allen Schichten von der Anwendung bis zur physikalischen Schicht sowohl eine garantierte Bandbreite als auch eine konstante Verzögerung anbieten.

Einer Implementierung solcher Protokolle stehen vor allem zwei Hindernisse entgegen: Die Belastung eines Rechners mit anderen Prozessen und die Protokolle heutiger Netze.

Heutige Workstations werden üblicherweise nicht von mehreren Personen gleichzeitig benutzt; deswegen stellt die leistungsfähige Prozessor-Hardware nach unseren Erfahrungen auf der Seite des X-Servers keinen Engpaß mehr dar. Wichtig ist allerdings, daß beim Scheduling der Prozesse auf der Workstation auf Echtzeitanforderungen mit konstanter Verzögerung Rücksicht genommen wird [ADH91].

Bei den Protokollen ist die Sache wesentlich schwieriger. Dies beginnt bereits beim Netzzugangsprotokoll: Auf einem CSMA/CD-basierten Netz kann weder Verzögerung noch Datenrate einer Verbindung garantiert werden. Ein anderes in lokalen Netzen verbreitetes Zugangsprotokoll, der Tokenring, garantiert immerhin eine maximale Verzögerung. Dies gilt ebenfalls für das token-basierte Hochgeschwindigkeitsnetz FDDI. Mit diesen Protokollen kann in bestimmten Grenzen ein isochroner Dienst durch Puffern verfrühter Datenpakete simuliert werden. Aber auch die heutigen Transportprotokolle sind für isochrone Datenübertragung ungeeignet. So benutzen beispielweise sowohl TCP als auch ISO TP4 eine Fehlerkorrektur durch Zeitschranken und Wiederholung. (Timeout and Retransmission). Derart wiederholte Pakete haben natürlich eine andere Verzögerung als erfolgreich übertragene Pakete. Erst mit dem zukünftigen DQDB-Protokoll oder dem kanalvermittelten Breitband-ISDN der Telekom für glasfaserbasierte Netze steht eine isochrone Datenübertragung zur Verfügung [Kil91]. Parallel dazu werden auch neue Transportprotokolle entwickelt [HS91].

4.3 Simulation isochronen Datenflusses auf einer herkömmlichen Transportschnittstelle

Im folgenden wird untersucht, inwieweit isochroner Datenfluß in der Internetprotokollwelt möglich ist. Als Basis dient dabei das UDP. Dabei ergeben sich durch zweierlei Einschränkungen Probleme: Zum ersten ist die Verzögerung für jedes einzelne UDP-Paket weder konstant noch beschränkt und zum zweiten steht die volle Bandbreite des Netzes im allgemeinen höchstens zeitweise zur Verfügung. Die Filmqualität kann somit nicht konstant bleiben. Es stellt sich nun die Frage, wie Qualitätseinbußen möglichst gering gehalten werden. Beispielsweise macht es wenig Sinn, ein verlorengegangenes Teilbild neu anzufordern, da es durch das nächste Bild ohnehin sofort wieder verdeckt wird. D.h. der Verlust solcher Datenpakete sollte übergangen werden. Die veränderliche Verzögerung kann durch Pufferung teilweise aufgefangen werden. Ist die Bandbreite dagegen zeitweise zu gering, so kann nur durch eine niedrigere Bildfrequenz der Bildlauf aufrecht erhalten werden. Es gibt aber auch Datenpakete, deren Verlust unbedingt bemerkt und korrigiert werden muß. Das ist zum einen der Header eines Films oder Bildes und zum anderen die Änderung der Farbtabelle. Außerdem dürfen die Anfragen des Clients an den Server nicht verloren gehen.

Um alle diese Anforderungen effizient erfüllen zu können, wurde ein eigenes hybrides Protokoll aufbauend auf dem UDP-Protokoll entwickelt. Dieses Protokoll sieht nur für einen Teil der Pakete Bestätigungen durch den Empfänger vor. Datenpakete mit Anfragen des Clients an den Server werden meist implizit durch die Antwort des Servers bestätigt (z.B. Start/Stop). Schwieriger ist es hingegen, zum Beispiel, zu erkennen, ob die Anfrage „Slower/Faster" angekommen ist. Dies wird daher vom Server explizit bestätigt.

In herkömmlichen Netzarchitekturen gibt es Flußkontrollmechanismen, wie zum Beispiel die „Sliding Window"-Technik, um zu verhindern, daß ein schneller Sender einen langsamen Empfänger überflutet. Bei der isochronen Übertragung macht eine Flußkontrolle durch Rückstau keinen Sinn mehr; vielmehr müssen sowohl Sender als auch Empfänger mit genau derselben Datenrate arbeiten: Sie müssen synchronisiert werden [Ste90]. Für die Filmdarstellung über ein Netz kommt noch dazu, daß diese Synchronisation auch noch mit der Realzeit in Gleichtakt gebracht werden muß; es sollen beispielweise genau 25 Bilder pro Sekunde gezeigt werden, oder auch eine andere benutzerdefinierte Bildfrequenz, z.B. bei Zeitraffer. Dazu muß sowohl beim Bewegtbild-Server als auch beim Bewegtbild-Client ein Synchronisationsmechanismus implementiert sein. Die Synchronisation ist erheblich komplizierter als eine Flußkontrolle im asynchronen Netz.

5 Integration von Bewegtbildern in ein Window-System

Dem Benutzer sollten Bewegtbilder als Teil des gewohnten Bildschirms erscheinen, d.h. der Film sollte in einem Window neben anderen Applikationen ablaufen. Die Bedienung des Bewegtbildfensters sollte in der gewohnten Weise erfolgen. Dies wird aber von heutigen Window-Systemen wie z.B. dem X-Window-System, nicht unterstützt. Die Integration eines Bewegtbildfensters kann prinzipiell auf drei Arten gelöst werden:

- Das Window-System kann mittels geeigneter Hardware umgangen werden (z.B. im Blue-Box-Verfahren).
- Da Window-Systeme die Ausgabe von Rasterbildern erlauben, könnten Einzelbilder in schneller Folge an das Window-System geschickt werden.
- Das Window-System wird um einen Service für Bewegtbildabläufe erweitert.

Wie Eingangs erwähnt ist es das Ziel des X-MOVIE-Projektes, ohne spezielle Hardware auszukommen; deshalb wird die erste Variante nicht weiter in Betracht gezogen. Die beiden Softwareansätze werden im folgenden kurz skizziert.

5.1 Schnelles Übertragen von Einzelbildern

Da jedes Window-System Funktionen zur Darstellung von Rastergraphiken zur Verfügung stellt, können Bewegtbilder prinzipiell durch schnelles Übertragen von Einzelbildern an das Window-System dargestellt werden [Lof90]. Im Falle des X-Systems bedeutet dies allerdings einen hohen Aufwand, da der X-Server ein eigener Prozeß ist. Zur Ausgabe eines Bildes sind daher folgende Schritte erforderlich (Client und Server befinden sich auf einem UNIX-Rechner): Kopieren des Bildes vom Prozeßraum des Clients in den Kernel, Prozeßumschaltung vom Client zum Server, Kopieren in den Prozeßraum des Servers, Kopieren in den Framebuffer des Videoadapters und schließlich die Prozeßumschaltung zum Client. Diese Optimierung setzt allerdings voraus, daß Client und Server auf einem Rechner laufen. Werden die Bilder nun vom Client auf einem dritten Rechner abgeholt, so gehen die Bilder sogar doppelt übers Netz.

5.2 Bewegtbildservice als Erweiterung des Window-Systems

Ein konsequenterer Ansatz ist die Integration eines digitalen Bewegtbildfenster als graphisches Primitiv des Window-Systems. Das bedeutet im Falle des X-Window-Systems die Übernahme der Bewegtbildkontrolle durch den X-Serverprozeß. Dieser Prozeß ist dann für den Verbindungsauf- und -abbau zu einem Bildserver sowie für die Kontrolle des Filmlaufs verantwortlich. Diese Vorgehensweise bietet gegenüber der oben erläuterten Lösung mehrere Vorteile:

- Durch die größere Mächtigkeit des X-Servers ist der Kommunikationsaufwand zwischen X-Client und X-Server deutlich geringer.
- Der Code zur Bewegtbilddarstellung muß nicht zu jedem X-Client gebunden werden.
- Da der X-Server besser an die gegebene Hardware angepaßt werden kann als ein Client, ist eine höhere Performance, also eine bessere Bildqualität zu erreichen.
- Die Manipulation des Bewegtbildfensters kann sowohl für den Benutzer als auch für den Programmierer in gewohnter Weise erfolgen (öffnen, schließen, zoomen, usw.).

Der Prototyp des X-MOVIE-Systems basiert auf der letztgenannten Lösung. Dazu wurde die Programmierschnittstelle des X-Window-Systems (die X-Library) um die folgenden Aufrufe erweitert [Kel91]:

XMListMovies(): Diese Anforderung an den X-Server liefert eine Liste verfügbarer Filme oder Bilder.

XMFreeMovieInfo(): Dient zum Freigeben vorher belegten Speichers.

XMOpenMovie(): Eröffnen einer Verbindung.

XMPlayMovie(): Startet einen Film im gewünschten Window.

XMShowSinglePicture(): Dient zum einfacheren Anzeigen eines einzelnen Bildes.

XMUpdateMovie(): Ändert die bei **PlayMovie()** übergebenen Parameter.

XMStopMovie(): Hält einen (laufenden) Film an.

XMDestroyMovie(): Gibt alle für den Film belegten Resourcen (Speicher, Kommunikationskanäle) frei.

6 Erfahrungen

Das System X-MOVIE wird zur Zeit auf Workstations des Typs IBM-PS/2 Modell 80 unter AIX als Clients und einer DEC-5400 unter Ultrix als Bewegtbild-Server entwickelt. Die PS/2-Rechner sind mit dem Bildschirmadapter 8514a ausgestattet. Ein erster Prototyp wird zur Zeit getestet.

Als Übertragungsprotokoll wurde UDP/IP verwendet, nachdem bei einem Vorversuch mit TCP/IP eine Datenübertragungsrate von nur ca. 0.15 Megabit/sec. festgestellt wurde. Diese unerwartet niedrige Datenrate wurde nur bei einer Datenübertragung von der DEC-5400 auf eine PS/2-Workstation gemessen. Mit dem ungesicherten UDP-Protokoll konnten dagegen Übertragungsraten von ungefähr 2 Megabit/sec. erreicht werden. Dabei gingen zwischen 10 und 20 Prozent der Pakete (=Bildteile) verloren. Dies beeinflußt die Bildqualität jedoch kaum. Diese geringe Verlustrate wurde allerdings erst erreicht, nachdem der Server gezielt abgebremst wurde, um zu verhindern, daß der Client überflutet wird.

Ein zweiter Testlauf, bei dem eine zweite PS/2-Workstation als Server mit einem Token-Ring angeschlossen wurde, ergab ungefähr die gleichen Ergebnisse. Allerdings mußte hier der Server nicht abgebremst werden.

In einem dritten Test wurden Client und Server auf derselben Workstation installiert. Es ist jetzt keine echte Parallelarbeit zwischen Client und Server mehr möglich, da beide auf demselben Prozessor ablaufen. Der Prozessor wird durch den häufigen Prozeßwechsel und die Kommunikation zwischen Client und Server stark belastet. Dadurch geht die Datenübertragungsrate auf unter ein Megabit pro Sekunde zurück [Sut90, Zim90].

Die Erfahrungen ergaben also, daß eine physikalische Datenrate von 10 Megabits pro Sekunde ausreichend wäre; jedoch sind die heute verfügbaren Protokolle für die Bewegtbildübertragung völlig ungeeignet. Erst durch die in der Normung befindlichen Hochgeschwindigkeitsnetze, insbesondere DQDB und B-ISDN über ATM, wird in näherer Zukunft der für die digitale Filmübertragung benötigte isochrone Dienst verfügbar werden.

In weiteren Leistungstests konnten wir feststellen, daß das langsamste Glied in der Kette der physikalischen Übertragungsmittel die Ethernetkarte darstellt. Nach dem Einbau einer anderen Karte müßte der neue Engpaß bei einem der folgenden Bauteile gesucht werden: der Magnetplatte des Servers, auf der die Filme gespeichert sind; dem Bus des Servers; dem Bus des Clients (Microchannel); dem Hauptspeicher des Clients, in dem die Protokollsegmente zwischengespeichert und zu einem Gesamtbild zusammengefügt werden; dem Prozessor des Clients; dem Graphikadapter 8514a;

7 Zusammenfassung und Ausblick

Das System X-MOVIE ist eine Testumgebung für Bewegtbilddarstellung in einem Netz aus Unix-Workstations. Es basiert auf Standard-Hardware und benutzt herkömmliche Netztechnik zur Übertragung und herkömmliche Graphik-Adapterkarten zur Darstellung von Bewegtbildsequenzen.

In dieser Arbeit wurden zunächst die wichtigsten Darstellungsformate für Standbilder und Bewegtbilder vorgestellt; dabei wurde auch auf Kompressionsalgorithmen kurz eingegangen. Da die gängigen Adapterkarten für Standbilder mit der Farbtabellentechnik (Color Lookup Table) arbeiten, wurden zwei Algorithmen vorgestellt, mit denen eine effiziente Aktualisierung der Farbtabelle während des Filmablaufs in Echtzeit ermöglicht wird.

In einem weiteren Kapitel wurde auf die speziellen Probleme bei der Übertragung von Bewegtbildsequenzen über ein Netz eingegangen. Es zeigte sich, daß herkömmliche Netzarchitekturen hierzu ungeeignet sind, und zwar sowohl auf Grund ihres Netzzugangsprotokolls als auch auf Grund ihres Transportprotokolls. In verbindungsorientierten Transportprotokollen, wie zum Beispiel TCP, wirkt sowohl die Flußkontrolle durch „Sliding Window" als auch die Fehlerkontrolle durch Zeitschranke und Übertragungswiederholung einem isochronen Datenfluß entgegen. In verbindungslosen Transportprotokollen, wie zum Beispiel UDP, gibt es keine Gewähr für die zuverlässige Aktualisierung der Farbtabelle, so daß allmählich immer stärker abweichende Fehlfarben entstehen können. Daher wurde ein hybrides Protokoll entwickelt und aufbauend auf UDP prototypisch implementiert, das sowohl die ungesicherte als auch die gesicherte Übertragung im gemischten Betrieb zuläßt.

Zur Darstellung der Filme auf einem fensterorientierten Bildschirm wurden drei Alternativen gegenübergestellt, von denen die volle Integration des Bewegtbilddienstes in das Window-System, sowohl auf der Client- als auch auf der Server-Seite als beste Lösung ausgewählt wurde. Eine Erweiterung des X-Window-Systems für den Bewegtbildservice wurde kurz skizziert.

Das System X-MOVIE ist als Labormodell gedacht, als Experimentierfeld für alternative Architekturen und Algorithmen. Insofern erhebt es nicht den Anspruch auf Vollständigkeit oder gar Produktqualität. In einer ersten Version konnte gezeigt werden, daß in einer Standard-Umgebung aus Unix-Workstations und Ethernet ohne spezielle Hardware bereits Farbfilme mit 10 Bildern/s in einer Größe von 128×128 Pixel übertragen und dargestellt werden können.

In den nächsten Schritten ist eine Erweiterung von X-MOVIE in verschiedene Richtungen vorgesehen. Zum einen wird das Ethernet-LAN durch ein FDDI-LAN ersetzt werden, so daß das Netz keinen Engpaß für größere Bildfenster und höhere Bildfrequenzen mehr darstellt. Weitere Untersuchungen sollen dann die Möglichkeiten einer annähernd isochronen Übertragung durch Pufferung bei den hohen Geschwindigkeiten von FDDI eruieren. Längerfristig wäre ein Anschluß an ein DQDB- oder B-ISDN-Netz wünschenswert, da in diesen Netzen Bitströme isochron übertragen werden können. Parallel zu den neuen Netzanschlüssen wird die Implementierung der Erweiterungen von X-Client und X-Server um die Filmfunktionen durchgeführt, um Erfahrungen mit einer integrierten Bewegtbilddarstellung aus der Sicht des Programmierers und des Benutzers gewinnen zu können.

Bisher ist in X-MOVIE kein Mechanismus zur Synchronisation mit der Realzeit vorhanden; die Filmgeschwindigkeit wird allein durch die Leistungsfähigkeit des schwächsten Gliedes in der Übertragungskette vom Server zum Client bestimmt. Verschiedene Synchronisationstechniken werden zur Zeit auf ihre Eignung für X-MOVIE untersucht.

Letztendlich ist auch bisher keine Tonübertragung und -ausgabe möglich; hierzu würde auch spezielle Audio-Hardware in den Workstations benötigt. In einer späteren Phase des Projekts, wenn einmal vollständig isochrone Datenströme über das Netz übertragen werden können, soll X-MOVIE um eine Audio-Komponente ergänzt werden.

Danksagung

Wir danken Prof. A. Schmitt, W. Leister und A. Stößer vom Institut für Betriebs- und Dialogsysteme der Universität Karlsruhe für die Überlassung der mit dem VERA-Raytracing-System erzeugten Filmausschnitte.

Dem Europäischen Zentrum für Netzwerkforschung der IBM in Heidelberg und insbesondere den Mitarbeitern des dortigen Multi-Media-Labors danken wir für die vertrauensvolle Zusammenarbeit.

Literatur

[ADH91] D.P. Anderson, L. Delgrossi und R.G. Herrtwich. Process Structure and Scheduling in Real-Time Protocol Implementations. In *Kommunikation in Verteilten Systemen,* Mannheim. Informatik-Fachberichte, Springer-Verlag, 1991.

[Bru89] T. Brunhoff. *VEX: Video Extension to X, Version 5.5.* Textronix, Inc., 1989.

[CDF+86] G. Campbell, T.A. DeFanti, J. Frederikson, S.A. Joyce, A.L. Lawrence, J.A. Lindberg und D.J. Sandin. Two Bit/Pixel Full Color Encoding. *Computer Graphics,* 1986.

[Com88] D. Comer. *Internetworking with TCP/IP.* Prentice-Hall International Editions, 1988.

[Gal91] Didier Le Gall. MPEG:A Video Compression Standard for Multimedia Applications. *Communications of the ACM,* 34(4):46–58, 1991.

[HS91] L. Henckel und H. Stüttgen. Transportdienste in Breitbandnetzen. In *Kommunikation in Verteilten Systemen,* Mannheim. Informatik-Fachberichte, Springer-Verlag, 1991.

[Her91] R.G. Herrtwich. Zeitkritische Datenströme in verteilten Multimedia-Systemen. In *Kommunikation in Verteilten Systemen,* Mannheim. Informatik-Fachberichte, Springer-Verlag, 1991.

[Jon88] O. Jones. *Introduction to the X-Window-System.* Prentice Hall, 1988.

[Kel91] R. Keller. Erweiterung des X11-Servers zur Bewegtbilddarstellung. Diplomarbeit, Universität Mannheim, 1991.

[Kil91] U. Killat. B-ISDN und MAN: Konkurrierende Netztechnologien. In *Kommunikation in Verteilten Systemen (Tutorial),* Mannheim. Deutsche Informatik-Akademie, 1991.

[LD87] L. F. Ludwig und D. F. Dunn. Laboratory for Emulation and Study of Intergrated and Coordinated Media Communication. *Proc. of ACM SIGCOMM 87,* Stowe, Vermont, Seiten 283–291, Aug. 1987.

[LMS91] W. Leister, H. Müller und A. Stößer. *Fotorealistische Computeranimation.* Springer-Verlag, (Heidelberg, Berlin), 1991.

[Lof90] G. Loff. Konzeption, Entwurf und Implementierung eines netzweiten Videodienstes. Diplomarbeit, Universität Karlsruhe, 1990.

[Lut89] A. Luther. *Digital Video in the PC Environment.* McGraw Hill Book Company, New York, 1989.

[MD89] W. E. Mackay und G. Davenport. Virtual Video Editing in Interactive Multimedia Applications. *Communications of the ACM,* 32(7):802–810, Juli 1989.

[Pin90] M. Pins. *Analyse und Auswahl von Algorithmen zur Datenkompression unter besonderer Berücksichtigung von Bildern und Bildfolgen.* Dissertation, Universität Karlsruhe, 1990.

[SGN88] R. W. Scheiffler, J. Gettys und R. Newman. *X-Window-System, C-Library and Protocol Reference.* Murray Printing Company, 1988.

[Ste90] R. Steinmetz. Synchronisation Properties in Multimedia Systems. *IEEE Journal on Selected Areas in Communications,* 8(3):401–412, 1990.

[Sut90] S. Sutter. *Datenformate bei der Bewegtbildübertragung über ein lokales Netz.* Studienarbeit, Lehrstuhl für Praktische Informatik IV, Universität Mannheim, 1990.

[Tin89] M. Tinker. DVI Parallel Image Compression. *Communications of the ACM,* 32(7):844–851, Juli 1989.

[Zim90] R. Zimmermann. *Übertragungsprotokolle für die Bewegtbildübertragung über ein lokales Netz.* Studienarbeit, Lehrstuhl für Praktische Informatik IV, Universität Mannheim, 1990.

ODIN und MARS: Werkzeuge zur Archivierung multimedialer Dokumente

H.-J. Appelrath, H. Eirund
Universität Oldenburg
Fachbereich Informatik
Postfach 2503
D-2900 Oldenburg

Zusammenfassung

In diesem Beitrag werden zwei Problembereiche bei der Archivierung multimedialer Dokumente diskutiert, für die die hier vorgestellten Projekte ODIN (Optical Disc Information Systems) und MARS (Multimedia Archiving and Retrieval System) Lösungsansätze verfolgen.

In ODIN stehen die Entwicklung von Werkzeugen zur Verbesserung konventioneller Ansätze der Dokumentenarchivierung (Retrieval, Navigation, konsistentes Update von Thesaurus und Deskriptoren) sowie die Integration von schnellen Bildspeichersystemen im Vordergrund.

In MARS wird ein Ansatz zur Repräsentation der Dokumentaussagen durch semantisch begründete, inhaltsorientierte Strukturen verfolgt, die die Grundlage für das Retrieval bilden. Insbesondere Anwendungen multimedialer Dokumente in Büroumgebungen werden hier untersucht.

1. Einleitung

Fragestellungen bei der Verwaltung *multimedialer Dokumente* - das sind elektronische Dokumente, die auch nicht-textuell dargestellte Informationen beinhalten können (z.B. Rasterbild, Graphik, Sprachanmerkung) - kann man sich unter verschiedenen Gesichtspunkten der Informationstechnik nähern, wie z. B.

(i) *(elektro-)technischen Aspekten*, die die Realisierung der physischen Kommunikation, der Entwicklung geeigneter Protokolle, der Erzeugung und Präsentation multimedialer Dokumente in geeigneten Umgebungen etc. umfassen;

(ii) *medialen und visuellen Aspekten*, die neue Einsatzweisen von multimedialen Informationen in Anwendungen herausstellen, insbesondere durch Hypermedia-Systeme mit hoher Benutzerinteraktion;

(iii) *strukturellen und datenbanktypischen Aspekten*, wie z.B. Entwurf von Inhaltsrepräsentationen, Auswahl von Zugriffsverfahren, Abfrageunterstützung, Optimierung der Abfragebearbeitung oder Einsatz spezieller Speichermedien.

In dieser Arbeit werden die beiden Systeme ODIN und MARS vorgestellt, in denen einige der unter (iii) genannten Aspekte näher untersucht werden, speziell im Hinblick auf die mit der Archivierung multimedialer Dokumente verbundenen Fragen der Ablage und des differenzierten Zugangs zu Dokumenten.

Abb. 1 zeigt die beiden Blickwinkel, unter denen dabei das zentrale Problem der Archivierung multimedialer Dokumente betrachtet wird.

Aus ODIN-Sicht reduzieren sich multimediale Dokumente auf Indexe/ Referenzen (i.a. als Adressen auf optischen Speichermedien) zu diesen Dokumenten und die Dokumente spielen "nur" bezüglich ihrer differenzierten Deskribierung sowie ihrer algorithmischen Besonderheiten etwa beim Retrieval oder Navigieren eine Rolle. Hingegen ist die MARS-Sicht insbesondere auf die Inhalte und Inhaltsstrukturen (semantische Einheiten) multimedialer Dokumente gerichtet.

Die Schwerpunkte in ODIN, dessen Retrievalansatz auf der Inhaltsrepräsentation durch Deskriptoren basiert, liegen in drei Bereichen:

(a) Verfahren zur konsistenten Erweiterung von Thesauri und Dokumentdeskribierungen,

(b) Techniken und Werkzeuge zur Einbindung multimedialler Speichermedien, hier insbesondere Videodisk-Systeme und CD-ROM zur Bildspeicherung, und

(c) Werkzeuge zur strukturierten Präsentation der Ergebnismenge durch Erzeugung und Ausführung von Navigationsprogrammen.

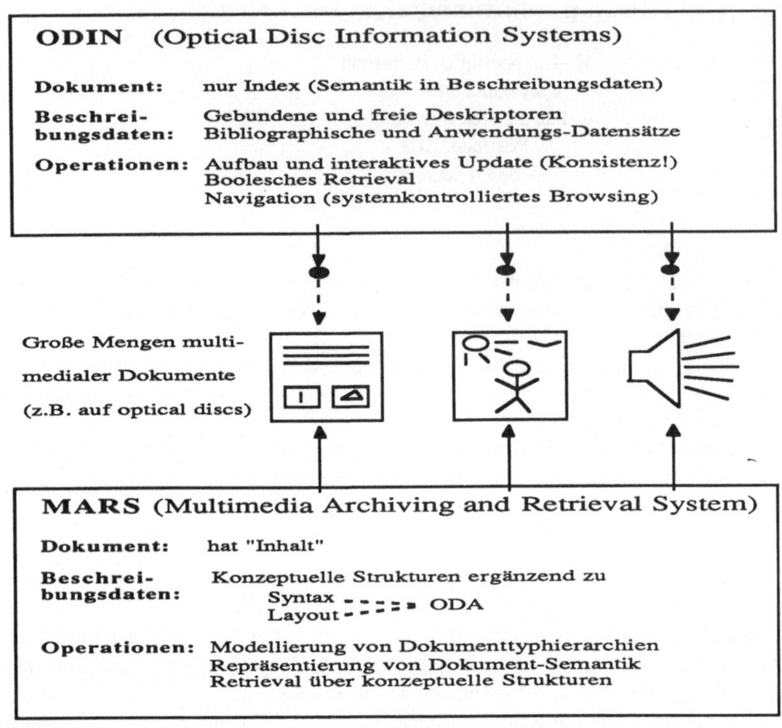

Abb.1 ODIN- und MARS-Sicht auf multimediale Dokumente

Im MARS-Projekt liegt der Schwerpunkt auf der Entwicklung eines Modells, in dem die zentralen Aussagen des Dokuments durch eine komplexe Struktur beschrieben werden können, die auch multimediale Inhaltsteile als Werte aufnehmen kann. Über Generalisierung dieser Strukturen gelangt man zu Dokumenttypen, die den Prozess der Dokumentmodellierung sowie die Abfrageformulierung und Bearbeitung unterstützen. Besondere Beachtung wird der Möglichkeit zur flexiblen Erweiterung der Typwelt sowie Einbindung neuer Inhaltsmedien geschenkt.

Die weitere Arbeit gliedert sich wie folgt: In Kapitel 2 und 3 werden die wesentlichen Konzepte von ODIN und MARS beschrieben. Kapitel 4 stellt typische Anwendungsszenarios für beide Systeme vor, und Kapitel 5 schließt mit einer Zusammenfassung und einem kurzen Ausblick.

2. Das System ODIN

ODIN [Appe87] ist ein Software-Werkzeugkasten, der die Realisierung von Informationssystemen für multimediale Dokumente erlaubt. Mit ODIN realisierte Informationssysteme enthalten u.a. mindestens folgende Objektklassen und Beziehungstypen:

Dokument bzw. Dokumentindex:

Video-, (Raster-)Bild-, Text-, Graphik-, Audio- oder multimediales Dokument in digitalisierter Form bzw. ein Verweis auf ein solches Dokument z.B. als Adresse eines Dokumentspeichers, etwa einer Optischen Platte.

Term:

Wort/ Phrase eines Diskursbereichs.

Inhaltsorientierter Deskriptor:

Term, der einem Dokument als Inhaltscharakterisierung zugeordnet ist.

Wir unterscheiden dabei (logisch) zwischen

- *gebundenen* Deskriptoren, die aus einem Thesaurus stammen (s.u.) und deren Anbindung an ein Dokument gewissen Konsistenzregeln unterworfen ist, und
- *freien* Deskriptoren, die nicht aus einem Thesaurus stammen und auch - bis auf natürliche Einschränkungen wie Eindeutigkeit der Bezeichnung - keinen Konsistenzregeln gehorchen müssen.

Bibliographiedeskriptorsatz:
i.a. komplexer, strukturierter Bibliographiedatensatz, der einem Dokument "bibliographische" Informationen (z.B. Erstellungsdatum und -ort, Autor, technische Angaben) zuordnet.

Anwendungsdeskriptorsatz:
i.a. komplexer, strukturierter Anwendungsdatensatz, der je nach Anwendung ein Dokument ergänzend mit anwendungsbezogenen Informationen beschreibt.

Thesaurus:
Beziehungsnetz in Form eines markierten Graphen, der Deskriptoren (= Knoten) in Beziehung (= Kanten) setzt (z.B. in Oberbegriff- oder Synonym-Beziehung).

Relevanz-Beziehung:
Verweis von einem inhaltsorientierten (gebundenen oder freien), bibliographischen oder anwendungsbezogenen Deskriptor auf ein Dokument.

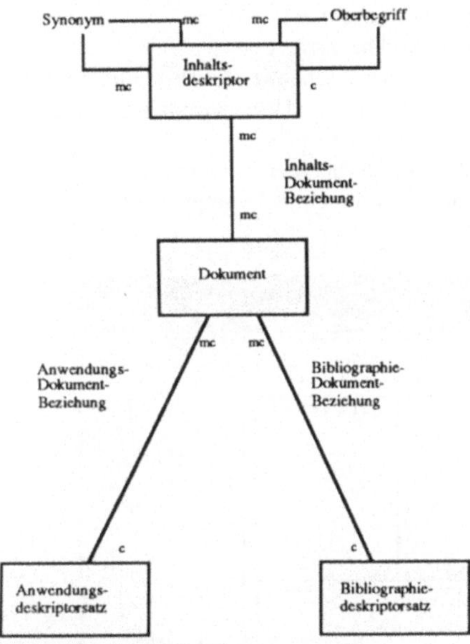

Abb. 2 ODIN-Datenschema

Abb. 2 zeigt einen Ausschnitt des für ODIN resultierenden Datenschemas in einem vereinfachten ER-Modell, das die oben aufgeführten Objektklassen und die Beziehungen noch einmal graphisch repräsentiert und dabei auch die Assoziationstypen (hier nur c: "kein oder ein" und mc: "kein, ein oder mehrere") präzisiert, wobei die Beziehungstypen im Thesaurus hier (aber nicht prinzipiell) auf Oberbegriff- und Synonym-Beziehung beschränkt sind.

ODIN-Systemkonzeption und -Realisierung

ODIN unterstützt *duale* Multimedia-Arbeitsumgebungen, d.h., daß
- die multimedialen Dokumente selbst vollständig auf optischen Speichermedien (bisher realisiert: Bildplatte, WORM, CD-ROM) liegen und
- die Beschreibungsdaten dazu (Deskriptoren, Bibliographie- und Anwendungsdatensätze) und die Zugriffspfade (zu den Dokumenten) im Rechner - und damit änderbar - und nicht wie bei anderen Systemen auf der Optischen Platte verwaltet werden.

ODIN zielt auf die Entwicklung von Werkzeugen, die Unterstützung beim
- Retrieval von Dokumenten und Dokumentbeschreibungsdaten (inhaltsbezogen und administrativ)
- Update von Dokumentbeschreibungsdaten mit Berücksichtigung von Konsistenzanforderungen
- Navigieren auf vordefinierten Pfaden innerhalb der gesamten Dokumentmenge

bieten (siehe auch [Appe87].

Dabei sind die Werkzeuge unabhängig von dem zugrundeliegenden Rechner, den eingesetzten Speichermedien für die Verwaltung multimedialer Dokumente und speziellen Anwendungsklassen konzipiert und implementiert.

Wie erwähnt unterstützen die ODIN-Werkzeuge keine Dokumenterkennung und -manipulation, denn die Semantik eines Dokuments liegt nur in den ihm zugordneten Beschreibungsdaten und, da nicht in die Dokumente "hineingeschaut" wird (keine Bildinterpretation, Text- oder Sprachanalyse), reicht den Werkzeugen zur Bearbeitung ein (eindeutiger) Index zur Repräsentation eines Dokuments.

Diese Vorüberlegungen führten zu der in Abb. 3 skizzierten ODIN-Systemarchitektur. Wir differenzieren in dieser Systemarchitektur drei Ebenen: die Basisfunktions-, die Werkzeug- und die Applikations-Ebene mit jeweils einer Reihe teilweiser umfangreicher, vollständig in MODULA-2 implementierter Software-Komponenten, deren Funktionalität im folgenden skizziert werden soll.

Die Ebene der Basisfunktionen (hellgrau in Abb. 3 eingezeichnet) umfaßt:
- HOST, eine Modula-2-Bibliothek, die vom zugrundeliegenden Rechner isoliert.
- das GridFile [Niev84] als Datenverwaltungssystem, wobei ein Zugriff auf anwendungs- neutrale Daten nur über ein Modul BasicDataHandler möglich ist, das unabhängig vom verwendeten Datenverwaltungssystem definiert ist. Damit ist z.B. auch die Integration eines Standard-SQL-Datenbanksystems möglich.
- Treiber zur Kapselung des Massenspeichers, auf dem alle Dokumente abgespeichert werden.
- Fantasy [Lorek88], ein User Interface Management System, das die einfache Implementierung interaktiver, graphik-basierter Systeme unterstützt.

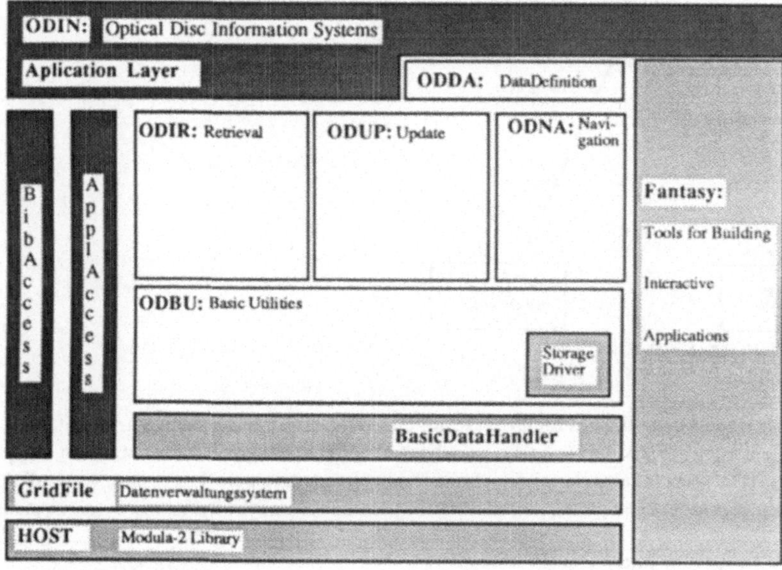

Abb. 3: ODIN-Systemarchitektur

Die gesamte Basisfunktions-Ebene - und damit natürlich auch die darauf aufbauende Werkzeug- und Applikations-Ebene - ist portabel, d.h. bis auf einen geringen, gut isolierten, rechnerabhängigen Teil, einen je nach Dokumentspeicher zu reimplementierenden Treiber sowie eine Schnittstelle zum Datenverwaltungssystem ist die gesamte Software auf anderen Rechnern und für andere Speichermedien ablauffähig.

Die W e r k z e u g - E b e n e (weiß in Abb. 3 eingezeichnet) beinhaltet die Komponenten

ODBU

- exportiert Basisfunktionenen für alle Werkzeuge, z.B. Mengenoperationen für Dokumente oder Deskriptoren.

ODDA

- erlaubt den Aufbau eines Thesaurus mit Ober-Unter-Begriffen und Synonymen und
- führt bei der Überführung der extern definierten Daten in eine Datenbank einfache Plausibilitätsprüfungen durch.

ODUP

- bietet Werkzeuge zur interaktiven Aktualisierung aller ODIN-Objektklassen und
- garantiert dabei die Einhaltung differenzierter Konsistenzbedingungen (inbesondere bei Update-Operationen bezüglich des Thesaurus).

ODIR

- bietet Werkzeuge zum Aufbau komplexer boolescher Abfragen und
- exportiert Funktionen zur effizienten Auswertung dieser Abfragen.

ODNA

- erlaubt eine einfache Erstellung von Navigationsprogrammen durch große Dokumentmengen und
- bietet eine komfortable Interaktion mit diesen Programmen für unterschiedliche Benutzerklassen.

Zur A p p l i k a t i o n s - E b e n e (dunkelgrau in Abb. 3 eingezeichnet) gehören zum einen die unter Nutzung von Werkzeugen der Werkzeug-Ebene geschriebenen Anwendungsprogramme selbst, aber auch applikationsabhängige Zugriffsmodule zu anwendungsbezogenen (ApplAccess) und bibliographischen Informationen (BibAccess).

Updatekonzept

Die ODUP-Datenschnittstelle exportiert zu allen Objektklassen und Beziehungsarten Funktionen, um konkrete Objekte und Beziehungen einfügen, ändern und löschen zu können.
BasicDataHandler isoliert alle auf diesem Modul aufbauenden Software-Bausteine, insbesondere also von dem verwendeten DB-System. Alle Funktionen, die Daten aus der DB lesen bzw. manipulieren, sind so definiert, daß sie weder eine Aussage über das verwendete Datenmodell noch über das benutzte DB-System machen. Somit ergibt sich eine maximale Flexibilität bezüglich des Einsatzes verschiedener DB-Systeme, und die Portabilität des gesamten Softwaresystems wird entscheidend erhöht. In der aktuellen Realisierung wird das GridFile-System zur Datenverwaltung verwendet.

Die Update-Operation auf der Ebene von ODUP kann zu umfangreichen Konsistenzüberprüfungen und eventuell notwendigen Folge-Updates führen (wie z.B. das Löschen eines Deskriptors aus einem Thesaurus).

Nachfolgend soll zunächst eine Trigger-Sprache als ein mögliches Modellierungswerkzeug für komplexe Konsistenzbedingungen vorgestellt werden. Notwendigkeit und Nutzen solcher Konsistenzregeln anhand einiger Bespiele werden in [Appe90] illustriert.

Konsistenzsicherung durch Triggermechanismen

Das in ODIN-Triggerkonzept zur Konsistenzsicherung basiert auf folgendem Sprachkonstrukt:

```
TRIGGER <name>
        [BEFORE, AFTER] <update>
          IF  <condition> /* optional */
             THEN <action>.
```

<name>	:	eindeutiger Triggerbezeichner
<update>	:	[INSERTION, DELETION] Objekt
<condition>	:	boolescher Ausdruck
<action>	:	Ausdruck, der mit booleschen Operatoren verknüpfte Update-Operationen, FOR-ALL-Kontrollstrukturen und REJECT beinhalten kann

Trigger werden während des Abarbeitungsprozesses wie folgt interpretiert: Auslöser eines Triggers ist ein <update> (Einfügen oder Löschen eines Objekts). Das Schlüsselwort BEFORE bzw. AFTER entscheidet, ob der

Trigger vor oder nach der Ausführung des <update> relevant ist, d.h. ob der <condition>-Teil sich auf den Zustand der DB vor oder nach Ausführung des <update> bezieht. Ist ein Trigger relevant für die gerade auszuführende Operation, so wird der <condition>-Teil des Triggers überprüft und, falls dieser erfüllt ist (d.h. true liefert), "feuert" der Trigger und der <action>-Teil kommt zur Ausführung.

Alle relevanten Trigger werden nach der Reihenfolge ihres Auftretens überprüft, d.h. sind zwei Trigger relevant, so wird zuerst der Trigger getestet, der in der Aufschreibe-Reihenfolge an erster Stelle steht. Im <action>-Teil können Folge-Updates auftreten, die wiederum die Überprüfung aller relevanten Trigger erfordern. Wenn im <action>-Teil eines Triggers, welcher zur Ausführung kommt, das Schlüsselwort REJECT steht, so wird die weitere Überprüfung der Trigger abgebrochen und das <update> wird nicht durchgeführt. Bereits durchgeführte Folge-Updates werden dann rückgängig gemacht. Nach Aufruf der Update-Operation befindet sich die DB also im gleichen Zustand wie vor dem Versuch, sie zu ändern.

Retrievalkonzept

Die zentrale Aufgabe der Tools aus ODIR ist die Bereitstellung einer Menge von Dokumenten aus einer i.a. sehr großen Dokumentmenge auf eine vom Benutzer vorgegebene Abfrage. ODIR unterscheidet dabei die beiden aufeinanderfolgen Schritte *Grobrecherche* (mit der Bestimmung der Kardinalität der Ergebnismenge) und *Feinrecherche* (mit der intellektuellen Überprüfung und manuellen Nachbereinigung der Ergebnismenge durch den Benutzer).
Diese Zweistufigkeit hilft, um iterativ:

- bei zu großer Ergebnismenge durch eine Verschärfung der Abfrage die Menge zu reduzieren, bis sie eine vom Benutzer akzeptierte, d.h. eine für die von ihm für die Feinrecherche eingeplante Zeit angemessene Größenordnung erreicht hat, bzw.
- bei zu kleiner Ergebnismenge durch eine Ausweitung der Abfrage die Menge zu vergrößern, um eine dem Informationsbedürfnis angepaßte Dokumentmenge zu erhalten.

Die ODIR-Retrievalsprache erlaubt boolesche ("und", "oder", "aber nicht") Verknüpfungen von Deskriptoren (sowohl gebundene, d.h. aus dem Thesaurus stammende als auch freie, d.h. nicht aus dem Thesaurus stammende Terme), bibliographischen Informationen (z.B. Autor eines Textes, Datum einer Bilderstellung) und anwendungsbezogenen Daten (z.B. Patientendaten zu einem Röntgenbild). Das iterative Fortschreiben der Abfrage entspricht dem Erweitern (OR), Einschränken (AND), bzw. Reduzieren (NOT) der jeweils gefundenen Dokumentmenge. Ergänzend werden die dabei realisierten Mengenoperatoren Vereinigung, Durchschnitt und Differenz auch als solche angeboten, um z.B. die Ergebnisse zweier Abfragen zu verknüpfen, indem die resultierenden Dokumentmengen vereinigt, geschnitten bzw. mengentheoretisch subtrahiert werden.

Neben dem eigentlichen Retrieval sind direkte Abfragen nach allen ODIN-Objektklassen möglich. Da das zugrundeliegende GridFile-System entsprechende Abfragetypen unterstützt, sind auch nur teilweise spezifizierte Abfragen ("partial match queries") erlaubt, z.B die Suche nach Bildern eines Monats (ohne den Tag anzugeben) oder nach Patientendaten nur aufgrund des Familiennamens.

Systemkontrollierte Navigation durch Dokumentmengen

Zielsetzung von ODNA ist die einfache Programmierung interaktiver Durchläufe (Pfade, durch Navigationsprogramme kontrolliert) durch Dokumentmengen, die durch - vor allem nur gelegentliche - Benutzer über Menüs ausgewählt und verfolgt werden können. Ein Pfad entspricht i.a. einer erwarteten Benutzerklasse bzw. eines vermuteten Benutzungskontextes. Die zur Programmierung verwendete Navigationssprache (ODNATalk) bietet Konstrukte wie z.B. Sequenz, Sprung, Schleife, Mehrfachverzweigung und Makros und erlaubt eine einfache Programmierung von interessanten Dokumentfolgen, die z.B. unter sinnfälligen Bezeichnungen abgelegt und jederzeit abrufbar sind.

Trotz vorgegebener Pfade ist eine ausreichende Interaktivität durch Menüangebote gegeben, die ein Verlassen der Pfade und eine Rückkehr (wichtig bei "Verlaufen") ermöglichen.

Abb. 4 zeigt im linken Teil Konstrukte von ODNATalk, im mittleren Teil eine graphische Repräsentierung von Pfaden durch eine Dokumentmenge und im rechten Teil ein ODNATalk-Programm.

Navigationssprache ODNA-Talk:
Audio: De/ Aktivierung von Audio-Spuren
DISF: Einblenden eines Dokumentindex
MENU: Angebot eines Menüs mit Titel
PROMPT: Menü-Auswahl
SFNR: Sprung zu einem Dokumentindex
WAIT: Wartet Zeitspanne (unterbrechbar)
(ca. 30 Befehle, einschließlich Befehle für spezielle Speichermedien wie CD-ROM, WORM, Bildplatte)

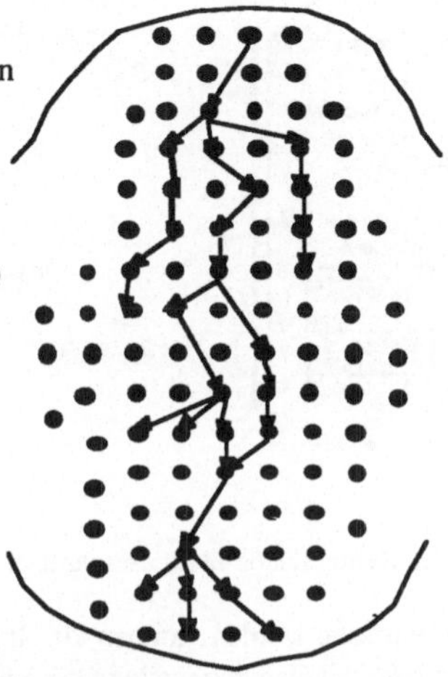

```
......
AUDIO(1,Y)
DISF(Y)
LOOP(10)
    MENU (Muecken)
        PROMPT(Entwicklung)
            SFNR(23119)
            WAIT(7, Y)
            SFNR(23121)
        PROMPT(Stech-Muecke)
            SFNR(23136)
            WAIT(5, N)
            SFNR(23145)
        PROMPT(Brutstaetten)
            SFNR(23156)
        PROMPT(Ende)
            EXIT
    END. *MENU
END. *LOOP
.......
```

Abb. 4: Navigationssprache, -pfade und -programm

Das ODNA-Werkzeug ist für Anwendungsszenarien geeignet, die Ausbildungs- oder Präsentationscharakter ("points of sale") haben und sich durch viele, häufig wechselnde und i.a. "naive" Benutzer auszeichnen.

Für diese Zielgruppe ist ein durch ODIR bereitgestelltes, reines Retrievalangebot zu anspruchsvoll, da entsprechenden Benutzern häufig ein angemessener Begriffsapparat (Terme des Diskursbereichs) nicht verfügbar ist, boolesche Verknüpfungen bezüglich ihrer Wirkung nicht gut genug abgeschätzt und das zielorientierte, iterative Expandieren von Queries nicht sicher beherrscht wird. Die Situation läßt sich anschaulich vergleichen mit einem ungeübten Benutzer, dem eine (Ergebnis-) Menge von Dokumenten durch einen Experten "aufbereitet" wird.

Mit dem Navigations-Werkzeug ODNA, ergänzt durch das Retrieval-Werkzeug ODIR, lassen sich sehr rasch Anwendungssysteme realisieren, die relevante Dokumente bestimmen, unterschiedliche Benutzerklassen gezielt durch Dokumentmaterial führen und einen Wechsel zwischen Retrieval und Navigation unterstützen. In ODNA können Navigationsprogramme sowohl durch (manuelles) Programmieren wie auch durch Beobachten einer benutzergesteuerten Auswahl von Dokumenen (Monitoring) erzeugt werden.

3. Das Projekt MARS

Der Inhalt multimedialer Dokumente kann abhängig von der Art der Anwendung aus verschiedenen Sichten betrachtet werden:

(a) Bezüglich der Layout-Präsentation auf einem Ausgabemedium mit Seiten, positionierten Bildern, Textblöcken etc. (Standards z.B. PostScript oder ODA(-Layoutstruktur))
(b) Bezüglich seiner syntaktischen Strukturierung, bestehend aus Paragraphen mit Titelzeile, Kapiteln etc. (z.B. in SGML oder ODA(-Logische Struktur)) und
(c) Bezüglich der Darstellung seiner semantischen Aussage.

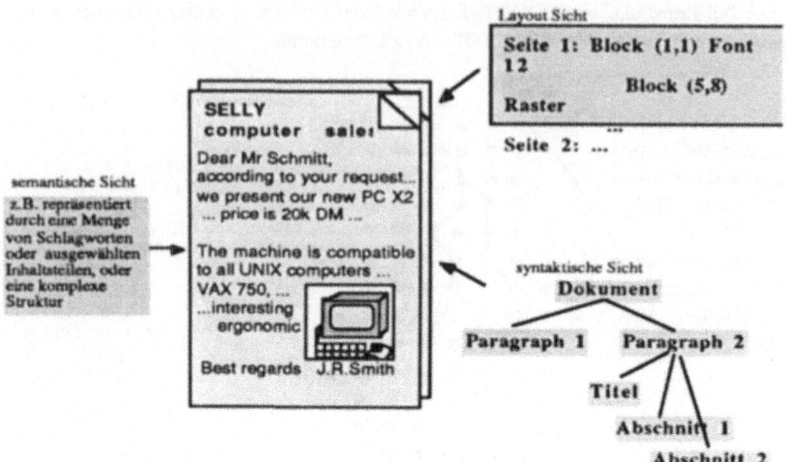

Abb. 5: Inhaltssichten: Layout Sicht, Syntaktische Sicht, Semantische Sicht

Alle diese Sichten, die in Abb. 5 umrissen werden, können als Grundlage für die Spezifikation des Zugriffs dienen. Während das Layout von Bildern oder die Kapiteleinteilung bei Büchern für das Wiederauffinden von "Bild"- oder "Buch"-Dokumenten interessant sein kann, ist ein Zugriff über die (in unterschiedlichen Abstraktionsebenen repräsentierbare) "semantische Sicht" am universellsten einsetzbar. Die einfache und flexible Modellierung dieser Sicht ist Gegenstand des MARS-Modells.

Der Einsatz multimedialer Dokumente in modernen Büroanwendungen fordert die Integration der Retrieval-Möglichkeiten vieler für eine semantische Sicht möglichen Repräsentationsformen: Schnelle Wortsuche in ausgezeichneten Kontexten, Abstraktion von Inhaltsteilen (z.B. Bildanteile) auf Konzepte, Dokument-Qualifikation und Faktensuche über bewerteten Eigenschaften. Die folgende Beispielanfrage (in natürlichsprachlicher Form) mit Bezug auf verschiedene Dokumentrepräsentationen motiviert diese Vielfalt:

"finde alle Dokumente von 1988,	<- Bezug auf **verdeckten Wert** (nicht im Dokumenttext)
in denen ein Computer angeboten wurde,	<- **Abstrahieren** auf konzeptuelle Aussage "Computer-Angebot"
dessen Preis unter 3000 $ liegt und	<- Zugriff über **Wert** aus dem Inhalt
als ergonomisch beschrieben wird"	<- **Wortsuche** in speziellem Kontext

Spezielle Erweiterungen von DB-Systemen für die Unterstützung von Dokumentenretrieval werden z.B. in [Ston83] vorgeschlagen, wobei für das System INGRES spezielle Textoperatoren eingeführt werden. System-R bietet variabel lange Felder, die auch Bilder als Werte aufnehmen können. Eine spezielle Ausrichtung auf die Dokumentenablage-Anwendung bieten Formularsysteme (z.B. [Tsic82], [Yao84]), bei denen auch NF^2-basierte DBS zum Einsatz kommen (z.B. in [Kapp85]). Auf der anderen Seite werden Informationssysteme, die sich auf Konzeptzugriff abstützen, um (bibliographische-) Attribute erweitert. In [Barb85] und [Cons86] werden inhaltsbeschreibende Konzepte zu konzeptuellen Strukturen aggregiert und Wertzuordnungen zu bestimmten Konzepten zugelassen. Neben atomaren Wertebereichen können Textwerte vergeben werden, die dann die Wortsuche in bestimmten Kontexten ermöglichen. [Eiru88] beschreibt Anwendungen auf der Basis konzeptueller Strukturen.

Das MARS-Dokumentenmodell

Das dem MARS-System zugrunde liegende Dokumentenmodell baut auf dem zuletzt genannten Ansatz auf und unterstützt eine flexible Modellierung einer sich dynamisch ändernden "Dokumentwelt", wie sie insbesondere im Büroeinsatz vorzufinden ist. Wie für Bücher (als Dokumente in Bibliothekssystemen) können in Büro-Dokumenten Aussagen identifiziert werden, die zu Begriffen abstrahiert werden können. Solche Begriffe, deren Bezeichner eine "natürliche" Bedeutung vermitteln, werden im MARS-Modell *Konzepte* genannt (für Bücher: Schlag- / Stichworte).

Nicht nur die Existenz bestimmter Konzepte (in den Beispielanfragen: Computerangebot, Preis, Beschreibung), sondern auch ihre Bewertung durch die Zuordnung von Werten aus einem festgelegten Wertebereich geben den für das Wiederauffinden der Dokumente relevanten Inhalt der Aussage des Dokuments wieder. Die Wertebereiche sind dabei nicht auf atomare Wertebereiche beschränkt, sondern umfassen verschiedene, in multimedialen Dokumenten anzutreffende Informationsmedien wie Text, Rasterbild, Graphik etc.. Konzepte mit zugeordneten Wertebereichen werden *Attribut-Konzepte* genannt.

Die Formulierung von Anfragen über Konzept-Werten mit üblichen Vergleichsrelationen (<, >, = etc.) ist für atomare Wertebereiche (z.B. Absender: String, Preis: ganze Zahlen, Datum: Zahl) sinnvoll. Für die weiteren Wertebereiche sind spezielle, für das Informationsmedium spezifische Vergleichsrelationen erforderlich (z.B. für Text: Wortmatch [Falo85], im Beispiel oben: Beschreibung - Textteil; für Rasterbild: Ähnlichkeitsmaße [Hach87]; für Graphik: räumliche Beziehungen zwischen den Graphikobjekten [Chan87]).

Das MARS-Modell ermöglicht sowohl Abstraktion auf Konzepte wie auch Wertzuordnung zu Konzepten. Vergleichsrelationen sind definiert für atomare Wertebereiche und Text. Weitere spezifische Vergleichsrelationen können als Vergleichsoperatoren definiert werden. Die Darstellung komplexer Aussagen durch Konzepte wird dadurch vereinfacht, daß sie als *Aggregation* ihrer Teilinformationen dargestellt werden, die wiederum auf Konzepte abstrahiert werden können (z.B. aggregiert sich ein Angebot aus einer Produktbeschreibung und einem Preis). In Dokumenten, die gleiche allgemeine Informationen beinhalten (z.B. ein Angebot), kann eine spezielle Charakteristik eines Konzeptes durch eine *Spezialisierung* ausgedrückt werden (z.B. Computer-Angebot). Mit Hilfe der Beziehungen können Konzepte und deren Werte in ihrem Kontext betrachtet werden. Zu einem Dokument bilden die Konzepte mit ihren Beziehungen (Aggregations- und Spezialisierungs-Beziehungen) einen benannten baumförmigen Graphen, in dem die Konzeptbeziehungen die Kanten darstellen. Dieser Graph wird die *konzeptuelle Struktur* oder auch *Dokumentstruktur* genannt (in Abb. 6 wird eine konzeptuelle Struktur eines Dokuments dargestellt: Aggregation und Spezialisierung durch Kasten bzw. Kreis, Wertzuordnungen durch Pfeile auf markierte Inhaltsteile).

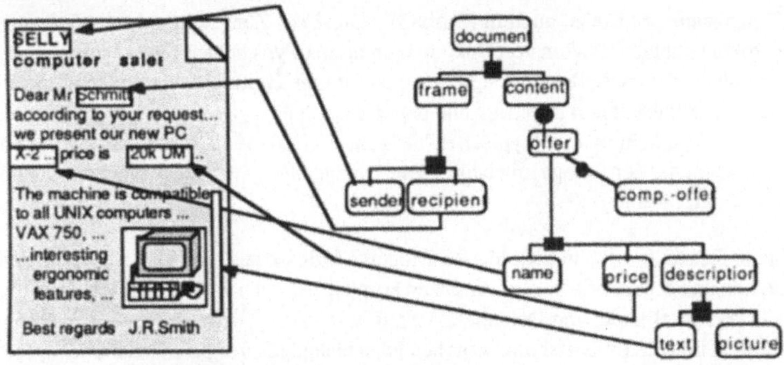

Abb. 6: Konzeptuelle Struktur eines Dokuments

Typisierung von Dokument-Repräsentationen

Dokumente werden gemäß ihrer in konzeptuellen Strukturen repräsentierten Eigenschaften zu *Dokumentklassen* zusammengefaßt. Gemeinsame Eigenschaften von Dokumenten einer Klasse lassen sich in *Dokumentschemata* (i.f. auch: *Dokumenttypen* oder bei eindeutigem Kontext: *Typen*) beschreiben. Dokumenttypen sind also Beschreibungen von Dokumenten mit ähnlichen Aussagen. Ein Dokumenttyp wird durch eine konzeptuelle Struktur, die *Typstruktur*, dargestellt. Die Gesamtheit aller für eine Anwendungsumgebung definierten Typen bildet den *Typkatalog* .

Sowohl Typstrukturen als auch Dokumentstrukturen werden durch Anwenden von *Verfeinerungsoperatoren* ausgehend von einem fest vorgegegebenen Typen hergeleitet, dessen Struktur dann als Schablone dient (Wertzuordnung, Einfügen innerer und äußerer Knoten). Die Möglichkeit der Erzeugung einer gegebenen konzeptuellen Struktur S_1 aus einer gebenen Struktur S_2 durch Anwenden einer Folge von Verfeinerungsschritten legt die Verfeinerungsrelation "(S_1 ist) *feiner* (als S_2)" fest (in der Literatur auch mit "is_a"-Relation bezeichnet, hier um die entsprechenden Verfeinerungsmöglichkeiten erweitert). Gemäß dieser Verfeinerungsrelation werden alle

Typen des Typkatalogs in einem gerichteten azyklischen Graphen angeordnet (Verband) und die Dokumentmenge bzgl. dieser Typen klassifiziert. Die zugeordneten Typen eines Dokumentes sind dabei die *am besten passenden* Typen, wobei die Dokumentstruktur feiner als die Typstrukturen sind und kein feinerer Typ existiert, für den dieses ebenfalls gilt. In Abb. 7 wird das Dokument aus Abb. 6 gemäß seiner Struktur dem Typ "OFFER-TYPE" des Typkatalogs zugeordnet.

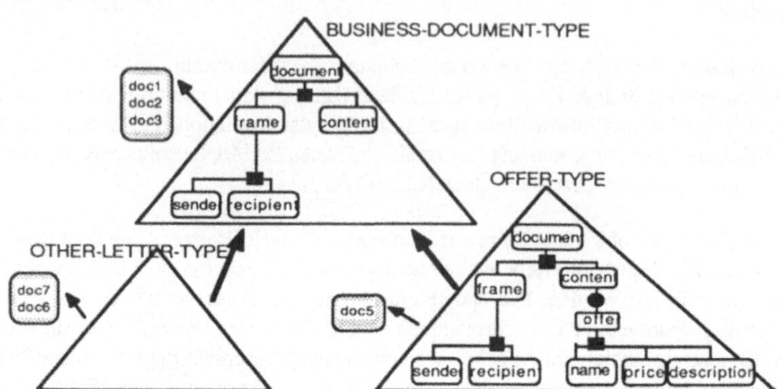

Abb. 9: Geordneter Typkatalog mit Dokumenten-Klassen und konzeptuellen Strukturen

Die Typisierung dient hauptsächlich drei Zwecken:

(a) Typen unterstützen die einfache und konsistente Manipulation von Dokumentstrukturen durch Ausnutzung der Vererbung der Typstruktur auf zu erzeugende Dokumentstrukturen. Diese werden durch die Zuordnung von Werten zu Attributkonzepten und ggf. weiteren Verfeinerungen aus einer Typstruktur erzeugt. Dabei müssen sie nicht eine identische Struktur des Types aufweisen, sondern können feinere strukturelle Merkmale aufweisen (*schwache Typisierung*, zu finden auch in [Barb85], [Grei87]). Zum Zeitpunkt der Typdefinition müssen daher nicht alle Dokumenteigenschaften von Dokumenten dieses Typs bekannt sein.

(b) In den Typstrukturen werden Eigenschaften der ihnen zugeordneten Dokumente repräsentiert. Bei der Anfrageformulierung kann diese Information abgefragt und benutzt werden.

(c) Das Wiederauffinden von Dokumenten mit "typischen" Eigenschaften wird durch die Zuordnung der Dokumente zu Typen zum Zeitpunkt der Ablage unterstützt. Die Typzugehörigkeit selbst kann zur Qualifikation der gesuchten Dokumente benutzt werden.

Wir können an dieser Stelle feststellen, daß in dem MARS-Modell Einflüsse aus dem klassischen Information-Retrieval, den (Text-)DB-Systemen und objekt-orientierten Ansätzen konvergieren. Dies wird deutlich durch

- die Abstraktion auf Konzepte und die Bildung von Dokumentklassen
- die Bewertung von Eigenschaften und die Einbeziehung wertebereichsabhängiger Vergleichsrelationen
- die einfache Modellierung durch Strukturierung, Typisierung und Repräsentation eines Benutzerobjekts in einem Modellobjekt.

Retrieval-Funktionalität in MARS

Die Anfragemöglichkeiten in MARS nutzen die Information der konzeptuellen Strukturen von Typen und Dokumenten. Sie erlauben die Spezifikation von relevanten Dokumenten durch explizite Benennung von Konzepten, wobei die Qualifikation von bewerteten Konzepten durch entsprechende Wertvergleiche eingeschlossen ist. Weiterhin soll die Möglichkeit zur Festlegung allgemeiner "typischer" Aussagen durch Spezifizieren von Dokumenttypen, die Klassen von ähnlich beschriebenen Dokumenten bezeichnen, eingeschlossen sein. Die Einbeziehung von benutzerdefinierten Dokumentmengen (*Kollektionen*) in die Abfrage erlaubt die Eingrenzung des Suchbereichs auf die Dokumentmenge, die aus einem Mengenausdruck über Kollektionen definiert wird. Weiterhin erlaubt die Sicherung eines Abfrageergebnisses in einer entsprechenden Kollektion die iterative Verfeinerung einer Abfrage.
Ein einfaches Beispiel eines entsprechenden Suchausdrucks ist:

```
SELECT DOCUMENTS  FROM TYPE Business-Document-Type
                  OF COLLECTION AktuellerVorgang AND AlteAbfrage
                  WHERE offer AND description CONTAINS "ergonomic"
```

Bei der Konzeption der Systemfunktionen wurde insbesondere zwei Optimierungsproblemen Beachtung geschenkt. Die bei der Verwaltung des Typkatalogs auftretenden Konzistenzfragen und Optimalitätsfragen bei der aufwandskritischen Reklassifikation nach einer Änderung des Typkatalogs wurden untersucht. Ein Problem, das bei der Abarbeitung obiger Abfragen auftritt, ist die Entwicklung geeigneter Optimierungsstrategien. In Archivsystemen für multimediale Dokumente können in der Abarbeitung einer Abfrage (wie oben) viele Zugriffsverfahren involviert sein, deren optimale Auswahl und Reihenfolge der Benutzung durch deren unterschiedliche aufwandsbeeinflussenden Charakteristika a priori nicht festlegbar ist. Für MARS kommt ein in [Eiru91] vorgestelltes Verfahren zum Einsatz. Dort wird ein universelles, nicht an eine bestimmte Konfiguration von Verfahren gebundenes Optimierungsschema vorgestellt, das anhand der gegebenen Verfahrenscharakteristika und Zustandsinformationen zu einer Abfrage eine optimierte Bearbeitung steuert.

4. Anwendungsszenarien für Dokumenten-Archivsysteme

Die Ansätze von MARS sind dort effektiv einsetzbar, wo überschaubare Mengen von Aussagen in Dokumenten leicht identifizierbar sind, die für Klassen von Dokumenten typisch sind. Solche Szenarien finden sich häufig in der Büroumgebung (z.B. Angebote, Mahnschreiben, Einladungen etc.). Hier findet sich auch häufig der Bedarf für eine Unterstützung, Dokumente über bestimmte, den Aussagen zugeordnete Inhaltsteile zu qualifizieren (Preis eines Produktes, Wortvorkommen in bestimmten Textteilen, Form eines Bildes). Anwendungsumgebungen, in denen die zu betrachtende "Dokumentwelt" noch nicht endgültig beschrieben werden kann, profitieren von der Möglichkeit des "schwachen Typkonzeptes" und der flexiblen Anpassung der "Typwelt" in MARS. Die Unterstützung bei der Integration neuer Zugriffsstrukturen für den Wertzugriff und deren Einbindung in die Abfrageoptimierung schafft aber auch Vorteile bei der Konfigurierung in Anwendungen mit einer eher statischen "Dokumentwelt" (z.B. Formularverwaltung).

In Anwendungen, in denen die zu verwaltenden Dokumente keine hinreichend homogene Welt beschreiben (z.B. Monographien oder Artikel in Bibliotheken) wird eine über die rein bibliographischen Angaben hinausgehende Generalisierung schwierig. Hier werden Werkzeuge zur Unterstützung der Deskribierung und des Thesaurus-management benötigt, wie sie in ODIN entwickelt wurden. ODIN-Anwendungsbeispiele sind u.a. "CH-MED" (Differenziertes Retrieval einer Bildplatte mit über 40.000 medizinischen Farbbildern), "Landkreis Celle" (Retrieval und Navigation von Bildern, Vidoes und Texttafeln aus dem Landkreis Celle für die CEBIT 88) und "ODIN-RAD" (Retrieval multimedialer Dokumente und administrativer Daten in der Radiologie einer großen Klinik mit differenzierten digitalen bildgebenden Systemen) [Appe90].

Bei Dokumenten, in denen die Repräsentation einer Vielzahl von Beziehungen zwischen den Teilaussagen für das Retrieval wichtig wird (z.B. räumliche Beziehungen zwischen Teilbildern in Nur-Bild-Dokumenten) reichen die Modellierungsmöglichkeiten des MARS-Modells nicht aus und bieten kaum Vorteile gegenüber dem Ansatz in ODIN. Hier kommen wohl eher Formalismen zum Einsatz, in denen die räumlichen Beziehungen einzelner Bestandteile in topologischen Bildern repräsentiert werden können (z.B. markierte Graphen bei [Rabi87], Linearisierung 2-dimensionale Bilder durch Projektion in [Chan87]) oder durch eine formalisierte textuelle Beschreibung (z.B. räumliche Präpositionalphrasen in [Meye91]).

Der Einsatz der Navigationswerkzeuge ist sinnvoll u.a. in kooperativen Prozessen (z.B. Aufbereitung einer Ergebnismenge durch Spezialisten, für weitere individuelle, benutzerverantwortete Suche, z.B. Zusammenstellung von Dokumenten zu Präsentations- oder Ausbildungszwecken; kooperative Suche und Ergenispräsentation).

5. Zusammenfassung und Ausblick

Mit den Systemen ODIN und MARS wurden zwei Realisierungen vorgestellt, die auf unterschiedlichen Inhaltsrepräsentationen beruhen (Deskriptoren bzw. konzeptuelle Strukturen). Die Speicherung der Dokumentinhalte selbst steht dabei nicht im Vordergrund. In ODIN können Videodisk-Systeme oder CD-ROM mit nichtänderbarem Dokumentbestand benutzt werden. MARS betrachtet in dem ersten Prototypen nur Textdokumente, die außerhalb des Systems verwaltet werden.

Zur laufenden Entwicklung und Integration eines Multimedia-Editors (mit Text, Graphik, Rasterbild) profitiert MARS von dem in ODIN entwickelten Oberflächenentwicklungs-Tool XFantasy [Boles91] und der Geräteunabhängigen Schnittstelle zum Speichersystem. Die Weiterentwicklung des Tools zur Erzeugung von Navigationsprogrammen zur Aufbereitung von Ergebnismengen ist in Vorbereitung.

Danksagung

Wir danken Helmut Lorek und den Mitgliedern der Projektgruppe TEXIDOS für die Unterstützung im bei der Realisierung von ODIN und MARS. Den Gutachtern gilt unser Dank für erfreulich zahlreiche und nützliche Hinweise zur Überarbeitung dieses Beitrags.

Literatur

Appe87 H.-J. Appelrath, *ODIR und ODNA: Retrieval und Navigation von Laser-Bildplatten*, Interner Bericht Nr. 75 des Instituts für Informatik der ETH Zürich, 1987

Appe90 H.-J. Appelrath, H. Lorek, *Softawre-Werkzeuge für multimediale optische Speichermedien*, In: Betriebliche Informations- und Kommunikationssysteme Band 13, E: Schmidt Verlag, 1990

Barb85 F. Barbic and F. Rabitti, *The Type Concept in Office Document Retrieval*, Proc. 11th Conf. on VLDB, Stockholm 1985

Boles91 D. Boles, H. Lorek, *XFantasy - Konzeptpapier*, Interner Bericht des FB Informatik der Universität Oldenburg, 1991

Chan87 S. Chang, Q. Shi, C. Yan, *Iconic Indexing by 2-D strings*, IEEE Transactions on Pattern Analysis and Machine Intelligence, Vol. 9/3,1987

Cons86 P. Constantopoulos, H. Eirund, K. Kreplin, F. Rabitti, C. Thanos et al., *Office Document Retrieval in MULTOS*, Proc. 3rd ESPRIT Tech. Week, North-Holland 1986.

Eiru88 H. Eirund, K. Kreplin, *Knowledge Based Document Classification Supporting Integrated Document Handling* , ACM Conf. on OIS, Palo Alto, 1988

Eiru91 H. Eirund, *Abfrageoptimierung in Archivsystemen für multimediale Dokumente*, Proc. BTW 91, IFB 270, Springer-Verlag, 1991

Falo85 C. Faloutsos, *Access Methods for Text*, Computing ACM Comp. Surveys, Vol. 17/1, 1985

Grei87 G. Greiter, *Zu zentralen Design-Entscheidungen beim Entwurf der Datenbank POINTE/PVS*, Proc. BTW 87, Darmstadt, IFB 136, Springer-Verlag, 1987

Hach87 K. Hachimura, *A Prototype PACS with the Capability of Retrieval by Image Data* , Proc. Computer Assited Radiology, Springer-Verlag, 1987

Kapp85 G. Kappel, A.M. Tjoa, R.R. Wagner, *Form flow systems based on NF^2-relations*, in BTW85, Karlsruhe, IFB 94, Springer-Verlag, 1985

Lorek88 H. Lorek, Fantasy: *Methoden und Werkzeuge für die Erstellung graphischer Benutzerschnittstellen*, in: K. Kansy, P. Wißkirchen (Hrsg.): Graphik im Bürobereich, GI-Fachgespräch, Bad Honnef, 1988

Meye91 K. Meyer-Wegener, *Multimedia-Datenbanksysteme*, Teubner Verlag, 1991

Niev84 J. Nievergelt, H. Hinterberger, K.C. Sevcik, *The Grid File: An Adaptable, Symmetric Multikey File Structure* , ACM TODS 9 / 1, 1984

Rabi87 F. Rabitti, P. Stanchev, *An Approach to Image Retrieval from Large Image Databases*, Proc. ACM-SIGIR Aonf., New Orleans, 1987

Ston83 M. Stonebraker et al., *Document Processing in a Relational Database System* , ACM TOIS, 1 / 2, 1983

Tsic82 D. Tsichritzis, *Form Management*, CACM 24 / 7, 1982

Yao84 S.B. Yao, A.R. Hevner, Z. Shi, D. Luo, *FORMMANAGER: An Office Forms Management System* , ACM TOOIS 2 / 3, 1984

A Concept for a Flexible High Performance Transport System

Martina Zitterbart, Burkhard Stiller
Institut für Telematik
Universität Karlsruhe
D-W7500 Karlsruhe
Tel.: +49 / +721 / 608-3983
FAX: +49 / +721 / 606097
E-Mail: Stiller%i70vca at iravcl.ira.uka.de

Abstract

Transport systems as they are known today will not be suited for the environment of high speed networks and future applications as e.g. multi-media applications. First of all, concepts to overcome the protocol processing bottleneck are necessary. They lead to a design of a transport oriented component within a network as a conglomeration of protocol functions without internal layering. This results in an impressive reduction of functions to be realized. Moreover, the transport service interface must be designed in a way that allows an easy and efficient support of various service requirements. Therefore, the transport service interface can be classified into four transport service classes, which are derived from the application requirements on performance and functional criteria. Each service class is associated with an individual protocol stack. The paper discusses these topics with special regards to flexible high speed transport systems. The flexibility, which is the possibility to adapt the transport system to the specific application dynamically, will be obtained by using an underlying multi-processor architecture, where the processing of different functions in parallel will be supported.

1. Introduction

Recently, a lot of high performance applications, like tele-robotics, video transmission and in general multi-media applications, came into fashion. This evolution has been supported by the possibility to use available high speed networks as local backbones, e.g. the FDDI-ring. In this case the bandwidth of the underlying network is no longer the performance bottleneck in computer networks. Now ***protocol processing inside network nodes*** becomes the limiting factor, because of the used protocols. They had been developed for lower data transmission rates. With regard to this, current research efforts take place in the design and evaluation of new concepts, architectures, and protocols for future high speed transport systems being the fundamentals for the mentioned high performance applications. For this reason, it is important to develop an efficient transport system, which reduces current overheads concerning the protocol itself and the runtime environment (system overhead).

This paper presents a concept for a high speed transport system, which is based on a simplified communication model consisting of three logical components. Thereby, it has been considered that processing of the protocol state machine within a layer is dominated by processing overhead caused by layer interfacing and system functions (e.g. process switching and buffer management). That is the motivation for the concept described throughout this paper. Additionally, the use of parallel architectures for protocol processing may be more efficient. Therefore, the concept was developed in a way, where it is possible to support parallel processing. The main advantages of our approach are the flexible dedication to totally different applications, the reduction of processing overhead, which will result in higher throughput, less processing delays, and if desired in the achievable compatibility to the ISO/OSI-Basic Reference Model at the component interfaces.

Section 2 summarizes a short overview of related work. The following section 3 presents the simplified communication model. The flexible transport system is discussed in sections 4 and 5. Section 4 focuses on the protocol functions of the transport oriented component, whereas section 5 presents design issues of the transport service interface. Additionally, in section 6 some implementation aspects of the flexible transport oriented component are put together. Section 7 concludes the paper.

2. Related Work

Research on high speed transport systems can be subdivided into several parts dealing with the transport protocols themselves, an efficient implementation of those protocols on suitable architectures as well as with the transport service interface. Some approaches are based on the development of new so-called high speed protocols that use adapted protocol mechanisms for dedicated protocol functions as e.g. connection handling, acknowledgement strategies, and flow control machanisms. Examples can be found in /ChWi89/, /Ches89/, and /Wats89/. In /DDKM90/ an excellent overview thereof is given. Section 5 will focus on the third point mentioned above. The transport service interface will be discussed in detail. Within the second point mentioned special network adapters /KaCh88/, VLSI-Implementations and parallel protocol implementations e.g. on Transputer networks /Zitt89/ are considered. The use of protocol inherent parallelism combined with the development of a well suited multi-processor architecture seems to be a promising approach to lead to performance increase (e.g. /GKWW89/ or /Zitt91/). Additionally, the decomposition of the strongly layered architecture of communication protocols may improve the performance of protocol implementations because of the reduction of layer interfacing. One approach of delayering has been done by /Haas91/. /MeTa90/ and /ClTe90/ have been dealing with a more generic consideration of removing certain layer boundaries and decomposition into functions. The next section describes a communication model of a similar approach.

Additionally, multi-media systems in general are an actual research field that is closely bound up with investigations in high speed networking. Especially multi-media applications make eminent demands on the performance of underlying networks and requested transport services. The papers /HSSt90/ and /Salm89/ discuss concepts for the design of high speed transport systems and transport service interfaces with special attention to multi-media applications. Collections of application requirements may be found in /Lidi90/, /HSSt90/, /Stil90/, /WFMS89/ and /WrTo90/. Some of them rely on the defined CCITT-Service-Categories /CCIT88/.

3. Simplified Communication Model

The ISO/OSI-Basic Reference Model subdivides communication protocols into seven layers that are called application oriented (layers 7-5) and transport oriented layers (layers 4-1). Each layer is transparent to each other and communicates via well defined service interfaces, which leads to functional overhead. As an example, layers 3 and 4 are allowed to segment and reassemble protocol data units. Furthermore, processing overhead is added due to the communication accross the layer interfaces, even if a special buffer-management will be used.

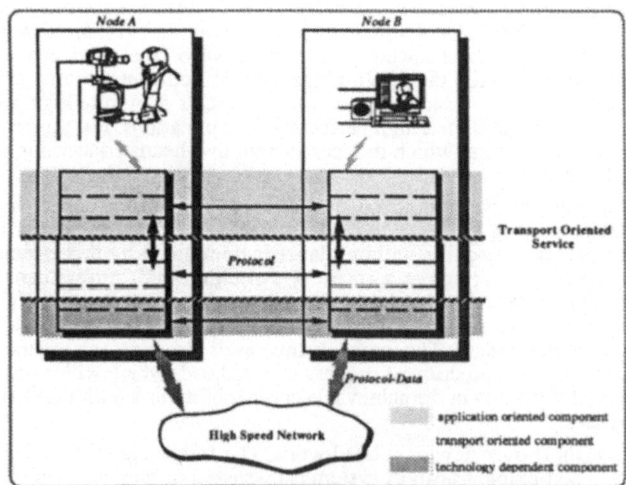

Figure 1 3-component model of communication

A simplified communication model, the so-called **3-component-model** /Zitt91/ being not as fine grained as the ISO/OSI-Basic Reference Model, will be the basis for the discussion of a **flexible transport system** throughout this paper. It consists of three components being responsible for the technology dependent, the transport oriented and the application oriented functions (cf. figure 1). Inside these components no separate layers are distinguished. Each component can be seen as a conglomeration of protocol functions, which are necessary to fulfill the required service at the specific interface. At the different service interfaces in figure 1 the thickness of the arrows associates the achievable maximum throughput being degraded in orders of magnitude by protocol processing within the communication nodes. The discussed delayering is more appropriate than the solution in /Haas90/ and is more specific than the general approaches of /MeTa90/ and /ClTe90/. On one hand, combining the layers 1 to 3 is not useful, because the media access control schemes are dependent on the underlying network and the routing- and relaying-functions are absolutely independent on the network. Additionally, regarding the unition of layers 4 to 6 a similar argument counts; the transport layer includes more functions that are similar to the network layer functions than to the presentation layer functions. There is no reason to mix application dependent presentation information with the transmission depending information like maximum supportable segment sizes or flow control parameters. On the other hand, the delayering in /MeTa90/ has been done for just two parts: the application process and a whole communication machine. The interiors of the communication machine are not adjusted to the ISO/OSI-Basic Reference Model and do not allow the explicit relating to functions separeted from OSI protocols.

The **technology dependent component** comprises the layers 1 and 2a of the ISO/OSI-Basic Reference Model and provides high performance at the service interface. The reason for combinig these one and a half layers is their common medium dependency. Defining and fixing an interface - that is compatible to the ISO/OSI-Basic Reference Model - above layer 2a allows the exchange of any underlying network. Therefore, it is no problem to use a HSLAN, e.g the FDDI-ring, or any imaginable network.

The **transport oriented component** includes functions of layers 2b up to 4 and is responsible for correct data transport through the network. The main difference contrary to the seperate layers lies in the internal organization of this component. The functions themself are considered as basic protocol elements (cf. section 4). This transport oriented component has to place a reliable and flexible transport service at the application interfaces disposal. In the following we will focus especially on the transport oriented component.

Finally the **application oriented component** includes layers 5 to 7 of the ISO/OSI-Basic Reference Model. Throughout this paper this component will not be regarded in detail.

4. Protocol Functions of the Transport Oriented Component

As stated above, the transport oriented component considered in this paper consists of a conglomeration of protocol functions without further subdivision into protocol layers. This set of functions comprises the **protocol functions** of the OSI layers 2b up to 4. They also may integrate additional functions of improved protocols, e.g. multi- and/or broadcasting as well as synchronization. Note, functions being necessary for the implementations of service interfaces are required for the communication with the technology dependent and the application oriented component only, but no longer inside the transport orientied component itself due to lack of internal layering. The functions of OSI protocols of the transport oriented component are categorized into six different domains that are further developments and some specializations of /MeTa90/:

- connection management,
- protocol data processing,
- routing,
- service interface,
- system functions, and
- network management functions.

The first category **connection management** functions includes besides connection set-up and release functions all functions and functionalities that influence any kind of connection or virtual channel. For that reason it is appropriate to tie them up into one categorie. Especially the flow control affects the data transmitted on a specific connection. For realizing this function the sequence number is needed. Distinguish the resequencing function which needs a sequence number also to work correctly. In fact, the sequence number is a necessary precondition for these two functions. Any (de-) multiplexing and synchronization influences the actual connection, where the second function is very important to transmit two or more streams of data which are related in one valid manner only. Any **protocol data processing** function manipulates the transmitted information, e.g. the packet header and/or trailer. These functions include for instance PDU-(de-)composition, packet header analysis, lifetime control, segmentation, reassembling, padding. Additionally the functions concerning the reactions onto different states of a protocol machine, e.g. acknowledgement, duplicate recognition, packet retransmission, and QOS-maintenance, belong to the second category. Obviously, all addressing mechanisms and relaying functions are separated in the

routing category. In detail, the route PDU function, source routing, record routing, addressmanagement, and multi- and/or broadcasting belong to the routing category. The **service interface** functions are used to communicate with the technology dependent and the application oriented component. These are send/receive SDU to/from down/up. **System functions** are heavily implementation dependent like the buffer- and timer-management. The **network management** function will not be regarded any further. Moreover, protocol functions of enhanced high speed protocols like VMTP or XTP are not discussed, because in general they are very similar to OSI protocol functions (cf. /Stil90/ and /ZiGe90/).

The suitability of this functional approach has to be proved by real implementations, which have to be evaluated under performance aspects. In /Zitt91/ parallel protocol implementations of transport oriented protocols are presented that follow the explicite separation of protocols into their functions directly. According to this, a multi-processor architecture for such an implementation has been developed. Besides the functional approach an important additional factor arises by using multi-processor architectures: the parallelism. Further performance increases are possible, while exploiting the parallelism of necessary functions; e.g. checksum calculation could be done in parallel to packet header analysis. Prototypical implementations of transport oriented OSI-protocols on Transputer networks have been done and show promising performance results. Because of the adaptability and the achieved performance, this underlying architectural concept seems to be well suited for a realization of the described transport oriented component.

In the following section design issues and concepts of a service interface for the transport oriented component as well as the combination of well suited protocol stacks are discussed.

5. Transport Service Interface

To characterize the interface of the flexible transport system to the application oriented component detailed application requirements on the service have to be investigated. Therefore, first of all in section 5.1 a subdivision of requirements on the flexible transport system has to be given. Subsequently, the resulting transport service classes, classified by qualitative and quantitative requirements, are shown and discussed in section 5.2. Some general remarks on the transport oriented component will be found in the last subsection.

5.1 Requirements on the Flexible Transport System

To give a structural overview, the **requirements** are divided into the following two categories:

- performance requirements and
- functional requirements.

The distinction of performance and functional requirements is straightforward, even if the distinctive marks are not completely independent from each other. But due to a more systematic discussion of these requirements and their distinction, it is well suited in this context. In particular, the quantity of requested functionalities as well as the selected protocol mechanisms influence the performance of the transport system. Later on, performance criteria are used to define the transport service class characteristics (cf. figure 4). Finally, the functional criteria and the protocol functions itself specify the functionality of each transport service class (cf. table 3).

5.1.1 Performance Criteria

To obtain a general view of the main **performance criteria**, which give a description of the application/transport interface take notice of figure 2.

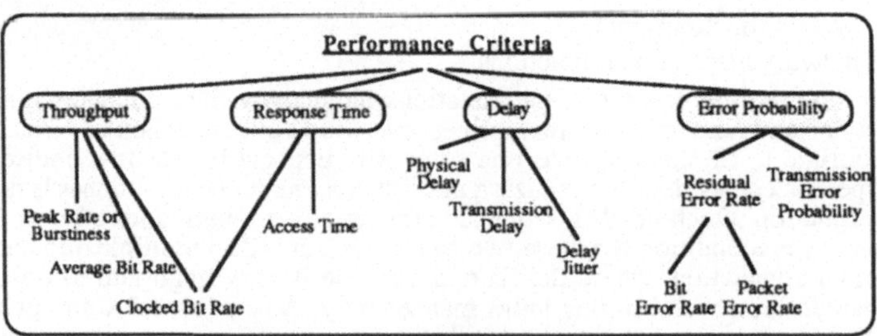

Figure 2 Performance criteria

The classification into four main categories has been done as follows:

- throughput,
- response time,
- delay, and
- error probabilities.

In detail, the clocked bit rate will be mentioned. It has to be calculated by dividing the to be transmitted amount of data by the response time. This number describes the actual needed throughput. Therefore, the necessary bandwidth for an application is heavily dependent on its timing requirements. Second, the peak rate defines the maximum needed throughput for a specific application. A third remark will be done; the delay jitter is the most important factor for transmitting isochronous information. Further criteria are discussed in /Stil90/. Some of these criteria will be found as quality of service parameters in ISO documents, cf. /ISO86/. Others, like the burstiness or the delay jitter, are not yet included in any standard. But it is obvious that future standards have to pay attention to these important criteria and that the QOS-parameters have to be requested and/or set by the application itself.

5.1.2 Functional Criteria

A survey of **functional criteria** for a transport system is presented in figure 3, where three main categories are distinguished:

- service type,
- error protection, and
- data-stream-control.

The importance of that distinction will be pointed out in section 5.2. Especially some of these functions have to be requested by the application. Others are absolutely necessary. The decomposition of functional criteria has been done up to an adequate level. The leaves of the developed graph represent functions or functionalities used in protocols. Any further decomposition into different mechanisms, e.g. rate based or window-size based flow control, would not be useful at this point.

The service type of the transport system considering connection oriented and connectionless services depicts an important distinctive mark. Regarding the error protection category the packet discarding belongs to error detection and recovery, because a detected and non treatable error inside a packet will cause the discarding of this packet. Finally, this results in the detection of a lost packet whereupon another recovery facility will be started.

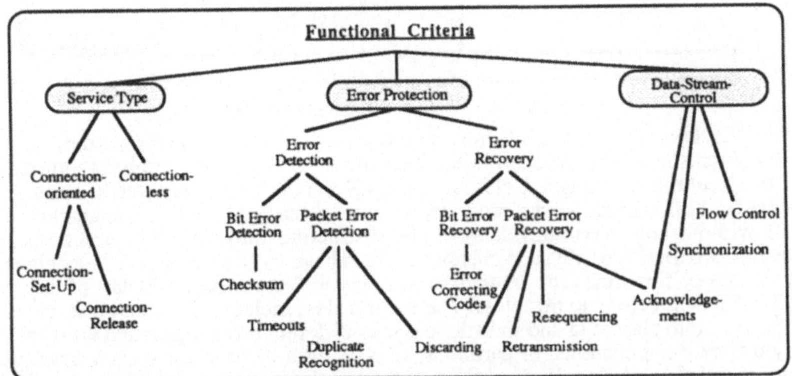

Figure 3 Functional criteria

The last category comprises protocol functions, which are responsible for the control of a single data stream and the synchronization between two or more of them. Their existence in the actual transport system may be choosen explicitly with a service request at the transport service interface. They include protocol functions, which may be designed within different mechanisms, e.g. acknowledgement (e.g. cumulative or selective) and flow control mechanisms (e.g. sliding window, interpacket gap). The overlapping of error protection and data stream control results in the acknowledgement function, because on one hand the data stream will be influenced by transmitting acknowledgements; on the other hand it is a packet error recovery function. Other approaches to disjoin flow control and error protection - especially acknowledgement strategies - have been proposed using for instance two different sequence numbers.

5.2 Transport Service Classes

The description of a various number of applications on the strength of performance and functional oriented criteria leads towards four **transport service classes**, identified by typical requirements as summarized in the three-dimensional figure 4, where the x-y-plane delineates the different transport service classes depending on the criteria (x-, y-, and z-achsis). This figure points out that the delay jitter and the real-time capability were applied as the most important and in a way general and equal classification criteria. The distinction into these four classes were driven by a detailed requirement study of existing applications. Only the essential criteria for the application itself and the important requirements for the protocol stack were regarded. For that reason, the useful combinations of requirements were choosen only. For each block at the lowest level in the figure 4 all necessary requirements are specified qualitatively as they are useful for the future protocol stack design (cf. table 3). The development of these transport service classes has been done, because of the non-suitability of the CCITT-service categories for specifying a set of functions for the associated protocol stacks.

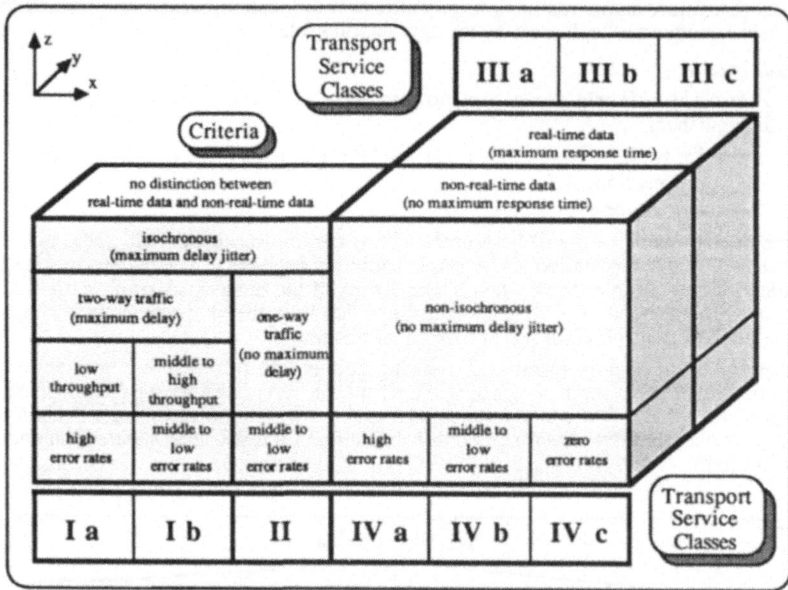

Figure 4 Transport service classes

The distinction of the non-isochronous information into real-time and non-real-time data only is sufficient. Any kind of delay thresholds are included in the maximum response time implicitly in case of real-time data. Finally, the isochronous transport classes are divided into a two-way traffic **class I**, specified by maximum delays for both transmission directions, because of the conversational character, and a one-way traffic **class II** without any maximum delays. The distinction into different directions results in the existence of one or two logical or physical channels, which have to be maintained. The delays in class I are considered as a sum of two times the transmission delay and the response delay of any receiver. Any conversational traffic will belong to this class. The second class includes distribution services /CCIT88/. A subdivision of class I into classes Ia and Ib is done due to different throughputs and error rates, which may be required by different applications. In particular, class Ia will be used for speech transmission mainly. The reason for one subclass of class II only is the existence of applications, which need various amounts of throughput, but always need middle to low error rates.

The concentration on the response time and non-isochronous data results in the **class III**, where real-time data are included. These data have to comply with a stricktly observed maximum response time. The distinction into three smaller subclasses is done due to different error rate requirements. The throughput requirements vary in a small throughput interval only and, therefore, it is not necessary to divide the class III - and class IV also - into more throughput oriented subclasses. Some kind of robotic control data or monitor data may belong to this class.

Last, the **class IV** combines all non-isochronous and non-real-time data. Any limiting value for the response time is unnecessary. The distinction of high, low, and zero error rates results in the subclasses IVa, IVb, and IVc argumented as for the subclasses of class III.

Determining exact **threshold values** is quite difficult or almost impossible, because the values are highly application dependent and the order of magnitude may be fixed only. Considering the compression of data results in smaller amounts of data but in larger delays, caused by (re-)calculating the original/compressed data. One attempt to give some value intervals for the thresholds is discussed in /Stil90/ circumstantially using limits, which were collected from various papers. Other approaches may be found in /HSSt90/ and /Salm89/. An association of applications to the defined transport service classes is given in table 2. Note, the classification of any specific application in more than one class with different requirements, e.g speech or text, is not contradictory because the lowest threshhold values are sufficient - better values do not allow a necessarily needed better quality! Therefore, it is possible to satisfy any application requirement located in a lower subclass in a higher subclass also, e.g. uncompressed graphic may be transmitted in class IVc also. That is the reason of defining minimal thresholds.

Transport Service Classes	I a	I b	II	III a-c	IV a	IV b	IV c
Appli-cations	speech	speech, music, slow-motion-images, interactive video, and tele-conferencing [conver-sational]	speech, music, slow-motion-images, video, and HDTV [distri-bution]	real-time data (text, data, formula, table, graphic and, photo like in classes IVa-c)	text, formula, table, graphic, and photo (uncom-pressed)	text, photo, and graphic (compressed)	text and data

Table 2 Association of applications to transport service classes

The collection of protocol functions (cf. table 1), which are necessary for the realization of each transport service class, will be shown in table 3 in their specific arrangement. This should be considered in relation to the transport service classes (cf. figure 4) and their basic requirements given by the included applications (cf. table 2). Some functions, in particular the multiplexing and demultiplexing function, are missing, because any protocol stack may be activated n-times in parallel while needing more than one connection. The split and recombine functions had been left out, because they may limit a more efficient protocol processing (e.g. immediate parsing the protocol header). Any kind of source and record routing may be regarded as a specific route-PDU function. Thus, they are not listed as separate functions in table 3 anymore. The sequence number is the absolutely necessary precondition for the acknowledgement function and the flow control.

The minimal set of functions for one single transport service class, represented by "x" in table 3, may be enhanced by some optional functions. Functions marked by "o(t)" have to be selected by the flexible transport system itself knowing the network details like maximum delay to a specific receiver and in general the quality of the underlying network. The "o(a)" functions will be choosen by an explicit request of the application. This may happen in case of a desired better transmission quality without changing the selected class. Finally, a non-used function is marked by a hyphen "-".

The functions to be used in general in every class are the PDU-composition and PDU-decomposition, the packet header analysis, the packet discard function, the QOS-maintenance, the route and relay PDU, and the service interface functions. Any kind of management functions, like address, timer, buffer, and network management will not be regarded in detail, but they are needed. The security and multi- and broadcasting functions are always application dependent and the padding is dependent on the transport oriented component. Considering the QOS-maintenance function in detail the QOS-parameters differ in maintaining connection less services, connection oriented services, or isochronous traffic /Stil90/ (cf. remarks in chapter 5.1.1 and 5.3).

No.	Function \ Transport Service Class	I		II	III			IV		
		a	b		a	b	c	a	b	c
1	Connection Set-Up	x	x	x	-	-	-	-	x	x
2	Connection Release	x	x	x	-	-	-	-	x	x
3	Resequencing	-	o(a)	o(a)	-	o(a)	x	o(a)	o(t)	x
4	Sequence Number	x	x	x	-	o(a)	x	o(a)	x	x
5	Flow Control	x	x	x	-	-	-	-	x	x
8	Synchronization	o(a)	x	x	o(a)	o(a)	o(a)	o(a)	o(a)	o(a)
9	PDU-Composition	x	x	x	x	x	x	x	x	x
10	PDU-Decomposition	x	x	x	x	x	x	x	x	x
11	Packet Header Analysis	x	x	x	x	x	x	x	x	x
12	Lifetime Control	o(t)	o(t)	o(t)	x	x	x	x	o(t)	o(t)
13	Segmentation	o(t)	o(t)	o(t)	x	x	x	x	x	x
14	Reassembly	o(t)	o(t)	o(t)	x	x	x	x	x	x
15	Discard PDU	x	x	x	x	x	x	x	x	x
16	Padding	o(t)	o(t)	o(t)	o(t)	o(t)	o(t)	o(t)	o(t)	o(t)
17	Security	o(a)	o(a)	o(a)	o(a)	o(a)	o(a)	o(a)	o(a)	o(a)
18	QOS-Maintenance	x	x	x	x	x	x	x	x	x
19	Acknowledgement	-	o(a)	-	-	o(a)	x	o(a)	o(a)	x
20	Checksum Calculation	-	o(a)	x	o(a)	o(a)	x	o(a)	o(a)	x
23	Packet Retransmission	-	o(a)	-	-	o(a)	x	o(a)	o(a)	x
24	Duplicate Recognition	-	o(a)	o(a)	o(a)	o(a)	x	o(a)	o(a)	x
25	Expedited Data	-	-	-	x	x	x	-	-	-
26	Route PDU	x	x	x	x	x	x	x	x	x
27	Relay PDU	x	x	x	x	x	x	x	x	x
31	Multi-Broadcasting	o(a)	o(a)	o(a)	o(a)	o(a)	o(a)	o(a)	o(a)	o(a)
32	Send SDU up	x	x	x	x	x	x	x	x	x
33	Send SDU down	x	x	x	x	x	x	x	x	x
34	Receive SDU from up	x	x	x	x	x	x	x	x	x
35	Receive SDU from down	x	x	x	x	x	x	x	x	x

Table 3 Protocol functions of transport service classes

The segmentation and reassembling of packets will be transport optional in classes I and II because often isochronous packets are smaller than non-isochronous packets. But depending on the underlying network, e.g. an ATM-network, these functions will be required. For that reason, in classes III and IV segmentation and reassembling are marked "x". Still it has to be decided whether the segmentation and reassembling function belong to the transport oriented or the technology dependent component only. The expedited data function is useful in the real time class III only, e.g. transmitting an emergency stop initiated by the application. In every other class a special management packet will be send as a normal data packet to solve appearing application problems. The synchronization is the most interesting function. It has to be utilized for classes Ib and II, because of the internal combination of speech and moving-images as one transmitting unit. Special internal synchronizing aspects may be satisfied with the same synchronizing function, marked as application optional in classes Ia, III and IV. Synchronization between different transport service classes will be mentioned in subsection 5.3.

The isochronous data transmission (classes I and II) needs a connection in case of a maximum delay jitter that cannot be guaranteed by the connectionless service. Furthermore the importance of low and zero error rates claim a connection oriented service for class IVb and IVc. The exception is class IIIb and IIIc because in general control packets for e.g. some numeric-controlled machine are independent of each other and carry information which fits in a packet with a maximum length. In this case no state information of any connection has to be maintained and less overhead is achievable due to the absence of connection set-up and release times. For that reason the whole class III provides the connection less service. The reliability will be increased using additional error protection mechanisms (functions 19, 20, 23, and 24).

Depending on the service type some special functions have to be used. In the connection oriented case the flow control will be needed. The lifetime control is optional; only the transport oriented component has to decide whether the underlying network does not guarantee the delay in any case. No flow control will be used and a lifetime control has to be used in the connection less service. The sequence number is dependent on the presence on the flow control and resequencing. For that reason, the functionality is marked "x" in the connection oriented service classes and denoted the same as the resequencing function in the connection less service classes.

Finally, the error protection functions (3, 19, 20, 23 and 24) have to be regarded. They vary depending on the used service type, the required error rates, and the application itself. Class Ia does not need any error protection functions. Concerning speech traffic packet retransmission has to be avoided. The class Ib includes conversational traffic that may be supported by any error protection function, where the transport oriented component will request the resequencing function only. Others may be requested by the application itself. In the distribution traffic class II the error protection has to be limited to a necessary checksum calculation and an optional duplicate recognition supported by the resequencing function. Any acknowledgement and packet retransmission is impossible, because of the one-way traffic and on that account useless. The resequencing is dependent on the application like in class Ib.

The whole real-time data class III includes the possibility for expedited data traffic. The class IIIa supports the optional checksum calculation and duplicate recognition only, because higher error rates are tolerable. Furthermore class IIIb includes the optional acknowledgement, packet retransmission, and the resequencing with the sequence number. In the real-time class IIIc all error protecting functions are undispensable.

The functions of class IVa are equal to those of class IIIb unless the expedited data function. Class IVa provides a connection less service because for transmitting short information it is totally sufficient to send single packets. No connection set-up delay and connection state maintaining overhead exists. The class IVb supports a connection oriented service and every error protection function is optional. the resequencing function may be requested by the transport oriented component only (see above). Finally, the class IVc uses every error protection function.

5.3 General Remarks

The transport component has to support the transmission of data depending on the QOS-parameters negotiated between the transport service interface and the application. This component does not have to decide what kind of alternatives concerning the functionalities of a protocol stack will be used to satisfy the application requirements in case of the mismatch of QOS-parameters. For that reason, the application itself is responsible for taking steps against it, which will be the change of QOS-parameters or the abort of the connection. This might be seen as a loss of transparency between the transport oriented component and the application. In the sense of pure OSI this will be right, but the must of an efficient transport component requires the exchange of QOS-parameters and demands the possibility to turn on or off functions explicitely.

An example concerning class IIIb: Transmitting a table, for instance including the expected room occupation of a hotel, in real time, it makes no difference in which sequence the packets will arrive. Any checksum calculation will be totally unnecessary, because of the non-importance of a single pixel inside the table. Instead transmitting a x-ray the checksum calculation has to be done, because each pixel may show important information. The difference between these two data streams will be known by the application only and, therefore, the application has to request the desired functionalities.

The approach of reducing the transparency does not challenge the ISO/OSI-BasicReference Model as a model to understand the complexity of data communication, but it will improve some details to make an efficient implementation feasible.

Concerning the problem of synchronization we propose the following: A separate **synchronization** function is needed to guarantee the correct run of events; that is the timing of packets with different but relating infomation content, e.g. moving-images and music. To our opinion this function should be located right above the transport oriented component, because on one hand the application itself determines the use of any synchronization and on the other hand that kind of function is not responsible for a reliable transport of data and, therefore, this function is not part of the transport oriented component. For instance the maximum delay for a specific data stream will be satisfied by the transport oriented component, but any relation between two or more of them is clearly application dependent.

6. Implementation Aspects on the Transport Oriented Component

Based on the general discussion of the design of the transport oriented component within the simplified communication model and the detailed investigations on the concepts for a transport service interface, a so-called **flexible transport oriented component** can be developed. A generic model of the flexible transport oriented component is shown in figure 5, where the application oriented and technology dependent components are not depicted explicitly. The circles inside the transport oriented component represent a protocol function within a protocol stack. It can be seen that different protocol stacks are supported concurrently for the transport services requested by different data streams of the application (e.g. classes Ia, II and IVc). The protocol stacks are configured according to the service request of the application, which may also include optional functions.

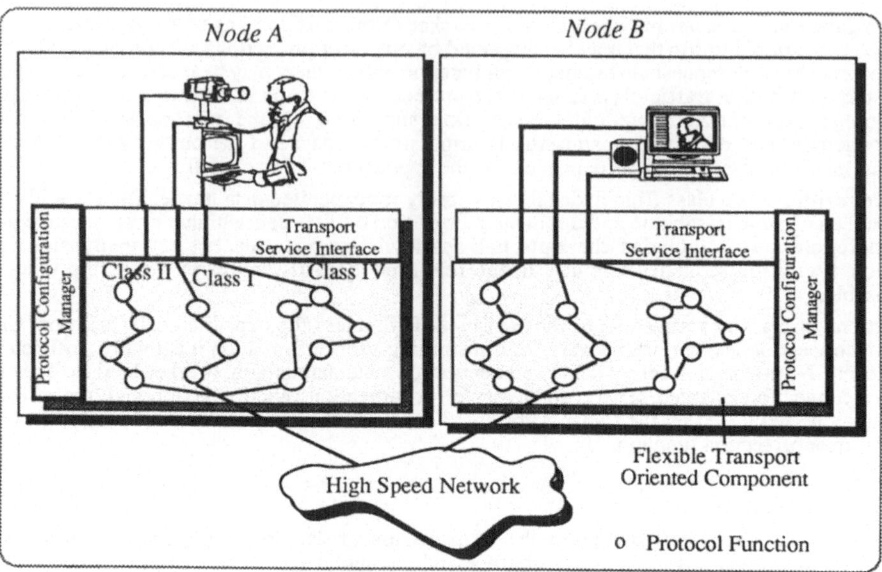

Figure 5 Generic model of a flexible transport oriented component

Thereby, the basic idea is to provide a flexible configuration of protocol stacks dependent on the requested transport service class as shown in table 3. Multiple protocol stacks supporting different service classes, even multiple entities of the same protocol stack may exist simultaneously. For the realization of such an approach especially multi-processor architectures will be well suited, because they allow parallel processing of multiple protocol stacks running on different processors. Moreover, parallelism within a protocol stack may also be provided as mentioned in section 2. Transputer networks will be a well suited implementation environment for a flexible transport oriented component. Especially the ability of an easy adaption of the configuration of the transputer network and the efficient support of parallel and quasi-parallel programming have to be pointed out in this context. The Connectionless Internetwork Protocol and the Transport Protocol Class 4 were implemented using a parallel version of the C language /Zitt91/. This implementation focused on the explicit separation of protocol functions using an underlying Transputer architecture. The interconnection of these protocols and the integration of the Logical Link Control Protocol will be the next step to reach the whole transport oriented component. At this point the quality of service parameters of the TP4 have to be enhanced to satisfy the mentioned possible requests. Finally, the **Protocol Configuration Manager** to support the dynamic configuration of protocol functions has to be developed and implemented. This results in an efficient and flexible benefit regarding the performance of the transport oriented component.

7. Conclusion

The paper has presented a concept for the design of a flexible transport system being well suited for advanced applications in high speed networks. The essential design issues are the reduction of the number of layers and, therefore, the reduction of interfacing overhead, which results in the definition of service classes, associated with an individual protocol stack each. This leads to a flexible design of the protocol stacks being dependent on the requested service. With the idea of parallel protocol processing in mind, first of all it will be possible to use different processors for each requested service and, moreover, the use of the inherent parallelism inside the protocol stacks on multi-processor architectures will be possible. Thus, an efficient implementation of the concepts presented within this paper will be practicable and delineates a progress in the realization of high-speed protocols.

Acknowledgements

The authors would like to thank Torsten Braun who contributed to the presented work, especially to the parallel aspects within communication protocols.

References

/Ches89/ G.Chesson; *XTP/PE-Design*; in "Protocols for High Speed Networks", Editors: H.Rudin, R.C.Williamson, pages 27-33, North-Holland, 1989.

/ChWi89/ D.R.Cheriton, C.L.Williamson; *VMTP as the Transport Layer for High Performance Distributed Systems*; IEEE Communications Magazine, pages 37-44, June 1989.

/CCIT88/ CCITT-Study-Group XVIII (Broadband Task Group), Part C of the Report of the Seoul Meeting (January 25 to February 5, 1988), CCITT COM XVIII-R55(C)-E, February 1988.

/ClTe90/ D.D.Clark, D.L.Tennenhouse; *Architectural Considerations for a New Generation of Protocols*; in Proceedings of ACM SIGCOM 1990, Philadelphia, Pa, USA, September 1990.

/DDKM90/ W.Doeringer, D.Dykemann, M.Kaiserswerth, B.Meister et.al.; *A Survey of Leigth-Weigth-Transport-Protocols for High-Speed Networks*; IBM Research Divison Zürich/Rüschlikon, Dokument 0.16, January 1990.

/GKWW89/ D.Giarrizo, M.Kaiserswerth, T.Wicki, R.C.Williamson; *High-Speed Parallel Protocol Implementations;* in "Protocols for High-Speed Networks", Editors: H.Rudin, R.Williamson, IFIP, pages 303-321, North Holland, 1989.

/Haas91/ Z.Haas; *A Protocol Structure for High Speed Communication over Broadband ISDN;* IEEE Network, January 1990.

/HSSt90/ D.Hehmann, M.Salmony, H.J.Stüttgen; *Transport services for multi media applications on broadband networks*; in "Computer Communications", special issue: MM-Communications, Vol 13 no. 4 , pages 197-203, Mai 1990.

/ISO86/ Information Processing System, OSI Connection Oriented Transport Srevice Definition, ISO Std 8072, 1986

/KaCh88/ H.Kanakia, D.R.Cheriton; *The VMP Network Adapter Board (NAB): High Performance Network Communications for Multiprocessors*; ACM SIGCOM 1988.

/Lidi90/ W.P.Lidi; *Data Communication Needs*; IEEE Network Magazine, pages 28-33, March 1990.

/MeTa90/ H.E.Meleis, A.N.Tantawy; *High Performance Networking and the Modular Communiction Machine (MCM) Approach,* in Second IEEE Workshop on Future Trends of Distributed Computing Systems, Cairo, Egypt, October 1990.

/Salm89/ M.Salmony; *On OSI-Based Transport Systems for Future Applications over High Speed Networks*; PhD-thesis, Tiergartenstraße 15, 6900 Heidelberg, 1989.

/Stil90/ B.Stiller; *Entwurf und Bewertung einer flexiblen Transportkomponente für Multimedia Anwendungen*; Diplomarbeit, Universität Karlsruhe, Institut für Telematik, Germany, October 1990.

/Wats89/ R.W.Watson; *The Delta-t Transport Protocol: Features and Experience*; in "Protocols for High-Speed Networks", Editors: H.Rudin, R.Williamson, IFIP, pages 3-18, North Holland, 1989.

/WFMS89/ M.Wernik, G.Fitzpatrick, R.McNamara, M.Seguin; *Supporting Mulimedia Applications in Asynchronous Transfer Mode Networks*; Multimedia 1989, paper 6.2, Ottawa, Canada, 20.-23. April 1989.

/WrTo90/ D.J.Wrigth, M.To; *Telecommunication Applications of the 1990s and their Transport Requirements*; IEEE Network Magazine, pages 34-40, March 1990.

/ZiGe90/ M.Zitterbart, W.Gerteis; *OSI-Protocols for high speed networks?*; ICCC ′90, New Dehli, India, 5.-9. November 1990.

/Zitt89/ M.Zitterbart; *A Multiprocessor Architecture for High Speed Network Interconnections*; IEEE INFOCOM 89, Ottawa, Canada, April 1989.

/Zitt91/ M.Zitterbart; *High Speed Transport Components;* IEEE Network, January 1991.

A
Guide For Advanced Broadband Multimedia Applications

The BERKOM Reference Model Application-Oriented Layers

UWE HOLZMANN - KAISER ECKHARD MOELLER
GERD SCHÜRMANN KARL HEINZ WEISS

GESELLSCHAFT FÜR MATHEMATIK UND DATENVERARBEITUNG M.B.H.
FORSCHUNGSZENTRUM FÜR OFFENE KOMMUNIKATIONSSYSTEME
HARDENBERGPLATZ 2
D-1000 BERLIN 12

Abstract

This paper gives an introduction to the BERKOM Reference Model providing profile recommendations for the application-oriented layers of the OSI Reference Model. The BERKOM Reference Model is a framework categorising communication standards and in addition to communication, data and information structures and aspects of the application environment.

Integrated services, which are provided by distributed applications, have a lot of common functionality and an application developer should be supported in respect to the selection of suitable combinations of standards and application profiles. Additionally, similar services should use the same profiles to provide interoperability. Using the BERKOM Reference Model, many distributed applications, especially in the area of multimedia, which have common functionality can use products based on the appropriate standards - if available at all.

KEYWORDS: Communication, Data and Informationstructures, Distributed Applications, Multimedia Representation Types, Open Distributed Processing

1 Introduction

The BERKOM project has been launched to promote the development of future services and end systems for broadband ISDN, so that various kinds of applications may profit from this new technology. A first version of the Reference Model was developed by working groups within BERKOM and issued in 1987 [1]. The present version of the Reference Model is structured into two parts, one part called *BERKOM Reference Model Transport-oriented layers*, [2] and a second part called *BERKOM Reference Model Application-oriented layers*[1] [3] being described in this paper and there abbreviated *RM*.

In order to provide services for communication between arbitrary communication partners in an open systems environment, future developments have to be based on international standards. An analysis of relevant standards [4] in the fields of communication, application environment, and information and data structures and interchange formats has shown that these developments have not been adequately coordinated. In several areas, an overlapping of functionality exists, while other areas have not even been considered within the standards. Going beyond the definition of base standards, a number of application profiles have been developed within CCITT and numerous regional workshops, which further restrict the range of functionality of the base standards within certain fields of application. For this reason, future products which are based upon the same standard may not even have the same functionality.

For that reason an objective of the Reference Model is to give a comprehensive description for the structuring of the integrated services provided by distributed applications. It is a framework for existing standards which also supports the identification of functional areas where no standards are available.

The RM categorizes the standards and specific profiles into

- information and data structure profiles,
- communication profiles for OSI upper layers and
- profiles for the application environment.

The communication profiles must be suitably combined with the transport system profiles defined in the Reference Model Transport-oriented layers to get a recommendation encompassing the seven OSI layers.

Distributed applications require identification, localization and management support [5]. These related functions, information and data structures and communication protocols are grouped in the RM under the overall term *administration*. The RM introduces functions and data structures required for such administrative support. Because administration aspects are integrated in most applications and related to all the categories listed above, they are included in the discussion throughout this paper, and are not described in a separate chapter.

[1]This paper was written within the framework of the BERMMD project, granted by the BERKOM program contract #1013 F. The BERKOM Reference Model Application-oriented layers is part of the results of the BERMMD project at GMD/Fokus in cooperation with the Fhg-AGD, Darmstadt, Institut für Schiffs- und Meerestechnik TU Berlin and Konrad-Zuse-Zentrum für Informationstechnik Berlin (ZIB).

2 Description of the Model

Integrated services are provided by distributed applications, which can be based on the information and data structure standards and profiles together with the communication and application environment standards and profiles. These profiles provide the basis for any distributed application in the scope of RM. They are called *application profiles* and have to be combined with the transport service provided by the Reference Model for the lower layers.

Presented in Figure 1 are the three categories of profiles upon which distributed applications and integrated services are based. Suitable profiles selected from each category can be used to develop specific distributed applications. One main reason for the distinction of communication on the one hand and information and data structure on the other hand is motivated by the separate development of standards in these areas in the past. Additionally, many standards cannot be grouped into one of these two categories, e.g., some communication-specific standards have been developed, which also possess aspects going beyond communication. These are categorized as an application environment. The three parts of the reference model, which are described in more detail in the sections below, are:

- Information and data structures,

  The information and data structures comprise *general structuring capabilities*, *common* representation types and *application-specific* representation types. The representation types character text, audio/speech, geometric graphics, moving pictures and raster graphics are classified as representation types for presentation. The representation types for presentation, which are used to present information in many different application areas and representation types for administration, such as management and directory are classified as common representation types.

  In addition to these common representation types, many applications require the interchange of specific information. Some of this *application-specific* information is not intended to be humanly perceptible, and therefore must undergo further processing before being presented through a common representation type - if possible at all (e.g. product definition data).

- Communication

  The communication aspect (with the meaning of interconnection) can be classified in

  - OSI application profiles,
  - Advanced application profiles encompassing light weight protocols for isochronous communication and multipoint communication, such as multicasting,
  - Industrial profiles such as FTP and Telnet (only mentioned for completeness),
  - Administration profiles encompassing directory and management.

- Application Environment

  The functionality needed for the development of distributed applications in different application areas in addition to communication aspects and information and data structures is classified according to aspects as identified in the area of open distributed processing [6] : *processing, storage, user access, identification and localization, management, security.*

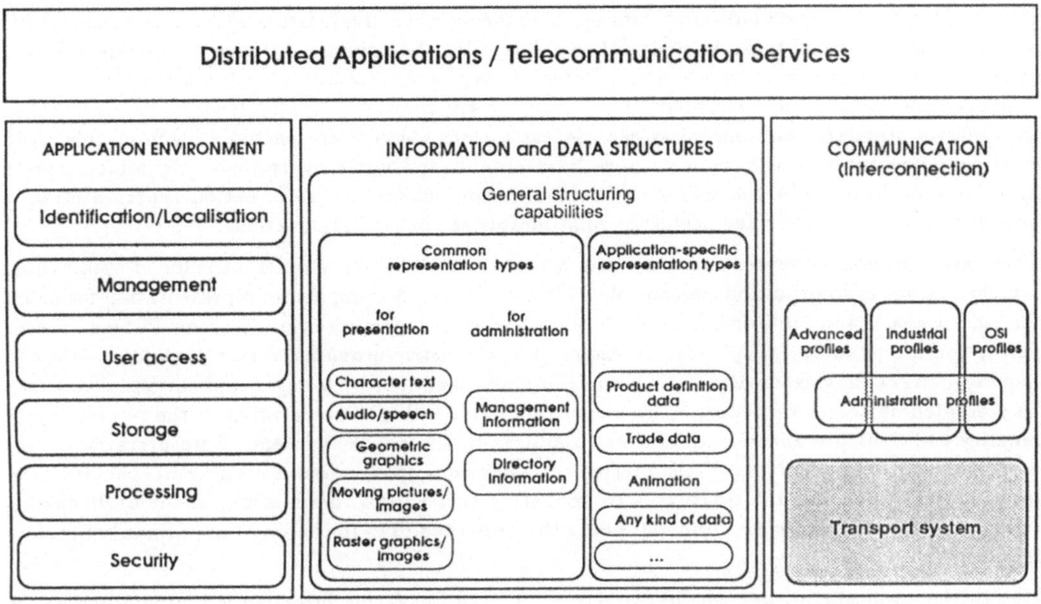

Figure 1 : Scheme for Application-oriented Profiles of the BERKOM Reference Model

Standards and profiles from each of those parts can be combined to fulfill specific application requirements. Some application profiles are data structure independent. They are transparent in relation to data structures. Some application profiles need only profiles from the communication part.

2.1 Information and Data Structures

Before the first part of the reference model is outlined in more detail, a few remarks regarding general aspects of information as related to data are necessary for clarity. *Information* may be considered as any kind of knowledge about things, facts, concepts, etc. of a UoD (universe of discourse) that is exchangeable among (not necessarily human) users. It provides people or machines with new facts about the real world. *Data* may be defined as physical symbols (representation forms) used to represent information in a formalized manner suitable for communication, interpretation or processing by humans or automatic means [7], [8], [9][2]. Thus, exchangeable information necessarily has a representation form (data) to make it communicable – and different data may be used to represent the same information, such as written text or speech – but it is the interpretation of this representation (the meaning) that is relevant in the first place. Any method of representing information has it merits, but is incomplete. If it is to remain understandable, it must omit details (of the UoD). The problem is to decide what details are to be left out in any given communication; this decision determines what representation forms can or should be used [9].

[2]For a discussion of the slight differences of various definitions see, e.g., [10].

On the other hand, the definition of data suitable for representing information of a particular UoD is by no means a trivial task. Various data models, with the primary purpose to provide a formal means of representing information and a formal means of manipulating such representations [11], have been developed in different fields of application. An analysis of IT standards [4] has shown that the different standards contain many representation forms which are partially incompatible. And moreover, structural (static) aspects have been emphasized while operational (dynamic) aspects have been neglected. In the following a classification scheme for information representation is introduced which is particular useful for *multimedia* (see below) applications.

There are a number of representation forms which are used in many different fields of application: representations of information perceptible to human users. Among these, representation forms intended for visual and auditory perception play a major role (representation forms for, e.g., tactile perception are possible). Typically, *character text*, *geometric graphics*, *raster graphics*, *audio* and *moving pictures*[3] in various degrees of complexity are mandated by today's applications. They may be presented on screen or paper, or by means of a loudspeaker. Presentation at the receiver's side requires sufficient presentation directives to adhere to the sender's intend. Interpretation of the corresponding data is subject of the presentation device, implemented according to specified rules for interpretation. This information is to be distinguished from the meaning of the exchange between sender and receiver, i.e., understanding the presented data in the way the (human) originator intended.

Due to their widespread use, the above five representation forms intended for human perception are classified as part of *common* representation forms, they are also called representation *types* in the following. Representation forms for the administration of distributed applications, such as management and directory information representation types, are also regarded as common representation types because they may be used by many different applications. Globally relevant information such as names and addresses will be stored using the standardized *Directory System* [12], e.g. identification and localization. Management Information is represented through logical units called *Managed Objects*. A *Management Information Base* conceptually contains all the Managed Objects of one OSI end system.

In addition to these common representation types, many applications require the exchange of specific information, which cannot at all or not without further processing be represented as common representation types. Examples are information for robot control, product definition data, data for animation and simulation applications or trade data. They are called *application-specific* representation types. The distinction between the two classes should be understood as a justification of the basic needs of many applications. Further developments in technology and/or standardization may very well change representation types classified as application-specific today to become common in the future.

Historically, the above representation types for human perception required different (physical) *media* to carry the information, e.g., character text, geometric and raster graphics on paper, audio on magnetic tape, moving picture on film. The term "multimedia" was used when presentation on different physical media was involved. Today, the terms multimedia information, representation (data) or presentation are still used, but with the meaning of a composition out of the five different representation types. "Hypermedia" data is an extension of multimedia data, allowing the human user to "navigate" through the information, based on predefined links between distinct parts of the information. In order to serve application-specific needs, integrated processing of common and application-specific representation types is required. Thus, the application-specific representation types are included in the definition of the term "multimedia".

[3]By means of a unified approach raster graphics and moving pictures are considered as special types of a more general type 'digital image'. The separate listing of the two types corresponds to current terminology.

In addition to specific representation types, *general structuring capabilities* are of importance, i.e., construction of structures out of representation types, such as multimedia document structures or directory name structures and structures of management information, where tree structures are a typical example. In case of standardized document structures the Office Document Architecture [13] and the Standard Generalized Markup Language (SGML) [14], emerging from the office and publishing environment, play a major role. Both standards are considered as relevant document interchange formats, although with its layout description and the Document Application Profiles (DAPs) under development, ODA has at present the advantage of allowing document interchange without prior agreement between the communicating parties. However, with the development of the Document Style Semantics and Specification Language (DSSSL) [15] and the Standard Page Description Language (SPDL) [16], the use of this standard will be less restricted to closed user groups, as is currently the case (at least for paper-oriented presentation). In addition, the standardization efforts of the Multimedia and Hypermedia information coding Expert Group (MHEG) in ISO/IEC JTC1/SC2 [17] need to be taken into consideration in further developments of the Reference Model. Other descriptions of general structuring capabilities mentioned above can be found in the Directory Standard (*Directory Information Model* [18]) and the set of OSI Management Standards (*Structure of Management Information (SMI)* [19]. Terms such as *Directory Information Tree (DIT), entry, attribute, managed object, containment tree, etc.* are introduced and defined in these standards. These concepts together with ODA and SGML concepts, their commonalities and differences should be considered for general modelling techniques.

One aspect of the general structuring capabilities, the presentation structures, and in particular the *synchronization* of the presentation in multimedia applications, is emphasized. That is, the presentation of multimedia data requires simultaneous and/or sequential presentation of several representation types at the same or at different locations.

The current version of the Reference Model issues recommendations for the common representation types, including the administration representation types, and some of the application-specific representation types, as well as for some of the structuring aspects.

2.2 Communication

In a distributed environment, applications composed of a number of components are split among the different nodes of systems. To interact, distributed application-processes require well defined sets of procedures along with the representation types required to carry out a specific type of application. The functions common to many applications, either defined in standards and profiles or still not defined, have to be identified. They describe those communication functions which can be used by many applications.

The **OSI application profiles** can be derived from combinations of communication standards and profiles. The communication protocols are defined without extension of the capabilities and definitions of the OSI Basic Reference Model [20].

The **advanced application profiles** encompass real-time communication support and multipoint communication support, such as multicasting. Real-time communication profiles can be used for example for real-time audio and real-time moving picture communication and the approach taken in the RM is shown in section 3. From the service point of view this is an OSI extension, and from the protocol point of view, the protocols are reduced in their complexity (light-weight protocols). Advanced application profiles supporting multi point connections, such as multicasting are currently not part of the RM.

Application-oriented **industrial profiles** are based on 'industrial standards' such as the suite of protocols developed by the ARPAnet community, which have raised to the level of a standard by the US Department of Defense. The latter are known as TCP/IP (Transmission Control Protocol / Internet Protocol). IP is the equivalent of the (upper sub-layer of the) Network Layer in the OSI Reference Model and TCP is roughly the equivalent of the Transport layer (Class 4) in the OSI Model. But TCP/IP also stands for a full stack of protocols which have been defined for the use on top of TCP/IP like TELNET, FTP, SMTP and X-Windows for advanced workstation functions.

In the area of **Administration Profiles** based on OSI Directory and Management the following profiles can be identified. The X.500/ISO-9594 standard defines the information model and the protocols to be used when developing a standardized directory service [12], [21]. The EWOS Profiles are being produced to provide guidelines for consistent interworking between different implementations of X.500

The OSI Management standards are concerned with the monitoring and controlling of resources involved in an OSI communication activity [22], providing the framework for further OSI management standardization. The Systems Management Functions specify standard mechanisms for a broad range of management activities [23], [19], [24]. The CMIS / CMIP standards are concerned with the set of common services and protocols for transmitting and manipulating management information [25], [26].

There are also **Non-OSI-Management-Profiles** available. The Simple Network Management Protocol (SNMP) has been developed for TCP/IP-based internets [27]. Although designed for use over TCP/IP it is currently being implemented experimentally over the OSI Transport Service [28]. The Telecommunications Management Network (TMN) is being defined by CCITT. It covers a range of functions needed to manage a telecommunications network [29].

2.3 Application Environment

In addition to communication-specific aspects, as described above, and data structure aspects (see section 2.1), telecommunication-specific characteristics, such as terminal characteristics of telematic services, which are beyond the scope of OSI, are within the scope of the reference model.

The functionality needed for the development of distributed BERKOM applications in different application areas is classified according to ODP aspects[4] [30], although there is some overlapping functionality between the different classes. The classification depends on its main characteristics.

Application environment profiles encompass non-communication specific aspects such as filing and retrieval, security, presentation of multimedia information and terminal characteristics. The functions listed under each aspect are part of the conceptional work within the scope of RM. These aspects, structuring the functionality, are briefly described below.

- Processing
  - generation or manipulation

    supports the distributed creation, manipulation, processing and deletion of the contents and structures of multimedia information.

[4] The communication aspect of ODP has been treated separately in the RM. It offers transmission or distribution supporting several different forms of communication for interchange of information between two or more users e.g.: (*conversation, multicast, broadcast*).

- Replication control, Transaction processing, Concurrency control and recovery

- Storage
  - filing and retrieval of multimedia information
    deals with different storage aspects such as short or long term storage, keywords and search strategies for multimedia information (Hypermedia Links).

- User Access
  - presentation of multimedia information
    has to cover the simultaneous presentation of several common representation types at the same or at different locations.
  - internationalisation
    describes the possibilities for cultural adaption.
  - multimedia operations
    encompasses all aspects of conversions between different media, if necessary at all.

- Identification and localization
  - Identification: global or context-specific unique object name
  - Localization: name / address mapping the name resolution process must be supported, such as name address resolution (directory).
  - Trading: Supports the finding of appropriate services for a given set of requirements.

- Management
  - Defining Policy, Configuration Management
  - Quality of Service Management, Accounting, Monitoring

- Security
  covers the areas of authentication, authorization, audit and cryptographic support facilities etc.

2.4 Relationship between Information and Data Structures, Communication and Application Environment

These three parts, as described in the previous sections, are not independent from each other. There is firstly the relationship between information/data structures and communication, secondly the relationship between communication and application environment and last but not least the relationship between information/data structures and application environment to be described.

The relationship between information/data structures and communication is given in the OSI context on the level of the application and presentation layer. Information/data structures are related to the application layer in respect to semantics and to the presentation layer in respect to the transfer syntax (representation of data). If an application is modelled adequately, from communication point of view AEs (application entities) or ASEs (application service elements) must be selected or developed. The definition of AEs and transfer syntaxes are influenced by the information/data structures of the RM. Two different approaches are possible. Either the ASEs and the presentation service are defined according to the information/data or the latter are more or less transparent to the ASEs and the presentation layer.

The relationship between communication and application environment can be described by means of the ODP aspects, which are high-level categories of useful groupings of functionality in respect to open distributed processing. The aspects are interrelated and offer no definite boundaries for classifying functionality. The OSI Reference Model has been defined for one aspect - the communication aspect - which enables the distribution of objects in open distributed processing. Because of the importance of the communication aspect and the given OSI Reference Model the communication part is regarded seperately in the RM. All non-communication specific aspects, which are by definition not part of the OSI Reference Model are called application environment (see section 2.3).

The relationship between information/data structures and application environment can be described by means of the functionality of the different aspects. As already mentioned in section 2.3 the processing aspect encompasses, for example, the generation and manipulation of multimedia structures, and the storage aspect the filing and retrieval of multimedia structures. The functionality classified by the different aspects and the data structures are very strongly related to each other and it depends on the modelling approach for each aspect how the information/data structures are defined. In many cases the definition of functions within the various aspects are governed by the data structures they are intended to act upon.

3 Examples of Specific Profiles

The follwing two examples show profiles out of the set of predefined profiles of the BERKOM Reference Model. The first profile can be derived from combinations of the communication and application environment standards and profiles and from the data structure standards and profiles. The profile is characterized in such a way that the communication protocols are defined without extension of the capabilities and definitions of the OSI Basic Reference Model.

Electronic Mail Profiles

The electronic mail functionality is supported by the Message Handling System (MHS). Figure 2 describes components which can be used for constructing MHS applications.

Advanced Real-time Profile for Audio/Video

The Advanced Real-time Communication Profiles are embedded into the OSI Reference Model and based as much as possible on existing and emerging standards or recommendations. As opposed to OSI profiles, building blocks of the Advanced Real-time Communication Profiles are neither defined in detail nor approved and registered by respective standardisation bodies. Consequently, the advanced profiles are outlined to stimulate and/or influence the standardisation process in order to capture appropriate support for real-time audio and video communication, multi-point configurations and conference control. The advanced profile is intended to be used to derive specific profiles through a selection process that determines the set of chosen building blocks for a particular field of application in the area of real-time audio and video communication.

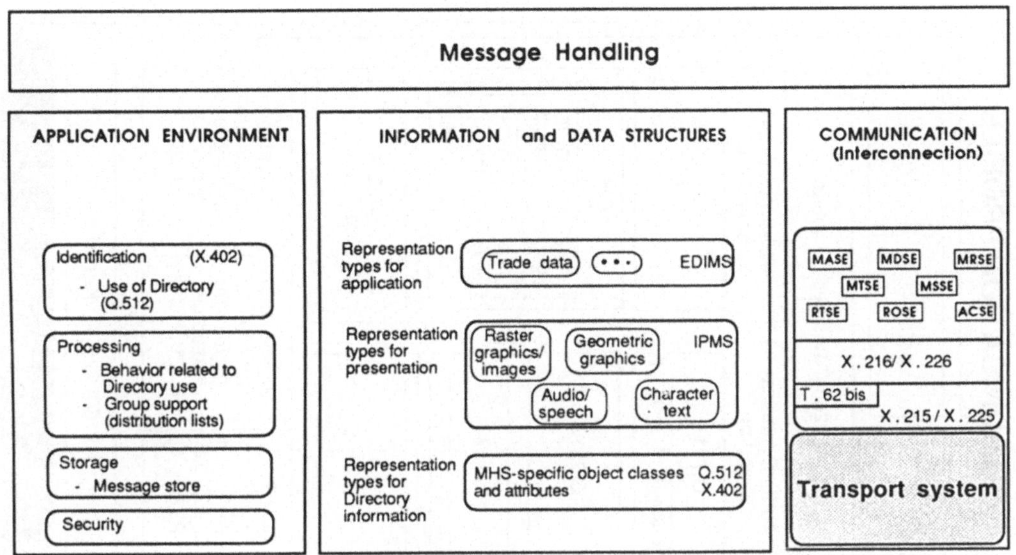

Figure 2 : Application profile: Electronic Mail

As the only layer in the OSI Reference Model that directly provides services to the application-processes, the application layer necessarily provides all OSI services directly usable by application-processes. Considering that it is the purpose of the application layer to serve as a window between correspondent application processes and that several future applications will use real-time audio and/or video communication, multi-point configurations and/or conference control facilities the following application service elements as parts of application entities are intended.

Continuous Media Service Element (CMSE) for real-time audio and video communication. The major difference to existing service elements is the fact, that audio and video data reflect a portion of the original signal over a certain period of time. The impact of time need to be considered with the provision of facilities to support the handling of timed data streams containing timed data items.

Multi point Communication Service Element (MCSE) for multicast services and protocols to support group communication. The MCSE provides services to create, modify, and delete group; add/delete group member, and multiple confirmation modes.

Conference Control Service Element (CCSE) for conferencing support. It is the sole role of the CCSE to facilitate establishment, modification and termination of various types of conferences, e.g. audiovisual- or multimedia conferences.

The definition of the three advanced ASEs (CMSE, MCSE, CCSE) has to be done by standardisation bodies and is therefore at the time being only hypothetical.

As far as audio and/or video transmission in real-time is concerned appropriate Quality of Service (QoS) parameter need to be defined. Those parameters, used by underlying service providers, capture: communication mode i.e. unidirectional, bidirectional synchronous/asynchronous, capabilities of I/O equipment (transmission rate, coding type), delay characteristics, error characteristic etc.

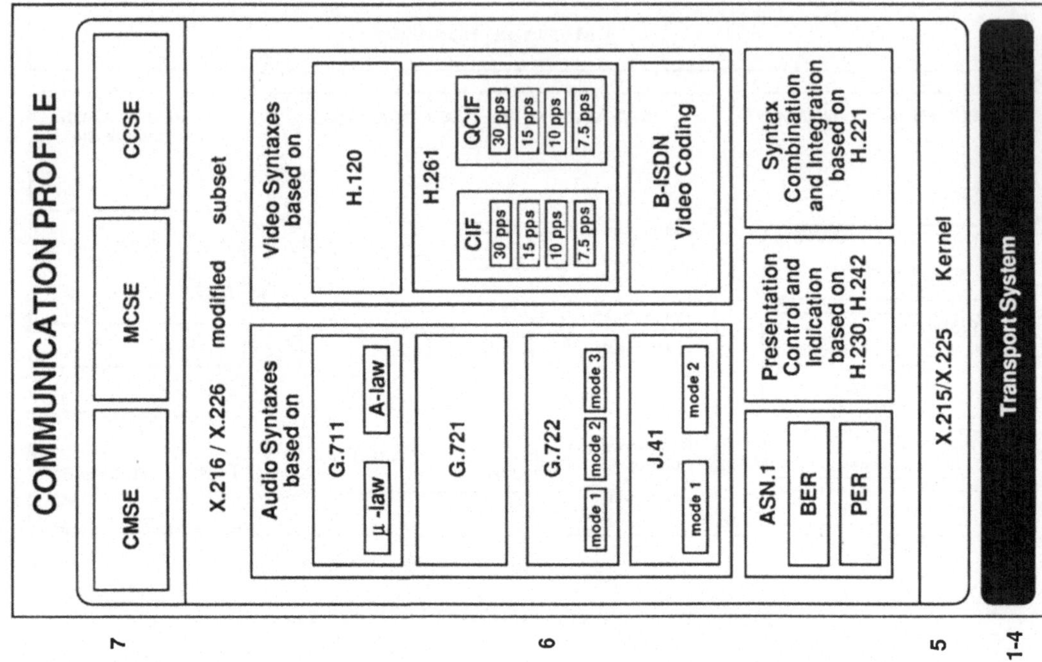

Figure 3 : Advanced Communication Profile for Real-time Audio and Video

The approach outlined before uses bearer service capabilities being supported in the BERKOM Reference Model Transport-oriented layers and providing service primitives and Quality of Service (QoS) parameters. Supported features and characteristics of data formats, such as audio or video are reflected in parameters used in bearer communication services (layer 3 - network layer). Parameters required by the bearer services are provided by using the QoS parameters of transport service primitives. The QoS parameters are passed down from the application layer to the transport layer. During the data transmission phase, the transport service provider may select an advanced transport profile to achieve the required QoS.

4 Conclusion

The presently available BERKOM Reference Model Application-oriented layers contains short-term and medium-term profile recommendations, which should serve as a basis for the development of applications within this framework. The profile recommendations are based on the OSI Basic Reference Model, existing CCITT telematic services, distributed applications already standardized within OSI, and other international standards and profiles in the areas of communication and data structures for information interchange. Hereby, existing telematic services are taken into consideration. The telematic services cover complete service descriptions including communication protocols, interchange formats, and terminal characteristics. The present version of the Reference Model concentrates, however, upon communication, management and information/data structures aspects. In addition to these profile recommendations, areas are identified for which the available standards do not yet offer any solutions. For some of these identified areas, solutions are offered, such as combinations of profiles for isochronous and anisochronous communication.

The disadvantage of standards which do not complement each other in most cases, but which are of overlapping functionality, is not solved by the current approach of the RM. Attention must also be paid to the fact that the developer finds no support for the concept of *how* (in the sense of design guidelines or architectural concepts) to build an integrated service as a distributed application. The current approach can be regarded as a first step to an architecture providing generic or common functions which can be combined according to general rules. In order to eliminate, in the long term, the inadequacies of the presently available standards, an architecture for distributed processing (including homogeneous concepts, generic functions and common data and information facilities) has to be developed in addition to the current Reference Model approach.

The architecture should permit the modelling of distributed applications and the integration of services independent of available technological constraints and transmission mechanisms. The approach to such general concepts is not, however, part of this report. An early version of a general telecommunication model was produced as a discussion paper in 1989 [31]. Further work in this direction is now being carried out as part of national and international standardization, where the development of such an architecture and such concepts is being undertaken primarily within the framework of Open Distributed Processing [6] and the Framework for the Support of Distributed Applications [32]. The impact on future versions of the Reference Model depends first of all on the stabilization of such concepts in international standardization, on the requirements of the RM user and will also be influenced by the results of the current research programs of the European Community, especially by the following projects: *Integrated Systems Architecture* (ISA), *RACE Open Services Architecture* (ROSA) and the *Communication Systems Architecture* (CSA).

References

[1] "Berkom-Referenzmodell", ed. BERKOM, DETECON Technisches Zentrum Berlin, Germany (Juni 1987), Version 01.

[2] "BERKOM Reference Model – Lower Layers", ed. DETECON and GMD-FOKUS, DETECON Technisches Zentrum Berlin, Germany (February 1991), Version 2.1.

[3] "BERKOM Reference Model II – Application-Oriented Recommendations", ed. GMD-FOKUS, DETECON Technisches Zentrum Berlin, Germany (September 1990), Version 3.1.

[4] "Analyse relevanter Normen und Empfehlungen zur verteilten Bearbeitung multimedialer Informationen im ISDN-B", ed. GMD-FOKUS, DETECON Technisches Zentrum Berlin, Germany (May 1989), Version 2.0, BERKOM Dokumentation Band V (in German).

[5] G. Schürmann and U. Holzmann-Kaiser, "Distributed Multimedia Information Handling and Processing", *IEEE Network*, vol. 4, no. 6 (November 1990).

[6] ISO/IEC JTC1/SC21 N4883-4888, "Information processing systems – Open Systems Interconnection (OSI) – Open Distributed Processing (ODP)" (June 1990).

[7] ISO 2382-1, "Data processing – Vocabulary", *Part 1: Fundamental Terms* (1984), 2nd edition.

[8] ISO/TR 9007, "Information processing systems – Concepts and terminology for the conceptual schema and the information base" (1987).

[9] J.J. van Griethuysen and D.A. Jardine, *Information Modelling with INFOMOD, Volume 1, Concepts and Information Struture*, Philips International B.V. (October 1989).

[10] S. Lehert and E. Moeller (eds.), *Data and Information Modelling*, Proc. of the BERKOM Workshop, Annelsbach - Hoechst/Odenwald, July 9-13, 1990,GMD Report no. 196, R. Oldenbourg Verlag, München (to be published).

[11] C.J. Date, *An Introduction to Database Systems*, vol. II, Addison-Wesley (1985).

[12] CCITT Recommendation X.500, "Open Systems Interconnection (OSI) - The Directory - Overview of Concepts, Models and Services" (1989).

[13] ISO 8613, "Information Processing - Text and Office Systems - Office Document Architecture (ODA) and Interchange Format" (1989).

[14] ISO 8879, "Information Processing - Text and Office Systems - Standard Generalized Markup Language (SGML)" (1986).

[15] ISO/CD 10179, "Information Technology - Text Composition -Document Style Semantics and Specification Language (DSSSL)" (1989).

[16] ISO/CD 10180, "Information Technology - Text Composition - Standard Page Description Language" (1989).

[17] ISO/IEC JTC1/SC2/WG12, "Coded Representation of Multimedia and Hypermedia Information", Version 3 (November 1990).

[18] CCITT Recommendation X.501, "Open Systems Interconnection (OSI) - The Directory - Models" (1989).

[19] ISO/DIS 10165, "Information processing systems — Open Systems Interconnection (OSI) — Structure of Management Information" (June 1990).

[20] ISO 7498, "Information processing systems - Open Systems Interconnection (OSI) - Basic Reference Model" (First Edition 1984).

[21] ISO 9594, "Information processing systems - Open Systems Interconnection (OSI) - The Directory" (1989).

[22] ISO 7498-4, "Information processing systems - Open Systems Interconnection (OSI)", *Part 4: Management Framework* (1989).

[23] ISO/DIS 10040, "Information processing systems — Open Systems Interconnection (OSI) — Systems Management Overview" (June 1990).

[24] ISO/DIS(Part1-7) 10164, "Information processing systems — Open Systems Interconnection (OSI) — Systems Management Functions".

[25] ISO 9595, "Information processing systems — Open Systems Interconnection (OSI) — Common Management Information Service Definition" (January 1990).

[26] ISO 9596, "Information processing systems — Open Systems Interconnection (OSI) — Common Management Information Protocol Specification" (January 1990).

[27] K. McCloghrie and M. Rose, "A Simple Network Management Protocol (SNMP)" (May 1990), RFC 1157.

[28] "SNMP over OSI", ed. M. Rose (June 1990), RFC 1161.

[29] CCITT Recommendation M.30, "Principles for a Telecommunications Management Network", Melbourne (November 1988).

[30] ISO/IEC JTC1/SC21 N4883 Working Document, "Working Document on Topic 4.1 - Requirements for Structures and Functions" (June 1990).

[31] "BERKOM Referenzmodel II - Verallgemeinertes Telekommunikationsmodell", ed. GMD-FOKUS, DETECON Technisches Zentrum Berlin, Germany (May 1989), Version 2.0.

[32] CCITT SQ VII/Q19, "Framework for the Support of Distributed Applications (DAF)", SG VII Plenary Geneva (July 1989).

Programmieren multimedialer Anwendungen

Bei interaktiven Systemen tritt die multimediale Anwendung immer mehr in den Vordergrund. Bei deren Programmierung wird neben der textuellen Vorgehensweise mit Hilfe einer Programmiersprache zunächst das Visuelle Programmieren immer mehr eingesetzt, im Kleinen durch Mosaik- und Puzzle-Ansätze, im Großen CASE-Techniken. Nicht nur der Ablauf im Rechner und die dabei angesprochenen Daten sind zu programmieren, auch die mehr oder minder multimediale Benutzungsumgebung (auch evtl. in der Verknüpfung zu einem Hypermedia-System) muß gestaltet werden. Für letztere Aufgabe ist das Programmieren durch Vormachen ein interessanter und vielversprechender Weg. Bei multimedialen Anwendungen könnten, auch müßten, audio-visuelle Techniken Bedeutung erlangen, beim Programmieren durch Vormachen auch das Einbeziehen von Information, die auf der multi-medialen Peripherie (CD-Platten, Video-Platten, BTX, ISDN-Dienste, Tele-Peripherie u.ä.) vorliegt bzw. dorthin gelangen soll. In verteilten Anwendungen muß die Kommunikation und Kommunikationskontrolle mitprogrammiert werden. Wie bei allen interaktiven Anwendungen ist auch und insbesondere bei multimedialen Anwendungen ergonomischen Anforderungen Rechnung zu tragen. Das gilt für die Benutzung und für die Programmierung, letztere geeignet integriert in programmiererfreundliche Software-Entwicklungstechniken.

Koordinator: Prof. Dr. H.-J. Hoffmann, TH Darmstadt

A MULTIMEDIA MULTIUSER STRATEGIC PLANNING SYSTEM

Lisa Neal, Michael McCormick, and David Temkin
EDS Center for Machine Intelligence
One Cambridge Center, Suite 408
Cambridge, MA 02142 USA

Abstract

We report on the design and development of a multimedia, multi-user system to support business strategic planning. The system can be used by a single user as a tutorial system or to prepare for or review a meeting, or a group of managers can use the system with a single computer that is projected on a screen. However, to utilize the full capabilities of the system, a computer-supported meeting room, such as CMI's Capture Lab, is required. In a computer-supported meeting room, the system allows individual and group work as part of the planning and consensus-building processes. Video, voice, and animation are incorporated to educate users in strategic planning concepts. To account for the expense and availability of equipment, the system is configurable so that a range from full multimedia capabilities to voice, text, and simple animation is allowed.

Introduction

The business strategic planning system is designed to guide groups of managers through strategic planning meetings. The system uses the strategic planning methodology of Professor Arnoldo Hax, the Alfred P. Sloan professor at the M.I.T. Sloan School of Management. The system also incorporates Professor Hax's expertise, using artificial intelligence techniques to provide evaluation and feedback to its users.

The strategic planning system is designed to encourage discussion and critical thought through a structured planning methodology and allows for recording of the group's process and decisions. It will be used in the Capture Lab, CMI's computer-supported meeting room. It will have a multimedia educational component that will be used by individuals as a strategic planning tutorial and by groups who need to learn about the strategic planning process as part of their annual workshops.

In addition to supporting the annual strategic planning process, later extensions to the system will be used on an ongoing basis. All too often strategic planning meetings result in massive documents that are shelved until the following year's meeting. Strategic planning needs to be a dynamic process. Our system will provide on-line access to the strategic plans developed during the face-to-face meetings and give the individual managers the ability to easily update and revise the plans as well as monitor their progress over time. Later development will support corporate and functional strategic planning as well.

Strategic Planning

Strategic planning is a structured process which helps a business determine how to achieve a long-term sustainable advantage over their competitors. The strategic planning process is a disciplined and well-defined organizational effort aimed at the complete specification of a firm's strategy and the assignment of responsibilities for its execution [Hax 91]. The dual purposes of an annual strategic planning meeting are to develop a plan and to develop better communication between a group of managers. The plan consists of an examination of where the business currently is, where they would like to be in the future, which is typically defined as five years hence, and how they plan to get there. The process of defining a plan culminates in the definition and evaluation of specific action programs at the business level. Improved communication results from the shared education about strategic planning, the common framework and semantics, and the consensus that is reached.

Goals

The strategic planning process is a paper-based one which traditionally has no reliance on the use of technology. Our goals in developing the system for business strategic planning were to determine how the introduction of technology could improve the process. Our overall goals were to provide education, to guide users through the process, to provide as much feedback and evaluation as possible, and to enhance and improve group communication. Our long term goals include supporting all phases of strategic planning (corporate and functional, in addition to business) and to provide ongoing strategic management support. This will result in a planning environment, rather than an infrequently used system, which can be used throughout an organization.

While much of our guidance in designing the system came from Professor Hax, many of the initial ideas arose from observing actual strategic planning workshops. This provided an opportunity to understand how the process worked without the introduction of technology, and to evaluate the ways in which it was effective and the ways in which technology could provide assistance and meet user's needs. One of the clear strengths of the workshop was the tight integration of education and engaging in actual planning activities: groups were educated about a facet of planning and immediately applied it. Another advantage was the lecture style, personality, and in-depth expertise of the seminar leader, Professor Hax. While the integration of education seemed possible to replicate, especially with the use of multimedia, the latter provided a significant challenge.

There were readily observable problems as well. Some were the difficulties groups had once engaged in a planning activity; these included forgetting the definitions of terminology and concepts and forgetting what they were supposed to be doing when engaged in a specific activity. In the latter case, it was because of a lack of understanding of the activity or not understanding what information they should be entering in a table or chart. For these types of problems, human intervention solved their problems, but it seemed clear that computer support could have provided the needed information. Also, groups frequently shuffled through stacks of paper to find information that they had written down earlier; this could be avoided by displaying needed information at the appropriate time.

Overall, we were concerned with how we could introduce technology to solve existing problems while not introducing new ones, and whether technology could add to the process such that the use of our system, while different from being guided through strategic planning by a human expert, provided significant benefits to its users. This involved serious consideration of how to extend and stretch the capabilities of paper onto the computer. Because the system will be used by managers, who may not have strong computer skills, we are concerned with introducing technology such that it is easy to learn, use, and understand, and provides an appropriate level of guidance.

Education

The system is designed for managers who may or may not know about strategic planning but who are unlikely to be familiar, at least during their initial system use, with the planning methodology of Professor Hax. Hence, education is an important component of our system. The integration of education into the process was very effective with groups, and we had the goal of supporting that in our design. We considered that the coupling could be even tighter than in the seminars we observed, which involved a lecture and discussion period of one to two hours followed by a similar time period for engaging in the activity. Also, we wanted to add educational components such that information is provided to prevent users from getting stuck in the way we observed they frequently did in a seminar.

We investigated the use of video as the main mechanism to provide educational support, with the use of voice, animation, and text as well, when appropriate. In providing multimedia, we are developing essentially low-, medium-, and high-end systems so that, depending on the hardware available, the system is configurable to take advantage of the as much multimedia as possible. Our rationale is that we do not want the strategic planning system to necessitate enormous expenditures.

Video was determined to be effective in other types of systems [Laurel 91] and we thought that, for the amount of material to be covered, would be more compelling than text. Video also had the advantage of capturing the personality of the human expert and reinforcing the notion that the system incorporated his methodology and expertise. Since we planned to provide feedback and evaluation as well, this reinforcement seemed important for feedback to be regarded seriously.

The educational components of the system were determined to be overviews, an introduction and summary for each stage of the process, examples, and help balloons. The overviews would consist of a videotaped session in which new material was covered such as an initial overview of business strategic planning and an overview for determining the mission of the business. Animation supplements the videotaped segments by highlighting points and by illustrating regions in tables which are being referred to in the video. The video segments are accessible through"Introduce" buttons, which are labelled by the type of material that selecting them will invoke and by the playing time (i.e. ten minute video). This is so that the users know what will happen when selecting a button and what time commitment is needed.

Most of the individual windows in the system contain introduce and conclude buttons, which enable users to see (in a live video window), hear (through digitized sound), or read (if neither of the others is available or desired) an explanation of what they should be doing at that particular point in the process. The introduce buttons, at points at which there is no need for an overview of new concepts, will consist of a brief video or voice-over to provide information to aid in accomplishing a specific task. For example, if a group is developing a list of current products, the "Introduce" button will give them guidance to complete the task, which would inform them of what they are doing, what is important about it, and what they need to keep in mind, such as the proper level of aggregation.

The "conclude" buttons will have a similar purpose. They will be used when a task has been completed, and will inform the users of what they have done, some simple feedback and evaluation where appropriate, what the next task is, and when they will revisit the information they just entered. For example, if a group just developed a list of current products, the "Conclude" button will remind them that if the list contains over twenty items they may want to rethink their aggregation. In addition, they will be told that the next step is to develop a list of current markets, but that they will soon revisit this list to augment it with a list of future products.

Examples will be included at each step; for instance, when developing the list of current products, the users could view a list of products for another business. Initially, the selection of form-based examples will be static, with one example available for each task. Our long term goal is to provide a library of examples, built up from use of the system. Within an organization, this will allow users to look at an example which is relevant to their business or one with which they are familiar. This mechanism can also be used to provide access to the strategic plans developed in previous sessions. Case-based reasoning techniques may later be used to provide access to examples deemed relevant to the group's business.

Examples have been shown to be effective in other domains [Neal 89] and should prevent some of the problems encountered by groups who did not know what type of information to input. However, we do not want the examples to mislead or limit the creativity of users. In the latter case, we fear that an example will be viewed as "perfect" and hence a group would try to replicate it or orient it to their business, without thinking creatively about their own business. If we determine through testing that this happens more frequently than the use of examples as an aid, we will remove them from the system or enrich the library of examples so the users can compare and contrast them.

The last mechanism which we are incorporating to provide education is help balloons. By activating a feature called *balloon help* [Apple Computer, 1991], users of the system can receive information on general Macintosh features as well as features specific to the strategic planning system. Balloon help works by tracking the mouse pointer (cursor), and providing explanation for the designated item. If the user wants to know the meaning of choosing a cell in a table, the system responds by displaying the help balloon which provides the meaning of that cell. Work can continue while help balloons are displayed, and the feature may be activated or deactivated by pointing at the small balloon with the question mark in it near the upper right of the screen. The same mechanism will provide information about the strategic planning system as well as about other applications and the Macintosh in general. It is a standard feature beginning with the Macintosh Operating System version 7.0. The use of this mechanism is consistent with our goal of providing a system that is easy for people to use and consistent with existing applications.

Process Support

The strategic planning system provides an orderly way of looking at strategy. The integration of education and process means that knowledge gained from the educational components can be immediately applied and that assistance is proved. The tasks users are asked to complete primarily involve filling in tables and charts as a group, which typically involve the creation of lists or the assessment of the attractiveness of items in a list based on a set of criteria.

When the strategic planning system is opened, two windows will appear. The first, labelled "Planning Overview," and appearing on the left, continuously provides an overview of the entire business planning process and highlights the group's current position within the process. The overview also provides a mechanism for navigation, by the selection of any item displayed. For the sake of space and to maintain a better picture of where the group is, the overview condenses all but the current portion being worked on.

Overlapping the Overview window is the "Introduction to Business Strategic Planning" window. This window is meant as an introduction to the system, which allows users to select the introduce button to view a video describing the overall process of business strategic planning. The video does not have to be viewed; rather, the user is free to ignore it or view it as often as desired. Users are encouraged to select the "Introduce" and "Conclude" buttons before and after filling out each window throughout the entire system. In this way they can receive the maximum amount of guidance and information to guide them through the process.

Following the introduction to Business Strategic Planning, the user presses the "Next Step" button in the Overview window. This button advances the process and updates the overview window, and is used to allow a group to step through the entire strategic planning process. When the highlighted label in the overview window has been condensed, it is automatically expanded when selected so that each of the individual steps of the process are visible to the user. Each of the sub-categories corresponds to at least one stage of planning and one window.

In each window, the "New Topic" and "New Subtopic" buttons can be used to enter hierarchical text in the window. This feature is common to all of the locations in the strategic planning system which support user-entered outlines or lists of information, and is similar to a new feature incorporated in the System 7 "Finder" or file manager. Triangles to the left of the category names indicate that those categories contain one or more items. Since the user can organize their items hierarchically, it is possible to "collapse" categories so that subcategories are not visible. When a triangle faces rightward rather than downward, it indicates that there are additional items which may be viewed by expanding the outline. Pressing on any triangle will expand that topic.

Other activities involve filling in attractiveness matrices. These display ratings, as indicated by a legend, and read the lists from the earlier windows for the axes, yet allow editing and regrouping of the lists in order to attain the proper level of aggregation for this process. Any editing of these lists will not affect the contents of the windows from which the information was taken. Those windows, which may be recalled at any time through the overview window, are affected only by editing them in their own window. Selecting a matrix label (vertical or horizontal) copies the item's text to the square in the upper left of the window, in which standard Macintosh editing features may be used to edit the text. As the text is edited, the label is updated. This way, long labels can be fully displayed and vertically oriented items can more easily be altered.

For any activity in which it is useful to view previously entered information rather than for information to be directly copied, the user is presented with the old window or windows, along with a new window in which the new task is being undertaken. An example of this is in determining the challenges that have emerged from the current to the future or providing a summary.

Most of the tasks users are asked to complete involve highly structured information. However, there is a great deal of information, such as the rationale behind decisions [MacLean 90], which should not be lost. Hence, we introduce a mechanism for capturing unstructured information: we allow users to essentially attach Post-It™ notes at any location when using the system. This notes capability means that users do not lose the flexibility of paper.

Notes are small yellow windows, and, as with other windows, note windows may contain hierarchical outlines which may be collapsed or expanded at will. Note windows may be affixed to any other window in the system, and later retrieved for reference. This is accomplished using the "drop box" on the upper left of the window (which is a "see-through" hole). The user places a note into position into the window which is visible through the drop box. The note can be retrieved by double-clicking (pointing with the mouse and pressing the button twice rapidly) on the miniature yellow document icon in the window. If the place where a note is attached is an outline, then if it is collapsed, expanded, or rearranged, notes may be moved or hidden, depending on their location in the window. Likewise, when lists are copied into tables in new windows, no notes are copied; they remain situated solely their original window.

Group Interaction

The purposes on a planning session are to produce a plan and to improve the communication between members of a group. The latter is accomplished through shared education and terminology, the process of reaching consensus on business strategy, and "buying in" to the plan, which increases the likelihood of it being effectively used within the organization. One of our goals, therefore, is to use technology to enable more effective group work and to enhance communication abilities within a group. This is done in part through design of the system to help a group work better together and reach consensus. One method is through the incorporation of "groupware" tools for brainstorming, voting, ranking, and decision-making. Another is through the use of the Capture Lab, both for the facility it offers and for the group dynamics feedback available to groups [Losada 90]. The strategic planning system, however, is designed so that it can be used by a single user as a tutorial or for meeting preparation or review, or by a group with a single computer with a shared, projected display.

The Capture Lab [Elwart-Keys 90], CMI's computer-supported meeting room, consists of an ergonomically-designed conference room equipped with nine Apple Macintosh personal computers. Eight are set into the meeting table and are for individual use, and they are linked together with the additional, shared computer which can be controlled by any meeting participant. The shared machine's display is rear-projected onto a screen at one end of the meeting room, so that all participants can view the same documents. The room is equipped with a variety of audiovisual equipment which projects onto the same screen and there is also an observation room which is used for recording the dynamics of groups meeting in the Capture Lab.

A planning session is conducted with the strategic planning system running on the shared computer. Capabilities are provided for individual or group work on the private computers. Although the Capture Lab will provide an optimal environment for the Strategic Planning System, the system will work in less elaborate configurations, although certain capabilities may not be available, depending on the specific hardware and software environment.

While all of the previous descriptions of the system have been of the public screen (a screen which is visible to everyone present, and which is operated by one person at a time), in he Capture Lab, the private computers can be used by individual participants for individual work, such as educational purposes or trying out ideas, or for brainstorming or voting. Similarly, the private computers can be used by sub-groups who are working together. To allow this type of communication and collaboration, the standard Macintosh desktop is supplemented by a miniaturized version of the public screen. Within this window is a small representation of the current contents of the public screen. It allows users to copy information to and from the public screen, for private (or sub-group) annotation and modification, and then return it to the public screen for consideration by the entire group.

When a user drags a copy of a window out of the public screen window on their individual machine, it remains shrunken. Upon release, the window expands to full size and acquires a label which is the original label appended with a designation of its owner. At this point, any modifications can be made. Typical modifications include revisions of ratings or annotations using notes. Like the note windows, this window has a drop box in its upper left for the purpose of returning it to the public screen by positioning the revised window over the reduced public screen representation. When a window has been returned to the public screen, it appears in front of the overview window, but behind everything else, so as not to disturb any work currently underway on the public screen. This allows individuals to show their work for comment and/or merging, and to take it back for revision if necessary. It also provides a mechanism for brainstorming by individuals or subgroups.

More traditional tools for the purposes of brainstorming, voting, and ranking are available in the Capture Lab. We plan to integrate them into our system, and will test there effectiveness with groups. While such tools have been shown to aid in group processes in a number of ways (lessening of a hierarchy amongst a group, for instance), we have considered that they may be detrimental to the progress of a group of managers doing strategic planning. The reasons are that group members may feel compelled to contribute when they would not have vocally, that the lessening of the hierarchy among the group might be detrimental to their working together since the hierarchy is well-established and is likely to remain so, and that, if brainstorming is done anonymously, contributors may worry that they will not receive credit for a worthwhile contribution. Hence we would like to incorporate and test the effectiveness of such tools.

Future Directions

There are a number of areas which are being pursued currently and others that will be examined in the future. One area of current work is the incorporation of artificial intelligence techniques to provide feedback and diagnostic information. The goal is to understand as much as possible of the information input by users, using contextual information and natural language understanding, and to determine inconsistencies, incomplete information, etc. and provide intelligent and tailored presentations of data. Ultimately, we would like to provide feedback much as a human expert working with a group would in response to information.

Areas that will be examined in the future include the advantages of alternate input methodologies, the effectiveness of meeting tools incorporated into the system and the effectiveness, over the short- and long-term, of group dynamics feedback. In addition, there are a number of questions we would like to answer through observation of the use of the strategic planning system. Foremost are the differences and relative effectiveness of the use of our system compared to the traditional (non-computer-based) strategic planning process, and the differences between the use of our system in a computer-supported meeting room and its use on a single computer which is projected. Through these questions we hope to better understand the strategic planning process, the incorporation of technology, and the interactions between groups working in a computer-supported meeting room.

Conclusions

The strategic planning system described here is under development. The design of the system has involved careful consideration of how to best support managers who may not be technically proficient, consistency with existing Macintosh applications, how to educate and assist users most effectively in the strategic planning process through the use of multimedia, and how to best support improved communication and consensus-building in a group.

There was a significant challenge in translating a paper-based process into a computer application. It necessitated an in-depth understanding of the goals at every stage of the process in order to understand what the paper forms were trying to accomplish and which parts were fixed and which parts were flexible, and then to go on to creatively develop a design which met the goals supported by the paper-based process yet, at the same time, took advantage of the flexibility and capabilities of the computer. For example, notes support the annotations and scribbling in margins common to the use of paper, yet are more elegant, are multifunctional, and can be hidden at will. Likewise, topics and subtopics in tables, which can be hidden, expanded, and edited, allow the viewing of large amounts of information and make it easy to keep details hidden in order to maintain an overall picture. The incorporation of multimedia for educational purposes attempts to replicate the interaction with a human expert, and may capture some of a human's effectiveness when working with a group but lacks a human ability to adapt and orient information easily to a group's needs and abilities.

System development will be completed in Fall, 1991, at which point system testing and evaluation will commence. In addition to determining the effectiveness of our design decisions, we hope to answer many of the above questions. As the system becomes used within organizations, we will acquire more data, especially with its use over many years for strategic planning.

References

Apple Computer (1991). *Inside Macintosh, Volume VI.* Cupertino: Apple Computer.

Elwart-Keys, M., Halonen, D., Horton, M., Kass, R. & Scott, P. (1990) User interface requirements for face to face groupware. *Proceedings of Conference on Human Factors in Computing Systems,* CHI '90, Seattle, WA (pp. 295-301). Association for Computing Machinery, New York.

Hax, A. & Majluf, N. (1991). The Strategy Concept & Process. New Jersey: Prentice Hall.

Laurel, B. (1991). *Computers as Theatre.* New York: Addison-Wesley Publishing Company.

Losada, M., Markovitch, S. (1990). GroupAnalyzer: A system for dynamic analysis of group interaction. *Proceedings of The 23rd Annual Hawaii International Conference on System Sciences,* HICSS '90, Kailua-Kona, HA (pp. 101-110). IEEE Computer Society, Los Alamitos.

MacLean, A., Carter, K., Lövstrand, L., Moran, T. (1990). User-tailorable systems: pressing the issue with buttons. *Proceedings of Conference on Human Factors in Computing Systems,* CHI '90, Seattle, WA (pp. 175-182). Association for Computing Machinery, New York.

Neal, L. R. (1989). A System for Example-Based Programming. *Proceedings of Conference on Human Factors in Computing Systems,* CHI '90, Seattle, WA (pp. 63-68). Association for Computing Machinery, New York.

Ein Visualisierungswerkzeug für die Wartung modularer Programme

Jürgen Lucas[1]

Abstract

Die vorliegende Arbeit analysiert die Probleme, die das Verstehen eines unbekannten Programms bereitet, zu dem schlimmstenfalls keine oder nur veraltete Dokumentation vorhanden ist. Hierzu werden die in Programmen vorkommenden Größen klassifiziert und unter dem Gesichtspunkt von Datenfluß, Kontrollfluß und Definitionszusammenhang untersucht. Die Analyse berücksichtigt besonders die Probleme bei großen, modularen Programmen. Als Ergebnis wird ein interaktives Werkzeug vorgestellt, das es erlaubt, die Größen und ihre Beziehungen untereinander zu analysieren und so das Programm zu verstehen. Zur Informationsdarstellung werden graphische Mittel ebenso verwendet wie Markierungen im Programmtext. Der Programmtext wird durch die Anwendung dieses Werkzeugs (VAMP) zu einem Hypertextsystem.

1 Einführung

In der Wartung von Programmen werden Änderungen an existierenden Programmen entworfen und durchgeführt. Diese Änderungen können der Fehlerverbesserung, der Anpassung an eine sich ändernde Umgebung oder dem Einbau neuer Funktionen dienen [Sch87], [Par86]. Änderungen werden aber auch als Präventivmaßnahme vorgenommen, um etwa erkannte Unsauberkeiten in der Implementierung zu beseitigen.

Ein Schlüsselproblem hierbei ist das Verstehen eines unbekannten Programms. Im ungünstigsten Fall [LSW90] gibt es keine Dokumentation, nur der Programmtext steht zur Verfügung. Zwar fehlt die Dokumentation nicht immer, aber auch eine aktuelle Dokumentation kann nicht alle Details eines Programms darstellen. Im Zweifelsfalle ist also auch hier der Programmtext entscheidend.

Es erheben sich Fragen beispielsweise nach dem Typ einer Variablen, nach den Seiteneffekten einer Prozedur oder nach den Verwendungsstellen einer Variablen. Die richtige und vollständige Beantwortung dieser Fragen ist unabdingbar für das Erfassen der Bedeutung einer Programmgröße, für das Verstehen des Programms. Sie ist somit notwendige Bedingung für das korrekte Einbringen einer Änderung am Programm.

Gegenstand der vorliegenden Arbeit ist der Entwurf eines Werkzeuges, das in dieser Situation den Programmierer angemessen unterstützt. Sie ist wie folgt gegliedert:

Nach einem Überblick über vorhandene Arbeiten zum Thema *Verstehen eines Programms* und über existierende Ansätze, die sich mit diesem Problem beschäftigen, wird eine Klassifikation der Größen in Programmen und der ihnen eigenen Probleme in Bezug auf das Erfassen ihrer Bedeutung vorgenommen. Im Abschnitt 4 wird dann die grundlegende Strategie von VAMP [2] vorgestellt und motiviert. Abschnitt 5 beschreibt die implementierte Benutzungsschnittstelle, in Abschnitt 6 werden Erfahrungen mit VAMP berichtet. Eine Zusammenfassung beschließt die vorliegende Arbeit.

2 Andere Arbeiten

Schon recht früh wurde erkannt, daß bestimmte Konstrukte von Programmiersprachen das Verständnis von Programmen behindern. In [Dij68] wird bei der Untersuchung des Sprungbefehls auf die Abweichungen des Kontrollflusses vom textuellen Ablauf hingewiesen, wodurch der Programmierer den Zusammenhang und den Überblick verliert. Ähnliche Überlegungen wurden in [WS73] für globale Variablen angestellt.

[1]Institut für Betriebs- und Dialogsysteme, Universität Karlsruhe, Am Fasanengarten 5, W-7500 Karlsruhe.
Die Arbeit wurde gefördert von der Siemens AG Abt. ZFE IS SOF

[2]Verstehen und Analyse Modularer Programme

Theorien für das Verstehen von Programmen werden unter anderem von [Bro83], [Luk80], [SP88] vorgestellt. In [Sen83] wird das Verstehen eines Programms als ein rekursiver Vorgang verstanden, der schrittweise einen Ausschnitt aus einem Programm identifiziert, seine Grenzen bestimmt und die Beziehungen, die diese Grenzen überschreiten, herausfiltert. Durch das Erfassen der Bedeutung dieser Beziehungen läßt sich die Bedeutung des Ausschnittes erfassen. Dieses Verständnis resultiert in einem internen Modell, das sich der Programmierer von der Bedeutung des Ausschnittes macht. Diesem Modell muß dann eine abstrahierte, zusammengefaßte Bedeutung zugeordnet werden, die dann zum Verstehen der übergeordneten Abschnitte verwendet wird.

Seit Anfang der achtziger Jahre wird das Verstehen von Programmen durch eine Reihe von Werkzeugen für den Programmierer in der Wartungsphase unterstützt.

Hierbei sind die folgenden Strategien zu unterscheiden:

- Typographische Ansätze betonen die Lesbarkeit von Programmen, um das Verstehen zu erleichtern [BM90], [OC90]. Durch das Verbessern des Schriftbildes wird versucht, das Erfassen wesentlicher Größen und struktureller Zusammenhänge zu erleichtern.
- Dokumentatoren versuchen, aus dem vorliegenden Quelltext durch automatische Analysen Informationen über die Definition und Verwendung von Größen zu beschaffen. Diese Information wird dann in schriftlicher Form ausgedruckt. Teils wird ein zusätzlicher Bericht über das Programm erzeugt [Con88], [BA86], teils wird die Information in den Programmtext eingefügt [LSW90]. Zu den Erfahrungen mit derartigen Werkzeugen sei auf [LSW90] verwiesen.
- Interaktive Analysewerkzeuge [Lin84],[CR86],[CNR90],[CER90] versuchen, die Nachteile statischer Dokumentatoren in Bezug auf Eindringtiefe und Informationsfülle zu beseitigen. Es werden in einem Analyseschritt Informationen aus dem Programm extrahiert und in einer Datenbank gespeichert. Hierzu werden Abfragemöglichkeiten bereitgestellt, über die sich der Programmierer die Information beschaffen kann, die ihn gerade interessiert. Die Anfragen und Analyseanforderungen werden über eine Kommandosprache abgewickelt. Die Leistungsfähigkeit und der Funktionsumfang derartiger Analysewerkzeuge hängen von der Genauigkeit und vom Umfang des internen Datenmodells ab, das die Analyseergebnisse aufnimmt.
- Graphische Mittel wie in INCENSE [Mye83], PegaSys [MH86] oder VIFOR [RDLS88] versuchen, Datenstrukturen, Datenfluß und Kontrollfluß durch Diagramme darzustellen. Einen Überblick über den Einsatz graphischer Hilfsmittel in der Programmierung geben [Rae85] und [Win90].
- Integrierte Programmierumgebungen wie PECAN [Rei84] erlauben es zumindest für kleine Sprachen, zwischen mehreren Ansichten eines Programms zu wählen. In PECAN sind dies vor allem ein syntax-gerichteter Editor und eine graphische Darstellung, die am Vorbild von Nassi-Shneiderman-Diagrammen orientiert ist.
- Programm-Browser[3] sind als erste Versuche zu verstehen, die Prinzipien von Hypertext [Con87] auf Programme anzuwenden. Klassische Vertreter sind etwa die SMALLTALK-80 Umgebung [Gol84] oder CEDAR [Tei85]. Neuere Ansätze werden in [SSSW86] und in [Cle89],[Cle88],[Cag90] vorgestellt. Auch derartige Browser erleichtern das Verständnis von Programmen. Allerdings fehlt in diesem Bereich zur Zeit noch eine gründliche Durchdringung der Probleme des Programmverstehens. Daher sind die Beziehungen, die untersucht werden können, nicht ausreichend und nicht allgemein genug. Auch die Interaktion läßt zu wünschen übrig.

Das hier vorgestellte Werkzeug VAMP ist prinzipiell dem Browsing-orientierten Ansatz zuzuordnen. Von existierenden Werkzeugen dieses Gebietes unterscheidet es sich durch die konsequente Verfolgung des Hypertext-Prinzips und durch die stark erweiterte Funktionalität. Das interne Modell und die Bedienung ermöglichen weiterhin die konsistente Behandlung und Analyse von modulübergreifenden Zusammenhängen. Die erweiterte Funktionalität ist auf das Verstehen von Programmen hin entwickelt worden. Durch diesen Funktionsumfang zeigt VAMP auch Eigenschaften eines interaktiven Analysewerkzeuges, das es mit einer besonders entworfenen Benutzungsschnittstelle erlaubt, das Verhalten und die Bedeutung von Programmgrößen zu untersuchen.

[3] Im deutschen Sprachgebrauch wird zur Zeit der Versuch unternomen, für derartige Werkzeuge den Begriff *Stöberer* einzuführen.

3 Größen und Beziehungen in Programmen

In VAMP unterscheiden wir als Größen die Klassen *Module, Konstanten, Typen, Prozeduren, Funktionen, Variablen, Verbundfelder, Parameter, Marken und Makros.* Die Probleme, die sich beim Versuch ergeben, ein unbekanntes Programm zu verstehen, lassen sich in drei Klassen einteilen:

- *Datenfluß*

  Fragen bezüglich des Datenflusses sind zum Beispiel:

  Welchen Wert kann eine Größe an einer gegebenen Programmzeile annehmen? Von welchen anderen Größen wird ihr Wert beeinflußt? Welche anderen Größen werden von ihr beeinflußt?

  Diese Fragen haben für Variablen, Parameter und Verbundfelder Bedeutung, aber auch für Funktionen, die ja einen Ergebniswert annehmen.

- *Kontrollfluß*

  Fragen bezüglich des Kontrollflusses sind zum Beispiel:

  Welche Prozeduren rufen eine Prozedur P auf, welche ruft sie selber auf? Wo ist eine Marke definiert, und von wo kann sie angesprungen werden?

- *Definitionszusammenhang*

  Dieser Zusammenhang ist in Verbindung mit Datentypen von Interesse. Typische Fragen sind zum Beispiel:

  Welche Größen werden mit Hilfe eines Typs T definiert, welche dienen zur Definition von T? Welchen Typ hat eine Variable v?

In großen, modular aufgebauten Programmen sind die Import-Beziehungen zwischen verschiedenen Moduln von großer Bedeutung. Die verschiedenen Beziehungen lassen sich also unterteilen: zum einen handelt es sich um Beziehungen, die lokal sind, also innerhalb eines Moduls liegen. Zur anderen Gruppe gehören die Beziehungen, die Modulgrenzen überschreiten.

4 Der VAMP - Ansatz

Aufgrund der Vielzahl der in Programmen vorliegenden Beziehungen ist es erforderlich, daß der Programmierer je nach Interessenlage die für ihn wichtigen Informationen abrufen kann. Hierbei darf er nicht durch andere Informationen, die zur fraglichen Zeit nebensächlich sind, abgelenkt werden. Statische Dokumentationswerkzeuge wie PasDok [LSW90] oder FORTVER [Con88] scheiden somit aus, denn diese Aufgabe ist nur im Dialog zu bewältigen.

Die Dialoggestaltung ermöglicht es, direkt über den Programmtext als Arbeitsgrundlage Anfragen an eine interne Datenbasis zu richten. Hierzu klickt der Benutzer eine ihn interessierende Größe im Programmtext an.

Ausgabeseitig wird eine problemangemessene Darstellung der Ergebnisse der Anfragen geliefert. Grobe, intermodulare Zusammenhänge werden durch Graphen visualiert, während die Details der innermodularen Beziehungen mit Hilfe des Programmtextes dargestellt werden. Hierzu werden im Programmtext Markierungen vorgenommen.

Zu jeder im Programm vorkommenden Größe können ihre Definition und alle sie beeinflussenden Beziehungen ermittelt werden.

Der sich ergebende Systementwurf läßt sich wie folgt beschreiben (vgl. Abb. 1): Durch statische Analyse der syntaktischen und semantischen Zusammenhänge im Programm wird eine Datenbasis geladen. An diese Datenbasis kann der Programmierer mittels einer Fenster-Oberfläche Anfragen an einen Kommando-Interpretierer richten. Dieser Kommando-Interpretierer beschafft sich die erforderliche Information aus der Datenbasis und stellt sie unter Zuhilfenahme der Fenster-Oberfläche dar.

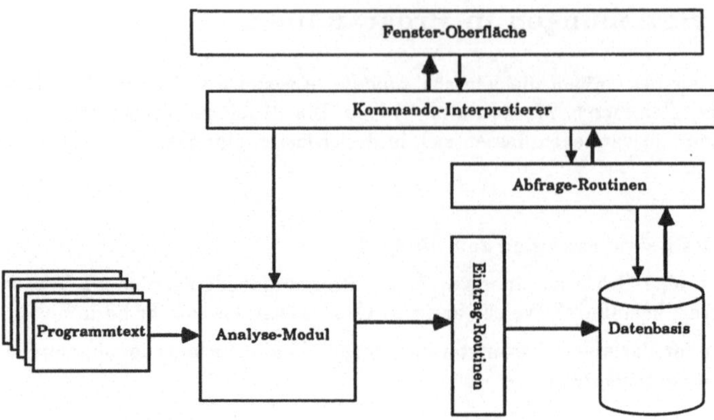

Abbildung 1: VAMP : Der Systementwurf

Um die Funktionen anbieten zu können, wurde ein einheitliches internes Programmodell entwickelt, das die oben angeführten Beziehungen repräsentiert. Das Modell ist an bekannte Datenmodelle aus Compilern [PD83] und an Programmabhängigkeitsgraphen [Ott78], [Mos90] angelehnt und wurde in Richtung des Datenflusses und des Definitionszusammenhangs erweitert. Das so entwickelte interne Modell erlaubt in einfacher Weise das schnelle Ermitteln von Beziehungen aller drei oben genannten Klassen.

Da VAMP große Programme behandeln können soll, ist weiterhin eine in Bezug auf den Speicherbedarf sparsame Datenstruktur nötig, die den schnellen Zugriff über den Programmtext auf die im Datenmodell gespeicherten Größen erlaubt. Durch eine Kombination von statischen und dynamischen Datenstrukturen konnte erreicht werden, daß die Zugriffszeit, die benötigt wird, um zu einer Mausklick-Position im Programmtext die ausgewählte Größe zu ermitteln, unabhängig von der Programmgröße ist. Sie ist umgekehrt proportional zur mittleren Bezeichnerlänge und nimmt im wesentlichen mit steigender Bezeichnerlänge ab. Der Speicherbedarf der Datenstruktur wächst im wesentlichen linear mit der Zahl der Referenzen. Hinzu kommt noch ein vernachlässigbarer Verwaltungsaufwand, der mit O(l) zu Buche schlägt, wenn l die Länge des Programmtextes ist. Das entworfene Modell ist sprachunabhängig, verschiedene prozedurale Programmiersprachen (Modula-2, Pascal, FORTRAN, COBOL, C, Ada) können darauf abgebildet werden.

5 Die Benutzungsschnittstelle

Wesentlich für die Leistung und Akzeptanz ist die Konzeption einer einfachen und problemangemessenen Benutzungsschnittstelle. Zu ihrer Implementierung wird das Fenstersytem X benutzt. Um die geforderten Eigenschaften zu erbringen, werden die folgenden 4 Fensterklassen konzipiert:

1. Das Hauptfenster (vgl. Abb. 2):

   Es dient dazu, die übergreifende Modul-Struktur im Programm zu verdeutlichen. Es soll gleichzeitig als globales Steuerinstrument für die Analyse dienen. Dieses Fenster wird dem Programmierer als Ausgangspunkt für seine Analysen angeboten. Über das Hauptfenster startet der Programmierer die Analyse und greift auf den Programmtext zu. Ferner kann er in diesem Fenster die Schnittstellen zwischen den verschiedenen Moduln ermitteln.

2. Das Textfenster (vgl. Abb. 3):

   Das Textfenster ist das grundlegende Element der Programmanalyse im Kleinen. In einem Textfenster kann der Programmtext eines Moduls dargestellt werden. Über das Textfenster kann der Programmierer auf einzelne Größen im Programm zugreifen, die ihm für die Lösung einer Frage relevant erscheinen. Hierbei kann er bottom-up vorgehen und sich zu einer bestimmten Stelle im Programm, die ihn interessiert, alles beschaffen, was an dieser Stelle von Bedeutung ist. Der Zugriff auf eine Größe im Programm erfolgt durch das Anklicken der Größe im Text per Maus. Als Ergebnis werden alle Auftretensstellen der angewählten Größe farbig markiert, ferner öffnet sich ein Bezeichnerfenster zu dieser Größe.

3. Das Bezeichnerfenster (vgl. Abb. 4):

   Zu jeder angeklickten Größe öffnet sich ein Bezeichnerfenster. Jedes Bezeichnerfenster ist farblich auf die Markierung der zugehörigen Größe abgestimmt. Über das Bezeichnerfenster kann der Programmierer systematisch alle Aufretensstellen der Größe ansteuern, indem er entsprechende Kommando-Knöpfe (≫ und ≪) bedient.

   Über das Bezeichnerfenster kann der Programmierer sich zusätzliche Information zu der fraglichen Größe beschaffen. Als erste Information zu der Größe wird die Definitionsstelle gezeigt. Ferner gibt es je nach Art der Größe eine Reihe von weiteren Analysemöglichkeiten, die durch Kommandoknöpfe bedient werden können. Diese Analysemöglichkeiten entsprechen den verschiedenen Beziehungen, die zu der fraglichen Größe gehören. Beispielsweise kann zu einer ausgewählten Prozedur p eine Liste aller Prozeduren ausgegeben werden, die p aufrufen.

4. Das Listenfenster (vgl. Abb. 5):

   Das Listenfenster wird immer dann eingesetzt, wenn das Ergebnis einer Anfrage eine Menge von Größen ist, die eventuell aus verschiedenen Moduln stammen können. Über das Listenfenster kann dann zu einzelnen Größen gezielt Information beschafft werden, indem die Größe in der Liste angeklickt wird. Daraufhin öffnet sich ein Bezeichnerfenster zu dieser Größe. Beispiele für den Einsatz des Listenfensters sind Import- und Exportlisten eines Moduls, Schnittstellen zwischen Moduln und die Liste aller Variablen, die mit Hilfe eines Typs definiert werden.

Ablauf der Arbeit mit VAMP

Als erstes erhält der Programmierer eine Übersicht über das Gesamtprogramm. Dieses setzt sich im allgemeinen aus verschiedenen Moduln zusammen, die via import/export-Klauseln Informationen austauschen. Dieser intermodulare Zusammenhang wird in Form eines Graphen dargestellt. Die Module werden als Knoten dargestellt, die Beziehungen zwischen Moduln als Kanten. Ein Modulgraph mit 12 Moduln ist in Abb. 2 wiedergegeben. Durch das Selektieren der Knoten und Kanten kann der Programmierer in die Analyse der Module bzw. der zwischen ihnen vorliegenden Beziehungen einsteigen.

- Durch das Selektieren eines Knotens erhält der Programmierer als Ausgangsinformation den Programmtext des betreffenden Moduls, so wie er programmiert wurde.
- Durch das Selektieren einer Kante erhält der Programmierer eine Liste aller Größen, die die Schnittstelle zwischen den betreffenden Moduln bilden.

In einem Programmtext kann der Programmierer die zugrundeliegende Struktur ermitteln, indem er Größen auswählt und dann die zugehörigen Beziehungen erforscht. Das Auswählen einer Größe im Programm erfolgt durch das Anklicken der Größe mit einem graphischen Eingabegerät (Maus o.ä.). Als Ergebnis werden alle Auftretensstellen dieser Größe farbig markiert. Außerdem wird ein Bezeichnerfenster geöffnet, das die Definitionsstelle der Größe enthält. Das Bezeichnerfenster bietet die Gelegenheit, die zu dieser Größe in Beziehung stehenden Größen zu ermitteln. Der Programmierer erreicht dies durch das Bedienen eines entsprechenden Kommandoknopfes.

Abbildung 3 zeigt den Programmtext des Moduls `Polygons` nach dem Anklicken der Größe `PolyTree`, Abbildung 4 das Bezeichnerfenster zu `PolyTree`, in dem die Definitionsstelle angegeben ist. Mittels der Knöpfe am rechten Fensterrand kann der Programmierer weiteres über den Typ `PolyTree` und seine Benutzung erfahren. Durch Anklicken des Knopfes **Definiert** werden alle Größen ermittelt, die mit Hilfe von `Polytree` definiert wurden. Diese Größen werden mit einer zweiten Farbe im Text markiert, Zu dieser Markierung öffnet sich ein Fenster, in dem eine Erklärung zu den markierten Größen steht. Auch dieses Fenster ist farblich auf die Markierung abgestimmt, so daß die Zuordnung von Markierung und zugehörigem Fenster

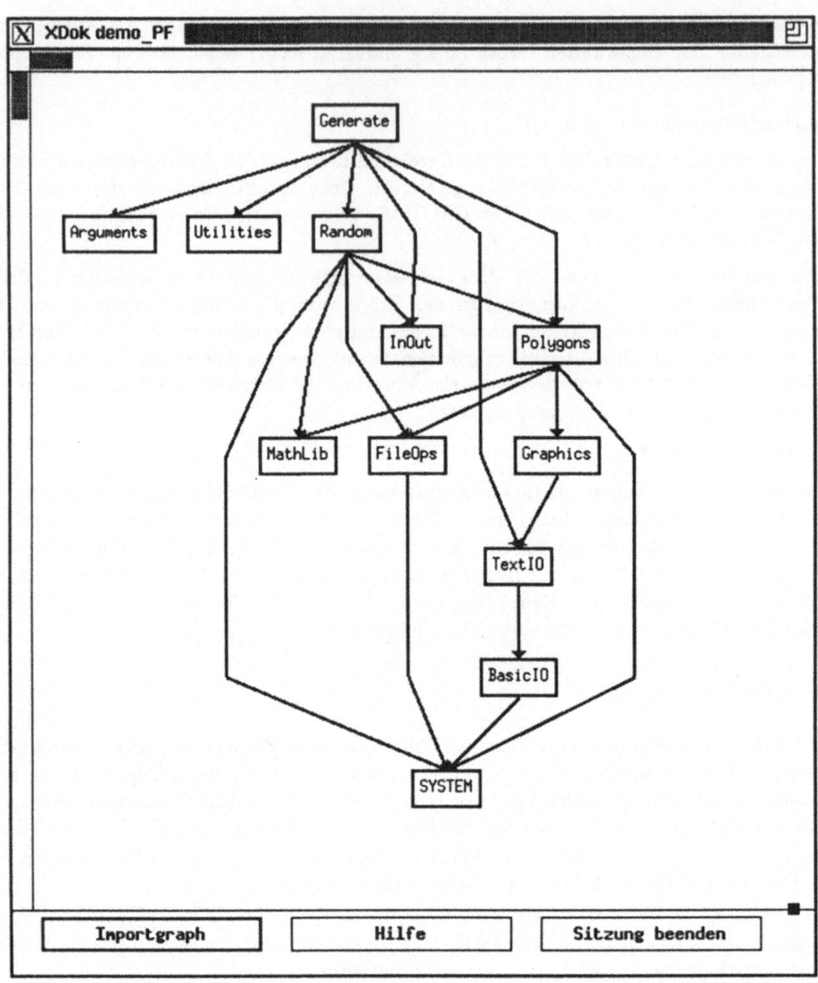

Abbildung 2: Ein Modulgraph

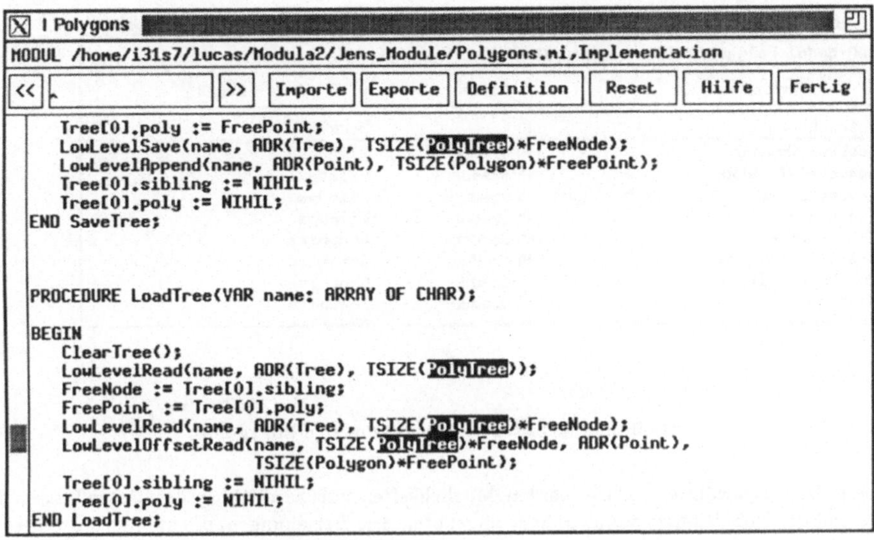

Abbildung 3: Programmtext mit markiertem `PolyTree`

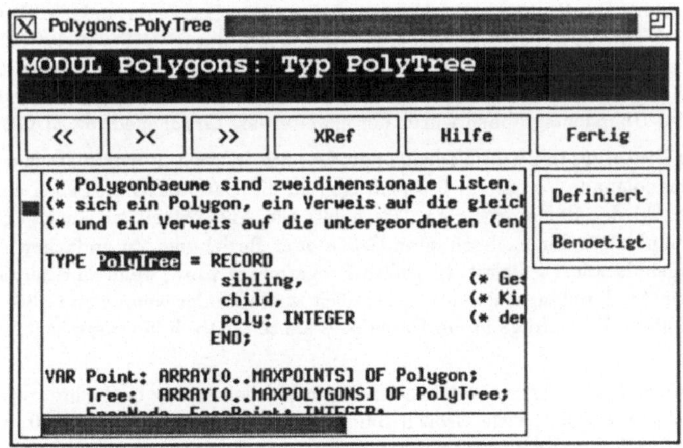

Abbildung 4: Bezeichnerfenster zu `PolyTree`

IM Polygons

IMP-Modul Polygons: Liste aller IMPORTE

Sortierung | Hilfe | Fertig

Bezeichner	Art	Herkunft
LowLevelAppend	Prozedur	FileOps
LowLevelOffsetRe	Prozedur	FileOps
LowLevelRead	Prozedur	FileOps
LowLevelSave	Prozedur	FileOps
ClearWindow	Prozedur	Graphics
DrawBlackPolygon	Prozedur	Graphics
DrawWhitePolygon	Prozedur	Graphics
SetSize	Prozedur	Graphics

Abbildung 5: Listenfenster: Importe zu `Polygons`

Je nach Art der ausgewählten Größe werden Möglichkeiten zur Ermittlung des Datenflusses, des Kontrollflusses und der Definitionszusammenhänge angeboten. Die Beziehungen, die untersucht werden können, sind intra- wie auch intermodular zu verstehen. Es werden also z.B. alle Aufrufstellen zu einer Prozedur in allen Moduln als Ergebnis einer Anfrage geliefert. Neben dem direkten Markieren nachgefragter Größen im Programmtext gibt es - bei vielen zu markierenden Größen - die Möglichkeit, die Größen zunächst in einer Liste (vgl. Abbildung 5) auszugeben. In dieser Liste kann der Benutzer dann Größen per Maus auswählen und so im Programmtext markieren lassen.

6 Erfahrungen mit VAMP

Das Programmodell und die Benutzungsoberfläche mit Teilen der in dieser Arbeit konzipierten Leistungsmerkmale konnten als Prototyp implementiert werden. Das entstandene Programmierwerkzeug VAMP läuft auf SUN-3 und SUN-4 (Sparc) Arbeitsplatzrechnern unter UNIX. Zielsprache dieses Prototyps ist Modula-2. Der Prototyp benutzt als Hilfsmittel zur Gestaltung der Benutzungsschnittstelle das Fenstersystem X (Version 11) und die ATHENA Widgets. Ein C-Frontend wird gegenwärtig implementiert. Zur Gestaltung des Modul-Übersichtsgraphen wurde der erweiterbare Grapheneditor EDGE [PT90] verwendet.

Eine Validierung des Prototyps mit verschiedenen mittelgroßen Modula-2 Programmen ergab in Hinblick auf die Antwortzeiten das folgende Verhalten: Die Anfragen an die Datenbasis bezüglich einer Größe g dauern proportional zur Anzahl der Referenzen r(g). Das heißt: Die Antwortzeit zu einer gegebenen Anfrage ist proportional zur Komplexität der nachgefragten Größe oder Beziehung. Zu einfachen Anfragen in kleinen Moduln wird das Ergebnis sofort geliefert. In großen Programmen kann, wenn zu einer Größe viele Referenzstellen existieren, die Analyse länger dauern. Bei großen Moduln oder wenn eine Größe viele Bezugsstellen hat, hat sich jedoch nicht die Anfrage in die Datenbasis als zeitkritisch herausgestellt, sondern das Färben vieler Textstellen.

Es können beliebig viele Text-, Listen- und Bezeichnerfenster gleichzeitig betrachtet werden. Dadurch wird einerseits die Möglichkeit eröffnet, viele Größen simultan zu analysieren. Andererseits verführt die Benutzungsschnittstelle dadurch zum uneingeschränkten Öffnen von Fenstern, wodurch leicht die Übersicht verloren gehen kann.

Eine Beseitigung dieser Gefahr wäre etwa durch Restriktion der Fensterzahl oder durch automatisches Aufgeben von offenen Fenstern zu erreichen, wenn eine obere Schranke überschritten wird. Dies wäre aber nur auf Kosten der Flexibilität möglich. Diese Flexibilität ist aber genau einer der Vorteile von VAMP gegenüber anderen Werkzeugen, die eine beschränkte Anzahl fest positionierter Fenster anbietet. Vorbehaltlich weiterer Erfahrungen aus dem praktischen Einsatz halten wir eine derartige Einschränkung des Benutzers für nicht angemessen.

7 Zusammenfassung und Ausblick

VAMP dient zur Unterstützung eines Programmierers beim Verstehen eines unbekannten Programms. Hierzu erhält er eine auf seine Probleme zugeschnittene Benutzungsschnittstelle, die es ihm erlaubt, je nach Bedarf Information über Größen im Programm abzurufen. Zu diesen Größen kann er die Beziehungen, die auf die Größe Einfluß nehmen, verfolgen. Das Abrufen der Information erfolgt durch das Anklicken der fraglichen Größe im Programmtext oder durch das Bedienen von Funktionsknöpfen, die bestimmte Analysen ausführen. Die Darstellung der Ergebnisse erfolgt je nach Anfrage durch farbige Markierung im Programmtext, durch Präsentation entsprechender Programmtextstücke oder – zwecks Abstraktion – in Form von Listen oder Graphen.

Als nächster Schritt soll die Leistung von VAMP in einer breit angelegten Praxisstudie untersucht werden. Ziel hierbei ist es, den Zeitvorteil zu ermitteln, der beim Erledigen einer Wartungsaufgabe an einem unbekannten Programm mit Unterstützung durch VAMP erreicht wird.

Literatur

[BA86] M. Ben-Ari. Foreet: A Tool for Design and Documentation of Fortran Programs. *Software – Practice & Experience*, 16(10):915 .. 924, 1986.

[BM90] R. Baecker und A. Marcus. *Human Factors and Typography for More Readable Programs.* Addison Wesley, Reading,MA, 1990.

[Bro83] R. Brooks. Towards a theory of the comprehension of computer programs. *International Journal of Man-Machine Studies*, 18(6):542 – 554, 1983.

[Cag90] M.R. Cagan. The HP SoftBench Environment: An Architecture for a New Generation of Software Tools. *Hewlett-Packard Journal*, 41(3):36 – 47, Juny 1990.

[CER90] J.R. Cordy, N.L. Eliot, und M.G. Robertson. TuringTool - A User Interface to Aid in the Software Maintenance Task. *IEEE Transactions on Software Engineering*, 16(3):294 – 301, March 1990.

[Cle88] L. Cleveland. A User Interface for an Environment to support Program Understanding. In *Proc. Conference on Software Maintenance*, Seite 86 – 91, 1988.

[Cle89] L. Cleveland. A program understanding support environment. *IBM Systems Journal*, 28(2):324 – 344, 1989.

[CNR90] Y.-F. Chen, M.Y. Nishimoto, und C.V. Ramamoorthy. The C Information Abstraction System. *IEEE Transactions on Software Engineering*, 16(3):325 – 334, March 1990.

[Con87] J. Conklin. Hypertext: An Introduction and Survey. *Computer*, 20(9):17 .. 41, September 1987.

[Con88] R. Conradi. Experience with Fortran Verifier. Technical report, Norwegian Institute of Technology, Trondheim, Norway, 1988.

[CR86] Y.-F. Chen und C.V. Ramamoorthy. The C Information Abstractor. Technical Report UCB/CSD 86/300, University of California, Berkeley, Juny 1986.

[Dij68] E.W. Dijkstra. Go To Statement Considered Harmful. *Communications of the ACM*, 11(3):147 – 148, March 1968.

[Gol84] A. Goldberg. *Smalltalk-80: The Interactive Programming Environment.* Addison-Wesley, Reading,MA, 1984.

[Lin84] M.A. Linton. Implementing Relational Views of Programs. In *Proceedings of the ACM SIGSOFT/SIGPLAN Software Engineering Symposium on Practical Software Development Environments*, May 1984.

[LSW90] J. Lucas, A. Schmitt, und J.F.H. Winkler. Automatic Documentation of Modular Programs using PasDok. Technical Report 29, Universität Karlsruhe, Fakultät für Informatik, Karlsruhe, Germany, 1990.

[Luk80] F.J. Lukey. Understanding and debugging programs. *International Journal of Man-Machine Studies*, 12(2):189 – 202, February 1980.

[MH86] M. Moriconi und D.F. Hare. The PegaSys System : Pictures as formal Documentation of large Programs. *ACM Transactions on Programming Languages and Systems*, 8:524 – 546, 1986.

[Mos90] L.E. Moser. Data Dependency Graphs for Ada Programs. *IEEE Transactions on Software Engineering*, 16(5):498 – 509, May 1990.

[Mye83] B.A. Myers. INCENSE : A System for Displaying Data-Structures. *Computer Graphics*, 17(3):115 – 125, July 1983.

[OC90] P.W. Oman und C.R. Cook. Typographic Style is More than Cosmetic. *Communications of the ACM*, 33:506 – 520, May 1990.

[Ott78] K.J. Ottenstein. *Data-Flow graphs as an Intermediate Program Form.* Dissertation, Computer Science Dept., Purdue University, Lafayette, IN, 1978.

[Par86] G. Parikh. *Handbook of Software Maintenance.* Wiley-Interscience, New York,NY, 1986.

[PD83] S. Pemberton und M.C. Daniels. *Pascal Implementation, Part 1: The P4-Compiler; Part 2: Compiler and Assembler.* Ellis Horwood Ltd., 1983.

[PT90] F.N. Paulisch und W.F. Tichy. EDGE : An extendible directed Graph Editor. *Software – Practice & Experience*, 20(S1):S1/63 .. S1/88, Juny 1990.

[Rae85] G. Raeder. A Survey of current graphical Programming Techniques. *IEEE Computer*, 18:11 – 25, August 1985.

[RDLS88] V. Rajlich, N. Damaskinos, P. Linos, und J. Silva. Visual Support for Programming-in-the-Large. In *Proc. Conference on Software Maintenance*, Seite 92 – 99, 1988.

[Rei84] S.P. Reiss. PECAN : Program Development Systems, that support multiple Views. In *Proc. 7. ICSE*, Seite 324 – 333, 1984.

[Sch87] N.F. Schneidewind. The State of Software Maintenance. *IEEE Transactions on Software Engineering*, 13:303 – 310, March 1987.

[Sen83] H.-E. Sengler. *Ein Modell des Verstehens von Programmen und seine Anwendung beim Entwurf der Programmiersprache GRADE.* Dissertation, Universität Hamburg, 1983.

[SP88] E. Soloway und J. Pinto. Designing Documentation to Compensate for Delocalized Plans. *Communications of the ACM*, Seite 1259 – 1267, November 1988.

[SSSW86] B. Shneiderman, P. Shafer, R. Simon, und L. Weldon. Display Stretegies for Program Browsing: Concepts and Experiments. *IEEE Software*, 3(3):7 – 15, May 1986.

[Tei85] W. Teitelman. A Tour through CEDAR. *IEEE Transactions on Software Engineering*, 11:285 – 302, March 1985.

[Win90] J.F.H. Winkler. Visualisierung in der Software-Entwicklung. In A. Reuter, Hrsg., *GI – 20. Jahrestagung I*, page 40 .. 72. Springer Verlag, October 1990.

[WS73] W. Wulf und M. Shaw. Global Variables Considered Harmful. *SIGPLAN Notices*, 8(2):28 – 34, February 1973.

Dynamische Icons:

Anforderungen, Entwurfs- und Darstellungshilfmittel

Oliver Bernhardt
TH Darmstadt

Gabriele Rohr
IBM Entwicklungslabor Böblingen

Problemstellung

Grafische Benutzungsschnittstellen haben in letzter Zeit immer mehr an Bedeutung für die Mensch-Rechner Interaktion gewonnen. Waren es zuerst Anwendungen, die eher einen technisch unerfahrenen Anwenderkreis bedienten (Textverarbeitung, Dokumentenverwaltung), so hat sich das jetzt auch immer mehr auf komplexe, stärker technisch orientierte Anwendungen ausgedehnt (Netzwerkanwendungen, Konfigurationen usw.). Der primäre Vorteil grafischer Benutzungsschnittstellen liegt in der Strukturierung komplexer Information. Visuelle, bildliche Information kann vom Menschen eher parallel verarbeitet werden (ROHR 1988, 1990a). Sie hilft überall dort, wo entweder strukturelle Zusammenhänge analysiert werden müssen (GERSTENDÖRFER & ROHR 1987) oder ganz allgemein ein Überblick über eine große Menge von Informationen gehalten werden muß.

Grafische Benutzungsschnittstellen haben darüberhinaus noch einen weiteren Vorteil: Sie erlauben abstrakte Funktionen in einer bildlichen Objektwelt abzubilden und diese zu manipulieren, d.h., sie können über eine Metapher einen verständlicheren Umgang mit einem komplexen abstrakten System herstellen. Im Gebrauch von Metaphern liegen aber auch grössere Probleme: Die Auswahl einer adäquaten Metapher und ihre konsistente Verwendung (siehe dazu auch CARROLL et al. 1988 und ROHR 1990b). Inadäquate Metaphern führen den Benutzer eher in die Irre, als daß sie hilfreich die Planung seiner Aktionen auf entsprechenden Objekten unterstützen.

Die Realisierung von Anwendungsfunktionen in direkt manipulativen Aktionen auf grafischen Objekten erfordert aber immer eine Metapher (ROHR 1990b). Diese kann für unterschiedliche Anwendungsbereiche sehr verschieden aussehen, d.h., Metaphern können je nach Anwendungsbereich unterschiedlich adäquat sein. Sorgfältige Untersuchungen auf der Benutzerseite sind dazu notwendig. Das erfordert eine Trennung der Funktionen des Entwurfs einer Benutzerschnittstelle von der der

Anwendungsentwicklung, wobei es eine klar und allgemein definierte Schnittstelle zwischen der Anwendung, der grafischen Objektwelt und Werkzeugen zu ihrer Erstellung geben muß. Erst das ermöglicht einen iterativen Entwurf der grafischen Oberfläche mit Untersuchungen an den Endbenutzern für die optimale Gestaltung, unabhängig von der konkreten Implementierung auf Seiten der Anwendungsentwicklung.

Ein besonders kritischer Bereich für solche Art von Untersuchungen ist, da relativ neu, der Bereich von Anwendungen mit Funktionen der Prozeß-Überwachung Prozeß-Steuerung. Visualisierung von Prozessen mit Hilfe grafischer Objekte, d.h., Metaphern für ihren Entwurf, sind bisher wenig untersucht, wenn man einmal von vereinfachter Verwendung klassischer Instrumenten-Anzeige absieht. Diese Instrumente-Icons entsprechen meistens sehr spezifischen Metaphern, können nicht verändert werden und sind häufig in der Schnittstelle an sehr spezielle Anwendungen angepaßt. Anwendungen auf die die jeweilige Metapher nicht paßt und/oder die mehr dynamische Eigenschaften eines Objektes zu visualisieren haben, können diese Icons nicht verwenden, sondern müssen andere komplett neu und wiederum speziell implementieren.

Um breitere Gestaltungsmöglichkeiten zu haben, inklusive von Untersuchungen am Endbenutzer über die Angemessenheit der verwendeten Metapher, sollte es möglich sein unterschiedlichste Icons mit beliebigen dynamischen Eigenschaften zu konstruieren, zu testen und in ein Anwendungsprogramm oder möglicherweise mehrere entsprechende einzuhängen. Sollten ganze Metaphernsysteme entworfen und konstruiert werden, die dann für ganze Gruppen von Einzelanwendungen gültig wären, d.h., deren dynamische grafische Objekte, "dynamische Icons", standardisiert wären, so müßten diese dynamischen Icons auch über Bibliotheken zugreifbar sein, um nicht ständig neue Re-implementierungen zu erfordern. Dafür benötigt man aber auch eine allgemeine, standardisierte Schnittstelle für diese Art von grafischen Objekten, an die sich jede beliebige Anwendung anpassen kann. Hierzu müssen die Eigenschaften solcher dynamisch ansteuerbaren grafischen Objekte, die wir im weiteren immer "dynamische Icons" nennen werden, genau analysiert werden, im Hinblick darauf, was für ihren Aufbau und die funktionalität ihrer Konstruktion und ihrer Darstellungsschnittstelle erforderlich ist. (Wir verwenden hier den englishsprachigen Begriff "Icon" und nicht den deutschen "Piktogramm", weil der englischsprachige weiter gefaßt ist und auch Strukturen enthalten kann, z.B. "Iconic Interfaces", während der Begriff "Piktogramm" eher für bildliche Stellvertreter von isolierten Objekten und Funktionen verwendet wird.)

Dynamische Icons können als grafische Objekte betrachtet werden, die aus einzelnen Komponenten zusammengesetzt sind, von denen mindestens eine Komponente mindestens eine dynamische Eigenschaft hat, d.h., daß ein grafischer Parameter der Komponente, wie Höhe oder Breite, dynamisch durch ein Programm verändert werden kann. Dabei reflektiert die dynamische grafische Veränderung dynamische Attributwerte-Änderungen eines Anwendungssobjektes. FAIRCHILD et al. (1989) sprechen hierbei von "Automatic Icons".

Es geht darum, die Anwendung mit einem möglichst einfach zu handhabenden und allgemein (generisch) verwendbaren Hilfsmittel (Werkzeug) für sowohl die Konstruktion als auch die On-Line Darstellung solcher Icons zu Unterstützen. Es liegen bereits einige Arbeiten vor (z.B. CROOK & AULETTA 1986, FAIRCHILD et al. 1989, HSIA & AMBLER 1988, MYERS 1987), die sich mit diesem Problem oder Teilaspekten davon befassen. Allerdings findet man unserer Meinung nach wenig wirklich generische Ansätze darunter. Die meisten unterstützen ein paar wenige, sehr spezifisch ausgewählte Aspekte dynamischer Icons. Unser Ansatz besteht darin, aus der allgemeinen Struktur solcher grafisch dynamischen Objekte und aus den funktionellen Anforderungen der Umgebung für die Anwendungsentwicklungs abgeleitet, ein möglichst einfaches, allgemein verwendbares Unterstützungspaket zu entwickeln.

Allgemeine Struktur dynamischer Icons

Dynamische Icons, also grafische Objekte mit dynamischen Eigenschaften, sind in gewisser Weise aus unterschiedlichen Komponenten zusammengesetzte Objekte. Die Komponenten können als grafische Primitive betrachtet werden, d.h., sie bestehen aus einem Form-Merkmal (Rechteck, Ellipse, Polygon) und einer Füllfarbe.

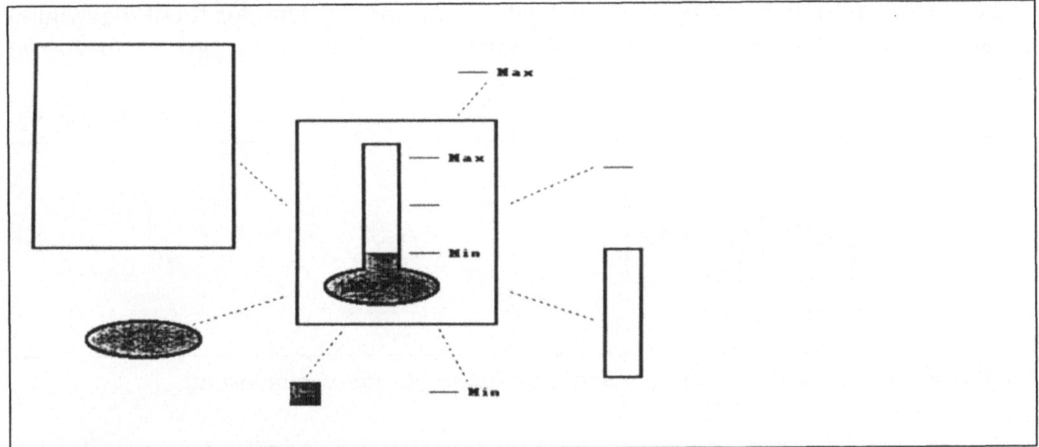

Abbildung 1. Aus Komponenten zusammengeztztes Grafisches Objekt

Die veränderlichen Parameter dieser Komponenten sind für die Formmerkmale die Höhe und die Breite und die Farbstufe für die Füllfarbe. Dynamische Eigenschaften können auf diese Parameter von Komponenten des Gesamtobjektes gelegt werden, indem ein Minimalwert, ein Maximalwert und die Anzahl der Stufen dazwischen und ihre Skala (linear, logarithmisch, exponential, usw.) für die Ausdehnung von Höhe und Breite definiert werden. Für die Füllfarbe wäre es analog die Bestimmung der Ausgangsfarbe, der Endfarbe und der Übergangsstufen dazwischen.

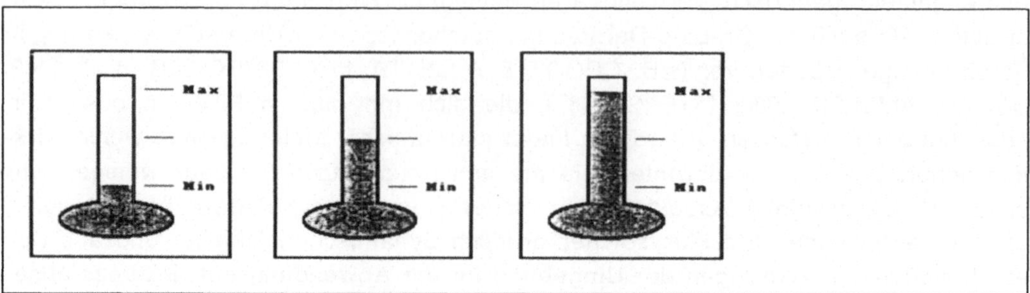

Abbildung 2. Dynamische Veränderung auf dem Parameter Höhe

Eine besondere Rolle spielt die dynamische Veränderung der Position einer Komponente, da für aus Komponenten zusammengesetzte Objekte die Position bedeutungstragend ist. Die Position ist immer im Bezug zum Gesamtobjekt zu sehen. Bei der Positionsänderung muß dieser Aspekt berücksichtigt werden, wenn außer der rein horizontalen oder vertikalen Verschiebung andere Positionsveränderungsformen vorgenommen werden sollen. In gewisser Weise spielen solche Grenzen allerdings auch in der Höhen- und Breiten-Veränderung eine Rolle. Die Maxima müssen in sinnvoller Beziehung zum Gesamtobjekt stehen. Analog verhält es sich auch mit einem weiteren dynamischen Parameter, dem der Richtungswinkel Veränderung (siehe nächste Abbildung). Auch er muß in einen sinnvollen Bezug zum Gesamtobjekt gebracht werden.

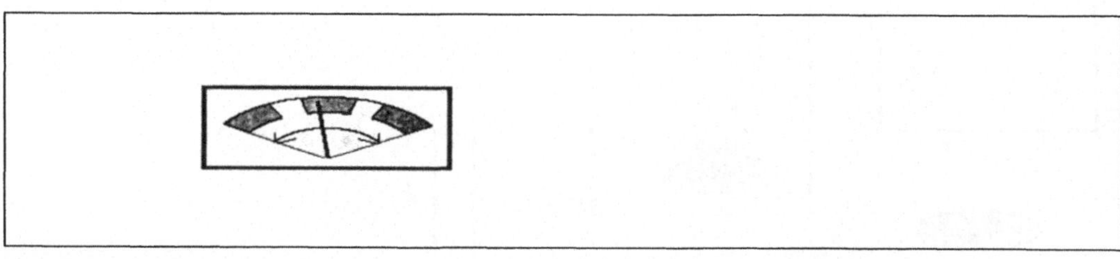

Abbildung 3. Beispiel für eine dynamische Ausrichtungs-Veränderung

Ein Spezialfall der dynamischen Veränderung ist das Wegblenden oder Hinzuschalten einer oder mehrerer Komponenten als Indikator für binäre Zustandsänderungen. Sie kann auch als Markierung eines Objektes angesehen werden.

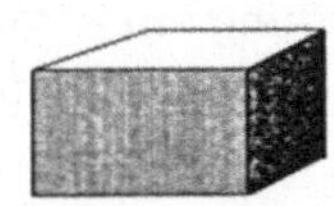
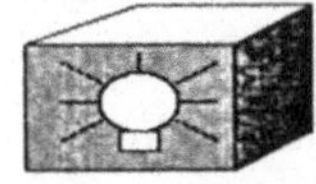

Abbildung 4. Beispiel für eine dynamische binäre Veränderung

Dynamische Icons können ganz allgemein nicht nur mehrere dynamische Komponenten besitzen, sondern jeder Komponente könnten auch mehrere dynamische Eigenschaften gleichzeitig zugewiesen werden.

Hilfsmittel für Design und Programm-Anbindung

Will man Unterstützungsfunktionen für die Konstruktion und den Gebrauch dynamischer Icons entwickeln, muß man zwei Funktionsbereiche der Anwendungsentwicklung unterscheiden: Das grafische Design der Benutzungsschnittstelle und die Programm-Entwicklung. Jeder der beiden Bereiche benötigt eine spezifische Unterstützung.

Bei Anwendungen mit Funktionen zur Prozess-Überwachungs und Prozeß-Kontrolle muß der Designer der Benutzerschnittstelle vorrangig die dem Endbenutzer verständlichen dynamischen Icons und der darin veränderlichen Parameter mit deren Grenzen und Stufen entwerfen, konstruieren und auf ihre Verständlichkeit hin testen. Der Programm-Entwickler dagegen benötigt eine Schnittstelle zur Objekt-Darstellung, die es ihm ermöglicht, sich ändernde Attributwerte im Anwendungsobjekt schnell und einfach auf das Darstellungsobjekt weiterzugeben. Dabei interessieren ihn weniger die konkreten grafischen Parameter, sondern das zu benutzende dynamische Icon und dessen anzusteuernde Attribute. Er sollte damit auf die vom Benutzerschnittstellen-Designer konstruierten und beim Benutzer ausgetesteten dynamischen Icons direkt zugreifen können.

Folglich benötigt der Designer der Benutzerschnittstelle einen Editor, der es ihm ermöglicht, beliebige grafische Objekte aus Basiskomponeneten zusammenzusetzen, selektiv Komponenten als dynamisch zu deklarieren, den grafischen Parameter (Höhe, Breite, Position, Farbe, etc.) für die dynamische Änderung auszuwählen, im Dialog Minimum und Maximum und die Anzahl der Stufen dazwischen zu definieren. Zusätzlich sollte auch ein grafisches Feedback über die definierte Dynamik gegeben werden, die es dem Designer erlaubt die gewählte Entscheidung zur Erstellungszeit zu testen und zu korrigieren.

Der Programm-Entwickler benötigt demgegenüber den Zugriff auf die abgelegten Daten des vom Designer konstruierten dynamischen Icons über einen Satz Darstellungsfunktionen. Für den Programm-Entwickler ist es nur wichtig, die Zuordnung von Darstellungsobjekt zu Anwendungsobjekt zu kennen und die dynamischen Attribute, die für die zu betrachtenden Attributwerte der Anwendung in Frage kommen. Die Darstellungsfunktionen müssen dem Programm-Entwickler ermöglichen, Minimum und Maximum des darzustellenden Wertebereichs zu Beginn zu definieren, um dann die Darstellung des dynamischen Objektes mit nur dem aktuellen Wert für das definierte Attribut anzustoßen.

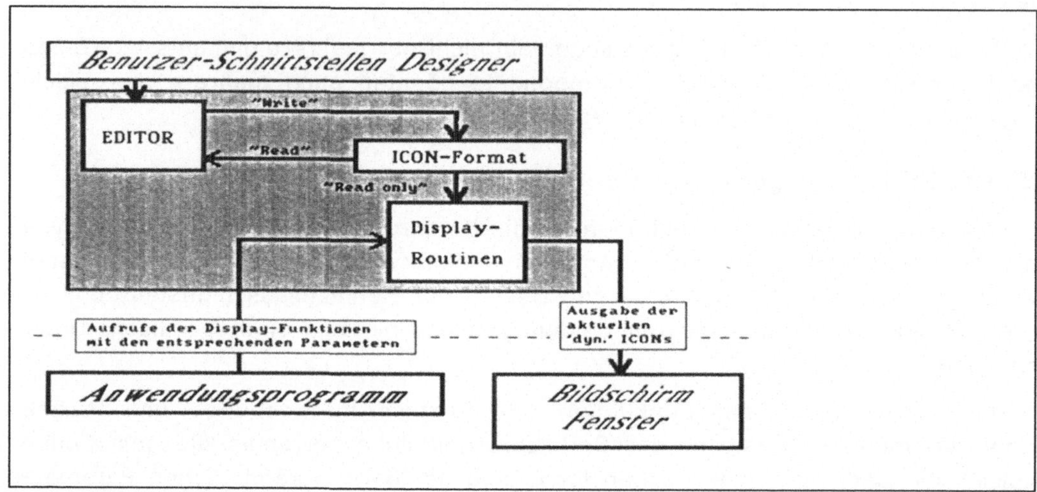

Abbildung 5. Struktur der Unterstützungsfunktionen

Der Editor für den Anwendungs-Designer

Der Editor bietet zur direkt manipulativen Erstellung von grafischen Objekten aus grafischen Komponenten über die üblichen Zeichenprimitive (Rechteck, Ellipse, Polyline, Füllfarbe, etc.) mit Editierfunktionen, wie sie auch in üblichen Zeicheneditoren gefunden werden, hinaus noch Möglichkeiten für spezielle Farbmischungen. Die wichtigste zusätzliche Funktion ist aber, daß eine mit einem Primitiv gezeichnete Komponente, immer als eine Komponente registriert ist, so daß man sich das Gesamtobjekt sequentiell komponentenweise ansehen kann. Man hat die Möglichkeit eine beliebige Komponente mit einem eindeutigen Namen zu markieren und über diesen Namen auch wieder aufzurufen. Diese, ursprünglich zum Editieren von Komponenten gedachte Funktion wurde nun benutzt, um darauf einen Dialog zu eröffnen, der es ermöglicht dynamische Eigenschaften einer solchen Komponente zu definieren. Die nächste Abbildung zeigt den Aufbau einer dynamischen Icon-Struktur am Beispiel einer Rechteck-Komponente.

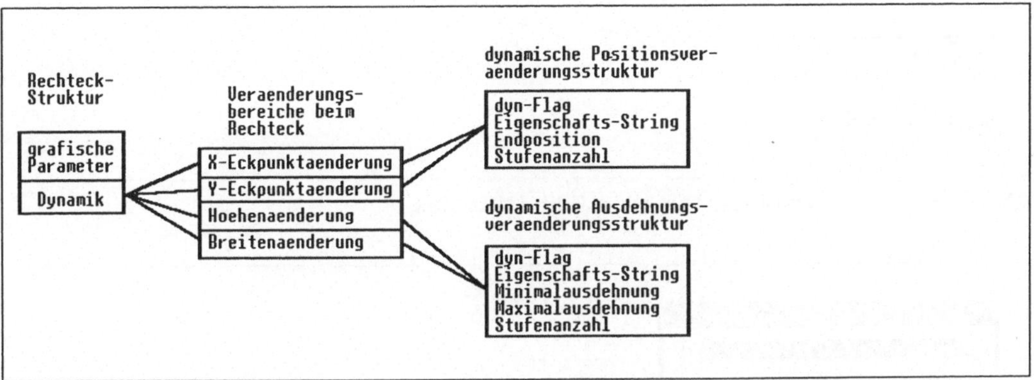

Abbildung 6. Dynamische Icon-Struktur am Beispiel einer Rechteck-Komponente

Der Benutzer kann einen grafischen Parameter wie Höhe, Breite, Position, etc. auswählen und erhält einen Dialog (siehe nächste Abbildungen), der ihn über die maximalen Grenzen der Dynamik bezogen auf das Gesamtobjekt informiert und den Eintrag von Minimalwert, Maximalwert und der Anzahl der Stufen erlaubt. Zur Kontrolle der eingetragenen Werte kann der Benutzer einen Stufentest laufen lassen, der dynamisch alle definierten Stufen vom Min- bis zum Maxwert für die ausgewählte Komponente darstellt, oder nur Minimal- oder Maximal-Wert anzeigen lassen. Diese Funktionen sind besonders wichtig, um den sinnvollen Bezug der dynamischen Änderung zum Gesamtobjekt zu testen. Die so erstellten Werte können mit dem Objekt abgespeichert werden.

Darstellungsfunktionen für den Programm-Entwickler

Die Darstellungsfunktionen greifen auf die abgespeicherten Werte zu. Der Benutzer der Darstellungsfunktionen muß den Namen des grafischen Objektes kennen und den oder die Namen der entsprechenden dynamischen Attribute.

Er hat dann drei Schritte zu vollziehen bis er das grafische Objekt mit dem oder den aktuellen dynamischen Wert darstellen kann:

- **Initialisierung**: Das dynamische Icon mit dem oder den entsprechenden Attributen initialisieren;
- **Eichung**: Das jeweilige Icon-Attribut mit dem entsprechenden Maximalwert und Minimalwert der Anwendungsfunktion eichen;
- **Aktuelle Darstellung**: Das Icon mit dem entsprechenden aktuellen Wert für das jeweilige Attribut darstellen.

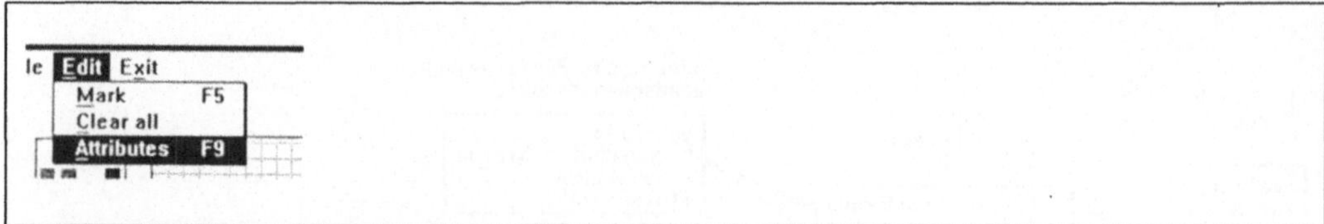

Abbildung 7. Auswahl der Funktion Edit Attributes

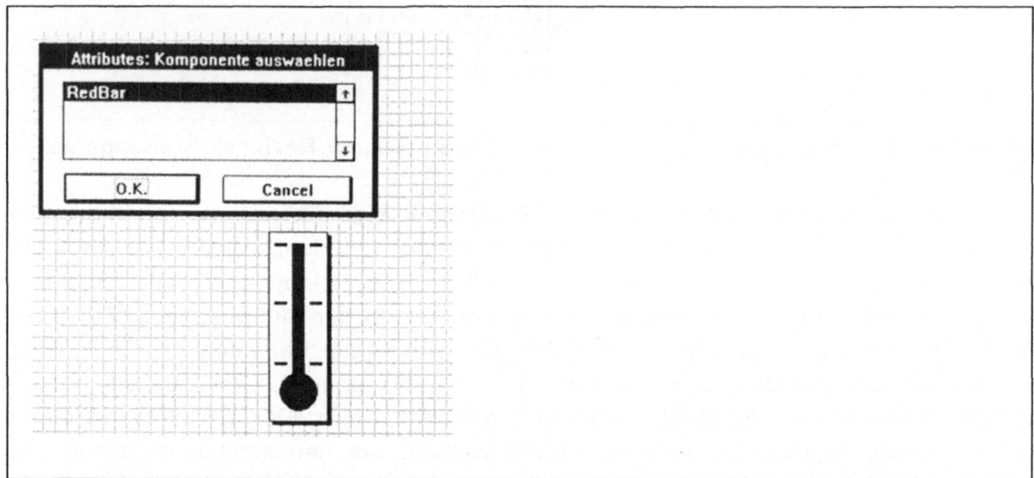

Abbildung 8. Auswahl der Komponente

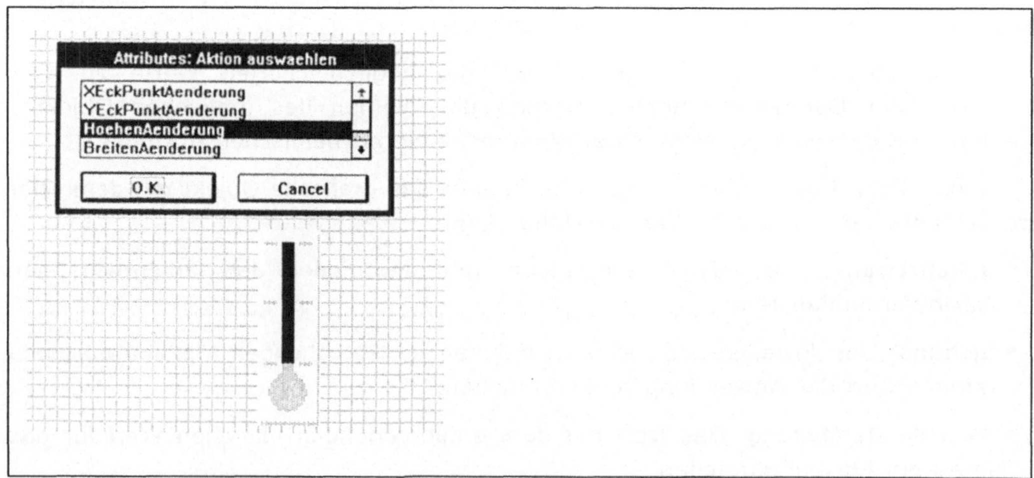

Abbildung 9. Auswahl der Art der dynamischen Veränderung

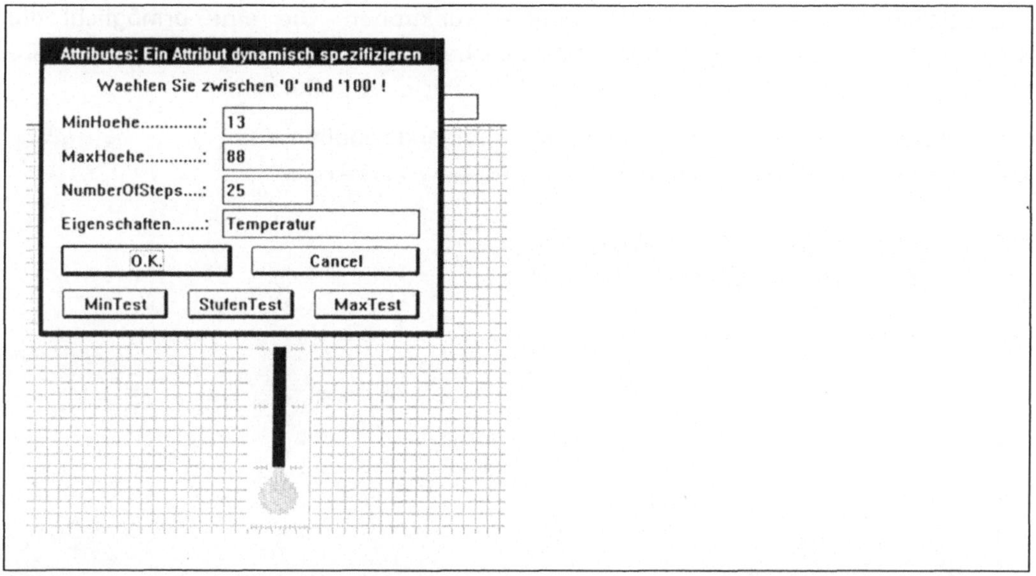

Abbildung 10. Definition der Max-, Min- und Stufenwerte und Eigenschaftsname

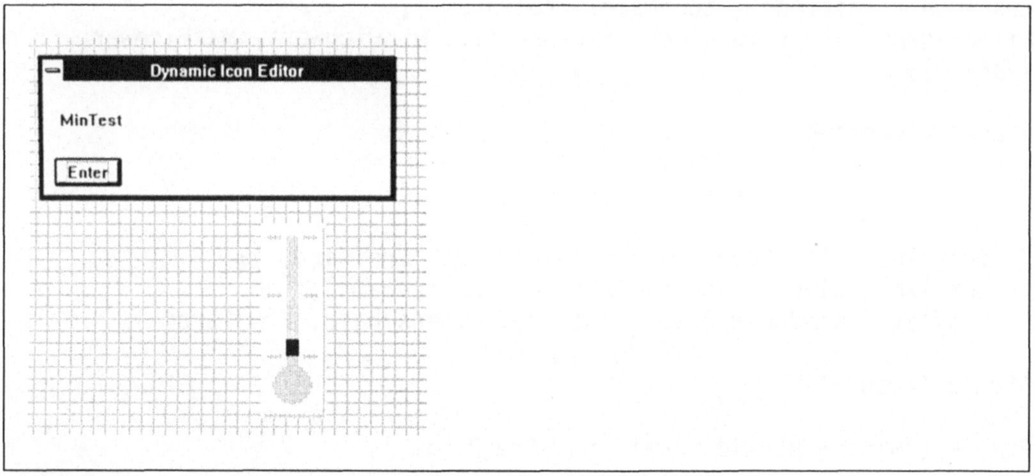

Abbildung 11. Ausführung des Min-Test

Die aktuelle Darstellungsfunktion bezieht jeden aktuellen Meßwert der Anwendung grafisch auf das definierte und geeichte Minimum und Maximum und ordnet ihn der am besten passenden der definierten grafischen Stufen zu (siehe Abb. 12). Bei jeder Änderung des jeweiligen Meßwertes muß die aktuelle Darstellungsfunktion mit dem neuen Meßwert für das entsprechende Icon mit dem entsprechenden Attribut wieder aufgerufen werden.

Zusätzlich existieren noch zwei weitere Funktionen. Die eine ermöglicht die Größe des dargestellten gesamten dynamischen Icons zu verändern, die andere ermöglicht das gesamte Icon wieder vom Bildschirm zu entfernen.

Der Aufbau eines Programmes sieht dann folgendermaßen aus:

```
/**************************************************/
/*                                                */
/*     Allgemeiner Aufbau eines Programms,        */
/*     das dynamische Icons verwendet.            */
/*                                                */
/**************************************************/


   :                    :                    :
   :                    :                    :
   :                    :                    :
   :                    :                    :
   :                    :                    :


TestIcon = InitIcon ( IconName, FreiParam );
EicheEigenschaft ( TestIcon, Eigenschaft, MinEichwert, MaxEichwert, Fre
iParam );

'Schleifenanfang'

   --- Erzeugen eines zu setzenden Wertes ---

   SetValue ( TestIcon, Eigenschaft, Wert, FreiParam );
   SetSize ( Window-Handle, TestIcon, FreiParam );
   Display ( Window-Handle, TestIcon, Eigenschaft, FreiParam );

'Schleifenende'

Remove ( Window-Handle, TestIcon, FreiParam );


   :                    :                    :
   :                    :                    :
   :                    :                    :
   :                    :                    :
   :                    :                    :

/**************************************************/
```

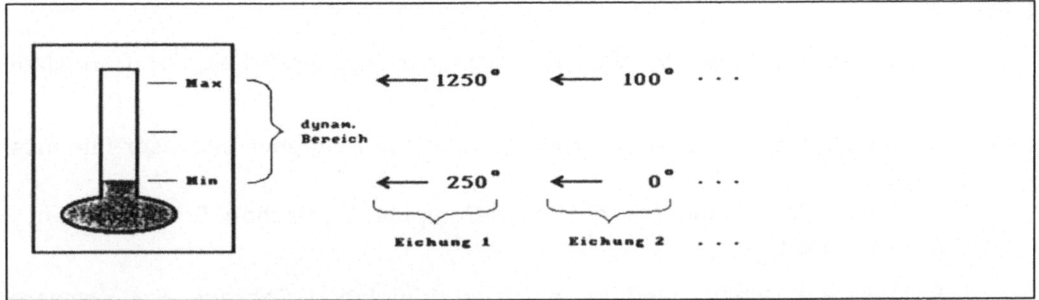

Abbildung 12. Zuweisung von Min/Max - Werten der Anwendung zu grafischen

Realisierung

Es wurde eine prototypische Implementierung vorgenommen, die vorerst die Definition der dynamischen Parameter Höhe, Breite und Position pro Komponente eines konstruierten dynamischen Icons in einem entsprechenden Editor ermöglicht. Die entsprechenden Darstellungsroutinen wurden ebenfalls implementiert und an einem kleinen Anwendungsprogramm getestet. (Für mehr Detail siehe BERNHARDT 1991.)

Die Implementierung erfolgte mit der Programmiersprache C auf dem Betriebssystem OS/2 unter zuhilfenahme der OS/2 Presentation Manager (TM) Routinen auf einem IBM PS/2 Model 70.

Es zeigte sich eine gute Handhabbarkeit sowohl des Editors mit der Definition der dynamischen Attribute (sehr schnelle Konstruktion der unterschiedlichsten dynamischen Icons mit unterschiedlichen dynamischen Komponenten mit jeweils unterschiedlichen dynamischen Attributen) als auch der dynamischen Darstellungfunktionen (sehr leichter Austausch von dynamischen Icons für die Darstellung).

Abschließende Bemerkungen und Ausblick

Obwohl die unterstützenden Darstellungshilfsmittel primär für den Anwendungsbereich der Prozeß-Überwachung entworfen wurden, sind sie so allgemein gehalten, daß sie bei Erweiterung um eine zusätzliche Ebene auch für die Konstruktion von "Graphic Controls", d.h. interaktive grafisch dynamische Einstellung von Wertebereichen, verwendet werden können. Diese zusätzliche Ebene müßte für die Zeigerabfrage (z.B. Maus-Position und Maustasten-Status) enthalten.

Die Schnittstellen sind so allgemein generisch entworfen, daß eine Reihe von Erweiterungen auch im Bezug auf die zu steuernden dynamischen Parameter möglich sind (Richtungsänderung von Komponenten, z.B. Zeiger, usw.). Weiterhin könnte auch eine Koppelung mit akustischen Parametern vorgenommen werden.

Literatur

O. Bernhardt: Ein dynamischer Icon-Manager. Diplom-Arbeit, TH Darmstadt (unveröffentl.) 1991

J.M. Carroll, R.L. Mack, W. A. Kellogg: Interface Metaphor and User Interface Design. In:
M. Helander (Ed.): Handbook of Human-Computer Interaction. Elsevier Science Publishers 1988

R.P.Cook, R.J.Auletta: StarLite, A Visual Simulation Package For Software Prototyping. ACM Sigplan Notices, January 1987, pp. 102-110, 1986

K.Fairchild, G.Meredith, A.Wexelblat: A Formal Structure for Automatic Icons. Interacting with Computers, vol.1 no.2, pp. 131-140, 1989

M. Gerstendörfer, G. Rohr: Which task in which representation on what kind of interface. In:
H.-J. Bullinger, B. Shackel (Eds.): Human-Computer Interaction - CHI '87. North Holland: Amsterdam 1987

Y.-T.Hsia, A.L.Ambler: Construction and Manipulation of Dynamic Icons. IEEE Conference on Coputer Languages, pp. 78-83, 1988

B.A.Myers: Creating User Interfaces by Demonstration. Dr.Ph. Dissertation, Technical Report CSRI-196, Toronto, May 1987

G. Rohr: Menschliche Informationsverarbeitung. In:
H. Balzert, U. Hoppe, R. Oppermann, H. Peschke, G. Rohr, N. Streitz: Einführung in die Software-Ergonomie. DeGruyter: Nürnberg 1988

G. Rohr: Mental concepts and direct manipulation: Drafting a direct manipulation query language. In:
D. Ackermann and M.J. Tauber (Eds.): Mental Models and Human-Computer Interaction 1. North-Holland: Amsterdam 1990 (a)

G. Rohr: Ikon-Techniken für Benutzerschnittstellen komplexer Anwendungssoftware. In:
A. Reuter (Ed.): GI - 20. Jahrestagung, Informatik auf dem Weg zum Anwender. Springer (Informatik Fachberichte): Heidelberg 1990 (b)

Pictorial Janus: Eine vollständig visuelle Programmiersprache und ihre Umgebung

Kenneth M. Kahn, Vijay A. Saraswat
Xerox Palo Alto Research Center
3333 Coyote Hill Road, CA 94304, USA
Email: <name>@parc.xerox.com

Volker Haarslev
Universität Hamburg, Fachbereich Informatik
Bodenstedtstr. 16, W-2000 Hamburg 50
Email: haarslev@informatik.uni-hamburg.de

Dieser Beitrag beschreibt einen Ansatz zur visuellen Programmierung, der auf wenigen grundlegenden, visuellen Formalismen basiert. Als Fallbeispiel stellen wir die konkurrente, logische Sprache Pictorial Janus und ihre visuelle Programmierumgebung vor. Die Struktur und der Zustand von Programmen werden in Pictorial Janus als sog topologisch vollständige Visualisierungen dargestellt. Sogar Programmabläufe werden in Form von graphischen Animationen visualisiert. Die zugehörige Sprachumgebung bietet Benutzern wichtige Werkzeuge wie einen visuellen Parser, Transformationen zwischen Bild- und Textrepräsentationen, ein visuelles Prüfsystem sowie ein graphisches Programmanimationssystem.

1 Einleitung

Im Bereich der visuellen Programmiersprachen und Systeme gibt es bereits über einen längeren Zeitraum vielfältige Aktivitäten. Für einen taxonomischen Überblick verweisen wir auf [10], der auch zahlreiche Beispiele vorstellt. Andere Übersichtsartikel zu diesem Thema sind in [3], [14] und [4] zu finden. Ein wichtiges Ziel visueller Programmiersprachen besteht darin, den Einsatz textueller Elemente und ihrer eindimensionalen (1-D) Anordnung – wie bisher in traditionellen Sprachen bekannt – bewußt zu reduzieren und durch systematischen Gebrauch von graphischen Repräsentationen zu ersetzen. Visuelle Programmiersprachen zeichnen sich insbesondere dadurch aus, daß sie Graphiken und Bilder (2-D) zur Programmdarstellung nutzen. Die Vorteile einer 2-D gegenüber einer 1-D-Repräsentation können insbesondere bei Sprachen zur Programmierung konkurrenter Systeme voll zum tragen kommen, wo Programme aus vielen kleinen Teilen (Prozessen) bestehen, die alle konkurrent zueinander ausgeführt werden.

Dieser Beitrag stellt deshalb einen Ansatz vor, der versucht, diese Vorteile konsequent auszunutzen und dabei den extremen Standpunkt vertritt, auf den Gebrauch textueller Elemente in einer speziellen, aber allgemein einsetzbaren Programmiersprache weitgehend zu verzichten. Als Fallbeispiel dient die Sprache *Janus* [13], die zur Programmierung hochgradig paralleler, verteilter Systeme entworfen wurde. Die vollständig visuelle Repräsentation von Janus, im weiteren als *Pictorial Janus* bezeichnet, basiert auf wenigen visuellen Formalismen, die sog. *topologisch vollständige* Programmvisualisierungen ermöglichen. Pictorial Janus ist in eine *visuelle Programmierumgebung* integriert, die vielfältige Hilfsmittel zur Programmentwicklung anbietet. Programme können durch Handzeichnungen, Graphikeditoren oder auch Texteditoren erstellt und Programmabläufe interaktiv durch *visuelle Prüfsysteme* (visual debugger) inspiziert oder gar in Form qualitativ hochwertiger Graphikanimationen dargestellt werden.

Dieser Ansatz zeichnet sich gegenüber anderen Ansätzen insbesondere dadurch aus, daß in Pictorial Janus Programme und Programmabläufe vollständig visuell dargestellt werden. Die Programmvisualisierungen beschreiben einzig durch ihre Erscheinungsform – und nicht durch den Vorgang ihrer Erzeugung – den vollständigen Zustand eines Programmablaufs.

Dieser Beitrag gliedert sich in die folgenden Abschnitte. Der nächste Abschnitt stellt das „Original" Janus kurz vor. Abschnitt 3 beschreibt die wichtigsten der Sprache Pictorial Janus zugrundeliegenden Entwurfsprinzipien. Abschnitt 4 verdeutlicht diese Prinzipien, indem er Beispielelemente der piktoriellen Syntax vorstellt. Im nächsten Abschnitt zeigen wir anhand eines kleinen Programmbeispiels die wesentlichen Konzepte der piktoriellen Programmdarstellung und die damit verbundenen Möglichkeiten zur Programmanimation. Im Abschnitt 6 erläutern wir die Grobkonzeption der Programmierumgebung und einige ihrer Komponenten. Abschnitt 7 beschreibt das Verhältnis zwischen der textuellen und piktoriellen Variante von Janus. Anschließend gehen wir kurz auf verwandte Ansätze ein. Wir beschließen diesen Beitrag mit einer Zusammenfassung und einem Ausblick auf zukünftige Forschungsaktivitäten.

2 Janus

Die Sprache *Janus* (und somit auch ihr „Zwilling" *Pictorial Janus*) gehört in die Familie der konkurrenten, einschränkungsorientierten (concurrent constraint) Programmiersprachen [12] und basiert auf dem Konzept sog. flacher, bewachter Hornklauseln (flat guarded horn clauses). Sie bietet weiterhin asynchronen, gepufferten Nachrichtenaustausch, rekursive Aktivierung von konkurrent zueinander ausgeführten Agenten, undeterministische Auswahl von Regeln (clauses) und „private", der „Umgebung" verborgene Kommunikation zwischen Agenten.

Janus stützt sich auf ein Einschränkungssystem über der Domäne unbeschränkter Bäume und ungeordneter Mengen (bags). Für Janus gilt weiterhin das Prinzip, einschränkungsorientierte Programmierung anzubieten, aber aus Gründen der Einfachheit und klaren Semantik auf eine Komponente zur automatischen Lösung von Einschränkungen zu verzichten ("concurrent constraint programming without constraint-solving"). Für eine ausführliche Erörterung dieser Problematik sei auf [13] verwiesen. Beim Entwurf von Janus wurde darauf geachtet, alle Möglichkeiten zur äußerst effizienten Implementation (vergleichbar mit konkurrenten C-Programmen) zu erhalten. Beispielsweise benötigen Janus-Programme zur Laufzeit keine Speicherbereinigung (garbage collection), die sonst für Programme vergleichbarer konkurrenter Sprachen immer notwendig ist. Die Konzeption von Janus ermöglicht, das Auftreten nicht mehr referenzierter, temporärer Speicherstrukturen bereits zur Übersetzungszeit zu erkennen und diese zur Laufzeit ohne Suchaufwand zu deallozieren und erneut zu verwenden.

Durch syntaktische Einschränkungen für das Auftreten logischer Variablen in Regeln wird es möglich, ungebundene (non-ground) Variablen als unidirektionale, Punkt-zu-Punkt Nachrichtenkanäle aufzufassen. Für jede ungebundene Variable kann es maximal ein *Lese-* (asker) und ein *Schreibrecht* (teller) geben. Das Besitzen eines Schreibrechts bedeutet, daß für diese Variable eine Einschränkung formuliert (eine Nachricht gesendet) werden darf. Ein Leserecht für eine Variable gestattet, diese auf die Gültigkeit von bereits formulierten Einschränkungen (auf den Eingang einer Nachricht) zu überprüfen. Sofern diese Variable ungebunden ist, wird deshalb die Überprüfung der erfragten Einschränkungen bis zur Bindung der Variablen (zum Empfang einer Nachricht) aufgeschoben, d.h. die Ausführung des entsprechenden Agenten suspendiert. Lese- und Schreibrechte für Variablen können ebenfalls als Einschränkungen interpretiert werden (Bestandteil von Nachrichten sein). Die bereits zur Übersetzungszeit überprüfbaren Restriktionen zum Auftreten logischer Variablen garantieren, daß in Janus-Programmen zur Laufzeit niemals inkonsistente Einschränkungen für Variablen formuliert werden können.

Die oben angedeutete Eigenschaft von Janus, das Erfragen und Setzen von Einschränkungen für logische Variablen auch operational als Empfangen und Versenden von unvollständigen Nachrichten über unidirektionale Nachrichtenkanäle zu deuten, hat sich bei der Entwicklung von Pictorial Janus als sehr vorteilhaft erwiesen. Nachfolgend werden wir uns deshalb überwiegend auf die operationale Deutung von Janus beziehen.

Ein *Janus-Programm* besteht somit aus einem Netzwerk von Agenten. Das Verhalten eines *Agenten* wird durch eine Menge von Regeln definiert. Diese *Regeln* beschreiben Vorbedingungen zur Ausführung eines Agenten. Sollten gleichzeitig *Vorbedingungen* mehrerer Regeln eines Agenten erfüllt sein, so wird eine dieser Regeln auf undeterministische Weise ausgewählt. Eine derartig ausgewählte Regel definiert somit das aktuelle Verhalten eines Agenten auf die von ihm empfangenen Nachrichten. Ein Agent kann sich selbst wieder durch ein neues Teilnetzwerk aus Nachrichten, Kanälen und Agenten ersetzen. An Datenstrukturen bietet Janus Listen, Reihungen (arrays), Tupel und ungeordnete Mengen (bags). Für eine weitergehende Beschreibung von Janus (nachfolgend auch als *Textual Janus* bezeichnet) sei auf [13] verwiesen.

3 Entwurfsprinzipien von Pictorial Janus

Pictorial Janus repräsentiert einen Ansatz, der versucht, die Programmierung konkurrenter Prozesse ausschließlich mithilfe *vollständig visueller Formalismen* durchzuführen [8]. Die folgenden Kriterien definieren den Begriff der *vollständigen Visualisierung*:

1. *Bilder sind Programme*: Bilder stellen einzig durch ihre Erscheinungsform – und nicht durch den Vorgang ihrer Erzeugung – ein eindeutiges und vollständiges Beschreibungsmittel dar.

2. *Bilder sind Zustandsbeschreibungen*: Bilder erfassen vollständig und eindeutig jeden Aspekt der Programmausführung.

3. *Einheitlichkeit*: Programme und ihre Ausführung werden mit denselben visuellen Formalismen dargestellt.

4. *Bildliche Alternativen zu Text*: Jede Benutzung von Text (mit Ausnahme von textuellen Daten) muß bildliche Alternativen ermöglichen.

5. *Objekt-kohärente Animation*: Die Ausführung von Programmen kann durch eine Bildfolge derart dargestellt werden, daß die Identität von Bildelementen und ihre zeitliche Transformation leicht zu verfolgen sind.

6. *Visuell erfaßbare „Ausführungsregeln"*: Die Zustandsübergänge des zugrundeliegenden Ausführungsmodells sollen durch entsprechende Bildmanipulationen visualisiert werden. Vom Betrachter nur schwer nachvollziehbare Bildtransformationen sind zu vermeiden.

Pictorial Janus erfüllt sogar eine noch weitergehende Bedingung, da die Programme topologisch vollständig sind. Eine Visualisierung sei *topologisch vollständig,* wenn sie zusätzlich die beiden folgenden Eigenschaften besitzt:

7. *Invarianz unter Rotation*: Die Bedeutung eines Bildes ist invariant gegenüber jeglicher Rotationstransformation.

8. *Invarianz unter Verzerrung*: Bilder können in ihrer Gesamtheit beliebig verzerrt werden, um bestimmte Merkmale hervorzuheben, ohne daß ihre Bedeutung verändert wird.

4 Elemente der piktoriellen Syntax

Abbildung 1 zeigt eine Zusammenstellung wesentlicher Teile der piktoriellen Syntax. Fast alle Elemente von Janus – Agenten, Regeln, Anschlüsse (ports), einfache Funktionen, Einschränkungen (constraints), Konstanten, Reihungen* (arrays) und Mengen* (bags) – werden als geschlossene Konturen dargestellt. Aufgrund der topologischen Vollständigkeit konnte die graphische Repräsentation der Sprachelemente derart gewählt werden, daß diese unabhängig von Form-, Größen- und Farbattributen sind. Diese Attribute können somit zur besseren graphischen Gestaltung der Programme verwendet werden, ohne die Bedeutung der Programme zu verändern.

Ein *Anschluß* (port) besteht aus einer eine leere Fläche umschließenden Kontur. Ein *interner* (*externer*) Anschluß befindet sich innerhalb (außerhalb) einer anderen Kontur und berührt diese Kontur von innen

* In Abbildung 1 nicht dargestellt.

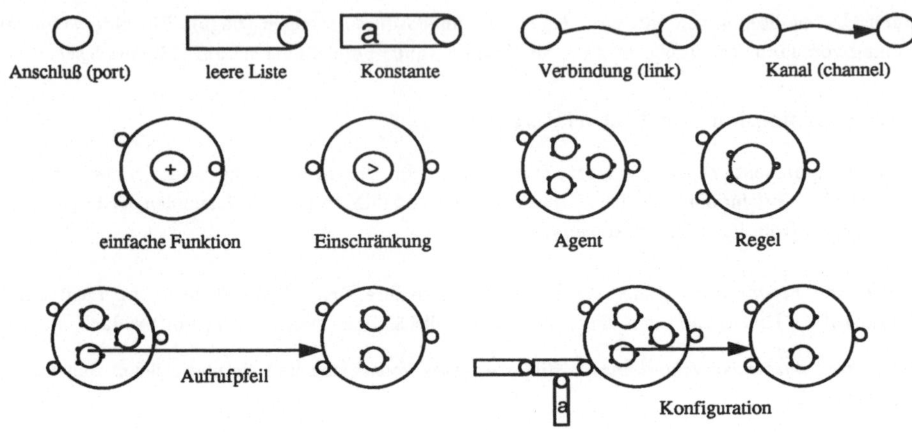

Abbildung 1: Beispielelemente der piktoriellen Syntax

(außen). Ein interner Anschluß dient als einziger Verknüpfungspunkt (Repräsentant) für die Kontur, in der er enthalten ist. Externe Anschlüsse werden implizit im Uhrzeigersinn – beginnend mit der 6:00 Uhr'-Position – angeordnet. Eine *leere Liste* besteht aus einer Kontur, die nur einen internen Anschluß umfaßt. Kopf bzw. Rest einer nicht-leeren Liste werden mithilfe von externen Anschlüssen dargestellt (siehe Abb. 3). Eine Konstante entspricht in der Darstellung einer leeren Liste, allerdings umschließt ihre Kontur zusätzlich eine Zeichenkette oder Zahl. Eine *Verbindung* (link) verknüpft zwei Anschlüsse und wird als ungerichtete Linie dargestellt. Ein *Kanal* verbindet ebenfalls zwei Anschlüsse, er besteht aber aus einer gerichteten Linie und repräsentiert ein Senderecht. Eine *Funktion* besitzt einen internen Anschluß, ein Funktionssymbol und externe Anschlüsse. Eine *Einschränkung* (constraint) hat externe Anschlüsse sowie ein Relationssymbol.

Ein *Agent* hat keinen internen Anschluß, aber eine beliebige Anzahl von externen Anschlüssen. Er enthält entweder Regeln (die sein Verhalten definieren) oder einen Aufrufpfeil (zu einem anderen Agenten) oder eine Namen (label) (der einen Agenten bezeichnet). Eine *Regel* entspricht in der Darstellung einem Agenten. Sie unterscheidet sich von einem Agenten dadurch, daß sie immer innerhalb eines Agenten sein muß. Ein *Aufrufpfeil* (call arrow) beginnt innerhalb der Kontur eines Agenten und berührt mit der Pfeilspitze die Kontur eines Agenten. Eine *Konfiguration* stellt einen Programmzustand dar. Sie besteht aus einer beliebigen Anzahl von untereinander verbundenen Agenten, deren externe Anschlüsse teilweise mit Datenelementen verknüpft sind.

Eine Regel besitzt immer dieselbe Anzahl von externen Anschlüssen wie der Agent, zu dem sie gehört. Alle Elemente, die mit den externen Anschlüssen einer Regel verknüpft sind, definieren die *Vorbedingungen* (ask devices) dieser Regel. Nur wenn ihre Vorbedingungen erfüllt sind, kann sie angewendet werden. Alle Elemente innerhalb der Regelkontur definieren ihren *Rumpf* (body devices). Abbildung 2 stellt diesen Sachverhalt anhand eines Beispiels dar. Eine Regel beschreibt durch ihre Vorbedingungen und ihren Rumpf die linke und rechte Seite einer Graphtransformation. Eine wesentliche Innovation der Syntax von Pictorial Janus besteht nun darin, die sonst übliche Links-Rechts-Beziehung bei Graphtransformationen durch eine Außen-Innen-Beziehung zu ersetzen. Diese Konvention erweist sich insbesondere bei der Animation des Programmablaufs als sehr vorteilhaft.

5 Beispiel zur Programmdarstellung und Animation

Abbildung 3 zeigt eine Beispielkonfiguration mit einem Agenten **append**, dessen linker unterer Anschluß mit einer Liste verknüpft ist. Dieser Agent verkettet zwei übergebene Listen und erzeugt als Ergebnis eine neue Liste. Er besitzt zwei Regeln, die sein Verhalten definieren. Die oberere Regel beschreibt das

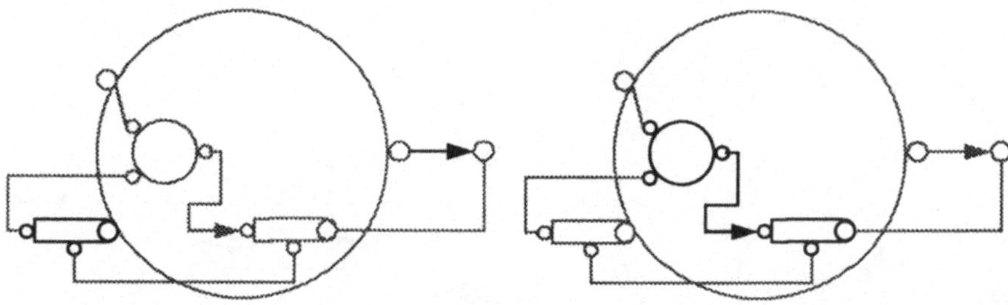

Abbildung 2: Vorbedingungen (links) und Rumpf (rechts) der unteren Regel von **append**

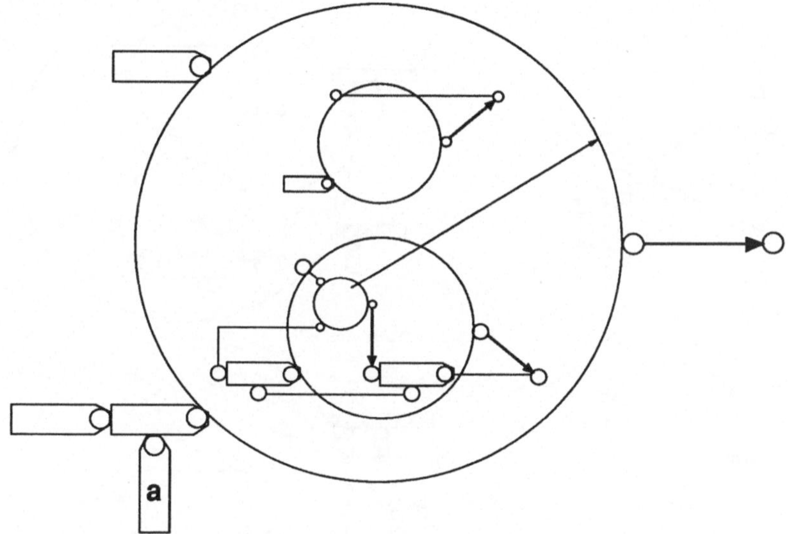

Abbildung 3: Konfiguration mit einem Agenten **append**

Verketten einer leeren Liste mit einer beliebigen anderen Liste. Die Vorbedingungen dieser Regel bestehen aus einer leeren Liste, die mit dem linken unteren Anschluß, und einem Senderecht, das mit dem rechten Anschluß verbunden sein muß. Der linke obere Anschluß ist leer und gestattet somit den Anschluß einer beliebigen Liste. Als Ergebnis wird die am linken oberen Anschluß anliegende Liste mit dem am rechten Anschluß anliegenden Senderecht verknüpft.

Die untere Regel innerhalb des Agenten beschreibt den (rekursiven) Fall, daß an das Ende einer nichtleeren Liste eine andere Liste angehängt werden soll. Die Vorbedingung unterscheidet sich von der ersten Regel nur dadurch, daß am linken unteren Anschluß eine nicht-leere Liste (im weiteren als Eingangsliste bezeichnet) verlangt wird. Der Rumpf dieser Regel besteht aus einem rekursiven Aufruf von **append**, der durch die Verwendung eines Aufrufpfeils spezifiziert ist. Die Eingangsdaten des linken oberen Anschlusses sowie der Rest der Eingangsliste werden dabei unverändert weitergegeben. Der Regelrumpf erzeugt weiterhin ein neues Listenelement, desses Kopf aus dem Kopf der Eingangsliste besteht. Für den Rest des neuen Listenelements wird ein Senderecht erzeugt und mit dem rechten Anschluß des rekursiven Aufrufs verknüpft. Das neue Listenelement wird über seinen internen Anschluß mit dem am rechten Anschluß der Regel anliegenden Senderecht verbunden.

Abbildung 4 zeigt eine Folge von Konfigurationen (von links nach rechts und oben nach unten geordnet) mit dem Agenten **append**. Das erste Bild (links oben) zeigt die Anfangskonfiguration mit den zu verkettenden

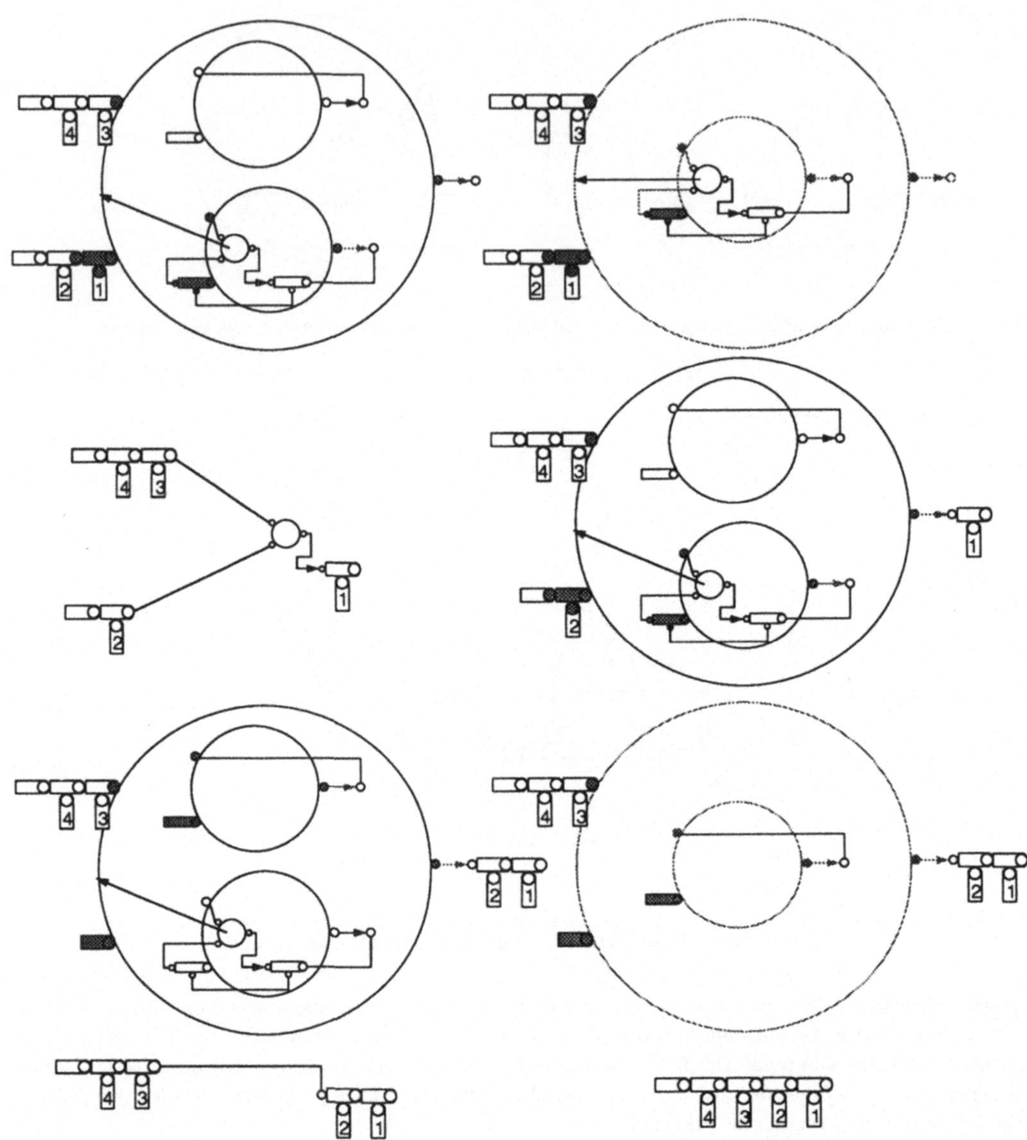

Abbildung 4: Ausschnitte einer Programmanimation mit **append**

Listen [1,2] und [3,4]. Nur die untere Regel von **append** kann aufgrund der Vorbedingungen (grau schraffiert) angewendet werden. Im zweiten Bild (rechts oben) wurde der Agent (Umrisse in Grau) durch den Rumpf der angewendeten Regel ersetzt. Das dritte Bild (mitte links) zeigt die nach der Resolvierung bereinigte Konfiguration. Die linke untere Liste wurde um ihr erstes Element (1) verkürzt, welches jetzt den Anfang der neu erstellten Ergebnisliste (mit dem rechten Anschluß verknüpft) bildet. Im vierten Bild wurde der Agent wieder an seine normalen Größe (analog zu Bild 1) angepaßt. Das fünfte Bild zeigt den Zustand nach der Abarbeitung des Elements 2. Aufgrund der leeren Liste kann nur noch die obere Regel angewendet werden. Im sechsten Bild wurde diese Regel angewendet. Als Ergebnis ihrer Resolvierung (siebtes und achtes Bild) wird an die Ergebnisliste die zweite Eingangsliste ([3,4]) angehängt sowie die

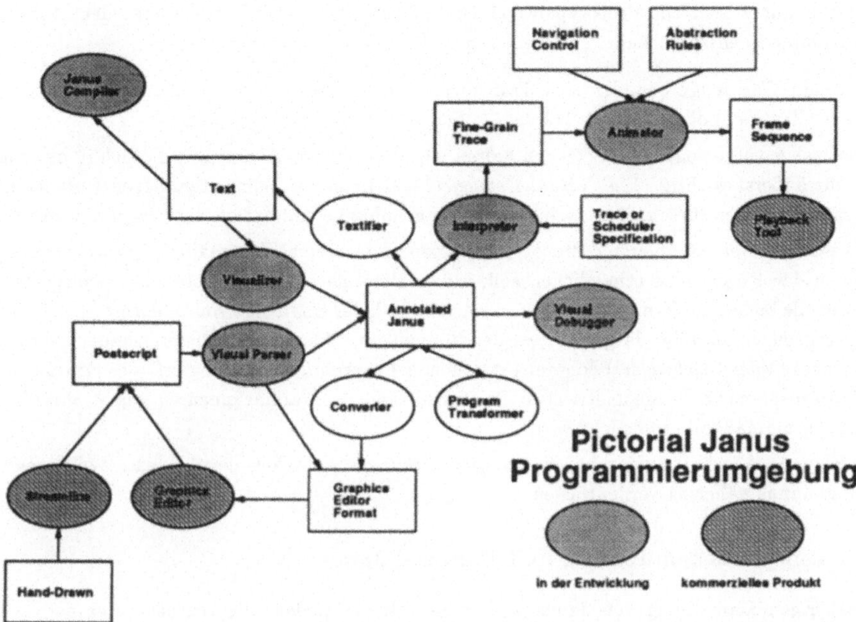

Abbildung 5: Gesamtstruktur der visuellen Programmierumgebung

Verbindung zur Länge Null geschrumpft (Normaldarstellung). Bei dieser Art der Programmanimation ist die oben erwähnte Außen-Innen-Beziehung der Bildransformationen von entscheidender Bedeutung.

6 Programmierumgebung

Eine visuelle Sprache erscheint nur dann sinnvoll und praktisch einsetzbar, wenn sie durch eine Programmierumgebung unterstützt wird. Dies erscheint umso wichtiger, da z.Z. noch keine allgemein gesicherten Erkenntnisse zur Generierung visueller Programmierumgebungen existieren. Abbildung 5 zeigt die (geplante) Programmierumgebung von Pictorial Janus. Alle Daten wurden als Rechtecke und alle Prozesse bzw. Programme als Ovale repräsentiert.

Wir möchten im Rahmen dieses Beitrags nur auf einige wichtige Teile dieser Programmierumgebung näher eingehen. Die zentrale Rolle spielt eine erweiterte Form von Janus. '*Annotated Janus*' ist als interne Repräsentation zu verstehen, die um geometrische Information zur visuellen Darstellung erweitert wurde. Teile der Umgebung bestehen aus (kommerziell) verfügbaren Werkzeugen (*Streamline, Graphics Editor, Playback Tool*) oder benutzen entsprechende Sprachen (*Postscript*™). Programme können auf verschiedene Arten erstellt werden:

1. *Handzeichnung*: Handzeichnungen von 'Pictorial Janus'-Programmen können in eine Postscript-Darstellung überführt, durch einen als Prototyp verfügbaren '*Visual Parser*' analysiert und als 'Annotated Janus' gespeichert werden.

2. *Graphikeditor*: Es ist geplant, kommerziell verfügbare Graphikeditoren zu verwenden oder zu modifizieren, die 'Pictorial Janus'-Programme im gewünschten Postscript-Format erzeugen können.

3. *Text*: Der traditionelle Weg, Programme zu erstellen, soll ebenfalls weiterhin unterstützt werden.

Dafür befindet sich ein Prototyp für einen '*Visualizer*' in der Entwicklung, der 'Textual Janus'-Programme in entsprechende 'Pictorial Janus'-Programme überführen kann.

Als ergänzende Werkzeuge befinden sich Transformationsprogramme in der Planung, die aus '*Annotated Janus*' Text (*Textifier*) oder Graphik (*Graphics Editor Format*) erzeugen können.

Neben den bisher vorgestellten Werkzeugen halten wir zwei weitere Werkzeuge, die sich in der Entwicklung befinden, für äußerst wichtig. Ein '*Visual Debugger*' soll Benutzer beim interaktiven Entwicklungsprozeß von Programmen unterstützen. Dies kann durch eine graphische Animation der Programmabläufe erfolgen.

Weiterhin existiert bereits ein erster Prototyp für einen '*Interpreter*', der qualitativ hochwertige graphische Animationen der Programme ermöglichen soll, indem detaillierte Ablaufprotokolle bereitgestellt werden. Diese Protokolle bilden die Grundlage für einen '*Animator*', der eine Folge von Bildern erzeugt, die dann mit Standardwerkzeugen als Film dargestellt werden können. Die Abtrennung dieses Animationsprozesses (im Gegensatz zum '*Visual Debugger*') von der eigentlichen Programmausführung erfolgte bewußt, da dadurch u.a. eine Vorausplanung der geometrischen Positionen von Programmelementen gemäß ihrer zukünftigen Rolle im Programmablauf ermöglicht wird.

Der nachfolgende Abschnitt versucht, die Wechselwirkungen zwischen den beiden „Zwillingen“ Pictorial und Textual Janus näher zu verdeutlichen.

7 Verhältnis zwischen Textual und Pictorial Janus

In der bisherigen Darstellung von Janus wurde bewußt vermieden, die textuelle Repräsentation dieser Sprache näher zu erläutern, um eine rein bildhafte Beschreibung von Janus zu ermöglichen. Nachfolgend stellen wir am Beispiel des Agenten `append` die Beziehung zwischen Textual und Pictorial Janus kurz dar. Die textuelle Version von `append` läßt sich analog zur Abbildung 3 durch die folgenden Klauseln definieren (die Hornklauselsyntax von Textual Janus ähnelt der von Prolog):

```
append([],X2,!X4) :- X4=X2.
append([X4|X5],X2,!X3) :- X3=[X4|X7], append(X5,X2,!X7).
```

Das Symbol '`!`' im Kopf der Klauseln bedeutet, daß für `X4` bzw. `X3` ein Sende- bzw. Schreibrecht vorliegen muß, während es im Rumpf ein Schreibrecht für `X7` weiterreicht.

Die textuelle bzw. piktorielle Darstellung von `append` benutzt einige Abkürzungen und Vereinfachungen, die die automatische Übersetzung zwischen den beiden Repräsentationen der Janus-Programme erschweren. Deshalb wird eine sog. *normalisierte Darstellung* erzeugt, bei der alle fragenden (ask) und setzenden (tell) Einschränkungen explizit formuliert worden sind (die Syntax für eine normalisierte Regel lautet: `<head> ':-' <asks> '|' <tells> <body>`).

```
append(X1,X2,X3) :- X1=[], X3=!X4 | X3=X2.
append(X1,X2,X3) :- X1=[X4|X5], X3=!X11 |
                    X11=[X6|X7], X6=X4, X8=X5, X9=X2, X10=!X7, append(X8,X9,X10).
```

Zu dieser normalisierten textuellen Darstellung existiert ebenfalls eine piktorielle Darstellung, die sich aus einer weitgehenden 1:1-Abbildung ergibt (siehe Abbildung 6). Der ein Senderecht darstellende Pfeil wird in der normalisierten piktoriellen Darstellung durch einen Term mit dem '`!`'-Operator ersetzt, da es sich bei diesem Pfeil um eine Abkürzung handelt.

Unsere Untersuchungen zur automatischen Transformation von Textual zu Pictorial Janus haben ergeben, daß es sich dabei in erster Linie um ein Planungs- und Layoutproblem handelt. Die Zuweisung von Position und Größe einzelner Elemente von Pictorial Janus hängt sehr davon ab, wie ihre Wechselwirkung zu anderen Elementen (z.B. Verknüpfung mit '='-Operatoren) aussieht. Weiterhin haben unsere Untersuchungen bestätigt, daß es sehr wichtig ist, diese Elemente nach ästhetischen Gesichtspunkten anzuordnen. Ein weiteres Kriterium besteht darin, die Wechselwirkungen bei der Ausführung von Programmen zu berücksichtigen. Idealerweise sollten Elemente derart angeordnet werden, daß ihre Position während der Programmausführung gar nicht oder nur noch sehr geringfügig verändert werden muß.

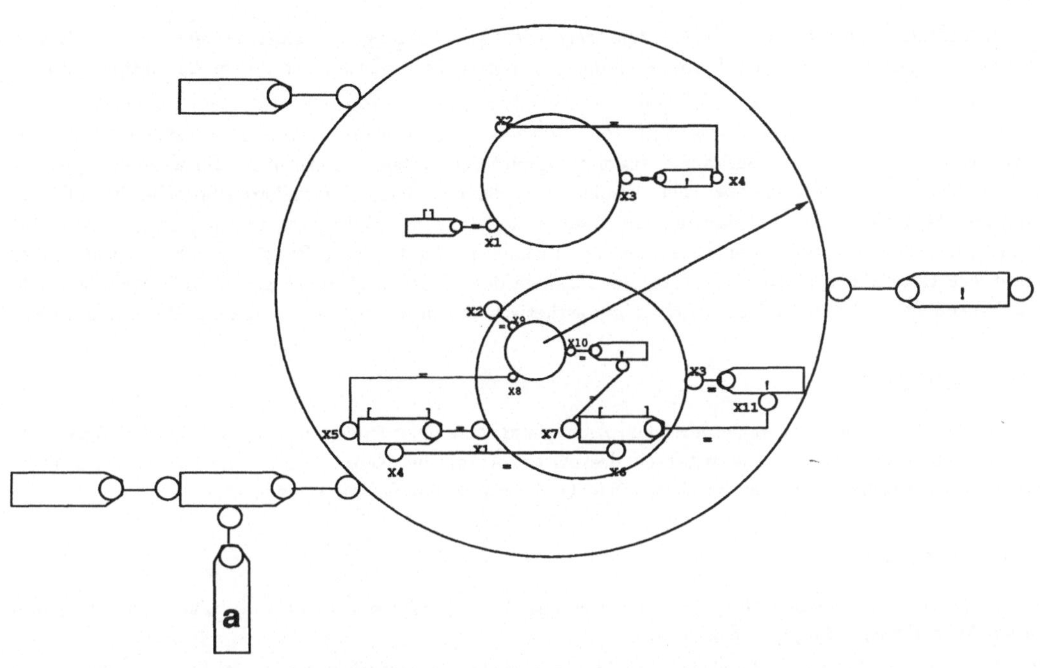

Abbildung 6: Normalisierte Form von append

8 Verwandte Arbeiten

Wir gehen nachfolgend auf einige Systeme ein, die gewisse Gemeinsamkeiten mit Pictorial Janus haben. Zuerst ist die Gruppe der Systeme zu nennen, die visuelles Programmieren für ähnliche Sprachen anbieten. Es gibt Vorschläge für Parlog [11], 'Connection Graphs' [1] und 'Interaction Nets' [9]. Diese Ansätze basieren allerdings noch in erheblichem Umfang auf textuellen Notationen. Das PICT/D-System [7] versucht dagegen, auf die Verwendung von Text völlig zu verzichten. Der Sprachumfang gestattet allerdings nur die Berechnung von einfachen numerischen Ausdrücken. Prograph [5] ist eine funktionale, datenflußorientierte Sprache, die auf Macintosh-Rechnern verfügbar ist. Sie implementiert elementare Konstrukte, die über Datenleitungen miteinander verbunden und als Prozeduren zusammengefaßt werden können. Weiterhin wird eine an Smalltalk angelehnte objektorientierte Erweiterung angeboten. Zur Programmierumgebung gehören ein graphischer Editor sowie ein visuelles Prüfsystem.

Für den Bereich der Programmanimation sind vor allem Balsa-II [2] und das Nachfolgesystem Tango [15] zu nennen. Diese Systeme können für konventionelle, textuelle Sprachen eingesetzt werden. Die zu animierenden Programme werden um Prozeduraufrufe ergänzt, die interessante Ereignisse signalisieren. Es können allerdings nur Zustandswechsel von Programmabläufen visualisiert werden. Möglichkeiten zur Visualisierung der Programmstruktur oder der zugrundeliegenden Ausführungsarchitektur sind im Gegensatz zu Pictorial Janus nicht vorhanden.

9 Zusammenfassung und Ausblick

Wir haben in diesem Beitrag einen Ansatz zur vollständigen Visualisierung konkurrenter, einschränkungsorientierter Programme vorgestellt und dessen Tragfähigkeit am Beispiel der Sprache Pictorial Janus exemplifiziert. Wesentliche Merkmale dieses Ansatzes sind die Verwendung topologisch vollständiger Visualisierungen, die die Struktur von Programmen und den Zustand ihrer Ausführung einzig durch ihre Erscheinungsform – und nicht durch den Vorgang ihrer Erzeugung – vollständig und eindeutig beschreiben.

Die zugehörige Sprachumgebung bietet Benutzern wichtige Werkzeuge wie einen visuellen Parser, Transformationen zwischen Bild- und Textdarstellungen, ein visuelles Prüfsystem sowie ein Animationssystem.

Unser momentanes Hauptaugenmerk gilt der Vervollständigung unserer Prototypen, um eine voll einsatzfähige Programmierumgebung zu erhalten. Nächste Schritte werden sein, die Tragfähigkeit unseres Ansatzes bei der visuellen Programmierung umfangreicherer Systeme zu erproben. So wurden bspw. der visuelle Parser, der Animator sowie der 'Visualizer' in der kommerziell verfügbaren Sprache Strand™ [6] implementiert, die große Ähnlichkeiten mit Janus besitzt und deren Programme mit geringem Aufwand nach Janus portiert werden können. Durch den 'Visualizer' könnten diese Programme dann leicht in ihre piktorielle Darstellung überführt werden. Im Rahmen des 'Visualizers' finden gegenwärtig weitergehende Untersuchungen zur automatischen Erzeugung ästhetisch angeordneter 'Pictorial Janus'-Programme statt.

Danksagungen

Wir danken Ralf Möller für seine hilfreichen Anmerkungen zu einer früheren Fassung dieses Beitrags sowie den anonymen Gutachtern. Die Mitarbeit des dritten Autors bei Xerox PARC wurde durch ein DAAD-Forschungsstipendium vergeben durch den NATO-Wissenschaftsausschuß ermöglicht.

Literaturverzeichnis

1. A. Bawden, Connection Graphs, In Proceedings *ACM Conference on Lisp and Functional Programming,* ACM Press, 1989, pp. 258–265.

2. M.H. Brown, Algorithm Animation, ACM Distinguished Dissertations Series, MIT Press, 1988.

3. S.K. Chang, Visual Languages: A Tutorial and Survey, *IEEE Software* 4, 1 (Jan. 1987), 29–39.

4. S.K. Chang (Hrsg.), Principles of Visual Programming Systems, Prentice Hall, Englewood Cliffs, New Jersey, 1990.

5. P.T. Cox, F.R. Giles, T. Pietrzykowski, Prograph: a step towards liberating programming from textual conditioning, In: Proceedings, *1989 IEEE Workshop on Visual Languages,* Rome (Italy), Oct. 4-6, IEEE Computer Society Press, 1989, pp. 150–156.

6. I. Foster, S. Taylor, Strand: New Concepts in Parallel Programming, Prentice Hall, Englewood Cliffs, New Jersey, 1990.

7. E.P. Glinert, S.L. Tanimoto, PICT: An Interactive Graphical Programming Environment, *IEEE Computer* **17**, 11 (Nov. 1984), 7–25.

8. K.M. Kahn, V.A. Saraswat, Complete Visualizations of Concurrent Programs and their Executions, In: Proceedings, *1990 IEEE Workshop on Visual Languages,* Skokie/IL, Oct. 4-6, 1990, IEEE Computer Society Press, 1990, pp. 7–14. Siehe auch *Technical Report* SSL-90-38 [P90-00099], Xerox Palo Alto Research Center, 1990.

9. Y. Lafont, Interaction Nets, In: Proceedings, *ACM Conference on Principles of Programming Languages,* January 1990, pp. 95–108.

10. B.A. Myers, Taxonomies of visual programming and program visualization, *Journal of Visual Languages and Computing* **1**, 1 (Mar. 1990), 97–123.

11. G. Ringwood, Pictures and pixels, *New Generation Computing* **7** (1989), 59–80.

12. V.A. Saraswat, Concurrent Constraint Programming, MIT Press, forthcoming, 1990.

13. V.A. Saraswat, K.M. Kahn, J. Levy, Janus: A step towards distributed constraint programming, Technical Report SSL-90-51 [P90-00112], Xerox Palo Alto Research Center, 1990. Auch erschienen in Proceedings, North American Conference on Logic Programming, Austin, Texas, October 1990.

14. N.C. Shu, Visual Programming, Van Nostrand Reinhold Company, New York, 1988.

15. J.T. Stasko, Simplifying Algorithm Animation with Tango, In: Proceedings, *1990 IEEE Workshop on Visual Languages,* Skokie/IL, Oct. 4-6, 1990, IEEE Computer Society Press, 1990, pp. 1–6.

Eine offene Entwurfsumgebung für multimediale Bedienoberflächen

Alex Jarczyk, Matthias Schneider-Hufschmidt
Thomas Kühme, Peter Witschital, Joachim Grollmann
Siemens Zentralabteilung Forschung und Entwicklung
ZFE IS KOM 3
Otto-Hahn-Ring 6, D-W8000 München 83,
Fed. Rep. of Germany

Keywords

User Interface Design, Multimediale Bedienoberflächen, Direkte Manipulation, Direkte Komposition, Design Environment, Entwurfsumgebungen.

Zusammenfassung

Dieser Beitrag beschreibt Konzepte und Lösungsansätze zur Einbindung multimedialer Aspekte in Entwurfsumgebungen für Bedienoberflächen. Anhand der Entwurfsumgebung SX/Tools werden charakteristische Eigenschaften aufgezeigt, die eine Integration multimedialer Interaktionstechniken ermöglichen und Gestaltungsmechanismen über den Entwurfszeitpunkt hinaus verfügbar machen.

Bedienoberflächen werden mit SX/Tools nach dem Prinzip der *Direkten Komposition* entworfen. Wir erläutern die Konsequenzen, die sich aus diesem Entwurfsprinzip ergeben. Dazu gehören die Erweiterbarkeit der Entwurfsumgebung, ihre Integrationsfähigkeit für neue Interaktionstechniken und verschiedene Interaktionsstile, und die Möglichkeit, dem Benutzer Hilfsmittel zur Modifikation existierender Bedienoberflächen in die Hand zu geben.

1 User Interface Management Systeme

Der Entwurf von graphischen Bedienoberflächen für interaktive Systeme ist eine der aufwendigsten Tätigkeiten bei der Entwicklung von Software-Systemen. Bedienoberflächen haben häufig eine komplexe Struktur und spiegeln sowohl Eigenschaften der Anwendung als auch technische Eigenschaften der verwendeten Hard- und Software wider.

Um Software-Entwickler beim Entwurf von Bedienoberflächen effizient zu unterstützen, wurden schon viele Anstrengungen unternommen. In den letzten Jahren entwickelten sich User Interface Managament Systeme [Pfaff 85] zum Stand der Technik. Diese Systeme erfüllen zwei Aufgaben: Zum einen unterstützen sie den Software-Entwickler beim Entwurf seiner Oberfläche (der User Interface Design System (UIDS)-Aspekt), zum anderen stellen sie eine Laufzeitumgebung zur Verfügung, in der die entwickelten Bedienoberflächen eingesetzt werden können. Die Entwicklung von Bedienoberflächen mit Hilfe eines UIMS umfaßt die Definition der statischen Eigenschaften, der Verbindung zur Anwendung und die des Verhaltens der Oberflächen, soweit dies auf der Ebene der Bedienoberfläche möglich ist. Eine wichtige Eigenschaft eines UIMS ist die strikte Trennung von Applikationssystem und Bedienoberfläche, welche die unabhängige Entwicklung beider Komponenten erlaubt. Allerdings muß der Software-Entwickler auch bei der Nutzung eines UIMS Details der Struktur und Funktionalität seiner Anwendung kennen, um ergonomisch sinnvolle Bedienoberflächen entwerfen zu können.

User Interface Management Systeme bieten eine Reihe von Vorteilen. Sie erlauben die einheitliche Gestaltung von Bedienoberflächen für mehrere Anwendungen. Auch die Nutzung eines einheitlichen Interaktionsstils wird durch Ihren Einsatz gewährleistet. Zum zweiten erlauben sie die Verbindung einer Anwendung mit verschiedenen Bedienoberflächen. Dies ermöglicht eine Anpassung von Systemen an verschiedene Hardware-Voraussetzungen oder individuelle Benutzerwünsche. Mit einem UIMS kann der Entwurf der Bedienoberfläche zu einem sehr frühen Zeitpunkt an einem Prototypen evaluiert werden, um Rückmeldungen der späteren Benutzer rechtzeitig einfließen zu lassen. Dieser prototypische Entwurf kann unabhängig von der zu entwickelnden Applikation durchgeführt werden. Schließlich bieten User Interface Managament Systeme die Möglichkeit, existierende Komponenten von Bedienoberflächen wiederzuverwenden.

Die Erweiterung eines solchen UIMS um multimediale Basiselemente wie Text, Grafik, Audio, Animation und Video erlaubt einerseits die einfache Integration dieser neuen Medien in das Gesamtkonzept, andererseits ist es möglich verschiedenste Multimedia-Player und Editoren mit Hilfe solcher UIMS zu erstellen. Dabei sollte aber beachtet werden, daß Realzeitanforderungen, Synchronisation und neue Visualisierungskonzepte (z.B. Layertechnik für verschiedene Video- und Animationskanäle) das Basismodell eines solchen UIMS erheblich erweitern.

Auch mit Hilfe eines UIMS bleibt der Entwurf einer Bedienoberfläche eine komplexe Aufgabe. Das notwendige Wissen, über das der Entwickler verfügen muß, läßt sich in drei Kategorien einteilen: Wissen über das Verhalten und Aussehen von guten Bedienoberflächen, Wissen über den Entwurfsprozeß und Wissen über die zugrundeliegende Anwendung. Fehlendes Wissen in einer dieser Kategorien führt zu nicht optimalen Bedienoberflächen. Da der Software-Entwickler selten ein Experte auf dem Gebieteder Applikation ist, sind schwerwiegende Entwurfsfehler zu erwarten.

Ein Verbesserung der Qualität von Bedienoberflächen kann man dadurch erreichen, daß man die Entwurfsumgebung den späteren Benutzern zur Verfügung stellt. Diese können dann ihre Bedienoberflächen selbst an ihren persönlichen Arbeitsstil und an geänderte Anforderungen der Applikation anpassen. In diesem Fall wird die Designaufgabe zum Teil den Benutzern übertragen. Wenn sie ihre Oberflächen anpassen wollen, müssen sie lernen, die Entwicklungsumgebung zu benutzen und gute Oberflächen zu gestalten. Dieses zusätzlich notwendige Wissen kann durch einen Entwurfstil mit dem Namen *Direkte*

Komposition verringert werden, bei dem jedes Objekt einer Bedienoberfläche Wissen darüber enthält, wie es im Entwicklungsprozeß verwendet werden kann [Kühme et al. 91, Schneider-Hufschmidt 91].

2 Das Prinzip der direkten Komposition

Unter dem Begriff *Direkte Komposition* [Kühme et al. 91] versteht man die vollständige Anwendung des Prinzips der Direkten Manipulation [Shneiderman 83, Hutchins et al. 86] auf den Entwurf und die Entwicklung graphischer Bedienoberflächen. Direkte Komposition wird charakterisiert durch einen objekt-orientierten Ansatz und ermöglicht die Defintion und Entwicklung von Bedienschnittstellen ohne spezielle Werkzeuge. Direkte Komposition basiert auf einem elementaren Objektmodell. Bedienschnittstellen werden durch Aggregation von solchen elementaren Objekten definiert. Das Objektmodell enthält sowohl die Beschreibung des interaktiven Entwurfs als auch die Beschreibung des Interaktionsverhaltens eines Objekts. Die Werkzeuge, die für den Entwurf und die Modifikiaton eines Interaktionsobjekts benötigt werden, können rekursiv mit den Mitteln der direkten Komposition erzeugt werden. Das Interaktionsverhalten eines Gesamtsystems kann als die Vereinigung der elementaren Interaktionsmechanismen der Bildschirmobjekte verstanden werden.

Das Aussehen und Verhalten von Schnittstellenobjekten kann durch interaktive Techniken definiert werden. Neue Objekte können aus exisitierenden durch Kopieren und Modifizieren erzeugt und neue Bedienoberflächen durch das Zusammensetzen solcher Objekte *komponiert* werden. Der Entwickler einer Bedienoberfläche kommuniziert direkt mit deren Objekten. Er benötigt dazu keine speziellen Werkzeuge. Wenn Werkzeuge notwendig werden, weiß das Objekt, welche sinnvoll eingesetzt werden können. Da die Bedienschnittstellenwerkzeuge auch mit Mitteln direkter Komposition definiert wurden, ist das Aussehen und das Verhalten dieser Werkzeuge mit dem anderer Bildschirmobjekte vergleichbar.

Objekte umfassen die Vereinigung ihrer Interaktionstechniken, die zum Entwurf und zu ihrer Nutzung notwendig sind, d.h. sämtliche Aspekte der Konstruktion, Manipulation und Visualisierung sind immer vorhanden. Je nach Zeitpunkt sind unterschiedliche Aspekte von Bedeutung, wobei der Benutzer einer Bedienoberfläche auch die konstruktiven Elemente jedes Objekts zur Verfügung hat und somit seine Oberflächen auch während der Nutzung modifizieren und testen kann.

3 Anforderungen an Entwurfsumgebungen für multimediale Bedienoberflächen

In Abschnitt 4 wird eine Entwurfsumgebung, die auf dem Prinzip der Direkten Komposition basiert, beschrieben. Es wird gezeigt, wie multimediale Interaktionstechniken in die Enturfsumgebung integriert werden können. Um das Design der Entwurfsumgebung SX/

Tools [Hermann 90] zu verstehen, ist es notwendig, die Anforderungen zu kennen, die von industriellen Anwendungen an Bedienschnittstellenwerkzeuge gestellt werden.

Bedienoberflächen in industriellen Anwendungen müssen auf *Hard- und Softwarestandards* bzw. verbreitete Systeme aufbauen. Die *Trennung von Applikation und Oberfläche* ist eine wichtige Anforderung, weil dadurch der evolutionäre Übergang von alphanumerischen auf graphische Schnittstellen ermöglicht wird. Außerdem kann man durch diese Trennung homogene Schnittstellen für mehrere Anwendungen und benutzerangepaßte Oberflächen für eine Anwendung schaffen.

Für die Nutzung multimedialer Interaktionstechniken sind die Eigenschaften der *Offenheit* und *Erweiterbarkeit*, der *Interaktionsfähigkeit* und der *Endbenutzer-Modifizierbarkeit* von zentraler Bedeutung. Auf diese Eigenschaften soll im folgenden näher eingegangen werden.

Erweiterbarkeit und Offenheit. Verschiedene Bereiche der Produktionsautomatisierung, Wartungsunterstützung (z.B. multimediale Leitwarten), Roboter-Kontrollsysteme, Nachrichtenschnittplätze, etc. stellen ähnliche Anforderungen bezüglich des Entwurfs der multimedialen Benutzeroberfläche. Für die Visualisierung und die Kontrolle der spezifischen Applikationen sollte sich die Aufgabe des Benutzeroberflächen-Entwicklers auf die Aggregation einzelner schon vorhandener visueller Basisobjekte zu einem homogenen Ganzen beschränken, ohne speziellen Programm-Code schreiben zu müssen.

Integrationsfähigkeit. Die prototypische Benutzeroberfläche eines Systems wird einerseits aus applikationsspezifischen Visualisierungsobjekten, multimedialen Basiselementen und "standard" Bedienoberflächen-Objekten wie Menüs, Fenster, Scrollbars, Masken und Icons zusammengesetzt, andererseits muß eine solche Bedienoberfläche möglichst viele verschiedene externe (multimediale) Applikationen integrieren können, um schon investierten Programmieraufwand effektiiv nutzen zu können. Damit ergeben sich für die zu erstellenden Bedienoberflächen-Objekte, z.B. für einen Time-Line-Editor, zwei Möglichkeiten: Der Editor wird mit Mitteln des generischen Bedienoberflächen-Werkzeugs "komponiert" und ist später an Veränderungen adaptierbar oder aber ein schon existierender Editor wird eingekapselt als integriertes Element in die "Gesamtkomposition" eingebunden.

Endbenutzermodifizierbarkeit. Oft wird eine ausgelieferte Benutzeroberfläche nach kurzer Zeit den veränderten Anforderungen nicht mehr gerecht oder der Entwurf ist noch nicht abgeschlossen und muß deshalb modifiziert werden [Fischer und Girgensohn 90]. Dies Ist nur möglich wenn der Endbenutzer die Werkzeuge für die Bedienoberflächen-Modifizierung selbst an die Hand bekommt und diese dann auch ohne großen Aufwand einsetzen kann. Durch ein solches Vorgehen ist der Endbenutzer in der Lage, einerseits den Entwurf seinen eigenen Vorlieben und andererseits den eben erwähnten veränderten Systemanforderungen anzupassen.

Realzeitverhalten und multimediale Komposition. Die Anforderungen an multimediale Bedienoberflächen berühren nicht nur die verschiedenen Medien allein (z.B. Videofenster mit Zooming, Rotation, etc.), sondern auch deren Komposition im zeitlichen und visuellen

Ablauf. Die akkurate Synchronisation der verschiedenen zeitlich abhängigen Medien wie Audio, Video und Animation, etc. ist für die visuelle und akustische Präsentation der Bedienoberfläche von großer Bedeutung für deren Akzeptanz. Aus diesem Grunde muß ein multimediales UIMS die Verknüpfung von den sich entwickelnden Standards (z.B. QuickTime) mit den tatsächlichen Realzeitanforderungen berücksichtigen.

4 SX/Tools: direkte Komposition von multimedialen Bedienoberflächen.

SX/Tools ist eine homogene, erweiterbare Entwurfsumgebung für Bedienoberflächen, mit der sich vollständige Bedienschnittstellen für industrielle Anwendugssysteme definieren lassen. Der Schnittstellenentwurf folgt dem Prinzip der direkten Manipulation. Jedes Objekt in SX/Tools enthält Informationen darüber, wie es im Designprozeß eingesetzt werden kann und wie es als Teil einer Anwendungsschnittstelle verwendet wird.

Mit SX/Tools werden Bedienoberflächen durch Kopieren und Zusammensetzen vorhandener Elemente erzeugt. Der Benutzer muß dabei keinen Code schreiben, sondern kann die Komposition mit Hilfe eines Zeigeinstruments direkt am Bildschirm vornehmen. Schnittstellenkomponenten können in Toolboxen gesammelt und in späteren Designs verwendet werden. Diese Toolboxen enthalten anwendungsspezifische Schnittstellenelemente. Basis-Toolboxen existieren für Graphik, Bedienelemente, Fenster und Menüs. Der Entwickler kann seine spezifischen Toolboxen hinzufügen, die beispielsweise Elemente zur Visualisierung von Robotern oder Ampelanlagen enthalten. Toolboxen können permanent gespeichert und in späteren Designprozessen wiederverwendet werden.

SX behandelt alle Interface-Objekte einheitlich. Das System unterscheidet nicht nach Bedienelementen, Graphik, Fenster oder ähnlichen Kategorien. Deswegen benötigt der Entwickler keine unterschiedlichen Techniken zur Festlegung des Aussehens und Verhaltens von Oberflächen-Objekten.

SX erlaubt die Definition von statischen und dynamischen Eigenschaften von Bildschirm-Objekten. Statische Eigenschaften werden direkt-manipulativ (z.B. Größe, Position) oder mit Hilfe von Eigenschaftsfenstern (z.B. Farben) definiert. Das Prinzip der Direkten Komposition findet auch bei diesen Eigenschaftsfenstern Anwendung. Jede Eigenschaft weiß. wie sie am besten definiert werden kann und benutzt ein entsprechendes Eigenschaftsfenster. Verschiedene Eigenschaftsfenster für Text, Selektion und Farbauswahl existieren.

Für die Definition dynamischer Eigenschaften wurden verschiedene Mechanismen vorgeschlagen, darunter State-Transition-Diagrams, kontextfreie Grammatiken und ereignis-basierte Techniken [Green 86]. SX/Tools verwendet eine einfache, ereignis-basierte Sprache [Witschital 91]. Der Vorteil dieser Vorgehensweise ist, daß die Reaktion auf eintreffende Ereignisse lokal an den einzelnen Objekten definiert werden kann und somit wieder dem Prinzip der direkten Komposition folgt. Das Verhalten das Gesamtsystems kann durch die Vereinigung der lokalen Verhaltensmuster beschrieben werden. Die vom

Entwickler definierten Programme werden nicht compiliert, sondern als Eigenschaften bei den Objekten definiert. Es ist deshalb nicht notwendig, nach jeder veränderung des Systems durch den Entwickler den Bedienschnittstellen-Code erneut zu compilieren und zu binden.

SX ist ein *erweiterbares* System. Neue Interface-Objekte können durch *Spezialisierung* oder durch *Aggregation* vorhandener Objekte erzeugt werden. Spezialisierung wird durch explizites Programmieren realisiert. Dabei werden die Vererbungsmechanismen des objektorientierten Konzeptes ausgenutzt. Neue Objekte erhalten alle Eigenschaften vorhandener Objekte und können neue Fähigkeiten definieren. Beispielsweise kann ein Bildschirmbereich zur Darstellung von Video-Einblendungen viele Eigenschaften und Fähigkeiten von Bitmap-Fensterbereichen verwenden. Dazu gehören Methoden zum Verschieben, Vergrößern und ähnliche Aktionen. Andere Eigenschaften muß der Baustein selbst definieren, wie Methoden zum Starten und Stoppen der Video-Einblendung.

Aggregation wird vom Entwickler mittels direkter Manipulation durchgeführt. Vorhandene Element werden zu neuen Objekten zusammengesetzt, die wiederum als Elemente in anderen aggregierten Objekten verwendet werden können. Aggregierte Objekte besitzen wie primitive Objekte eine Beschreibung ihres Verhaltens und ihrer statischen Eigenschaften. Mit dieser Technik läßt sich beispielsweise die Visualiserung eines Video-Recorders durch die Aggregierung eines Video-Bildschirmbereichs mit einer Sammlung von Bedienelementen zur Steuerung des Video-Recorders realisieren.

SX integriert neue Schnittstellenobjekte und verschiedene Interaktionsstile. Durch die Realisierung von SX als Server, der sich mit beliebigen Anwendungen koordinieren kann, lassen sich neue Werkzeuge einfach in die Entwicklungsumgebung integrieren. So kann beispielsweise ein Time-Line-Editor als separates Modul verwendet werden, das für die Darstellung auf dem Bildschirm die Elemente der SX-Umgebung verwendet. Andere Werkzeuge zur Aufgabe von akustischer Information lassen sich ebenso integrieren. Vorhandene Systeme wie z.b Desktop-Publishing-Systeme (Framemaker) können aus SX-Umgebungen verwendet werden, indem man sie mittels dazu definierter Dynamik-Skripten ansteuert.

Vorhandene Bedienoberflächen-Elemente, die mit anderen Werkzeugen erstellt wurden, sollen in SX wiederverwendbar sein. Die Integration verschiedener Widget-Sets (wie Xm oder OLIT) wird durch eine Definition entsprechender SX-Klassen ermöglicht. Dadurch ist ein evolutionärer Übergang von Motif- oder OpenLook-Oberflächen zur SX/Tools-Umgebung möglich.

SX ermöglicht die Anpassung der Bedienoberflächen durch den Endbenutzer. In SX/Tools wird nicht zwischen der Design- und der Laufzeit-Umgebung unterschieden. Das bedeutet, daß der Endbenutzer in der Lage ist, Oberflächen seinen spezifischen Wünschen und Fäigkeiten anzupassen. Außerdem ist er dadurch in der Lage, seine Bedienschnittstelle zu modifizieren, wenn veränderte Anforderungen die Modifiaktion des Anwendungssystems erforderlich machen [Brown et al. 90].

Die Eigenschaft der Endbenutzer-Modifizierbarkeit ist gerade für multimediale Oberflächen von zentraler Bedeutung. Zum einen ergeben sich mit der großen Zahl von möglichen Interaktionstechniken neue Anforderungen, Oberflächen individuell anzupassen, zum anderen kann man heute nicht voraussehen, welche Interaktionsmöglichkeit in zukünftigen Systemen Verwendung finden.

Die Realisierung von SX/Tools wurde auf der Basis von C++ und dem X-Window-System durchgeführt. Diese Entwurfsentscheidung wurde getroffen, um industriellen Quasi-Standards zu entsprechen. Das System ist deswegen auch einfach auf verschiedene-Rechner zu portieren. Der derzeit realisierte Interaktionsstil entspricht dem Motif-Look&-Feel [Motif 90a, Motif 90b], andere Interaktionsstile lassen sich aber wie oben beschrieben in die Entwurfsumgebung einbringen.

SX wurde bislang als ein Prototyp realisiert und in verschiedenen Beispielanwendungen (Medizientechnik, Anlagentechnik) erprobt. Bedienschnittstellen, die mit SX entwickelt wurden, zeigen ein hohes Maß an Homogenität der Interaktion. Die Integration der oben beschriebenen Multimedia-Fähigkeiten wird derzeit durchgeführt. Eine vollständige Integration von SX mit OSF-Motif und eine ergonische Evaluierung der Entwicklungsumgebung wird in den nächsten Entwicklungsschritten durchgeführt.

5 Abschließende Bemerkungen

Entwurfsumgebungen für Bedienschnittstellen werden auch in Zukunft sowohl den Dialog zum Benutzer als auch den Entwurf von Schnittstellen verbessern müssen.

Die Dialogseite wird dabei an der *Integrationsfähigkeit für multimediale Basiselemente* und deren speziellen *Kompositionsfähigkeiten* gemessen, sowie an der Fähigkeit, *Dialogkonzepte von Hypermediasystemen* repräsentieren zu können. Die Qualität der Entwurfsumgebung kann durch *alternative Repräsentationen der Script-Sprache* und die Fähigkeit zur *Dokumentation von Designentscheidungen* verbessert werden.

Die Fähigkeit, *Hyper*-Links in SX zu repräsentieren, ist durch die Erweiterbarkeit der Objekte mittels ihrer Eigenschaftsfenster gegeben. Das Konzept, einen Stack wie in HyperCard zu generieren, ist von der UIMS-Seite ebenfalls ohne Probleme zu realisieren, allerdings müssen von der Applikationsseite her Datenbankaspekte berücksichtigt werden. Neben den lokalen Hyper-Dialogkriterien sind außerdem Browsing-Mechanismen notwendig.

Die Definition von Dynamik-Skripten findet in der derzeitigen SX-Version durch das textuelle Erstellen von Code-Segmenten statt. Alternativen in Form von graphischen Darstellungen der Skripten und die direkt-manipulative Erstellung können die Modifikation von Bedienoberflächen auch für unerfahrene Anwender erleichtern.

Der Entwurf einer Bedienoberfläche in industriellen Anwenndungsbereichen ist eine immer komplexer werdende Aufgabe. Die Dokumentation von Design-Entscheidungen für solche Bedienoberflächen in Form von elektronischen Design-Rationale-Dokumenten könnte dem Designer oder besonders einem Team die Lokalisierung einer Designentscheidung und deren kontextuelle Beziehungen direkt an der Bedienoberfläche erleichtern. Damit sind bei einem Redesign schnell die relevanten Fakten, Argumente und Meinungen zur Hand, welche normalerweise nur sehr schwierig und mühsam rekonstruiert werden können. Zukünftige Bedienoberflächenentwurfsumgebungen müssen deshalb nicht nur den Entwurf selbst, sondern auch die Entscheidungen, die hinter dem Entwurf stehen, dem Designer zur Verfügung stellen.

Literatur

[Brown et al. 90]
D. Brown, P.Totterdell und M. Norman: *Adaptive User Interfaces.* Academic Press, Computer and People Series, 1990

[Fischer und Girgensohn 90]
G. Fischer und A. Girgensohn: *End-user Modifiability in Design Environments.* In Proceedings of CHI '90,ACM press, 1990, pp 183-190.

[Hermann 90]
R. Hermann: *SX/Tools - Externe Architektur Spezifikation.* Internal Paper, Siemens AG, 1990.

[Hutchins et al. 86]
Edwin L. Hutchins, James D. Hollan und Donald A. Norman, *Direct Manipulation Interfaces.* In D. A. Norman und S. W. Draper (eds.), *User Centered System Design*, Erlbaum, Hillsdale, 1986, pp. 87-124.

[Kühme et al. 91]
Th. Kühme, G. Hornung und P. *Witschital: Conceptual models of the design process of direct manipulation user interfaces.* To appear in : Proceedings of HCI International '91, Stuttgart, FRG, September 1991

[Motif 90a]
Open Software Foundation: *OSF/Motif Programmer's Guide.* Prentice Hall, 1990

[Motif 90b]
Open Software Foundation: *OSF/Motif Style Guide.* Prentice Hall, 1990

[Pfaff 85]
Günther E. Pfaff (ed.), *User Interface Management Systems.* Proceedings of the Workshop on User Interface Management Systems held in Seeheim, FRG, November 1-3, 1983, Springer, Berlin, 1985.

[Schneider-Hufschmidt 91]
M. Schneider-Hufschmidt: *Designing User Interfaces by Direct Composition - Prototyping Appearance and Behavior of User Interfaces,* to appear in the proceedings of the NATO-workshop on User-centered Requirements for Software Engineering Environments, Chateau Bonas, France, September 1991.

[Shneiderman 83]
Ben Shneiderman, *Direct Manipulation: A Step Beyond Programming Languages,* IEEE Computer 16(8), August 1983, pp. 57-69.

[Witschital 91]
P. Witschital: *SX/Tools Dialogbeschreibungssprache.* Internal Paper, Siemens AG, 1991

MOTIFATION
ein User Interface Development System

Peer Griebel
Manfred Pöpping
Gerd Szwillus

Universität - GH - Paderborn
Fachbereich Mathematik/Informatik
Postfach 1621
D-4790 Paderborn
F R Germany

MOTIFATION ist ein Programm zum interaktiven Entwurf von graphischen Benutzerschnittstellen, basierend auf dem X Window System und dem OSF/Motif Widget Set. Es hilft dabei, viele der Probleme zu überwinden, die beim Benutzen dieses sehr leistungsfähigen Graphikpaketes auftauchen. MOTIFATION wird bereits in seiner jetzigen Version von mehreren studentische Gruppen eingesetzt. Das Programm wurde im Bootstrapping-Verfahren entwickelt: nur die allerersten Versionen wurden per Hand programmiert, spätere Versionen wurden dann mit MOTIFATION selber erzeugt.

SCHLÜSSELBEGRIFFE
Interface Builder, OSF/Motif, UIDS, UIMS, visuelles Programmieren

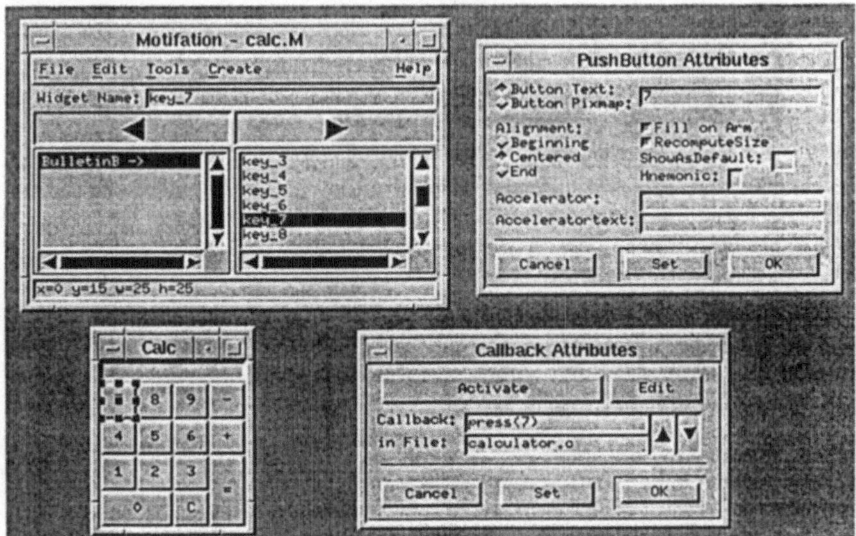

Abbildung 1: Beispiel eines Arbeitsschrittes bei der Verwendung von MOTIFATION

1. EINFÜHRUNG

Mit einer zunehmenden Verbreitung von graphischen Benutzeroberflächen wächst das Bedürfnis, leistungsfähige Hilfsmittel einsetzen zu können, die das Erstellen von Programmen unter diesen Benutzeroberflächen erleichtern. Ursache für dieses Bedürfnis sind eine Reihe von Problemen, die im folgenden aufgeführt werden sollen.

- o Der Programmierer muß sich in eine Vielzahl verfügbarer graphischer Objekte einarbeiten. Er muß lernen, welche Aufgaben sie erfüllen, welches Verhalten sie haben und wie sie aussehen.
- o Aufgabe, Verhalten und Aussehen sind keine statischen Merkmale, sondern können durch eine Vielzahl von Parametern variiert werden.
- o Bei der Erzeugung graphischer Elemente und der Einstellung dazugehöriger Attribute müssen teilweise Reihenfolgebedingungen berücksichtigt werden. Es können auch Abhängigkeiten existieren, so daß Objekte bzw. Attribute sich gegenseitig beeinflussen.
- o Der notwendige C-Quellcode wird schon für einfache Benutzerschnittstellen sehr umfangreich, und der Programmierer verliert leicht die Übersicht.
- o Der Programmierer gewinnt kein Gefühl für das Aussehen der Schnittstelle, da er gezwungen ist, eine "graphische Realität" textuell zu beschreiben. Er muß sich durch "Versuch und Irrtum" an das gewünschte Design herantasten.
- o Generell - nicht nur bei Verwendung von Toolkits - besteht die Gefahr einer zu engen Mischung von Programmcode für Applikation und Benutzerschnittstelle. Daraus resultieren eine unklare Programmschnittstelle zwischen diesen Komponenten und eine verringerte Portierbarkeit auf andere Rechnerplattformen. Eine klare Schnittstelle ist auch Voraussetzung für gute Wartbarkeit bzw. Adaptierbarkeit. Änderungen an der Oberfläche sollten ohne größere Änderungen an der Applikation durchgeführt werden können (und umgekehrt).

Das UIDS MOTIFATION wurde entwickelt, um dem Programmierer die Möglichkeit zu geben, OSF/Motif unter weitgehender Vermeidung der oben angegebenen Probleme einzusetzen.

2. ALLGEMEINE EIGENSCHAFTEN VON MOTIFATION

Innerhalb graphischer Benutzerschnittstellen unterscheiden wir zwischen statischen und dynamischen Anteilen.

- o Die dynamischen Anteile umfassen die anwendungsspezifischen Darstellungen eines Programmes. Eigenschaften wie Anzahl, Position, Aufbau und innere Struktur der Darstellungen sind hochgradig abhängig von Daten der Anwendung und dementsprechend auch veränderlich. Beispiele für dynamische Anteile wären etwa Werkstückzeichnungen (CAD-Programm) oder Projektnetzpläne (Projektplanungswerkzeug).
- o Die statischen Anteile dagegen sind von den Daten der Applikation weitgehend unabhängig. Sie dienen meist zur Steuerung des Programms und zur Kommunikation zwischen Benutzer und Applikation. Der Zeitpunkt des Auftretens dieser Dialoganteile ist meist abhängig von Bedingungen in der Benutzerschnittstelle oder der Applikation, aber das Aussehen ist im allgemeinen festgelegt. Statische Anteile wären etwa die Menüs oder Dateiauswahl-Fenster von interaktiven Systemen.

Die Abgrenzung dieser beiden Anteile einer Benutzerschnittstelle kann nicht präzise sein. Der Übergang ist fließend. MOTIFATION konzentriert sich auf die interaktive graphische Spezifikation der statischen Anteile einer Benutzerschnittstelle.

MOTIFATION bietet dem Benutzer eine Arbeitsweise wie bei objektorientierten Zeichenprogrammen an. MOTIFATION arbeitet dabei nach dem WYSIWYG-Prinzip: der Interface Designer hat bereits während des Entwurfs einen unmittelbaren Eindruck des optischen Erscheinungsbildes der graphischen Oberfläche. MOTIFATION unterstützt die graphischen Objekte (Widgets) von OSF/Motif. Motif ist ein Graphikpaket basierend auf MIT's X Window System und X Toolkit.

Der folgende Überblick von Features soll einen ersten Eindruck von der Leistungsfähigkeit von MOTIFATION ermöglichen:

- o Interaktives Plazieren aller von OSF/Motif angebotenen graphischen Objekte
- o Größen- und Lageveränderung durch direkte Manipulation unterstützt durch zusätzliche Anzeige von Koordinaten
- o Ausrichtungsfunktionen
- o Zwischenablage
- o Verwaltung von häufig benutzen Dialogelementen in Bibliotheken
- o strukturierte Attributboxen zum Einstellen der objektspezifischen Attribute wie Farbe oder Font, bei unmittelbarer Anzeige der Auswirkungen am Bildschirm
- o Tabgroup-Recorder: ein OSF/Motif-Mechanismus zur Definition der Benutzerführung innerhalb einer Dialogbox
- o Auflösungsunabhängigkeit (Resolution Independence): zur automatischen Anpassung der geometrischen Größe der Benutzerschnittstelle in Abhängigkeit eines bestimmten Zeichensatzes bzw. zur leichten Adaptierbarkeit von Programmen auf Monitoren verschiedener Auflösung
- o Spezifikation graphischer Zwänge (Constraints), die Objekte untereinander ausüben.
- o Anbindung der Applikation an graphische Objekte über Zuordnung auszuführender Operationen gesteuert von Ereignissen
- o Erzeugung einer textuellen Dokumentation der Benutzerschnittstelle unter verschiedenen Sichten
- o nachträgliche Modifizierbarkeit der Benutzerschnittstelle des spezifizierten Systems
- o automatische Erzeugung aller notwendigen C-Code-Dateien aus der interaktiven Spezifikation
- o Erzeugung eines entsprechenden Makefiles zur Erstellung des lauffähigen Gesamtprogrammes

Bei der Erzeugung des Benutzerschnittstellen-Codes mit MOTIFATION ist automatisch gewährleistet, daß eine klare Trennung zwischen Applikation und Benutzerschnittstelle existiert. Die Benutzeroberfläche und die Anwendungsroutinen stehen in separaten Dateien. Eine Kommunikation zwischen der Applikation und der graphischen Benutzeroberfläche findet über die vom Interface Designer vergebenen Objektnamen und entsprechenden, vordefinierten Zugriffsfunktionen statt. Die Benutzeroberfläche kommuniziert mit der Applikation über sogenannte Callbacks. Es handelt sich dabei um einen X Toolkit Mechanismus zur Definition der Reaktionen auf Ereignisse.

MOTIFATIONs Benutzerschnittstelle wurde nach dem Boot-Strapping-Verfahren erstellt, das heißt zur Entwicklung einer neuen Benutzerschnittstelle wurde jeweils eine ältere Version von MOTIFATION zur Hilfe genommen. Dadurch konnte MOTIFATIONs Leistungsfähigkeit bereits intensiv getestet werden. MOTIFATION wird zur Zeit von mehreren studentischen Arbeitsgruppen eingesetzt.

3. BEGRIFFSBESTIMMUNG UND VERGLEICH: UIDS - UIMS

3.1 Charakterisierung von MOTIFATION

Unserer Meinung nach ist MOTIFATION kein User Interface Management System (UIMS). Wir verwenden den allgemeineren Begriff User Interface Development System (UIDS). Myers [Myers 1989] fordert für ein UIMS neben Komponenten für die äußere Gestaltung der Benutzerschnittstelle (Präsentationskomponente) und einem Mechanismus zur Anbindung der Applikation (Applikationsschnittstelle) eine explizite Instanz zur Dialogsteuerung (Dialogkomponente), um es gegenüber Toolkits oder Frameworks abzugrenzen.

MOTIFATION besitzt keine Dialogkomponente in diesem Sinne: Wann genau zum Beispiel ein Fenster sich öffnet, etwa um eine anwendungsspezifische Fehlermeldung zu zeigen, wird in der Applikation abhängig von semantischen Bedingungen programmiert. Das Wissen um diese semantischen Bedingungen ist innerhalb der Spezifikation mit MOTIFATION nicht vorhanden.

MOTIFATION unterstützt allerdings die Dokumentation dynamischer Verbindungen über sogenannte Softlinks (siehe Abschnitt 5.2). Dieser Mechanismus ist unverbindlich - die angegebenen Verzweigungen werden weder erzwungen noch ihre Realisierung innerhalb der Applikation überprüft. Er dient dazu, Verbindungen zwischen Dialogelementen zu dokumentieren, um bei der separaten Programmierung der Anwendung diese Verbindung nicht zu vergessen.

Myers unterscheidet weiterhin "language based", "automatic creation" und "graphical specification" UIDS. MOTIFATION baut auf dem "graphical specification"-Ansatz auf, verzichtet also auf eine syntaktische Spezifikation der Dialoge.

Daß MOTIFATION kein UIMS ist, stellt unserer Meinung nach keinen schwerwiegenden Mangel dar. Generell geht der Trend zu Direct-Manipulation-Schnittstellen, die möglichst wenig oder keine globale Syntaxkomponente besitzen. Hartson [Hartson 1989] beschreibt eindrücklich, daß bei Direct-Manipulation-Schnittstellen eine klare Trennung der drei Komponenten Input, Computation und Display nicht realistisch ist.

Vielmehr beschreibt man Benutzerschnittstellen in Begriffen von aktiven Objekten, die "lokale" Miniaturdialoge mit dem Benutzer führen. Auch OSF/Motif folgt diesem Trend: Jedes Widget besitzt nicht nur die Eigenschaft, Benutzereingaben zu verwalten (Input), sondern es berechnet automatisch neue Zustände (Computation) und zeigt diese auch an (Display). Ein UIDS läßt die Verschmelzung dieser Komponenten zu. Die bei einem UIMS geforderte Dialogkomponente wirkt sich bei Direct-Manipulation-Schnittstellen eher hinderlich aus.

3.2 Andere Systeme

Ein Programmsystem, daß MOTIFATION ähnlich ist, ist das UIMS TeleUSE von TeleSoft. Im Vergleich zu TeleUSE erscheint uns die Bedienung von MOTIFATION einfacher und intuitiver. TeleUSEs Editor (VIP), der für das Zusammenstellen der graphischen Benutzerschnittstelle zuständig ist, benutzt den Umweg über sogenannte Templates. Der Interface Designer muß sich deshalb nach dem Ändern von Attributen erst eine Repräsentation des Templates erzeugen lassen, bevor er das wirkliche Aussehen vor Augen hat.

TeleUSE erzeugt anders als MOTIFATION keinen C-Quellcode, sondern einen eigenen Zwischencode, der bei der Ausführung des Programmes interpretiert wird. Problematisch ist, daß die Größe der erzeugten Programme bereits bei kleinsten Anwendungen sehr umfangreich ist: Das "Hello-World-Programm" ist bei TeleUSE um den Faktor sechs größer als bei MOTIFATION.

Im Gegensatz zu MOTIFATION besitzt TeleUSE als UIMS eine explizite Dialogkontrollkomponente. Sie wird in einer auf C basierenden Sprache D spezifiziert. Neben der Tatsache, daß der Interface Designer eine weitere Sprache erlernen muß, ist die Beschreibung selbst einfacher Dialoge aufwendig.

Ein weiterer Unterschied ist, daß TeleUSE nicht auf die Verwendung des OSF/Motif Widget Set fixiert ist. Es können vielmehr auch andere Toolkits eingebunden werden. Dies ist prinzipiell natürlich ein Vorteil, da man nicht unbedingt an OSF/Motif gebunden ist.

Von van der Zanden und Myers wird das System JADE [Zanden 1990] zur Erzeugung von Benutzeroberflächen vorgestellt. Die sehr interessante Idee dieses Ansatzes, Look&Feel-Unabhängigkeit zu garantieren, scheitert an der unterschiedlichen Vielzahl der graphischen Objekte verschiedener Toolkits. Es kann nicht auf spezielle Eigenschaften und Vorzüge einzelner Toolkits eingegangen werden. Nur die Schnittmenge der Eigenschaften der von JADE unterstützten Benutzeroberflächen kann Look&Feel unabhängig realisiert werden. Zu der FileSelectionBox in OSF/Motif (siehe nächstes Kapitel) existiert in den meisten anderen Benutzeroberflächen kein entsprechendes Objekt. JADE gehört in die Klasse der "automatic creation"-UIDS und ist damit nicht in allen Punkten mit MOTIFATION vergleichbar.

4. PROGRAMMIEREN MIT OSF/MOTIF

OSF/Motif ist ein auf dem X Window System / X Toolkit aufbauendes Widget Set. Ein Widget Set kann am ehesten als eine Sammlung von Objekten beschrieben werden, die als Interaktionsobjekte in Benutzerschnittstellen auftauchen. Sie besitzen also eine graphische Gestalt und Funktionalität zur Kommunikation mit dem Benutzer.

Einige der vielen von OSF/Motif angebotenen Widgets sollen hier kurz erwähnt werden.

- Das Label-Widget dient zur Darstellung von statischen Texten als nicht änderbare Ausgabe an den Benutzer.
- Über das Text-Widget kann textuelle Eingabe vom Benutzer eingeholt werden. Das Text-Widget hat dabei die Funktionalität eines kleinen Editors.
- PushButton-Widgets sind Objekte, die Ereignisse erzeugen (und damit eine Funktion aktivieren), falls der Bediener sie anwählt.
- Das BulletinBoard dient als Behälter für weitere Objekte und hat kein Aussehen, ist also für den Bediener unsichtbar.
- In einer ScrolledList kann eine Liste von Textzeilen dargestellt werden. Wenn der sichtbare Teil der Liste kleiner ist, als die gesamte Liste, erscheint automatisch ein Scrollbar, mit dessen Hilfe man sich in der Liste bewegen kann.
- In der FileSelectionBox wird eine Liste der Dateien eines Verzeichnisses angezeigt. Der Benutzer kann eine Datei wählen, um eine Aktion wie beispielsweise laden oder speichern auszuführen.

Weitere leistungsfähige Objekte, wie sie aus anderen Benutzeroberflächen bekannt sind, werden angeboten wie Menüs und Scrollbars. In Abbildung 1 sind im oberen linken Fenster (Browser) zwei ScrolledLists zu sehen. Bei dem Text "Widget Name:" handelt es sich um ein Label. Rechts daneben befindet sich ein Text-Widget.

Bei der Erzeugung einer Benutzeroberfläche unter OSF/Motif werden die einzelnen Widgets hierarchisch angeordnet. Typischerweise besteht eine Anwendung aus einer Shell, auf der ein BulletinBoard erzeugt wird. Das BulletinBoard ist seinerseits Vater von diversen Labels, PushButtons, Text-Widgets etc. Die Hierarchie betrifft nicht nur die Sichtbarkeit, sondern auch semantische Aspekte. Beispielsweise können bestimmte Widgets ihren Söhnen ein graphisches Aussehen und andere Einstellungen aufzwingen. Folglich existieren Einschränkungen bezüglich der Widgettypen, die als Sohn eines Widgets erlaubt sind.

Das Verhalten und Aussehen der verschiedenen Objekte wird über Attribute gesteuert. Zu den Attributen gehören auch die sogenannten Callback-Attribute, bei denen es sich um C-Funktionen handelt, die bei Auftreten bestimmter Ereignisse aufgerufen werden sollen. Aufgrund dieser Architektur entfällt bei der Programmierung die explizite Einbindung von Programmcode zum Abfangen von Ereignissen (Window-Main-Loop). Dieser Anteil wird von MOTIFATION bzw. dem Widget Set übernommen.

5. PROGRAMMIEREN MIT MOTIFATION: DER ENTWURF EINER BENUTZERSCHNITTSTELLE

5.1 Creator, Browser und Workbench

Beim Programmstart werden als erstes die beiden Dialogelemente "Browser" und "Creator" sichtbar.

- o Der Creator dient dazu, ein neues Widget in die bislang erstellte Hierarchie einzufügen. Diese Einfügungen - genau wie andere Aktionen - geschehen immer an einem bestimmten Arbeitspunkt, dem aktuell selektierten Widget.
- o Jedes Widget hat einen eindeutigen Namen. Beim Erzeugen über den Creator wird von MOTIFATION automatisch ein eindeutiger Name generiert, der jedoch nachträglich verändert werden kann. Dieser Name wird benutzt, wenn aus dem Applikationscode heraus auf dieses Widget zugegriffen werden soll.
- o Mit dem Browser, ähnlich dem Smalltalk-Browser, kann man die Hierarchie durchwandern, inspizieren und Teile davon selektieren.
- o Die graphische Repräsentation der Hierarchie wird ständig in der sogenannten Workbench (Arbeitsfläche) angezeigt.

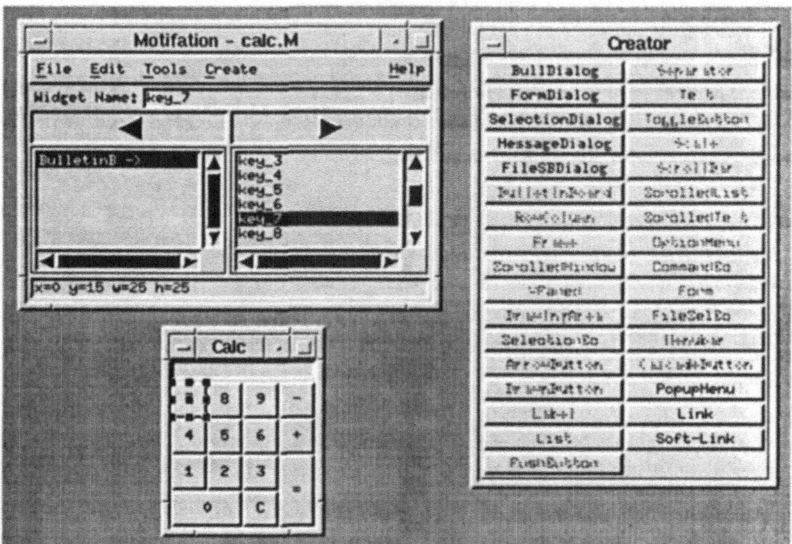

Abbildung 2: Taschenrechner mit Creator

Die obige Abbildung 2 zeigt eine typische Situation beim Erstellen einer Widgethierarchie: Aktuell ist das Widget mit dem Namen key_7 vom Typ PushButton selektiert. Auf diesem Widget könnten alle im Creator (rechts oben) schwarz angezeigten Widgets erzeugt werden - zum Beispiel ein PopupMenu oder ein Softlink. Softlinks sind nicht Bestandteil von OSF/Motif. Sie werden jedoch von MOTIFATION wie Widgets behandelt, um die Erstellung des User Interfaces zu erleichtern (siehe auch 5.2). Auf der Workbench unterhalb des Browsers wird das graphische Erscheinungsbild der bearbeiteten Widgethierarchie, in diesem Fall der Taschenrechner, dargestellt.

Der Creator zeigt zu jedem Zeitpunkt eine Liste der Widgets, die als Sohn des aktuell selektierten Objektes erzeugt werden können. Ein neues Widget wird im Browser und auf der Workbench sofort dargestellt. Um ein Objekt zu selektieren, kann man es entweder im Browser über seinen Namen oder auf der Workbench über seine graphische Darstellung mit der Maus anwählen. Das selektierte Widget erhält wie von objektorientierten Zeichenprogrammen gewohnt, eine Resize-Box. Über diese kann das Objekt interaktiv vergrößert und verschoben werden.

Zu jedem Widget existieren eine Vielzahl von Attributboxen. Man ändert die Attribute des aktuell selektierten Widgets, indem man die Werte in der entsprechenden Attributbox editiert. Die Eingaben werden auf Konsistenz geprüft und ihre Auswirkungen unmittelbar auf der Workbench sichtbar. Dem Interface Designer werden dabei nur Änderungen erlaubt, die im aktuellen Zusammenhang möglich sind.

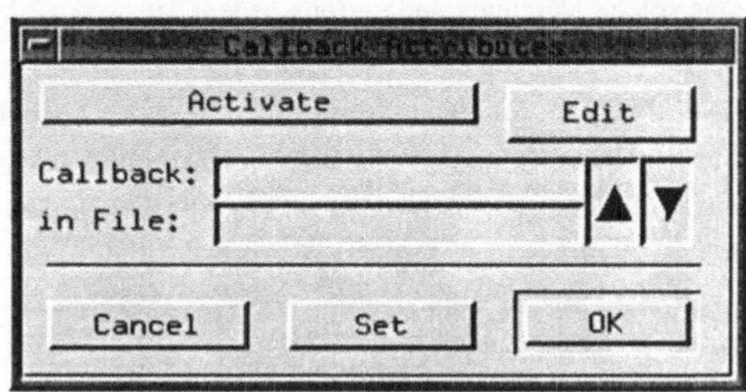

Abbildung 3: Callback-Attributbox

Eine spezielle Attributbox ist die Callback-Attributbox. In ihr werden die Verbindungen zwischen der Oberfläche und der Applikation hergestellt. Dies geschieht, indem der Interface Designer angibt, welche C-Funktionen bei Auftreten eines bestimmten Ereignisses aufgerufen werden sollen. Zusätzlich wird vermerkt, in welcher Datei die Funktionen zu finden sind.

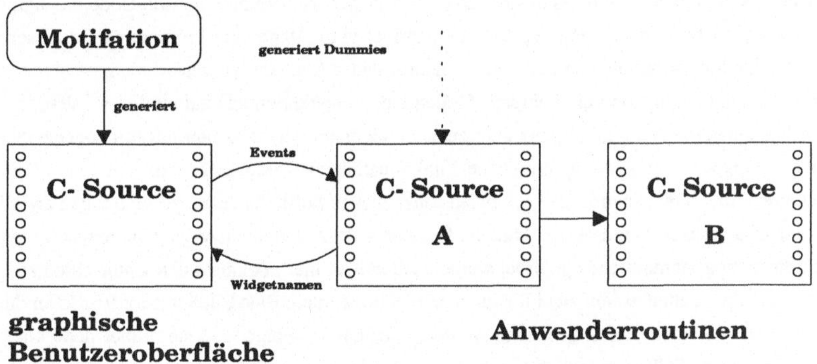

Abbildung 4: Dreiteilung des C-Quellcodes

MOTIFATION unterscheidet drei Arten von C-Quellcode-Dateien.

- o Der die Präsentationskomponente ausmachende C-Quellcode wird komplett von MOTIFATION verwaltet. Der Benutzer macht Änderungen hierin nur über MOTIFATION.
- o Der Applikations-Quellcode ist vollkommen unabhängig von der Benutzeroberfläche. Die Programmierung dieses Teils wird deshalb von MOTIFATION nicht unterstützt.
- o Dazwischen liegt das Modul, das die Schnittstelle zwischen der Oberfläche und der Applikation bildet.

In diesem Modul befinden sich die Funktionen, die als Callbacks auf Ereignisse reagieren. MOTIFATION generiert hierzu leere Funktionsrümpfe (Dummies), falls die entsprechenden Funktionen noch nicht ausformuliert sind. Dem Programmierer bleibt die Aufgabe, diese Rümpfe mit seinen Routinen zu füllen. Da MOTIFATION die genaue Kenntnis der notwendigen Quelldateien hat, wird gleichzeitig ein entsprechendes Makefile generiert. Mit den generierten Funktionen zur Erzeugung der Oberfläche und den Dummies liegen alle notwendigen Funktionen vor, um ein lauffähiges Programm zu übersetzen.

Der Programmierer muß also keine Zeile von Hand programmiert haben und hat trotzdem ein lauffähiges Programm, das unabhängig von der Entwicklungsumgebung übersetzt und ausgeführt werden kann.
Die oben aufgeführte Aufteilung der Anwenderroutinen in die Teile A und B wird von MOTIFATION nicht erzwungen. Einerseits ist es günstig, eine enge Kopplung zwischen der graphischen Benutzeroberfläche und den Anwenderroutinen herzustellen. Dies ist beispielsweise bei Direct-Manipulation-Schnittstellen oder sehr kleinen Anwendungen sinnvoll. Aus softwaretechnologischer Sicht ist eine solche Mischung andererseits bedenklich, so daß MOTIFATION die Trennung unterstützt.
Der größte Vorteil der Aufteilung aus Abbildung 4 besteht jedoch in der Wiederverwendbarkeit der Applikationsroutinen (Teile A und B). Eine nachträgliche Veränderung der graphischen Benutzeroberfläche (z.B.: Hinzufügen eines Menüeintrages zu einem PopupMenu) kann vollkommen separat durchgeführt werden. Da MOTIFATION über die Applikationsroutinen aus Teil A informiert ist, werden diese bei einer Codegenerierung nicht überschrieben.
Zusätzlich zu den generierten C-Quellcode-Dateien kann sich der Interface Designer eine strukturierte Beschreibung der graphischen Benutzeroberfläche unter verschiedenen Gesichtspunkten in Textform erzeugen lassen. Er erhält hierüber eine gute Dokumentation über die Verbindung zwischen der Oberfläche und den Anwenderroutinen. Auch können die verwendeten maschinenabhängigen Ressourcen wie Farben, Zeichensätze oder Bitmaps aufgezählt werden.

5.2 Spezielle Eigenschaften

Die oben beschriebene Widgethierarchie von OSF/Motif in einer Applikation wird durch MOTIFATION erweitert.

- Für eine Dialogbox, die als Sohn eines PushButtons erzeugt wurde, wird automatisch Code erzeugt, der dafür sorgt, daß diese Dialogbox beim Aktivieren des Buttons geöffnet wird. Ohne eine Programmzeile zu schreiben, legt der Interface Designer hiermit schon einen Teil der Semantik seiner Applikation fest.
- Soll ein und dieselbe Dialogbox von mehreren Stellen einer Applikation aus aufgerufen werden, so ist es nicht notwendig, mehrere Kopien dieser Dialogbox zu erzeugen. Vielmehr verweist man nur mit sogenannten Links auf die gewünschte Dialogbox. Die oben beschriebene Funktionalität, die eine Dialogbox auf einem Button hat, bleibt dadurch erhalten. Im Unterschied zu den normalen Links dienen Softlinks dazu, Verbindungen zwischen Anwenderroutinen und Dialogboxen herzustellen. Dies ist besonders dann vonnöten, wenn Dialogboxen sporadisch in Abhängigkeit eines Programmzustands geöffnet werden sollen. Um hier nochmals den Unterschied zwischen Link und Softlink deutlich zu machen, sollen zwei Beispiele angeführt werden. Eine FileSelectionBox kann durch ein Link an den Button "Save" in einem Menü gebunden werden. Dies hat zur Folge, daß sie immer dann erscheint, wenn der Button aktiviert wird. Sollten jedoch keine Daten vorhanden sein, die gespeichert werden können, so kann als nächstes eine Dialogbox mit der entsprechenden Fehlermeldung angezeigt werden. Eine solches Verhalten kann in MOTIFATION durch Anlegen eines Softlinks ausgedrückt werden, der den "Save"-Button und die Fehlermeldungs-Dialogbox verbindet.
- MOTIFATION stellt ein Clipboard zur Verfügung. Die Wiederverwendbarkeit von Teilbäumen der Widgethierarchie wird über diesen Mechanismus unterstützt. Man kann damit auch Bibliotheken anlegen, in denen man Teile von Benutzerschnittstellen in speziellen Dateien abspeichert. Häufig benötigte Teile einer Oberfläche sind somit schnell zur Hand, und die Standardisierung von Benutzerschnittstellen wird unterstützt.

5.3 Beispiel

Der Entwurf eines kleinen Taschenrechners soll als Beispiel dienen, um zu verdeutlichen, wie mit MOTIFATION Objekte erzeugt und plaziert werden und wie die Anbindung der Callbacks vor sich geht.

Schritt 1: Über den Menüeintrag NEW wird ein leeres Fenster angelegt; damit ist eine neue Widgethierarchie kreiert, die zu Beginn lediglich aus der Wurzel, dem Hauptfenster besteht.

Schritt 2: Auf diesem wird mit Hilfe des Creators ein BulletinBoard plaziert. Wie schon oben erwähnt, ist das BulletinBoard ein Container-Widget und dient hier zur Aufnahme der Taschenrechnertasten und des Displays.

Schritt 3: Der Interface Designer erzeugt darauf ein Text-Widget, das als Display des Taschenrechners dienen soll, und ein PushButton zur Aufnahme einer Taste des Taschenrechners. In Abbildung 5 ist der aktuelle Stand des Taschenrechners dargestellt.

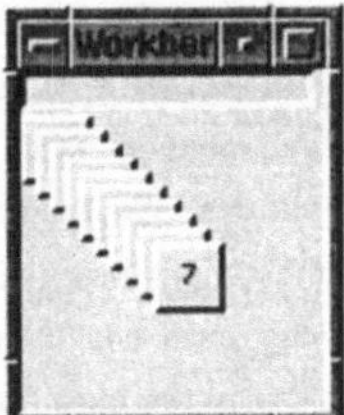

Abbildung 5, 6: Während des Taschenrechnerentwurfs

Schritt 4: Das Text-Widget und der PushButton werden nacheinander selektiert und über die Resize-Box in die richtige Größe und Position gebracht. Für den PushButton wird die Beschriftung '7' vorgenommen.

Schritt 5: Mit dem Clipboard ist es möglich, mehrere Kopien von dem PushButton zu erzeugen. Sie haben damit zwar schon die richtige Größe, müssen aber noch auf die richtige Position gebracht werden. Abbildung 6 zeigt die Workbench mit den jeweils um einen kleinen Betrag verschobenen PushButtons. Durch diese Verschiebung ist es einfacher, die einzelnen Widgets zu selektieren.

Schritt 6: Für jeden Button wird über die PushButton-Attributbox die Beschriftung eingestellt und über die Callback-Attributbox die Anbindung an die eigentliche Applikation hergestellt.

Somit ist die Erzeugung der Benutzeroberfläche für den Taschenrechner abgeschlossen. Den jetzigen Zustand sieht man in Abbildung 1. Es ist zu sehen, daß bei Aktivierung des Buttons 7 die Funktion press() mit dem Argument 7 in der Datei calculator.c aufgerufen wird.

Da nun die Arbeit an der Benutzeroberfläche beendet ist, kann sie abgespeichert und der Code generiert werden. Nun müssen noch die Anwendungsroutinen erstellt werden, also etwa Funktionen zum Verwalten der Tastendrucke und zur Berechnung des Rechenergebnisses. Damit ist der Taschenrechner funktionsfähig. Die gesamte Erstellung dieser Benutzerschnittstelle dauert etwa 30 Minuten.

5.4 Erfahrungen und Ausblicke

Die statischen Anteile graphischer Benutzeroberflächen lassen sich durch MOTIFATION sehr schnell erzeugen. MOTIFATION wird bereits von mehreren Studenten eingesetzt, die die Praxistauglichkeit des Programmsystems bestätigen. MOTIFATION wurde im Bootstrapping-Verfahren entwickelt, wodurch rechtzeitig Probleme erkannt und beseitigt werden konnten. Schon früh wurden die Dialogelemente von MOTIFATION durch MOTIFATION selbst erzeugt und erweitert.

Die klare Trennung (Abbildung 4) des C-Quellcodes hat sich als besonders positiv ausgezeichnet, weil in der Praxis eine häufige Veränderung der Benutzerschnittstelle üblich ist. Die Veränderungen können sehr bequem und schnell mit MOTIFATION durchgeführt werden. Ebenfalls positiv wird die Tatsache bewertet, daß nur reiner C-Quellcode generiert wird. Aufgrund der eigenen Erfahrungen und der von außen herangetragenen Anregungen, sind folgende Verbesserungen und Erweiterungen geplant:

- Der Benutzer soll leistungsfähigere Hilfsmittel erhalten, um die Attribute einzustellen. Im Augenblick ist es notwendig, Widgetnamen generell über die Tastatur einzugeben. In Zukunft soll es möglich sein, Widgets durch Anklicken mit der Maus auszuwählen.
- Weiterhin planen wir, die einfache Erstellung verschiedener nationaler Versionen der Benutzerschnittstelle einer Applikation komfortabel zu unterstützen.
- MOTIFATION unterstützt noch nicht sämtliche Attribute der OSF/Motif Widgets. Es ist daher notwendig, weitere Attributboxen zur Verfügung zu stellen.
- Desweiteren planen wir eine weitergehende Konsistenzprüfung, die auch den Applikationscode beinhalten soll. Bei Änderungen an einem Widgetnamen soll der Interface Designer dann beispielsweise darüber informiert werden, daß er diesen Widgetnamen in seinem Applikationscode bereits benutzt hat.
- Eine weitergehende Unterstützung im dynamischen Teil einer Applikation wird angestrebt. Graphische Ausgaben einer Applikation werden von OSF/Motif im Augenblick nicht oder nicht ausreichend unterstützt. Es ist daher notwendig, über die von OSF/Motif bereitgestellten Mechanismen hinaus weitere Funktionen und Widgets zur Verfügung zu stellen. So wäre etwa ein "Icon"-Widget denkbar, das die Programmierung von Direkt-Manipulations-Objekten, ähnlich den Piktogrammen für Dateien des Macintosh Desktops, erlaubt. Allgemeinere Überlegungen gehen in die Richtung der Integration einer leistungsfähigen Verwaltung für die graphischen Objekte der Anwendung in MOTIFATION.
- Darüber hinaus muß die Erweiterung bzw. Modifikation der Oberfläche während der Laufzeit möglich werden. Wir denken dabei an Widgets bzw. Widgethierarchien, deren Aussehen und Eigenschaften schon mit MOTIFATION eingestellt werden können, deren Erzeugung jedoch erst durch die Applikation in beliebiger Anzahl erfolgt.
- Weiterhin planen wir einen Mechanismus, über den die Anwählbarkeit von Widgets in Abhängigkeit vom Programmzustand gesteuert werden kann. Beispielsweise soll der Cut-Button im Edit-Menü eines Programmes nicht anwählbar sein, wenn vom Benutzer nichts selektiert wurde. Der Interface Designer soll die Möglichkeit bekommen, solche oder ähnliche Programmzustände mit MOTIFATION zu spezifizieren. Diese Spezifikationen sollen in den generierten C-Quellcode einfließen und somit Auswirkungen auf die Oberfläche haben.

LITERATURANGABEN

Hartson, R.: *User-Interface Management Control and Communication*, **IEEE Software**, January 1989, pp. 62-70

Myers, B.: *User Interface Tools: Introduction and Survey*, **IEEE Software**, January 1989, pp. 15-23

Open Software Foundation: **OSF/Motif Reference Manual**, Revision 1.0, Open Software Foundation, 11 Cambridge Center, Cambridge, 1989

Telesoft Inc.: **TeleUSE Reference Manual**, Version 1.0, Teknikringen 1989

Van der Zanden, B.; Myers B.: *Automatic, Look-and-Feel Independant Dialog Creation for Graphical User Interfaces*, **Proceedings of the CHI '90 Conference on Human Factors in Computing Systems**, Seattle, Washington, April 1990, pp 27-34

Webster, B.F.: **The NeXT Book (The Interface Builder)**, Addison-Wesley, July 1989

Synchronisation der Präsentation von Multimedia-Objekten - Modell und Beispiele -

PETRA HOEPNER

GESELLSCHAFT FÜR MATHEMATIK UND DATENVERARBEITUNG
FORSCHUNGSZENTRUM FÜR OFFENE KOMMUNIKATIONSSYSTEME
HARDENBERGPLATZ 2
1000 BERLIN 12

Die Präsentation von Multimedia-Objekten umfaßt die gleichzeitige und/oder sequentielle Präsentation von verschiedenen Repräsentationstypen (z.B. Text, Grafik, Bild, Audio und Video). Die zeitlichen Relationen zwischen verschiedenen Aktionen müssen spezifiziert und durch Anwendung geeigneter Synchronisationsmechanismen realisiert werden. Dieser Beitrag stellt ein Synchronisationsmodell und eine auf Pfadausdrücken basierende Notation für die Beschreibung von Synchronisationsbeziehungen vor. Eine äquivalente Darstellung durch Petri-Netze wird eingeführt und anhand von Beispielen erläutert.

1 Einführung

Der vorliegende Beitrag wurde im Rahmen des Projekts "Multi-Media-Dokumente im ISDN-B" (BERMMD), einem Teilprojekt des BERKOM Projekts, erstellt. Die Entwicklung des "BERKOM Reference Model - Application-Oriented Layers" [1] führte unter anderem zu der Anforderung, die Synchronisation für die Präsentation von Multimedia-Objekten zu modellieren.

Wird ein Multimedia-Objekt präsentiert, so müssen sowohl zeitunabhängige Repräsentationstypen wie Text, Grafik, Bild, als auch zeitabhängige Repräsentationstypen wie Audio und Video dargestellt werden. Das bedeutet, daß verschiedene, zeitlich in Beziehung stehende Aktionen in einer definierten Anordnung auszuführen sind. Die Spezifikation und Überwachung dieser Ordnung von Aktionen wird im allgemeinen als Synchronisation bezeichnet.

In den folgenden Abschnitten wird ein Synchronisationsmodell und eine auf Pfadausdrücken (path expressions) basierende Notation vorgestellt, die eine Beschreibung der Synchronisation von Aktionen unterstützen.

2 Modell für die Synchronisation von Multimedia-Objekten

Die Präsentation eines Multimedia-Objekts besteht aus einer Menge von Aktionen, die in einer zeitlichen Relation zueinander stehen, d. h. synchronisiert werden müssen. Diese zeitlichen Relationen sollen durch eine geeignete Notation spezifiziert werden.

2.1 Aktionen

Eine **Aktion** (action) wird im "Open Distributed Processing (ODP) Reference Model" wie folgt definiert [2]: "An action is the representation of something which happens." Eine Aktion besteht aus atomaren Einheiten, die als **Ereignisse** (events) bezeichnet werden [2]. Für diese Ereigisse kann eine partielle zeitliche Ordnung definiert werden [3].

Im Rahmen der Synchronisation sind nur die Ereignisse relevant, die in einer Relation zu Ereignissen anderer Aktionen stehen. Diese Ereignisse werden **Synchronisationspunkte** genannt. Aktionen werden an ihren Synchronisationspunkten synchronisiert, indem bestimmte Synchronisationsmechanismen angewendet werden.

Jede Aktion wird durch zwei zeitliche Ereignisse begrenzt: **Startzeitpunkt** und **Endzeitpunkt**. Start- und Endzeitpunkt können sowohl als absolute Zeitpunkte bezüglich eines zeitlichen Referenzsystems, als auch durch einen zeitlichen Bezug zu Ereignissen anderer Aktionen definiert werden. Start- und Endzeitpunkte fungieren somit als Synchronisationspunkte.

In Abhängigkeit von der Existenz von Synchronisationspunkten zwischen dem Start- und Endzeitpunkt einer Aktion werden zwei Arten von Aktionen unterschieden: (1) **Atomare Aktionen** (atomic actions) enthalten keine Synchronisationspunkte, (2) **zusammengesetzte Aktionen** (composed actions) enthalten Synchronisationspunkte. Zusammengesetzte Aktionen werden aus atomaren oder zusammengesetzten Aktionen gebildet. Die an einer zusammengesetzten Aktion beteiligten Aktionen müssen an ihren Synchronisationspunkten synchronisiert werden, um die vorgeschriebene Ausführungsanordnung zu erfüllen. Zusammengesetzte Aktionen werden daher in atomare Aktionen zerlegt, so daß die partiellen Aktionen nur zu Start- und Endzeitpunkten mit anderen Aktionen synchronisiert werden müssen. Eine Zerlegung einer zusammengesetzten Aktion in atomare Aktionen erfolgt dabei auf einer logischen Abstraktionsebene und ändert nicht die Semantik der ursprünglichen Aktion.

2.2 Zeitliche Relationen zwischen Aktionen

Allen definiert in [4] dreizehn zeitliche Relationen zwischen zwei Zeitintervallen: *before, meets, during, overlaps, starts, finishes* and *equal*, sowie deren inverse Relationen (außer equal). Diese Relationen beschreiben alle möglichen zeitlichen Beziehungen zwischen *zwei* Zeitintervallen. Die linke Seite der Abbildung 1 veranschaulicht diese Relationen; die Zeitintervalle werden durch die Aktionen A und B repräsentiert.

Um allgemeine zeitliche Relationen zwischen Aktionen für die Präsentation zu definieren, werden die 'leeren' Zeitintervalle in Allen's zeitlichen Beziehungen durch eine Zeitgeber-Aktion T (timer) ersetzt. Eine zeitlich versetzte Aktion kann somit als zusammengesetzte Aktion modelliert werden, die aus einer Folge von Zeitgeber-Aktion und versetzter Aktion besteht. Die rechte Seite der Abbildung 1 zeigt die entstandenen allgemeinen zeitlichen Relationen zwischen den Aktionen. Eine Folge von Aktionen wird dabei als *sequentiell* bezeichnet, gleichzeitige Aktionen (mit gleichem Startzeitpunkt) werden *parallel* genannt. Der Endzeitpunkt einer zusammengesetzten Aktion, die aus *parallelen* Aktionen besteht, kann einerseits definiert werden als der Endzeitpunkt der zeitlich zuerst beendeten Aktion (*parallel-erste*) oder andererseits als der Endzeitpunkt der zeitlich zuletzt beendeten Aktion (*parallel-letzte*).

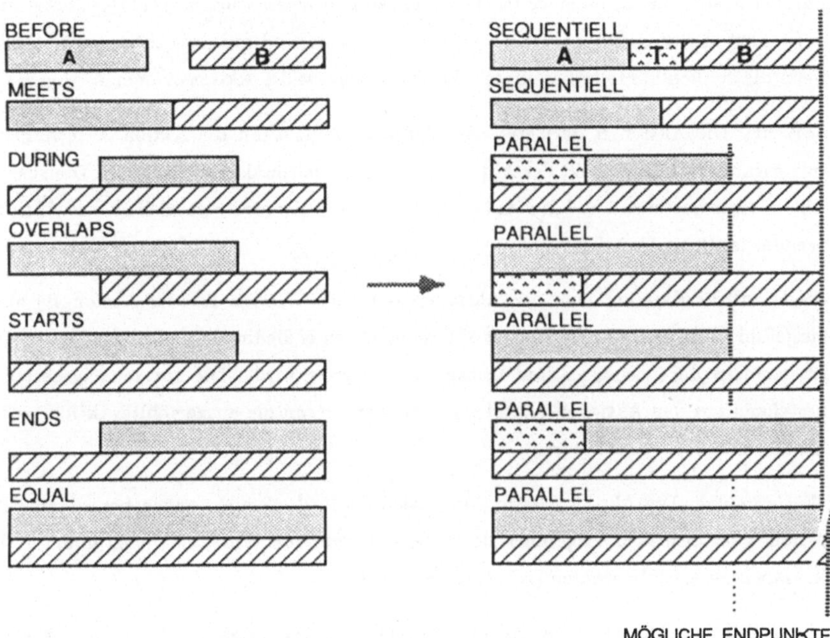

Abbildung 1 : Zeitliche Relationen

2.3 Spezifikationsnotation basierend auf Pfadausdrücken

Campbell und Habermann führten 1974 **Pfadausdrücke** (path expressions) für die Synchronisation von Aktionen auf Prozedurebene ein [5]. Ein Pfadausdruck besteht aus Aktionen und Pfadoperatoren, die einen **Pfad** für erlaubte Aktionen bzw. eine ausführbare Anordnung von Aktionen beschreiben. In einem Pfadausdruck spezifizieren die **Pfadoperatoren** (path operators) die Synchronisation der Aktionen. Pfadoperatoren bestimmen somit die Flexibilität und Mächtigkeit von Pfadausdrücken hinsichtlich ihrer Synchronisationseigenschaften.

Betrachtet man die ursprünglichen 'Aktionen auf Prozedurebene' allgemein als Aktionen (wie im Abschnitt 2.1 beschrieben), dann können Pfadausdrücke auch die für die Beschreibung der Synchronisation von Aktionen im Sinne der Präsentation von Multimedia-Objekten verwendet werden. Da die Synchronisationseigenschaften durch die Pfadoperatoren definiert werden, muß eine geeignete Menge von Pfadoperatoren ausgewählt und deren Semantik spezifiziert werden. Basierend auf den allgemeinen zeitlichen Relationen aus Abschnitt 2.2 werden folgende Pfadoperatoren mit aufsteigender Priorität eingeführt:

$A \wedge B$ *Parallel-Letzte*: Die Aktionen A und B werden zum gleichen Startzeitpunkt gestartet und gleichzeitig ausgeführt. Die zusammengesetzte Aktion (hier bestehend aus den Aktionen A und B) terminiert, wenn alle beteiligten Aktionen (hier Aktionen A und B) beendet sind.

$A \vee B$ *Parallel-Erste*: Die Aktionen A und B werden zum gleichen Startzeitpunkt gestartet und gleichzeitig ausgeführt. Die zusammengesetzte Aktion (hier bestehend aus den Aktionen A und B) terminiert, wenn die zeitlich kürzeste beteiligte Aktion (hier entweder Aktion A oder Aktion B) beendet ist.

$A \,;\, B$ *Sequenz*: Die Aktion B darf erst ausgeführt werden, wenn die Aktion A terminiert ist. Dabei entspricht der Endzeitpunkt der Aktion A dem Startzeitpunkt der Aktion B. Die zusammengesetzte Aktion (hier bestehend aus den Aktionen A und B) terminiert, wenn die letzte Aktion der Sequenz beendet ist (hier Aktion B).

$A \mid B$ *Selektion*: Es darf nur eine der Aktionen (entweder Aktion A oder Aktion B) ausgewählt und ausgeführt werden. Die Selektion ist abhängig von einer Bedingung, die nicht Teil des Pfadausdrucks ist, sondern durch eine externe Instanz ausgewertet wird. Die zusammengesetzte Aktion (hier bestehend aus den Aktionen A und B) terminiert, wenn die ausgewählte Aktion (entweder Aktion A oder Aktion B) beendet ist.

A^{i*} *Wiederholung*: Die Aktion A wird i-mal wiederholt. Ist i nicht angegeben, dann wird die Aktion A null- oder mehrfach ausgeführt; die genaue Anzahl von Wiederholungen muß durch eine externe Instanz bereitgestellt werden (z.B. vom Benutzer).

$n{:}A$ *Gleichzeitigkeit*: Die Aktion A darf bis zu n-mal gleichzeitig ausgeführt werden. Wenn n=1 (default) ist, dann schließen sich die Ausführungen von A gegenseitig aus (mutual exclusive). Wenn $n=\infty$ ist, dann ist keine Ausführung oder jede beliebige nebenläufige Ausführung von A erlaubt.

Die Definition von **Aktionen für die Präsentation** von Multimedia-Objekten ist ein Teil der Spezifikation, wird jedoch nicht in dem Synchronisationsmodell beschrieben. Eine eindeutige Namensgebung für Aktionen ist erforderlich, da eine Aktion nur durch ihren Namen in einem Pfadausdruck identifiziert wird. Gleiche Namen identifizieren dieselbe Aktion. Gleichartige Aktionen, die mehrfach in einem Pfadausdruck auftreten, werden durch verschiedene Namen unterschieden. Eine Präsentationsbeschreibung kann aus mehreren Pfadausdrücken bestehen.

BEISPIEL:

path A ; ((B $\wedge$ C) $\vee$ (D $\wedge$ E)) ; F^* end
path B ; G end

BESCHREIBUNG: Die Aktion A wird gestartet; zum Endzeitpunkt von Aktion A werden genau die vier Aktionen B, C, D und E gestartet. Aktion F wird ausgeführt sobald die Aktionen B und C oder sobald die Aktionen D und E beendet sind. Aktion F kann null- oder mehrfach, jedoch nicht mehrfach gleichzeitig ausgeführt werden. Soll außerdem nach der Aktion B eine Aktion G ausgeführt werden (Aktion G steht in keiner zeitlichen Relation zu den restlichen Aktionen), dann wird ein zweiter Pfadausdruck angegeben (Aktion B wird jedoch nur einmal ausgeführt).

3 Darstellung des Synchronisationsmodells mit Petri-Netzen

Petri-Netze sind eine Modellierungstechnik für die Darstellung von nebenläufigen Systemen und deren Synchronisation. Sie wurden erstmals 1962 von C.A. Petri eingeführt und seitdem unter verschiedenen Aspekten modifiziert und erweitert. Die Basiskonzepte von Petri-Netzen werden zum Beispiel in [6] und [7] beschrieben und hier als bekannt vorausgesetzt. Die Beschreibung von Pfadausdrücken durch Petri-Netze wurde in [8] eingeführt.

Petri-Netze dienen unter anderem einer visuellen Darstellung der zeitlichen Relationen zwischen Aktionen. Dies wird durch eine spezielle Anordnung von Stellen und Transitionen erreicht. Transitionen repräsentieren dabei die Aktionen der Pfadausdrücke. Die Synchronisation von Aktionen wird in den Pfadausdrücken durch die Pfadoperatoren spezifiziert. Die Semantik der Pfadoperatoren (wie in Abschnitt 2.3 beschrieben) wird nachfolgend durch Petri-Netze dargestellt. (Das Petri-Netz für den Pfadoperator "Parallel-Erste" enthält eine positive natürliche Zahl n, die der Anzahl der beteiligten parallelen Aktionen entspricht (hier ist n=2 für die Aktionen A und B).)

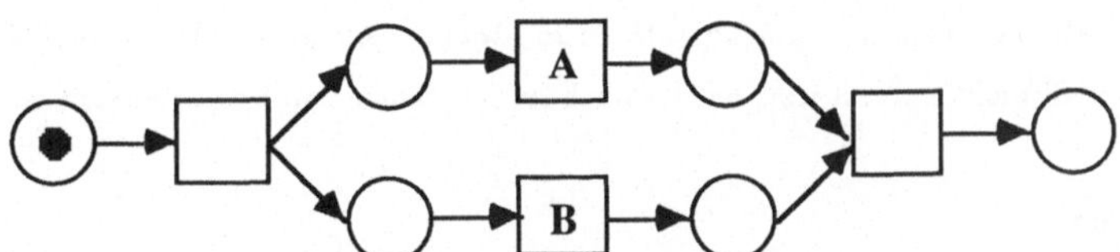

Abbildung 2 : Pfadoperator "Parallel-Letzte ($\wedge$)"

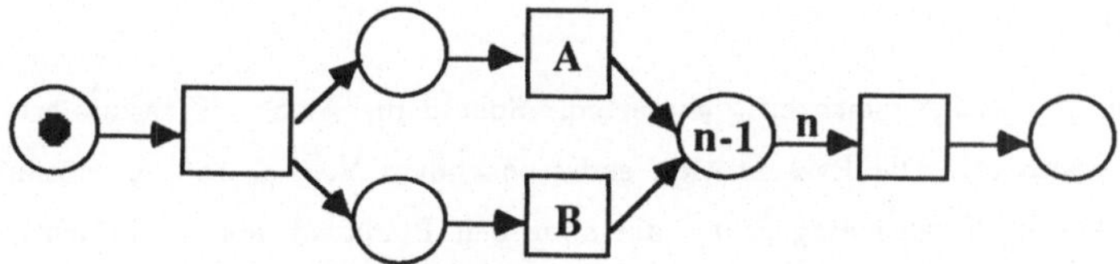

Abbildung 3 : Pfadoperator "Parallel-Erste ($\vee$)"

Abbildung 4 : Pfadoperator "Sequenz (;)"

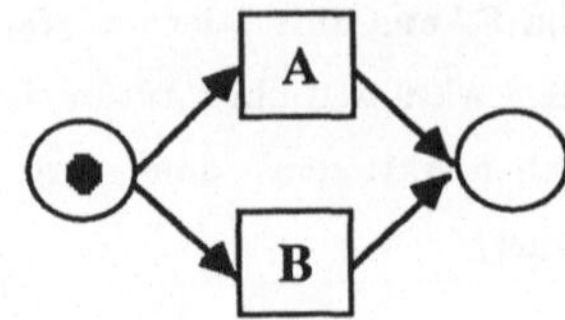

Abbildung 5 : Pfadoperator "Selektion (|)"

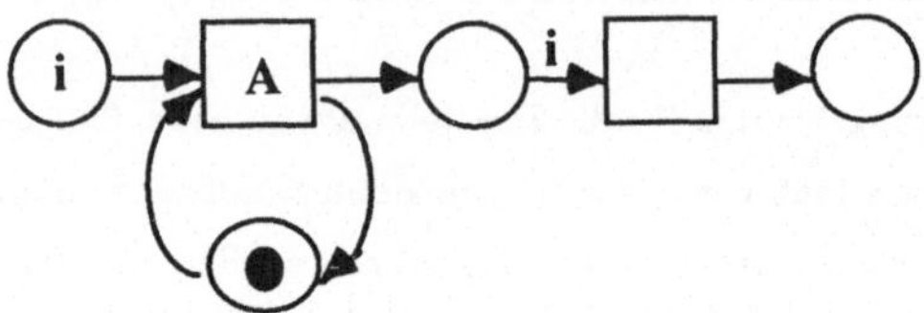

Abbildung 6 : Pfadoperator "Wiederholung (A^{i*})"

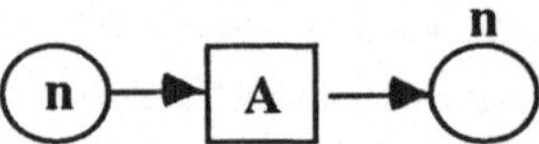

Abbildung 7 : Pfadoperator "Gleichzeitigkeit (n:)"

4 Beispiele von Multimedia-Objekten und deren Modellierung

Die folgenden Präsentationsbeispiele für Multimedia-Objekte wurden aus [9] und [10] ausgewählt. Einen wesentlichen Aspekt der Synchronisationsmodellierung bildet die Identifikation von atomaren Aktionen beziehungsweise die Zerlegung von zusammengesetzten in atomare Aktionen. Dieser Aspekt wird durch das Synchronisationsmodell nicht unterstützt, wird jedoch in den Beispielen beschrieben.

4.1 Videosequenz

Eine Videosequenz (V) soll vorgeführt werden. Bei Bild 1025 sollen ein Text (TX) und eine Audiopräsentation (A) beginnen. Das Warten auf Bild 1025 (T) wird durch eine Echtzeit-Analyse der Videosequenz realisiert.

Die Aktionen V und T werden gleichzeitig gestarted. Sobald die Aktion T terminiert, werden die Aktionen TX und A gestartet. Die Präsentation endet, nachdem V, TX und A terminiert sind. (Eine zusammengesetzte Aktion X wird eingeführt, die zwar den Pfadausdruck vereinfacht, nicht jedoch die Ausführungsreihenfolge beeinflußt.)

path X := (T ; (A $\wedge$ TX)) end

path V $\wedge$ X end

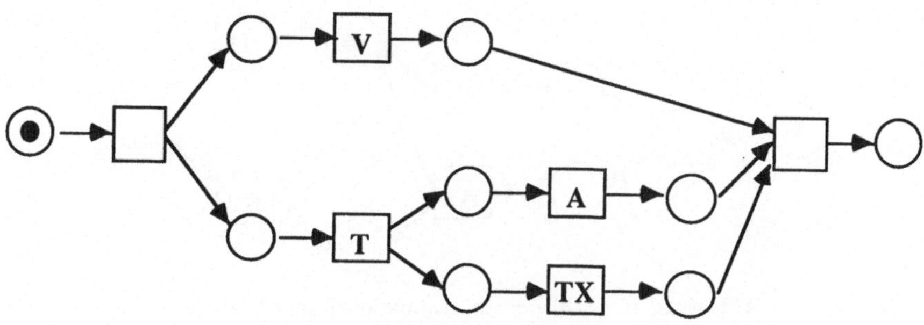

Abbildung 8 : Videosequenz

4.2 Graphische Darstellungen mit gesprochenen Kommentaren

Zeitabhängige graphische Darstellungen (G1, G2) und gesprochene Kommentare (K1, K2) sollen synchronisiert werden. Ein Kommentar und die dazugehörige graphische Darstellung beginnen zum gleichen Zeitpunkt. Sobald der Kommentar und die graphische Darstellung beendet sind, soll der nächste Kommentar und die dazugehörige graphische Darstellung präsentiert werden. Die graphischen Darstellungen können zeitlich länger dauern als die Kommentare und umgekehrt.

path (G1 $\wedge$ K1) ; (G2 $\wedge$ K2) end

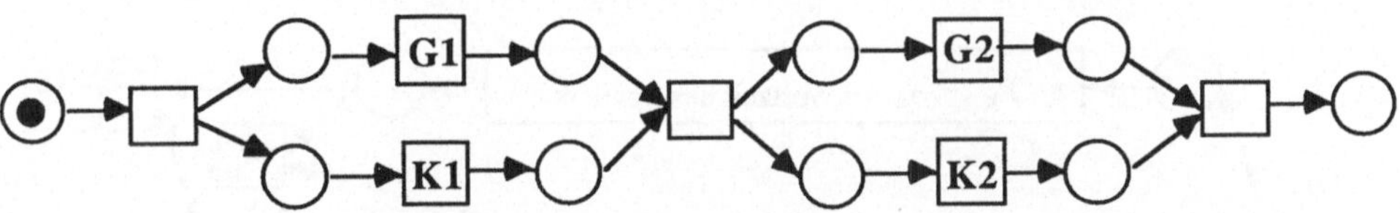

Abbildung 9 : Graphische Darstellungen mit Kommentaren

4.3 Sequenz von Animationen und Bildern

Zwei Animationen (A1, A2) starten zum gleichen Zeitpunkt und werden gleichzeitig präsentiert. Nachdem die erste Animation terminiert (entweder A1 oder A2), wird Bild 1 gezeigt (B1). Nachdem die zweite Animation terminiert, wird Bild 2 gezeigt (B2). (Bemerkung: Die Aktionen A1 und A2 sind in beiden Pfadausdrücken vorhanden, werden jedoch nicht zweimal präsentiert (siehe dazugehöriges Petri-Netz und Abschnitt 2.3).)

path (A1 $\vee$ A2) ; B1 end

path (A1 $\wedge$ A2) ; B2 end

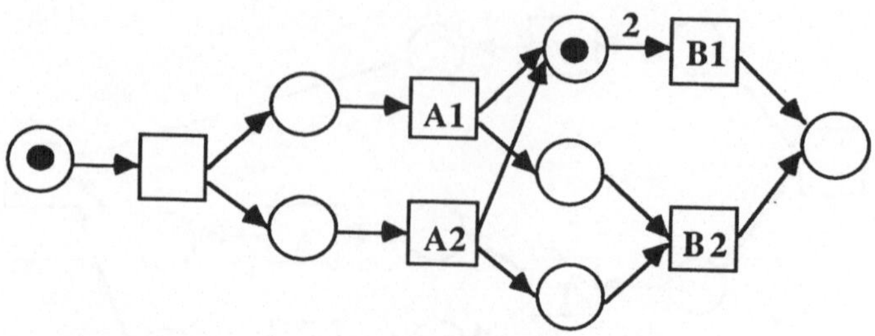

Abbildung 10 : Sequenz von Animationen und Bildern

4.4 Frage/Antwort-Spiel

Einem Spieler werden eine Reihe von Fragen (F_i) gestellt, die er jeweils während einer Minute beantworten soll (das Warten auf die Antwort wird durch die Aktion T_X repräsentiert, die Antwort auf die Frage F_i entspricht der Aktion A_i). Alle 5 Sekunden ertönt eine Glocke (G), die den Ablauf der Zeit wiedergibt (Aktion T5 wartet jeweils 5 Sekunden). Wenn keine Antwort gegeben wurde, soll eine Tonsequenz (E) das Ende der Antwortzeit anzeigen und die nächste Frage wird gestellt.

path X := $((T5 ; G)^{12*} ; E)$ end
path $(F_i ; ((T_X ; A_i) \vee X))^*$ end
path $A_i \mid X$ end

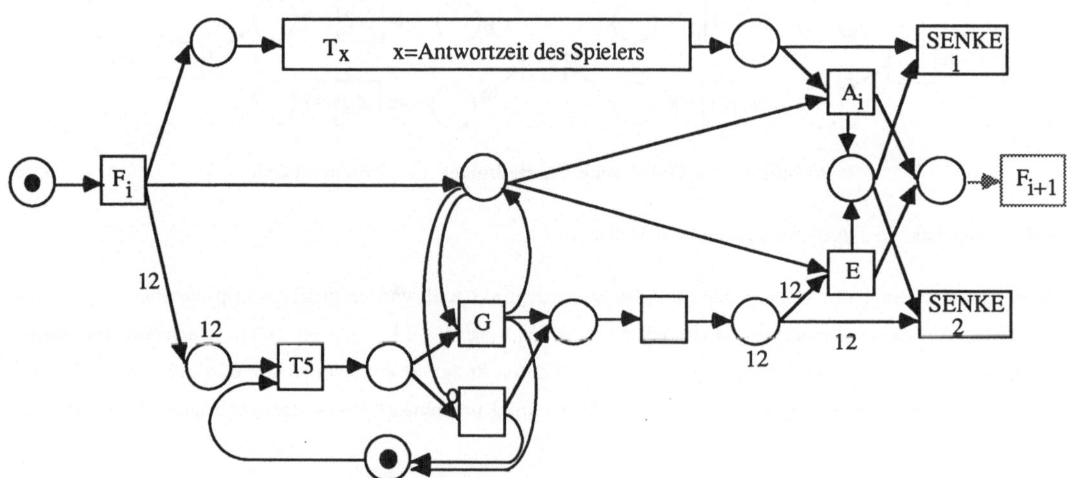

Abbildung 11 : Frage/Antwort-Spiel

5 Realisierungsaspekte

Die vorliegende Modellierung der Synchronisation von Aktionen für die Präsentation von Multimedia-Objekten ist als ein mögliches 'ODP Computational Model' für die Synchronisation zu verstehen. Es dient somit nicht als System- oder Implementationsbescheibung, sondern spezifiziert systemunabhägig ein idealisiertes Synchronisationsverhalten, wie es für den Austausch von Präsentationsabläufen in heterogenen Systemumgebungen notwendig ist.

Nicht beschrieben wurde ein 'ODP Engineering Model', welches die Mechanismen umfaßt, die das 'Computational Model' realisieren. Ein 'Engineering Model' integriert nicht nur synchronisationsspezifische, sondern auch ablaufsteuerungsspezifische Komponenten, um die Präsentation zu steuern. Dabei kann ein Präsentationsskript erstellt werden, das durch Interpretation oder Übersetzung in eine spezielle Konfiguration von Prozessen überführt wird und entsprechende Synchronisationsmechanismen implementiert. Eine mögliche Skriptsprache kann dabei direkt auf Pfadausdrücken basieren oder indirekt deren Semantik integrieren (zum Beispiel Esterel [11]). Ebenfalls sind Fehlererkennungs- und Fehlerbearbeitungsmechanismen erforderlich.

Die Erstellung eines Synchronisationsskripts sollte durch eine geeignete Benutzerschnittstelle unterstützt werden, zum Beispiel eine graphische Schnittstelle [12], eine Petri-Netz-orientierte Schnittstelle [13], oder eine Pfadausdruck-orientierte Schnittstelle.

Die Ausführung des Skripts erfordert eine zeitabhängige Bereitstellung der Präsentationsinformationen, das heißt in verteilten Systemen ist die medienbezogene, (echt)zeitabhängige Übermittlung von Datenstömen und deren Synchronisation notwendig.

Eine andere Anwendung des vorgestellten Modells auf ODA-Dokumente (ODA - Open Document Architecture [14]) wurde in [15] vorgestellt. Die ODA Layout-Struktur wurde durch zusätzliche Attribute so erweitert, daß die zeitlichen Relationen zwischen ODA Objekten hinsichtlich der Präsentation spezifiziert werden können.

6 Zusammenfassung

Das vorgestellte Synchronisationsmodell stellt eine Beschreibungsmethode für die Synchronisation von Aktionen im Rahmen der Präsentation von Multimedia-Objekten zur Verfügung. Synchronisation wird dabei anwendungs- und implementationsunabhängig spezifiziert und dient somit als Grundlage für verschiedenartige Anwendungen. Das Modell dient u.a. als Basis für eine synchronisationsunterstützende Präsentationsumgebung. Pfadausdrücke sind dabei nicht als Benutzerschnittstelle vorgesehen, sondern als Spezifikationsmethode für deren Leistungsumfang.

Danksagungen

Ich danke allen Kollegen im Projekt "Multi-Media-Dokumente im ISDN-B" (BERMMD) für ihre Diskussionsbereitschaft. Ebenfalls danke ich Herrn Prof. Dr. H.-J. Hoffmann für die hilfreichen Kommentare hinsichtlich der Modellierung mit Petri-Netzen.

Literaturverzeichnis

[1] "BERKOM Reference Model II – Application-Oriented Layers", DETECON, Technisches Zentrum Berlin, Voltastr.5, 1000 Berlin 65, Germany (March 1991), Version 3.2.

[2] ISO/IEC JTC1/SC21/WG7 N315, "Working Document - Recommendation X9yy | Basic Reference Model of Open Distributed Processing", *Part II : Descriptive Model* (December 1990).

[3] L. Lamport, "Time, Clocks, and the Ordering of Events in a Distributed System", *Communications of the ACM*, vol. 21, no. 7, pp. 558–565 (July 1978).

[4] J.F. Allen, "Maintaining Knowledge about Temporal Intervals", *Communications of the ACM*, vol. 26, no. 11, pp. 832–843 (November 1983).

[5] R.H. Campbell and A.N. Habermann, "The Specification of Process Synchronization by Path Expressions", *Lecture Notes in Computer Science No. 16, Operating Systems*, ed. G. Goos and J. Hartmanis, pp. 89–102, Springer-Verlag (1974).

[6] J.L. Peterson, "Petri Nets", *ACM Computing Surveys*, vol. 9, no. 3, pp. 223–252 (September 1977).

[7] E.V. Krishnamurthy, *Parallel Processing, Principles and Practice*, Addison-Wesley Publishing Company, Syndey (1989).

[8] P.E. Lauer and R.H. Campbell, "A Description of Path Expressions by Petri Nets", in: *Conference Record of the Second ACM Symposium on Principles of Programming Languages*, pp. 95–105 (1975).

[9] ISO/IEC JTC1/SC18 N2028, "Information on French Proposal on Audiovisual Interactive Applications" (September 1989).

[10] K.-D. Engel, "Zeitliche Synchronisation bei der Präsentation von Multimedia-Informationsobjekten" (June 1990), GMD-FOKUS Internal Report.

[11] G. Berry, P. Couronne, and G. Gonthier, "Synchronous Programming of Reactive Systems: An Introduction to Esterel", INRIA, Institut National de Recherche en Informatique et en Automatique, Technical Report No. 647 (1987).

[12] R. Ogawa, H. Harada, and A. Kaneko, "Scenario-based Hypermedia: A Model and a System", in: *Hypertext: Concepts, Systems and Applications; Proceedings of the First European Conference on Hypertext*, ed. N. Streitz, A. Rizk, and J. Andre, pp. 38–51, Cambridge University Press (November 1990).

[13] P.D. Stotts and R. Furuta, "Hierarchy, Composition, Scripting Languages, and Translators for Structured Hypertext", in: *Hypertext: Concepts, Systems and Applications; Proceedings of the First European Conference on Hypertext*, ed. N. Streitz, A. Rizk, and J. Andre, pp. 180–193, Cambridge University Press (November 1990).

[14] ISO 8613, "Information Processing – Text and Office Systems – Office Document Architecture (ODA) and Interchange Format" (1989).

[15] P. Hoepner, "Synchronizing the Presentation of Multimedia Objects - ODA Extensions", Eurographics multimedia workshop 1991 (April 1991).

Konzeption für eine Sprache zur Beschreibung von Transport- und Darstellungseigenschaften multimedialer Objekte

Gerold Blakowski
Institut für Telematik, Universität Karlsruhe, Zirkel 2, 7500 Karlsruhe, Email: blakowski@ira.uka.de

Zusammenfassung

Viele verteilte, multimediale Anwendungen überschreiten den Bereich eines lokalen Netzes und basieren nicht mehr auf homogenen Arbeitsstationen. Die Entwicklung verteilter Anwendungen in einer solchen heterogenen Umgebung ist sehr komplex. Ziel des MODE-Projektes (Multimedia Objects in a Distributed Environment) ist es, die Entwicklung dieser Anwendungen durch ein geeignetes Objektmodell und durch auf diesem Modell basierende Objekttransport- und Darstellungsdienste zu unterstützen. Aufgabe des Diensterbringers ist es, den Prozeß des Objekttransports und der Objektdarstellung für beliebige benutzerdefinierte Medien zu automatisieren. Für diesen Automatisierungsvorgang verwendet der Diensterbringer Information über die Anwendung und die zur Verfügung stehenden verteilten Ressourcen und Basisdienste. Zur Vermittlung der zur Automatisierung relevanten Anwendungsinformation wird die Anwendungsbeschreibungssprache AMDL-MODE verwendet. MODE wird im Rahmen von NESTOR, einem Kooperationsprojekt zum Erstellen einer Autoren/Lernerumgebung basierend auf vernetzten multimedialen Arbeitsstationen, eingesetzt. Kooperationspartner sind die Universitäten Karlsruhe und Kaiserslautern, das Digital Equipment Corporation CEC Karlsruhe sowie andere Forschungseinrichtungen. In diesem Artikel wird ein Überblick über MODE und die Konzepte von AMDL-MODE gegeben.

1 Einleitung

Die verteilte Verarbeitung multimedialer Information vergrößert nicht nur das Potential existierender Anwendungsbereiche, sondern erschließt auch völlig neue. Viele dieser multimedialen Anwendungen [13] überschreiten den Bereich eines lokalen Netzes, und es kann bei der Entwicklung solcher Anwendungen nicht mehr von der Homogenität der Arbeitsstationen und zugrundeliegenden Netzwerke ausgegangen werden. Beispiele hierfür sind das entfernte Lernen im Rahmen des computerunterstützten Unterrichts [3] und Computerunterstützung der Zusammenarbeit räumlich weit entfernter Personen [19] durch Telekonferenzsysteme.
Die Erstellung verteilter, multimedialer Anwendungen in einer Umgebung heterogener Arbeitsstationen, die heterogen vernetzt sind, ist durch den hohen Ressourcenbedarf bei der Verarbeitung und dem Transport [25] der multimedialen Daten besonders schwierig. Ziel des MODE (Multimedia Objects in a Distributed Environment) Projektes ist, den Bedarf nach Entwurfs- und Laufzeitunterstützung für Anwendungen in solchen Umgebungen zu decken. Die Unterstützung soll dabei so flexibel sein, daß sie bei allen CCITT-Dienstkategorien [7] wirksam wird. Typische verteilte Anwendungen, die von MODE unterstützt werden sollen, sind verteilte Ein- bzw. Mehrbenutzeranwendungen mit multimediabasierten Dialogen und Konferenzen, Broad- oder Multicasting multimedialer Information, Retrieval auf verteilt gespeicherten Daten und Erfassung von multimedialen Daten. Das computerunterstützte Lernen vereinigt zum Beispiel alle diese Anwendungsformen.

Insbesondere soll MODE alle Medien digital handhaben, nicht auf eine feste Medienmenge eingeschränkt sein und eine über die Formatkonvertierung oder Formatdefinition hinausgehende Unterstützung der multimedialen Verarbeitung in verteilten, heterogenen Systemen ermöglichen.

In diesem Sinne erweitert MODE die Funktionalität vieler existierender Ansätze zur Unterstützung der verteilten Verarbeitung multimedialer Information [2, 8, 9, 16, 18], die feste Medientypen unterstützen, sich auf hybride, d.h. gemischte Digital- und Analognetze abstützen, die Unterstützung auf homogenen Arbeitsstationen und/oder Netzwerken anbieten oder Darstellungsaustauschformate definieren. Bei der Erbringung der Unterstützung soll MODE jedoch die verfügbaren Funktionalitäten, wie sie sich aus obigen Ansätzen oder anderen Projekten zur Entwicklung von Basisdiensten [1, 5, 6, 17, 23] für multimediale Anwendungen ergeben, nutzen.

Zentrale Komponenten von MODE sind das MODE-MODEL, ein Objektmodell für verteilte, multimediale Information, das hinsichtlich der Handhabung multimedialer Information in verteilten, heterogenen Umgebungen entworfen wurde und die MODE-SERVICES, verteilte Objekttransport- und Darstellungsdienste, die vom MODE-SERVER zur Verfügung gestellt werden.

Die Unterstützung nicht vordefinierter Medien erfolgt dadurch, daß dem Anwender ermöglicht wird, innerhalb von MODE, unter Nutzung des MODE-MODELs, die für den Transport und die Darstellung relevanten

Eigenschaften seiner Objekte und deren Bedarf an den verteilten Ressourcen zu beschreiben. Er kann sie so der automatisierten Verarbeitung durch den MODE-SERVER zugänglich machen. Es wird also nicht von einer Menge vordefinierter, hierarchisch angeordneter Medienklassen ausgegangen, wie zum Beispiel im MHEG-Vorschlag [11].

Aufgabe des MODE-SERVERs ist es, bei einer konkreten Dienstanforderung einen Ablaufplan zur Erbringung des Dienstes zu erstellen und auszuführen. Dabei sollen die verfügbaren, auf den heterogenen Arbeitsstationen und heterogenen Netzwerken verteilten Ressourcen sowie die existierenden Basisdienste effizient genutzt werden, der Transport und die Darstellung der Information im Rahmen der bei einer Dienstanforderung vorgegebenen Qualitätsmargen, wie zum Beispiel Darstellungsqualitätsmargen und Kostenrahmen an die verfügbaren Ressourcen angepaßt werden und die Synchronisations- und die sich aus der Anwendungskategorie ergebenden Interaktionszeitbedingungen eingehalten werden.

Neben der gesteigerten Effizienz durch die Nutzung der Dienste wird der Anwendungsprogrammierer durch die homogene, heterogenitäts- und netzwerktransparente Schnittstelle von der Berechnung der Abläufe, dem Berücksichtigen der verfügbaren verteilten Ressourcen, der Auswahl von Basisdiensten mit geeigneten Dienstqualitäten und der Zeitplanung und zeitlichen Koordination des Ablaufs entlastet.

Weitere wichtige Ziele sind die einfache, automatisierte Handhabung verschiedener Darstellungsformen und Kompressionsarten, die einfache Integration von Objekten in mehrere synchronisierte Darstellungen, die Wiederverwendbarkeit von und einfache Erweiterbarkeit um Anwendungsmedien und insbesondere auch die leichte Portabilität auf neue Zielumgebungen und die einfache Integration neuer Systemkomponenten.

Um dem MODE-SERVER die notwendige Information zum Erstellen des Ablaufplans zu übermitteln, werden drei Sprachen eingesetzt. AMDL-MODE (**A**pplication **M**edia **D**escription **L**anguage for MODE) zur Beschreibung der zur Diensterbringung relevanten Anwendungsinformation, NDL-MODE (**N**etwork **D**escription **L**anguage for MODE) zur Beschreibung der Netzwerke und ihrer Ressourcen und WDL-MODE (**W**orkstation **D**escription **L**anguage for MODE) zur Beschreibung der Arbeitsstationen und ihrer Ressourcen. Zusätzlich steht noch SDL-MODE (**S**ynchronisation **D**escription **L**anguage for MODE) zur Verfügung. Diese wird zur Beschreibung synchronisierter Darstellungen [10, 22] verwendet.

Im folgenden werden in den Kapiteln 2 und 3 das MODE-MODEL und die MODE-SERVICEs und ihre für AMDL-MODE relevanten Eigenschaften erläutert. In Kapitel 4 werden dann die Anforderungen und Konzepte von AMDL-MODE beschrieben. Den Abschluß bildet Kapitel 5 mit einer Zusammenfassung und einem Ausblick.

2 Objektmodell

Ein objektorientierter Ansatz wurde gewählt, da die Entwicklung verteilter objektorientierter Systeme gezeigt hat, daß das objektorientierte Programmierparadigma, kombiniert mit der Behandlung von Objekten als Verteilungseinheiten, gut zur Entwicklung verteilter Anwendungen geeignet ist [21]. Objektorientierung im MODE-MODEL basiert auf Objekten, Klassen und Klassenvererbung, wie in [24] beschrieben. Das MODE-MODEL ist unabhängig von und anwendbar auf das von der Anwendung konkret verwendete Objektsystem beziehungsweise die verwendete objektorientierte Programmiersprache, solange diese dem obigen Begriff von Objektorientierung entsprechen.

Das MODE-MODEL erlaubt es, Klassen eines oder mehrere der Attribute *Informations-*, *Darstellungs-* und *Transportklasse* zu geben:

Informationsklassen: Objekte von Informationsklassen (*Informationsobjekte*) sind die grundlegenden Informationseinheiten.

Darstellungsklassen: Um Informationsobjekte darzustellen, wird ein Objekt oder eine Folge von Objekten einer geeigneten Darstellungsklasse (*Darstellungsobjekte*) *generiert*. Darstellungsobjekte beinhalten alle zur Darstellung und eventuellen Interaktion (bei *interaktiven Darstellungsobjekten*) mit dem Benutzer notwendigen Daten und Methoden und können unabhängig von Informationsobjekten existieren. Zusätzlich können mehrere Darstellungsobjekte zu einem Darstellungsobjekt *zusammengefaßt* werden, zum Beispiel um Kommunikationskosten zu sparen oder eine implizite Synchronisation durchzuführen. Nach der Darstellung wird das Darstellungsobjekt gelöscht.

Transportklassen: Wenn Informations- oder Darstellungsobjekte transportiert werden sollen, werden sie für die Dauer des Transports in Objekte geeigneter Transportklassen (*Transportobjekte*) *konvertiert*. Transportobjekte beinhalten alle zum Transport notwendigen Daten und Methoden. Die Wiederherstellung des ursprünglichen Objektes ist die *Rekonvertierung*. Die Konvertierung und Rekonvertierung können dabei eine Kompression und Dekompression implizieren.

Abbildung 1 zeigt die möglichen Übergänge, die für die im weiteren besprochenen Darstellungs- und Transportdienste relevant sind. Weitere Übergänge unter der Kontrolle des MODE-SERVERs sind möglich, zum Beispiel die Übergabe von Informationsobjekten an Sammelobjekte im Rahmen von Informationserfassungs-

und Objekterzeugungsdiensten. Natürlich hat die Anwendung selbst auch noch die beliebige Manipulationsmöglichkeit von Objekten außerhalb der Kontrolle des MODE-SERVERs.

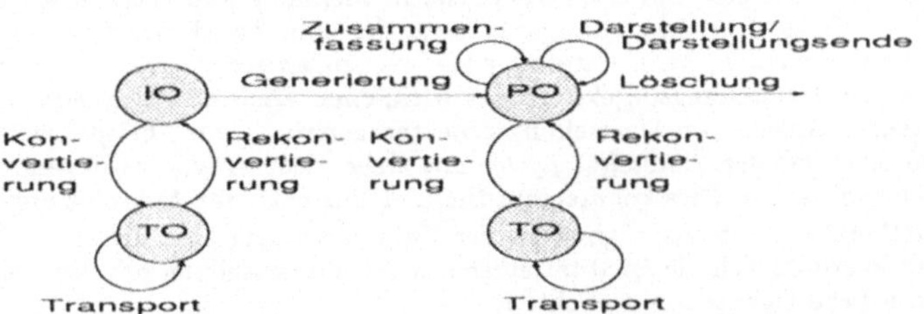

Abbildung 1: Zustandsübergang für Informationseinheit (IO = Informationsobjekt, TO = Transportobjekt, DO = Darstellungsobjekt)

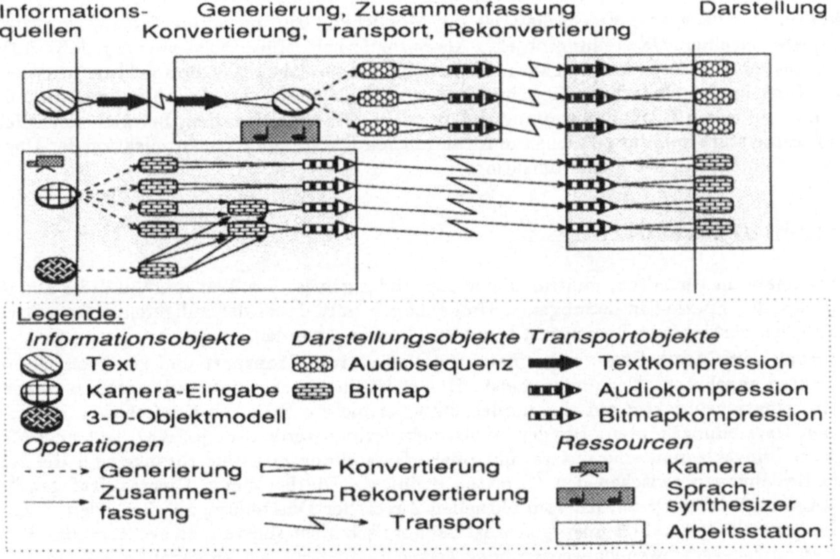

Abbildung 2: Beispiel eines MODE-FLOW-GRAPHs

Die Attribuierung von Klassen ist nicht zwingend abhängig von einer gemeinsamen Vererbungsstruktur. Dies erlaubt es auch, einer Klasse mehrere Attribute zuzuordnen, ohne daß das verwendete Objektsystem auf multipler Vererbung beruht. Durch die Nutzung der mehrfachen Attribuierung einer Klasse oder auch gemeinsamer Unterobjekte ist es nicht zwingend notwendig, daß bei einer Generierung oder Konvertierung ein Kopiervorgang für die gesamte Objektinformation stattfindet.

Der Mechanismus des Transports und der Darstellung von Objekten, welcher einen der zentralen Aspekte in einer verteilten, heterogenen Umgebung darstellt, läßt sich in MODE als sogenannter MODE-FLOW-GRAPH modellieren. Abbildung 2 veranschaulicht den Ablauf einer synchronisierten Darstellung am Beispiel eines MODE-FLOW-GRAPHs, der sich für eine synchronisierte Darstellung ergibt, bei der die entfernte Wiedergabe einer Kameraeingabe zeitweise durch ein 3-D-Objekt überblendet wird und zu der eine von einem Sprachsynthesizer aus einem Text erzeugte Audiosequenz wiedergegeben wird. Die zeitweise Überblendung spiegelt sich dabei in den nur auf einer Auswahl der Bitmaps stattfindenden Zusammenfassungen wieder.

Im folgenden werden einige für AMDL-MODE relevante Aspekte der Generierungen, Konvertierungen und Zusammenfassungen erläutert.

Eine wichtige Eigenschaft multimedialer Systeme ist die n:m Relation zwischen der gespeicherten Information und den möglichen Darstellungsformen. Eine Informationsform kann in mehreren Darstellungsformen auftreten und eine Darstellungsform kann von mehreren Informationsformen benutzt werden. Im MODE-MODEL wird diese wichtige Beziehung dadurch modelliert, daß ein Informationsobjekt Darstellungsobjekte verschiedener Darstellungsklassen erzeugen kann und eine Darstellungsklasse von mehreren Informationsklassen zur Darstellung verwendet werden kann.

Einzelne Darstellungsobjekte haben immer eine endliche Ausdehnung und entsprechen einer Darstellungseinheit. Bezüglich der Menge von Darstellungsobjekten, die zum Zweck der Darstellung von einem Informationsobjekt generiert werden, kann unterschieden werden zwischen einer *Einzeldarstellung*, bei der zur Darstellung genau ein Darstellungsobjekt erzeugt wird, einer *endlichen Darstellungsfolge*, bei der eine von vornherein begrenzte Anzahl von Darstellungsobjekten generiert wird, beispielsweise die Einzelbilder eines Videoclips und einer *offenen Darstellungsfolge* mit einer nicht im voraus bestimmbaren Anzahl von generierten Darstellungsobjekten. Dies könnte zum Beispiel eine entfernte Mikrofoneingabe sein.

Eine wichtige Darstellungsqualitätsausprägung ist der Zusammenhang zwischen den Objekten einer Darstellungsobjektfolge. So erhöht sich die Qualität einer Bewegtbilddarstellung mit der Anzahl der Bilder pro Sekunde, bis eine natürliche Obergrenze erreicht ist.

Bei den Konvertierungen existiert eine n:m Konvertierungsrelation zwischen der Menge der Objekte von Informations- und Darstellungsklassen einerseits und den Transportklassen andererseits. Diese ist darin begründet, daß eine Komprimierungsform von verschiedenen Informations- oder Darstellungsobjekten verwendet werden kann und umgekehrt ein Darstellungs- oder Transportobjekt auf verschiedene Arten komprimiert werden kann.

Eine bereits oben angedeutete Eigenschaft ist die Möglichkeit zur Zusammenfassung von mehreren Darstellungsobjekten zu einem Darstellungsobjekt. Als Besonderheit ist hier zu vermerken, daß ein Darstellungsobjekt durchaus mit mehreren anderen zusammengefaßt werden kann. Ein Beispiel hierfür ist eine Graphik, die mehrere Einzelbilder einer Videosequenz überlagert und bereits vor dem Transport mit den Bitmaps der Einzelbilder zu neuen Bitmaps zusammengefaßt wird. Wie schon aus dem Beispiel ersichtlich, hängt die Zusammenfassung stark von der zeitlichen und räumlichen Darstellungssynchronisation der Darstellungsobjekte ab und ist eine Form der Synchronisation.

3 Laufzeitunterstützung

Eine Hauptaufgabe in verteilten, multimedialen Anwendungen ist der Transport und die Darstellung multimedialer Objekte. Speziell in heterogenen Umgebungen muß dabei der auftretende Ressourcenbedarf im Verhältnis zu den verfügbaren Ressourcen besonders beachtet werden.

Deshalb werden vom MODE-SERVER zur Laufzeit Dienste zum Transport und zur Darstellung von Informationsobjekten angeboten. Der umfassendste Dienst ist die synchronisierte Darstellung einer Menge von verteilten Informationsobjekten auf einem oder, zum Beispiel im Falle von Konferenzen oder Broadcasting, auf mehreren Darstellungsknoten. Bei der Dienstanforderung werden für jedes Objekt eine Menge von erlaubten Darstellungsklassen, eine untere und obere Darstellungsqualitätsschranke und die zeitlichen und räumlichen Beziehungen zwischen den Objekten definiert. Die bei einem Dienstaufruf gewählte Darstellungsqualität hängt dabei wesentlich vom aktuellen Zweck der Darstellung ab. Bei jedem Dienstaufruf von der gleichen Darstellungsqualität auszugehen ist ökonomisch nicht sinnvoll, da eventuell durch übersteigerte Qualität Ressourcen verschwendet werden oder Zeitbedingungen nicht eingehalten werden können. Andererseits kann aber eine hohe Darstellungsqualität wichtiger sein als die Einhaltung von Zeitbedingungen, die sich aus Interaktionszeitbeschränkungen, das heißt der Zeit zwischen Dienstaufruf und erster Darstellung, ergeben, zum Beispiel bei der Übertragung von Röntgenbildern. Es muß also möglich sein, bei einer Dienstanforderung die Mindestdarstellungsqualitätsanforderung und ihr Verhältnis zu Zeitbedingungen flexibel zu definieren.

Die Aufgabe des MODE-SERVERS ist es, einen MODE-FLOW-GRAPH zu konstruieren, dessen Darstellungsobjekte auf den Endknoten die Qualitäts- und Darstellungsklassenanforderungen erfüllen und der den Synchronisations- und Interaktionszeitbedingungen genügt. Der gewählte MODE-FLOW-GRAPH soll dabei die verteilten Ressourcen möglichst effizient nutzen. Es können bezüglich der Effizienz innerhalb des durch die Dienstanforderung und verfügbaren Ressourcen vorgegebenen Rahmens zwei gegenläufige Optimierungsziele unterschieden werden: Einerseits eine Optimierung der Ressourcennutzung in Richtung möglichst gute Darstellungsqualität, andererseits eine Optimierung in Richtung möglichst geringer Bedarf an bestimmten Ressourcen. Ein Einsatzbeispiel für letzteres ist die Minimierung der Benutzung öffentlicher Weitverkehrsnetze, um Kosten zu sparen. Ist kein gültiger MODE-FLOW-GRAPH zu konstruieren, so muß der MODE-SERVER den Auftrag zurückweisen.

Die Auswahl geeigneter Generierungs-, Zusammenfassungs- und Konvertierungsmethoden mit ausreichender resultierender Darstellungsqualität, die Replikation und Mehrfachverwendung von Darstellungsobjekten, die Sequenzierung und Auswahl der Ausführungsorte der Methoden, die Zeitplanung als Grundlage der Synchronisation und die Auswahl geeigneter existierender Basisdienste sind elementare Unteraufgaben bei der Konstruktion des MODE-FLOW-GRAPHS. Die Auswahl der Basisdienste betrifft die eigentliche Wahl unter den Diensten und die geeignete Auswahl an Dienstqualitäten mit eventueller Ressourcenallokation.

Beispielsweise bieten verschiedene Netzwerke unter Umständen verschiedene Transportdienstqualitäten zur Unterstützung der multimedialen Datenübertragung [5, 26] an. Der MODE-SERVER muß dann die für die Übertragung der Objekte geeigneten Transportdienstqualitäten auswählen und bei unzureichender Dienstqualität geeignete Kompensationsverfahren, zum Beispiel Pufferung, anwenden.

Um die notwendigen Entscheidungen im Rahmen der Berechnung eines MODE-FLOW-GRAPHs zu treffen, benutzt der MODE-SERVER Anwendungs-, a priori-Verarbeitungs-, Umgebungs- und Synchronisationsinformation, die auf sprachlichen Beschreibungen, ergänzt durch zur Laufzeit von der Umgebung erfaßter Information beruht.

Zu der **Anwendungsinformation** gehört die klassenbasierte Information über die existierenden Klassen, Generierungen, Kon- und Rekonvertierungen, Zusammenfassungen und deren Kosten, die Darstellungsqualität von generierten Darstellungsobjekten sowie die Verfügbarkeit von Klassen und Methoden auf den heterogenen Arbeitsstationen. Diese Information wird mittels AMDL-MODE beschrieben.

A priori-Verarbeitungsinformation über zukünftige Verarbeitungsanforderungen ergibt sich bei der synchronisierten Darstellung einer Menge von Objekten sowie aus vorhersehbaren Anwendungsabläufen. Zum Beispiel können bei einem Hypermedianetz Voraussagen gemacht werden, welches der nächste darzustellende Knoten ist. Ebenfalls besteht die Möglichkeit, daß der MODE-SERVER-MANAGER Information über Dienstabfolgen sammelt, um vorzeitige Übertragungen vorzunehmen.

Die **Umgebungsinformation** umfaßt ein Modell der Arbeitsstationen und der Netzwerke. Die sprachliche Beschreibung notwendiger Information über die Arbeitsstationen erfolgt mittels der Sprache WDL-MODE. Sie muß Mechanismen zur Verfügung stellen, um die Ressourcen und Basisdienste von Arbeitsstationen zu beschreiben. Es wird dabei unterschieden zwischen statischen Ressourcen, die prinzipiell einem Benutzer exklusiv zugeordnet werden und dynamischen Ressourcen, die von Benutzern geteilt werden. Letztere werden noch unterteilt in Ressourcen mit absoluter Kapazität, zum Beispiel Plattenspeicher, und dynamische Ressourcen, die eine gewisse Leistung pro Zeiteinheit bringen, zum Beispiel CPU oder I/O-Interfaces. Letztere müssen dann noch daraufhin betrachtet werden, ob sie realzeitfähig verwaltet werden.

Das Modell der Netzwerke [15] berücksichtigt unter anderem den zu erwartenden Durchsatz auf Netzwerkebene, Mehrbenutzerkommunikationsfähigkeiten, Transportdienstqualitäten (zum Beispiel Realzeitfähigkeit oder Isochronität) und aktuelle Auslastung sowie Information über die Topologie und Internetzwerkcharakteristiken. Hier wird zur Beschreibung die Sprache NDL-MODE verwendet. Sie muß ebenfalls die Möglichkeit zur abstrakten Beschreibung von Ressourcen besitzen.

Die **Synchronisationsinformation** wird mittels SDL-MODE vermittelt. Dies ist eine separate Sprache, um die Synchronisation der Darstellung einzelner Informationsobjekte zu einer gemeinsamen Präsentation zu ermöglichen. Sie basiert auf dem MODE-MODEL und berücksichtigt die besonderen Eigenschaften der verteilten Umgebung. Bestandteile von SDL-MODE sind die Beschreibung der involvierten Objekte, welche Darstellungsformen genutzt werden können, welches die Standardqualitätsschranken sind und wie die zeitlichen und räumlichen Abhängigkeiten sind. Die Struktur ist hierarchisch, so daß synchronisierte Beschreibungen selbst wieder Elemente der Synchronisation sein können. Im Rahmen der Beschreibung können Verzögerungsrestriktionen und zugehörende Ausnahmebehandlungen beschrieben werden. Dem Benutzer stellt sich SDL-MODE als graphischer, sichtenorientierter Editor dar. Die Synchronisation erfolgt über Synchronisationspunkte, die auf Referenzpunkte innerhalb eines Darstellungsobjektes oder einer Darstellungsobjektfolge abgebildet werden. Die Ausführung erfolgt über einen lokalen Synchronisator, der für die Synchronisation auf dem Darstellungsknoten verantwortlich ist und einen globalen Synchronisator, der für die Synchronisation auf den anderen Knoten und während des Transports zuständig ist und dessen Verantwortung es ist, die Anlieferung der Darstellungsobjekte an den Darstellungsknoten so zu koordinieren, daß der lokale Synchronisator seine Aufgabe erfüllen kann.

Insgesamt ergibt sich folgender Zusammenhang der Sprachen: Mit AMDL-MODE werden die zur Ausführung einer Methode benötigten Ressourcen beschrieben, mit WDL-MODE und NDL-MODE wird Information über die in der verteilten Umgebung existierenden Ressourcen und Basisdienste gegeben, die zur Laufzeit um vom MODE-SERVER erfaßte Information ergänzt wird. Bei einem Dienstaufruf muß dann der MODE-SERVER unter Berücksichtigung der Optimierungsziele einen Abgleich zwischen den Zeitbedingungen und Darstellungsqualitätsanforderungen mit ihrem resultierenden Ressourcenbedarf und -verbrauch und den aktuell zur Verfügung stehenden Ressourcen vornehmen.

MODE basiert also auf einer Trennung der eigentlichen Basisinformation und der Information über die Darstellungsformen, Darstellungsqualitäten und Synchronisationsbeziehungen. Eine Transformation in die kombinierte Darstellungsform- und Synchronisationsinformation, wie zum Beispiel im MHEG-Vorschlag [11] oder bei Multimediadokumentarchitekturen [9] ist möglich, wobei jedoch Information über variante Darstellungsformen und Darstellungsqualitätsmargen verlorengeht und eine Restriktion auf die dort vorgegebenen Medien erfolgen muß. Eine auf diesen Einschränkungen basierende Objektmodellierung kann auch in MODE erfolgen, wobei jedoch dann ein Teil der Funktionalität des MODE-SERVERs nicht genutzt wird.

Neben der Auswahl eines geeigneten MODE-FLOW-GRAPHs hat der MODE-SERVER die Aufgabe, den MODE-

Flow-Graph auszuführen. Hierzu muß der Mode-Server, entsprechend der zeitlichen Koordination des Mode-Flow-Graphs die Generierungen, Konvertierungen, Rekonvertierungen und Zusammenfassungen, den eigentlichen Transport und die Darstellung initiieren und dabei die Parametrisierung der verwendeten Basisdienste vornehmen, zum Beispiel bei realzeitfähigen Komponenten die Allokation der Ressourcen.

Die Initiierungen erfolgen dadurch, daß der lokale Mode-Server-Manager mittels einer zur Anwendung gebundenen Dienstanbindungskomponente (Mode-Server-Binding) zum geeigneten Zeitpunkt entsprechende Methoden auf den Anwendungsobjekten anstößt. Die eigentliche Ausführung der Methoden findet jeweils in den Anwendungsinstanzen statt. Ebenso findet der initiierte Transport von Transportobjekten mit Hilfe der Dienstanbindung immer direkt zwischen Anwendungsinstanzen statt.

Das Mode-Server-Binding bildet die operationale Schnittstelle zur Anwendung. Es ist der Teil vom Mode-Server, der abhängig von dem von der Anwendung verwendeten Objektsystem ist. Der Mode-Server-Manager ist vom von der Anwendung verwendeten Objektsystem unabhängig und kommuniziert mit dem Mode-Server-Binding über eine Interprozeßkommunikationsschnittstelle. Die Einbettung in die Umgebung veranschaulicht Abbildung 3.

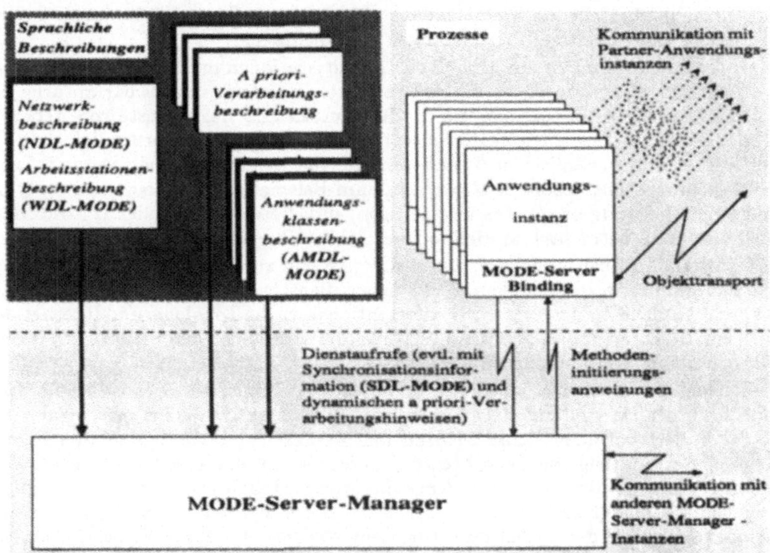

Abbildung 3: Einbindung einer lokalen Mode-Server - Instanz

4 AMDL-Mode

Um einen Mode-Flow-Graph zu planen, benötigt der Mode-Server Anwendungsinformation über die existierenden Klassen, ihre Verfügbarkeit auf den Arbeitsstationen, den Ressourcenbedarf der Informations-, Darstellungs- und Transportobjekte, die Migrationseigenschaften der Objekte, die existierenden Generierungs-, Transport- und Zusammenfassungsmethoden mit Parameterbeschreibung, Ressourcenbedarf und -verbrauch und resultierenden Darstellungsqualitäten.

In den folgenden Unterkapiteln wird auf diejenigen Konzepte von AMDL-Mode eingegangen, die hierfür notwendig sind.

4.1 Typisierung der Klassen

Innerhalb von AMDL-Mode können Anwendungsklassen eines oder mehrere der Attribute Informations-, Darstellungs- und Transportklassen zugeordnet werden. Dies kann durch eine rein textuelle Zuordnung der Attribute zu den entsprechenden Klassennamen erfolgen.

Bei der Definition einer Informationsklasse können Oberklassen angegeben werden. Methodendefinitionen der Oberklassen werden dann für die definierte Klasse übernommen, sofern sie nicht überschrieben werden. Es wird also keine strikte Vererbung erzwungen. Bei Verwendung der multiplen Vererbung ist es notwendig, eindeutige Methodennamen oder Auflösungshinweise zu geben, da die Vorgabe eines Suchalgorithmus die Sprachunabhängigkeit einschränken würde.

4.2 Verfügbarkeit von Klassen und Methoden auf Rechnern

MODE fordert nicht, daß jeder Knoten jede Klasse und Methode zur Verfügung stellt. So können zum Beispiel einzelne Darstellungsformate oder Komprimierungsverfahren nur auf einer Teilmenge der Rechner verfügbar sein. AMDL-MODE muß deshalb die Möglichkeit beinhalten zu beschreiben, auf welchen Rechnern welche Klassen mit welchen Methoden der Klasse verfügbar sind.

4.3 Ressourcenbedarf der Informations-, Darstellungs- und Transportobjekte

Der Ressourcenbedarf von Objekten muß in die Planung eines MODE-FLOW-GRAPHs einbezogen werden. Hierzu müssen sowohl statische Ressourcenanforderungen für alle Objekte einer Klasse sowie Methoden, über die die speziellen Eigenschaften eines Objektes abgefragt werden können, wie zum Beispiel dessen Größe, beschrieben werden können. Die Beschreibung der Ressourcen bezieht sich auf die Ressourcendefinitionen, die in den Sprachen WDL-MODE und NDL-MODE gegeben worden sind. Generell können hier Mechanismen verwendet werden, wie sie für die Konfigurationsverwaltung [20] in verteilten Systemen entwickelt worden sind.

4.4 Migrationseigenschaften von Informationsobjekten

Ein Teil der Migrationsbeschränkungen ergibt sich bereits indirekt über den Ressourcenbedarf eines Objektes. Es sollte jedoch für den Anwender möglich sein, Klassen oder einzelne Objekte für eine Migration zu sperren. Dies kann notwendig sein, um ein Objekt fest an eine Ressourceninstanz zu binden, zum Beispiel an die Kamera eines bestimmten Rechners, um internen Abarbeitungsbedingungen der Anwendung, wie zum Beispiel die Fixierung eines Objektes an den Knoten seiner Bearbeitung zu genügen oder um Sicherheitsaspekte zu beachten.

Hierzu sollen Sprachkonstrukte zur Verfügung gestellt werden, die es gestatten, daß einer Klasse eine feste Migrationsbedingung zugeordnet werden kann oder eine Methode angegeben werden kann, über die der MODE-SERVER die aktuelle Migrationsbedingung abfragen kann.

4.5 Beschreibung der Generierungsmethoden

Die Beschreibung der Generierungsmethoden ist eine der wichtigsten Aufgaben von AMDL-MODE. Zuerst muß bei einer Generierungsmethode die anbietende Informationsklasse und die resultierende Darstellungsklasse beschrieben werden. Anschließend soll eine derartige Beschreibung der Parameter einer Generierungsmethode ermöglicht werden, daß der MODE-SERVER eine Zuordnung der Parameterauswahl zur erzeugten Darstellungsqualität des Darstellungsobjektes, dem Ressourcenbedarf und -verbrauch für die Ausführung der Methode sowie dem zu erwartenden Ressourcenbedarf für das erzeugte Objekt vornehmen kann.

Diese Angaben ermöglichen dem MODE-SERVER, einen Abgleich zwischen den zur Verfügung stehenden Ressourcen und der vom Benutzer bei einem Dienstaufruf geforderten Darstellungsqualität unter Berücksichtigung der vorgegebenen Optimierungsziele vorzunehmen.

Zur Beschreibung der Methodenparameter wird folgender grundlegender und erweiterbarer Beschreibungsmechanismus vorgeschlagen. Die Parameter $par_1, ..., par_n$ einer Generierungsmethode werden mit ihren Typen und Wertebereichen beschrieben. Es wird also von einem typisierten Ansatz ausgegangen, der im Bedarfsfall auf typfreie Methodenparameter abgebildet wird. Anschließend können folgende Funktionen definiert werden:

- Lineare Funktionen, die in Abhängigkeit des Wertes eines Eingabeparameters par_i die zu erwartende Darstellungsqualität $DQ(par_i)$ angeben. Dies sollte sowohl über eine funktionelle Definition als auch über die Angabe von Stützstellen ermöglicht werden. Diese Funktionen erlauben eine abstraktere, flexiblere und homogenere Darstellungsqualitätsbeschreibung als die einfache Identifikation der Darstellungsqualität mit dem Parameter. Zum Beispiel ist es möglich, alle Darstellungsqualitäten auf Werte zwischen 1 und 10 abzubilden und dann bei einem Dienstaufruf zu verlangen, daß alle Darstellungsobjekte zumindest der Darstellungsqualität 5 entsprechen.
- Lineare Funktionen, die in Abhängigkeit einer angegebenen Darstellungsqualität den zu wählenden Wert des Parameters par_i angeben.
- Funktionen, die in Abhängigkeit aller Parameter eine Darstellungsqualität angeben.
- Funktionen, die den Wert eines Parameters aus anderen Parametern berechnen.
- Funktionen, die Angaben über den (eventuell geschätzten) Ressourcenbedarf in Abhängigkeit von den Eingabeparametern machen. Der Ressourcenbedarf sollte in Einheiten angegeben werden, welche in Bezug stehen zu einer Ressourcendefinition, die im Rahmen von WDL-MODE oder NDL-MODE gegeben wurde.

Ein einfaches Beispiel für eine Methodendefinition:

Informationsklasse: Videosequenz

Darstellungsklasse: Bitmap

Generierungsmethode:
Videosequenz.generiere_Einzelbild(Farbtiefe,X_Auflösung,Y_Auflösung)

Wertebereiche:

Farbtiefe	Integer: 2-24
X_Auflösung	Integer: 10-640
Y_Auflösung	Integer: 8-512

Darstellungsqualitäten DQ:
DQ(Farbtiefe) = Farbtiefe / 2
DQ(X_Auflösung) = { 10 → 1, 640 → 10 }

Parameterverhältnisse:
Y_Auflösung = 4/5 * X_Auflösung

Ressourcenbedarf für Generierung:
CPU_Bedarf= Farbtiefe * X_Auflösung * Y_Auflösung * empirischer_CPU_Bedarfswert

Ressourcenbedarf für generiertes Objekt:
Speicherbedarf= Farbtiefe * X_Auflösung * Y_Auflösung + konstanter_Verwaltungsaufwand

Eine Dienstanforderung kann nun die Darstellung eines Einzelbildes als Bitmap mit DQ(Farbtiefe) > 8 und DQ(X_Auflösung) > 2 sein. Hier würde der Schwerpunkt auf eine gute Farbdarstellung gelegt sein. Umgekehrt könnte durch einen Dienstaufruf mit DQ(Farbtiefe) = 1 und DQ(X_Auflösung) > 8 ein Schwarz-Weiß-Bild mit hoher Auflösung angefordert werden.

Soll die Generierungsparameterisierung sich nur auf die Angabe der gewünschten Qualität beschränken, so ist dies durch Verwendung eines einzelnen Parameters mit der funktionellen Zuordnung DQ(Parameter) = Parameter möglich.

AMDL-MODE sollte des weiteren ermöglichen zu definieren, daß der Aufwand für eine spezifische Generierung beim generierenden Objekt nachgefragt wird. Anforderungen an statische Ressourcen, wie zum Beispiel die Existenz eines Spracherzeugers auf einem Knoten, lassen sich über parameterlose Ressourcenfunktionen beschreiben.

Betrachtet man Generierungsfolgen, so ist der Zusammenhang der Generierung einzelner Darstellungsobjekte zu berücksichtigen. Deshalb muß eine Generierungsmethodendefinition um Sprachkonstrukte zur Beschreibung der Folgen mit ihrem zeitlichen Zusammenhang und der resultierenden Darstellungsqualität erweitert werden:

Wertebereich:

Zeitparameter	Integer:	6..30

Zeitzusammenhang (Generierungen pro Zeiteinheit):
Zeitparameter / Sekunde

Darstellungsqualität:
DQ(Zeitparameter) = Zeitparameter/ 2

Die Festlegung, ob einzelne Darstellungseinheiten fehlen dürfen oder Beschränkungen bezüglich Jitter vorliegen, wird bei der Dienstanforderung gemacht und hängt vom konkreten Zweck der Dienstanforderung ab. Soll die von einem Animationsprogramm erzeugte Bildsequenz auf einem entfernten Knoten auch gespeichert werden, sollte kein Bild verlorengehen. Zur reinen Darstellung ist der Verlust eines Bildes im allgemeinen tragbar.

4.6 Beschreibung der Konvertierungsmethoden

Zur Beschreibung der Konvertierungsmethoden kommen ähnliche Mechanismen wie bei den Generierungsmethoden zum Einsatz, wobei jedoch der Darstellungsqualitätsaspekt wegfällt. Der zu beschreibende Komprimierungseffekt wird durch den Mechanismus der Ressourcenbedarfsbeschreibung abgedeckt. Gleiches gilt für den zu erwartenden Aufwand für die Komprimierung und den eventuellen Bedarf an Komprimierungshardware. Das folgende Beispiel zeigt auch, wie Objektmethodenaufrufe in die Beschreibung integriert werden können.

Darstellungsklasse:	Text
Transportklasse:	Komprimierter_Text

Konvertierungsmethode:
Text.zu_Komprimierter_Text()

Ressourcenbedarf für Kon- und Rekonvertierung:
CPU_Bedarf_Konvertierung = Text.Größe() * empirischer_CPU_Bedarfswert
CPU_Bedarf_Rekonvertierung = Text.Größe() * empirischer_CPU_Bedarfswert

Ressourcenbedarf für Transportobjekt:

Speicherbedarf = Text.Größe() * 0.4

Aufgrund dieser Beschreibungen kann der MODE-SERVER entscheiden, ob und wenn ja welche Komprimierungsverfahren bei einem konkreten MODE-FLOW-GRAPH verwendet werden. Es gilt, ein günstiges Verhältnis zwischen dem hohen Ressourcenbedarf und Zeitverbrauch für die Komprimierung und den Vorteilen durch den dadurch erzielten Komprimierungseffekt zu ermitteln, wie zum Beispiel Verringerung des Ressourcenbedarfs an Bandbreite. Eine wichtige Rolle bei dieser Abwägung spielt das Optimierungsziel.

4.7 Beschreibung der Zusammenfassungsmethoden

Darstellungsobjekte haben als Parameter die Objekte, mit denen sie sich zusammenfassen können, sowie die notwendige Synchronisationsinformation zur Zusammenfassung. Die Wertebereiche der Objektparameter sind die Klassen, aus denen das Parameterobjekt entstammen muß. Die Wertebereiche der Synchronisationsinformation definieren das räumliche oder zeitliche Verhältnis von Objekten.

Der Bedarf an Ressourcen für die aktuelle Zusammenfassung und der Ressourcenbedarf des zusammengeführten Objektes lassen sich wiederum mit den Ressourcenbeschreibungsmechanismen erfassen:

```
Darstellungsklasse:   Bitmap

Zusammenfassungsmethode:
    Blende_Text_ein(Text,X_Position,Y_Position,Größe)

Wertebereiche:
    Textobjekt        Klasse Text
    X_Position        Integer: 1..1000
    Y_Position        Integer: 1..1000
    Größe             Integer: 1..50

Ressourcenbedarf für Zusammenfassung:
    CPU_Bedarf_Zusammenfassung = Bitmap.Größe() * empirischer_CPU_Bedarfswert

Ressourcenbedarf für zusammengefaßtes Objekt:
    Speicherbedarf= Bitmap.Größe()
```

Die aktuelle Belegung der Parameter X_Position, Y_Position und Größe entnimmt der MODE-SERVER der Synchronisationsinformation.

5 Zusammenfassung und Ausblick

Durch die Benutzung von AMDL-MODE kann der MODE-Anwender die Objekte seiner Anwendung bezüglich ihrer Transport- und Darstellungseigenschaften und dem damit zusammenhängenden Ressourcenverbrauch beschreiben.

Basierend auf einer solchen Objektbeschreibung kann der MODE-SERVER Transport- und Darstellungsdienste für die Anwendungsobjekte anbieten. Zusätzlich zur Anwendungsinformation verwendet der MODE-SERVER a-priori Verarbeitungsinformation und Umgebungsinformation über die Arbeitsstationen und die Netzwerke. Aufgrund dieser Information und der dazu entwickelten Auswertungsalgorithmen kann der MODE-SERVER Transport- und Darstellungsabläufe festlegen und initiieren, die den Transport und die Darstellung der Information im Rahmen der bei einer Dienstanforderung vorgegebenen Dienstqualitätsmargen an verfügbare Ressourcen anpassen. Hierdurch kann, je nach Optimierungsziel, eine Darstellungsqualitätsverbesserung bzw. eine Kostensenkung erreicht werden.

Neben der gesteigerten Effizienz durch die Nutzung der Dienste, wird die Entwicklung der Anwendungen in heterogenen Umgebungen wesentlich vereinfacht durch die homogene, heterogenitäts- und netztransparente Schnittstelle, wie sie von den MODE-SERVICEs angeboten wird, die den Anwendungsprogrammierer von der Berechnung der Abläufe, dem Berücksichtigen der verfügbaren verteilten Ressourcen und der Zeitplanung und zeitlichen Koordination des Ablaufs entlastet. Das MODE-MODEL Konzept der Darstellungsobjekte erleichtert zusätzlich die Handhabung von verschiedenen Darstellungsformen eines Informationsobjektes und die Integration der Darstellung eines Informationsobjektes in eine synchronisierte Darstellung. Das Transportobjektkonzept verbirgt Kompressions- und Transporteigenschaften von Objekten. Insgesamt erhöht das Klassenattributskonzept des MODE-MODELs die Wiederverwendbarkeit von Darstellungs- und Komprimierungsverfahren und erlaubt die einfache Erweiterung um neue Informations-, Transport- und Darstellungsformen. Wichtig ist ebenfalls die hohe Portabilität der Anwendungen auf neue Umgebungen beziehungsweise die einfache Erweiterung der Umgebung durch die Mechanismen zur Beschreibung des Ressourcenbedarfs und der Verfügbarkeit von Klassen auf den Knoten.

Ein Prototyp des MODE-Systems befindet sich in der Entwicklung. Er wird in NESTOR (Networked Systems for TutORring) [4, 12, 14], eine Autoren/Lernerumgebung, basierend auf vernetzten, multimedialen

Arbeitsstationen, integriert werden. NESTOR ist ein Kooperationsprojekt der Universitäten Karlsruhe und Kaiserslautern, dem Digital Equipment Corporation CEC Karlsruhe sowie anderen Forschungseinrichtungen. Zur Entwicklung des MODE-SERVER-MANAGERs wird C++ verwendet. NESTOR als Anwendung und das zugehörige MODE-SERVER-BINDING sind Smalltalk-80 basiert. In [14] kann auch zusätzliche Information über hier weniger ausführlich diskutierte Aspekte von MODE gefunden werden.

6 Danksagung

An dieser Stelle danke ich Herrn Prof. Dr. Max Mühlhäuser für zahlreiche gewinnbringende Diskussionen und Kommentare bei der Entstehung der vorliegenden Arbeit.

Literatur

[1] D. Anderson, R. Herrtwich, C. Schaefer; *SRP: A Resource Reservation Protocol for Guaranteed-Performance Communication in the Internet*; TR-90-006, International Computer Science Institute, Berkeley, USA, Feb. 1990

[2] B. Arons, C. Binding, K. Lantz, C. Schmandt; *The VOX Audio Server*; 2nd IEEE COMSOC International Multimedia Communications Workshop, Quebec, Apr. 1989

[3] G. Blakowski, T. Rüdebusch, J. Schaper; *NESTOR - die Architektur einer fortgeschrittenen Autoren-/Lernerumgebung*; Workshop "Informatik in der Aus- und Weiterbildung", Dresden, Feb. 91

[4] G. Blakowski, K. Coyle, J. Dirnberger, M. Dürr, M. Mühlhäuser, B. Neidecker-Lutz, M. Richartz, T. Rüdebusch, J. Schaper, F. Spanachi, P. Tallet, I. Varsek; *NESTOR: Requirements and Architecture*; Interner Bericht Nr. 13/89, Universität Karlsruhe, Aug. 1989

[5] B. Butscher; *A Flexible Transport Service in the BERKOM Broadband Environment*; Protocols for High-Speed Networks, IFIP Workshop, Zürich, Mai 1989

[6] D.Ferrari, D. Verma; *A Schema for Real-Time Channel Establishment in Wide-Area Networks*; IEEE Journal on Selected Areas in Communication, Vol. 8, Nr. 3, S. 368-379, Apr. 1990

[7] D. Hehmann, M. Salmony, H. Stüttgen; *High-Speed Transport for Multi-Media Applications*; Protocols for High-Speed Networks, IFIP Workshop, Zürich, Mai 1989

[8] M. Hodges, R. Sasnett, M. Ackermann; *A Construction Set for Multimedia Applications*; IEEE Software, Jan. 1989

[9] R. Hunter, P. Kaijser, F. Nielsen; *ODA: A Document Architecture for Open Systems*; Computer Communications, Vol. 12, Nr. 2, S. 69-79, Apr. 1989

[10] T. Little, A. Ghafoor; *Synchronisation and Storage Models for Multimedia Objekts*; IEEE Journal on Selected Areas in Communication, Vol. 8, Nr. 3, S. 413-426, Apr. 1990

[11] *ISO/IEC JTC1/SC2/WG12 MHEG working document "S", Version 3*; Nov. 1990

[12] M. Mühlhäuser;*Hypermedia-Techniken für 'Instructional Tool Environments' - Anforderungen und Ansätze*; Proceedings GI 91, Darmstadt, Okt. 1991, in: GI - 21. Jahrestagung, Informatik-Fachberichte, Springer-Verlag, Berlin, etc., 1991

[13] N.Naffah; *Multimedia Applications*; Computer Communications, Vol. 13, Nr. 4, S. 243-249, Mai 1990

[14] *NESTOR - A New Approach to Course Authoring and Learning*; Erscheint als interner Bericht, Universität Karlsruhe, 1991

[15] G. Parulkar, J. Turner; *Towards a Framework for High-Speed Communication in a Heterogeneous Environment*; IEEE Network Magazine, Vol. 4, Nr. 2, S. 19-27, März 1990

[16] A. Poggio, J. Garcia Luna Aceves, E. Craighill, D. Moran, L. Aguilar, D. Worthington, J. Hight; *CCWS: A Computer-Based, Multimedia Information System*; IEEE Computer, S. 93-103, Oct. 1985

[17] R. Popescu-Zeletin; *From Broadband ISDN to Multimedia Computer Networks*; Computer Networks and ISDN-Systems, Vol. 18, Nr. 1, S. 47-54, Nov. 1989

[18] R. Pretty, K. Zoehner; *MMA - A System for Multimedia Communication*; 2nd IEEE COMSOC International Multimedia Communcations Workshop, Quebec, Apr. 1989

[19] T. Rüdebusch, M. Mühlhäuser; *Ein Unterstützungssystem für Gruppenarbeit in verteilten Systemen*; in W. Effelsberg, H. W. Meuer, G. Müller (Hrsg.), Kommunikation in verteilten Systemen, Proceedings GI/ITG-Fachtagung, Informatik-Fachberichte, Springer-Verlag, Berlin, etc., S. 464-478, 1991

[20] A. Schill, G. Blakowski; *Configuration Management for Distributed Object-Oriented Applications*; Proceedings International Telecommunication Symposium, Rio de Janeiro, Brasilien, Sep. 1990

[21] A. Schill; *Objects and Distribution: Advantages, Problems and Solutions*; Int. TOOLS Conf., Paris, Nov. 1989

[22] R. Steinmetz; *Synchronisation Properties in Multimedia Systems*; IEEE Journal on Selected Areas in Communication, Vol. 8, Nr. 3, S. 401-412, Apr. 1990

[23] D. Verma, H. Zhang, D. Ferrari; *Guaranteeing Delay Jitter Bounds a Packet-Switching Networks*; First International Workshop on Network and Operating System Support for Digital Audio and Video, TR-90-062, International Computer Science Institute, Berkeley, California, USA, Nov. 1990

[24] P. Wegner; *The Object-Oriented Classification Paradigm*, in B. Shriver, P. Wegner (Hrsg.), Research Directions in Object-Oriented Programming, Computer System Series, MIT Press, Cambrige, etc., S. 479-560, 1987

[25] D. Wright, M. To; *Telecommunication Applications of the 1990s and their Transport Requirements*; IEEE Network Magazine, Vol. 4, Nr.2, S. 34-40, März 1990

[26] M. Zitterbart, B. Stiller; *A Concept For a Flexible High Performance Transport System*; Proceedings GI 91, Darmstadt, Okt. 1991, in: GI - 21. Jahrestagung, Informatik-Fachberichte, Springer-Verlag, Berlin, etc., 1991

Objektorientierte graphische Benutzerschnittstellen für Hypermedia

Informationstechnische Systeme werden in zunehmendem Maße mit anwendungsorientierten, wirklichkeitsgetreuen Bedienoberflächen ausgestattet. Graphische Darstellungs- und Interaktionstechniken, erweitert um hypermediale Techniken wie Animation, Bild, Video und Akustik, bilden daher ein Forschungsthema, das die Zukunft der Informationstechnik maßgeblich mitbeeinflußt.

Koordinatoren: Dr. J. Röhrich, FhG-IITB, Karlsruhe

Dr. P. Wißkirchen, GMD, St. Augustin

Graph-Layout für eine objektorientierte Benutzungsoberfläche

Ein Vergleich verschiedener Graph-Layout–Verfahren für den Browser des Hypertextsystems CONCORDE

K.Laue M.Hofmann H.Langendörfer

TU Braunschweig
Institut für Betriebssysteme
und Rechnerverbund
Bültenweg 74-75
D-3300 Braunschweig

Tel.: (0531) 391 3249
FAX: (0531) 391 4577
UUCP: hofmann@infbs.uucp
EARN: hofmann@dbsinf6.BITNET

1 Einleitung und Motivation

Hypertextsysteme sind Informationssysteme, deren Inhalt durch Einheiten repräsentiert wird, die *Knoten* genannt werden. Diese Knoten werden durch *explizite Verweise* miteinander verbunden. Auf diese Weise wird Benutzern die Möglichkeit gegeben, Beziehungen zwischen Entitäten beliebiger Art auszudrücken [HCL89]. Im Gegensatz zu anderen Informationssystemen wie bspw. relationalen Datenbanken werden in Hypertextsystemen *Objektinstanzen* zueinander in Beziehung gesetzt, nicht etwa Objekttypen oder Relationen.

Neben dieser eher strukturellen Eigenschaft weisen Hypertextsysteme eine weitere Besonderheit auf, wenn man sie mit anderen Informationssystemen vergleicht: Sie sind ausschließlich *interaktive Systeme*; die Gestaltung der Benutzungsoberfläche ist daher bei ihrer Entwicklung sehr wesentlich. Zu dieser Gestaltung gibt es inzwischen eine Fülle von Material, siehe u.a. [AMY88], [Foss88], [HoLa90a], [Walk90]. Die mit ihnen verwalteten Objekte und Datenstrukturen werden i.allg. über eine *objektorientierte graphische Oberfläche* dem Benutzer zugänglich gemacht; er navigiert von Knoten zu Knoten durch die direkte Manipulation der Oberfläche.

Für die Gestaltung der Benutzungsoberfläche nennt Walker in [Walk90] drei gebräuchliche Metaphern: die des *elektronischen Buchs* [YHMD88], die des *persönlichen Assistenten* [FrCo89] und die einer *räumlichen netzwerkartigen Darstellung* [Hala87]. Gerade letztere Metapher erfordert die Realisierung objektorientierter graphischer Oberflächen und bildete in jüngster Zeit die Basis für eine Reihe von Systementwicklungen.

In vielen derartigen auf der Netzwerkmetapher beruhenden Systemen wird die Präsentation des Hypertexts in einem graphischen Browser nur einmal (nach Anstarten des Systems) berechnet. Dem Benutzer wird dann das gesamte Netz oder bestimmte Ausschnitte davon dargestellt. Diese Methode besitzt die gravierenden Nachteile, daß

- beliebige nicht in Beziehung stehende Objekte zufällig nah beieinander abgebildet werden können [Foss88];

- das entstehende Gewirr von Knoten und Verweisen die kognitive Beanspruchung des Benutzers erhöht statt ihn zu entlasten [UtYa89].

Wir beschreiben nun in dem folgenden Papier den Prototyp unseres Hypertextsystems CONCORDE, dessen graphischer Browser auf der Netzwerkmetapher basiert. Die genannten Nachteile werden in diesem System durch folgende Eigenschaften vermieden:

1. Der graphische Browser präsentiert nur die Knoten, die mit einem Objekt, das für einen Benutzer gerade im Mittelpunkt seines Interesses steht (focus of interest), verbunden sind. Dabei muß die Beziehung nicht direkt sein; auch weiter entfernt liegende Knoten und Verweise können bei Existenz entsprechender Verbindungen abgebildet werden.

2. Dargestellt wird nur bis zu einer gewissen Entfernung vom Interessenmittelpunkt. Diese Entfernung ist vom Benutzer interaktiv einstellbar. Im Unterschied zu den „Fish-eye views" von Furnas [Furn86] haben wir jedoch die Darstellung von Nachbarschaft und das Filtern durch Prioritätsvergabe strikt getrennt.

3. Eine Reihe von Graph-Layout–Algorithmen wurde analysiert, von denen ein geeigneter letztlich zur Präsentation des Hypertextgraphen implementiert wurde. Während in vielen anderen Systemen das Graph-Layout dem Benutzer überlassen wird (bspw. im ursprünglichen Ansatz von Intermedia [UtYa89]), was den Benutzer zusätzlich belastet, übernimmt bei uns das System selbst das Layout.

Dieser zuletzt genannt Punkt ist für die Handhabbarkeit (usability) des Systems sehr entscheidend, da nur eine geeignet entflechtete Präsentation des Hypertexts eine objektorientierte Darstellung im Browser sinnvoll macht. Wir konzentrieren uns daher bei der folgenden Beschreibung auf die Diskussion der untersuchten Algorithmen; Näheres zur Funktionalität der Oberfläche findet sich in [HoLa88] und [HoLa90a].

In Abschnitt 2 wird die Präsentation der Oberfläche von CONCORDE dargestellt und einige Randbedingungen erläutert, denen der Hypertextgraph unterliegt. In Abschnitt 3 werden einige Kriterien aufgeführt, deren Einhaltung zu einer „schönen Graphpräsentation" und damit zu einer guten Handhabbarkeit der Oberfläche beitragen. Abschnitt 4 beschreibt die getesteten und implementierten Algorithmen, Abschnitt 5 einige Vorzüge, die die Smalltalkumgebung, mit der wir CONCORDE implementierten, bei der Realisierung dieser Untersuchung bot. In Abschnitt 6 werden die Ergebnisse diskutiert, die die Implementierung der Algorithmen und ihr Test ergab. Dabei wurde auf die Einhaltung der Kriterien und auf das Laufzeitverhalten der einzelnen Methoden geachtet. In den abschließenden Abschnitten werden die Ergebnisse der Arbeit zusammengefaßt und ein Überblick über die verwendete Literatur gegeben.

2 Oberfläche von CONCORDE

Dieser Abschnitt beschreibt die Benutzungsoberfläche von CONCORDE. CONCORDE ist ein Hypertextsystem für aktive Anwendungen, das Benutzern neben dem Zugriff auf eine allgemein zugängliche Knoten- und Verweismenge das Halten individueller Objekte gestattet. Es bietet ferner typisierte Knoten und Verweise sowie Komplexobjekte. Ein Prototyp von CONCORDE wurde im April 1990 auf der ersten deutschen Hypertextkonferenz in Darmstadt vorgeführt. CONCORDE ist in Smalltalk-80 implementiert; die Vorteile, die dies gerade für die Oberflächenentwicklung hat, werden beispielhaft in Abschnitt 5 diskutiert. Näheres zur Architektur und zum Datenmodell finden sich in [HoLa90b], zu möglichen Anwendungen in [HSL90]. Bild 1 zeigt das Arbeitsfenster des derzeitigen Prototypen von CONCORDE. Der Smalltalk-Philosophie gemäß können mehrere dieser Fenster geöffnet sein; dies ist insbesondere beim Filtern von Information bei Vergleichsvorgängen sehr vorteilhaft. Andererseits wurde dafür gesorgt, daß jede Funktion, auch das Erschaffen neuer Verweise, mit nur einem geöffneten Arbeitsfenster durchgeführt werden kann.

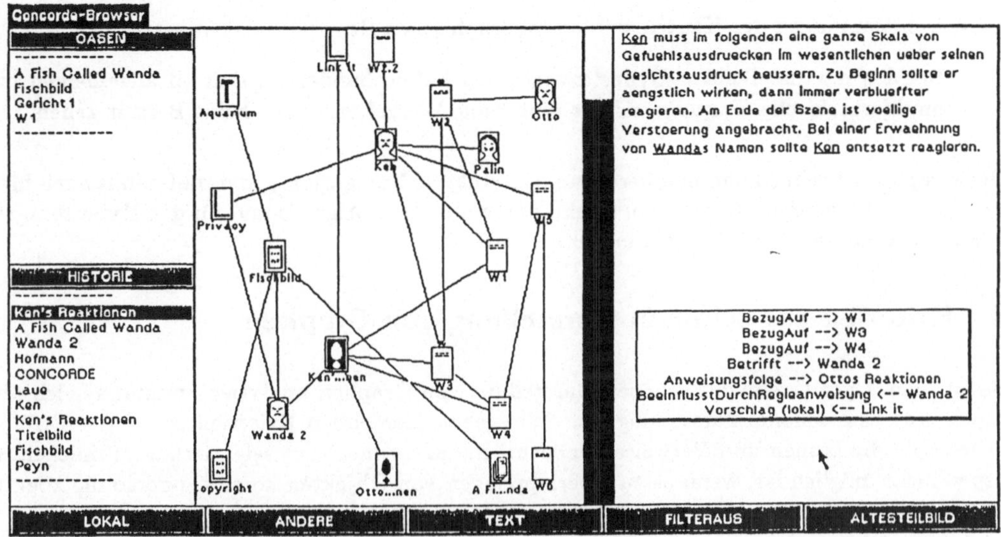

Bild 1: Arbeitsfenster des derzeitigen Prototypen von CONCORDE.

Das Arbeitsfenster teilt sich in vier Teilfenster, einen senkrechten Balken und eine Schalterleiste am unteren Rand. Das mittlere Fenster ist das Browserfenster. Hier werden Knotenmengen abgebildet. Ein ausgewählter Knoten (focus of interest) wird mit seinen direkt verbundenen Knoten (Nachbarn) und, je nach Auswahl durch den Benutzer, noch weiter entfernten Knoten gezeigt. Im allgemeinen erweist sich eine Umgebung, die um den ausgewählten Knoten alle die Knoten darstellt, die drei oder weniger Verweise entfernt sind, als sehr geeignet für die Navigation. Jeder Knoten und jeder Verweis ist selektierbar; dies ist typisch für eine objektorientierte graphische Oberfläche. Der Benutzer hat die Möglichkeit, andere Knoten anzuwählen und sich deren Umgebung anzusehen. Das große rechte Fenster zeigt den Inhalt des aktuellen, im Browserfenster invertiert dargestellten Knoten an; die zwei kleinen linken Teilfenster dienen verschiedenen CONCORDE-Funktionalitäten. Die Schalterleiste ermöglicht dem Benutzer, Veränderungen an den Standardeinstellungen der Fenstereigenschaften vorzunehmen. Der Benutzer kann z.B. die maximale Entfernung eines noch sichtbaren Knotens vom aktuell angewählten auswählen. Der Balken wird für Veränderungen an der Qualität der Graph-Layouts benötigt; er wird später noch näher erläutert.

Ein Graph in CONCORDE besitzt einige sehr wichtige Randbedingungen; der Graph ist so programmiert, daß zunächst nur ein Netzwerk aus Ikonen und Linien im Browser sichtbar ist; bei der Navigation eventuell störende Einzelheiten werden erst nach dem Anwählen mit der Maus dargestellt (etwa die Informationen zu den einzelnen Verweisen oder Knoten).

- Verweise zwischen einem Knotenpaar werden als einfache Linien und Knoten selber als Rechteckikonen mit Namen dargestellt. Die Ikonen sind so gestaltet, daß es für jede Knotenklasse eine eigene Ikone gibt.

- Das Browserfenster bietet maximal 63 Ikonen Platz. Es ist in ein 7×9–Gitter unterteilt; jedes Gitterfeld kann überdeckungsfrei eine Ikone samt Namen darstellen.
- Zwischen zwei Ikonen wird höchstens eine Linie angezeigt. Sie kann mehrere Verweise repräsentieren; wählt der Benutzer diese Linie durch einen Mausklick an, so erhält er Hintergrundinformationen über die Verweise, bspw. ihre Zahl und Richtung.
- Die Richtungen der Kanten werden im graphischen Browser nicht angezeigt.
- Bei den darzustellenden Hypertextnachbarschaften handelt es sich im allgemeinen nicht um hierarchische Graphen, d.h. es muß keine Wurzel und auch keine Blätter geben.

Der Graph wird sehr häufig neu berechnet (interaktive Navigation!) und muß somit auch häufig neu dargestellt werden. Aus diesem Grund darf ein Layout–Algorithmus für den Hypertextgraph möglichst wenig Rechenzeit verbrauchen.

3 Kriterien zur schönen Darstellung von Graphen

Bevor man Algorithmen zur schönen Darstellung von Graphen analysiert, muß man sich überlegen, was eine schöne Darstellung ist. Wir geben hier einige Kriterien an, deren Erfüllung helfen soll, die Ikonen und Verweise übersichtlich anzuordnen, gut selektierbar zu machen, was bspw. nicht möglich ist, wenn es zu Überlappungen von Objekten kommt, und so die kognitive Beanspruchung des Benutzers zu senken.

In der Graphtheorie wird unter einem „schönen Graph" meist ein planarer Graph verstanden [Bran88]. Für ein interaktives System mit objektorientierter graphischer Benutzungsoberfläche genügt jedoch dieser „Schönheitsbegriff" nicht; Graphen sind hier im Unterschied zur Mathematik keine Verbindungen von dimensionslosen Punkten mit Geraden, sondern von Ikonen, die eine Fläche beanspruchen, mit Linien, die u.U. einige Pixel breit sein können. Folgende Kriterien sind für die Erfüllung der genannten Ziele wichtig:

1. *Ikonen dürfen sich nicht gegenseitig überdecken.* Wenn eine Ikone von einer anderen überlappt wird, ist sie nicht mehr zu sehen und somit auch nicht mehr anzuwählen.
2. *Alle Ikonen müssen vollständig im Fenster liegen.* Eine Ikone, die nur zum Teil zu sehen ist, ist schwer anzuwählen. Namen, die nur teilweise sichtbar sind, kann man nicht mehr lesen. Es müssen auch alle Zielikonen von Verbindungen sichtbar sein.
3. *Verbindungen dürfen sich nicht teilweise überdecken (Steigungsproblem).* Wenn sich zwei Verbindungen teilweise überdecken, können sie nicht mehr unterschieden und auch nicht ausgewählt werden. Das bedeutet, daß sich der Winkel zwischen zwei Verbindungsgeraden um einen gewissen Betrag unterscheiden muß.
4. *Verbindungen dürfen nicht durch Ikonen verlaufen.* Wenn eine Verbindung durch den Namen einer Ikone verläuft, ist der Name nur noch schwer zu lesen. Wenn eine Verbindung durch eine Ikone selbst verläuft, ist schwer zu erkennen, ob die Verbindung an der Ikone endet oder durch sie verläuft.
5. *Der Bildschirm soll ausgenutzt werden.* Wenn der gesamte Bildschirm ausgenutzt wird, ist die Wahrscheinlichkeit von Überlappungen geringer. Der Graph wirkt insgesamt nicht so „gedrängt".

6. *Kanten dürfen sich nicht kreuzen.* Dieses Kriterium ist zwar von untergeordneter Wichtigkeit, wir versuchen aber trotzdem – wenn möglich – Planarität zu erreichen.

Neben den hier erwähnten Kriterien muß bei der Bewertung von Algorithmen, die ein Graph-Layout für die Darstellung eines Graphen vornehmen, auch ihre Komplexität und somit die *Rechenzeit* einfließen.

4 Beschreibung der einzelnen Algorithmen

Bereits vor der Analyse der untersuchten Algorithmen (siehe Abschnitt 6) sei bemerkt, daß die ersten beiden in Abschnitt 3 erwähnten Kriterien von allen Algorithmen eingehalten werden. Dies liegt bei Kriterium 1 an der Einteilung des Browserfensters in ein Gitter und bei Kriterium 2 daran, daß wir nur zusammenhängende Nachbarschaften von Knoten darstellen, d.h. eine Menge von Objekten, die Beziehungen zueinander besitzen.

Bei der Beschreibung der analysierten Algorithmen verwenden wir im folgenden für Ikonen den Begriff „Karte"; Karten sind die Bezeichnung für die Hypertextknoten des CONCORDE–Datenmodells. Die untersuchten Algorithmen lassen sich in vier Kategorien einteilen: *Entfernungsalgorithmen, Algorithmen zur hierarchischen Darstellung, heuristische Ansätze* und *andere Algorithmen.* Zu der letzten Kategorie gehört die Visualisierung von ER–Diagrammen.

ER–Diagramm: Die Methode zur Berechnung des Layouts von ER–Diagrammen [TaBT83] gliedert sich in mehrere Teilschritte, die z.T. sehr rechenzeitaufwendig sind. Dieser Algorithmus ist für ein interaktives System viel zu langsam; nach ersten Untersuchungen wurde daher auf eine nähere Analyse verzichtet.

Entfernungsalgorithmen: Mehrere Entfernungsalgorithmen wurden von uns untersucht. Sie arbeiten alle mit einer Entfernungsmatrix, in der jeweils der Verwandtschaftsgrad von zwei Objekten eingetragen ist. Der Verwandtschaftsgrad von zwei Karten wurde dabei von uns als abhängig von der Länge des kürzesten Wegs definiert, der zwischen diesen beiden Karten verläuft. Die einzelnen Algorithmen versuchen nun, die Karten entsprechend dieser „Wunschentfernung" zu plazieren. Das heißt, direkt miteinander verbundene Karten werden nah beieinander plaziert und Karten, zwischen denen mehrere Verbindungen liegen, werden weiter voneinander weg plaziert. Es ist unmittelbar einsichtig, daß eine ideale Plazierung entsprechend der „Wunschentfernung" nicht immer möglich ist.

Party–Planner: Die Idee zu diesem Algorithmus stammt von Dewdney [Dewd87]. Die Karten bekommen zunächst alle willkürlich einen Platz im Fenster zugewiesen. Danach wird für jeweils eine Karte der tatsächliche Abstand zu den anderen Karten mit dem „Wunschabstand" auf der aktuellen Position berechnet. Dieser Wert wird mit den Werten verglichen, die sich für die Karte auf ihren acht Nachbarpositionen ergäbe. Die Karte wird dann auf der Stelle positioniert, wo dieser Wert am kleinsten ist, d.h. dort, wo sich die geringste Differenz zwischen Wunsch und Realität ergibt. Das wird mehrfach für alle Karten iteriert (5–10 mal ist ausreichend, da die Laufzeit sonst zu lang ist). Dieser Algorithmus wurde in einer ersten Version des Prototypen implementiert.

Distanzanalyse: Eine mathematische Lösung der Plazierung der Karten nach der Entfernungsmatrix ist in [PiTs88] beschrieben. Die Matrix wird transformiert, und die beiden größten Eigenwerte werden berechnet. Das Produkt der zugehörigen Eigenvektoren mit der Wurzel des Eigenwertes ergibt die Koordinaten der einzelnen Karten.

Landkarte: Dieser Algorithmus von Spencer [Spen86] plaziert die Karten schon bei der Initialisierung möglichst nach ihren Entfernungen. An diese Initialisierung schließen sich dann noch mehrere Iterationen an, die die Positionen der Karten optimieren sollen.

Hierarchische Algorithmen: Obwohl die Graphen bei vielen Hypertextsystemen selten rein hierarchische Strukturen aufweisen, haben wir dennoch einen sehr schnellen Algorithmus von Robins [Robi87] getestet. Ein sehr großes Problem lag in der dabei nötigen Definition einer künstlichen Knotenhierarchie für einen eigentlich nicht–hierarchischen Graphen und speziell in der Behandlung der Zyklen, die in unseren Hypertexten existieren.

Nach Festlegung einer solchen „Pseudohierarchie" erfolgt die Berechnung der x– und y–Koordinaten nun unabhängig voneinander. Die x–Koordinate von Karte N wird wie folgt berechnet: Wenn N ein Blatt ist, bekommt sie die zuletzt vergebene x–Koordinate plus 1. Ist N kein Blatt, werden zuerst die x–Koordinaten der Kinder berechnet, und N wird dann der Mittelwert dieser Koordinaten zugewiesen. Berechnung der y–Koordinate: Ist N die Wurzel, bekommt sie die y–Koordinate 1. Hat N dagegen Eltern, werden zuerst deren y–Koordinaten berechnet, und N bekommt den größten Wert plus 1.

Ein anderer aufwendigerer Algorithmus, der eine bessere Behandlung von Zyklen vorsieht, wird in [RDMMST87] beschrieben. Bei diesem Algorithmus werden durch Kartenikonen verlaufende Verbindungen durch Knicken der Linien vermieden. Dieser Algorithmus ist sehr zeitintensiv.

Heuristische Lösungsansätze: Einige heuristische Algorithmen wurden von uns entwickelt, die die vorgegebenen Kriterien einhalten sollen. Bei diesen Algorithmen werden die Knoten zuerst nach der Anzahl ihrer Nachbarn sortiert. Die Knoten mit den meisten Nachbarn werden zuerst behandelt.

Feste Positionierung: Eine sehr einfache und sehr schnelle Methode der Berechnung des Graph–Layouts ist die feste Positionierung. Die Positionen im Gitter werden sortiert, und die Karten werden der gebildeten Reihenfolge gemäß auf diese Positionen gesetzt. Die Reihenfolge der Karten ergibt sich aus der Anzahl der Verbindungen. Das bedeutet, daß Karten, an denen viele Verbindungen ankommen oder abgehen, auf „gute" Positionen plaziert werden. Die Heuristik besteht in dieser Methode in der Zuordnung zu den für „gut" befundenen Positionen.

Mittel–Plazierung: Die Karte mit den meisten Nachbarn wird auf die mittlere Position des linken Rands des Gitters gesetzt, ihre Nachbarn werden um sie herum plaziert. Danach werden die Nachbarn der bereits gesetzten Karten plaziert. Sie werden jeweils auf den Mittelwert der Positionen ihrer bereits gesetzten Nachbarn plaziert oder – wenn diese Position schon mit einer anderen Karte belegt ist – auf einen Platz in der Nähe der errechneten Position. Das wird für alle Karten durchgeführt. Nach Plazierung aller Karten wird der gesamte Graph noch in die Mitte geschoben und vergrößert. Das ist sehr wichtig, da der Graph durch die Anfangsposition (1,4) sehr „linkslastig" ist.

Regel–Algorithmus: Hier sind die Positionen auf dem Bildschirmgitter geordnet. Die ersten fünf vergebenen Positionen bilden ein Pentagramm. Für jede weitere Karte wird dann eine gewichtete Summe der aufgetretenen Übertretungen der einzelnen Kriterien auf der gerade betrachteten Position berechnet. Liegt diese Summe unter einem vorgegebenen Schwellwert, wird die Karte auf diese Position gesetzt, sonst wird die nächste Position der Reihenfolge ausprobiert. Findet sich keine Position, auf der der Fehlerwert unter dem Schwellwert liegt, wird die Karte auf die Position gesetzt, auf der sich der geringste Fehler ergab.

Um die Rechenzeit gering zu halten, sehen wir *kein Backtracking* vor. Das bedeutet, daß einmal plazierte Karten nicht mehr umgesetzt werden. Das kann dazu führen, daß die n–te Karte, deren ideale Position m wäre mit $m < n$, auf eine schlechtere Position plaziert wird, da die rechnerische Idealposition bereits besetzt ist. Die Rechenzeit ist sehr abhängig von dem gewählten Schwellwert. Je kleiner er ist, desto „schöner" wird das Layout, desto länger dauert aber auch dessen Berechnung. Das Layout hängt außerdem von der Wertigkeit der Verletzung der einzelnen Kriterien ab. Beides kann in CONCORDE durch den Benutzer interaktiv beeinflußt werden: die Wertigkeit der Kriterien durch Anwählen eines Menüs und der Schwellwert durch den Balken zwischen Browser- und Inhaltsfenster.

5 Vorteile der objektorientierten Programmierung beim Vergleich von verschiedenen Layout–Algorithmen

In diesem Abschnitt wird beschrieben, wie Eigenschaften der objektorientierten Programmierung, speziell der Vererbung, beim Vergleich der im vorigen Abschnitt beschriebenen Algorithmen genutzt werden. Zunächst wurden zwei Klassen **Graph** und **Graphenknoten** als Unterklasse von **Object** eingerichtet, die die für alle Algorithmen notwendigen Variablen und Methoden enthalten. Die Klasse **Graph** beschreibt dabei den gesamten Graphen mit seinen Knoten und Verbindungen und enthält in ihren Instanzvariablen u.a. die darzustellenden Ikonen, die Anzahl der abzubildenden Ikonen und die Dreiecksmatrix sämtlicher Verbindungen. Ferner enthält diese Klasse Methoden zur Manipulation dieser Instanzvariablen. Weiterhin gibt es eine Methode, die ausgehend von der Karte, um die das Bild aufgebaut werden soll und der maximalen Entfernung, die gesamte darzustellende Kartenmenge berechnet und in der Variable *knoten* ablegt. Zur Manipulation der Koordinaten gibt es noch die zwei Methoden SCHIEBEINMITTE, die den Graph in der Mitte des Browserfensters zentriert und VERGROESSERE, die dafür sorgt, daß der Graph das Browserfenster gut ausnutzt.

Graphenknoten beschreibt einen einzelnen Knoten und seine Position auf dem Bildschirm. Für jeden einzelnen Layout–Algorithmus werden dann jeweils zwei Klassen eingerichtet, die unter **Graph** bzw. **Graphenknoten** gehängt werden. Die Unterklasse von **Graph** muß die Nachricht BERECHNEKOORDINATEN verstehen, die an eine Instanz von ihr geschickt wird. Diese Methode sieht das Berechnen der Koordinaten nach dem jeweiligen Algorithmus und das Zeichnen der Ikonen und Verbindungen vor. Es ist nun nicht mehr notwendig, die bereits implementierten Methoden wie z.B. das Berechnen der darzustellenden Kartenmenge neu zu schreiben, da sie an die Unterklassen vererbt werden. Bei Implementieren eines neuen Layout–Algorithmus mit einer prozeduralen Programmiersprache müßten diese Methoden und Variablen kopiert und dann geeignet modifiziert werden. Wir werden nun die beiden Oberklassen und (als Beispiel) die zwei Klassen des hierarchischen Algorithmus kurz vorstellen. Die neue Klasse **Baum** wird unter die Klasse **Graph** gehängt. Sie besitzt drei weitere Instanzvariablen, die die Reihenfolge der darzustellenden Knoten nach ihrer Hierarchisierung festlegen.

Die neuen Methoden sind BERECHNEKOORDINATEN, die die Berechnung der x–und y–Koordinaten veranlaßt und die Karten zeichnet, BERECHNEXKOORDINATENVON: AKNOTEN, die nach dem beschriebenen Algorithmus die x–Koordinate für AKNOTEN berechnet und BERECHNEY-KOORDINATEVON: AKNOTEN, die entsprechend die y–Koordinate berechnet.

Die Klasse **Knoten** wird unter **Graphenknoten** gehängt. Sie besitzt ebenfalls drei eigene Instanzvariablen, die die Beziehungen der Knoten nach ihrer Hierarchisierung ausdrücken.

Wenn man nun einen anderen Layout–Algorithmus im System verwenden will, muß die Nachricht zum Berechnen der Positionen und Zeichnen der Ikonen nur an eine Instanz der Klasse des jeweiligen Algorithmus geschickt werden. Das erlaubt den leichten Austausch von Layoutalgorithmen, falls applikationsspezifische Beschränkungen dies notwendig erscheinen lassen. Eine Anwendung mit vielen Entscheidungsbäumen macht etwa eine darstellung des Hypertexts durch den hier beschriebenen hierarchischen Algorithmus vorteilhaft; der in der ersten Prototypversion implementierte Party-Planner oder die derzeit verwendeten heuristischen Ansätze sind somit leicht auswechselbar.

6 Vergleich der Algorithmen

Um die Wirkung der verschiedenen Algorithmen vergleichen zu können, wurden unterschiedlich große Kartenmengen erzeugt. Zusätzlich wurden für gleiche Kartenmengen die Anzahl der Verbindungen variiert, so daß der Einfluß des Verbindungs-Karten–Verhältnisses auf die verschiedenen Algorithmen untersucht werden konnte. Wir haben dann jede Kartenmenge von jedem Algorithmus darstellen lassen und die Verletzungen der einzelnen Kriterien gezählt. Außerdem wurde die Zeit gemessen, die die Algorithmen zur Berechnung der Graph–Darstellungen benötigten.

Die Kriterien 1 und 2 werden von allen Algorithmen eingehalten, da der Bildschirm in ein Gitter eingeteilt wird und die berechnete Knotenmenge immer vollständig auf dem Bildschirm dargestellt wird (Nachbarschaftsprinzip).

Kriterium 3 (Vermeiden von Verweisüberdeckungen) wird am besten vom Regel–Algorithmus eingehalten. Recht gute Ergebnisse liefert auch noch die Mittel–Plazierung. Überraschend gut schneidet die feste Plazierung ab, sie führt zu weniger Überlappungen von Verbindungen als sämtliche Entfernungsalgorithmen. Der hierarchische Algorithmus ist sehr gut für ein Verbindungs-Karten–Verhältnis von annähernd 1:1, arbeitet sonst aber sehr schlecht.

Das vierte Kriterium (Vermeiden der Überkreuzung von Ikonen durch Verweise) wird ebenfalls am besten vom Regel–Algorithmus eingehalten. Der Party–Planner ist am schlechtesten, was für uns eine große Enttäuschung darstellte, da wir diesen Algorithmus in der ersten Version unseres Prototypen implementiert hatten.

Kriterium 5 (gute Ausnutzung des Bildschirms) wird von allen Algorithmen bis auf den Party–Planner recht gut eingehalten, da bei allen anderen Algorithmen von uns zusätzlich nach dem Berechnen der Positionen noch ein zentrierendes Justieren und anschließendes Vergrößern vorgenommen wird.

Die Ergebnisse des sechsten Kriteriums (Planarität des Graphen) zeigen erwartungsgemäß eine deutliche Verschlechterung des Ergebnisses bei großen Karten– und Verbindungsmengen. Bemerkenswert sind die Werte des hierarchischen Algorithmus bei einem Verhältnis von Karten zu Verbindungen von etwa 1:1.

Nach dem Vergleich sind die Ergebnisse des Regel–Algorithmus also die besten. Da die Wahl des Schwellwertes einen großen Einfluß auf das Ergebnis hat, wurde auch der Einfluß des Schwellwerts näher untersucht. Dabei konnte festgestellt werden, daß der Algorithmus noch recht gute Bilder liefert, wenn man einige Verletzungen von weniger wichtigen Kriterien wie etwa die der Planarität in Kauf nimmt.

Für eine abschließende Bewertung muß auch die Rechenzeit berücksichtigt werden. Die Rechenzeit wurde mit einer Smalltalk–Methode gemessen, die u.a. die Zahl der verschickten Botschaften berücksichtigt, die von großem Einfluß auf die Laufzeit sind. Deren Einheit sind sog. „tallies“. Erwartungsgemäß benötigt die feste Positionierung die wenigste Zeit. Nur unwesentlich langsamer ist die Mittel–Plazierung, deren Layout in den meisten Fällen trotzdem recht gut ist. Der Party–Planner ist ebenfalls recht schnell (weswegen wir ihn in der ersten Version des Prototypen von CONCORDE auch implementiert hatten). Die anderen Entfernungsalgorithmen (Landkarte,

Distanzanalyse) benötigen noch mehr Rechenzeit als der Regel–Algorithmus. Die Rechenzeit beim Regel–Algorithmus hängt sehr stark von dem gewählten Schwellwert ab. Der Regel–Algorithmus mit dem geringsten Schwellwert, der zu den schönsten Bildern führt, benötigt sehr viel mehr Rechenzeit als mit einem Schwellwert, der einige Verletzungen zuläßt. Wenn man mehr Verletzungen zuläßt (wie einige durch Karten verlaufende Verbindungen oder sich fast überlappende Verbindungen), wird der Algorithmus sogar schneller als der Party–Planner oder der hierarchische Algorithmus und liefert trotzdem noch bessere Bilder.

Für die zweite Prototypversion von CONCORDE haben wir uns daher entschlossen, den Graph-Layout–Algorithmus zu reimplementieren. Der Regel–Algorithmus wurde mit verschiedenen Schwellwerten implementiert und zusätzlich die Mittel–Plazierung eingebunden. Zur Auswahl der verschiedenen Algorithmen wird der erwähnte senkrechte Balken benutzt, der in zehn verschiedene Stufen eingeteilt ist. Je höher das schwarze Rechteck ist (d.h. je mehr der Balken ausgefüllt scheint), desto besser wird das Layout, desto länger braucht allerdings auch seine Berechnung. Die obersten acht Teile des Balkens entsprechen dem Regel–Algorithmus mit verschiedenen Schwellwerten. Ganz oben befindet sich Schwellwert 0, der Schwellwert steigt mit sinkender Balkenhöhe. Für die unterste Stufe des Balkens haben wir die feste Positionierung implementiert, für die vorletzte Stufe die Mittel–Plazierung, die an dieser Stelle schneller als der Regel–Algorithmus, in der Graphqualität aber etwa gleich gut ist.

7 Zusammenfassung

In dieser Arbeit wurde eine Benutzungsoberfläche für ein Hypertextsystem diskutiert, das – im Gegensatz etwa zu elektronischen Büchern – auf der Metapher einer räumlichen netzwerkartigen Darstellung („spatial metaphor“ [Walk90]) beruht. Kennzeichen dieser Oberfläche ist, daß in dem von ihr angebotenen *graphischen Browser* immer eine Menge von semantisch zusammmenhängenden Knoten des Hypertexts samt der zugehörigen Verweise präsentiert werden. Da der Benutzer bei der Navigation durch den Hypertext in dieser Umgebung seinen Schwerpunkt von einem Knoten auf andere Knoten verlagert (inkrementelle Navigation [Foss88], [HoLa90a]), benötigt CONCORDE einen schnellen Graph-Layout–Algorithmus. Dieser muß zudem noch einen „schönen“ Graph erzeugen, da nur ein „schöner Graph“ die kognitive Beanspruchung des Benutzers gering hält. Effekte, die erhöhte Beanspruchung hervorrufen können und deshalb vermieden werden sollten, sind etwa schwer zu selektierende Ikonen und schwer voneinander unterscheidbare Verweise.

Wir stellten daher eine Reihe von Kriterien auf, deren Einhaltung einen entsprechend „schönen Graph“ erzeugt. Neben diesen Kriterien ist für *interaktive Systeme*, wie Hypertextsysteme sie darstellen, eine schnelle Antwortzeit sehr wichtig. Wir analysierten unter diesen Gesichtspunkten eine Reihe von Algorithmen (Entfernungsalgorithmen, hierarchische Algorithmen, heuristische Methoden und andere) auf ihre mögliche Verwendbarkeit für das Graph-Layout in Hypertextsystemen, die die bewußte Netzwerksmetapher verwenden. Das Ergebnis der Analyse führte konkret für unser System zu einer Reimplementierung des Graph-Layout–Algorithmus; wir ersetzten die vorher verwendete Methode Party-Planner durch von uns entwickelte heuristische Methoden.

Die Präsentation des Hypertexts in CONCORDE wurde dadurch stark verbessert. Der Benutzer hat die Möglichkeit, interaktiv die Ergebnisse des Graph-Layouts zu beeinflussen. Dies geschieht über die Veränderung einer Markierung (Balken) und mit Menüfunktionen. Diese Einflußmöglichkeit ordnet sich zwanglos in unsere Entwurfsphilosophie ein, dem Benutzer eine reichhaltige Menge an Funktionalität zur Verfügung zu stellen und ihm damit genügend Flexibilität für die Entwicklung eines persönlichen Arbeitsstils zu ermöglichen.

8 Literaturverzeichnis

[AMY88] R.Akscyn, D.McCracken, E.Yoder. *KMS: A Distributed Hypermedia System for Managing Knowledge in Organizations.* CACM, Vol.31, No.7, Juli 1988, pp.820–835

[Bran88] F.J.Brandenburg. *Nice Drawings of Graphs and Trees are Computationally Hard.* Tech. Bericht, Univ. Passau, September 1988

[Dewd87] A.K.Dewdney. *Computer-Kurzweil.* Spektrum der Wissenschaft, Dezember 1987, S.6–9.

[DJAY83] C.G.Davis, S.Jajodia, P.Ann-Beng, R.T.Yeh (eds). *Entity-Relationship Approach to Software Engineering.* Proc. of the 3rd International Conference on ER-Approach, Oktober 1983.

[Foss88] C.L.Foss. *Effective Browsing in Hypertext Systems.* Proc. RIAO'88, M.I.T. Cambridge (MA), März 1988, pp.82–98

[FrCo89] M.Frisse, S.Cousins. *Information Retrieval from Hypertext: Update on the Dynamic Medical Handbook Project.* Proc. Hypertext'89, Pittsburgh, November 1989, pp.199–212

[Furn86] G.W.Furnas. *Generalized Fisheye Views.* Proc. CHI'86, Boston (MA), April 1986, pp.16–23

[Hala87] F.G.Halasz. *NoteCards: An Experimental Environment For Authoring And Idea Processing.* Proc. BTW'87, GI–Fachtagung, Darmstadt, April 1987, pp.56–67

[HCL89] M.Hofmann, R.Cordes, H.Langendörfer. *Hypertext/Hypermedia.* Informatik-Spektrum, Vol.12, No.4, August 1989, pp.218–220

[HoLa88] M.Hofmann, H.Langendörfer. *Interaktives Navigieren im Browser des Hypertextsystems CONCORDE.* Proc. Benutzerschnittstellen - Werkzeuge und Techniken zur Gestaltung interaktiver Systeme, GI–Fachtagung, Darmstadt, November 1988

[HoLa90a] M.Hofmann, H.Langendörfer. *Browsing as Incremental Access of Information in the Hypertext System CONCORDE.* Proc. Interactive Communication, Paris, Mai 1990, pp.17–35

[HoLa90b] M.Hofmann, H.Langendörfer. *User Support by Typed Links and Local Contexts in a Hypertext System.* Proc. Workshop Intelligent Integrated Information Systems (III), Schloß Tuczno (Polen), September 1990, pp.106–124

[HSL90] M.Hofmann, U.Schreiweis, H.Langendörfer. *An Integrated Approach of Knowledge Acquisition by the Hypertext System CONCORDE.* Proc. ECHT'90, Versailles, November 1990, pp.166–179

[PiTs88] X.Pintado, D.Tsichritzis. *An Affinity Browser.* In D.Tsichritzis (ed.). *Active Object Environments.* Université de Genève, Centre Universitaire d'Informatique, Genf 1988. S.51–60.

[RDMMST87] L.A.Rowe, M.Davis, E.Messinger, C.Meyer, C.Spirakis, A.Tuan. *A Browser for Directed Graphs.* Software - Practice and Experience, Vol. 17(1), Januar 1987, S.61–76.

[Robi87] G.Robins. *The ISI grapher - a portable tool for displaying graphs pictorially.* Proceedings of Symboliikka, August 1987, S.44–60.

[Spen86] R.Spencer. *Similarity Mapping.* BYTE, August 1986, S.85–92.

[TaBT83] R.Tamassia, C.Batini, M.Talamo. *An Algorithm for Automatic Layout of Entity Relationship Diagrams.* in [DJAY83], S.421–439.

[UtYa89] K.Utting, N.Yankelovich. *Context and Orientation in Hypermedia Networks.* ACM ToIS, Vol.7, No.1, Januar 1989, pp.58–84

[Walk90] J.Walker. *User Interface Metaphors in Hypertext.* Tutorial 7, ECHT'90, Versailles, November 1990

[YHMD88] N.Yankelovich, B.J.Haan, N.Meyrowitz, S.M.Drucker. *Intermedia: The Concept of a Seamless Information Environment.* IEEE Computer, Vol.21, No.1, Januar 1988, pp.81–96

Objektorientierte Graphik-Ausgabe für Benutzerschnittstellen-Werkzeuge

Gregor Lux

Zentrum für Graphische Datenverarbeitung e.V.
Wilhelminenstr. 7
D-6100 Darmstadt
E-mail: lux@zgdvda.uucp

1. Einführung

Die Benutzerschnittstellen hypermedialer Anwendungen umfassen in der Regel einen großen Anteil direkt-manipulativer Interaktionen zur Durchführung navigierender und ändernder Operationen auf graphisch dargestellten Informationen /GlSt-90/. Wichtig sind speziell kontinuierliche Interaktionen wie Verschieben (*dragging*) oder Dehnen (*stretching*) /FWC-84/. Gerade in Bereichen wie Hypermedia, in denen die graphischen Darstellungen hochvernetzte Anwendungsstrukturen repräsentieren, müssen die Benutzeroperationen oft unter Einhaltung anwendungsspezifischer Randbedingungen auf den Darstellungen durchgeführt werden. Solche Randbedingungen können z.B. das geometrische Ausrichten von Elementen sein, das Verbot von Überlappungen oder das Nachziehen von Verbindungslinien.

Während der letzten Jahre wurden eine Reihe von Werkzeugen zur Entwicklung von graphischen Benutzerschnittstellen vorgestellt. *User Interface Toolkits* wie OSF/Motif /OSF-89/ und MacApp /Schm-86/ erleichtern die Programmierung einer begrenzten Menge häufig verwendeter graphischer Dialogelemente wie Fenster, Rollbalken, Menüs, Buttons und Formulare.

Die Funktionalität zur graphischen Ausgabe solcher Toolkits ist beschränkt auf wenige zweidimensionale Elemente mit geringer Komplexität. Sie ist üblicherweise unter Verwendung einer graphischen Basisschnittstelle wie X Window System oder QuickDraw implementiert. Diese Schnittstellen bieten rasterorientierte Zeichenfunktionen zur schnellen Ausgabe von Primitiven wie Linien, Kreisen und Pixelmatrizen. Die meisten Toolkits geben diese Schnittstelle dann auch weiter an ihre Anwendungen, um deren anwendungsspezifische graphische Ausgabe zu realisieren. Dies ist ausreichend für Anwendungen, die keine oder nur sehr rudimentäre graphische Ausgabe zusätzlich zu den vordefinierten Dialogelementen benötigen. Hypermediale Anwendungen benötigen jedoch wesentlich komfortablere graphische Schnittstellen zu ihrer effektiven Unterstützung.

Andererseits bieten graphische Systeme wie GKS /ISO-85/ oder PHIGS /ISO-88/ ihren Anwendungen eine mächtige graphische Ausgabe-Funktionalität wie z.B. die interne Speicherung und Verwaltung anwendungsdefinierter graphischer Elemente. Auf diesen sind beliebige geometrische Transformationen möglich und zur Verbindung zwischen Ein- und Ausgabe werden Funktionen zur Identifikation dieser Elemente angeboten. Darüber hinaus erlaubt PHIGS hierarchische Bildbeschreibungen über die mögliche Definition strukturierter Ausgabeelemente. Diesen Vorteilen stehen jedoch eine Reihe von Nachteilen gegenüber:

(1) Graphische Systeme bieten keine ausreichende Unterstützung für die oben beschriebenen kontinuierliche Interaktionen auf den anwendungsspezifischen graphischen Elementen an. Da sie sehr häufige auftreten, sollten sie vom Werkzeug und nicht von der Anwendung

gesteuert werden. Dieses Problem wurde von graphischen Benutzerschnittstellen-Werkzeugen wie THESEUS /HLM-87/, InterViews /LVC-89/ und PROMETHEUS /Ehmk-90/ angegangen, welche die höhere Ausgabefunktionalität graphischen Systeme in ihre Modelle integrierten und zusätzlich kontinuierliches Verschieben auf isolierten graphischen Elementen erlauben, was relativ einfach zu realisieren ist. Die Anwendung weiterer Interaktionstechniken ist jedoch bis heute noch Sache des Anwendungsprogrammierers. Dies macht die Entwicklung solcher Interaktionen zu einer ziemlich komplizierten und teuren Arbeit.

(2) Im Gegensatz zu den objektorientierten Architekturen und Programmierschnittstellen der meisten modernen Toolkits, die dadurch ein hohes Maß an Erweiterbarkeit und Flexibilität offerieren, werden die Programmierschnittstellen der graphischen Systeme solchen Erfordernissen nicht gerecht. Es ist z.B. nicht möglich, neue anwendungsdefinierte graphische Elementtypen mit typspezifischen Interaktionstechniken in den Systemkern zu integrieren. Diese generische Funktionalität ist jedoch essentiell für einen Dialogentwerfer.

(3) Eine letzte Beschränkung der existierenden graphischen Systeme und Normen betreffend den Bereich der Benutzerschnittstellen ist ihre mangelhafte Kompatibilität und Anpaßbarkeit an Fenstersysteme wie z.B. X Window System. Die daraus resultierenden Probleme zusammen mit möglichen Lösungen sind beschrieben in /Lux-90/.

Zur Vereinfachung der Programmierung anwendungsspezifischer graphischer Interaktionen und zur Unterstützung der Definition neuer Interaktionstypen wurde das objektorientierte Benutzerschnittstellen-Werkzeug THESEUS++[1)] entwickelt, welches speziell zur Entwicklung graphisch-interaktiver Anwendungen entworfen wurde. Auf dem Bereich der Eingabe bietet THESEUS++ ein hierarchisches Eingabemodell an, das in /Hübn-91/ genauer beschrieben ist.

Die folgenden Abschnitte dieses Aufsatzes beschreiben das Ausgabemodell von THESEUS++. Ziel des Modells ist ein Beitrag einerseits zur Überwindung von Beschränkungen heutiger *User Interface Toolkits* wie OSF/Motif bei der Beschreibung graphischer Ausgabe, andererseits die Beseitigung von Schwächen graphischer Systeme wie GKS und PHIGS bei der Realisierung direkt-manipulativer Benutzerschnittstellen.

2. Architektur des Modells

Das Modell basiert auf einer objektorientierten Architektur und bietet die Funktionalität konventioneller graphischer Systeme in objektorientierter Weise an. Darüber hinaus wurden die Ausgabeelemente von Fenstersystemen und Toolkits in das Systemmodell integriert.

Die Grundidee des Systems zur Unterstützung direkt-manipulativer Interaktionen ist in Abb. 1 skizziert. Die erste Art des Zugriffs durch die Anwendung steuert die Editier-Schleife, innerhalb derer die Struktur der Darstellung definiert und geändert wird. Dies beinhaltet die Erzeugung der graphischen Elemente mit ihren Geometrien und Attributen sowie die Definition graphischer Beziehungen zwischen den graphischen Elementen und ihren definierenden Parametern.

Die zweite Art des Zugriffes durch das Eingabesystem steuert die direkte Manipulations-Schleife. Sie wird benötigt zur Änderung des Layouts der Elemente in der Darstellung, was durch die Änderung ihrer Werte und Zustände geschieht.

[1)] **THESEUS++ wurde entwickelt im Verbundprojekt STONE, einem offenen und erweiterbaren Kern für Software-Entwicklungsumgebungen, der zur Zeit entwickelt wird. Partner sind neben dem ZGDV Darmstadt die TU Berlin, die GMD Birlinghoven, die TU Dresden, die TH Ilmenau, das FZI Karlsruhe, die GMD Karsruhe, das IITB Karlsruhe und die Universität Rostock. Die Arbeiten werden gefördert vom Bundesminister für Forschung und Technologie (BMFT) unter dem Kennzeichen ITS 8902 D/5.**

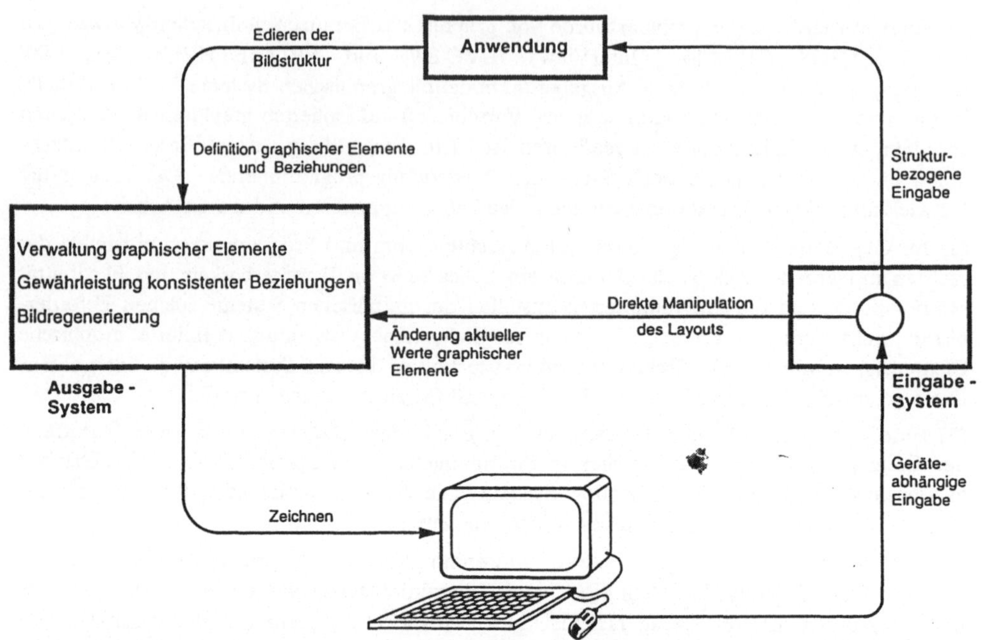

Abb. 1: Zugriff auf das Ausgabesystem bei direkter Manipulation

Das Ausgabesystem verwaltet die Datenstrukturen der graphischen Elemente unter gleichzeitiger Sicherstellung ihrer inneren Beziehungen, wenn die Elementwerte während der Interaktionen geändert werden. Desweiteren steuert es das kontinuierliche Neuzeichnen auf dem Bildschirm.

Mit dieser Aufgabenverteilung wird die sinnvolle Separierung der Anwendung von THESEUS++ unterstützt. Letzteres verwaltet (als Kombination von Ein- und Ausgabesystem) die direkte Manipulations-Schleife völlig lokal. Der Anwendungsprogrammierer wird damit weitgehend von der Programmierung geometrischer Algorithmen unter Wahrung von Restriktionen während der kontinuierlichen Layout-Änderungen entlastet.

Abb. 2 zeigt die Kernklassen des Modells. Der Kern enthält zwei voneinander separierte Klassenhierarchien, die die fundamentalen Konzepte des Modells realisieren, und die in den folgenden Abschnitten beschrieben werden. Die *GraphicalObject*-Hierarchie repräsentiert die graphischen Elemente, während die *Constraint*-Hierarchie die graphischen Beziehungen und Restriktionen enthält.

3. Graphische Objekte

Graphische Objekte werden gleichermaßen von Anwendungs- und Dialogprogrammierern zum Entwurf anwendungsspezifischer Darstellungen und graphischem Feedback für Dialogelemente genutzt. Sie werden definiert und arrangiert innerhalb zweidimensionaler Koordinatensysteme.

Im Gegensatz zu den Segmenten und Strukturen traditioneller graphischer Systeme wie GKS und PHIGS stellt ein graphisches Objekt des Ausgabemodells eine unabhängige Einheit im objektorientierten Sinn mit eigenem lokalen Zustand dar. Jedes Objekt gehört zu einer Klasse, deren spezielle Methoden beim Empfang korrespondierender Botschaften anderer

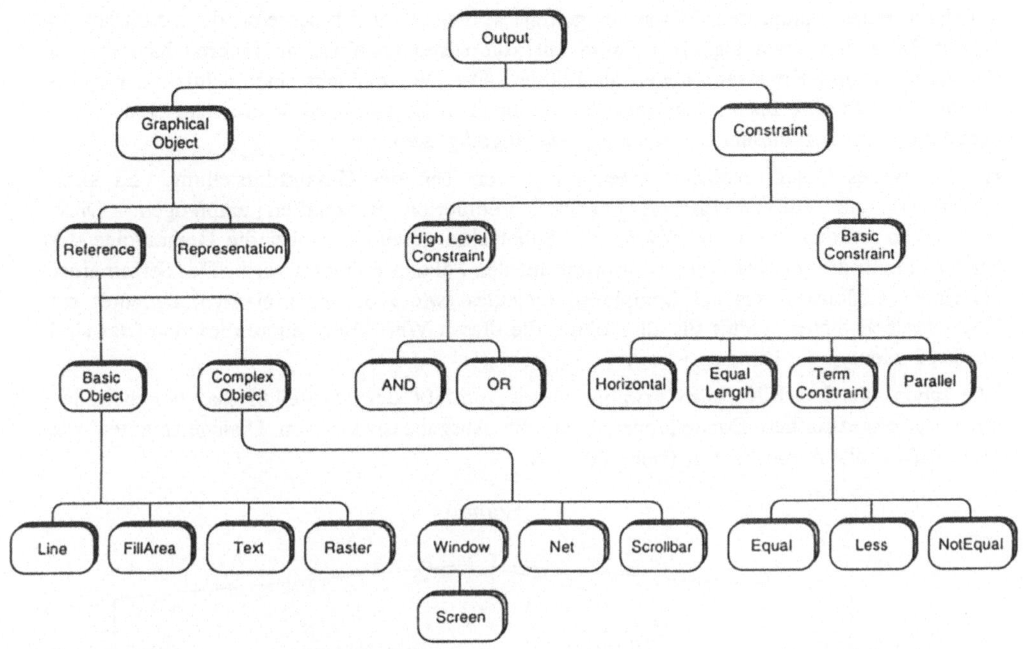

Abb. 2: Kernklassen des Ausgabemodells

Objekte ausgeführt werden.

Allen Klassen graphischer Objekte gemeinsam sind Methoden zum Sichtbar- und Unsichtbarmachen, zur geometrischen Transformation, zur Kontrolle graphischer Attribute, zum Klippen und zum Kopieren. Sie sind zusammen mit ihren zugehörigen Daten definiert in der abstrakten Klasse[1)] *GraphicalObject* und werden dann in spezifischer Weise innerhalb der konkreten Unterklassen implementiert.

Zur Definition und zur Strukturierung von Darstellungen stellt das Modell die drei folgenden Konzepte zur Verfügung:

Basisobjekte

Die Klasse *BasicObject* stellt die elementaren Ausgabe-Elemente des Modells zur Verfügung. Sie verwaltet dazu ein Geometrie-Objekt. Jede Unterklasse von *BasicObject* (die selbst wiederum eine abstrakte Klasse darstellt) fügt klassenspezifische Daten wie Bitmaps oder Textstrings hinzu. Jede dieser Unterklassen beschränkt ihre geerbte allgemein definierte Geometrie sowie die Attribute zu spezialisierten Klassen. Z.B. interpretiert ein *Area*-Objekt nur geschlossene Pfade und Füll-Attribute.

Teilestrukturen

Zum Aufbau hierarchischer Bilder können Teilestrukturen auf graphischen Objekten durch Verwendung der Klasse *ComplexObject* konstruiert werden. Ein komplexes Objekt ist rekursiv definiert als eine Aggregation beliebiger graphischer Objekte. Da jedes Objekt nur

[1)] Eine abstrakte Klasse kann nicht instantiiert werden und wird nur zur Definition von Daten und Operationen verwendet, die von Unterklassen geerbt und implementiert werden.

innerhalb eines komplexen Vaters eingefügt werden kann, beschreibt die resultierende Objekt-Hierarchie einen logischen Baum mit komplexen Objekten als inneren Knoten und Basisobjekten und Repräsentationen als Blätter. Das Wurzelobjekt einer sichtbaren Objekt-Hierarchie muß eine Instanz der speziellen komplexen Objektklasse *Screen* sein. Jede sichtbare Einheit muß in diesem *Darstellungsbaum* eingefügt sein.

Ein komplexes Objekt realisiert einen separierten Teil der Gesamtdarstellung. Es kann unabhängig alle auf graphischen Objekten definierten Botschaften empfangen. Diese Botschaften werden dann rekursiv an die Subobjekte gesendet, wobei die Reihenfolge des Sendens durch die relative Ausgabepriorität auf den Söhnen gesteuert wird. Das Setzen eines einzelnen Attribut-Wertes des komplexen Objektes (wie z.B. der Linientyp) bedeutet ein Überschreiben dieses Wertes für alle Söhne, die diesen Wert interpretieren können (das sind für den Linientyp die Linienobjekte).

Wie schon für den Bildschirm erwähnt wurde, umfaßt der Darstellungsbaum neben den anwendungsspezifischen Darstellungen auch die Ausgabeaspekte von Dialogelementen wie Menüs, Rollbalken und Fenster (siehe Abb. 3).

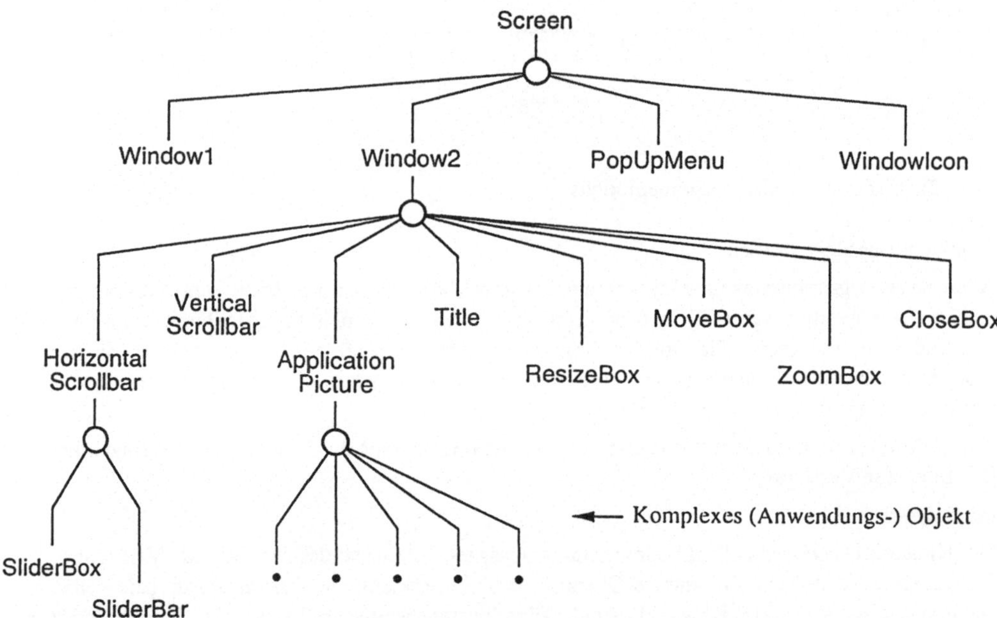

Abb. 3: Darstellungsbaum mit Dialogelementen

Sichtenstrukturen

Die Klassen *Reference* und *Representation* werden zur Definition von Sichtenstrukturen verwendet. Das Konzept erlaubt verschiedene Ansichten und Verwendungen eines (ggf. hochkomplexen) graphischen Objektes innerhalb verschiedener Darstellungsumgebungen. Ein Beispiel zeigt die Abb. 4, wo verschiedene Ansichten eines Netzdiagrammes in unterschiedlichen Fenstern presentiert werden.

Eine Repräsentation (Klasse *Representation*) stellt ein graphisches Objekt ohne eigene interne Struktur dar. Diese erbt es von seinem Referenzobjekt (Klassen-Hierarchie *Reference*). Die Repräsentation verwaltet lediglich die Abweichung von dieser Struktur in Form einer geometrischen Transformation (in Relation zum Koordinatensystem der Referenz) und dem

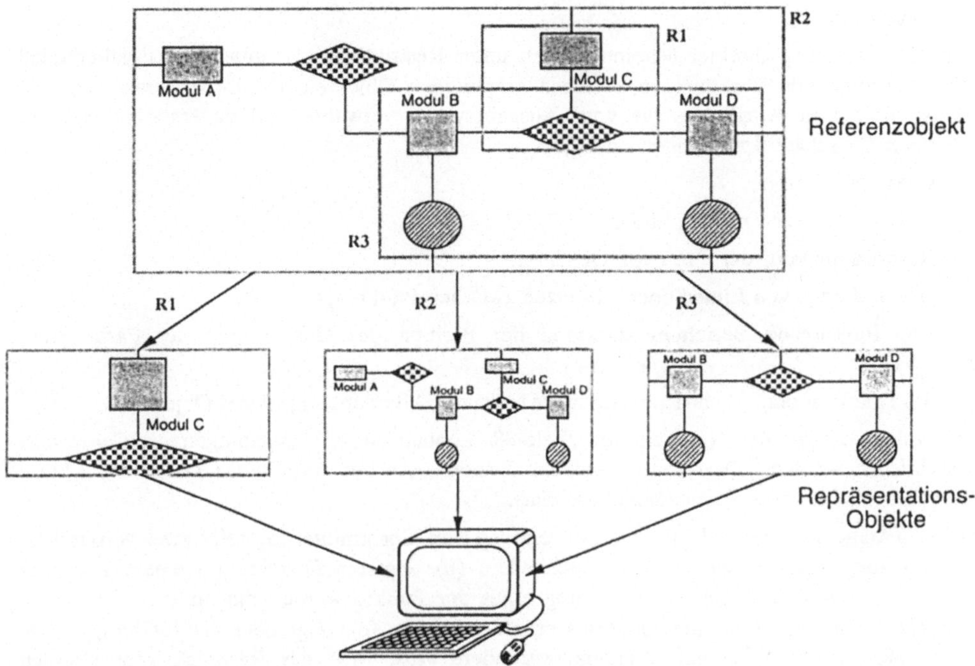

Abb. 4: Verschiedene Repräsentationen eines Netzdiagrammes

optionalen Überschreiben graphischer Attribute.
Die Abhängigkeit zwischen Referenz und Repräsentation ist unidirektional: die Änderung der internen Struktur der Referenz bewirkt die Änderung all ihrer Repräsentationen, umgekehrt hat die Änderung des Zustandes einer Repräsentation keine Auswirkung auf ihre Referenz.

Ein Forschungsansatz, der mit dem Modell graphischer Objekte vergleichbar ist, wurde realisiert im System GEO++ /Wißk-90/. Es wurde speziell zur objektorientierten Beschreibung der Modellierungsfunktionen von PHIGS entwickelt. GEO++ unterscheidet ebenfalls zwischen Teilestrukturen und einer mehrfachen Darstellung graphischer Elemente, ähnlich wie dies in THESEUS++ der Fall ist. Allerdings ist die Beschreibung der Anwendungsschnittstelle völlig unterschiedlich, z.B. verwendet GEO++ z.B. im Gegensatz zu THESEUS++ nicht explizit das Vererbungsprinzip objektorientierter Systeme.

Das Problem der Einbettung von Teilestrukturen in objektorientierte Umgebungen und Sprachen wurde beschrieben von Blake and Cook /BlCo-87/. Ihr Ansatz sieht Datenkapselung für die Teile einer Hierarchie vor, die nur über Vermittlung ihrer Väter zugegriffen werden sollten, d.h. die Botschaften nicht selbständig empfangen und entsprechende Methoden abarbeiten sollten. Im hier beschriebenen Modell werden auch Teile als unabhängige Objekte angesehen, die direkt zugegriffen werden können. Wenn allerdings globale Auswirkungen durch die Abarbeitung einer Methode zu erwarten sind, z.B. durch eine Löschoperation, sendet das ausführende Subobjekt (d.h. das Teil) dem Vater eine entsprechende Botschaft. So werden inkonsistente Zustände der Gesamtstruktur vermieden.

4. Constraints

Zur Unterstützung direkter Manipulationen unter Restriktionen erlaubt das Ausgabemodell die Definition von Beziehungen zwischen graphischen Objekten und ihren Parametern wie Geometrien und Attributen, die vom Ausgabesystem verwaltet und sichergestellt werden. Beispiele für solche Beziehungen sind:

- verknüpfte Punkte,
- horizontale oder vertikale Linien,
- Restriktionen auf der Form von Objekten,
- gleiche Länge von Linien oder Distanzen zwischen Punkten,
- eine funktionale Beziehung zwischen der Position des Cursors und der Farbe eines Objektes
- das Erlauben oder Nicht-Erlauben einer teilweisen Überlappung zweier Objekte.

In Anlehnung an die Notation von /Lele-88/ können solche Beziehungen als *Constraints* beschrieben werden. Da wir ausschließlich Beziehungen auf graphischen Daten betrachten, sprechen wir hier von *graphischen Constraints*.

Abb. 5 zeigt drei Beispiele für die Anwendung von Constraints: (a) zeigt zwei verbundene Linien, die einen rechten Winkel umschließen, (b) zeigt ein Quadrat mit einem auf einer Linie fixierten Punkt: bei der Änderung eines der Punkte werden die anderen Punkte so geändert, daß die Form des Quadrats erhalten bleibt. (c) zeigt neun Verbindungspunkte eines Rechtecks: Wenn ein Eckpunkt verändert wird, wird das Rechteck kontinuierlich gedehnt (*stretching*), während das Bewegen des Mittelpunktes ein kontinuierliches Verschieben (*dragging*) bedeutet. In dieser Weise kann ein spezifisches Interaktionsverhalten für ein graphisches Objektes deskriptiv beschrieben werden.

Zur Beschreibung der graphischen Constraints durch die Anwendung wurden die folgenden Entwurfs-Entscheidungen getroffen:

- Constraints werden als Klassen repräsentiert, um die komplizierte Funktionalität zur Constraint-Sicherstellung zu kapseln. Ein zweiter Grund für die Beschreibung der Constraints in Form von Klassen besteht in der vereinfachten Möglichkeit zur Definition neuer Constraint-Klassen.
- Das Modell beinhaltet Kompositionsmechanismen zum Aufbau höherer Constraint-Instanzen auf der Basis von existierenden Constraints.
- Das Modell beschreibt Constraints bidirektional, d.h. eine Relation auf einer Menge von Objekten wird für beliebige Änderungen aller in der Relation enthaltenen Objekte sichergestellt.

Das Modell sieht zwei abstrakte Hauptklassen graphischer Constraints vor:

Basis-Constraints (Klasse *BasicConstraint*) sind solche Constraints, die nicht über andere Constraints parametrisiert werden. Die Term-Constraints (Klasse *TermConstraint*) sind dabei die elementarsten Constraints. Sie werden zusammengesetzt aus arithmetischen Termen, die wiederum über *Atomen* als elementaren Datentypen der graphischen Objektklassen definiert sind. Solche Atome sind z.B. reellwertige Skalare, Indizes innerhalb von Attributlisten, Punkte und Transformationsmatrizen. Zwei arithmetische Terme können dann durch Operatoren wie *Equal*, *Less* und *NotEqual* zu einem Term-Constraint verknüpft werden. Auf der Basis solcher Constraints können dann weitere Unterklassen von *BasicConstraint* wie *Horizontal* realisiert werden, die dann über nicht-atomare Datentypen wie z.B. Geometrien parametrisiert werden können.

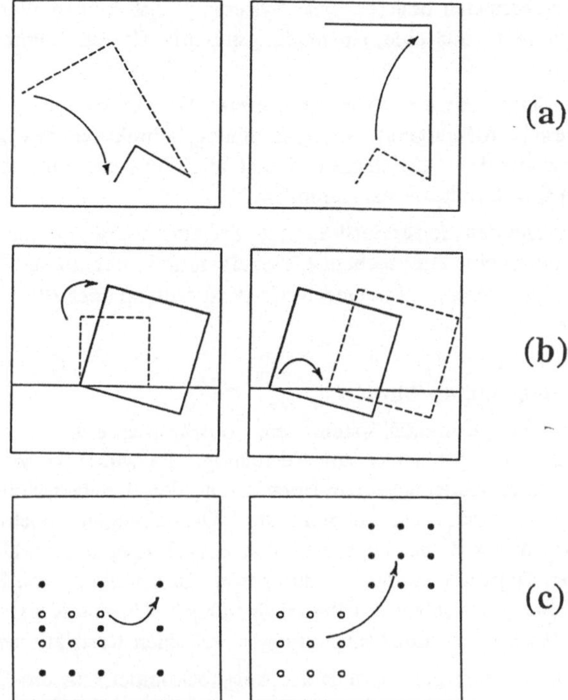

Abb. 5: Beispiel für den Einsatz graphischer Constraints

Höhere Constraints (Klasse *HighLevelConstraint*) sind Baumstrukturen, deren innere Knoten wiederum höhere Constraints, die Blätter elementare Constraints darstellen. Ein höherer Constraint verknüpft seine Söhne durch einen der beiden logischen Operatoren ODER und UND. Ein typisches Beispiel für einen ODER-Constraint ist die Beschreibung mehrerer erlaubter Bereiche für die Position eines graphischen Objektes.

Die *Constraint-Sicherstellung* (engl. *Constraint Satisfaction*) ist vor dem Anwendungsprogrammierer vollständig verborgen /Hopp-90/. Dazu werden die Constraints intern auf ein System linearer Gleichungen über *real*-Variablen abgebildet. Der Algorithmus ist realisiert durch numerische Relaxationsverfahren. In THESEUS++ werden zur Zeit zwei Methoden verwendet, zum ein Minimierungsverfahren über den Defektvektor des Gleichungssystems, zum anderen das aus der numerischen Mathematik bekannte Standardverfahren SOR (siehe z.B. /Finc-77/). Beide Verfahren lösen sowohl lineare, als auch nicht-lineare Systeme (allerdings ist die Konvergenz des SOR-Verfahrens nicht garantiert) unter zufriedenstellender Performanz beim Feedback auch während kontinuierlicher Interaktionen.

Der Algorithmus zur Sicherstellung muß nach jeder Zuweisung abgearbeitet werden. Eine solche Zuweisung kann atomar sein, d.h. ein einzelner skalarer Wert wird einer einzelnen Variablen zugewiesen, jedoch besteht in vielen Fällen eine Zuweisung aus einer Menge von Zuweisungen (z.B. nach einer geometrischen Transformation).

Constraint-Techniken wurden bisher schon verwendet in einigen graphischen Editoren für spezielle Anwendungen wie ThingLab /Born-79/ oder Juno /Nels-85/. Im Gegensatz zum hier präsentierten Ausgabemodell besitzen diese Systeme jedoch keinen Werkzeug-Charakter und konzentrieren sich nicht auf eine Anwendungsschnittstelle zur komfortablen Beschreibung ihrer Funktionalität.

Auf der anderen Seite besitzen constraint-basierte Toolkits wie Coral /SzMy-88/ oder MEL /Hill-90/ nicht dieselbe Mächtigkeit zur Beschreibung bidirektionaler Constraints wie das hier vorgestellte Ausgabemodell. Das letztere bietet allerdings ebenfalls höhere Constraints zur Kombination von Constraints aus existierenden.

Die meisten existierenden constraint-basierten Systeme repräsentieren Constraints in einer impliziten Weise durch eine entsprechende Verzeigerung innerhalb der Objekte, die durch die Constraints verknüpft werden. Im Gegensatz dazu werden hier die Constraints explizit als Klassen verwaltet.

5. Zusammenfassung und Ausblick

Das vorgestellte Ausgabemodell stellt ein objektorientiertes Beschreibungsmittel für Anwendungs- und Dialogentwickler zur Verfügung. Es wurde konzipiert als erweiterbarer graphischer Kern eines Werkzeuges zur Entwicklung der Benutzerschnittstelle für graphisch-interaktive Anwendungen, bei denen die Darstellungen hochvernetzte Strukturen repräsentieren, wie dies z.B. bei hypermedialen Anwendungen der Fall ist. Neben strukturierten graphischen Objekten wurden Constraints zur Beschreibung von Beziehungen zwischen den Geometrien und graphischen Attributen der Objekte in das Kernsystem eingebettet, um kontinuierliche Interaktionen unter anwendungsspezifischen Restriktionen zu unterstützen.

Eine Pilot-Implementierung des Modells als Ausgabekomponente des Dialog- und Graphik-Werkzeuges THESEUS++ wurde durchgeführt in der Sprache C++ und auf der Basis einer Graphik-Erweiterung des OSF/Motif-Toolkits. Das System wurde implementiert auf VaxStation 2000 und Sun SPARC. Eine detailliertere Beschreibung von Modell und Implementierung ist zu finden in der Arbeit /Lux-91/.

Die zukünftige Arbeit wird sich auf die folgenden Schwerpunkte konzentrieren:

(1) Zunächst ist ein interaktives Entwicklungs-Werkzeug oberhalb von THESEUS++ zur Definition neuer Klassen von Interaktionstechniken geplant.

(2) Das Modell soll zur Behandlung dreidimensionaler Graphik erweitert werden. Obwohl die aktuelle Version des Modells auf zwei Dimensionen beschränkt ist, sollte eine 3D-Erweiterung keine grundsätzlichen konzeptionellen Probleme mit sich bringen.

(3) Eine weiterer Arbeitspunkt umfaßt die Integration von Datentypen wie Audio und Video.

6. Literatur

/BlCo-87/ Blake, E.; Cook, S.: On Including Part Hierarchies in Object-Oriented Languages, with an Implementation in SMALLTALK. European Conference on Object-Oriented Programming, Proceedings, 1987, S. 41-50.

/Born-79/ Borning, A.: ThingLab - A Constraint-Oriented Simulation Laboratory. PhD Theses, Stanford University, 1979.

/Ehmk-90/ Ehmke, D.: PROMETHEUS - A System for Programming Graphical User Interfaces. Esprit/Eurographics International Workshop on "User Interface Management Systems and Environments", Lisbon, 1990.

/Finc-77/ Finckenstein, F. v.: Einführung in die numerische Mathematik. München, Wien: Carl Hanser 1977.

/FWC-84/ Foley, J. D.; Wallace, V. L.; Chan, P.: The Human Factors of Computer Graphics Interaction Techniques. IEEE Computer Graphics and Applications, November 1984, S. 13-45.

/GlSt-90/ Gloor, P.A.; Streitz, N.A. (Hrsg.): Hypertext und Hypermedia. Informatik Fachberichte 249. Springer 1990.

/Hill-90/ Hill, R. D.: A 2-D graphics system for multi-user interactive graphics based on objects and constraints. Eurographics Workshop on Object Oriented Graphics, Königswinter, 1990.

/HLM-87/ Hübner, W., Lux-Mülders, G., Muth, M.: Designing a system to provide graphical user interfaces: the THESEUS approach. In: Marechal, G. (ed.): EUROGRAPHICS'87, Proceedings, S. 309-322. Amsterdam, New York, Oxford, Tokyo: North-Holland 1987

/Hopp-90/ Hopp, J.: Implementierung graphischer Constraints. Diplomarbeit, Technische Hochschule Darmstadt, 1990.

/Hübn-91/ Hübner, W.: Entwurf Graphischer Benutzerschnittstellen. Springer, 1991.

/ISO-85/ International Organization for Standardization (ISO): Information Processing Systems - Computer Graphics - Graphical Kernel System (GKS) - Functional Description. ISO International Standard (IS) 7942. New York: ISO, 1985.

/ISO-88/ International Organization for Standardization (ISO). Information Processing Systems - Computer Graphics - Programmer's Hierarchical Interactive Graphics System (PHIGS) - Functional Description. ISO International Standard (IS) 9592. New York: ISO, 1988.

/Lele-88/ Leler, W.: Constraint Programming Languages. Addison Wesley 1988.

/Lux-90/ Lux-Mülders, G.: Integration graphischer Normen in Fensterumgebungen. Informatik Spektrum, Band 13, Heft 3, 1990 S. 151-162.

/Lux-91/ Lux, G.: Graphische Ausgabe für Werkzeuge zur Entwicklung interaktiver Benutzerschnittstellen. Dissertation (in Vorbereitung), Technische Hochschule Darmstadt, 1991.

/LVC-89/ Linton, M. A.; Vlissides, J. M.; Calder, P. R.: Composing User Interfaces with InterViews. IEEE Computer, Vol. 22, No. 2, 1989, S. 8-22.

/Nels-85/ Nelson, G.: Juno, a constraint based graphics system. Volume 19, Number 3, 1985. ACM SIGGRAPH Proceedings 1985, S. 235-243.

/OSF-89/ Open Software Foundation: OSF Motif Toolkit External Specifications. 1989.

/Schm-86/ Schmucker, K. J.: Object-Oriented Programming for the Macintosh. Hayden, Hasbrouck Heights (NJ), 1986.

/SzMy-88/ Szekely, Pedro A.; Myers, Brad A.: A User Interface Toolkit Based on Graphical Objects and Constraints. SIGPLAN Notices, November '88, Vol. 23 No. 11. Proceedings 3rd Annual Conference on Object-Oriented Programming Systems, Languages and Applications, (OOPSLA) '88, S. 36-45.

/Wißk-90/ Wißkirchen P.: Object-Oriented Graphics. Springer, 1990.

Natürliche Sprache und Computer-Animation als Komponenten eines Multi-Media Systems

U. Harke, U. Leiner, M. Niemöller, K. Zünkler, J. Grollmann
Siemens AG
Zentralabteilung Forschung und Entwicklung
W-8000 München 83, Otto-Hahn-Ring 6
Fax.: +49 - 89 - 636 - 48000

Zusammenfassung

In dem Multi-Media-Projekt der Siemens Zentralabteilung Forschung und Entwicklung wird ein interaktives Computeranimationssystem mit einer um die Eingabe natürlicher Sprache erweiterte Bedienoberfläche erstellt. Ausgehend von der Analyse der heutigen Situation wird die Forderung nach besserer Bedienbarkeit insbesondere der Werzeuge der Computer-Animation begründet. Der erste Prototyp besteht aus einer virtuellen Welt, die verschiedene taktile Eingabewerkzeuge kennt, und aus einem fortgeschrittenen Spracherkennungssystem. Die Objekte der virtuellen Welt folgen physikalischen Bewegungsgesetzen und können interaktiv manipuliert werden.

Abstract

The objectives of the multi-media project currently carried out at Siemens Corporate R&D laboratories are outlined. The longterm goal is to develop a computer animation system with a speech dialog interface. We sketch the experiences made in a production of an animation with tools available today. Based on that experience research was done on the extension of animation tools by physics-based models and on the improvement of the user interface by providing a command-word recognizer leading to a first prototype. Objects in a virtual world can be manipulated interactively and follow physical laws in their kinematical behaviour.

Einleitung

Die traditionellen Eingabemittel bei Computern sind Tastatur und Maus. Nicht nur im Hinblick auf Multi-Media Systeme ist diese Beschränkung unzureichend. Mit der Erweiterung der Einsatzfelder für die elektronische Datenverarbeitung steigen die Anforderungen an die Bedienoberflächen für Rechner im Hinblick auf Benutzungsfreundlichkeit und problemgerechte Bedienoberfläche. Der Dialog mit dem Rechner ist den menschlichen Interaktionsformen anzupassen. Neben dem Sprachdialog sind fotorealistische Festbilder und Bewegtbildsequenzen zur Darstellung komplexer Objekte und deren Dynamik in Bedienoberflächen als Interaktionsformen einzubringen. Die zwischenmenschliche Kommunikation basiert hauptsächlich auf Sprache und visueller Information, die insbesondere beide simultan wahrgenommen werden können. Beim Entwurf zukünftiger Bedienoberflächen sollte daher dem natürlichen Bedürfnis des Menschen, Dinge durch Zeigen, Berühren und Bewegen zu identifizieren und Anweisungen „einfach auszusprechen", entgegengekommen werden.

Auf drei Gebieten lassen die Kommunikationsfähigkeiten der meisten Rechner heute noch besonders viel zu wünschen übrig:

- Verstehen von kontinuierlich gesprochener Sprache
- Verarbeitung und Erstellung von visueller 3D-Information (Handskizzen, Bilder, Filme etc.)
- Ergonomische und standardisierte Bedienoberflächen.

Auf diesen drei Gebieten wurden in den letzten Jahren erhebliche Fortschritte gemacht. Sie werden jedoch jedes für sich und getrennt von den anderen bearbeitet.

Ein grundlegendes Anliegen unseres Projektes ist die Zusammenführung dieser Techniken, wobei in einem ersten Prototypen Überlegungen zur Standardisierung nicht im Vordergrund standen. Das Fernziel ist es, eine interaktive Animationsumgebung mit taktilen Eingabemitteln und mit Hilfe von fließend gesprochener Sprache bedienen und steuern zu können.

Auf dem Weg dahin haben wir bisher folgende Stationen erreicht: Zunächst wurde eine anspruchsvolle Animation (Videofilm) erstellt. Dabei wurden heutige Animationssysteme im Gebrauch erprobt und der gesamte Erstellungsprozeß studiert. Im nächsten Abschnitt werden die dabei erkannten Probleme und gewonnenen Erfahrungen besprochen. Im zweiten Schritt entwickelten wir ein kleines Animationssystem, das es erlaubt, eine virtuelle Welt zu erzeugen und mit den Objekten zu arbeiten. Dies ermöglicht uns, zwei wesentliche Probleme der heutigen Computer-Animation anzupacken: die Simulation von dynamischem Verhalten der virtuellen Körper sowie die Verbesserung der Bedienoberflächen.

Zukünftige Arbeiten werden es ermöglichen, Animation auf Task-Level und somit auf hohem semantischem Niveau zu steuern. Dies betrifft insbesondere die Steuerung von simulierten menschlichen Bewegungen (human animation) über Sprache. Dabei werden Befehle in Form von ganzen Sätzen gegeben, die in Inhalt und Art Anweisungen in zwischenmenschlichen Dialogen ähnlich sind. Typische Beispiele hierfür sind: „Lege Schalter XY auf Stellung 5." oder „Gehe durch den Raum zur Werkbank".

Heutige Erstellung von Computer-Animation

Im technisch-wissenschaftlichen Bereich ist die Anwendung von Computer-Animation noch relativ neu. Eine virtuelle Kamera, die den Betrachter z.B. durch das Innere eines virtuellen Hauses führt, oder die Beobachtung der komplizierten Bewegungen eines ebenfalls simulierten Montageroboters, bieten neue Möglichkeiten zur Optimierung technischer Systeme. Unsere Arbeiten zielen auf dem Gebiet der Computer-Animation vor allem auf die Simulation der Bewegungsabläufe von Menschen und Robotern (human animation). Damit wird es möglich sein, neue Arbeitsplätze im voraus ergonomisch günstig zu gestalten, oder die technische Realisierbarkeit von Arbeitsabläufen zu prüfen. Heute ist die Bedienung solcher Systeme noch recht kompliziert. Neue Interaktionsformen, insbesondere die Kommunikation in natürlicher Sprache, versprechen hier, neue Dimensionen zu erschließen.

Vor der technischen Erstellung einer Animation, muß ein Dokument vorliegen, das in der Informatik als Spezifikation und Pflichtenheft bezeichnet würde. Es handelt sich um das Drehbuch, welches den Inhalt, zeitlichen Ablauf und die Choreographie der Animation vorschreibt. Zum Produkt führen von da aus die drei technischen Schritte:

- Modellierung der Objekte,
- Bewegungsdefinition,
- Visualisierung.

Daran schließt sich die Nachbearbeitung (postprocessing) an, die das Produkt in das gewünschte Format, z.B. Videoband, bringt.

Modellierung

Modellieren in diesem Zusammenhang bedeutet die Eingabe, Beschreibung, Strukturierung, Verknüpfung und visuelle Darstellung von dreidimensionalen Objekten. Bei komplexen und fein strukturierten Objekten ist der Aufwand an Manpower hierfür erheblich. Solid Modelling erzeugt starre, nichtverformbare Objekte mit geschlossenen Oberflächen. Rotationssymmetrische Objekte sind leicht durch ein Polygon, welches das Profil beschreibt, zu definieren. Beim Sweeping definiert man eine 2-D Fläche, die, entlang einer geraden oder gekrümmten Strecke bewegt, einen 3-D Körper definiert. Das Constructive Solid Modelling baut komplexe Objekte aus mathematisch einfach zu beschreibenden Elementen, wie Block, Zylinder, Kugel und Kegel, durch deren Verküpfung mit booleschen Operationen auf. Anspruchsvoller wird die Technik zur Beschreibung von Freiformflächen. Dies kann man z.B. dadurch erreichen, daß man den Querschnitt beim Sweeping verändert. Zur Behandlung weicher Gegenstände und in der Medizin gemessener 3-D Daten, setzt man sogenannte "Voxels" ein. Diese stellen ein Volumenelement (in Anlehnung an pixel - picture element) mit Eigenschaften wie Farbe, Transparenz etc. dar. Es existieren erst wenige Lösungen, die bei der Darstellung von Gasen und Flüssigkeiten benötigt werden, wie z.B. particle based animation.

Bewegungsdefinition

Man unterscheidet die drei elementaren Arten der dreidimensionalen Animation: Keyframes [Gom 85], parametrisierte Keyframes und Algorithmen, auf die dann höhere Stufen aufbauen (Task-Level Animation). Die zweite Art, also die parametrisierte Interpolation zwischen Keyframes, ist für die Darstellung der Bewegungen des menschlichen Körpers geeignet. Ein simplifiziertes Skelett definiert dabei die Gelenke und deren Freiheitsgrade. Um aber Kommandos der Art „Gehe zur Tür" verarbeiten zu können, ist Task-Level Animation notwendig. Hier sind Aufgabenplanung und mögliche Interruptbehandlungen zu berücksichtigen. Derzeit gibt es hierfür in nur wenigen Instituten [Bad87] Ansätze. Zeltzer [McK89] verfolgt einen verwandten Ansatz mit „autonomous agents", die quasi aus eigenem Antrieb einem Ziel zusteuern.

Die heute verbreiteste Animationsart ist die Keyframe-Animation. Der Mensch muß dabei in mühseliger, handwerklicher Arbeit Details festlegen. Eine gewisse Entlastung ist die Interpolation (inbetweening), so daß nicht jedes einzelne Bild neu erstellt werden muß. Von einem Keyframe zum nächsten können Parameter, wie Positionen, Lage, Formen und Farben der Objekte verändert werden. Lichtquellen können ein- oder ausgeschaltet, ihre Position und Lichtstärke verändert werden. Die Hauptaufgabe des Animators ist es festzulegen, innerhalb welcher Framesequenz und somit zu welcher Zeit die Änderungen stattfinden sollen. Schließlich gehört auch die Animation der virtuellen Kamera dazu. Dies bedeutet, Position, Blickwinkel und Blickpunkt (Fokus) der Kamera in der zeitlichen Abfolge zu beschreiben.

Eine weitere algorithmische Animationsmethode benutzt kinematische Regeln, sowie inverse Kinematik zusammen mit Randbedingungen (Beschränkungen) in einem Simulationssystem. Die Bewegungen von Körpern werden durch Lösung der physikalischen Bewegungsgleichungen beschrieben. Dies führt zu Bewegungen, die für den Beobachter natürlich erscheinen. Hinzu kommt hierbei noch die Kollisionsvermeidung, eine Methode, die in der Robotik viel eingesetzt wird.

Visualisierung

Visualisierung ist das Erstellen eines am Bildschirm sichtbaren Bildes für einen Rahmen (Frame). Dieser Vorgang wird auch Rendern genannt. Hier kann die Berücksichtung von Eigenschaften wie Schattierung, Transparenz von Objekten, und Spiegelungseffekte die Bildqualität wesentlich beeinflussen.

Die Animation oder Bewegungdefinition eröffnet neben den 3 räumlichen Dimensionen beim Modellieren die vierte Dimension, die Zeit. Um bei der interaktiven Erstellung einer Animationssequenz das Ergebnis, nämlich die schnelle Abfolge von Frames und damit den

Bewegungseffekt zu sehen, enthalten solche Animationssysteme einfache "Renderer". Diese sind deshalb sehr schnell, da sie sich darauf beschränken, bunte Drahtmodelle der Objekte zu zeigen. Bestenfalls werden die Oberflächen der Objekte mit sehr simplem sogenannten "Solid Rendering" ausgefüllt. Dadurch bekommt der Anwender schon einen guten Eindruck über die Lage der Objekte zueinander.

Diese recht bescheidene Qualitätsstufe des Renderns ist in den Arbeitsprozessen der Modellierung und der Bewegungsdefinition ausreichend. Anspruchsvoller ist man bei dem Rendern des Endergebnisses. Hier kann man bei gegebener Auflösung und Farbqualität des Monitors bis zur Qualität von fotorealistischen Bildern gelangen. Dieser Schritt findet i. a. in länger laufenden Hintergrundprozessen statt. Jeder Rahmeninhalt muß einzeln visualisiert werden. Bei längeren Sequenzen ist dies nicht nur sehr CPU-intensiv, sondern produziert auch Daten von enormer Menge.

Begründet in dieser Notwendigkeit, zwischen Geschwindigkeit und Qualität zu wählen, liegt die Forderung an moderne Animationssysteme, einerseits interaktives Previewing in niedriger Auflösung und andererseits hochauflösendes Rendering in der Stapelverarbeitung anzubieten.

Postprocessing

In der Regel kann weder das Rendern selbst noch das Lesen der dazu notwendigen Daten von einer Festplatte in Echtzeit geschehen. Deshalb ist das Übertragen der Daten in ein Videoformat, und zwar Rahmen für Rahmen, unumgänglich.

Durch die geeignete Vertonung mit Text, Musik und / oder Umweltgeräusche kann die Aussagekraft eines Filmes wesentlich ergänzt werden. Dies ist in der Regel unverzichtbar.

Trip to the Insights

Um den Status Quo der Computer-Animation zu beurteilen und Schwächen der heutigen Werkzeuge aufzudecken, haben wir eine kurze Animation "Trip to the Insights" erstellt. Der Inhalt des Filmes zeigt eine Projektion der langfristigen Ziele unseres Vorhabens. Eine Anwenderin arbeitet mit einem futuristischen, sehr komfortablen Animationssystem. Die interaktive Konstruktion eines Hauses beginnt mit 2D-Grundrißzeichungen und wächst Stockwerk um Stockwerk in die Höhe. Durch natürlichsprachliche Eingaben können Objekte geändert oder ausgetauscht werden, der Blickwinkel verändert werden. Man kann in das Haus eintreten und sich darin umschauen. Durch diesen Film können wir die dargestellte Zielumgebung mit der heutigen Arbeitsumgebung anschaulich vergleichen. Darüberhinaus regt das Projekt fruchtbare Diskussionen zwischen Fachleuten auf den drei Gebieten Sprachverarbeitung, Bedienoberflächen und Computer-Animation an. Nicht zuletzt demonstriert der Film die Nutzbarkeit der Computer-Animation in Architektur, Produktion, Simulation und vielen anderen Bereichen.

Für den etwa dreiminütigen Film wurden mehr als 3000 Frames in VHS-Qualität erzeugt. Auf Workstations war die durchschnittliche Rendering-Zeit ca. 3 Minuten je Rahmen, was sich zu 150 Stunden aufaddiert. Die höhere UMATIC-Highband-Qualität erfordert einen etwa sechsfachen Rechenaufwand. Es ist unbestritten, daß Rendern außerordentlich teuer ist, wenn fotorealistische Qualität erreicht werden soll. Auch mit sehr leistungsfähigen Rechnern der Zukunft wird Echtzeit-Rendern für interaktive Arbeit nur bei reduzierter Qualität möglich sein.

Mit Ausnahme des Renderns sind alle Arbeitsgänge der Animationserstellung sehr interaktiv und fordern erheblichen Aufwand an Arbeitskraft. Während einige Modellierer bereits graphische Bedienoberfächen haben und die Maus als Eingabegerät kennen, gestatten die meisten Animationssysteme lediglich Eingaben über die Tastatur. Der Anwender, insbesondere der Neuling, bekommt zu wenig Orientierungshilfen und verliert sich leicht im virtuellen dreidimensionalen Raum. Die Verbesserung der Bedienoberfächen würde zu erheblich besseren Produkten führen und die Technik der Computer-Animation, die heute noch einer relativ kleinen Gruppe von Experten vorbehalten ist, einer breiteren Anwendung zuführen. Die Spezifizierung von Bewegung durch Keyframe-Animation bedarf gut ausgebildeter Leute mit einem ausgeprägten räumlichen Vorstellungsvermögen und Sinn für natürliche Bewegungsabläufe und Effekte. Besonders schwierig ist es, Bewegungsabläufe nicht kantig sondern natürlich aussehen zu lassen, als ob die Körper physikalischen Gesetzen gehorchten. Hierzu bedient man sich am besten der Simulation, deren Algorithmen aber ihrerseits wieder sehr CPU-intensiv sein können.

Die softwaretechnische Aufteilung der drei oben beschriebenen Schritte auf drei unterschiedliche Softwarepakete ist historisch bedingt und muß überwunden werden. Die Kommunikation zwischen den Paketen geschieht über Datenfiles. Dies impliziert eine strenge Trennung der Arbeitschritte und hohen Zeitaufwand beim Wechsel von einer Phase in eine andere. Mit einem integrierten System kann der Benutzer effektiver arbeiten, da Iterationsschritte leichter durchführbar sind und keine Wiederholungen implizieren.

Prototyp für ein Animations-System mit Spracheingabe

Angeregt durch die im vorangegangenen Abschnitt diskutierten Erkenntnisse, wurden Arbeiten auf den drei Gebieten Simulation von Bewegungsdynamik, Darstellung und Interaktion mit einer virtuellen Welt, sowie Integration von Sprache in die Bedienoberfläche begonnen, die alle in einem gemeinsamen Prototypen mündeten. SESAM (Siemens Environment for Speech controlled AniMation) ist der erste Prototyp einer angestrebten, interaktiven Animationsumgebung mit einer multimedialen Bedienoberfläche.

Virtuelle Welt und Bewegungssimulation
SESAM enthält eine einfache 3D-Welt, in der verschiedene vorgefertigte Objekte erzeugt werden können, indem auf deren Beschreibungen in einer Datenbasis zurückgegriffen wird. Neben den üblichen geometrischen Strukturen, die durch Eckpunkte und Polygone beschrieben werden, haben unsere Objekte auch physikalische Eigenschaften, die durch Massenpunkte und Randbedingungen oder Beschränkungen von Freiheitsgraden definiert sind [Nie90]. Den Objekten können Attribute wie Elastizität, Anfangsgeschwindigkeit, Gravitation und die Art der Darstellung zugeordnet werden, die vom Animator während der Sitzung interaktiv verändert werden können. Die freie Bewegung der Körper wird durch die Lagrangefunktion der pysikalischen Bewegungsgleichungen beschrieben, die die Einschränkungen und Randbedingungen mitberücksichtigt. Die Lösung dieser Gleichungen erfolgt durch einen neuen approximativen, iterativen Algorithmus [Nie90]. Dadurch, daß in der Approximation nur diejenigen Parameter berücksichtigt werden, die einer Veränderung unterliegen, wird der Algorithmus echtzeitfähig. Die Bewegungen akzeptiert der Beobachter als vollkommen natürlich, wenn auch, um das Echtzeitverhalten zu erreichen, z. B. der Drehimpulserhaltungssatz nicht immer vollständig erhalten ist.

Benutzungsoberfläche
Die Funktionalität des Animationssystems für die virtuelle Welt in SESAM ist soweit möglich getrennt gehalten von Funktionen des Dialogteiles. Der Dialog kann interaktiv durch den Benutzer geführt werden, oder durch Lesen von Skriptdaten aus einer Datei selbständig ablaufen. Seine Aufgabe ist es, die Dialogoberfläche (s. Bild1) aktuell zu halten. Dies

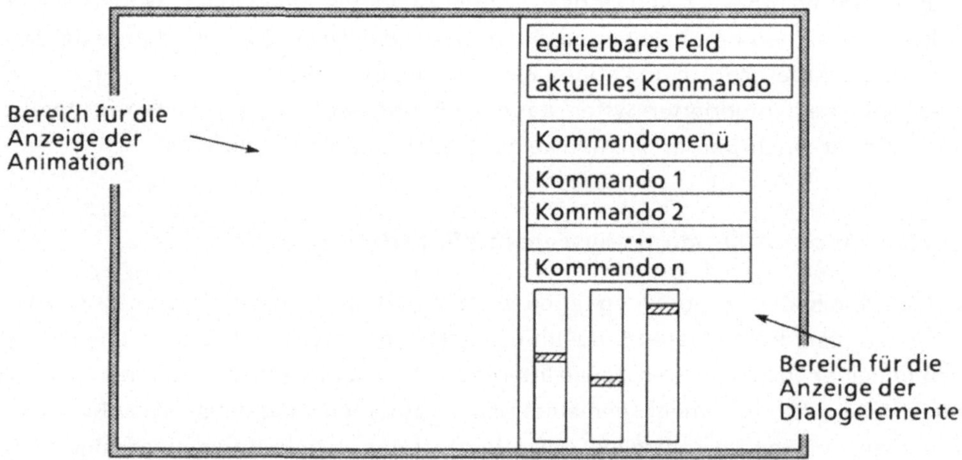

Bild 1: Bedienoberfläche des Systems

beinhaltet, Feedback nach erfolgten Eingaben zu geben, indem die dann gültigen Kommandos, das zuletzt ausgewählte Kommando und Fehlermeldungen angezeigt werden. Neben Menüs stehen auch andere interaktive Eingabemöglichkeiten zur Verfügung, wie editierbare Felder, Schieberegler (slider), 3D-Kraftmoment-Sensor (Geometry Ball) und, wie weiter unten noch ausgeführt wird, die natürliche Sprache. Unabhängig davon, mit welchem Interaktionsmedium ein spezielles Kommando eingegeben wurde, reagieren der Dialogteil und das Animationssystem in identischer Weise.

Unsere Oberfläche erlaubt die simultane Eingabe über verschiedene Eingabekanäle. Der Animator kann zu jedem Zeitpunkt die für ihn und die Situation optimale Methode benutzen: Tastatur, Menüauswahl und der Wechsel der Sliderfunktionalität durch Mausklick. 3D-Eingabegeräte wie die Sensor-Kugel erleichtern die Orientierung und die Bewegung von Objekten im virtuellen Raum. Die natürlichere und ergonomische Kommunikationsform ist die Sprache. Die hier vorgestellte Verknüpfung von verschiedenen Eingabegeräten mit der Eingabe in natürlicher Sprache wertet die Bedienoberfläche erheblich auf, da die Interaktion mit dem Animationssystem dadurch wesentlich verbessert wird.

Spracheingabe

Im ersten Prototypen unserer multi-medialen Bedienoberfläche ist ein sprecherunabhängiger Worterkenner integriert, der kontinuierliche Hidden-Markov-Modelle für Phoneme benutzt. Jedes erkannte Wort ist mit Hilfe eines Ausspracheiexikons aus einer Sequenz von Phonemen zusammengesetzt. Der Worterkenner kann zu mehreren alternativen Sequenzen (Hypothesen) kommen und wählt daraus diejenige aus, die die Wahrscheinlichste ist und keinen speziellen Kriterien widerspricht. Die Erkennungsrate ist in gewissem Maße abhängig von dem konkreten Kommandovorrat und liegt bei 99.5% [Akt91]. Mit speziellen HW-Boards kann der Erkennungsprozeß in Echtzeit ablaufen. Es handelt sich dabei um zwei Boards mit A/D-Wandler, drei Signalprozessoren und einem ASIC-Chip für das vector-matching. Diese Boards sind als Teil des Projektes SPICOS [Zue90] für das Erkennen von kontinuierlicher Sprache entstanden. Durch das Hinzufügen eines dritten Boards wird man in der näheren Zukunft Echtzeit für das Erkennen kontinuierlich gesprochener, natürlicher Sprache erreichen.

Software- und Hardwaretechnische Aspekte

Die Dialogkomponente ist der Monitor für die Ein-/Ausgaben. Sie wurde mit den Standard-Werkzeugen LEX und YACC in UNIX erstellt. Die Eventschleife prüft ständig ab, ob Eingaben vorliegen (s. Bild 2). Diese Eingaben können derzeit über die physikalischen Geräte Tastatur, Maus oder Geometry-Ball sowie über Spracheingabe ausgelöst werden. Sie werden interpretiert und in eine geräteunabhängige Form gebracht. Die der Dialog-Grammatik entsprechenden Kommandosequenzen werden in Aktionen des Animationssystems übersetzt. Diese Sequenzen sind die einzige Schnittstelle zwischen dem Dialogpaket und dem Animations-

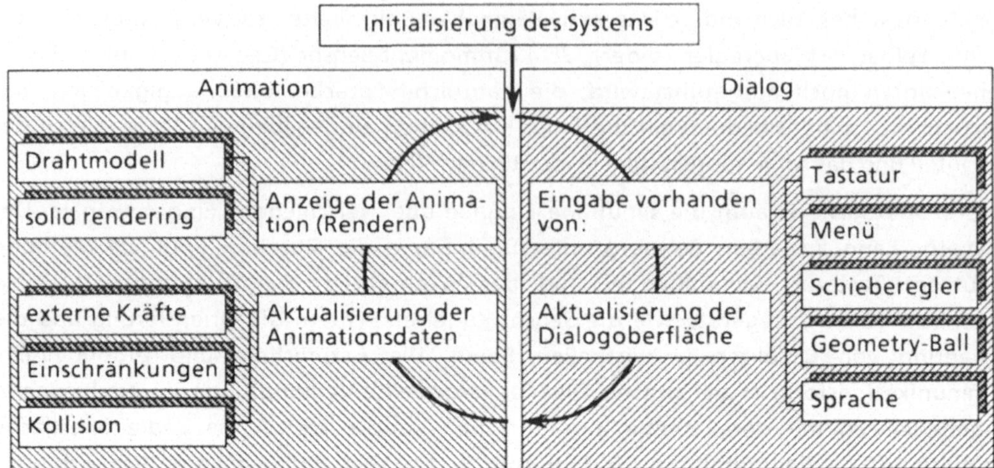

Bild 2: Ablaufdiagramm

system. Liegen keine Eingaben vor, fährt das Animationssystem in der Berechnung der laufenden Prozesse ohne Einfluß von außen fort.

Der Prototyp ist innerhalb der letzen zwölf Monate entstanden. Die verwendete Programmiersprache ist C. Die Animation und die Kontrolle des Benutzerdialoges laufen auf einer Silicon Graphics 4D 50G unter Verwendung der Graphikbibliothek ab. Der Spracherkennungsteil läuft auf einer Sun 4 Workstation, die mit den beiden speziellen Sprachboards am VME-Bus ausgestattet ist. Die Sun 4 und die Silicon Graphics sind über TCP/IP gekoppelt.

Die Trennung des Dialogteils von der eigentlichen Anwendung sowie die Verwendung des UNIX Werkzeugs LEX für die Interpretation der Eingaben haben erwiesen, daß eine Integration von neuen Eingabewerkzeugen innerhalb des Systems leichter möglich ist als bei herkömmlicher Programmierung. Die Realisierung der Dialog-Grammatik innerhalb von YACC ermöglicht zusäztlich eine schnelle Erweiterung und Umstellung der Dialoggestaltung.
Der Dialogteil kann nicht nur die über Tastatur eingegebenen Daten interpretieren, sondern auch Inhalte aus Dateien lesen. Daher ist es möglich, z.B. selbstablauffähige Demonstrationen aus Dateien abzurufen.

Erste Erfahrungen im Gebrauch der integrierten Oberfläche

Es hat sich gezeigt, daß die Spracheingabe die Handhabung des komplexen Systems beschleunigt, insbesondere dann, wenn ein Wechsel der Hand von einem Gerät zum anderen (etwa von Maus zu Tastatur) dadurch vermieden werden kann. Die Bedienung des Geometry-Balls wird anfangs als gewöhnungsbedürftig betrachtet. Anwender sind jedoch schon nach kurzer Zeit mit dem Gerät vertraut. Es bietet eine besonders gute Möglichkeit, sich innerhalb eines Raumes zu orientieren, sowie die Lage von Objekten im Raum zu verändern.

Eine gute Rückmeldung von Informationsdaten (etwa Koordinaten, Drehachse) an den Benutzer würde den Einsatz dieses Geräts noch mehr unterstützen. Als positiv hat sich das Angebot der Schieberegler herausgestellt. Da diese für unterschiedliche Aktionen (z.B. Ändern von Farbe, Drehen eines Objektes) verschiedene Wertebereiche liefern, geben sie durch zusätzliche Anzeige der aktuellen Werte eine gute Rückkopplung für den Anwender. Die Kombination verschiedener Eingabemedien innerhalb einer Dialogoberfläche kommt sowohl dem unerfahrenen Anwender als auch dem Experten zugute.

Zukünftige Arbeiten

In Zukunft werden Forschungsarbeiten bezüglich der Animation selbst und der multimedialen Benutzungsoberfläche fortgesetzt. Im Animationsbereich werden wir die Funktionalität von SESAM durch die konventionelle Keyframe-Technik erweitern. Sowohl die Simulation der Bewegungsdynamik als auch die Key-Frame-Methode haben ihre Vorteile und Anwendungsgebiete. Die Zusammenführung wird dem Anwender beides anbieten.

Animation auf dem bisher diskutierten recht elementaren Niveau genügt nicht den Anforderungen für verschiedene Anwendungen wie z.B. der Human Animation. Hierbei handelt es sich um Task-Level Animation auf hohem Niveau [Bad 89]. Hier liegt auch die wahre Rechtfertigung für die Integration der Eingabe von fließend gesprochenen Sätzen in die Bedienoberfläche. Die typischen Befehle sind derart, daß sie den Sätzen der zwischenmenschlichen Kommunikation äußerst nahe kommen.

Die Sprache ist ein Medium mit der Möglichkeit, ein interaktives System zu erstellen, das einen Menschen oder Roboter, der Befehle und Aufgaben durchführt, ziemlich nahe simulieren kann. Der Mensch gibt seine Befehle oder ganze Handlungspläne in natürlicher Sprache ein und verändert oder bewegt Objekte der virtuellen Welt direkt durch taktile Eingabegeräte. Das System analysiert die Aktionspläne und bildet sie auf elemetare Befehle ab. Unsere nächsten Schritte werden untersuchen, wie das System für die Erkennung fließend gesprocher Sprache [Akt89] mit einem Task-Level Animationssystem verknüpft werden kann.

Zusammenfassung

Die Erstellung von Computer-Animationen ist sehr aufwendig sowohl was Rechenzeit als auch insbesondere die menschliche Arbeitszeit betrifft. Die wesentliche Verbesserung der Bedienbarkeit solcher Animationssysteme wird die Techniken einem größeren Nutzerkreis eröffnen.

Die Bereitstellung von Animationssystemen auf sehr hohem Task-Level in Verbindung mit verschiedenen Eingabemedien insbesondere durch fließend gesprochene Sätze, wird ein anschauliches Beispiel liefern, wie der Umgang mit dem Computer sich künftig vereinfachen wird. So werden die dem Menschen zur Verfügung stehenden Kommunikationskanäle besser genutzt und angesprochen.

Literatur

[Akt89] Aktas A., Höge H.: Real-time recognition of subword units on a hybrid multi-DSP/ ASIC based acoustic front-end; Proc. of the IEEE International Conference on Acoustics, Speech and Signal Processing, pp. 101 - 103, Edinburgh, May 1989

[Akt91] Aktas A., Zünkler K.: Speaker independent continous HMM-Based recognition of isolated words on a real-time multi-DSP System; Eurospeech 91, Genova, Sept. 1991

[Bad 89] Badler N.I.: Communication Tasks and their Performance via Animation and Natural Language; Technical Report University of Pennsylvania, June 1989

[Gom 85] Gomez J.: TWIXT: A 3D Animation System Comput. & Graphics Vol.9 No. 3, pp. 291 - 298

[McK89] M. McKenna, D. Zeltzer: Dynamic Simulation of Autonomous Legged Locomotion, Tch.Report The Media Lab, Massachusetts Institute of Technology, Cambridge, MA, January 1989

[Nie 90] Niemöller M., Leiner U.: Approximative, Dynamic Simulation of Objects in Computer Animation Systems; Proceedings of Computer Graphics 90 Conference, 6.-8. Nov. 1990, London

[Zue90] Zünkler K.: Speech-understanding systems: The communication technology of tomorrow; in: Schwärzel and Mizin (Eds.): Advanced Information Processing, pp. 227 - 251, Springer 1990

Interaktive Schnittstellengestaltung in einem objektorientierten Rahmen

Thomas Berlage

GMD, Postfach 1240, D-5205 Sankt Augustin 1

Zusammenfassung

Als Bestandteil des Systems GINA (Generische Interaktive Anwendung) wurde ein Interface Builder entwickelt, ein Werkzeug, mit dem interaktiv graphische Benutzeroberflächen auf Basis von OSF/Motif erstellt werden können. Der Interface Builder setzt konsequent Techniken der direkten Manipulation zur Beschleunigung des Definitionsprozesses und zur Unterstützung des explorativen Lernens ein. Er unterstützt besonders die dynamischen Layout-Fähigkeiten von Motif. Der Interface Builder erleichtert den Umgang mit dem Form-Widget durch ein einfacheres Linealmodell, das intern auf die Motif-Mechanismen abgebildet wird. Außerdem kann er aus einer ungenauen Lagevorgabe automatisch Parameter für ein dynamisches Layout der Komponenten generieren.

1. Einführung

Das System GINA (Generische Interaktive Anwendung) wird in der GMD als User Interface Management System (UIMS) zur Erstellung einheitlicher Benutzerschnittstellen für das Leitprojekt "Assistenz-Computer" entwickelt (Spenke90b). Der Schwerpunkt des Systems liegt auf der vereinfachten Erstellung graphischer, direkt-manipulativer Oberflächen.

Um portabel zu sein und möglichst viele Standardkomponenten auszunutzen, setzt der Assistenz-Computer auf dem Betriebssystem UNIX, dem X Window System (Scheifler86, McCormack88) und dem OSF/Motif-Toolkit (Berlage91) auf. Motif ist eine Bibliothek von Schnittstellenbausteinen, die von der Open Software Foundation, einem Industriekonsortium, entwickelt wurde. GINA unterstützt die Programmiersprachen C++ und Common Lisp mit CLOS.

Ein Ziel von GINA ist es, möglichst viele Teile einer Anwendung interaktiv zu spezifizieren (insbesondere den Fensteraufbau) und nur die unbedingt notwendigen Teile zu programmieren.

GINA besteht aus zwei wesentlichen Teilen: der eigentlichen *generischen Anwendung* (Spenke-90a), einer Klassenbibliothek, und dem *Interface Builder*, einem Werkzeug zur graphischen Komposition von Motif-Oberflächen. Die Klassenbibliothek verwendet das Konzept einer *Rahmenanwendung* ähnlich wie in MacApp (Schmucker86, Wilson90) oder ET++ (Weinand89), um Gemeinsamkeiten zwischen Anwendungen vorgefertigt zur Verfügung zu stellen. Der Interface Builder erzeugt als Ergebnis Code für GINA.

Eine interaktive graphische Benutzerschnittstelle wird in einem objektorientierten System aus einzelnen Schnittstellenobjekten zusammengesetzt, die in Motif *Widgets* genannt werden. Widgets sind rechteckige Bereiche mit einem bestimmten Aussehen und Verhalten (Abb. 1). Ein ganzes Anwendungsfenster besteht aus einer Anzahl entsprechend parametrisierter Widgets.

Nicht für alle Schnittstellenobjekte einer Anwendung können Motif-Widgets eingesetzt werden, zum einen, weil Motif-Widgets immer rechteckig sind, und zum anderen, weil es nicht ganz einfach ist, neue Widgetklassen zu programmieren, wenn die existierende Funktionalität nicht ausreicht. Daher werden eigene Graphiken oder Objekte mit unregelmäßiger Form innerhalb von Zeichenflächen (Views) unter Benutzung der X-Zeichenoperationen realisiert. Die Programmierung solcher *View-Objekte* wird durch eine Reihe von GINA-Klassen unterstützt. Dabei finden Konzepte ähnlich wie in Unidraw (Vlissides90) oder im Garnet Toolkit (Myers90b) Anwendung.

Im Gegensatz zu vielen Systemen aus der Forschung hat Motif eine sehr viel höhere Komplexität. Widgetklassen haben bis zu 70 verschiedene Parameter, deren komplette Dokumentation mehrere hundert Seiten benötigt. Um die interaktive Gestaltung einer Motif-Schnittstelle

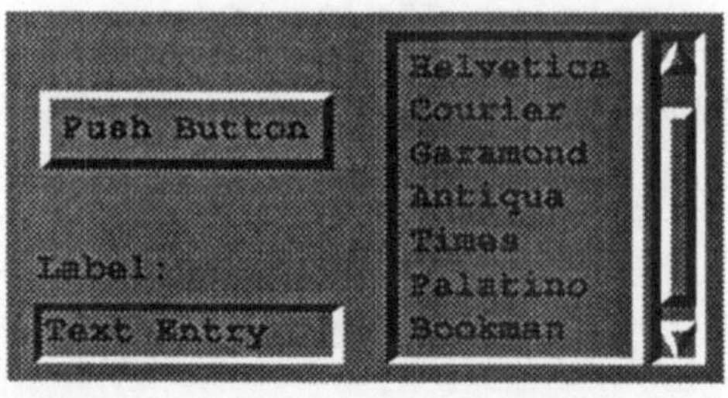

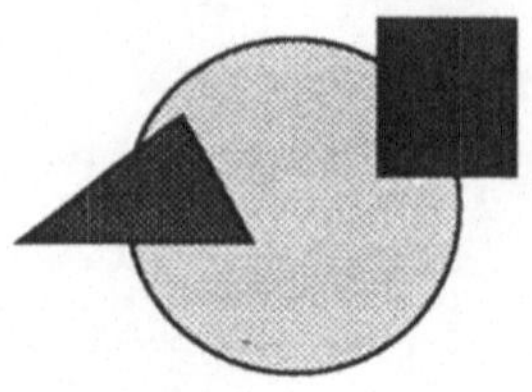

Abb. 1: Vergleich Widgets – View-Objekte. Widgets (links) sind rechteckige Standardbausteine. View-Objekte (rechts) können beliebige Formen und anwendungsspezifisches Aussehen und Verhalten besitzen.

trotz dieser Schwierigkeiten möglichst einfach zu machen, verwendet der GINA Interface Builder eine ganze Reihe von Maßnahmen mit den beiden folgenden Schwerpunkten:

- Das explorative Arbeiten soll erleichtert werden.
- Der Umfang der notwendigen Spezifikationen soll reduziert werden.

Die Möglichkeit des explorativen Arbeitens senkt den Lernaufwand erheblich. Weil sehr viel einfacher alternative Lösungen ausprobiert werden können, wird die optimale Lösung schneller erreicht. Die folgenden Punkte, die das Prinzip der direkten Manipulation charakterisieren, tragen dazu bei:

- Alle Bedienungsmöglichkeiten müssen an der Benutzerschnittstelle des Interface Builders sichtbar sein (zum Beispiel darf es keine geheimen Tastenoptionen geben, deren Wirkung nicht auch anders zugänglich ist). Ein Beispiel dafür ist der *Magnet-Mechanismus*, der in Abschnitt 3 beschrieben wird und der es ermöglicht, Operationen auf Objekten anzubieten, die nicht direkt anwählbar sind.
- Jede Benutzeraktion muß eine unmittelbare Rückkopplung über seine Auswirkungen produzieren. Je früher diese Rückkopplung erfolgt, desto schneller können Alternativen geprüft werden. Der Interface Builder produziert in der Regel eine vollständige Rückkopplung bei jedem einzelnen Tastendruck und jeder Mausbewegung.
- Alle Änderungen müssen zuverlässig rückgängig gemacht werden können. Der Interface Builder erlaubt dies für beliebig viele Schritte zurück, ein Merkmal, das von der GINA Klassenbibliothek bereitgestellt wird.

In Abschnitt 3 wird beschrieben, wie sich diese Prinzipien in der tatsächlichen Benutzung des Interface Builders niederschlagen.

Motif unterscheidet sich von vielen anderen Systemen dadurch, daß die Fensteraufteilung dynamisch durch eine Hierarchie von Manager-Widgets realisiert wird (dies gilt mehr oder weniger für alle Toolkits, die auf den X Toolkit Intrinsics (McCormack88) basieren). Abschnitt 4 beschreibt, wie der Interface Builder diesen Mechanismus unterstützt.

Der GINA Interface Builder reduziert den Umfang der für eine bestimmte Schnittstelle nötigen Spezifikationsschritte auf verschiedene Weise:

- Es werden einige Bausteine als Einheiten angeboten, die sich aus mehreren Motif-Widgets zusammensetzen (zum Beispiel Radio-Button-Groups oder Menüs). Diese Kombinationen sind bereits in der Motif-Implementation und dem Motif Style Guide (OSF90) vorgesehen.
- Die komplexen Layout-Parameter des Form-Widgets von Motif werden in einem *Linealmodell* dargeboten, das einfacher zu verstehen und zu spezifizieren ist. Dieses Modell wird in den Abschnitten 5 und 6 detailliert beschrieben.
- Anfangswerte für die Layout-Parameter von Manager-Widgets werden automatisch aus der ungefähren Anordnung der Komponenten abgeleitet. Auf diese Weise kann ein dynamisch veränderliches Layout aus einem statischen Beispiel gewonnen werden (Abschnitt 7).

2. Vergleichbare Systeme

Eine Reihe von Interface Buildern wurde schon in der Literatur beschrieben. Wegweisende Vorbilder sind zum Beispiel DialogEditor (Cardelli88) oder der NeXT Interface Builder

(Webster89). Der GINA Interface Builder unterscheidet sich von den meisten Systemen dadurch, daß er auf einem Standard-Toolkit aufsetzt und keinen eigenen Satz von Schnittstellenobjekten definiert. In vergleichbarer Situation sind Druid (Singh90b), das auch auf Motif basiert, sowie OSU (Lewis89), das die Macintosh Toolbox benutzt.

Die grundlegende Interaktionsmöglichkeiten der Motif-Widgets sind durch die vordefinierten Parameter der einzelnen Widgetklassen festgelegt. Es besteht keine einfache Möglichkeit, wie in Peridot (Myers90a) oder Lapidary (Myers89) neue Interaktionstechniken zu definieren. Dies ist teilweise auch nicht erwünscht, da die Standardobjekte in allen Anwendungen gleich aussehen und das gleiche Verhalten zeigen sollen (OSF90).

Systeme unterscheiden sich auch darin, wie sie die Fensteraufteilung beschreiben. Frühere Systeme, zum Beispiel vu (Singh90a), unterstützen nur eine statische Spezifikation wie in einem Graphikeditor. Neuere Modelle, die das Layout dynamisch anpassen können, zum Beispiel stretching (Cardelli88), glue (Vlissides90) oder beliebige Constraints (Myers90b), lassen sich nicht ohne weiteres auf Motif übertragen. Dazu wäre eine neue Manager-Widgetklasse erforderlich. Der GINA Interface Builder bietet daher ein Linealmodell an, das auf die Motif-eigenen Fähigkeiten abgebildet werden kann. Ein vergleichbares Linealmodell, aber mit allgemeinen Constraints, wird in (Hudson90b) beschrieben. Das Linealmodell in GINA ist jedoch stärker spezialisiert, weil es nur Fähigkeiten abdecken muß, die durch das Form-Widget vorgegeben sind.

Die Idee, Layout-Parameter aus einem Beispiel abzuleiten, wurde durch die Arbeitsweise von Peridot (Myers90a) motiviert. Der GINA Interface Builder überträgt die Idee auf die speziellen Parameter der Motif-Widgets. Die Entscheidungen des Systems müssen jedoch nicht wie in Peridot vom Benutzer bestätigt werden, weil sie anschließend sehr einfach direkt-manipulativ wieder geändert werden können.

Mittlerweile sind Interface Builder für Motif auch als kommerzielle Produkte erhältlich, zum Beispiel TeleUSE (Larsson90). Der GINA Interface Builder hat ihnen die einfache Bedienung, die konsequente Verwendung direkter Manipulation und die spezielle Unterstützung des Layout-Prozesses voraus.

3. Umgang mit dem Interface Builder

Jedes zusammengesetzte Fenster einer Anwendung wird auf einer großen Arbeitsfläche entworfen (Abb. 2). Die einzelnen Widgets können mit der Maus aus einer Palette "gegriffen" und an einer bestimmten Stelle der Arbeitsfläche "fallengelassen" werden. Dieser "Drag and Drop"-

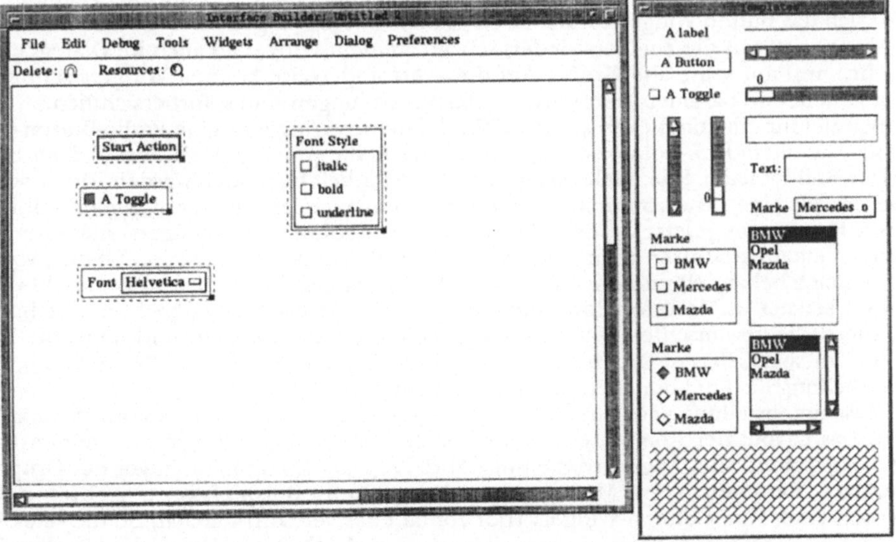

Abb. 2: Arbeitsfenster des Interface Builders mit Palette. Einige Widgets wurden bereits plaziert.

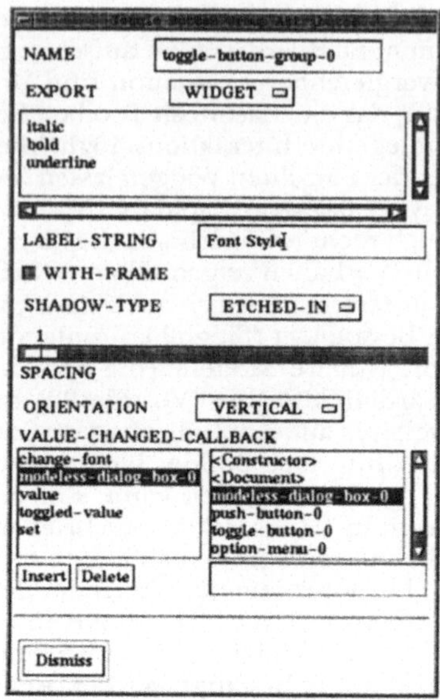

Abb. 3: Ressourcedialoge für ein Push-Button-Widget und eine Toggle-Button-Group.

Mechanismus wird auch im NeXT Interface Builder verwendet (Webster89).

Die Widgets können zunächst beliebig verschoben werden. Außerdem sind sie nicht nur Bilder, sondern voll funktionsfähig, das heißt Button-Widgets können gedrückt, Scroll-Bar-Widgets betätigt und Einträge von List-Widgets selektiert werden. Ein separater Testmodus ist damit überflüssig. Um die Betätigung der Widgets von Metaoperationen zu unterscheiden, hat jedes Widget einen schmalen Rand, in dem Metaoperationen wie etwa das Verschieben ausgelöst werden können.

Die Einstellung der Widget-Parameter, der sogenannten *Ressourcen*, geschieht über nichtmodale Dialoge. Für jedes Widget steht ein eigener Dialog zur Verfügung (Abb. 3). Änderungen haben sofortige Auswirkung auf das Widget; wenn also der Text eines Buttons geändert wird, verlängert sich das Button-Widget auf der Arbeitsfläche mit jedem getippten Zeichen. Der Menüeintrag "Undo" entfernt die Zeichen wieder, sowohl auf dem Widget als auch im Dialog, und das Widget schrumpft auf seine alte Größe. Auf diese Art und Weise können Widgets sehr schnell zusammengestellt und parametrisiert werden; alle Auswirkungen sind sofort ersichtlich.

Ein Beispiel für ein komplexeres, von GINA definiertes Widget ist eine Radio-Button-Group, die in Motif aus einem Row-Column-Widget und einer Reihe von Toggle-Button-Widgets besteht. Im Interface Builder kann eine solche Gruppe als eine Einheit behandelt werden. Die möglichen Werte werden als eine Liste von textuellen Einträgen angegeben, die dann dynamisch in eine Anzahl von Buttons umgesetzt werden. Auf diese Weise können alle Widgets, die eine 1-aus-n bzw. m-aus-n-Semantik besitzen (zum Beispiel das List-Widget oder ein Option-Menü), vom Programm aus gleich behandelt und ohne Programmänderung gegeneinander ausgetauscht werden.

Obwohl Fenster in Motif top-down aufgebaut werden müssen, ermöglicht es der Interface Builder, zuerst die gewünschten primitiven Komponenten auszuwählen und dann die Zusammenstellung zu spezifizieren. Dies ist die natürlichere Reihenfolge für die Entwicklung, da die Zusammenstellung die Funktionalität nur wenig beeinflußt.

Zur Zusammenstellung wird ein Teil der Widgets selektiert und mit einem *Manager-Widget* gruppiert. Das Layout der Komponenten wird dann vom Manager-Widget und seinen Layout-Parametern bestimmt, allerdings wird die ungefähre Lage der Komponenten vor der Gruppierug dazu benutzt, Parameter des Managers zu bestimmen. Beispielsweise richtet sich die Orientierung eines Row-Column-Widgets (horizontal oder vertikal) danach, ob die selektierten Widgets eher horizontal oder vertikal angeordnet sind (Abb. 4). Besonders wichtig ist dies für die zahlreichen Parameter des Form-Widgets (siehe Abschnitt 7).

Der Benutzer kann das Geometrieverhalten des Manager-Widgets direkt ausprobieren, indem er das Widget vergrößert und verkleinert und dabei die Änderung des Layouts beobachtet. Die Layout-Parameter des Manager-Widgets, zum Beispiel der Abstand zwischen den Komponenten, werden ebenfalls über einen Dialog angegeben.

Der Interface Builder ist selbst eine mit GINA realisierte Anwendung. Entsprechend dem Dokumentmodell von GINA kann eine Fensterdefinition in einer Datei abgespeichert und später wieder geladen werden.

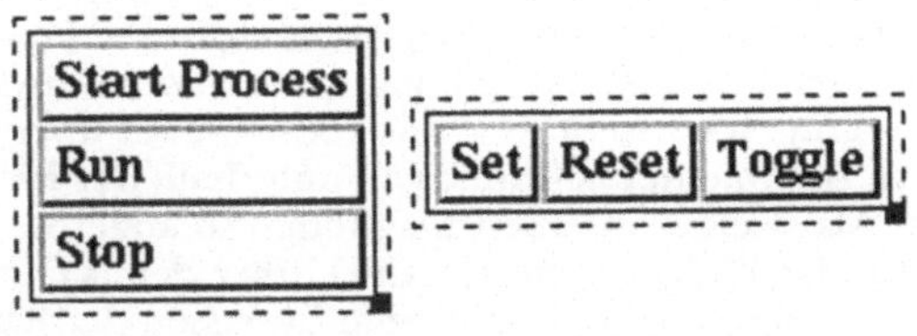

Abb. 4: Zwei Row-Column-Widgets in verschiedener Orientierung.

Als Ergebnis erzeugt der Interface Builder Code für GINA, der eine neue Fensterklasse definiert. Die Klasse stellt eine Konstruktorfunktion bereit, mit der beliebig viele Instanzen dieser Fensterklasse erzeugt werden können. Für Änderungen zur Laufzeit sind alle Bestandteile des Fensters als Variablen der Fensterinstanzen zugreifbar. Der Code wird wahlweise in Lisp oder C++ erzeugt, passend zur GINA-Version in der jeweiligen Sprache.

4. Verwaltung der Widget-Hierarchie

In Motif wird die Fensteraufteilung durch eine Hierarchie von Manager-Widgets bestimmt. Viele andere Systeme erlauben nur feste Koordinaten oder Constraints zwischen Objekten auf einer Ebene. Manager-Widgets enthalten eine Anzahl von anderen Widgets als Komponenten. Jedes Fenster bildet eine solche Widget-Hierarchie (Abb. 5). Das Layout der Komponenten hängt von der Strategie der gewählten Managerklasse ab. Beispielsweise sorgt das Row-Column-Widget dafür, daß alle seine Komponenten links und rechts bündig in einer Spalte angeordnet werden. Die Gesamtbreite richtet sich nach der breitesten Komponente.

Das *Geometrie-Management*, die dynamische Anpassung von Lage und Größe der Widgets in der Hierarchie, verfolgt zwei Ziele:

- Die Widgets bestimmen untereinander die optimale Anfangsgröße des gesamten Fensters. Diese Größe ist abhängig von den verwendeten Schriftgrößen oder Texten (die in X vom Benutzer vorgegeben werden können und aufgrund von Umgebungsbedingungen oft auch müssen).
- Der Benutzer kann ein Fenster beliebig vergrößern oder verkleinern. In diesem Fall soll eine sinnvolle Reaktion erfolgen, beispielsweise daß variable oder weniger wichtige Komponenten verkleinert werden, während wichtige Bedienungselemente sichtbar bleiben.

Um diese Fähigkeiten von Motif richtig zu nutzen, muß die geeignete Hierarchie von Manager-Widgets mit den geeigneten Layout-Parametern ausgewählt werden. Der GINA Interface Builder erleichtert den Aufbau dieser Hierarchie. Widgets können zwar auch frei positioniert werden, diese Positionen werden jedoch nur als Anhaltspunkt genommen, um daraus automatisch Parameter für die Manager-Widgets zu berechnen.

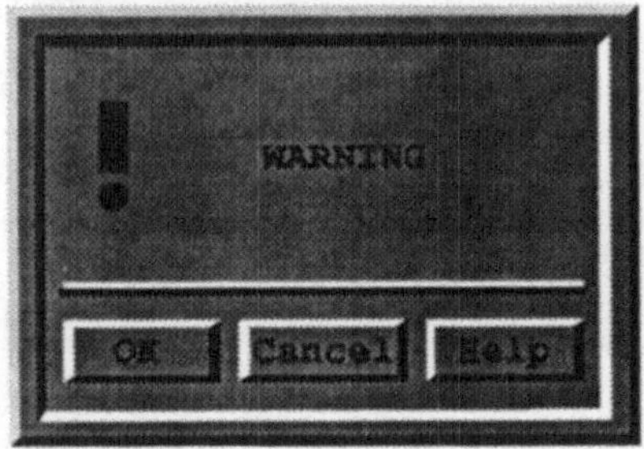

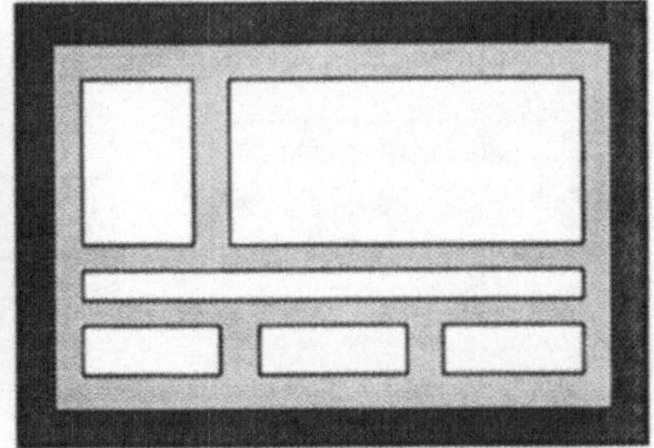

Abb. 5: Widget-Hierarchie für ein einfaches Dialogfenster.

Sobald Widgets in einem Manager zusammgefaßt werden, sind sie in der Hierarchie eingespannt. Nur die Widgets auf oberster Ebene können im Interface Builder unabhängig bewegt und selektiert werden. Dies ist eine direkte Folge davon, daß wirkliche Widgets verwendet werden, um die Fensteraufteilung exakt zu demonstrieren.

Eingeschlossene Widgets können aber über Hilfsmittel wie den *Magneten* angesprochen werden. Der Magnet befindet sich unterhalb der Menüleiste (s. Abb. 2). Beim Anklicken ändert sich der Mauszeiger in einen Magneten, solange die Maustaste gedrückt bleibt. Beim Loslassen wird dann das Widget entfernt (Cut), über dem sich der Magnet gerade befunden hat. Ähnlich funktioniert die *Lupe*, nur wird beim Loslassen der Parameterdialog des Widgets gezeigt.

Mit diesem Magnet-Mechanismus lassen sich verschiedene Operationen auf einem Objekt ausführen, ohne zwischen Maustasten unterscheiden zu müssen oder Umschalttasten der Tastatur hinzuzuziehen. Allerdings wird in Umkehrung der üblichen Reihenfolge zuerst die Aktion und dann das Objekt gewählt. Der gleiche Mechanismus wird auch für kontextsensitive Hilfe verwendet. Ein Fragezeichen kann gegriffen werden, um Informationen über ein bestimmtes Element der Schnittstelle abzurufen. Wesentliches Merkmal ist, daß der durch Auswählen der Aktion betretene Modus automatisch beim Loslassen der Maustaste wieder verlassen wird.

Zur Unterstützung beim Zielen wird das Widget, über dem sich der Zeiger befindet, durch einen blinkenden Rahmen hervorgehoben. Aus Performance-Gründen wird diese Hilfe aber erst gezeigt, wenn die Maus ruht oder sich nur noch langsam bewegt.

Auch das Einfügen neuer Widgets in die Hierarchie ist möglich, indem man sie passend fallenläßt. Auch hier wird beim Anhalten der Maus eine passende Rückkopplung erzeugt.

Umbauten in der Hierarchie können auch in einer baumartigen Ansicht der Hierarchie vorgenommen werden. Hier können ganze Unterbäume angefaßt und umgehängt werden.

5. Ein Linealmodell für das Form-Widget

Zur Realisierung komplexer Layoutstrategien dient in Motif das Form-Widget. Es ermöglicht es, für jede Seite jeder Komponente ein *Attachment* zu spezifizieren. Ein Attachment ist ein gerichteter Constraint: diese Seite richtet sich nach etwas anderem. Es stehen acht verschiedene Arten von Attachments zur Verfügung. Abb. 6 zeigt ein Beispiel und das resultierende Verhalten.

Die Spezifikation eines Layouts durch Attachments ist recht mühsam, schon allein durch die große Anzahl von Einzelheiten (9 verschiedene Attachments im Beispiel von Abb. 6). Außerdem sind leicht unsinnige oder widersprüchliche Spezifikationen möglich, als Reaktion fällt das Form-Widget dann in sich zusammen.

Um diese Probleme zu vermeiden, verwendet der GINA Interface Builder ein weniger detailliertes Modell aus *Linealen*, das dann intern auf Attachments abgebildet wird. Durch die Zusammenfassung mehrerer Attachments zu einem Lineal sinkt der Spezifikationsaufwand, die Spezifikation wird übersichtlicher und leichter zu verstehen. Da die unsinnigen Fälle ausgeschlossen sind, ist das Linealmodell nicht so allgemein wie die zugrundeliegenden Attachments, aber alle praxisrelevanten Fälle sind enthalten. Ein ähnliches Modell, allerdings mit allgemeinen Constraints, wird in (Hudson90b) beschrieben.

Ein Lineal verbindet gleiche Seiten mehrerer Widgets, die alle auf einer Linie liegen sollen. Entsprechend den verschiedenen Typen von Attachments gibt es verschiedene Linealtypen. Es werden drei wesentliche Typen unterschieden:

- Lineale mit einem festen Abstand zu einer Seite des Form-Widgets, zum Beispiel 12 Punkte zum linken Rand.

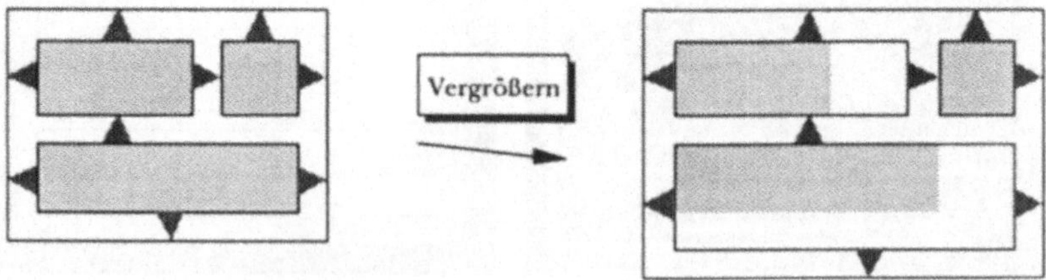

Abb. 6: Beispielattachments eines Form-Widgets und das resultierende Verhalten beim Vergrößern.

- Lineale mit einer relativen Position innerhalb des Form-Widgets, zum Beispiel auf einer Höhe von 30 Prozent der Gesamthöhe.
- Lineale deren Position von der Seite eines anderen Widgets, des *Führers*, abhängt.

Eines der Widgets entlang eines Lineals kann *variabel* sein, das heißt, dieses Widget wird vergrößert, wenn das Form-Widget größer gezogen wird.

Abb. 7 zeigt die Linealkonfiguration für das Beispiel von Abb. 6. Lineale sind als gestrichelte Linien dargestellt. Das Dreieck am linken oder oberen Ende symbolisiert den Typ des Lineals (zu welcher Seite der Abstand konstant gehalten wird). Das variable Widget wird durch einen grauen Balken entlang des Lineals gekennzeichnet.

Sobald ein Widget an einem Lineal liegt, sind die Positionen aller seiner Seiten festgelegt, und zwar aus folgenden Gründen:

- Das Lineal hält den Abstand zwischen seinen Widgets konstant.
- Jedes Motif-Widget hat eine bevorzugte Größe, die es einnimmt, wenn keine Beschränkungen vorliegen.

Eine Ausnahme bildet der Führer eines Lineals: da er einen Bezugspunkt für die anderen Widgets am Lineal darstellt, muß er selbst noch durch ein anderes Lineal positioniert werden. Im GINA Interface Builder können nur Konfigurationen erzeugt werden, in denen alle Widgets festgelegt sind.

6. Direkte Manipulation der Lineale

Das Linealmodell kann im Interface Builder als alternative Sichtweise auf ein Form-Widget gewählt werden (Abb. 7). Technisch geschieht dies durch Überlagerung eines Views, in dem die Linealsicht dargestellt und manipuliert wird. Die Linealsicht ist orthogonal zu den sonstigen Operationen, das heißt, auch in der Linealsicht kann das Form-Widget verschoben oder vergrößert werden.

Änderungen des Layouts können nun durch direkte Manipulation in der Linealsicht vorgenommen werden. Die Änderungen werden in Attachments umgesetzt und an das Form-Widget weitergereicht. Die neue Form-Konfiguration wird dann in die Linealsicht übernommen. Die Funktion des Form-Widgets wird also nicht simuliert, die Linealsicht ist immer konsistent mit dem tatsächlichen Form-Verhalten.

Durch Anklicken des Typsymbols kann der Typ eines Lineals umgeschaltet werden. Durch Anfassen dieses Symbols wird das Lineal mit allen Widgets verschoben. Ein Widget kann verschoben werden, um den Abstand zu seinem oberen oder linken Nachbarn zu verändern, dabei sind Überlappungen verboten.

Die Zuordnung einer Seite zu einem Lineal kann geändert werden, indem man ein Kästchen an der jeweiligen Seite anfaßt und verschiebt. Ist das Ziel ein anderes Widget, werden diese

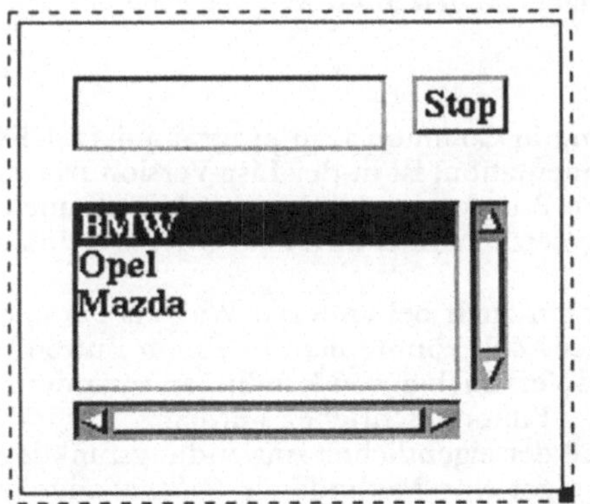

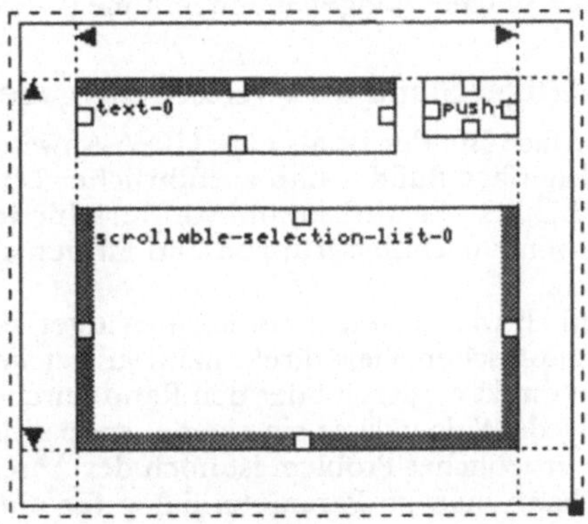

Abb. 7: Normale Ansicht und Linealsicht eines Form-Widgets.

beiden Widgets mit einem neuen Lineal verbunden, falls möglich. Ansonsten wird ein neues Lineal an der Zielposition erzeugt. Auf diese Weise werden keine besonderen Operationen zum Erzeugen und Löschen von Linealen benötigt (ein Lineal wird gelöscht, wenn es keine Widgets mehr besitzt).

Durch die Technik des *semantic snapping* (Hudson90a) wird sichergestellt, daß keine unzulässigen Konfigurationen entstehen können. Zum Beispiel werden zirkuläre Abhängigkeiten zwischen Führern verhindert. Die graphische Rückkopplung während des Verschiebens zeigt immer exakt die Position, an der ein neues Lineal erzeugt würde. Wenn der Mauszeiger in ein anderes Widget wandert, wird das neue Lineal an der Position dieses Widgets gezeigt. Ist der Zeiger außerhalb von Widgets, oder ist die Verbindung mit einem Widget nicht erlaubt, folgt die Rückkopplung genau der Mausposition.

7. Bestimmung des Anfangszustandes

Selbst mit dem Linealmodell ist die Spezifikation eines dynamischen Layout nicht ganz einfach. Daher versucht der Interface Builder, eine Konfiguration aus einer ungefähren Plazierung der Komponenten abzuleiten. Diese Konfiguration kann dann gegebenenfalls noch interaktiv abgeändert werden.

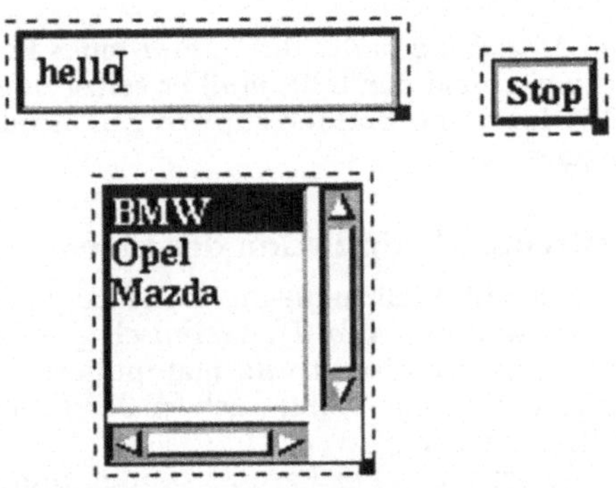

Abb. 8: Eine mögliche Ausgangslage die zur Konfiguration in Abb. 7 führt.

Zunächst ordnet der Benutzer alle gewünschten Komponenten ungefähr an (Abb. 8). Beim Gruppieren zu einem Form-Widget wertet der Interface Builder die Lage der Komponenten sowie deren Typen und Parameter aus und leitet daraus eine Linealkonfiguration ab, die dann in dem Form-Widget installiert wird.

Aus der in Abb. 8 gezeigten Lage produziert der Interface Builder automatisch die in Abb. 7 gezeigte Konfiguration. Er erkennt, daß Text und Button an der oberen Kante und Text und Scrolled-List an der linken Kante ausgerichtet werden sollen. Er erkennt außerdem, daß Scrolled-Lists typischerweise in beiden Richtungen variabel sein sollen, einzeilige Texte nur horizontal, und Buttons gar nicht.

In vielen Fällen produziert der Interface Builder automatisch die richtige Konfiguration, in vielen anderen Fällen müssen nur geringfügige Änderungen vorgenommen werden.

Der Interface Builder benutzt einen grundsätzlich recht einfachen Algorithmus. Zuerst werden alle Komponenten grob in Zeilen und Spalten eingeteilt. Diese Zeilen und Spalten werden von links oben anfangend nach und nach in Lineale verwandelt, bis alle Komponenten festgelegt sind. Eine Heuristik entscheidet schließlich aufgrund der Klassen der Komponenten und ihrer Parameter, welches Widget an einem Lineal als variabel ausgewählt wird.

8. Derzeitiger Stand und weitere Aufgaben

Der Interface Builder ist als eine GINA-Anwendung in Common Lisp programmiert. GINA, inklusive Interface Builder und ausführlicher Dokumentation, ist in der Lisp-Version frei von der GMD erhältlich. Es wird sowohl von verschiedenen Projektgruppen in der GMD als auch international von Forschungsgruppen und Firmen eingesetzt. Parallel wird eine GINA-Version in C++ entwickelt.

Ähnlich wie derzeit beim Form-Widget könnten auch bei anderen Widgets Parameter in einem graphischen View direkt manipuliert werden. Z.B. könnte man in einem Button-Widget den Text direkt editieren oder den Rand durch das Verschieben von Randlinien verändern. Dazu muß für jede Widgetklasse ein eigener spezialisierter Editor geschrieben werden.

Ein erhebliches Problem ist noch der Anschluß der eigentlichen Anwendungsfunktionalität. Derzeit kann man im Parameterdialog für ein Widget eine Methode als *Callback* angeben, die dann programmiert werden muß. In Zukunft soll der Interface Builder um graphische Programmiermöglichkeiten erweitert werden, wie sie zum Beispiel in (Harbert90) beschrieben sind. Dies würde dann Rapid Prototyping ohne Programmierung erlauben.

Sobald Anwendungsfunktionalität angeschlossen wird, ist die Fensterdefinition nicht mehr unabhängig, sondern Programmcode und Fensterdefinition müssen konsistent gehalten werden. In Zukunft soll der Interface Builder diese Abhängigkeiten verwalten. Dann kann ein Endbenutzer gefahrlos das Fenster eines existierenden Programms verändern, indem er den Interface Builder in einem Modus verwendet, in dem nur Änderungen erlaubt sind, die das Anwendungsprogramm nicht betreffen.

9. Schlußfolgerungen

Wie bei Verwendung eines Standard-Toolkits zu erwarten, werden bei der Benutzung des GINA Interface Builders Designprobleme und einzelne Programmfehler von Motif deutlich. In solchen Fällen ist beim Anwender noch zusätzliches Wissen über Motif notwendig.

Das Konzept, die Resultate aller Änderungen sofort beobachten zu können, hat hohe Akzeptanz bei den Anwendern gefunden. Die Motif-Dokumentation braucht nur noch wenig in Anspruch genommen zu werden. Durch das unbegrenzte und vor allen Dingen verläßliche Rückgängigmachen wird der Lerneffekt wesentlich gesteigert.

Die Kombination der automatischen Generierung von Anfangswerten mit der leichten Änderbarkeit ist im Mittel wesentlich schneller als die komplette Eingabe aller Parameter. Trotzdem hat der Benutzer nicht das Gefühl, er habe geringere Kontrolle über das Anwendungsprogramm, wie dies bei vollständig automatischen Verfahren leicht vorkommt.

Ein wesentliches Problem ist noch der Anschluß der Anwendungsfunktionalität. Gerade für Rapid Prototyping fehlen Funktionen, da man hier eigentlich ganz ohne Programmierung in GINA auskommen möchte. Auf diesem Gebiet ist noch weitere Forschungsarbeit zu leisten.

Literatur

[Berlage91] Thomas Berlage
OSF/Motif und das X Window System
Addison-Wesley, Bonn, 1991

[Cardelli88] Luca Cardelli
Building User Interfaces by Direct Manipulation
Proceedings of the ACM SIGGRAPH Symposium on User Interface Software (Banff, Alberta, Canada, Oct. 1988), pp. 152-166

[Harbert90] Andrew Harbert, William Lively, Sallie Sheppard
A Graphical Specification System for User-Interface Design
IEEE Software 7, 7 (July 1990), pp. 12-20

[Hudson90a] Scott E. Hudson
Adaptive Semantic Snapping - A Technique for Semantic Feedback at the Lexical Level
Proceedings of CHI Conference on Human Factors in Computing Systems (Seattle, WA, April 1-5, 1990), pp. 65-70

[Hudson90b] Scott E. Hudson, Shamim P. Mohamed
Interactive Specification of Flexible User Interface Displays
ACM Trans. Inf. Syst. 8, 3 (July 1990), pp. 269-288

[Larsson90] Leif Larsson
Creating Interaction Primitives
In User Interface Management and Design, Proceedings of the Workshop on User Interface Management Systems and Environments (Lisbon, Portugal, June 4-6, 1990), eds. D. A. Duce, M. R. Gomes, F. R. A. Hopgood, J. R. Lee, Springer Verlag: Berlin, pp. 273-293

[Lewis89] T. G. Lewis, Fred Handloser III, Sharada Bose, Sherry Yang
Prototypes from Standard User Interface Management Systems
IEEE Computer, May 1989, pp. 51-60

[McCormack88] Joel McCormack, Paul Asente
An Overview of the X Toolkit

Proceedings of the ACM SIGGRAPH Symposium on User Interface Software (Banff, Alberta, Canada, Oct. 1988), pp. 46-55

[Myers89] Brad A. Myers, Brad Vander Zanden, Roger B. Dannenberg
Creating Graphical Objects by Demonstration
Proceedings of the ACM SIGGRAPH Symposium on User Interface Software and Technology (Williamsburg, VA, Nov. 13-15, 1989), pp. 95-104

[Myers90a] Brad A. Myers
Creating User Interfaces Using Programming by Example, Visual Programming, and Constraints
ACM Trans. Program. Lang. Syst. 12, 2 (April 1990), pp. 143-177

[Myers90b] Brad A. Myers, Dario Giuse, Roger B. Dannenberg, Brad Vander Zanden, David Kosbie, Ed Pervin, Andrew Mickish, Philippe Marchal
Garnet: Comprehensive Support for Graphical, Highly Interactive User Interfaces
IEEE Computer, Nov. 1990, pp. 71-85

[OSF90] Open Software Foundation
OSF/Motif Style Guide, Release 1.1
Prentice-Hall, 1990

[Scheifler86] Robert W. Scheifler, Jim Gettys
The X Window System
ACM Trans. Graph. 5, 2 (1986), pp. 79-109

[Schmucker86] Kurt Schmucker
Object-Oriented Programming for the Macintosh
Hayden Book Company, New Jersey, 1986

[Singh90a] Gurminder Singh
Vu: Visual User-Interface Design
The Visual Computer 6 (1990), Springer Verlag, pp. 230-241

[Singh90b] Gurminder Singh, Chun Hong Kok, Teng Ye Ngan
Druid: A System for Demonstrational Rapid User Interface Development
Proceedings of the ACM SIGGRAPH Symposium on User Interface Software and Technology (Snowbird, Utah, Oct. 3-5, 1990), pp. 167-177

[Spenke90a] Michael Spenke, Christian Beilken
An Overview of GINA - the Generic Interactive Application
In User Interface Management and Design, Proceedings of the Workshop on User Interface Management Systems and Environments (Lisbon, Portugal, June 4-6, 1990), eds. D. A. Duce, M. R. Gomes, F. R. A. Hopgood, J. R. Lee, Springer Verlag: Berlin, pp. 273-293

[Spenke90b] Michael Spenke, Thomas Berlage
GINA - Ein objektorientiertes Integrationsmodell für die Benutzerschnittstelle des Assistenz-Computers
GMD Jahresbericht 1989, Sankt Augustin 1990, pp. 90-100

[Vlissides90] John M. Vlissides, Mark A. Linton
Unidraw: A Framework for Building Domain-Specific Graphical Editors
ACM Trans. Inf. Syst. 8, 3 (July 1990), pp. 237-268

[Webster89] Bruce F. Webster
The NeXT Book
Addison-Wesley, 1989

[Weinand89] André Weinand, Erich Gamma, Rudolf Marty
Design and Implementation of ET++, a Seamless Object-Oriented Application Framework
Structured Programming (1989), Springer-Verlag New York

[Wilson90] David A. Wilson, Larry S. Rosenstein, Dan Shafer
Programming with MacApp
Addison-Wesley, 1990

Objektorientierte Dialogspezifikation

Rainer Götze
Universität Oldenburg
Fachbereich Informatik
Postfach 2503
D-2900 Oldenburg
Arbeitsgruppe "Informatik-Systeme"[1]

Kurzfassung

Eine objektorientierte Dialogspezifikation beschreibt den statischen Aufbau ("Layout") und das dynamische Verhalten einer Benutzerschnittstelle in Form einer Menge von dialogspezifischen Objekten sowie den Beziehungen und die Kommunikation zwischen diesen Objekten. In dieser Arbeit werden die Konzepte einer objektorientierten Dialogspezifikationssprache beschrieben. Diese Sprache unterstützt insbesondere die Spezifikation von interaktiven, graphischen Benutzerschnittstellen und damit auch die Entwicklung von Hypermedia-Benutzerschnittstellen. Die Beschreibung der ODISS (Objektorientierte DIalogSpezifikationsSprache) genannten Sprache beinhaltet die wesentlichen Sprachkonstrukte und deren informelle Semantik.

1. Einleitung

Die Software-Architektur moderner Anwendungssysteme weist in zunehmendem Maße eine Aufteilung in die Benutzerschnittstelle und den anwendungsspezifischen Berechnungsteil auf. Dies gilt auch für Hypermedia-Systeme [Con87], deren hochgradig interaktive, graphische Benutzerschnittstellen durch

- eine fensterorientierte Darstellung der Informationen, z.B. Texte, Graphiken und Bilder,
- eine graphische Darstellung struktureller Informationen, z.B. logisch vernetzte Struktur multimedialer Informationsknoten (Hypermedia-Netzwerk), und
- direkte Manipulation graphischer Objekte, z.B. Navigieren in einem Hypermedia-Netzwerk durch Aktivieren von Verbindungen zwischen Informationsknoten

gekennzeichnet sind. Der Begriff Benutzerschnittstelle bezieht sich dabei sowohl auf die graphische Oberfläche einer Anwendung (Benutzungsoberfläche) als auch die Softwaremodule für die Kommunikation mit dem Benutzer.

Für die Entwicklung dieser Art von Benutzerschnittstellen steht eine Reihe von Werkzeugen zur Verfügung, die sich in User Interface Toolkits (UIT, [My89]), User Interface Management Systeme (UIMS,[Ka82],[Pf85]) und User Interface Development Systeme (UIDS,[Har89],[Pf85]) aufteilen lassen. UIT bieten dem Anwendungsprogrammierer eine evtl. erweiterbare Bibliothek von Interaktionskomponenten, aus denen graphische Benutzerschnittstellen mit Hilfe einer Programmierschnittstelle aufgebaut werden können. Die Struktur und der Dialogablauf einer mit einem UIT entworfenen Bentuzerschnittstelle werden explizit programmiert und damit auf einem sehr niedriegen Abstraktionsniveau spezifiziert. UIMS ermöglichen dagegen eine abstrakte Spezifikation von Benutzerschnittstellen in textueller oder graphischer Form, die in eine interne, ausführbare Repräsentation übersetzt wird. Die Entwicklung von Benutzerschnittstellen wird dadurch übersichtlicher und erheblich beschleunigt. UIDS können als Weiterentwicklung der UIMS betrachtet werden und unterstützen eine interaktive Entwicklung von Benutzerschnittstellen.

Ziel dieser Arbeit ist die Unterstützung einer objektorientierten Entwicklung von interaktiven, graphischen Benutzerschnittstellen durch ein entsprechendes UIMS. Hypermedia-Benutzerschnittstellen stellen nur einen Spezialfall der hier betrachteten graphischen Benutzerschnittstellen dar. In objektorientierten Hypermedia-Benutzerschnittstellen werden z.B. die Informationsknoten durch Objekte und die Verbindungen zwischen ihnen durch Kommunikationsverbindungen zwischen den Objekten modelliert.

---

[1]gefördert durch die Stiftung Volkswagenwerk (Az. 210-70631/9-13-14/89)

Während die interaktive Entwicklung der statischen Struktur ("Layout") bereits von einigen Systemen (z.B. [My86]) ermöglicht wird, ist eine vollständig interaktive Entwicklung des dynamischen Verhaltens einer Benutzerschnittstelle sehr schwierig zu realisieren. Aufgrund dieser Probleme wird mit dieser Arbeit eine textuelle Spezifikation von Benutzerschnittstellen angestrebt. Die dafür entwickelte *Dialogspezifikationssprache* ODISS ermöglicht die objektorientierte Beschreibung der Struktur und der Dynamik von graphischen Benutzerschnittstellen.
Die Notwendigkeit der Entwicklung einer objektorientierten Dialogspezifikationssprache beruht auf dialogspezifischen Anforderungen, denen die Objektmodelle objektorientierter Programmiersprachen nicht genügen und die auf folgenden Eigenschaften beruhen:

- Objekte reagieren nicht nur auf Nachrichten in Form von Methodenaufrufen, sondern auch auf die vom Benutzer erzeugten Ereignisse (*externe Events*).
- Externe Events besitzen im Gegensatz zu Methodenaufrufen kein Zielobjekt, sondern müssen an alle daran interessierten Objekte verteilt werden (*Eventverteilung*).
- Benutzerschnittstellen unterliegen oft bestimmten Constraints, das sind Einschränkungen der zulässigen Objektzustände, Konfigurationen und Dialogabläufe.

Die Dialogspezifikationssprache ODISS und ihr Laufzeitsystem berücksichtigen diese Anforderungen und unterstützen insbesondere die

- objektorientierte Spezifikation des dynamischen Verhaltens einer Benutzerschnittstelle, die
- objektorientierte Spezifikation von Constraints mit Hilfe eines speziellen Kommunikationskonzeptes (*Eventreceiving*) und die
- objektorientierte Eventverteilung durch das Laufzeitsystem.

Um die umfangreiche Entwicklungsarbeit erweiterbarer, objektorientierter User Interface Toolkits auszunutzen, soll ODISS auf einem solchen Toolkit aufbauen und dessen UI-Komponenten als Präsentationskomponenten der ODISS-Objekte verwenden. Die in einem Toolkit integrierte Eventbehandlung soll dabei von einer durch das ODISS-Laufzeitsystem realisierten objektorientierten Eventverteilung, der sogenannten hierarchischen Eventverteilung, ersetzt werden.

In dieser Arbeit werden zunächst die Anforderungen an eine Spezifikation dynamischen Verhaltens in Benutzerschnittstellen dargestellt. Darauf aufbauend werden das ODISS-Objektmodell und die ODISS-Sprachkonstrukte zur Spezifikation dynamischen Verhaltens vorgestellt. Die Beschreibung des ODISS-Objektmodells beinhaltet dabei eine Vorstellung anderer objektorientierter Ansätze zur Dialogmodellierung. Abschliessend wird die vom Laufzeitsystem realisierte Eventverteilung erläutert.

2. Dynamisches Verhalten von Benutzerschnittstellen

Als dynamisches Verhalten einer Benutzerschnittstelle werden die Reaktionen auf Events und die Veränderungen zur Laufzeit (z.B. Erzeugen und Löschen von Informationsknoten eines Hypermedia-Netzwerkes, Navigieren entlang eines Pfades) bezeichnet. Eine objektorientierte Beschreibung dynamischen Verhaltens einer Benutzerschnittstelle kann durch das Verhalten einzelner Objekte, das Verhalten einer Gruppe von Objekten und das globale Verhalten der Benutzerschnittstelle klassifiziert werden:

- Verhalten einzelner Objekte:
  - Reaktionen auf die vom Benutzer erzeugten Events, z.B. direktes Feedback auf Manipulation graphischer Objekte,
  - Ereignisfolgen, die ein Objekt durchlaufen darf, z.B. zulässige Eventverarbeitungsfolgen (objektspezifisches Verhalten),
  - Einschränkungen (Constraints) der gültigen Zustandstransitionen einzelner Objekte.
- Verhalten einer Gruppe von Objekten:
  - zulässige Eventverarbeitungsfolgen mit Beteiligung von mehrerer Objekten, z.B. Navigieren durch ein Hypermedia-Netzwerk,

- Einschränkungen (Constraints) der Beziehungen zwischen mehreren Objekten, z.B. Integritätsbedingungen eines Hypermedia-Netzwerkes.
- globales Verhalten einer Benutzerschnittstelle:
  - Änderungen des globalen Zustandes einer Benutzerschnittstelle (Konfiguration), z.B. Änderungen der Menge der sichtbaren Objekte und zulässigen Aktionen.

Diese Klassifikation beschreibt gleichzeitig die Anforderungen an eine Sprache zur Spezifikation des dynamischen Verhaltens einer Benutzerschnittstelle. ODISS und das zugrundeliegende Laufzeitsystem enthalten folgende (Sprach-) Konstrukte, die diesen Anforderungen genügen sollen und im Rahmen dieser Arbeit erläutert werden:

- *Eventverarbeitung* zur Spezifikation der objektspezifischen Reaktion auf Events,
- *Eventreceiving* zur Kommunikation zwischen den Objekten,
- *Eventsequenzen* zur Spezifikation zulässiger Eventverarbeitungsfolgen eines Objektes,
- *Objektinvarianten* zur Spezifikation von Constraints,
- *Kommunikationsgraph* zur effizienten Eventverteilung und Konfigurationsverwaltung.

3. Das ODISS-Objektmodell und andere objektorientierte Ansätze

Das ODISS-Objektmodell stellt ein für die Dialogspezifikation erweitertes Objektmodell dar, das in mehrere Abschnitte aufgeteilt ist und der Spezifikation von Objektklassen und Objekten dient. Die wichtigsten Erweiterungen betreffen dabei die Spezifikation dynamischen Verhaltens. Die Spezifikation einer ODISS-Objektklasse hat folgende Form (Kennzeichnung optionaler Teile analog zur EBNF in [Wi85]):

```
OBJECTCLASS <IDENTIFIER>
  SUPERCLASSES <IDENTIFIERLIST>
  [ WIDGETCLASS <IDENTIFIER> ]
  [ ATTRIBUTES
    ... ]
  [ METHODS
    ... ]
  [ EVENTHANDLING
    ... ]
  [ EVENTSEQUENCES
    ... ]
  [ INVARIANT
    ... ]
END
```

Eine ODISS-Objektklassendefinition beginnt mit der Einordnung der neuen Objektklasse in die Klassenhierarchie. Aufgrund von möglicher Mehrfachvererbung (engl.: multiple inheritance, [Me88]) kann im Abschnitt "SUPERCLASSES" eine Liste von Oberklassen angegeben werden. Für Objektklassen, deren Instanzen eine graphische Repräsentation besitzen, muß aufgrund der Verwendung eines objektorientierten Toolkits eine sogenannte *Widgetklasse* angegeben werden. Die Abschnitte "ATTRIBUTES" und "METHODS" dienen wie in objektorientierten Programmiersprachen der Beschreibung der Struktur (Attribute) und des potentiellen Verhaltens (Methoden) der Objekte dieser Klasse.

Das zulässige dynamische Verhalten der Instanzen einer Objektklasse kann in den Abschnitten "EVENTHANDLING" (Reaktionen auf Events und Nachrichten), "EVENTSEQUENCES" (zulässige Eventverarbeitungsfolgen) und "INVARIANT" (Constraints) spezifiziert werden.

Dynamisches Verhalten von Objekten kann in klassenspezifisches und objektspezifisches Verhalten aufgeteilt werden. Während klassenspezifisches Verhalten das gemeinsame Verhalten der Objekte einer Klasse beschreibt, können besondere Eigenschaften einzelner Objekte durch das objektspezifische Verhalten beschrieben werden. Aus diesem Grund können in ODISS Objektspezifikationen klassenspezifisches Verhalten analog zur Spezialisierung in einer Klassenhierarchie lokal erweitern. Die Spezifikation eines ODISS-Objektes hat folgende Form:

```
OBJECT <IDENTIFIER>
  CLASS <IDENTIFIER>
  [ ATTRIBUTVALUES
    ... ]
  [ EVENTHANDLING
    ... ]
  [ EVENTSEQUENCES
    ... ]
  [ INVARIANT
    ... ]
  [ PARENTWIDGET <IDENTIFIER> ]
  [ COMMUNICATION <IDENTIFIERLIST> ]
END
```

In einer ODISS-Objektspezifikation muß zunächst die Klasse des Objektes angegeben werden. Im Abschnitt "ATTRIBUTVALUES" werden die Initialwerte der Attribute spezifiziert. Objektspezifisches Verhalten kann analog zum klassenspezifischen Verhalten in den Abschnitten "EVENTHANDLING", "EVENTSEQUENCES" und "INVARIANT" spezifiziert werden. Der Abschnitt "PARENTWIDGET" dient dem Aufbau einer Widgethierarchie des zugrundeliegenden Toolkits und damit der Spezifikation der statischen Struktur einer Benutzerschnittstelle. Der Abschnitt "COMMUNICATION" enthält eine Liste der Objekte, an die Events und Nachrichten weitergeleitet werden können und dient damit der Eventverteilung.

Die Objektklassen von ODISS lassen sich in folgende Kategorien einteilen:

- Objektklassen mit graphischer Repräsentation besitzen eine zugeordnete Wigetklasse des zugrundeliegenden Toolkits.
- Metaobjektklassen besitzen keine graphische Repräsentation und dienen der Dialogkontrolle und Spezifikation von Constraints.
- Die Schnittstelle zu den physikalischen Eingabegeräten besteht aus einer fest konfigurierten Menge von Objektklassen, den sogenannten Geräteklassen [An82, Hü90].
- Die Schnittstelle zur Applikation (berechnender Teil des Anwendungsprogramms) wird ebenfalls durch Objektklassen modelliert [Si86].

Der hier vorgestellte Ansatz basiert auf der in [Ja86] vorgestellten Spezifikationssprache für direktmanipulative Benutzerschnittstellen, die eine Beschreibung von Benutzerschnittstellen durch eine Menge miteinander kommunizierender Interaktionsbjekte ermöglicht. Das dynamische Verhalten dieser Interaktionsobjekte und die Kommunikation zwischen ihnen werden durch den Objekten zugeordnete Zustandstransitionsdiagramme beschrieben. Die Interaktionsobjekte besitzen prozedurale Schnittstellen zur Präsentation und Applikation. Das ODISS-Modell verwendet dagegen kontextfreie Grammatiken zur Beschreibung objektspezifischen Verhaltens und modelliert sowohl die Präsentationskomponente als auch die Applikationsschnittstelle der Interaktionsobjekte analog dem MVC-Modell von Smalltalk [Go84] als eigenständige Objekte.

Das PAC-Modell [Cou87] baut wie das ODISS-Modell auf dem MVC-Modell auf, aber faßt die Präsentations-, Dialogkontroll- und Applikationsschnittstellenkomponente in den Interaktionsobjekten zusammen. Während die Dialogkontrollkomponente im PAC- und MVC-Modell primär der Kommunikation zwischen der Präsentations- und Applikationsschnittstellenkomponente dient, ermöglichen die entsprechenden Komponenten das ODISS-Modells außerdem eine objektspezifische Beschreibung des Dialogablaufs. Damit werden Konzepte des DIWA-Modells [Vo90] aufgegriffen, das eine Erweiterung des PAC-Modells um objektspezifische Dialogablaufbeschreibungen darstellt.

Ein weiteres Modell für die Modellierung und Implementierung graphischer Benutzerschnittstellen stellt das in [Hü90] vorgestellte Interaktionsmodell dar. Die Interaktionsobjekte dieses Modells integrieren ebenfalls die verschiedenen Komponenten eines Benutzerschnittstellenobjektes, aber mit einer vom PAC- und MVC-Modell abweichenden Aufteilung. Die Dialogkontrolle erfogt dabei durch hierarchische Anordnung der Interaktionsobjekte zur Modellierung komplexer Events. Die im ODISS-Modell verwendeten Dialogablaufbeschreibungen durch kontextfreie Grammatiken sind flexibler und ermöglichen in Verbindung mit Eventreceiving ebenfalls die Beschreibung von komplexen Events.

Die objektorientierte Modellierung der Applikationsschnittstelle im ODISS-Modell wurde bereits im GWUIMS [Si86] realisiert. Während im GWUIMS jedoch die Schnittstellen der durch das Seeheim-Modell [Pf85] implizierten Schichtenarchitektur durch Objekte modelliert werden, bindet das ODISS-Modell die Funktionalität dieser Schichten an die einzelnen Objekte, ohne dabei jedoch die Selbständigkeit der Präsentations- und Applikationsschnittstellenobjekte aufzuheben.
Einen weitereren Ansatz für eine objektorientierte Entwicklung von graphischen Benutzerschnittstellen stellen die in [Sch86] und [Wei89] vorgestellten "Application Frameworks" dar. Diese Werkzeuge bestehen aus einer Klassenhierarchie und einem objektorientierten Programmrahmen, der den anwendungsunabhängigen Teil einer Benutzerschnittstelle realisiert. Die Definition einer konkreten Benutzerschnittstelle erfolgt durch eine objektorientierte Erweiterung dieses Programmrahmens. Im Gegensatz zum ODISS-Modell bieten diese Werkzeuge jedoch keine Möglichkeit, Dialogablaufbeschreibungen explizit zu spezifizieren.
Zusammenfassend betrachetet, kombiniert das ODISS-Modell bekannte Konzepte, die bereits in anderen Ansätzen verwendet wurden, mit neuen Konzepten. Diese Kombination zeichnet sich dadurch aus, daß sie konsequent objektorientiert ist (Anlehnung an das MVC-Modell), die Modellierung anderer Ansätze teilweise ermöglicht und eine höhere Flexibilität bietet.

4. Eventverarbeitung und Eventreceiving

Die Kommunikation zwischen ODISS-Objekten wird zum einen durch die vom Benutzer erzeugten externen Events und zum anderen durch intern erzeugte Events, den *Nachrichten*, realisiert. Während externe Events kein explizites Zielobjekt besitzen (siehe Eventverteilung), sind Nachrichten i.a. direkt an ein bestimmtes Objekt gerichtet. Dagegen wird in objektorientierten Toolkits (z.B. X Toolkit, [Ny90]) ein Event oft als direkt an ein Objekt gerichtete Nachricht betrachtet. Da das ODISS-Laufzeitsystem jedoch eine eigene Eventverteilung realisieren wird, betrachten wir Events als ungerichtet. In den Fällen, in denen externe und interne Events jedoch identisch behandelt werden, wird in den folgenden Abschnitten der Begriff Event als gemeinsamer Oberbegriff verwendet.

Als Eventverarbeitung wird die Reaktion eines Objektes auf bestimmte Events oder Nachrichten bezeichnet. Reaktionen auf Events und Nachrichten werden im Abschnitt "EVENTHANDLING" des ODISS-Objektmodells in Form von sogenannten *Eventhandlern* spezifiziert. Eventhandler werden durch eine Liste von Events und Nachrichten gefolgt von einer Liste von Aktionen (z.B. Versenden von Nachrichten, Aufruf von Methoden) spezifiziert. Empfängt ein Objekt ein Event dieser Liste, so werden die spezifizierten Aktionen ausgeführt. Die Beispiele in dieser Arbeit behandeln alle die Verschiebung graphischer Objekte in einer direktmanipulativen Benutzerschnittstelle. Für die Realisierung dieser Interaktion wird zunächst die notwendige Eventverarbeitung definiert:

```
OBJECTCLASS GraphObject;
   ...
   ATTRIBUTES
      float x,y;
   METHODS
      void ChangePosition (float x, float y);
   EVENTHANDLING
      ButtonDown:        Presentation.Highlight(); Presentation.DrawFrame();
      MouseMove(x,y):    Presentation.MoveFrame(x,y); ChangePosition(x,y);
      ButtonUp:          PresentationDeleteFrame(); Presentation.DeHighlight();
                         ChangePosition(x,y); Application.NewPosition (x,y);
      ...
```

Durch *Eventreceiving* (vgl. Eventsharing in [So90]) werden Objekte über die Eventverarbeitung bestimmter anderer Objekte informiert. Eventreceiving ist wie das Message-Passing eine asymmetrische Kommunikation, die aber im Gegensatz zum letzteren vom Empfänger ausgeht. Ein Objekt o_1 bekundet sein Interesse an der Verarbeitung eines Events e durch ein Objekt o_2 durch einen Eventausdruck der Form $o_2.e$ in der Eventliste eines seiner Eventhandler.

Nach der Verarbeitung des Events *e* durch das Objekt o_2 wird das Objekt o_1 entsprechend informiert und führt die für $o_2.e$ spezifizierte Eventverarbeitung aus.

Eventreceiving wird durch interne Nachrichten realisiert und unterstützt die im Rahmen objektorientierter Programmierung oft hervorgehobene Erweiterbarkeit dadurch, daß neue Kommunikationsverbindungen nur die Änderungen des Empfängerobjektes erfordern. Die Erweiterung einer Dialogspezifikation um neue Objekte läßt die Spezifikation bereits vorhandener Objekte i.a. unverändert. Beim Message-Passing müssen dagegen Spezifikationen von Senderobjekten um entsprechende Anweisungen zum Versenden von Nachrichten erweitert werden.

Das Eventreceiving kann auch auf Methoden erweitert werden, in dem die Ausführung einer Methode *m* durch ein Objekt *o* als Pseudoevent dieses Objektes betrachtet wird. Objekte, die an der Ausführung dieser Methode interessiert sind, können dieses Interesse durch einen Eventausdruck der Form *o.m* darstellen und werden nach Ausführung der Methode *m* durch Objekt *o* durch eine Nachricht informiert.

Eventreceiving kann sowohl auf einzelne Objekte als auch auf eine ganze Objektklasse bezogen werden:

- Objektbezogenes Eventreceiving hat die Syntax *o.e* ; dieses Event tritt ein, wenn das Objekt *o* das Event *e* verarbeitet hat.
- Klassenbezogenes Eventreceiving hat die Syntax *C.e* ; dieses Event tritt ein wenn ein beliebiges Objekt der Klasse *C* oder eine ihrer Unterklassen das Event *e* verarbeitet hat.

Die Berücksichtigung der Unterklassen beim klassenbezogenen Eventreceiving beschreibt ein *hierarchisches Eventreceiving*, das besonders bei der Formulierung von klassenbezogenen Constraints nützlich sein kann. Aufgrund der Unabhängigkeit einer Objektklassendefinition von konkreten Objekten (Instanzen), darf in der entsprechenden Spezifikation nur klassenbezogenes Eventreceiving verwendet werden, während in der Spezifikation eines Objektes beide Formen des Eventreceivings zulässig sind.

Durch die Erweiterung des Eventreceivings auf Methoden (s.o.) können Attributabhängigkeiten ([Ba86], [Sz88]) relativ einfach modelliert werden. Dabei ist der Wert eines Attributes abhängig von den Werten anderer Attribute, so daß nach Änderungen von deren Werten der Wert dieses Attributes neu berechnet werden muß. Um die notwendigen Neuberechnungen einfach zu spezifizieren, muß sichergestellt werden, daß die Änderung von Attributen nur durch speziell dafür vorgesehene Methoden der entsprechenden Objekte erfolgen können. Durch Eventreceiving auf diesen Methoden werden dann Objekte, deren Attributwerte von Attributwerten anderer Objekte abhängen, über die relevanten Attributwertänderungen informiert. Als Beispiel soll die Speicherung der Positionen graphischer Objekte durch einen Windowmanager beschrieben werden. Nach der Verschiebung eines graphischen Objektes (Änderung der Koordinatenattribute) muß die von einem Windowmanager verwaltete Liste der Positionen aller graphischen Objekte aktualisiert werden. Für die Modellierung dieser Attributabhängigkeit wird klassenbezogenes Eventreceiving verwendet. Der Zugriff auf das Objekt, das durch seinen Methodenaufruf das Eventreceiving ausgelöst hat, erfolgt dabei über den Klassennamen (siehe Parameter des Aufrufs von ChangeList).

```
OBJECTCLASS WindowManager;
  ...
  ATTRIBUTES
    ListOfObject List;
    ...
  METHODS
    ChangeList(GraphObject* go, float x, float y);
    ...
  EVENTHANDLING
    GraphObject.ChangePosition(x,y) : ChangeList (GraphObject,x,y);
    ...
```

5. Constraints

In Benutzerschnittstellen beschreiben Constraints ([Bo87],[Ba86],[Sz88]) Einschränkungen der zulässigen Objektzustände, Konfigurationen und dynamischen Dialogabläufe. Für ihre Spezifikation existieren verschiedene Alternativen, z.B. Aussagenlogik, Prädikatenlogik oder temporale Logik, die sich in ihrer Ausdrucksmächtigkeit und im Aufwand für die Überprüfung der Einhaltung deutlich voneinander unterscheiden. Die in ODISS verwendeten *Objektinvarianten* [Me88] erlauben die Spezifikation von objektspezifischen Constraints auf den Objektzuständen im Prädikatenkalkül. Um auch Constraints auf den Zustandstransitionen von Objekten (Änderung der Attributwerte) beschreiben zu können, kann den Attributnamen innerhalb von Ausdrücken das Schlüsselwort "OLD" vorangestellt werden [Me88]. Damit wird dann der Wert dieser Attribute unmittelbar vor der Zustandsänderung spezifiziert. Dieser Ansatz stellt eine passive ereignisgesteuerte Constraint-Überwachung dar. Die in [Ma89] beschriebene Constraint-Programmierumgebung ThingLab II ist wesentlich mächtiger, da sie Hierarchien von Constraints unterstützt und einen integrierten Algorithmus zur aktiven Constraintbefriedigung durch Ausführung von Methoden besitzt.

Objektinvarianten und damit auch Constraints können für einzelne Objekte und für Objektklassen angegeben werden. Die Überwachung von Constraints, die mehrere Objekte betreffen, erfolgt durch die Objektinvarianten von Metaobjekten (Objekte ohne graphische Repräsentation). Dabei kommunizieren die beteiligten Objekte durch Event-receiving mit den Metaobjekten. Als Beispiel für diesen Fall soll die Spezifikation einer unteren Grenze für den Abstand zwischen zwei Objekten *X* und *Y* betrachtet werden. Aufgrund des objektbezogenen Eventreceivings hat eine Verschiebung eines der beiden Objekte zur Folge, daß die entsprechenden Koordinaten durch das Metaobjekt aktualisiert werden. Als Folge dieser Eventverarbeitung wird die Invariante des Metaobjektes und damit die Einhaltung des Constraints überprüft.

```
OBJECTCLASS DistanceConstraint;
   ...
   ATTRIBUTES
      float Xx, Xy;
      float Yx, Yy;
   EVENTHANDLING
      X.Create, X.ChangePosition(x,y): Xx = x; Xy = y;
      Y.Create, Y.ChangePosition(x,y): Yx = x; Yy = y;
      ...
   INVARIANT
      Distance(XCoordinates,YCoordinates) ≥ 50;
      ...
```

6. Eventsequenzen

Während in dem Abschnitt "EVENTHANDLING" die Verarbeitung einzelner Events durch die Objekte spezifiziert wird, können in einem Abschnitt names "EVENTSEQUENCES" die zulässigen Eventverarbeitungsfolgen eines Objektes in Form von kontextfreien Produktionen in EBNF spezifiziert werden. Ein Objekt kann genau die Eventsequenzen verarbeiten, die der unter "EVENTSEQUENCES" aufgeführten Grammatik entsprechen. Events, die zu einer Abweichung von dieser Grammatik führen, werden von dem entsprechenden Objekt nicht verarbeitet, auch wenn in dem Abschnitt "EVENTHANDLING" eine Verarbeitung dieses Events spezifiziert wurde.
Durch die Benennung von Eventsequenzen können komplexe Events (vgl. [Hü90]) definiert werden. Eine komplexes Event tritt genau dann ein, wenn seine definierende Eventsequenz von dem entsprechenden Objekt bearbeitet wurde. Komplexe Events können wie elementare Events verwendet werden, so daß die Eventverarbeitung des Abschnitts "EVENTHANDLING" sowohl elementare vom Benutzer erzeugte Events als auch komplexe Events betrifft. Damit ermöglichen Eventsequenzen die Eventbehandlung auf verschiedenen Abstraktionsebenen durch Zusammenfassung bereits definierter Events zu komplexeren Events.

Als Beispiel für die Spezifikation zulässiger Eventsequenzen soll hier wieder das Verschieben graphischer Objekte betrachtet werden. Diese Interaktion erfolgt durch ein ButtonDown-Event, beliebig viele Move-Events und ein abschließendes ButtonUp-Event. Die Verarbeitung der einzelnen Events, z.B. Speicherung der aktuellen Position und Zeichnen eines Rechteckrahmens, wird bereits im Abschnitt "EVENTHANDLING" spezifiziert. Die beschriebene Eventsequenz kann dann wie folgt spezifiziert werden:

```
OBJECTCLASS GraphObject;
  ...
  EVENTHANDLING
  ...
  EVENTSEQUENCES
    ObjectMove(x,y) = ButtonDown { MouseMove(x,y) } ButtonUp;
    ...
```

Ein Objekt, für das keine Eventsequenzen spezifiziert werden, kann beliebige Eventsequenzen verarbeiten, so daß der Abschnitt "EVENTSEQUENCES" als Einschränkung des zulässigen dynamischen Verhaltens von Objekten betrachtet werden kann.

Analog zur Spezifikation von Constraints kann auch die Spezifikation von zulässigen Eventsequenzen durch Verwendung von Metaobjekten auf eine Gruppe von Objekten erweitert werden. Im Abschnitt "EVENT-SEQUENCES" dieser Metaobjekte werden die entsprechenden Sequenzen festgelegt. Die beteiligten Objekte kommunizieren durch Eventreceiving mit den entsprechenden Metaobjekten, die i.a. keine graphische Repräsentation besitzen. Mit diesem Konzept können dann auch komplexe Interaktionen wie z.B. SEQUENCE, AND, OR und REPEAT (vgl. [Hü90]) modelliert werden:

```
Sequence =    Event1 Event2 Event3
Or =          Event1 | Event2 | Event3
Repeat =      { Event1 }
And =         Event1 Event2 Event3 |
              Event2 Event1 Event3 |
              ...
              Event3 Event2 Event1
```

7. Eventverteilung

Der Begriff "Eventverteilung" bezeichnet die Verteilung von Events an alle daran interessierten Objekt der Benutzerschnittstelle. Die Eventverteilung ist notwendig, weil die vom Benutzer erzeugten externen Events nicht wie Nachrichten direkt an Objekte gerichtet sind. Sie wird als objektorientiert bezeichnet, falls die Empfänger der Events Objekte sind.

In dem ODISS zugrundeliegenden Laufzeitsystem soll eine *hierarchische Eventverteilung* realisiert werden, die Flexibilität bei der Konzeption der Eventverteilung gewährleistet. Diese Flexibilität wird dadurch ausgedrückt, daß die hierarchische Eventverteilung die Modellierung anderer Ansätze zur Eventverteilung (z.B. [An82],[Gr86], [Hü90]) ermöglicht.
Bei der hierarchischen Eventverteilung werden die Objekte in einer Polyhierarchie (gerichteter azyklischer Graph) verwaltet, die im folgenden als *Kommunikationsgraph* bezeichnet wird. Events werden von den Objekten entlang der gerichteten Kanten dieses Graphen an Unterobjekte weitergeleitet. Dabei kann ein Objekt durchaus mehrere Vorgängerobjekte besitzen, von denen Events an dieses Objekt weitergeleitet werden. Der initiale Zustand des Kommunikationsgraphen ergibt sich direkt aus der Dialogspezifikation (Abschnitt "COMMUNICATION"). Seine Verwaltung (z.B. Einfügen und Entfernen von Objekten) erfolgt durch die Methoden eines speziellen Objektes, des sogenannten *Kommunikationsmanagers*. Die Objekte des Kommunikationsgraphen können aktiviert und deaktiviert werden. Deaktivierte Objekte reagieren nicht auf das Empfangen von Events und leiten diese auch nicht weiter. Der Kommunikationsgraph besitzt ein ausgezeichnetes Objekt, das die vom Benutzer erzeugten Events direkt von den Geräteobjekten erhält und als *Einstiegspunkt* für die Verteilung der Events an die aktiven Objekte des von diesem Objekt ausgehenden Teilgraphen dient. Für die Aktivierung von Subdialogen, z.B. in Dialogboxen, können auch interne Knoten des Kommunikationsgraphen als Einstiegspunkt deklariert werden (vgl. Abb. 1).

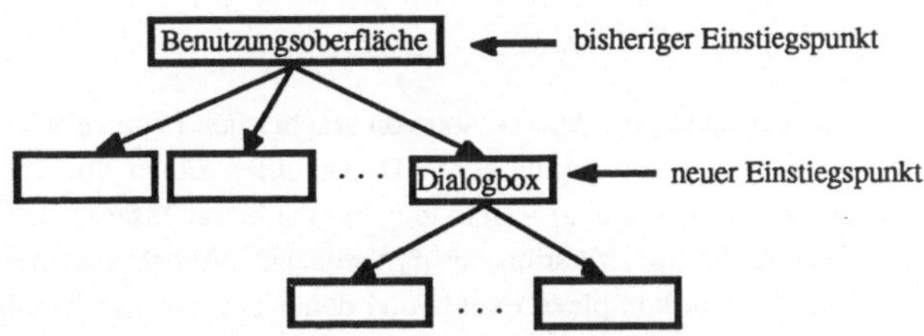

Abb. 1 Einstiegspunkt in den Kommunikationsgraphen

Damit könnte dann z.B. spezifiziert werden, daß alle Events zwischen dem Öffnen und Schließen einer Dialogbox nur die Komponenten dieser Dialogbox betreffen können. Durch die blockierende Eigenschaft deaktivierter Objekte, deren Unterobjekte nicht mehr über das Eintreten neuer Events informiert werden, wird bei der hierarchischen Eventverteilung ein Effizienzgewinn erzielt, dem jedoch der zusätzliche Verwaltungsaufwand für den Kommunikationsgraph und der damit verbundene erhöhte Speicherplatzbedarf gegenübergestellt werden muß.

Eine Änderung der Konfiguration der Benutzerschnittstelle wird durch eine Änderung des Kommunikationsgraphen (Aktivierung und Deaktivierung von Objekten, Einfügen und Entfernen von Objekten, Änderung des aktuellen Einstiegspunktes) modelliert. Der Kommunikationsgraph stellt damit ein Konzept zur Modellierung globalen Vorhaltens einer Menge von Objekten dar.

Abb. 2 zeigt einen Ausschnitt des Kommunikationsgraphen einer Benutzerschnittstelle. Der Einstiegspunkt in diesen Kommunikationsgraph ist das Objekt "Benutzerschnittstelle", das im Sinn einer Top-down-Sicht die gesamte Benutzerschnittstelle repräsentiert. Die Dialogbox-Objekte sind nur dann aktiv, wenn der entsprechenden Subdialog ausgewählt wurde. In diesen Fällen leiten sie alle Events an ihre ebenfalls aktiven Unterobjekte weiter. Die Button-Objekte in den Menüs werden erst durch die Auswahl des entsprechenden Menüs aktiviert und können deshalb in allen anderen Fällen keine Events empfangen. Entsprechend können die graphischen Objekte nur dann Events verarbeiten, wenn sowohl das Viewport-Objekt (Sichtfenster auf eine Fläche) auch als das Canvas-Objekt (Zeichenfläche für graphische Objekte) aktiv sind.

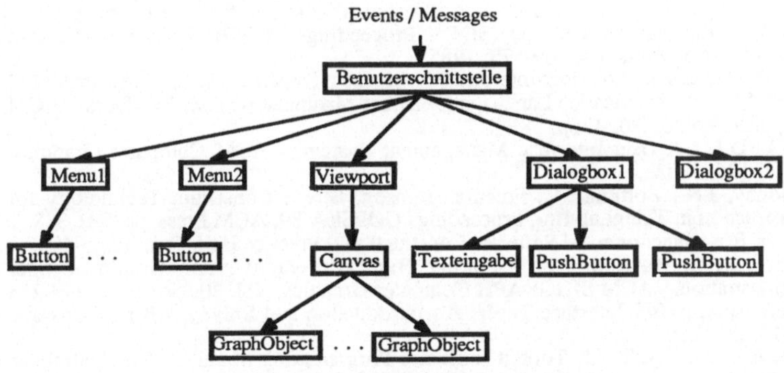

Abb. 2 Kommunikationsgraph einer Benutzerschnittstelle

Die Struktur eines Kommunikationsgraphen ist unabhängig von der hierarchischen Anordnung der Objekte im eingebetteten Toolkit. Während letztere den statischen Aufbau von Benutzerschnittstellen beschreiben, dient der Kommunikationsgraph der Verteilung von Events und der Modellierung von Konfigurationsänderungen zur Laufzeit. Im ODISS-Objektmodell werden sie deshalb auch getrennt spezifiziert (Abschnitte "PARENTWIDGET" und "COMMUNICATION") obwohl, wie im Beispiel gezeigt, durchaus ähnliche Strukturen möglich sind.

8. Zusammenfassung

Die Sprache ODISS dient der objektorientierten Spezifikation von graphischen Benutzerschnittstellen und damit u.a. der Entwicklung von Hypermedia-Benutzerschnittstellen. Dabei unterstützen die Sprachkonstrukte Eventverarbeitung, Eventsequenzen, Objektinvarianten in Verbindung mit der hierarchischen Anordnung der Objekte im Kommunikationsgraph insbesondere die Spezifikation des dynamischen Verhaltens einer Benutzerschnittstelle. Durch Eventsequenzen können außerdem komplexe Events und damit eine Eventbehandlung auf verschiedenen Abstraktionsebenen definiert werden. Objektinvarianten und Eventreceiving als neue Form der Kommunikation zwischen Objekten ermöglichen außerdem die Spezifikation von Constraints, die sich auf einzelne Objekte, auf Objektklassen und durch Verwendung von Metaobjekten auch auf eine Gruppe von Objekten beziehen können.
Die hierarchische Anordnung der Objekte im Kommunikationsgraph erlaubt eine effiziente und flexible Eventverteilung. Durch Manipulationen des Kommunikationsgraphen kann außerdem das globale Verhalten einer Benutzerschnittstelle in Form von Konfigurationsänderungen modelliert werden.

Literatur

[An82] Anson, E. : The Device Model of Interaction. - ACM SIGGRAPH Computer Graphics Vol. 16, No. 3, pp. 107-114, 1982.

[Ba86] Barth, P.S. : An Object-Oriented Approach to Graphical Interfaces. - ACM Transactions on Graphics Vol. 5, No. 2, pp. 142-172, 1986.

[Bo87] Borning, A., Duisberg, R., Freeman-Benson, B., Kramer, A., Woolf, M. : Constraint Hierarchies. - Proceedings OOPSLA´87, ACM Press, pp. 48-60, 1987.

[Cou87] Coutaz, J. : PAC, an Object Oriented Model for Dialog Design. - Proceedings Human-Computer Interaction - INTERACT´87, North-Holland, pp.431-436, 1987.

[Con87] Conklin, J. : Hypertext: An Introduction and Survey. - IEEE Computer Vol. 20, No.9, pp. 17-41, 1987.

[Go84] Goldberg, A. : Smalltalk-80: The Interactive Programming Environment. - Addison-Wesley, Reading, Mass., 1984.

[Gr86] Green, M. : A Survey of Three Dialogue Models. - ACM Transaction on Graphics Vol. 5, No. 3, pp. 245-275, 1986.

[Hü90] Hübner, W. : Entwurf Graphischer Benutzerschnittstellen. - Srpinger-Verlag, Berlin, 1990.

[Har89] Hartson, H.R., Hix, D. : Human-Computer Interface Development: Concepts and Systems for Its Management. - ACM Computing Surveys, Vol. 21, No.1, pp. 5-92, 1989.

[Hay85] Hayes, P.J., Szekely, P.A., Lerner, R.A.: Design Alternatives for User Interface Management Systems Based on Experience with COUSIN. - Proceedings SIGCHI´85 Human Factors in Computing Systems, ACM Press, pp. 169-175, 1985.

[Hi90] Hix, D. : Generations of User-Interface Management Systems. - IEEE Software 9 (1990), pp. 228-234.

[Ja86] Jacob, J.K. : A Specification Language for Direct-Manipulation User Interfaces. - ACM Transactions on Graphics, Vol. 5, No. 4, pp. 283-317, 1986.

[Ka82] Kasik, D.J. : A User Interface Management System. - ACM Computer Graphics Vol 16, No. 3, pp. 99-106, 1982.

[Ma89] Maloney, J.H., Borning, A., Freeman-Benson, B.N. : Constraint Technology for User-Interface Construction in ThingLab II. - Proceedings OOPSLA´89, ACM Press, pp. 381-388, 1989.

[Me88] Meyer, B. : Object-oriented Software Construction. - Prentice Hall, Cambridge, 1988.

[My86] Myers, B.A., Buxton, W. : Creating Highly-Interactive and Graphical User Interfaces by Demonstration. - ACM SIGGRAPH Computer Graphics, Vol. 20, No. 4, pp. 249-258, 1986.

[My89] Myers, B.A. : User-Interface Tools: An Introduction and Survey. - IEEE Software Vol. 6, No. 1, pp. 15-23, 1889.

[Ny90] Nye, A., O´Reilly, T. : X Toolkit Intrinsics Programming Manual. - The Definitive Guide to the X Window System (Volume Four), O´Reilly & Associates, 1990.

[Pf85] Pfaff, G.E. (Hrsg.): User Interface Management Systems. - Springer-Verlag, 1985.

[Sch86] Schmucker, K.J. : MacApp: An Application Framework. - Byte, August 1986, pp. 72-75.

[Si86] Sibert, J.L., Hurley, W.D., Bleser, T.W. : An Object-Oriented User Interface Management System. - ACM Computer Graphics Vol. 20, No. 4, pp. 259-268, 1986.

[So90] Sousa, J.P., Sernadas, C., Sernadas, A. : An Object-Oriented Specification Tool for Graphical Interfaces. - Computer & Graphics, Vol 14, No. 1, pp. 29-40, 1990.

[Sz88] Szekely, P.A., Myers, B.A. : A User Interface Toolkit on Graphical Objects and Constraints. - Proceedings OOPSLA´88, ACM Press, pp. 25-30, 1988.

[Wei89] Weinand, A., Gamma, E., Marty, R. : Design and Implementation of ET++, a Seamless Object--Oriented Application Framework. - Structured Programming, Vol. 10, No. 2, pp. 63-87, 1989.

[Wi85] Wirth, N.: Programmieren in Modula-2. - Dritte korrigierte Überarbeitung, Springer-Verlag, 1985.

Design und Realisierung einer objektorientierten Bedienoberfläche für die Multimedia-Kommunikation

T. Töpperwien T. Weidenfeller

TELENORMA GmbH (Bosch Telecom)
Vorentwicklung
Kleyerstr. 94
6000 Frankfurt am Main 1
Tel. 069/266-3372 bzw. 3042
email: {toepper,weidi}@tnevo.uucp

1 Einleitung

Multimedia-Kommunikation zeichnet sich dadurch aus, daß in einer netzartigen Systemumgebung neben zeitunabhängigen Komponenten (z.B. Text, Graphik, Image) auch zeitabhängige Daten (z.B. Video, Audio) verwaltet, integriert, dargestellt und ausgetauscht werden.

Eine Systemumgebung zur interaktiven Multimedia-Kommunikation ist gekennzeichnet durch:

- physikalische Übertragungsprotokolle unterschiedlicher Bandbreite und Reichweite,
- multimediale Dienste, Dienstelemente und Datentransferprotokolle und
- modulare Softwaregestaltung und Hardwarearchitektur (Baukasten) für multimediale Endgeräte rsp. Workstations.

Für die Integration unterschiedlicher Dienste sowie dem damit verbundenen Softwaredesign ist eine objektorientierte Entwicklungsumgebung eine mögliche Lösung. Um der einfachen Bedienbarkeit unterschiedlicher Dienste und Dienstelemente gerecht zu werden, eignen sich spezielle Werkzeuge zum Entwurf und der Realisierung von adaptiven Bedienoberflächen, die jedoch ein einheitliches 'look and feel' gewährleisten müssen. Durch Verwendung von objektorientierten Werkzeugen zur Gestaltung von Bedienoberflächen ergeben sich folgende Vorteile:

- **Rapid Prototyping**
  Der Entwickler kann das entworfene Konzept schneller in sichtbare Ergebnisse umwandeln und eventuelle Fehler in der Konzeptarbeit beseitigen bzw. Alternativen im Entwurf effizient testen.

- **Wartung und Austauschbarkeit**
  Durch Datenkapselung wird ein modulares Konzept erzwungen, das die Integration weiterer bzw. die Aktualisierung bestehender Dienste zuläßt.

- **Wiederverwendbarkeit**
  Generell benötigte Funktionalitäten werden von den Ober- auf die Unterklassen vererbt und stehen somit dort zur Verfügung.

Im folgenden Abschnitt werden wir eine Klassifizierung und Untersuchung einiger graphischer Entwicklungsumgebungen vornehmen und diese anhand von Beispielen illustrieren. In unserer Entwicklungsumgebung, so zeigte unsere Analyse, war InterViews in der Version 2.6 ein geeigneter Kandidat, um multimediale Anwendungen in der Präsentation zu unterstützen. Wir werden in Abschnitt drei genauer auf die Ideen und das Konzept der InterViews-Bibliothek eingehen. Abschnitt vier befaßt sich dann mit der Vorgehensweise bei der Programmierung in InterViews anhand eines Beispieles, bevor wir im letzten Abschnitt ein Resumee über InterViews ziehen und allgemein noch fehlende Konzepte aufzählen, die teilweise in anderen Werkzeugen verwirklicht werden.

2 Objektorientierte Werkzeuge für die Entwicklung von Bedienoberflächen

Im Gegensatz zum TOS von Atari und dem Apple-Betriebssystem sind die graphischen Komponenten für UNIX-Rechner erst recht spät entwickelt worden und nicht Bestandteil des Betriebssystems. Dieses ist wohl auch der Grund, weshalb der Markt im Bereich der Entwicklungswerkzeuge für graphische Oberflächen bei UNIX-Rechnern stark expandiert und dadurch momentan unüberschaubar geworden ist.

Als Basis für graphische Bedienoberflächen unter UNIX dient häufig das X-Window-System des MIT. So bauen auch die beiden bekanntesten, von zwei unterschiedlichen Interessengruppen entwickelten Oberflächen auf diesem System auf. Zum einen ist dies OSF/Motif von der Open Software Foundation (OSF), zum anderen Open Windows bzw. Open Look, das von AT&T in Zusammenarbeit mit Sun Microsystems entwickelt worden ist. Zu beiden Systemen existieren diverse Bibliotheken oder Entwicklungsumgebungen zur Erstellung von Benutzerapplikationen. Um eine Übersicht zu gewinnen, benutzen wir die folgenden Kriterien für eine Klassifizierung:

1. Das Werkzeug bietet eine fast objektorientierte Struktur von nicht objektorientierten graphischen Basisfunktionen an.

2. Das Werkzeug stellt eine Klassenhierarchie unter Verwendung von graphischen Basisfunktionen zur Verfügung.

3. Das Werkzeug besitzt eine eigene (objektorientierte) Hochsprache zur Manipulation von graphischen Objekten.

4. Graphische Objekte sind Bestandteile der Entwicklungsumgebung.

Als zusätzliche wünschenswerte Erweiterungen bieten manche Werkzeuge noch die Möglichkeit, die Oberfläche interaktiv über Maussteuerung zu entwerfen.

Unter die erste obengenannte Rubrik fällt die Xt-Intrinsic-Bibliothek [16], die von der Grundidee das baukastenförmige Zusammenstellen von komplexen graphischen Objekten erlaubt. Weitere Vertreter dieser Kategorie sind XView und das Open Look Intrinsic Toolkit (OLIT), die sich zwar durch eine Klassenstruktur auszeichnen, denen aber der für objektorientierte Konzepte typische Vererbungsmechanismus fehlt. Die Attribute der mit diesen Werkzeugen erzeugten Objekte können durch einfache Prozeduraufrufe geändert werden. Durch Verwendung eines interaktiven Entwicklungswerkzeugs kann der Quellcode für XView automatisch erstellt werden, der wahlweise als C- oder C++-Code vorliegen kann. In einer weiteren Version dieses Werkzeugs wird es möglich sein, auch den OLIT-Quellcode in den obengenannten Wirtssprachen zu erzeugen.

Werkzeuge der zweiten Kategorie sind etwa Screens++ [9] und InterViews, wobei es sich bei Screens++ um eine Bibliothek zur reinen Tastatursteuerung der Oberfläche handelt; es können keine Graphiken eingebunden werden. InterViews hingegen unterstützt die Fähigkeiten eines graphischen Terminals voll.

Zur dritten Gruppe gehören der Dialog-Manager von ISA [6], TeleUSE [15] von Telesoft und die Kombination aus Datenbank und Entwicklungsumgebung 4GL [7] von INGRES.

Der Dialog-Manager kann systemübergreifend eingesetzt werden, da er nicht nur zu Oberflächen aus der UNIX-Welt, sondern auch zu solchen aus der PC-Welt kompatibel ist. Im Gegensatz zu TeleUSE besitzt er noch eine Schnittstelle zu SQL-Datenbanksystemen.

Die Leistungsmerkmale von TeleUSE sind:

1. interaktives Entwerfen der Oberfläche

2. eigene Hochsprache für die Manipulation graphischer Objekte

3. C-Schnittstelle

4. OSF/Motif basiert

5. Testumgebung für die Hochsprache zur effizienteren Entwicklung von Bedienoberflächen

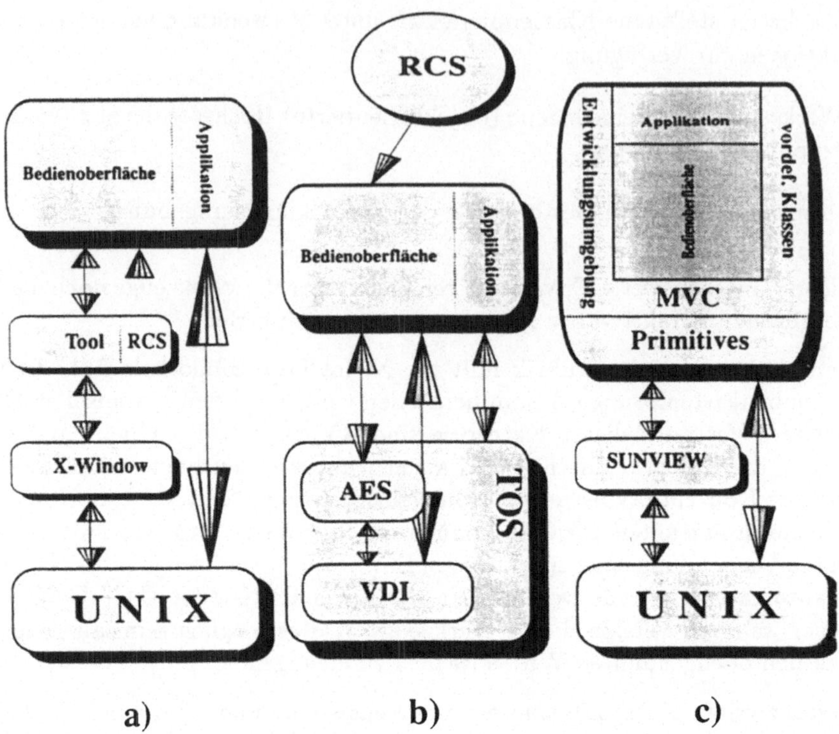

Abbildung 1: Die verschiedenen Architekturen der Entwicklungsumgebungen für graphische Oberflächen

Für 4GL gelten zusätzlich zu den oben genannten Punkten 1-3 noch folgende:

- Paßt sich dem 'look and feel' der jeweils unterliegenden Oberfläche (Open Look bzw. OSF/Motif) an
- Verwaltet die graphischen Objekte in einer Datenbank mit den damit verbundenen Annehmlichkeiten wie Verteiltheit und Transaktionskonzept

Zur vierten Kategorie gehören in sich abgeschlossene Entwicklungsumgebungen, in denen ein Oberflächenkonzept schon Bestandteil der Sprache ist. Ein Vertreter dieser Kategorie ist die Umgebung von Smalltalk-80. Hierbei ist – wie in Abbildung 1c[1] zu sehen – die Applikation in die Smalltalk-80-Umgebung eingebettet.

Sie ist aber im Gegensatz zu den Architekturen weiterer unter UNIX angebotener Entwicklungsumgebungen (Abbildung 1a) und denen des Ataris (Abbildung 1b)[2] nicht 'stand alone'. Dieses führt zu einem Geschwindigkeitsverlust; ein gerade bei Bedienoberflächen nicht zu vernachlässigender Aspekt. Weiterhin ist auch die obengenannte wünschenswerte Komponente des interaktiven Designs der Oberfläche, wie z.B. das Resource Construction Set (RCS), nicht vorhanden.

---

[1]MVC = Model-View-Controler

[2]AES = Application Environment System, VDI = Virtual Device Interface

Ein wesentliches Ziel bei der Multimedia-Kommunikation ist eine multimediale Bedienoberfläche, die sowohl in das modulare Konzept zukünftiger Kommunikationsendgeräte integrierbar ist, als auch die Möglichkeit bietet, zwischen unterschiedlichen multimedialen Diensten und Informationstypen umzuschalten [4], [14].

Unsere Entwicklungsumgebung besitzt derzeit folgende Randbedingungen:

- SUN4-Architekturen
- Kommunikation der Rechner untereinander über Ethernet oder Breitbanderweiterungen in privaten Nebenstellenanlagen
- UNIX (SunOS 4.1)
- C++-Entwicklungsumgebung
- X-Window-System
- Open Look

Diese Umgebung sowie die Untersuchung unterschiedlicher Werkzeuge hat zunächst den Einsatz von InterViews als Oberflächenentwicklungswerkzeug impliziert, da InterViews eine C++-Klassenstruktur anbietet und unabhängig von den unterliegenden Bedienoberflächen ist.

3 Das Konzept von InterViews

InterViews ist ein objektorientiertes Werkzeug zur Entwicklung von Oberflächen für Benutzerapplikationen, das an der Universität Stanford entwickelt worden ist. Von den ca. 30000 Zeilen Quellcode der Version 2.6 basieren nur ca. 6% auf der X-Window-Bibliothek [8], wodurch allerdings InterViews-Applikationen unabhängig von beiden obengenannten Bedienoberflächen (Open Windows, OSF/Motif) sind, aber wie alle X-Applikationen problemlos unter ihnen laufen können. Sie fallen aber unter diesen Bedienoberflächen durch ein eigenes 'look and feel' auf.

InterViews besitzt in der Version 2.6 einen hierarchischen Klassenbaum (siehe Abbildung 2), dessen Wurzelklasse *Interactor*[3] ist, von der alle graphischen Objekte abgeleitet werden.

Konkrete Objekte wie z.B. *PushButtons* sind meistens Unterklassen abstrakter Oberklassen (in diesem Beispiel *Button*), von denen häufig keine Instanzen gebildet werden, da sie selbst nur allgemeine Funktionalitäten auf ihre Unterklassen vererben.

Komplexe graphische Objekte können nach dem Baukastenverfahren aus weniger komplexen zusammengebaut werden. Hierbei übernimmt immer das in der Hierarchiestufe

[3]In der InterViews-Version 3.0 haben sich Änderungen ergeben.

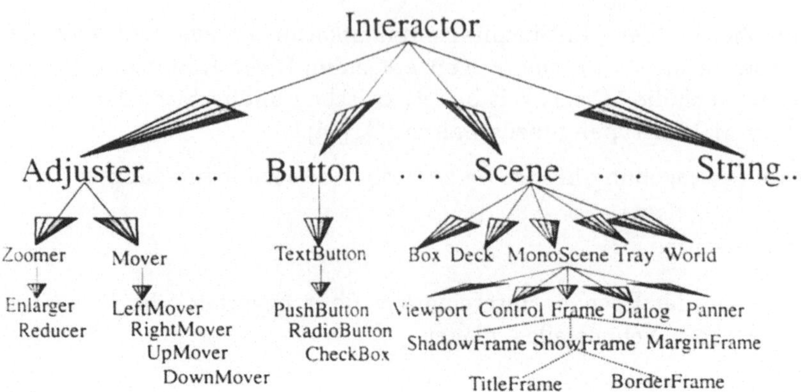

Abbildung 2: Teil des InterViews-Klassenbaumes

höher stehende Objekt die Koordination seiner Unterklassenobjekte. InterViews stellt im Gegensatz zum X-Window-System schon gewisse, häufig benötigte komplexere Objekte zur Verfügung, wie z.B. den *Slider*, die dann nicht vom Programmierer aus den Basiskomponenten zusammengesetzt werden müssen. Weiterhin existieren für diese vordefinierten, komplexeren Objekte Methoden, die deren Manipulation gewährleisten, wie z.B. das Verschieben der Marke auf dem *Slider*. Diese beiden Merkmale unterstützen ein Rapid Prototyping.

In InterViews existieren 'Behälterklassen', deren Instanzen graphische Objekte (*Interactors*) aufnehmen und strukturiert darstellen können. Die Strukturierungsmittel reichen von einer *Box*, die ihre Bestandteile von oben nach unten bzw. von rechts nach links anordnet, bis zu einem *Tray*, bei dem sich die Bestandteile überlappen können oder einem *Viewport*, der in Verbindung mit einem *Slider* Teilansichten seiner Bestandteile liefern kann.

InterViews stellt ein methoden-basiertes Konzept zur Interaktion mit den verschiedenen graphischen Objekten bereit. Hierzu wird die Klasse *Event* verwendet, die die unter dem X-Window-Systems gelieferten Events aufbereitet und auf die zugehörigen Objekte verteilt. Diese Objekte haben die Möglichkeit, sich einen *Sensor* zu definieren, den sie dann mehr oder weniger sensibilisieren können. Dieser *Sensor* übernimmt für sie dann die Klassifizierung der einkommenden Ereignisse, wobei er nur in interessante und uninteressante unterscheidet. Jedem graphischen Objekt ist auch noch eine Darstellungsform zugeordnet; z.B. stellt sich ein *Button* einfach gesprochen durch ein Rechteck mit innen liegendem

Text dar.

Zur Verdeutlichung der oben erwähnten Konzepte von InterViews wird im nächsten Abschnitt näher auf die Vorgehensweise bei der Programmierung eingegangen. Dieses geschieht anhand eines Beispiels.

4 Realisierung mit InterViews

Unter den in Abschnitt 2 gegebenen Zielsetzungen sowie unter Verwendung von InterViews ist eine objektorientierte interaktive Bedienoberfläche für eine Multimedia-Kommunikation entwickelt worden. Dieser erste Prototyp ist auf der CeBIT '91 präsentiert worden.

Ein spezieller Teilaspekt – die Steuerung einer Videoquelle – dieser Multimedia-Kommunikation soll die Anwendung von InterViews veranschaulichen. Zur Steuerung der Videoquelle ist ein Fenster programmiert worden, das eine Reihe von Buttons beinhaltet, deren Funktionalität im wesentlichen mit der einer Fernbedienung für einen Fernseher zu vergleichen ist. Es existieren z.B. Buttons, die die Helligkeit, den Kontrast und die Sättigung regeln oder auch ein Freezing oder Grabbing der Videoinformation erlauben. Der folgende Auszug aus dem Quellcode-Listing soll das Erzeugen von InterViews-Basisobjekten und deren Integration in komplexere Objekte verdeutlichen. Weiterhin zeigt das Beispiel auch, wie komplexere Objekte in in der Hierarchie höherstehende Objekte eingefügt werden können:

```
static PropertyData pro[] = {
     {"*font","9x15bold"},
     {"background","White"},
     {"*PushButton.foreground","Blue"},
     {"*PushButton.background","Green"},
     {nil}
};
static OptionDesk options[] = {
     . . .
};
```

In den Feldern *pro* und *options* werden die Voreinstellungen für die Oberfläche abgelegt. Hierbei werden u.a. die Schriftart, die Hintergrundfarbe und die Farbe der Buttons (Knöpfe) festgelegt.

```
main(int argc, char* argv[]) {
     World *world = new World("Video",pro,options,argc,argv);
                                                   . . .
     ButtonState *aState = new ButtonState();
                                                   . . .
     PushButton *aButton1 = new PushButton("Text1",aState,0);
                                                   . . .
```

```
HBox *box1 = new HBox(
                                        new HGlue(),
                                        aButton1,
                                        new HGlue()
                              );
                                    . . .
Frame *frame = new Frame (
                                        new VGlue(),
                                         box1,
                                         . . .
                                );
                                    . . .
world->InsertApplication(frame);
do {
    world->Read(event);
    if (event.eventType == DownEvent) {
        if (event.target == aButton1) {
            /*Aktion, die mit aButton1 assoziiert ist,
               anstarten    */
        };
                                    . . .
    }
}
} /*end main*/
```

Der Aufruf von *new World(...)* erschafft das Top-Level-Windows, in dem alle weiteren Objekte visualisiert werden. Hier werden auch die Voreinstellungen für die Oberfläche übernommen.

Das Erzeugen eines Knopfes erfolgt durch die Aufrufe *new ButtonState(...)* und *new PushButton(...)*. Der Status eines Knopfes ist nur bei gegenseitiger Abhängigkeit (*Radio-Button*) wichtig, da über ihn das Umschalten von einem Knopf auf den anderen koordiniert wird.

Anschließend können die einfachen Objekte (z.B. Buttons) zu beliebig kompliziert verschachtelten Objekten zusammengefaßt werden. Diese geschieht durch die Aufrufe *new HBox(...)* und *new Frame(...)*. Bis zu diesem Zeitpunkt ist auf dem Bildschirm noch nichts passiert; es wurde lediglich das Layout der Oberfläche bestimmt. Für diese Aufgabe benutzen andere Systeme Werkzeuge für das interaktive Design.

Erst durch den Aufruf *InsertApplication(...)* erscheint das Control-Panel auf dem Bildschirm. In der folgenden *do-Schleife* erfolgt die Event-Behandlung, die in diesem Fall auf das Drücken einer Maustaste reagiert und die Aktion auslöst, die mit dem Zielobjekt assoziiert ist. Eine weitere Möglichkeit der Event-Behandlung wäre das sofortige Weiterreichen des Events an das betreffende Objekte. wobei jedes Objekt selbst weiß, wie es mit diesem Event umzugehen hat. Das Ergebnis des auszugsweise angedeuteten Programms ist das Control-Panel in Abbildung 3a. Zur Verdeutlichung der Unterschiede

a)

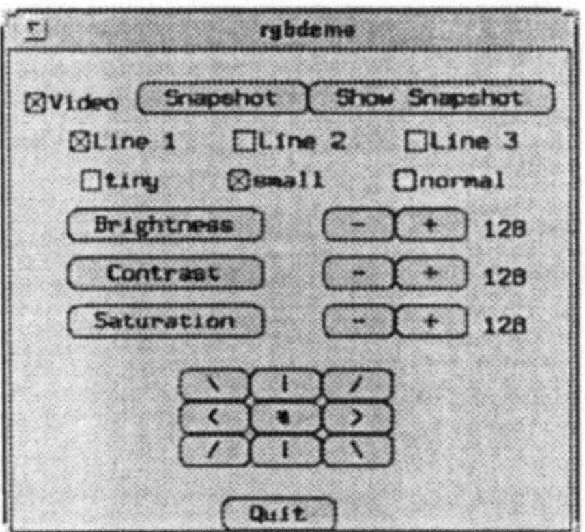

b)

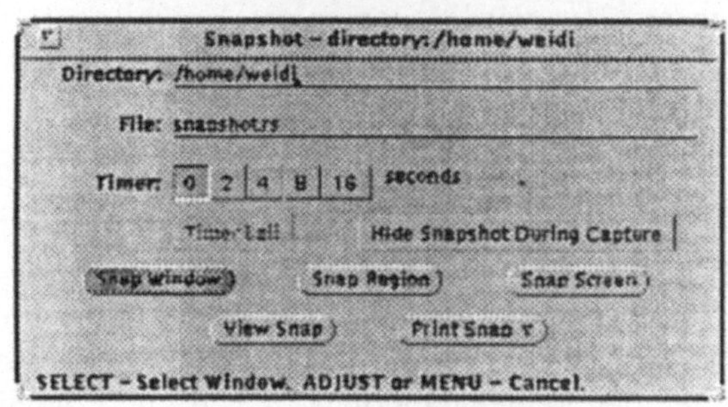

Abbildung 3: Das 'look and feel' von InterViews (a) und Open Look (b)

im 'look and feel' zwischen InterViews und Open Look dient ein Control-Panel aus der Open-Look-Oberfläche (in Abbildung 3b).

Wie man sieht, arbeitet InterViews nur mit einfachen optischen Mitteln (zwei-dimensionale Knöpfe), während unter Open Look die drei-dimensionalen Effekte z.B. durch Änderung des Schattenwurfes unterstützt werden. In Abbildung 3b erkennt man diesen Effekt bei den *Radio-Buttons* für den *Timer*.

5 Zusammenfassung, offene Probleme und Ausblick

Durch die Erfahrungen, die wir mit InterViews gemacht haben, ergibt sich folgendes Resumee:

+ Rapid Prototyping ist effizient möglich, da dem Programmierer schon zusammengesetzte graphische Elemente mit den dazugehörigen Methoden zu deren Manipulation an die Hand gegeben werden (z.B. Slider).

+ Es existieren schon vordefinierte Objekte hoher Komplexität (z.B. ein Editor) und InterViews-Applikationen (z.B. IDraw), die der Programmierer in seine Applikationen einbinden kann.

- InterViews besitzt eine gute objektorientierte Struktur.
- Durch die starke Kapselung von InterViews hat der Programmierer nicht mehr die Möglichkeit, Unschönheiten, die in InterViews vorhanden sind, zu umgehen.
- InterViews hat ein eigenes 'look and feel', das im Gegensatz zu den beiden Standard-Oberflächen OSF/Motif und Open Look spartanisch ist.
- Eine Dokumentation ist praktisch nicht vorhanden; man muß das Wissen aus den mitgelieferten Beispielprogrammen extrahieren.

Ein weiterer Schritt zur Integration von Medien in die multimediale Kommunikation ist die Videointegration in die Oberfläche, die aber derzeit noch von keiner Oberflächenentwicklungsumgebung unterstützt wird. Eine solche Komponente (VEX genannt) soll in nächster Zeit in das X-Window-System eingeführt werden; aber auch bei der Andrew-Umgebung [12] gibt es Ansätze hierzu.

Die von uns untersuchte InterViews-Version 2.6 besitzt im Gegensatz zu XView bzw. OLIT kein interaktives Werkzeug zum Designen von Oberflächen. In der neuen Version 3.0 existiert ein solches Werkzeug, das sich IBuild nennt.

Ein weiterer Nachteil von InterViews ist das eigene 'look and feel', das – wie oben erwähnt – gegenüber dem 'look and feel' von Open Look bzw. OSF/Motif spartanisch ist und in Verbindung mit einer der beiden Oberflächen wie ein Fremdkörper wirkt. Da wir aber mit Open Look arbeiten, mußten wir eine Entwicklungsumgebung wählen, deren Applikationen sich in die Oberfläche einpassen lassen. Aber auch am look and feel' von InterViews wird in Stanford gearbeitet und in einer der nächsten Versionen soll eine Anpassung an OSF/Motif erreicht werden.

Literatur

[1] G. Cockroft, L. Hourvitz; *NeXTstep: Putting JPEG to Multiple Uses.* Communication of the ACM, April 1991

[2] R. Cordes, M. Hofmann, H. Langendörfer; *Layered Object-oriented Techniques Supporting Hypermedia and Multimedia Applications*, Proc. WOODMAN89, May 1989, pp. 286-296

[3] R. Cordes; *On the Way to hypermedia and multimedia Services and Terminals*, Conf. on New Interactive Communication Tools, Paris, May 1990

[4] R. Cordes, T. Kummerow; *Multimedia Communication and Information Management based on Available and Emerging Standards*, accepted paper: IEEE-COMSOC, International Workshop on Telematics 91

[5] K. Golm, E. Schnell; *Kommunizierende Objekte*, iX, April 1989,pp.82

[6] Informationssysteme für computerintegrierte Automatisierung; *Informationsmaterial*, Stuttgart

[7] Ingres; *Informationsmaterial*, Frankfurt

[8] M. Linton, J.M. Vlissides, P.R. Calder; *Composing User Interfaces with InterViews*, Computer, Feb. 1989, pp.8

[9] Oasys; *Informationsmaterial*, Distr.: Xcc. Karlsruhe

[10] J. Rosenberg, M. Sherman, A. Marks, J. Akkerhuis, *Multi-media Document Translation: ODA and the EXPRES Project*, Springer Verlag, 1991

[11] W.R. Scheifler, J. Gettys, R. Newman; *X Window System*, DEC Press, 1988

[12] M. Sherman, W.J. Hansen, M. McInerny. T. Neuendorfer; *Building Hypertext on Multimedia Toolkit: An Overview of Andrew Toolkit Hypermedia Facilities*, Proc. of the European Conf. on Hypertext, Paris, Nov. 1990

[13] B. Stroustrup; *The C++ Programming Language*, Addison-Wesley, 1986

[14] K. Süllow, R. Cordes; *Einbeziehung von Hypermediatechniken in die multimediale Kommunikation* Proc. Workshop Hypertext/Hypermedia 90, Darmstadt, April 1990

[15] Telesoft; *Informationsmaterial*, Distr.: Integrierte Informationssysteme, Konstanz

[16] D. Young; *X Window Systems; Programming and Applications with Xt*, Prentice-Hall 1989

Hypertextsysteme-Architekturen, Werkzeuge und Anwendungen

In Verbindung mit Informationssystemen, die nicht als reine Datenbanken oder Retrievalsysteme zu klassifizieren sind sowie im Bereich des elektronischen Publizierens, des rechnergestützten Unterrichts sowie in einigen Gebieten der Kulturproduktion wird momentan die Entwicklung von Hypertextsystemen forciert. Hypertext wird dabei als ein Informationsmedium angesehen, das Informationsportionen logisch in nicht-sequentieller Form miteinander verbindet. Erste verfügbare Produkte zeigen neben einem weiten Anwendungsspektrum eine Vielzahl neuer Möglichkeiten und Synergien, die aus der Verbindung von Hypertextsystemen mit bekannten Informatikkonzepten wie Datenbanken, Expertensystemen oder des Software-Engineerings erwachsen können. Im Rahmen dieses Fachgesprächs sollen sowohl interessante Anwendungen - speziell interdisziplinäre und ingenieurwissenschaftliche Systeme - auch neuartige Werkzeuge oder zukünftige Architekturen und Konzeptionen vorgestellt werden.

Koordinator: Dr. R. Cordes, Telenorma, Frankfurt a.M.

Hypertext:
Bestandsaufnahme, Trends und Perspektiven

Norbert A. Streitz

Institut für Integrierte Publikations- und Informationssysteme (IPSI)
Gesellschaft für Mathematik und Datenverarbeitung (GMD)
D - 6100 Darmstadt

e-mail: streitz@darmstadt.gmd.de

Zusammenfassung

In diesem Beitrag wird - nach einer kurzen Einführung in die Hypertextproblematik - eine Bestandsaufnahme vorgenommen, in der die bisherige Entwicklung auf dem Gebiet der Hypertextsysteme kritisch beleuchtet werden soll. Dazu werden vor dem Hintergrund ausgewählter Anwendungen aktuelle Trends und ihre Defizite untersucht. Schließlich werden Konsequenzen und Perspektiven aufgezeigt für eine dem innovativen Potential adäquaten Realisierung der Hypertext-Idee.

1 Einleitung

Die Konzepte Hypertext und Hypermedia sind im Sprachgebrauch keine unbekannten Größen mehr. Wo man auch hinschaut - auf Tagungen und Workshops, bei Messen und Kongressen, bei Anwendungsprojekten - Hypertext und Hypermedia sind überall anzutreffen. Das mag einerseits eine Modeerscheinung sein wie sie auch in der Informatik anzutreffen ist, andererseits steckt aber wohl mehr dahinter. Für diese, die zweite Position, gibt es wiederum zwei Argumentationsschienen. Entweder ist das Konzept selbst sehr innovativ und ermöglicht neue Anwendungen oder es stößt eine notwendige Weiterentwicklung auf existierenden Gebieten an, die nun mit Hypertext eine neue und adäquatere Plattform vorfindet. Es ist damit an der Zeit, eine erste Bestandsaufnahme vorzunehmen. Allerdings kann dieser Beitrag in dem hier gegebenen Rahmen nur Schlaglichter setzen und erhebt keinen Anspruch auf eine erschöpfende Behandlung des Themas.

2 *Konzept und Terminologie*

Bevor man sich über die Möglichkeiten von Hypertext unterhält, sollte man sich zunächst darüber verständigen, was unter Hypertext zu verstehen ist. Das Konzept Hypertext/Hypermedia begegnet uns in vielen Ausprägungen und der Begriff in vielerlei Interpretationen. Für die einen ist es die Einbeziehung von "elektronischen Referenzen" in einen Text: z.B. die Möglichkeit, von einem Wort in einem Text auf ein Wörterbuch direkt und on-line Zugriff zu haben oder auf die in den Literaturzitaten genannten Artikel direkt verweisen zu können, Querbezügen zwischen allen gleichen, bzw. ähnlichen Begriffen in einem Dokument nachgehen zu können. Andere verwenden diese Begriffe, obwohl sie eigentlich (nur) Multimedia meinen, d.h. eine vielfältige Mischung und Abfolge von Text, Grafik, Animation und Video, z.B. im Rahmen einer multimedialen Präsentation. Darüber hinaus gehen die Vorstellungen derjenigen, die bei Hypermedia vor allem an ihre Verwendung im Unterricht, bzw. allgemeiner in Lehr-/Lernsituationen denken. Wieder andere denken dabei an neue Möglichkeiten des Schreibens als Externalisieren von Gedanken, wobei die Nicht-Linearität von Hypertext den assoziativen Wissensstrukturen des menschlichen Gedächtnisses entgegenkommt. Viele Interpretationen sind vornehmlich durch den Verwendungszweck geprägt: Hypertext als Medium zur Repräsentation von Argumentationen (z.B. im juristischen Bereich oder bei der Erfassung und Diskussion von Designentscheidungen); Hypertextsysteme als Werkzeuge zur Erstellung und Präsentation technischer Dokumentationen; Hypertext als Mittel zur Vermittlung von Informationen in Museen, als elektronische Enzyklopädie oder allgemeiner als das Medium elektronischer Publikationen überhaupt. Dazu gehören dann auch die Vorstellungen desjenigen, der den Begriff "hypertext" geprägt hat - Ted Nelson - und der Hypertext einerseits als "nonsequential writing" charakterisiert (Nelson, 1987 a) und andererseits von einem "docuverse" - einem Universum miteinander verknüpfter elektronischer (Hyper)Dokumente - spricht (Nelson, 1987 b). Die Verweise sollen dabei über den einzelnen Rechner hinaus Dokumente in einem lokalen oder globalen Netzwerk von Rechnern verknüpfen können. Und schließlich: Hypertext als *das* Medium zur Unterstützung von Gruppenarbeit durch den Computer (CSCW = Computer Supported Cooperative Work). Das Interessante an dieser Liste ist, daß die genannten Interpretationen irgendwie alle zutreffen, aber jede nur eine bestimmte Sichtweise eines allgemeinen und einfachen generischen Konzeptes darstellen, für das die Idee der nicht-linearen Verknüpfung von Informationen zentral ist.

Das Hypertext-Konzept wurde bereits mehrfach beschrieben. Wir verweisen hierzu z.B. auf die Ausführungen der Hypertext-Pioniere Bush (1945), Engelbart (1963) und Nelson (1965, 1987 a,b) und auf die danach erfolgten Charakterisierungen, wie sie von Conklin (1987), Halasz (1988), Fiderio (1988), Smith & Weiss (1988), Nielsen (1990) gegeben wurden. Darüberhinaus erscheint uns die Unterscheidung zwischen Hyper*text* und Hyper*media* sinnvoll und notwendig (Streitz, 1990), die wir deshalb hier noch einmal (etwas überarbeitet) wiederholen und damit gleichzeitig unsere Sichtweise deutlich machen:

- **Hypertext ⇒ struktureller Aspekt**

  Mit Hypertext meinen wir eine Kategorie von (elektronischen) Dokumenten, deren definierende Merkmale mit sog. nicht-linearen Netzwerkstrukturen und assoziativen Verweisketten - innerhalb und zwischen Dokumenten (bzw. Objekten) - bei denen Zyklen möglich sind, am besten beschrieben werden. Dabei verwenden wir eine verallgemeinerte Vorstellung des Begriffs "Dokumente" und meinen damit Medien zur Externalisierung, Repräsentation, Kommunikation, Präsentation und Rezeption von Wissen. Traditionelle, klassische Dokumente, wie z.B. gedruckte Bücher, sind durch im Prinzip hierarchische Organisationsstrukturen charakterisiert, die eine sequentielle Produktion, Präsentation und Rezeption von Informationen nahelegen[1]. Damit ist nicht gesagt, daß der Text - aus textlinguistischer oder kognitionswissenschaftlicher Sicht - keine internen Strukturen hat, die Netzwerkcharakter aufweisen. Sie sind implizit vorhanden, werden aber nicht explizit kommuniziert. Damit ist auch einer Gründe genannt, warum unterschiedliche Leser bei ihrer Textanalyse zu unterschiedlichen Ergebnissen kommen können. Dieser Umstand macht es u.a. auch für die maschinelle Textanalyse und Übersetzung so schwierig, satzübergreifende Strukturen zu erkennen und adäquat auszuwerten. Für die Charakterisierung von Hypertexten ist besonders wichtig, daß die zum Einsatz kommenden Verweisketten als "machine-supported links" nur in elektronischen Dokumenten realisierbar sind. (Es würde den Rahmen dieser Charakterisierung sprengen, auf Realisierungsmöglichkeiten wie Mehrfenstersysteme, Aktivieren maussensitiver Bereiche zum Verfolgen von Verweisen, etc. einzugehen, obwohl gerade erst diese Techniken dem Hypertext-Konzept den Durchbruch ermöglichten.) Andererseits bedeutet es aber nicht, daß ein "normales" elektronisches Dokument, in dem ich z.B. mit einer Suche-Finde-Funktion zu einem anderen Teil des Dokumentes "springen" kann, auch ein Hyper(text)-Dokument ist. In einem Hypertext muß die modulare Struktur aus Knoten (= "nodes"), die die Informationseinheiten enthalten, und Kanten (="links"), die die Verbindungen darstellen, explizit vom Autor erzeugt worden sein. Damit ist sie kommunizierbar, als "node-link"-Struktur navigierbar und kann vom Rezipienten nachvollzogen werden. Diese Netzstrukturen sind insbesondere dann explizit, wenn die Verknüpfungen zwischen den Informationseinheiten, z.B. über eine grafische Präsentation, auch als Netz visualisiert werden.

Hypermedia ⇒ multi-medialer Aspekt

Enthalten die Knoten eines dem Hyperdokument zugrunde liegenden Netzes multimediale Inhalte, dann sprechen wir von Hypermedia. Dabei ist an über Text und einfache Strichzeichnungen hinausgehende Medien wie z.B. Ton (Geräusche, Sprache, Musik), komplexe (objektorientierte) Grafiken, Stand- und Bewegtbilder, Video, Animationen und Simulationen gedacht. Diese können entweder allein oder in Kombination in Erscheinung treten. Will man diese Inhalte von Knoten effizient unterstützen, so sind entsprechende Speichermedien notwendig, wie z.B. Bildplatte, CD-

---

[1] Das bedeutet natürlich nicht, daß man lineare Bücher nicht auch nicht-linear lesen kann. Schließlich kann man nicht daran gehindert werden, in der Mitte des Buches zu beginnen, Seiten zu überspringen, einer Fußnote, einer Literaturquelle oder einem Querverweis nachzugehen. Es ist jedoch wichtig festzuhalten, daß die primäre Intention des Autors und die sequentielle Präsentation und Organisation - z.B. über fortlaufende Kapitelnummern und Seitenzahlen - eine lineare Rezeption intendieren. Wir nehmen an dieser Stelle Nachschlagewerke und Lexika aus, die viele Parallelen zur Hypertext-Idee aufweisen, aber nicht das maschinengestützte Verfolgen von Verweisen realisieren und auch keine die Einheiten übergreifende Kohärenz erfordern.

ROM, DVI ("digital video interactive"). Es ist festzustellen, daß die multimedialen Aspekte das zunächst eher strukturell innovative Hypertext-Konzept mit zusätzlicher Attraktivität versehen und zu seiner Verbreitung entscheidend beitragen werden. Aber hier gilt: nicht alles, was unterschiedliche Medien verwendet und damit *multi*medial ist, ist auch gleich *hyper*medial.

3 *Bestandsaufnahme: Eine Auswahl*

Für eine Bestandsaufnahme der gegenwärtigen Forschung und Entwicklung auf dem Gebiet von Hypertext und Hypermedia ist es hilfreich, zwischen der Basistechnologie und der Integration dieser Basistechnologie mit anderen Komponenten der Informationstechnik zu unterscheiden. Dabei ist keine Auflistung verfügbarer Systeme oder eine umfassende Darstellung existierender Entwicklungen intendiert; diese würde auch einen anderen Rahmen erfordern. Vielmehr soll eine Auswahl von Beobachtungen vorgestellt werden, die dann als Basis für eine Trendbeschreibung verwendet werden kann.

3.1 Basistechnologie

Zur Basistechnologie von Hypertext zähle ich die Systemfunktionalität, die es - zunächst einem Autor, aber auch einem aktiven Leser - erlaubt, Knoten anzulegen und diese mit Hilfe von "Links" miteinander zu verknüpfen, so daß ein Netzwerk von Informationseinheiten entstehen kann. Inzwischen gibt es eine Vielzahl von Hypertextsystemen, die diese Grundfunktionalität anbieten. Sie laufen auf den unterschiedlichsten Hardware- und Software-Plattformen (z.B. Macintosh mit Mac-OS, Workstations mit Unix, PCs mit MS-DOS) und bieten in den meisten Fällen die Möglichkeit, sowohl Text als auch Grafik als Inhalte der Knoten zu verwenden. Dies sei im weiteren hier vorausgesetzt. In welcher Weise diese Systemfunktionalität verwendet wird, liegt natürlich in der Hand des jeweiligen Autors. Er kann bei den klassischen hierarchischen Strukturen bleiben und damit das Grundmuster des gedruckten Buches beibehalten. Mehr im Sinne der zugrunde liegenden Hypertextidee ist es aber, von den Möglichkeiten nichtlinearer, assoziativer Verknüpfungen Gebrauch zu machen. Hiermit soll deutlich gemacht werden, daß nicht alles, was mit einem Hypertextsystem erzeugt wird, auch notwendigerweise ein *Hyper*dokument ist. Die Qualität der am Ende stehenden Produkte wird somit sowohl durch die Basistechnologie der Hypertextsysteme als auch durch die Bereitschaft der Autoren, sich auf diese neuen Strukturierungsmöglichkeiten einzulassen, bestimmt. Bei der Basistechnologie spielen die "Typen"-Vielfalt und die Eigenschaften der Knoten- und Link-Objekte eine ausgezeichnete Rolle.

In Hinblick auf die **Art der Knoten** ist die Unterscheidung in sog. "card sharks" vs. "holy scrollers" (Raskin, 1987) ein wichtiges Differenzierungsmerkmal. Als "card sharks" werden dabei diejenigen Hypertextsysteme bezeichnet, bei denen ein Knoten eine feste "Größe" hat, wie z.B. die "Karte" in HyperCard (Goodman, 1987) oder der "frame" in KMS (Akscyn et al., 1988). Innerhalb eines solchen Rahmens können nun Inhalte in Form von Grafik, Text etc. plaziert sein. Die Begrenzung auf eine feste Größe wird naturgemäß als Beschränkung empfunden. Aber sie gibt auch immer wieder einen Anstoß, sich mit der Frage nach einer sinnvollen Segmentierung der zu vermittelnden Information und der Art der dann notwendigen Verknüpfung auseinanderzusetzen. Auf der anderen Seite dieser Dimension stehen Systeme wie Guide (Brown, 1987) und Intermedia (Meyrowitz, 1986), bei denen die Knoten im Prinzip beliebige Dokumente sind, deren Inhalte über die Länge eines Bildschirms hinausgehen können. Sie werden deshalb - wie in fensterorientierten Texteditoren üblich - meistens nur ausschnittsweise in einem Fenster angezeigt und können dort z.B. mit Hilfe eines Rollbalkens "gescrollt", bzw. durchblättert werden. In diesen Fällen hat der Autor die Freiheit, das Ausmaß der Inhalte in einem Knoten selbst zu bestimmen. Andererseits ist dies mit der Gefahr verbunden, daß oft sehr lange Texte "am Stück" in einen Knoten geschrieben werden und man sich damit von einer der Grundideen von Hypertext sehr weit entfernen kann. Es wird sicherlich noch lange umstritten bleiben, wieviel Informationen in einen Knoten sollen. Einerseits ist dies eine Entscheidung, die zur Freiheit des Autors gehört, andererseits gibt es z.Zt. noch keine eindeutigen Erkenntnisse zu dieser Frage und schließlich kann die Antwort nicht unabhängig von der jeweiligen Anwendung gegeben werden.

Entscheidender als die Eigenschaften der Knoten sind aber die Möglichkeiten der Erstellung und Verfolgung von **Verbindungen/ Verknüpfungen zwischen den Knoten.** Sie nehmen in Form der "Links" die zentrale Rolle für die Beurteilung eines Hypertextsystems ein. Links können z.B. Referenzen auf ein Zielobjekt darstellen, ohne daß das Objekt "weiß", daß auf es verwiesen wird, und man dem Objekt auch nicht "ansehen" kann, daß auf es verwiesen wird. Für viele Fälle benötigt man aber eine bidirektionale Verbindung zwischen zwei Knoten, bzw. Regionen in den Knoten, derart, daß die Tatsache der Verbindung auch beiden Knoten "bekannt" ist. Den ersten Fall findet man bei NLS/Augment (Engelbart, 1963), während der zweite nur selten, aber z.B. in Intermedia realisiert wurde. Über diese Basisentscheidung hinaus ist die weitere Funktion von Links zu klären. Dies betrifft z.B. das Ausmaß an Informationen, das mit einem Link verbunden sein soll: Name nur als "label", gerichtete Links, Stichworte oder Liste von Eigenschaften ? Was soll als Ausgangsbereich ("source") und Zielbereich ("destination") von Links zugelassen werden und wie sollen sie realisiert werden? Üblich sind entweder "Knoten --> Knoten" Links (gIBIS; Begemann & Conklin, 1988) oder "Punkt/Region --> Knoten" Links (NoteCards; Halasz et al, 1987, Halasz, 1988) oder auch Mischformen wie bei HyperCard. Seltener sind Links zwischen beliebigen Regionen in den jeweiligen Knoten (z.B. span-to-span Links bei Intermedia). Die Frage der Realisierung der verschiedenen Möglichkeiten beinhaltet die Frage nach dem Typ des "Link-Anchors". Die bestehenden Systeme beantworten diese Frage sehr unterschiedlich: als Anker werden Worte, Grafiken, Regionen einer Grafik oder der gesamte Knoten (z.B. eine Karte) verwendet. Die Detailentscheidung ist zwar anwendungsabhängig zu gestalten, aber das System sollte eine Bandbreite an Möglichkeiten ambieten, von denen der Autor/Designer des Hyperdokumentes dann Gebrauch machen kann. In welcher Weise soll dem Benutzer das Vorhandensein eines Ankers - insbesondere für den Ausgangsbereich - angezeigt werden? Es geschieht oft durch den sog. "Link Marker", aber es gibt auch Systeme/ Dokumente, in denen im Prinzip jedes Wort oder allgemeiner jedes Objekt als Ausgangspunkt für einen Link (z.B. auf ein Lexikon) erlaubt ist. In einigen Fällen, z.B. HyperCard, ist es ein "Button" (auf Deutsch ungeschickt als "Taste" übersetzt), meistens als Umriß des maussensitiven Bereiches markiert, der zum Aktivieren und Verfolgen des Links durch Mausklick vorgesehen ist. In anderen Fällen deutet ein "Marker" auf das Vorhandensein eines maussensitiven Bereichs, der aber erst nach einer Aktion des

Benutzers erkennbar oder beim Bewegen des Cursors durch Wechsel seiner Form deutlich wird und sehr viel größer als der Marker selbst sein kann. Diese Fragen stellen Teilaspekte der umfassenderen Fragestellung der Präsentation des Hyperdokumentes gegenüber dem Leser dar. Damit impliziert das Design eines Hyperdokumentes die Gestaltung einer Benutzungsoberfläche für ein interaktives System und die damit verbundenen bekannten Probleme.

Bei den in der Literatur anzutreffenden Definitionen wird meistens nur von "links", Verweisen, Verknüpfungen, etc. geprochen. Damit sind Verweise im Sinne von "zeigen auf" gemeint, dem nach erfolgten Aktivieren des Ausgangspunktes ("source") das "Anzeigen" des Inhaltes des Knotens folgt, auf den der "link" zeigt. Diese auch "points_to" genannten Verweise stellen aber nur elementare Ausprägungen des Hypertext-Konzeptes dar. Interessanter - und für die wirklich innovativen Anwendungen unbedingt notwendig - sind sog. "getypte" Verweise ("typed links"). Diese tragen einen Bezeichner ("label"), der ihre Bedeutung kennzeichnet. Dabei ist wiederum zu unterscheiden, ob dieser "label" nur eine "angeheftete" Textmarke ist oder ob mit ihm auch eine spezifische im Rechner repräsentierte Semantik verbunden ist. Diese könnte sich z.B. so ausdrücken, daß bestimmte Verweise nur in definierten Kontexten sichtbar und aktivierbar sind; daß bei der Aktivierung mit diesem "link" assoziierte und für ihn spezifische Operationen - z.B. eine Tonfolge oder eine Animation - ausgeführt werden, oder daß ein Link vom Typ "support" sich selbst daran "hindert", zwischen zwei Knoten vom Typ "claim" und "rebuttal" eingefügt zu werden (Streitz et al, 1989). Die Möglichkeit getypter Knoten und Links und das automatische Überprüfen der Verträglichkeit von Knoten- und Linktypen bringt eine sehr große Flexibilität beim Systemverhalten mit sich. Damit ist eine entscheidende Stärkung der Mächtigkeit der Repräsentationsstrukturen verbunden.

Ein wichtiger Aspekt bezieht sich auf das Ausmaß der Unterstützung, das ein gegebenes Hypertextsystem für die Erstellung und Verwaltung von **zusammengesetzten Objekten** bietet. Diese auch als "containers", "organizers", "composites" oder "complex objects" bezeichneten Objektklassen ermöglichen das Zusammenfassen und Organisieren von Knoten und Links zu größeren Einheiten. Beispiele für "container": "notefiles" in NoteCards, die zum Ablegen eines Netzwerkes verwendet werden; "framesets" in KMS, in denen Mengen von "frames" zusammengefaßt werden, und die "stacks" (Stapel) bei HyperCard zur Organisation von Karten. Die "folders" in Intermedia, in denen "documents" gespeichert werden, übernehmen die Funktion von "organizers". "Composites" ermöglichen das Behandeln von zusammengesetzten Strukturen (z.B. Teilnetze) als "first class objects", wie sie z.B. in HyperBase (Schütt & Streitz, 1990) realisiert und im Dexter Hypertext Reference Model (Halasz & Schwartz, 1990) vorgesehen sind.

Weitere Themen, die zur Basistechnologie gehören, aber aus Platzgründen hier nicht behandelt werden, beziehen sich auf "views", das Erstellen von "virtual entities", die auf erst zur Laufzeit dynamisch konstruierten Links basieren, die Möglichkeiten für "overviews" mit Hilfe eines Inhaltsverzeichnisses, mit einem graphischen Browser (Diagramme, Netzdarstellungen), "fish-eye views", "guided tours", etc.

Bei der Klassifizierung von Hypertextsystemen hat sich die Unterscheidung zwischen **Autoren-Systemen** und **Browsing-Systemen** als nützlich erwiesen (Streitz el al, 1989). Natürlich muß jedes Hypertextsystem beide Aspekte unterstützen. Dies kann aber auf unterschiedliche Art und Weise geschehen und die Komponenten werden für den Benutzer unterschiedlich zugänglich sein. So ist es von Bedeutung, ob man innerhalb eines Systems zwischen Autorenrolle und Leserrolle schnell umschalten kann (z.B. in HyperCard) oder ob man zwischen zwei Systemen wechseln muß (Concordia und Document Examiner; Walker, 1987). Nimmt man eine Trennung vor, muß es eine "preview"-Möglichkeit für den Autor geben. Diese Problematik ist sehr eng mit der nächsten Frage verknüpft: der Qualität von Hyperdokumenten.

Betrachtet man die Ergebnisse der unterschiedlichsten Bemühungen, Hyperdokumente zu erzeugen, dann läßt die Qualität in vielen Fällen zu wünschen übrig. Teilweise sind die Dokumente zu sehr an klassischen hierarchischen Dokumentstrukturen orientiert, teilweise gefallen sie sich in einer wenig motivierten oder durchdachten Aneinanderreihung von Multimedia-Effekten. Sind sie andererseits sehr hypertextartig, dann bieten sie oft viele Probleme beim Benutzen durch den Leser (die viel zitierte Desorientierung; Conklin, 1987). Die **Qualität von Hyperdokumenten** ist von einer Reihe von Faktoren abhängig. Einmal natürlich - wie bei jedem Dokument - von der Kompetenz des Autors. Weiterhin von der Qualität des Systems (Hardware/Software), mit dem das Hyperdokument dem Leser präsentiert wird. Aber ganz entscheidend ist das Wissen um die zusätzlichen Bedingungen, die ein Hyperdokument erfüllen muß, und die Art und Weise, in der das Beachten und Einhalten dieser Bedingungen von dem System, besonders von der Autorenkomponente, auch unterstützt wird. Dieses zusätzliche hypertextspezifische Wissen wird unter dem Stichwort "new rhetoric" (Landow, 1987) thematisiert. Wir kommen darauf in Kapitel 4 zurück.

3.2 Integration von Hypertextsystemen mit anderen Informationssystemen

Obwohl Hypertextsysteme bisher vor allem als isolierte Systeme verwendet werden, muß man sich der Frage stellen, wie sie in die Landschaft der existierenden Informationssysteme passen und welche Rolle sie dort übernehmen können und sollen. Dies ist umso notwendiger, als man einerseits zwar die innovativen Möglichkeiten von Hypertextsystemen nutzen, andererseits aber auf die gewohnten Unterstützungsmöglichkeiten nicht verzichten möchte. Dies gilt insbesondere für die Nutzung existierender Datenbestände. Vor diesem Hintergrund ist es notwendig, daß ein Hypertextsystem Daten auf unterschiedliche Arten importieren und exportieren kann. Damit stellt sich die Frage nach Austauschformaten und zwar sowohl zwischen verschiedenen Hypertextsystemen als auch zwischen einem Hypertextsystem und anderen Textverarbeitungssystemen, Grafikprogrammen, Multimedia-Umgebungen, etc.

Viele Hypertextsysteme erlauben es inzwischen, zuvor mit anderen Programmen erstellte Text- und Grafikdaten zu importieren - mal einfacher, mal komplizierter. Der umgekehrte Weg - und da vor allem die saubere Trennung in entsprechende Bestandteile - ist oft nicht möglich. Das Dexter Hypertext Reference Model (Halasz & Schwartz, 1990) bereitet den Boden für einen Austausch von Dokumenten zwischen verschiedenen Hypertextsystemen vor. Dazu gibt es auch schon erste Umsetzungen, z.B. wurde der Austausch zwischen HyperCard und NoteCards sowie zwischen KMS und Intermedia realisiert.

Insbesondere dann, wenn die Datenbestände groß werden - und das ist bei realistischen Anwendungen sehr schnell der Fall - ist es unumgänglich, eine Datenbank zur Verwaltung der Hypertextobjekte zu verwenden. Damit stellt sich das Problem der Integration, bzw. des Anschlusses eines existierenden DBMS an ein

Hypertextsystem. Es gibt zur Zeit kaum Realisierungen, die schon in großem Maßstab eingesetzt werden. Oft werden Hypertextsysteme mit einer selbst geschriebenen Datenverwaltung kombiniert. Diese Thematik stellt ein sehr aktives Forschungsgebiet dar, wobei hier wiederum zwischen den sog."hypermedia engines", die in Kombination mit einem Filesystem (z.B. UNIX) als "backends" fungieren, und der Kombination einer "hypermedia engine" mit einem DBMS unterschieden werden muß. Die in diesem Zusammenhang viel zitierte HAM ("hypertext abstract machine") von Tektronix (Campbell & Goodman, 1988) stellt zwar ein entsprechendes Modell vor, benutzt aber selbst kein DBMS. Dies tut dagegen HyperBase (Schütt & Streitz, 1990), das eine objektorientierte, mehrbenutzerfähige persistente Verwaltung von Hypertextobjekten in Kombination mit einer relationalen Datenbank (Sybase) realisiert.

Eine weitere Frage betrifft die Integration von Lexika, von existierenden Information Retrieval-Systemen und von wissensbasierten Komponenten (Haake & Schütt, 1990). Schließlich werden die Anforderungen an die ästhetische Qualität und Lesbarkeit von Hyperdokumenten auf dem Bildschirm anspruchsvoller. In dem Maße, in dem Hypertext das Experimentierstadium verläßt und zu einer Alternative für das Printmedium werden soll, benötigt man Darstellungen, die ein belastungsarmes Lesen am Bildschirm ermöglichen. Dies erfordert neben den entsprechenden Hardware-Voraussetzungen die Integration von Layoutkomponenten traditioneller Publikationssysteme in Hypertextumgebungen.

Die Art und das Ausmaß der Integrationsfragen sind zu einem großen Teil auch von der gewählten Anwendung abhängig. Die z.Zt. dominanten Anwendungen finden sich in den folgenden Bereichen: Elektronisches Publizieren mit einem Schwerpunkt bei der technischen Dokumentation (Gebrauchsanweisung, Reparaturanleitung) und bei interaktiven Enzyklopädien und annotierbaren historischen oder juristischen Quellenmaterialien, interaktive Demonstrations- und Präsentationssysteme für ein weites Themenspektrum (von der vortragsbegleitenden interaktiven Dia-Show bis zum Museum), Aus- und Weiterbildung (z.B. Englisch- und Biologie-Kurse an der Universität, Schulungsmaterial für neue technische Systeme), CAD, Software-Engineering, Bürokommunikation, Auskunftssysteme, Problemerforschungssysteme (für eine Person und ein oder mehrere Probleme) und Argumentations- und Diskussionsunterstützungssysteme (eine Gruppe von Personen und ein oder mehrere Probleme). Im Fall der Unterstützung von Gruppenarbeit mit Hypertextsystemen kommen zusätzliche Fragestellungen hinzu, wie z.B. Versionsverwaltung, Zugriffsrechte, etc.

Trends und Perspektiven

Es ist sicherlich deutlich geworden, daß es eine Vielzahl von Hypertextsystemen gibt, die sich aber auf einer breiten Palette unterschiedlicher Funktionalität anordnen lassen. Sie existieren auf unterschiedlichen Plattformen (Software/ Hardware), und es gibt sie auch in unterschiedlichen Preislagen - von umsonst bis sehr teuer. Es stellt sich schnell die Frage, ob man so viele unterschiedliche Systeme braucht und - wenn ja- wozu? Einerseits stellen die meisten die zuvor benannte Grundfunktionalität - verschiedene Informationseinheiten/ Objekte auf unkomplizierte Weise miteinander zu verknüpfen - bereit. Andererseits unterscheiden sie sich in dem Ausmaß an zusätzlicher Funktionalität, um Anforderungen der Benutzer und der

Anwendungen nachzukommen. Dabei handelt es sich oft um Möglichkeiten, die außerhalb eines Hypertextsystems als selbstverständlich verlangt werden und dort auch oft vorhanden sind. Es verwundert daher nicht, wenn sich ein gewichtiger Teil der aktuellen Diskussion mit der Frage beschäftigt, in welcher Weise man Hypertextfunktionalität haben kann und zugleich auch diejenigen Vorteile und Möglichkeiten, die man bisher schon gewohnt ist. Hierzu gibt es nun im wesentlichen zwei Standpunkte. Die erste Position ist: man baut ein Hypertextsystem und es verfügt über eine datenbankbasierte Objektverwaltung, weist Schnittstellen/Austauschformate (z.B. SGML, ODA) auf, verfügt über hochwertige Layoutqualität, integriert eine intelligente IR-Schnittstelle und wissensbasierte Komponenten. Der Ausgangspunkt ist aber ein Hypertextystem, das um die genannten Aspekte angereichert wird. Die andere Position besteht darin, die Vielfalt der existierenden Systeme als gegeben zur Kenntnis zu nehmen und mit ihnen zufrieden zu sein. Dem Fehlen von Hypertextfunktionalität wird nun dadurch Rechnung getragen, daß man diese Grundfunktionalität anwendungsunabhängig derart bereitstellt, daß im Prinzip beliebige Anwendungen sie nutzen können. Meyrowitz (1989) propagiert die Existenz eines Basis-Linkservice, der innerhalb des "desktop" als nächste Ebene der Integration angesehen werden kann. So wie man in den meisten Systemen mit "select", "cut", "copy", und "paste" zwischen verschiedenen Anwendungen Informationen austauschen kann, soll man mit "select", "start link" und "complete link" Informationen über Anwendungen hinweg miteinander verknüpfen können. Diese "navigational links" sollen dabei bidirektional sein und ein einfaches Navigieren ermöglichen. Intermedia realisiert diese bidirektionalen Links zunächst innerhalb eines Hypertextsystem, geht aber darüber hinaus mit "warm linking" (dem Austauschen des Inhaltes des Ankers über einen Link mit "pull" und "push") und "hot linking" (der automatischen Synchronisation bei Veränderungen der Inhalte eines als "master" fungierenden Ankers mit seinem Gegenüber). Von einigen Firmen (SUN, DEC) wird in ihren Umgebungen nun auch ein Basis-Linkservice angeboten, der das Setzen und Verfolgen von Links zwischen unterschiedlichen Anwendungen erlaubt. Allerdings handelt es sich dabei immer nur um Navigations-Links.

Bei dem Konzept des anwendungsunabhängigen Link-Service ist anzuerkennen, daß die Basisoperationen des Desktops mit dem Hypertext-Konzept verbunden werden. Dies ist eine unbedingt begrüßenswerte Integration und Erweiterung. Sie berücksichtigt aber keine getypten Links, die unserer Meinung nach das Hypertext-Konzept erst richtig umzusetzen gestatten. Außerdem geht man dabei von einer sehr eingeschränkten Vorstellung dessen aus, was ein Hyperdokument sein kann oder soll. Hyperdokumente sollten meiner Meinung nach mehr leisten als Informationsausschnitte in existierenden normalen Dokumenten nur im Sinne einer Referenz miteinander zu verbinden. Damit sind wir wieder bei der ersten Position. Um komplexe und anspruchsvolle Hyperdokumente zu realisieren und den Lesern zu präsentieren, bedarf es dedizierter Hypertextsysteme. Diese Forderung ist eng verknüpft mit der Problematik einer neuen Hypertextrhetorik. So vertreten wir am GMD-IPSI z.B. die These, daß die Orientierungs-und Navigationsproblematik ("getting lost in hyper space") keine dem Hypertext-Konzept "angeborene" Eigenschaft ist, sondern daß sie zu einem großen Teil aus den unzureichenden Erfahrungen von Autoren mit dem Design von Hyperdokumenten resultiert. Weder Autoren noch Leser können auf eine langjährige Tradition oder Erfahrung mit Hyperdokumenten zurückgreifen wie wir es bei Printmedien tun. Dies zeigt sich insbesondere dann, wenn man existierende lineare Dokumente und Hyperdokumente in

Bezug auf ein entscheidendes Qualitätsmerkmal, dem der Kohärenz, vergleicht. Es zeigt sich, daß ein Großteil der Desorientierung durch das Fehlen von Kohärenz verursacht wird. Diese herzustellen ist aber für den Autor eines Hyperdokuments zur Zeit sehr schwierig, da ihm für diese Autorenproblematik keine Unterstützung von Seiten der Hypertextsysteme angeboten wird. Der am GMD-IPSI verfolgte Lösungsansatz basiert darauf, ein benutzer- und verwendungsorientiertes Dokumentmodell für Hyperdokumente zu definieren und in Verbindung damit sog. Designobjekte und zugehörige Operationen durch das System zur Verfügung zu stellen. Diese geben dann einen Rahmen ab, innerhalb dessen sich der Autor kreativ entfalten kann. Details zu der Kohärenzproblematik und dem skizzierten Lösungsvorschlag finden sich bei Haake, Hannemann & Thüring (1991) und Thüring, Haake & Hannemann (1991).

Hypertextsysteme repräsentieren sicherlich den Beginn der Entwicklung einer neuen Generation von Informations- und Publikationssystemen und dies sowohl mit Bezug auf die Möglichkeiten für eine benutzerorientierte Gestaltung der Mensch-Computer-Schnittstelle als auch für die aufgabenzentrierte Bereitstellung von Funktionalität zur Unterstützung einer Vielzahl von Arbeitstätigkeiten. Aber die Umsetzung dieses Anspruchs wird in vielen Fällen noch Wünsche offen lassen und nicht immer von den im Hypertext-Konzept inhärent angelegten Möglichkeiten Gebrauch machen. So gilt es einerseits, eine Reihe noch ungelöster technischer Probleme zu bewältigen und andererseits mehr Erfahrung mit entsprechenden Realisierungen von Hyperdokumenten zu sammeln.

Literatur

Akscyn, R., McCracken, D. Yoder, E. (1988). KMS: A distributed hypermedia system for managing knowledge in organizations. *Communications of the ACM,* 31 (7), 820 - 835.

Begemann, M. & Conklin, J. (1988). The right tool for the right job. *BYTE* , 13 (10), 255 - 266.

Brown, P. (1987). Turning ideas into products: The Guide system. In *Proceedings of the HYPERTEXT'87 Workshop.* (pp. 33 - 40). Chapel Hill, NC. (November 13 - 15, 1987).

Bush, V. (1945). As we may think. *Atlantic Monthly,* 176 (1), 101-108.

Campbell, B. & Goodman, J.M. (1988). HAM: A general purpose hypertext abstract machine. *Communications of the ACM,* 31 (7), 856 - 861.

Conklin, J. (1987). Hypertext: An introduction and survey. *IEEE Computer Magazine,* 20 (9), 17-41.

Engelbart, D. (1963). A conceptual framework for the augmentation of man's intellect. In P. Howerton (Ed.), *Vistas in information handling* (pp. 1-29). Spartan Books.

Fiderio, J. (1988). Hypertext: A grand vision. *BYTE* 13 (10), 234 - 244 (October 1988).

Goodman, D. (1987). *The complete HyperCard book.* New York: Bantam Books.

Haake, J., Hannemann, J. & Thüring, M. (1991). Ein Ansatz zur Organisation von Hyperdokumenten. In H. Maurer (Hsrg.), *Proceedings der Tagung Hypertext/Hypermedia '91* (S. 119 - 134). Informatik-Fachberichte 276. Heidelberg: Springer Verlag.

Haake, J. & Schütt, H. (1990). Eine Systemarchitektur für ein wissensbasiertes Hypertext-Autorensystem. In P. Gloor & N. Streitz (Hrsg.), *Hypertext und Hypermedia: Von theoretischen Konzepten zu praktischen Anwendungen.* (S. 65-78). Informatik-Fachberichte 249 . Heidelberg: Springer.

Halasz, F.G. (1988). Reflections on Notecards: Seven issues for the next generation of hypermedia systems. *Communication of the ACM*, 31, 836-852.

Halasz, F.G., Moran, T. & Trigg, R. (1987). NoteCards in a nutshell. In *Proceedings of the ACM CHI+GI'87 Conference* (pp. 45 - 52) Toronto, Canada (April 5 -9, 1987).

Halasz, F. G. & Schwartz, M. (1990). The Dexter Hypertext Reference Model. In *Proceedings of the NIST Hypertext Standardization Workshop*, Gaithersburg, MD. (January 16-18, 1990).

Landow, G. (1987). Relationally encoded links and the rhetoric of hypertext. In: *Proceedings of the HYPERTEXT '87 Workshop* (pp. 331 - 343). Chapel Hill, NC.(November 13 - 15, 1987).

Meyrowitz, N. (1986). Intermedia: The architecture and construction of an object-oriented hypermedia system and applications framework. In *Proceedings of the OOPSLA '86 Conference* (pp. 186-201). Portland, OR (September 29 - October 2, 1986).

Meyrowitz, N. (1989). Hypertext - Does it reduce cholesterol, too? Keynote address at the *ACM-Conference HYPERTEXT '89* . Pittsburgh, PA. (November 5-8, 1990).

Nelson, T. (1965). A file structure of the complex, the changing, and the indeterminate. In: *Proceedings of the 20-th National ACM-Conference* (pp. 84 - 100). Cleveland, OH.

Nelson T. (1987 a). *Literary machines.* Edition 87.1 (Ted Nelson Eigenverlag).

Nelson T. (1987 b). All for one and one for all. In *Proceedings of the HYPERTEXT '87 Workshop.* (pp. v-vii). Chapel Hill, NC. (November 13 - 15, 1987).

Nielsen, J. (1990). *Hypertext and Hypermedia.* Boston: Academic Press.

Raskin, J. (1987). The hype in hypertext: A critique. In *Proceedings of the HYPERTEXT '87 Workshop.* (pp.325 - 330). Chapel Hill, NC. (November 13 - 15, 1987).

Schütt, H. & Streitz, N. (1990). Hyperbase: A hypermedia engine based on a relational database management system. In A. Rizk, N. Streitz & J. André (Eds.), *Hypertext: Concepts, systems, and applications.* (pp. 95 - 108) Cambridge : Cambridge University Press.

Smith, J. & Weiss, S. (1988). An overview of hypertext. *Communications of the ACM*, 37(7), 816 - 819.

Streitz, N.A. (1990). Hypertext: Ein innovatives Medium zur Kommunikation von Wissen. In P. Gloor & N. Streitz (Hrsg.), *Hypertext und Hypermedia: Von theoretischen Konzepten zu praktischen Anwendungen.* (S. 10-27) . Informatik-Fachberichte 249. Heidelberg: Springer.

Streitz, N.A., Hannemann, J. & Thüring, M.(1989). From ideas and arguments to hyperdocuments: Travelling through activity spaces. In *Proceedings of the ACM-Conference HYPERTEXT '89* (pp. 343 - 363). Pittsburgh, PA. (November 5-8, 1990).

Thüring, M., Haake, J., & Hannemann, J. (1991). What's Eliza doing in the Chinese Room ? Incoherent hyperdocuments and how to avoid them. Paper accepted and to be published in *Proceedings of the ACM-Conference HYPERTEXT '91* . San Antonio, Tx. (December 15-18, 1991).

Walker, J. (1987). Document Examiner: Delivery interface for hypertext documents. *Proceedings of the HYPERTEXT '87 Workshop.* (pp. 307-324). Chapel Hill, NC. (November 13 - 15, 1987).

Ein Bewertungsschema für HyperText-Systeme und ein vergleichender Überblick über HyperTies, HyperPad, HyperCard und Guide

Ch. Meyer, M. Rauterberg, M. Strässler
Institut für Arbeitspsychologie (IfAP)
Eidgenössische Technische Hochschule Zürich
Nelkenstrasse 11, CH-8092 Zürich

Das hier vorgestellte Bewertungsschema soll primär dazu dienen, dem nicht professionellen Hypertextanwender (-Autor und -Leser) einen systematischen Überblick über Eigenschaften von Hypertextsystemen zu geben. Das Bewertungsschema enthält Eigenschaften, welche bei nichttrivialen Hypertextentwicklungen empfehlenswert, bzw. unabdingbar sind. Wie man an der Schwerpunktsetzung zwischen Benutzersicht und Autorensicht erkennen kann, werden die Probleme des Hypertextautors und eine adäquate Unterstützung seitens der Entwicklungsumgebung noch zu sehr unterschätzt. Erst wenn entsprechend komfortable Entwicklungsumgebungen zur Verfügung stehen, werden Hypertextsysteme die ihnen zustehende Rolle einnehmen können.

1 Einführung

Der Begriff 'Hypertext' hat viele Facetten, und es gibt verschiedene Ansätze, sich ihm zu nähern. Jedoch trifft man stets, wenngleich in verschiedenen Varianten, auf die beiden folgenden Grundaspekte von Hypertexten:

Die **Struktur**: Jeder Hypertext ist ein *beliebiges Netzwerk* von strukturierten, informationstragendenen *Knoten*; Verbindungspunkte im Innern der Knoten setzen diese über *Querverweise* zueinander in Beziehung.

Der **Zugriff** auf die Struktur: Jedes Hypertextsystem ermöglicht neben dem gewöhnlichen Direktzugriff auf einzelne Knoten auch die *Navigation* im Hypertextnetz, wodurch man jede gewünschte Knoteninformation mittels einer Wanderung entlang einem geeigneten Pfad des Querverweis-Netzwerks erreicht.

Diesen beiden Aspekten stehen zwei Grundkonzepte zur technischen Umsetzung auf Computern gegenüber. **Datenorientierte Hypertextsysteme**: die Knoten sind *Informationsspeicher*, die nur gelesen, also nicht verändert werden können; Navigation im Hypertext entspricht einem Suchzugriff auf eine netzwerkartig organisierte Datenbank. **Objektorientierte Hypertextsysteme**: die Knoten sind *Zustandsautomaten*, d.h. nach aussen abgeschlossene Objekte, die aus einem Informationsspeicher (Datenstrukturen) sowie aus *Transformationsregeln* für diesen Speicher (Operationen auf den Datenstrukturen) bestehen. Die Transformationen werden ausgelöst, indem die Bildschirmdarstellungen der Objekte manipuliert werden und die betreffenden Knotenobjekte sich darauf gegenseitig *Nachrichten* mit entsprechenden Transformationsanforderungen zusenden. Dabei muss zwischen den Nachrichtenpfaden und den Netzwerkpfaden kein Zusammenhang bestehen. Navigation im Objektnetzwerk dient zur Inspektion des Objektzustandes, d.h. des Speicherinhaltes, und zur Auslösung von Nachrichtenflüssen.

2 Grundlegende Konzepte

2.1 Daten- versus Objekt-Orientierung

Das **datenorientierte Hypertextkonzept** ist ein neuer Lösungsansatz für ein altes Problem. In herkömmlichen Texten reiht sich Satz an Satz: die sichtbare Form des Textes, seine typographische Organisation ist der Inbegriff der *Sequenz*. Die Welt, die mit dem Text beschrieben werden soll, ist jedoch meist nicht von solch sequentieller Natur, sondern besitzt im günstigen Fall eine *hierarchische*, im allgemeinsten Fall jedoch eine beliebige *netzartige*, also *zirkuläre* Beziehungsstruktur zwischen den zu beschreibenden Objektbereichen.

Im herkömmlichen Text gibt einzig die in ihm enthaltene Information, der Textinhalt, Aufschluss über die strukturellen Beziehungen in der dargestellten Wirklichkeit — die *äussere Form* hingegen ist nicht sehr aussagekräftig. Zum Teil lässt sich dieser Mangel mit Methoden der *Informationstypographie* beheben

(Seitenraster, Schriftwahl und Schriftmischung, Titelgestaltung, Auszeichnung von Textpassagen, Farben, graphische Elemente, Illustrationen). Mit rein typographischen Hilfsmitteln lässt sich aber höchstens eine *sequentiell dargestellte hierarchische Organisation* erreichen, wie sie auch diesem Aufsatz zugrunde liegt: Kapitel gliedern sich in Abschnitte, diese in Absätze bzw. einzelne Punkte. Aber schon ein umfangreicher Fussnotenapparat oder gar eine echt nicht-hierarchische Kapitelstruktur lassen diese Bemühungen am sequentiellen Grundprinzip des gedruckten Textes scheitern.

Erheblich näher an die Lösung des Problems, die Form des Textes mit den abzubildenden Beziehungsstrukturen in Übereinstimmung zu bringen, kommt man mit der Methode der 'Strukturbilder' oder **Blockdiagramme**. Der sequentielle Text wird in Teilsequenzen, d.h. in Textblöcke zerlegt und diese mit einfachen graphischen Elementen wie Striche, Pfeile und Rahmen zweidimensional zueinander in Beziehung gesetzt. Bei grösseren Textmengen versagt aber auch dieser Ansatz, zwar keineswegs prinzipiell, jedoch aus rein praktischen Gründen.

Hingegen wird auf *Computersystemen* das Blockdiagramm im Rechnerspeicher repräsentiert und bei Bedarf auszugs- oder blockweise in gewohnter Darstellung auf einen Bildschirm gebracht. Das ist bereits im wesentlichen, was als 'Hypertext' bezeichnet wird: eine Menge von gespeicherten Textfragmenten (Kapitel, Absätze), die durch elektronische Verbindungen zu einer Beziehungsstruktur verbunden werden. Deshalb auch die Bezeichnung: ein 'Hypertext' ist ein '**Über-Text**', eine Textsammlung mit einer Überstruktur. Darüberhinaus umfasst das Hypertextkonzept nicht nur Information in Form von Schrift, sondern auch in Form von Bildern und bewegten Bildsequenzen, d.h. Animation. Kommt noch Ton und Film hinzu, so spricht man meist nicht mehr von 'Hypertext', sondern von '*Hypermedia*'.

Objektorientierte Hypertextsysteme sind dedizierte objektorientierte Programmiersysteme, die für die Programmierung von Interaktion im Knotennetzwerk ausgelegt sind. Ein objektorientierter Hypertext ist demnach ein Programm, bestehend aus untereinander vernetzten Knotenobjekten, welche die eigentliche Information in ihren Datenspeichern halten und bei Erhalt von entsprechenden Aufforderungen transformieren können. Die Knotenobjekte werden aus vorgegebenen Grundobjekten aufgebaut, z.B.Textfelder zur Informationsvermittlung oder Bedienungsfelder zur Auslösung von Transformationen, und auf dem Bildschirm graphisch dargestellt.

Das datenorientierte Konzept ist ein Spezialfall des objektorientierten, wo auf die Knoteninhalte ausser den Navigationsoperationen keine Transformationen wirken, die 'Datenobjekte' also unveränderlich sind. Trotz dieser scheinbaren Einschränkung verkörpert das datenorientierte Konzept das eigentliche Grundprinzip von Hypertexten; die Objektorientierung ist in diesem Sinne eine nicht unbedingt notwendige, aber ausserordentlich leistungsfähige und flexible Erweiterung. Natürlich sind in konkreten Anwendungsfällen beliebige Übergangsformen zwischen den beiden Positionen denkbar, d.h datenorientierte Hypertexte können auch objektorientierte Komponenten aufweisen, indem sie auch Knoten oder Knotenbestandteile enthalten, die vom Benutzer transformiert werden können.

2.2 Informationsstrukturierung

Allgemein betrachtet ist Hypertext eine *Methode zur Wissensrepräsentation*, d.h. ein Werkzeug zur gezielten Erschliessung von Wissen in Form strukturierter begrifflicher und anschaulicher Information über einen Ausschnitt aus unserer Innen- und Umwelt. Das Wissen soll in Form von Text und Bild in einer Maschine nachgebildet werden und gezielt abrufbar sein. Man könnte Hypertext zudem als einen graphentheoretischen Ansatz zur Informationsstrukturierung einordnen. Der allgemeinste Graph, das Netzwerk, ist gut geeignet, um das Beziehungsnetz zwischen in sich geschlossenen Informationseinheiten zu modellieren und damit Wissen auf einen Graphen abzubilden. Entsprechend der Terminologie der Graphentheorie werden die abgeschlossenen Informationseinheiten Knoten genannt, die Kanten des Graphen jedoch als Verweise ('Link') bezeichnet.

Ein Hypertext unterscheidet sich jedoch von einem gewöhnlichen Graphen darin, dass ein Hypertextknoten kein null-dimensionaler Punkt ist, sondern eine zwei-dimensionale (Bildschirm)- Fläche mit innerer Text-Bild-Struktur. Die Anfangs- und Endpunkte der Verweise, welche im folgenden als Verweispunkte bezeichnet werden, können im Knoten beliebig angeordnet sein. Die Verweise führen aus dem Inneren des Ausgangsknotens heraus in den Zielknoten. Dabei kann es sich auch um den gleichen Knoten handeln, denn in der Regel enthält ein Knoten eine Vielzahl von Verweispunkten, und diese können auch im Knoteninnern untereinander vernetzt sein. Im folgenden wird bei graphentheoretischen Interpretationen des Hypertextnetzes stets von der inneren Knotenstruktur abgesehen und innere Verweise als Schlingen im Graph betrachtet.

Die Informationsstruktur von Hypertexten beruht demnach auf drei Grundelementen:

2.2 a Der **Knoten** ist die abgeschlossene Informationseinheit des Hypertextes. Jeder Knoten ist durch einen Namen, einen Titel oder sonstige Klassifizierungsinformation eindeutig identifiziert und kann eine beliebige Sequenz von Text, Bild und Animation enthalten. Die Informationsmengen in den Knoten können sehr unterschiedlich sein, von einem einzigen Wort über einen Bildschirminhalt bis zu einem ganzen Buch; sinnvollerweise wird man jedoch den Knotenumfang auf ein Kapitel oder einen inhaltlich abgeschlossenen Abschnitt festlegen. Im Unterschied zu einem gedruckten Text sind die Knoteninhalte bei objektorientierten Hypertexten prinzipiell, bei datenorientierten in Sonderfällen (objektorientierte Bestandteile, z.B. Textfelder für Anmerkungen und Notizen) veränderbar.

2.2 b Jeder Knoten kann eine beliebige Anzahl **Verweispunkte**, d.h. Anfangs- und Endpunkte von Verweisen enthalten. Verweispunkte können natürlich auch geschachtelt werden, z.B. kann ein Satz, der als Anfangspunkt für einen Verweis dient, ein Stichwort enthalten, das ebenfalls als Anfangspunkt für einen weiteren Verweis verwendet wird. Als Verweispunkte können beliebige Knotenbestandteile dienen, im besonderen:

2.2 ba STICHWÖRTER, vergleichbar mit Querverweisen in Lexika, aber auch Wortgruppen und ganze Sätze.

2.2 bb in sich geschlossene KNOTENTEILE, wie Fussnoten, Absätze und Kapitel; auch der Knoten als Ganzes kann Verweispunkt sein (vgl. Punkt 2.2 d).

2.2 bc BILDKOMPONENTEN, die z.B. in der Art einer Explosionszeichnung dargestellt und ausgewählt werden können.

2.2 bd KONTROLLFELDER (z.B. Dialogboxen, etc.) zur Interaktion mit dem Knoten bzw. dem Knotennetzwerk; solche Simulationen von 'Bedienungskonsolen' mit Ein- und Ausgabefeldern sind für objektorientierte Hypertexte von grundlegender Bedeutung.

2.2 c **Verweise** verbinden genau je zwei Verweispunkte miteinander, innerhalb desselben Knotens oder ausserhalb. Navigation in einem Hypertext bedeutet, dass man im aktuellen Knoten einen Anfangspunkt auswählt und über den Verweis automatisch zum Endpunkt in einem neuen Knoten gelangt und hier erneut einen Anfangspunkt auswählt. Jeder Verweis ist wie ein Knoten eindeutig identifiziert und gehört einem bestimmten Typ an. Der **Verweistyp** ist bestimmt durch:

2.2 ca RICHTUNG. Analog zur gerichteten Kante in einem Graphen werden beim gerichteten Verweis Anfangs- und Endpunkt unterschieden, beim ungerichteten können die zugeordneten Verweispunkte sowohl Anfangspunkt als auch Endpunkt sein. Von jedem Anfangspunkt geht genau ein Verweis aus; in einen Endpunkt können beliebig viele gerichtete und, falls sonst kein Verweis wegführt, genau ein ungerichteter Verweis einlaufen.

2.2 cb GEWICHTUNG. Analog zu einem bewerteten Graphen kann einem Verweis ein Zahlengewicht zugeordnet werden, z.B. je nach Wichtigkeit oder Detailliertheitsgrad der zugehörigen Verweispunkte.

2.2 cc ORDNUNG. Je nachdem, ob der Verweis auf ein übergeordnetes Stichwort, einen untergeordneten Begriff, auf ein verwandtes Thema, ein Gegenbeispiel oder eine zugeordnete Fussnote verweist, hat er den Typ einer Überordnung, Unterordnung, Nebenordnung, Gegenordnung oder Zuordnung. Zusammen mit der Gewichtung ist die Ordnung von grosser Bedeutung für die Informationswiedergewinnung (siehe Kap. 2.3).

2.2 cd DARSTELLUNGSFORM. Wird ein Verweis aktiviert, d.h. ein Anfangspunkt auf dem Bildschirm angewählt, so gibt es verschiedene Möglichkeiten, den Endpunkt auf der zweidimensionalen Bildschirmfläche in sichtbaren Bezug zum Anfangspunkt zu setzen:

cda Einschub. Ist der Endpunkt ein Textabschnitt oder ein Bild, das den Ausgangstext oder das Ausgangsbild inhaltlich und formal passend erweitert, so wird der Endpunkt in der Darstellung des aktuellen Knotens an der entsprechenden Stelle eingefügt. Der Anfangspunkt wird sozusagen 'aufgefaltet' und sein Inhalt, der Endpunkt, frei gelegt. Ist ein Hypertext rein hierarchisch organisiert, so kann er ausschliesslich mit (in sich geschachtelten) Einschüben dargestellt werden.

cdb Liegt der Endpunkt in einem gänzlich anderen inhaltlichen Umfeld als der Ausgangspunkt oder hat eine unvereinbare äussere Form (z.B. bei Verweisen von Bildern heraus auf Text), so ist ein Einschub nicht sinnvoll:

cdba *Ersetzung*. Der Knoten, der den Endpunkt enthält, ersetzt den aktuellen Ausgangsknoten. Diese Darstellung muss gewählt werden, wenn auf der Darstellungsfläche, d.h.auf dem Bildschirm nicht mehr genügend Platz verfügbar ist.

cdbb *Paralleldarstellung*. Der Knoten, der den Endpunkt enthält, wird parallel zum aktuellen Knoten dargestellt. Diese Darstellungsform ist wegen der Übersichtlichkeit gegenüber der Ersetzung vorzuziehen. Ist ein Hypertext netzartig organisiert, so muss im allgemeinen eine dieser beiden Darstellungsformen verwendet werden.

cdc Überlagerung. Hat der Endpunkt den Charakter einer Fussnote oder Anmerkung, also einer erläuternden,nur kurzzeitig benötigten Hilfsinformation, so wird diese vorübergehend dem aktuellen Knoten in der Darstellung überlagert und nach Gebrauch wieder entfernt. Bei allen Darstellungsformen ist zu beachten, dass nur die Darstellung der aktuellen Knoten transformiert wird, nicht aber die Knoteninhalte an sich.

Bisher war nur von einem einzigen, gesamthaften Netzwerk die Rede. Analog zu gedruckten Publikationen lässt sich die Vernetzung jedoch in weitere **Strukturebenen** aufspalten:

2.2 d Das **Knotennetz** besteht aus denjenigen Verweisen, die ausschliesslich ganze Knoten sowohl zum Anfangs- als auch zum Endpunkt haben. Das Knotennetz repräsentiert die thematische Gliederung des im Hypertext gespeicherten Wissens, es ist sozusagen das Inhaltsverzeichnis des Hypertextes. Analog zur Schachtelung von Kapiteln in konventionellen, hierarchischen Inhaltsverzeichnissen können Teilnetze des Knotennetzes zu Überknoten zusammengefasst und als ein einzelner Knoten behandelt werden.

2.2 e Das **Verweisnetz** besteht aus allen Verweisen, die keinen ganzen Knoten zum Anfangs- und Endpunkt haben. Das Verweisnetz bestimmt die detaillierte Beziehungsstruktur des Hypertextes und entspricht z.B. dem System von Querverweisen in Lexika oder den kapitelübergreifenden Hinweisen in Lehrbüchern. Es wäre denkbar, das Verweisnetz nochmals aufzuspalten, und zwar in abgeschlossene, untereinander unabhängige Verweisnetzebenen, die sich alle auf dasselbe Knotennetz beziehen, aber z.B. für verschiedene Benutzerkreise bestimmt sind.

2.2 f Der **Thesaurus** erlaubt die direkte Lokalisierung aller Verweispunkte, die den Charakter eines Stichwortes haben, also nur aus einzelnen Zeichen, Wörtern oder kurzen Wortverbindungen bestehen. Ausserdem hat er die Funktion eines Glossars, enthält also Kurzerklärungen zu den verzeichneten Stichwörtern. Der Thesaurus besteht aus zwei Komponenten:

2.2 fa ALPHABETISCHES STICHWORTVERZEICHNIS (Index). Bei der automatischen Ableitung dieses Verzeichnisses müssen orthographische Abweichungen, unterschiedliche Flexionen und Numeri der Stichwörter auf die entsprechenden Grundformen zurückgeführt werden.

2.2 fb SYSTEMATISCHES STICHWORTVERZEICHNIS. Dieses Verzeichnis lässt sich zum Teil aus dem Ordnungstyp der Verweise herleiten. Die Stichwortbegriffe lassen sich nach folgenden Beziehungsgruppen klassifizieren:
fba sinngleiche (synonyme) und sinnverwandte Begriffe, gegensinnige (antonyme) Begriffe;
fbb übergeordnete (hyperonyme) und untergeordnete (hyponyme) Begriffe;
fbc sachverwandte Begriffe.

2.3 Informationswiedergewinnung

'Information Retrieval' , d.h. Wiedergewinnung von Information, ist in allen Informationssystemen eine zentrale Tätigkeit. Es gibt im wesentlichen zwei Verfahren zur Informationswiedergewinnung. Das **deskriptive Verfahren**: Die gewünschte Information wird in einem iterativen Ablauf immer genauer umrissen, bis eine ausreichende Genauigkeit der Beschreibung erreicht ist und die gesuchte Information abgerufen werden kann. Zur Beschreibung werden mengentheoretische, boolesche und prädikatenlogische Operatoren eingesetzt, die auf die sogenannten *Deskriptoren* angewendet werden. Die Menge der Deskriptoren stellt eine kompakte Beschreibung der zu durchsuchenden Textmenge dar; vor allem handelt es sich dabei um Stichwörter und Begriffe, die mit Zeichenkettenoperationen zugegriffen werden können. Das deskriptive Verfahren ermöglicht somit einen Direktzugriff, der von der konkreten Netztopologie unabhängig ist, jedoch entscheidend von der zur Verfügung stehenden Deskriptorenmenge bestimmt wird. Dabei muss die inhaltliche Struktur sowohl der gesuchten Information als auch ihres Umfeldes ungefähr bekannt sein, da sich die Beschreibung darauf bezieht und nur so eine Lokalisierung möglich ist. Der deskriptive Ansatz eignet sich für die systematische, flächendeckende Suche in unbeschränkten Themenbereichen bei fortlaufender Einengung des Themenkreises. Für die oberflächliche Durchsicht und die Übersichtsgewinnung in unbekannten Bereichen ist diese Methode ungeeignet. Das deskriptive Verfahren ist die klassische Methode zur Informationswiedergewinnung in herkömmlichen Datenbanken. In Hypertextsystemen wird es bei Zugriffen auf Thesauri benötigt (vgl. Kap. 2.2 f).

Das **navigierende Verfahren**: Obwohl diese Methode auch bei herkömmlichen Datenbanken, z.B. relationalen, eingesetzt werden kann, erreicht sie ihre volle Leistungsfähigkeit und den grössten Kontrast zur deskriptiven Methode erst mit dem Hypertext-Konzept, da die Netzwerkstruktur optimal mit der anschaulichen Vorstellung von 'Navigation' übereinstimmt. Deshalb wird im folgenden die ganz spezielle, hypertextspezifische Variante des navigierenden Verfahrens beschrieben.

Es handelt sich um eine *exploratorische* Methode. Ausgehend von einem Einstiegspunkt in das Netzwerk bewegt sich die Suche entlang eines frei wählbaren Verweispfades durch die Knoten des Netzwerkes. In jedem Knoten muss entschieden werden, welcher Verweispunkt als nächster ausgewählt wird; bei mehreren Möglichkeiten müssen die Endpunkte mittels geeigneter Deskriptoren zur Auswahl gestellt werden. Im Gegensatz zum deskriptiven Verfahren, wo ein bestimmtes Vorwissen über die abzufrufende Information notwendig ist, kann man sich beim selektiven Vorgang der Navigation auf den inhaltlichen Kontext des umgebenden Knotens und des bisher beschrittenen Pfades abstützen und so problemlos auch in unbekanntes Gebiet vorstossen. Navigation ermöglicht, ganz im Unterschied zur Deskription, dass Umfang und Beziehungsstruktur der Information, die irgendwo unsichtbar in den Tiefen des Systems lagert, dem Benutzer offenbar und fassbar werden.

Das navigierende Verfahren ermöglicht im wesentlichen zwei verschiedene **Zugriffsarten** auf das Netzwerk:

Suchen nach einem unklaren Begriff oder einem nicht näher bestimmten Themenbereich: Befindet man sich in einem unbekannten Wissensgebiet ohne begriffliche Anhaltspunkte, so kommt man nur mit einer navigierenden Methode zum Ziel. Bei bekanntem Begriff oder klar umrissenem Thema ist der deskriptive Zugriff über Thesauri dagegen einfacher und effizienter. Man kann deshalb zuerst über den Thesaurus eine Groblokalisierung vornehmen, anschliessend mit navigierendem Suchen das Ziel genau ansteuern.

Sichten eines ganzen Bereichs: Soll ein unbekannter Teil des Verweisnetzes gezielt oder spielerisch erforscht werden, so ist das Sichten oder 'Browsen', wie es auch genannt wird, eine sehr intuitive und effiziente Methode. Navigierendes Sichten ist die eigentliche Stärke des Hypertext-Konzepts, denn es entspricht genau der gewöhnlichen Methode, ohne bestimmte Anhaltspunkte auf gedruckte Informationssammlungen zuzugreifen: man durchstöbert die Seiten, liest hier etwas und dort, folgt einem Querverweis, merkt sich interessante Stichwörter. Bei komplizierten, hochgradig vernetzten, d.h. zyklischen Strukturen ist es jedoch sehr schwierig, einen grösseren Bereich systematisch und flächendeckend zu sichten.

Bei der Navigation, sowohl beim Suchen als auch beim Sichten, ist es sehr wichtig, während der Wanderung durch das Verweisnetz die Übersicht zu behalten und den eigenen Standort bestimmen zu können (deshalb auch die Bezeichnung 'Navigation'). Wie die Erfahrung zeigt, können schon einfache zyklische Strukturen ungemein verwirren. Deshalb ist mit der Informationsgewinnung untrennbar stets auch die Übersichtsgewinnung verbunden. Grundprobleme bei der Gewinnung von Übersicht sind: Reduktion der Informationsmenge, Verringerung der Strukturkomplexität, Visualisierung komplizierter Beziehungsgeflechte.

3 Die beiden Sichten auf HyperTexte

3.1 Die Benutzersicht: Interaktion mit Hypertexten

Der Benutzer eines Hypertextes entspricht dem herkömmlichen Leser von gedruckten Publikationen. Ein Buch zu benutzen bereitet dem geübten Leser keine Probleme. Allein dadurch, dass er das Buch in der Hand hält, seine Seitenzahl ermittelt und die Schriftgrösse in Betracht zieht, kann er den Informationsgehalt des Werkes abschätzen. Auch über die Struktur ist er sich völlig im klaren: das Buch ist physisch eine Folge von numerierten Seiten, logisch gemäss dem hierarchischen Inhaltsverzeichnis organisiert und über ein Stichwortverzeichnis direkt zugreifbar. Die Orientierung zu verlieren ist nicht möglich, da die Stelle, an der das Buch aufgeschlagen ist, bereits rein optischen Aufschluss über die ungefähre Leseposition gibt. Hingegen sieht sich der Leser eines Hypertextes einer ganzen Reihe von Problemen gegenüber. Der Bildschirm bietet dem Benutzer stets nur einen kleinen Ausschnitt dar; die gesamte zur Verfügung stehende Information ist in ihrem Umfang und in ihrer physischen Struktur anschaulich nicht fassbar. Die neuen Möglichkeiten zur Informationswiedergewinnung führen zu einem Orientierungsverlust, der durch eine künstliche Visualisierung der logischen Organisation und der Position im Netzwerk ausgeglichen werden muss. Ein Hypertextsystem muss daher gewissen Anforderungen genügen. An erster Stelle kommen sicher Forderungen bezüglich **Leistungsfähigkeit und Ergonomie der Hardware.** Wichtig sind vor allem: eine möglichst intuitive, direkte Schnittstelle zum Rechner, da die Navigation ein höchst interaktiver Vorgang ist; eine möglichst angenehme Informationsaufnahme ab Bildschirm, da man in Hypertexten sehr oft auch längere Texte lesen muss; möglichst kurze Antwort- und Reaktionszeiten, besonders bei Massenspeicherzugriffen, da Hypertext sehr grosse Datenmengen mit sich bringen kann. Ebenso wichtig sind Forderungen bezüglich **Leistungsfähigkeit und Ergonomie der Software**. Neben allgemeinen Anforderungen wie Betriebseffizienz und Zuverlässigkeit sind speziell folgende Punkte zu beachten:

3.1.a **Interaktion**. Das Dialogsystem deckt die interaktiven Aspekte des Hypertextsystemes ab und bestimmt damit unmittelbar die ergonomischen Qualitäten von Hypertext. Eine Kommandooberfläche ist für die Zwecke von Hypertext ungeeignet und für Gelegenheitsbenutzer unzumutbar; besonders bei einer ausgesprochen interaktiven Tätigkeit wie Navigation (siehe Punkt b) ist die direkte Manipulation mittels Zeigens auf die Verweispunkte unbedingte Voraussetzung.

3.1.b **Navigation**. Das Hypertextbetriebssystem ist für die Navigation zuständig. Im einzelnen sind folgende Punkte zu betrachten:

3.1.ba MODELLVORSTELLUNG DES BENUTZERS. Der Mensch hat allgemein die hochentwickelte Fähigkeit, Vorgänge und Sachverhalte, die ihm nicht oder nur teilweise einsichtig sind, zu abstrahieren und sich eine Modellvorstellung zu bilden. Von der Qualität dieses gedanklichen Modells hängt direkt die Qualität des sachbezogenen Verhaltens und Handelns ab. Das gilt im besonderen für die Benutzer von Software-Systemen, da die Programme unsichtbar in der Maschine ablaufen, die Datenmengen meist sehr kompliziert strukturiert sind und die Maschinenlogik der Alltagserfahrung widersprechen kann. Auch bei Hypertext muss der Benutzer dazu angeleitet werden, sich eine klare, adäquate Modellvorstellung von der vorliegenden Netzstruktur und den darauf wirkenden Operationen zu bilden. Vor allem müssen gedankliche Strukturkonflikte verhindert werden, da sie die Modellbildung und somit die rationelle Bedienung erschweren. So muss z.B. der Unterschied zwischen Knotennetz, Verweisnetz und Thesaurus dem Benutzer jederzeit präsent sein, denn jede der drei Strukturebenen ermöglicht einen eigenen gedanklichen Zugang zum Hypertext, was die Navigationsstrategie des Benutzers erheblich beeinflusst.

3.1.bb FORTBEWEGUNG IM NETZWERK. Informationswiedergewinnung in einem Hypertext erfordert Fortbewegung in dessen Informationsnetzwerk, d.h. Suchen entlang geeigneter Verweispfade. Bei Verzweigungsalternativen müssen leicht handzuhabende Entscheidungshilfen angeboten werden. Falls der Pfad in einer Sackgasse endet, sollte dem Benützer ein brauchbarer Mechanismus zur Wegrückverfolgung, zum 'Backtracking', zur Verfügung stehen. Um rationell arbeiten zu können, müssen häufig benutzte Zielpunkte und Pfade markiert werden können.

3.1.bc ORIENTIERUNG IM NETZWERK. Da der Bildschirm nur einen eng beschränkten Ausschnitt der Netzwerklandschaft darstellen kann, ist der Benutzer stets gefährdet, die Orientierung zu verlieren. Wichtigster Grundsatz ist: der Benutzer muss jederzeit seinen Standort im Netzwerk bestimmen und notfalls zu einem ihm wohlbekannten Ausgangspunkt zurückkehren können. Damit die Position bestimmt werden kann, muss in der Bildschirmdarstellung eine Informationsreduktion erfolgen. Dazu wird mit Hilfe der Verweisgewichtungen und -ordnungen (Kap. 2.2 c) die strukturelle Information des betreffenden Netzwerkausschnittes in Übersichtsdarstellungen soweit reduziert, bis charakteristische Strukturmerkmale zutage treten, die der Benutzer mit seiner Modellvorstellung identifizieren und lokalisieren kann.

3.1.bd SICHERN UND WEITERVERWENDEN VON ARBEITSRESULTATEN. Will man nicht nur im Datenbestand herumstöbern, so müssen komfortable Möglichkeiten vorhanden sein, Informationsfragmente aus dem Netzwerk dauerhaft zu extrahieren, zu ordnen und weiterzuverwenden. Dabei ist zu beachten, dass weder das bestehende Netzwerk verändert wird noch ein neues Netz vom Benutzer konstruiert werden soll. Die interessanten Knotenteile werden lediglich aus dem Netzwerk herauskopiert und in konventioneller Weise sequentiell, allenfalls hierarchisch zusammengestellt, formatiert und nach Bedarf ausgedruckt.

3.1.c **Anpassungsfähigkeit**. Ein weiteres Mass für die ergonomische Qualität bilden die Möglichkeiten, die dem Benutzer offenstehen, um einen gegebenen Hypertext seinen Bedürfnissen individuell anzupassen.

3.1.ca ANPASSUNG DES DIALOGSYSTEMS. Soll Hypertext einer breiten Benutzerschicht zugänglich werden, muss das Dialogsystem sehr anpassungsfähig sein. Das heisst: eine ausführliche, schrittweise Dialogführung für den Anfänger, eine knappe, abgekürzte für den Fortgeschrittenen. Im Idealfall ist das Dialogsystem komplett nach individuellen Ansprüchen reorganisierbar.

3.1.cb ANPASSUNG DER NETZSTRUKTUR. Das Hypertextkonzept lässt veränderliche Knoteninhalte zu (Objektorientierung), aber es verbietet, dass der Benutzer die Vernetzungsstruktur modifiziert. Solchen Anpassungswünschen sind deshalb enge Grenzen gesetzt. Denkbar wäre, selten verwendete Netzwerkteile zu entfernen bzw. das wirklich benötigte Teilnetzwerk aus dem Gesamtnetz herauszulösen, was einen entsprechend modularen Netzentwurf voraussetzt (siehe Kapitel 3.2).

Nachstehend werden aus den obigen Betrachtungen die Schlüsse gezogen und **konkrete Realisierungsvorschläge** bezüglich der entsprechenden Systemkomponenten aufgelistet:

3.1.d Bedienungsschnittstelle

3.1.da Der BILDSCHIRM muss unbedingt Grossformat besitzen, mindestens A4-hoch, aber noch besser A3-quer. Eine grosse Darstellungsfläche ist eine einfache und sehr effiziente Massnahme, einen Teil der Desorientierungsprobleme vollständig zu lösen. Ausserdem muss der Bildschirm eine exzellente Darstellungsqualität besitzen, d.h. er muss den Text in hoher Auflösung, randscharf, flimmer- und reflexfrei in schwarzer Schrift auf papierweissem Hintergrund abbilden. Einen Hypertext zu benutzen heisst auch vor allem zu lesen, und deshalb dürfen bezüglich Bildschirmqualität keine Kompromisse eingegangen werden, da sonst die Gebrauchstauglichkeit von Hypertexten ernsthaft gefährdet ist. Farbbildschirme können im Bereich der Orientierung sehr gute Dienste leisten, allerdings wirkt sich ein übermässiger oder unzweckmässiger Einsatz von Farbe eher kontraproduktiv aus. Als Pendant zum Bildschirm muss der DRUCKER gleich hohe Qualitätsansprüche erfüllen, denn er verhilft der gewonnenen Information erst zu einer dauerhaften, computerunabhängigen Repräsentation.

3.1.db Als EINGABEGERÄTE eignen sich die herkömmliche Tastatur und Zeigegeräte, wie Maus und Steuerkugel. Die Tastatur wird hauptsächlich zur Dateneingabe verwendet und dient nur optional zur Verkürzung von direktmanipulativen Sequenzen. Für öffentlich zugängliche Hypertextanwendungen sind auch berührungsempfindliche Bildschirme geeignet, da sie den intuitivsten, direktesten Zugriff auf die dargestellten Objekte ermöglichen.

3.1.dc Die BEDIENUNGSOBERFLÄCHE muss den Punkten 3.1.da und 3.1.db Rechnung tragen und deshalb für direkte Manipulation und Fenstertechnik ausgelegt sein. Die direkte Manipulierbarkeit ergibt sich aus der Verwendung von Zeigegeräten, die Fenstertechnik aus den grossen Bildschirmformaten und den hypertextspezifischen Erfordernissen: jeder Knoten lässt sich als Fenster darstellen und zu anderen Fenstern des Bildschirms in Beziehung setzen (2.2 cdb). Ausserdem sind Kontrollfelder (2.2 bd) und überlagernde Meldungen (2.2 cdc) integrale Bestandteile von direkt manipulierbaren Fensteroberflächen, welche damit als einziger Typus von Bedienungsoberflächen alle darstellungsmässigen Anforderungen erfüllen.

3.1.e Das **Dialogsystem** muss von allen Möglichkeiten der direkt manipulierbaren, fensterorientierten Bedienungsoberfläche Gebrauch machen: durchgehende Menüsteuerung, konsequente Fensterdarstellung, graphische Repräsentation der zu manipulierenden Objekte. Das Dialogsystem muss ausserdem möglichst einfach und möglichst umfassend an die sehr unterschiedlichen Bedürfnisse der einzelnen Benutzer anpassbar sein. Dazu bieten sich an: Parametermenüs zur Feinabstimmung des Systemverhaltens, Tastaturmakros zur Verkürzung von langen Aktionsfolgen, Ausblenden von nicht benötigten Menüpunkten bzw. Direktzugriff auf häufig benötigte, selbstdefinierbare Menüs zur Zusammenstellung von wichtigen Funktion und Tastaturmakros. Im Endeffekt ist das Dialogsystem vollständig reorganisierbar; um dies zu ermöglichen, ist es naheliegend, das Dialogsystem in der flexiblen Form eines objektorientierten Hypertextes zu implementieren.

3.1.f Hypertextbetriebssystem

3.1.fa ZUGRIFFSFUNKTIONEN. Dazu gehören alle Funktionen, die nötig sind, um in suchender oder sichtender Weise auf einen Hypertext zuzugreifen.

faa Direktzugriff. Ist man mit dem vorliegenden Hypertext bereits vertraut, hat sich also bereits eine Modellvorstellung des Netzwerks und des darin enthaltenen Wissens gebildet, so wird man eine bestimmte Position direkt ansteuern oder zumindest in unmittelbarer Umgebung einsteigen:

faaa *Zugriff über Einstiegspunkte*. Das 'elektronische Buchzeichen', also eine speziell definierte Position in einem bestimmten Knoten, ist der schnellste Weg, eine gewünschte Information wieder aufzufinden. Allerdings sollte die Zahl der Einstiegspunkte beschränkt sein, analog den richtigen Buchzeichen.

faab *Zugriff über den Thesaurus*. Im einfachsten Fall entspricht dies dem gewohnten Nachschlagen in einem alphabetischen oder thematischen Stichwortverzeichnis. Im allgemeinen Fall wird auf den Thesaurus mit dem in Kapitel 2.3 beschriebenen deskriptiven Verfahren zugegriffen.

faac *Zugriff über ein benutzerdefiniertes Stichwortverzeichnis*. Dies entspricht dem Thesauruszugriff, mit dem Unterschied, dass das dabei verwendete Verzeichnis neu erzeugt wurde, sei es automatisch mittels statistischer Auswertungen durch das System oder durch explizite Anweisung des Benutzers.

fab <u>Navigierender Zugriff.</u> Der navigierende Zugriff entspricht nichts anderem als einer Folge von Sprüngen von Verweispunkt zu Verweispunkt. Die Funktionalität des Sprunges wird bestimmt durch:

faba *Operationelle Funktionen.* Dazu gehören Grundfunktionen wie Zielauswahl bei mehreren Möglichkeiten und Pfadkorrektur durch Rücksprung um mehrere Schritte oder an einen Ausgangspunkt. Soll der beschrittene Weg später nochmals wiederholt werden, z.B. als 'Einführungstour' für einen ungeübten Benutzer, so muss der Pfad fortlaufend protokolliert und nach Wunsch dauerhaft gespeichert werden können.

fabb *Informatorische Funktionen.* Darunter fallen: Anzeige von Art und thematischer Zuordnung des angewählten Verweispunktes, Anzeige des Verweistyps, Hinweis auf allfällig vorhandene Synonyme, Hyponyme etc. Damit ausserdem der Benutzer sofort bemerkt, wenn er seinen Weg kreuzt oder gar in einen Zyklus geraten ist, müssen die bereits besuchten Knoten mit einer temporären Markierung versehen werden.

3.1.fb ORIENTIERUNGSFUNKTIONEN. Sich Orientieren heisst, sich in einer Darstellung zu lokalisieren. Deshalb geht es in den folgenden Punkten um die Visualisierung der verschiedenen Hypertextstrukturen:

fba <u>Knotendarstellung.</u> Der Inhalt des Knotens ist durch die zweidimensionale Sequenz des Textes bestimmt. Deshalb ist der Knoten ausschliesslich mit den Mitteln der Typographie übersichtlich zu gestalten. Besonders bei hochauflösenden Grossformatbildschirmen ist den typographischen Regeln und Gesetzmässigkeiten besondere Beachtung zu schenken. Leider werden oft die typographischen und graphischen Möglichkeiten entweder gar nicht wahrgenommen oder aber bis zum Exzess ausgereizt.

fbb <u>Thesaurusdarstellung.</u> Ist man im Verlauf der Navigation bei einem Stichwort angelangt, dessen Bedeutung oder thematische Zuordnung unklar ist, so müssen die entsprechenden Thesauruseinträge angezeigt werden. Dies geschieht am besten mittels überlagernder Hinweise, die anschliessend wieder entfernt werden.

fbc <u>Netzdarstellung</u>. Die Visualisierung des Knoten- und Verweisnetzes ist ein Kernproblem von Hypertextsystemen. Gewiss lässt sich mit rein numerischen, statistischen Beschreibungen bereits eine gewisse Übersicht verschaffen: Anzahl Knoten, Anzahl Verweise, Durchschnitt der Verweise pro Knoten, durchschnittliche Anzahl einlaufender Verweise pro Verweispunkt usw. Dazu gehören auch kompliziertere graphentheoretische Aussagen wie Anzahl und Grösse der Zyklen im Netzgraphen, durchschnittliche und maximale Pfadlängen. Für die Orientierung im Detail wird man jedoch das Netz in expliziter Form, textmässig oder graphisch, darstellen müssen.

fbca *Knotennetzdarstellung.* Ist das Knotennetz im wesentlichen hierarchisch, so lässt es sich rein textmässig darstellen, mit Einrückungen usw. Im allgemeinen Fall jedoch ist das Knotennetz nur graphisch in der Art eines Blockdiagramms darstellbar; die einzelnen Blöcke enthalten dabei den Knotentitel und allenfalls weitere Information. Um auch bei grossen Knotennetzen die Übersicht zu behalten, muss der aktuelle Netzausschnitt in verkleinerter Form in Zusammenhang mit seiner Umgebung oder im Gesamtzusammenhang betrachtbar sein; zudem müssen Netzteile temporär in Überknoten zusammengefasst werden können. Das System sollte automatisch diejenige Darstellung des Netzwerkgraphen wählen, die möglichst planar, d.h. überkreuzungsfrei ist.

fbcb *Verweisnetzdarstellung.* Während für das Knotennetz eine Blockdiagrammdarstellung noch einigermassen realistisch ist, kann das Verweisnetz nur in einfachen Fällen mit Pfeilen und Kästchen visualisiert werden. Im allgemeinen scheitert dieses Vorhaben jedoch in einem undurchdringlichen Netzgewirr. Ausserdem zeigt sich ein Effekt, der bereits bei der Darstellung des Knotennetzes zu Schwierigkeiten führen kann: die optische Wirkung eines grossen Blockdiagramms, das im wesentlichen nur aus linearen Elementen, nämlich Linien und Text besteht, ist zu wenig differenziert, als dass es als charakteristische Form wahrgenommen werden könnte. Deshalb muss zumindest für grosse Verweisnetze eine alternative Form der Visualisierung gefunden werden. Es ist denkbar, dass mit der Metapher der 'Landkarte' und den Methoden der Topographie und Kartographie ein neuer Zugang zur Visualisierung von Verweisnetzen gefunden werden könnte. Der Benutzer könnte durch die verschiedenen Gegenden des Hypertextlandes schweifen, Planquadrat um Planquadrat erforschen und bekannte Punkte auch direkt auf dem Luftweg erreichen.

3.1.fc AUSGABEFUNKTIONEN

fca <u>Extraktion</u> nach einem Suchvorgang. Recherchen in Hypertexten sollen nicht nur zur 'Instant-Information' dienen, sondern auch dauerhafte Resultate hervorbringen. Zu diesem Zweck kann man die interessierenden Knoten oder Knotenteile aus dem Netzwerk extrahieren, d.h. kopieren, in eine beliebige lineare Folge einreihen, formatieren und abspeichern bzw. ausdrucken.

fcb Zusammenfassung während des Sichtens. Dieser Vorgang ist ähnlich zur Extraktion, aber mit folgenden Unterschieden: es wird fortlaufend extrahiert, also während der Wanderung durch das Netzwerk, und es werden nur kleine Knotenteile, meist nur Stichworte kopiert. Am Schluss wird die automatisch erstellte Liste der extrahierten Knotenabschnitte nach Wunsch umgestellt und weiterverarbeitet.

fcc Linearisierung eines Hypertexts. Soll ein Hypertext in Buchform umgesetzt werden, so muss das Netzwerk automatisch oder manuell in eine Sequenz aufgelöst werden. Die Linearisierung kann durch die Extraktion von Knoten des Netzwerkes erfolgen. Die Formatierung ist von grosser Wichtigkeit; sie gibt dem gedruckten Hypertext die erforderliche typographische Konsequenz, fügt die Fussnotenverweispunkte am richtigen Ort ein und ergänzt automatisch alle anderen Verweispunkte mit einem konventionellen Querverweis zur entsprechenden Seite. Zum Schluss werden ebenfalls automatisch das Inhaltsverzeichnis und der Thesaurus erzeugt.

3.2 Die Autorensicht: Konstruktion von Hypertexten

Der in Analogie zum Buchautor benannte Hypertextautor ist der Konstrukteur und Produzent des Hypertextes. Die gesamten Ausführungen dieses Kapitels gelten zwar ganz allgemein für Hypertexte, jedoch richtig bedeutsam werden sie erst für kompliziert strukturierte Hypertexte mit grossem Informationsgehalt. Die Konstruktion von Hypertexten ist ein technischer Vorgang. Die dabei auftretenden Probleme und die entsprechenden Methoden zur Lösung dieser Probleme haben sehr viele Gemeinsamkeiten mit der Softwaretechnik, bzw. dem Software-Engineering.

Problemanalyse. Wie auch bei einem Softwareprodukt muss bei einem Hypertextprojekt zuerst das gestellte Problem analysiert, d.h. Art, Umfang und Interaktivität des Wissens bestimmt werden, das durch den Hypertext vermittelt werden soll. Das Pflichtenheft konkretisiert dann die Informationsbedürfnisse, die der Hypertext gegenüber dem Benutzer befriedigen soll. Wie bei herkömmlicher Software sind in dieser Phase nicht so sehr hypertexttechnische Kenntnisse von nöten, sondern vor allem fachtechnische und produktbezogene. Hingegen müssen nun alle folgenden Konstruktionsschritte in eine integrierte Produktionsumgebung eingebettet werden, wie dies für den Fall von herkömmlicher Software gilt. Dies ist eine unbedingte Voraussetzung, denn die meisten potentiellen Hypertextautoren werden kaum mit den elementaren und unabdingbaren Prinzipien der Softwaretechnik vertraut sein. Auf den Hypertextautor bezogen heisst das: Er muss dazu angehalten werden, sich bereits in einem frühen Entwurfstadium die Strukturierung seines Hypertexts zu überlegen und ihn nicht ziellos zu einem Netzgewirr wuchern zu lassen. In dieser Phase wird die Gebrauchstauglichkeit des Hypertextes entscheidend vorbestimmt.

Systemstrukturierung bedeutet in Hypertext, die Struktur des Knotennetzes zu entwerfen. Allgemein gelten bei der Netzkonstruktion folgende zwei Grundsätze: möglichst einfache Knotennetzstruktur und möglichst lose gekoppelte Netzknoten. Der erste Grundsatz verlangt, dass das darzustellende Gesamtthema mit einem minimalen Vernetzungsaufwand in die Einzelthemen, in die Kapitel aufgelöst wird, was sich direkt und sehr förderlich auf die Übersichtlichkeit auswirkt. Der zweite Grundsatz entspricht eigentlich dem ersten, nur dass nicht das Knotennetz, sondern das Verweisnetz betrachtet wird: Die Netzknoten enthalten in sich geschlossene Informationseinheiten, sozusagen 'Informationsmodule'. Vermeidet man unnötige Abhängigkeiten, d.h. Verweise zwischen den Knoten, so vereinfacht sich das Verweisnetz und die Orientierung wird unmittelbar erleichtert. Daraus folgt, dass man der Desorientierung nicht nur mit ausgefeilten Navigationsoperationen begegnen muss, sondern auch mit einer handlichen, problemangemessenen Netzstruktur. Diese an sich selbstverständliche Erkenntnis muss nun die Produktionsumgebung in den Entwurfsprozess einbringen, indem der Hypertextautor in seinen Strukturentscheidungen unterstützt und geführt wird, das heisst, indem der Autor stets die Übersicht über den Entwurfsprozess behält und angehalten wird, nur im tatsächlichen Bedarfsfall kompliziertere Netzstrukturen zu benützen.

Im folgenden werden einige der wesentlichen Eigenschaften für den Autor von Hypertexten aufgeführt.

3.2.a **eingebettete Programmiersprache**: da der Hypertextautor sich während der Erstellungsphase ständig in der zu schaffenden Hypertextstruktur aufhält, ist es wichtig, dass der Wechsel zwischen dem Ausführungmodus und dem Entwicklungsmodus rasch von statten gehen kann. Dieser Umstand hat auch zur Folge, dass der Hypertextautor auf dieselben Unterstützungs- und Hilfsmittel angewiesen ist, wie der spätere Hypertextleser.

3.2.aa Die EINBINDUNG EXTERNER PROZEDUREN ist deshalb nützlich, weil es oftmals für spezielle Aspekte nicht die entsprechenden Ausdrücke in der jeweiligen Hypertextprogrammiersprache gibt.

3.2.ab Ein Hypertextsystem, welches den IMPORT VON TEXTEN nicht zumindest teilweise automatisch unterstützt, ist für echte Hypertextanwendungen ungeeignet. Hierzu zählt generell der Import von vorhandenem Datenmaterial.

3.2.b Da heutzutage mit die wichtigsten Güter 'Informationen' sind, sollte ein Hypertextsystem als Informationssystem gegen unbefugten Gebrauch mittels eines **Passwortsystems** schützbar sein.

3.2.c Bei grossen Hypertexten wird man um ein kooperatives Arbeiten und Entwickeln nicht umhin kommen. Eine wesentliche Voraussetzung dafür ist die **Netzwerkfähigkeit**.

3.2.d Je nach konkretem Verwendungszusammenhang des Hypertextes ist es nötig, eine **Runtime-Version** erstellen zu können. Dies ist z.B. bei reinen Auskunftssystemen gegeben. Wenn jedoch auf dem Hypertext selbst fortlaufend Veränderungen vorzunehmen sind, so muß die Entwicklungsumgebung zur Verfügung stehen.

4 Fazit

Je nach Verwendungszweck des zu erstellenden Hypertextes wird die Bewertung der zur Auswahl stehenden Hypertextsysteme unterschiedlich ausfallen. Das allen Ansprüchen gerechtwerdende und preiswerte Hypertextsysteme gibt es bisher (noch) nicht. Die wohl wichtigste Unterscheidung ist die Wahl zwischen einem *daten-* und einem *objekt*-orientierten Hypertextsystem. Während das daten-orientierte Hypertextsystem die Erstellung von Hypertexten mit vorwiegendem Textcharakter dadurch optimal unterstützt, dass es die Verweispunkte unmittelbar an den semantischen Kontext koppelt, wird demgegenüber bei den hier untersuchten objekt-orientierten Hypertextsystemen primär nur eine rein syntaktische Lösung für die Kopplung zwischen Verweispunkten und semantischem Kontext angeboten. Dies wirkt sich unmittelbar auf den Entwicklungsaufwand aus. So können zwar bei HyperCard auch Texte grundsätzlich mit Verweispunkten ausgestattet werden, dennoch müssen bei einer Änderung der Textstruktur alle Verweispunkte 'von Hand' nachjustiert werden; dies gilt z.B. bei HyperCard auch für die Verweispunkte mit visuellem semantischen Kontext. Bei grossen Hypertexten z.B. mit HyperCard sind Änderungen ohne entsprechende automatische Unterstützung seitens der Entwicklungsumgebung praktisch undurchführbar; deshalb erscheint es notwendig, die Konzepte von Guide in HyperCard einzubringen. Die Stärke von objekt-orientierten Hypertextsystemen liegt sicherlich darin, dass die von den Objekten ausgesandten Nachrichten den Hypertext zum 'Leben erwecken'. Dies führt dann unmittelbar zu verstärktem Einsatz von Bildern, bis hin zu Animationseffekten (siehe 'Hypermedia'-Systeme).

Hypertextsysteme nehmen eine Schlüsselstellung zwischen traditionellen Datenbanksystemen, modernen Textverarbeitungssystemen und Systemen zur Verwaltung semantischer Netzwerke ein. Erst wenn eine entsprechende automatische Unterstützung des Hypertext-Lesers und -Autors zur Gewährleistung ihrer Orientierung in den verschiedenen Netzstrukturen eines Hypertextes gegeben ist, wird die Entwicklung und Nutzung von Hypertexten einen breiten Einsatz erfahren können. Diese automatische Unterstützung setzt voraus, dass das Wissen über die hypertext-spezifischen Strukturen in das Hypertextsystem implementiert ist.

5 Ein vergleichender Überblick

Im folgenden werden die wesentlichsten Beschreibungskategorien aus dem zuvor dargestellten Bewertungsschema hergenommen, um vier am Markt erhältliche Hyptertextsysteme mit einander zu vergleichen. Das Produkt 'Guide' lag sowohl in der Version für den Macintosh, als auch in der Windows-Version für MS-DOS Rechner vor. HyperPad ist das entsprechende Gegenstück auf MS-DOS Maschinen zu dem bekannten HyperCard auf Macintosh-Rechnern.

		Programm-Name	Hyper-ties Ms-DOS	Hyper-Pad Ms-DOS	Guide Windows	Guide Macintosh	Hyper-Card Macintosh
Symbol	Bedeutung	Versions-Nr.	*	1.0	2.00	2.0	2.0
++	gut unterstützt		10	12	24	24	19
+	teilweise unterstützt		3	3	1	1	9
(+)	"von Hand" behelfsmäßig möglich		3	5	4	4	8
leer	nicht vorhanden		30	31	22	22	15

		Betriebssystem des Rechners	MS-DOS	MS-DOS	MS-DOS Windows	Macin-tosh	Macin-tosh
		Programm-Name	Hyper-ties	Hyper-Pad	Guide	Guide	Hyper-Card
Kapitel	Bewertungsbegriffe	Versions-Nr.	*	1.0	2.00	2.0	2.0
2.1	DATENORIENTIERT		++		++	++	
2.1	OBJEKTORIENTIERT			++			++
2.2a	Knoten		++	++	++	++	++
2.2b	**Verweispunkte**						
2.2ba	Stichwörter, Wortgruppen		++		++	++	(+)
2.2bb	beliebige Textbereiche				++	++	(+)
2.2bc	Bilder (als Ganzes)			++	++	++	++
2.2bc	Bildteile						+
2.2bd	Kontrollfelder, Dialogboxen			++			++
2.2c	**Verweisattribute**						
2.2ca	gerichtet		++	++	++	++	++
2.2ca	ungerichtet			(+)			(+)
2.2cb	Gewichtung						
2.2cc	Ordnung				++	++	(+)
2.2cd	**Darstellung eines Verweises**						
2.2cda	Einschub				++	++	
2.2cdba	Ersetzung		++	++	++	++	++
2.2cdbb	Paralleldarstellung				++	++	+
2.2cde	Überlagerung				++	++	+
	Strukturebenen						
2.2d	Knotennetz		++	++	++	++	++
2.2e	Verweisnetz						
2.2f	**Thesaurus**						
2.2fa	alphabetisches Stichwortverzeichnis		++	(+)	(+)	(+)	(+)
2.2fb	**systematisches Stichwortverzeichnis**						
2.2fba	nach Synonymen / Antonymen		(+)	(+)	(+)	(+)	(+)
2.2fbb	nach Über-/ Unterordnungen		(+)	(+)	(+)	(+)	(+)
2.2fbc	nach sachverwandten Begriffen		(+)	(+)	(+)	(+)	(+)
2.3	deskriptiver Zugriff		++	+	+	+	+
2.3	navigierender Zugriff		++	++	++	++	++
3.1d	**Bedienungsschnittstelle**						
3.1da	Grossformatdarstellung (A3)				++	++	++
3.1da	Positivdarstellung				++	++	++
3.1da	Farbdarstellung			++			
3.1da	hochauflösende Grafikdarstellung				++	++	++
3.1da	verschiedene Schriftarten und -grössen				++	++	++
3.1db	Mauseingabe			++	++	++	++
3.1dc	Fenstertechnik				++	++	++
3.1e	Menüs		+	++	++	++	++
3.1e	selbstdefinierbare Menüs						++
3.1fa	**Zugriffsfunktionen**						
3.1faaa	über Einstiegspunkte						
3.1faab	über Thesaurus						
3.1faac	über benutzerdef. Stichwortverzeichnis						
3.1faba	Backtracking			+	++	++	+
3.1faba	Pfadprotokollierung		++				+

Kapitel	Bewertungsbegriffe					
	Betriebssystem des Rechners	MS-DOS	MS-DOS	MS-DOS Windows	Macin-tosh	Macin-tosh
	Programm-Name	Hyper-ties 1)	Hyper-Pad	Guide	Guide	Hyper-Card
	Versions-Nr.	*	1.0	2.00	2.0	2.0
	Informationsfunktionen					
3.1fabb	Anzeige Verweistyp	+		++	++	+
3.1fabb	Markierung besuchter Punkte					
3.1fbc	**Netzdarstellung**					
3.1fbca	Knotennetzdarstellung					
3.1fbcb	Verweisnetzdarstellung					
3.1fc	**Ausgabefunktionen**					
3.1fca	Extraktion		+			+
3.1fcb	Zusammenfassung					
3.1fcc	Linearisierung			++	++	
3.2	**Funktionen für Autoren**					
3.2a	eingebettete Programmiersprache	*	++			++
3.2aa	Einbindung externer Prozeduren	*				++
3.2ab	automatischer Import von Texten	*				
3.2b	Passwortschutz	*				++
3.2c	netzwerkfähig	*				
3.2d	Runtime-Version möglich	+		++	++	+

++ gut unterstützt: * nur Runtime-Version "Hands-On!" untersucht;
+ teilweise unterstützt:
(+) 'von Hand' behelfsmässig möglich, jedoch unpraktikabel;
leer nicht vorhanden.

6 Weiterführende Literatur

Barrett E. (1988): Text, ConText, and HyperText - writing with and for the computer. Cambridge: MIT Press.

Conklin J. (1987): Hypertext: An Introduction and Survey. IEEE Computer. Sept. 87, pp.17-40.

Fiderio J. (1988): A Grand Vision. Byte. Oct. 88, pp. 237-244.

Frisse M. (1988): From Text to Hypertext. Byte. Oct. 88, pp. 247-253.

Goodman D. (1988): The complete HyperCard Handbook. (2nd edition), New York: Bantam Books.

Gloor P. & Streitz N. (1990; Hrsg.): Hypertext und Hypermedia. Heidelberg New York: Springer Verlag.

Hofmann M., Cordes R. & Langendörfer H. (1989): Hypertext/Hypermedia. Informatik-Spektrum. Band 12 Heft 4, Aug. 89, S. 218-219.

Hypertext'87 Papers. Technical Report no 88-013. Department of Computer Science, The University of North Carolina at Chapel Hill, USA, 1988.

Marchionini G. & Shneiderman B. (1988): Finding Facts vs. Browsing Knowledge in Hypertext Systems. IEEE Computer. Jan. 88, pp.70-80.

McAleese R. (1989; ed.): Hypertext: theory into practice.Osney Oxford: Blackwell Scientific Publ.

Monk A. (1989): The Personal Browser: a tool for directed navigation in hypertext systems. Interacting with Computers. vol. 1 no. 2, pp.190-196.

Nelson T. (1967): Getting It Out of Our System. in: Information Retrieval: A Critical Review. (Schechter G., ed.), Washington: Thompson Books.

Nielson J. (1990): Trip Report: Hypertext'89. SIGCHI Bulletin. vol. 21 no. 4, pp. 52-61.

Pollitt A.S. (1989): Information Storage and Retrieval Systems - origin, development and applications. Chichester: Ellis Horwood.

Ranade J. (1989): On-Line Text Management - Hypertext and other Techniques. New York: Intertext Publ.

Shneiderman B. (1987): Designing the User Interface. Reading, Mass.: Addison-Wesley Publ.

Shneiderman B.& Kearsley G. (1989): Hypertext Hands-On! Reading, Mass.: Addison-Wesley Publ.

Stepno B. (1989): A HyperCard for the PC. Byte. Sept. 89, pp.189-192.

EIN OFFENES HYPERMEDIASYSTEM FÜR GRAPHISCHE APPLIKATIONEN

Thomas Kirste

Zentrum für Graphische Datenverarbeitung e.V. (ZGDV)
D–6100 Darmstadt, Wilhelminenstr. 7
Tel.: (06151) 155–241
email: kirste@zgdvda.uucp

Abstract

Offene Hypermediasysteme bieten neben der Unterstützung der Organisation und Manipulation heterogener Datenobjekte wie Text, Rasterbilder, Vektorbilder, Video, Animation und Ton vor allen Dingen eine leichte Erweiterbarkeit auf neue Datenobjekte und Tools. In dieser Arbeit wird eine Basisarchitektur für derartige Systeme vorgestellt. Ausgehend von einer Anforderungsdefinition an diese Systeme werden die Aufgabenbereiche in einem offenen Hypermediasystem definiert und in eine konzeptionelle Architektur eingeordnet. Auf der Basis dieser Architektur werden die einzelnen Komponenten (Massenspeicherverwaltung, Objektverwaltung, Sitzungsverwaltung und Präsentations–/Interaktionsverwaltung) sowie die Beschreibungselemente (Informationsobjekte, Funktionen, Bereiche, Bindungen, Ereignisse und Aktionen) eines offenen Hypermediasystems für graphische Applikationen definiert und deren Interaktion beschrieben. Anschließend wird eine Basisrealisierung des vorgestellten Konzeptes, das HyperPicture–Toolkit, vorgestellt.

1 Einführung

In der vorliegenden Arbeit wird eine Architektur für offene Hypermedia–Systeme für graphische Applikationen vorgestellt und eine partielle Realisierung dieser Architektur beschrieben.

Unter "Hypermedia" versteht man inzwischen vor allem die Integration verschiedenster Aspekte in den Bereichen Hardware, Benutzerschnittstellen, multimedialen Datentypen und Datenorganisation (z.B. durch Verweise). Als zentrale Eigenschaft von Hypermedia, der Kern, kann deshalb die Integration über die Achsen 'Datentypen', 'Applikationstools' und 'Systemdienste & –Resourcen' gesehen werden (Abb. 1). Jedes Hypertext/Hypermediasystem füllt hier ein mehr oder weniger großes Volumen aus (zur Unterscheidung zw. Hypertext und Hypermedia siehe [15]).

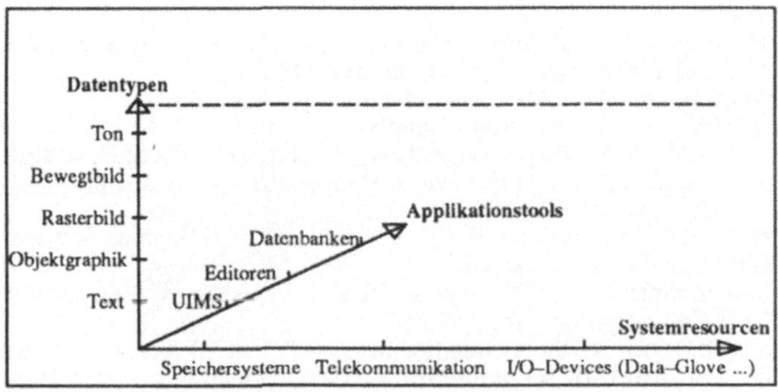

Abb. 1 – Der Integrationsraum offener Hypermediasysteme

Das Ziel eines "offenen Hypermediasystems" ist es, den gesamten Integrationsbereich potentiell abdecken zu können. Ein offenes Hypermediasystem muß daher Mechanismen bieten, um neue Datentypen, Tools und Systemdienste in das bestehende System integrieren zu können.

Eine charakteristische Anforderung an ein offenes Hypermediasystem für graphische Anwendungen, auf dessen Basis beispielsweise Rendering–Systeme, Bildarchive, Bildverarbeitungssysteme und multimediale Informationssysteme realisiert werden sollen, ist die Bewältigung der folgenden Aufgaben:

Organisation von heterogenen Datenobjekten – Objekttypen: z.B. Szenendefinition, Geometrie, Textur, Bild und Text; Objektverknüpfungen: z.B. Komplexe Objekte wie *Szene*, bestehend aus Szenendefinition, Geometrie, mehreren Texturen und Rasterbildern; freie Verknüpfungen wie z.B. Anmerkungen; Objektmengen: z.B. benutzer–, applikations–, und projektspezifisch.

Präsentation der Datenobjekte – Darstellung von Bildern, Szenendefinitionen, Geometrien etc.; interaktive Manipulation der Darstellung (z.B. Drehen einer Geometrie, Kontrasterhöhung eines Bildes).

Transformation & (interaktive) Manipulation der Datenobjekte – Editieren von Szenendefinitionen, Geometrien und Anmerkungen; Transformation von Szenendefinitionen in Geometrien, von Geometrien und Texturen in Rasterbildern; Erzeugung von Anmerkungen und freien Verknüpfungen, Verfolgung von Verknüpfungen.

Speicherung von großvolumigen Datenobjekten – z.B. hochaufgelöste Rasterbilder, Bildsequenzen zu einer Szenendefinition.

Ein offenes Hypermediasystem muß damit für den Anwender in einem integrierten System sämtliche Mechanismen anbieten, um die Datenobjekte – interaktiv – zu organisieren und zu manipulieren. Es stellt dadurch eine High–level Schnittstelle zu den verschiedenen 'klassischen' Systemdiensten dar. Ein derartiges System kann deshalb von Benutzerseite als vollständiger Ersatz der üblichen Systemschnittstellen (z.B. Kommandosprache) und Systemdienste gesehen werden [11].

Bei einer Systemerweiterung (z.B. auf Animation) kommen weitere Datenobjekte, Manipulations– und Interaktionsmechanismen hinzu. Über die o.g. Anforderungen hinaus benötigt ein offenes Hypermediasystem aus diesem Grunde Beschreibungswerkzeuge, um neue Objektklassen, Manipulationsmechanismen sowie Interaktions– und Präsentationsmethoden definieren zu können. Daraus folgt, daß sämtliche Mechanismen innerhalb eines offenen Hypermediasystems generische Mechanismen sein müssen, deren Funktionalitäten bezüglich bestimmter Objektklassen, Anwendungen usw. durch explizite Beschreibungen definiert werden. Zusätzlich werden die Anforderungen an die Speicherverwaltung auf großvolumige *und* isochrone Datenobjekte verschärft.

2 Das HyperPicture–Konzept

Das Zentrum für Graphische Datenverarbeitung e.V. (ZGDV) hat mit HyperPicture eine Basisarchitektur für ein offenes Hypermediasystem für graphische Anwendungen entwickelt. Ziel war es, ein Toolkit zu realisieren, mit dem Informationssysteme auf der Basis multimedialer Datentypen (Rasterbild, Bewegtbild, Ton etc.) entwickelt werden können. Bei der Entwicklung des HyperPicture–Konzeptes wurde dabei die Manipulation von heterogenen Datenobjekten als wichtige Teilfunktion des Systems verstanden. Daraus folgte, daß insbesondere für die Beschreibung der Relationen zwischen multimedialen Objekten ein gegenüber dem konzeptionell statischen Verweismodell erweiterter Mechanismus benötigt wurde. Daneben wurde eine Detaillierung der Konzepte "Anker" bzw. "Auslöser" notwendig, um komplexere Funktionalitäten der Benutzerschnittstelle einfach beschreibbar zu machen.

2.1 Die Architektur des HyperPicture–Konzeptes

Die Problematik der Interpretation unterschiedlichster Objektinhalte wie Rasterbilder, Video, Text, wird im HyperPicture–Konzept – genauso wie in den Hypertext–Modellen [5][6][9] – durch eine Trennung von abstrakter Identität und konkretem Inhalt eines Objektes gelöst [1].

Innerhalb eines objekttypunabhängigen Kernsystems werden *abstrakte* Objekt–Identitäten als Synonyme für Objekte verwendet. Dadurch lassen sich die Mechanismen des Kernsystems zur Verarbeitung der verschiedenen Objekte und Objektstrukturen unabhängig von der tatsächlichen, typspezifischen Bedeutung des *Inhal-*

---

1. Daraus folgt, daß die Identität eines Objektes nicht von seinem Inhalt abhängt, eine Eigenschaft, die bei der Manipulation des Inhalts vor allen Dingen für Integritätsüberprüfungen nicht unproblematisch ist. Eine Lösung wie in XANADU[10], sämtliche Versionen eines Objektes verfügbar zu halten, d.h. die Zuordnung zwischen Identität und Inhalt eineindeutig zu gestalten, ist für multimediale Dokumente aufgrund deren sehr hohen Speicherplatzbedürfnisse nicht möglich.

tes eines Objektes formulieren. Um das Kernsystem herum sind die inhaltsspezifischen Mechanismen (ISM) angeordnet, die spezielle Werkzeuge für den Zugriff und die Manipulation der Inhalte von Objekten anbieten. Ein Beispiel ist die Definition eines Teiles eines Objektes. Da im allgemeinen der Inhalt eines Objektes beliebig strukturiert sein kann wäre eine Sprache notwendig, mit der jeder beliebige Teil eines jeden beliebigen Objektes beschreibbar wäre, um für ein typunabhängiges System Bereiche von Objekten definierbar zu machen. Sehr viel einfacher ist es, die Verantwortung für die *Interpretation* der Bereichsspezifikation an eine inhaltsspezifische Komponente zu delegieren und auf der Ebene des Basissystems nur mit der abstrakten *Identität* des Bereichs zu arbeiten.

Eine zweite Strukturierungsebene ist die Trennung der Verantwortlichkeiten für das *dynamische* und das *strukturelle* Verhalten des Systems. Das dynamische Verhalten steuert und beschreibt die Interaktion des Benutzers mit dem System, während das strukturelle Verhalten die Transformation von Datenobjekten ineinander und die Mechanismen für den Aufbau von Informationsstrukturen definiert.

Diese Strukturierungsebenen des HyperPicture–Konzeptes – die Trennung in abstrakte und inhaltliche Ebene sowie in Dynamik– und Strukturebene erlaubt eine einfache Zuordnung der Komponenten *Massenspeicherverwaltung* (Storage Management, STM), *Objektverwaltung* (Object Management, OM), *Sitzungsverwaltung* (Session Management, SM) sowie *Präsentation– und Interaktion* (PIM),

zu den Strukturebenen:

	dynamisch	strukturell
abstrakt	SM	OM
inhaltsbezogen	PIM	STM

Die verschiedenen Komponenten haben dabei die folgenden Aufgabenbereiche:

Massenspeicherverwaltung – Diese Komponente ist unter anderem verantwortlich für die Zuordnung der Datenobjekte zu den einzelnen Speicherpools, den Zugriff auf den Inhalt der Datenobjekte, die zeitsynchrone Wiedergabe von Datenobjekten und die Cache–Steuerung.

Objektverwaltung – Die Objektverwaltung ist verantwortlich für die Speicherung und Organisation der abstrakten Elemente des HyperPicture–Konzeptes, sie bildet die *Abstract Machine* im Sinne der HAM [3].

Sitzungsverwaltung – Die Sitzungsverwaltung steuert die Koordination und Synchronisation der verschiedenen an einer Sitzung beteiligten Präsentations– und Interaktionskomponenten (PIMs) und ist verantwortlich für die Verarbeitung der abstrakten Ereignisse, die Abarbeitung von Aktionen sowie die Ausführung von Funktionen.

Präsentation und Interaktion – Durch diese Komponente erfolgt die typspezifische Darstellung von Datenobjekten, die Steuerung der Interaktion des Benutzers mit den Datenobjekten, die Interpretation von Ereignis– und Bereichsspezifikationen, die Erzeugung von Ereignissen und die Bereitstellung von Standardmechanismen für die Objektmanipulation.

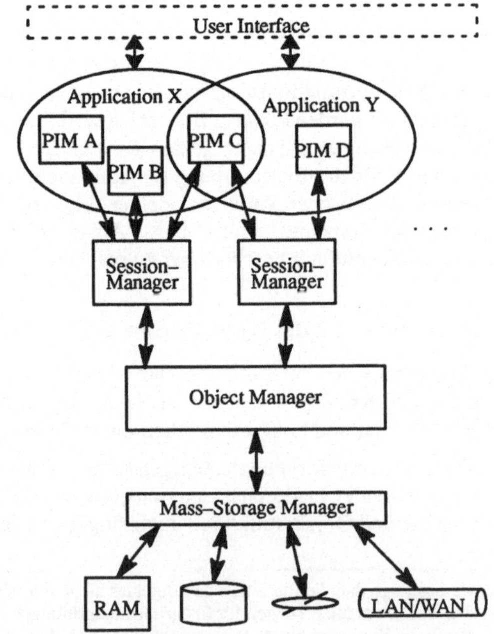

Abb. 2 – Die Architektur des HyperPicture–Konzeptes

In HyperPicture sind Objektverwaltung, Sitzungsverwaltung und insbesondere die jeweiligen PIMs als eigene Prozesse realisierbar. Die Trennung der Daten– und Sitzungsverwaltung ermöglicht eine parallele Verarbeitung ihrer jeweiligen Aufgaben, die einerseits I/O–intensiv (Objektverwaltung), andererseits berechnungsintensiv (Sitzungsverwaltung) sind. Dies gilt insbesondere bei mehreren parallel arbeitenden Benutzern (Sitzungen). Die Formulierung der PIMs als separate Prozesse ermöglicht eine einfache Erweiterung des Systems durch neue PIMs und den Austausch durch verbesserte/erweiterte PIMs, ohne daß das Kernsystem (Objekt– und Sitzungsverwaltung) davon betroffen wird[2].

2.2 Objekte und deren strukturelles Verhalten

Das strukturelle Verhalten des Systems wird auf der Basis von 'Informationsobjekten' und 'Funktionen' auf diesen Informationsobjekten beschrieben. Zusätzlich werden 'Bereiche' und 'Bindungen' verwendet, um spezielle Organisationsstrukturen einfach realisieren zu können.

Informationsobjekte – Informationsobjekte im HyperPicture–Konzept sind Behälter für Anwendungsdaten. Für die Beschreibung der Struktur und des Verhaltens eines Informationsobjektes wird ein Klassensystem verwendet. Jedes Informationsobjekt gehört einer bestimmten Klasse an, d.h., ist von einem bestimmten eindeutigen Typ ("Rasterbild", "Animation"). Die Objektklasse bestimmt die dem System bekannte Struktur des Informationsobjektes, also die Instanzvariablen und ihre Typen, z.B. Szene = { Szenendefinition: Text, Geometrie: Vectors, Version: Integer ... }.

Zu jeder Objektklasse gehört in der Regel zumindest ein inhaltsspezifischer Mechanismus, die Präsentations– und Interaktionsmethode (PIM). Die PIM wird verwendet, um ein Objekt[3], d.h. insbesondere seinen Inhalt, bei dessen Aktivierung dem Anwender in einer geeigneten Form darzubieten und die Interaktion mit dem Objekt zu ermöglichen.

Bereiche – Neben der Referenzierung vollständiger Objekte ist es sinnvoll, Teile eines Objektes näher spezifizieren zu können, beispielsweise um aus einem Rasterbild einen Ausschnitt für die Verwendung als Textur zu definieren oder um für einen Teil eines Objektes eine Referenz auf ein anderes Objekt bzw. Objektteil zu definieren.

Da die Art der Bereichsspezifikation für ein Objekt vom Inhaltstyp des Objektes abhängt, obliegt die Interpretation der Bereichsspezifikation dem jeweiligen ISM. Für das System ist lediglich die Identität des Bereichs von Interesse. Die ISM sind insbesondere auch dafür verantwortlich, daß bei einer identitätserhaltenden Manipulation des Objektinhalts die Bereichsspezifikationen entsprechend modifiziert werden.

Funktionen – In HyperPicture werden Funktionen zur Datenmanipulation, "active links" und statische Verweise im gemeinsamen Konzept der "Funktion" integriert.

Das HyperPicture–Konzept kennt als einzigen Mechanismus für den Zugriff auf Daten Funktionen, die sowohl für die Datenstrukturierung (im Sinne abstrakter Datentypen) eingesetzt werden, als auch für die Datenmanipulation. Abstrahiert man von realisierungstechnischen Überlegungen, dann ist es völlig unerheblich, ob Informationen zu einem Objekt in einer Datenstruktur gespeichert oder durch einen Funktionsaufruf errechnet werden. Relevant ist hier einzig, daß man ausgehend von einem bestimmten Objekt weitere Informationen erhalten möchte, die eine bestimmte Funktion von diesem Objekt sind[4]. Damit läßt sich ein binäres Hyperlink von x nach y mit Hilfe einer objektwertigen Funktion von Objekten, 'jump', durch **jump: x |–> y** beschreiben.

In HyperPicture stellt der Mechanismus für die Deklaration von Funktionen und die Applikation von Funktionen auf Argumente eine zentrale Systemkomponente dar. Der Anwender kann jederzeit das System durch die Definition neuer Funktionen erweitern. Die kanonische Aktivität in HyperPicture ist damit (zumindest aus Systemsicht) nicht die Verweisverfolgung wie in typischen Hypertext/Hypermedia–Systemen, sondern die Funktionsapplikation.

---

2. Die einfache Austauschbarkeit von PIMs erlaubt es dem Benutzer, z.B. unter verschiedenen Browsern denjenigen auszuwählen, der seine Anforderungen am besten erfüllt – ähnlich wie es im Fenstersystem X möglich ist, den Window–Manager nach rein persönlichen Präferenzen auszuwählen.
3. Im Folgenden wird für 'Informationsobjekt' der Begriff 'Objekt' synonym verwendet.
4. Dieses funktionale Datenmodell ([12], [13]) wird z.B. auch im IRIS–Datenbankystem [4] eingesetzt..

Bindungen – Der oben vorgestellte am funktionalen Datenmodell orientierte Ansatz wirft jedoch die folgenden Probleme auf:

a) einige Funktionen sind bezüglich des für die Auswertung notwendigen Aufwandes sehr viel kostspieliger als die Speicherung des Ergebnisses (z.B. Rendering eines photorealistischen Bildes).

b) es gibt Funktionen, für die sich kein endlicher Algorithmus angeben läßt, d.h. der Aufwand zur Berechnung ist unendlich (z.B. die Funktion "Gegenposition:Behauptung -> Wiederlegung").

Es ist deshalb sinnvoll (a) und notwendig (b), einen Mechanismus einzuführen, mit dessen Hilfe individuelle Abbildungsvorschriften für eine Funktion explizit deklariert werden können, und der zusätzlich das Ergebnis einer Funktionsauswertung speichert. In HyperPicture existiert hierfür der Bindungsmechanismus, durch den für eine Funktion zu einem Argumenttupel das Resultattupel explizit definiert werden kann.

Bei der Anwendung einer Funktion auf ein bestimmtes Argumenttupel wird zunächst überprüft, ob für diese Funktion das entsprechende Tupel gebunden ist. Ist dies der Fall, so ist das Ergebnis bereits bekannt und kann direkt zurückgeliefert werden. Andernfalls muß der Funktionskörper ausgewertet werden, um das Funktionsergebnis zu erhalten[5]. Das Verhalten einer Funktion gegenüber dem Bindungsmechanismus kann sowohl bezüglich der Auswertung des Funktionskörpers, als auch bezüglich der Erzeugung von Bindungen attribuiert werden. Auf der Basis dieses Bindungsverhaltens kann beispielsweise die Funktion 'annotate' als automatisch bindende Funktion implementiert werden, wobei die Annotation in einem nur bei nichtvorhandener Bindung auszuwertenden Funktionskörper erzeugt wird.

Die Datendefinitions- und Datenmanipulationssprache – Für die Beschreibung von Objektklassen und Funktionen wird eine Datendefinitionssprache und eine Datenmanipulationssprache benötigt. Neben speziellen Konstrukten für die Bedienung der Mechanismen von HyperPicture, muß die Datenmanipulationssprache die Mächtigkeit einer vollständigen Programmiersprache bereitstellen, um beliebige Funktionen auf der Objektmenge ausdrücken zu können.

2.3 Das dynamische Verhalten

Neben der Beschreibung der Datenobjekte und der Funktionen auf diesen Datenobjekten ist es notwendig, das Verhalten des Systems gegenüber äußeren Vorkommnissen gegenüber definieren zu können. Diese Definitionsmöglichkeit des dynamischen Verhaltens wird sowohl für die Beschreibung der Benutzerschnittstelle (Welche Eingabeoperationen sind erlaubt und welche Systemaktionen werden dadurch ausgelöst ?), als auch für die Beschreibung des Objektverhaltens anderen Ereignissen gegenüber benötigt (Was passiert, wenn das Objekt modifiziert wird oder wie wird auf den Ablauf eines Zeitintervalls reagiert ?)

Das dynamische Verhalten wird mit Ereignissen und Aktionen beschrieben. Im Bereich Hypertext/Hypermedia wird eine derartige Trennung zwischen der Spezifikation eines Ereignisses und der dadurch – eventuell – ausgelösten Handlung in der Regel nicht vorgenommen. Hier sind 'Ereignis' und 'Handlung' (und auch 'Bereich') implizit und in spezifischer Ausprägung im Begriff des 'Ankers' enthalten. Ein Anker ist ein Bereich, dessen Aktivierung (z.B. durch Mausklick) zu einem Ereignis führt, durch den das System zu der Handlung veranlaßt wird, ein Link zu aktivieren. In HyperPicture wurde diese direkte Verknüpfung der Konzepte 'Ereignis' und 'Handlung' (sowie 'Bereich' – s.o.) im Konzept 'Anker' aufgelöst. Die Einzelkonzepte sind dadurch direkt verfügbar, so daß eine sehr flexible Definition des Schnittstellenverhaltens ermöglicht wird.

Ereignisse – Ereignisse sind der Basiselemente, durch die der Anwender oder das Betriebssystem mit HyperPicture kommuniziert.

Ereignisse können von den unterschiedlichsten Vorfällen erzeugt werden, etwa vom Ablauf eines Zeitintervalls, durch das Löschen einer Datei oder durch einen Mausklick auf einem bestimmten Bereich. Im allgemeinen läßt sich die Menge der Ereignisquellen und die Arten der erzeugten Ereignisse nicht vorherbestimmen. Aus diesem Grund wird auch für Ereignisse eine Trennung in abstrakte Identität und konkrete Spezifikation vorgenommen, wie es für Bereiche und Objekte der Fall ist.

Für jedes Objekt (und jede Klasse) wird eine Menge von Ereignissen verwaltet, die zu dem Objekt (oder einer Instanz der Klasse) erzeugt werden können. Bei der Aktivierung des Objektes werden die konkreten Ereignis-

5. Funktion und Bindung erlauben dadurch intensionale und extensionale Definition von Prädikaten im Sinne Woods [16].

spezifikationen von der verantwortlichen PIM[6] interpretiert. Die PIM sendet beim Eintreten eines der definierten Ereignisse dessen abstrakte Identifikation an des Kernsystem. Ereignisse in HyperPicture bilden dadurch einen von anderen Ereignismodellen (etwa X–Windows) vollkommen unabhängigen, eigenständigen Mechanismus, der nicht auf eine bestimmte Menge von Ereignistypen (wie eben X) beschränkt ist.

Aktionen – Ereignisse legen fest, welche Benutzer– und Betriebssystemaktionen von HyperPicture erkannt werden. Ausgehend hiervon benötigt das System Informationen darüber, was geschehen soll, wenn ein bestimmtes Ereignis eintritt. Anders gesagt, Ereignisse definieren nur die 'lexikalischen Elemente' der Interaktion. Für die Definition der 'Syntax' und der 'Semantik' der Interaktion wird deshalb ein weiterer Mechanismus benötigt.

HyperPicture setzt hierfür die "Aktionsdefinition" ein. Eine Aktionsdefinition besteht aus einer Bedingung (Condition) und einer Handlung (Action), die bei Eintreten der Bedingung ausgeführt werden soll[7]. Die wichtigste Eigenschaft dieser Form der Verhaltensbeschreibung ist, daß einzelne Aktionen beliebig aus der Menge der definierten Aktionen entfernt oder hinzugefügt werden können. Dadurch läßt sich das Schnittstellenverhalten leicht modifizieren. Dies bedeutet auch, daß das Schnittstellenverhalten nicht an einer Stelle vollständig definiert werden muß, sondern daß eine verteilte Form der Definition verwendet werden kann, z.B. kann jede Objektklasse den sie betreffenden Teil der Schnittstelle für sich definieren[8]. Zusätzliche Datentypen lassen sich dadurch auch auf der Schnittstellenseite leicht integrieren.

Im einfachsten Fall besteht eine Aktionsbeschreibung aus einem Ereignis als Bedingung und einem parametrisierten Funktionsaufruf als Handlung. Typischerweise ist eine Bedingung jedoch komplexer und besteht aus einer Liste von Zuständen und Ereignissen, die gelten müssen, um die Handlung eintreten zu lassen. Dadurch läßt sich das Schnittstellenverhalten an den jeweiligen Zustand der Applikation anpassen. Durch die Aktionsbeschreibung kann das Verhalten eines PIMs in weiten Bereichen beeinflußt werden, ohne das PIM selbst zu modifizieren. Es bietet sich dadurch z.B. die Möglichkeit, das PIM als funktional leeres User–Interface zu implementieren und unterschiedliche Funktionalitäten durch entsprechende Aktionsbeschreibungen zu bestimmen.

Wie Ereignisse können Aktionen sowohl für Klassen als auch für individuelle Informationsobjekte definiert werden. Mit Aktionen und Message–Events kann der übliche Methodenbegriff realisiert werden. Dies wird vom System bei der Aktivierung von Objekten ausgenutzt[9].

2.4 Verweise in HyperPicture

Gegenüber klassischen Hypertextsystemen sind mit der Erzeugung und der Traversierung von Verweisen in HyperPicture recht komplexe Funktionsabläufe verbunden:

Verweis–Erzeugung – Zur Erzeugung eines Verweises müssen neben einer Verweisfunktion und der entsprechenden Bindung die Bereiche, Ereignisse und Aktionen definiert werden, durch die der Verweis ausgelöst werden soll. Im einzelnen sind (exemplarisch) die folgenden Definitionen notwendig um einen binären Verweis, der durch einen Mausklick auf einem rechteckigen Rasterbereich des Quellobjektes auslösbar sein soll, zu erzeugen. Der Verweis hat als Ziel einen – hier nicht näher angegebenen – Bereich im Zielobjekt:

---

6. Beliebige Ereignisquellen ('Ereignisdämonen') können auf der Basis spezieller PIMs erzeugt werden. Diese PIMs müssen keine Datenobjekte darstellen, sondern werden lediglich verwendet, um Ereignisse zu erzeugen. Von Systemseite wird dadurch die Einführung eines separaten Ereignisquellenkonzeptes unnötig.

7. Damit sind Aktionsdefinitionen prinzipiell Produktionssysteme. Produktionssysteme bieten einerseits die Möglichkeit, adaptives Verhalten zu modellieren, andererseits bilden sie einen formalen Apparat, der zur Überprüfung gewisser Eigenschaften des Schnittstellenverhaltens ähnlich wie Petrinetze [14] eingesetzt werden kann.

8. Dieser Mechanismus der verteilten Schnittstellendefinition ist nicht mit 'Vererbung' im objektorientierten Sinn zu verwechseln. Hier können für *individuelle* Objekte separat die jeweilig benötigten Teile des Verhaltens definiert werde. Erst das *Zusammenspiel* der Objekte bildet die Benutzerschnittstelle. Im Gegensatz hierzu der Mechanismus der Vererbung: Hier erbt ein Objekt das für seine Superklassen definierte Schnittstellenverhalten.

9. Tatsächlich wird durch Aktionen und Messages das *Objektverhalten* modellierbar, wodurch die Auswirkungen von Objektmanipulationen auf das System beschrieben werden kann [2]. Beispielsweise können hierdurch automatische Integritätsprüfungen bei Modify–Ereignissen formuliert werden.

```
Create extent(ob#1): xt#1 = "pixelarea 10 10 100 100"
Create event(ob#1): ev#1 = "mouseclick" xt#1
Create action(ob#1): ev#1 -> jump(ob#1,xt#1)
Create extent(ob#2): xt#357 = ...
Create binding(jump): ob#1 xt#1 -> ob#2 xt#357
```

Verweis–Auslösung – Die in Hypertext übliche Aktion der Verweis–Auslösung wird von HyperPicture auf den folgenden Aktionsablauf abgebildet:

	Preprocessing		Evaluation	Postprocessing	
Execution Agent	PIM	SM	SM/DM	SM	PIM
Activity	Event →	Action–Exec. →	Function–Appl. →	Activation →	Display
Definition	Event–Descr.	Action–Descr.	Function–definition + Binding	Action–Descr. (activate)	Contents (+ selected extent)

2.5 Applikationen auf HyperPicture

Die wichtigste Eigenschaft des HyperPicture–Konzeptes besteht darin, vollständige Applikationen allein auf der Basis der HyperPicture–Beschreibungsmechanismen für Datenobjekte, Funktionen und Schnittstellenverhalten definieren zu können. Eine Applikation auf HyperPicture besteht im einfachsten Fall aus einer Menge von Objektklassendefinitionen, einer Menge von Funktionsdefinitionen sowie den klassenspezifischen Ereignis– und Aktionsdefinitionen (analog der "Stackware" in HyperCard [1]). Zusätzlich kann eine Menge von Datenobjekten für den Aufbau der Benutzungsschnittstelle verwendet werden. Alle o.g. Definitionen können ohne Modifikation des Systems, zur Laufzeit, in das System geladen und verfügbar gemacht werden.

2.6 Einordnung in das Dexter–Referenz–Modell

Das HyperPicture–Konzept läßt sich in das Dexter–Referenz–Modells [6] einordnen. Es stellt unter dieser Einordnung eine Detaillierung und Konkretisierung des Dexter–Modells dar, insbesondere bezüglich der Datenmanipulation und der Definition des dynamischen Verhaltens. Im HyperPicture–Konzept werden als zusätzliche Elemente *Ereignisse* und *Actions* für die Beschreibung des Objektverhaltens eingeführt. Weiterhin wird die im Dexter–Modell enthaltene *Resolver–Function* durch das HyperPicture–Funktions/Bindungs–Konzept explizit gemacht und für die Datenmanipulation erweitert. Die Elemente *Presentation Specification* und *Accessor Function* des Dexter–Modells sind im HyperPicture–Konzept noch nicht ausformuliert.

3 Die Realisierung von HyperPicture

3.1 Ziel der HyperPicture Entwicklung

Im ZGDV wurde ein Experimentalsystem für die Organisation und Manipulation von multimedialen Datenobjekten auf der Basis des HyperPicture–Konzeptes entwickelt [8]. Neben der Untersuchung der einzelnen Aspekte einer Realisierung des HyperPicture–Konzeptes wird das System verwendet, um die Einsetzbarkeit der einzelnen Elemente von HyperPicture für die Realisierung verschiedener Anwendungen im Bereich graphischer Informationssysteme zu demonstrieren und vor allen Dingen zu evaluieren; wie etwa für Rasterbildarchive, Satellitendatenretrievalsysteme oder Röntgenbildanalyse– und Dokumentationssysteme.

HyperPicture ist gemäß der o.g. Architektur realisiert, als System zur Verwaltung und Bearbeitung heterogener Daten, die auf optischen Speichermedien abgelegt und mit Verweisen strukturiert werden[10]. Eine graphische Benutzungsoberfläche auf der Basis von X–Windows und OSF/Motif ermöglich die interaktive Manipulation der Daten und unterstützt den einfachen Zugriff auf die gespeicherten Dokumente.

Das HyperPicture–Toolkit unterstützt zur Zeit die Datentypen Text, Rasterbild, Vektorgraphik, Animation und Video, wobei als primären Datenstrukturierungs– und Manipulationsmechanismus das HyperPicture–Funktions/Bindungskonzept Verwendung findet. Zusätzlich wird das klassische Konzept des Hyperlinks, des Verweises, an der Benutzerschnittstelle direkt angeboten.

Das HyperPicture–Basissystem besteht aus zwei Schichten:

- Das Objekt–Management–System umfaßt die Komponenten STM, OM und SM. Es ist eine applikationsunabhängige Hyperlinkmaschine für die Verarbeitung von Anfrage– und Änderungsaufträgen an die Datenmenge.
- Für die einzelnen Objekttypen sind individuelle Präsentations– und Interaktionsmodule vorhanden. Sie enthalten einen Browser, der als eine spezielle PIM für Objekte angesehen werden kann.

In HyperPicture werden, ähnlich wie im CONCORDE–System [7], ein globaler sowie private und private temporäre Datenräume unterstützt. Für jedes aktive (dargestellte) Objekt kann interaktiv definiert werden, welche Datenräume von hier aus sichtbar sind und welcher Datenraum für die Erzeugung von neuen Bindungen und Objekten verwendet werden soll.

Der Browser – Zur Unterstützung des Anwenders bei der Orientierung im Netzwerk bietet HyperPicture wie einige andere Hypersysteme einen Browser an, von dem Ausschnitte der Netzstruktur mit den einzelnen Dokumenten und den definierten Bindungen graphisch dargestellt werden. Der für die HyperPicture–Toolkit derzeit implementierte Browser wird in der Anwendung des Rasterbildarchivierungssystems eingesetzt.

Die Objekte werden im Browser als Icons, symbolische Darstellungen oder verkleinerte Rasterbilder (Bitmap oder Pixmap) dargestellt. Ein Objekt kann sowohl durch ein individuelles Icon im Browser repräsentiert werden, als auch durch ein für den jeweiligen Objekttyp definiertes Standard–Icon. Durch geeignete Verwendung individueller Icons kann schon im Browser auf den Inhalt eines Objekts geschlossen werden, ohne das Objekt selbst aktivieren zu müssen. Bei komplexen Objekten wie Rasterbildern kann damit bereits im Browser der Bildinhalt erkannt werden, wodurch der zeitliche Aufwand für die Suche nach Bildern mit bestimmtem Inhalt erheblich veringert wird.

Die objekttypspezifischen PIMs – Die PIMs des HyperPicture–Systems sind einfache Beispiele für das PIM–Konzept. Alle PIMs erlauben die interaktive Auslösung, Definition und Löschung von Verweisen. Zusätzlich sind objekttypspezifische Funktionen implementiert, wie etwa die Veränderung der Bildgröße und die Manipulation der Farbtabelle für graphisch dargestellte Objekte oder das Edieren von Textobjekten[11].

Das jeweilig von einer PIM darzustellende Informationsobjekt wird in einem Fenster auf dem Bildschirm angezeigt, wobei beliebig viele Objekte gleichzeitig aktiv sein dürfen. Der Benutzer kann mit den dargestellten Datenobjekten durch die Aktivierung von Links, die Applikation von Funktionen oder die Auswahl spezieller Funktionen des Anzeigemoduls interagieren.

Bei der interaktiven Definition eines Verweises – ausgelöst durch ein "Link To ..."–Kommando im Befehlsmenü eines Anzeigemoduls – hat der Benutzer zunächst die Möglichkeit, den Auslöserbereich auf dem Objekt zu definieren und danach das Zielobjekt für den Verweis zu bestimmen[12]. Als Bereiche werden bisher Zeichenketten (Text), rechteckige und polygonal begrenzte Gebiete (Raster & Vektor) sowie zeitvariante rechteckige Gebiete (Animation &Video) von den jeweiligen PIMs unterstützt. Auch bei der einrichtung von

10. Die Arbeiten am HyperPicture–Projekt des ZGDV wurden im Dezember 1989 begonnen. Als Basissystem wurde eine HP 9000/360 Workstation mit HP–UX Betriebssystem eingesetzt. Die ersten lauffähigen Versionen wurden auf der CeBit '90 und anläßlich des Workshops Hypertext/Hypermedia '90 präsentiert.
11. Die Textobjekt–PIM ist ein Beispiel für eine inhaltsverändernde ISM, die Bereichsdefinitionen entsprechend der Inhaltsmanipulation aktualisieren muß.
12. Das Ausgangsobjekt des Verweises ist implizit dasjenige, in dessen Fenster das "Link To ..."–Kommando ausgelöst wurde.

Verweisen sind sämtliche Navigationsfunktionen des Browsers verfügbar. Nach der Zieldefinition erfolgt die optional Angabe des Namens und die Festlegung des Typs (d.h., die der resultierenden Bindung unterlagerte Funktion) des Verweises. Danach steht der Verweis im System für die Aktivierung zur Verfügung.

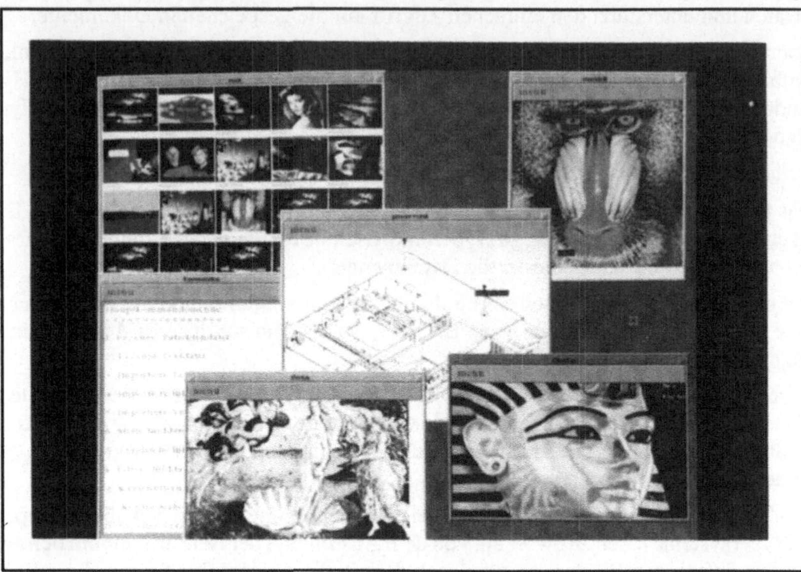

Abb.3 – Das HyperPicture Bildarchiv

3.2 SpacePicture – Ein Beispiel einer HyperPicture Anwendung

SpacePicture ist ein Satellitenbildretrievalsystem, das auf dem HyperPicture–Toolkit aufsetzt. Ziel der Entwicklung ist es, ein interaktives Retrievalsystem für den Zugriff auf hochaufgelöste Satellitenbilder (z.B. 6000 x 5000 Pixel, 7 x 8 Bit/Pixel) bereitzustellen, das insbesondere die geographisch–orientierte Suche nach Bilddaten zuläßt. Als Massenspeichermedium wird hierbei das Optical–Disc Jukebox–System EPOCH–I mit einem Festplatten–Cache eingesetzt.

Ein zentrales Merkmal des SpacePicture–Systems aus der Benutzersicht besteht darin, die geographische Lage der Satellitenbilder auf der Erdoberfläche als Suchschlüssel für das Retrieval zu benutzen. Der Systembenutzer kann graphisch–interaktiv ein bestimmtes Gebiet auf der Erdoberfläche markieren und sich danach vom System alle in diesem Gebiet befindlichen Satellitenbilder, soweit sie dem System bekannt sind, heraussuchen lassen.

Neben der geographischen Lage eines Satellitenbildes können außerdem zusätzliche Suchmerkmale definiert werden, um die Ergebnismenge einer Suche genauer zu spezifizieren. Zu den sekundären Suchmerkmalen gehören das Erstellungsdatum des Bildes und der Sensortyp, mit dem das Bild aufgezeichnet wurde.

Die Objekttypen in SpacePicture sind die 'Karten', 'Satellitenbilder' und 'Mengen' (Suchergebnisse). Für Karten und Suchergebnisse ist die PIM 'Map–Browser' verantwortlich, auf Satellitenbilder wird mit der PIM 'Quick–Look–Display' zugegriffen.

4 Ausblick

Das HyperPicture–Konzept bietet in seiner jetzigen Version eine leicht erweiterbare Plattform für offene Hypermediasysteme. Es stellt die Mechanismen für die Beschreibung von Objekten, deren Organisation, Manipulation und Präsentation/Interaktion auf einer abstrakten Ebene zur Verfügung, so daß keine Beschränkung auf bestimmte Ausprägungen von Objekten, Funktionen und Interaktionsformen vorliegt. Die grundsätzliche Einsetzbarkeit dieser Konzepte wurde durch die Implementierung des HyperPicture–Toolkits und der SpacePicture–Applikation demonstriert.

Das Fernziel ist, das HyperPicture–Konzept zu einem vollständigen Basissystem – analog zu der heute üblichen Kombination Betriebssystem+Datenbanksystem+User–Interface – für multimediale Daten weiterzuentwickeln. Als DDL/DML soll hierbei der speziell für HyperPicture entwickelte LISP–Dialekt *HCL* ('HyperPicture Control Language') eingesetzt werden.

Einige wichtige Aspekte werden vom jetzigen Stand des HyperPicture–Konzeptes hierfür noch nicht befriedigend behandelt. Dies gilt vor allen Dingen für die Beschreibungselemente des HyperPicture–Konzeptes (Objekte, Bereiche, Funktionen, Bindungen, Ereignisse und Aktionen). Hier sind exakte formale Definitionen ihres Verhaltens und ihrer Struktur zu entwickeln. Analog sind die Schnittstellen zwischen PIM/Session–Server und Session–Server/Object–Management sind zu formalisieren und zu detaillieren.

Danksagung

Der Autor möchte sich an dieser Stelle besonders bei Herrn Dr. W. Hübner bedanken, dessen Kommentare und Anmerkungen bei der Erstellung dieser Arbeit sehr hilfreich waren. Bei der Implementierung des HyperPicture–Toolkits und der SpacePicture–Applikation wurde der Autor unterstützt von Susanne Boll, Rolf Knittel, Rafael Kozlowski, Jochen Weiss und Jürgen Werkmann.

Die Arbeiten an HyperPicture werden unterstützt von Hewlett–Packard, DisCom und DLR.

A Literatur

[1] Apple Computer Inc. HyperCard Script Language Guide: The HyperTalk Language. *Addison–Wesley*, 1988.

[2] Baumann, P. Die Spezifikation informationsverarbeitender Systeme mit Abstrakten Objekttypen, *Tagungsband der BTW–Fachtagung*, Kaiserslautern, 6.–8. März, 1991.

[3] Campbell, B., Goodman, J.M. HAM: A general purpose hypertext abstract machine. *Communications of the ACM*, 31(7), 1988, 856–861.

[4] Fishman, D.H. et al. Overview of the Iris DBMS, *Technical report HPL–SAL–89–15*, Hewlett–Packard Company, 1989.

[5] Furuta, R., Stotts, P.D. The Trellis Hypertext Reference Model. *Proceedings of the NIST Hypertext Standardisation Workshop*, Gaithersburg, Maryland, January 16–18, 1990.

[6] Halasz, F., Schwartz, M. The Dexter Hypertext Reference Model. *Proceedings of the NIST Hypertext Standardisation Workshop*, Gaithersburg, Maryland, January 16–18, 1990.

[7] Hofmann, M., Cordes, R., Langendörfer, H., Lübben, E., Peyn, H., Süllow, K., Töpperwien, T. Vom lokalen Hypertext zum verteilten Hypermediasystem. In Gloor, P., Streitz, N. (Hrsg.): *Hypertext und Hypermedia: Von theoretischen Konzepten zu praktischen Anwendungen*, Reihe Informatik Fachberichte Band 249, Springer, 1990, 28–42.

[8] Kirste, T., Hübner, W. HyperPicture – ein Archivierungs– und Retrievalsystem auf optischen Speichermedien. In Gloor, P., Streitz, N. (Hrsg.): *Hypertext und Hypermedia: Von theoretischen Konzepten zu praktischen Anwendungen*, Reihe Informatik Fachberichte Band 249, Springer, 1990, 144–148.

[9] Lange, D.B. A Formal Model of Hypertext. *Proceedings of the NIST Hypertext Standardisation Workshop*, Gaithersburg, Maryland, January 16–18, 1990.

[10] Nelson, T. Replacing the Printed word: A complete literary system. In Lanvington, S.H. (Hrsg.): *Proceedings of the IFIP Congress 1980*, North–Holland, 1013–1023.

[11] McCracken, D.L., Akscyn, R.M. Experience with the ZOG human–computer interface system. *Int. J. Man–Machine Studies*, 21, 1984, 293–310.

[12] Shipman, D. W. The Functional Data Model and the Data Langage DAPLEX, *ACM Transactions on Database Systems*, 6(1), 1981, 140–173.

[13] Sibley, E.H., Kershberg, L. Data architecture and data model considerations. *Proceedings of the AFIPS National Computer Conference*, Dallas, Texas, 6/1977, 85–96.

[14] Stotts, P.D., Furuta, R. Petri–Net–Based Hypertext: Document Structure with Browsing Semantics. *ACM Transactions on Information Systems*, 7(1), 1989, 3–29.

[15] Streitz, N.A. Hypertext: Ein innovatives Medium zur Kommunikation von Wissen. In Gloor, P., Streitz, N. (Hrsg.): *Hypertext und Hypermedia: Von theoretischen Konzepten zu praktischen Anwendungen*, Reihe Informatik Fachberichte Band 249, Springer, 1990, 10–27.

[16] Woods, W.A. What's in a Link: Foundations for Semantic Networks, in Bobrow, D.G., Collins, A.M. (Eds.): *Representation and Understanding: Studies in Cognitive Science*, Academic Press, New York, 1975.

The Multi Media Jotter

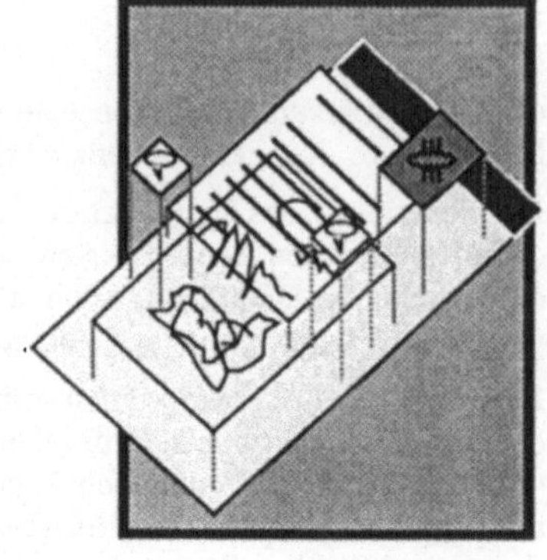

Arno Spinner *
Hans Schuttenbelt
Philips Innovation Centre
Weißhausstraße
D - 5100 Aachen, West Germany

*seit 01.07.1991 bei
ERICSSON Eurolab Deutschland GmbH

Abstract

In this paper a concept for the highly interactive creation and reading of multi media documents will be described. The concept is based on the existence of a central carrier integrating single media sheets into one document. The power of the editor comes from a limited but dedicated functionality, not from a wide range of sophisticated, but most often not needed, functionality (like with desk top publishing systems). This intentionally restricted functionality makes the editor and the produced documents easy to understand and to use for everybody. This makes the editor an innovative product within the multi media world of our daily life. To stress this fact the editor is open to be integrated into office environments. One of the main result is that faxes, voice messages, scanned images, recorded phone conversations, e-mail, etc. are visualized and modified by the editor in exactly the same way as documents created by the editor itself.

The paper describes the design principles and the user interface of the first fully implemented prototype of the editor named multi media jotter. Additionally its daily usage is demonstrated. This paper does not elaborate in detail on the jotter's implementation in an object oriented prototyping environment. The jotter was developed as a substantial part of the PICA project (a Personal Information and Communication Appliance) of the Philips Innovation Centre Aachen.

Keywords: multi media (editors), office automation, communication and information systems

The reader will find in section 1 some motivation of our work and the basic concepts of the jotter. Media specific editing facilities are described in section 2 and in section 3 some remarks on implementation are given. A typical document of daily usage of the jotter is shown in section 4. Section5 discusses the acceptance by the potential users. The last section tells in which directions we plan to continue our work.

1. Introduction

Our daily life is "multi media": during a phone call we make some notes on a piece of paper. We annotate a received fax for later discussions and we like to show some nice photographs when we tell our colleagues that we had a nice business trip. The list of these situations, in our private life and at work, is endless. As a consequence office systems and home appliances, which try to integrate the information usage of a computer and the communication aspects of devices like phone and fax, need multi media editors. These editors must be easy to handle for casual users.

We call our multi media editor a ***jotter***, since it is not meant as a full blown multi media desktop publishing tool to be used by experts. In our concept no text paragraphs, automatic reformatting, numbering of pictures, header pages etc. exist. But the jotter tries to help people to perform simple but essential and powerful tasks such as: record a phone call, make handwritten annotations in a document, put voice stickers on a discussion paper or scan a map and use graphic operators to describe a route and add a spoken greeting.

The fact that it is restricted is exactly its strength, since it is assumed that the use of simple mechanisms satisfies the daily multi media needs of the non-expert user. We will call the documents ***jottings***.

In the first prototype the jotter provides the following functionality:

- general functions (for all media):
  load, store, print document
  cut, copy, paste, to top, to bottom of selections
  iconize, open, opaque, transparent of single media content units
- together with media specific functionality as:
  - text: cursor positioning, word and range selection
    small set of typefaces, sizes and styles
    word wrap and explicit newlines
  - voice: record, play, "forward/rewind", fast play and insert,
    cursor positioning, range selection
  - drawings: lines, arrows, ovals, arcs, rectangles, polygons and free hand drawings
    different widths, greys and fill patterns
  - images: import from file or (hand)scanner or from selected parts of the screen
    rectangular selection, object driven selection
  - postscript: import encapsulated postscript files, read - only

In our opinion the jotter fills one of the most interesting gaps in the range of existing tools in the world of multi/hyper media: he satisfies the trade off between a document in which information is expressed in the best suited media and a simple to use product.

1.1 Basic Concepts

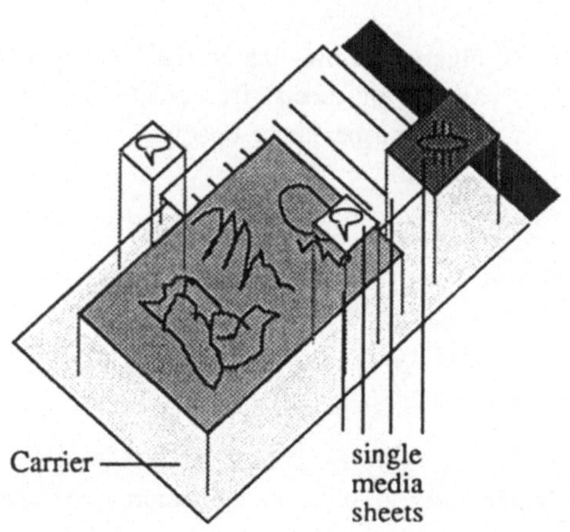

FIGURE 1.

A jotting is a document of one or more pages. Every ***page*** carries one or more single media elements. Therefore we call the page a ***carrier***. The single media elements are called ***sheets***. Sheets have a top to bottom relation and can be presented iconized or opened. When activated an iconized sheet opens and reveals its content e.g. the image, the drawing or the text. Iconized voice sheets are able to play their recorded message. The contents of a sheet are only editable when opened. A sheet is by default transparent and opened and one sheet can only be presented on one carrier.

The number of sheets of the same media type per carrier is not restricted. In this concept one carrier can carry a collection of sheets, iconized or opened, integrating a lot of multi media information on one page. The creator of the jotting will decide which sheets are most important and therefor to be displayed open. Other sheets, of minor importance, will be displayed iconized. The reader will decide what iconized sheets to open, depending on the information he gets from the other, opened, sheets. The reader can easily add sheets to the collection, already integrated on the carrier, with his own multi media jotter..

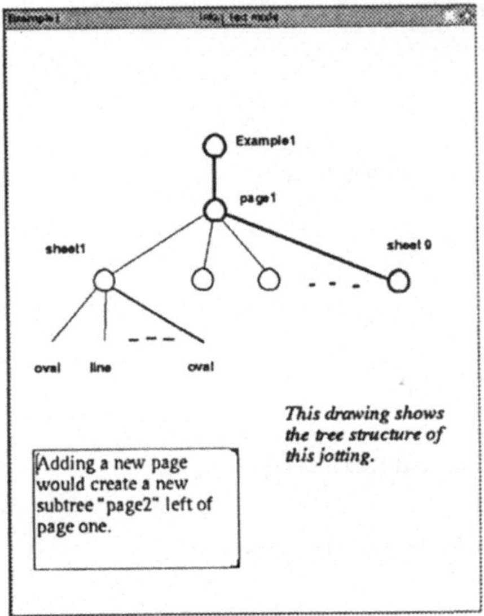

This concept is similar for all sheets of all media. Both reader and creator can easily operate with the jottings since the concept is transparent and consistent over all media.

The document structures of jottings are trees. The tree in figure2 shows only the upper part of the document's own structure. E.g. "sheet9" (this is the boxed sheet) is the root of a subtree containing its position, font information etc. and the text.

The thick path in the tree visualizes the current selection tupple: (Example1, page1, sheet 9). Selected sheets are boxed with handles to move and resize them. All jotter commands refer to the current selection.

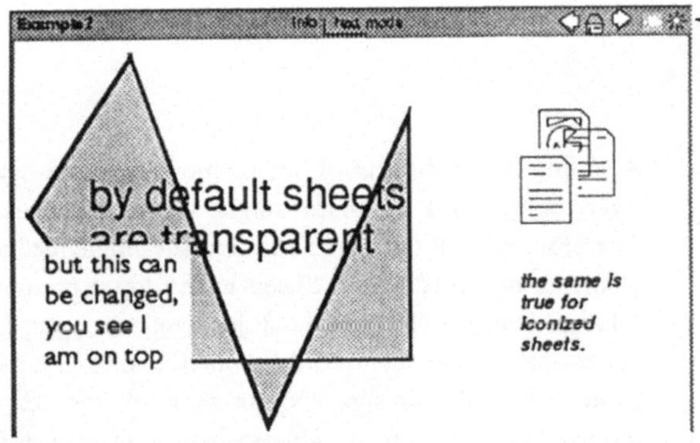

FIGURE 3. sheet configurations

From left to right: title, status information, goto previous, first and next page, miniview and quit.

Figure 3 shows the media independent aspects of sheets: free positioning and size, transparent or opaque, iconized or opened.

2. Dedicated Editing

The user can use a toolbox, menus or shortcuts to work with the jotter. To illustrate the features we focus on the toolbox (see figure 4).

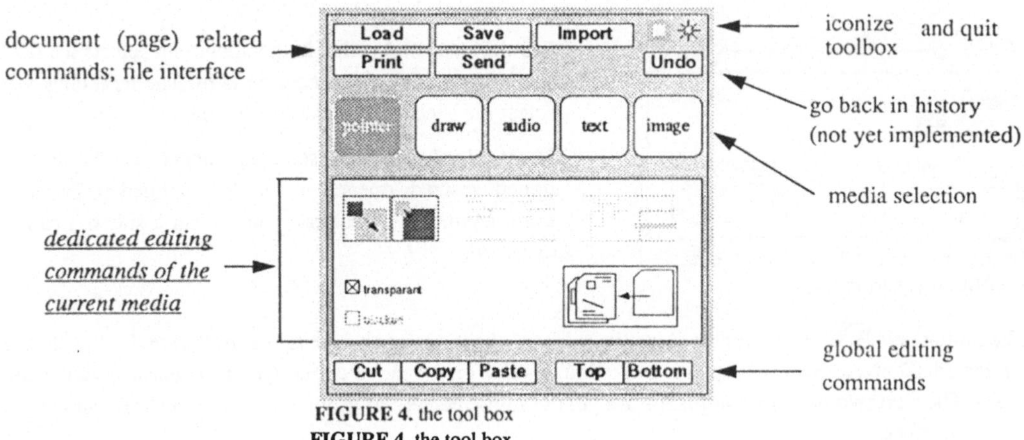

FIGURE 4. the tool box
FIGURE 4. the tool box

Before we will discuss the dedicated editing we have to tell what is global and/or common.

- Save: Saves a snapshot of the jotting.
- Load: Loads a jotting. If the name isn't known in the users "jotting domain"then a new.

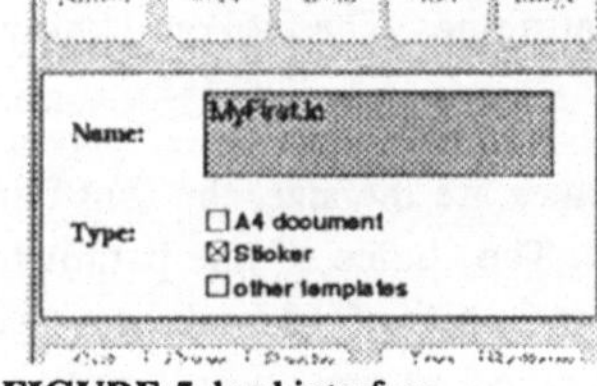

FIGURE 5. load interface

jotting will be created. In the current implementation the user can choose between two templates: a many paged A4 Jotting and a single paged small (yellow) sticker. ("other templates" are not implemented.)

- Print: Sends the document to the printer service. The jotter converts the internal jotting format to a postscript file.
- Send: The document is converted into a postscript file and sent to a chosen application. In our office environment (see section 4) the document can be sent to a multi media mail application. E.g. that uses a translation from a postscript file to a fax G3 data format to send it as a fax.
- Import: A file browser will be activated to select files containing text, audio or bitmaps. The selected file is used to create a new sheet of that media.

Cut...Bottom: These commands are related to the current selection.

2.1 Pointer Mode

In pointer mode the user is able to browse through the sheets of a page, to manipulate some sheet parameters and to add new pages. It is not possible to edit the content of a sheet or some part of it.

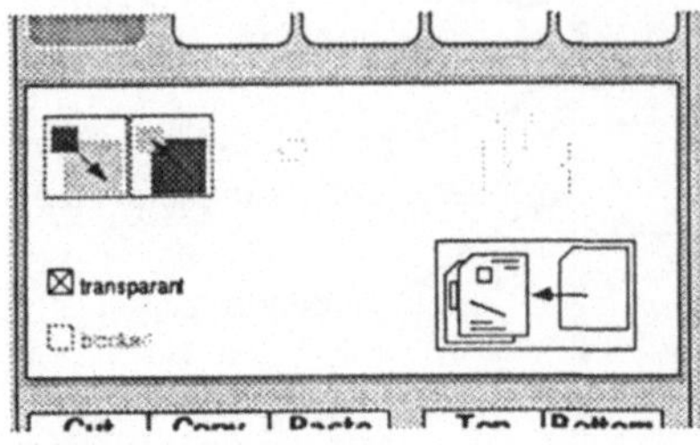

FIGURE 6. pointer mode

- ***Browsing***: a single click in the document selects the (top) sheet where the cursor location is in. The next click will select the sheet underneath it and so on.
- ***Sheet Manipulations***: selected sheets can be repositioned, resized, opened or closed, changed to be (non) transparent. All global commands (cut a sheet...) can be executed.

The button pair is to open or close the selected sheet. Instead of using the toolbox a double click on a selected, iconized sheet will open it and a double click on the move corner (right top corner) will close it again. The next two pairs of commands are yet not implemented: to group and ungroup sheets and to glue sheets together.

2.2 Text

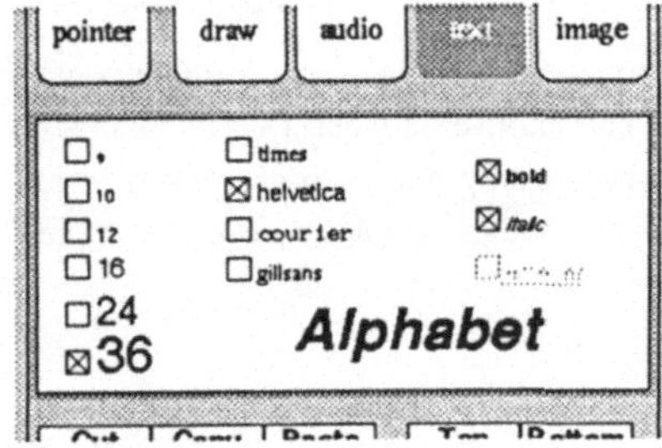

FIGURE 7. text mode

The text capabilities are very simple. All text in one sheet has the same font attributes. The tool only knows about words and lines.

The text attributes are the size, the font family and yes/no of bold and italic. The choice of one parameter will change the text in the sheet (and the feed back box "Alphabet") immediately. Changing the font size doesn't result in an automatic adaptation of the sheet size.

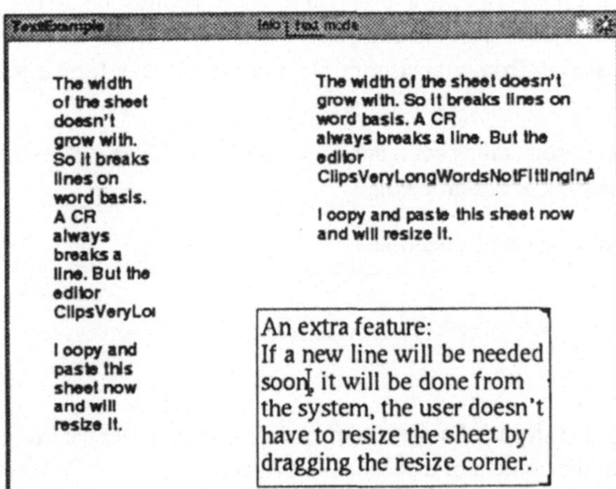

FIGURE 8. line breaking and automatic sheet resize

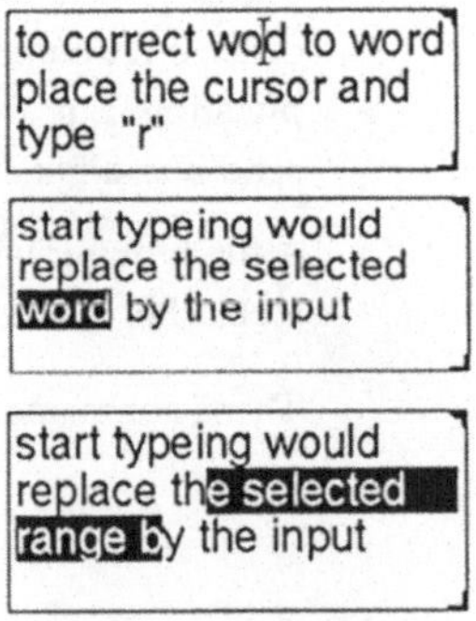

to correct wod to word place the cursor and type "r"

start typeing would replace the selected word by the input

start typeing would replace the selected range by the input

FIGURES 8b: as usual: a cursor; word and range selections

The text editor doesn't support direct functionality of:

- scrolling / paging
- hyphenation, right aligned or centered text
- catalogue of types of paragraph or bullets, etc.
- find / search / replace
- spelling checking
- markers, anchored frames
- headers / footers / page numbering

But with the supplied building blocks (sheets) users have the functionality of:

- headers and footers (independent text sheet (with smallest font))
- columns (independent text sheets, but no overlap)
- (not yet implemented are "master pages": jotting templates)

2.3 Graphics

The PostScript based drawings are objects of drawing sheets. Like a page only knows its sheets a drawing sheet only knows its draw objects and their bottom to top order. Position, type and other attributes are only known by the object itself. The attributes are type (line, ... , free hand drawing), the indication if the draw object is bordered and filled and the grey values of the border and the filled area (see figure 9).

Before the user creates a draw object he has to define its type. There are polymorphic types like "ovals/arcs": ◎. Selecting the "hot spots" changes the type to: ◔. Other examples are "lines/arrows" and "closed/open polygons".

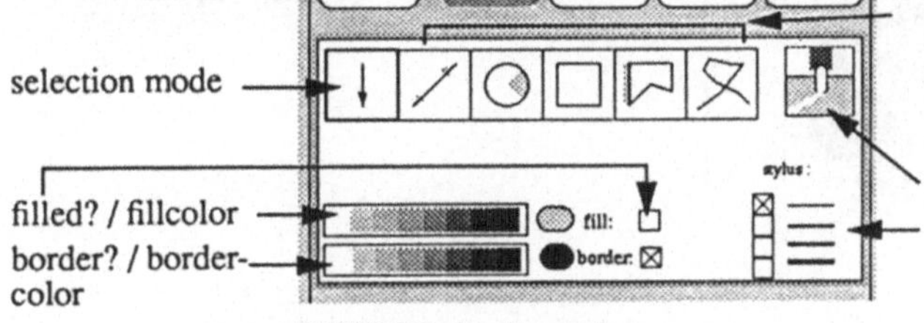

types from left to right: lines and arrows, ovals and arcs, rectangles, opened and closed polygons, free hand drawings.

eraser

stroke width

FIGURE 9. drawing mode

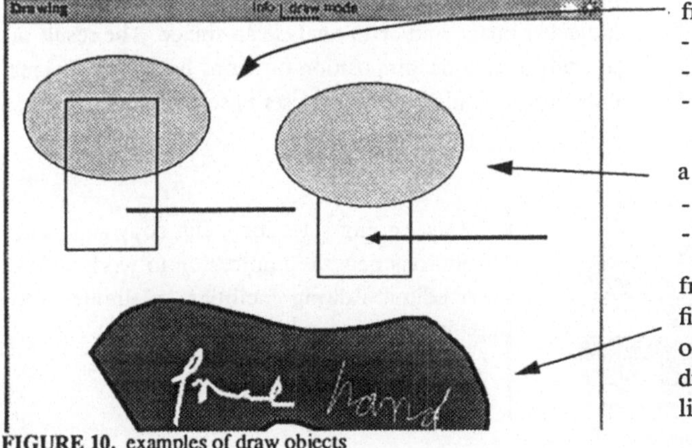

from top to bottom:
- a line
- a non-filled rectangle
- a filled oval

a copy of the 3 objects and
- line changed to arrow
- rectangle to bottom

free hand drawing, filled and closed; on top two free hand drawings with different line width

FIGURE 10. examples of draw objects

2.4 Images

Beside graphics there is a mode named "image mode" that handles bitmaps. A bitmap might be imported from a raster file or from a scanner or handscanner. In our office environment a received fax (G3 data format) is converted to a jotting, where each page has exactly one image sheet.The image mode can be driven in a simple mode and in a sophisticated mode. Let's first have a look at the basic version:

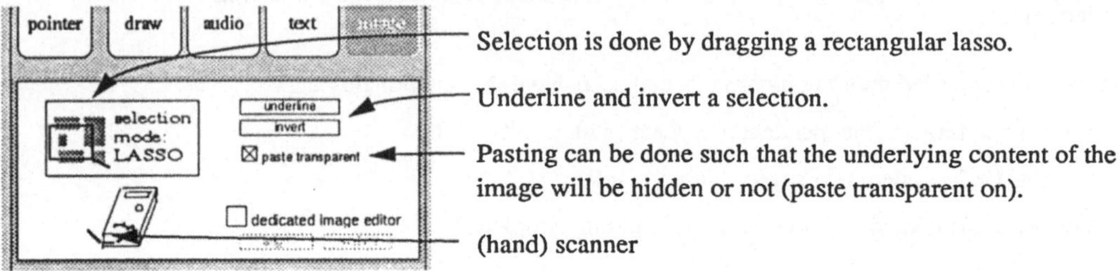

The (hand)scanner button starts the scanning operation. In our environment we make use of a simple handscanner. Contrast and brightness can be controlled by some buttons on the device itself. The "ok button" on the device will finish the scan operation. The result is a new sheet containing the scanned image.

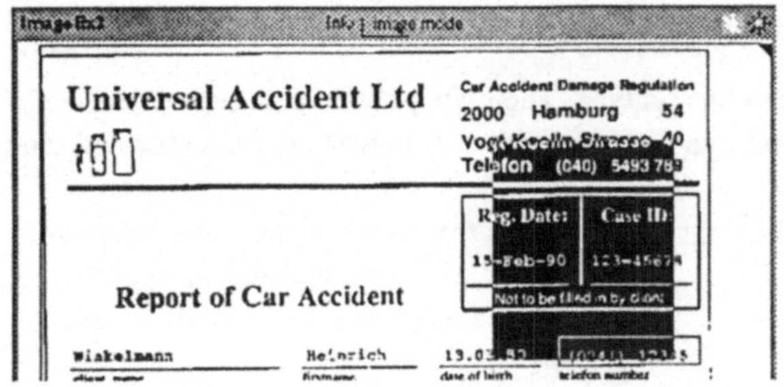

FIGURE 12. selection by rectangular lasso

Copy (cut) will put the selection into the clipboard (and clear it). To paste it the user will see a rectangle (its bounding box). After having positioned this rectangle the content will be copied from the clipboard.

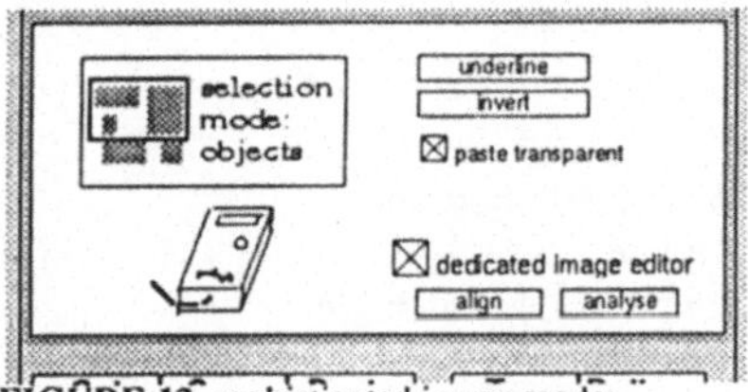

FIGURE 13. sophisticated image mode:

As a feasibility study we integrated some Philips research work on optical formulae recognition [Greef 89]. In the sophisticated mode the user is able to ask the system to align a slop sided image and/or to analyse an image. The result of the analysis is a decomposition on word basis. For images containing text this enables the user to select on word basis.

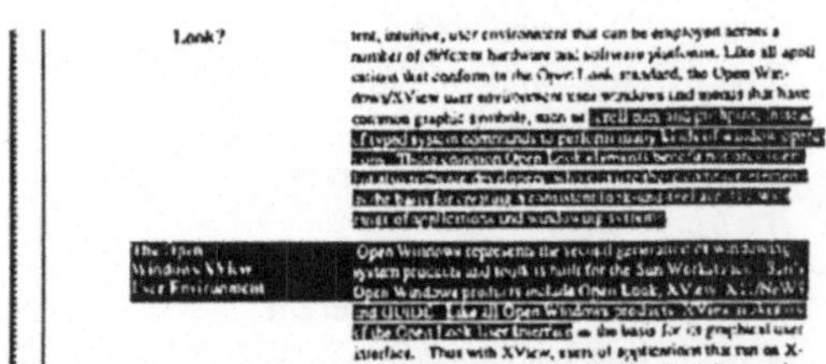

FIGURE 14. text selection in sophisticated image mode

The user doing selections, cut, copy and paste operations gets the impression to work with a text editor. Editing facilities are limited (no text input from keyboard etc.).

2.5 Audio

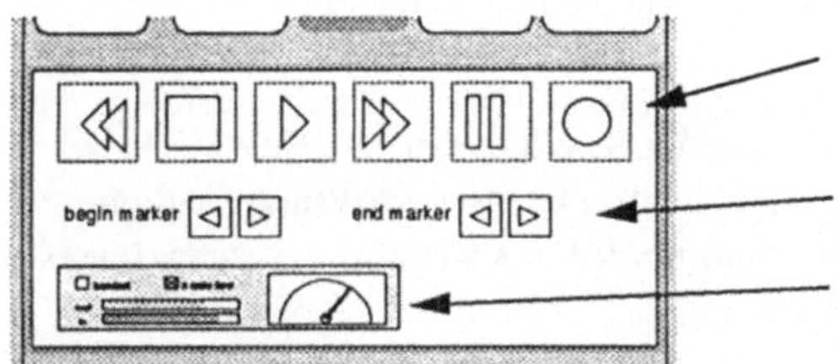

from left to right: play fast backwards, stop, play, play fast forward, pause, record

modify the range of selections (see figure 17)

access to the audio panel to control the microphone sensitivity, the loudness of the loudspeakers etc.

FIGURE 15. audio mode

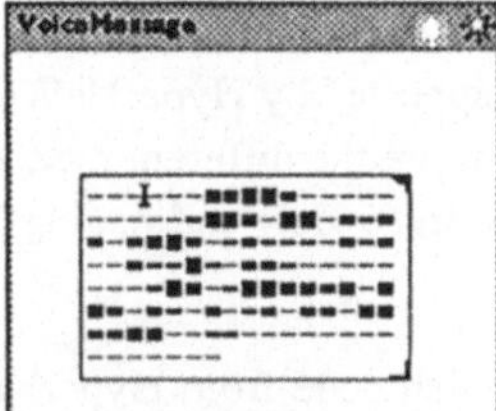

FIGURE 16. a/b: stickers with closed and opened voice sheet

Figure 16a) shows a voice sheet in iconized state. It is possible to play and record them in this state. Only an open sheet leads to full editing functionality.

This functionality can be compared with the jotter's text editing. Instead of a text font there is a block font indicating the amplitudes. There is a cursor like a text cursor. New input (by recording) is inserted at its position. As doing it with text the user can select some part and use the cut, copy and paste operations. This makes it an audio editor, being much more than a cassette recorder. The cursor can be placed at a special position by a mouse click. But the user can make use of the feature, that the cursor moves during play. Pressing stop will leave the cursor at the current position.

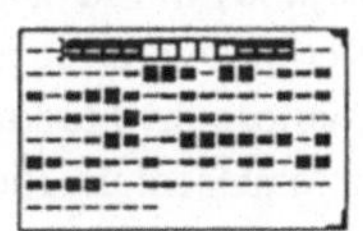

FIGURE 17. modifying the range of a selection

If there is a some part selected the command is restricted to that selection. E.g. the user wants first listen to a selection before he decides to cut it. The selection range can additionally be modified by the 4 buttons in the toolbox (see figure 15).

2.6 PostScript - Only

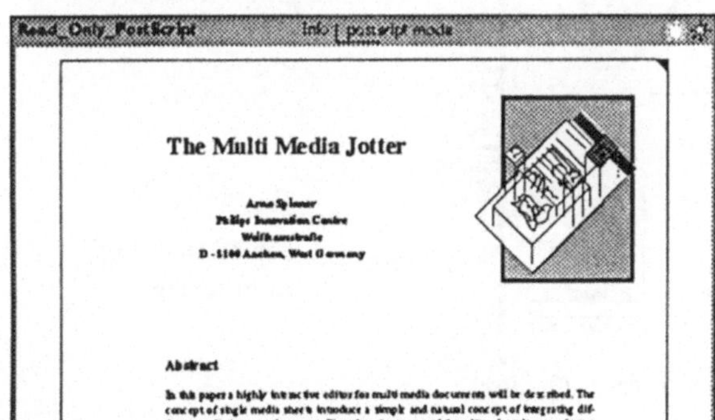

Read_Only_PostScript

Info | postscript mode

The Multi Media Jotter

Abstract

FIGURE 18. import a page of a FrameMaker document

"PostScript-Only" is a read only medium to import postscript files. The result is a new sheet in which the postscript file is executed.

This can be used to have visual effects or to import pages of documents from other editors which have a PostScript description.

3. Implementation

The jotter's prototype is implemented on a SUN sparc station. The code is written in C and in HyperNeWS [Hoff 89].

HyperNeWS is a (user interface) prototyping environment based on NeWS [Arden 89]. NeWS is based on PostScript [Adobe 90]. It extends PostScript by adding a Smalltalk like class mechanism and offering building blocks like canvases, lightweight processes and events needed in a windowing system. It uses a client server model, serving the UI needs of (remote) clients.

Like Apple's HyperCard [Goodm 88] the stack model is the basic architecture of HyperNeWS application. The UI of an application (and possibly the whole application) consists of one or more windows called stacks. A stack consists of cards and cards contain objects like buttons, sliders or text fields.

Jottings can easily be mapped to the stack structure supported by HyperNeWS: the stack is the document, the cards are its pages and the objects held by the cards are the different media sheets. Objects held by the stack are not part of the document. These are the quit, mini-view and paging buttons and the status information and document name fields.

The intension of the implementation was to use as much code from HyperNeWS as possible. Therefore text sheets and audio sheets are modifications of the HyperNeWS text objects and the code for graphic objects is derived from the HyperNeWS drawing tool.

4. Integration into an Office System

The jotter is integrated in an office system named PICA being developed (as a prototype)by the advanced development groupof the Philips Innovation centre in Aken. Therefore there are interfaces to electronic mail, to fax, ISDN phone applications etc. It is integrated in the general workspace philosophy (desktop) and in the common look and feel of all the applications. Because it isn't our intension to present the PICA system figure 19 should give the reader an impression of a typical very short, but expressive jotting

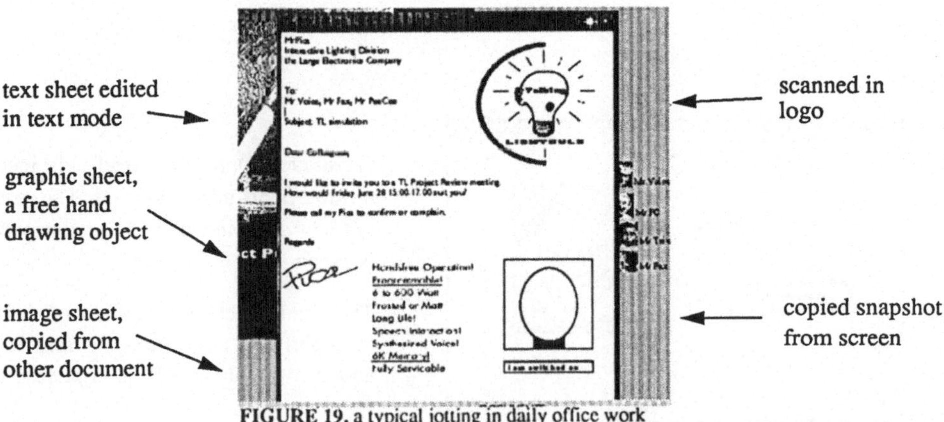

FIGURE 19. a typical jotting in daily office work

Using the "Send" command (see figure 4) the user is able to send the document e.g. as a fax to someone else. If the receiver also has this office system it can be sent as a jotting, being editable for the receiver.

In the future we will be conform to the ODA/ODIF standard. Than the receiver needs "only" an arbitrary ODA editor. To give another example the jotter is used to record greetings for the answering machine and to listen (and annotate) received voice messages. The jotter can be shared during "telephone conferencing" by the partners.

5. Ergonomic Appraisal

We asked an external group of human factor experts of Philips CID (corporate industrial design) to judge the user interface of the PICA prototype with respect to the acceptance of the potential end users. The work is mainly based on the draft standard ISO 9241 Part 10 "Dialogue principles" and on an accepted checklist. The aim was to determine to what extent the PICA prototype allows the user to complete certain routine office tasks to provide him with support in his work. Related to the jotter the tasks were:

- creating multi-media jottings
- reviewing and altering jottings

The analysis of these tasks was structured by following aspects of user interface design:

- controllability of procedure
- intuitiveness of procedure and actions
- conformity between expectations and results of actions
- error tolerance of system, procedures and individual actions
- necessity of individual actions
- clarity and content of display
- help and tutorial support

A lot of the criticism that came out is related to the fact that the jotter is a prototype:

- Performance will be gained simply by replacing the prototyping environment by a real development environment.
- A three button mouse seems to be too complicated for the user. It will be replaced either by a two or one button mouse or by a stylus. (The jotter uses logic events.)
- Error tolerance and some help facilities can simple be added.

In the analysis the toolbox was replaced by a menu. The consequences where a lack of feedback on the ongoing operation and a lack of intuitive help by graphic symbols of the toolbox. The toolbox is an essential part of the jotter's user interface design and helps the jotter to reach the goal to be a self explaining tool.

But the "how to learn it" needs more attention. We think the novice user will not need a reference manual if we add some carefully designed templates showing the features and a tutorial (may be a hyper jotting). Last not least error tolerance will help the user not to be afraid of making mistakes. But in general there is a problem: a human being learns to handle all the media during a lot of years and gains experience in a natural way. Suddenly he has to learn a technical system that tries to deal and integrate these media.

The positive results were the reduced but sufficient complexity and the integration of the different media. But the user has to accept the idea that the jotter does not replace a document editor. E.g. he has to write a report with a dedicated sophisticated word processor; the note telling the reader to inform him about his opinion could be created with the jotter.

6. Future Work

As mentioned this paper described the implementation of the first prototype. We now work on a real implementation. The human factors experts didn't investigate the graphic aspects, because one of the goals is to adapt the jotter to the OpenLook [Sun 89] user interface specification. To reach this goal we changed the development platform from HyperNeWS to tNt (the NeWS toolkit [Sun 90]), that offers the building blocks for OpenLook.

Another goal is to learn from the ergonomic analysis and implement the aspects mentioned above. Additionally there are some small topics that can be done better or have to be added (grouping mechanism etc.).

We will not change the strong structure of the jottings but we will implement a layer that converts the internal representation such that jottings are documents fulfilling the ODA/ODIF standard. But we will not develop an ODA editorm and therefore we will not be able to produce or understand each ODA document.

And we did not investigate in hyper aspects, but we believe that the usage of hyper links will be very fruitful. E.g. reading a message (jotting) could be guided by the links, or selecting a page or object could start to play some audio. One can dream up simple planning capabilities, agents etc.. For the novice user it would be nice to have guided tours through the usage of the jotter. Analogous to teachware using hypertext technology the trainee should be able to have control over the session according to his interests.

literature:

[Adobe90] Adobe Systems Incorporated: "POSTSCRIPT language, Tutorial and Cookbook / Reference Manual", Addison-Wesley Publishing Company, Inc., 16th Printing, July 1990

[Arden 89] Arden, Gosling, Rosenthal: "The NeWS Book An Introduction to the Networked Extensible Window System", Springer Verlag, Technical Publishing Series, 1989

[Greef 90] de Greef: "Multi resolution Document Analysis", Philips NatLab Eindhoven (NL), draft of technical note, 1989

[Hoff 89] Hoff, Abu-Hakima, Niblett: "HyperNeWS1.3 User Manual", The Turing Institute, 1989

[Phil 90] Philips Innovation Centre Aachen: "The June 1990 Project Status Presentation", internal report, july 1990

[Sun 89] Sun Microsystems, Inc.: "Open Look™ User Interface Style Guide", august 1989

[Sun 90] Sun Microsystems, Inc.: "The NeWS Toolkit Reference Manual", august 1990

[Goodm 88] Goodman: "The complete HyperCard Handbook", Bantam Books, October 1988

Architektur und Realisierung eines Multimedia-Dialogmanagers

Klaus-Peter Fähnrich
Fraunhofer-Institut für Arbeitswirtschaft und Organisation (IAO)
Institut für Arbeitswissenschaft und Technologiemanagement (IAT),
Universität Stuttgart

Abstract

The notion of interactive multi-media systems is introduced. A brief review of user interface management systems is given and the ISA Dialog Manager with its main features is presented. Possible architectures for the extension of user interface management systems towards multimedia user interface management systems are discussed. The current implementation status of the Multimedia Builder, an extension of the ISA Dialog Manager and its future directions are outlined. The Multex system - a multimedia user interface management system based on the DIAMANT user interface management system is discusses as example of a multi-modal user interface management system.

Keywords: Interactive Multi-media Systems, Multi-modal Systems, User Interface Management System, Speech Input/Output, Live-Video, Animation.

1. Einleitung

Die letzte Entwicklung von wesentlicher Bedeutung im Bereich der Benutzerschnittstellentechnologie waren die sog. graphischen, direkt manipulativen Benutzerschnittstellen /1/, die momentan den Standard bei Unix-Workstations sowie nach der Einführung von Windows 3.0 auch bei PC's bilden. Für diese Technologien haben sich mittlerweile mit Microsoft Windows 3.0, X Windows mit OSF-Motif und dem Presentation Manager Industriestandards herausgebildet. Weiterhin entwickelt wurden hochstehende Entwicklungswerkzeuge (User Interface Management Systeme).

Der nächste wichtige Schritt bei der technologischen Weiterentwicklung wird von sogenannten Multimedia-Systemen erwartet. Sie sollen neuartige Anwendungsbereiche bei der Produktvermarktung bzw. bei Lerntechnologien oder Informationssystemen erschließen. Sie sollen anderen bisher unzulänglichen Anwendungssystemen zu breiterem Durchbruch verhelfen (z.B. Diagnosesysteme bzw. Wartungssysteme oder Schadstoffdatenbanken) und sie sollen existierende Applikationen attraktiver oder effizienter machen, wie z.B. durch neuartige Trainings- und Hilfesysteme bzw. Zugangssysteme.

Auch bei Multimedia-Systemen werden, wie bei den direkt manipulativen Benutzerschnittstellen, Entwicklungswerkzeuge benötigt, um eine effiziente Entwicklung zu gewährleisten.

2. Interaktive Multimedia-Systeme

Multimedia-Systeme werden gemeinhin dadurch definiert, daß sie mehrere unterschiedliche statische und dynamische Medien zu Eingabe- und/oder Ausgabezwecken an der Benutzerschnittstelle kombinieren. Zu den statischen Medien gehören neben Text und Daten Rastergraphik und Vektorgraphik. Zu den dynamischen Medien werden Ton und Sprache, Animation sowie Bewegtbild (Video) gerechnet. Am weitesten gereift sind momentan Systeme, die neben Daten und Text bzw. den übliche Dialogobjekten wie Pushbuttons, Menüs und Eingabefeldern Rastergraphik sowie Ton bzw. Sprachausgabe integrieren. Weiterhin wird an den hard- und softwaremäßigen Voraussetzungen zur Integration von Video gearbeitet. Sprachausgabe (bis hin zu natürlichsprachlichen Systemen) ist technisch und von der Benutzung her noch sehr aufwendig. Animationstechniken bis hin zu aufwendigen 3-D Simulationen sind in der Entwicklung. Ein weiterer Schritt besteht in der Entwicklung sog. "virtueller Realitäten".

Unter interaktiven Multimedia-Systemen werden in Abgrenzung zur Mischung unterschiedlicher Medien zur reinen Präsentation Multimedia-Systeme verstanden, bei denen Darstellung und Abläufe durch den Benutzer steuerbar sind, und Daten bzw. Informationen vom Benutzer z.B. zum Zwecke der Anwendungssteuerung manipuliert werden. Kann die Interaktionstechnik (bzw. der Interaktionsmodus) vom Benutzer angepasst werden, so spricht man auch von multimodalen Systemen /2/.

Interaktive Multimedia-Systeme sind klar von Hypertext-Systemen abzugrenzen. Unter Hypertext versteht man gemeinhin die nicht sequentielle Organisation von Textdokumenten mit netzartiger Zugriffsstruktur. Hypermedia bezeichnet dabei die netzartige Verknüpfung von Multimedia-Komponenten mit interaktiver Suche (Browsing) mit unterschiedlichen Medien.

3. User Interface Management Systeme

Für die unterschiedlichen Standardbetriebssysteme MS-DOS, OS/2 und Unix haben sich mit Microsoft Windows, Presentation Manager bzw. X Windows und Motif jeweils unterschiedliche Fenstersysteme und Oberflächenbaukästen etabliert.

User Interface Management Systeme besitzen darüber hinaus ein Dialogmanagement-System mit portabler Dialogbeschreibungssprache, Layout-Editoren und Regeleditoren (zum Editieren der meist regelbasierten, ereignisorientierten Dialogbeschreibung), ein interpretatives Simulationssystem sowie Schnittstellen zur Anbindung von Datenbanken bzw, Applikationsroutinen sowie Austauschformate zum Austausch der Dialogbeschreibungen. Sie reduzieren den Aufwand bei der Implementierung von Benutzerschnittstellen gegenüber der Implementation in einer Programmiersprache unter Verwendung von Oberflächenbaukästen mindestens um einen Faktor 3 bis 5.

Von kommerzieller Bedeutung sind bisher die Systeme TeleUse /3/, UIMX /4/ sowie ISA Dialog Manager /5/. Dabei bietet UIMX interpretatives C zur Ablaufsteuerung und schränkt damit die Zielgruppe auf C-Programmierer ein. Der ISA Dialog Manager unterstützt als einziges User Interface Management System alle drei vorgenannten Industriestandards sowie alphanumerische Terminals durch ein sog. Generic Window System Interface und ermöglicht die Entwicklung portabler Anwendungen unter Beibehaltung des systemtypischen "look and feel".

4. Der Multimedia-Builder: Ein UIMS für interaktives Multimedia

Der Multimedia-Builder ist eine Weiterentwicklung des ISA Dialog Manager. Im folgenden werden nach einer kurzen Einführung in den ISA Dialog Manager mögliche Architekturen für Erweiterungen diskutiert, die momentane Realisierung beschrieben sowie die kurzfristigen Weiterentwicklungen aufgezeigt.

4.1 Der ISA Dialog Manager

Der ISA Dialog Manager ist ein UIMS, daß:

- o für C (C++), Fortran und Cobol-Umgebungen die Entwicklung von portablen, von der Applikation und der Datenverwaltung getrennten Benutzerschnittstellen ermöglicht;

- o dabei für OSF-Motif, Windows 3.0 sowie den Presentation Manager und Alpha-Windows (ein Fenstersystem für alphanumerische Terminals) eine einheitliche Dialogbeschreibung realisiert;
- o das Generieren zur Laufzeit, die Zusammenfassung von Objekten zu neuen Objekten, das Zusammenfassen von Objekten und Modellen zu neuen Modellen sowie die Übernahme von Attributwerten entlang einer Vererbungshierarchie realisiert;
- o und darüberhinaus alle Komponenten eines UIMS (Interpretatives Simulieren, Layout- und Regeleditoren, Applikationsschnittstelle, Austauschformate) implementiert hat.

Dieses UIMS wird schrittweise für Multimedia-Anwendungen zu einem Multimedia-UIMS erweitert.

4.2 Zur Architektur von Multimedia UIMS

Grundsätzlich existieren drei Möglichkeiten, eine Multimedia-Erweiterung von User Interface Management Systemen vorzunehmen:

- o die Einbindung über Application-Calls bzw. Service-Calls entsprechen den Konventionen der Applikationsschnittstelle;
- o die Vergabe von neuen Attributen für bestehende Objekte mit entsprechenden Veränderungen im Kern des User Interface Management Systems;
- o die Einbindung von neuen Objekten im Kern des User Interface Management Systems, die über die Regelsprache mit den bereits existierenden Objekten verknüpft werden.

Die erste Möglichkeit stellt die Einbindung mit möglichst geringen Konsequenzen für den Rest des Systems dar. Es ergeben sich keine Veränderungen im Kernsystem, der Regelsprache oder in den Editoren. Die Einbindung erfolgt also wie bei einer normalen Applikation. Der einzige Unterschied besteht darin, daß durch Interprozeßkommunikation Synchronisationsfunktionen ermöglicht werden. Der eindeutige Nachteil besteht darin, daß dem Anwendungsprogrammierer die Dienstleistungen des Systemkerns wie Model-Inheritance oder Portabilität vorenthalten werden. Für Anwendungen, bei denen Multimedia eher eine Zusatzleistung zu konventionellen, direktmanipulativen Benutzerschnittstellen darstellt (bei denen der Multimedia-Anteil also eher klein ist), stellt die erste Möglichkeit momentan sicherlich die gegenwärtig zu bevorzugende Realisierungsmöglichkeit dar. Auf Dauer werden sich allerdings Lösungen entsprechend der zweiten oder dritten Möglichkeit etablieren.

Dabei unterscheiden sich Möglichkeit zwei und drei konzeptionell nicht; die Anlage der Regelsprache ermöglicht die Simulation einer Attributierung für vorhandene Objekte mit Hilfe der Modellbildung. Dies erfordert natürlich vom

Anwendungsprogrammierer ein entsprechendes Verständnis der Möglichkeiten der Regelsprache. Andererseits wird bei Möglichkeit drei eine unnötige konzeptionelle Einschränkung vermieden.

Für die Medien "Still-Video" und "Live-Video" wurden bzw. werden neue Objekte eingeführt. Dabei ist zu unterscheiden zwischen Präsentationsobjekten (Multimedia) und Dialogobjekten (Interaktives Multimedia). Faßt man Still-Video und Live-Video als reine Präsentationsobjekte auf, so stellt sich die Definition der entsprechenden Objekte mit ihren Attributen als relativ unproblematisch dar. Sollen Still-Video und Live-Video jedoch als dialogfähige Objekte aufgefaßt werden, so steht man momentan erst am Anfang der Überlegungen. Besondere Bedeutung erlangt hier der Umgang mit einer impliziten oder expliziten Zeitdarstellung, der bei bisherigen Dialogobjekten nicht in dieser Form bekannt ist. Weiterhin werden Verfahren der Mustererkennung benötigt, um Teilobjekte ohne explizite Markierung vor Beginn der Laufzeit (sensitive regions) einmalig und in ihrer zeitlichen Entwicklung identifizieren zu können. Für die Sprach- oder Tonausgabe ist eine Repräsentation als Präsentationsobjekt ausreichend.

4.3 Still-Video

Die Einbindung von Still-Video ist bereits in der gegenwärtigen Implementierung des ISA Dialog Managers als Object "image" vorhanden. Das Dialogobjekt "image" ist dabei den anderen in einem Fenster auftretenden Dialogobjekten (wie z.B. Buttons, Listboxes, etc.) gleichgestellt. Dabei werden die dargestellten (Farb-) Bilder als Dateien folgender Formate unterstützt: GIF, TIFF, XWD, CCITT T.6 (Fax Gruppe 4). Das GIF steht als universelles Format auf allen Plattformen zur Verfügung; die übrigen Formate sind plattformspezifisch; es stehen jedoch Konvertierungs-Werkzeuge zwischen diesen und einer Vielzahl weiterer Formate zur Verfügung. Die Erstellung kann über Scanner, Video-Kamera, Dateiaustausch oder Wandlung der Daten aus Fremdprogrammen (z.B. über Postscript) erfolgen.

4.4 Die Audio-Schnittstelle

Momentan steht eine SONY NEWS-Workstation mit integriertem Audio-Interface (A/D und D/A Wandlung von maximal zwei Kanälen mit Abstastfrequenzen zwischen 8 und 38 kHz; verschiedene Kompressionsverfahren) zur Verfügung. Es liefert damit Qualitäten zwischen ISDN-Telefonqualität und CD-Qualität in Stereo. Als Basissoftware werden Bibliotheken zur Verarbeitung / Verwaltung des Audiodateiformates mitgeliefert.

In der momentanen Version wird der Sound-Daemon als eigenständiger Prozess aufgesetzt, der sich über eine Interprozesskommunikatikon (TCP/IP) mit den Dialog Manager koordiniert. Auf seiten des Dialog Manager ist eine Multi-Media Extension implementiert. Die Extensions werden aus der Regelsprache in funktionaler Notation angesteuert. Die dabei zur Verfügung gestellten Funktionen umfassen: Starten und Beenden eine Aufnahme bzw. Wiedergabe, Setzen und

Ändern diverser Attribute (Digitalisierungsrate, Kompressionsverfahren, Kanalzahl, etc.).

In Zukunft ist angedacht, ein eigenständiges Audio-Objekt in den Kern einzubringen, um damit eine hardware-unabhängige Schnittstelle anzubieten. Das Abspielen erfolgt dann wie folgt über die Regelsprache:

```
// Objekt-Definition
sound SoundObject4711
{
    .file "soundfile.snd";
    .volume 80;
    .balance -10;
}


// Regel zum Auslösen der Wiedergabe
on InfoButton select
{
    SoundObject4711.play := true; // Wiedergabe beginnen
    this.bgc := ColorLilaBlassBlau; // Farbe des Buttons ändern
}


// Regel nach Beendigung der Audio-Wiedergabe
on SoundObject4711 finish
{
    this.bgc := this.model.bgc; // Original-Farbe restaurieren
}
```

Ein entsprechendes Definitionsfenster zur Festlegung der Objekt-Attribute ist in den Editor einzufügen. Werkzeuge zum Erfassen und Modifizieren von Ton- und Sprachsequenzen müssen nicht im Kernsystem realisiert werden, können aber effizient mit Hilfe des UIMS implementiert werden.

4.5 Live-Video

Gegenwärtig wird auf der SONY NEWS Workstation mit einem "true color framebuffer" (NWB-254P) gearbeitet, der das direkte Einmischen eines (NTSC- oder PAL-) Video-Signals erlaubt. Dabei kann dieses Video-Signal von verschiedenen Quellen wie z.B. Video-Recorder (SONY U-Matic, VHS), Bildplatte (SONY Analog Draw mit 72.000 Einzelbildern bzw. 2 * 24 Minuten Film im Direktzugriff) oder Kamera stammen. Im gegenwärtigen System wird zur Applikationsentwicklung ein SONY U-Matic Recorder eingesetzt, da dieser eine Steuerung über RS 232 Schnittstelle erlaubt. Entsprechende Video-Adapterkarten sind auch für PC´s verfügbar (Screen Machine).

Softwareseitig wird die genannte Konfiguration im X Windows Server über eine Video Extension unterstützt. Dies ermöglicht, das in Echtzeit anstehende Videosignal in einem X-Fenster darzustellen; dabei werden Funktionen wie das Skalieren des Videosignals in das X-Fenster direkt in Hardware realisiert und vom X-Server entsprechend kontrolliert. Für die PC-Welt wird gegenwärtig eine Treiberarchitektur, die entsprechende Leistungsmerkmale beinhaltet, unter Beteiligung namhafter Soft- und Hardwareanbieter spezifiziert.

Die Einbindung in den Dialog Manager erfolgt wie beim Audio-Interface. Zu realiserende Funktionen sind z.B. Positionierung auf einen bestimmten Frame, Abfragen des aktuellen Frames und Starten und Beenden des Abspielen.

Momentan laufen im X Consortium Bestrebungen, eine Video Extension in die X11 Protocol Specification aufzunehmen. Damit wäre eine standardisiertes API für X Windows in bezug auf die Integration von Live-Video hergestellt. Für den PC-Bereich sind wie bereits oben ausgeführt entsprechende Entwicklungen im Gange. Auf seiten des UIMS werden entsprechend den Ausführungen in 4.4 ein oder mehrere neue Video-Objekte eingeführt.

5. Multex: Ein multimodales Multi-Media UIMS

5.1 Das DIAMANT - UIMS

Ziel bei der Entwicklung des User Interface Systemes DIAMANT, die bereits 1989 abgeschlossen war, war die prototypische Erprobung eines objektorientierten User Interface Management Systems /6/. Dabei standen die Ziele Erweiterbarkeit in Bezug auf Oberflächenobjekte, dynamische Objekterzeugung zur Laufzeit, Unterstützung multimedialer Benutzerschnittstellen und Flexibilität bei der Anbindung von Anwendungen im Vordergrund. DIAMANT enthält eine objektorientierte Dialogbeschreibungssprache, auf die im weiteren eingegangen wird. Weitere Komponenten sind ein Oberflächeneditor und ein Laufzeitkern. Der Laufzeitkern stellt Klassen für die interne Kommunikation zwischen Dialogobjekten und für die externe Interprozesskomunikation bereit. Damit unterstützt DIAMANT sowohl die Verteilung der Anwendung als auch die Integration neuer E/A Medien, was für eine Multimedia-Erweiterung von Wichtigkeit ist. Als graphischer Oberflächenbaukasten wurde InterViews verwendet. InterViews wurde dabei um einige Bausteine erweitert.

Bei der Entwicklung der Dialogbeschreibungssprache UIDL des User Interface Management Systems DIAMANT wurde wie beim ISA Dialog Manager ein ereignisorientiertes Dialogmodell zugrunde gelegt. Dieses hat sich als besonders geeignet für die Beschreibung graphisch-interaktiver direkt-manipulativer Dialoge erwiesen. Ein Dialog besteht in DIAMANT aus einer Menge von Ereignisbehandlern (Event Handler). Ein Ereignisbehandler verfügt über eine Menge von Regeln, wobei jede Regel die Reaktion auf genau einem Ereignistyp beschreibt. Ereignistypen sind für die vordefinierten Ereignisbehandler festgelegt. Bei neuen Ereignisbehandlern können auch eigene Ereignistypen definiert werden. Quellen der Ereignisse können sein:

- o Ein Oberflächenobjekt: Jedes Oberflächenobjekt hat einen Ereignisbehandler, an den die Oberflächenereignisse weitergegeben werden und der die Darstellung des Objektes steuern kann;
- o Ein anderer Ereignisbehandler;
- o Ein externer Prozeß.

Das Versenden von Ereignissen ist damit das Kommunikationsmodell der UIDL für die interne und externe Komunikation. Die Reaktion auf ein Ereignis kann darin bestehen, daß:

- o neue Ereignisse versendet werden, z.B. um den Zustand von Ereignisbehandlern oder Ihrer Oberflächenobjekte zu verändern;
- o neue Ereignisbehandler erzeugt werden oder alte gelöscht werden;
- o lokale Variable des Ereignisbehandlers verändert werden;
- o und Anwendungsfunktionen aufgerufen werden.

Die UIDL ist objektorientiert ausgelegt. Ereignisbehandler sind Instanzen von Klassen, wobei die Regeln als Methoden aufgefasst werden. Zwischen den Klassen kann es Vererbungsbeziehungen geben (einfache Vererbung). Damit lassen sich die Vorteile objektorientierter Programmierung auch auf der Dialogebene nutzen.

Eine Anwendung der Klassenbildung ist die Definition von Modellen oder Vorlagen, die das Befolgen von Richtlinien (Style Guides) für die Oberfläche unterstützen.

DIAMANT wurde auf Sun-Workstations unter SunOS in C++ implementiert. Verwendung von C++ lag schon insofern nahe, als DIAMANT auf InterViews aufsetzt. Die Vorteile einer objektorientierten Implementierung liegen darin begründet, daß die gesamten objektorientieren Basismechanismen von C++ für die Objektverwaltung benutzt werden können. Dies schränkt natürlich auch den Anwendungsbereich des Systemes in entsprechender Weise für eine kommerzielle Nutzung ein. Da dieser aber bei dem Prototypensystem nicht im Vordergrund steht, war eine entsprechende objektorientierte Implementierung gerechtfertigt.

Die Klassen für die Ereignisbehandler werden vom Übersetzer der UIDL in C++ Klassen übersetzt. Der Laufzeitkern enthält einen Ereignisverteiler, der je eine Warteschlange für lokale (innerhalb der Dialogsteuerung) auftretende und globale (von der Oberfläche oder anderen Prozessen kommende) Ereignisse verwaltet. Außerdem verwaltet er eine Liste aller existierenden Ereignisbehandlerinstanzen. Die Ereignisverteilung besteht darin, zunächst die lokale und dann die globale Warteschlange abzuarbeiten, da lokale Ereignisse Priorität haben. Jedes Ereignis wird dabei an den zugehörigen Ereignisbehandler weitergegeben.

5.2 Die Erweiterung von DIAMANT zu Multex

Das System Multex ist einerseits ein Multimedia User Interface Management System und namensgleich mit einem implementierten Anwendungssystem im Bereich Maschinendiagnose und -wartung. Ausgehend von der Diagnose-Expertensystem-Shell IDS /7/, die am IAO entwickelt wurde und für Anwendungen in der Maschinendiagnose eingesetzt wird, wurde ein Multimedia Expertensystem aufseztend auf dem prototypischen Multimedia UIMS implementiert. Dabei enthält das System die folgenden Komponenten: Die Laufzeitkomponente des Expertensystems wurde durch einen Medienmanager erweitert. Die Aufgabe des Medienmanagers ist die Steuerung und Synchronisation der angeschlossenen Medien: synthetische Sprachausgabe, Einzelworterkenner, Live-Video, Animation, Grafik und Text. Der Medien-Manager wurde dabei in der Beschreibungssprache UIDL von DIAMANT implementiert.

Der Anwendungsteil eines Systems kann von der Benutzerschnittstelle auf zwei Arten angesprochen werden. Die erste Möglichkeit besteht im Aufruf einer in C++ oder in C geschriebenen Anwendungsfunktionen aus einer Regel heraus. Als zweite Möglichkeit kann die Anwendungs als eigener Prozeß implementiert werden, welcher mit der Benutzerschnittstelle über die Interprozeßkommunikationsmöglichkeiten von DIAMANT komuniziert. Auf diese Art und Weise werden vom Medien-Manager sowohl die verschiedenen Medien, als auch die eigentliche Anwendung (des Expertensystem) angesprochen. Das folgende Beispiel zeigt, wie Erläuterung zu einer Animationsequenzen gegeben werden.

```
// Ereignis MESSAGE von Animation
on "MESSAGE" with (text) from "animation"
{
    if ((TALK == "on") && (any_window == "open"))
        // Text an Prozess "talk" schicken
        send ("talk", "speak", text);
}


// Ereignis Objekt selektiert von Animation
on "SELECTED" with (object) from "animation"
{
    if (object == "Ruecklicht")
        send (explainer, "show", "Rucklichtdata");
    else if (object=="Schaltung")
        send (explainer, "show", "Schaltungdata");
    else // Undefined object??
        send (explainer, "sing song", "doodel doedel");
}
```

Der Laufzeitkern von DIAMANT übernimmt die Aufgabe, die Verbindung zu den Prozessen ("talk" und "animation") aufzubauen. Falls ihm diese Prozesse bisher nicht bekannt sind, sucht er automatisch in einer Datei nach einem Eintrag, auf welchen Rechner unter welchem Port er den jeweiligen Prozeß finden kann. Danach baut der Laufzeitkern die Verbindung auf und verwaltet die gesamte Komunikation mit diesen Prozessen.

Auf dieselbe Art wurde das Expertensystem eingebunden. Es steuert einen einfachen Frage-/ Antwortdialog. Es erfragt Zustände der Maschine und gibt Anweisungen, wie diese Zustände zu überprüfen sind, beziehungsweise wie ein gefundener Fehler zu beheben ist. Dabei berücksichtigt es die vorhandenen Werte von Maschinensensoren und benutzt Heuristik-Funktionen (um den Aufwand zur Fehlersuche und Reparatur möglichst minimal zu halten.

Obwohl Multex und der Multimedia Builder intern in der Architektur viele Ähnlichkeiten aufweisen, standen bei der Entwicklung von Multex andere, weitergehende Ziele im Vordergrund:

- o zum einen sollte eine größere Palette von Medien (also auch synthetische Sprachausgabe sowie Animation) integriert werden;

- o Es sollten höherwertige Dialogphänomene untersucht werden, die Multex zu einem multimodalen, interaktiven User Interface Management System machen.

Bei den zusätzlichen Medien wurde auf eine am Institut entwickelte deutsche "Text-To-Speech"- Karte zurückgegriffen, die von der Leistungsfähigkeit und vom Preis-Leistungsverhältnis her keine Alternative zu der im Multimedia Builder verwendeten Sprachausgabetechnologie darstellt. Weiterhin integriert wurde ein sprecherunabhängiger Einzelworterkenner mit einem großen Wortschatz. Weiterhin wurden animierte Bitmaps und animierte Grafiken verwendet, für die eigene Entwicklungswerkzeuge geschaffen wurden.

Das System ermöglicht die alternative Verwendung unterschiedlicher Medien unter Benutzerkontrolle oder in Abhängigkeit von internen Zuständen des angeschlossenen Applikationssystems. Es kennt unterschiedliche Dialogmodi für verschiedene Einlernzustände des Benutzers. In grafischen Animationen ermöglicht es die Referenzierung auf Teilobjekte, um in der Art von Hyper-Media Systemen zusätzliche Informationen zu den referenzierten Objekten zu erhalten. Die unterschiedlichen Medien können je nach Arbeitsaufgabe und Zustand des Dialoges singulär oder parallel (redundant) dargestellt werden. Aufgrund der Implementation sind dabei die entsprechenden Prozesse jeweils unabhänig voneinander ansprechbar.

6. Zusammenfassung

Die Erfahrungen mit den Systemen Multex und Multimedia Builder lassen sich wie folgt zusammenfassen:

- o Die Erweiterung konventioneller, marktgängiger User Interface Management Systeme zu Entwicklungswerkzeugen für interaktives Multimedia unter Einschluß der Medien Still-Video, Live-Video und Sprachausgabe bzw. Tonausgabe ist möglich. Dabei existieren kurzfristige Realisierungsmöglichkeiten, die eine schnelle Erweiterung des Systems zulassen. Es wurden allerdings auch mittelfristige Weiterentwicklungsmöglichkeiten zu einer systemkonformen Integration aufgezeigt und erprobt.

- o Experimentell wurden die Integration weiterer Techniken im Bereich vollsynthetische Sprachausgabe, Spracheingabe, sprecherunabhängige Spracheingabe, Animationstechniken und sogar Gestikeingaben erprobt. Hier ist in der nächsten Zeit zumindest im Bereich Animation mit entsprechenden Kommerzialisierungen zu rechnen.

o Es wurden Versuche unternommen, Leistungsmerkmale von multimodalen interaktiven Multimedia Systemen zu definieren. Hier werden sich Umsetzungen weit langsamer als im vorgenannten Bereich vollziehen.

7. Literatur

/1/ Ziegler, J.E., Fähnrich, K.P. (1988). Direct Manipulation. In: Helander, M. (Ed.) (1988). Handbook of Human-Computer Interaction. Amsterdam: Elsevier, 123-133.

/2/ H.-J. Bullinger, K.-P. Fähnrich (1984). Symbiotic Man Computer Interfaces And The User Assistant Concept. In: G. Salvendy (Hrsg.): Human Computer Interaction. Elsevier 1984.

/3/ Marmolin, H. (1991). The TELEUSE Dialog Management System. Erscheint in: Bullinger, H.-J., Proc. HCI International 91, Stuttgart: Elsevier 1991.

/4/ Mikes, S. (1991). Two Gooey Builders that Stick. UNIX World, Jan. 1991, 105-112.

/5/ Raether, C. (1990). ISA-Dialogmanager - Ein Entwicklungswerkzeug zur Erstellung portabler, graphischer Benutzeroberflächen. In: Bullinger, H.-J. (Hrsg.). Software-Ergonomie in der Praxis. IAO-Forum. Berlin, Heidelberg: Springer, 43-56.

/6/ Trefz, B. Ziegler, J. (1989). DIAMANT - Ein User Interface Management System für graphische Benutzerschnittstellen. In: Maaß, S.; Oberquelle, H. (1989). Software-Ergonomie`89. Stuttgart: Teubner, 264-273.

/7/ Koller, F. Ziegler, J.(1989). MULTEX - Eine multimediale Benutzeroberfläche für ein Expertensystem zu Maschinendiagnose. In: Paul, M (Hrsg.). GI-19. Jahrestagung I. Berlin, Heidelberg: Springer, 96-107.

Hypermedia-Techniken für 'Instuctional Tool Environments' - Anforderungen und Ansätze

Max Mühlhäuser

Universität Kaiserslautern, FB Informatik, AG Telematik
Erwin-Schrödinger-Str., D-6750 Kaiserslautern
[+49] (631) 205-2992, Fax ...-2640, max@informatik.uni-kl.de

1 ÜBERBLICK

Für die 90er Jahre wird vielfach das Zusammenwachsen von Informations- und Kommunikationstechnologie sowie Unterhaltungselektronik prognostiziert. Für die dabei entstehenden *kooperativen* (verteilten, Gruppenarbeit unterstützenden) *multimedialen* Systeme wird häufig als zentrales Software-*Grundkonzept 'Hypertext/Hypermedia'* [9] propagiert und als erstes wichtiges *Anwendungsfeld* die *Aus-/Weiterbildung*. Der Begriff Hypermedia ist aber noch nicht konsolidiert: es fehlt an einem allgemein akzeptierten Grundverständnis, an breit eingeführten Standards, an einer soliden theoretischen Untermauerung. Heutige Hypermedia-Systeme erweisen sich meist auch als viel zu eingeschränkt, um als *die* zentrale Komponente großer Softwaresysteme verwendet zu werden.

In diesem Beitrag wird berichtet über den Versuch, Hypermedia-Konzepte und –Techniken zu entwickeln, welche möglichst tragfähig und breit einsetzbar sind und welche die Wiederverwendbarkeit und formale Handhabbarkeit der Hypermedia-Dokumentverarbeitung verbessern. Die vorgestellten Arbeiten entstanden im Rahmen des Projektes *Nestor*. In Nestor wird an einem sogenannten 'Instructional Tool Environment' gearbeitet, d.h. einer Entwicklungsumgebung im Bereich der computergestützten Aus- und Weiterbildung, welche (auf der Basis von kooperativen multimedialen Systemen, s.o.) Werkzeuge, Informationen und Prozesse (Abläufe) zu integrieren gestattet.

Nestor wird vorgestellt, wichtige Hypermedia-Konzepte beschrieben. Der Teilaspekt 'Prozeßunterstützung durch *generische Hypermedia-Navigation*' wird näher betrachtet.

2 PROJEKT NESTOR

Nestor ist ein Gemeinschaftsprojekt mehrerer deutscher Universitäten (Karlsruhe, Kaiserslautern, Freiburg) und des Forschungszentrums CEC von Digital Equipment. Es basiert auf einer umfassenden Gesamtarchitektur für 'Instructional Tool Environments', welche im Zuge des Projektes fortgeschrieben wird (der Begriff 'instructional' umfaßt dabei unterschiedlichste Formen der Informationsvermittlung, z.B. Aus-/Weiterbildung, Fernunterricht, Dokumentations-Logistik, Online-Hilfen für Softwaresysteme). Besonderes Gewicht wird gelegt auf die Unterstützung von Multimedia-Information und von Kooperation (Teams von Autoren und/oder Lernenden, verteilt auf ggf. entfernte Arbeitsstationen). In dieser Gesamtarchitektur identifiziert Nestor besonders untersuchenswerte Teilaspekte, die dann in einer der Nestor-'Projektschienen' ('Tracks') bearbeitet werden:

- Research Track: in diesem werden längerfristige dedizierte Forschungsarbeiten angesiedelt; ihre Ergebnisse fließen ein in den
- Development Track: hier steht weniger wissenschaftliches Forschungsinteresse im Vordergrund als softwaretechnische Machbarkeitsnachweise (Prototypen).

Als grundlegende Prämissen dienen

1. *Hypermedia* und *Objekt-Orientierung* als wesentliche, zu integrierende Paradigmen.
2. Ein Szenario *kooperierender multimedialer* Arbeitsstationen,
3. *Prozesse* oder Abläufe *(instruktionelle Prozesse, Autorenprozesse,* s.u.), für welche *Prozeßmodellierung und -steuerung* zu unterstützen ist.

Aus diesen Prämissen ergeben sich direkt die wesentlichen Arbeitsfelder im Bereich der Hypermedia-Konzepte und -Techniken:

1. Konzepte für ***o***bjekt***o***rientierte ***H***yper***m***edia-Systeme *('OOHM'-Grundlagen).*
2. 'OOHM'-Erweiterungen bezüglich
   a. Kooperationsunterstützung
   b. Multimedia-Unterstützung
   c. Prozeßunterstützung (für instruktionelle Prozesse und Autorenprozesse)

Dabei zeigt sich, daß eine sinnvolle Erweiterung von Hypermedia gemäß 2a-c erst durch die Integration und Erweiterung von Objektorientierung und Hypermedia möglich wird.

3 GRUNDLEGENDE HYPERMEDIA-KONZEPTE

3.1 OOHM-Grundlagen

Ein Blick auf jüngere Literatur zum Thema Hypermedia läßt den Schluß zu, daß einerseits die Zahl verfügbarer Hypertext- und Hypermedia-Systeme immer weiter zunimmt, daß andererseits jedoch die theoretische Fundierung des Gebiets vergleichsweise langsam vorankommt.

So zeichnet sich trotz Standardisierungsbemühungen weder ein breit akzeptiertes Grundverständnis eines Modelles ab, welches über die Begriffe 'Knoten' und 'Link' weit hinausginge, noch eine Formalisierung als Grundlage besserer Informatik-technischer Unterstützung. *Beispiele* für die Folgen sind: Koexistenz von Systemen mit ausschließlich statischen Links und solchen mit ausschließlich dynamischen Links, Bearbeitung

(Erstellung, Programmierung, Wartung...) von Hypertexten praktisch ausschließlich 'instanzorientiert' ("a single node or link at a time") ohne Wiederverwendbarkeit oder automatische Konsistenzprüfungen u.v.a.m.

Im OOHM-Konzept wird nicht nur ein Hypermedia-System mit einer objektorientierten Programmiersprache implementiert, sondern die beiden Konzepte werden verschmolzen. Das führt zunächst zu folgenden wichtigen Eigenschaften:

- strenge objektorientierte *Typisierung* (inkl. Vererbung, Polymorphie, Aggregation) für Knoten *und* Links als Grundlagen für Wiederverwendbarkeit, Konsistenzprüfung, etc.
- keine Einschränkung der *Präsentation* von Knoten z.B. auf 'eine Bildschirmseite', Möglichkeit der Unterordnung unter ein UIMS (User Interface Management System).
- statische *und dynamische* Knoten (statisch z.B. für zu präsentierende Hypermedia-Information, dynamisch z.B. für ein auszuführendes Animationsprogramm), statische *und* dynamische Links (Dynamik als dynamische Berechnung des Zielknotens oder zur Verknüpfung des Link-Überganges mit auszuführenden Aktionen)
- *n-äre Links* (mehrere Quell-, mehrere Zielknoten), bei denen z.B. die o.g. Dynamik für Multimedia-Synchronisation ausgenutzt werden kann

Die Nestor-spezifische Weiterentwicklung des OOHM-Paradigmas ergibt weitere entscheidende Vorteile:

- Den *'dualen Ansatz'* [2], durch den auf die Einführung expliziter Objektklassen 'Knoten' und 'Link' verzichtet werden kann, wodurch existierende objekt-orientierte Systeme unverändert einer Hypermedia-Defaultinterpretation unterzogen werden können
- *Optional* jedoch volle Flexibilität der *Spezialisierung:* unterschiedliche Knoten- und Link-Grundtypen können im selben System gemischt werden, Browser-Funktionen (wie 'Anzeigen eines Knotens', 'Link-Übergang' etc.) können als Bestandteil der Knoten / Links, also informationsspezifisch, implementiert werden, etc
- *Persistente* Speicherung, entweder durch Abbildung auf herkömmliche Datenbankmodelle (objekt-orientiert, relational), oder durch dedizierte objekt-orientierte Unterstützung, welche effizient konkurrierenden Zugriff im verteilten System unterstützt
- Hypermedia-spezifische Versionskontrolle (in Entwicklung)
- Kontextabhängigkeit (Unterschiedliches Verhalten in unterschiedlichem Benutzungskontext; in Entwicklung)

3.2 Kooperationsunterstützung

Das CSCW-Paradigma [4] (computer supported cooperative work) hat in den vergangenen Jahren zunehmend Beachtung gefunden. Ein Blick auf CSCW-Systeme aus der Literatur zeigt, daß diese meist für sehr *spezifische* Anwendungsdomänen (joint editing, conferencing, decision support, ...) geschrieben wurden; die wenigen existierenden *allgemeineren* Ansätze der CSCW-Unterstützung decken längst nicht alle Mechanismen und Strukturen ab, die im Bereich Gruppenarbeit anzutreffen sind.

Besonders interessant erscheint CSCW im Kontext von Hypermedia, da Hypermedia-Systeme entwickelt wurden, um die Handhabung großer und komplex vernetzter Informationsmengen zu unterstützen - die Erstellung, Manipulation und Akquisition solcher komplexer Informationsmengen hat der Natur nach Teamarbeits-Charakter, die meisten gängigen Hypermedia-Systeme sind aber eher auf 'single-user'-Betrieb vorbereitet.

Gewünscht ist also ein generischer CSCW-Ansatz auf Hypermedia-Basis: in Nestor wird daher in folgenden Schritten vorgegangen: a) Identifikation und Klassifikation generischer CSCW-Bausteine/Aktionen im Rahmen eines umfassenden Modellierungskonzeptes, b) Implementierung von Bausteinen und zugehörigem Unterstützungssystem ('Group Interaction Environment') c) Integration der CSCW-Konzepte mit dem OOHM-Konzept, d) Implementierung der Integration (dieser Schritt ist noch in Arbeuit [7]).

Es zeigt sich, daß beim Einsatz moderner Fenstersystem-basierter Arbeitsstationen elementare Systemunterstützung für CSCW-Anwendungen fehlt, welche bei einfachen alphanumerischen Arbeitsstationen selbstverständlich ist: so ist es z.B. in solchen Arbeitsstationen ein Problem, die Ausgabe einer Anwendung auf Datei zu 'konservieren' (für späteres 'Replay') oder die Ausgabe einer Anwendung auf mehrere 'Displays' gleichzeitig zu lenken. Entsprechende Systemerweiterungen wurden im Nestor-'Development-Track' (s.o.) entwickelt.

3.3 Multimedia-Unterstützung

Während frühe Multimedia-Systeme im wesentlichen einen 'Computeranschluß' für Multimedia-Speicher- und -Display-Spezialperipherie schufen, setzt sich zunehmend die Einsicht durch, daß für den breiten Einsatz von Multimedia die Daten als digitalisierte, vom System 'standardmäßig' verwaltete und verarbeitbare Medien vorliegen müssen. Mit diesem Schritt taucht jedoch eine große Zahl neuer Probleme auf, von denen Nestor im 'Research-Track' folgende untersucht:

1. Die netzwerk-, lokations- und gerätetransparente Unterstützung von Multimedia-Objekten durch einen 'Präsentationsdienst', welcher Anforderungen an Multimedia-Objekte (z.B. Darstellung eines solchen Objektes an einer bestimmten Arbeitsstation) optimierend befriedigt aufgrund von statischer Information (derzeitiger Ort des Objektes, Information über Transportwege, Kompressions-Verfahren und -Stationen, etc.) und dynamischer Information (aktueller Netzzustand, prognostiziertes Anwendungsverhalten etc.) [1].
2. Synchronisationsmodell und Synchronisationsunterstützung zur Integration mehrerer Einzelmedien in Multimedia-Objekte (jeweils modelliert als Knoten)

Die Objektorientierung des OOHM-Ansatzes erleichtert hier wesentlich die Transparenz des 'Präsentationsdienstes', die Verwendung n-ärer Links ist eine wichtige Grundlage für das Synchronisationsmodell.

Auch im Bereich Multimedia-Unterstützung wird im Nestor Development Track ergänzend an besserer Systemunterstützung gearbeitet, vor allem um für die Bearbeitung und Darstellung zeitabhängiger Medien (Audio, Video) 'Standard'-Arbeitsstationen anstelle spezialisierter 'Multimedia-Arbeitsstationen' verwenden zu können (z.B. Dekompression von Video in Software mittels Standard-Prozessoren, Darstellung von Video mittels Standard-Graphiksubsystemen, Anschluß von Audio an Standard-Peripheriebus).

3.4 Prozeßunterstützung

Im Development Track wird Prozeßunterstützung für dedizierte Autorenprozesse (zum Begriff s.u.) aufbauend auf dem OOHM-Grundsystem implementiert.

Als grundlegende Hypermedia-Erweiterung zur Prozeßunterstützung wird im Research Track an der 'generischen Hypermedia-Navigation' gearbeitet. Der Begriff *Navigation* steht hierbei für das computerunterstützte 'Durchschreiten' eines Hypermedia-Dokumentes durch den Benutzer, also für eine spezifische Abfolge von Knoten-Darstellungen und Link-Übergängen [10]. 'Rechnergestützt' wiederum heißt in Nestor, daß zur Ausführungszeit 'Navigationsprogramme' (sog. PreScripts) den Benutzer nach jeder Knoten-Darstellung bei der Auswahl des nächstfolgenden Link-Überganges unterstützen. Zentral und neu ist hierbei der Ansatz zur *wiederverwendbaren, generischen* Navigation.

4 PROZESSUNTERSTÜTZUNG: GENERISCHE HYPERMEDIA-NAVIGATION

4.1 Prozeßunterstützung im Kontext der computerunterstützten Aus-/Weiterbildung

Die wichtigsten Arten von Prozessen, die im Zusammenhang mit 'Instructional Tool Environments' betrachtet werden müssen, seien wie erwähnt bezeichnet als *'instruktionelle Prozesse'* und *'Autorenprozesse'* (die Terminologie ist nicht einheitlich).

Instruktionelle Prozesse bestimmen die Aufbereitung und Organisation von Lehrstoff und bestimmen als Bestandteil computerunterstützter Unterrichts-Software ('courseware') die Vermittlung von Lehrstoff; sie werden bestimmt durch drei wesentliche CAI-Bereiche:

1. Übergreifende *Strategien* wie 'component display theory', 'progressive deepening' etc. [6], die den *Gesamtzusammenhang* zwischen Präsentationen, Tests etc. herstellen.
2. Sogenannte 'Lernermodelle', welche z.B. in 'intelligenten tutoriellen Systemen' (ITS) [5] mit KI-Methoden unterstützt werden. Es lassen sich unterscheiden:
   a. *Strategieabhängige Lernermodelle,* welche die o.g. Strategien mit dem Basiswissen und Lernfortschritt des Lernenden in Beziehung setzen und
   b. *Domänenabhängige Lernermodelle,* welche lehrstoffspezifisches Metawissen (richtig erlernte Konzepte, Fehlannahmen, ...) beim Lernenden diagnostizieren und 'lokale' Regeln, z.B. zur Beseitigung von Falschwissen, angeben.
3. Allgemeine lokale *Präsentationsregeln* (simplifizierte Beispiele: 'verlange spätestens alle 90 Sekunden eine Eingabe', 'benutze Rot für zentrale Leitthesen').

Aufgrund der (i.a. ohne Lernermodellierung arbeitenden) viel zu starren frühen Courseware-Entwicklungen (in den 60er Jahren) und aufgrund der trotz enormem Entwicklungsaufwand oft unvollkommenen ITS-Entwicklungen (ab ca. Mitte der 70er Jahre) sind computerunterstützte instruktionelle Prozesse heute etwas in Verruf geraten; vielfach wird geraten, durch Hypermedia-Einsatz lediglich eine gute Aufbereitung und Strukturierung des Lernstoffes zu gewährleisten und Navigations-Entscheidungen gänzlich dem Lernenden zu überlassen. Dem steht aber z.B. die Erfahrung des 'Verlorenseins' in großen Hypermedia-Dokumenten gegenüber ('getting lost in Hyperspace') sowie im Falle von Nestor die Chance, aufgrund der Wiederverwendbarkeit (s.u.) instruktionelle Prozesse zunehmend ausgefeilt unterstützen zu können. Hilfreich kann hier zudem eine *vorschlagende* (statt 'diktierende') Navigationskomponente sein.

Autorenprozesse bezeichnen - analog Softwareentwicklungsprozessen in Softwareproduktionsumgebungen - den Vorgang, der von der ersten Idee für eine Courseware-Entwicklung bis zu deren Einsatz führt, ja weitergehend bis zum Rückfluß von Erfahrungen und Änderungswünschen auf nachfolgende Versionen. Das OOHM-Konzept in Nestor erlaubt es dabei, die für einen Autorenprozeß (also einen Courseware-Entwicklungsprozeß) benutzten Werkzeuge (Storyboard-Editor, Hypermedia-Lernstoff-Editor, ...) und die im Prozeß entstehenden und verwendeten Entwicklungsdaten (Lernziel-Sammlung, Stoffgliederung, Projektplan, Courseware selbst, ...) allesamt als Hypermedia-Knoten aufzufassen und Beziehungen zwischen diesen (zeitliche und kausale Zusammenhänge, Erzeugung/Verwendung, Zugriffsberechtigungen) als Links darzustellen. Dann können Benutzer und Benutzergruppen (Autoren, Medienexperten, Projektleiter, Verbesserungen vorschlagende Lernende, ...) im Sinne von Navigationsunterstützung durch ein solches Hypermedia-Netz aus Werkzeugen und Dokumenten durchgeführt werden.

4.2 Anforderungen

Als wesentliche Anforderungen an das PreScript-Konzept ergeben sich:

- *Computerunterstützte Prozeßmodellierung* allgemein: instruktionelle Prozesse und Autorenprozesse (auch Softwareentwicklungsprozesse allgemein) liegen bis heute im wesentlichen in Büchern oder als Expertenwissen vor; höchstens konkrete Inkarnationen (Anwendung eines bestimmten instruktionellen Prozesses innerhalb einer domänenspezifischen ITS-Software, selten auch Unterstützung eines spezifischen Autorenprozesses) werden 'fest codiert' unterstützt. Gefordert ist ein formales Prozeßmodell (generische Navigationsvorschrift), das sowohl für die Mensch-Mensch-Kommunika- tion (anstelle von Büchern oder Expertenwissen) als auch für die Mensch-Maschine-Kommunikation eingesetzt werden kann.
- *Flexibilität* im Grad der Benutzerführung: es sollte z.B. möglich sein, abhängig vom

gewählten instruktionellen Prozeß die Navigation *nicht einzuschränken* (exploratives Lernen), *einzuschränken* oder gar *völlig zu determinieren* (letzteres wenn die Pfadauswahl, z.B. durch Simulation, ganz von einer instruktionellen Strategie bestimmt wird).

- *Wiederverwendbarkeit:* einerseits sollte z.B. derselbe instruktionelle Prozeß (als PreScript) auf viele Hypermedia-Netze anwendbar sein, andererseits sollten mehrere instruktionelle Prozesse alternativ auf dasselbe Hypermedia-Netz anwendbar sein; Hierarchisierung (ein PreScript ruft zur Navigation innerhalb eines Subnetzes ein anderes PreScript auf) sollte ebenfalls unterstützt werden.

4.3 Grundprinzip

PreScripts sind Navigationsvorschriften für eine 'Familie von Hypermedia-Dokumenten' (s. auch [3]). Diese 'Familie' wird beschrieben durch ein *generisches* Hypermedia-Netz. Dieses generische Hypermedia-Netz beschreibt in Anlehnung an die Theorie der *Graph-Grammatiken* [8], wie ein 'gültiges' Hypermedia-Dokument (Mitglied der 'Familie) *konstruiert* wird. Neben *Navigations*vorschriften gibt es also auch *Konstruktions*vorschriften. Folglich besteht ein PreScript aus zwei Teilen:

- dem *Konstruktionsteil,* der die Regeln zur Erstellung eines 'gültigen' Hypermedia-Netzes (d.h. eines Mitgliedes der 'Familie') auf Graph-Grammatik-Basis beschreibt,
- dem *Navigationsteil,* der die Navigation generisch für alle Netze der 'Familie' regelt.

Die Zuordnung eines PreScripts zu einem konkreten Hypermedia-Netz wird als *Mapping* bezeichnet. Z.B. wird durch das Mapping ein bestimmter instruktioneller Prozeß (ggf. mit einer Menge von Regeln, welche Lernermodellierung und lokale Präsentationsregeln festlegen) einem bestimmten Lehrstoff zugeordnet, es entsteht also spezifische Courseware. Zu beachten ist, daß nur dieser Mapping-Vorgang nicht-wiederverwendbar ist, während alle anderen Teile (PreScripts, d.h. instruktionelle Prozesse sowie konkretes Hypermedia-Netz, d.h. Lehrstoff) generisch und wiederverwendbar sind.

4.4 Konstruktion

Die Grundbausteine des Konstruktionsteils sind, wie für Grammatiken üblich, 'Terminale' und 'Nichtterminale' (in unserem Falle für Knoten; Links sind in PreScripts immer Terminale) sowie 'Option' und 'Iteration'. Der Konstruktionsteil besteht aus einer Menge von Ersetzungsvorschriften, also zulässigen 'Expansionen' nichtterminaler Knoten in detailliertere Subnetze, also in zusammenhängende Graphen aus Knoten (gekennzeichnet durch den Typ), Metaknoten (ebenfalls durch einen Typ gekennzeichnet) und Links (optional mit Typkennung versehen). Iteration und Option werden unterstützt durch

- Iterationen von Link/Zielknoten-Paaren 'seriell' (das Paar kann zu einer beliebig langen Sequenz von Links und Zielknoten des angegebenen Typs expandiert werden) oder 'parallel' (beliebig viele Links des angegebenen Typs, mit angehängtem Zielknoten, dürfen am *selben* angegebenen Startknoten beginnen).

- optionale Angabe von Minima, Maxima und Formalparameter für Zahl der Iterationen
- 'Option'-Vermerk '[0..1]' für Links und für Link/Knoten-Paare

Nichtterminale Knoten heißen Metaknoten; sie können

- im selben Konstruktionsteil in Ersetzungsvorschriften beschrieben sein,
- durch ein separates PreScript beschrieben werden,
- direkt durch ein kohärentes Hypermedia-Subnetz instanziiert werden (-> innerhalb dieses Subnetzes wird dann nur 'exploratives Lernen' ohne Navigation unterstützt)

4.5 Navigation

Die Navigationsvorschriften sind (in der derzeitigen Fassung von PreScripts) an die Theorie der *Zustandsübergangsmodelle* angelehnt. In der derzeitigen Version werden Übergangs-Tabellen (ähnlich denen in Automaten) verwendet, worin erlaubte Übergänge (Quellknoten→Link→Zielknoten) beschrieben sind durch folgende Tabelleneinträge: *Quellknotenselektor* (als Selektor kann dienen: ein Typ-Ausdruck, ein Instanz-Ausdruck oder eine allgemeine Regel), *Zielselektor* (entweder als Linkselektor oder als Zielknotenselektor), *Bedingung* (objektorientiert als Methode spezifiziert, dadurch beliebige Mächtigkeit) und *Aktion* (analog, bei Auswahl des Überganges auszuführen).

4.6 Mapping

Der Mapping-Prozeß kann im Prinzip auf verschiedene Weisen erfolgen:

- Das Hypermedia-Netz kann *neu* erstellt werden. Dann kann der Konstruktionsteil des PreScripts wie bei einem syntaxgestützten Editor direkt den Erstellungsvorgang treiben, die Zuordnung der Knoten-, Link- und Metaknotentypen zum konkreten OOHM-Netz ergibt sich automatisch bei diesem Vorgang.
- Ein existierendes OOHM-Netz kann zunächst manuell so erweitert werden, daß es dem Konstruktionsteil genügt (durch Hinzufügen von Knoten/Links oder durch Vererbung zusätzlicher Typen an existierende Knoten/Links). Die Zuordnung der Elemente des Konstruktionsteils zum konkreten Netz muß (in Nestor, auf absehbare Zeit) ebenfalls manuell erfolgen. Automatisches Zerteilen ('Parsing') des OOHM-Netzes soll erst zukünftig unterstützt werden; da die erwähnte Erweiterung existierender Netze i.a. ohnehin erfolgen muß, erscheint die manuelle Zuordnung zumutbar.
- In einer zukünftigen Version ist vorgesehen, daß Konstruktionsregeln teilweise erst nach Start des Navigationsvorganges angewendet werden, so daß auch die dynamische Manipulation von Knoten und Links (Erzeugung, Veränderung) während der Navigation durch PreScripts kontrolliert werden kann. Dies ist eine wesentliche Voraussetzung für die umfassende Unterstützung von Autorenprozessen (wo Entwicklungsdaten im Verlauf der Werkzeugbenutzung, also zur Navigationszeit, erzeugt werden).

4.7 Beispiel

Die nachfolgende Abbildung zeigt ein sehr einfaches beispielhaftes PreScript in graphischer Notation. Ellipsen stehen für Metaknoten, beschränkte Iterationen werden in der Form y:[u...o] gekennzeichnet, wobei *u* für das Minimum und *o* für das Maximum an Iterationen steht und *y* eine Variable kennzeichnet, die die tatsächliche Zahl (beim Mapping) aufnimmt. Wie erwähnt steht u=0, o=1 für 'optional'. Das Beispiel zeigt eine sehr primitive 'Familie' von Hypermedia-Netzen für *Tutorien:*

- Auf einen terminalen Knoten 'In' (*in*troduction) folgt eine Sequenz von *1+x* Metaknoten des Typs 'Short' (*short* topics); diese sollen eine kurze Einführung in jedes Kapitel des Tutoriums geben (*'x'* ≥ 0: es gibt mindestens *einen Short*-Metaknoten).
- Der letzte *Short* Metaknoten führt zu einem terminalen Knoten 'End' (Evaluation).
- Vom Knoten 'In' führen auch *y* Links zu Metaknoten des Typs *Main* (main topic). Von dort können optional Links zu jedem anderen Knoten führen (Der Linktyp 'pr' für 'prerequisite' soll anzeigen, daß der Quellknoten dann notwendiges Vorwissen für den Zielknoten enthält).

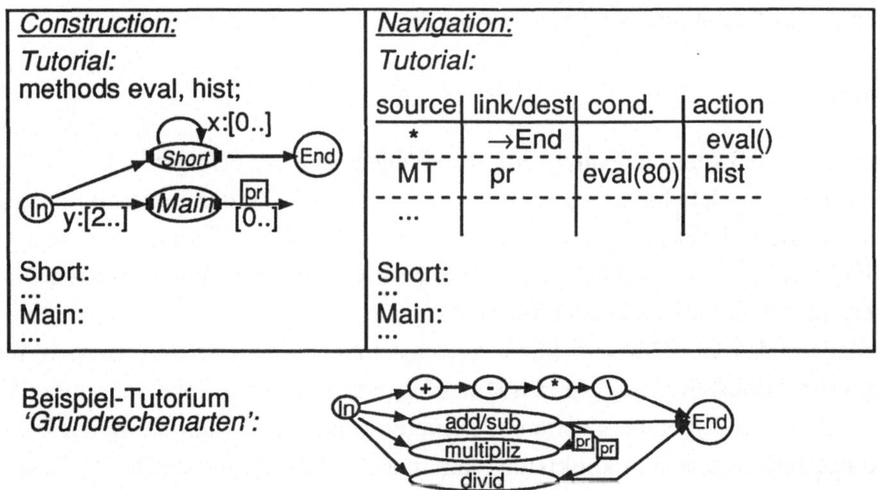

Beispiel-PreScript für Tutorien, Beispiel-Tutorium 'Grundrechenarten'

In einem vollständigen Beispiel würden dann Ersetzungsregeln für *Short* und *Main* folgen. Im Beispiel ist noch angedeutet, daß die beiden objektorientierten Methoden 'eval' und 'hist' definiert seien. In der Abbildung ist außerdem ein primitives Beispiel-Netz angedeutet: Ein Tutorium über die Grundrechenarten. Dort ist auch ersichtlich, daß das Netz zusätzliche, nicht im PreScript vorgesehene Links enthalten kann.

Was die *Navigation* betrifft, ist das Beispiel folgendermaßen zu interpretieren:

- von jedem Knoten aus (*) kann der Lerner, ohne weitere Bedingung, jederzeit zum

'End'-Konten navigieren (d.h. 'exit' ist jederzeit erlaubt), beim Übergang wird die Methode 'eval' ausgeführt.

- von jedem Metaknoten vom Typ *Main* aus darf der Lerner nur dann zu einem Knoten, der den derzeitigen als Vorwissen voraussetzt, wenn das Ergebnis der 'eval'-Methode mindestens bei 80 liegt (soll andeuten, daß der Stoff des aktuellen Knotens zu 80% beherrscht wird); die 'Lern-Historie' wird beim Übergang mittels 'hist' aktualisiert

5 ZUSAMMENFASSUNG

Der Weg zu kooperativen Multimedia-Anwendungen führt über eine adäquate Softwaretechnik. Ein leistungsfähiges Hypermedia-Konzept mit Unterstützung wiederverwendbarer Software kann dazu eine wesentliche Grundlage sein. Im Beitrag wurden als Vorschlag in diese Richtung die Hypermedia-Konzepte und -Techniken von Nestor angesprochen, der Navigationsansatz wurde etwas vertieft. Die vorgestellten Konzepte werden weiterentwickelt und miteinander integriert. Kontextabhängigkeit, umfassende Versionskontrolle und Unterstützung für manipulierende Navigation sind dabei die wichtigsten nächsten Schritte.

6 LITERATUR

1. Blakowski, G.: Supporting Multimedia Information Presentation in a Distributed, Heterogeneous Environment. Proc. 2nd IEEE WS Future Trends of Distributed Comp.Sys., Kairo Sept. 1990
2. Dürr, M., Lang, S.: Hypertext and Object-Orientation: The Dual Approach. Proc. BTW'91, Kaiserslautern, März 1991.
3. Furuta, R. et al: A Spectrum of Automatic Hypertext Constructions. Hypermedia 1 (1989), 179-195
4. Kraemer, K., King, J.: Computer-Based Systems f. Cooperative Work and Group Decision Making. ACM Comp. Survey 20 (1988), 115-146
5. Mandl, H., Lesgold, A.: Learning Issues for Intelligent Tutoring Systems. New York: Springer 1988
6. Reigeluth, C.M.: Instructional Theories in Action. Hillsdale, NJ: Lawrence Erlbaum Ass. 1987
7. Rüdebusch, T., Mühlhäuser, M.: Ein Unterstützungsssystem für Gruppenarbeit in verteilten Syst. IFB267, 464-478, Berlin: Springer 1991
8. Schürr, A. et al: Introduction to Progress, an Attribute Graph Grammar Based Spec. Lang. LNCS 411, Berlin: Springer 1989, 151-165
9. Smith, J.B., Weiss, S.F.: Hypertext. CACM 31 (1988), 816 - 819
10. Zellweger, P.T.: Scripted Documents: A HM Path Mechanism. Proc. Hypertext'89 ACM SIGCHI (1989), 1-14

HERMES – EIN HYPERMEDIASYSTEM FÜR DIE BETRIEBSWIRTSCHAFTLICHE AUSBILDUNG

Eric Schoop
Lehrstuhl für Betriebswirtschaftslehre und Wirtschaftsinformatik,
Prof. Dr. R. Thome, Universität Würzburg
Neubaustr. 66, W-8700 Würzburg

1 Zusammenfassung

Computergestützten Lehr-/Lernsystemen haftet der Ruch exorbitant hoher Erstellungskosten bei gleichzeitig zu geringer Anwenderflexibilität an. Das am Lehrstuhl für Betriebswirtschaftslehre und Wirtschaftsinformatik an der Universität Würzburg entwickelte Hypermediasystem ***HERMES*** ist der Versuch, mit vergleichsweise einfachen Mitteln in einer selbst konzipierten Entwicklungsumgebung in kurzer Zeit Wissensbausteine aus einem großen Fachgebiet unter Beteiligung vieler Autoren so aufzubereiten, daß eine sinnvolle Ergänzung klassischen Lehrmaterials für alternative Anwendungsmöglichkeiten (Informieren, Wiederholen, Vorbereiten, Lernen) entsteht. Vorrangige Erklärungsobjekte sind dabei dynamische Zusammenhänge, deren zeitliche Abläufe und gegenseitige Beeinflussungen durch den kombinierten Einsatz verschiedener Darstellungsformen (Text, Grafik, Standbild, Sprache, Animation, demnächst Video) besser beschrieben werden können, als dies in Lehrbüchern möglich wäre. Im Mittelpunkt der Systementwicklung steht der mündige Anwender, der am Bildschirm mit grafischer Oberfläche über die Maus jederzeit aktiv entscheiden kann, was er wann, wie oft, in welcher Reihenfolge und wie detailliert durchsehen möchte. Dies wird erreicht durch die permanente Verfügbarkeit alternativer Navigationskonzepte durch das Themengebiet, die von dem vom Autor eines Beitrags vorgegebenen, sequentiell zu verfolgenden Lernpfad ("guided tour") bis zur effizienten Nutzung grafischer Orientierungshilfen ("graphical browser") reichen, welche außerhalb des eigentlichen Stoffgebietes liegen und eine Vogelperspektive auf die Zusammenhänge erlauben. Technisches Grundkonzept ist der Aufbau einer multimedialen Datenbank und deren Organisation als Hypertext unter einer direkt manipulierbaren, grafischen Benutzeroberfläche. Eine erste Systemversion mit ca. 100 Megabyte multimedialer Informationen ist seit Juni 1991 auf CD ROM Datenträger verfügbar.

2 Ausgangssituation

2.1 Das BWL-Studium heute

Die Überlast in der universitären Ausbildung im Fach Betriebswirtschaftslehre verursacht seit Jahren Kapazitätsengpässe, die ein ausschließliches Beibehalten herkömmlicher Veranstaltungsformen (Vorlesungen, Seminare im Frontalunterricht) zunehmend in Frage stellen, soll die Qualität der Lehre aufgrund der Massensituation nicht bleibenden Schaden erleiden. Selbst an kleineren Fakultäten wie in Würzburg sind Seminargrößen von mehr als 100 Teilnehmern im Hauptstudium der Betriebswirtschaftslehre keine Seltenheit mehr, und Vorlesungen für das Grundstudium werden mittlerweile bei 800 und mehr Studienanfängern pro Semester in Ermangelung geeigneter, großer Hörsäle täglich zwischen 16 und 20 Uhr in der Mensa des naturwissenschaftlichen Campus abgehalten. Selbst bei größtem Bemühen auf Seiten der Lehrenden wie auch der Lernenden kann hier kaum mehr von akademischem Studium im Sinne selbständiger, kritischer Auseinandersetzung mit der angebotenen Thematik im Rahmen von Diskussionen, Gruppenarbeiten oder praktischen Übungen gesprochen werden. Es muß daher nach Alternativen der Vermittlung insbesondere von Basiswissen gesucht werden. Die hohen Studentenzahlen, die Knappheit an Räumen und Lehrpersonal sowie die immer schnellere Zunahme des geforderten Theorie- und Praxiswissens, insbesondere auch im Umgang mit modernen Informationstechnologien, legen es nahe, zentrale Veranstaltungen durch zeitlich und räumlich entkoppelte, dezentrale Wissensaufnahme an rechnergestützten Selbstlernarbeitsplätzen zu ergänzen, die auch außerhalb der regulären Veranstaltungszeiten zugänglich sind. Hier können die betrieblichen Führungskräfte von morgen bei entsprechender systemseitiger Unterstützung individuelle Lern- und Arbeitsstile üben, sich aktiv in interessierende Zusammenhänge einarbeiten und gleichzeitig die notwendigen Kenntnisse im Umgang mit der rechnergestützten Informationsverarbeitung erwerben, die sie nach dem Studium am Arbeitsplatz erwartet.

2.2 Erklärungsobjekte der Betriebswirtschaftslehre

„Gegenstand betriebswirtschaftlicher Forschung sind die Grundlagen, Abläufe und Auswirkungen menschlicher Entscheidungen in allen Funktionsbereichen und auf allen hierarchischen Ebenen einer Betriebswirtschaft" [HEIN80, Sp. 697]. Inhalt der betriebswirtschaftlichen Lehre, also Erklärungsobjekte, sind damit ganz unterschiedliche Problemtypen, die gemäß Bild 1 zusammengefaßt werden können [HART91, S. 50]. Für die Wissensvermittlung auf konventionelle Weise wie auch in computergestützten Lehr-/Lernprogrammen sollten sie so dargestellt werden, daß die Zusammenhänge leicht nachvollziehbar sind. Dies bedarf der Wahl jeweils besonders geeigneter Präsentationsformen.

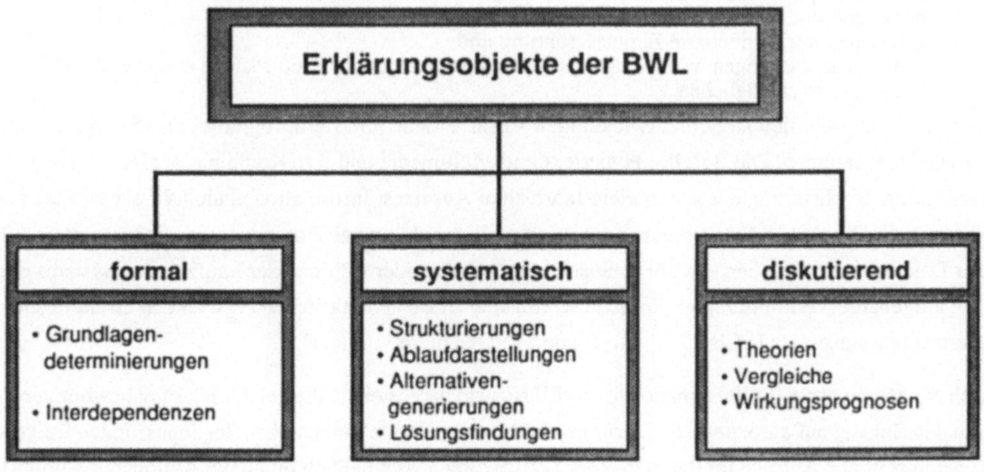

Bild 1: Erklärungsobjekte der BWL

2.3 Hypermedia: Neue Möglichkeiten der Wissensdarstellung

Die Präsentation von Informationen mit Hilfe des Computers kann prinzipiell visuell, auditiv oder audiovisuell erfolgen. In die erste Gruppe gehören die Darstellungsformen

- Linearer Text,
- Standbilder/-grafiken,
- Grafikfolgen (schrittweiser Aufbau komplexer Zusammenhänge),
- Kombinationsformen (Mischung von Grafiken mit ein-/ausblendbaren Zusatztexten).

Rein auditive Präsentationsformen sind Sprache und alle nichtsprachlichen akustischen Effekte (Geräusche, Musik, Signaltöne). Die audiovisuelle Darstellung von Informationen kombiniert die beiden anderen Gruppen und kommt damit der Realität, dem Unterricht durch einen Lehrer, der spricht, schreibt, skizziert und gestikuliert, am nächsten. Die medientechnische Umsetzung umfaßt die Bereiche

- Trickfilm (Animation),
- Dokumentarfilm und
- Spielfilm,

in der Regel auf Basis von Videosequenzen.

Aus pädagogischer Sicht kann es jedoch nicht genügen, Technik vorrangig als Lehrer-Ersatz und damit als Substitut eines deskriptiven, ausschließlich präsentierenden Frontalunterrichts einzusetzen. "Learning by doing" ist vielmehr die zentrale Zielvorstellung der Pädagogik. Das aktive Durcharbeiten von Fallbeispielen in der Gruppe oder Rollenspiele als Beispiele moderner Lehrkonzepte müssen bei Nachahmung mittels Film oder Computer in einem möglichst hohen Anteil interaktiver Systemkomponenten ihren Niederschlag finden. Vorlesungen, Seminare und praktische Übungen finden ihre technischen Pendants in Lehrfilmen, Lernprogrammen auf Basis konventioneller Autorensysteme und Simulationen im Sinne computergestützter Unternehmensplanspiele und Fallstudien [THOM91a, THOM91b].

In den letzten Jahren führen Weiterentwicklungen bei Hard- und Software wie

- direkt manipulierende Interaktionsmittel (Maus, Joystick oder Touchscreen),
- hochauflösende Bildschirme und entsprechende Rechenleistung am Lerner-Arbeitsplatz,
- grafische, objektorientierte Benutzerführung und
- Datenstrukturierungen, welche die Verwaltung unterschiedlicher Formate in einem einheitlichen System ermöglichen,

die drei sich ursprünglich isoliert entwickelnden Ansätze Lehrfilm, Lernprogramm und Simulation/Planspiel wieder zusammen. Die Begriffe Hypertext und Multimedia und ihre Kombinationsform Hypermedia stehen für die Realisierung des schon viele Jahre alten Ansatzes, Informationen nicht linear zu verwalten, sondern vielmehr als ein Netz voneinander prinzipiell unabhängiger Komponenten (nodes) unterschiedlicher Datenformate zu sehen, die über statisch vorgegebene oder während der Laufzeit des Systems dynamisch aufgebaute Verknüpfungen (links) untereinander in Beziehung stehen (vgl. zu den Grundlagen von Hypertext beispielsweise [SCHO91a] und die dort referenzierte Literatur).

Durch die Möglichkeit, Teilinformationen am Bildschirm darzustellen, die bei Aktivierung bestimmter Verweise den Zugang auf assoziierte Ergänzungen, alternative Darstellungen oder Gegenpositionen freigeben, und deren zeitliche Abfolge im Rahmen einer Lernsitzung ausschließlich durch die Aktionsreihenfolge des Benutzers, also seine Interessen und Prioritäten, bestimmt wird, werden enorme Systemflexibilität und Darstellungsvielfalt des gespeicherten Wissens erreicht. Die neuartige, selbst zu beeinflussende Umgebung fördert, eine entsprechende „Aufmachung" des Systems durch Design, Dialogformen und abwechslungsreiche, themenorientierte Präsentation vorausgesetzt, die intrinsische Motivation des Lernenden [ADER90, S. 236]. Damit kann auch eine grundsätzlich positive Veränderung des Lernverhaltens im Sinne höherer Selbstbestimmung, aktiverer Wissensaufnahme und eigener Beteiligung durch Kommentieren einhergehen [JONA90].

Doch stehen diesen prinzipiellen Vorteilen von Hypermedia-Ansätzen für die Nutzung als Lernsysteme auch Nachteile gegenüber, die sich neben den allgemeinen Kritikpunkten an Hypertextkonzepten ("cognitive overhead" und "disorientation" [CONK87]) insbesondere auf das Fehlen von Leitfäden im Sinne einer Vorstrukturierung des Stoffes durch den Lehrer/Autor sowie eine begleitende Lernerfolgskontrolle und gegebenenfalls adaptive Rückkopplungen konzentrieren (vgl. [SCHO91b]).

3 Projekt

Vor obengenanntem Hintergrund war es eine der wesentlichen selbstgestellten Aufgaben des nachfolgend beschriebenen Projektes ***HERMES***, die beiden gegensätzlichen Aspekte des Hypertext- und des Lernprogrammkonzeptes in einem gemeinsamen Lehr-/Lernsystem über betriebswirtschaftliche Fragestellungen zu bündeln, um den Hauptforderungen einfacher, intuitiver Benutzbarkeit und hoher Funktionalität unter einer ansprechenden Oberfläche sowohl bei Erstbenutzern wie auch bei fortgeschrittenen Anwendern gerecht zu werden und damit eine gute Akzeptanz und einen erfolgreichen Einsatz des Systems zu erreichen.

3.1 Rahmen

Die ***HERMES*** zugrundeliegende Idee manifestierte sich im August 1989, als für das 11. internationale Forum der EAIR (European Association for Institutional Research) in Trier unter dem Namen „BWL Info" eine Demonstrationsanwendung eines hypermediagestützten Informationssystems über betriebswirtschaftliche Fragestellungen entwickelt und vorgestellt wurde [SCHO89, S. 168, SCHO90, S. 261]. Vor dem Hintergrund der sich stetig verschlechternden Studiensituation im Fach Betriebswirtschaftslehre erfuhr das System in den darauffolgenden Monaten im Rahmen eines explorativen Prototypings hinsichtlich Design, Dialoggestaltung und Funktionalität umfangreiche Modifikationen und mehrere Neuaufwürfe, bis im Sommer 1990 der Rahmen für die derzeitige Version fixiert wurde. Daran schlossen sich die Entwicklung von Hilfswerkzeugen und die eigentliche inhaltliche Ausarbeitung.der ersten Beiträge, die seit Juni 1991 auf CD ROM Datenträger erhältlich sind. Jährliche Neuauflagen im Sinne optischer, funktionaler und inhaltlicher Verbesserungen sowie thematischer Erweiterungen sind geplant.

HERMES wird ausschließlich am Lehrstuhl für Betriebswirtschaftslehre und Wirtschaftsinformatik der Universität Würzburg entwickelt, ist als wissenschaftliches Forschungs- und Anwendungssystem ausgelegt und wird vollständig aus Eigenmitteln finanziert.

3.2 Gestaltung

Der angestrebte Umfang des Systems sowie die Struktur des Projektes – geringe Mittel, kleines Team, hohe Grundauslastung durch die Lehre – ließen von Anfang an nur ein Gemeinschaftsprojekt vieler Autoren unter Federführung des Lehrstuhls bei strenger zeitlicher Strukturierung gemäß einem in Bild 2 dargestellten Phasenschema zu. Entsprechend großen Anteil nahmen während der gesamten Laufzeit rein organisatorische Aufgaben ein wie die Vermittlung pädagogischer Grundlagen, die Festlegung einheitlicher Interaktions- und Darstellungsprinzipien, die thematische Betreuung, die Strukturierung der Beiträge und deren Verknüpfungen untereinander.

3.2.1 Vorarbeiten

Die Autoren nehmen neben den inhaltlichen Ausarbeitungen auch deren Umsetzung in multimediale Hypertexte selbst vor. In den ersten Projektmonaten wurden zahlreiche Hilfswerkzeuge sowie die Beschreibung der Vorgehensweise weitgehend parallel zu den ersten inhaltlichen Ausarbeitungen erstellt. Zur Angleichung unterschiedlicher Kenntnisstände der Autoren im Umgang mit Computer und Programmen erfolgte zunächst ein einwöchiger Blockkurs, der anschließend während der gesamten Projektlaufzeit durch begleitende Tutorien und Nachschulungen ergänzt wurde.

3.2.2 Erstellung der Beiträge

Das Erstellen inhaltlicher Ausarbeitungen kann grob in die zwei Bereiche Deskription und Simulation aufgeteilt werden. Die deskriptiven Arbeiten zur näheren Themenbeschreibung werden vorgenommen mit Hilfe des Werkzeuges ***HERMES***-Autor , das ein systemgestütztes Anlegen und Verwalten von Verknüpfungen, den Aufbau graphischer Übersichten im Sinne eines hierarchischen Browsers, sowie das Anlegen und Verwalten mehrerer Lernpfade ermöglicht, ohne dem Autor hierfür Programmierleistung abzuverlangen. Auch die Einbindung externer Dokumente wie Sprache, Animationen, Geschäftsgraphiken, künftig auch Videoausschnitte, ist auf sehr einfache Weise möglich. Gleichzeitig wird eine systemeinheitliche Oberfläche vorgegeben. Das Stichwort Simulation steht für das Projektziel, nicht nur ein Thema in Form multimedialer, hoch vernetzter Informationsstrukturen zu erarbeiten, sondern darüber hinaus Lernenden auch die Möglichkeit zu geben, in jedem Bereich anhand von einfachen Fallstudien oder Planspielen interaktiv Zusammenhänge erarbeiten zu können. Während die ersten Beiträge überwiegend deskriptiv ausgelegt sind, befassen sich laufende Ausarbeitungen verstärkt mit ihrer Ergänzung um integrierte Fallstudien.

3.2.3 Redaktionelle Überarbeitung

Während der Arbeitsschwerpunkt der Autoren im Anlegen der lokalen, einzelthemenbezogenen Informationsknoten sowie im Aufbau der innerthematischen Beziehungen liegt, werden in der anschließenden redaktionellen Überarbeitung die Beiträge untereinander verknüpft (interthematische Bezüge). Neben ausführlichen Tests werden die Beiträge um Fragen zur Lernstoffkontrolle ergänzt, umfangreiche Konsistenzprüfungen vorgenommen, gegebenenfalls zwischenzeitliche Änderungen der Benutzeroberfläche nachgezogen und anwenderorientierte Funktionen, die in der Entwicklungsumgebung nicht zur Verfügung stehen, implementiert. Die Redaktionsphase soll gemäß Projektplan zweimal jährlich erfolgen, umspannt jeweils mehrere Wochen und bildet die Sammelstelle für das Zusammenfassen aller in der Zwischenzeit fertiggestellten einzelnen Beiträge. Ergebnis der Überarbeitungen ist die jeweils nächste Systemversion.

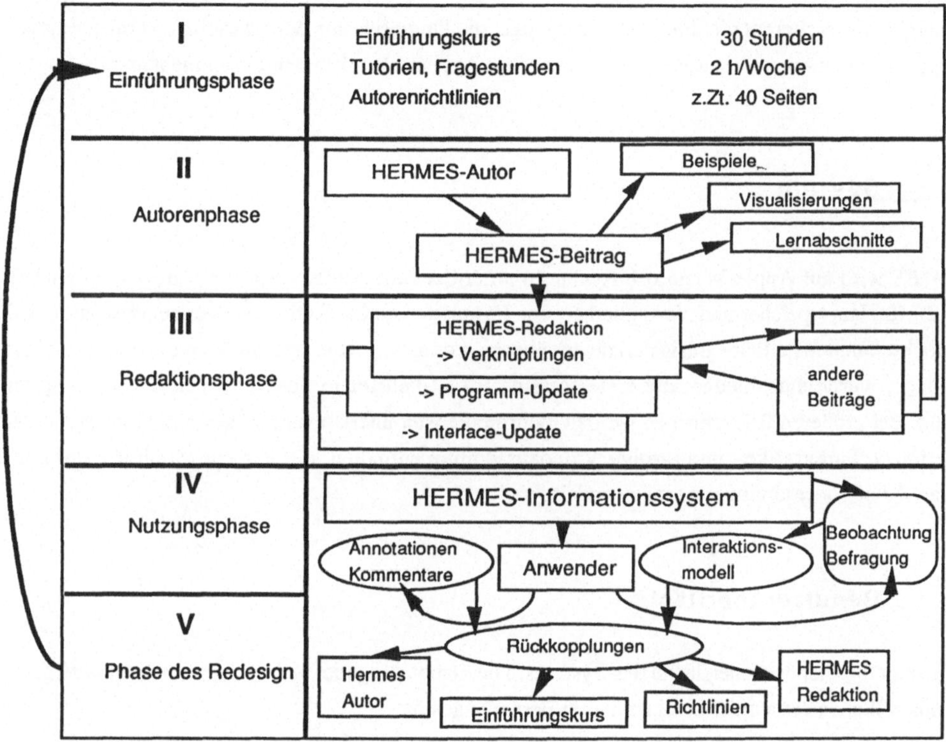

Bild 2: ***HERMES***-Projektphasen

3.2.4 Anwendungseinsatz, Untersuchungen und Redesign

Die erste ***HERMES***-Version ist den Studenten der Betriebswirtschaftslehre an der Universität Würzburg ab Juni 1991 an etwa 70 Stunden in der Woche frei zugänglich. In den nächsten Monaten sollen Funktionalität, Akzeptanz und Erfolg des Systems analysiert werden. Im Mittelpunkt stehen die Aufgaben,

- über Fragebogen und mit Hilfe der im System integrierten Anmerkungsfunktionen subjektive Benutzerangaben zu erfassen und auszuwerten,
- durch persönliche Beobachtung konkretes Benutzerverhalten bei der Lösung von Aufgaben festzustellen und

- mittels systemgesteuerter Routinen allgemeines Benutzerverhalten für spätere Auswertungen zu protokollieren.

Neben persönlichen Einschätzungen des Systems soll insbesondere ermittelt werden, wie die vorhandenen Navigationsalternativen in der Praxis genutzt werden, ob Anwender beispielsweise im Zuge steigender Benutzerkenntnisse ihr persönliches Zugriffsverhalten auf schon bekannte oder neu zu suchende Informationen ändern oder systematisches, auf Oberfläche beziehungsweise Interaktionsmodell zurückzuführendes Fehlverhalten feststellbar ist. Die Untersuchungsergebnisse werden im Sinne des Prototyping als Rückkopplung in die Weiterentwicklung der Werkzeuge, die Durchführung der tutoriellen Veranstaltungen und die Ergänzung der Autorenrichtlinien einfließen. Bild 2 veranschaulicht den Zusammenhang.

4 System

HERMES wird auf Apple Macintosh Rechnern unter der Basis-Software HyperCard 2.0 entwickelt und ist ab 2 MB Hauptspeicher und Verfügbarkeit von Festplatte und CD ROM Laufwerk einsetzbar. Um die an den Hochschulen installierte Basis gerade der älteren, kompakten Systeme mit kleinem Monitor nicht auszugrenzen, wurde entschieden, die erste Version im 9"-Fenster in Schwarz/Weiß-Darstellung zu entwickeln. Bei größeren Bildschirmen werden mehrere Fenster überlappend dargestellt. Künftige Versionen werden auch Farbgrafiken und farbige Videoausschnitte enthalten, die je nach Monitor automatisch in Schwarz/Weiß- oder Grauwerte umgewandelt werden.

4.1 Benutzeroberfläche

Bild 3 zeigt die Standardoberfläche des Systems. Die Hauptnavigationsleiste liegt am rechten Fensterrand, wird durch Ikonen verdeutlicht und ist in drei Bereiche geteilt:

- Zurück in die globale Inhaltsübersicht bzw. zum letzten Informationsknoten,
- Hierarchieübersicht mit der Richtung zum Oberkapitel bzw. zu den Unterkapiteln,
- Lernpfad als vom Autor vorgeschlagene Abarbbeitungsreihenfolge in einem von drei einstellbaren Schwierigkeitsgraden mit den Alternativen „rückwärts" und „vorwärts".

Am unteren Bildschirmrand befindet sich eine Menüleiste mit weiterführenden Funktionen. Diese sind – ausgehend vom zentralen Navigationselement Rechtspfeil – von rechts nach links mit abnehmender Zugriffswahrscheinlichkeit angeordnet. Die Taste Beispiel verweist auf dynamische Darstellungen, die den aktuellen Kontext durch alternative Präsentation vertiefen, oder Fallbeispiele. Test führt zu Kontrollfragen, die den Lernstoff des Kapitels überprüfen. Die Taste daneben zeigt die kapitelbezogenen Literatur-Hinweise. Die weiteren Funktionen der Menüleiste erlauben bei Selektion den Zugriff auf entsprechende Unterverzeichnisse. Sie beinhalten im einzelnen:

- <u>Hilfe</u> mit Benutzungshinweisen, Standortauskünften und Navigationsvorschlägen,
- <u>Suche</u> über Inhaltsverzeichnisse, Thesaurus (Schlagworte) und im Volltext (Stichworte),
- <u>Liste</u> zur Zusammenstellung und Speicherung individueller Memory-Listen, mit deren Unterstützung der Anwender seine eigenen Kurzpfade durch das System erstellt und abarbeitet,
- <u>Notiz</u> zum Anlegen privater oder öffentlicher Anmerkungen, also Randnotizen, Ergänzungen oder Kritik an der aktuellen Aussage (ein Hilfsmittel für das gemeinsame Erarbeiten von Stoffzusammenhängen in kleinen Lernergruppen = "collaborative learning"),
- <u>Druck</u> der verschiedenen Verzeichnisse, Kapitel oder Bildschirmseiten,
- <u>Modus</u> zum Einstellen eines Lernpfades (kurzer Überblick, allgemeine Darstellung oder vertiefende Themenausführung) und
- <u>Info</u> mit Auskünften über Projekt, System und Autoren.

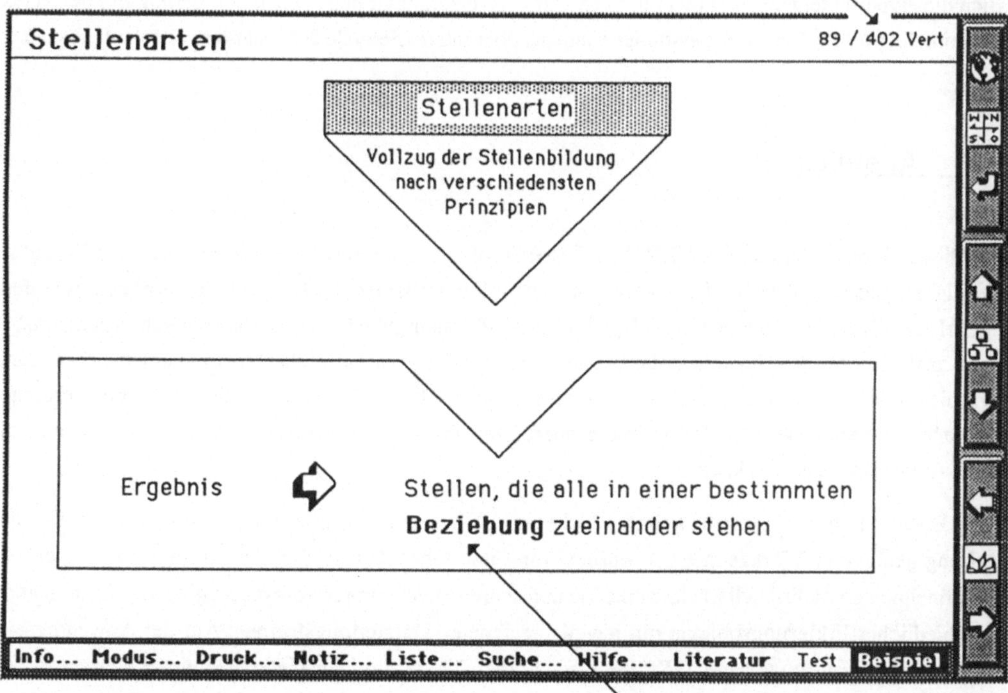

Bild 3: ***HERMES*-Oberfläche**

4.2 Orientierungs- und Navigationsalternativen

In prinzipiell schwach strukturierten Informationsbeständen, wie sie typisch sind für Hypermediasysteme, ist es notwendig, spezielle Orientierungshilfen anzubieten. In ***HERMES*** zeigt die in der oberen rechten

Ecke des Fensters sichtbare Anzeige die aktuelle Position im eingestellten – linearen – Lernpfad an (vgl. Bild 3). Eine hierarchische Orientierungs- und Navigationsmöglichkeit mit Standortanzeige bietet der jederzeit über die vertikalen Pfeiltasten erreichbare grafische Browser als kapitelbezogener Hierarchiebaum. Es wird jeweils die aktuelle Hierarchiestufe angezeigt, und über den Objekten zugeordnete weitere Verweiselemente können die Ebenen beliebig gewechselt werden. Jedes als grafisches Objekt repräsentierte Kapitel kann durch Aktivieren mit der Maus erreicht werden. Diese beiden Standardhilfen des Systems werden durch von den Autoren zusätzlich eingebrachte grafische Strukturdarstellungen der Beiträge mit Anzeige der schon abgearbeiteten Kapitel ergänzt.

Lernen kann, insbesondere im fortgeschrittenen Fall, als repetitiver Vorgang einer gezielten Informationsaufnahme aufgefaßt werden. Neben der Verfügbarkeit konventioneller Such- und Zugriffsmethoden des Information Retrieval (Schlagwortsuche mit bool'scher Verknüpfung der Deskriptoren, Volltextsuche und Zugriff über Inhaltsverzeichnisse) und der Nutzung eines umfangreichen Glossars bietet das System zusätzlich die Hypertexteigenschaft einer aktiven Verweisverfolgung durch Objektaktivierung mit der Maus, ergänzt um die Möglichkeit, sich persönliche Indices über interessierende Informationsseiten aufzubauen.

5 Ausblick

Die vorliegende erste Version des ***HERMES***-Systems soll nicht nur kurzfristig die beschriebene Engpaßsituation in der betriebswirtschaftlichen Ausbildung an bundesdeutschen Hochschulen verbessern helfen. Längerfristig soll an dem Umsetzungsbeispiel der Lehr-/Lernaufgabe in der Betriebswirtschaftslehre insbesondere analysiert werden, inwieweit die Organisation von Informationen als Hypertext unter Verwendung unterschiedlicher Medien und deren Präsentation in einer leicht handhabbaren, grafischen Benutzerumgebung die bisher noch zu geringe Akzeptanz rechnergestützter Informationssysteme gerade auch bei eher ungeübten Anwendern steigern kann.

Bei der Entwicklung des Systems konnten bisher schon Erfahrungen gewonnen werden hinsichtlich der Erstellung geeigneter Werkzeuge, der Aufbereitung von Oberflächen, der Vermittlung von Hypertext-Grundprinzipien an traditionell linear denkende und schreibende Autoren sowie bezüglich der Kombination unterschiedlicher Erklärungsobjekte mit jeweils geeigneten Darstellungsformen. Auf der Anwenderseite stehen größere Erfahrungen mit ***HERMES*** noch aus. Diesbezügliche Untersuchungen werden in den nächsten Monaten vorrangig betrieben, um möglichst schnell Erkenntnisse für die geplanten Weiterentwicklungen zu gewinnen und schon in die nächste Version im Frühjahr 1992 wieder einfließen zu lassen. Aufgrund des momentan beobachtbaren technologischen Fortschritts im Bereich der Datenkompression ist außerdem geplant, als weiteres Darstellungsmedium Videoausschnitte in digitalisierter Form aufzunehmen. Funktionale Verbesserungen betreffen in besonderem Maße den bisher noch nicht genügend berücksichtigten Bereich der Lernstoffkontrolle. Hier wird ein erfolgsorientiertes, lerneradaptives Systemverhalten hinsichtlich Fragenauswahl und Erklärungskomponenten angestrebt. Auf Basis von Prolog in Verbindung mit Hypercard finden derzeit erste Prototypen-Entwicklungen statt.

6 Literatur

[ADER90] Aders, A./Ansel, B.:
Hypertext für den Unterricht – Eine kritische Standortbestimmung, S. 235-248 in: Gloor, P./Streitz, N. (Hrsg.): Hypertext und Hypermedia: Von theoretischen Konzepten zur praktischen Anwendung, Springer, Berlin-Heidelberg 1990.

[CONK87] Conklin, J.:
Hypertext: An Introduction and Survey, S. 17-41 in IEEE Computer, Bd. 20, September 1987.

[HEIN80] Heinen, E.:
Betriebswirtschaftslehre, Sp. 697-703 in: Sellien, R./Sellien, H. (Hrsg.): Gablers Wirtschafts-Lexikon, 10., neu bearbeitete Auflage, Gabler, Wiesbaden 1980.

[HART91] Hartenstein, H.:
Darstellungsformen von Erklärungsobjekten der BWL für das Hypermedia-Informationssystem ***HERMES***, Diplomarbeit am Lehrstuhl für BWL und Wirtschaftsinformatik der Universität Würzburg, 1991.

[JONA90] Jonassen, D.H./Grabinger, S.R.:
Problems and Issues in Designing Hypertext/Hypermedia for Learning, S. 3-25 in: Jonassen, D.H./Mandl, H. (Hrsg.): Designing Hypermedia for Learning, NATO ASI Series, Series F, Vol. 67, Springer, Berlin-Heidelberg 1990.

[SCHO89] Schoop, E.:
Der Einsatz multimedialer Informationssysteme für die betriebswirtschaftliche Aus- und Weiterbildung, S. 161-169 in: Roithmayer, F. (Hrsg.): Der Computer als Instrument der Forschung und Lehre in den Sozial- und Wirtschaftswissenschaften, Schriftenreihe der Österreichischen Computer Gesellschaft, Bd. 50, Oldenbourg, Wien-München 1989.

[SCHO90] Schoop, E.:
HERMES – Botschafter eines neuen Ausbildungskonzeptes für die Betriebswirtschaftslehre, S. 258-262 in: Gloor, P./Streitz, N. (Hrsg.): Hypertext und Hypermedia: Von theoretischen Konzepten zur praktischen Anwendung, Springer, Berlin-Heidelberg 1990.

[SCHO91a] Schoop, E.:
Hypertext: Organisation schlecht strukturierbarer Information, S. 20 - 25 in technologie & management 1/91.

[SCHO91b] Schoop, E.:
Entwicklung einer Hypermedia-Autorenumgebung für die Erstellung interaktiver Lehr-/Lernsysteme, in: Wallmannsberger, J. (Hrsg.): Hypertext – State of the Art, Oldenbourg, Wien - München 1991 (in Vorbereitung).

[THOM91a] Thome, R.:
Hypermedia als Basis für Selbstlernsysteme, S. 20-23 in technologie & management 2/91.

[THOM91B] Thome, R.:
Hypermedia: Lehrer Lämpels Nachfolger, in Wirtschaftsinformatik 3/91 (in Vorbereitung).

Elektronisches Publizieren von Normdokumenten mit SGML

B.W. Alheit, F. Häfemeier, W. Hübner, H.H. Rath

Zentrum für Graphische Datenverarbeitung e.V. (ZGDV)
W-6100 Darmstadt, Wilhelminenstraße 7
Tel.: 06151 / 155 231
FAX: 06151 / 155 299

Abstract

Diese Arbeit befaßt sich mit der elektronischen Handhabung und dem Austausch von technischen Dokumenten, insbesondere von technischen Regelwerken (Normen). Grundlegende Aufgaben im Anwendungsgebiet Technische Dokumentation sind dabei die systemunabhängige Erstellung strukturierter Dokumente, die Festlegung und Einhaltung einer einheitlichen Dokumentstruktur, der elektronische Austausch, die dezentrale Weiterverarbeitung und Aktualisierung, Archivierung und Retrieval sowie Hypertext-Konzepte. Nach einer Einführung in die Problematik des Normenwesens werden die verschiedenen Austauschformate (insbesondere die Standard Generalized Markup Language SGML) für technische Dokumente vorgestellt und das entwickelte Werkzeug INES zur elektronischen Handhabung von Normdokumenten beschrieben.

1 Einführung

1.1 Handhabung von technischen Regelwerken

Technische Regelwerke und Normen sind derzeit sicher die am weitesten verbreitete Form des technischen Dokuments. In den Betriebsabläufen der Unternehmen bilden Technische Normen und Regelwerke in vielfältiger Weise die Grundlage für fast alle Produkte bzw. Dienstleistungen. Zur Zeit sind in Deutschland ca. 28.000 Normen, Norm-Entwürfe und Regelwerke in der Nutzung. Durch Neuausgaben und Änderungen werden hiervon jährlich 3.000 bis 4.000 neu herausgegeben. Dies ist eine erhebliche Belastung für die Normennutzer und bedeutet gerade in größeren Firmen, daß kaum ein aktueller konsistenter Stand in allen Abteilungen und bei allen Zulieferern sichergestellt werden kann.

Das Deutsche Institut für Normung (DIN) und der Verband Deutscher Elektrotechniker (VDE) verbreiten ihre Standards derzeit ausschließlich in Papierform. Gleiches gilt für Firmen- und Werknormen.

Künftig sollen Normdokumente elektronisch erstellt und publiziert werden, in den Firmen in elektronischer Form zur Verfügung stehen und zu Werknormen weiterverarbeitet werden können. Bereits auf internationaler Ebene sollen von der Internationalen Organisation für Standardisierung (ISO) Normdokumente elektronisch zur Verfügung gestellt werden, die dann zu den nationalen Instituten übertragen und dort weiterverarbeitet werden (Abbildung 1).

Das Zentrum für Graphische Datenverarbeitung e.V. (ZGDV) wurde beauftragt, die konzeptionellen und technischen Voraussetzungen hierfür auf nationaler Ebene zu schaffen.

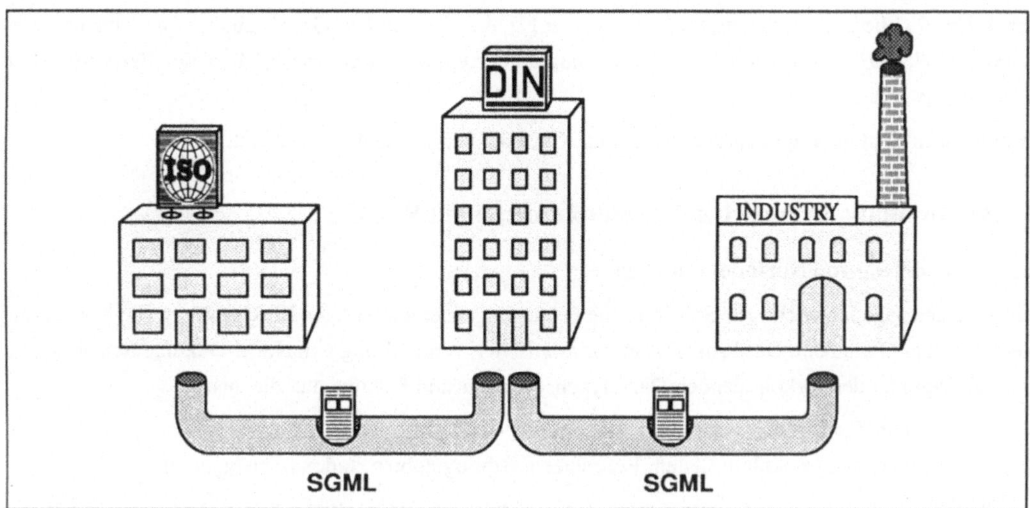

Abbildung 1: Internationaler Normenaustausch

Die wichtigsten Aufgaben sind dabei,

(1) die Struktur dieser Normdokumente in einem systemunabhängigen Format festzulegen und
(2) geeignete Werkzeuge (Editoren, Filter) zu entwickeln,

sodaß die elektronische Publikation dieser technischen Normen gewährleistet ist [16].

Als Beschreibungs- und Austauschformate wurden weit verbreitete Standards herangezogen:

- SGML [14] für Textelemente und die Gesamtstruktur des Dokuments
- CGM [15] für Graphiken
- CCITT FAX Group 4 [10] für Rasterbilder

Die Möglichkeiten und Grenzen von SGML (Standard Generalized Markup Language) wurden in diesem Projekt erkennbar.

Als Ergebnis liegt für DIN- und VDE-Normen eine SGML-Dokumenttypdefinition vor [8] [19]. Sie wird derzeit in einem eigenen Normvorhaben in einen DIN-Standard für den Austausch von Normdokumenten überführt [9].

1.2 Normen-Editor INES

Als zweiter Schritt wurde auf der Basis des verfügbaren Editors „Technical Publishing System TPS" das Werkzeug INES (Integrierter Normen-Editor auf der Basis von SGML) entwickelt, mit dem Normdokumente WYSIWYG-gerecht erstellt werden können. Durch INES ist sichergestellt, daß das erstellte Dokument der vorgegebenen SGML-Dokumenttypdefinition entspricht und damit DIN-Norm-gerecht ist. Durch entsprechende Filter ist eine Überführung von bzw. nach SGML-Notation und damit der standardisierte Austausch möglich.

Seit März 1991 liegt ebenfalls der INES-Editor vor [2], den die Interleaf GmbH übernommen hat und vermarktet. DIN, VDE und eine Reihe von Industrieunternehmen, wie AEG, Bosch, Mercedes-Benz, SEL oder Siemens setzen diese Möglichkeiten ein.

Im folgenden wird über die zugrundeliegenden Konzepte und Ergebnisse berichtet.

2 Beschreibung technischer Normdokumente mit SGML

2.1 Bestandteile von Normdokumenten

Der erste und zugleich wichtigste Schritt ist die Festlegung einer eindeutigen Beschreibung von Normen als elektronisches Dokument. Das Problem der elektronischen Verarbeitung von Norm-Dokumenten besteht in der Heterogenität der vorkommenden Datentypen. Normen sind Dokumente, die aus

- strukturiertem Text,
- Inhaltsverzeichnis, Deckblatt, Listen, Referenzen, Bibliographien, Index, Anhängen etc.,
- Tabellen,
- mathematischen und chemischen Formeln,
- Liniengraphiken,
- Rastergraphiken

bestehen.

2.2 Mögliche Austauschformate

Die Integration heterogener Datentypen zu Normdokumenten und deren elektronischer Austausch verlangt den Einsatz von standardisierten Beschreibungs- und Austauschformaten. Die Definition muß alle in Normen möglichen Inhalte abdecken, das Dokument eindeutig als Norm kennzeichnen, den Austausch auf elektronischem Wege gestatten und gleichzeitig die Weiterverarbeitung der Norm und ihres Inhalts nicht einschränken. Das Beschreibungsformat sollte insbesondere die Möglichkeit der EDV-Unterstützung in allen Phasen eines Normenlebenszyklus, von der Ausschußarbeit über den Vertrieb bis zum Austausch zwischen Normenanwendern (Werknormen) fördern.

Aus diesen Anforderungen ergibt sich, daß die Normen beim Empfänger in bearbeitbarer Form vorliegen müssen. Damit scheiden Formate, welche die Layoutinformation, aber nicht mehr die logische Struktur enthalten (z. B. Seitenbeschreibungssprachen, Pixeldaten), als Austauschformat aus. Im Dokumentenbereich sind im heutigen Normungswerk zwei Alternativen zu finden, die normierte Formate beschreiben:

- ODA/ODIF (Office Document Architecture / Office Document Interchange Format [17] [18]) beinhaltet sowohl Struktur als auch Layoutbeschreibungen.
- SGML (Standard Generalized Markup Language [14]) ist ein Standard zur Definition von Dokumenstrukturen, der den Bereich des Layouts nicht abdeckt.

Bei der Analyse der Anforderungen im Normenbereich wurde festgestellt, daß die damals existierende Fassung von ODA/ODIF nicht die vollständige Funktionalität bietet, die benötigt wird, um das Dokument „Norm“ abzubilden. Insbesondere in den Bereichen Tabellen, Formeln und Graphiken bestanden noch Lükken, die erst durch laufende Aktivitäten vervollständigt werden.

SGML bietet dagegen zwar alle Möglichkeiten, die für die Normenbearbeitung notwendig sind, erfordert aber im Layout-Bereich zusätzliche Absprachen/Konventionen zwischen den Benutzern. In der definierten

Zielgruppe Normenwesen ist diese Festlegung getroffen. Die DIN 820 [7] legt fest, wie Normen bzgl. Layout und Struktur auszusehen haben. Eine zukünftige Einbindung des SGML-Formates in den ODA/ODIF-Rahmen ist möglich.

2.3 Austauschformat SGML

Die Kodierung von Dokumenten in SGML [4] [11] erfolgt im Klartext-Format, so daß beim elektronischen Datenaustausch keine Probleme auftreten können. Auch die nationalen Zeichensätze (z. B. deutsch, französisch, skandinavisch) sind in einer Standardform darstellbar.

Der SGML-Standard erlaubt es, praktisch beliebig komplexe logische Dokumentstrukturen zu definieren; was mit Hilfe sogenannter Tags, z. B. <Kapitel>, <Absatz> oder <Fussnote>, die in den Text eingefügt werden können, um die einzelnen Elemente voneinander zu trennen, geschieht.

Über eine Grammatik wird der Zusammenhang der Elemente (die Struktur) definiert, z. B. daß Absätze nur in Kapiteln vorkommen können. Diese Grammatik nennt man Dokumenttypdefinition (DTD).

Für DIN-Normen ist eine DTD „DIN-Norm" festgelegt worden. Sie enthält eine komplette Festlegung der logischen Struktur der Normen, basierend auf den Regeln nach DIN 820 (Gestaltung von Normen, abgedruckt in [7]). Die DTD umfaßt die Typen Normen, Vornormen, Norm-Entwürfe, Beiblätter, Norm-Entwürfe zu Beiblättern sowie deren Verwendung als Werknorm. Um die Austauschbarkeit auch international sicherzustellen, wurde frühzeitig eine Harmonisierung der Arbeiten zwischen DIN und ISO angestrebt.

Zu jeder DTD benötigt man zusätzlich Informationen, wie das Layout der Elemente aussieht. Dies soll in Zukunft über den ISO-Standard „Document Style Semantics and Specification Language (DSSSL)", der zur Zeit als DIS [13] vorliegt, spezifiziert werden können. Daher obliegt es momentan noch jeder Nutzergruppe, die sich auf eine DTD für ihre Dokumente geeinigt hat, auch eine Layout-Festlegung zu treffen und auf ihren Systemen speziell zu realisieren. So wurden zum Beispiel im Rahmen des Deutschen Forschungsnetzes (DFN) Layout-Definitionen für mehrere DTDs vorgenommen und für die gebräuchlichsten Formatierer wie TeX, troff und DCF realisiert.

Für die Dokumenttypdefinition „DIN-Norm" gibt es bzgl. Layout eine einheitliche Vorgabe durch die DIN 820 [7]. Diese DIN-Norm legt fest, wie Normdokumente auf Papier formatiert sein müssen.

2.4 SGML und Hypertext

Das folgende Beispiel zeigt einen Ausschnitt aus einer DIN-Norm im SGML-Format:

```
<DINNORM TOC=YES><FRONTM><STANDARD>
<DINNO>17440<NOFIELD>Teil 1
<TITLE><GROUPTIT>Metalle<GENRLTIT>Nichtrostende St&auml;hle
<MLTITLE>Stainless steels
<REPLACES>Ersatz f&uuml;r Ausgabe 07.80
<DATE>Januar 1991<DKNO>699.14.018.8 : 620.1
<RESPONS>Normenausschu&szlig; Eisen und Stahl (FES) im DIN Deutsches
Institut f&uuml;r Normung e. V.
<BODY>
<SECTION ID=SEC>Begriffe und Definitionen
<P>Zur Veranschaulichung der festzulegenen Begriffe siehe auch Bild
<FIGREF REFID=FIG1>. Die Begriffe beruhen auf der <CIT>DIN 4711</CIT>.
```

```
...
<FIG FLOW WIDTH="COLUMN" ID=FIG1>
<FIGCOMM>Ma&szlig;e in mm
<ARTWORK NAME="BILD_1">
<FIGCAP>Dicke bei einfachen Schmiedest&uuml;cken
</FIG>
...
<SECTION>Anwendung
<P>Die in Abschnitt <HDREF REFID=SEC> auf Seite <HDREF REFID=SEC PAGE=YES>
festgelegten Begriffe und Definitionen gelten f&uuml;r die gesamte Norm.
...
</DINNORM>
```

Der Text ist mit SGML-Markierungen (Tags) ausgezeichnet (z. B. `<DINNORM>`, `<FRONTM>`). Die Namen und die erlaubte Strukturierung dieser Tags sind in der dazugehörenden DTD festgelegt; der Bezeichner eines Tags nennt sich Element (z. B. `DINNROM`, `FRONTM`). Nach einem Tag kann entweder ein Subtag (z. B. `<TITLE><GRROUPTIT>`) oder auch Text (z. B. `<GROUPTIT>Metalle`) folgen; daraus ergibt sich eine Dokumenthierarchie, die man als Baum darstellen kann (Abbildung 2). Für gewisse Elemente ist es zudem notwendig, daß ihr Ende durch ein Ende-Tag gekennzeichnet wird (z. B. `</FIG>`, `</DINNORM>`). Durch Attribute (z. B. `TOC=YES`) können Tags mit verschiedenen Eigenschaften belegt werden; auch die möglichen Attribute der einzelnen Elemente sind in der DTD festgelegt. Um Sonderzeichen, wie z. B. Umlaute im SGML-Format, das auf 7-bit-ASCII basiert, darstellen zu können, bedient man sich der Entities (z. B. `ä ü`). Diese sind zu standardisierten Sonderzeichensätzen [14] zusammengefaßt und müssen dem Formatierer bekannt sein, damit dieser das dazugehörende Zeichen darstellen kann.

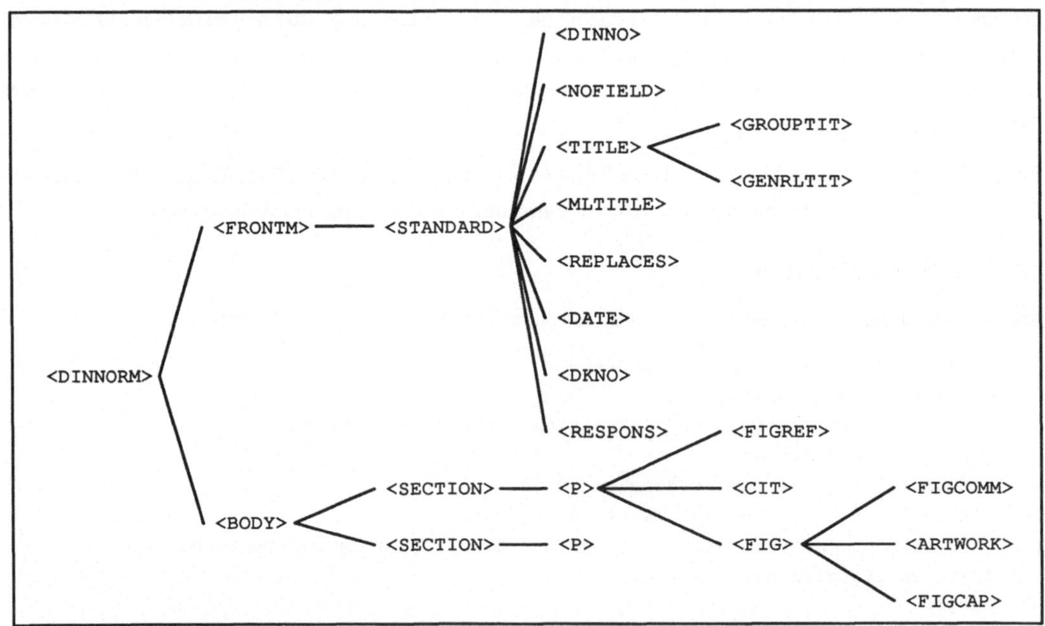

Abbildung 2: SGML-Dokument als Strukturbaum (Texte wurden weggelassen)

Das Beispiel deutet auch an, daß im SGML-Format Hypertext-Eigenschaften realisiert werden können, die der SGML-Standard anbietet. Mittels den auf den Attributen basierenden SGML-Referenzen lassen sich Verweise zwischen verschiedenen Tags herstellen. Das Ziel eines Verweises erhält einen eindeutigen Bezeichner (z. B. ID=SEC) auf den von der oder den Quellen verwiesen werden kann (REFID=SEC). Durch diesen eindeutigen Bezeichner ist die Verbindung zwischen der Quelle und dem Ziel hergestellt, sodaß einerseits Querverweise wie „siehe auch Abschnitt 5" automatisch vom Formatierer erzeugt werden können und andererseits ein Hypertext-System diesen Verweisen folgen kann, um von der Quelle zum Ziel zu gelangen.

Für Normdokumente sind Querverweise auf Überschriften, Abbildungen, Tabellen, Fußnoten, Bibliografielisten, Aufzählungen und mathematische Formeln in der DTD spezifiziert wurden. Zusätzlich zu den Verweisen innerhalb eines Dokumentes kann der Inhalt ausgewählter SGML-Elemente (z. B. <CIT>DIN 4711</CIT>) als Verweis auf ein anderes Dokument angesehen werden. Somit ist es mit dem Einsatz geeigneter Werkzeuge leicht möglich, aus einer Normendatenbank ein Hypertext-System für Normen zu generieren.

2.5 Graphik-Austauschformate

Die in den Normdokumenten enthaltenen Vektor- und Rasterbilder werden nicht in SGML-Notation abgelegt und ausgetauscht, sondern als separate Dateien in speziellen Graphiknotationen: „Computer Graphics Metafile (CGM)" [3], [15] für Vektorbilder und „CCITT FAX Group 4" [10] für Rasterbilder. Im Dokument wird an der entsprechenden Stelle auf die Datei, die das betreffende Bild enthält, mittels eines im SGML-Standard vorgesehenen Mechanismus (general entity) verwiesen.

2.6 Begleitende Arbeiten

Die Ergebnisse wurden als DIN-Fachbericht 27 „Rechnergestützte Dokumentbearbeitung von Normen; Austausch- und Bearbeitungsformat in SGML" [8] veröffentlicht.

Die SGML-DTD und die Layoutvorgaben für die einzelnen SGML-Elemente werden derzeit vom DIN im NABD 13.2 (Normenausschuß Bibliotheks- und Dokumentationswesen, Arbeitsausschuß 13 „Elektronisches Publizieren", Unterausschuß 2 „Dokumenttypdefinition Norm") zur Norm erhoben, die bereits als Entwurf [6] vorliegt. In gleicher Weise wie die existierende Norm DIN 820 vorschreibt, wie Normdokumente auf Papier auszusehen haben, wird diese Norm die elektronische Beschreibung von Normdokumenten vorschreiben.

Zeitgleich wurde in den USA zur Erstellung, Bearbeitung und Austausch technischer Dokumente im Rahmen von CALS (Computer-Aided Acquisition and Logistics Support) vom Department of Defence (DoD) eine SGML-DTD entwickelt [5].

3 Integrierter Normen-Editor auf der Basis von SGML: INES

In einem weiteren Schritt wurde ein Werkzeug entwickelt, um den theoretischen Ansatz praktisch zu erproben, zu vervollständigen und die Anwendbarkeit zu gewährleisten. Als Resultat entstanden der Editor INES (Integrierter Normen-Editor auf der Basis von SGML) [12] zum Erzeugen und Bearbeiten von Normen im spezifiziertem Standardformat und Umsetzungsprogramme für die Austauschformate SGML, CGM und CCITT/4.

3.1 WYSIWYG-Editor Interleaf TPS

Zur Erstellung und Bearbeitung einer Norm in SGML-Form wurde der Editor so entwickelt, das der Benutzer nach dem WYSIWYG-Prinzip (What-You-See-Is-What-You-Get) arbeitet und eine Kontrolle des Layouts jederzeit möglich ist (Abbildung 3).

INES basiert auf dem „Technical Publishing System (TPS)" (Version 4) der Firma Interleaf. TPS ist eines der leistungsfähigsten interaktiven Dokumentenverarbeitungssyteme, das nach dem WYSIWYG-Prinzip arbeitet. Es integriert neben der reinen Textverarbeitung Spezialeditoren für Tabellen, mathematische Formeln und Gleichungen, Linien- und Rastergraphiken, Geschäftsgraphik, Schraffuren und Muster. Darüberhinaus werden Eingabefilter für die Dateiformate verschiedener Textverarbeitungsysteme und Graphik- bzw. CAD-Pakete angeboten. TPS ist weitgehend systemunabhängig.

Die Auswahl von TPS als Basis dieser SGML-Anwendung wurde getroffen, da die derzeit angebotenen voll SGML-fähigen Dokumentensysteme die Anforderungen der Normdokumente nicht vollständig erfüllen konnten. Die speziellen Satzregeln für DIN-Normen, sowie die Komplexität der Inhalte insbesondere im Bereich Tabellen, Formeln und Graphiken übersteigen die Leistungsfähigkeit vieler Systeme.

3.2 Aufbau und Format der TPS-Dokumente

TPS-Dokumente sind ähnlich wie in SGML aus generischen Komponenten aufgebaut. Für jede Komponente existiert eine Stammvorgabe, die Name, Attribute und die voreingestellten Gestaltungsmerkmale festlegt. Das eigentliche Dokument besteht aus Instanzen dieser Komponenten, wobei die Gestaltung der einzelnen Instanzen gegenüber der Stammvorgabe individuell geändert werden kann. Die Komponenteninstanzen werden sequentiell aneinandergereiht, eine hierarchische Struktur, wie es der SGML-Standard vorsieht, wird derzeit nicht unterstützt.

TPS-Dokumente können in einem offenen, lesbaren Datenformat, dem sogenannten Interleaf-ASCII-File-Format abgelegt werden. Dieses bietet Anwendern die Möglichkeit, TPS-Dokumente mit eigenen Programmen weiterzuverarbeiten oder zu erzeugen. Theoretisch wäre es möglich, TPS-Dokumente statt mit TPS mittels eines ASCII-Texteditors zu schreiben. Dieses offene Dokumentenformat ist die Basis aller TPS-Applikationssysteme wie zum Beispiel die oben erwähnten Umsetzungsprogramme. Auch INES benutzt das Interleaf-ASCII-File-Format für die Filterung in das SGML-Format.

3.3 INES-Systemarchitektur

Für die Anwendung im Normenbereich wurden spezielle Programme entwickelt, die als Eingabe- und Ausgabekonverter die Abbildung von Normdokumenten in SGML-Notation in das ASCII-File-Format und umgekehrt vornehmen [1]. Dabei wird intern die Dokumenttypkonformität durch einen SGML-Parser geprüft. Die SGML-Fähigkeiten von INES sind als sogenanntes Pre- bzw. Postprocessing von TPS realisiert. Abbildung 4 zeigt im Überblick die INES-Systemarchitektur.

Ein Parser prüft zunächst das SGML-Dokument in Hinsicht auf die DTD und erweitert deren Elemente mit Hilfe des Eingabefilters um die TPS-Layout-Informationen. Nach diesem Schritt steht das Dokument zum Editieren in TPS-Format bereit.

Nach Abschluß des Editiervorganges entfernt ein zweiter Filter, der Ausgabefilter, die Layout-Informationen und erzeugt so das SGML-Format. Der Parser überprüft dieses Dokument; bei aufgetretenen Fehlern kann der Benutzer das Dokument korrigieren und den Ausgabefilter erneut starten.

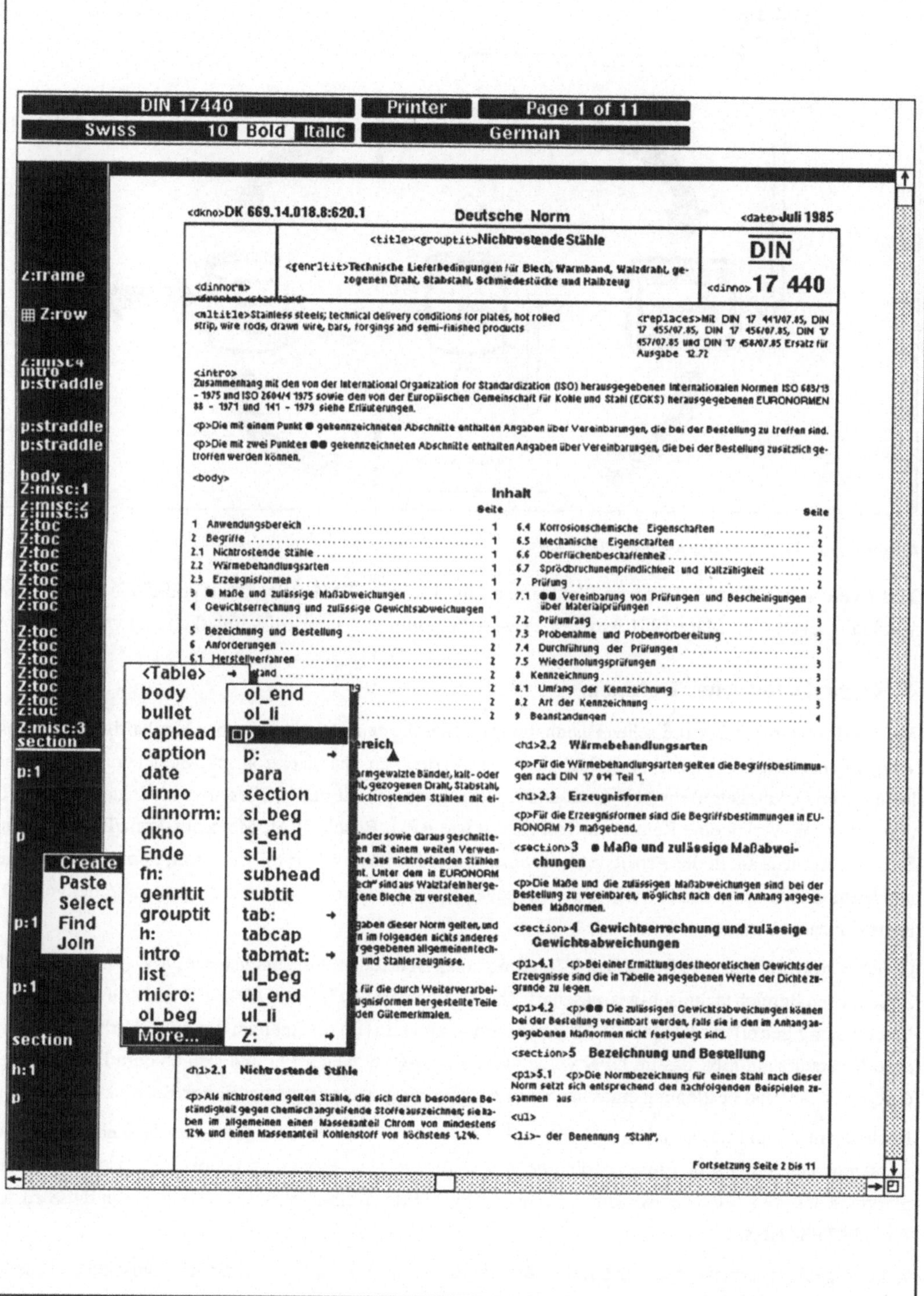

Abbildung 3: Benutzerschnittstelle des TPS-INES-Editors

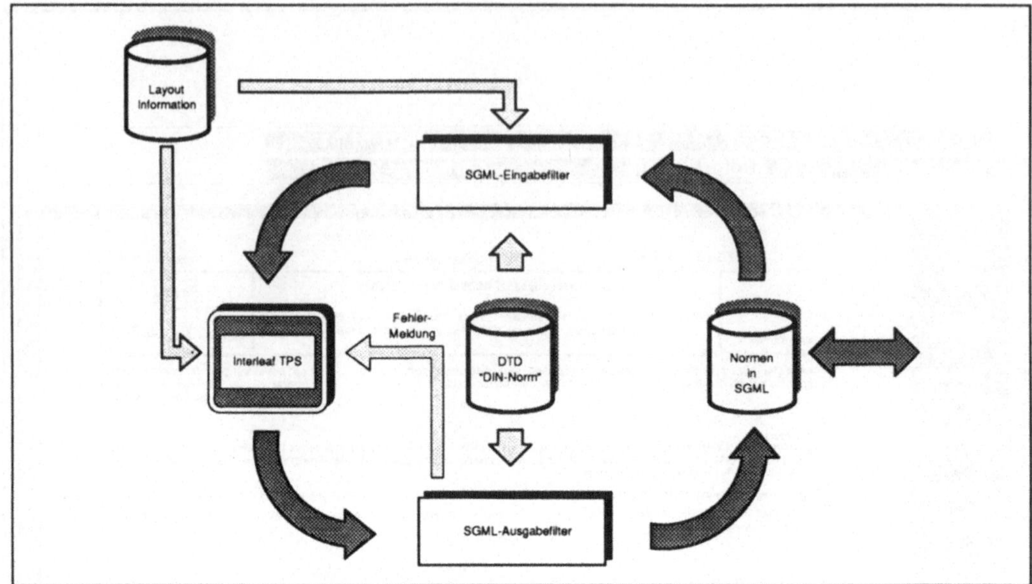

Abbildung 4: Die Architektur von TPS-INES

TPS-INES wurde für die Hardware-Plattformen Sun-3, Sun-4, IBM RS/6000, Apollo/WS-30 und VAX/VMS entwickelt. Seit März 1991 wird der TPS-INES-Editor von Interleaf vermarktet und vertrieben.

4 Organisatorischer Rahmen

Mit dem Problem der elektronischen Handhabung und des Austausches von Normen beschäftigte sich bereits seit geraumer Zeit der Arbeitskreis „Süd" der ESHD-Arbeitsgruppe (Elektronische Speicherung und Handhabung von Dokumenten) im ANP (Ausschuß „Normenpraxis") des DIN. Diesem gehören neben dem DIN und dem VDE-Verlag eine Reihe von Großfirmen wie AEG, Bosch, IBM, Mercedes-Benz/Daimler-Benz, SEL und Siemens an. In der Arbeitsgruppe wurde dieses Projekt initiiert mit dem Ziel, Formate festzulegen und Methoden zu entwickeln, um Normen zwischen Firmen elektronisch auszutauschen und/oder vom DIN übernehmen zu können.

Gleichzeitig verfolgte der „Verein zur Förderung eines Deutschen Forschungsnetzes e.V. (DFN)" ähnliche Ziele für den Bereich technisch-wissenschaftlicher Dokumente. Zwischen DFN, DIN und ZGDV wurde eine Vereinbarung getroffen, daß das ZGDV der Arbeitsgruppe ESHD im Rahmen des DFN-Vorhabens technische Unterstützung gibt, das im DFN vorhandene Know-How weitergibt und sich an dem ESHD-Projekt durch Auswahl und Festlegung eines neutralen Austauschformates sowie der Editor-Entwicklung beteiligt.

Nachdem in einem Pilotbetrieb von INES (1. 3. 1990 – 28. 2. 1991) bei ausgewählten Teilnehmern des Arbeitskreises die praktische Einsetzbarkeit geprüft wurde, zeigte die Interleaf GmbH Interesse an einer Vermarktung des vom ZGDV entwickelten Systems. Dies führte zu einer Weiterentwicklung von INES zu dem Produkt TPS-INES.

Zukünftige Aufgaben werden sich auf die Weiterentwicklung des Editors und die Archivierung und das Retrieval von Normdokumenten konzentrieren.

5 Danksagung

Die vorgestellten Arbeiten wurden in Kooperation mit der DIN-Gruppe „ESHD AK Süd", unter Vorsitz von U. Sälzer, durchgeführt. Wir möchten an dieser Stelle allen Teilnehmern für die Unterstützung danken. Weiterhin bedanken wir uns bei Herrn Kuhlmann und Herrn Pandikow für ihre Beiträge an der INES-Entwicklung.

Die Autoren danken der Interleaf GmbH für Förderung und Unterstützung dieser Arbeit sowie dem „Verein zur Förderung eines Deutschen Forschungsnetzes e. V. (DFN)", der das Projekt ebenfalls gefördert hat (unter der Kennziffer TK 558 - VA 007).

6 Literatur

[1] Alheit, B.W.; Häfemeier, F.; Hübner, W.; Kuhlmann, H.; Pandikow, M.: *INES - Integrierter Normeneditor auf der Basis von SGML*, Zentrum für Graphische Datenverarbeitung, Bericht 43/90, Darmstadt, 1990.

[2] Alheit, B.W.; Häfemeier, F.; Hübner, W.; Rath, H.H.: *INES – Ein Werkzeug zum Bearbeiten und elektronischen Austausch multimedialer Norm-Dokumente*, Working Proc., GI-Arbeitstagung „Multimediale elektronische Dokumente", Heidelberg, 1990.

[3] Arnold, D.B.; Bono, P.: *CGM and CGI*, Springer Verlag, 1988.

[4] Bryan, M.: *SGML: an author's guide to the Standard Generalized Markup Language*, Addison-Wesley, 1988.

[5] Department of Defense: *Markup Requirements and Generic Style Specification for Electronic Printed Output and Exchange of Text (MIL-M-28001)*, Department of Defense, 1987.

[6] Deutsches Institut für Normung (Hrsg.): *DIN-Entwurf 33900 – Rechnergestützte Dokumentenbearbeitung von Normen*, Beuth Verlag, 1982.

[7] Deutsches Institut für Normung (Hrsg.): *DIN 820 – Grundlagen der Normungsarbeit des DIN*, Beuth Verlag, 1982.

[8] Deutsches Institut für Normung (Hrsg.): DIN-Fachbericht 27, *Rechnergestützte Dokumentbearbeitung von Normen, Austausch- und Bearbeitungsformat in SGML*, Beuth Verlag, 1990.

[9] Deutsches Institut für Normung (Hrsg.): *Rechnergestützte Dokumentbearbeitung von Normen*, Normentwurf, Beuth Verlag, 1991.

[10] *FAX Group 4 (CCITT T.6)*, Norm für Rastergraphiken.

[11] Goldfarb, C.F.: *The SGML Handbook*, Oxford University Press, 1990.

[12] Hübner, W.; Rath, H.H.: *Electronic Handling of Standard Documents with a SGML-Based Editor TPS-INES*, International Markup 91, Lugano, 1991.

[13] International Standards Organization: *Document Style Semantics and Specification Language (DSSSL)* (Draft International Standard ISO/IEC 10179), ISO, 1991.

[14] International Standards Organization: *Information Processing – Text and Office Systems - Standard Generalized Markup Language (SGML)* (ISO 8879), ISO, Genf, 1986.

[15] International Standards Organization: *Metafile for Transfer and Storage of Picture Description Information (CGM)*, ISO 8632/1-4, 1987.

[16] Kuhlmann, H.: *Elektronische Handhabung von Dokumenten*, Proceedings der IFAN'89, 26. Konferenz Normenpraxis, 6. Internationale Konferenz, DIN, 1989.

[17] *ODA (DIS 8613)*, Standard zur Beschreibung der elektronischen Repräsentation von Dokumenten aus dem Bereich der Bürokommunikation.

[18] *ODIF (DIS 8613)*, Standard-Format für den Austausch von formatierten Dokumenten aus dem Bereich der Bürokommunikation.

[19] Sälzer, U.; Hübner, W.; Kuhlmann, H.: *Elektronische Handhabung von Dokumenten*, DIN-Mitteilungen 69 Nr. 9, S. 492-494, 1990.

Echtzeitkommunikationssysteme

Zeitgerechtes und vorhersehbares Verhalten sind die Charakteristika des Echtzeitbetriebes. Mithin werden von den "Echtzeitkommunikationssystemen" in verteilten Rechenanlagen zeitlicher Determinismus bei der Abwick-lung von Kommunikationsdiensten, transaktionenorientierte und zeitgerechte Verarbeitung zur Verwaltung verteilter Betriebsmittel und, um in Ausnahmesituationen Rekonfigurationsmaßnahmen durch-führen zu können, systemtransparente Bereitstellung von Diensten erwartet. Da derzeit verfügbare Kommunikationssysteme die genannten Anforderungen noch nicht hinreichend erfüllen, werden für das Fachgespräch Beiträge zu den Themen Kommunikationsarchitekturen für Echtzeitanwendungen, Standards für Echtzeit-kommunikationsdienste und -realisierungen sowie Test- und Unterstützungshilfsmittel für verteilte Echtzeit-anwendungen und zur Modellierung und Simulation letzterer vorgestellt.

Koordinator: Prof. Dr. W. A. Halang, Universität Groningen

Echtzeitkommunikationssysteme

Eine Einführung in die Problembereiche und Lösungsansätze

Prof. Dr. Helmut Rzehak
Universität der Bundeswehr München
Werner-Heisenberg-Weg 39
8014 Neubiberg

Zusammenfassung:
Ausgehend von den Anforderungen für zeitkritische Anwendungen wird erläutert, wodurch der Zeitbedarf im Kommunikationssystem verursacht wird, und an Hand von Ergebnissen aus Messungen und Analysen dargelegt, in welchen Größenordnungen dieser derzeit bei lokalen Netzen liegt. Es wird diskutiert, welche Verbesserungen in der Zukunft zu erwarten sind. Die Aufgaben des Kommunikationssystems bei der Koordination der Anwendungsinstanzen und bei der Rekonfiguration in einem verteilten System werden behandelt.

1. Begriffe, Anforderungen

1.1 Kommunikation in Echtzeitsystemen

Für ein Echtzeitsystem wird definitionsgemäß gefordert, daß das zeitliche Verhalten des Systems innerhalb vorgegebener Grenzen vorhersehbar ist. Dabei spielen die beiden nachfolgenden Aspekte eine besondere Rolle:

- die Ausführung eines Algorithmus innerhalb eines relativ engen Zeitfensters, und
- die zeitlichen und logischen Abhängigkeiten zwischen der Ausführung von Algorithmen (Rechenprozessen), deren Zeitfenster sich überlappen, d.h. die idealisiert nebeneinander gleichzeitig (nebenläufig) ablaufen.

Den ersten Punkt bezeichnet man auch als Forderung nach garantierten Antwortzeiten. Die zeitlichen und logischen Abhängigkeiten resultieren aus Abhängigkeiten der Abläufe innerhalb und außerhalb des Rechensystems [1], [2].

Kommunikation soll hier im Sinne von Kooperation zwischen Partnerinstanzen verstanden werden und sich nicht auf den reinen Nachrichtenaustausch beziehen. Das Kommunikationssystem umfaßt aus der Sicht des Anwenders alles, was am Datenaustausch zwischen den Anwendungsinstanzen beteiligt ist. Das ISO-Referenzmodell für die Kommunikation in offenen Rechnernetzen sieht vor, daß die Kommunikationsdienste von der Anwendungsschicht den Anwendungsinstanzen zur Verfügung gestellt werden. Da diese Dienste gewöhnlich durch das Betriebssystem hindurchgereicht werden, ist dieses an der Kommunikation mit beteiligt. Die Zeitverhältnisse bei der Kommunikation werden durch die Eigenschaften des Betriebssystems mit beeinflußt.

Unter einem "Echtzeitkommunikationssystem" soll nicht einengend verstanden werden, daß der Zeitbedarf für den Nachrichtenaustausch innerhalb vorgegebener Grenzen vorhersehbar ist, sondern daß das Kommunikationssystem für Echtzeitanwendungen geeignet ist. Da ein Teil des Nachrichtenaustausches zur Koordinierung der Anwendungsinstanzen dient, muß ein Echtzeitkommunikationssystem auch Funktionen hierfür unter Berücksichtigung der zeitlichen und logischen Abhängigkeiten zur Verfügung stellen.

Das Kommunikationssystem muß Rechensysteme von verschiedenen Herstellern miteinander verbinden können. Man spricht daher auch von offenen Systemen. "Offen" meint aber auch, daß einzelne Knoten während des Betriebes hinzugefügt und entfernt werden können, und daß Anwendungsinstanzen hinzugefügt, entfernt oder neu im Netz verteilt werden können (inkrementelle Erweiterbarkeit).

Die nachfolgenden Ausführungen beziehen sich auf lokale Rechnernetze, da für zeitkritische Anwendungen nur diese von Bedeutung sind.

1.2 Verteilte Systeme

Ohne auf die verschiedenen Aspekte einer Verteilung von Komponenten einzugehen sei hier ein "Verteiltes System" durch die nachfolgenden zwei Merkmale gekennzeichnet:

- Die Teilsysteme verfügen über keinen gemeinsamen Speicher.
- Zwischen den Teilsystemen liegt eine gewisse räumliche Distanz, so daß der Zeitbedarf für den Nachrichtenaustausch keinesfalls vernachlässigbar ist.

Dies bedeutet, daß ein Teilsystem - im nachfolgenden auch Knoten genannt - keine aktuelle Sicht des globalen Zustands haben kann, und folglich für zeitkritische Anwendungen a priori kein aktueller Wert einer für alle verbindlichen globalen Zeit zur Verfügung steht. Es können vielmehr nur die lokalen Zeitgeber mit begrenzter Genauigkeit synchronisiert werden. Der Aufwand kann hierfür erheblich werden [3]. Da der Zeitbedarf für den Nachrichtenaustausch nicht konstant ist, muß die zeitliche Reihenfolge von Ereignissen aus der Sicht zweier Knoten nicht übereinstimmen [4]. Es liegt also ein zeitlicher Nichtdeterminismus vor, der aber nicht zu einem unbestimmten Verhalten des Systems führen darf. Dies ist von großer Bedeutung für die Koordination von verteilten Anwendungsinstanzen.

1.3 Anwendungsbereich

Zu den Echtzeitanwendungen ist insbesondere die Steuerung technischer Prozesse zu zählen. Speziell für den Fertigungsbereich ist das "Manufacturing Automation Protocol (MAP)" [5] als Standard für ein einheitliches Kommunikationssystem entwickelt worden. Die Echtzeiteigenschaften von MAP-Netzen sind daher von besonderem Interesse. Verteilte Systeme mit Echtzeitbedingungen findet man z.B. zunehmend auch in Echtzeitsimulationen, wobei derzeit noch maßgeschneiderte Lösungen dominieren.

2. Zeitverhältnisse in Kommunikationssystemen

2.1 Ursachen für den Zeitbedarf

Um einen Überblick über die Bearbeitungszeiten einer Nachricht zu gewinnen, ist in Bild 2.1.1 eine typische Implementierung des Kommunikationssystems in einem Knoten dargestellt. Für den Mediumzugriff (MAC-Schicht) wird ein VLSI-Baustein verwendet, während die übrigen Protokollinstanzen (Schichten) einen Prozessor zur Bearbeitung benötigen. Dadurch können diese nicht gleichzeitig aktiv sein und es ergibt sich ein sehr komplexes Warteschlangen-Netz. Die Zuteilung des Prozessors an die Protokollinstanzen entspricht der Prozessorzuteilung in einem Betriebssystem für Mehrprogrammbetrieb.

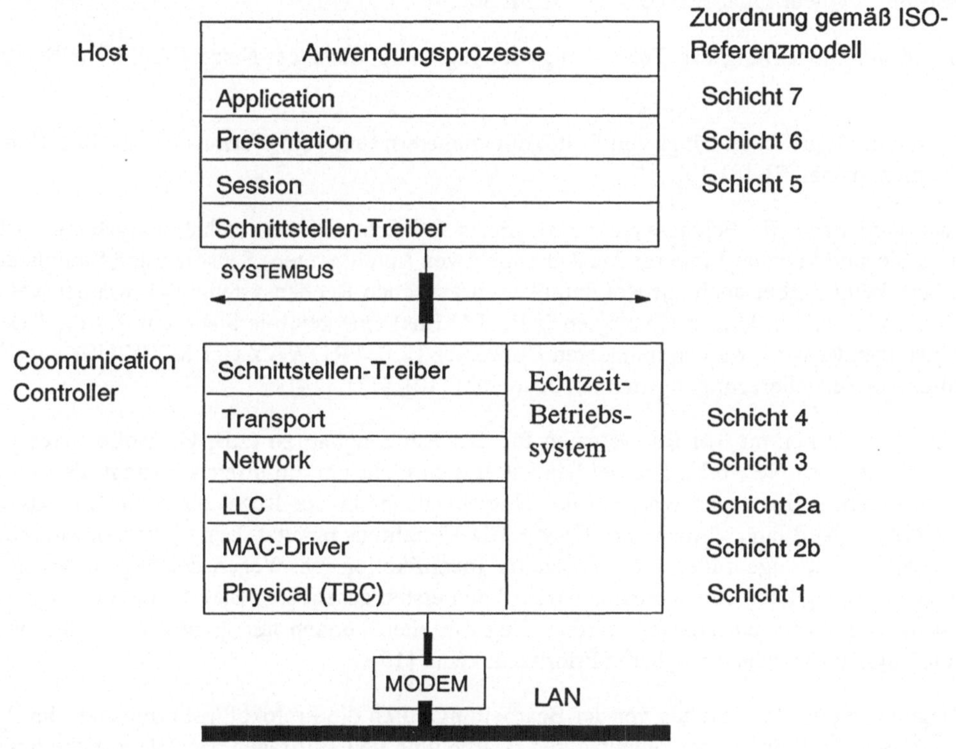

Bild 2.1.1 Typischer Aufbau eines Kommunikationskontrollers

Der Zeitbedarf für die Kommunikation entsteht durch

- die Übertragungszeit auf dem physikalischen Medium,
- die Wartezeit auf die Zuteilung des Übertragungsmediums [6],
- die Bearbeitungszeit für die Protokollabwicklung und
- Wartezeiten, wenn eine Protokollinstanz die Bearbeitung nicht sofort übernehmen kann.

Die Übertragungszeit und die Bearbeitungszeit für die Protokollabwicklung sind gut abschätzbar. Für ein vorhersehbares Zeitverhalten muß jedoch auch

- das Zugriffsverfahren auf das Übertragungsmedium deterministisch, und

- die Wartezeiten bei der Auftragsbearbeitung in den einzelnen Protokollschichten begrenzt sein.

Unter den standardisierten Zugriffsverfahren für lokale Rechnernetze sind die deterministischen Verfahren prinzipiell für Echtzeitanwendungen geeignet:

- Verfahren mit Sendeberechtigungsmarken (Token):
  - Token-Bus; IEEE 802.4 bzw. ISO 8802, Teil 4
  - Token-Ring; IEEE 802.5 bzw. ISO 8802, Teil 5

- ein Verfahren für regionale Glasfaser-Netze mit besonderer Eignung für synchronen Verkehr (Distributed Queue Dual Bus (DQDB); IEEE 802.6)

- eine Weiterentwicklung des Token-Rings für regionale Glasfaser-Netze (FDDI; ANSI-Standard).

Die maximale Zugriffszeit hängt von Protokollparametern und vom Netzausbau ab. Ihre Ermittlung ist nicht trivial [7], [8], [9].

Daneben sind noch die Feldbussysteme als Basis für Echtzeitkommunikationssysteme zu betrachten. Sie sind in erster Linie für das Ankoppeln von (intelligenten) Sensoren und Stellgliedern konzipiert, können aber auch zur Kommunikation zwischen Rechensystemen verwendet werden [10], [11]. Während in älteren Konzepten (z.B. IEC-Bus) eine zentrale Steuerung für die Buszuteilung verwendet wird, werden in neueren Konzepten (z.B. PROWAY C; PROFIBUS) Zugriffsverfahren mit Sendeberechtigungsmarken bevorzugt [12], [13], [14].

Eine Sonderrolle kommt den Bussystemen für den Fahrzeugbau zu (z.B. Controller Area Network; Vehicle Area Network). Hierbei handelt es sich nicht um ein offenes System, da sich die kommunizierenden Instanzen während der Nutzungsdauer in der Regel nicht ändern. Als Zugriffsverfahren wird eine Variante des CSMA/CD-Verfahrens benutzt. Den einzelnen Zugreifern wird eine Priorität zugeordnet, die gleichzeitig Identifikation ist. Wegen der begrenzten räumlichen Ausdehnung kann eine Kollision bereits beim ersten gesendeten Bit erkannt und zugunsten der höchsten Priorität entschieden werden. Zugriffszeiten können nur garantiert werden, wenn man die Zugriffswünsche der höheren Prioritäten kennt [15].

Die Begrenzung der Wartezeiten vor der Bearbeitung durch die Protokollinstanzen stellt im Prinzip das klassische Problem der zeitgerechten Bearbeitung von Aufträgen dar. Die Randbedingungen sind allerdings weniger überschaubar, weil

- die Zeiten für Bearbeitungsende und -beginn der Protokollinstanzen (Prozesswechsel-Zeiten) im Vergleich zu der Bearbeitungszeit für einen Auftrag nicht vernachlässigbar sind,

- die Bearbeitbarkeit von Aufträgen vom Zustand der Partnerinstanz im entfernten Knoten abhängen kann,

- zur guten Ausnutzung des Übertragungsmediums genügend viele Aufträge vor der MAC-Schicht warten sollten, und

- ankommende Nachrichten möglichst schnell "nach oben" weitergegeben werden sollen, damit kein Rückstau entsteht.

In Anlehnung an Strategien zur zeitgerechten Auftragsbearbeitung in Betriebssystemen könnte daran gedacht werden, jeder Nachricht eine Auslieferungszeit mitzugeben und die Bearbeitungsreihenfolge entsprechend der nächsten anstehenden Auslieferungszeit zu wählen. Ein solches Vorgehen müßte mit den oben genannten Randbedingungen in Einklang gebracht werden. Besondere Probleme entstehen dabei durch die Abhängigkeiten von der Partnerinstanz (Flussregelung, Sicherung der Übertragung).

Leichter zu implementieren sind Prioritätsklassen für die Nachrichten, zumal verschiedene Zugriffsverfahren (z.B. Token-Bus und Token-Ring) diese bereits vorsehen. Man kann so die zeitkritischen und weniger zeitkritischen Nachrichten voneinander trennen, so daß man nur die Wartezeiten der zeitkritischen Nachrichten begrenzen muß. Die nicht zeitkritischen Nachrichten werden nur dann bearbeitet, wenn das Kommunikationssystem keine zeitkritischen Nachrichten zu übertragen hat.

2.2 Einflußgrößen und Meßergebnisse

Um konkrete Anhaltspunkte für den Zeitbedarf zu geben, sind in Bild 2.2.1 die Antwortzeiten für einen Client-Server Aufgabenzyklus in einem MAP-Netz (Carrier-Band, 5 Mbit*s^{-1}) dargestellt.

Eine detaillierte Diskussion findet sich in [9], weitere Untersuchungen in [16] bis [23]. Zusammenfassend ergibt sich:

- Die Übertragungszeit auf dem Medium liefert nur einen geringen Beitrag für den Zeitbedarf (ca. 3,4 ms in Bild 2.2.1).

- Die Zeit für die Protokollbearbeitung und die Wartezeiten bis zur Bearbeitung bestimmen im wesentlichen den Zeitbedarf. Hierbei sind die Flußregelungsmechanismen des Transportprotokolls von erheblicher Bedeutung.

- Ab einer Grenzlast wird die Antwortzeit beliebig groß, da mehr Aufträge pro Zeiteinheit ankommen, als das Kommunikationssystem verarbeiten kann (Instabilität). Für die Transportschicht liegt der maximale Durchsatz für eine Transportverbindung deutlich unter der Grenze, die durch die Übertragungsrate auf dem Medium gegeben ist.

- In dem Netz, an dem die Messungen zu Bild 2.2.1 durchgeführt wurden, konnte auf der Transportschicht nur 20% der Übertragungskapazität bei der Kommunikation zwischen zwei Knoten ausgenutzt werden. Die Kommunikationskontroller stellen somit einen Engpaß dar, der auf heutiger technologischer Basis nicht um Größenordnungen verbessert werden kann.

- Für zeitkritische Anwendungen wird daher vielfach die Verwendung einer verkürzten Kommunikationsarchitektur vorgeschlagen (PROWAY C; PROFIBUS; Mini-MAP), bei der die Anwendungsschicht unmittelbar auf der LLC-Schicht aufsetzt.

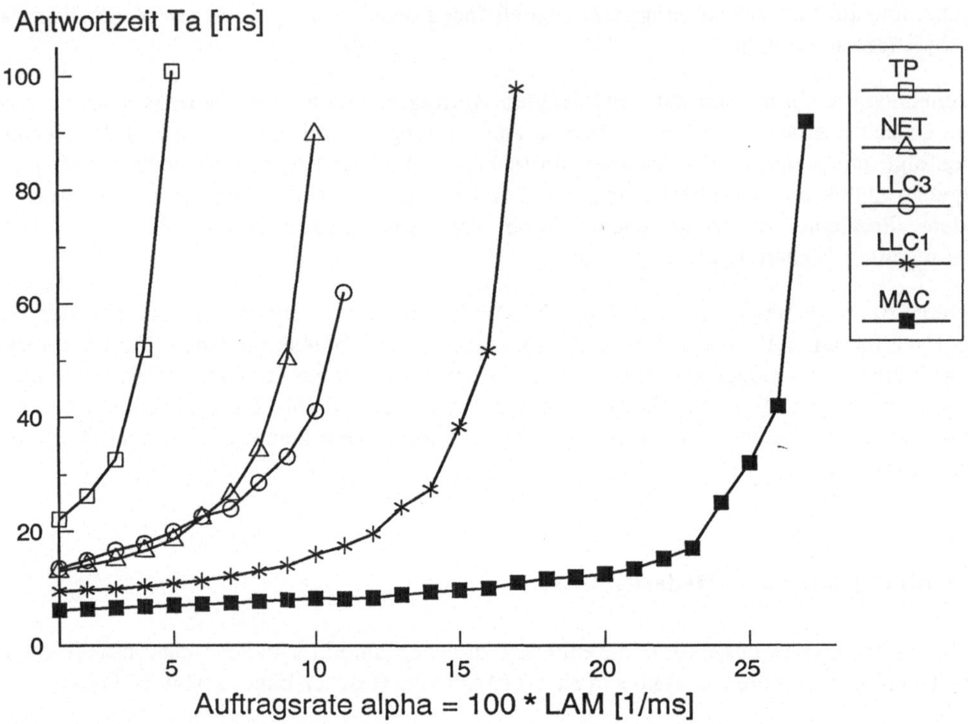

Bild 2.2.1 Gemessene Antwortzeiten als Funktion der Netzlast

Leider werden in den vorliegenden Normen der ISO-Kommunikationsarchitektur einzelne Funktionen der LLC-Schicht von den darüberliegenden Schichten nicht unterstützt. Hierzu gehören Broadcast-Dienste und auch die Prioritätsklassen des Zugriffsverfahrens. Bisher existiert kein Konzept zur angemessenen Berücksichtigung der Nachrichtenprioritäten bei der Protokollausführung in den darüberliegenden Schichten. Dadurch sind die verschiedenen Prioritätsklassen für die Anwendungsinstanzen nicht sichtbar, und eine Begrenzung der Wartezeiten für zeitkritische Kommunikationsaufträge ist nicht gegeben. Für eine verkürzte Kommunikationsarchitektur lassen sich hier leicht Lösungen finden, da die Prioritäten in der LLC-Schicht noch sichtbar sind.

2.3 Vermindern des Zeitbedarfs

Auf Grund der dargestellten Ergebnisse kann man die Möglichkeiten zum Vermindern des Zeitbedarfs grob bewerten:

- Schnellere Prozessoren im Kommunikationskontroller

  Hierdurch wird die Zeit für die Protokollbearbeitung reduziert, und indirekt auch die Wartezeiten durch die Flußregelungsmechanismen. Auch wenn die Verbesserung nicht mit dem selben Faktor zu erwarten ist, so dürfte dies der bedeutendste Beitrag sein. Natürlich müssen Speicher und VLSI-Bausteine für das Zugriffsverfahren auch entsprechend schneller werden.

- Höhere Übertragungsrate auf dem Medium [24]
  Diese läßt sich gegenüber den heute in LANs üblichen Werten noch deutlich steigern. Allerdings ist der Einfluß auf den Zeitbedarf vergleichsweise klein, so daß eine Steigerung nur zusammen mit schnelleren Kommunikationskontrollern sinnvoll ist.

- Mehrere Prozessoren im Kommunikationskontroller [25] bis [28]
  Hierzu muß die Protokollabwicklung in parallel ausführbare Teile zerlegt werden. Sieht man für jede Schicht einen eigenen Prozessor vor, so können diese nicht parallel an einem Kommunikationsauftrag arbeiten. Es kann jedoch in jeder Schicht gleichzeitig ein Auftrag bearbeitet werden. Dadurch können sich die Wartezeiten verringern. Untersuchungen zeigen, daß dieser Effekt erst bei unrealistisch hoher Fenstergröße eintritt. Zerlegungen in parallele Automaten für einzelne Schichten sind untersucht worden, ergeben jedoch keinen hohen Grad an Parallelität. Es ist daher nicht zu erwarten, daß viele Prozessoren in einem Kommunikationskontroller eingesetzt werden können.

- Spezielle VLSI-Bausteine
  Diese könnten zu einer erheblichen Beschleunigung der Protokollabwicklung führen. Unter wirtschaftlichen Gesichtspunkten können VLSI-Bausteine nur für Teilaufgaben entwickelt werden, die in verschiedenen Schichten gleichartig auftreten.

- Verkürzte Kommunikationsarchitekturen [29], [30]
  Hierdurch werden die Aufgaben bei der Protokollabwicklung reduziert. Es ergeben sich Einschränkungen beim Einsatz dieser Architekturen.

- Optimierung der Netzauslegung [7], [8], [9]
  Verschiedene Protokollparameter haben einen erheblichen Einfluß auf die Echtzeiteigenschaften des Kommunikationssystems. Die Werte können jedoch nicht ohne Berücksichtigung anderer Anforderungen und Randbedingungen gewählt werden. Es ergibt sich damit ein komplexes Optimierungsproblem.

Die Verbesserung der Zeitverhältnisse im Kommunikationskontroller bedarf großer Anstrengungen. Es ist nicht zu erwarten, daß diese Verbesserungen mit der Steigerung der Übertragungsraten Schritt halten, so daß der Kommunikationskontroller ein Engpaß bleiben wird und damit entscheidend die Echtzeiteigenschaften des Kommunikationssystems bestimmt.

3. Koordination der Anwendungsinstanzen

Die Koordination von Anwendungsinstanzen (gelegentlich auch Synchronisation genannt) kann aus zwei Gründen erforderlich sein [1],[4]:

- Die Instanzen sind kausal voneinander abhängig und deshalb in einer bestimmten Reihenfolge auszuführen.

- Die Instanzen stehen in Konkurrenz zueinander und können nicht konfliktfrei gleichzeitig ausgeführt werden. Der Konflikt entsteht durch die gemeinsame Benutzung von internen oder externen Betriebsmitteln durch die konkurrierenden Instanzen.

Beide Punkte betreffen klassische Anforderungen an Echtzeitbetriebssysteme. Die Lösungen für ein verteiltes System stellen Anforderungen an das Kommunikationssystem. Entsprechende Dienste sind der Anwendungsschicht zuzuordnen. Am Beispiel von MMS (Manufacturing Message Specification; DIN ISO 9506), der für die Fertigungsautomatisierung vorgesehenen Norm für die Anwendungsschicht [31], sollen die Probleme dargestellt werden.

Die kausale Abhängigkeit berücksichtigt man durch Starten bzw. Anhalten und Fortsetzen der abhängigen Instanz. Dies kann explizit (ProgramInvokation Management in MMS) oder implizit, ausgelöst durch Ereignisse (Event Management), erfolgen. Beim Verknüpfen von Ereignissen, die über das Netz gemeldet werden, ist der in Abschnitt 1.2 erwähnte zeitliche Nichtdeterminismus zu berücksichtigen.

Für die Koordinierung von konkurrierenden Instanzen kennt man einen objektorientierten Ansatz, indem man die Kontrolle über eine unzulässige mehrfache Benutzung dem benutzten Objekt zuordnet. Bei einem tätigkeitsorientierten Ansatz ordnet man die Kontrolle den benutzenden Instanzen zu. Der objektorientierte Ansatz ergibt dann ein geschlossenes Konzept, wenn man im Sinne eines abstrakten Datentyps die Zugriffsprozeduren mit dem Objekt zusammenfaßt. Handelt es sich aber um ein verteiltes Objekt, so stößt man auf Schwierigkeiten.

MMS sieht die klassische Semaphor-Lösung vor (Semaphor Management). Zur Realisierung wird ein Semaphorverwalter verwendet, der als ein Objekt der Anwendungsschicht anzusehen ist, und beim Einrichten des Semaphors einem Knoten fest zugeordnet wird. Durch die Operation TakeControl fordert eine Instanz die Berechtigung zur Benutzung des durch das Semaphor geschützten Betriebsmittels an, und der Verwalter erteilt sie (bei freiem Semaphor) durch Senden einer Berechtigungsmarke. Durch RelinquishControl wird diese nach Beendigung der Benutzung wieder zurückgegeben, wodurch das Semaphor wieder frei wird (vgl. Bild 3.1). Es kann festgelegt werden, daß bis zu n parallele Benutzungen möglich sind (zählendes Semaphor mit n Berechtigungsmarken). Die Dienste des Semaphorverwalters sind als unteilbare Aktionen zu implementieren. Prinzipiell entspricht TakeControl der P-Operation und RelinquishControl der V-Operation. Dem Semaphorverwalter ist jedoch der Besitzer einer Berechtigungsmarke bekannt und der Besitz kann zeitlich terminiert werden. Dies erleichtert die Auflösung von Ausnahmesituationen. Zusätzlich kann dieser bei der Vergabe der Berechtigungsmarke eine benutzerspezifische Prozedur ausführen, womit Schutzmechanismen gegen inkorrekte Benutzung von Betriebsmitteln eingerichtet werden können.

Semaphore gelten bei der Programmierung als zu fehlerträchtig und Ausfälle während des Betriebs sind schwer beherrschbar. Auch mit den in MMS definierten erweiterten Funktionen ist ein Wiederaufsetzen nach einem Ausfall des Semaphorverwalters nicht möglich.

Im Datenbankbereich wurde das Transaktionen-Konzept für die Koordinierung von konkurrierenden Instanzen entwickelt. Insbesondere zum Wiederaufsetzen nach Ausnahmesituationen sind Konzepte zur transaktionenorientierten Datenverarbeitung auch für Echtzeitsysteme von Interesse. Solche Konzepte müssen die folgenden Randbedingungen berücksichtigen:

- An einer Transaktion können mehrere Instanzen im Netz beteiligt sein (verteilte Transaktion).

- Es gelten zusätzliche Terminierungsbedingungen (z.B. Ablauf einer Zeit).

- Eine Transaktion kann möglicherweise nicht zurückgesetzt werden, weil sie die Umgebung der Rechensysteme bereits beeinflußt hat.

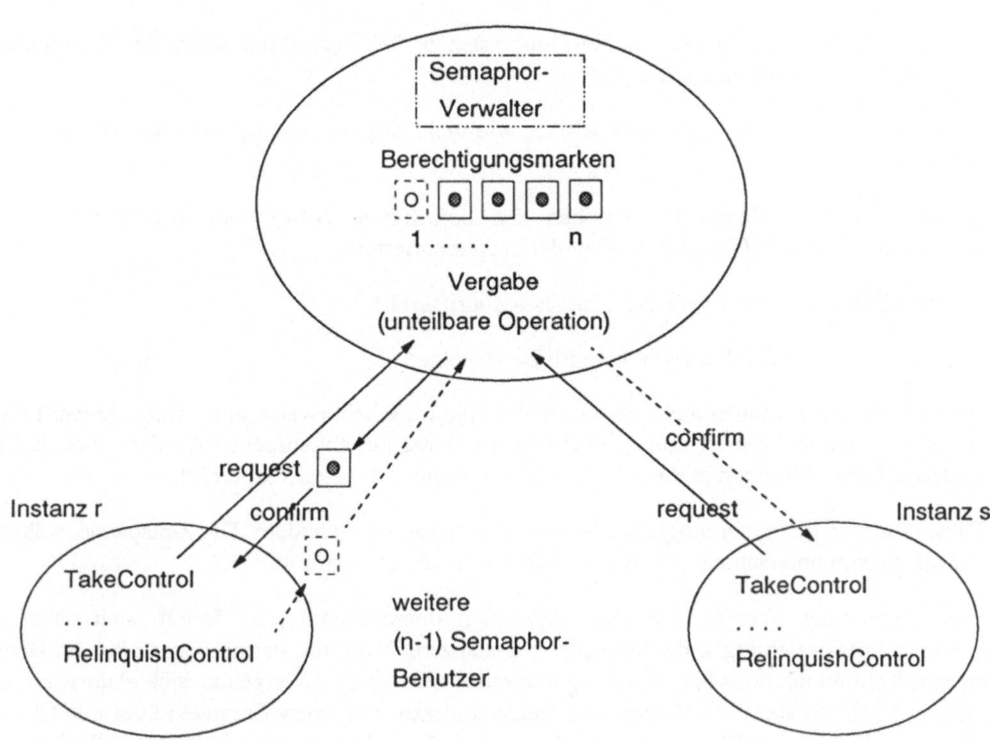

Bild 3.1 Aktionen des Semaphorverwalters

In der ISO wird derzeit eine Norm der Anwendungsschicht für verteilte Transaktionen ausgearbeitet. Diese orientiert sich an den Verhältnissen in Weitverkehrsnetzen und berücksichtigt nicht die Randbedingungen für Echtzeitanwendungen. Eine detaillierte Diskussion der Probleme findet sich in [32].

4. Replizierte Daten und Fehlertoleranz

Da ein defektes System seine Aufgaben nicht zeitgerecht erfüllen kann, muß ein Echtzeitsystem eigentlich fehlertolerant sein. Unterscheidet man jedoch zwischen den Anforderungen im Normalbetrieb und den Anforderungen zur Vermeidung von Gefährdungen bei Teilausfällen, so kann es ausreichen, wenn Aufgaben des defekten Teilsystems von einem anderen übernommen werden, bzw. die verbleibende Funktionalität für ein geordnetes Einstellen des Betriebs ausreicht. Es genügt, wenn das System rekonfigurierbar ist. Neben anderen Problemen (vgl. [33], [34]) ergibt sich die Notwendigkeit, Datenbestände eines defekten Knoten noch verfügbar zu haben. Dies kann nur dadurch erreicht werden, daß die Daten mehrfach (repliziert) im System gehalten werden. Replizierte Daten können darüber hinaus bei überwiegend lesendem Zugriff eine deutliche Leistungssteigerung durch verkürzte Zugriffszeiten bewirken.

Die Verwaltung von replizierten Daten stellt prinzipiell einen Sonderfall der transaktionenorientierten Datenverarbeitung dar [32] und kann daher von den Anwendungsinstanzen erledigt werden, wenn entsprechende Dienste zur Verfügung stehen. Die Verwaltung durch das Kommunikationssystem bietet jedoch einige Vorteile:

- Die Anwendungsinstanz muß nicht wissen, wie viele Kopien angelegt sind und wo diese sich befinden.

- Es bedarf keiner Absprache zwischen den betroffenen Anwendungsinstanzen über das Verwaltungsprotokoll (Unabhängigkeit der Programmierung).

- Der Zugriff kann optimiert werden (kürzeste Zugriffszeit).

Zu berücksichtigen sind dabei zwei wesentliche Aspekte:

- Wenn Schreiboperationen auf eine einzelne Kopie (möglicherweise nur vorübergehend) nicht ausgeführt werden können, soll nicht die ganze Operation abgebrochen werden. Dies darf jedoch nicht dazu führen, daß die defekte Kopie für immer inkonsistent bleibt.

- Es ist grundsätzlich nicht möglich alle Kopien gleichzeitig zu ändern. Die Zeitspanne, während der die Kopien unterschiedliche Werte liefern würden, ist zu minimieren.

Ist als Folge eines Fehlers eine Rekonfiguration durchzuführen, so ändern sich dabei die Kommunikationsbeziehungen der betroffenen Instanzen. Wenn die Partnerinstanz bei der Benutzung der Kommunikationsdienste explizit bekannt sein muß, so ergeben sich Adressierungsprobleme. MAP [5] sieht das Führen von Verzeichnissen (Directory Services) über alle Objekte im Netz vor. Bricht eine Kommunikationsbeziehung ab, so könnte der arbeitsfähige Partner über diese Dienste eine neue Partnerinstanz erfragen. Eine Rekonfiguration würde dann in zwei Schritten ablaufen:

- Die Einträge in den Verzeichnissen (Directories) werden geändert.

- Die neuen Kommunikationsbeziehungen bauen sich einzeln auf, sobald ein Kommunikationsauftrag nicht ausgeführt werden kann.

Von einem Echtzeitverhalten kann so kaum die Rede sein. Zudem führt das Sichern der Verzeichnisse gegen Ausfälle auf die Notwendigkeit zur replizierten Datenhaltung. MAP bietet hierfür keine direkten Lösungen an.

Eine bessere Lösung ist die Kommunikation über einen PORT [33]. Dieser ist ein vom Kommunikationssystem verwaltetes Objekt. Er wird einer Instanz fest zugeordnet, die ihre Kommunikationsaufträge stellvertretend für die Partnerinstanz an diesen richtet. Über Anweisungen an das Kommunikationssystem wird festgelegt, welche PORTs zu einer Kommunikationsbeziehung gehören. PORTs können Attribute tragen (Kontext für die Kommunikation). Ihre Bezeichner müssen global eindeutig sein. Zur Rekonfiguration müssen die Zuordnungen der PORTs zu Kommunikationsbeziehungen neu festgelegt werden. Die einzelne Instanz ist davon nicht betroffen. Das Konzept ist bisher noch nicht in der Normung von Kommunikationsprotokollen berücksichtigt. Der Kommunikationsmechanismus in Mehrrechner PEARL (DIN 66 253 Teil 3) baut darauf auf.

5. Ausblick

Kommunikationssysteme mit hohen Übertragungsraten werden heute vielfach eingesetzt. Der Zeitbedarf für einen einzelnen Kommunikationsauftrag beträgt jedoch ein Vielfaches des Zeitbedarfs für die Übertragung auf dem Medium, und die Übertragungskapazität kann von einem einzelnen Kontroller nicht genutzt werden. Dies gilt insbesondere für offene Netze mit standardisierten Protokollen. Die Maximalwerte für den Zeitbedarf, die von verschiedenen Protokollvarianten garantiert werden können, liegen beträchtlich höher als teilweise erwartet. Verbesserungen sind durch leistungsfähigere Mikroprozessoren möglich, jedoch dürften die Verbesserungsfaktoren in kurzer Zeit kaum eine Größenordnung erreichen.

In den standardisierten Protokollen sind derzeit auch keine Konzepte für eine verteilte Datenverarbeitung aufgenommen. Die zur Zeit laufenden Arbeiten in den Normungsgremien berücksichtigen nicht die Randbedingungen in Echtzeitsystemen. Diese etwas negative Bilanz sollte in dem Sinne verstanden werden, daß es noch viel zu tun gibt, bis verteilte Echtzeitsysteme technologischer Standard sind.

Literatur

[1] Herrtwich, R.G.; G. Hommel: Kooperation und Konkurrenz; Springer-Verlag 1989

[2] Millner, R.: Communication and Concurrency; Prentice Hall, London 1989

[3] Hartlmüller, P.: Wahrung der Konsistenz wesentlicher systeminterner Daten in fehlertoleranten Realzeitsystemen; Dissertation; Universität der Bundeswehr München, 1988

[4] Rzehak, H.: Distributed Systems for Real Time Applications Using Manufacturing Automation as an Example; in "Real-Time Systems, Engineering and Applications", M. Schiebe, S. Pferrer (Eds.), Kluwer Verlag, 1991

[5] Manufacturing Automation Protocol Specification, Version 3.0; European MAP Users Group, 1989

[6] Schwartz, M.: Telecommunication Networks; Addision Wesley, 1987

[7] Jäger, R.: Leistungsanalyse und Verbesserung der Realzeiteigenschaften bei LANs mit dem Tokenbus-Mediumzugriffsverfahren mit Prioritäten; Dissertation; Universität der Bundeswehr München, 1990

[8] Gorur, R.M.; A.C. Weaver: Setting Target Rotation Times in an IEEE Token Bus Network; IEEE Trans. on Industrial Electronics, Vol. 35, No. 3, Aug. 1988

[9] Rzehak, H.; A.E. Elnakhal; R. Jäger: Analysis of Real-Time Properties and Rules for Setting Protocol Parameters of MAP Network; The Journal of Real-Time Systems, Vol. 1, 1989, S. 219-239

[10] Kaminski, M: Protocols for Communication in the Factory; IEEE-Spectrum, April 1986

[11] Offene Kommunikation im Feldbusbereich, VDI-Bericht 728, VDI-Verlag 1989

[12] Process Communications Architecture; Draft Standard ISA-DS 72.03 - 1988; Instrument Society of America

[13] Jayasumana, A.P.; G.G. Jayasumana: On the Use of IEEE 802.4 Token Bus in Distributed Real-Time Control Systems; IEEE Trans. on Industrial Electronics, Vol. 36, No. 3, Aug. 1989

[14] Schümmer, M.: Bewertung von Polling-Verfahren in realzeitkritischen Fertigungsumgebungen; PEARL 89 - Workshop über Realzeitsysteme, Informatik-Fachbereichte Nr. 231, Springer-Verlag

[15] Gupta, S.: CAN Facilities in Vehicle Networking; SAE Technical Paper Series No. 900695

[16] Elnakhal, A.E.; H. Rzehak: Der Zeitbedarf für Kommunikationsaufträge in MAP-Netzen; PEARL 89 - Workshop über Realzeitsysteme, Boppard, 7.-8. Dez. 1989; Informatik-Fachberichte, Springer-Verlag Bd. 231, S. 40-52

[17] Janetzky, D.; K.S. Watson: Token Bus Performance in MAP and PROWAY; IFAC Workshop on Distributed Computer Control Systems, Mayschoss (FRG), 30.9. - 2.10.1986

[18] Janetzky, D.; K.S. Watson: Performance Evaluation of MAP Token Bus in Conjunction with LLC Protocols; Presented at the NBS/IEEE Workshop on Factory Communication, Geithersburg, Maryland, 1987

[19] Ciminiera, L.; A. Valenzano: Acknowledgement and Priority Mechanisms in the 802.4 Token Bus; IEEE-Trans. on Industrial Electronics 35, pp. 307-316, 1988

[20] Strayer, W.T.; A.C. Weaver: Performance Measurements of Motorola's Implementation of MAP; 13th Local Computer Networks Conference, Minneapolis, MN, (c) IEEE, pp. 216-221, 1988

[21] Marathe, M.V.; R. A. Smith: Performance of a MAP Network Adapter; IEEE Network, Vol. 1, No. 3, pp. 82-89, May 1988

[22] Svobodova, L.: Measured Performance of Transport Service in LANs; North-Holland Computer Networks and ISDN Systems 18, pp. 31-45, 1989/90

[23] Wu, Z.D.; E.B. Spratt: The Performance Analysis of a Local Area Network from a User Point of View; IEEE Trans. Communication, pp. 370-377, 1987

[24] Rudin, H.; R.C. Williamson: Protocols for High-Speed Networks; North-Holland, 1989

[25] Dietasch, H.; R. Ulrich: Ein OSI-Kommunikationswerk auf Transputer-Basis für den Einsatz in der dezentralen Prozeßrechentechnik; Prozeßrechensysteme '91, Informatik-Fachberichte Nr. 269, Springer-Verlag, Berlin 1991

[26] Zitterbart, M.: A Parallel Architecture for Transport Systems and Gateways; Proc. Kommunikation in verteilten Systemen, Stuttgart 1989, Informatik-Fachberichte Nr. 205, Springer-Verlag

[27] Zitterbart, M.: OSI-Internetzwerkprotokoll auf Transputer-Netzwerken; ITG/GI-Fachtagung: Architektur von Rechensystemen, VDE-Verlag, 1990

[28] Zitterbart, M.: High-Speed Transport Components; IEEE Network Magazine, Jan. 1991

[29] Rzehak, H.: Three-Layer Communication Architecture for Real Time Applications; Proceedings of the EUROMICRO'90 Workshop on Real Time, June 6-8, 1990; IEEE Computer Society Press, pp. 224-228

[30] Elnakhal, A.E.; H. Rzehak: Untersuchungen zur Implementierung einer verkürzten Kommunikationsarchitektur (Mini-MAP-Konzept); GI/GMA-Fachtagung "Prozeßrechensysteme '91", Berlin, 26.-27.2.1991; Informatik Fachberichte Bd. 269, S. 238-250, Springer-Verlag

[31] Withnell, S.; W. van Puymbroeack (Eds.): Communications for Manufacturing; Proc. of the Open Congress, Stuttgart, 4. bis. 7. Sept. 1990; Springer-Verlag 1990

[32] Stieger, K.: Randbedingungen für Protokolle zur transaktionsorientierten Datenverarbeitung in verteilten Realzeitsystemen; Beitrag für das Fachgespräch Echtzeitkommunikationssysteme auf der GI-Jahrestagung vom 14.-18.10.1991 in Darmstadt

[33] Stoll, J.: Anwendungsorientierte Techniken zur Fehlertoleranz in verteilten Realzeitsystemen; Dissertation; Universität der Bundeswehr München, 1987; Diese Arbeit erschien auch als Buch im Springer-Verlag, Informatik Fachberichte Bd. 236, März 1990

[34] Echtle, K.: Fehelrtoleranzverfahren; Studienreihe Informatik, Springer Verlag 1990

Randbedingungen für Protokolle zur transaktionsorientierten Datenverarbeitung in verteilten Realzeitsystemen *

Klaus Stieger
Universität der Bundeswehr München
Fakultät für Informatik - Inst. 3.3
Werner-Heisenberg-Weg 39
8014 Neubiberg

Zusammenfassung: Die transaktionsorientierte Datenverarbeitung ist bisher eine Domäne der (verteilten) Datenbankmanagementsysteme, wohingegen der Begriff zeitgerecht in angestammten Gebieten der Realzeitdatenverarbeitung zu Hause ist, die die Automatisierung technischer Prozesse zum Ziel hat. In diesem Beitrag wird untersucht, inwieweit die Konjunktion zwischen den beiden Attributen *transaktionsorientiert* und *zeitgerecht* gerechtfertigt ist. Nach einer Begriffsklärung werden die Randbedingungen für eine transaktionsorientierte Verarbeitung unter Realzeitbedingungen näher untersucht und am Beispiel des 2-Phasen-Commit-Protokolls auf notwendige Modifikationen eines Protokolls zur Gewährleistung der Atomizität eingegangen.

1. Einleitung

Beim Entwurf verteilter Datenverarbeitungs-Systeme können drei unterschiedliche Zielsetzungen verfolgt werden. Neben der Leistungssteigerung durch Vervielfachung der Betriebsmittel und dem Aspekt der Fehlertoleranz steht die inkrementelle Erweiterbarkeit im Vordergrund der Diskussion. Je nach Anwendungsgebiet wird der eine oder andere Aspekt stärker zum Tragen kommen. In Systemen zur Automatisierung technischer Prozesse ist bedingt durch den Einsatz von "intelligenten" Komponenten die räumliche Verteilung a priori gegeben. Und obwohl in diesem Anwendungsgebiet viele spezialisierte Geräte zum Einsatz kommen, kann man die vorhandene Redundanz der Komponenten für fehlertolerantes Verhalten und zur Leistungssteigerung des Gesamtsystems nutzen, auch wenn keines der beiden Ziele als Hauptargument für die Verteilung der Komponenten angeführt wird. Die inkrementelle Erweiterbarkeit spielt bei der Automatisierung technischer Prozesse eine große Rolle.

1.1 Modell eines verteilten Systems

In diesem Beitrag wird von verteilten Systemen ausgegangen, die aus lose gekoppelten, autonomen Komponenten (d.h. physischen und logischen Betriebsmitteln) bestehen, denen dynamisch Teilaufgaben zur kooperativen Lösung einer Gesamtaufgabe zugeordnet werden. Die Verteilung betrifft grundsätzlich die Daten (Objekte), die Verarbeitung (Operationen auf den Objekten) und die Kontrolle (Koordinierung) [Enslow 78].

In Anlehnung an das Basis-Referenzmodell der ISO [ISO 7498] wird im Folgenden der Begriff "Instanz" ('entity') verwendet, der sowohl für ein ausführbares Programmodul (Software) als auch für ein Gerät (Hardware) - die Unterscheidung ist für die Definition ohne Bedeutung - stehen kann. Ein verteiltes

* Dieser Beitrag wurde von der Deutschen Forschungsgemeinschaft im Rahmen des Projekts *Kommunikationsprotokolle für verteilte Transaktionensysteme in lokalen Rechnernetzen zur Fertigungsautomatisierung* gefördert.

System besteht dann aus N (N > 1) Instanzen, zwischen denen eine Beziehung bestehen muß. Diese besteht darin, daß die Instanzen gewisse Dienstleistungen ('services') anbieten, die von anderen Instanzen - durch den Austausch von Nachrichten[1)] - genutzt werden können. Eine einzelne Dienstleistung ist als Folge von Operationen auf den von einer Instanz verwalteten Objekten zu betrachten und drückt damit die logische Zusammengehörigkeit von Objekten und den auf den Objektmengen definierten Operationen aus. Der Vorteil liegt in der großen Transparenz. So muß ein Dienstbenutzer ('service user'), weder die ausführende Instanz kennen noch irgendwelche sonstigen Details; z.B. den Knoten auf dem diese Instanz gerade abläuft. Es ist Aufgabe des Diensterbringers ('service provider') den jeweils aktuellen Kontext zu kennen. Die Nutzung der Dienstleistung erfolgt unter Einhaltung von Regeln, die das kooperative Zusammenwirken der Instanzen gewährleisten. Im allgemeinen Fall wird eine Instanz ihrerseits auf die Dienstleistung anderer Instanzen zurückgreifen, wodurch hierarchische Beziehungen zwischen den Instanzen entstehen. Im Sinne der inkrementellen Erweiterbarkeit eines verteilten Systems und aus Gründen der Fehlertoleranz (Rekonfiguration) wird weiter angenommen, daß Instanzen dynamisch generiert werden können.

Die Interaktionen der einzelnen verteilten Instanzen innerhalb einer typischerweise hierarchischen Struktur ist einerseits durch größtmögliche Autonomie und andererseits vom kooperativen Zusammenwirken im Sinne einer gemeinsamen Aufgabenstellung gekennzeichnet. Die Autonomie der Instanzen bezieht sich auf die lokale Aufgabenstellung, die der Gesamtaufgabenstellung dienen muß, aber auch dann - eventuell im eingeschränkten Umfang - erbracht werden muß, wenn die Kommunikation mit Partnern, die für das Zusammenwirken Voraussetzung ist, gestört ist. Stichworte hierzu sind: 'graceful degradation' oder 'fail safe'-Verhalten.

Eine zentrale Anforderung an die Systeme zur Automatisierung technischer Prozesse ist die Einhaltung von Zeitbedingungen bei der Ausführung einzelner Aktionen. Sie hat entscheidenden Einfluß auf die Kommunikation, die für das kooperative Zusammenwirken der Instanzen unerläßlich ist. Wählt man eine Betrachtungsweise wie sie im Basis-Referenzmodell der ISO [ISO 7498] vorgeschlagen wird, so wird die Kommunikation zum Zweck des Datenaustausches und der Koordination zwischen Anwendungsprozessen auf die Nutzung von entsprechenden Diensten zurückgeführt. Die Berücksichtigung von Zeitbedingungen beim Aufruf einzelner Dienste ist bisher bestenfalls andeutungsweise erkennbar.

1.2 Transaktionsorientierte verteilte Systeme

Die Anwendungsprozesse werden im allgemeinen mehrere - oft auch viele - Dienstaufrufe zur Bearbeitung der Objekte benutzen. Zwischen den einzelnen Dienstaufrufen besteht ein logischer Zusammenhang. Da in verteilten Systemen im allgemeinen mehrere Instanzen beteiligt sind, und da mit dem Versagen einzelner Instanzen gerechnet werden muß, braucht man ein Bewertungskriterium, das über den Erfolg oder Mißerfolg der Bearbeitung Auskunft gibt. Wenn, wie in diesem Beitrag, die Koordinierung der Objekte im Vordergrund steht, bietet sich der Begriff der *Konsistenz* an. Eine Folge von Bearbeitungsschritten wird dann als korrekt angesehen, wenn sie einen konsistenten Datenbestand wieder in einen konsistenten überführt. Darunter ist zu verstehen, daß für die Beziehung zwischen Objekten, die als Abstraktion realer Objekte interpretierbar ist, gewisse Regeln oder Bedingungen gelten, die bei der Ausführung zu berücksichtigen sind.

Eine Lösung des Problems, die ihren Ursprung im Bereich der Datenbankmanagementsysteme (kurz DBMS) hat, ist das Konzept der Transaktion [Eswaran et al. 76] [Gray et al. 75]. Bei Transaktionen wird von dem Ansatz ausgegangen, daß die damit verbundene Folge von Dienstaufrufen die Konsistenz ver-

---

1) Es wird angenommen, daß die Kommunikation über ein lokales Netzwerk (LAN) abgewickelt wird. Diese Voraussetzung wird noch dahingehend verschärft, daß für eine Klasse von Nachrichten eine obere Schranke der Übertragungszeit angenommen wird (vgl. Abschnitt 4).

letzen kann, wenn sie nicht vollständig ausgeführt wird. Folglich fordert man die *Atomizität*, d.h. eine Folge von Dienstaufrufen soll (bezogen auf die transaktionsgebundenen Objekte) keinerlei Effekte zeigen, wenn sie nicht vollständig ausgeführt werden kann. Das heißt aber auch, daß die eigentliche Anforderung an das System bestehen bleibt. Ein Buchungsvorgang wird deswegen nicht hinfällig und muß zu einem späteren Zeitpunkt erneut gestartet werden. Die Vorstellung, daß eine Transaktion keinerlei Effekte zeigt impliziert eine bestimmte Recovery-Strategie ('backward recovery'). Bei der Automatisierung technischer Prozesse kann diese Strategie verhängnisvolle Folgen haben, so daß in jedem Fall eine vorwärts gerichtete Strategie im Sinne eines 'fail-safe'-Verhaltens anzustreben ist. Dieses Vorgehen kann nicht ohne Auswirkungen auf die Konsistenzdefinition bleiben.

Weiter ist zu berücksichtigen, daß mehrere Anwendungsprozesse gleichzeitig Dienste aufrufen. Daher muß sichergestellt sein, daß auch in Konfliktsituationen die Objekte konsistent verarbeitet werden. Man fordert daher von Transaktionen die *Isolation*, das heißt, daß alle Zwischenergebnisse einer Transaktion, die inkonsistent sein können, allen anderen Anwendungsprozessen verborgen bleiben.

Die vierte Eigenschaft, die üblicherweise von Transaktionen gefordert wird, ist die *Dauerhaftigkeit*. Darunter versteht man, daß Ergebnisse aus konsistenzerhaltenden Operationen vom Versagen beliebiger Instanzen unberührt bleiben.

Zusammenfassend kann man sagen, daß eine Transaktion aus der Sicht eines Anwendungsprozesses die vier Eigenschaften Atomizität, Konsistenz, Isolation und Dauerhaftigkeit garantiert. Sie werden gern unter dem Akronym ACID (von den englischen Begriffen 'atomicity', 'consistency', 'isolation' und 'durability') zusammengefaßt. Es ist jedoch anzumerken, daß die durch die Aufzählung suggerierte Gleichwertigkeit dieser Eigenschaften nicht zutrifft. Das Ziel der transaktionsorientierten Verarbeitung ist die Erhaltung der Konsistenz. Die drei anderen Eigenschaften dienen nur dazu dieses Ziel zu erreichen, insbesondere unter dem Aspekt der parallelen Ausführung mehrerer Transaktionen und der Fehlertoleranz. Das setzt voraus, daß neben der Steuerung nebenläufiger Aktivitäten, für jede Transaktion auch eine geeignete Fehlerbehandlung vorgesehen ist. Beide Aspekte sind überwiegend im Bereich der Datenbankmanagementsysteme gründlich untersucht worden [Bernstein et al. 87], [Cellary et al. 88], [Weikum 88].

1.3 Verteilte Realzeitsysteme (Zum Begriff der Zeit in verteilten Systemen)

Wenn man von Realzeitsystemen spricht - ob verteilt oder nicht - muß man sich über den Unterschied der drei Attribute

- zeitabhängig,
- zeitgerecht und
- gleichzeitig

im Klaren sein. Das mag selbstverständlich klingen, ist es aber nur solange man nicht das beliebte Lehnwort 'real-time' dafür einsetzt. Insbesondere da 'real-time' auch mißbräuchlich für "sehr schnell" benutzt wird.

Die *Zeitabhängig*keit beim Erbringen einer Dienstleistung bezieht sich auf die Fähigkeit einer Instanz, dieses nicht nur auf der Basis von vorhandenen Objekten und Meßgrößen an sich zu tun, sondern deren zeitlichen Bezug untereinander und zur Gesamtsituation mit einzubeziehen. Dabei ist es zunächst unerheblich, in welcher Zeit eine Instanz diesen Dienst erbringt.

Dagegen wird in diesem Beitrag mit dem Attribut *zeitgerecht*, man kann auch sagen termingerecht oder rechtzeitig, grundsätzlich eine Anforderung bezeichnet, die dem System (oder einer einzelnen Instanz) von außerhalb vorgegeben wird. Der entscheidende Punkt dabei ist, daß die Einhaltung dieser zeitlichen

Bedingungen essentiell für das (im Sinne der Spezifikation) korrekte Funktionieren des Systems ist. Für sie ist der Begriff *Realzeitbedingung* angebracht.

Geht man der Frage nach, welche Auswirkungen eine Verletzung von Realzeitbedingungen hat, so muß es ein wesentlicher Bestandteil der Spezifikation sein, auch für diese Fälle deterministische Reaktionen vorzusehen. Gelegentlich kann man in diesem Zusammenhang die Unterscheidung in harte und weiche Realzeitbedingungen finden [Halang 89]. Gemäß dieser Klassifikation sind bei der Verletzung von harten Realzeitbedingungen schwerwiegende Schäden in der Umgebung der Instanz die Folge, wohingegen die Verletzung von weichen Realzeitbedingungen unter Inkaufnahme höherer Kosten tolerierbar ist.

Schließlich ist noch zu berücksichtigen, daß die Anforderungen aus der Umgebung *gleichzeitig* bestehen können. Das trifft auch zu, wenn die Rechenleistung nicht verteilt ist, d.h. wenn ein zentraler Prozeßrechner über Sensoren mit einem technischen Prozeß verbunden ist. In verteilten Systemen ist ferner zu berücksichtigen, daß einerseits der Gleichzeitigkeit durch das Vorhandensein mehrerer Instanzen Rechnung getragen werden kann, daß andererseits die Koordination der Instanzen dadurch erschwert wird, daß keine Instanz die Gleichzeitigkeit von zwei Ereignissen beurteilen kann [Lamport 78].

Die Forderung nach zeitgerecht erbrachter Dienstleistung impliziert, daß eine Dienstleistung innerhalb eines - oft engen - Zeitintervalls erbracht werden muß. Eine Verletzung (Unter- bzw. Überschreiten) von Zeitschranken ist in jedem Fall als Versagen des Diensterbringers der entsprechenden Abstraktionsebene anzusehen. Es ist ein wichtiger Aspekt innerhalb der hierarchischen Gliederung eines verteilten Systems, möglichst bereits auf der nächsthöheren Ebene eine Fehlerbehandlung vorzusehen, die eine Fehlerausprägung verhindern oder einschränken kann. Das kann bedeuten, daß die geforderte Leistung "etwas später" oder eine verminderte Leistung rechtzeitig erbracht wird ('graceful degradation'). Die Strategie ist es, die kausale Kette Fehlerursache, Fehlerausprägung und Versagen auf einer Ebene, was auf der nächst höheren als Fehlerursache registriert wird, zu unterbrechen. Zur Definition der Begriffe Fehlerursache ('fault'), Fehlerausprägung ('error') und Versagen ('failure') sei auf [Stoll 90] verwiesen.

Ein Beispiel aus der Kommunikation ist die wiederholte Übertragung von Datenpaketen (SDUs, PDUs). Erst wenn sich innerhalb einer vorgegebenen oberen Zeitschranke kein Erfolg einstellt, macht sich das Versagen der Übertragung auf der nächst höheren Schicht bemerkbar.

2. Transaktionsorientierte verteilte Realzeitsysteme

In diesem Abschnitt werden die Eigenschaften von transaktionsorientierten verteilten Systemen unter der Maßgabe der zeitgerechten Terminierung diskutiert. Insbesondere interessiert dabei, inwieweit die ACID-Eigenschaften unter dieser Randbedingung garantiert werden können.

Der zunächst allgemein gehaltene Begriff Konsistenz ist zu präzisieren, da er "der einzige völlig anwendungsspezifische unter den vier konstitutiven Begriffen [ist]. 'Konsistenz' muß von Fall zu Fall aus der vom Benutzer gesehenen Bedeutung der manipulierten Daten heraus definiert werden." [Austen et al. 89]

2.1 Unterschiede im Anforderungsprofil

Transaktionensysteme sind bisher fast ausschließlich im Zusammenhang mit Datenbankmanagementsystemen zu finden[2)] und orientieren sich im Funktionsumfang an den Anwendungen dieses Bereichs. So sind auch alle Untersuchungen über Transaktionensysteme, sowie Prototypen von Transaktionensystemen oder kommerziell eingesetzte Transaktionensysteme praktisch auf DBMS-Anwendungen zugeschnitten. Der klassische Anwendungsfall ist die Kontenbuchung, debit_credit, [Gray 78].

2) Dabei ist weder eine Verteilung der Daten noch der Bearbeitung gefordert.

Der Zugriff auf die Datenbestände erfolgt aus der Sicht des Anwendungsprozesses gewöhnlich nicht (zumindest nicht unmittelbar) mit Hilfe einer Datenbank-Abfragesprache sondern mittels eines weitgehend vorgegebenen Dialogs, der zwischen einem Anwendungsprogramm und dem Benutzer an einem Terminal abläuft. Man bedient sich dabei meistens vorgegebener Bildschirmmasken für die Ein- und Ausgabe und spricht von einem Teilhabersystem (englisch 'teleprocessing system'). Diese Bezeichnung führt wie auch der Begriff 'transaction processing system' zur Abkürzung TP-System.

2.2 Datenmodell

TP-Systeme im herkömmlichen Sinn bieten den Anwendungsprozessen einen relativ kleinen Satz von Zugriffsprozeduren an. Im Fall von Endbenutzern erfolgt die Dateneingabe über Masken, was einer Parametrierung der Prozeduren entspricht. In jedem Fall findet ein strenges Wechselgespräch zwischen dem TP-System und Anwendungsprozeß, bzw. zwischen Koordinator und Agenten statt. Diese Festlegung hat auch Eingang in die Normung gefunden [ISO 10 026]. Die Folge ist, daß dem Anwendungsprozeß theoretisch beliebig lange Wartezustände "zugemutet" werden können. Das heißt, die Rate mit der Objekte ihre Werte ändern können wird ausschließlich vom TP-System kontrolliert.

Dagegen werden sich Daten, die von Sensoren stammen, mit für den technischen Prozeß typischen Raten oder auch spontan ändern. Es ist nun Aufgabe des TP-Systems, mit dem technischen Prozeß "Schritt zu halten". Das betrifft sowohl die reine Datenaufzeichnung, als auch Ausgaben an den technischen Prozeß. Es sei angemerkt, daß nicht jede Änderung eines Objekts bereits essentielle Folgen haben muß und eventuell nicht jede Änderung registriert werden muß, d.h., der Verlust einzelner Meßwerte kann durchaus tolerierbar sein. Dagegen muß eine Änderung des selben Objekts dann berücksichtigt werden, wenn applikationsspezifische Grenzwerte über- bzw. unterschritten werden.

Ein weiterer Unterschied ist der, daß in traditionellen TP-Systemen mit vergleichsweise einfachen Mitteln erreicht werden kann, daß eine konsistente Variante eines Objekts, im allgemeinen die, die vor Beginn der Transaktion gültig war, neben der noch vorläufigen, "neuen" Variante existieren kann. Es liegt an der Art der Objekte, die als Datensätze auf "stabilem Speicher" stehen. Es muß nur erreicht werden, daß die vorläufige Variante "unsichtbar" bleibt. Ob man den lesenden Zugriff auf die alte Variante zu diesem Zeitpunkt noch erlaubt, ist unter diesem Gesichtspunkt zweitrangig. Ist dagegen das Objekt die Abstraktion eines technischen Prozesses, die durch Meßwerte physikalischer Größen repräsentiert wird, so ist die "alte" Beschreibung, obwohl konsistent im herkömmlichen Sinn, wertlos, da sie eben nicht mehr das Abbild des technischen Prozesses ist.

Oft ist die Situation gegeben, daß die aktuelle Variante - unter einem modifizierten Konsistenzbegriff - die "bessere" ist. So hat das zugrundeliegende Datenmodell entscheidenden Einfluß auf den Konsistenzbegriff und damit die Mechanismen, die beim Recovery eingesetzt werden können. 'Forward recovery' kann beispielsweise eine Strategie sein.

2.3 Recovery

Das Ziel des Recovery ist es, durch geeignete Maßnahmen, sei es durch Behebung von Fehlerursachen oder durch Maskierung der Fehlerausprägung, das Versagen einzelner Dienste oder Instanzen zu verhindern und somit ein fehlertolerantes Verhalten des Gesamtsystems zu erreichen.

Bereits in Anschnitt 2.2 wurde darauf verwiesen, daß es aus kausalen Gründen nicht zweckmäßig ist, sich auf zwei Abbilder ("alt" und "neu") eines technischen Prozesses zu stützen. Insbesondere muß man sich vergegenwärtigen, daß das Zurücksetzen ('backward recovery') keine Operation ist, die selbstverständlich durchgeführt werden kann. Jedoch ist nicht auszuschließen, daß Recovery-Maßnahmen durch inverse

Operationen erreichen, daß der Zustand vor der Transaktion hergestellt werden kann, wenn der beabsichtigte Zustand nach der Transaktion nicht erreicht werden kann. Beispielsweise kann man einen angelaufenen Motor wieder abschalten. Dagegen ist für ein bereits gestartetes Flugzeug die Maßnahme für das 'backward recovery' und das reguläre Ende des Flugs immer die Gleiche: das Flugzeug muß landen. Allenfalls in der Wahl des Flugplatzes bestehen noch Freiheiten.

Grundsätzlich besteht neben dem 'backward recovery' noch die Möglichkeit durch 'forward recovery' zu einem konsistenten Zustand zu kommen. Diese Maßnahme ist wie folgt definiert:

"Vorwärtsbehebung versetzt Komponenten in einen konsistenten Zustand, ohne auf Zustandsinformation zurückzugreifen, die in der Vergangenheit zum Zweck der Fehlertoleranz (d.h. als Rücksetzpunkt) abgespeichert wurde." [Echtle 90]

Es ist mit einer Ausnahme keine allgemeine Lösung für die Vorwärtsbehebung bekannt. Ein konsistenter Zustand der durch Vorwärtsbehebung erreicht werden soll, wird daher meistens anwendungsbezogen formuliert sein müssen. Die Ausnahme, die zu einer allgemeinen Lösung herangezogen werden kann, ist im Zusammenhang mit der Automatisierung technischer Prozesse oft anwendbar. Es handelt sich um den Absoluttest, das heißt, die Ausprägung eines Objekts (z.B. der Wert einer Variablen) kann, sofern sie in (direktem) Zusammenhang mit dem technischen Prozeß steht, durch erneute Messung wiedergewonnen werden. Ob dabei auf die in verteilten Systemen vorhandene dynamische Redundanz zurückgegriffen wird, oder ob ein transienter Fehler (z.B. vorübergehende Störung einer Kommunikationsverbindung) durch wiederholten Zugriff ('retry') maskiert werden kann, ist in diesem Zusammenhang von untergeordneter Bedeutung. In beiden Fällen hat der zurückliegende Zustand keinen Einfluß mehr auf das Erreichen des neuen konsistenten Zustands.

Das Zurücksetzen einzelner Instanzen auf einen Initialzustand ist, auch wenn die Ausdrucksweise das verschleiert, ein Fall von Vorwärtsbehebung, da der Initialzustand von keinem anderen, also auch keinem zurückliegenden Zustand, abhängt.

Jedenfalls sind für Transaktionen mit Zeitschranken auch die Recovery-Maßnahmen bei der Einhaltung der Zeitschranken zu beachten.

2.4 Prioritäten

Können die Anwendungsprozesse, die ein traditionelles TP-System nutzen, als gleichwertig im Sinne von gleichwichtig angesehen werden, so ist die Situation bei der Automatisierung technischer Prozesse grundsätzlich anders. Anwendungsprozesse werden Anfragen an ein TP-System mit unterschiedlicher Priorität stellen. Erschwerend kommt hinzu, daß sich die Prioritäten im Laufe der Missionszeit beliebig oft ändern können.

3. 2-Phasen-Commit-Protokoll zur Validierung von verteilten Transaktionen unter Berücksichtigung von Zeitschranken

Transaktionen sind ein sehr anwenderfreundlicher Mechanismus, der dem Benutzer auch in Fehlerfällen ein deterministisches Ergebnis liefert. Dabei wird stillschweigend vorausgesetzt, daß "beliebig viel" Zeit zur Behebung von Fehlern zur Verfügung steht. Es wird im Folgenden untersucht, inwieweit dieser Mechanismus auch unter Realzeitbedingungen genutzt werden kann. Dabei soll hier nicht auf den anwendungsspezifischen Teil einer Transaktion eingegangen werden, sondern auf den allgemein formulierbaren Teil, der die Atomizität gewährleistet.

3.1 Das klassische 2-Phasen-Commit-Protokoll - 2PC

Das einfachste und bekannteste Protokoll, das die Atomizität garantiert, ist das 2-Phasen-Commit-Protokoll (2PC). Es sind mehrere Varianten bekannt, z.B. [Bayer et al. 84], [Bernstein et al. 87]. Im wesentlichen werden die folgenden Schritte durchlaufen, wobei die ersten beiden Schritte der ersten Phase, der vote-Phase, zugerechnet werden, und die Schritte drei und vier zur decision-Phase zählen:

1) Eine ausgezeichnete Instanz - Koordinator genannt - fordert alle an der Transaktion beteiligten Instanzen - die Agenten - zur "Stimmabgabe" auf, indem er das Dienstprimitiv PREPARE_req[3)] verwendet.
2) Nach[4)] dieser Aufforderung müssen die Agenten ihre "Aktionen" beenden und danach das Ergebnis der Teiltransaktion aus ihrer - d.h. aus lokaler - Sicht dem Koordinator durch die Dienstprimitive READY_req bzw. ABORT_req bekanntgeben. Beide Dienstprimitive enthalten das "Versprechen", die durch die Teiltransaktion berührten Objekte in einen konsistenten Zustand zu überführen.
3) Der Koordinator sammelt alle "Stimmen" ein und entscheidet über den Erfolg oder Mißerfolg der Transaktion. Für den Erfolg muß immer ein einstimmiges Votum aller Beteiligten vorliegen. Diese Entscheidung teilt er allen Agenten mit (COMMIT_req bzw. ABORT_req).
4) Die Agenten, die mit READY gestimmt haben, warten auf die Entscheidung des Koordinators und führen dementsprechend die Transaktion zu Ende.

Es ist hinlänglich bekannt, daß dieses Protokoll nur in einer "perfekten" Umgebung eingesetzt werden kann, da davon ausgegangen wird, daß alle beteiligten Instanzen die Dienstprimitive in dem Sinne zuverlässig abarbeiten, daß sie die geforderten Aktionen in endlicher Zeit erbringen. Diese Annahme ist unrealistisch. Bedingt durch die Verteilung ist der Austausch von Nachrichten notwendig, die nicht zuverlässig übertragen werden. Ferner können Instanzen ausfallen, so daß eine geforderte Dienstleistung nicht oder, betrachtet man fehlertolerante Systeme mit redundanten Instanzen, zu spät erbracht wird.

Zur Auflösung der potentiell unendlichen Wartezustände wird gewöhnlich eine Zeitüberwachung ('timeout') der einzelnen Schritte verwendet. Für den Fall des Überschreitens der Zeitgrenze sind spezielle Aktionen ('timeout actions') vorzusehen.

Im klassischen 2PC gibt es drei Stellen, an denen die Instanzen warten müssen: jeweils zu Beginn der Schritte zwei, drei und vier. In [Bernstein et al. 87] wird im Hinblick auf die Behandlung eines 'timeout' folgendes vorgeschlagen:

Im zweiten Schritt - das Warten auf PREPARE_ind - sollen sich die Agenten autonom für ABORT_req entscheiden. Im dritten Schritt, der Koordinator wartet auf READY_ind bzw. ABORT_ind der Agenten, soll der Koordinator, da er zu diesem Zeitpunkt noch keine Entscheidung getroffen hat und auch keiner der Agenten zu diesem Zeitpunkt die Transaktion über den perform(commit)-Zustand beendet haben kann, für ein ABORT_req entscheiden, falls die 'timeout'-Schranke erreicht wird.

Im letzten Schritt warten die Agenten auf die Entscheidung des Koordinators. Im Gegensatz zu den vorigen Schritten kann in diesem Fall keine "unilaterale" Entscheidung getroffen werden. Es sind besondere Maßnahmen ('termination protocol') erforderlich um den Wartezustand aufzulösen. Für 2PC sind keine allgemeinen Lösungen bekannt, so daß es als blockierendes Verfahren zu betrachten ist, insbesondere wenn ein Versagen des Koordinators vorliegt.

---

3) Die Bezeichnung der Dienstprimitive erfolgt in Anlehnung an [ISO 9804.3] bzw. [ISO 9805.3], wobei die Dienstprimitve ROLLBACK_xxx durch ABORT_xxx ersetzt sind, da sie begrifflich mit einer 'forward recovery'-Strategie nicht in Einklang zu bringen sind. Das allgemeinere RECOVER ist in der CCR-Norm mit einer anderen Semantik belegt.

4) Es gibt gute Gründe das Wort "nach" durch "spätestens nach" zu ersetzen, die in Abschnitt 3.2 angegeben sind. Diese Änderung gegenüber dem Protokoll wie es in der angegebenen Literatur beschrieben ist, wurde auch in CCR nach [ISO 9804.3] und [ISO 9805.3] gemacht.

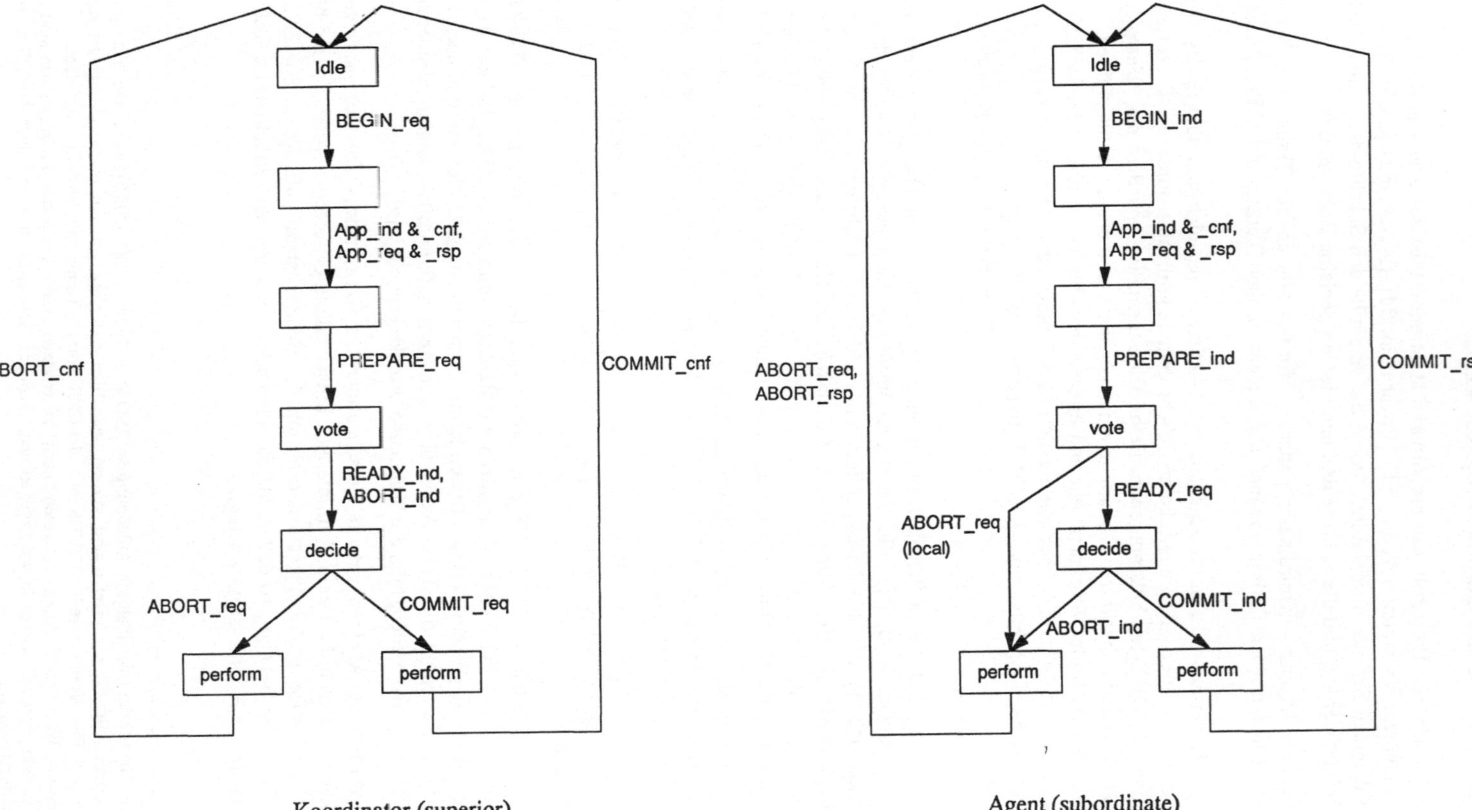

Figur 1: Zustandsübergangsdiagramm für das 2-Phasen-Commit-Protokoll (App_ind, App_cnf, App_req und App_rsp stehen für Dienstprimitive, die die anwendungsbezogene Modifikation von Transaktionsgebundenen Objekten betreffen)

3.2 Diskussion der Wartezustände im 2PC

Es soll nun gezeigt werden, daß die unilaterale Entscheidung für ein ABORT_req in den Schritten zwei und drei "unglücklich" wenn nicht gar nachteilig ist. Der Nachteil liegt darin, daß durch sie beispielsweise verhindert wird, daß alle Instanzen bis auf Eine die Transaktion regulär abschließen können und nur dieser eine Agent in einem Ausnahmezustand "landet". Das ist im Sinne einer 'graceful degradation'-Strategie ungünstig. Selbstverständlich hängt es von der jeweiligen Anwendung ab, inwieweit diese Strategie sinnvoll ist.

Der Haupteinwand ist aber, daß bei dieser Lösung zwei semantisch sehr unterschiedliche Ereignisse auf ein und denselben Zustand zurückgeführt werden. Es bietet sich also an, die zwei semantisch unterschiedlichen Ereignisse ABORT_req und 'timeout' auch durch zwei Zustände perform(abort) und Exception darzustellen.
Der Unterschied mag zunächst etwas diffizil erscheinen, er ist aber äußerst wichtig.

Durch die Unterscheidung der zwei Zustände perform(abort) und Exception wird berücksichtigt, daß der perform(abort)-Zustand mit einem durch die Transaktion oder einen Anwendungsprozeß aufgerufenen Dienstprimitiv (ABORT_req) erreicht wird, wohingegen das ohne weitere Semantik behaftete 'timeout'-Ereignis auf ein Versagen des darunterliegenden Systems hindeutet, das mit dem anwendungsbezogenen Ablauf der Transaktion in keinem Zusammenhang steht. Somit wird erreicht, daß die beiden typischen Gründe aus denen eine Transaktion nicht beendet werden kann - 'transaction aborts' und 'system crashes' [Weikum 88] - wohl unterschieden werden. In beiden Fällen führt das zu Recovery-Aktionen. Während beim Recovery aufgrund eines ABORTs die Aktionen anwendungsbezogen sind (meistens als Rücksetzen der Transaktion bezeichnet), muß ein 'crash recovery' auf das jeweilige System bezogen werden [King et al. 90]. Typische Fehlerursachen sind: Versagen des Kommunikationssystems oder einzelner Instanzen, Verklemmungen und Überlastsituationen. Es wird in diesem Beitrag die vereinfachende Annahme gemacht, daß das Aufspüren der Fehlerursachen lokal durch einen 'timeout'-Mechanismus überwacht wird.

Es sei noch angemerkt, daß, geeignete Recovery-Maßnahmen vorausgesetzt, sowohl eine Anwendungstransaktion ein 'crash recovery' unbeschadet überstehen kann, wie auch 'transaction aborts' aufgrund von Programmierfehlern vom darunterliegenden System toleriert werden.

Es wurde bereits in der Fußnote 4) hingewiesen, daß ein Agent von sich aus, d.h. ohne Aufforderung durch den Koordinator, READY_req ausführen kann. Sicher muß er, wenn man diese Lösung anvisiert, über den nötigen Kontext verfügen. In [ISO 9804.3] ist diese Variante ausdrücklich vorgesehen. Etwas lax ausgedrückt, ist das vielleicht die "letzte Chance" die Transaktion noch zu retten. Ein zweckmäßiges (aber nicht hinreichendes) Kriterium dafür ist sicher, daß ein Agent dies tut, solange er aus lokaler Sicht die Transaktion (gerade) noch regulär beenden kann. Unterscheidet man dabei noch zwischen perform(ready) und perform(abort), so kann der Zeitpunkt für die ABORT-Entscheidung vor dem für die READY-Entscheidung liegen. Der Fall ist gar nicht unwahrscheinlich, denn je nach Art der Recovery-Maßnahmen können diese wesentlich aufwendiger sein, als eine praktisch abgeschlossene Aktion für den READY-Fall, die vom Koordinator "nur" noch nicht bestätigt ist. Zur Abschätzung dieses Zeitpunkts (ohne die Unterscheidung READY/ABORT) siehe Abschnitt 4.

Ein weiteres Problem ist, daß das 2PC in der vorgestellten Ausprägung von einer Zeitüberwachung der einzelnen Schritte ausgeht, nicht aber für eine Strategie geeignet ist, die eine Transaktion innerhalb eines fest vorgegebenen Zeitintervalls durchführt.

4. Definition eines zeitgerechten 2PC - T2PC

Im diesem Abschnitt wird ein 2PC-Protokoll mit einer oberen Schranke für die Ausführungszeit diskutiert, d.h. es wird eine obere Zeitschranke für 2PC "als ganzes" und nicht nur mit einer Zeitüberwachung der einzelnen Schritte betrachtet. Dabei wird vor allem darauf eingegangen, daß sowohl im COMMIT- wie im ABORT-Fall noch Aktionen "sicher und rechtzeitig" zu Ende geführt werden müssen. Was zweckmäßigerweise durch eine Bestätigung dieser Aktion überwacht wird, wozu die Dienstprimitive COMMIT_rsp, bzw. ABORT_rsp dienen.

Will man erreichen, daß ein 2PC - unabhängig vom Ausgang - zu einem bestimmten Zeitpunkt T beendet ist, überlegt man zweckmäßigerweise von T ausgehend, wann die einzelnen Phasen abgeschlossen sein müssen.

Annahmen: Es gibt eine obere Schranke für die Nachrichtenübertragung D_m und für die Zeit bis eine Nachricht aus der (Empfangs-)Warteschlange entnommen wird D_q; - bzw. für die Ausführungszeit der einzelnen Dienstprimitive D_{se} - ferner gibt es eine globale Übereinkunft hinsichtlich T. Die zweite Voraussetzung kann dadurch erreicht werden, daß in den einzelnen Instanzen (bis auf die Drift e) synchronisierte lokale Zeitgeber existieren, was gleichbedeutend ist mit der Existenz eines virtuellen globalen Zeitgebers. Über die Möglichkeiten zur Implementierung von synchronisierten Zeitgebern sei auf [Hartlmüller 88] verwiesen. Abschätzungen einer oberen Schranke für die Ausführungszeit von Dienstprimitiven kann man in [Rzehak et al. 89] und [Rzehak, Elnakhal 91] finden.

Ferner wird vorausgesetzt, daß der Koordinator einen Zustandsvektor verwaltet, der für jeden Agenten einen Eintrag enthält, der den Zustand des Agenten aus der Sicht des Koordinators widerspiegelt. Initialisiert wird dieser "globale" Zustandsvektor mit 'exception'.

Zunächst muß man sich darüber im Klaren sein, daß das Versenden der COMMIT- bzw. ABORT-Nachrichten - besser: daß der Aufruf der COMMIT- bzw. ABORT-Dienstprimitive - nicht die letzte Aktion in der Transaktion ist. Die Instanzen müssen noch - und zwar bis zum Zeitpunkt T - ihren Teil der Transaktion erfolgreich ausführen, d.h. entsprechend der Entscheidung des Koordinators das Ergebnis der Transaktion "endgültig machen", wozu beispielsweise auch gehört, daß alle gehaltenen Sperren freigegeben werden. Ausgehend vom Endzeitpunkt T und unter den genannten Annahmen kann man die Zeitpunkte festlegen, zu denen die einzelnen Phasen bzw. Schritte abgeschlossen sein müssen.

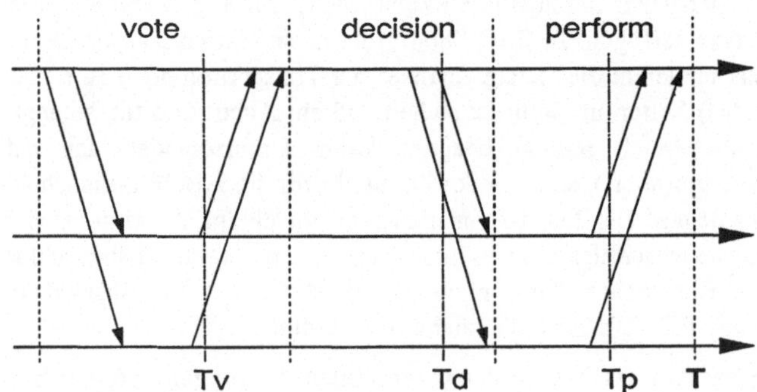

Figur 2: Zeitlicher Verlauf des T2PC

- Der Endzeitpunkt für die Performance-Phase ergibt sich aus der Sicht eines Agenten gemessen mit seiner lokalen Uhr (d.h. seinem Zeitgeber)

  $T_p := T - D_{se} - e$

- Der Zeitpunkt, bis zu dem der Koordinator seine Entscheidung gefällt haben muß, ist
  $T_d := T_p - D_{se} - e - D_p$
  wobei D_p die Zeit ist, die die Agenten im fehlerfreien Fall längstens benötigen, um die lokalen Aktionen auszuführen, die zum Abschluß der Transaktion notwendig sind.
- Der späteste Zeitpunkt, bis zu dem die Agenten ihre Stimme abgegeben haben müssen, ist
  $T_v := T_d - D_{se} - e - D_d$
  wobei D_d die Zeit ist, die der Koordinator im fehlerfreien Fall längstens benötigt, um die Entscheidung über den Ausgang der Transaktion zu fällen.

Die drei Zeitpunkte T_p, T_d und T_v sind "Zwischenendzeitpunkte" ('intermediate deadlines') für die einzelnen Phasen des 2PC. Es ist zu überlegen, was zu geschehen hat, wenn eine dieser Realzeitbedingungen, und damit auch der Endzeitpunkt der Transaktion, nicht eingehalten werden kann. Fürs erste wird dieser Fall einfach als Ausnahmefall (Exception) betrachtet und als neuer Zustand ins Übergangsdiagramm eingefügt. Das hat allerdings Konsequenzen für den Ausgang der Transaktion: Wenn das klassische 2PC terminiert, haben entweder alle beteiligten Instanzen über den perform(commit)- oder alle Instanzen über den perform(abort)-Zustand den Idle-Zustand erreicht. Die Terminierung ist aber nicht garantiert. Im Gegensatz dazu wird beim T2PC die Terminierung durch Zeitüberwachung erzwungen. Dafür muß man in Kauf nehmen, daß das System eventuell in einem Zustand ist, in dem alle Instanzen im Exception-Zustand sind, oder einige über den perform(commit)-Zustand die Transaktion beendet haben und einige noch im Exception-Zustand sind, bzw. einige über den perform(abort)-Zustand die Transaktion beendet haben und einige noch im Exception-Zustand sind.

Es soll jedoch weiterhin eine "funktionelle Konsistenz" in dem Sinne gewährleistet sein, daß auf keinen Fall einige Instanzen über den perform(commit)-Zustand und einige über den perform(abort)-Zustand die Transaktion beenden. Das heißt mit anderen Worten: Das klassische 2PC garantiert die Einstimmigkeit aber nicht die Terminierung. 2PC mit Zeitüberwachung garantiert die Terminierung, die Einstimmigkeit aber nur für die Instanzen, die das 2PC termingerecht beenden.

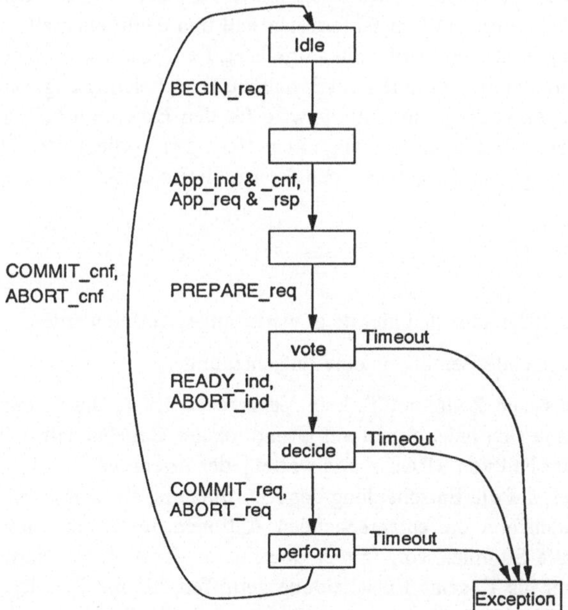

Figur 3: Zustandsübergangsdiagramm des Koordinators für T2PC

5. Korrektheit des T2PC

5.1 Kriterien

Die Bewertungskriterien für die Korrektheit des T2PC sind weitgehend mit denen des 2PC (1-4a) nach [Bernstein et al. 87] identisch. Für die Situation beim Erreichen der Zeitschranken sind zusätzliche Kriterien angegeben (4b - 5):

1. Alle beteiligten Instanzen, die zu einer Entscheidung (COMMIT oder ABORT) kommen, treffen die gleiche Entscheidung.
2. Keine Instanz darf eine einmal getroffene Entscheidung zurücknehmen.
3. Eine Transaktion wird nur dann mit COMMIT beendet, wenn alle Beteiligten mit READY gestimmt haben.

4a. Wenn im fehlerfreien Fall alle Beteiligten mit READY stimmen, dann ist die Entscheidung COMMIT.

4b. Im fehlerfreien Fall treffen alle Beteiligten eine Entscheidung.

4c. Im fehlerfreien Fall beenden alle Beteiligten ihre Teiltransaktion entsprechend der Entscheidung des Koordinators bis zum Zeitpunkt T.

4d. Im fehlerfreien Fall ist zum Zeitpunkt T der Zustand aller Agenten im Zustandsvektor des Koordinators registriert.

5. Zum Zeitpunkt T haben die Agenten entweder ihre Teiltransaktion abgeschlossen (mit COMMIT oder ABORT), das heißt sie sind im Idle-Zustand, und der Koordinator hat in seinem Zustandsvektor für jeden Beteiligten einen Eintrag, der diesen Zustand widerspiegelt, ansonsten ist der Eintrag im Zustandsvektor 'exception'.

Zur Erläuterung kann dazu noch folgendes gesagt werden:

Bedingung 1 garantiert die Konsistenz, aber nicht die Terminierung. Bedingung 2 ist nicht unumstritten. So wird beispielsweise von H.-R. Wiehle ein rücknehmbares READY vorgeschlagen [Austen et al. 89]. Bedingung 3 sorgt für die Einstimmigkeit im fehlerfreien Fall und damit ebenfalls die Konsistenz. Unter 4 sind minimale Anforderungen für den Erfolgsfall aufgelistet: Bedingung 4a ist eine schwächere Version der Umkehrung von 3. 4b impliziert zusammen mit 3, daß die Entscheidung zugunsten des ABORT-Falls ausfällt, solange kein Versagen vorliegt, andernfalls wird für den Exception-Fall entschieden. 4c garantiert die Terminierung bis zum Zeitpunkt T solange kein Versagen vorliegt und 4d besagt, daß 4c dem Koordinator bekannt ist. Bedingung 5 garantiert einen deterministischen Zustand zum Zeitpunkt T.

5.2 Nachweis

1-4a sind identisch mit dem 2PC und sollen hier nicht weiter angezweifelt werden.

4b - Im fehlerfreien Fall treffen alle Beteiligten eine Entscheidung:

Unter der Annahme, daß bis zum Zeitpunkt T_v kein Versagen vorliegt, d.h., jeder der Agenten hat ein PREPARE_ind erhalten, kann sich jeder Agent auf Grund lokaler Gegebenheiten für READY_req bzw. ABORT_req entscheiden. Im Fall von READY_req wartet jeder Agent bis $T_v = T_d - D_{se} - e - D_d$, bezogen auf seinen lokalen Zeitgeber, auf die Entscheidung des Koordinators, andernfalls kann der sich sofort für den ABORT-Fall entscheiden und die entsprechenden Aktionen ausführen. Zum Zeitpunkt $T_v + D_{se}$ liegen beim Koordinator alle Stimmen vor. Immer noch unter der Voraussetzung, daß kein Versagen vorliegt, hat der Koordinator bis T_d seine Entscheidung getroffen, die bis $T_d + D_{se} + e$ bei den Agenten eingetroffen ist, die mit READY gestimmt haben. Damit sind alle Beteiligten zu einer Entscheidung gekommen.

4c - Im fehlerfreien Fall beenden alle Beteiligten ihre Teiltransaktion entsprechend der Entscheidung des Koordinators bis zum Zeitpunkt T:

Wenn ein Agent mit ABORT gestimmt hat, so hat er das bis zum Zeitpunkt T_v getan. Er benötigt zur Ausführung D_p, wobei definitionsgemäß gilt:

$$T_v = T_p - D_p - 2D_{se} - 2e - D_d \Rightarrow T_v < T_p - D_p - D_{se} - e$$

Hat ein Agent dagegen mit READY gestimmt, so wird er die Entscheidung des Koordinators bis $T_d + D_{se}$ erhalten haben. Die verbleibende Zeit genügt um die entsprechenden Aktionen auszuführen und die Bestätigung (COMMIT_rsp, bzw. ABORT_rsp) an den Koordinator zu schicken (vgl. Figur 2).

4d - Im fehlerfreien Fall ist zum Zeitpunkt T der Zustand aller Agenten im Zustandsvektor des Koordinators registriert.

Wieder unter der globalen Voraussetzung, daß kein Versagen vorliegt, folgt diese Aussage unmittelbar aus der vorherigen.

5. - Zum Zeitpunkt T haben die Agenten entweder ihre Teiltransaktion abgeschlossen (mit COMMIT oder ABORT), das heißt sie sind im Idle-Zustand, und der Koordinator hat in seinem Zustandsvektor für jeden Beteiligten einen Eintrag, der diesen Zustand widerspiegelt, ansonsten ist der Eintrag im Zustandsvektor 'exception'.

Für den Fall, daß kein Versagen vorliegt, trifft 4d zu. Für die in Abschnitt 3.2 genannten Ausfälle ergibt sich aus der Tatsache, daß für alle Beteiligten zu Beginn des T2PC der Eintrag im Zustandsvektor des Koordinators mit 'exception' initialisiert wurde, und dieser Eintrag nur überschrieben wird, wenn ein Agent rechtzeitig das T2PC beendet hat. Hier kommt zum Tragen, daß die lokalen Zeitgeber zu Beginn des T2PC synchronisiert waren und bis zum Zeitpunkt T nicht mehr als um die Drift e auseinanderlaufen.

6. Schlußbemerkungen

In diesem Beitrag wurden die verschärften Anforderungen untersucht, die bei der transaktionsorientierten Verarbeitung unter dem Aspekt der rechtzeitigen Terminierung bestehen. Es wurde am Beispiel der 2-Phasen-Commit-Protokolls gezeigt, daß eine "funktionelle Konsistenz" für die korrekten Instanzen gewährleistet werden kann. Die rechtzeitige Behandlung des Ausnahmezustands gilt noch als ein ungelöstes Problem [Stankovic 88]. Es ist jedoch denkbar, daß mit anwendungsspezifischen (pragmatischen) Maßnahmen etwa mit "Notlösungen", die einer 'graceful degradation'-Strategie folgen, ein 'fail-safe'-Verhalten des Systems erreicht werden kann. Dazu müssen lokale Aktionen eingeplant sein, die globalen Regeln folgend, eine determinierte Behandlung des Ausnahmezustands durchführen [Davidson et al. 89]. Die Verwendung von redundanten Instanzen wird im allgemeinen notwendig sein [Burkes, Teiber 90].

In [ISO 9804.3] sind - ohne jedoch den Aspekt der Rechtzeitigkeit zu betonen - ebenfalls autonome Entscheidungen einzelner Instanzen außerhalb des Protokolls vorgesehen. Sie werden als *heuristische Entscheidungen* bezeichnet.

Weder 2PC noch T2PC können den Ausfall des Koordinators tolerieren. In weiteren Untersuchungen werden das 4PC untersucht, das den Ausfall des Koordinators toleriert. Ferner sollen die Eigenschaften anderer Validierungsprotokolle, insbesondere von Zeitstempel- und 'Voting'-Verfahren, hinsichtlich der rechtzeitigen Terminierung untersucht werden. Dabei ist allerdings zu bedenken, daß bei den optimistischen Verfahren, die für den Normalfall (COMMIT) optimiert sind, Recovery-Aktionen wesentlich aufwendiger sind, und somit der Zeitpunkt für die Entscheidung für ABORT vor dem für COMMIT liegen wird. Auf diese Problematik wurde bereits im Abschnitt 3.2 hingewiesen.

Literatur

[Austen et al. 89] Austen, M.W.; J.M. Janas; H.-R. Wiehle: Über das ISO Norm-Projekt zur verteilten Transaktionsverarbeitung: Stand und technische Alternativen. ITG/GI-Fachtagung: Kommunikation in verteilten Systemen. Stuttgart, Feb. 1989, S. 99-114, Informatik-Fachberichte Nr. 205, Springer-Verlag, 1989

[Bayer et al. 84] Bayer, R.; K. Elhardt; W. Kießling; D. Killar: Verteilte Datenbanksysteme. Informatik-Spektrum, Band 7, Heft 1, Feb. 1984, S. 1-19

[Bernstein et al. 87] Bernstein, P.A.; V. Hadzilacos; N. Goodman: Concurrency Control and Recovery in Database Systems. Addison-Wesley, 1987

[Burkes, Teiber 90] Burkes, D.L.; R.K. Treiber: Design Approaches for Real-Time Transaction Processing Remote Site Recovery. IEEE Proceedings of COMPCON Spring 90, Los Alamitos, CA, Feb. 26 - March 2, 1990, pp 568-572

[Cellary et al. 88] Cellary, W.; E. Gelenbe; T. Morzy: Concurrency in Distributed Database Systems. North-Holland, 1988

[Davidson et al. 89] Davidson, S.B.; I. Lee; V. Wolfe: Language Constructs for Timed Atomic Commitment. IEEE Proceedings of 19th International Symposium on Fault-Tolerant Computing, 1989, pp 470-477

[Echtle 90] Echtle, K.: Fehlertoleranzverfahren. Springer-Verlag (Studienreihe Informatik), 1990

[Enslow 78] Enslow, P.H. Jr.: What is a "Distributed" Data Processing System? IEEE Computer, Vol. 11, No. 1, Jan. 1978 pp 13-21

[Eswaran et al. 76] Eswaran, K.P.; J.N. Gray; R.A. Lorie; I.L. Traiger: The Notions of Consistency and Predicate Locks in a Database System. Communications of the ACM, Vol. 19, No. 11, Nov. 1976, pp 624-633

[Gray et al. 75] Gray, J.N.; R.A. Lorie; G.R. Putzulo; I.L. Traiger: Granularity of Locks and Degrees of Consistency in a Shared Database. Research Report RJ1654, IBM, Sep. 1975

[Gray 78] Gray, J.N.: Notes on Data Base Operating Systems. Operating Systems: An Advanced Course. R. Bayer, R.M. Graham, G. Seegmüller (eds.) Lecture Notes in Computer Science, No. 60, Springer-Verlag, 1978, pp 393-481

[Halang 89] Halang, W.: Schwerpunkte der internationalen Forschung im Bereich Echtzeitsysteme. PEARL 89 - Workshop über Realzeitsysteme, Boppard, Dez. 1989, S. 1-12, Reihe Informatik-Fachberichte Nr. 231, Springer-Verlag, 1989

[Hartlmüller 88] Hartlmüller, P.: Wahrung der Konsistenz wesentlicher systeminterner Daten in fehlertoleranten verteilten Realzeitsystemen. Dissertation, Universität der Bundeswehr München, 1988

[ISO 7498] ISO 7498, Information Processing Systems - Open Systems Interconnection - Basic Reference Model. 1984

[ISO 9804.3] ISO/IEC DIS 9804.3, Information Processing Systems - Open Systems Interconnection - Service definition for the Commitment, Concurrency and Recovery service element. 1989

[ISO 9805.3] ISO/IEC DIS 9805.3, Information Processing Systems - Open Systems Interconnection - Protocol specification for the Commitment, Concurrency and Recovery service element. 1989

[ISO 10 026] ISO/IEC DIS 10026, Information Processing Systems - Open Systems Interconnection - Distributed Transaction Processing - Part 1: Model; Part 2: Service Definition; Part 3: Protocol Specification. Oct. 1989

[King et al. 90] King, R.P.; N. Halim; H. Garcia-Molina; C.A. Polyzois: Overview of Disaster Recovery for Transaction Processing Systems. IEEE Proceedings of the 10th International Conference on Distributed Computing Systems, Paris, France, May 28 - June 1, 1990, pp 286-293

[Lamport 78] Lamport, L.: Time, Clocks, and the Ordering of Events in a Distributed System. Communications of the ACM, Vol. 21, No. 7, July 1978, pp 558-565

[Rzehak et al. 89] Rzehak, H.; A.E. Elnakhal; R. Jäger: Analysis of Real-Time Properties and Rules for Setting Protocol Parameters of MAP Network. Real-Time Systems, Vol. 1, No. 3, 1989, pp 221-241

[Rzehak, Elnakhal 91] Rzehak, H.; A.E. Elnakhal: Untersuchungen zur Implementierung einer verkürzten Kommunikationsarchitektur (Mini-MAP-Konzept). Prozeßrechensysteme 91, Berlin, Feb. 1991, S. 238-250, Reihe Informatik-Fachberichte Nr. 269, Springer-Verlag, 1991

[Stankovic 88] Stankovic, J.A.: Misconceptions About Real-Time Computing: A Serious Problem for Next-Generation Systems. IEEE Computer, Vol. 21, No. 10, Oct. 1988, pp 10-19

[Stoll 90] Stoll, J.: Fehlertoleranz in verteilten Realzeitsystemen - Anwendungsorientierte Techniken. Reihe Informatik-Fachberichte Nr. 236, Springer-Verlag, 1990

[Weikum 88] Weikum, G.: Transaktionen in Datenbanksystemen - Fehlertolerante Steuerung paralleler Abläufe. Addison-Wesley Verlag (Deutschland) GmbH, Bonn, 1988

Die PROFIBUS-Anwendungsschicht

K. Bindbeutel, A. Funke, M. Katz, G. Biwer, K. Bender, Karlsruhe

Zusammenfassung

Ausgehend von dem Grundgedanken einer offenen Kommunikation im Feldbereich und dem Konzept des ISO/OSI-Basisreferenzmodells wird die Einordnung des PROFIBUS in die Schichtenstruktur besprochen. Anschließend wird genauer auf die Anwendungsschicht eingegangen, und zwar mit den Schwerpunkten Aufbau, Verbindungstypen und Dienstekonzept. Ein Beispiel verdeutlicht den Ablauf bei dem Auslesen einer Prozeßvariablen aus einem abgesetzten Gerät. Abschließend wird auf einige Leistungsaspekte der PROFIBUS Schicht 7 eingegangen.

1. Offene Kommunikation im Feldbereich

Der ständig steigende Automatisierungsgrad in der Fertigungs- und Prozeßtechnik macht den Aufbau von hierarchisch gegliederten Kommunikationsstrukturen erforderlich. Auf der untersten Ebene dieser Hierarchie stehen die Feldbusse. Sie stellen die Verbindungen zwischen Feldgeräten (i.a. Aktoren und Sensoren) und mit übergeordneten Steuereinheiten (SPS, PC, ...) her. Für ein solches Kommunikationssystem gibt es eine Reihe von Anforderungen, wie z.B.

- Übertragung von vielen kurzen Nachrichten
- kurze, garantierbare Übertragungszeiten (möglichst Echtzeitfähigkeit)
- Störsicherheit (z.B. Explosionsschutz)
- geringe Anschaltkosten

Der Grundgedanke der "Offenen Kommunikation" ist es, herstellerspezifische Anwendungssysteme, bei denen in der Regel keine Kompatibilität untereinander besteht, miteinander kommunizieren zu lassen. Um dieses Ziel zu erreichen, sind umfangreiche Normungsanstrengungen erforderlich. Diese betreffen ausschließlich die Kommunikation, d.h. die Art und Weise, wie die Nachrichten ausgetauscht werden sollen. Welche konkrete Bedeutung (Semantik) die ausgetauschten Informationen haben, ist abhängig von der Anwendung, d.h. es bedarf weiterer Absprachen der beteiligten Anwendungen über den Inhalt der ausgetauschten Information.

2. Das Schichtenmodell des PROFIBUS

Moderne Kommunikationssysteme sind nach dem ISO/OSI-Referenzmodell gestaltet, das einen 7-schichtigen Aufbau der Funktionalität empfiehlt. Jede einzelne Schicht erstreckt sich global über das Kommunikationssystem und legt jeweils ein zwischen den angeschlossenen Geräten vereinbartes schichtspezifisches Protokoll fest. An der Schnittstelle zur nächsthöheren Schicht (bei der Schicht 7 ist es die Schnittstelle zum Anwendungssystem) werden *Dienste* bereitgestellt. Über sie verfügt der Anwender über ein transparentes Kommunikationssystem, mit dessen Hilfe er den Datentransport zwischen den einzelnen Stationen auf einfache Weise in einer der Automatisierungsaufgabe entsprechenden Form bewerkstelligen kann. Das PRO-

FIBUS-Kommunikationssystem gliedert sich in Anlehnung an das OSI-Referenzmodell (Bild 1) in die Schichten 1 (Übertragungsschicht), 2 (Sicherungsschicht) und 7 (Anwendungsschicht). Die beiden unteren Schichten sind in [DIN91a], die Anwendungsschicht in [DIN91b] spezifiziert. Aus Effizienzgründen sind die Schichten 3-6 nicht explizit ausgefüllt. Ein Teil von deren Funktionalität wird von der Schicht 7 mit übernommen, und zwar durch das LLI (*Lower Layer Interface*). Dessen Hauptaufgaben sind die Umsetzung aller Schicht 7-Dienste auf Schicht 2-Dienste, sowie die Verwaltung der Verbindungen im Netz (Aufbau, Abbau, Überwachung). Die eigentliche Schicht 7 wird von dem FMS (*Fieldbus Message Specification*) in Form einer Protokollmaschine realisiert. Die Anpassung der Eigenschaften eines realen Gerätes, wie z.B. Variablentypen, Mechanismen zum Umgang mit Programmen, etc. an die von FMS geforderten abstrakten Strukturen wird vom ALI (*Application Layer Interface*) vorgenommen. Dies ist nicht Bestandteil der Norm, sondern muß vom Anbieter eines Feldgerätes implementiert werden.

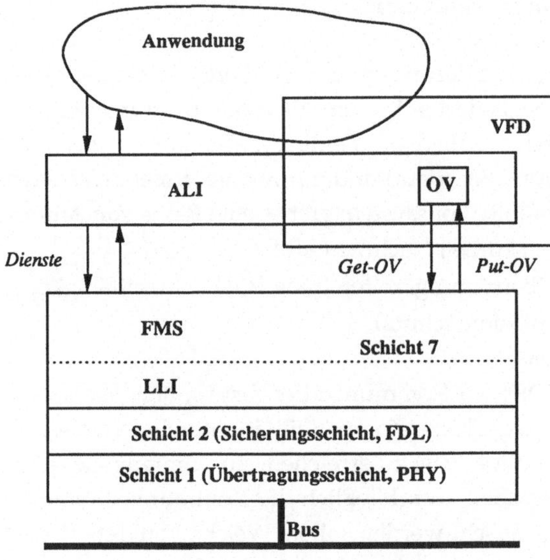

Bild 1: Schichtenstruktur

3. Die PROFIBUS Schicht 7

Die Anwendungsschicht (Schicht 7 der OSI-Hierarchie) öffnet sich direkt dem Anwender, bzw. dem Anwendungsprozeß eines Gerätes. Die einem Anwender dargebotene Abstraktion einer bestimmten Klasse von realen Feldgeräten nennt man allgemein *virtuelles Feldgerät* (Virtual Field Device, VFD). Jede virtuelle Geräteklasse zeichnet sich durch die Festlegung bestimmter Klassenmerkmale aus, die die gemeinsame Verständigung erlauben. Bei PROFIBUS werden diese Klassenmerkmale durch den Teil eines Anwendungsprozesses repräsentiert, der durch PROFIBUS-Dienste sicht- und manipulierbar ist. Dieser besteht aus Objekten, auf die eine Reihe von Diensten angewendet werden können.
Objekte sind die Gegenstände, über die Information ausgetauscht wird. Sie bilden die Basis einer gemeinsamen Verständigung über ein Kommunikationssystem hinweg. Jeder Anwen-

dungsprozeß besitzt eine Reihe von Objekten, wie z.B. Variablen, Speicherbereiche, Programme, usw. Alle die Objekte, die von der Kommunikation her sicht- und erreichbar sind, heißen *Kommunikationsobjekte*.

3.1. Zugriffsmechanismus auf Objekte bei PROFIBUS

Der Zugriff auf die Kommunikationsobjekte erfolgt über spezielle Kurzadressen, in PROFIBUS *Indizes* genannt, die eine schnelle, nur wenige Oktetts lange Nachricht über den Bus erlauben. Diese Indizes sind beim jeweiligen Feldgerät im sogenannten *Objektverzeichnis* aufgeführt. Dieses Objektverzeichnis beinhaltet die Beschreibungen aller bei diesem Gerät über den Bus zugänglichen Objekte. Die Definition der Objektbeschreibung erfolgt bei demjenigen Busteilnehmer, bei dem das Objekt real existiert (*Source-OV*). Die Kommunikationspartner halten sich jeweils eine vollständige oder teilweise Kopie der Objektbeschreibungen (*Remote-OV*) ihrer Kommunikationspartner. Somit kann jeder Busteilnehmer ein Source-OV für lokal existierende Kommunikationsobjekte und eine oder mehrere Remote-OV besitzen, so wie es in Bild 2 dargestellt ist. Zum Auslesen und Verändern eines Source-Objektverzeichnisses über den Bus werden spezielle Management-Dienste zur Verfügung gestellt. Die Beschreibung eines Objekts besteht u.a. aus dem Index, Namen, Datentyp, evtl. der Länge des Objekts.

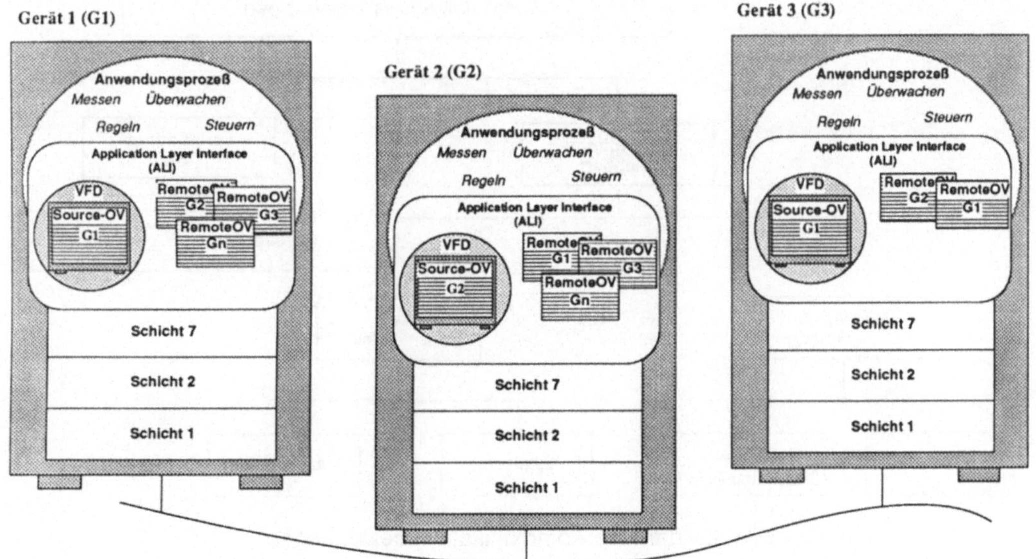

Bild 2: Source- und Remote-Objektverzeichnisse

Die Objekt-Beschreibungen können zwischen den Kommunikationsteilnehmern einmalig oder mehrmalig ausgetauscht werden. Zum eigentlichen Nutzdatenaustausch ist nur noch die Angabe des Index im Dienstauftrag erforderlich.

5. Kommunikationsbeziehungen

a) Kommunikationsbeziehungsliste

Aus Sicht des Anwenders findet die Kommunikation mit den Anwendungsprozessen der Kommunikationspartner über logische Kanäle statt. Diese logischen Kanäle zu den Kommunikationspartnern werden in der Projektierungsphase in der sogenannten *Kommunikationsbeziehungsliste* (KBL) definiert, die jeder Kommunikationspartner für seine Kommunikationsbeziehungen besitzt. Das bedeutet, Kommunikationsbeziehungen werden unabhängig von ihrer Nutzung festgelegt und durch das Netzmanagement in die Schicht 7 geladen.

b) Verbindungslose und verbindungsorientierte Kommunikationsbeziehungen

Grundsätzlich wird bei PROFIBUS zwischen verbindungslosen und verbindungsorientierten Kommunikationsbeziehungen unterschieden (Bild 3). Bei **verbindungsloser** Kommunikation werden die Daten in Pakete verpackt und über einen „öffentlichen" Übertragungsweg an mehrere Adressaten (Multi-/Broadcast) gleichzeitig verschickt. Bei dieser Verbindungsart, vergleichbar mit der eines Rundfunksenders, gibt es keinen zugehörigen Rückkanal. Dem Anwender stehen hier naturgemäß nur die unbestätigten Dienste für Multi- bzw. Broadcast zur Verfügung.

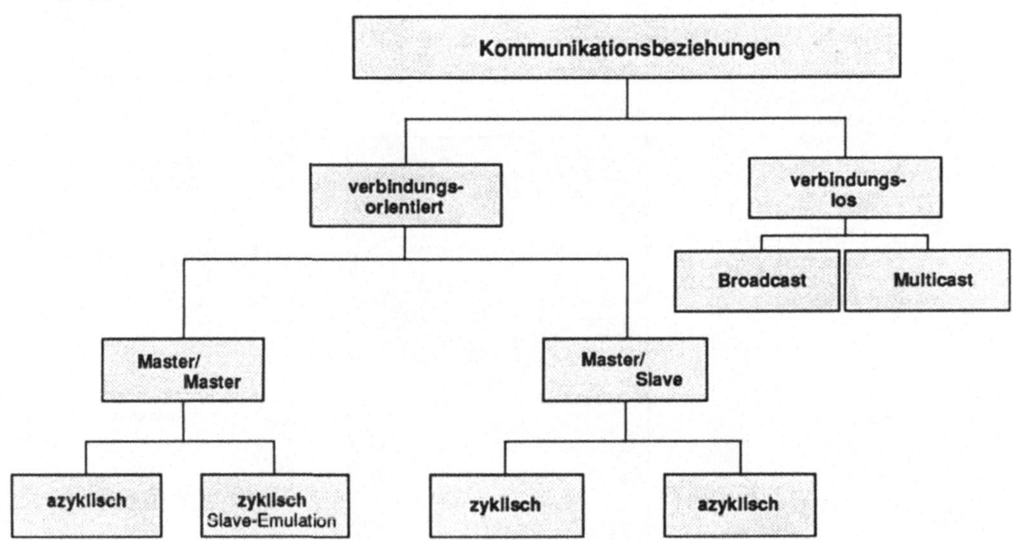

Bild 3: Arten von Kommunikationsbeziehungen

Bestätigte Dienste müssen über **verbindungsorientierte** Kommunikationsbeziehungen übertragen werden, da der Rücklauf der Quittung aus Adressierungsgründen einen Rückkanal erfordert. Im Unterschied zur verbindungslosen Kommunikation muß vor der Übertragung von Daten die Verbindung in einer Verbindungsaufbauphase initialisiert und nach Beendigung der Datenübertragung wieder freigegeben werden. Die Initialisierung erfordert den Austausch und die Vereinbarung von Qualitätsmerkmalen, unter anderem die Sende- und Empfangs-Puffergrößen, und die Art der unterstützten Dienste.

PROFIBUS stellt dem Anwender eine Vielzahl von Kommunikationsbeziehungsarten mit unterschiedlichsten Merkmalen zur Verfügung. In Bild 3 werden schematisch alle unterstützten Varianten aufgezeigt.

c) Offene und definierte Verbindungen

Neben den oben aufgeführten Merkmalen verbindungslos und verbindungsorientiert, gibt es für verbindungsorientierte Kommunikationsbeziehungen noch die Unterscheidungskriterien „Definiert" und „Offen". Bei definierten Verbindungen wird der Kommunikationspartner bereits in der Projektierungsphase festgelegt. In den drei Phasen Verbindungsaufbau, Datentransferphase und Verbindungsabbau liegen die Schicht 2-Adressen bereits fest. Dadurch ist der von der Schicht 2 zur Verfügung gestellte Zugriffsschutz bereits bei der Initialisierung gewährleistet. Bei offenen Verbindungen werden Schicht 2-Adressen erst während der Verbindungsaufbauphase eingetragen. Anschließend verhält sich eine offene Verbindung wie eine definierte.

d) Zyklischer und azyklischer Datenverkehr

Bei Master/Slave-Kommunikationsbeziehungen wird auf der Slave-Seite grundsätzlich von einfachen Geräten ausgegangen, die bezüglich der Schicht 7-Funktionalität bis auf Alarmmeldungen lediglich ein Server-Verhalten aufweisen. Für ein Prozeßabbild wird auf der Master-Seite ein schneller zyklischer Datenaustausch zu Sensoren benötigt. Aktoren müssen vielfach zyklisch aktualisiert werden, damit ein Ausfall der Kommunikationsbeziehung vom Master erkannt wird (Fail-Safe, Redundanzsteuerung usw.). Einfache Ein-/Ausgabegeräte, Barcode-Leser, einfache Regler müssen azyklisch kommunizieren können, wenn der Anwendungsprozeß des Masters im Spontanbetrieb (Laden von Parametern) arbeitet oder die Häufigkeit des Datenaustauschs gering ist (alle 24 Stunden, Prozeßanlauf...). PROFIBUS unterstützt deshalb zur effizienten Abwicklung von Datentransporten beide Methoden des Datenverkehrs.

6. Dienste der Schicht 7

a) Dienstbegriff

Aus der Sicht eines Anwendungsprozesses erbringt das Kommunikationssystem eine Dienstleistung (wie ein Treiber). Diese Dienstleistung besteht darin, den Benutzern eine Reihe von Diensten zur Verfügung zu stellen, mit deren Hilfe genau spezifizierte Aufgabenstellungen erledigt werden können. Bei PROFIBUS erfüllt sie die Aufgabe, dem Benutzer eine transparente und offene Kommunikation zwischen Anwendungsprozessen zu ermöglichen.

b) Bestätigte und unbestätigte Dienste

Grundsätzlich wird zwischen bestätigten (*confirmed*) und unbestätigten (*unconfirmed*) Diensten unterschieden. Bestätigte Dienste sind, aufgrund der zuzuordnenden Bestätigung zum abgesandten Dienstauftrag, nur bei eindeutigen, sogenannten One-to-One-Kommunikationsbeziehungen erlaubt. Unbestätigte Dienste können auch bei den Kommunikationsarten Multi- bzw. Broadcast verwendet werden, d.h. bei One-to-Many oder One-to-All-Kommunikationsbeziehungen (Bild 4).

Diesen verschiedenen Dienstarten stehen Kommunikationsprotokolle für verbindungslose und verbindungsorientierte Kommunikationsbeziehungen zur Verfügung. Das Protokoll für verbindungsorientierte Kommunikationsbeziehungen erfordert zu Beginn der Kommunikation einen Verbindungsaufbau. Nach erfolgreicher Einrichtung einer Verbindung wird in die Datentransferphase übergegangen. Zur Beendigung der Datenübertragungsphase kann die Verbindung mit Hilfe des Abort-Dienstes abgebaut werden.

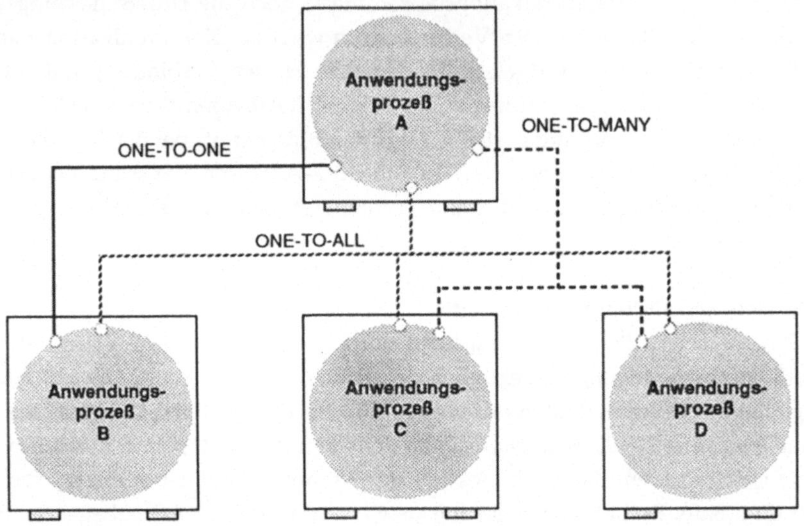

Bild 4: Kommunikationsbeziehungstypen

c) Dienstgruppen

Generell können die PROFIBUS-Kommunikationsdienste in die folgenden drei Funktionsgruppen eingeteilt werden:

- Anwendungsdienste
- Verwaltungsdienste
- Netzmanagementdienste

Zu der ersten Dienstgruppe zählen alle Dienste, mit deren Hilfe auf Kommunikationsobjekte eines Anwendungsprozesses zugegriffen werden kann. Das sind zum Beispiel die Dienste „Read" zum Lesen bzw. „Write" zum Schreiben von Variablen oder etwa Dienste zum Laden, Starten und Stoppen von Programmen.

Die Dienste der zweiten Gruppe werden zwar auch vom Anwender angestoßen, berühren aber nur Objekte bzw. die Kommunikationsbeziehungen eines virtuellen Feldgeräts. Ein Objekt ist zum Beispiel die Beschreibung eines Kommunikationsobjekts im Objektverzeichnis. Ein möglicher Dienst auf dieses Objekt ist „Get-OV". „Get-OV" liest einen Eintrag des Objektverzeichnisses. Zum Einrichten und Abbrechen einer Verbindung auf einer Kommunikationsbeziehung werden die Dienste „Initiate" und „Abort" verwendet.

Die Netzmanagementdienste dienen zum Aufrechterhalten des technischen Betriebs eines Netzes. Hier gibt es z.B. Dienste, die die Größe von Nachrichtenpuffern festlegen, Schichten

zurücksetzen und die Liste der möglichen Kommunikationsbeziehungen laden und verändern. Das Netzmanagement erstreckt sich über alle Schichten von PROFIBUS.
Es müssen nicht alle Dienst-Gruppen von jedem PROFIBUS-Gerät unterstützt werden. Beim Verbindungsaufbau geben sich die Partner gegenseitig die unterstützten Dienste an.

d) Dienstprimitiven

Ein Dienst wird an den Schnittstellen zu den Diensterbringern (Schichten 7 und 2 in PROFIBUS) durch eine genau festgelegte Folge von Dienstprimitiven erbracht. In der Spezifikation der Norm versteht man unter einer Dienstprimitive ein abstraktes, implementierungsunabhängiges Element einer Interaktion zwischen Dienstanforderer und Diensterbringer (Bild 5).

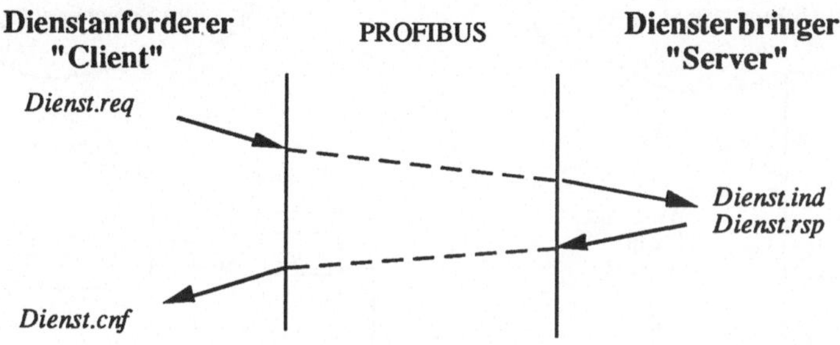

Bild 5: Dienstprimitive

Ein bestätigter Dienst wird auf folgende Dienstprimitive an den Verbindungsendpunkten abgebildet:

Dienstprimitiv	*Kürzel*	*Bedeutung*
Request	.req	Dienstanforderung des lokalen Benutzers
Indication	.ind	Anzeige der eingetroffenen Dienstanforderung beim partnerseitigen Diensterbringer
Response	.res	Antwort des partnerseitigen Diensterbringers
Confirmation	.cnf	Bestätigung des Erhalts der Dienstanforderung beim lokalen Diensterbringer

Bei unbestätigten Diensten entfallen die Dienstprimitive Response und Confirmation.

7. Beispiel

Ein Anwendungsprogramm, lauffähig auf einem PC, hat die Aufgabe, über das Kommunikationssystem PROFIBUS die Öl-Temperatur in einem Kessel zu messen (Bild 6).
Vorausgesetzt wird in diesem Beispiel eine bereits eingerichtete azyklische Verbindung zu dem Slave, dessen Anwendungsprozeß mit der Meßaufgabe betraut ist; die entsprechende Kommunikationsreferenz (KR) soll die 7 sein. Außerdem soll der Einfachheit halber bereits das Objektverzeichnis (OV) aufgebaut sein und eine Kopie als Remote-Objektverzeichnis dem ALI des PC zur Verfügung stehen.
Da PROFIBUS aus Echtzeitgründen bei der Adressierung mit Indizes anstatt mit Namen arbeitet, muß der Name des Meßwerts „Temp_Öl" in den Index mit Hilfe des Remote-Objekt-

verzeichnisses übersetzt werden. Dies kann zur Laufzeit oder statisch, je nach Anforderung an die Anwendung, geschehen. Die Verwaltung der Objektverzeichnisse obliegt dem ALI. In unserem Fall soll Temp_Öl dem Index 112 entsprechen.

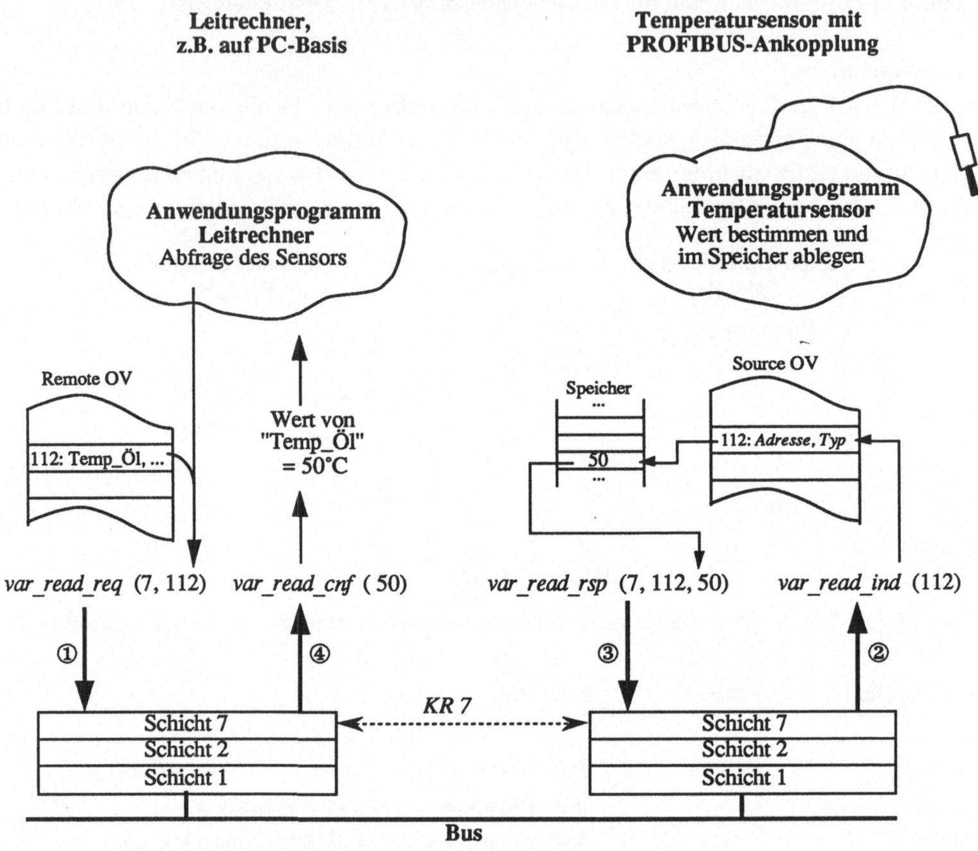

Bild 6: Beispielanwendung

Es muß ein Leseauftrag für eine Einfachvariable mit dem Index 112 und dem Dienst *„Read.request <index>"* vom Anwendungsprogramm über das Application Layer Interface an die Schicht 7/(FMS) übergeben werden (Bild 6, ①). Durch die Angabe der Kommunikationsreferenz und des Index ist die gewünschte Variable eindeutig adressiert.
Im ALI des Slaves wird bei der ankommenden *Var_Read.indication* (Bild 6, ②) über das Objektverzeichnis die Adresse und der Datentyp (z.B. Integer) der Einfachvariable bestimmt, um sie auszulesen. Entspricht die Datendarstellung im realen Gerät nicht dem im FMS festgelegten Integer-Datentyp, ist eine Konvertierung erforderlich. Die Antwort des Leseauftrags enthält nur den Meßwert, hier im Beispiel ist das der Wert 50; die Einheit Grad Celsius ist dem Anwender über das Objektverzeichnis bekannt.
Nach dem Auslesen der Temperatur durch den Anwender wird das Unterprogramm *var_read_rsp (...)* (Bild 6, ③) zum Absenden der Response (Antwort) aufgerufen. Beide Unterprogramme gehören organisatorisch in den Ablaufbereich des ALI.

Ist die Response-PDU beim Dienstanforderer angekommen, muß analog zur *Var_Read.indication* eine *Var_Read.confirmation* (Bild 6, ④) an das ALI abgesetzt werden. Im ALI wird anhand der Auftragskennung dieser Confirmation der zugehörige Leseauftrag identifiziert und in eine vom Anwendungsprozeß zur Verfügung gestellte Datenstruktur der Meßwert 50 eingetragen. Dadurch steht dem Anwendungsprozeß der aktuelle Wert der Öltemperatur zur Verfügung.

8. Leistungsaspekte der Schicht 7

Von einem Feldbus werden im Gegensatz zu anderen Kommunikationssystemen verschiedene Eigenschaften erwartet. Er muß echtzeitfähig sein, d.h. ein bestimmtes Datenaufkommen in deterministisch kurzer Zeit ($\leq$ 10 ms) mit hoher Zuverlässigkeit transportieren. Das Datenaufkommen ist geprägt durch zahlreiche, kurze (wenige Byte) und sich häufig periodisch wiederholende Nachrichten. Ein Feldbusnetz weist oft eine Kommunikationsstruktur mit wenigen zentralen Stationen auf, die mit mehreren ihnen zugeordneten Teilnehmern kommunizieren.

Viele potentielle Feldbusanwender stellen die Frage, ob sich ein Feldbus eine komplexe, noch dazu objektorientierte Schicht 7 unter Leistungsgesichtspunkten überhaupt leisten könne und ob die PROFIBUS Schicht 7 wegen ihrer hohen Funktionalität zu einem deutlichen Leistungsabfall führen würde. Mit derartigen Fragestellungen befaßt sich die Leistungsanalyse.
Das Ziel einer Leistungsanalyse ist die Ermittlung von Abhängigkeiten zwischen *Einflußgrössen* (Systemkonfiguration, Netz- und Lastparameter) und *Leistungsgrößen* [Fun90]. Zu letzteren gehören:

- Laufzeit
  Übertragungszeit zwischen zwei definierten Punkten des Kommunikationssystems
- Reaktionszeit
  Zeit vom Absenden einer Anforderung bis zum Eintreffen der Antwort
- Durchsatz
  Datenübertragungsfluß an einem definierten Punkt des Kommunikationssystems
- Listenbearbeitungszeit
  Zeitspanne in der eine Station sämtliche ihr zugeordneten Stationen einmal abfragt

Prinzipiell lassen sich die gesuchten Leistungsgrößen über Messung in realen Netzen oder durch Simulation von Netzmodellen bestimmen. Für die diesem Beitrag zugrundeliegenden Untersuchungen wurde ein Meßansatz gewählt, weil damit die Anwendersicht erhalten bleibt ("black-box-Methodik"; in das Kommunikationssystem wird nicht eingegriffen) und sich hochgenaue Aussagen über die interessierenden Leistungsgrößen letztlich nur so gewinnen lassen. Die Grundlage der Messungen bildet ein Leistungsmeßnetz, bei dessen Entwicklung auf eine hohe Flexibilität bei der Nachbildung realer Anlagen geachtet wurde.
Das Leistungsmeßnetz besteht aus 5 zum Teil optionalen Komponenten, die auch mehrfach vorhanden sein können (vgl. Bild 7). Das Meßwerkzeug führt die Messungen in Zusammenwirken mit dem Meßobjekt durch und wertet die Meßdaten aus. Es hat sich gezeigt, daß aussagekräftige Leistungsgrößen realistische Lastprofile voraussetzen. Mindestens ein Lastgene-

rator erzeugt diese Buslast, die Lastabsorber wieder korrekt vernichten. Ein Busmonitor beobachtet die Abläufe auf dem Medium.

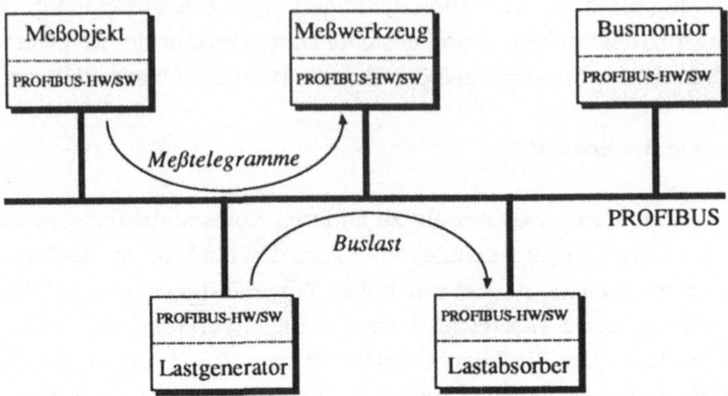

Bild 7: Leistungsmeßnetz

Die durchgeführten Untersuchungen führten zu folgenden tendenziellen Ergebnissen:

- die Busparameter (optimales Einstellen) sind wesentlich für die Leistungsfähigkeit des Gesamtnetzes
- kein erkennbarer Einfluß der Ausbaustufe einer Implementierung (Mindestdienste ... alle Dienste) auf die untersuchten Leistungsgrößen
- sehr deutliche Leistungssteigerung der untersuchten Produkte auf 8051-Basis im Vergleich zu den ersten Prototypimplementierungen
- kurze, effiziente Übertragung (Optimierung durch Indexadressierung, ...)
- optimierte Übertragung bei zyklischen Diensten, da das LLI die Kommunikationsabwicklung überwacht

Zusammenfassend läßt sich festhalten, daß die untersuchten PROFIBUS-Produkte die vorgegebenen Anwendungen aus Gebäudeleit- und Fertigungstechnik gut bewältigen. Auch harte zeitliche Anforderungen werden eingehalten. Die Schicht 7 hat nicht zu den befürchteten drastischen Leistungseinbußen geführt. Für ein offenes Kommunikationssystem, das ein derart breites Anwendungsspektrum abdeckt, weist PROFIBUS im Gegenteil sehr gute Leistungswerte auf.

Literatur

[Ben90] Klaus Bender (Hrsg.) - PROFIBUS - Der Feldbus für die Automation; Hanser 1990

[DIN91a] DIN 19245 Teil 1; Beuth-Verlag, Berlin, 1991

[DIN91b] DIN 19245 Teil 2; Beuth-Verlag, Berlin, 1991

[Fun90] A. Funke - Leistungsuntersuchungen am PROFIBUS; VDI-Berichte Nr. 855, VDI-Verlag 1990

[KaBB90] Katz, Biwer, Bender - Die PROFIBUS-Anwendungsschicht; Automatisierungstechnische Praxis - atp, R. Oldenbourg Verlag, Band 31, Heft 12

Die ISO-Transaktionsverarbeitung als Grundlage für den Nachrichtenaustausch in Verteiltem PEARL

Gabriele Dobler
Ulrich Bohnert
Peter Holleczek

Universität Erlangen-Nürnberg
Regionales Rechenzentrum
Martensstraße 1, D-8520 Erlangen
Telefon: (09131) 85-7031

Zusammenfassung:

Dienste zur Unterstützung verteilter Anwendungen müssen hohe Anforderungen in Bezug auf Kommunikation und Synchronisation erfüllen. Die Standardisierung eines Dienstes für verteilte Transaktionsverarbeitung eröffnet nun die Möglichkeit, solche Anwendungen auf heterogene Systeme zu verteilen und die bisherigen Individuallösungen abzulösen.

Ziel des hier vorgestellten Projekts 'Verteilte Anwendungen im DFN' ist die Bereitstellung eines Dienstes für verteilte Transaktionsverarbeitung und die Entwicklung einer Beispielanwendung.

Der ISO-TP-Dienst ist nur funktional festgelegt und damit unabhängig von der Implementationssprache einer verteilten Anwendung definiert. Mit der Abbildung des Botschaftskonzeptes von Verteiltem PEARL wurde eine Sprachschnittstelle für TP geschaffen, die das Prozeßkonzept von PEARL für eine verteilte PEARL Anwendung erhält und die Verarbeitung von Zeitbedingungen unterstützt. Um ISO-TP in andere Programmiersprachen einbinden zu können, die keine Unterstützung verteilter Anwendungen beinhalten, wird eine prozedurale Pascal-Schnittstelle entwickelt. Mit Hilfe dieser beiden Schnittstellen lassen sich zum einen nicht-PEARL-Programme als Quasi-Prozeß unter verteiltem PEARL betrachten. Zum anderen wird für verteilte Anwendungen, deren Implementation unter Verwendung mehrerer Programmiersprachen erfolgte, die optimale Unterstützung erreicht (z.B. Vektor-Fortran für arithmetische Berechnungen, PEARL für Prozeß-Kontrolle).

Als Beispiel wird eine Anwendung aus dem Bereich der Hochenergiephysik entwickelt. Sie soll die Online-Auswertung von Experimentdaten ermöglichen und eine möglichst rasche Einflußnahme auf die Parametrisierung des Experiments gewährleisten.

Die Implementierung des TP-Protokolls wird für einen IBM-Großrechner mit dem Betriebssystem IBM-VM durchgeführt. Für die Realisierung der TP-Protokoll-Maschine wird die Programmiersprache PEARL eingesetzt, da sie sich als Strukturierungsmittel für die unabhängige Realisierung der Dialoge eignet. Teile der Anwendung werden ebenfalls für diese Anlage entwickelt. Als zweiter Rechner soll eine VAX-VMS zum Einsatz kommen, welcher für die Datenaufnahme vom physikalischen Experiment eingesetzt wird.

1. Einleitung

Dienste zur Unterstützung verteilter Anwendungen müssen hohe Anforderungen in Bezug auf Kommunikation und Synchronisation erfüllen. In der Praxis treten immer häufiger Anwendungen auf, die eine Verteilung der Verarbeitung auf verschiedene Systeme erzwingen oder aus Effizienzgründen nahelegen. Die Standardisierung eines Dienstes für verteilte Transaktionsverarbeitung eröffnet nun die Möglichkeit, solche Anwendungen auf heterogene Systeme zu verteilen und die bisher existierenden Individuallösungen zu ersetzen.

2. Modellvorstellungen

Nachfolgend werden zwei Konzepte zur Unterstützung verteilter Anwendungen vorgestellt und die Unterschiede aufgezeigt.

2.1. Der Dienst 'Transaction Processing' (TP)

Für die verteilte Verarbeitung unter Verwendung von ISO-TP läßt sich durch echte Parallelität von Programmen sowie durch Einsatz spezieller Hardware eine Geschwindigkeitssteigerung erreichen, die dem Geschwindigkeitsverlust durch Kommunikation gegenüber steht.

Ziel der Entwicklung des TP-Dienstes /ISO88/ ist die Unterstützung verteilter Transaktionen. Das zentrale Konzept für die Gewährleistung der Datenkonsistenz /SFB182,89/5/ ist durch die Forderung gegeben, daß eine Transaktion in einer verteilten Anwendung den 4 sogenannten ACID-Eigenschaften (atomicity, consistency, isolation, durability) gehorchen muß. Diese sind in /Weg86/ beschrieben.

Unter einer Transaktion ist ein abgeschlossener, nicht unterbrechbarer Verarbeitungsschritt zu verstehen, welcher von einem konsistenten Zustand ausgehend in einem neuen konsistenten Zustand endet, wobei seine Auswirkungen nachträglich nicht mehr beeinflußt werden können. Als Kommunikationsmittel zwischen den an der Transaktion beteiligten Programmen wird der Dialog eingesetzt, der wahlweise als halb-duplex- oder voll-duplex-Verbindung aufgebaut werden kann.

Als Voraussetzung für die Benutzung des TP-Dienstes gilt:

- Es muß berücksichtigt werden, daß die Transaktionsprogramme durch Dialoge nur zu einer Baumstruktur verknüpft werden können (siehe Bild 1).

- Alle Programme, die über transaktionsunterstützende Dialoge kommunizieren, sind Bestandteil eines Transaktionsbereichs. Ein Transaktionsbereich ist immer ein Teilbaum eines Dialoges und wird als Transaktionsbaum bezeichnet.
- Jedes durch einen Dialogaufbau dem Baum neu hinzugefügte Transaktionsprogramm befindet sich im Initialzustand.
- TP beinhaltet keine Unterstützung bei der Verarbeitung von Zeitbedingungen. Ereignisse, die zeitabhängig sind, unterliegen der lokalen Entscheidung einzelner Programme der verteilten Anwendung.

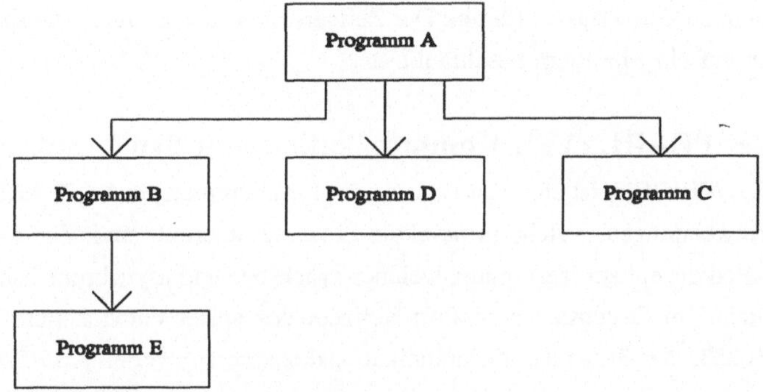

Bild 1: Beispiel eines Dialogbaumes für verteilte Transaktionsverarbeitung

Der TP-Dienst setzt sich zusammen aus Diensten für

- Aufbau, Abbau, Abbruch von Dialogen,
- asynchronen Datentransfer,
- Synchronisation über einen Dialog,
- Fehlermeldungen,
- Anforderungen über Übergabe der Sendeberechtigung,
- Erzeugen von baumübergreifenden Sicherungspunkten,
- Rücksetzen einer Transaktion.

Der asynchrone Datentransfer hat eine sogenannte "send and pray-" oder "no wait send-" Semantik entsprechend dem in /SFB182,89/5/ beschriebenen Kommunikationskonzept. Hinter dem asynchronen Datenaustausch verbirgt sich die Möglichkeit, anwendungsspezifische Dienste (z.B. Remote Operation Service - ROS) zu integrieren.

Für Synchronisation wird ein Handshake-Mechanismus angeboten. Der Initiator des Handshake-Verfahrens kann den betroffenen Dialog bis zum Eintreffen der Handshake-Antwort nur für Fehlermeldungen benutzen. Dieser Dienst kann vom Handshake-Initiator für die Zeitüberwachung eingesetzt werden.

Das Erzeugen konsistenter Zustände geschieht mit Hilfe eines zwei-Phasen-Sperr-Protokolls und bewirkt, daß sich eine verteilte Anwendung in eine Folge von Transaktionen unterteilen läßt. Ein konsistenter Zustand entspricht dem Abschluß einer Transaktion aus einer Transaktionsfolge. Alle in die Transaktion einbezogenen Programme können während der ersten Phase des zwei-Phasen-Sperr-Protokolls eine Zeitüberwachung durchführen und als Ergebnis der Zeitüberwachung oder anderer lokaler Ereignisse das Zurücksetzen der Transaktion erzwingen.

Das Zurücksetzen einer Transaktion ist gleichbedeutend mit dem Wiederaufsetzen auf den letzten Sicherungspunkt.
Dieser Dienst kann zu Zeitverlusten führen. Der Zeitverlust wird aber durch die Existenz von Sicherungspunkten auf ein Minimum beschränkt.

2.2. Verteiltes PEARL (VP), Kommunikation und Synchronisation

Verteiltes PEARL /FHKK83/ ist eine Erweiterung der Programmiersprache PEARL /DIN66253/. Es unterstützt Anwendungen, welche in parallele Prozesse unterteilt sind. Zur Kommunikation steht ein Botschaftskonzept zur Verfügung, welches synchrone und asynchrone Nachrichtenübertragung ermöglicht. Im Gegensatz zum Port-Konzept des später entwickelten "Mehrrechner-PEARL" /DIN66253, Part 3/ ist das Kommunikationskonzept von verteiltem PEARL nicht verbindungsorientiert.

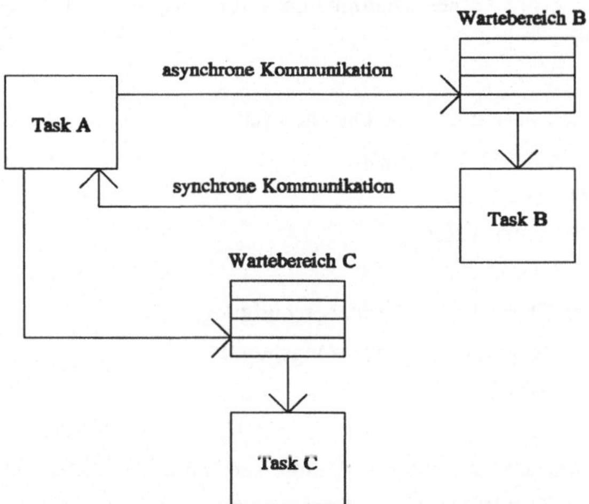

Bild 2: Beispiel einer verteilten Anwendung in Verteiltem PEARL

Die Definition eines Wartebereichs für die Botschaften von einem Sendeprozeß zu einem Empfangsprozeß erzwingt asynchrone Kommunikation, solange im Wartebereich Platz für die Einlagerung einer Botschaft ist. Davon unabhängig kann in der Gegenrichtung synchrone Kommunikation erfolgen, wenn für diese Richtung kein Wartebereich definiert wurde (siehe Bild 2). Die Funktionalität des Botschaftskonzeptes wird in Mehrrechner-PEARL durch die Definition der Protokolle No-Wait-Send, Blocking-Send und Send-Reply zwischen Sende- und Empfangstask abgedeckt.

Außer dem bisher vorgestellten Botschaftskonzept beinhaltet Verteiltes PEARL nichtdeterministische Kontrollanweisungen (guarded statements). Ein guarded statement setzt sich aus mehreren guards (Wächtern) zusammen, welche exklusiv ausgeführt werden. Jeder guard besteht aus einem Bedingungsteil und einem Reaktionsteil. Letzterer kommt nur dann zur Ausführung, wenn alle im Bedingungsteil enthaltenen Bedingungen erfüllt sind.

Ein guarded statement kann von einem Timeout-Zweig abgeschlossen werden (guarded region) oder einen Default-Zweig enthalten (guarded command). Im Timeout-Zweig wird die maximale Wartezeit einer Task für den Fall, daß kein guard ausführbar ist, angegeben. Ein guarded region ohne Timeout-Zweig entspricht einem guarded region mit Timer-Wert unendlich. Wenn ein guarded command einen Default-Zweig enthält, wird für den Fall, daß kein guard sofort ausführbar ist, die Default-Anweisung durchgeführt. Ein guarded command ohne Default-Zweig wirkt wie eine leere Anweisung, wenn kein guard sofort ausführbar ist. Mit Hilfe der nichtdeterministischen Kontrollanweisungen läßt sich das Senden bzw. Empfangen beliebig vieler Botschaften synchronisieren. In Mehrrechner-PEARL ist die durch guarded statements erreichte Funktionalität nur zum Teil nachbildbar. Das gleichzeitige Senden an verschiedene Empfänger ist auf eine vervielfältigte Botschaft beschränkt. Der gleichzeitige Empfang mehrerer Botschaften ist in Mehrrechner-PEARL nicht möglich.

2.3 Unterschiede

Aus dem Vergleich der beiden Konzepte ergeben sich folgende Unterschiede:

- Verteiltes PEARL läßt beliebige Verbindungsstrukturen zwischen Tasks zu. In TP können Transaktionsprogramme nur zu einer Baumstruktur verbunden sein.
- TP-Dialoge sind in zwei Richtungen einsetzbar, während eine PEARL-Botschaft einer gerichteten Verbindung entspricht.
- Die Anzahl der Dialoge zwischen zwei Transaktionsprogrammen ist auf einen Dialog begrenzt. Die Anzahl der Botschaften zwischen zwei Tasks ist nicht beschränkt.
- In Verteiltem PEARL ist synchrone Kommunikation möglich. TP bietet einen Dienst ohne Benutzerdaten für die Synchronisation zweier Dialogpartner an.
- Das Erzeugen von und Rücksetzen zu baumweit gültigen Sicherungspunkten wird nur von TP unterstützt. Der Anwendungsprogrammierer von Verteiltem PEARL muß solche Sicherungspunkte und Rücksetzmechanismen ohne Systemunterstützung realisieren.
- Verteiltes PEARL bietet zeitlich bedingte Kommunikation und Synchronisation explizit an. ISO-TP klammert zeitliche Aspekte aus und überläßt damit die zeitabhängige Reaktion dem Einzelknoten eines Dialogbaumes.

3. Das Projekt

Die Realisierung des TP-Dienstes ist Gegenstand eines vom DFN-Verein geförderten Projekts am RRZE.

3.1. Sinn und Zweck

Ziel dieses Projektes ist der Einsatz des TP-Dienstes für eine Anwendung aus der Hochenergiephysik. Bei der Durchführung von Experimenten aus diesem Bereich fallen am Ort des Experiments hohe Datenraten an, welche nach einer Vorverarbeitung mit Datenreduktion durch ein rechenintensives Analyseverfahren ausgewertet werden. Die für die Analyse nötige Rechenkapazität steht nur an Großrechnern zur Verfügung. Die Kontrolle und Beeinflussung des Meßverlaufs ist durch schnelle Aufbereitung der Analyseergebnisse möglich.

Für die anfallenden Aufgaben ist eine Verteilung auf geeignete Systeme unerläßlich. Der TP-Dienst bietet die geeignete Unterstützung für die Koordination und Kontrolle des geschilderten Ablaufs.

3.2. Betriebssystem-Einbettung der ISO-TP-Protokollmaschine

Der TP-Dienst ist in die Schicht 7 des ISO-Referenzmodells einzuordnen und setzt Dienste unterlagerter Schichten voraus.

Die Realisierung des TP-Protokolls erfolgt zunächst für eine IBM 3090-Anlage mit VM-Betriebssystem. Hier stehen die Netzwerkschichten 1-3 (X.25) und eine Transportschicht für Tranportklasse 0 (OTSS) zur Verfügung. Als zweites System bietet sich ein DEC-Rechner mit VMS-Betriebssystem an, welcher ebenfalls mit den Netzwerkschichten 1-3 (X.25) und einer Transportschicht für Transportklasse 0 (VOTS) ausgestattet ist. Die Installation des TP-Protokolls erfolgt als Service-Maschine, um die Unabhängigkeit der Transaktionsprogramme von der Implementationssprache des TP-Protokolls zu erreichen. Als Implementationssprache für das TP-Protokoll wurde PEARL ausgewählt, da das Prozeßkonzept die unabhängige Verwaltung von TP-Dialogen unterstützt. Die beiden Betriebssysteme VM und VMS werden vom PEARL-Betriebssystem überlagert.

3.3. Schnittstellen für die ISO-TP-Protokollmaschine

Als Schnittstelle wurde zum ersten Verteites PEARL bereitgestellt. Dies bietet den Vorteil, daß eine TP-Anwendung durch ein syntaktisches Hilfsmittel für zeitabhängige Kommunikation bzw. Synchronisation unterstützt wird. Für die Abbildung von verteiltem PEARL auf die ISO-TP-Funktionalität kommen zwei Möglichkeiten in Betracht.

Die Möglichkeiten 1 und 2 in Bild 3 bieten einer PEARL-Task die Schnittstelle von Verteiltem PEARL für TP an. Diese beiden Realisierungen unterscheiden sich in der Art der Abbildung. Möglichkeit 1 bewirkt eine Umsetzung zwischen dem Dienst für Verteiltes PEARL und dem TP-Dienst. Bei der Möglichkeit 2 wird das Protokoll für Verteiltes PEARL mit den genannten Einschränkungen dem TP-Protokoll unterlagert.

Bei der Abbildung sind folgende Punkte zu beachten:

- Eine Anwendung für Verteiltes PEARL, welche auf der Funktionalität von TP aufbaut, muß der in TP geforderten Baumstruktur gehorchen.
- Ein Dialog wird als Träger aller Botschaften zwischen zwei Tasks definiert.
- Synchrone Botschaften sind nur dann zulässig, wenn sie auf eine Folge von TP-Diensten abgebildet werden.
- Der Anwendungsprogrammierer muß sicherstellen, daß das Erzeugen von Sicherungspunkten und das Zurücksetzen auf einen Sicherungspunkt alle Tasks des Teilbaumes erfaßt.

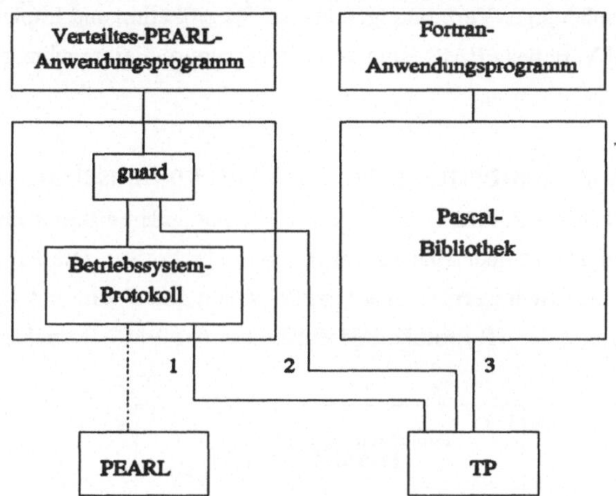

Bild 3: Möglichkeiten für die Anbindung von Tp-Anwendungsprogrammen an TP

Für die Implementation einer zweiten, prozeduralen Schnittstelle wurde die Programmiersprache Pascal ausgewählt. Diese Lösung entspricht dem Zweig 3 in Bild 3. Bei der Verwendung der Pascal-Schnittstelle muß die Verarbeitung von Zeitbedingungen im Anwendungsprogramm semantisch erfaßt werden. Die prozedurale Pascal-Schnittstelle ist in verschiedene Programmiersprachen einbindbar.

Durch Bereitstellen dieser beiden Schnittstellen wird die Kommunikation verteilter PEARL-Tasks mit "nicht-PEARL-Prozessen" ermöglicht.

3.4. Gesamt-Konfiguration

Die Auswahl der beteiligten Systeme wurde auf die Anforderungen der physikalischen Anwendung abgestimmt. Es ergibt sich daraus eine typisch heterogene Hardware-Konfiguration mit räumlicher Trennung. Der Einsatz von standardisierten Protokollen gemäß der Idee von OSI gewährleistet die Kommunikation für solche Anwendungen.

4. Implementation der Protokollautomaten und Betriebssystem-Einbettung

4.1. Programmierumgebung

Als Spezifikationsmethode für die Entwicklung der TP-Service-Maschine wurde PASS (Parallel Activities Specification Scheme) /FLEI84/ eingesetzt. PASS benutzt als Strukturierungsmittel den Begriff des Prozesses. Als Kommunikationsmittel der Prozesse sind sowohl gemeinsame Objekte als auch Botschaften vorgesehen. Bei der Implementierung wurden PASS-Prozesse in PEARL-Tasks umgesetzt. PASS und PEARL bieten durch das Prozeß- bzw. Task-Konzept Vorteile für die Verwaltung unabhängiger Verbindungen bei der Spezifikation und Implementierung von Protokollen. PASS und Verteiles PEARL sind Teil einer Programmierumgebung für verteilte Anwendungen /HA89/.

4.2. Betriebssystemeinbettung der TP-Protokollmaschine unter VM

Die Kommunikation einer PEARL-Benutzermaschine mit anderen Benutzermaschinen über dem VM-Betriebssystem erfolgt über eine Pascal-Programmbibliothek. Die Pascalprozeduren bilden die Schnittstelle zu den Benutzermaschinen der Anwendung und zum unterlagerten OTSS (siehe Bild 4). Die Portierung der TP-Protokoll-Maschine auf das VMS-Betriebssystem ist geplant.

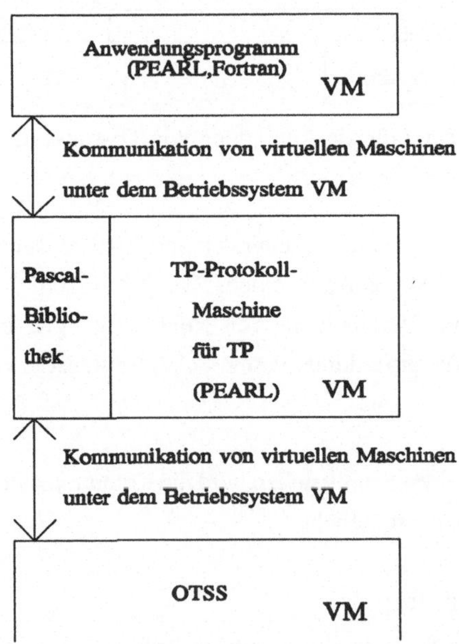

Bild 4: Kommunikation der TP-VM mit anderen VM-Benutzermaschinen

5. Implementation einer exemplarischen Anwendung

Als Anwendung wurde die Auswertung eines physikalischen Experiments gewählt, das im Rahmen eines Projekts des Physikalischen Instituts der Universität Erlangen am CERN durchgeführt wird (siehe Bild 5).

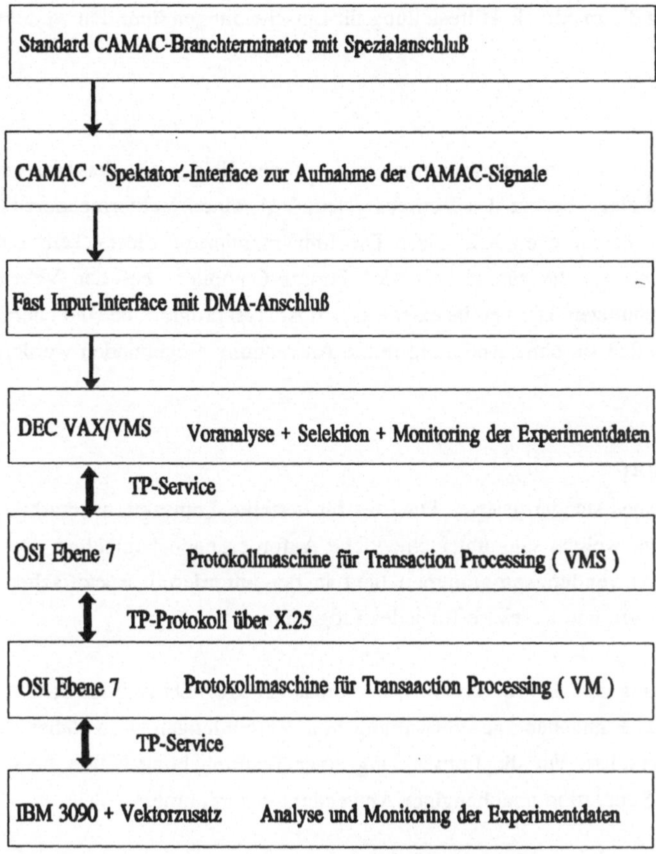

Bild 5: Erfassung und Verarbeitung der Experimentdaten

5.1. VMS

Für die Datenaufnahme am Ort des Experiments wird ein DEC-Rechner mit VMS-Betriebssystem eingesetzt. Die Meßdaten werden durch Vorverarbeitung auf diesem System auf die physikalisch relevanten Ereignisse reduziert. Für die Analyse werden die Daten über einen Zugang zum öffentlichen Netz an die IBM-Anlage der Universität Erlangen übertragen. Die Darstellung der Analyseergebnisse soll graphisch erfolgen, um für die Überwachung und Einflußnahme auf den Experimentverlauf einen möglichst umfassenden Einblick zu gewährleisten. Die Histogramme werden im Hauptspeicher gehalten und asynchron gesichert. Dadurch sind zum einen 1 Hz Raten für das Schreiben von Histogrammen, zum anderen Antwortzeiten im Sekundenbereich möglich. Der Zugriff auf die Histogramm-Datenbank und die graphische Wiedergabe können entweder als zweiter Prozeß oder als seperater TP-Knoten implementiert werden. Die Ergebnisse der Analyse werden graphisch dargestellt. Dies erlaubt den größtmöglichen Überblick über den Experimentverlauf und damit die maximale Hilfestellung für Entscheidungen über den weiteren Verlauf des Experimentes.

5.2. VM

Der rechenintensive Teil der Anwendung wird auf der IBM 3090 der Universität Erlangen ausgeführt. Die Anlage besitzt das Betriebssystem VM, einen Vektorrechnerzusatz und einen Zugang zu einem öffentlichen X.25-Netz. Die Implementierung dieses Teils der Anwendung erfolgt in Fortran, um die Fähigkeiten des Fortran-Compilers bei der Vektorisierung von Programmen auszunutzen. Die bereits existierenden Analyseprogramme sind ebenfalls in Fortran programmiert, so daß sie ohne Änderung in die Anwendung eingebunden werden können.

6. Diskussion

Die Bedeutung eines standardisierten Dienstes für verteilte Transaktionsverarbeitung zeigt sich bei Anwendungen, welche sehr unterschiedliche Anforderungen beinhalten. Der Vorteil einer Verteilung von Anwendungsprogrammen liegt in der anforderungsspezifischen Auswahl der geeigneten Hardware und Software für jedes Programm.

Für die Entwicklung der TP-Protokollmaschine bietet der Prozeßbegriff als Strukturierungsmittel große Vorteile. Die unabhängige Verwaltung von Verbindungen wird daher von PASS und PEARL gut unterstützt. Für die Entwicklung einer Testumgebung bieten PASS und PEARL Vorteile bei der Simulation unabhängiger Anwendungsprogramme.

Die Anwendung ermöglicht Leistungsanalysen in verschiedenen Situationen. Hohe Schreibraten auf Histogramme sind zu erwarten. Außerdem sind hohe Antwortraten notwendig, um den aktuellen Zustand des Experiments zu verfolgen. Die Erfahrung aus diesem Projekt ermutigt uns, TP auch für den Hochgeschwindigkeitsvorrechner im Datenaufnahmesystem einzusetzen.

Literatur

/HOAR 78/ C.A.R. Hoare: Communicating Sequential Processes; Comm. ACM, August 1978

/Dijk 75/ E.W. Dijkstra: Guarded Commands, Nondeterminacy and Derivation of Programs; Comm. ACM, August 1974

/SFB 182,89/5/ M. Fäustle, T. Ruf, P. Schlenk: Grundlage verteilter Systeme; Bericht 89/5 des Sonderforschungsbereichs 182, Erlangen, 1989

/Weg 86/ Klaus Meyer-Wegener: Transaktionssysteme; Dissertation, Universität Kaiserslautern 1986

/ISO88/ Information Processing System - Open System Interconnection - Distributed Transaction Processing, Second Draft Proposal Part 1-3, 9. Dec. 88, ISO/IEC DP 10026 1-3

/FHKK83/ A. Fleischmann, P. Holleczek, G. Klebes, R. Kummer: Synchronisation und Kommunikation verteilter Automatisierungsprogramme; Angewandte Informatik 7/83, 290-297

/FLEI84/ A. Fleischmann: Ein Konzept zur Darstellung und Realisierung von verteilten Prozeßautomatisierungssystemen; Dissertation, Universität Erlangen-Nürnberg, 1984

/HA89/ P. Holleczek, C. Andres: A Programming Environment for Distributed Realtime Applications; IEEE/acm - Hawaii International Conference on System Sciences, 1989

Interprozeßkommunikation und verteiltes Filesystem in einem Echtzeit-Netzwerksystem

Dipl.-Ing. H. Husmann

Universität Hannover
Institut für Regelungstechnik
Appelstr. 11, 3000 Hannover 1

1. Zusammenfassung

Die immer komplexer werdenden Steuerungs- und Regelungsaufgaben in der Automatisierungstechnik lassen sich mit einem Zentralrechner oft nicht mehr sinnvoll lösen. Probleme sind die räumliche Ausdehnung der Prozesse und die mangelnde Leistungsfähigkeit eines einzelnen Rechners. In diesen Fällen bietet sich der Einsatz von verteilten Prozeßrechnersystemen an, die über lokale Netzwerke gekoppelt sind. In diesem Verbund werden neben sehr leistungsfähigen Prozeßführungseinheiten auch einfache Frontend-Datenerfassungsrechner und unabhängige Prozeßvisualisierungsstationen enthalten sein.

Über Feldbusse, deren Normung recht weit fortgeschritten ist, kann eine einfache, herstellerunabhängige Ankopplung von Sensoren, Aktoren und speicherprogrammierbaren Steuerungen erreicht werden. Für die Kopplung von intelligenten Prozeßeinheiten sind die in der Feldbus-Norm Ebene 7 (ISO/OSI Schichtenmodell: Anwendungsschicht) angebotenen Dienste jedoch unzweckmäßig.

Als Alternative wird hier ein verteiltes Filesystem vorgestellt, über das ein durchgängiger Zugriff auf die I/O-Ressourcen im lokalen Netz möglich ist. Zusätzlich können virtuelle Kanäle eingerichtet werden, um eine Prozeßkommunikation über Stationsgrenzen hinweg zu erlauben. Dieses Konzept wurde im Institut für Regelungstechnik realisiert und hat sich im täglichen Einsatz bewährt. Am Beispiel der Steuerung und Überwachung einer Fermentation über das Netzwerk soll die Anwendung der Netzdienste gezeigt werden.

2. Anforderungen an ein lokales Netz in der Prozeßleittechnik

In der Prozeßleittechnik steht man vor der Aufgabe eine große Anzahl von Sensoren und Aktoren, die häufig im Feld räumlich weit verteilt sind, mit der Steuer- und Regelungshardware zu verbinden (s. Bild 2.1). Geht man dabei davon aus, daß der oder die Prozeßrechner in einer zentralen Leitwarte konzentriert sind, so bedeutet das eine sehr aufwendige und teure Verdrahtung über Spezialkabel.

Vernetzt man die Sensoren/Aktoren direkt über einen Feldbus, muß jeder Sensor/Aktor mit der Netzwerk Hard- und Software ausgestattet sein. Zur Zeit ist dies jedoch noch nicht mit vertretbaren Kosten realisierbar. Außerdem müßten dann Regelkreise über den Feldbus geschlossen werden, wodurch sich die Anforderungen an die Reaktionszeiten und die Datenrate des Feldbusses enorm erhöhen. Daher lohnt sich schon im Feld der Einsatz von Prozeßrechnern um die Verbindungen zu Sensoren/Aktoren kurz zu halten und vor Ort eine Datenvorverarbeitung vorzunehmen (s. Bild 2.1). Die eigenständigen Prozeßeinheiten können dann selbstständig Regelungsaufgaben übernehmen und müssen von der Leitebene nur noch mit Parametern versorgt werden.

Diese verteilten Systeme sind nie vollständig entkoppelt, so daß ein Datenaustausch zwischen allen Stationen über ein lokales Netz möglich sein muß. Die im Feld installierten Prozeßrechner sind häufig einer agressiven Umgebung ausgesetzt, die den Einsatz von mechanisch empfindlichen Massenspeichern und Druckern vor Ort verbietet. Daher sollten die Prozeßrechner Zugriff auf I/O-Ressourcen haben, die z.B. in einer Leitwarte den Netzteilnehmern zur Verfügung gestellt werden.

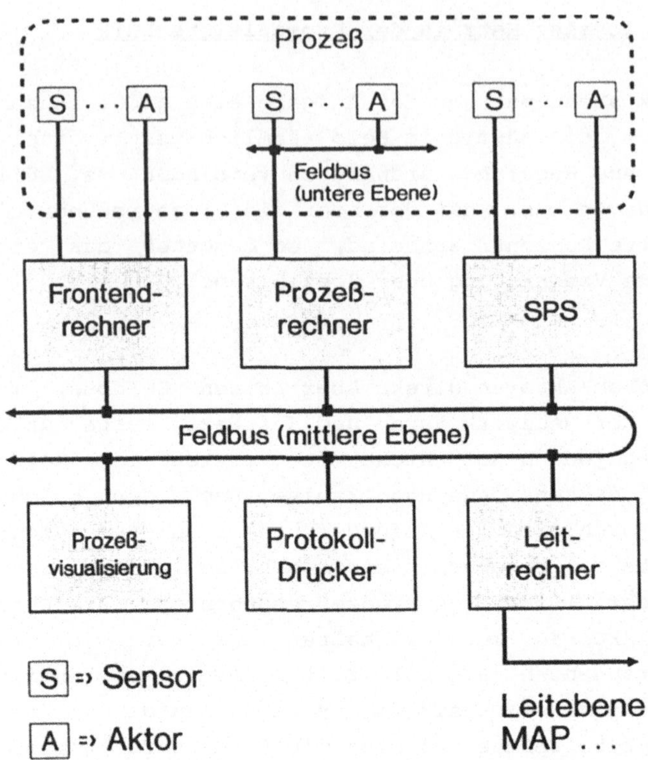

Bild 2.1 Hierarchisches Feldbussystem

2.1. Dienste der Anwenderschicht von Feldbussystemen

Die Normung der Anwenderschicht von Feldbussystemen (Ebene 7 des ISO/OSI Schichtenmodells) ist schon sehr weit vorangeschritten (z.B.: DIN 19245 Teil 2, PROFIBUS). Dort wird jedoch in erster Linie die Vernetzung von Sensoren/Aktoren und speicherprogrammierbaren Steuerungen mit evtl. mehreren Leitstationen unterstützt. Für die Rechner/ Rechner-Kopplung und damit gemeinsamer Nutzung der I/O-Ressourcen im Netz sind sie jedoch nicht ausgelegt. Die geringe Akzeptanz der bisher genormten Feldbus Anwenderschnittstellen zeigt, daß man deren Dienste auf eine einfache Master-Slave Kopplung von Sensoren/Aktoren mit einem Prozeßrechner beschränken sollte, um dann die verteilten Systeme über eine speziell für die Rechner/Rechner-Kopplung ausgelegte Anwenderschicht zu verbinden.

2.2. Dienste eines echtzeitfähigen verteilten Filesystems

Das I/O-System eines Prozeßrechners wird um alle im Netz enthaltenen I/O-Ressourcen erweitert. Das heißt, es entsteht ein verteiltes Filesystem, bei dem nicht mehr zwischen lokalen und auf einer anderen Netzwerkstation vorhandenen Files unterschieden wird. Dieses Konzept ist speziell auf die Rechner/Rechner-Kopplung zugeschnitten und gestattet einen Datenaustausch zwischen den Netzwerkteilnehmern auf Dateiebene. Der Zugriff auf externe Datenstationen sollte auch auf Hochsprachebene ohne Strukturänderung der Programme möglich sein.

Ein Zugriff auf das verteiltes Filesystem in einer Echtzeitumgebung darf, wie jeder andere I/O-Zugriff auch, nie zu einer Blockierung des eigenen Rechners oder der Zielstation führen. Ein zyklisch eingeplanter Regler darf z.B. nicht unzumutbar weit aus seinem Zeittakt gedrängt werden. Ein echtzeit-/multitaskingfähiges Betriebssystem ist daher unabdingbare Voraussetzung für die Implementierung eines verteilten echtzeitfähigen Filesystems. Daurch kann der Datenaustausch über das lokale Netz quasiparallel und interruptgesteuert im Hintergrund abgearbeitet werden, ohne die eigentlichen Regelungs- und Steuerungsaufgaben zu beeinträchtigen.

3. Netzwerkrealisierung am Institut für Regelungstechnik

3.1. Untere Protokollebenen des lokalen Netzes

Die Schichten eins und zwei des lokalen Netzes werden über ein an die DIN 19241 (PDV-Bus) angelehntes Protokoll abgewickelt. Als Übertragungsmedium dient eine verdrillte Zweidrahtleitung, an die die Prozeßeinheiten über sogenannte "Network Interface Units" (NIU) parallel an den Bus angeschaltet werden. Die Datenübertragung erfolgt seriell asynchron mit Pegeln nach RS 485 und einer Übertragungsrate von 48 kBaud.

Der PDV-Bus benutzt ein deterministisches Zugriffsverfahren mit zentraler Vergabe der Sendeberechtigung durch eine Leitstation. Dabei ist prinzipiell jede Station in der Lage, die Funktionen einer Leitstation zu übernehmen. Es ist dadurch im Betrieb des Netzes jederzeit möglich, beliebige Stationen abzuschalten oder neue einzuschalten, ohne eine laufende Datenübertragung zu stören.

Die Protokollsoftware unterstützt den Nachrichtendienst "Daten schreiben mit Quittung", so daß eine angeschlossene Prozeßeinheit Nachrichtenpakete codetransparent an eine andere Netzwerk Station verschicken kann. Die gesamte Datensicherung wird dabei von der NIU übernommen.

3.2. Netzwerkankopplung

Für die Ankopplung einer Prozeßeinheit an den PDV-Bus stehen drei Interface-Typen zur Verfügung (s. Bild 3.1):

1) VME-Bus Interface

Für Prozeßrechner, die einen VME-Bus Anschluß besitzen, steht eine Slaveprozessorkarte für die Ankopplung an den PDV-Bus zur Verfügung. Das Bus-Protokoll wird vollständig auf der Slave-CPU abgewickelt, so daß die Master-CPU entlastet wird. Ein 64 kByte großes dual-portet Ram dient als Schnittstelle zum Austausch der Nachrichtenpakete.

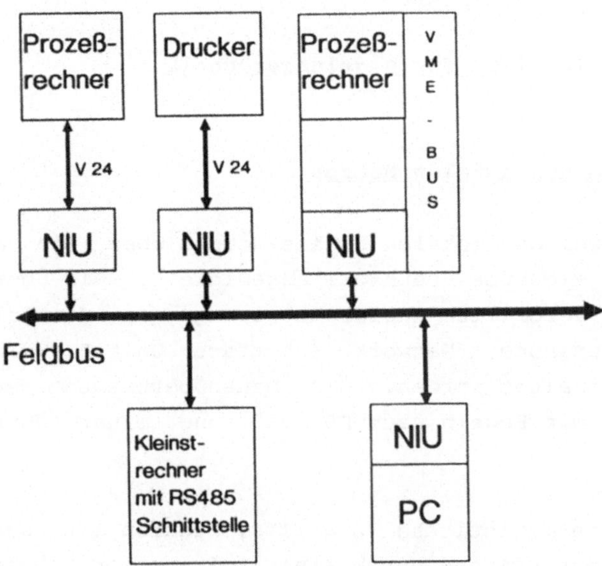

Bild 3.1 Ankopplung an den Feldbus

2) V-24 Interface

Ein mit einem Motorola 68008 ausgerüstetes Interface dient zum Anschluß von Prozeßeinheiten mit V-24 Schnittstelle. Über einen Schalter kann eine Software Option angewählt werden, bei der das Interface auch Teile der höheren Protokollebenen ausführt, um z.B. einen Drucker direkt am Interface zu betreiben.

3) Kleinstrechner mit RS 485 Schnittstelle

Bei den ersten beiden Interface-Typen wurde das PDV-Bus Protokoll von einem eigens dafür abgestellten Prozessor abgewickelt. Für einen einfachen Steuerrechner ist das nicht tragbar. Es zeigte sich jedoch, daß für die Abwicklung des PDV-Bus Protokolls maximal 20% der Rechenleistung des 68008 benötigt werden, so daß für Anwenderprogramme genügend Reserven verbleiben. Rechner, die mit einer RS-485 Schnittstelle und dem Betriebssystem RTOS-UH ausgestattet sind, können daher ohne Slave-Prozessor am PDV-Bus betrieben werden.

3.3. Der Netzwerk Filehandler

Auf die von den NIU's angebotene Schnittstelle setzt ein Filehandler auf, der den Anwenderprozessen folgende Dienste zur Verfügung stellt:

- Ein verteiltes Filesystem, das jedem Prozeß den Zugriff auf alle I/O-Geräte im Netz erlaubt. (s. 3.3.1.)

- Virtuelle Kanäle, über die Prozesse auf unterschiedlichen Stationen Daten austauschen können. (s. 3.3.2.)

- Über eine Kommandodatenstation ist es möglich, jeden Rechner im Netz fernzusteuern (remote control) oder über ein login von einer anderen Netzwerkstation zu bedienen (remote login). (s. 3.3.3)

- Ein Nachrichtensystem, mit dem der Nutzer eines Prozeßrechners von der Konsole aus Nachrichten an andere Stationen verschicken kann.

Der Filehandler ist in einen hardwareunabhängigen Protokollhandler und einem hardwareabhängigen Driver aufgeteilt. Um den Filehandler an einen anderen Interface-Typ anzupassen genügt es daher den relativ kleinen Driver zu ändern, während die Protokollmaschine unverändert bleibt. Auch der Übergang zu einem anderen Netzwerk ist somit problemlos möglich, solange der Dienst "Daten schreiben mit Quittung" unterstützt wird. Durch diese Struktur ist die Implementierung eines Gateways denkbar einfach, es müssen dann zwei Driver, jeweils einer pro Netz, eingerichtet werden.

Bei derart weitreichenden Möglichkeiten des Eingriffs über das Netz muß auch der Schutz einer Station vor unberechtigtem Zugriff gewährleistet sein. Der RTOS-UH Filehandler ist prinzipiell als offenes System konzipiert worden. Man kann jedoch in jeder Station über eine interne Liste bestimmen, welche I/O-Pfade für einen Netzzugriff freigegeben sind. Diese Liste ist selbstverständlich über das Netz nicht manipulierbar.

Der beschriebene Filehandler ist optimal an das Echtzeitbetriebssystem RTOS-UH angepaßt. Daneben wurde ein portabler Filehandler in der Programmiersprache "C" implementiert, um die unbedingt notwendige Verbindung zu anderen Betriebssystemen zu schaffen. Dadurch konnten z.B. UNIX- und MS-DOS-Rechner in das Netz integriert werden.

3.3.1. Das verteilte Filesystem aus der Sicht des Anwenders

Um auf I/O-Geräte einer fremden Netzwerkstation zuzugreifen wird dem Gerätebezeichner der Name dieser Station vorangestellt. Mit dem Befehl "LOAD /ST4/H0/PROG.EXE" wird z.B. das Programm PROG.EXE von der Festplatte H0 der Netzwerkstation ST4 geladen. Durch diese Erweiterung des Filesystems fügt sich der Zugriff auf Netzwerkstationen nahtlos in die Bedienung der lokalen Station ein. Auch der Zugriff von Hochsprachebene aus erfordert keine Sonderbehandlung von externe I/O-Geräten, da sie sich exakt so verhalten wie die lokalen.

3.3.2. Nutzung von virtuellen Datenkanälen

Neben dem Zugriff auf I/O-Ressourcen benötigen viele Anwendungen eine Möglichkeit der Kommunikation von Prozessen über das Netzwerk. Es lassen sich daher virtuelle Netzwerk-Kanäle einrichten, die eine unidirektionale Verbindung zwischen zwei Stationen darstellen. Auch für die Implementierung von Mehrrechner PEARL (DIN 66253 Teil 3) sind solche Kanäle zwingend notwendig. Die virtuellen Kanäle sind jedoch auch ohne Spracherweiterung auf Hochsprachebene nutzbar, da sie im Filehandler als Datenstationen implementiert sind. Sie sind über den Gerätenamen "/CH*" erreichbar (z.B.: /ST4/CHANEL1) und können mit den normalen Ein/Ausgabebefehlen (in PEARL z.B.: PUT, GET) angesprochen werden.

3.3.3. Fernbedienung/Fernsteuerung über das lokale Netzwerk

Das RTOS-UH Betriebssystem bietet durch die sogenannte "XC-Datenstation" die Möglichkeit Bedienbefehle automatisiert auszuführen. An die XC-Datenstation übergebene Texte werden ausgeführt, als wären sie von der Konsole aus eingegeben worden. Da auch diese Datenstation über das Netz erreichbar ist, können z.B einfache Frontendrechner programmgesteuert über das Netz fernbedient werden (remote control).

Die Bedienung einer Netzwerkstation kann über eine Login-Prozedur erreicht werden. Dabei richtet der Filehandler auf dem Zielrechner dynamisch einen neuen User ein, der nach dem Logout wieder aufgelöst wird. Nach dem Login bedient man die Zielstation, als wäre das eigene Terminal als Konsole angeschlossen.

3.3.4. Echtzeitfähigkeit des Filehandlers

Grundlage für einen echtzeitfähigen Filehandler ist selbstverständlich ein echtzeitfähiges Netzwerk. Bei der Realisierung am Institut für Regelungstechnik wurde ein PDV-Feldbussystem benutzt. Durch das deterministische Zugriffsverfahren ist eine garantierte maximale Zeitspanne angebbar, innerhalb der ein vom Filehandler an den PDV-Bus übergebenes Nachrichtenpaket, bei der Empfängerstation ankommt.

Daneben sind auf Filehandlerebene zusätzliche Vorkehrungen zu treffen, um die Echtzeiteigenschaften des Netzwerkes zu garantieren. Die Aufträge des Filehandlers sind nach der Auftrags-Priorität geordnet und es wird der lauffähige Auftrag mit der höchsten Priorität bearbeitet. Auf der Server-Seite richtet der Filehandler pro Auftrag eine Subtask mit der Priorität des initiierenden Auftrags ein, so daß auch ein Server-Auftrag prioritätsgerecht bearbeitet wird. Das bedeutet, daß der Prioritätsvergabe ein systemweites Konzept zugrunde liegen muß, wobei ein geeignetes CASE-Tool hilfreich sein kann.

4. Ein Anwendungsbeispiel

Am Institut für Regelungstechnik wurde im Rahmen eines Forschungsprojektes die Lysin-Fermentation in einem rechnergesteuerten Biofermenter untersucht. Als Steuerrechner dient ein mit einem Motorola 68008 Prozessor ausgerüsteter Einplatinenrechner (BDE 1000 der Fa. Gefec Hannover), der über ein serielles Interface in das Feldbus Netz eingebunden ist (s. Bild 4.1). Die Bedienung der BDE 1000 kann von jeder Netzwerkstation erfolgen, die die entsprechenden Zugriffsrechte besitzt. Die Fernbedienbarkeit erlaubt einen komfortablen Umgang mit dem Fermenter, der sich mehrere Kilometer entfernt in einem anderen Gebäude befindet. Da sich die BDE 1000 direkt neben dem Fermenter in einem biotechnologischen Labor befindet, verbietet sich der Einsatz von mechanisch anfälligen Massenspeichern. Daher wird die gesamte Steuer-, Regelungs- und Dokumentationssoftware vor Beginn einer Fermentation über das Netz geladen und gestartet.

Während der Laufzeit der Fermentation steuert und regelt die BDE 1000 eigenständig den Fermenter, und sammelt alle von der Instrumentierung gelieferten Prozeßdaten. Die Kommunikation mit einem aktiven Leitrechner und einer Prozeßvisualisierungsstation erfolgt über drei Datenkanäle. Über einen Befehlskanal erhält die BDE 1000 Befehle und neue Parameter, über einen Datenkanal werden die aktuellen Prozeßdaten an eine Visualisierungseinheit geschickt. Der dritte Kanal dient zum zyklischen Abruf der auch in der BDE 1000 gespeicherten Prozeßdaten.

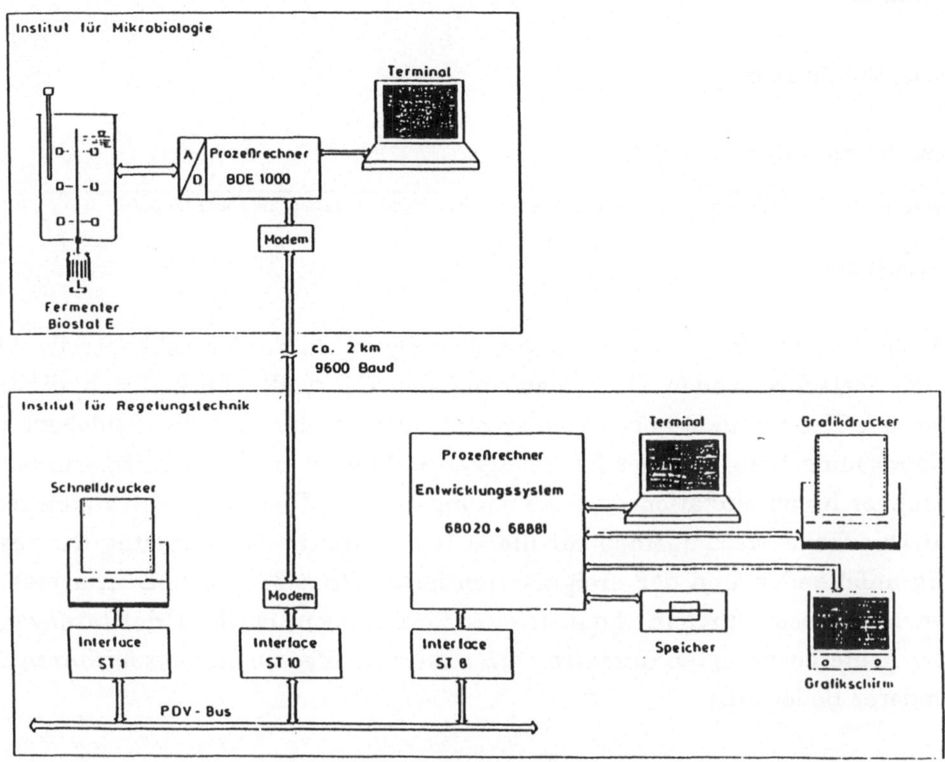

Bild 4.1 Überwachung und Regelung einer Lysin-Fermentation

Prioritätsmechanismen für Kommunikationssysteme der Prozeßkontrolle

Martine Schümmer
RWTH Aachen*)
Lehrstuhl für Informatik IV

*) inzwischen Philips Forschungslaboratorium Aachen, Weißhausstr., 5100 Aachen, Tel.: 0241-6003509

1. Einleitung

Die bisher sternförmige analoge Datenübertragung in der Prozeßkontrolle wird mit der fortschreitenden Digitalisierung der Kontrolltechnik durch digitale busförmige Kommunikationsmedien ersetzt. Insbesondere auf der Feldebene der Fertigungsumgebung und der Fahrzeugkontrolle wird zur Zeit die Spezifikation innovativer Kommunikationssysteme vorangetrieben. Diese Systeme bieten recht unterschiedliche Mechanismen für die echtzeitkritische Übertragung der regelmäßig anfallenden und der ereignisorientierten Prozeßdaten (z.B. Alarme). In Anbetracht dieser durchaus konträren Anforderungen, ist die Prioritätenvergabe der heute bevorzugten *dezentral organisierten Medienzugangsverfahren* von besonderer Bedeutung.

So unterscheiden Feldbussysteme in der Regel nur zwischen zwei (PROFIBUS) bis vier (MiniMAP) verschiedene Prioritäten für den Medienzugang. Alternative Systeme wie das japanische Ringkonzept IINET oder die Intra-Vehikel-Übertragungssysteme unterscheiden eine Vielzahl von Prioritäten. Der Hauptunterschied dieser Systeme zu den Feldbussystemen besteht in der Tatsache, daß die Prioritätenzuordnung in Intra-Vehikel-Systemen den Medienzugang einzelner Nachrichten regelt. Die Zuordnung in Feldbussystemen dient nur dazu, stationsbezogen die Sendereihenfolge der Datenpakete zu entscheiden. Die Prioritätenvergabe geschieht hier stationsintern vorrangig nach *Dringlichkeit* des Nachrichtentyps, während sie in Intra-Vehikel-Systemen eher *nachrichtenorientiert* (bzw. funktionsorientiert) vergeben wird.

Im folgenden werden die Konzepte und adäquaten Metriken für die Bewertung von prioritätsgesteuerten Medienzugangsverfahren für die Prozeßsteuerung mit Echtzeitanspruch vorgestellt. Darüberhinaus werden exemplarisch einige Ergebnisse von Performance-Analysen der jeweiligen Konzepte präsentiert und diskutiert. Ausführliche Darstellungen der Ergebnisse können in /SCH90b, HER90, REI90, RID90/ nachgelesen werden.

2. Metriken der Prozeßkontroll-Kommunikation

Neben den üblichen Leistungsmaßen sind in der Prozeßkontrolle spezielle Leistungsgrößen zu beachten.

Besonders wichtig ist die *Polling-Verzögerung* /SCH89a/, die den Abstand aufeinanderfolgender Abfragezyklen (synchroner Verkehr) wiedergibt. Abfragezyklen stellen in Kontrollsystemen die Hauptkommunikationsstruktur zwischen den Master-Stationen (Kontrollstationen) und den Slave-Stationen (Meßgeräte) dar. Dieser Zeitabstand unterliegt in Kontrollsystemen, wie in der Fertigung oder Fahrzeugkontrolle, der Forderung der *Rechtzeitigkeit.* So muß für eine Kontrollstation sichergestellt werden, daß der jeweils aktuelle Wert regelmäßig in anwendungsabhängigen Zeitabständen abgefragt wird und zur Weiterverarbeitung vorliegt.

Bei Systemen, die Prioritäten vergeben, muß insbesondere die Möglichkeit der Verzögerung hochpriorer Datenpaketen durch niederpriore untersucht werden (*Prioritätskonflikt*). Prioritätskonflikte können in zwei Ausprägungen vorliegen. Zum einen kann die Weiterleitung hochpriorer Datenpakete stationsintern durch niederpriore Datenübertragungen (z.B. bei Zeitscheibenverfahren) verhindert werden. Diese Konfliktsituation soll im folgenden als *Blockierung* definiert und als eigener Leistungsparameter untersucht werden, da stationsinterne Verzögerungen hochpriorer Daten durch adäquate Mechanismen, die allerdings einen Mehraufwand an Protokollroutinen erforderlich machen, verhindert werden könnten. Die stationsexterne Form des Prioritätskonflikts wird im folgenden nur im Leistungsmaß der Paketverzögerung mitbetrachtet, da es sich hierbei um protokollbedingte Verzögerungen handelt, die eine Neuorganisierung des Medienzugangsverfahrens erfoderlich machen würde. Insbesondere in den nachrichtenorierentierten Medienzugangsverfahren (Intra-Vehikel-Systeme und IINET) treten stationsinterne Prioritätskonflikte durch die große Anzahl verschiedener Prioritäten und die von Paketen unterschiedlicher Priorität gemeinsam benutzten Warteschlangen auf.

Ebenso wichtig wie diese meßbaren Größen ist die *Klassifizierung der Echtzeitfähigkeit* der Systeme. So stellt die Verwendung von Warteschlangensystemen und Prioritäten eine Einschränkung der Rechtzeitigkeitsbedingung dar. Um trotzdem Aussagen über die Qualität der Verfahren machen zu können, wird neben der Rechtzeitigkeitsbedingung im strengen Sinn eine abgeschwächte Form definiert. Während die *strenge Rechtzeitigkeitsbedingung* die Garantie einer oberen Grenze für den Zugang einer Nachricht auf das gemeinsame Medium verlangt, reduziert die abgeschwächte Echtzeitbedingung das Problem auf den Medienzugriff einer Station (Abb. 2.1). Die *abgeschwächte Rechtzeitigkeitsbedingung* ist demnach erfüllt, wenn ein oberes Zeitlimit für den Medienzugang einer Station angegeben werden kann.

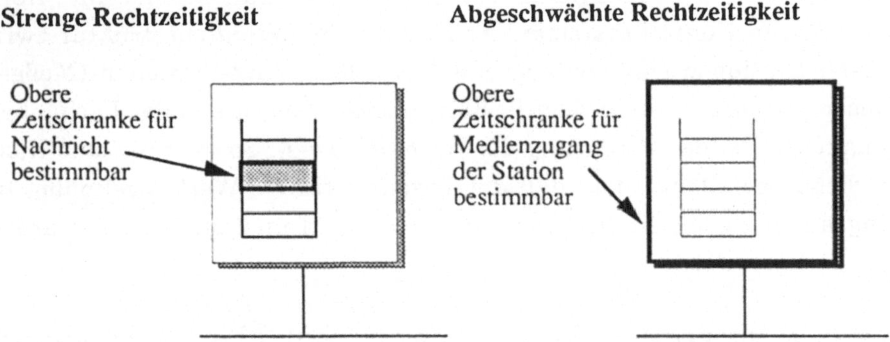

Abb. 2.1: Strenge und abgeschwächte Rechtzeitigkeitsbedingung

Die Anforderung der *Gleichzeitigkeit* der Bereitstellung von Daten für die Synchronisierung und Koordinierung der Geräte (z.B. Roboterarme, die gemeinsam ein Werkstück bearbeiten) soll hier nicht weiter betrachtet werden. Der/die interessierte Leser/in sei auf /SCH90a, SCE*89/ verwiesen.

3. Stationsinterne Vergabe nach Dringlichkeit in Token-Passing-Systemen

Feldbussysteme, wie z.B. der deutsche PROFIBUS und die Feldversion von MAP MiniMAP, verwenden auf MAC-Ebene Token-Bus-Verfahren, die nur eine stark begrenzte Anzahl an Prioritäten vorsehen (zwei /DIN88/ bis vier /IEEE84). Hier führt die Verwendung von *Timern* für die einzelnen Prioritätsklassen oder die Begrenzung der Paketlänge zur Erfüllung der *abgeschwächten Rechtzeitigkeitsbedingung*. Nach der Notation des IEEE 802.4 Standards werden die Timer für die Medienzugangskontrolle einer Station als Token-Holding-Timer (THT) bezeichnet. Um sich dynamisch der Last auf dem Netz anpassen zu können, errechnet sich die Einstellung der THT einer Station aus der Differenz zwischen der vorgegebenen Sollzeit für die Tokenumlaufzeit (target Token Rotation Time (TRT)) und der in der aktuellen Tokenrunde bereits verbrauchten Zeit.

Nach dem Tokenerhalt ist zu klären, in welcher Reihenfolge die sendebereiten Pakete auf das Medium gegeben werden sollen. Hier greift die Prioritätenvergabe von Token-Bus-Systemen. Die Einteilung in Prioritätsklassen wird nach der allgemeinen Funktioalitätstypisierung der Nachrichten in Alarme, Prozeßdaten, Managementdaten u.a. abgeleitet. Somit werden für die normale Prozeßkontrolle keine Prioritäten unterschieden. Im folgenden wird diese Zuordnung mit *Klassifizierung nach Nachrichtentyp* bezeichnet.

Für niederpriore Pakete kann keine Berechnung einer obere Schranke für die Verzögerungen eines Pakets angegeben werden, da sie bei entsprechender Last hochpriorer Pakete das Senderecht innerhalb der THT nicht erhalten (*Monopolisierung* durch Pakete höherer Priorität), falls nur ein Timer für alle Prioritätsklassen vorgesehen ist (z.B. im PROFIBUS). Systeme wie z.B. MiniMAP erlauben die Aufteilung der THT für die verschiedenen Prioritäten durch die Benutzung von einem *Timer pro Priorität*. Somit erfüllen diese Systeme die *Rechtzeitigkeitseigenschaft im strengen Sinn*, wenn die Timer adäquat eingestellt werden. Adäquat heißt hier, daß jede Station nach Tokenerhalt für mindestens ein Paket pro Priorität bei Bedarf übertragen kann. Dies geht bei Beibehaltung der Sollzeit für die Token-Rotationszeit auf Kosten der hochprioren Pakete.

Besonders wichtig ist in Produktions- und Fahrzeugkontrollsystemen, daß das *regelmäßige Polling* der Geräte rechtzeitig durchgeführt wird. Mit der Einführung von Prioritäten auf der MAC-Ebene ist die Möglichkeit gegeben, diesem Nachrichtentyp eine spezielle Priorität zuzuordnen und damit gegenüber anderen Nachrichten zu bevorzugen. In der Regel kann es jedoch nicht sinnvoll sein, den Polling-Paketen auf MAC-Ebene die höchste Priorität zuzuordnen, da diese für die Alarmnachrichten vorbehalten sein sollte. Aus diesem Grund werden den Produktionskontroll-Paketen i.a. die zweit-höchste Prioritätsklasse zugeteilt. Für diese Klasse gilt jedoch auch die oben erwähnte Monopolisierungsgefahr durch Pakete der höchsten Priorität.

Der Prioritätsvergabemechanismus des PROFIBUS bietet eine besondere Unterstützung von Pollingnachrichten, die in der folgenden Graphik abgebildet ist (Abb. 3.1).

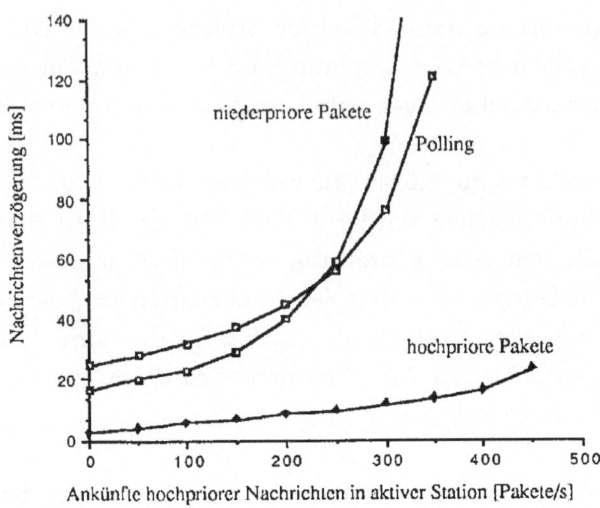

Abb. 3.1: Pollingverzögerung im Vergleich zu nieder und hochpriorer Paketverzögerung

Den im Abb. 3.1 dargestellten Simulationsergebnissen liegt ein Szenario von vier Master-Stationen, eine Datenrate von 500 Kbit/s, exponentiell verteilten Zwischenankunftszeiten, eine TRT-Einstellung von 10 ms und eine konstant angebotene Last an asynchronen, niederprioren Paketen von 13,19 Kbit/s pro Station zugrunde. Die Hauptlast auf dem Medium entsteht jedoch durch zyklische Wiederholung der Pollingaufträge, die stationär auf MAC-Ebene in einer Pollingtabelle abgelegt sind. Das Medienzugangsverfahren funktioniert nach stationsintern wie folgt: zunächst werden die hochprioren Datenpakete übertragen. Im Anschluß werden die Pollingaufträge abgewickelt, bevor die niederprioren Daten gesendet werden. Wird die Abarbeitung der Pollingtabelle durch den Timer unterbrochen können keine niederprioren Nachrichten übertragen werden. Die Pollingtabelle wird an der Abruchstelle bei erneutem Tokenerhalt weiterbearbeitet /REI*90/.

Die mittlere Polling-Verzögerung berechnet sich aus der Zeit, die zwischen zwei aufeinanderfolgen Ausführungen der Pollingtabellen vergeht, da für Pollingaufträge (synchron) keine Ankunftsrate angegeben werden kann. Die unterschiedliche Berechnung der Verzögerung ist bei der Bewertung der Ergebnisse zu beachten.

Insgesamt ist festzuhalten, daß die Steigerung der hochprioren Nachrichten sich direkt auf die Verzögerungen der Pollingnachrichten und damit indirekt auch auf die asynchronen Nachrichten niedriger Priorität auswirkt. Der Anstieg der Verzögerungszeiten der hochprioren Pakete verläuft ab der Ankunftsrate von 150 Paketen pro Sekunde viel moderater als jener der übrigen Pakete. Bis zu dieser Belastung werden die hochprioren Nachrichten in jeder Tokenrunde auf Kosten der Pollingnachrichten übertragen, ohne jedoch die Abarbeitung der Pollingtabelle um mehr als eine TRT zu verzögern. Dabei liegt die Verzögerung der niederprioren, asynchronen Pakete unter denen der Pollingverzögerungen, solange die niederprioren asynchronen Pakete alle ihre Sendewünsche nach jeder Abarbeitung der Pollingtabelle erfüllen können. Bei der Last von 300 hochprioren Paketen pro Sekunde werden die niederprioren, asynchronen Pakete bereits im Mittel durch die zweimalige Abarbeitung der Pollingtabelle verzögert. Dies wird dadurch deutlich, daß die mittlere Verzögerung dieser Pakete über der Verzögerung liegt, die sich aus Abarbeitung der Pollingtabelle plus Verzögerung bis zum nächsten Start einer Pollingtabellenabarbeitung ergibt.

4. Systemweit eindeutige Vergabe (nachrichtenorientiert)

Der zufällige Zugriff *(Random Access)* auf das gemeinsame Medium stellt die einfachste Form der dezentrale Organisation eines Buszugangs dar. Allerdings sind in echtzeitkritischen Anwendungsumgebungen Kollisionen durch den zufälligen Zugang oder die nicht nach einer vorgegebenen Zeit beendeten Kollisionsaufhebungsmechanismen untragbar. Deswegen werden für in Kommunikationssystemen für Kontrollumgebungen, die Random-Access-Verfahren anwenden, verschiedenen Nachrichtentypen jeweils Prioritäten zugeteilt, die eine Bestimmung einer *eindeutigen Rangfolge* aller gleichzeitig auf das Medium gegebener Datenpakete zuläßt. So werden in den Industriestandards für Intra-Vehikel-Kommunikationssyteme (wie z.B. CAN und Abus) bis zu 2023 verschiedene Prioritäten angeboten. Diese Art der Prioritätszuteilung erlaubt in Zusammenhang mit engen hardwäremäßigen Randbedingungen (Medienlänge kleiner 40m) und einer rezessiven bzw. dominanten Darstellung des Eins- bzw. Nullbits eine effiziente Regelung des Medienzugangs.

In ringförmigen Token-Passing-Systemen kann der Medienzugang ebenfalls nachrichtenorientiert erfolgen, indem das Token durch ein Prioritätsfeld erweitert wird, das die jeweils höchste Priorität aller vorliegenden Übertragungswünsche angibt. Zunächst überschreibt jede Station eine im Vergleich zu der bei ihr vorliegenden niedrigere Priorität im Tokenfeld. In einer zweiten Tokenrunde wird anschließend die Station, die das Senderecht erhält, bekanntgegeben. Ein solches Medienzugangsverfahren liegt dem japanischen Industrie-Standard IINET (Integrated Instrument Network) zugrunde.

In beiden Systemarten wird die Prioritätenvergabe nach der stationsübergreifenden prozeßrelevanten Dringlichkeit der Datenübertragung vorgenommen. Probleme ergeben sich allerdings durch die gleichzeitige Verwendung der Prioritäten für die Zuordnung des Senderechts. Nur der Nachrichtentyp mit der systemweit höchsten Priorität erfüllt die strenge Rechtzeitigkeitsbedingung. Alle anderen Nachrichtentypen können durch diesen Typ auf unbestimmte Zeit verzögert werden und erfüllen somit nicht einmal die abgeschwächte Rechtzeittigkeitsanforderung.

Ein weiterer Schwachpunkt der nachrichtenorientierten Prioritätsvergabe ist das stationsinterne Blockieren von höherprioren Datenpaketen, durch die Benutzung von Wartestrukturen, welche die Priorität der Daten nicht berücksichtigen (z.B.: FIFO-Warteschlangen). In der Regel können höherpriore Nachrichten nur durch anwendungsbedingte Unterbrechungsroutinen der höheren Ebenen aus einer Blockierungssituation befreit werden. Die gleichzeitig auftretenden Nebeneffekte, wie z.B. der Verlust der niederprioren Nachricht, sollten dazu führen solche Aktionen weitestgehend zu vermeiden. Sieht man von solchen Mechanismen ab, ergibt ein im folgenden dargestelltes Systemverhalten.

Die folgenden Ergebnisse beziehen sich auf das IINET-System, dem eine Datenrate von 1 Mbps, eine Medienlänge von 300 m, eine Signalgeschwindigkeit von 2/3 c, vier Master- und 25 Slave-Stationen sowie eine Sende- und Empfangspufferkapazität von 100 bzw. 10 Paketen zugrundeliegt.

Im dargestellten Versuch liegen zusätzlich zu den asynchronen Datenübertragungen Polling-Pakete zur Übertragung an. Das ursprüngliche Systemdesign erlaubt kein permanentes Polling, da durch den exhaustiv Service die Systemkapazität allein durch diese Übertragungen vollkommen ausgeschöpft würde. Um dies zu verhindern werden für jede Station Pollingtimer eingeführt, die das Polling zeitlich begrenzen. Den Pollingdaten (5 Byte Nutzdaten und 2 Byte Quittungsnutzdaten) wird von vier Prioritäten die vorletzte zugeordnet. Insgesamt treten neben den stets zur Übertragung anstehenden 15 Pollingaufträgen (Pollingtabelle) Daten von drei weitere Prioritäten auf (Priorität 1, 3, 4). Die Pakete jeder Priorität haben den gleichen Anteil an der asynchronen Gesamtlast, die Zwischenankunftszeiten der Pakete sind exponentialverteilt. Die Pakete der Priorität 1 stellen mit einer mittleren Länge von 110 Byte kleinere Datenblöcke dar. Die Pakete der Priorität 3 enthalten Synchronisations- und Managementinformationen, haben eine mittlere Länge von 10 Byte und werden als einzige als Multicast-Pakete gesendet. Alarme schließlich tragen die höchste Priorität (4) und enthalten im Mittel 7 Byte Nutzdaten.

Hier werden die Pollingtimer aller Master mit der Einstellung T_p=50 ms vorbelegt. Bei 4 Masterstationen und dieser Timervorgabe ergibt sich nach der oben entwickelten Formel eine grundlegende Systemauslastung von ρ = 0,8, welche alleine auf das Polling zurückzuführen ist. Damit werden innerhalb der Timervorgabe von 50 ms bereits 40 ms durch Polling belegt. Etwa 10 ms können noch von asynchronen Paketen genutzt werden, ohne daß eine Beeinflussung der Polling-Verzögerung durch diese Pakete eintritt. Da sich die asynchronen Pakete zu dem Pollingpaketen mischen wird sich die Polling-Bearbeitungszeit mit steigender asynchroner Ankunftsrate erhöhen.

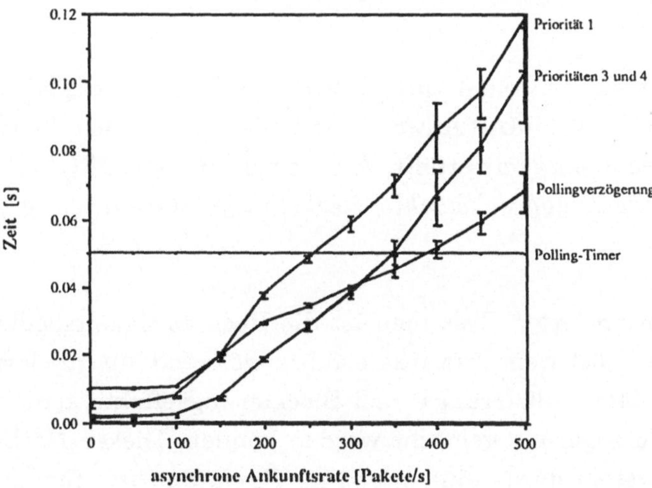

Abb.4.1: Polling-Bearbeitungszeit und Paketverzögerung bei steigender asynchroner Ankunftsrate

Erst ab einer asynchronen Ankunftsrate von etwa 300 Paketen/s steigt die Wartezeit der höherprioren asynchronen Pakete über die der Pollingdaten. Dieser Umstand erklärt sich aus der einzelnen Sende-Warteschlange der IINET-Sicherungsschicht, in welcher neu ankommende asynchrone Pakete unter höherer Last häufiger Polling-Pakete vorfinden. Da die Pakete der Pollingtabelle bei der hier betrachteten Polling-Abwicklung als Block übergeben werden, kann weiterhin davon ausgegangen werden, daß asynchrone Pakete so lange auf ihre Bearbeitung warten müssen, bis die gesamte Pollingtabelle gesendet worden ist. Die asynchronen Paketen der niedrigsten Priorität können auf jeden Fall erst dann gesendet werden, wenn kein Master mehr Polling-Pakete zu senden hat.

Die Betrachtungen können analog für die Intra-Vehikel-Kommunikationssysteme durchgeführt werden. Auch hier sind für ein regelmäßiges Polling zusätzliche Mechanismen einzuführen. Desweiteren bleibt anzumerken, daß durch die Blockierungssituationen ein unvorhersehbares Systemverhalten auftreten kann, dem nur durch Überdimensionalisierung (d.h. niedrige Auslastung) begegnet werden kann /HER90, SCH90b/.

5. Schlußbemerkungen

Insgesamt ist festzuhalten, daß das Prioritätenschema von IEEE für den Token Bus, wie es auch in MiniMAP Anwendung findet, die strenge Echtzeitbedingung im Sinne der Rechtzeitigkeit erfüllt, das Verfahren vom PROFIBUS jedoch nur die stationsbezogene *abgeschwächte Rechtzeitigkeitsbedingung* gewährleisten kann.

In nachrichtenorientierten Systemen ist die Rechtzeitigkeitsbedingung nur für Nachrichten der höchsten Priorität erfüllt, während für niederpriore Pakete durch zeitweise Monopolisierungs- und Blockierungseffekte keine Angaben über die maximale Verzögerung gemacht werden können. Diesen Nachteil versuchen Intra-Vehikel-System durch eine *adäquate Funktionalität* für die Fehlererkennung und -behebung auszugleichen. So bieten Intra-Vehikel-Systeme in der Regel besondere Mechanismen zur Anzeige von fehlerhaften Übertragungen, die Dank der besonderen physikalischen Realisierung kaum Zeitverzögerungen verursachen. Pilot- und Versuchsinstallationen haben gezeigt, daß die Systeme für ihre jeweiligen Anwendungsbereiche die benötigte Leistungskapazität (z.B.

durch Überdimensionalisierung) besitzen. Zusätzliche oder steigende Anforderungen (z.B. durch neue Technologien) können den Leistungsrahmen der Kommunikationssysteme sprengen. Ohne neue Lösungen kann bei steigenden Anforderungen nur auf das altbewährte Prinzip des *"Divide and Conquer"* zurückgegriffen werden.

Diese Philosophie des "Divide and Conquer" gewährleistet außerdem die *konzeptionelle Simplizität* der Systeme, wie sie für sicherheitsrelevante Systeme, wie z.B. Intra-Vehikel-Kommunikationssysteme, und für komplexe Umgebungen, wie in Produktionsumgebungen, gefordert wird. Die Aufteilung der Kommunikationssysteme erlaubt den Betrieb überschaubarer und damit beherrschbarer Subsysteme. So reduzieren sich Abhängigkeiten von Prioritätszuordnungen und Timer-Einstellungen in den Teilsystemen erheblich. Allerdings müssen funktionale Abhängigkeiten zwischen den Systemen in die Überlegungen miteinbezogen werden. Die Tatsache, daß effiziente Lösungen nur system- und anwendungsabhängig gefunden werden können, verursacht einen hohen Aufwand für das Systemdesign.

Literatur

/DIN88/ Messen Steuern Regeln, PROFIBUS - Process Field Bus, Teil 1 und 2, DIN Vornorm 19245, Jan. 1988

/HER90/ Hermanns O., 'Leistungsbewertung des alternativen Ringkonzepts für die Fertigungsumgebung IINET', Diplomarbeit an der RWTH Aachen, Lehrstuhl für Informatik IV, 1990

/IEEE 84/ IEEE Project 802, Local Area Network Standards, IEEE Standard 802.4, Token- Passing Bus Access Method and Physical Layer Specification, Revision F, Juli 1984

/REI*90/ Reinhardt W., Schümmer M., 'Leistungsbewertung des Feldbusstandards PROFIBUS', Tagungsband ECHTZEIT'90, Sindelfingen, S. 85-95, Juni 1990

/REI90/ Reinhardt W., 'Leistungsbewertung des Feldbusstandards PROFIBUS', Diplomarbeit an der RWTH Aachen, Lehrstuhl für Informatik IV, 1990

/RID90/ Ridder Ralph, 'Leistungsbewertung von MiniMAP als Kommunikationssystem der Fertigungsumgebung', Diplomarbeit an der RWTH Aachen, Lehrstuhl für Informatik IV, 1990

/SCE*89/ Scevarolli M., Simoncini L., 'Reliable multicast protocols for a token-ring architecture', Computer Systems Science and Engineering, Vol.4, No.2, pp.67-77, April 1989.

/SCH89a/ Schümmer M., 'Field Bus Systems - Communication Links of the Lowest Automation Level', Proc. of ISATA'89, Wiesbaden, pp. 1541-53, Nov. 1989.

/SCH89b/ Schümmer M.: 'Bewertung von Token-Passing-Verfahren in realzeitkritischen Fertigungsumgebungen', Tagungsband Workshop Realzeitsysteme - PEARL'89, Boppard, Springer Verlag, S. 25-39, Dez. 1989

/SCH90a/ Schümmer M., 'Multicasting in Realzeitumgebungen', Tagungsband ECHTZEIT'90, Sindelfingen, S. 55-64, Juni 1990

/SCH90b/ Schümmer M., ' Kommunikationssysteme in Echtzeitumgebungen - Modellierung und Bewertung von Fertigungs- und Intra-Vehikel-Kommunikationssystemen', Dissertation an der RWTH Aachen, Lehrstuhl für Informatik IV, Okt. 1990

Computerintegrierte Telefonie

In der ISO wird z.Zt. die Anlagenschnittstelle für CSTA, Computer Supported Telephone Applications, standardisiert. Die technische Grundlage ist die Übertragung aufbereiteter Informationen zwischen Koppelfeld- und Universalrechnern. Mit Hilfe von "proxy" und "mirror" werden zwischen der TK-Anlage und dem Rechner Kommandos und Statusmeldungen für eine Vielzahl neuer Applikationen ausgetauscht. Durch Verwendung eines Client/Server Konzeptes lassen sich Applikationen in einer heterogenen Betriebssystemwelt entwickeln. Die volle Netzwerkfähigkeit des Konzeptes ermöglicht ganz neue Ansätze für Applikationen.

Koordinator: L. Fahrenbach, Digital Equipment, München

CSTA: STANDARDARCHITEKTUR FÜR DIE FUNKTIONALE INTEGRATION VON RECHNER- UND VERMITTLUNGSSYSTEMEN

Jörg Eberspächer, Technische Universität München

1 EINFÜHRUNG

Der Begriff der Integration wird im Zusammenhang mit Kommunikationsnetzen in sehr unterschiedlicher Bedeutung verwendet. Zum einen versteht man darunter die **Transportintegration** in Form der "Leitungsintegration", der "Protokollintegration" oder der "Vermittlungsintegration". Zum anderen haben sich in den letzten Jahren Formen der **Funktionalen Integration** entwickelt in Gestalt der "Geräteintegration" und der "Anwendungsintegration". Unter Geräteintegration wird die Vereinigung verschiedener Funktionen in einem Gerät verstanden, wie sie in der Entwicklung von multifunktionalen Endgeräten zum Ausdruck kommt. Eine **Anwendungsintegration** erreicht man durch die Kopplung von Vermittlungs- und Verarbeitungsfunktionen auf der OSI-Anwenderschicht (Application Layer). Die unabhängigen Teilsysteme "Computer"("Host", "DV-System") und "Vermittlung" ("Switch") bilden dabei eine funktionale Einheit (Bild 1). Wesentlich ist dabei, daß vermittlungstechnische und datentechnische Verbindungen ("Calls" bzw. "Sessions") gemeinsam gesteuert und überwacht werden können. Konzepte und Architekturen für eine derartige Funktionale Integration für **Private Netze** (also unter Verwendung von Nebenstellenanlagen (Private Branch Exchange PBX) werden international unter verschiedenen Bezeichnungen diskutiert, standardisiert und realisiert:

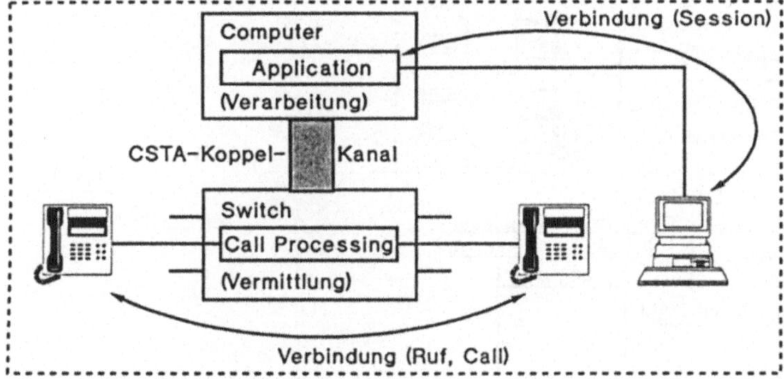

Bild 1: Funktionale Integration zwischen Vermittlungs- und Verarbeitungssystem

- CIT ("Computer Integrated Telephony", eine von der Firma Digital Equipment eingeführte Bezeichnung in Anlehnung an CIM (Computer Integrated Manufacturing), /1/.
- CSTA ("Computer Supported Telecommunications Applications") als Bezeichnung der Standardisierungsaktivitäten bei ECMA (European Computer Manufacturers Association, /2/).
- SCAI ("Switch-Computer Applications Interface"), eine Aktivität der T1S1.1 ISDN Networking Group bei ANSI (American National Standards Institute), /3/).

Daneben gibt es firmenspezifische Bezeichnungen wie ACL von Siemens (/4/), HCI von Mitel /1/.

Ähnliche Ziele verfolgt man im öffentlichen Netzbereich mit dem Konzept des "Intelligent Network" (IN). Es bestehen jedoch Unterschiede in Technik und Anwendung.
Auf Architektur- und Standardisierungsaspekte der im Rahmen von CSTA erarbeiteten funktionalen Integration von PBX und Hostrechnern soll im folgenden eingegangen werden.

2 Anforderungen an eine CSTA Architektur

2.1 Ein Anwendungsbeispiel

Im folgenden sollen die typischen Abläufe bei einer CSTA-Applikation anhand eines Beispiels beschrieben werden. Bild 2 zeigt ein Szenario "Steuern abgehender Verbindungen" ("Outbound Call Handling"). Bei dieser Anwendung geht es darum, daß ein in einem Hostrechner ablaufendes Programm selbsttätig Telefonverbindungen zwischen einem Teilnehmer A, z.B. einem Vertriebsbeauftragten (Agent A) und einem bestimmten Kunden B initiiert. Der Teilnehmer B kann dabei über das öffentliche Netz an die Vermittlungsstelle (PBX) angebunden sein. Der Agent A ist gleichzeitig an der Vermittlung und an einem Hostrechner angeschlossen. Er hat über sein Datenterminal im Rahmen einer "DV-Session" Zugriff auf den Hostrechner und somit auf die aktuellen Daten des Kunden.
Zur Durchführung seiner Aufgabe muß das Hostprogramm mit der Vermittlungsfunktion kommunizieren und Meldungen austauschen. In der Folge soll dann in diesem Beispiel das Telefongespräch mitsamt der Session zum Teilnehmer C (Agent C) transferiert werden.

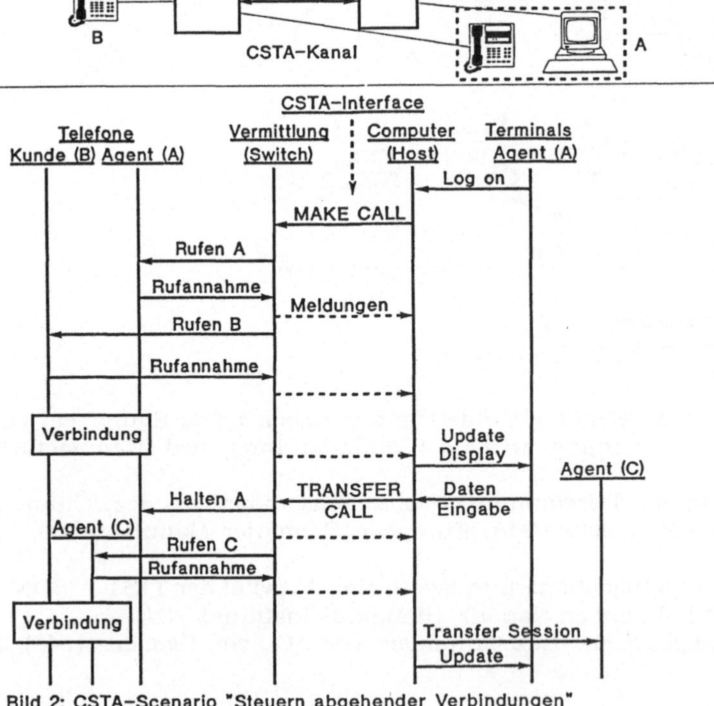

Bild 2: CSTA-Scenario "Steuern abgehender Verbindungen" (Outbound Call Handling)

Folgende Schritte werden durchgeführt:

- Der Agent A ruft das Applikationsprogramm auf dem Host auf (Log on)
- Die Applikation prüft in einem "MONITOR"-Modus, ob das Telefon von A frei ist. Wenn ja, schickt sie über den CSTA-Koppelkanal ein Kommando (Request) "MAKE CALL" (A, B) zur Vermittlungseinheit.
- Diese setzt daraufhin je einen Ruf an A (der Agent initiiert diesen Ruf also nicht selbst!) und anschließend an den Kunden B ab.
- Nach Rufannahme (z.B. Abheben) und Durchschaltung der Sprachverbindung werden die jeweiligen Ereignisse ("Events") von der Vermittlungsanlage an den Hostrechner und die Applikation gemeldet.
- Die Applikation kann nun dem Agenten A synchron mit der Durchschaltung der Sprachverbindung die aktuellen Daten des Kunden B auf dem Bildschirm bereitstellen, da sie die Zustandsmeldungen aus der Vermittlung in Echtzeit erhält.
- Während der Verbindung kann der Fall eintreten, daß der Agent A einen weiteren Agenten C zu Rate ziehen muß und z.B. das CSTA-Gespräch mit dem Teilnehmer B an C übergeben will. Zu diesem Zweck sendet die Applikation den CSTA-Befehl "TRANSFER CALL" an die Vermittlung.
- Während die Vermittlung die vermittlungstechnischen Prozeduren für den Transfer der Sprachverbindung durchführt, bereitet das Applikationsprogramm den Transfer der datentechnischen Sitzung zum Agenten C vor.
- Nach erfolgreichem "Übergeben" des Telefonrufs besteht daher nicht nur eine Sprachverbindung zwischen B und C, sondern C hat auch die aktuellen Daten des Teilnehmers B auf seinem Bildschirm.

2.2 Dienste

Aus diesem und anderen Anwendungsfällen lassen sich die drei wichtigsten Dienstarten ableiten, die von einer CSTA-Architektur zur Verfügung gestellt werden müssen:

1) Verbindungssteuerung (Call Handling)
   Dazu gehört das Auf- und Abbauen von Einzel- und Mehrfachverbindungen (Konferenz) ebenso wie die Modifizierung bestehender Verbindungen (Transferieren, Halten, Makeln).
2) Monitoring
   Darunter wird das Überwachen von Verbindungen bzw. Endgeräten und das Melden auftretender Ereignisse ("Events") zum Hostrechner verstanden.
3) Rechnersteuerung von Leistungsmerkmalen
   Möglich sind hierbei z.B. die Aktivierung einer Rufumlenkung oder eines automatischen Rückrufs bei besetztem Teilnehmer etc.

Bei 1) und 3) operiert die Applikation gewissermaßen "stellvertretend" für bestimmte Teilnehmer ("Proxy-Prinzip"): Die Initiative für den Verbindungsaufbau geht nicht, wie bei normalen Telefongesprächen, vom Sprachendgerät eines Teilnehmers aus, sondern vom Hostrechner. Bei 2) "spiegelt" das Vermittlungssystem Ereignisse bestimmter ausgewählter Ressourcen des Vermittlungssystems zur Host-Applikation ("Mirror-Prinzip").
Über diese klassischen Telefonfunktionen hinaus sind aber weitere CSTA-Dienste ("Services") denkbar und sinnvoll, z.B. Routing-Unterstützungs-Funktionen oder ein gegenseitiger Zugriff auf Datenbasen in Rechner und Vermittlungsanlage.

2.3 Sicherheit und Leistungsfähigkeit

Die enge Kopplung zwischen Rechner und Vermittlungssystem stellt hohe Anforderungen an die Systemsicherheit und zwar hinsichtlich der Zuverlässigkeit und der Integrität beider Systeme.
Die Initiierung von Vermittlungsvorgängen durch angeschlossene Rechner und die Übermittlung von Zustandsmeldungen kann zu einer erhöhten dynamischen Belastung der Vermittlungsanlage führen. Da Einschränkungen in der Dienstgüte für die Teilnehmer nicht toleriert werden können, muß dies bei der Auslegung der Vermittlungssteuerungen und Applikationen berücksichtigt werden.

2.4 Netzfähigkeit und Protokolle

Während in dem Prinzipbild (Bild 1) nur jeweils **ein** Vermittlungsknoten bzw. **eine** DV-Einheit betrachtet wurde, müssen in Wirklichkeit komplexere Netzkonfigurationen beherrscht werden (Bild 3). Eine CSTA-Architektur muß es erlauben, daß CSTA-Applikationsprogramme, die (im allgemeinen sogar verteilt) auf irgendwelchen Rechnern ablaufen, Vermittlungsvorgänge auch an solchen Knoten des Vermittlungsnetzes steuern bzw. überwachen können, die nicht unmittelbar mit einem Knoten des Rechnernetzes verbunden sind. Es muß möglich sein, die Kopplung nur über eine einzige Schnittstelle, den "CSTA-Koppelkanal" abzuwickeln (CSTA-Service Interface).
Sofern neue Protokolle für CSTA-Zwecke erforderlich sind, muß möglichst vermieden werden, die vermittlungstechnischen Protokolle an der Benutzer-Netz-Schnittstelle (Teilnehmer-Signalisierung) oder zwischen Vermittlungsknoten (Netzsignalisierung) CSTA-spezifisch zu modifizieren. Diese Protokolle sind weitgehend international genormt und nur sehr schwer änderbar.

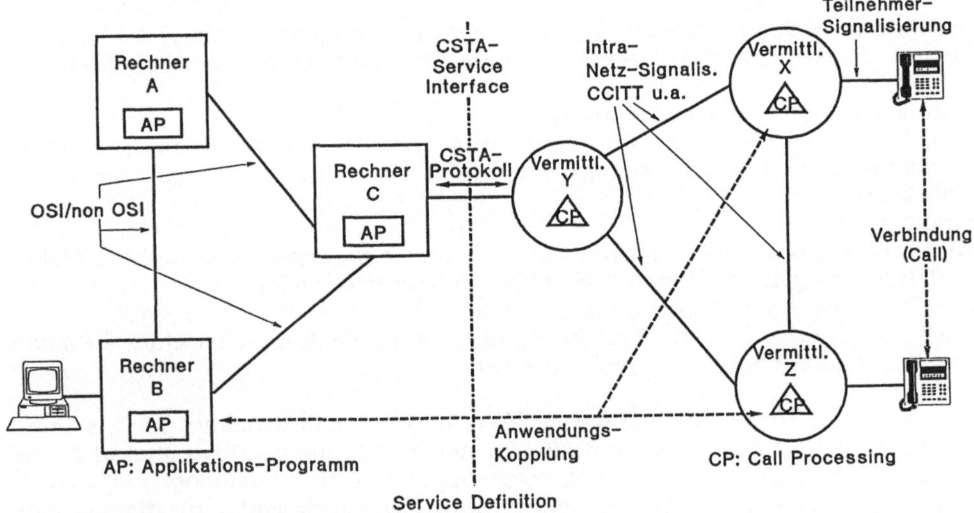

Bild 3: Anwendungskopplung in Rechner- und Vermittlungsnetzen

3 CSTA-Funktionen

3.1 Übersicht über CSTA-Services

Die oben genannten drei Dienstekategorien setzen sich aus einer Vielzahl von einzelnen, separat definierten "CSTA-Services" zusammen (Bild 4). Diese sind sehr von den derzeit im Vordergrund stehenden Telefon-Anwendungen geprägt und entsprechen weitgehend den heute in digitalen PBX-Systemen an der Teilnehmerschnittstelle verfügbaren Dienstmerkmalen.

3.2 Ein Service-Beispiel

Anhand des in Bild 5 graphisch dargestellten Service CONSULTATION_CALL werden Art und Umfang einer Service Definition erläutert. Der Service-Aufruf bewirkt, daß ein existierender, aktiver Ruf C1 zwischen zwei Teilnehmern (Devices) D1 und D2 in den Zustand "HALTEN (HOLD)" gebracht wird (vorübergehend inaktiv).
Anschließend wird, wiederum initiiert von der CSTA-Applikation, von der Seite eines der beiden Teilnehmer (z.B. D1) aus ein Rückfrage-Ruf (Consultation Call C2) zu einem dritten Teilnehmer D3 eingeleitet und, falls dieser frei ist und sich meldet, auch etabliert. Der Aufruf des Service CONSULTATON_CALL muß folgende Identifikationen beinhalten:

- aktiver Ruf C1;
- Teilnehmer D1, von dessen Terminal aus der Rückfrage-Ruf C2 ausgehen soll;
- Teilnehmer D3, der rückgefragt wird.

1. Switching Function Services
    - Make_Call Service
    - Clear_Call Service
    - Clear_Connection Service
    - Hold_Call Service
    - Retrieve_Call Service
    - Consultation_Call Service
    - Alternate_Call Service
    - Reconnect_Call Service
    - Transfer_Call Service
    - Conference_Call Service
    - Answer_Call Service
    - Call-Completion Service
    - Make_Predictive_Call Service
    - Divert_Call Service
    - Feature_Access Servece
2. Status Reporting Services
    - Snapshot Service
    - Monitor Service
3. Andere Services
    - Input/Output Services
    - Database Services

Bild 4: Übersicht geplanter CSTA-Services

a) Ausgangszustand

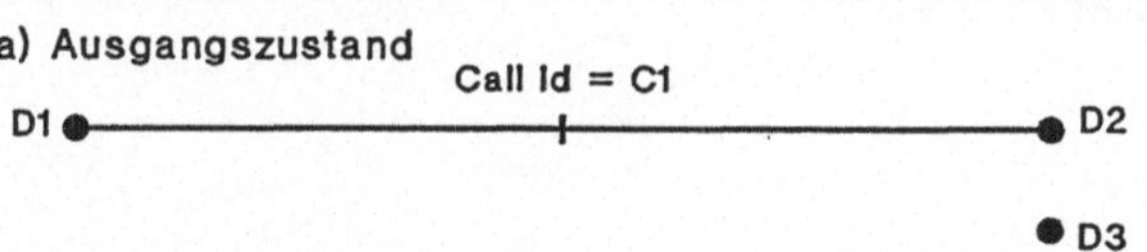

b) Ausführung der Operation CONSULTATION_CALL in Bezug auf T1

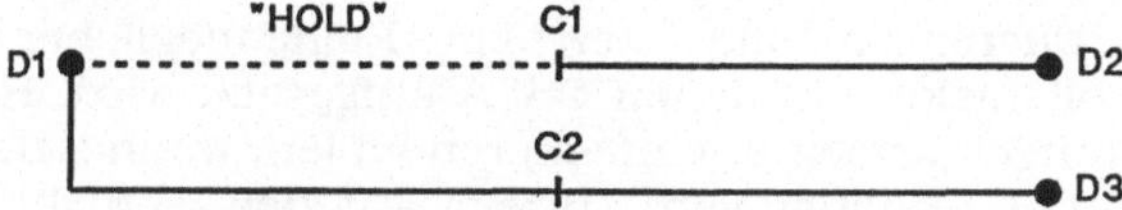

c) Wiederherstellung des ursprünglichen Zustands durch RECONNECT_CALL

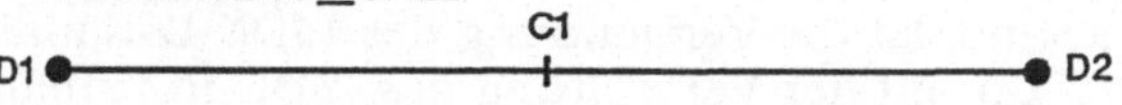

Bild 5: Ablauf der CSTA-Operationen (Services) CONSULTATION_CALL und RECONNECT_CALL

Durch das komplementäre Kommando RECONNECT_CALL kann die Applikation den ursprünglichen Zustand wiederherstellen. Die korrekte Durchführung des Service

kann durch Anwendung des MONITOR-Kommandos auf die beteiligten Endstellen D1,...D3 bzw. die Rufe C1, C2 überwacht werden.
Die konsistente und eindeutige Definition der CSTA-Services setzt eine exakte Definition und Modellierung der "Objekte" (z.B. Gerät, Ruf, Verbindung...) voraus, die mit den CSTA-Services ("Kommandos") zu beeinflussen und zu überwachen sind (siehe Kap. 5).

4 Protokollarchitektur

CSTA ist als eine Verteilte Schicht 7-Applikation zu betrachten und entsprechend den OSI-Prinzipien strukturiert /6/.

4.1 Application Layer

Die Struktur der Application Layer zeigt Bild 6. Die CSTA-spezifischen Application Service Elements (ASE) werden dabei kombiniert mit Common Service Elements, darunter ACSE (Association Control Service Element) zur Steuerung der "Verbindungen" zwischen den Application Service Elements und ROSE (Remote Operations) zur Ausführung von "entfernten Operationen". Wie auch für andere Application Layer Architekturen vorgeschlagen, läßt sich die Beziehung zwischen anforderndem Prozeß und ausführendem Prozeß durch ein Client-Server-Modell beschreiben (/7/).

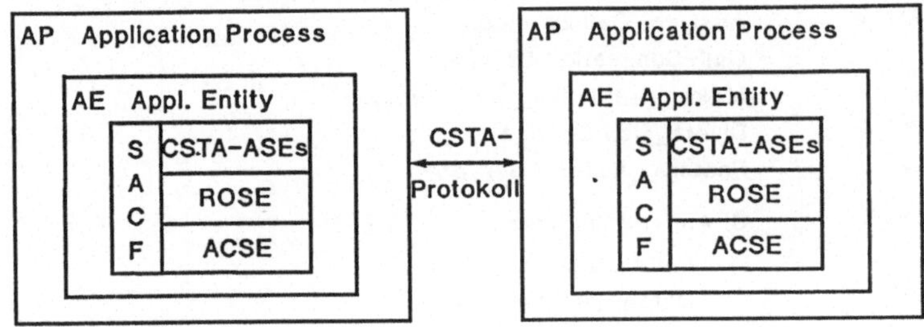

SACF Single Association Control Function
ROSE Remote Operations Service Element
ACSE Association Control Service Element

Bild 6: Struktur der Application Layer

4.2 Untere Schichten

Die Kommunikationsinfrastruktur von CSTA ist zwar aufgrund der OSI-Struktur relativ frei wählbar, doch gibt es einige Präferenzen. Bild 7 zeigt eine Reihe möglicher oder heute schon implementierter Protokollstacks, auf denen CSTA aufgesetzt wird. Heute schon verfügbare CSTA-Vorläuferlösungen setzen in einfacheren Fällen, wenn z.B. keine Multiplexfähigkeit des Koppelkanals benötigt wird, direkt auf den Schichten 2 (HDLC LAP B) auf. In anderen Fällen sind die Protokolle der Schichten 3 und 4, jedoch keine Session oder Presentation Layer vorhanden.
Eine in Zukunft wichtige Protokollvariante ist die Verwendung des ISDN-D-Kanal-Protokolls (Q.931 mit Ergänzungen /8/). Da mit der Verbreitung des ISDN in zunehmendem Maße auch Rechner mit ISDN-Anschlüssen ausgestattet sind, liegt es nahe, diese auch für CSTA zu nutzen. Zum Transport der CSTA-Schicht 7-Nachrichten gibt es dabei verschiedene Möglichkeiten /5/. Bild 8 zeigt eine Lösung unter Verwendung des "User-User-Signalling" im ISDN.
Hierbei werden die ACSE-Funktionen zum Schicht 7-Verbindungs-Management ebenso wie die Datentransfer-Operationen von ROSE auf D-Kanal-Meldungen abgebildet.

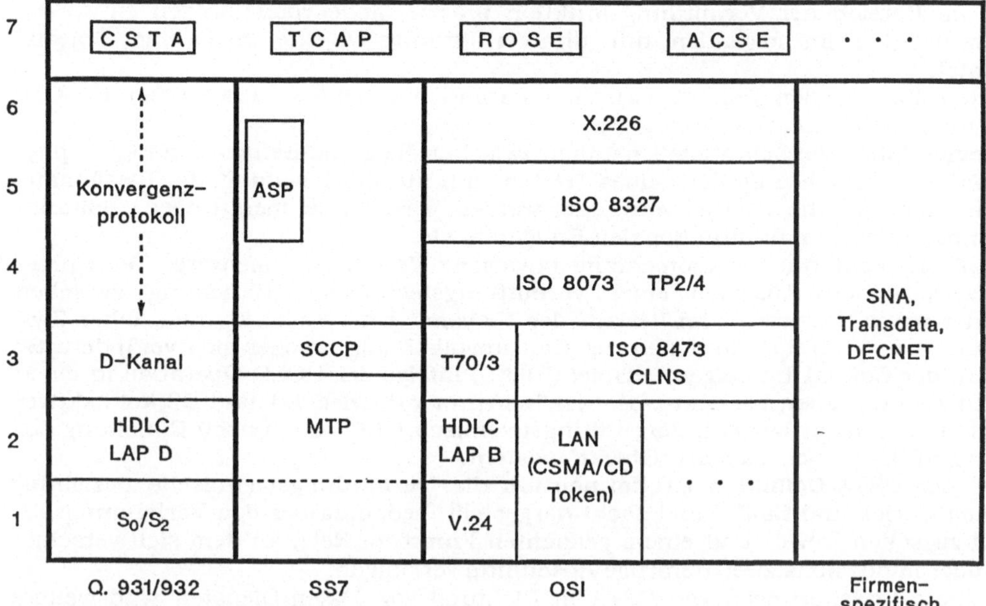

Bild 7: CSTA Protokoll Infrastruktur

	LAYER 7		Layer 3
1	A-ASSOCIATE . Req . Ind	⟶	SETUP
2	A-ASSOCIATE . Resp . Conf	⟶	CONNECT
3	Data Transfer:		
	RO-INVOKE . Req . Ind	⟶	USER INFO
	RO-RESULT . Req . Ind	⟶	USER INFO
4	A-RELEASE . Req . Ind	⟶	RELEASE
5	A-RELEASE . Resp . Conf	⟶	RELEASE COMPLETE

Bild 8: Verwendung des D-Kanal-Protokolls für CSTA (User-to-User-Signalling)

5 CSTA-Objekte und ihre Modellierung

5.1 CSTA-Objekte

Die Definition der CSTA-Services beruht auf ganz bestimmten Modellvorstellungen über das Vermittlungs- bzw. Rechnernetz, die darin definierten Objekte sowie die zulässigen Operationen in Bezug auf diese Objekte. Da die CSTA-Architektur bisher lediglich im Bereich der Vermittlungsfunktion präzise Modellvorstellungen entwickelt wurden, werden im folgenden nur die Vermittlungs-Objekte (Switching Objects) betrachtet.

Es wird unterschieden (Bild 9) zwischen folgenden Objekten Device ("Gerät"), Call ("Ruf") und Connection ("Verbindung").

Das **Device** bildet die Schnittstelleneinheit zwischen Netz und Benutzer. Es kann physisch definiert sein, wie im Falle eines Telefons mit nur einer Leitung; darüber hinaus können auch logische Devices verwendet werden, z.B. Teilnehmergruppen, Teilfunktionseinheiten eines multifunktionalen Endgeräts etc.

Der **Call** ist nicht nur die momentane physische Verbindung mehrerer Teilnehmer (Benutzer), sondern Ausdruck einer "vermittlungstechnischen Beziehung" zwischen zwei oder mehr Benutzern. Im Verlauf des "Lebens" eines Calls können daher Zwischenzustände auftreten, in denen der Call unvollständig ist oder sich verändert. So geht z.B. der Call C1 im obigen Beispiel (Bild 5) infolge der HOLD-Operation in einen anderen Zustand über, bei dem zwischen Teilnehmer (Device) D1 und D2 kein aktiver Transportkanal mehr besteht. Vermittlungstechnisch ist C1 auch nach Einleitung des Rückfrage-Rufs immer noch als "Objekt" existent.

Gemäß der CSTA-Definition ist **Connection** die "vermittlungstechnische Beziehung zwischen Device und Call". Vereinfacht dargestellt, bedeutet dies den Verbindungsabschnitt zwischen Device und einem gedachten Punkt im Netz, an dem sich verschiedene (aber mindestens zwei) derartige Abschnitte vereinigen.

Für spätere Erweiterungen von CSTA in Richtung von Daten-Diensten sind weitere Objekte denkbar wie Programmkomplexe, Datenbasen, DV-Terminals, Files usw.

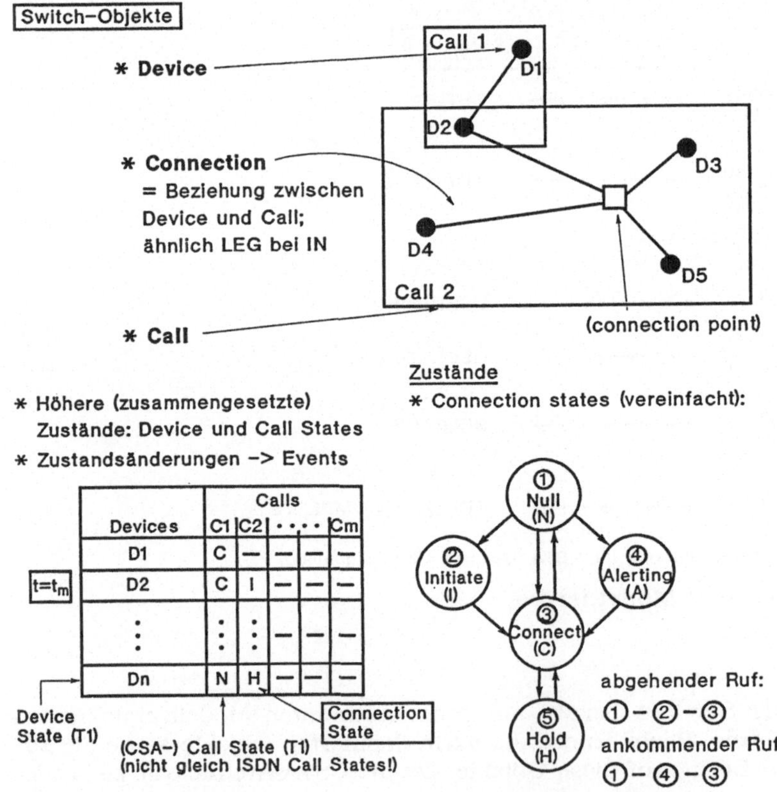

Bild 9: CSTA-Objekte

5.2 Zustände von Objekten

Eine exakte Modellierung von CSTA-Funktionen und Operationen ist unter Verwendung von Zuständen und Zustandsdiagrammen möglich.
CSTA baut hier auf einem zweistufigen Zustandsmodell auf. Die Connections bilden die Grundelemente und haben wenige, für die Anwendungen wesentliche Zustände (States). Die Connection-States ändern sich zu diskreten Zeitpunkten. Diese Änderungen führen zu Ereignissen (Events). Die Beziehungen zwischen Devices, Calls und Connections lassen sich gemäß Bild 9 für einen bestimmten Zeitpunkt $t=t_m$ in einer Tabelle darstellen. Die Spalten enthalten sämtliche Connections eines bestimmten Call Ci, und die Zeilen sämtliche Connections eines Device Dj. Bei Änderung eines "Connection State" ergibt sich eine modifizierte Tabelle zum Folgezeitpunkt $t=t_n$.
Im allgemeinen wird eine Applikation an "höheren" Zuständen interessiert sein, also z.B. am Zustand eines Call. In diesem Fall lassen sich durch geeignete Zusammenfassung von Connection States weitere Zustände definieren und, abgeleitet aus den Änderungen dieser Zustände, auch weitere Ereignisse (z.B. CALL_CONFERENCED).

6 Vergleich mit dem Intelligent Network (IN)

Angesichts der laufenden Spezifikations- und Entwicklungsarbeiten und der zahlreichen offenen Punkte ist ein Vergleich zwischen IN und CSTA nur eingeschränkt möglich. Die Grundkonzeptionen sind ähnlich, nämlich die funktionale Kopplung von DV-Systemen mit Vermittlungssystemen. In CSTA-Anwendungen steht das vorwiegend auf dem Hostrechner residente Applikationsprogramm im Mittelpunkt. Es bedient sich der Vermittlungsfunktionen einer PBX zur Ausführung einer Aufgabe. Beim IN geht es primär um die Vermittlungsaufgabe im Vermittlungsknoten (Service Switching Point SCP), der dabei durch die Kooperation mit einem Rechner (Service Control Point SCP) unterstützt wird. IN verwendet derzeit ein Call Modell, das sich vom CSTA-Modell darin unterscheidet, daß den Connections bzw. "Legs" eine größere Bedeutung zukommt als bei CSTA (/9/). Die Steueroperationen bei IN beziehen sich fast ausschließlich auf diese "atomaren" Objekte (z.B. "Join Leg"). Dies hat seinen Grund darin, daß beim IN eine "Intra-Netz-Perspektive" vorherrscht, bei der detaillierte "Eingriffe" in die vermittlungstechnischen Vorgänge erwünscht sind. Demgegenüber ist CSTA mehr aus einer "Extra-Netz-Perspektive" konstruiert, bei der man sich weniger für die Detailsteuerung von netzinternen Abläufen, als für die korrekte Ausführung der CSTA-Services als ganzes interessiert.
Die technischen Diskussionen zwischen den IN- und CSTA-Experten zeigen jedoch eine zunehmende Annäherung der Architekturkonzepte. Dies ist auch sehr wünschenswert, da eine Reihe von Anwendungen sowohl in privaten als auch in öffentlichen Netzen sinnvoll sind.

7 Stand der Normierungsaktivitäten

Die Standardisierung von CSTA begann im Herbst 1988 bei der ECMA (European Computer Manufacturers Association) als "Spin-off" der Arbeiten zur ISDN-Protokollnormierung in Privaten Netzen. Beteiligt ist die Mehrzahl aller bedeutenden Computer- und PBX-Hersteller aus Europa und Nordamerika. Als erstes Ergebnis wurde ein Architekturvorschlag erarbeitet und als ECMA Technical Report TR-52 im Juni 1990 veröffentlicht /2/. Gegenwärtig werden die Arbeiten bei ECMA fortgesetzt mit der Festschreibung konkreter Services und Protokolle für die o.g. telefonorientierten Dienste. Eine Erweiterung in Richtung "Computer Services" ist ebenfalls geplant.
Fast gleichzeitig mit dieser Aktivität bei ECMA startete eine Arbeitsgruppe beim amerikanischen ANSI-Komitee T1S1.1 mit der Definition einer SCAI-Architektur (Switch-

Computer-Application Interface). Zwischen beiden Komitees gibt es eine enge Liaison. Mit der Fertigstellung von implementierbaren Standards kann im Zeitfenster 1991/92 gerechnet werden.

8 Zusammenfassung

Die Kopplung von Vermittlungs- und Datenverarbeitungssystemen auf Anwenderebene hat aufgrund verschiedener internationaler Aktivitäten zur Entwicklung der CSTA-Architektur geführt. Dabei können von Applikationsprogrammen aus Funktionen in Vermittlungsanlagen gesteuert und überwacht werden.
Die Funktionalität der Dienste (Services) und Protokolle sowie die Modellierung der steuerbaren Objekte wurde anhand von Beispielen vorgestellt. Die Standardisierung der CSTA-Architektur ist noch im Gange. Erste Normvorschläge sind im Jahre 1992 zu erwarten. Auf dieser Basis werden dann auf verschiedensten Computersystemen flexible und PBX-unabhängige Applikationsschnittstellen (Application Programm Interfaces API) zur Verfügung gestellt werden können. Damit erschließen sich neue Felder für wirklich "integrierte" Systeme und Anwendungen.

9 Literatur

/1/ C.R.Strathmeyer: Voice in Computing: An Overview of Available Technologies. Computer, Aug. 1990, S. 10-15.

/2/ ECMA: ECMA TR/52 Computer-Supported Telecommunications Applications. Genf Juni 1990.

/3/ ANSI T1S1.1: Switch-Computer Applications Interface. Working Document-Version 8., Oct. 1990.

/4/ G. Baum: Öffnung der Telefonie. Vom Vermittlungssystem zum Anwendungsverbund. Telematica Stuttgart, 1990.

/5/ J. Eberspächer: CSTA in an ISDN Environment. Contribution to ECMA TC 32/TG6, Sophia Antipolis 1988.

/6/ ISO/DP 9545 Open System Interconnection - Application Layer Structure.

/7/ ISO 10031-1 Telecommunications - Distributed Office Applications Model Part 1: General Model.

/8/ CCITT: Rec. Q.931 Digital Subscriber Signalling System No. 1 (DSS1), Network Layer.

/9/ Bellcore: Advanced Intelligent Network Release 1 Baseline Architecture. SR-NPL-001555, Issue 1, Red Bank, NJ, Mar. 1990.